Numerical Methods in Laminar and Turbulent Flow

Numerical Methods in Laminar and Turbulent Flow

Editors:

C. Taylor

B. A. Schrefler

Proceedings of the Second International Conference held at Venice, 13th - 16th July, 1981

PINERIDGE PRESS

Swansea, U.K.

First Published, 1981 by
Pineridge Press Limited
91, West Cross Lane, West Cross, Swansea, U.K.

ISBN 0–906674–15–8

British Library Cataloguing in Publication Data

Numerical methods in laminar and turbulent flow
1.Fluid dynamics – Mathematics – Congresses
I. Taylor, C. II. Schrefler, B.
532'.051015'117 QA911
ISBN 0–906674–15–8

Printed and bound in Great Britain by
Robert MacLehose and Co. Ltd.
Printers to the University of Glasgow

PREFACE

The proceedings contains the papers presented at the International Conference on Numerical Methods in Laminar and Turbulent Flow held at Venice during the period JULY 13th-16th, 1981. Due to the high standard and the number of submitted abstracts, the final selection of contributed papers for presentation proved a difficult task. The organisers wish to express their thanks to the authors and extend their apologises to those whose papers were rejected.

In the short space of time since the first conference under the present heading was held at Swansea (U.K.), JULY 1978, significant advances have been made in the application of numerical methods to solve laminar and turbulent flow problems. This is adequately demonstrated in the current proceedings where a wide range of engineering problems are analysed. When such advances are made it is almost inevitable that further desirable avenues of research are uncovered. The present text is no exception and a large number of papers indicate that further fundamental as well as applied research is required. The papers should prove to be an asset to scientists and engineers engaged in the investigation of flow problems.

The papers have been reproduced directly from lithographs of the authors' manuscripts and the editors do not except responsibility for any erroneous comments or opinions expressed.

Finally, the editors wish to thank the Consiglio Nazionale Delle Ricerche (Italy) and the British Council for their support and sponsorship.

C. TAYLOR Department of Civil Engineering, University College, Swansea

B.A. SCHREFLER Universita di Padova, Padova, Italy

CONTENTS

PREFACE

SECTION 1 GENERAL VISCOUS FLOW

1. M. ISRAELI, V. REITMAN, S. SOLOMON and M. WOLFSHTEIN "On the marching solution of elliptic equations in viscous fluid mechanics" 3

2. M.D. OLSON and J.F. BALDWIN "Eigenvalue petubation of Navier Stokes equations by finite elements" 15

3. P.M. GRESHO, S.T.K. CHAN, R.L. LEE and C.D. UPSON "Solution of time-dependent, three dimensional incompressible Navier-Stokes equations via a finite elements" 27

4. R.L. SANI and B.E. EATON "On the solution of the time-dependent incompressible Navier-Stokes equations via a penalty Galerkin finite element method" 41

5. Y. LECIONTE and J. PIQUET "On the numerical solution of same types of unsteady incompressible viscous flow" 53

6. A. DANAEE and D.J. EVANS "The application of boundary value techniques in the solution of the Navier-Stokes equations" 65

7. L. QUARTAPELLE, F. VALZ-GRIS and M. NAPOLITANO "Intergral conditions on the vorticity:numerical alogorithms for the steady and unsteady Navier-Stokes equations" 79

8. R.K. ROUT, A. ECER and P. WARD — 91
"Investigation of solution of Navier-Stokes equations using a variational formulation"

9. W.C. HORAK and J.J. DORNING — 103
"A modal Green's tensor method for the efficient numerical solution of laminar flow problems"

10. F.J.K. IDERIAH — 113
"The numerical computation of laminar pipe flow"

11. M.B. BUSH and R.I. TANNER — 119
"The boundary element method applied to creeping motion of a sphere"

12. J.H. WANG and W.J. YANG — 129
"Numerical solutions for laminar flows in disc-type strrers"

13. T. THUNELL and L. FUCHS — 141
"Numerical solution of the Navier-Stokes equations by multi-grid techniques"

14. K. NANDARKUMAR and J.H. MASLIYAH — 153
"Laminar flow in curved moon shaped ducts"

15. L. FUCHS and H. ZHAO — 167
"Numerical simulation of three dimensional flows in ducts"

16. G. DHATT — 179
"Different finite element formulations for the Navier-Stokes equations"

17. J. RIEDLER — 191
"Quasi three-dimensional finite element solution for a twin extruder"

SECTION 2 TURBULENT FLOW

1. G. GOUESBET and A. BERLAMONT — 205
"Prediction of turbulent fields, including fluctuating velocity correlations and approximate spectra, by means of a simple second order closure scheme: the round free jet and developed pipe flow"

2. P.M. BAUER, F. HOISLBAUER, H. BARTOSCH and C. JAQUEMAR — 217
"Numerical solution of electromagnetically driven liquid metals"

3. A.G. HUTTON and R.M. SMITH — 229
"On the finite element simulation of incompressible turbulent flow in general two dimensional geometries"

4. C.W. RAPLEY — 243
"The prediction of local mean flow characteristics for fully developed turbulent flow in straight non-circular passages"

5. Z. MAZHAR and G.D. RAITHBY — 255
"A refined PUMPIN (Pressure Update by Multiple Path Integration) method for updating pressures in the numerical solution of the incompressible fluid flow equations"

6. W. SLAGTER and H.A. ROODBERGEN — 267
"Prediction of developing turbulent flow by the finite element method"

7. R.W. DAVIS and E.F. MOORE — 279
"The numerical simualtion of flow around squares"

8. O. KVERVOLD, A.K. RASTOGI and T. SØNTVEDT — 291
"Turbulent flow and turbulent mass transfer in partly closed pipes and channels"

9. M. BUFFAT, D. JEANDEL and J.F. BRISON — 303
"Improvements of the optimal control method to solve turbulent flows"

10. P. MELE, M. MORGANTI, A. Di CARLO and A. TATONE — 315
"Laminar to turbuent flow study by means of the F.E.M."

11. A. BERLEMONT and G. GOUESBET — 327
"Prediction of the behaviour of a cloud of discrete particles released in a fully developed turbulent pipe flow, using a non-discrete dispersive approach"

12. M.M.M. ABOU-ELLAIL and T.W. ABOU-ARAB — 339
"Prediction of turbulent mixing in confined co-axial reacting jets"

13. J.K. PLATTEN, J.M. LUIJKX and P. FLANDROY — 351
"On the transition to turbulence"

14. E.E. KHALIL and H.M.W. ASSAF 363
"Computer modelling of turbulent recirculating flows in engineering applications"

15. A. POLLARD 377
"On the calculation of the laminar and turbulent flow between parallel discs"

16. B.J. ALFRINK 389
"On the Neumann problem for the pressure in a Navier-Stokes model"

17. J.I. RAMOS 401
"A numerical study of confined turbulent jets"

18. I. GRANT, F.H. BARNES and T.H. LOON 413
"The prediction of the drag on structural beams"

19. F. BOYSAN and J. SWITHENBANK 425
"Numerical prediction of combined vortex flows"

20. V.K. GARG 439
"Finite difference solution for axially developing flow in an annulus"

21. C. TAYLOR, C.E. THOMAS and K. MORGAN 449
"F.E.M. and the two equation model of turbulance"

SECTION 3 BOUNDARY LAYER ANALYSIS

1. C. PATTERSON and M.A. SHEIK 463
"Regular boundary integral equations for fluid flow"

2. A. POSTAD, O. PADE, D. ANSHELOVITZ and M. WOLFSHTEIN 475
"Accurate solutions for laminar and turbulent boundary layers and very large pressure gradient parameters"

3. S.P. KLOTZ and R.L. STREET 487
"The numerical simulation of the turbulent boundary layers at a rough air-water interface"

4. C-C HSU and A. LIAKOPOULOS 497
"A finite element differential method for a class of compressible laminar boundary-layer flows"

5. A.R. WADIA and F.R. PAYNE 505
"On the OLP prediction of the unstable modes of the flat plate turbulent boundary layer"

6. S.P. KLOTZ and R.L. STREET 517
"On the simulation of the coupled laminar boundary layers at a smooth, phase changing gas-liquid interface"

7. V. GANESAN and B.H.L. GOWDA 531
"Numerical prediction of turbulent boundary layer developement on a two-dimensional curved wall"

SECTION 4 FLOW WITH HEAT TRANSFER

1. B.P. LEONARD 543
"A stable, accurate, economical and comprehendible algorithm for the Navier-Stokes and scalar transport equations"

2. Ph. ROUZAUD, J. CHINARDET, B. GAY and R. VERBIEST 555
"Thermal hydraulic calculations of wire-wrapped bundles using a finite element method. THESEE code"

3. P.L. VIOLLET 569
"On the computing of turbulent Heat and Mass Transfer for the computation of buoyancy affected flows"

4. J.A. RIEZES 581
"Developing compressible flow between parallel heated plates"

5. G. GRÖTZBACH 593
"Spatial resolution requirements for numerical simulation of internally heated fluid layers"

6. S. DEL GIUDICE and A. TROTTA 605
"Heat and mass transfer in tubular reactors: a finite element approach"

7. L. KATGERMAN 615
"A numerical solution of ribbon thickness formation during meltspinning"

8. L.A. BERTRAM 625
"Numerical 'turbulent transition', locus for continuous casting type flows"

9. A.M. GOORAY, C.B. WATKINS and W. AUNG 639
"Numerical calculations of turbulent heat transfer downstream of a rearward facing step"

10. M.S. KHADER and J.S. GOODLING 653
"Low Reynolds number heat transfer in cylindrical packed beds"

11. T. DROZD 665
"Momentum and heat transfer characteristics in turbulent flow"

SECTION 5 FREE SURFACE FLOW

1. G. LABADIE, J.P. BENQUE and B. LATTEUX 681
"A finite element method for the shallow water equations"

2. E.P. QUERNER 693
"The finite element method applied to turbulent open channel flow"

3. E.W. MINER, O.M. GRIFFIN, S.E. RAMBERG and M.J. FRITTS 705
"A numerical and experimental study of free surface wave flow over a half cylinder"

4. D.L.C. LAM 719
"Stratified flow models for Lake Eirie"

5. N. PRAAGMAN 731
"Numerical models for the computation of residual currents"

6. T.W. KAO 741
"Current and density structure in shelf waters due to fresh water discharge : a numerical Study"

7. A.D. JENKINS 753
"Determination of tidal residual currents in wide estuaries using finite element methods"

8. J. BENNER 763
"Some variants of the ICE technique"

SECTION 6 TURBOMACHINERY AND AIR FOIL FLOW

1. K.Y. SZEMA and C.H. LEWIS — 777
"Computation of three-dimensional hypersonic viscous flows over lifting bodies at high angles of attack"

2. A.S. ÜCER, I. YEĞEN and T. CETINLAYA — 789
"A finite element solution of compressible flow through cascades of turbo machines"

3. M.A. SERAG-ELDIN — 801
"Semi-elliptic computation of axi-symmetric transonic flows"

4. M.D. KIM, R.R. THAREJA and C.H. LEWIS — 813
"Hypersonic viscous flows in a streamline coordinate system"

SECTION 7 TWO PHASE FLOW

1. R.I. ISSA and A.D. GOSMAN — 827
"The computation of three-dimensional turbulent two-phase flows in mixer vessels"

2. W.H. LEE and V.L. SHAN — 841
"Numerical simulation for two-phase jet problem"

3. T. SPENDEL — 853
"Condensation heat transfer in two-phase thermo-syphons"

4. I.P. JONES and A.V. JONES — 863
"The numerical solution of simple one-dimensional multi-phase flows in shock tubes"

5. L. CONTRI and B.A. SCHREFLER — 877
"A microstructural multiphase material model for concrete"

6. L.G. NHAN and G. DE VAHL DAVIS — 893
"A numerical study of barbotage"

7. S. KIEDA, K. SUZUKI and T. SATO — 905
"Numerical study on flow behaviour and heat transfer in the vicinity of starting point of transpiration"

8. T. SIIKONEN 917
"An application of the method of characteristics for the simulation of single and two-phase flow"

9. A.C. BUCKINGHAM and W.J. SIEKHAUS 929
"Simulating interactions between turbulence and particles in erosive flow and transport"

SECTION 8 MASS TRANSPORT AND CONVECTION

1. W. SCHONAUER, K. RAITH and G. GLOTZ 943
"The selfadaptive solution of nonlinear 2-D boundary value problems in a rectangular domain"

2. Y. KRONZON, I. PARTOM and M. WOLFSHTEIN 955
"Numerical solution of the momentum equations in unsteady incompressible flow"

3. A. SATAKE and J.N. REDDY 965
"Natural convection between concentric (horizontal) circular cylinders by a penalty - finite element method"

4. K. RAMAKRISHNA, P.K. KHOSLA and S.G. RUBIN 981
"Laminar natural convection along vertical corners and rectangular channels"

5. E. LEONARDI, J.A. RIEZES and G. DE VAHL DAVIS 995
"Natural convection in a rotating annulus"

6. O. KVERNVOLD, A.K. RASTOGI and T. SONTVEDT 1007
"Diffusion controlled sea water corrosion of offshore pipelines under disbanded protecture coatings"

7. M. MIHELCIC, C. SCHROCK-PAULI, K. WINGERATH, H. WENZL, W. UELHOFF and A. VAN DER HART 1019
"Digital simulation of forced convection in a rotating fluid"

8. R.A. SKOP, M.L. MORRELL and G.A. KERAMIDAS 1031
"A comparison of explicit integration techniques for the advection equation"

9. S. PAOLUCCI and D.R. CHENOWETH 1045
"The effect of forced and free convection in the discharge of a pressurised gas"

10. L. ROBILLARD, P. VASSEUR and N.T. HUNG 1059
"Thermoconvective heat transfer in a rectangular cavity with constant wall cooling rate"

11. S.J.M. LINTHORST and C.J. HOOGENDOORN 1069
"Numerical calculation of the heat transfer by natural convection in a cubical enclosure"

12. P. LESAINT 1079
"Finite elements with upwinding for the diffusion-convection equation"

13. A. CAMPO and L.C. CHOW 1087
"Numerical solution for laminar forced convection in channels with combined wall suction and fluid axial conduction"

SECTION 9 NUMERICAL AND MATHEMATICAL CONCEPTS

1. J.J. PYUN 1099
"The second order particle-in-cell (PIC) computational method in the one-dimensional variable eulerian mesh system"

2. E.F. TORO and M.J. O'CARROL 1111
"On the nature of the stationary point in finite element determination of open channel flows"

3. M.M. GUPTA 1119
"A semi analytic method for viscous flows in the vicinity of singular corners"

4. M. LINDROOS 1129
"On a deferred convection procedure for determination of central difference solutions to the Navier-Stokes equations"

5. E.C.P. RANSOM,, J.H. BARNES and P.D. PHIPPS 1143
"An experimental investigation of the dispersion of a gas jet in a co-following stream of air"

6. G.A. KERAMIDAS and T.A. PAPATHEODOROU 1155
"Error and stability analysis of the finite element solution for the transport equation"

7. J. HAUSER, D. EPPEL nad F. TANZER 1165
"Solution of fluid flow problems by a direct spline interpolation method"

8. E-M. SALONEN, P. LEHTONEN and A. PRAMILA 1185
"Further u,υ - formulations for ideal fluid flow"

9. N.L. KALTHIA 1197
"Numerical techniques in a wedge flow of a power law fluid"

10. S.D. BAJPAI 1211
"The use of Bessel function and JACOBI polynomial in radial vibrations of a gas in an infinite cylindrical tube"

11. M.K. KADALBAJOO and K.K. BHARADWAJ 1125
"Symmetric marching technique (SMT) for the efficient solution of discretised poisson equation on non-rectangular regions"

12. R. PEYRET and B. REBOURCET 1229
"Numerical study of a jet in a stratified fluid"

13. D. AUBRY and D. ZANDVLIET 1241
"Finite element analysis of mixed convection applied to the storage of solar energy"

14. W.G. HABASHI and P.L. KOTIUGA 1253
"Finite element solution of subsonic and transonic cascade flows"

SECTION 1
GENERAL VISCOUS FLOW

ON THE MARCHING SOLUTION OF ELLIPTIC EQUATIONS IN VISCOUS FLUID MECHANICS

M. Israeli[(1)], V. Reitman[(2)], S. Solomon[(2)], M. Wolfshtein[(2)]

(1) Computer Science Dept., (2) Aeronautical Engineering Dept., Technion - Israel Institute of Technology, Haifa, Israel

SUMMARY

Marching solutions for the Poisson equation and for a set of linearized parabolized Navier-Stokes equations are considered. Enhancement of stability by high order derivatives is investigated, in order to find their influence on the accuracy and the extent of the region of integration. The results indicate that the addition of a mixed fourth order derivative damps instabilities and allows longer integration distance but may reduce the accuracy. Higher order mixed derivatives are more effective. Richardson extrapolation can be used to improve the accuracy of the stabilized solutions.

1. INTRODUCTION

In the endeavor to obtain solutions to more complex problems in fluid mechanics one is faced with the choice between general three-dimensional Navier-Stokes equations and simpler sets of equations. One important such simplification is the parabolization of the Navier-Stokes equations by neglecting the second derivatives (or diffusional fluxes) in the direction of the flow. The three-dimensional parabolic equations thus obtained are easier to solve than the full Navier-Stokes equations. However, this simplification is somewhat artificial because even the parabolized set of equations contains some ellipticity. This is best shown by the fact that the pressure is governed by an elliptic equation, be it a Poisson equation for the pressure in viscous flows, or a Laplace equation for the velocity potential in irrotational flows.

Thus, whenever the pressure field is not known beforehand, iterations on the pressure are required, and the elliptic pressure equation must be solved in each such iteration. It seems very useful to devise a method for the solution of elliptic equations by marching methods, in order to take full advantage of the parabolization of the equations. Such methods were proposed by Roache[1] and Davis and Rubin[2]. The present paper is devoted to a study of such methods and their incorporation in the parabolized equations.

2. THE PARABOLIZED NAVIER-STOKES EQUATIONS (PNS)

We are trying to find an efficient marching method for solving the PNS equations. In order to understand the difficulties we are starting with a very simple two-dimensional model.

The general equations for an incompressible fluid are:

$$\text{(continuity)} \qquad \operatorname{div}(V) = 0$$

$$\text{(Navier-Stokes)} \quad \frac{dV}{dt} + (V \operatorname{grad})V = -\frac{1}{\rho}\operatorname{grad}(p) + \frac{\mu}{\rho}\operatorname{div}\operatorname{grad}(V)$$

$$V = (u,v,w).$$

A simple steady two-dimensional model of these equations is:

$$\left.\begin{aligned} &\frac{\partial u}{\partial x} + \frac{\partial w}{\partial z} = 0 \\ &\bar{w}\frac{\partial u}{\partial z} = -\frac{1}{\rho}\frac{\partial p}{\partial x} + \frac{\mu}{\rho}\frac{\partial^2 u}{\partial x^2} \\ &\bar{w}\frac{\partial w}{\partial z} = -\frac{1}{\rho}\frac{\partial p}{\partial z} + \frac{\mu}{\rho}\frac{\partial^2 w}{\partial x^2} \end{aligned}\right\} \quad \text{The "Linearized Parabolized" set.}$$

where $\bar{w}$ is some uniform representative value of w.
We write these equations in a "marching form"

$$\frac{\partial w}{\partial z} = -\frac{\partial u}{\partial x} \tag{1.1}$$

$$\frac{\partial p}{\partial z} = -R\frac{\partial w}{\partial z} + \mu\frac{\partial^2 w}{\partial x^2} \tag{1.2}$$

$$\frac{\partial u}{\partial z} = -\frac{1}{R}\frac{\partial p}{\partial x} + \frac{\mu}{R}\frac{\partial^2 u}{\partial x^2} \quad \text{where} \quad R = \rho\bar{w}\ . \tag{1.3}$$

We consider these equations in the slab

$$0 \leqslant x \leqslant 1, \quad z \geqslant 0$$

and we prescribe

$$u(x,0),\ w(x,0),\ p(x,0),\ u(0,z),\ u(1,z),\ \frac{\partial w}{\partial x}(0,z),\ \frac{\partial w}{\partial x}(1,z).$$

By exploiting the parabolic form we construct a Crank-Nicolson scheme. First we consider a staggered grid (see Fig.1) such that we determine the values of u in the points designated by circles, the values of w, p in the points designated by crosses, and $p(x - \Delta x/2,\ z - \Delta z/2)$, $w(x - \Delta x/2,\ z - \Delta z/2)$, $u(x,z)$ are related to the same point (i,j). We associate to the system (1.1)-(1.3) the following difference scheme:

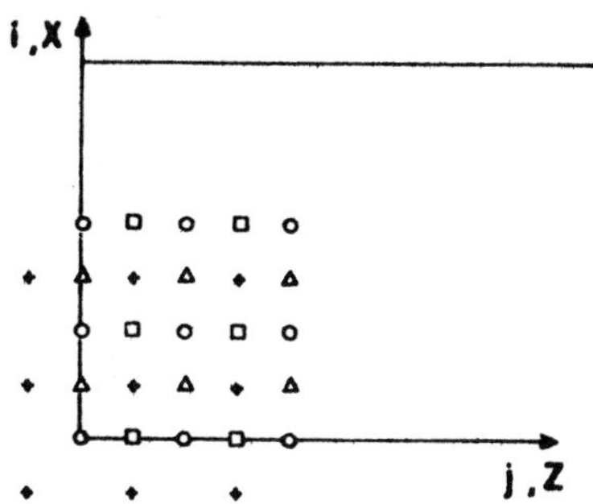

Figure 1: Finite Difference Mesh

$$\Delta_z w = -\Delta_x u$$

$$\Delta_z p = -R\Delta_z w + \mu\mu_z \delta_x^2 w$$

$$\Delta_z u = -\frac{1}{R} E_z \Delta_x p + \frac{\mu}{R} \mu_z \delta_x^2 u$$

where the first two equations are calculated at the points designated by triangles, and the third one is calculated at the points designated by squares. Explicitly, we have:

(2.1) $(w_{i,j+1} - w_{i,j})/\Delta z = -(u_{i,j} - u_{i-1,j})/\Delta x$

(2.2) $(p_{i,j+1} - p_{i,j})/\Delta z = -R(w_{i,j+1} - w_{i,j})/\Delta z +$

$+ \frac{\mu}{2} (w_{i+1,j+1} - 2w_{i,j+1} + w_{i-1,j+1} + w_{i+1,j} -$

$- 2w_{i,j} + w_{i-1,j})/\Delta x^2$

(2.3) $(u_{i,j+1} - u_{i,j})/\Delta z = -\frac{1}{R} (p_{i+1,j+1} - p_{i,j+1})/\Delta x +$

$+ \frac{\mu}{2R} (u_{i+1,j+1} - 2u_{i,j+1} + u_{i-1,j+1} + u_{i+1,j} - 2u_{i,j} +$

$+ u_{i-1,j})/\Delta x^2$.

The system of the equation (2.1)-(2.3) leads us to an obvious marching method in $j(z)$ direction. The equations (3.1), (3.2) allow us to advance explicitly from a "j" to a "j+1" level for w and p values:

(3.1) $w_{i,j+1} = w_{i,j} - (u_{i,j} - u_{i-1,j})\Delta z/\Delta x$

$$(3.2)\quad p_{i,j+1} = p_{i,j} - R(w_{i,j+1} - w_{i,j}) + $$
$$+ (w_{i+1,j+1} + \ldots + w_{i-1,j})\mu\Delta z/2\Delta x^2 .$$

For the u-values we have to solve a system of linear equations

$$(3.3)\quad u_{i+1,j+1} - (2+A)u_{i,j+1} + u_{i-1,j+1} =$$
$$(p_{i+1,j+1} - p_{i,j+1})A\Delta z/R\Delta x - u_{i+1,j} + (2-A)u_{i,j} - u_{i-1,j}$$

where $A = 2R\Delta x^2/\mu\Delta z$.

The coefficient matrix of this system is tridiagonal and can be solved by the Thomas algorithm. We have written a program which solves the discrete problem. For the boundary conditions we used an exact solution of the continuous problem (1.1-1.3) namely:

$$u = \bar{u}\, e^{-\pi z}\sin(\pi x)$$

$$w = \bar{w}\, e^{-\pi z}\cos(\pi x)$$

$$p = \bar{p}\, e^{-\pi z}\cos(\pi x)$$

with suitable constants $\bar{u}$, $\bar{w}$, $\bar{p}$. This exact solution will be used also for computing the absolute error. We have chosen $\Delta z = \Delta x = .1$. After 10 steps we reached the level $z = 1$ with an absolute-error of order $(\Delta x)^2$ for u, w and p. After more steps the errors increase exponentially. An explanation for this phenomenon is the elliptic character of the problem, namely p satisfies the Laplace equation $\partial^2 p/\partial x^2 + \partial^2 p/\partial z^2 = 0$. In a simple "computational experiment" we supplied the exact values of p and applied the parabolic equations (3.2), (3.3) for u and w. We then obtained an extremely accurate solution for u and w, and there was no evidence of error growth.

Following the above study we decided to analyze in more detail the marching methods for the Poisson equation.

3. MARCHING SOLUTION OF THE POISSON EQUATION AND THE LINEARIZED SYSTEM

Consider Poisson's equation in a rectangle:

$$(4)\quad \frac{\partial^2 p}{\partial x^2} + \frac{\partial^2 p}{\partial z^2} = f \quad \text{on } D = \{0 \leqslant x \leqslant a,\ \ 0 \leqslant z \leqslant c\}$$

$P\big|_{\partial D}$ given (∂D is the boundary of D).

A marching methods for this problem was proposed by Roache[1,3] and will be sketched here:

Let us consider a uniform grid defined by the points $(x_i, z_j) = (ih, jk)$ $i = 0, 1 \ldots M$, $j = 0,1 \ldots N$ where $h = a/M$

and $k = c/N$.
We associate to (4) the finite difference equation

(5) $(\delta_x^2 + \delta_z^2)p = f$

with

(5a) $p_{o,j}, p_{M,j}, p_{i,o}, p_{i,N}$ given.

Equation (5) leads to the following marching formula:

(6) $p_{i,j+1} = k^2 f_{i,j} + 2(1+\lambda)p_{i,j} - \lambda(p_{i+1,j} + p_{i-1,j}) - p_{i,j-1}$;
$i = 1,\ldots,M-1;\ j = 1,\ldots,N-1,\ \lambda = (k/h)^2$.

In order to apply it to marching in $z(j)$ direction we need $p_{i,1}$ $(i = 1 \ldots M-1)$. By choosing some values for $p_{i,1}$, say $p'_{i,1}$, we apply (6) and (5a) to obtain some solution $p'_{i,j}$. We define the errors $e_{i,j} = p_{i,j} - p'_{i,j}$ which satisfy the following formula:

(7) $e_{i,j+1} = 2(1+\lambda)e_{i,j} - \lambda(e_{i+1,j} + e_{i-1,j}) - e_{i,j-1}$

with

(7a) $e_{o,j} = e_{M,j} = e_{i,o} = 0$.

From Equation (7) we see that $(e_{i,1})$ and $(e_{i,N})$ are linearly related by:

(8) $e_{i,N} = \sum_n C_{i,n} e_{n,1}$ $\qquad i,n = 1,2,\ldots,M-1$

with the influence matrix C.

As $e_{i,N} = p_{i,N} - p'_{i,N}$, we can use (8) to find $e_{i,1}$, then $p_{i,1} = p'_{i,1} + e_{i,1}$ and a second application of (6) and (5a) allows us to find all $p_{i,j}$.

For the system (3) however, initial conditions for u,w,p determine the solution for all z. We note that for the full Navier-Stokes equations velocity boundary conditions are sufficient and the pressure is required only at one point. To apply the "Marching" procedure we can start with any reasonable pressure field, we also prepare the matrix C relating pressure perturbations on the initial and final line. We can avoid the large downstream pressure errors by using the matrix C to correct the initial profile. Any smooth downstream pressure can be used as the pressure level is immaterial. Derivative or extrapolation downstream pressure boundary conditions can be easily accommodated.

4. STABILITY ANALYSIS WITH PERTURBATION TERMS

We apply the Von-Neumann stability procedure. We consider the homogeneous variant of (5) with zero boundary conditions. The exact solution is

$$(9)\quad p_{i,j} = \mu^j \sin(iH)$$

where $H = m\pi h/a$, $m = 1,2,\ldots,M-1$ and μ is the amplification factor. We have:

$$\delta_z^2 p = (\mu - 2 + \mu^{-1})p/k^2 .$$

So we get the following equation for the amplification factor μ

$$(10)\quad \mu^2 - 2[1 + 2(\frac{k}{h})^2 \sin^2(\frac{H}{2})]\mu + 1 = 0$$

and clearly it has a root greater than 1

$$(11)\quad \mu = b + \sqrt{b^2 - 1}$$

with

$$b = 1 + 2(\frac{k}{h})^2 \sin^2(\frac{H}{2}) > 1 .$$

This is yet another manifestation of the fact that marching solution of an elliptic equation is unstable. It also shows that Roache's method is bound to cause exponential growth of small initial errors during the marching sweeps of Eq. (6). Therefore the total integration distance in the z-direction is limited by the resolution of the particular computer used. When the numbers are too large, Eq. (8) cannot be successfully applied.

The integration length may be increased if we choose a small $\lambda = k/h$ so that the greatest root will be closer to 1. But we choose another way: we propose to introduce a perturbation into the Eqs. (4) for which the discrete scheme will behave better. Namely, instead of Eq. (4) we consider:

$$(12)\quad \frac{\partial^2 p}{\partial x^2} + \frac{\partial^2 p}{\partial z^2} + \varepsilon h^2 \frac{\partial^4 p}{\partial x^4} - \alpha h^2 \frac{\partial^4 p}{\partial x^2 \partial z^2} = f$$

where ε, α, h are positive constants and h is the increment in x for the finite difference equation. By using the same method for the finite-difference equation:

$$(13)\quad \delta_z^2 p + \delta_x^2 p + \varepsilon h^2 \delta_x^4 p - \alpha h^2 \delta_z^2 \delta_x^2 p = f,$$

we find that the amplification factor μ is governed by:

$$\frac{\mu^2-2\mu+1}{\mu k^2} - \frac{4S^2}{h^2} + \varepsilon h^2 \frac{16S^4}{h^4} - \alpha h^2 \frac{\mu^2-2\mu+1}{\mu k^2}(-\frac{4S^2}{h^2}) = 0$$

where $S^2 = \sin^2(H/2)$.

The equation is simplified to:

$$(14) \qquad \mu^2 - 2\mu\left(1 + \frac{2\lambda^2 S^2(1-4\varepsilon S^2)}{1+4\alpha S^2}\right) + 1 = 0.$$

The amplification factor will satisfy $|\mu| = 1$ when

$$(15) \qquad -1 \leqslant 1 + \frac{2\lambda^2 S^2(1-4\varepsilon S^2)}{1+4\alpha S^2} \leqslant 1$$

or

$$-2 \leqslant \frac{2\lambda^2 S^2(1-4\varepsilon S^2)}{1+4\alpha S^2} \leqslant 0 .$$

The right hand side inequality is satisfied when

$$1 - 4\varepsilon S^2 \leqslant 0$$

i.e. $\varepsilon \geqslant \dfrac{1}{4S^2} = \dfrac{1}{4\sin^2 \frac{m\pi h}{2a}}$

and the largest value is obtained for $m = 1$, which gives

$$\varepsilon \geqslant \max_m \frac{1}{4\sin^2 \frac{m\pi h}{2a}} \approx \frac{a^2}{\pi^2 h^2} .$$

The left hand side inequality gives for such an ε,

$$1 + 4\alpha S^2 \geqslant \lambda^2 S^2 (4\varepsilon S^2 - 1) \quad \text{or}$$

$$\alpha \geqslant \max\left[-\frac{1}{4S^2} + \frac{\lambda^2}{4}(4\varepsilon S^2 - 1)\right] = -\frac{1}{4} + \frac{\lambda^2}{4}(4\varepsilon - 1)$$

i.e. $\alpha \geqslant \lambda^2 \varepsilon - \left(\dfrac{1+\lambda^2}{4}\right)$.

For $\varepsilon = \dfrac{a^2}{\pi^2 h^2}$ and $\lambda = 1$ we get $\alpha \geqslant \dfrac{a^2}{\pi^2 h^2} - \dfrac{1}{2}$.

These values may become too large to be reasonable, i.e. the scheme will be stable but the solution will be far from the solution of Equation (4), since (12) is not consistent now.

Nevertheless, for positive ε, α, the amplification factor μ will be closer to 1 and the marching process is expected to be improved.

The α term of (12) is implemented by the scheme:

$$(16) \qquad \delta_z^2 p + \delta_x^2 p - \alpha h^2 \delta_z^2 \delta_x^2 p = f.$$

This is an implicit marching scheme and requires the solution of a tridiagonal system of equations at each step.

5. HIGHER ORDER PERTURBATION TERMS AND RICHARDSON EXTRAPOLATION

With only a small extra computational effort we can replace the mixed fourth order derivative of Eq. (12) with a sum of higher order derivatives. Following is the analysis for the modified Poisson equation:

$$\frac{\partial^2 p}{\partial x^2} + \frac{\partial^2 p}{\partial z^2} - G^q \frac{\partial^{2+q} p}{\partial x^2 \partial z^q} = f . \tag{17}$$

The difference analogue of the last term is shifted backwards so that its computational star does not extend forward beyond that of the Laplacian (as in (16)).

Following our previous analysis we find for the error:

$$- 4S^2h^{-2} + (\mu-1)^2\mu^{-1}k^{-2}+4S^2G^qh^{-2}k^{-q}(\mu-1)^q\mu^{-q+1} = 0. \tag{18}$$

There are two ranges of interest; that of "long waves" when $S^2 = O(h^2)$, then $\mu = 1+O(k)$, the amplification is small and the first two terms balance, and that of short waves when $S^2 = O(1)$ then the first and last term balance and the second term is negligible. In this range we obtain the higher growth rates with the maximum at $S = 1$ (note that $S^2 < 1$):

$$((\mu-1)G/k\mu)^q = \mu^{-1} - \boxed{(h(\mu-1)/2\mu k\ S)^2} \leftarrow \text{small} \tag{19}$$

If $G = O(k)$ the perturbation term is very small, but $\mu-1 = O(1)$ and the growth rate is too large. If $G = O(1)$, $\mu-1 = O(k)$ but the equation becomes inconsistent with large truncation errors. If we require $G^q = O(k^2)$ we find: $\mu = 1+O(k^{1-2/q})$. It seems that here the choice $q= 4$ is most advantageous.

For $G^4 = k^2$ case we find $\mu = 1 + \sqrt{k}/(1-.75\sqrt{k})$ to be an approximation within 1% for the growth rate when $k/h \geqslant 1$.

It follows from the present analysis that for $q= 2$ an $O(h^2)$ perturbation term will imply $\mu-1 = O(1)$ which is not sufficient unless α is large with a corresponding large error. If we require $G^q = O(k^4)$ which is appropriate when one uses a 4-th order scheme for the Laplacian (the compact 9 point scheme is especially easy to implement in the present context) then $\mu = 1 + O(k^{1-4/q})$, here the choices $q = 6$ or 8 should be considered.

When marching over a distance z, the error growth will be $\mu^{z/k}$. Taking μ in the form $\mu = (1+(\beta k)^{1-\gamma})$ we find $\mu^{z/k} \cong e^{\beta z/(\beta k)^\gamma}$ as $k \to 0$. This implies (when (19) applies) that it pays to take a small γ for fixed k or a large k for fixed γ. These stability arguments should be subject to an accuracy investigation.

The accuracy can be improved by an a posteriori Richardson extrapolation, which is attractive for the marching methods, because it does not affect the stability. When a second order method is used for the Laplace operator we should choose an $O(k^2)$ perturbation term. The extrapolation will remove the leading terms of the truncation error as well as the leading terms in the error associated with the perturbation. When using the above mentioned compact 4-th order scheme with the corresponding $O(k^4)$ perturbation a Richardson extrapolation using a coarse net will again remove the $O(k^4)$ leading terms in the error expansion.

6. NUMERICAL TESTS AND CONCLUSIONS

The perturbation terms were implemented in a marching program and the results were compared with the exact solution

$$p(x,z) = \sin(\pi x)\,\mathrm{tgh}(\pi z) \ .$$

The ε term of (12) by itself was not satisfactory (which can be infered from 15). The α term can extend the maximum integration length z_M for which we can solve the Poisson equation with reasonable accuracy (Figure 2).

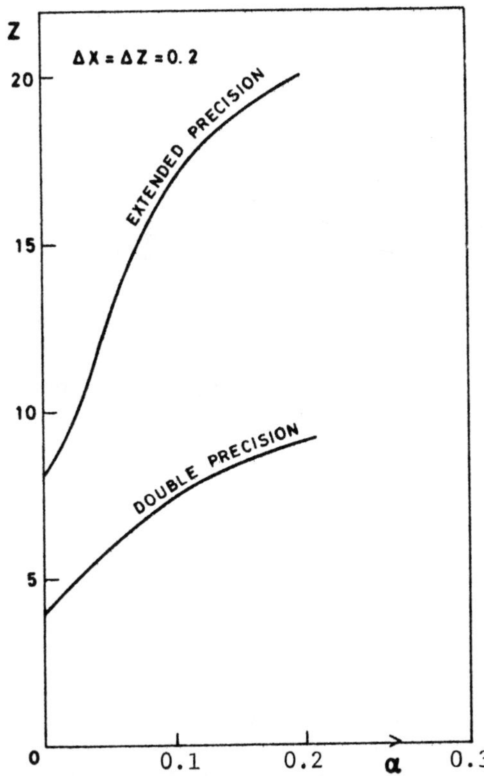

Figure 2: Maximum integration length vs damping parameter α, for double and extended precision (on IBM 370/168)

In Figure 3 we see the effect of varying $h(=\Delta x)$ and α on the accuracy of the solution for $z_M = 1$. Decreasing h improves the truncation error according to a h^2 law but eventually the round off errors become large. For fixed z_M increasing α allows us to use smaller h and get better results. Richardson extrapolation improves the results considerably. The extrapolated results are plotted against the h of the coarse grid.

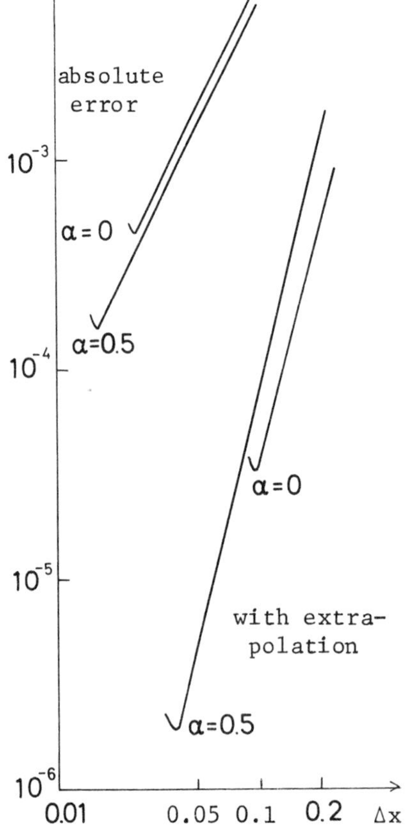

Figure 3: Maximum error in the field $0 \le x \le 1$, $0 \le z \le 1$ vs $\Delta x = \Delta z$, for $\alpha = 0$ and 0.5, before and after Richardson extrapolation (extended precision on IBM 370/168).

Figure 4 is similar to Figure 3 but with $z_M = 5$. Here we must use larger h for stability and the error increases correspondingly. On the same figure we have results for the perturbation term $-G^4 \partial^6 p/\partial x^2 \partial z^4$, (with $G^4 = \eta k^2$). The improvement is considerable as we get now with double precision results which are better than the previous results with extended precision.

We also applied Roache's method to the system (1.1-1.3) according to the explanation in paragraph 3, but without perturbation terms. The errors in the pressure were comparable to those obtained in Figures 3, and 4 with $\alpha = 0$, the errors in u and w were smaller. Work is now in progress on testing the perturbation terms and extrapolation as applied to the system (1).

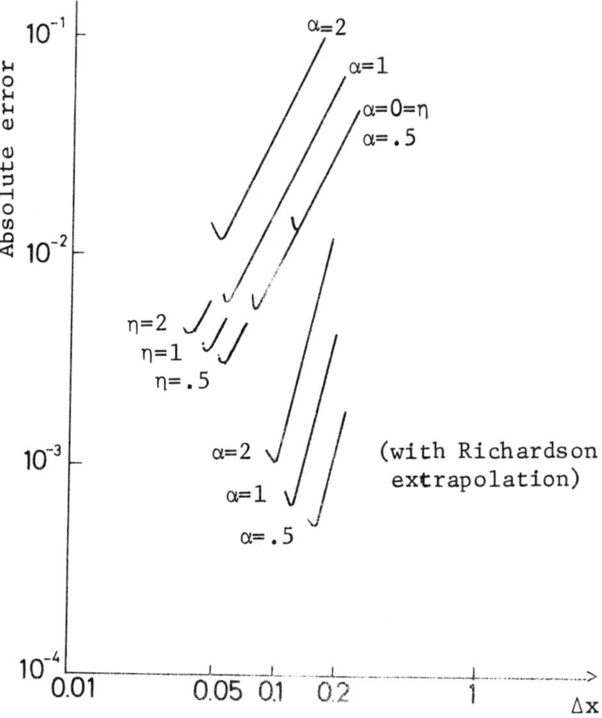

Figure 4: Maximum error in the field $0 \le x \le 1$, $0 \le z \le 5$ vs $\Delta x = \Delta z$ for various α, before and after Richardson extrapolation (extended precision on IBM 370/168). Also results with the damping term $-\eta k^2 \frac{\partial^6 p}{\partial z^4 \partial x^2}$ (double precision).

ACKNOWLEDGMENTS

The first author was partially supported by the National Science Foundation under grant NSF-ATM-78-17092.

REFERENCES

1. ROACHE, P.J. - Proc. 2nd Int. Conf. Numerical Methods in Fluid Dynamics, 1970, pp. 48-53.

2. DAVIS, R.T. and RUBIN, S.G. - Computers and Fluids, 8, 1, (1980).

3. ROACHE, P.J. - Marching Methods for Elliptic Problems: Parts 1, 2, 3. Numerical Heat Transfer, Vol. 1, (1978).

EIGENVALUE PERTURBATION OF NAVIER-STOKES EQUATIONS BY FINITE ELEMENTS

Mervyn D. Olson* and John F. Baldwin**

SUMMARY

Initial work towards formulating a finite element method for analysing the stability of laminar flow problems is presented. The starting point is the classical linear perturbation of the Navier-Stokes equations which is then discretised with finite elements. The procedure is illustrated firstly by formulating a finite element model of the one-dimensional Orr-Sommerfeld equation, which is then applied to the stability of plane Poiseuille flow. The results indicate that the method works well but it does require a fairly refined grid of elements to achieve accurate quantities. A two-dimensional representation is then formulated by perturbing the discrete stream function representation of the unsteady Navier-Stokes equations. A numerical check out of the method is then carried out on a two-dimensional discretisation of plane Poiseuille flow. Initial results indicate that the problem is very sensitive to boundary conditions and grid refinement.

1. INTRODUCTION

A variety of efficient numerical methods are now available for analysing viscous flow problems. In particular, steady two-dimensional flows governed by the Navier-Stokes equations can now be calculated routinely with finite difference or finite element methods, although the cost may be high for complicated flows or for higher Reynolds numbers. However, it is well-known that as the Reynolds number increases, laminar steady flows become unstable and go transient. The calculation of transient flows while feasible becomes very expensive. Further increases in Reynolds number lead to turbulence.

---

*Professor, **Graduate Student, Department of Civil Engineering, University of British Columbia, Vancouver, B. C., Canada.

The change from steady laminar flow to transient is considered to be an instability and when it happens is of prime importance in engineering design. So far predicting the critical Reynolds number for this instability has mainly been done on an empirical basis, although some simple configurations have been studied analytically, eg. two-dimensional Poiseuille flow.

The purpose of the present work is to develope a finite element method for analysing the two-dimensional instability problem. The linearized perturbation analysis of the Navier-Stokes equations is first reviewed. Then procedures are formulated for discretizing the perturbation equations with finite elements. As an illustration, the Orr-Sommerfeld equation for Poiseuille flow is discretised with one-dimensional finite elements. Sufficient results are calculated to compare with known results from the literature and to show convergence trends, etc. Then the general two-dimensional problem is presented. The Poiseuille problem is again analysed but now with a two-dimensional discretization. A variety of grids and Reynolds numbers are analysed to determine trends, accuracy requirements, boundary conditions and so on.

2. THEORY

The classical method of small disturbances (Cf. Schlicting [1]) is followed here. Since the present work is limited to two-dimensions, it is convenient to start with the stream function formulation of the Navier-Stokes equations

$$\frac{\partial(\nabla^2\psi)}{\partial t} = -\frac{\partial}{\partial x}\left(\frac{\partial\psi}{\partial y}\nabla^2\psi\right) + \frac{\partial}{\partial y}\left(\frac{\partial\psi}{\partial y}\nabla^2\psi\right) + \frac{1}{Re}\nabla^4\psi \tag{1}$$

where $\partial\psi/\partial y=u$, $\partial\psi/\partial x=-v$ are the x,-y velocities, respectively. Then it is assumed that there is a steady flow solution $\psi(x,y)$ of Eq. (1) that is perturbed by an unsteady component $\varepsilon(x,y,t)$. Then ψ in Eq. (1) is replaced by $\psi+\varepsilon$, and after expansion only linear terms in ε are retained. This then yields

$$\nabla^2\varepsilon_t = -\psi_y\nabla^2\varepsilon_x + \psi_x\nabla^2\varepsilon_y - \nabla^2\psi_x\varepsilon_y + \nabla^2\psi_y\varepsilon_x + \frac{1}{R_e}\nabla^4\varepsilon \tag{2}$$

where the subscripts x,y,t represent partial derivatives. Then because the coefficients are constant in time, Eq. (2) admits a solution exponential in time, say exp $[-\lambda t]$. Then Eq. (2) becomes

$$\frac{1}{R_e}\nabla^4\varepsilon - \psi_y\nabla^2\varepsilon_x + \psi_x\nabla^2\varepsilon_y - \nabla^2\psi_x\varepsilon_y + \nabla^2\psi_y\varepsilon_x + \lambda\nabla^2\varepsilon = 0 \tag{3}$$

Eq. (3) together with appropriate boundary conditions becomes an eigenvalue problem in which the aim is to find under what conditions the real part of λ changes from positive (stable perturbations) to negative (unstable).

2.1 Orr-Sommerfeld Equation.

The simplest stability problem is obtained when the steady flow is a function only of y, eg. Poiseuille flow. Then Eq. (3) reduces to (using $u = \psi_y$, $u' = du/dy$, etc.)

$$\frac{1}{R_e}\nabla^4\varepsilon - u\nabla^2\varepsilon_x + u''\varepsilon_x + \lambda\nabla^2\varepsilon = 0 \qquad (4)$$

The coefficients now are not functions of x and so the solution may be taken as

$$\varepsilon(x,y) = \phi(y)\, e^{i\alpha x} \qquad (5)$$

which represents a wave in the x direction of wave number α. Then Eq. (4) becomes

$$\frac{1}{R_e}(\phi^{iv} - 2\alpha^2\phi'' + \alpha^4\phi) + i\alpha[u''\phi - u(\phi''-\alpha^2\phi)] + \lambda(\phi''-\alpha^2\phi) = 0 \qquad (6)$$

This is the well-known Orr-Sommerfeld equation for flow stability.

2.2 Finite Element Representation

Consider the one-dimensional Orr-Sommerfeld equation first. A functional representation of Eq. (6) may be obtained in the form

$$I_1 = \frac{1}{2}\int_0^L \{\frac{1}{R_e}[(\phi'')^2 + 2\alpha^2(\phi')^2 + \alpha^4\phi^2] + i\alpha[(u''+\alpha^2 u)\phi^2 - 2u\bar{\phi}''\phi] - \lambda[(\phi')^2 + \alpha^2\phi^2]\}dy \qquad (7)$$

where the $\bar{\phi}''$ term is not varied when the first variation of I_1 is taken. Then setting $\delta I_1 = 0$ will reproduce Eq.(6) as the Euler equation of the calculus of variations. For the present application, all the boundary conditions are kinematic so they may be ignored here.

Eq. (7) is easily discretized with one-dimensional finite elements following standard procedures. Here we use a cubic element with C^1 continuity. The velocity coefficient u is assumed to vary quadratically within each element (thereby providing an exact fit for Poiseuille flow), and all matrices are calculated exactly. This then finally yields a discrete representation of the problem in the form

$$[K_1 + iK_2]\underline{\phi} = \lambda M\underline{\phi} \qquad (8)$$

where K_1, K_2, M are associated with the three square bracketted terms of Eq. (7), respectively. Note that K_1 and M are symmetric whereas K_2 is not. Further since Eq. (8) is complex, a complex eigenvalue routine had to be used for the calculations. The results follow in Sec. 3.

Now consider the two-dimensional problem of Eq. (3). This could be handled by a Galerkin or variational representation directly. However it seems easier to consider the discretized representation of the full Navier-Stokes problem of Eq. (1) and redo the perturbation directly on the discrete problem. That is Eq.(1) may be represented by the functional

$$I_2 = \iint [\frac{1}{2R_e} (\nabla^2\psi)^2 + (\overline{\psi_y \nabla^2\psi})\psi_x - (\overline{\psi_x \nabla^2\psi})\psi_y$$

$$+ \overline{\nabla\psi_t} \nabla\psi] \, dA \tag{9}$$

where the barred quantities are held fixed when the first variation is taken. Again setting the first variation $\delta I_2=0$ will reproduce Eq. (1) as Euler equation. There are questions about the boundary conditions but we ignore them for the moment. After suitable finite element discretization, we get the matrix equation

$$M \dot{\underline{\psi}} + [K + Q(\psi)]\underline{\psi} = 0 \tag{10}$$

where $\underline{\psi}$ is the vector of nodal variables. Then assuming $\underline{\psi}$ represents a steady solution about which we wish a perturbation $\underline{\varepsilon}$, we replace $\underline{\psi}$ by $\underline{\psi}+\underline{\varepsilon}$ in Eq. (10) and expand and linearize to get

$$M \dot{\underline{\varepsilon}} + [K + Q + \frac{\partial Q}{\partial \underline{\psi}}\underline{\psi}] \underline{\varepsilon} = 0 \tag{11}$$

where Q, $\partial Q/\partial\underline{\psi}$ are evaluated at the steady solution $\underline{\psi}$. Then assuming exponential time dependent $\underline{\varepsilon} = \underline{\phi} \exp [-\lambda t]$ yields the eigenvalue problem

$$[K + Q + \frac{\partial Q}{\partial \underline{\psi}}\underline{\psi}] \underline{\phi} = \lambda M \underline{\phi} \tag{12}$$

Again the question of interest is to find under what conditions the real part of λ changes sign. Note that the matrix on the left side of Eq. (12) is exactly the Jacobian matrix used in the Newton-Raphson process in solving the steady flow problem. Hence it is already available in our steady program and only the mass matrix M has to be added. Of course, the Jacobian matrix is unsymmetric and so a general eigenvalue solver for complex eigenvalues is necessary.

The solution procedure then is as follows. Our previous

program [2] using C^1 triangular elements is used to solve the steady flow for the problem of interest (for a particular finite element grid). Then the Jacobian matrix is recalculated for that solution and with the mass matrix goes to the eigenvalue solver for the stability calculation. The whole procedure is repeated for different values of Reynolds number to determine the stability boundary. Implicit in this procedure is the fact that the perturbation problem is being represented by the same C^1 triangular elements with the same gridwork as for the steady problem. Note however that the size of the perturbation problem is different because of the different boundary conditions.

3. NUMERICAL EXAMPLES

3.1 Orr-Sommerfeld: Poiseuille Flow

The one-dimensional finite element formulation of Eq. (8) was applied to Poiseuille flow in a channel of unit width. The u velocity was taken as $u=4y(1-y)$ with a maximum velocity equal to 1 at $y=0.5$. Therefore the Reynolds number based on channel width and maximum velocity was $Re=1/\nu$, where ν is the kinematic viscosity.

The unit channel was divided into equal length elements and homogenous boundary conditions of $\phi=\phi'=0$ were invoked on both walls. From solutions in the literature [3,4], the most unstable wave number was known to be about 2.0 Hence, α was set to 2.0 for all the calculations. Then eigenvalue results were calculated for a variety of Reynolds numbers and numbers of elements N.

The real part of the first eigenvalue (ie. the one with lowest real part) is plotted in Figure 1 versus Reynolds number for different N. Clearly the results are very sensitive to grid refinement, in that the low values of N did not show a change in sign at all in the range studied. This change was exhibited first by the N=10 grid but at rather low Re. Then finally the N=32 results showed it where expected. This result was verified with N=40 (not plotted).

This sensitivity is shown more clearly in Figure 2 where the real part of λ_1 is plotted versus N for a fixed Re=12,000. The behaviour at low N is very extreme with apparently about 15 elements required before final monotonic convergence is obtained. Of course, the problem is not self-adjoint, so monotonic behaviour could not be expected. Interestingly, Figure 1 is very similar to Dowell's [4] convergence plot with number of modes in a modal analysis

Our best result yielded a critical Re of 11,540 with N=40 for $\alpha=2$. This compares favourably to Grosch and Salwens' [3]

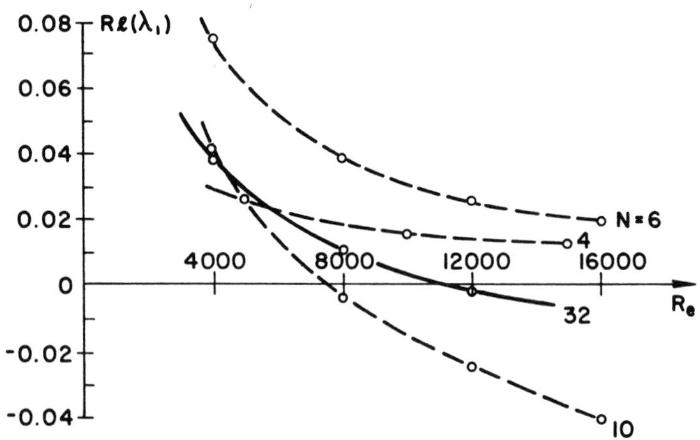

Figure 1 λ_1 versus R_e for different N

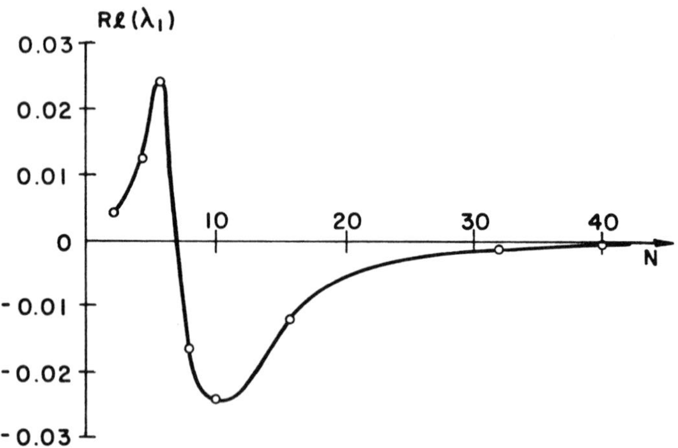

Figure 2 Convergence of λ_1 with N at Re=12000

result of 11,600 from their Table 1. Both methods gave the same imaginary part of λ_1 as 0.52 at these respective Reynolds numbers. The corresponding eigenvectors (real and imaginary parts) are plotted in Figure 3. The two results compare very well indeed thus effectively verifying each other since the methods are so vastly different.

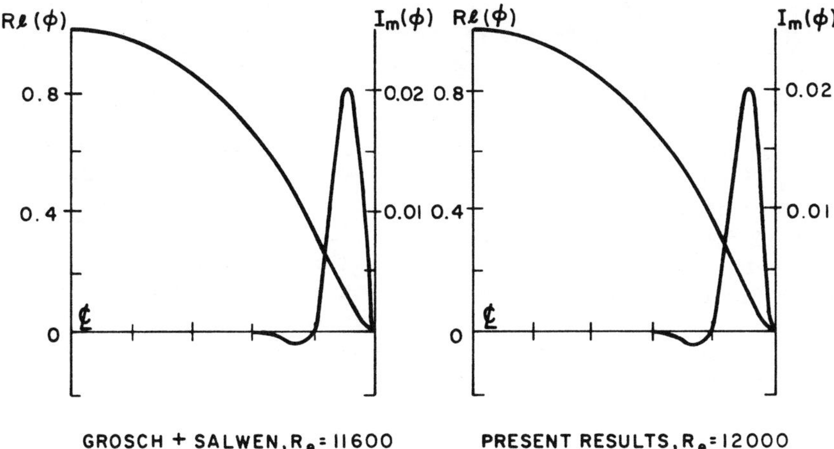

Figure 3 First Eigenvector for α=2.0

3.2 Two-Dimensional Formulation: Poiseuille flow.

A first check of the two-dimensional program was carried out by applying it to the same Poiseuille flow problem. However now the x dependence of the perturbation is not assumed a priori. That is, the problem is modelled by a two-dimensional finite element grid as eg. shown in Figure 4 (called 4x2). The x dimension was taken as 3.2 in an attempt to capture one full wave length, since from Sec. 3.1 the critical wave number is expected to be about 2.0. The perturbation boundary conditions were taken to be entirely kinematic ($\varepsilon=\varepsilon_y=0$) on the top and bottom edges, half kinematic ($\varepsilon=0$) on the left or upstream side and completely natural on the right or downstream side. The natural boundary conditions are a

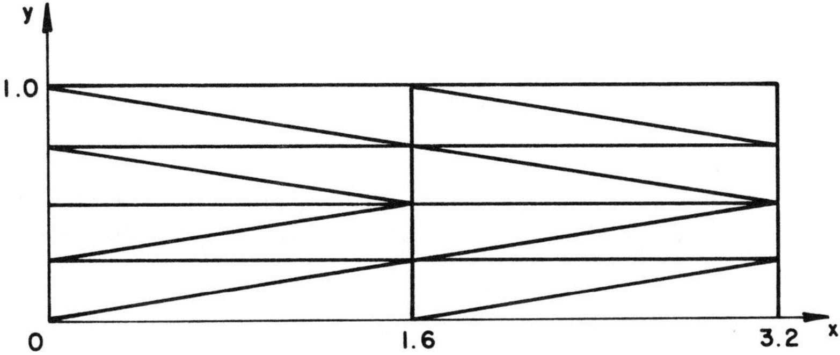

Figure 4 4x2 Grid for Poiseuille Flow

difficulty because they are not really known, whereas the ones in the program via the variational principle are zero traction which of course are incorrect for the perturbation. Eventually we hope to minimize their effect by adding more elements in the x direction.

So far calculations have been carried out with the 4x2 grid and a refined one of 8x2 for a variety of Reynolds numbers. Some preliminary results have also been obtained from a 4x4 grid of double length 6.4 in the x direction. In all cases, the exact steady solution is obtained in essentially one iteration (as expected since each element can reproduce the fully developed parabolic flow). Hence the results are only limited by the capability of the finite elements to represent the perturbation.

A sample of the eigenvalue results is shown in Table 1. They are ordered in ascending value of the real parts. Most of the eigenvalues are complex and occur in complex conjugate pairs as expected since the matrices are real. Unfortunately the lowest ones are real and negtive even down to very low Reynolds numbers (not shown). The negative value means instability. However we now feel that these are specious modes which have inadvertently crept in because of the incorrect boundary conditions. That is, the upstream condition of $\varepsilon=0$ is wrong for the travelling wave solution expected here. The proof for this is shown in Figure 5 which shows the 8x2, $\lambda=-1.47$ eigenvector for Re=12000 llustrated by equal increment contours. The plot shows that

Table 1. Eigenvalues for Poiseuille Flow

Re	FE Grid 4x2	FE Grid 4x4	FE Grid 8x2
8000	-1.51	-1.49	-1.44
	-0.39	-0.411	0.466±0.0711i
	0.161±0.0416i	0.0886±0.0516i	0.370±0.913i
	0.191±3.07i	0.151±0.516i	0.400±1.81i
12000	-1.49	-1.47	-1.47
	-0.476	-0.501	0.095±.0751i
	0.141±.0631i	0.083±0.048i	0.292±1.42i
	0.146±3.08i	0.127±3.18i	0.293±1.84i
16000	-1.48	-1.46	-1.48
	-0.52	-0.548	0.0568±.063i
	0.124±3.08i	0.080±0.046i	0.236±1.39i
	0.131±0.0703i	0.100±3.18i	0.236±1.86i

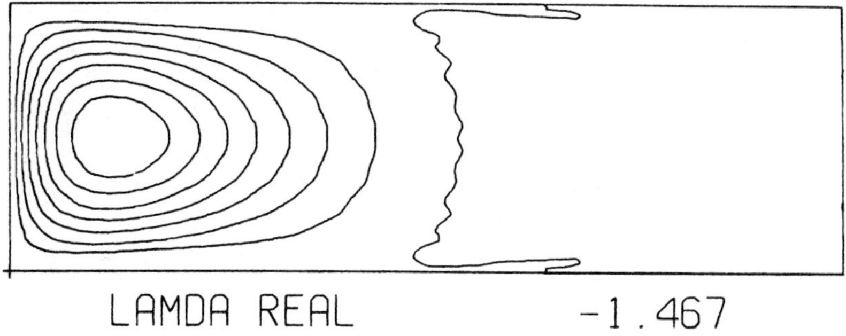

Figure 5 Re=12000 lowest mode from 8x2 grid

the mode decays rapidly to zero in the downstream direction and since λ_1 is real, it represents a non-oscillating standing wave.

So it appears that the complex eigenvalues are the ones of interest here. A study of the associated eigenvectors confirms this. For instance in Figures 6 and 7 we show the real parts of the eigenvectors associated with the first complex eigenvalues at Re=12000 from the 4x2 and 8x2 grids, respectively.

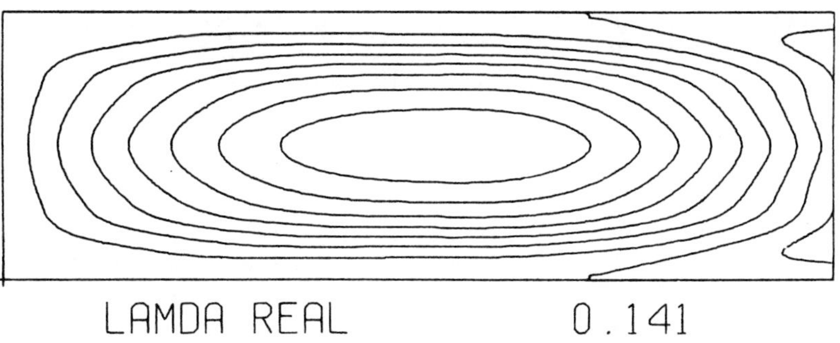

Figure 6 Real part of first complex mode from 4x2

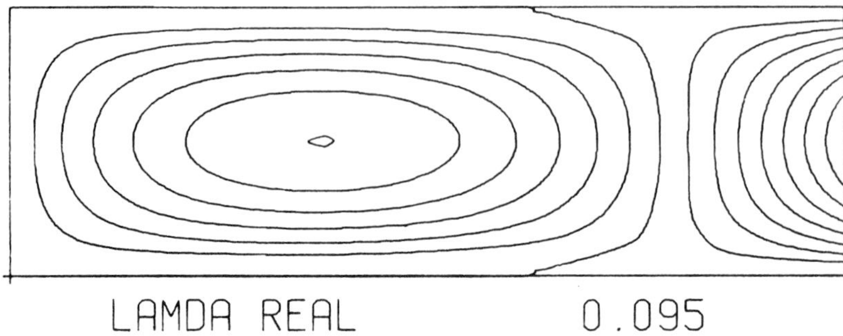

Figure 7 Real part of first complex mode from 8x2

Figure 6 shows one half wave in the x direction, whereas Figure 7 shows 1.5 half waves. Note that we expect two half waves or one full wave to be the critical travelling wave. Hence in that sense, these figures show the correct convergence with grid refinement. Unfortunately the real parts of the associated eigenvalues are still positive even at Re=16000 indicating the modes are stable. However the tabulated results indicate that these real parts are decreasing both with increasing Reynolds number and with grid refinement. These trends are in the correct direction. At this stage, we must conclude that the incorrect boundary conditions are stabilizing the flow and the flow domain is not large enough to make up for them. Unfortunatley the latter area is limited by computer capacity and cost. That is, our eigenvalue solver is of square form and finds all the eigenvalues. Present limits are about 200 variables. Alternately, the boundary conditions must be relaxed and this avenue is currently being persued.

4. CONCLUDING REMARKS

Progress has been made towards formulating a finite element method for analysing the stability of laminar flow problems. The modelling of the one-dimensional Orr-Sommerfeld equation as exemplified by the results for the Poiseuille flow problem was very successful. However it did show that a fine grid of finite elements was required to achieve good accuracy. Although not shown here, we found that a graded mesh with refinement near the walls worked best. This makes sense in that these regions contain the largest velocity gradients in the steady flow, and it is well known that it is from these gradients that the instability arises.

The two-dimensional formulation is straight forward and in principle should work well. Unfortunately there is a difficulty with boundary conditions in unconfined flows as exemplified by the application to Poiseuille flow. That is, the upstream and downstream boundary conditions are not known a priori and the imposition of incorrect ones has a marked influence on the results. It seems difficult to reduce their effect without resorting to large grids of elements. Hopefully a better solution can be found. Of course, this difficutly disappears in confined flow configurations and we hope to treat the cavity flow problem as an example of this.

5. REFERENCES

1. SCHLICHTING, H.- Boundary Layer Theory, McGraw Hill, 4th Ed., 1960, Chapt 16.
2. OLSON, M.D.- Variational-Finite Element Methods for Two-Dimensional and Axisymmetric Navier-Stokes Equations, Finite Elements in Flow Problems, Vol. 1., Eds. Zienkiewicz, O.C. et al, John Wiley, 1975.
3. GROSCH, C.E. and SALWEN, H. - The Stability of Steady and Time-Dependent Plane Poiseuille Flow, J. Fl. Mechs., Vol. 34(I), pp. 177-205, 1968.
4. DOWELL, E.H. - Non-Linear Theory of Unstable Plane Poiseuille Flow, J. Fl. Mechs., Vol. 38(II), pp. 401-414, 1969.

6. ACKNOWLEDGEMENT

The financial support of this work by the National Science and Engineering Research Council of Canada is gratefully acknowledged.

SOLUTION OF THE TIME-DEPENDENT, THREE-DIMENSIONAL INCOMPRESSIBLE NAVIER-STOKES EQUATIONS VIA FEM*

Philip M. Gresho, Stevens T. K. Chan, Robert L. Lee, and Craig D. Upson

Lawrence Livermore National Laboratory,
University of California
Livermore, California 94550

1. SUMMARY Our first venture into the area expressed in the title utilizes (we believe) the simplest techniques possible for solving the primitive ($\underline{u}$,p) equations: (1) the trilinear 8-node isoparametric "brick" element is utilized for the velocity approximation (also for temperature and concentration where relevant) with piecewise-constant approximation of the pressure, (2) the time-integration technique is basically the explicit forward Euler method, with the pressure obtained by solving the consistently derived Poisson equation. The advantages (speed and simplicity) and the disadvantages (stability limits on time-step size and the need to invoke mass lumping) of this technique are discussed and numerical examples presented.

2. INTRODUCTION This paper represents another progress report of our efforts to generate FEM simulations of small-scale phenomena in planetary boundary layer flows over complex terrain. We will describe and demonstrate our first three-dimensional, time-dependent code for solving the incompressible Navier-Stokes (and related) equations. As will become apparent, our first venture into affordable simulations of this type is based on simplicity, which involves taking certain liberties tending to cause our resulting algorithm to be (loosely) interpretable as a blend of finite elements and finite differences.

---

*This work was performed under the auspices of the U. S. Department of Energy by the Lawrence Livermore Laboratory under contract No. W-7405-Eng-48.

3. NAVIER-STOKES EQUATIONS AND SPATIAL DISCRETIZATION

The equations of motion and continuity for a constant property, incompressible Newtonian fluid are the Navier-Stokes equations, which can be written

$$\rho\left(\frac{\partial \underline{u}}{\partial t} + \underline{u}\cdot\nabla\underline{u}\right) = -\nabla p + \mu\nabla^2\underline{u} \ , \qquad (1a)$$

$$\nabla\cdot\underline{u} = 0 \ , \qquad (1b)$$

where $\underline{u} = (u,v,w)$ is the velocity, p is the pressure, ρ is the density and μ is the viscosity. Given appropriate initial conditions and boundary conditions [1], Eqn. 1 can be solved (in principle) for the velocity and pressure as functions of space and time.

If these equations are spatially discretized via the conventional Galerkin FEM, (see [1] for details), the analogous matrix system of coupled first order differential equations obtains:

$$M\dot{u} + [K + N(u)]u + Cp = f \ , \qquad (2a)$$

$$C^T u = g \ , \qquad (2b)$$

where $u(o) = u_o$ and $C^T u_o = g_o$. Here u is a global vector containing nodal values of u, v, and w, and p is a global pressure vector. The "force" vector, f, incorporates the effects of boundary conditions on the velocity which can be either essential (specified velocity) or natural (combinations of velocity derivatives and pressure; no boundary conditions are explicitly imposed on the pressure - see [1]). The vector g corresponds only to the specified velocities.

Since $C^T u = g$ for all time, $C^T \dot{u} = \dot{g}$, which can be combined with (2a) to yield the <u>consistent</u> Poisson equation for the pressure,

$$(C^T M^{-1} C)p = C^T M^{-1}[f - Ku - N(u)u] - \dot{g} \ , \qquad (3)$$

an equation implicitly contained in (2). The coefficient matrix for p is a discretized approximation to the Laplacian and Eqn. (3) is called the consistent Poisson equation in that it is the proper approximation to the continuum equation

$$\nabla^2 p = -\rho\nabla\cdot(\underline{u}\cdot\nabla\underline{u}) \ , \qquad (4)$$

(obtained via the divergence of (1a), using (1b)) in which the appropriate boundary conditions are automatically incorporated and the appropriate continuity equation is satisfied (i.e. (2a) and (2b) imply (3) and, conversely, (2a) and (3) imply (2b) - or rather, its time-derivative).

Thus far we have employed only the simplest element type in 3-D; the isoparametric 8-node brick. While this element is known to suffer from multiple "spurious pressure modes" under certain boundary conditions, we have devised an appropriate and apparently successful filter which recovers the physical pressures at the velocity nodes (see [2]).

4. TIME INTEGRATION METHOD Whereas our earlier 2-D codes were designed using a stable, implicit integration technique incorporating automatically varied time steps based on temporal accuracy [1], the size of the matrix problem in 3-D seems to currently preclude such techniques (at least for large problems - $>\sim 10^4$ nodes - and using direct solution techniques). Thus, our first 3-D code is based on the simplest possible first order explicit method (forward Euler) to the maximum extent possible (the pressure is always "implicit" in an incompressible fluid). When applied to (2a) it gives

$$u_{n+1} = u_n + \Delta t M^{-1}[f_n - Ku_n - N(u_n)u_n - Cp_n] \ , \quad (5)$$

where u_n is available and p_n is first obtained from (3), evaluated at time t_n.

The actual implementation is performed according to the following algorithm:

(1) Compute part of the acceleration vector (temporarily ignoring the pressure gradient):

$$a_n = M^{-1} \ [f_n - Ku_n - N(u_n)u_n]$$

(2) Solve the consistent Poisson equation for the pressure:

$$(C^T M^{-1} C)p_n = C^T a_n - \dot{g}_n$$

(3) Update the velocity, accounting for the pressure gradient:

$$u_{n+1} = u_n + \Delta t \ (a_n - M^{-1}Cp_n).$$

Several other features of this technique merit further comment:

(1) It appears hopeless to use the completely honest Galerkin FEM with respect to the mass matrix, M, since the inverse of the sparse and banded consistent mass matrix is a full matrix. The simplest viable approximation, and the one

we have used, replaces the consistent mass matrix by a diagonal lumped mass matrix using the row-sum technique [3], thus rendering operations with M^{-1} simple and fast. That this compromise can, however, reduce the accuracy for the simplest element was demonstrated in Gresho et al. [4]. Steady-state results are, of course, unaffected by the lumped mass approximation.

(2) We corrupt the Galerkin method in one other place - the advection terms, $\rho \underline{u} \cdot \nabla \underline{u} \sim N(u)u$. Rather than forming the true advection matrix (at element level),

$$N_{ij}^{e} = \rho \sum_{k} \int_{e} \phi_i \phi_k \underline{u}_k \cdot \nabla \phi_j \qquad (6a)$$

via full quadrature, we use the following simpler approximation:

$$N_{ij}^{e} = \rho \underline{u}_c \cdot \int_{e} \phi_i \nabla \phi_j \ , \qquad (6b)$$

where $\underline{u}_c$ is the centroid velocity (the average of the element nodal velocities). This shortcut has the effect of changing the global advection approximation from a complex stencil of 2/3 centered differences, 1/6 upwind differences, and 1/6 downwind differences (at least in 2-D on a regular grid; see [5]) to a simpler stencil of 1/2 centered, 1/4 upwind, and 1/4 downwind which is easily stated in words (for 2-D and 3-D): The average (centroid) velocity over an element is multiplied by the average gradient within the element and this result is averaged over the elements sharing the node in question. While we have not carefully evaluated the ostensible loss in accuracy caused by this approximation, preliminary tests indicate that it is small enough to make the "trick" cost effective.

(3) The "Laplacian" matrix, $C^T M^{-1} C$, is symmetric and invariant with time. Hence we have chosen to solve (3) via direct methods, using a recently developed profile (or skyline) method [6]. This matrix is formed and factored in a pre-processor code and stored on disk for later retrieval by the main code. During the time integration, each pressure update is obtained by reading the disk file and performing one forward reduction and back substitution.

(4) The explicit Euler method, while simple and fast, has one serious disadvantage (as do essentially all explicit schemes): it is unstable when the time steps are too large. The integration may become unstable if any one of several stability limits is exceeded: (1) the diffusive limit, (2) a linear advective-diffusive stability limit, and (3) certain types of nonlinear advection stability limits [7]. While we

are not able to predict these Δt limits a priori for the nonlinear FEM system on an arbitrary mesh, we have generally been reasonably successful by satisfying the following step-size restrictions, which come from analyzing the constant coefficient advection-diffusion equation via second-order centered finite differences on a grid with fixed Δx, Δy, and Δz:

(a) Diffusive limit:

$$\Delta t \leq \frac{1/2K}{1/\Delta x^2 + 1/\Delta y^2 + 1/\Delta z^2} \tag{7a}$$

(b) Linear advective-diffusive limit:

$$\Delta t \leq \frac{2K}{u^2 + v^2 + w^2}, \tag{7b}$$

where K is the diffusivity (μ/ρ for the Navier-Stokes equations). While the diffusive limit is well-known (e.g. Roache [8], there has been much confusion regarding the advective-diffusive limit (see Leonard [9] for 1-D, Hindmarsh and Gresho [10] for 2- and 3-D); in particular, it is important to note that there is no stability limit directly associated with a grid Reynolds number, $u\Delta x/K$ (it <u>may</u> exceed 2), as erroneously stated by Fromm [11] and promulgated by, e.g., Roache [8] (Dr. Roache is aware of this error, and will rectify it in his next edition). It is also noteworthy that Courant numbers do not appear directly; it turns out that satisfaction of (7) will always assure that the relevant Courant numbers (e.g. $u\Delta t/\Delta x$) are less than unity.

5. ANELASTIC EQUATIONS FOR DENSE GAS SIMULATIONS One of our current projects involves the simulation of the dynamics (gravitational spread and diffusion) of heavier-than-air gas releases into the environment. In particular, when LNG (liquefied natural gas) is spilled onto water, it quickly evaporates and a cold, dense cloud of natural gas (NG), typically methane at $\sim -160^{o}$C, is generated. Since the NG density is significantly greater than that of the ambient air (by $>\sim 50\%$), the use of the Boussinesq equations for modelling these buoyancy driven flows is felt to be inappropriate (see also [12]). We have therefore employed a new set of equations which we call the generalized anelastic equations (generalized from those originally presented by Ogura and Phillips [13]) on the premise that the important fluid dynamics are still basically incompressible and that acoustic waves are therefore unimportant and should (for computational efficiency) be filtered a priori. Our generalized anelastic model is thus given by

$$\frac{\partial(\rho \underline{u})}{\partial t} + (\rho \underline{u})\cdot\nabla\underline{u} = -\nabla p + \nabla\cdot(\rho \underline{\underline{K}}^m\cdot\nabla\underline{u}) + (\rho-\rho_0)\underline{g} , \quad (8a)$$

$$\nabla\cdot(\rho\underline{u}) = 0 , \quad (8b)$$

$$\frac{\partial T}{\partial t} + \underline{u}\cdot\nabla T = \nabla\cdot(\underline{\underline{K}}^T\cdot\nabla T) + (C_{pN} - C_{pA})/C_p\ (\underline{\underline{K}}^{\omega}\cdot\nabla\omega)\cdot\nabla T , \quad (8c)$$

$$\frac{\partial\omega}{\partial t} + \underline{u}\cdot\nabla\omega = \nabla\cdot(\underline{\underline{K}}^{\omega}\cdot\nabla\omega) , \quad (8d)$$

where p is now the pressure deviation from an isothermal system at rest (at T_0,ρ_0), T is the (mixture) deviation temperature, ω is the mass fraction of NG, $\underline{\underline{K}}$ is the (diagonal) eddy diffusivity tensor for momentum (m), energy (T), and mass (ω) transfer, $\underline{g}$ is gravitational acceleration, and C_{pN}, C_{pA} are the heat capacities of NG and air, respectively. The last term in (8c) accounts for energy transfer via interdiffusion of species with different enthalpies. We also need the mixture heat capacity,

$$C_p = \omega\ C_{pN} + (1-\omega)\ C_{pA} \quad (9a)$$

and the mixture density (ideal gas law)

$$\rho = \frac{PM}{RT_T} = \frac{P}{RT_T\left[\frac{\omega}{M_N} + \frac{(1-\omega)}{M_A}\right]}, \quad (9b)$$

where P is the total pressure (e.g. $P = p + P_0 - \rho_0 gz$, where P_0 = 1 atm.), $T_T = T_0 + T$, R is the gas constant, and M_N, M_A are the molecular weights of NG and air, respectively.

The approximate solution of (8) is obtained by a reasonably straightforward generalization of the techniques presented in the previous sections, in which T and ω are also approximated via trilinear piecewise polynomials. Other noteworthy features are: (1) in the momentum equations, the "principal variable" is the momentum ($\rho\underline{u}$) $\equiv$ $\underline{U}$ rather than the velocity, $\underline{u}$; i.e. $\underline{U} = \sum \underline{U}_j\phi_j$ etc., (2) the discretized energy and mass (NG) conservation equations are also integrated via the forward Euler method, after lumping the "mass" matrices, (3) the density (and heat capacity) is computed point-wise, at the nodes, but only after smoothing the pressure from the centroids to the nodes via element volume weighting (this also removes the spurious checkerboard pressure modes when present; see [2]), and (4) the (centroid) advection velocities in the energy and mass conservation equations are obtained by first dividing (point-wise) the momentum ($\underline{U}_i$) by the density (ρ_i) and then averaging the 8 nodal values.

Both the anelastic model and its (occasionally) ad hoc FEM approximation seem to be viable in that (1) sound waves are indeed precluded, (2) when $\Delta\rho/\rho<<1$, the results agree well with those in which the standard Boussinesq equations are employed.

6. NUMERICAL RESULTS In this section we present results from two examples. The first uses the isothermal version (Navier-Stokes) of the code and only the steady-state solution is of interest. In the second example, we use the anelastic version and present a realistic simulation of an actual LNG field experiment, including variable topography.

A. Lid-Driven Cavity The first non-trivial "verification run" was the three-dimensional lid-driven cubical cavity ($u = 1$ at all nodes on the top surface), using the graded mesh of 500 elements shown in Figure 1a (only one-half of the cavity was computed and symmetry boundary conditions were employed at $z = 0$; the node locations are shown along the edges). The code was run to steady state and the results, at Re = 100, are compared to some recent finite difference results [14, 15] in Figures 1b and 1c. These authors demonstrated adequate accuracy by comparison with others and with experiment, as well as by grid size variations. Our results appear to be nearly as good as theirs with only about 1/8 the number of nodes.

B. Simulation of a Heavy Gas Release Figure 2 shows the topography at the LNG experimental facility near China Lake, California. LNG is poured onto the spill pond (the zero elevation portion of Figure 2) from which it rapidly boils to form a dense cloud of NG which interacts with the pre-existing wind field (nominally left to right in the figure). A portion of the 6400 element discretization is shown in Figure 3, in which the mesh grading near the pond is displayed. The horizontal extent of the grid is the same as in Figure 2 (500 x 400 m) and the vertical extent is only 20 m. The discretization employs 40 elements in the flow direction (ranging from 6 to 40 m in length), 20 in the transverse direction (from 6 to 60 m), and only 8 in the vertical (from 1 m to 7 m). The total system comprises $\sim$ 45000 equations and is approximately what we consider the minimum size for meaningful simulations.

The initial conditions for the simulation were a steady isothermal wind field (3-4 m/sec at the top of the grid) and no NG vapor; we used 2 m^2/sec for the horizontal diffusivities and 0.4 m^2/sec for the vertical diffusivities, both prior to and during injection.

The known boil-off rate of LNG from water translates into a specified vertical injection velocity (w_0) of $\sim$ 0.1

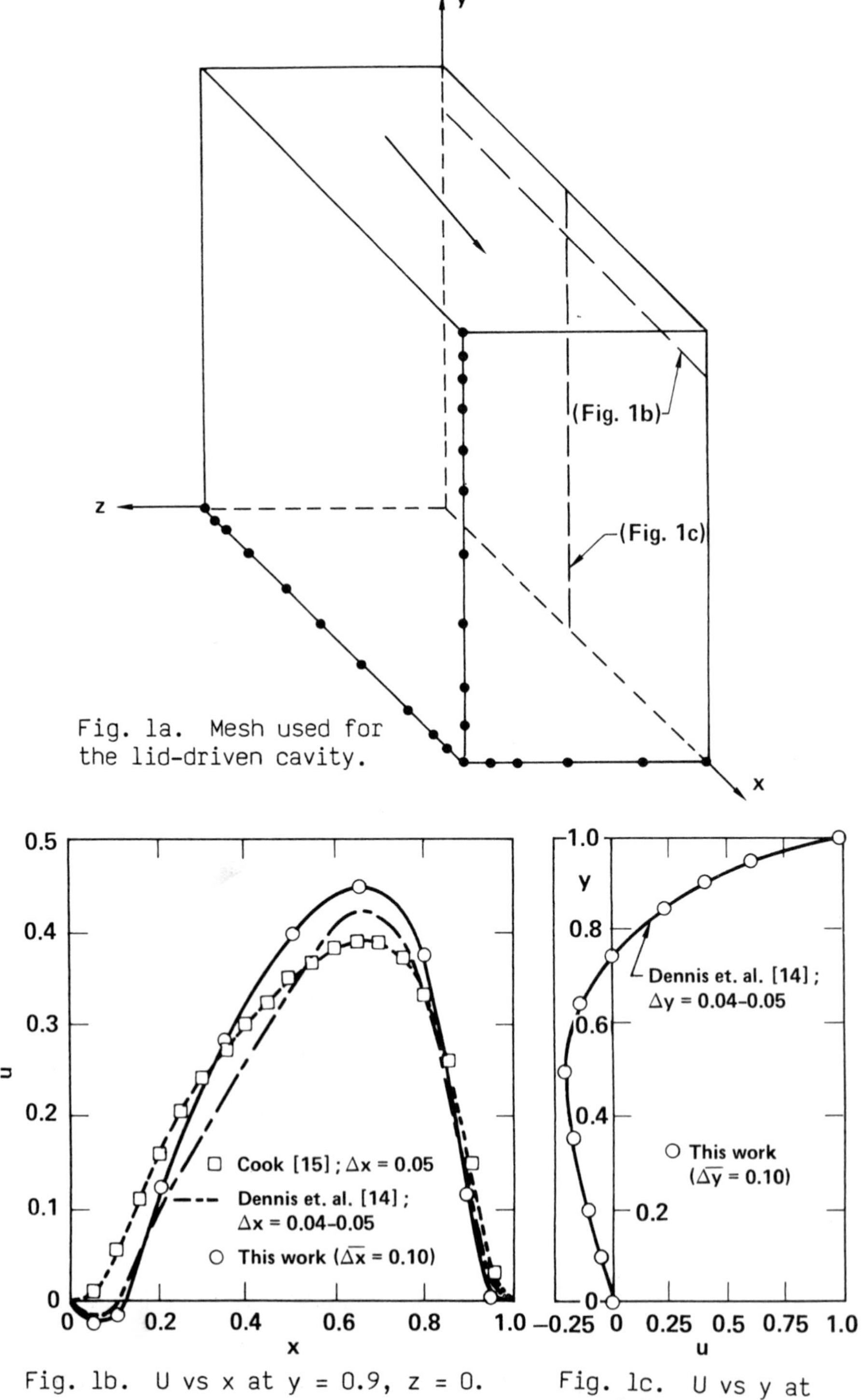

Fig. 1a. Mesh used for the lid-driven cavity.

Fig. 1b. U vs x at y = 0.9, z = 0.

Fig. 1c. U vs y at x = 0.5, z = 0.

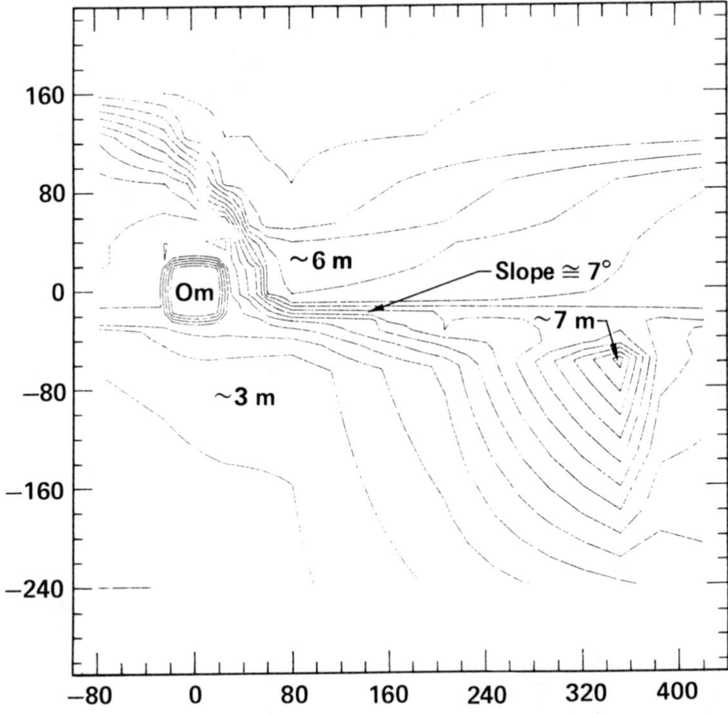

Fig. 2. Topography at China Lake LNG Spill Facility.

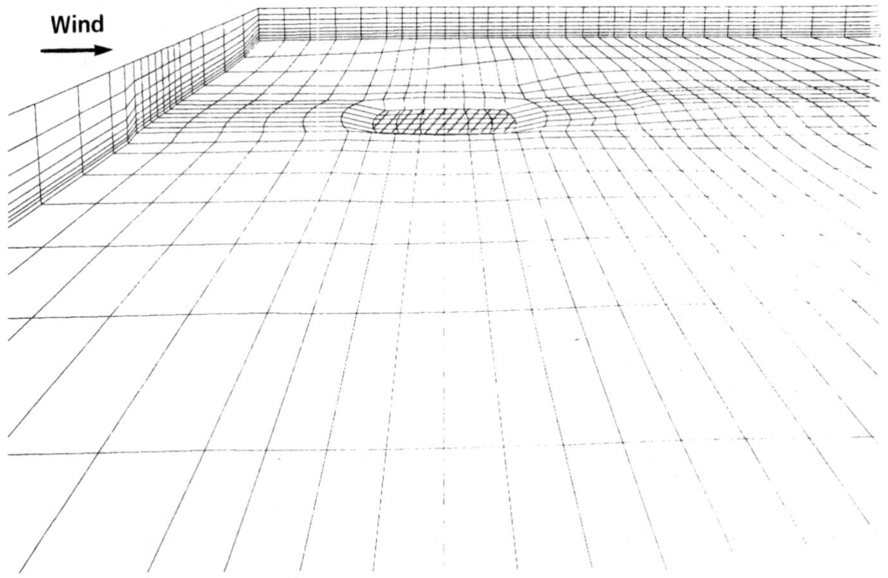

Fig. 3. View of a portion of the mesh (spill pond cross-hatched).

m/sec, which is the velocity boundary condition applied over the central 12 elements of the 30 elements comprising the spill pond. The temperature is specified at the LNG boiling temperature (-160°C) and the mass fraction boundary condition is $K^{\omega}\frac{\partial\omega}{\partial n} = w_0(1-\omega)$, which reflects a specified total mass flux of NG. Away from the source region, we used $\underline{u} = \underline{0}$ and $\partial T/\partial n = \partial\omega/\partial n = 0$ at the ground. The remaining boundary conditions employed were: specified $\underline{u}$, T, ω at the inlet plane, natural boundary conditions at the outlet, and symmetry conditions at the top and two lateral sides.

A sample of the numerical results is shown in Figures 4-6. The horizontal velocity and NG concentrations at the first "horizontal" plane of node points above terrain are shown in the vicinity of the pond at t = 100 sec in Figure 4. Both the gravitational spread (even in the upwind direction, which is also observed in the field) and the effect of the terrain shape on the motion of the cloud are clearly apparent (the NG is flammable in the range of 5-15%). Figure 5 shows results at the same time for a longitudinal cut through the pond centerline and along the nominal wind direction; again the influence of the dense cloud is apparent, this time in the form of a sort of rollup vortex in the upwind direction caused by the sharp horizontal density gradient at the cloud front. Finally, Figure 6 shows the third planar projection, this time transverse to the nominal wind and at 60 m downwind from the pond center (see Figure 2); this view is looking toward the pond so that the wind is nominally blowing toward you. The t = 0 vectors in Figure 6a show the considerable effect of the topography on the wind field in the absence of NG while those in Figure 6b again show the significant influence of buoyancy (via vorticity generation) on the flow field.

Overall, this early simulation agreed only qualitatively with the measured experimental results, which showed a thinner, more persistent cloud. Apparently we must use a much smaller value of vertical diffusivity (in lieu of a good turbulence model), at least in the NG cloud itself.

7. DISCUSSION The stability-induced time step (estimated at 0.4 sec) for the LNG simulation, combined with significant I/O (input/output) costs associated with this early version of the code, resulted in a fairly high total computational cost, with each time step requiring ∿ 7 sec CPU and ∿ 13 sec I/O on the CRAY-1. We are currently working toward reducing these costs in several ways, only one of which we mention here: 1-point quadrature. The simulation reported above used 2 x 2 x 2 Gauss-Legendre quadrature for all element matrices. This resulted in a disk file of about 2 mil-

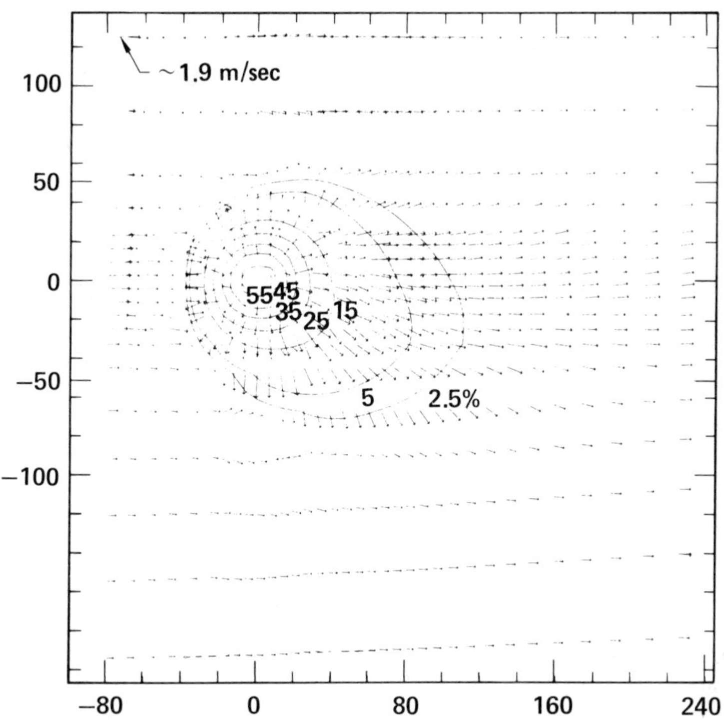

Fig. 4. Horizontal velocities and NG concentration at 1 m above terrain; t = 100 sec after the start of NG injection.

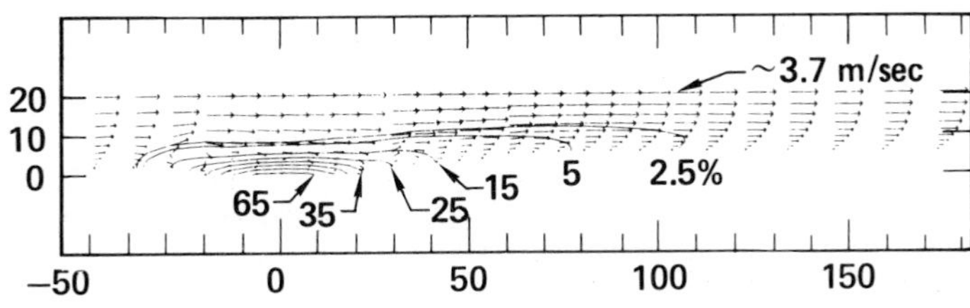

Fig. 5. Velocity and concentration in a longitudinal plane through the pond centerline; t = 100 sec.

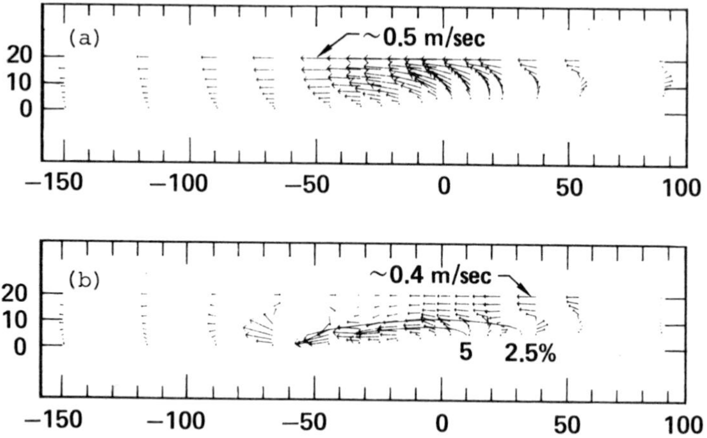

Fig. 6. Velocity and concentration in a transverse plane about 60 m downwind from the pond; (a) t = 0, (b) t = 80 sec.

lion words of storage for the element matrices (another million words were required to store the factored Laplacian matrix for pressure). If we can achieve reasonable accuracy via (vectorized) 1-point quadrature (see [4]), we will reduce the overall cost significantly.

8. SUMMARY/CONCLUSIONS We have designed and demonstrated a viable technique for approximating the solution of the time-dependent Navier-Stokes (and generalized anelastic) equations in three dimensions via the finite element method.

9. REFERENCES

1. GRESHO, P., R. L. LEE and R. L. SANI - On the Time-Dependent Solution of the Incompressible Navier-Stokes Equations in Two- and Three-Dimensions - in C. Taylor and K. Morgan, Recent Advances in Numerical Methods in Fluids, Pineridge Press Ltd., Swansea, U. K. (1980).

2. SANI, R. L., P. M. GRESHO, R. L. LEE, and D. F. GRIFFITHS - The Cause and Cure (?) of the Spurious Pressures Generated by Certain FEM Solutions of the Incompressible Navier-Stokes Equations - Int. J. Num. Meth. in Fluids, 1, No. 1 (to appear, 1981).

3. LEE, R. L., P. M. GRESHO, and R. L. SANI - Smoothing Techniques for Certain Primitive Variable Solutions of the Navier-Stokes Equations - Int. J. Num. Meth. Eng., 14, 1985-1804 (1979).

4. GRESHO, P. M., R. L. LEE, and C. D. UPSON - FEM Solution of the Navier-Stokes Equations for Vortex Shedding Behind a Cylinder: Experiments with the Four-Node Element - in Finite Elements in Water Resources, Proceedings of Third Inter. Conf., Univ. of Miss., USA (1980).

5. GRESHO, P. M. and R. L. LEE - Don't Suppress the Wiggles - They're Telling You Something! - Comp. and Fluids (to appear, 1981).

6. TAYLOR, R. L., E. L. WILSON and S. J. SACKETT - Direct Solution of Equations by Frontal and Variable Band, Active Column Methods - to appear in Proc. U.S. European Workshop on Nonlinear Finite Element Analysis in Structural Mechanics, Bochum, W. Germany, (July, 1980).

7. HIRT, C. W. - Heuristic Stability Theory for Finite-Difference Equations - J. Comp. Phys., 2, 339-355 (1968).

8. ROACHE, P. - Computational Fluid Dynamics, 2nd Ed., Hermosa Press, Albuquerque, New Mexico (1976).

9. LEONARD, B. P. - Note on the Von Neumann Stability of the Explicit FTCS Convective Diffusion Equation - Appl. Math. Mod., 4, p. 401 (October, 1980).

10. HINDMARSH, A. and P. GRESHO (in preparation).

11. FROMM, J. - Methods in Comp. Phys., 3, p. 345 (1964).

12. DALEY, B. and W. PRACHT - Numerical Study of Density - Current Surges - Phys. of Fluids, 11, 1, p. 15 (1968).

13. OGURA, Y. and N. PHILLIPS - Scale Analysis of Deep and Shallow Convection in the Atmosphere - J. Atm. Sci., 19, p. 173 (1962).

14. DENNIS, S., D. INGHAM, and R. COOK - Finite-Difference Methods for Calculating Steady Incompressible Flows in Three Dimensions - J. Comp. Phys., 33, p. 325 (1979).

15. COOK, R. N. - Ph.D. Thesis - Univ. Western Ontario, London, Ontario, Canada (1976).

ON THE SOLUTION OF THE TIME-DEPENDENT INCOMPRESSIBLE NAVIER-STOKES EQUATIONS VIA A PENALTY GALERKIN FINITE ELEMENT METHOD

Robert L. Sani and Brian E. Eaton
CIRES/NOAA and Department of Chemical Engineering
University of Colorado
Boulder, Colorado 80309/USA

and

Philip M. Gresho, Robert L. Lee and Stevens T. Chan
Lawrence Livermore National Laboratory
University of California
Livermore, California 94550/USA

1. SUMMARY The existence of an initial non-physical transient when using the penalty method in modeling the time-dependent incompressible Navier-Stokes equations is investigated theoretically and demonstrated numerically using the Galerkin finite element technique. A stable, variable step time integration scheme which can "overlook" the initial non-physical transient while using "reasonable-sized" time steps is described. Numerical examples illustrating the time integration scheme and concomitantly the difference in transient response of an incompressible fluid and its slightly compressible (penalty) analog are presented.

2. INTRODUCTION One of the methods which has been employed in modeling the time-dependent incompressible Navier-Stokes (and Boussinesq) equations is the penalty method in which the solenoidal constraint on the velocity field is weakened to $p = -\lambda\nabla\cdot u$ where $\lambda \gg 1$ is the penalty parameter. (See, for example, [1] and [2].) As a consequence, the infinite speed of propagation of a pressure signal in an incompressible fluid is replaced by an $O(\lambda)$ "diffusive" speed of propagation - a property which can lead to a non-physical transient. Since the penalty parameter is large, the initial non-physical transient is fast, occurring on a time scale of $O(1/\lambda)$, and reflects the slight compressibility inherent in the weakened solenoidal constraint. While the analogy with a "special" slightly incompressible fluid has been noted by others, it appears that the associated non-physical transient

has received little attention in the literature. Here we focus on the existence, characterization and remedy of the initial non-physical portion of the transient associated with a well-posed (in the incompressible sense) problem and also illustrate the physically uninteresting transient associated with an ill-posed problem.

3. BASIC EQUATIONS In the penalty technique the basic incompressible Navier-Stokes equations are replaced by a regularly perturbed system;

$$\frac{\partial \underline{u}}{\partial t} + \underline{u}\cdot\nabla\underline{u} = -\nabla p + \nu(\nabla^2\underline{u}+\nabla(\nabla\cdot\underline{u})), \tag{1}$$

$$\lambda\nabla\cdot\underline{u} = -p, \text{ where } \lambda >> \nu, \tag{2}$$

plus an initial condition, $\underline{u}\,(\underline{x},o) = \underline{u}_0$ and appropriate boundary conditions. It is noteworthy here that p is the pressure divided by the density and the initial velocity field, $\underline{u}_0$, which is restricted to be solenoidal in the incompressible case, seemingly is not so constrained here; however, as one of the numerical examples will clearly illustrate, a violation of this condition generally leads to a transient of little physical interest.

The characterization of the non-physical transient is best achieved by an analysis of the pressure equation implicitly associated with (1) and (2),

$$\frac{\partial p}{\partial t} + \underline{u}\cdot\nabla p = \lambda[(1 + \frac{2\nu}{\lambda})\nabla^2 p + \frac{p^2}{\lambda^2} + 2\left(\frac{\partial u}{\partial y}\frac{\partial v}{\partial x} - \frac{\partial u}{\partial x}\frac{\partial v}{\partial y}\right)], \tag{3}$$

which is obtained by forming the divergence of (1), using (2) where appropriate. Thus, in general the pressure satisfies a diffusion-dominated advection-diffusion equation with "velocity-and pressure-generated sources and sinks," and in particular, in the Stokes flow case (retaining only local acceleration terms) it satisfies

$$\frac{\partial p}{\partial t} = \lambda(1 + \frac{2\nu}{\lambda})\nabla^2 p, \tag{4}$$

a transient diffusion equation. The boundary conditions associated with (3) or (4) are given by (2) and are, in general, time-dependent; hence, the solution of (3) (or(4)) is closely coupled to the velocity solution, from (1). Since $\lambda(1 + 2\nu/\lambda) \sim \lambda >> 1$, singular perturbation analysis reveals that there is an initial fast transient with a time scale of $O(1/\lambda)$ associated with both (3) and (4) and represents a non-physical response induced by the weakened continuity constraint. When this spurious transient is completed, the pressure is in a quasi-equilibrium state, in which the left sides of (3) and (4) are replaced by zero. In contrast to (3) (and (4)), the pressure associated with an incompressible fluid satisfies, in general,

$$\nabla^2 p = -2\left(\frac{\partial u}{\partial y}\frac{\partial v}{\partial x} - \frac{\partial u}{\partial x}\frac{\partial v}{\partial y}\right) \quad , \tag{5}$$

and in Stokes flow,

$$\nabla^2 p = 0; \text{ both (5) and (6) are true for } \underline{\text{all}} \text{ time.} \tag{6}$$

4. GALERKIN FINITE ELEMENT EQUATIONS The finite element spatial discretization of (1)-(2) is performed by the Galerkin method and leads to a system of coupled ordinary differential equations,

$$\underline{\underline{M}}\dot{\underline{U}} + (\underline{\underline{K}} + \lambda\underline{\underline{B}})\underline{U} + \underline{\underline{N}}(\underline{U})\underline{U} = \underline{f}, \tag{7}$$

$$\underline{U}(o) = \underline{U}_o$$

for the vector of nodal velocities $\underline{U}(t)$. The associated vector of nodal pressures is obtained at any time from $\underline{P} = \lambda\underline{\underline{Q}}^{-1}\underline{\underline{C}}^T\underline{U}$, which is obtained from the discretized form of (2). The reader is referred to [3] for a detailed definition of the various finite element matrices except $\underline{\underline{Q}}$ which is defined here as the "pressure mass matrix" and $\underline{\underline{B}} \equiv \underline{\underline{C}}\underline{\underline{Q}}^{-1}\underline{\underline{C}}^T$. It is noteworthy, firstly, that the penalty technique is normally implemented using a C^{-1} (discontinuous) pressure approximation and a conforming C^0 velocity approximation which allows $\underline{\underline{B}}$, the penalty matrix, to be efficiently evaluated in many cases (but not all - see [4]) by reduced quadrature ([1],[2]) and allows efficient processing of the nodal pressure vector; and secondly, that certain element choices lead to unacceptable solutions ([5],[6]).

Since the penalty parameter, λ, is very large, it is clear that an explicit time integration algorithm applied to (7) (or its Stokes flow equivalent, which lacks the $\underline{\underline{N}}$-term) leads to time steps (Δt) of $O(1/\lambda)$ for stability - an unacceptable restriction imposed by the non-physical part of the transient. Even implicit algorithms which are non-dissipative can be severely limited in time step size; for example, if the trapezoid rule algorithm is chosen, undamped oscillations will occur if $\lambda\Delta t$ is much greater than unity. Consequently, the time integrator employed herein was designed with the capability of starting out in a constant, or variable, Δt backward Euler (BE) mode (an implicit, first-order, dissipative scheme) and switching to a variable Δt BE mode or trapezoid (TR) mode (an implicit, second-order, non-dissipative scheme) after a preassigned number of time steps. Using this algorithm one can either follow the <u>entire</u> transient to within a prescribed local time truncation accuracy via a variable step method exclusively, or follow only the <u>physical</u> part of the transient by initially taking a few BE steps at a constant $\Delta t >> (1/\lambda)$ and then switching to variable step BE or TR for efficiency. The initial non-physical portion of the transient appears as "high frequency signals" to the BE scheme and is therefore quickly damped by the numerical diffusion associated with this scheme. Consequently, this stable time integration strategy can "overlook" the initial non-physical transient while using reasonable-sized time steps, i.e., time steps not $O(1/\lambda)$, and can be up to second order accurate in the remaining (physical) portion of the transient while automatically varying Δt, based solely on temporal accuracy requirements. These features are illustrated in the next section.

5. NUMERICAL EXAMPLES We demonstrate the nature of the penalty transient with a simple but demonstrative example - developing linear Poisieulle flow, using two different boundary conditions, one legal and the other illegal (from the viewpoint of an incompressible flow simulation). The domain (see Fig. 1) is one unit high (H) and two units long and the

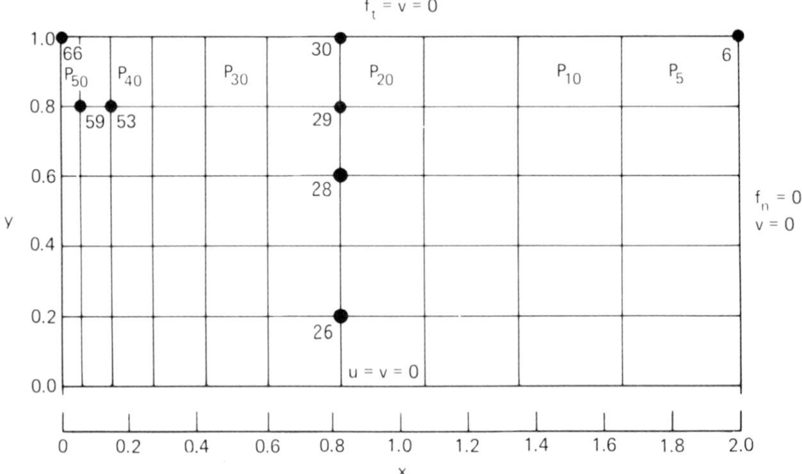

Fig. 1. Domain, FE mesh, and boundary conditions.

boundary conditions (BC's) at the inlet are: $v = 0$ and either $f_n = 4$, (the legal BC - an imposed normal force) or $u(y) = 2y(1-0.5y)$, (the illegal BC). The remaining BC's are shown in Fig. 1. The density and viscosity are 1.0 and the exact steady solution is $u(y) = 2y(1-.5y)$, $v = 0$, $p = 4(1-x/2)$. The domain is discretized using 10 x 5 four-node bilinear elements (with piecewise constant pressure) and is graded in the flow direction with the smallest elements at the inlet.

5.1. The Physical Solution Fig. 2 shows the velocity history at three y-locations and the timestep history (TR with $\varepsilon = .001$ required only 16 time steps; ε is the time truncation error control parameter, see [3]) using mixed-interpolation and (of course) the legal inlet BC. Here, $v = 0$ and $p = 4(1-x/2)$ for all time, and u is a function of y and t only. The physical transient is seen to be a simple, unidirectional and monotonically increasing velocity field with a time scale of $H^2/\nu = 1.0$.

5.2 The Penalty Transient Using the Legal BC For this case, the rapid penalty transient occurs on a time scale of

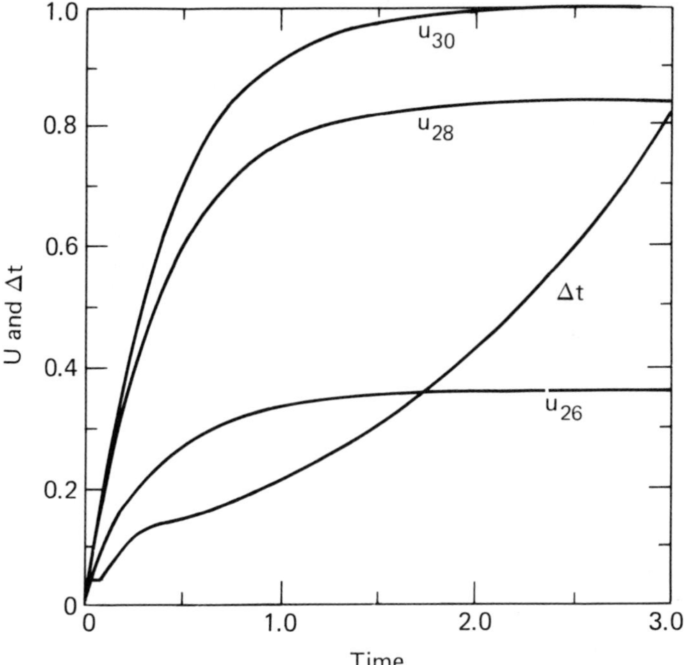

Fig. 2. Time histories and Δt vs t for the physical transient.

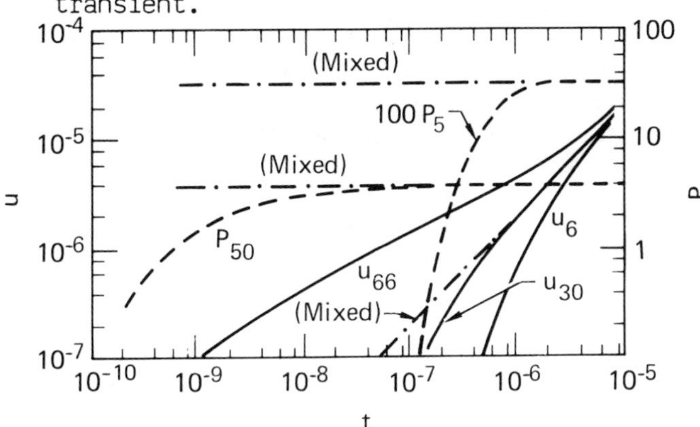

Fig. 3. Selected time histories during the penalty transient.

$O(1/\lambda) = 10^{-6}$ and several representative time history results are shown in Figs. 3 and 4. Figure 3 shows the horizontal velocities for three nodes at the top of the grid: the inlet (U_{66}, at x = 0), about half-way through the channel U_{30}, at x = .83), and the outlet (U_6, at x = 2). Also shown is the result using mixed interpolation (for any x), and it is seen that the penalty results converge to those using mixed interpolation by $t \simeq 10^{-5}$, at which time the penalty transient is essentially completed. Also shown are the pressures at the inlet (P_{50}, at x = .031, y = .9) and outlet (P_5, at x = 1.83, y = .9); these are also compared

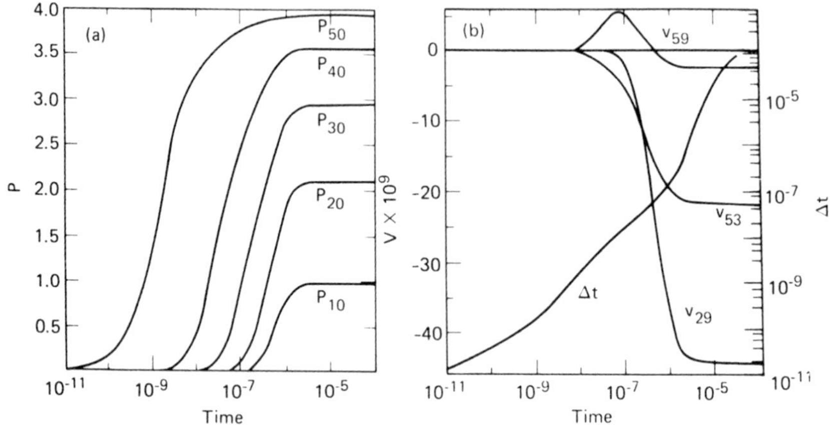

Fig. 4. Additional time histories, and Δt vs t, during the penalty transient.

with the corresponding mixed interpolation results which are invariant in time. The progression of the penalty transient through the grid is clearly displayed in Fig. 4 in the form of time histories of several vertical velocities and pressures. Again, by $t \simeq 10^{-5}$, the penalty transient is completed, the pressure is linear in the x-direction, and the vertical velocities are very small ($O(1/\lambda)$). Also shown is the timestep history - for this run we used the BE method with $\varepsilon = .001$; Δt grows by over seven orders of magnitude during this rapid transient and a total of 80 time steps were required to accurately follow the (spurious) solution.

Further insight into the nature of the penalty transient is revealed by displaying the velocity vectors and pressure contours; see Fig. 5. The solution goes through a clearly discernable "compression-wave transient" (not to be construed as physical, since the penalty "equation of state" is non-physical), which has progressed about half-way through the grid by $t \simeq 10^{-7}$. The results shown in Fig. 5c are essentially the same as those from a mixed interpolation calculation; hence, a continuation of the simulation will produce results which basically duplicate those of Fig. 2.

If the penalty transient is of no real interest (which is ostensibly the usual situation), it need not be computed - but <u>only</u> if a sufficiently dissipative time integration scheme is employed. We have done this using the BE method with fixed time steps (and no predictor equation) of a size large compared to the spurious time scale ($O(1/\lambda)$), but sufficiently small to accurately follow the physical transient. For the above example, the penalty transient is successfully damped out using just one timestep with $\Delta t = .01$; after this, we can switch to the automatic Δt option (using

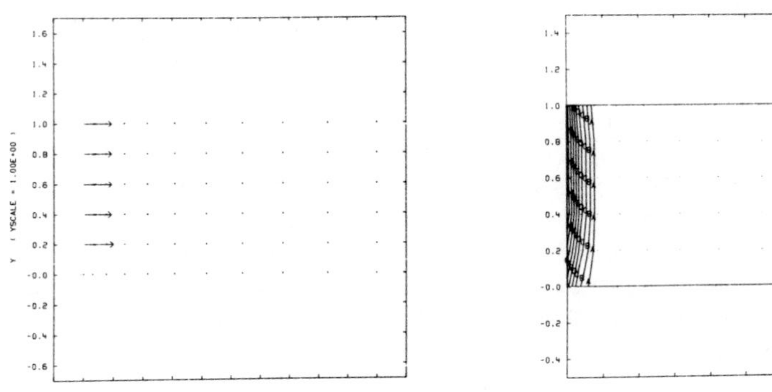

(a) $t = 10^{-9}$, $U_{max} = 1.2 \times 10^{-7}$, isobar $\Delta p = 0.16$.

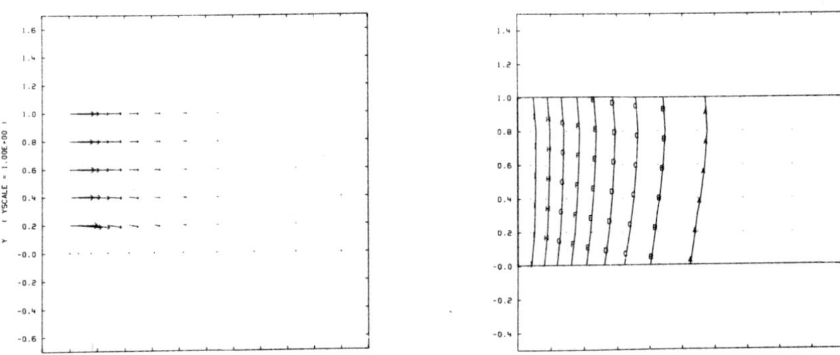

(b) $t = 2.5 \times 10^{-7}$, $U_{max} = 2.6 \times 10^{-6}$, isobar $\Delta p = 0.38$.

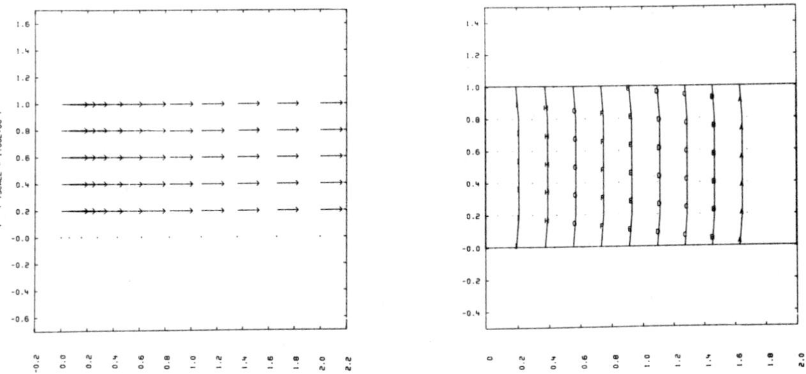

(c) $t = 1.3 \times 10^{-5}$, $U_{max} = 3 \times 10^{-5}$, isobar $\Delta p = 0.36$.

Fig. 5. Velocity vectors and isobars during the penalty transient.

either BE or TR) to efficiently compute the physical transient (a la Fig. 2).

5.3 <u>The Penalty Transient Using the Illegal BC</u> The specified inlet velocity BC (with zero initial velocity), while illegal for an incompressible flow (or using mixed interpolation), is not illegal when the penalty method is used. The ensuing penalty transient, although perhaps interesting, is nevertheless of no physical value. The next simulation was made to illustrate this situation, with the results summarized in Figs. 6 and 7. The time histories of one velocity

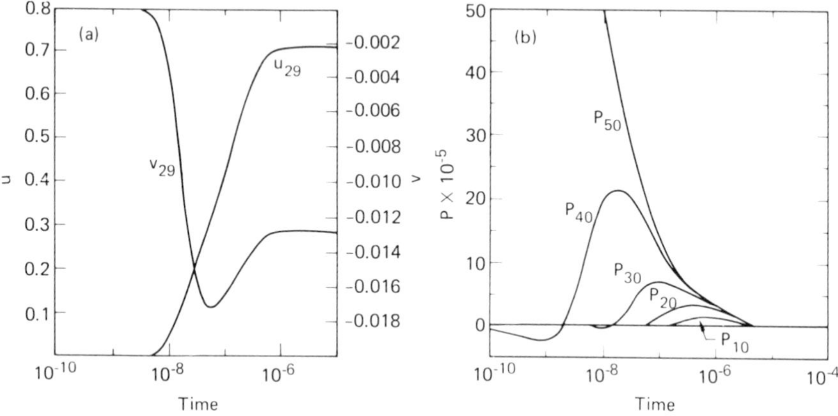

Fig. 6. Selected time histories during the spurious penalty transient.

node and several pressure "nodes" are shown in Fig. 6. (Δt vs t behaved similarly to that in Fig. 4b; BE required 94 time steps and TR required 43, both using ε = .001). The inlet pressure corresponding to this BC will actually tend to infinity at t = 0 as $\Delta x \to 0$

$(\lambda\frac{\partial u}{\partial x} \to \infty)$, but will decrease rapidly for t > 0 as

the "compression wave" moves through the fluid. These inordinately large pressures are, of course, a simple consequence of the penalty equations and the imposed BC. The initially negative values of the pressures are a consequence of using the consistent mass matrix; these wiggles are suppressed and the pressures always positive if mass lumping is invoked (see [7]). The apparent steady-state for t > ∿ 10^{-5} simply heralds the completion of the penalty transient and the overall solution is still quite far from that of steady Poisieulle flow (e.g., U_{29} is 0.96 for Poisieulle flow). The penalty transient is again seen quite clearly in the form of vectors and isobars - Fig. 7. The tremendous pressure gradients present at small time are quickly reduced as the flow accelerates through the grid. The final result (Fig. 7c) shows the now incompressible (to $O(1/\lambda)$) flow field at the conclusion of the penalty transient. This

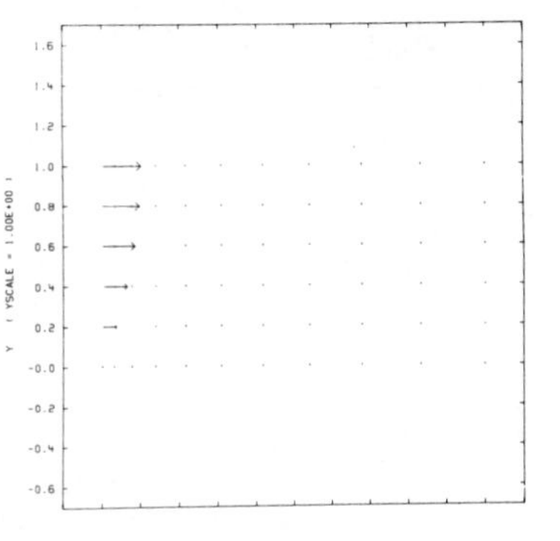
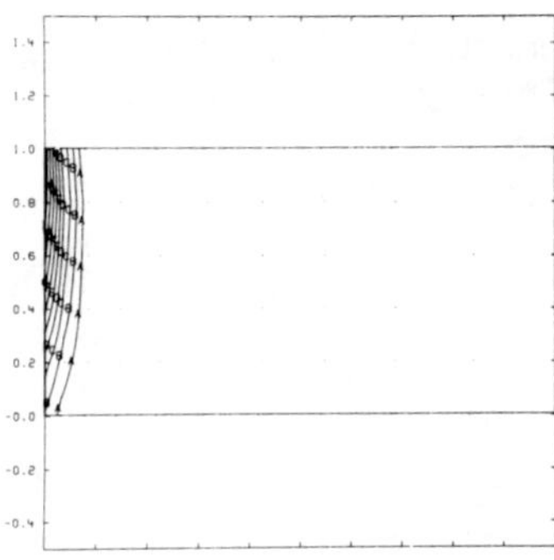

(a) $t = 10^{-10}$, $U_{max} = 1.0$, isobar $\Delta p = 1.5 \times 10^6$.

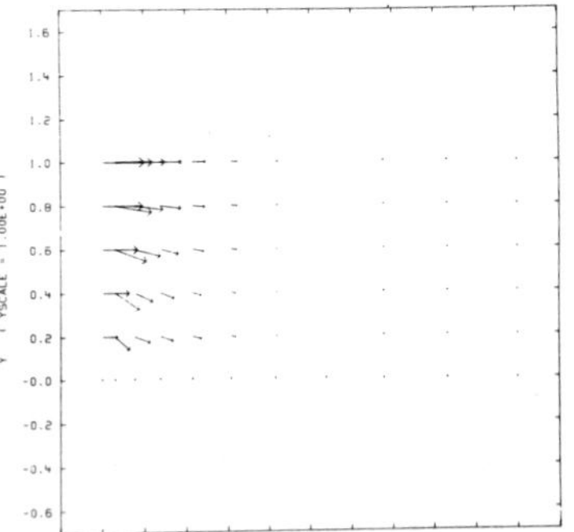
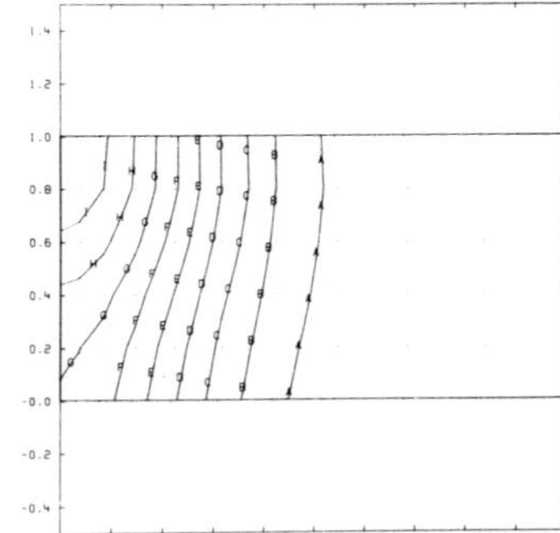

(b) $t = 10^{-7}$, $U_{max} = 1.0$, isobar $\Delta p = 1.37 \times 10^5$.

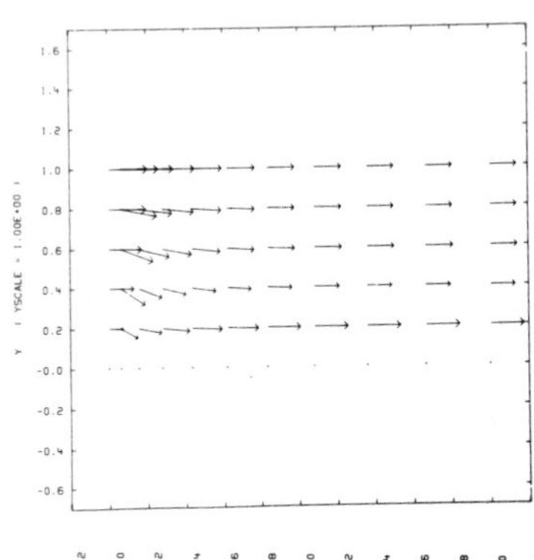
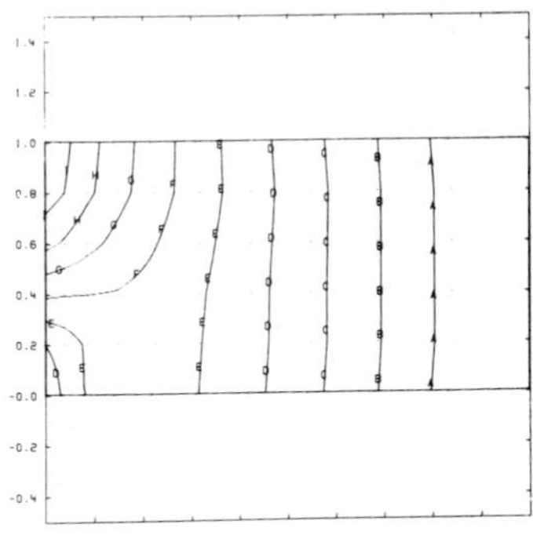

(c) $t = 10^{-4}$, $U_{max} = 1.0$, isobar $\Delta p = 1.73$.

Fig. 7. Velocity vectors and isobars during the penalty transient.

c

unusual velocity field is thus now legal in the sense that it is incompressible (weakly); further time integration simply recovers the steady Poisieulle solution on the physical time scale (O(1)).

6. SUMMARY and CONCLUSIONS There is an extraneous and spurious time scale associated with the penalty method approximation to the incompressible Navier-Stokes equations. This time scale is generally very short relative to any physically relevant time scales. For problems which are (incompressibly) well-posed, the spurious penalty transient can be bypassed by using a sufficiently dissipative time integration method. For problems which are ill-posed, the penalty transient disposes with the ill posedness and generates an (almost) incompressible "initial" velocity field; however, this modified velocity field is probably rarely of any physical interest.

7. REFERENCES

1. BERCOVIER, M. and M. ENGLEMAN - A Finite Element For Incompressible Viscous Flows. J. Comp. Phys., Vol. 30, 181-201, 1979.

2. HUGHES, T, J. R., W. T. LIU and A. BROOKS - Finite Element Analysis of Incompressible Viscous Flows by the Penalty Function Formulation. J. Comp. Phys., Vol. 30, 1-60, 1979.

3. GRESHO, P. M., R. L. LEE and R. L. Sani - On the Time-Dependent Solution of the Incompressible Navier-Stokes Equations in Two and Three Dimensions. Recent Advances in Numerical Methods in Fluids, Pineridge Press, 1980.

4. BERCOVIER, M., R. L. SANI, M. ENGLEMAN and P. M. GRESHO - in preparation.

5. R. L. SANI, P. M. GRESHO, R. L. LEE and D. F. GRIFFITHS - The Cause and Cure (?) of the Spurious Pressures Generated by Certain FEM Solutions of the Incompressible Navier-Stokes Equations, Part 1, Int. J. Num. Meth. Fluids, to appear 1981.

6. MALKUS, D. S. - Incompressible Finite Elements: The LBB Condition and the Discrete Eigenstructure, Int. J. Eng. Sci., to appear, 1981.

7. GRESHO, P. M. and R. L. LEE, - Don't Suppress the Wiggles - They're Telling You Something', to appear Computer and Fluids, 1981.

This work was performed under the auspices of the U. S. Department of Energy by the Lawrence Livermore Laboratory under contract No. W-7405-Eng-48. RLS and BEE would like to acknowledge support from both Lawrence Livermore National Laboratory and the U.S. Army Research Office (Grant DAAG29-79-6-0045).

ON THE NUMERICAL SOLUTION OF SOME TYPES OF UNSTEADY INCOMPRESSIBLE VISCOUS FLOW

LECOINTE, Y. & PIQUET, J.

Ecole Nationale Supérieure de Mécanique
1 rue de la Noe 44072 NANTES CEDEX

SUMMARY

This paper is devoted to the study of a new class of accurate compact discretization techniques for the Navier Stokes equations in their vorticity-stream function formulation both in inner problems (driven cavity, free convection in a rectangular enclosure) and outer problems (cylinder and airfoil).Mehrstellen methods are detailed with or without built-in upwinding which allows mesh Reynolds number restrictions to be avoided.

INTRODUCTION

Our aim is to discuss some difficulties connected with the efficiency of a new Navier-Stokes solver for the study of several aspects of unsteady viscous flow. The unsteady Navier-Stokes equations in their vorticity-stream function formulation are investigated with the help of "mehrstellen" techniques.

The Poisson equation is solved by a ADI method. A "mehrstellen" discretization is defer corrected so that h^6 accuracy on ψ is obtained. The vorticity equation is time-differenced with the help of a Beam & Warming type two-step, Δt^2 accurate, A-stable method which allows the splitting of the space operator so that unidimensionnal problems are solved with the help of distinct "mehrstellen" schemes which will be presented in this paper.

The first part of the paper is devoted to a study of the constructions and errors in the "mehrstellen" methods for a simple linear differential equation. In the second part, the method is applied to several test problems for the unsteady Navier-Stokes equations (e.g. the Pearson test, the driven cavity problem and a free convection problem). Further to this, some other results obtained around a cylinder are used to compare the performances of the algorithms.

CONSTRUCTION AND STUDY OF THE ERROR IN "MEHRSTELLEN" METHODS

"Mehrstellen" methods have a long history which goes back to early works of Collatz and Numerov. The method have been extensively used for boundary layer problems by Krause etAl (1976, and more recently by Ciment etAl (1978) and Berger etAl (1980). These techniques appear well suited to the modelling of diffusion -convection phenomena where two (or more) spatial operators of the form (1.1) are usually involved.

(1.1) $\ell u = \varepsilon u_{xx}+bu_x+cu = f$

In most applications and particularly in the context of large Reynolds number flow ε is much smaller than b and c.

"Mehrstellen" techniques (also called operator compact implicit techniques) as applied to (1.1) can be written in the following form:

(1.2) $\tau_j(U) = \varepsilon(r_j^+U_{j+1}+r_j^cU_j+r_j^-U_{j-1})/h^2- q_j^+f_{j+1}+q_j^cf_j+q_j^-f_{j-1} =0$

Such schemes lead to a tridiagonal system which is sol- -ved by factorization. They are at most formally fourth order accurate in the sense that the truncation error is

$$\tau_j(u) = \sum_{\nu=0}^{\infty} T_j^{[\nu]} u_j^{(\nu)} = 0(h^4) \quad \text{when } h\to 0 \text{ for fixed } \varepsilon.$$

The standard "Mehrstellen" scheme is obtained by writing $T_j^{[\nu]}= 0$, $\nu=0,1,\dots,6$ (see e.g. Schwarz etAl(1974)). Berger etAl, by writing the weaker condition $T_j^{[\nu]}=0$, $\nu=0,1,2$; $T_j^{[3]}$ and $T_j^{[4]}=0(h^4)$ have been able to formulate unrestricted schemes which do not exhibit a mesh Reynolds number limitation on h. Unfortunately, several arguments given in their work need to be corrected if one wants profitably to use their construction for viscous flow applications.

First, the diagonal dominance condition over r

(1.3) r_j^+, $r_j^- \geqslant 0$ and $-r_j^c \geqslant r_j^+ + r_j^-$ is a sufficient condition which ensures that the two roots of the characteristic polynomial associated with the homogeneous part of (1.2) are real, that one of these is greater than 1 and the other less than 1, whatever the sign of b (we limit the discussion to the case where $c\leqslant 0$ in (1.1)). The inequality on r_j^c in (1.3) needs $q_j^+c_{j+1}+q_j^cc_j+q_j^-c_{j-1}\leqslant 0$ which results either from the fact that q is non negative or from $c_j=0$ (when time dependent problems are considered)

Berger etAl (1980) also introduce a second important condition which should be written $b_j(b_jq_j^c-b_{j+1}q_j^+-b_{j-1}q_j^-)\geqslant 0$. This condition is a heuristic extension of an inequality allowing the time discretization to be unconditionnally stable, in any case where L_x^e and L_y^e (the space operators to be defined further) are of the form (1.1) with $c_j=0$. It allows also marginally the solvability in r in the sense that r_j^- (resp r_j^+) is sufficient to satisfy if $c_{j+1}\leqslant 2$ and $b_j\geqslant 0$ (resp $c_{j-1}+2\geqslant 0$ and $b_j\leqslant 0$).

An interesting property of these schemes is their built in monotonicity property which comes from the fact that they switch their form to a second order accurate upwind scheme when the mesh Reynolds number ρ becomes large. The conditions which allow q_j^-, q_j^c, q_j^+, r_j^-, r_j^+, to be non negative and lead to monotonicity and stability have been given by Berger etAl (1980) and they can be somewhat weakened : particularly $q_j^- \geqslant 0$ iff $p_2 \geqslant \pi_o^-$ where $24\pi_o^- = \langle p_1 - 3\rangle^3$ with $\langle\Phi\rangle = \frac{1}{2}(\Phi - |\Phi|)$; then the weakest sufficient condition over q_j^+ arises when $p_1 < 0$ if $p_4 \geqslant \pi_o^+$ where $24\pi_o^+ = \langle p_1\rangle^2$. Fourth order local accuracy results from the condition that r_j^- is a second order polynomial (which specifies p_4). Two sufficient conditions can then be obtained over r_j^+ either by specifying that $r_j^+ \geqslant r_j^-$ or by writing the non negativity of the second order polynomial obtained from r_j^+.

The aforementionned results have been applied to the test problem $\ell u \equiv u'' + bu' = 0$ $u(0) = U_o$, $u(1) = 0$. The local error can here be explicitly computed and written

(1.4) $\quad U_j - u(x_j) = U_o \; bR^4 \psi(R) \phi(x_j)/(1-e^{-b})^2 + 0(h^6)$ with

(1.5) $\quad \phi(x) = e^{-b}(1-e^{-bx}) - xe^{-bx}(1-e^{-b})$

$\psi(R) = (10p_1 - 10p_2 - 5p_3 - 3)/72$ iff $4p_4 = 2p_2 + p_3 - p_1$

It can be verified that the condition over p_4 which is necessary for h^4 local accuracy implies also that r_j^- is a second order polynomial.

APPLICATION TO THE NAVIER STOKES EQUATIONS.

Poisson Equation. The Poisson equation is considered in the following formulation:

(2.1) $\quad D(\psi) = A(\xi)\psi_{\xi\xi} + B(\xi)\psi_\xi + \psi_{\eta\eta} = \zeta$

The form (2.1) is well suited to airfoils as a result of a con--formal transformation into a unit circle combined with a stretching transformation in the radial direction (Mehta(1977)). This equation is solved using a Numerov η-discretization com--bined with a mehrstellen ξ-discretization in an optimized ADI method. The h^4 accurate obtained solution is then defer correc--ted in order to enforce h^6 accuracy of ψ with little extra cost. This method has been tested with the case $A(\xi) = \text{ch}\xi$, $B(\xi) = \text{sh}\xi$, $\zeta = \sin\eta \; \text{sh}\xi \; (2\text{ch}\xi - 1)$, the solution of which is $\psi = \sin\eta \, \text{sh}\xi$.

Error curves given in Fig. 1 show that while double precision is necessary for 10 points in each direction, it must be pointed out that only one correction is sufficient to increase significantly the accuracy even if it does not lead to a h^6 accurate result.

Time discretization algorithms. An approximate factorization of Beam & Warming /1980/ type is used. It rests upon the formulation of a linear multistep method and it is combined with a linearization of spatial operators. The result can be written :

(2.2) $|1-\omega\Delta t L_x^e|\bar{W} = \phi^{n+1,n}$; $|1-\omega\Delta t L_y^e| W^+ = \bar{W}$; $\zeta^{n+2} = W^+ + (1+\alpha)\zeta^{n+1} - \alpha\zeta^n$

(2.3) $\phi^{n+1,n} = \frac{\Delta t}{1+\xi}(L_x^e + L_y^e)\{1 + |\xi + \theta(\alpha-1) + 1/2|D^-\}\zeta^{n+1} + |\xi/(1+\xi) - \alpha|D^- \zeta^{n+1}$

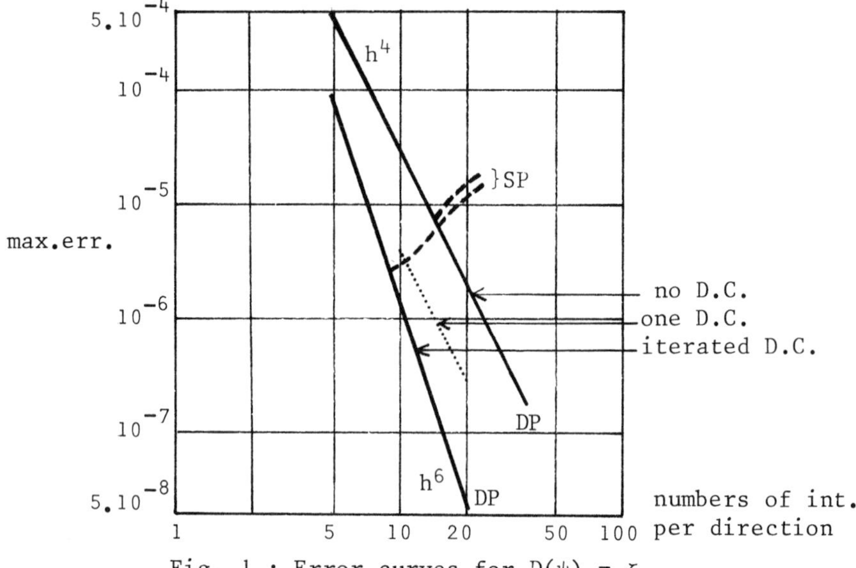

Fig. 1 : Error curves for $D(\psi) = \zeta$

where $D^{-}=1-E$; $E\zeta^{n}=\zeta^{n+1}$
L_x^e and L_y^e are extrapolated operators obtained with help of $\sigma_e(E)=(1+\Phi_e)E-\Phi_e 1$ where $\Phi_e=\theta+\Phi$ in order to ensure Δt^2 accurate linearization. $\omega=\theta/(1+\xi)$; α,ξ,θ being Beam & Warming parameters of the linear multistep method which will be A-stable iff $\xi\leqslant 2\theta-1$ $\xi\geqslant -1/2$; $-1\leqslant\alpha\leqslant 1$

Relationships (2.2) and (2.3) call for the following remarks : (i) the accuracy of this approximate factorization compares favourably with the accuracy of a standard ADI method, but its main advantage lies in the fact that a threedimensional extension is quite straightforward. (ii) the iteration over the vorticity ζ^{n+2} (unknown on the boundaries) will not need a re-calculation of the velocity field which is very time consuming and appears sometimes to lead to specific convergence problems. (iii) the aforementionned A-stability condition does not appear to be sufficient for adjacent points to the boundaries because of the boundary condition which is written on ζ. The Poisson problem is coupled with the vorticity equation in such a way that the V. Neumann analysis must be performed both on ψ and ζ (Bontoux etAl (1980)) and lead to a condition of explicit type.

Calculation of ϕ. It appears that the critical point with two dimensionnal "mehrstellen" techniques lies in the computation of ϕ. This problem arises because ϕ must be computed at points where ζ is known but not derivatives of ζ. Thus this will be the first step in the calculation. In the computational domain, the first derivatives are computed by solving :

$$(2.4)\qquad \zeta'_{i+1}+4\zeta'_i+\zeta'_{i-1}=3h^{-1}(\zeta_{i+1}-\zeta_{i-1})$$

This h^4 accurate hermitian formula is a very good smoother as it extinguishes Nyquist frequency, gives a good representation of long wave lengths and damps out shorter wavelength. It needs compact formulae for end points which are taken under one of the following forms :

(2.5a) $2\zeta'_i+\zeta'_{i+1}-(5\zeta_{i+1}-4\zeta_i-\zeta_{i-1})/2h=0$

(2.5b) $\zeta'_i+3\zeta'_{i+1}-(-17\zeta_i+9\zeta_{i+1}+9\zeta_{i+2}-\zeta_{i+3})/6h=0$

Second derivatives are calculated explicity with :

(2.6a) $\zeta''_i=2(\zeta_{i+1}-2\zeta_i+\zeta_{i+1})/h^2-(\zeta'_{i+1}-\zeta'_{i-1})/2h$

(2.6b) $\zeta''_{i+1}=(7\zeta_{i-1}+16\zeta_i-23\zeta_{i+1})/2h^2+(\zeta'_{i-1}+8\zeta'_i+6\zeta'_{i+1})/h$

Such relationships do not filter out short wavelengths but they are nevertheless used because they do not need more tridiagonal systems to be solved (Notice also that (2.6b) gives a h^4 boundary condition over ζ if ζ is replaced by ψ).

A by-product of these difficulties lies in the fact that "mehrstellen" techniques are as time consuming as hermitian methods but they are easier to code, and moreover, control of monotonicity with respect to mesh Reynolds number is also easier. Fig. 2 gives the impact of the computation of ϕ on the accuracy of ζ. If derivatives are calculated with five point formulae, the accuracy on ζ (Pearson (1965) test) is not very good. One decade better is obtained with (2.4), (2.6), with the h^3 accurate relationship (2.5a) for end points. Another decade better is obtained if (2.5a) is replaced by (2.5b) which is h^4 accurate and gives a loss of less than one decade relatively to an analytic computation of ϕ.

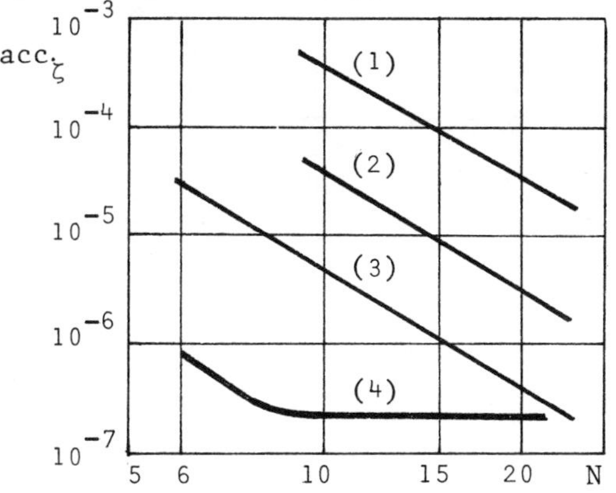

derivatives

(1) Calculated with 5 points

(2) with (2.4) (2.5a) (2.6)

(3) with (2.4) (2.5b) (2.6)

(4) analytically

Fig. 2 : Impact of the computation of ϕ on the accuracy of ζ.

Other filtering procedures. Hermitian schemes have also been used. They rest upon the following type of relationships :

(2.7) $S(\alpha,\beta,\gamma,\theta)\equiv KU_i+hLU'_i+h^2MU''_i=0$ where

$K=16\gamma I-8\gamma\mu-3\beta D_\circ;L=16\theta I+(3\beta-8\theta)\mu-(3\alpha-5\gamma)D_\circ;M=4\alpha I+(\alpha-\gamma)\mu+(4\theta-\beta)D_\circ$

$\mu U_i=U_{i+2}+U_{i-2}$;$D_\circ U_i=U_{i+1}-U_{i-1}$

which are combined with the ODE to be solved (see e.g.(1.1))

A compact block implicit method is obtained by considering, besides the equation to be solved, two formulae $S_1U_i=S_2U_i=0$ corresponding to two sets of coefficients $\alpha,\beta,\gamma,\theta$. It is always possible to form a pentadiagonal system for the unknowns U_i (Peyret(1978);Lecointe etAl(1980)),from which upwind hermitian schemes, for instance, can be easily written. One can also work with a semi-inverse procedure formulated in order to build in monotonicity. Let the characteristic polynomial of the pentadia--gonal matrix for unknowns be $P(q)=Aq^4+Bq^3+Cq^2+Dq+E=0$. Two of the roots q_0 and q_1 approximate the basic solutions of the dif--ferential problem.The two other roots are spurious and are partly responsible for the wiggles occurring as the mesh Reynolds number increases. These $2\Delta x$ oscillations can be prevented if we ensure the monotonicity of the underlined part of the solution: $U_j=\underline{C_0q_0^j+C_1q_1^j}+C_2q_2^j+C_3q_3^j$. Now, if $P(q)=(q-q_0)(q-q_1)P^+(q)$, where $P^+(q)=Aq^2+B^+q+C^+$. The roots of $P^+(q)$ will be of modulus less than 1 if $A=1+\gamma_1+\gamma_2$; $B=2(\gamma_2-1)$; $C^+=1-\gamma_1+\gamma_2$;$\gamma_1,\gamma_2\geqslant 0$. The deter-mination of the admissible zone in the $\gamma_1-\gamma_2$ plane is then straightforward and the result is given in fig3.

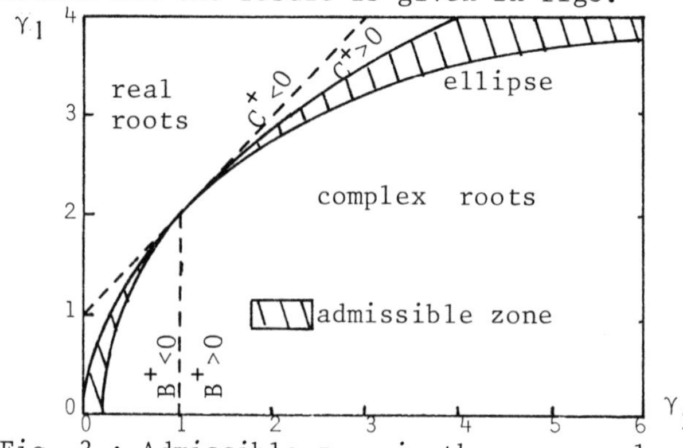

Fig. 3 : Admissible zone in the $\gamma_1-\gamma_2$ plane $-1<j<\sigma$

Then q_0 and q_1 are determined by exponential fitting by using the same type of method as proposed by El Mistikawy&Werle (1978). Such a procedure is known to be uniformly accurate for singular perturbation problems.

Endly, let us mention a negative result which bears some analogy with Dahlquist theorems. If one defines explicit,compact accurate filters with help of coefficients of tridiagonal ma--trices A_i to be determined by Taylor formula, the following definition is introduced:

(2.8) $\bar{U}_i=A_1U_i+hA_2U_i'+h^2A_3U_i''=U_i+2\sigma h^4U_i^{IV}/4!+2\theta h^6U_i^{VI}/6!$

If one then writes that no phase shift is allowed and that the Nyquist frequency $2\Delta x$ is completely damped out, one finds that the whole set of coefficients in A_i can be expressed with only two coefficients σ and θ. Now, if one assumes admissibility for the filter: $0\leqslant \bar{U}/U\leqslant 1$; it can be found that no values for σ and θ allow for admissibility.

NUMERICAL TESTS.

Pearson Test. We first have tested the Pearson Test which specifies an analytical unsteady solution of Navier-Stokes equations :

$$\psi=e^{-2\pi^2 t}\cos\pi y\cos\pi x \;;\; \zeta=-2\pi^2\psi \quad -\frac{1}{2}<x,y<\frac{1}{2}$$

Such a test gives a zero vorticity on the boundaries allowing for slip. Results given in fig. 2 show that the scheme used is h^4 accurate. The maximum error is in fact obtained along the first points in the vicinity of x=1/2 and y=1/2 and comes evidently from the evaluation of ϕ. It can also be verified that Δt^2 accurary is obtained but this test is in fact not significantbecause convection terms cancel each other in Pearson Test so that neither the vorticity boundary condition, nor the linearization procedure can be validated with this exemple.

Natural convection problem. Fig. 4 gives numerical results obtained in the steady limit with a "mehrstellen" scheme for Rayleigh and Schmidt number given by Ra=10000, Sc=100 with 10 points in each direction. Convergence is easily obtained.

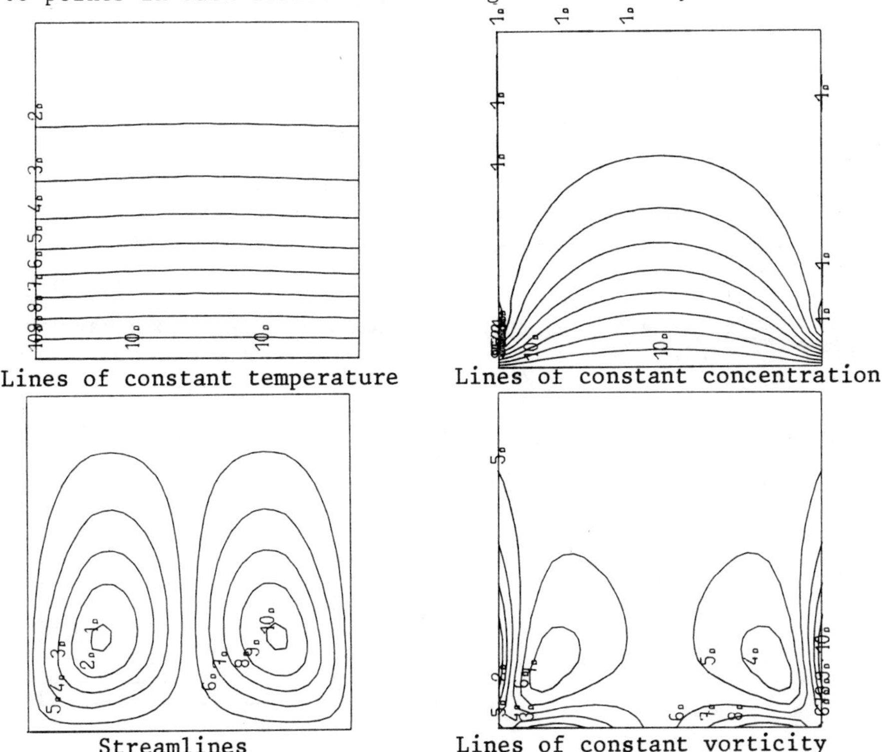

Lines of constant temperature — Lines of constant concentration

Streamlines — Lines of constant vorticity

Figure 4

In this example a zero flux condition is retained on vertical boundaries for temperature while concentration is imposed. Here again a specific uncentered mehrstellen h^3 accurate formula is necessary ; it can be written :

$$\frac{\varepsilon}{h^2}(r_j^+ T_{j+1} + r_j^c T_j) + \theta_j T'_{j+1} = q_j^+ f_{j+1} + q_j^c f_j + q_j^- f_{j-1}$$

A simple analysis performed in the model equation shows that the numerical solution remains $O(h^4)$. The classical double roll configuration is obtained.

Driven cavity. Numerical results which are given in the following concern only the steady limit and are obtained with a standard "mehrstellen" scheme. A h^4-accurate boundary condition on ζ of type (2.6b) has been used but no iteration on the value of ζ at the boundaries is performed at each time step consequen--tly numerical result is only Δt-accurate in the transient part.

For "mehrstellen" type algorithms, corner points must be excluded. Derivatives on the boundaries near these corner points need to be computed with the help of uncentered relationship like (2.5) and (2.6). In each direction an ordinary differential equation of the form (1.1) has to be integrated, and an uncentered "mehrstellen" formula is necessary ; it can be written :

$$\varepsilon(r_j^+\zeta_{j+1} + r_j^c\zeta_j + r_j^-\zeta_{j-1})/h^2 = q_j^c f_j + q_j^- f_{j-1} + q_j^= f_{j-2}$$

The coefficients $r_j^{+,c,-}$ and $q_j^{c,-,=}$ are computed by writing that such a scheme is h^3 accurate, while the numerical solution for a test problem remains in this case $O(h^4)$.

We shall now present briefly the case Re=100 for which it appears that the results are rather insensitive to the type of boundary conditions retained over ζ. Thirty points per direction are used.

Fig. 6 gives a schematic sketch of the square cavity flow and table 1 gives the corresponding vortex properties in the square cavity for t=9,1 and $\sup_{ij}|\zeta^{n+1}_{i,j} - \zeta^n_{i,j}| = 0{,}022$

Primary	LL	LR
x_{vc} = 0.616	x_1 = 0.03	x_2 = 0.95
y_{vc} = 0.738	y_1 = 0.03	y_2 = 0.095
ψ_{vc} = 0.1029	ψ_1 = -2.10^{-6}	ψ_2 = $-1.5\ 10^{-5}$
ζ_{vc} = -3.169	ζ_1 = 0.016	ζ_2 = 0.0215
	b_1 = 0.07	b_2 = 0.11
	d_1 = 0.07	d_2 = 0.13

Table 1 : Vortex properties in the square cavity

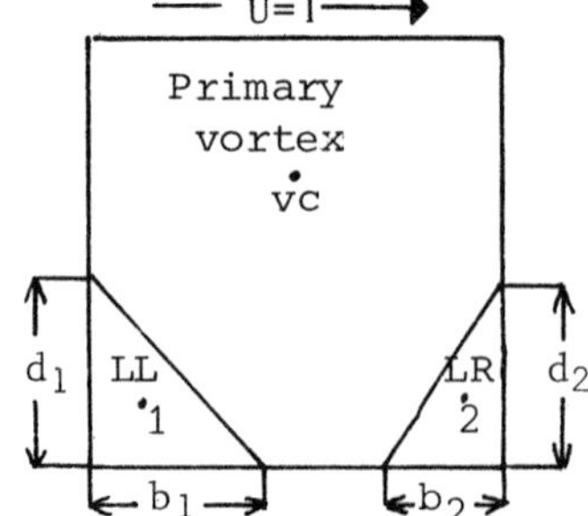

Fig.6 : Schematic Sketch of the square cavity flow

Fig. 7 and 8 give streamlines and lines of constant vorticity, they look very similar to those presented by other authors (see e.g. Tuann & Olson (1979)). Results have been obtained also for Re=500 but it must be noticed that it is necessary in this case

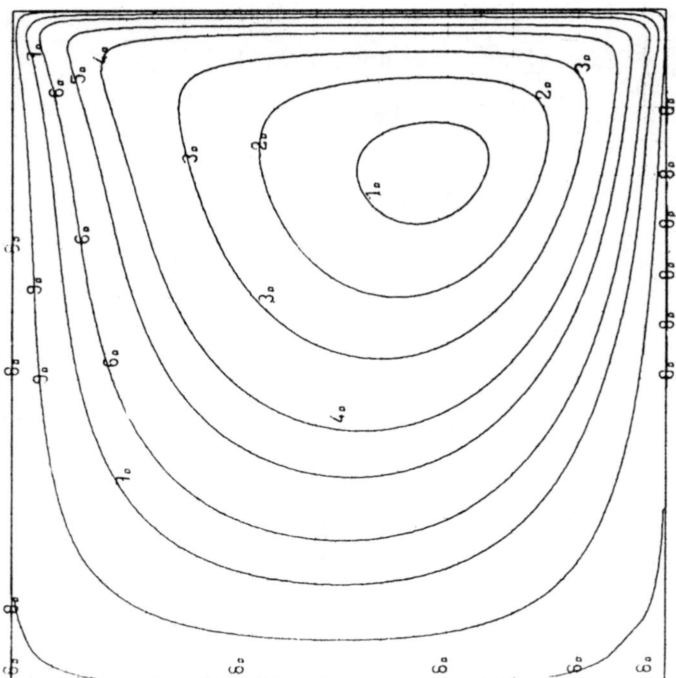

Fig. 7 : Streamlines for Re = 100

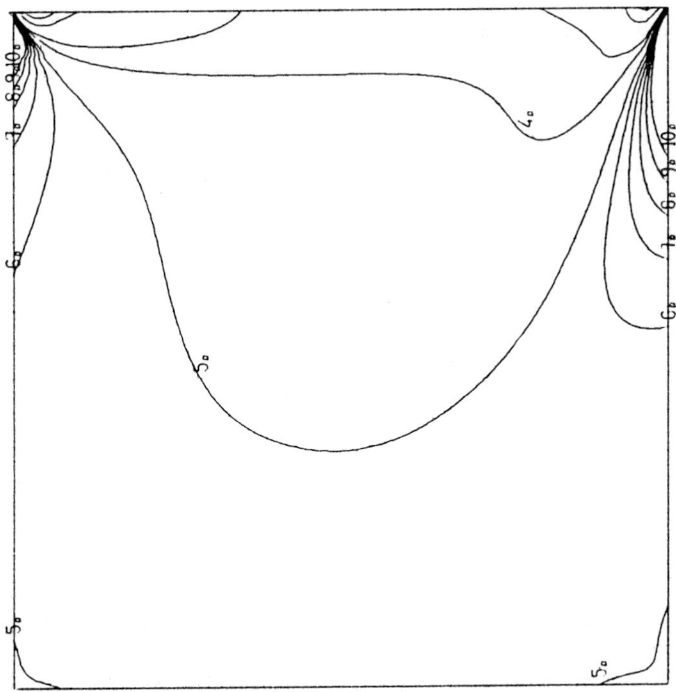

Fig. 8 : Lines of constant vorticity for Re = 100

to underrelax on the value of ζ on the driven boundary.

Fig. 9 gives u-velocity profiles along the vertical centreline.

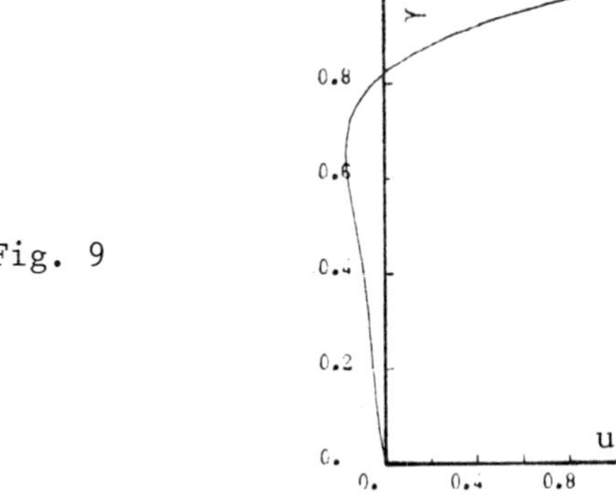

Fig. 9

Cylinder. Here we shall only present a low Reynolds number ($Re=U_\infty a/\nu=20$) symetrical flow. Numerical results are obtained with a standard "mehrstellen" scheme, but unrestricted OCI schemes or different boundary conditions on ζ lead to similar results (Woods condition, Hirsh h^3 accurate condition, a h^4 accurate condition of type (2.6b) have been used). No iteration on ζ is performed for any time step. No corner points problem arises here because of θ-periodicity. Irrotational flow is assumed at the outer boundary which is circular and taken at 10 radii. 30 points in each direction are taken and no symetry condition is written.

Fig. 10 gives the time evolution of the wall vorticity ζ_p with respect to θ. It appears that separation at $\theta=\pi$ does not appear immediately.

Fig. 11 give the time evolution of the recirculation length. It appears that good agreement is obtained with regard to Bouard & Coutanceau (1977).

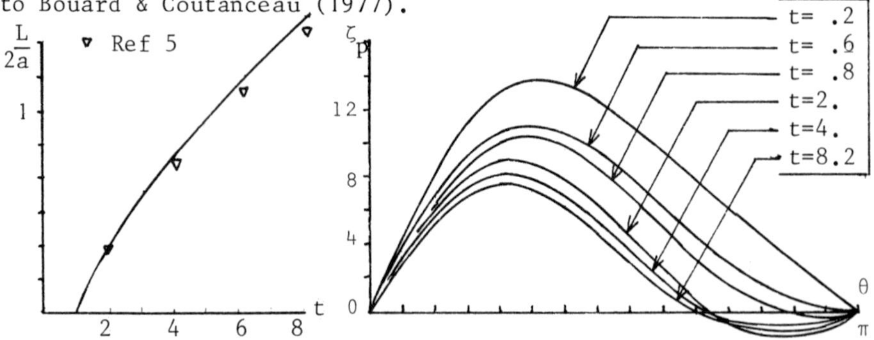

Fig. 11 : Time evolution of recirculation length

Fig. 10 : Time evolution of wall vorticity

Figs 12 and 13 give plots for streamlines and lines of constant vorticity in the steady limit.

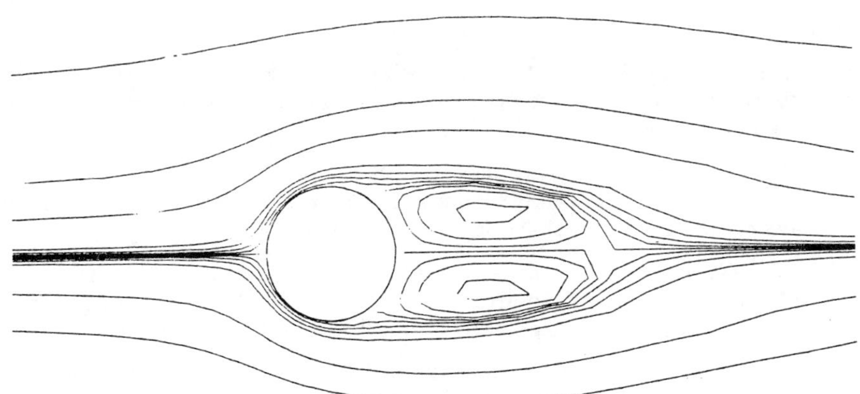

Fig. 12 : Streamlines for R_e=20

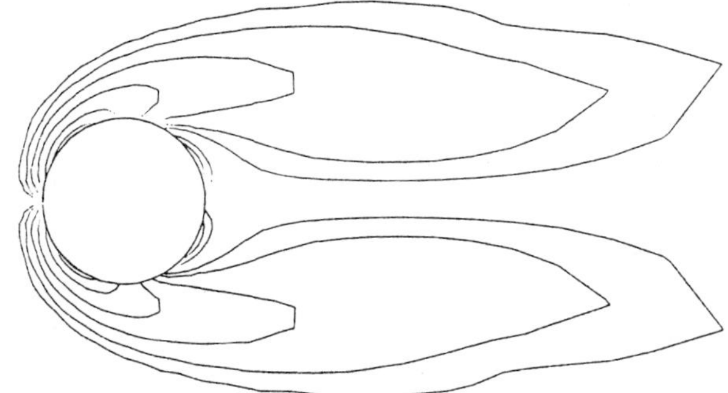

Fig. 13 : Lines of constant vorticity for R_e= 20

CONCLUSIONS

The results which are presented here concern only low Reynolds numbers. Systematic tests are being performed for higher Re than those considered here and for different schemes. Besides the standard mehrstellen and unrestricted OCI schemes which lead to a h^4 accurate vorticity, an upwind conservative scheme h^2 accurate on ζ, but h^4 accurate on ψ (Daube etAl (1979)) is also considered in order to compare them with spatial accuracy and time differencing methods. The Joukowski airfoil is also under investigation but this case is more difficult because of the severe restrictions on Δt due to the values of the jacobian near the (blunted) trailing edge.

ACKNOWLEDGMENTS

The authors are indebted to Mr. Visonneau for calculations performed on the cavity problem and to Dr. Ta P. Loc for valuable discussions during the course of the work. The financial support of this work through DRET contract 79-649 is also gratefully acknowledged.

REFERENCES

1- BEAM, R.M. & WARMING, R.F. (1980) "An implicit factored scheme for the compressible Navier-Stokes Equations II : the numerical ODE Connection" AIAA Paper 79-1146

2- BERGER, A.E ; SOLOMON, J.M.; CIMENT, M.; LEVENTHAL, S.H. & WEINBERG, B.C. (1980) "Generalized OCI Schemes for boundary layer problems" Maths. Comp. 35 p. 695

3- BONTOUX, P.; GILLY, B. & ROUX, B.(1980) "Analysis of the effect of boundary conditions on numerical stability of solutions of Navier Stokes Equations" Journ. Comp. Phys. 36 p.417

4- CIMENT, M.; LEVENTHAL, S.H. & WEINBERG, B.C.(1978) "The Operator Compact Implicit Method for Parabolic Equations" Journn. Comp. Phys. p.135

5- COUTANCEAU, M. & BOUARD, R.(1977) "Experimental Determination of the Main Features of the Viscous Flow in the Wake of a circular Cylinder in Uniform Translation II ; unsteady flow" J. of Fluid. Mech. 79 p. 257-272.

6- DAUBE, O. & TA P. LOC (1978) "Etude numérique d'écoulements instationnaires de fluide visqueux incompressible autour de corps profilés par une méthode combinée d'ordre $0(h^2)$ et $0(h^4)$" Journal de Méca. 17 pp. 651-678

7- EL MISTIKAWY, T.M. & WERLE, M.J.(1978) "Numerical method for boundary layers with blowing. The exponential box scheme" AIAA Journ. 16 p. 749-751

8- KRAUSE, E.; HIRSCHEL, E.H. & KORDULLA, W.(1976) "Fourth order "Mehrstellen" integration for three-dimensional turbulent Boundary layers". Computers and Fluids 4 p. 77

9- LECOINTE, Y. & PIQUET, J.(1980) "Examination of some numerical and asymptotical problems connected to the resolution of unsteady Navier Stokes Equations around Airfoils" Proc. Euromech 129

10-MEHTA, U.B.(1977) "Dynamic Stall of an Oscillating Airfoil" AGARD Conf. Proc No 227 Unsteady Aerodynamics

11-PEARSON, C.E.(1965) "A computational method for viscous flow problems" J. Fluid Mech. 21 p.611

12-PEYRET, R.(1979) "A hermitian finite Difference Method for the solution of the Navier-Stokes Equations" Proc. Ist Int Conf. Num. Methods in laminar and turbulent flows. Ed. Pentech Press.

13-SCHWARTZ, B.K.(1974) "The construction of finite difference analys of some finite element schemes" in Mathematical aspects of finite elements in Partial Differential Equations (Cde Boor Ed.) pp 279-312 Academic Press New York

14-TUANN, S.Y. & OLSON, M.D.(1979) "New finite element results for the square cavity" Computers & Fluids 7, 2, p. 123-135.

THE APPLICATION OF BOUNDARY VALUE TECHNIQUES IN THE SOLUTION OF THE NAVIER-STOKES EQUATION

A. Danaee[†] and D.J. Evans[††]

SUMMARY

In this paper the boundary value technique has been applied to solve the Navier-Stokes equations for the two dimensional, steady state, viscous, incompressible flow in a rectangular cavity.

The resulting finite difference equations using upwind differences are solved by successive line overrelaxation methods. The merit of the method is its fast convergence to obtain a solution for any required Reynolds number.

1. INTRODUCTION

The solution of scientific problems defined by partial differential equations presents such analytical difficulties that numerical methods appear to be the only reasonable means of solution within the foreseeable future.

Usual numerical methods of solution for the initial boundary-value type problems posed by parabolic partial differential equations have been the explicit and implicit (Crank-Nicolson) methods in which the solution is obtained by a point or line step-by-step process on a chosen network of lines over the given open domain.

An alternative strategy (the boundary value technique [1]) which has recently come into prominence is that of determining the steady state solution as $t \to \infty$. Then, with the assumption that this is the solution on row $t=T$ (large), we solve the resulting boundary value problem on the truncated closed

[†] Lecturer, Dept. of Mathematics, Isfahan University, IRAN.

[††] Professor, Dept. of Computer Studies, University of Technology Loughborough, Leicestershire, ENGLAND.

region by the iterative or direct methods normally associated with elliptic partial differential equations. It can be readily seen that this approach does not suffer from the row-to-row error accumulation which one associates with initial value techniques. This technique can be advantageous if the solution for large times is required.

2. PROBLEM FORMULATION

We investigate here the boundary value techniques for the two dimensional steady-state, viscous, incompressible flow in a rectangular cavity.

Consider a square cavity DABC as shown in Figure 1 within which a steady fluid motion is generated by sliding an infinitely long plate lying on top of the cavity. Suppose that all the variables are normalised so that the size of the cavity is the unit square and the sliding velocity is -1 ([2]), in the negative x-direction.

Let S be the square ABCD and denote its interior by R. On R the equation of motion to be satisfied are the two dimensional steady-state Navier-Stokes equations, namely:

$$\Delta\zeta + R_e \left(\frac{\partial\psi}{\partial x} \cdot \frac{\partial\zeta}{\partial y} - \frac{\partial\psi}{\partial y} \cdot \frac{\partial\zeta}{\partial x}\right) = 0, \qquad R_e \geq 0 \quad , \tag{1}$$

$$\Delta\psi = -\zeta, \tag{2}$$

where ψ is the stream function, ζ is the vorticity, and R_e is a non-negative constant called the Reynolds number.

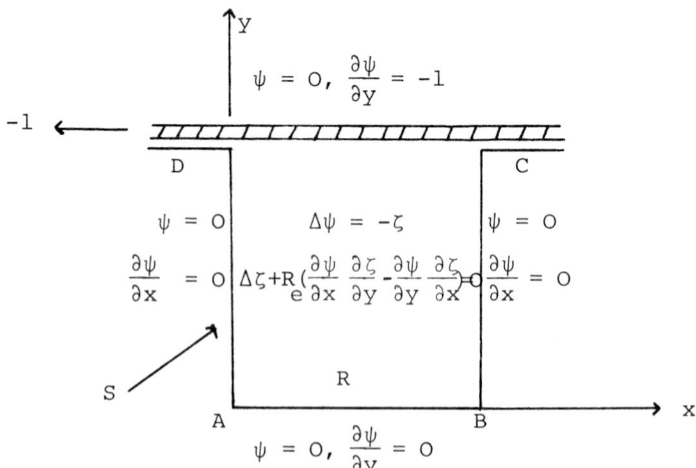

FIGURE 1: Cavity flow caused by a moving plate

If no fluid is squeezed out of the cavity below the moving plate, the fluid motion forms closed paths within the cavity. The surfaces DA,AB,BC and CD are then segments of the bounding stream lines designated by ψ=0, along which can be specified that the velocities normal to these four surfaces are zero. On the other hand, we require that the tangential velocity vanishes on all the surfaces except the top plate which is given as -1. Thus we obtain the four additional boundary conditions,

$$\frac{\partial\psi}{\partial x} = 0 \text{ on DA and CB}, \quad \frac{\partial\psi}{\partial y} = 0 \text{ on AB and } \frac{\partial\psi}{\partial y} = -1 \text{ on CD}.$$

The vorticity equation (1) can also be regarded as the asymptotic time limit of the non-steady, time dependent equations and the system (1)-(2) now becomes:

$$\Delta\psi = -\zeta \tag{3}$$

$$\frac{\partial\zeta}{\partial t} = \frac{1}{R_e}\Delta\zeta + \frac{\partial\psi}{\partial x}\cdot\frac{\partial\zeta}{\partial y} - \frac{\partial\psi}{\partial y}\cdot\frac{\partial\zeta}{\partial x}, \quad 0<x<1, 0<y<1 \text{ and } t>0 \tag{4}$$

which is elliptic-parabolic partial differential equation system. This new formulation can be solved by the marching procedures as well as by the boundary value technique which is the scope of this paper. In many problems of practical concern only the steady-state solution is the subject of interest.

3. FORMULATION OF THE FINITE DIFFERENCE EQUATIONS

Numerical solutions of the equations (3)-(4) have been the subject of many investigations. This study has been attempted to increase the understanding of optimum numerical solution techniques. Because of its simplicity and economy a boundary value technique for the problem (3)-(4) has been employed by Greenspan [2] where the generalized Newton iteration scheme has been applied to solve the system of non-linear equations corresponding to the non-linear vorticity equation (4). Other methods also have been attempted by Greenspan which are fully described in [1].

From the experience gained from the one-dimensional Burger's equation [3], we now concentrate our attention on the boundary value technique when successive block (line) over-relaxation schemes are used and when the Reynolds number R_e is chosen in the range 100-50,000. The method proceeds as follows.

Let D be the rectangular parallelpiped defined by $D=\{(x,y,t): 0\leq x\leq 1,\ 0\leq y\leq 1,\ 0\leq t\leq T\}$. Define R to be the interior and S to be the boundary of D. Using space grid size $\Delta x=\Delta y=h$ and time size Δt, we construct in the usual way by ordering the points <u>row-wise</u> the three dimensional set of interior grid points, denoted by $R_{h,\Delta t}$ and boundary grid points, denoted by $S_{h,\Delta t}$ of totality $(n\times n\times m)$ where $n=1/h$, $m=T/\Delta t$.

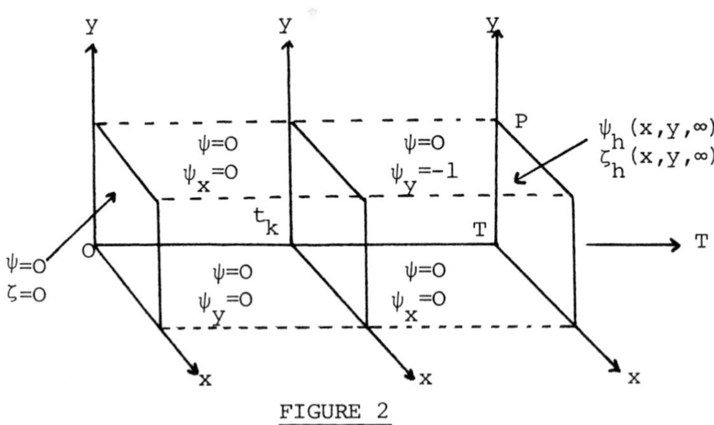

FIGURE 2

We observe that (3)-(4) is a coupled system of partial differential equations in ψ and ζ. But if ζ is known (3) is a linear elliptic equation in ψ, while if ψ is known, (4) is a linear parabolic equation in ζ. Thus, the initial guesses $\psi^{(0)}$ and $\zeta^{(0)}$ can be applied and we can construct a sequence of iterative solutions as follows:

Use $\zeta^{(0)}$ in (3) to produce $\psi^{(1)}$ whilst $\zeta^{(1)}$ can be obtained by solving the parabolic equation (4) in D. We shall remember that, the solution at t=T is chosen to be $\zeta=0$ (the steady-state solution which is the essence of the technique). Therefore $\zeta^{(1)}$ can be inserted in (3) to produce $\psi^{(2)}$ and then using $\psi^{(2)}$ we find $\zeta^{(2)}$ and so on. In this way, we can construct a sequence of discrete functions

$$\psi^{(1)},\psi^{(2)},\ldots\psi^{(\ell)}, \ \ldots \quad , \tag{5}$$

on $R_{h,\Delta t}$ and a sequence of discrete functions

$$\zeta^{(1)},\zeta^{(2)},\ldots\zeta^{(\ell)}, \ \ldots \quad , \tag{6}$$

on $R_{h,\Delta t}+S_{h,\Delta t}$ such that for some pre-assigned tolerances ε_1 and ε_2,

$$|\psi^{(\ell)}-\psi^{(\ell-1)}|<\varepsilon_1 \text{ and } |\zeta^{(\ell)}-\zeta^{(\ell-1)}|<\varepsilon_2 .$$

The discrete functions $\psi^{(\ell)}$ and $\zeta^{(\ell)}$ are then taken to be the numerical approximations to $\psi(x,y,t)$ and $\zeta(x,y,t)$ on the chosen grid.

For this purpose at each point of $R_{h,\Delta t}$ we obtain the difference equation,

$$\psi_{i+1,j,k}+\psi_{i-1,j,k}+\psi_{i,j+1,k}+\psi_{i,j-1,k}-4\psi_{i,j,k}=-h^2\zeta^{(0)}_{i,j,k} ,$$

$$\text{for } i=1,2,\ldots,n-1 \quad j=1,2,\ldots,n-1 \quad \text{and } k=1,2,\ldots,m-1. \tag{7}$$

where $\psi_{i,j,k}$ denotes $\psi^{(1)}_{i,j,k}$.

At the boundary grid points we set,

$$\psi(x,y,0) = 0, \quad \psi(x,y,T) = \psi(x,y,\infty) ,$$
$$\psi(0,y,t) = \psi(x,0,t) = \psi(x,1,t) \ \psi(1,y,t) = 0. \tag{8}$$

The solution $\psi^{(1)}(x,y,t_k)$ can now be obtained at every step by inserting (8) whenever necessary. Here we apply the SLOR iterative method to solve the stream equation (7) for $\psi^{(1)}$ [4].

Next step is generating the sequence $\zeta^{(1)}$ on $R_{h,\Delta t}$ which requires the values of ζ on $S_{h,\Delta t}$. The boundary conditions on the vorticity equation can now be obtained by central differencing equation (3), applying the boundary conditions for the stream function and enforcing the reflection condition at the boundary. These are as follows:-

at the surface DC (Figure 1),

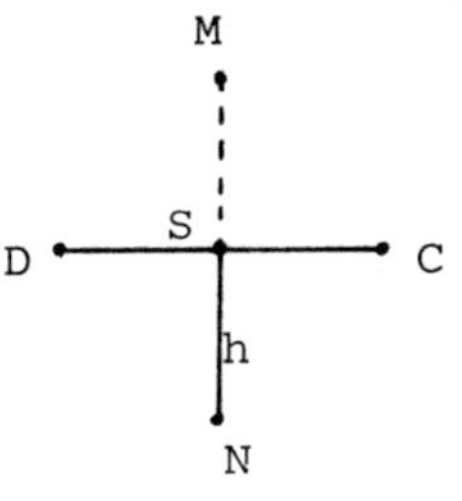

$$\frac{\psi_M-2\psi_S+\psi_N}{h^2} = -\zeta_S ,$$

$$\psi_S = 0 ,$$

$$\frac{\psi_M-\psi_N}{2h} = -1.$$

Combining these conditions results in

$$\frac{2\psi_N-2h}{h^2} = -\zeta_S . \tag{9}$$

Also, at the surface AB we have,

$$\frac{\psi_M-2\psi_S+\psi_N}{h^2} = -\zeta_S ,$$

$$\psi_S = 0 ,$$

$$\frac{\psi_M-\psi_N}{2h} = 0, \text{ or } \psi_M = \psi_N ,$$

or

$$\frac{2\psi_N}{h^2} = -\zeta_S . \tag{10}$$

At the surface AD and BC we can write

$$\frac{\psi_M - 2\psi_S + \psi_N}{h^2} = -\zeta_S ,$$

$$\psi_S = 0 \text{ and } \psi_M = \psi_N ,$$

or

$$\frac{2\psi_N}{h^2} = -\zeta_S . \qquad (11)$$

Now we can obtain the finite-difference replacement of the vorticity equation and insert the approximate boundaries (9)-(11) to generate the solution $\zeta^{(1)}$.

We apply the Richardson formula to solve the vorticity equation (4), i.e.

$$\frac{(\zeta_{i,j,k+1}-\zeta_{i,j,k-1})}{2\Delta t} = \frac{1}{R_e}\left[\frac{(\zeta_{i+1,j,k}-2\zeta_{i,j,k}+\zeta_{i-1,j,k})}{h^2} + \frac{(\zeta_{i,j+1,k}-2\zeta_{i,j,k}+\zeta_{i,j-1,k})}{h^2}\right] + \left\{\frac{\psi^{(1)}_{i+1,j,k}-\psi^{(1)}_{i-1,j,k}}{2h}\right\}$$
$$\left\{\frac{\zeta_{i,j-1,k}-\zeta_{i,j,k}}{h}\right\} - \left\{\frac{\psi^{(1)}_{i,j+1,k}-\psi^{(1)}_{i,j-1,k}}{2h}\right\}\left\{\frac{\zeta_{i,j,k}-\zeta_{i-1,j,k}}{h}\right\}, \qquad (12)$$

for $i=1,2,\ldots,n-1$
$j=1,2,\ldots,n-1$
and $k=1,2,\ldots,m-1$.

However, one might obtain better accuracy if the term $\frac{\partial\zeta}{\partial t}$ in (12) is replaced by

$$\frac{1}{2\Delta t}\{-3\zeta_{i,j,k-1}+4\zeta_{i,j,k}-\zeta_{i,j,k+1}\} . \qquad (12a)$$

The equation (12) can be solved for $\zeta^{(1)}_{i,j,k}$ at all the interior grid points of $R_{h,\Delta t}+S_{h,\Delta t}$ by a suitable iterative method (i.e. SLOR).

The presence of advective (or convective) terms in the vorticity equation causes some difficulties when one solves the iterative scheme such as (12).

If the terms $\frac{\partial\zeta}{\partial y}$ and $\frac{\partial\zeta}{\partial x}$ are replaced by central differences, some of the points $\zeta(x\pm h,y,t)$ and $\zeta(x,y\pm h,t)$ are outside $R_{h,\Delta t}$. To remedy this difficulty and avoid exterior points, central differences must be replaced by one-sided differences. Now as can be seen from (12), the terms $\zeta_{i,j,k}$ which arise from $\frac{\partial\zeta}{\partial y}$ and $\frac{\partial\zeta}{\partial x}$ play a different role for the iteration scheme, since they may appear with different sign with the similar term obtained from $\Delta^2\zeta$ and therefore the diagonal dominancy of the system may be reduced or even lost, which leads to the divergence of the iterative scheme. Therefore, we choose forward or backward difference replacements for these terms as follows:

if

$$A_{i,j,k} = \frac{\psi_{i+1,j,k}-\psi_{i-1,j,k}}{2\Delta x} \geq 0 \text{ then } \frac{\partial\zeta}{\partial y} \to \frac{(\zeta_{i,j,+1,k}-\zeta_{i,j,k})}{\Delta y}$$

else

$$\frac{\partial\zeta}{\partial y} \to \frac{(\zeta_{i,j,k}-\zeta_{i,j-1,k})}{\Delta y} \quad , \tag{13}$$

and if,

$$B_{i,j,k} = \frac{\psi_{i,j+1,k}-\psi_{i,j-1,k}}{2\Delta y} \geq 0 \text{ then } \frac{\partial\zeta}{\partial x} \to \frac{(\zeta_{i,j,k}-\zeta_{i-1,j,k})}{\Delta x}$$

else

$$\frac{\partial\zeta}{\partial x} \to \frac{(\zeta_{i+1,j,k}-\zeta_{i,j,k})}{\Delta x} \quad . \tag{14}$$

Thus, the term $\frac{\partial\psi}{\partial x} \cdot \frac{\partial\zeta}{\partial y} \cdot \frac{\partial\psi}{\partial y} \cdot \frac{\partial\zeta}{\partial x}$ can assume four different forms. It is said that in (12) one always employs "upwind differences" [5].

3. ITERATIVE METHOD OF SOLUTION

From the equations (7) it can be readily seen that by grouping the points (i,j,k) along the rows in the x- or y-direction in the domain $R_{h,\Delta t}$ then the totality of difference equations at all the grid points yields a large sparse (M×M) linear system of the form,

$$A\underline{\Psi} = \underline{b} \quad , \tag{15}$$

where

$$A = \begin{bmatrix} D & -F & & & 0 \\ -E & D & \ddots & & \\ & \ddots & \ddots & \ddots & \\ & & \ddots & \ddots & -F \\ 0 & & & -E & D \end{bmatrix}_{(m-1)\times(m-1)} ,$$

with

$$D = \begin{bmatrix} T & -U & & & 0 \\ -L & T & \ddots & & \\ & \ddots & \ddots & \ddots & \\ & & \ddots & \ddots & U \\ 0 & & & -L & T \end{bmatrix}_{(n-1)\times(n-1)} , \quad E = F = O ,$$

and

$$T = \begin{bmatrix} 4 & -1 & & & 0 \\ -1 & 4 & \ddots & & \\ & \ddots & \ddots & \ddots & \\ & & \ddots & \ddots & -1 \\ 0 & & & -1 & 4 \end{bmatrix}_{(n-1)\times(n-1)} , \quad U = L = I_{n-1} ,$$

where $\underline{\Psi} \equiv \psi_{i,j,k}$ for $i=1,2,\ldots,n-1$; $j=1,2,\ldots,n-1$, $k=1,2,\ldots,m-1$.

$\underline{b} \equiv \underline{\zeta}^{(\ell-1)}$ +contributions derived from the boundary conditions,

ℓ denotes the outer iteration index on the $\underline{\Psi}$ and $\underline{\zeta}$ solutions, and finally the order of the system $M=(m-1)(n-1)^2$.

In a similar row-wise grouping of the points (i,j,k) for equation (12) and (12a) we can derive another block structured linear system, of the form,

$$\tilde{A}\underline{\zeta} = \tilde{\underline{b}} \quad , \tag{16}$$

where

$$\tilde{A} = \begin{bmatrix} \tilde{D} & -\tilde{F} & & & 0 \\ -\tilde{E} & \tilde{D} & \ddots & & \\ & \ddots & \ddots & \ddots & \\ & & \ddots & \ddots & -\tilde{F} \\ 0 & & & -\tilde{E} & \tilde{D} \end{bmatrix}_{(m-1)\times(m-1)} ,$$

with

$$\tilde{D} = \begin{bmatrix} \tilde{T} & -\tilde{U} & & & 0 \\ -\tilde{L} & \tilde{T} & \ddots & & \\ & \ddots & \ddots & \ddots & \\ & & \ddots & \ddots & -\tilde{U} \\ 0 & & & -\tilde{L} & \tilde{T} \end{bmatrix}_{(n-1)\times(n-1)} \qquad \tilde{E} = 3I_{n-1}, \; \tilde{F} = I_{n-1},$$

and

$$\tilde{T} = \begin{bmatrix} \alpha_{1,j} & -2p_1 & & & 0 \\ -2p_1 & \alpha_{2,j} & \ddots & & \\ & \ddots & \ddots & \ddots & \\ & & \ddots & \ddots & -2p_1 \\ 0 & & & -2p_1 & \alpha_{n-1,j} \end{bmatrix}_{(n-1)\times(n-1)} \qquad \begin{aligned} \tilde{L} &= 2p_1 - \frac{\Delta t}{h} A^{(\ell)}_{i,j,k}, \\ \tilde{U} &= 2p_1 + \frac{\Delta t}{h} B^{(\ell)}_{i,j,k}, \end{aligned}$$

with $p_1 = \dfrac{\Delta t}{h^2 R_e}$, $\alpha_{i,j} = 4+8p_1 + \dfrac{\Delta t}{h^2}(|A_{i,j,k}|+|B_{i,j,k}|)$, $i=1,2,\ldots,n-1$,

where we define

$$A_{i,j,k} = (\psi_{i+1,j,k}-\psi_{i-1,j,k})/2h,$$

$$B_{i,j,k} = (\psi_{i,j+1,k}-\psi_{i,j-1,k})/2h, \text{ for } i,j=1,2,\ldots,n-1, \text{ and } k=1,2,\ldots,m-1,$$

$\underline{\zeta} \equiv \zeta_{i,j,k}$ for $i,j=1,2,\ldots,n-1$, $k=1,2,\ldots,m-1$,

and $\underline{b} \equiv$ boundary contributions derived from $\underline{\Psi}^{(\ell)}$.

Now, using the block structure of the matrix A, a block Jacobi iterative method can be expressed in the form,

$$D\underline{u}^{(s+1)} = (E+F)\underline{u}^{(s)}+b = \underline{f}^{(s)}, \text{for } 1\leq k\leq m-1, \tag{17}$$

where $\underline{u}$ can denote either $\underline{\Psi}$ the stream function or $\underline{\zeta}$ the vorticity function.

Further, use of the block structure of D, yields the row-line Jacobi method,

$$T\underline{u}^{(s+1)} = (L+U)\underline{u}^{(s)}+f^{(s)}, \text{ for } 1\leq j\leq n-1, \quad (18)$$

and the row-block successive overrelaxation method by,

$$\left.\begin{aligned} T\underline{\tilde{u}}^{(s+1)} &= L\underline{u}^{(s+1)}+U\underline{u}^{(s)}+\underline{f}^{(s)} \\ \underline{u}^{(s+1)} &= \underline{u}^{(s)}+\omega(\underline{\tilde{u}}^{(s+1)}-u^{(s)}) \end{aligned}\right\} , \quad \begin{aligned} &(19a) \\ &(19b) \end{aligned}$$

and

where the superscript s denotes the iteration cycle and ω the block overrelaxation parameter.

It can be observed that since the form of the matrix T in (15) and (16) is tridiagonal then the solution of (19a) requires the application of the Gauss elimination method to obtain the new iterative values along each row of the domain. However since T is diagonally dominant then no pivoting strategy is required and the solution can be obtained easily from a standard algorithm given in [4] before applying the principle of block overrelaxation in (19b).

The application of block overrelaxation methods to problems using the boundary value technique is given more fully in Evans [6] and shortage of space allows only a brief outline to be presented here. However, widely differing rates of convergence for the standard iterative methods such as point S.O.R., row block S.L.O.R. and column block S.L.O.R. can be achieved [6] and the reader is advised to be cautious and aware of this fact before selecting the method of solution.

4. NUMERICAL RESULTS

We have solved the system (1)-(2) using the boundary value technique as presented by the S.L.O.R. method for values of R_e in the range 100-50,000 with the space and time increments chosen to be $h=1/20$ and $\Delta t=1/5$ respectively. We have also chosen $T_\infty=5$ and set the boundary values at this level equal to zero.

In the accompanying Table 1 the number of iterations required to obtain an accuracy of 10^{-3} is given. The initial guesses for both systems were chosen to be equal to zero. Finally, the streamlines and equivorticity contours at t=1,2,3 and 4 for R_e=1000 are given in Figures 3-10.

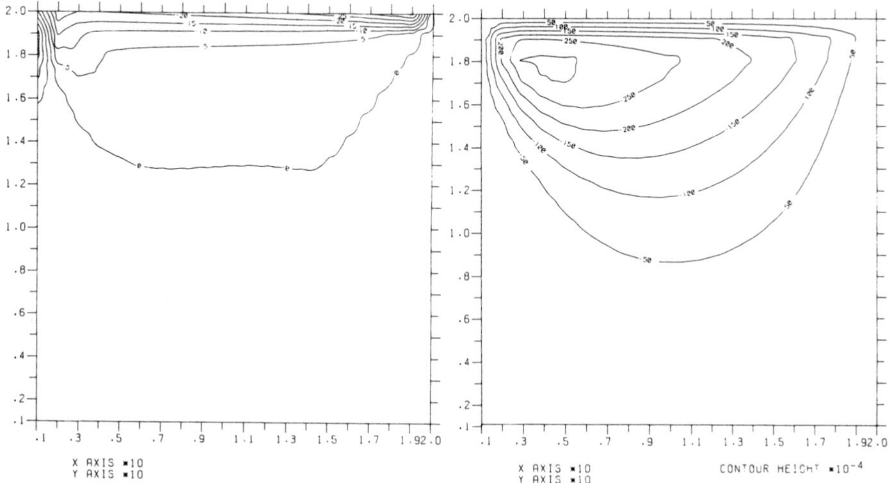

<u>FIG 3</u>: Equivorticity curves at t=1

<u>FIG 4</u>: Streamlines at t=1

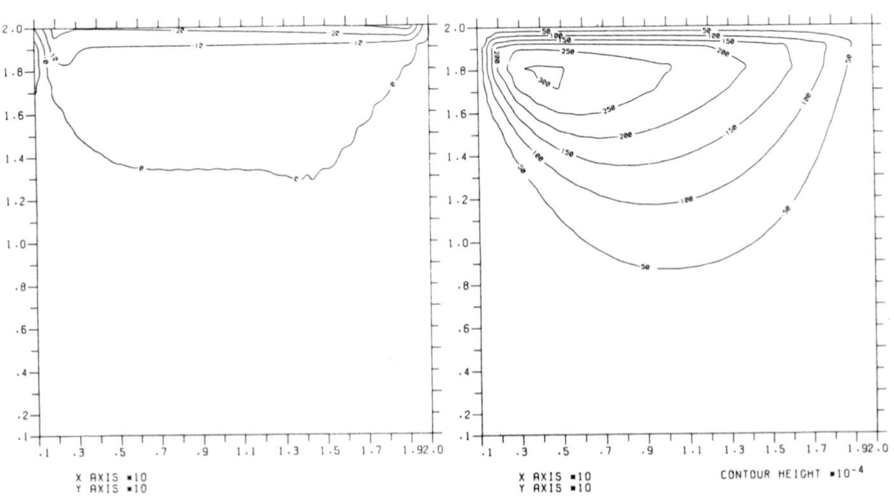

<u>FIG 5</u>: Equivorticity curves at t=2

<u>FIG 6</u>: Streamlines at t=2

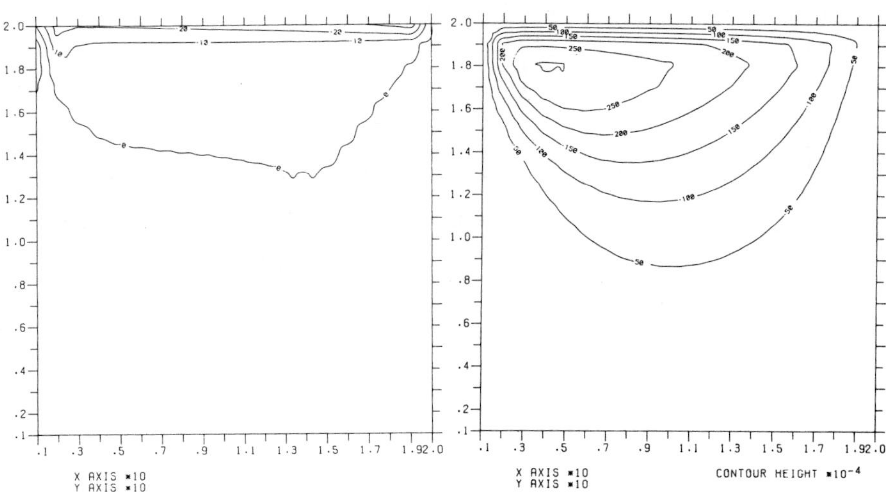

FIG 7: Equivorticity curves at t=3

FIG 8: Streamlines at t=3

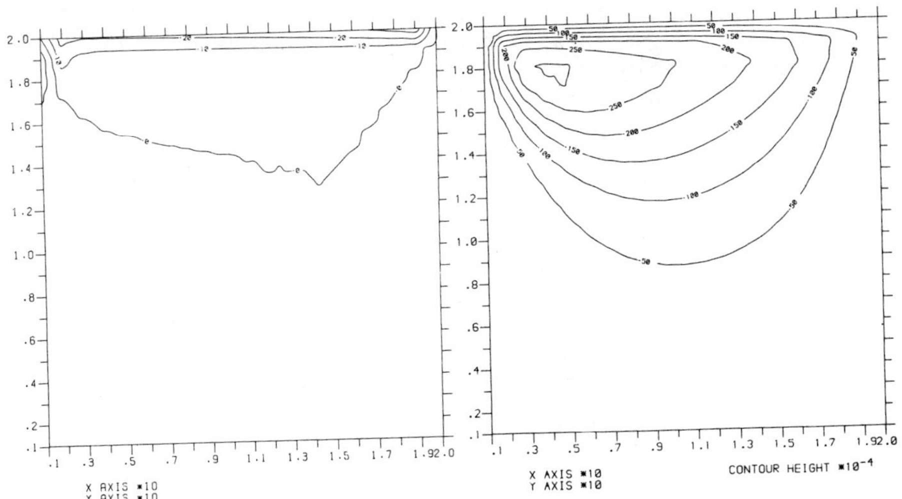

FIG 9: Equivorticity curves at t=4

FIG 10: Streamlines at t=4

R_e Number	Block overrelaxation parameter		S.L.O.R. iterations
	Stream Function	Vorticity Eqn.	
100	0.2	1.6	73
500	0.15	1	101
1000	0.1	1	100
5000	0.1	1	75
10,000	0.1	1	66
50,000	0.1	1	46

TABLE 1: Values of the block overrelaxation parameters and iteration numbers for various values of R_e.

As can be seen from the table, the values of the overrelaxation parameters were mostly 1 or less. Thus, the methods are of underrelaxation type.

5. CONCLUSIONS

From the results displayed in Table 1, it is clear that by using block overrelaxation methods in conjunction with boundary value techniques, we are able to obtain solutions to equations (1) and (2) more efficiently than the generalised Newton method suggested by Greenspan [2].

A further comparison was made with the results for $h=1/16$, $\Delta t=1/5$ and $R_e=100$ given by Smith et al [7] who applied the Hopscotch technique for the solution of the vorticity equation together with a fast direct method (Buneman) and also the S.O.R. method for the solution of the stream function. The number of iterations for an accuracy of $\varepsilon=10^{-3}$ was found to be 123 for the S.O.R. method and 62 iterations for the SLOR method. Although we could not obtain satisfactory convergence for $\varepsilon=10^{-4}$, the method seems to be comparable with the other iterative schemes suggested in [7] for the lower accuracy.

It is well known that by the standard methods the solution of the Navier-Stokes equation becomes much harder when R_e increases. In this paper however, we have shown that the reverse situation occurs when the boundary value technique is used. Although the mesh increments are quite large we were able to achieve an accuracy of 10^{-3}. However an attempt to obtain better accuracy failed.

Obviously one may obtain a better accuracy by increasing the mesh points, but in this case the amount of storage required may cause difficulties. In this case it seems more reasonable to use the approximate solution obtained by the above technique as the initial guess for a more accurate method e.g. A.D.I. (see [7]).

6. REFERENCES

[1] Carasso, A. and Parter, S.V. - An Analysis of Boundary Value Techniques for Parabolic Problems. Math.Comp. Vol.24, pp.315-340, 1970.

[2] GREENSPAN, D. - Discrete Numerical Methods in Physics and Engineering, Academic Press, New York, 1974.

[3] DANAEE, A. - A Study of Hopscotch Methods for Solving Parabolic Partial Differential Equations. Ph.D. Thesis, Loughborough University of Technology, 1980.

[4] VARGA, R.S. - Matrix Iterative Analysis, Prentice Hall Inc. Englewood Cliffs, N.J., 1962.

[5] FORSYTHE, G.E. and WASOW, W.R. - Finite Difference Methods for Partial Differential Equations, J. Wiley & Sons, New York, 1960.

[6] EVANS, D.J. - The Solution of Parabolic Partial Differential Equations by Boundary Value Methods. Greek Math.Bull. p.71-85, 1980.

[7] SMITH, R.E., and KIDD, A. - Comparative Study of Two Numerical Techniques for the Solution of Viscous Flow in a Driven Cavity. NASA SP-378, 1975.

INTEGRAL CONDITIONS ON THE VORTICITY: NUMERICAL ALGORITHMS FOR THE STEADY AND UNSTEADY NAVIER-STOKES EQUATIONS

L. Quartapelle, F. Valz-Gris, M. Napolitano (^)

SUMMARY

This paper describes algorithms for the solution of the incompressible Navier-Stokes equations in the vorticity/stream function representation. A distinctive feature of the proposed method is the use of exact vorticity conditions in the form of projection conditions of an integral character on the vorticity field. By means of them, an original formulation of the continuum problem with the equations in split form is introduced. The equations are discretized in time considering different linear approximations of the advective terms in the vorticity transport equation. Then, the process of spatial discretization is considered and a decomposition scheme for the solution of the integrally conditioned equations of the vorticity is proposed. The properties of the resulting algorithms for the approximate calculation of time-dependent and steady-state incompressible flows in two dimensions are numerically investigated in connection with the finite difference solution of the driven cavity problem. The results agree with published data and illustrate the potentialities of the present approach.

1. INTRODUCTION

The main difficulty in the approximate solution of the vorticity/stream function equations lies in the fact that the values of the stream function and of its normal derivative are prescribed on the boundary, whereas no boundary condition

(^) Luigi Quartapelle, Assistant Professor
Istituto di Fisica, Politecnico di Milano, Italy.
Fausto Valz-Gris, Assistant Professor
Istituto di Fisica, Università degli Studi di Milano, Italy
Michele Napolitano, Professor
Istituto di Macchine, Università di Bari, Italy.

is available for the vorticity. Therefore, most numerical schemes are faced with the problem of obtaining the correct vorticity boundary values. This is not the case if the vorticity and stream function equations are solved coupled together, even in the linear case (Stokes' problem), by an iterative approach [1] or a direct method [2,3]. On the other hand, if the solution of two uncoupled second-order problems is preferred, conditions for the vorticity equation which serve as an equivalente substitute for those attached to the stream function are mandatory.

Recently, it has been shown that proper conditions for the vorticity can indeed be formulated so as to allow the complete splitting, in the linear case at least, of the vorticity equation from the stream function one [4]. These conditions determine a projection of the vorticity field on the linear manifold of the harmonic functions in terms of the boundary data of the stream function and of its normal derivative. The aim of the present paper is to derive the discrete counterpart of the projection conditions on the vorticity and to present some numerical algorithms which rely upon these conditions for the solution of steady-state and evolution Navier-Stokes equations.

2. PROJECTION CONDITIONS ON THE VORTICITY

Let us consider the nonprimitive variables, vorticity ζ and stream function ψ, describing the two-dimensional motion of a viscous incompressible fluid. It is possible to formulate the basic equations for ζ and ψ of the continuum problem as a set of two split equations, letting aside the coupling due to the nonlinear advective terms, provided that the vorticity equation is supplemented by conditions of an integral character [4]. The equations for the dimensionless variables are

$$-\nabla^2\zeta/Re + \frac{\partial\zeta}{\partial t} + J(\zeta,\psi) = 0 \qquad (1)$$

$$\int dV\zeta\eta = -\int ds(b\eta - a\frac{\partial\eta}{\partial n}), \quad \forall\eta : \nabla^2\eta = 0 \qquad (2)$$

$$-\nabla^2\psi = \zeta \qquad (3)$$

$$\psi|_S = a \qquad (4)$$

where $a = \psi|_S$ and $b = \partial\psi/\partial n|_S$ are the boundary data that can be expressed in terms of the velocity prescribed on the boundary S of the domain V occupied by the fluid. Furthermore, $J(\zeta,\psi) = \partial(\zeta,\psi)/\partial(x,y)$ is the Jacobian. Conditions (2) state that the projection of the vorticity field on the linear manifold of the harmonic functions η is determined by the boundary data a and b. Thus, the conditions on ζ are independent of the values of ψ at interior points of V, and allow the splitting of the vorticity transport equation (1) from the stream function equation (3), as far as the imposition of the respective conditioning is concerned.

3. TIME-DISCRETIZED EQUATIONS AND ALGORITHMS

Starting from the new formulation of the continuum problem provided by equations (1) - (4), discrete approximations and numerical algorithms for the calculation of steady and unsteady flows can be derived. Let us discretize in time problem (1) - (4) by finite differences assuming a two-level time-stepping scheme (Euler differencing) in the vorticity equation. In compliance with the integral, and henceforth implicit, character of the vorticity projection conditions, we assume an always implicit treatment of the diffusion term, whereas the nonlinear advection term is taken into account with an increasing degree of implicitness depending on the linearization approximation which is considered. The discretized equations of each scheme are found as a particular case of the equations of the generalized scheme

$$(-\nabla^2+e)\zeta^{n+1} + Re\ J(\zeta^r,\psi^s) = e\zeta^n \tag{5}$$

$$\int dV\zeta^{n+1}\eta = -\int ds(b^{n+1}\eta - a^{n+1}\frac{\partial\eta}{\partial n}) \tag{6}$$

$$-\nabla^2\psi^{n+1} = \zeta^{n+1} \tag{7}$$

$$\psi^{n+1}|_S = a^{n+1} \tag{8}$$

where $e\equiv Re/\Delta t$, $a^{n+1}=a(t^{n+1})$ and $b^{n+1}=b(t^{n+1})$, $t^{n+1}=(n+1)\Delta t$. According to the values chosen for r and s we have four different time-integration schemes.

explicit scheme : $r = s = n$. Helmholtz equation (5) for the vorticity is supplemented by the integral conditions (6) and can be solved independently from Poisson equation (7) for the stream function supplied by Dirichlet boundary conditions (8).

ζ-implicit scheme : $r = n+1$, $s = n$. Helmholtz equation for the vorticity is still uncoupled from the stream function equation but now it contains the term $J(\zeta^{n+1},\psi^n)$ with first-order spatial derivatives of the unknown ζ^{n+1} (equation with a nonsymmetric linear operator).

<u>ψ-implicit scheme : r = n, s = n+1</u>. Equations (5) and (7) are now coupled (linearly) through $J(\zeta^n,\psi^{n+1})$ and therefore problem (5) - (8) constitutes a linear system of two coupled partial differential equations for the unknowns ζ^{n+1} and ψ^{n+1} with conditions of an integral type for ζ^{n+1} and of Dirichlet type for ψ^{n+1}.

<u>implicit scheme : r = s = n+1</u>. In this case problem (5) - (8) becomes a system of two equations nonlinearly coupled through $J(\zeta^{n+1},\psi^{n+1})$. As an iterative method of solution we can consider the following algorithm. The solution $(\zeta,\psi)^{n+1}$ is calculated as the limit of the sequence $(\zeta,\psi)^m$, m = 0,1,..., defined as follows. Start from $(\zeta,\psi)^{m=0} = (\zeta,\psi)^n$. Then, from ζ^n, a^{n+1}, b^{n+1} and $(\zeta,\psi)^m$, the new approximation $(\zeta,\psi)^{m+1}$ is obtained as the solution of

$$(-\nabla^2+e)\zeta^{m+1} + Re\ J(\zeta^p,\psi^q) = e\zeta^n \tag{9}$$

$$\int dV\zeta^{m+1}\eta = -\int ds(b^{n+1}\eta - a^{n+1}\frac{\partial\eta}{\partial n}) \tag{10}$$

$$-\nabla^2\psi^{m+1} = \zeta^{m+1} \tag{11}$$

$$\psi^{m+1}\Big|_S = a^{n+1} \tag{12}$$

The iterative process is terminated when $\|(\zeta,\psi)^{m+1}-(\zeta,\psi)^m\| < \varepsilon$ for some prescribed norm $\|\cdot\|$ and a small positive ε. According to the values chosen for p and q in (9) we have three algorithms for the iterative solution at each time step of the nonlinear equations of the implicit scheme :

explicit algorithm : p = q = m
ζ-implicit algorithm : p = m+1, q = m
ψ-implicit algorithm : p = m, q = m+1

Some comments about the numerical stability of the different schemes in the case that the spatial derivatives are discretized by means of centred second-order accurate finite differences. The stability analysis of the one-dimensional advection-diffusion equation, with an explicit treatment of the linearized advection and an implicit treatment of the diffusion, leads to the stability conditions $\Delta t\ u^2\ Re \leqslant 2$, u is the velocity. A similar result can be established for the two-dimensional equations of the explicit scheme, which is thus conditionally stable. The ζ-implicit and ψ-implicit schemes are both fully implicit within the respective linearization assumption. They are therefore expected to be unconditionally stable and are indeed found always stable up to the explored value $\Delta t = 10$ in the numerical tests on the driven cavity problem at Re = 100 and using a 9x9 mesh. A possible numerical instability at higher values of Δt could be called legitimately nonlinear. Finally, the implicit scheme is non-linearly implicit and it ought suffer not even from nonlinear instabilities. As a matter of fact, this scheme has been found numerically stable up to $\Delta t = 100$ for the same model problem as before.

Remark. Equations (5) - (8) of the implicit scheme, i.e. with $r = s = n+1$, once the superscript index n+1 is dropped out and setting $e = 0$, become the system of nonlinear equations for the steady state. Thus, the three iterative algorithms (9) - (12) for the solution of the time-dependent equations of the implicit scheme can also be used with $e = 0$ to determine the solution of the nonlinear equations of the stationary problem.

4. INTEGRALLY CONDITIONED VORTICITY EQUATIONS

As a consequence of the spatial discretization process by finite differences or finite elements, the vorticity projection conditions (6) become a finite number of linear algebraic equations that supplement the discretized counterpart of the elliptic equation for the vorticity. In fact, it can be easily shown that the manifold of the discrete harmonic functions contains exactly as many linearly independent functions as there are boundary points, namely the discrete harmonic functions η_α vanishing at all boundary points except for one

$$\nabla^2 \eta_\alpha = 0 \qquad\qquad \eta_\alpha|_S = \delta_{\alpha\beta} \qquad (13)$$

where $\delta_{\alpha\beta}$ is the Kronecker symbol and where the Laplace operator must be interpreted in the sense of the assumed spatial discretization. The integrally conditioned vorticity equations to be solved in the above schemes and algorithms, assume the typical form ·

$$(-\nabla^2 + e)\zeta = f \qquad (14)$$

$$\int dV \zeta \eta = -\int ds \left(b\eta_\alpha - a\,\frac{\partial \eta_\alpha}{\partial n}\right) \qquad (15)$$

where f, a and b are known, whereas ζ is the unknown variable. Instead of constructing the discretized counterpart of the operator of orthogonal projection onto the manifold of the discrete harmonics, we solve the linear system of algebraic equations (14) - (15) by means of the following decomposition scheme. The solution ζ is sought in the form

$$\zeta = \zeta_o + \sum_\alpha \lambda_\alpha \theta_\alpha \tag{16}$$

where ζ_o is the solution of

$$(-\nabla^2+e)\zeta_o = f \qquad \zeta_o|_S = 0 \tag{17}$$

and the functions θ_α are the solutions of the homogeneous Helmholtz problems with Dirichlet boundary conditions

$$(-\nabla^2+e)\theta_\alpha = 0 \qquad \theta_\alpha|_S = \delta_{\alpha\beta} \tag{18}$$

Therefore, $\lambda = \{\lambda_\alpha\}$ is the discrete approximation of the (unknown) trace of the vorticity, i.e. $\lambda = \zeta|_S$, and it is found by imposing that the vorticity expressed by (16) satisfies the integral conditions (15). We obtain the linear system of algebraic equations

$$A\lambda = c \tag{19}$$

where matrix A and vector c are defined, respectively, by

$$A_{\alpha\beta} = \int dV \theta_\alpha \eta_\beta \tag{20}$$

$$c_\alpha = -\int ds(b\eta_\alpha - a\frac{\partial\eta_\alpha}{\partial n}) - \int dV\zeta_o\eta_\alpha \tag{21}$$

By Green's theorem, from (13) and (18) it follows that matrix A is symmetric. The calculation of $A_{\alpha\beta}$ and c_α through expressions (20) and (21) is computationally inconvenient in that it requires to store the two manifolds of functions η_α and θ_α. However, the use of Green's theorem shows that $A_{\alpha\beta}$ and c_α can be evaluated in a different manner which avoids the above inconvenience at the expense of doubling the number of elliptic problems to be solved, using a technique suggested by Glowinski and Pironneau in a paper on the biharmonic problem [5]. This idea has been extensively used in the numerical experiments to be discussed in the next section.

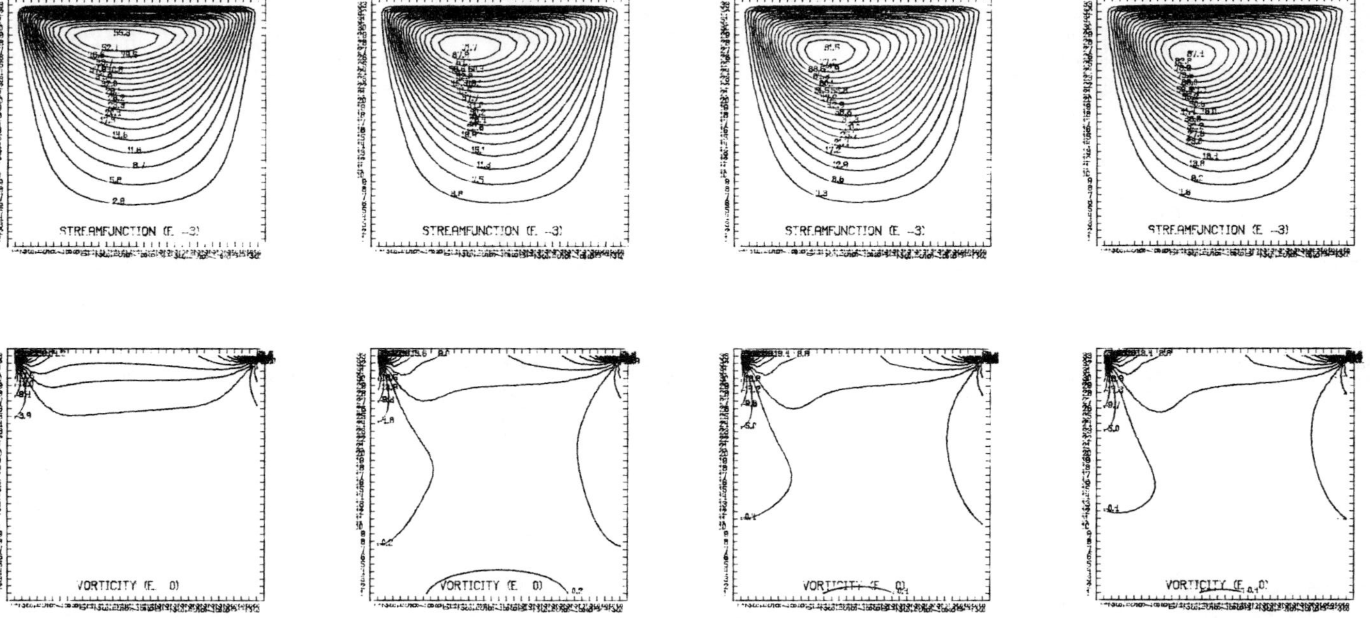

Figure 1. Time dependent flow in a square cavity produced by the impulsive motion of the upper wall. Re = 100, t = 0.5, 1, 1.5 and 2 (from left to right). Explicit scheme, $\Delta t = 0.05$ (Arakawa differencing of the advective terms, mesh 33x33).

5. NUMERICAL EXAMPLE AND DISCUSSION

For the numerical test of the proposed method the driven cavity problem has been chosen [6-7]. We have considered a time-dependent version of the classical steady-state problem assuming that the fluid is at rest initially and that the driving wall is set in motion impulsively at $t = 0$. Finite differences have been employed using centred approximations on a uniform mesh of 33x33 points. All the calculations have been performed using single precision arithmetic on a IBM 370/165.

To assess the consistency of the four schemes and of the three algorithms of the nonlinear scheme, we have calculated the transient solution at Re = 100 with a fixed time step $\Delta t = 0.05$. In the convergence test, we have relied upon the maximum norm taking $\varepsilon = 10^{-4}$, and we have used 2nd-order Arakawa differencing of the advective terms [8]. In all cases we have found numerical results coincident within the requested accuracy at all times. An example of this transient solution evaluated by the explicit algorithm is shown in Fig. 1 by the plots of the streamlines and equivorticity lines at $t = 0.5$, 1, 1.5 and 2. As a further test, we have solved the same problem with $\Delta t = 0.05$, 0.1, 0.5 using the implicit scheme/explicit algorithm. The transient solutions with the smaller values of Δt are almost coincident, that one corresponding to the larger value differs appreciably, but the steady-state solutions are anyway coincident in all three cases. Then, we have considered the problem at the moderate Reynolds number Re = 1000 using 4th-order accurate discrete approximation of the advective terms, as suggested by Ozawa [7]. The evolution of the flow field, obtained by the implicit scheme/explicit algorithm, is shown in Fig. 2 by plots of the vorticity, stream function, pressure and total pressure at $t = 5$, 10, 15. The pressure fields have been obtained by the procedure of line integration introduced by Burggraf [6] adapted to the case of the time-dependent equations. The plots detail the mechanism of formation of the main vortex and evidentiate the propagation of a pulse of the total pressure as the responsible of vortex acceleration. Finally, we have calculated the solution at Re = 1000 employing the ψ-implicit scheme with $\Delta t = 0.05$. A steady-state solution within an accuracy $< 5.10^{-4}$ is obtained at $t = 30$ in $n = 600$ iterations. The contour plots are shown in Fig. 3 and are in a fair qualitative agreement with the literature. The position of the main vortex centre with respect to the left lower corner of the cavity is $(x = 0.438,\ y = 0.594)$ where $\psi = 0.0683$ and $\zeta = 1.258$. The comparison with the values $(x = 0.453,\ y = 0.587)$, $\psi = 0.0756$ and $\zeta = 1.458$ reported by Ozawa [7] for a similar mesh (31x31) indicates that the computation must be pursued to reach the true steady-state solution.

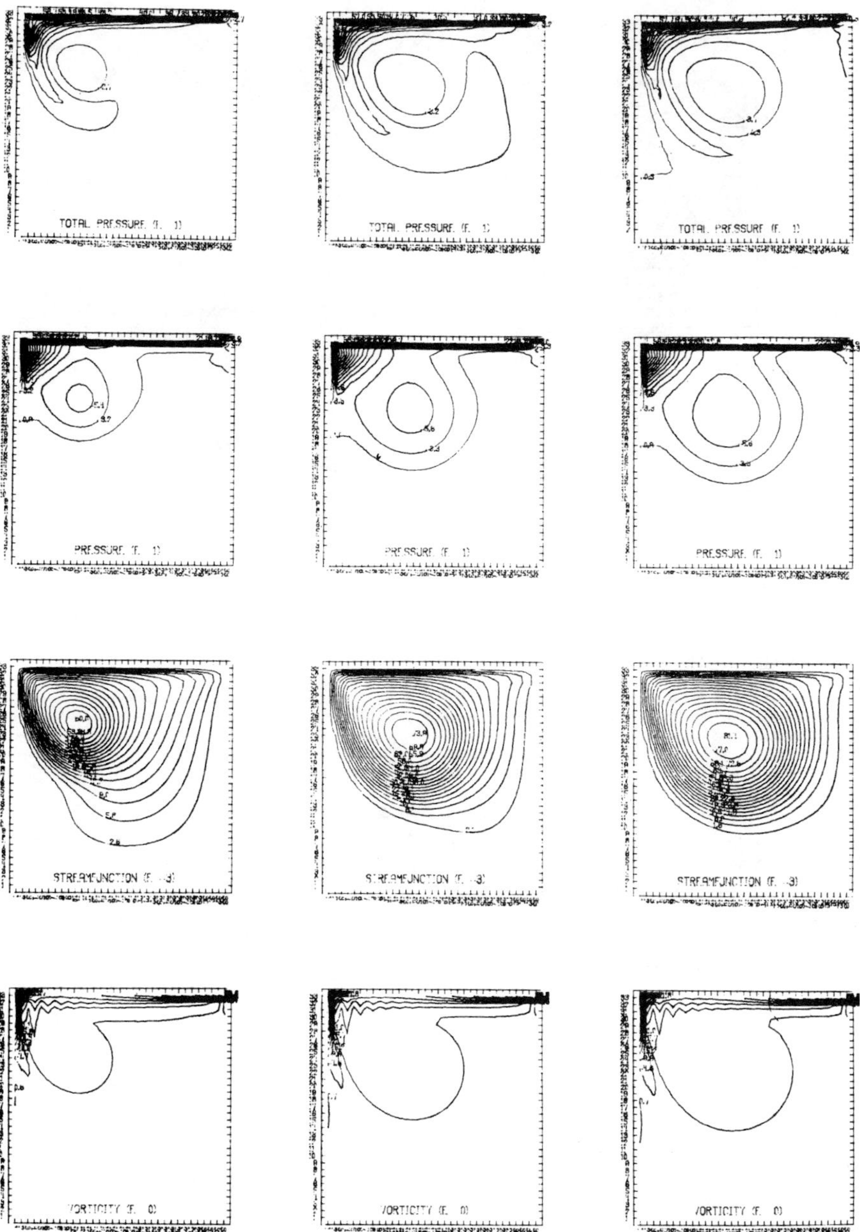

Figure 2 - Time-dependent flow in a square cavity produced by the impulsive motion of the upper wall. Re = 1000, t = 5,10 and 15. Implicit scheme/explicit algorithm with $\Delta t = 0.05$ and $\varepsilon = 10^{-3}$ (4th-order centred-difference approximation to the advective terms, mesh 33x33).

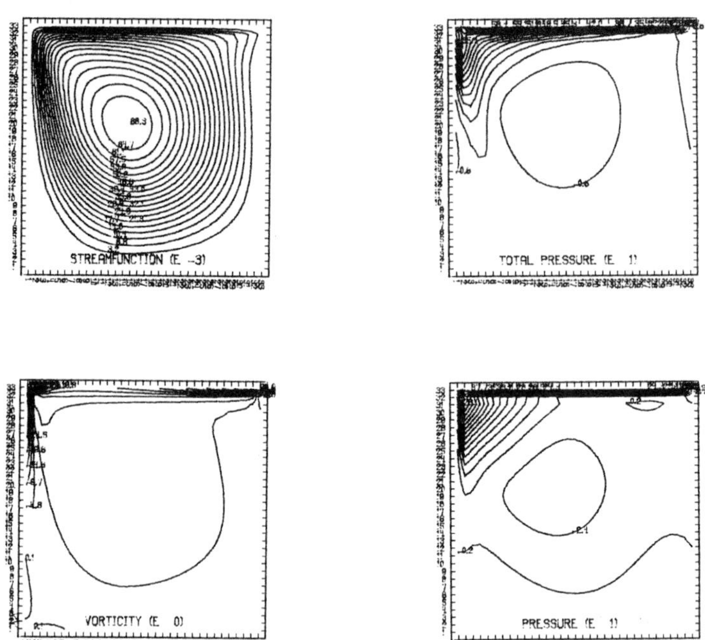

Figure 3. Flow pattern in a square cavity for Re = 1000 at t = 30. ψ-implicit scheme, Δt = 0.05. Steady-state solution within 5.10^{-4} (4th-order centred-difference approximation to the advective terms, mesh 33x33).

REFERENCES

1. Mc LAURIN, J.W. - A General Coupled Equations Approach for Solving the Biharmonic Boundary Value Problem, SIAM J. Numer. Anal., Vol. 11, pp.14-33, 1974.

2. FIX, G.J. - Hybrid Finite Element Methods, SIAM Review, Vol. 18, No. 2, pp. 460-480, 1976.

3. AXELSSON, O., and GUSTAFSSON, I. - An Iterative Solver for a Mixed Variable Variational Formulation of the (First) Biharmonic Problem, Comp. Meth. Appl. Mech. Eng., Vol. 20, pp. 9-16, 1979.

4. QUARTAPELLE, L. and VALZ-GRIS, F. - Projection Conditions on the Vorticity in Viscous Incompressible Flows, Int. J. Num. Meth. Fluids, Vol. 1, No. 1, 1981.

5. GLOWINSKI, R. and PIRONNEAU, O. - Numerical Methods for the Biharmonic Equation and for the Two-Dimensional Stokes Problem, SIAM Review, Vol. 21, No. 2, pp. 167-212, 1979.

6. BURGGRAF, O. - Analytical and Numerical Studies of the Structure of Steady Separated Flows, J. Fluid Mech., Vol. 24, No. 1, pp. 113-151, 1966.

7. OZAWA, S. - Numerical Studies of Steady Flow in a Two-Dimensional Square Cavity at High Reynolds Numbers. J. Phys. Soc. Japan, Vol. 30, No. 3, pp. 889-895, 1975.

8. ARAKAWA, A. - Computational Design for the Long-Term Numerical Integration of the equations of Motion: Two-Dimensional Incompressible Flow. Part I, J. Comput. Phys., Vol. 1, pp. 119-143, 1966.

INVESTIGATION OF SOLUTION OF NAVIER-STOKES EQUATIONS USING A VARIATIONAL FORMULATION

R.K. Rout, A. Ecer*, and P. Ward
Structural Dynamics Research Corporation
Milford, Ohio 45150 U.S.A.

1. SUMMARY

A variational formulation was developed for the analysis of two-dimensional, incompressible, and viscous flows[1]. The main objective of the present paper is to demonstrate the applicability of this approach for the solution of practical problems and in particular to investigate the introduction of boundary conditions to the Navier-Stokes equations through a variational formulation. The application of boundary conditions for typical internal and external flow problems are presented. Sample cases include flow around a cylinder and flow through a stepped channel.

Quadrilateral, bilinear isoparametric elements are utilized in the formulation. A single-step, implicit, and fully coupled numerical integration scheme is employed based on the variational principle. Presented results include sample cases with different Reynolds numbers for laminar and turbulent flows. Turbulence is modelled using a simple mixing length model. Numerical results show good agreement with existing solutions.

2. INTRODUCTION

The numerical solution of Navier-Stokes equations for incompressible, viscous flows requires the coupled solutions of two types of equations. The first equation is the condition of incompressibility:

$$u_x + v_y = 0. \tag{1}$$

---

* Professor, Purdue University at Indianapolis, Indianapolis, Indiana 46205.

This condition is a kinematic constraint on the velocity field (u,v). A second set of equations specifies the conservation of momentum:

$$u_t + u\, u_x + v\, u_y = - \frac{1}{p} p_x + \nu \nabla^2 u \tag{2}$$

$$v_t + u\, v_x + v\, v_y = - \frac{1}{p} p_y + \nu \nabla^2 v \tag{3}$$

These equations are nonlinear due to the presence of the convection terms. The solution of the above set of equations requires the treatment of the coupling between the two types of equations. Boundary conditions for u,v and consequently p have to be included consistently in the formulation. In practice, however, one observes that coupling between pressures and velocities produces difficulties both for the coupled numerical integration of the system and the introduction of the boundary conditions.

A second approach to the solution of Navier-Stokes equations is to introduce vorticity as a new variable and replace equations (2) and (3) with the vorticity transport equation:

$$v_x - u_y = \zeta \tag{4}$$

$$\zeta_t + u\, \zeta_x + v\, \zeta_y = \nu\, (\zeta_{xx} + \zeta_{yy}) \tag{5}$$

In this case, pressure is eliminated as a variable. A new kinematic relationship as defined in equation (4) is introduced and only one vorticity transport equation has to be integrated. Now, the problem reduces to one of obtaining a coupled solution of the velocities u,v and the vorticity ζ with the proper boundary conditions. In practice, again difficulties are encountered in obtaining full coupling between these equations, especially at the boundaries.

In this paper, the solution of Navier-Stokes equations are discussed based on a variational form. Using a Clebsch type of transformation:

$$u = \varphi_x + \beta_y - \eta\zeta_x \;, \tag{6}$$

$$v = \varphi_y - \beta_x - \eta\zeta_y , \tag{7}$$

a variational formulation for the solution of equations (1, 4, and 5) was presented in Reference 1. The details of the formulation are not repeated here. However, the importance of the proper treatment of the boundary conditions is presented in detail.

3. METHOD OF SOLUTION

A variational functional can be defined for the solution of Navier-Stokes equations in equations (1, 4, and 5) in terms of basic variables (φ, β, ζ, η) as defined in equations (6 and 7). The Lagrangian, L, for this case can be written as follows:

$$L = \tfrac{1}{2}(\varphi_x + \beta_y - \eta\zeta_x)^2 + \tfrac{1}{2}(\varphi_y - \beta_x - \eta\zeta_y)^2$$

$$- \beta\zeta + \nu\eta\nabla^2\zeta + \eta\zeta_t \qquad (8)$$

From the above Lagrangian, one can proceed to show that the minimization of the variational functional satisfies equations (1, 4, and 5) with the following natural boundary conditions on the boundary:

$$\int_C (u_n - p)\delta\varphi \, dC = 0 \qquad (9)$$

$$\int_C (u_t - q)\delta\varphi \, dC = 0 \qquad (10)$$

$$\int_C \eta^2 [u_n (\frac{\delta\zeta}{\eta}) - \nu \frac{\partial}{\partial n} (\frac{\delta\zeta}{\eta})] \, dC = 0 \qquad (11)$$

where n and t are defined in the normal and tangential directions to the boundary, C, respectively, and p and q are specified values of normal and tangential values on the boundary, C.

Bilinear, four-noded isoparametric elements were employed in the formulation. At each node of the element four variables (φ_i, β_i, ζ_i, η_i) were specified as unknowns. The resulting set of implicit nonlinear equations are written in matrix form as:

$$\underline{K}^{n+1} \, \underline{\varphi}^{n+1} = \underline{f}^{n+1} \qquad (12)$$

where $\underline{K}$ is an unsymmetric matrix and is a nonlinear function of the nodal variables. The equations are solved in a fully coupled fashion. The vector $\underline{f}$ includes the effects of the specified natural boundary conditions, p and q. To improve the efficiency of the solution scheme a constant coefficient matrix was employed in the following form[2]:

$$\underline{K}^o \, \tilde{\underline{\varphi}}^{n+k} = f^{n+1} + (\underline{K}^o - \underline{K}^{n+1}) \, \underline{\varphi}^n \qquad (13)$$

with

$$\underline{\varphi}^{n+1} = \omega\, \tilde{\underline{\varphi}}^{n+1} + (1 - \omega)\, \underline{\varphi}^{n} \qquad (14)$$

where $\underline{K}^0$ is the coefficient matrix formulated at the first time step and ω is a relaxation parameter.

The main advantages of the above formulation can be summarized in two parts:

a. The resulting formulation produces a fully coupled system for the solution of equations.

b. Most of the flow boundary conditions become natural boundary conditions.

In order to illustrate these advantages two sample problems were investigated using the finite element grids shown in Figures 1a and 1b. In the case of two-dimensional, symmetric flow around a cylinder shown in Figure 1a, the boundary conditions can be specified as follows:

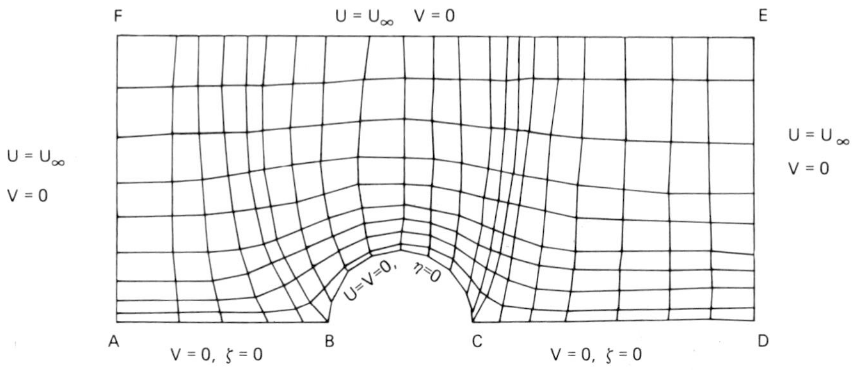

Figure 1a. Finite Element Model for Circular Cylinder

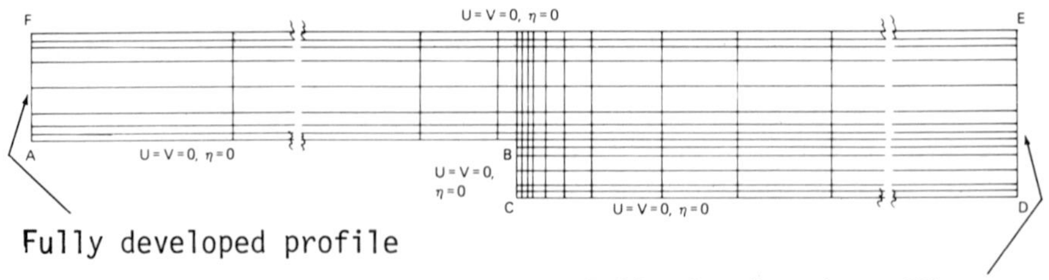

Figure 1b. Finite Element Model for Stepped Channel

- o On the free stream boundaries (D-E-F-A), u and v are specified as natural boundary conditions.

- o On the line of symmetry (A-B,C-D), v=0 is specified as natural boundary condition. ζ is specified as a forced boundary condition. β is specified as a forced boundary condition and then u on this line is calculated automatically.

- o On the cylinder surface (B-C), u=v=0 is specified as natural boundary conditions. Also, $\eta = 0$ is specified as a forced boundary condition. From equation (11), this corresponds to defining the rigid wall as a vorticity generating surface. Vorticity is automatically calculated from the kinematic relationship given in equation (4).

For the second sample problem of a channel with a backward facing step, inlet and outlet boundary conditions are specified as natural boundary conditions for velocities u and v. Vorticity specification is not required on these boundaries. Along the solid walls of the channel (A-B-C-D, and E-F), natural boundary conditions for (u=v=0) and the vorticity generating boundary condition (η=0) are specified.

The above examples illustrate the practicality of the present approach for coupling vorticity and velocities. As stated in the introduction previously, this has been a major difficulty in the solutions of Navier-Stokes equations. While only two of of the three basic variables (u, v, ζ) are needed to specify the boundary conditions uniquely, the coupled form of the governing equations as derived from the variational form provides the necessary coupling to determine the third variable.

5. MODELLING OF THE TURBULENT FLOW

A simple model was defined to account for the turbulence in the flow based on a mixing length theory[3]. The eddy viscosity near the wall region is defined as:

$$\mu_T = \rho \ell_p^{\,2} \left| \frac{\partial u}{\partial y} + \frac{\partial v}{\partial x} \right| , \tag{15}$$

where

$$\ell_p = 0.4\, y\, [1 - \exp\,(-y \sqrt{\left|\frac{\rho}{\mu}\frac{\partial u}{\partial y}\right|_w} /26)] \tag{16}$$

until the following value of ℓ_p is reached and subsequently used:

$$\ell_p = 0.07\,\delta \tag{17}$$

where δ is an arbitrary cutoff criterion based on the vorticity. For the wake region, the eddy viscosity is calculated from the following equation:

$$\mu_T = 0.001176\rho\delta \, |u_\delta - u_{\mathbb{C}}| \tag{18}$$

where u_δ and $u_{\mathbb{C}}$ are the velocities at the edge of the wake and its centerline, respectively.

6. DISCUSSION OF RESULTS

Several test cases for the solution of Navier-Stokes equations were analyzed based on the formulation presented above. These results were obtained for the flow around the cylinder and for the flow through a stepped channel using the finite element grids shown in Figures 1a and 1b. These cases are typical examples of external and internal flow problems with separation. Present results were compared with existing numerical and experimental results.

6.1 Flow Around a Cylinder:

Flow around a cylinder for Re = 200 was calculated using the developed numerical procedure. In this case, the Reynolds number is defined as $Re = U_\infty D/\nu$, where U_∞ is the free stream velocity, D is the diameter of the cylinder and ν is the kinematic viscosity. $\Delta t U_\infty/D=0.05$ was employed for numerical integration of the equations. Fifteen steps were required to reach a steady state solution. The obtained numerical results are shown in Figure 2(a-b).

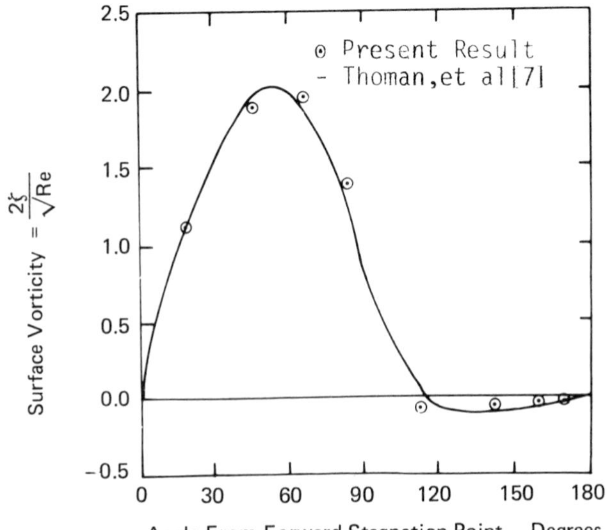

Figure 2a. Vorticity Distributions for **Circular Cylinder** (Re = 200)

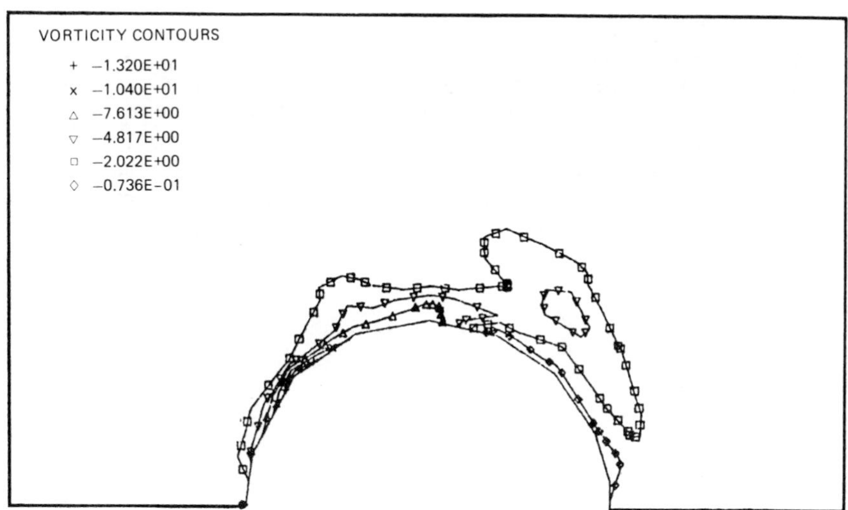

Figure 2b. Vorticity Contours Around Circular Cylinder (Re=200)

6.2 Flow in a Stepped-Channel:

Flow in a stepped-channel was analyzed using the finite element grid shown in Figure 1b. Fourteen elements were placed in the flow direction. Eight elements were placed across the duct in the upstream of the step and fourteen elements were placed in the downstream of the step as shown in the figure. Both laminar and turbulent flow cases were investigated for this problem. The flow Reynolds number is given as $Re = U_o h/\nu$, where U_o is the average velocity at the inlet and h is the step height. Laminar flow cases include Re=25, 73, and 229. Turbulent flow calculations were obtained for Re=3025. For the last two test cases, the grid was extended as shown in Table 1, while the number of elements were kept the same. The velocity profiles for all of the above

Table 1. Number of Iterations for Converged Solutions for Different Reynolds Numbers

Reynolds	△t	Number of Iterations	Inlet Length	Outlet Length
25	0.1	125	- 25.0	+ 25.0
73	0.1	170	- 25.0	+ 25.0
229	0.01	210	- 70.0	+ 70.0
3025	0.01	30*	- 70.0	+ 70.0

* Solution of Re=229 is used as an initial distribution for Re=3025

test cases are presented in Figures 3 (a-d) and compared with results by others[4,5,6]. Also, in Figure 4, variation of the length of the separated region is shown as a function of the Reynolds number and compared with previously published results. The presented results show good overall agreement. The time steps and the number of iterations to obtain convergence are also summarized in Table 1 for flows with different Reynolds numbers.

⊙ Present Finite Element Results
- Atkins, et al [5]
x Taylor, et al [4]

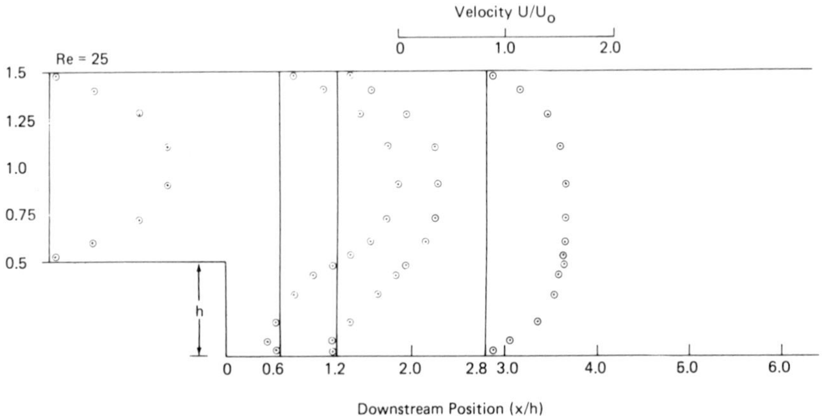

Figure 3a. Axial Velocity Profile for Stepped Channel (Re=25)

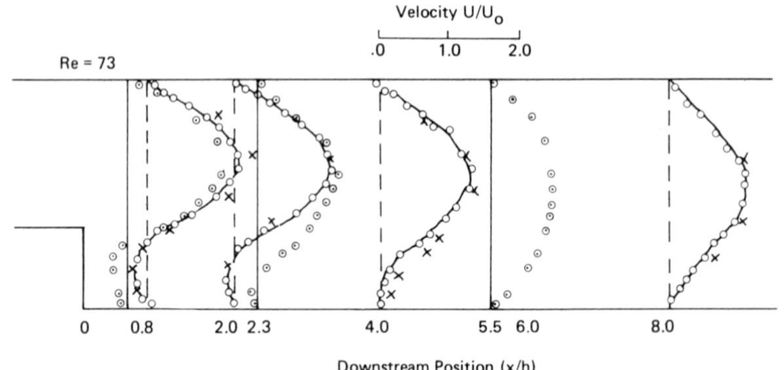

Figure 3b. Axial Velocity Profile for Stepped Channel (Re=73)

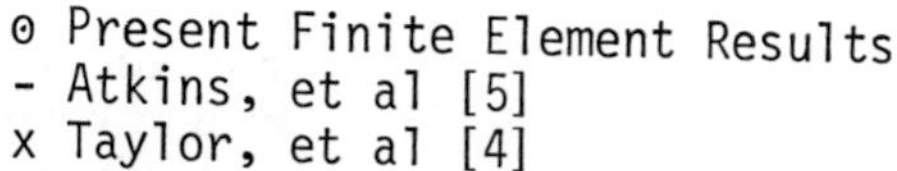

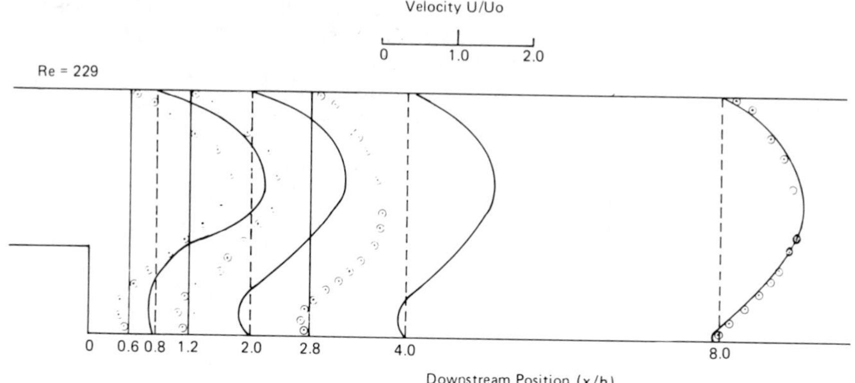

Figure 3c. Axial Velocity Profile for Stepped Channel (Re=229)

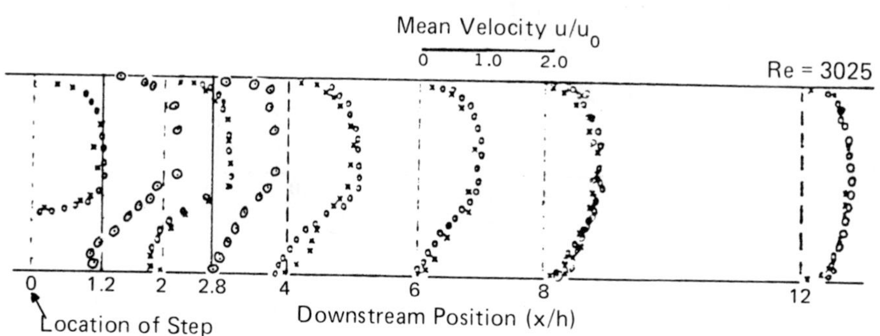

Figure 3d. Mean Velocity Profile for Turbulent Flow (Re=3025)

Finite Element Method:
- ■ Present results
- ⊗ Fully developed inlet profiles[4]
- △ Experimental inlet profiles[4]
- ☼ Fully developed inlet profiles[5]
- + Experimental inlet profiles[5]

Measurements:
- ● Dye tracer[6]
- o Laser anemometer[6]

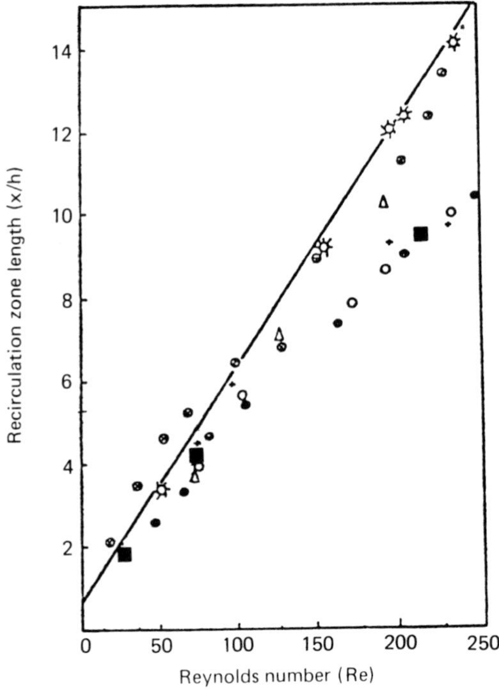

Figure 4. Variation of Separation Zone Length with Flow Reynolds Number

6.3 Convergence Characteristics with the Constant Coefficient Matrix

One numerical experiment was performed to determine the convergence characteristics of a solution scheme in which the coefficient matrix was updated periodically. This drastically improves the most costly part of a step, namely the decomposition of the global coefficient matrix.

Convergence results are shown in Figure 5. The inlet wall vorticity for Re=73 in the stepped channel

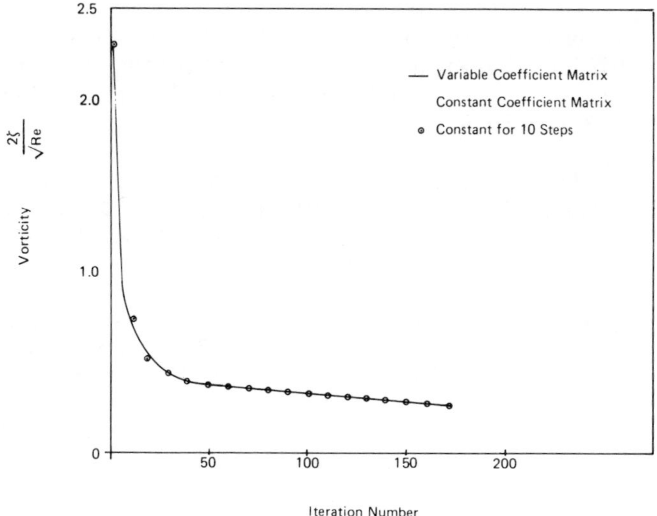

Figure 5. Comparison of Convergence of Constant and Variable Coefficient Matrix

example is plotted as a function of the number of time steps. Results for which the coefficient matrix was decomposed at every step are shown as a solid line. It can be seen from Figure 5 that the results obtained by decomposing the coefficient matrix in every step and every ten steps show negligible difference.

7. REFERENCES

1. ECER, A. - Variational Formulation of Viscous Flows, Int. J. for Numerical Methods in Engineering, Vol. 15, 1355-1361 1980.

2. AKAY, H.U. and ECER, A. - Transonic Flow Computations in Cascades Using Finite Element Methods, ASME 26th International Gas Turbine Conference, Houston, Texas, March 8-12, 1981, to be published at ASME Journal of Fluids.

3. DEIWERT, G.S. - Numerical Solution of High Reynolds Number Transonic Flows, AIAA Paper 74-603, presented at AIAA 7th Fluid and Plazma Dynamics Conference in Palo Alto, June 17-19, 1974.

4. TAYLOR, C., THOMAS, C.E., and MORGAN, K. - Confined Turbulent Flow Utilizing the Finite Element Method, Finite Element Methods for Convection Dominated Flows, Ed. by T.J.R. Hughes, AMD Vol. 34, ASME, 213-224 1979.

5. ATKINS, D.J., MASKEL, S.J., and PATRICK, M.A. - Numerical Predictions of Separated Flows, Int. J. for Numerical Methods in Engineering, Volume 15, 129-144 1980.

6. DENHAM, M.K. and PATRICK, M.A. - Laminar Flow Over a Downstream-facing Step in a Two-dimensional Flow Channel, Tans.Gnst.Chem. Engrs 52(4), 361-367 1974.

7. THOMAN, DAVID C., and SZEWCZYK, ALBIN A. - Time Dependent Viscous Flow over a Circular Cylinder, High-Speed Computing in Fluid Dynamics, The Physics of Fluids Supplement II, pp 75-86, 1969.

A NODAL GREEN'S TENSOR METHOD FOR THE EFFICIENT NUMERICAL SOLUTION OF LAMINAR FLOW PROBLEMS

By

W. C. Horak
Reactor Safety Division
Brookhaven National Laboratory
Upton, NY 11973

and

J. J. Dorning
Nuclear Engineering Program
University of Illinois
Urbana, IL 60801

SUMMARY

A coarse-mesh nodal method for the efficient numerical solution of incompressible laminar flow problems is developed using a transverse integration procedure followed by the introduction of locally-defined Green's tensors of the transverse-integrated in-node Navier-Stokes and mass conservation equations.

In applications to two-dimensional flow problems, including fully-developed flow, inlet flow, and modified driven cavity problems (driven cavites with inlet and outlet sections), this new nodal Green's tensor method is demonstrated to have very high accuracy even when applied on very large nodes. The high accuracy of this new method on very coarse meshes leads to a high computational efficiency (reduced computer time for fixed accuracy requirements).

1. INTRODUCTION

A new coarse-mesh nodal method, based *in part* on two concepts used in the development of high efficiency computational methods for neutron diffusion problems in fission reactors, [1,2] has been developed and applied to the numerical solution of laminar incompressible flow problems. In the first application to steady-state neutron diffusion problems, locally defined multi-dimensional Green's functions were used to convert differential equations applied locally over subdomains of the system to local multi-dimensional integral equations[1]. Later, this concept was extended to the solution of steady-state and transient heat-conduction problems[3,4]. When these local Green's function methods were improved by combining them with a transverse integration procedure for converting partial differential equations into sets of ordinary differential equations, computational efficiencies one-thousand times greater than conventional fully-accelerated fine-mesh finite difference method efficiencies were obtained for practical three-dimensional problems[2].

In the present method, *locally defined Green's tensors* of scalar arguments are used to convert a set of ordinary differential equations, obtained by integrating the partial differential equations of fluid flow over one independent variable, to a set of local one-dimensional integral equations. The "transverse-integrated" velocities and pressure are described by these local integral equations in terms of an integral of the source within the volume element and integrals of related quantities defined only on the surfaces of the volume element. A nonlinear algebraic equation set for the related surface quantities is then developed by evaluating the local integral equations at element surfaces. This nonlinear algebraic equation set is solved by a Newton-Raphson procedure. The surface quantities are then substituted into the local integral equations for the transverse-integrated velocities and pressure; these equations are solved on an element-by-element basis through the application of a local weighted residuals procedure.

2. FORMALISM

2.1 Transverse Integration

The development of the formalism for two-dimensional problems begins by decomposing the fluid system into L rectangular volume elements (nodes), $V^\ell = (-a_\ell \leq x \leq +a_\ell,\ -b_\ell \leq y \leq +b_\ell, \ell = 1,\ldots,L)$. The coupled mass conservation equation and Navier-Stokes equations are then integrated over one of the two independent variables (e.g., the x-variable) within the node to form

$$\begin{bmatrix} -\nu \dfrac{d^2}{dy^2} & 0 & 0 \\ 0 & -\nu \dfrac{d^2}{dy^2} & \dfrac{d}{dy} \\ 0 & \dfrac{d}{dy} & 0 \end{bmatrix} \begin{Bmatrix} \bar{u}^x(y) \\ \bar{v}^x(y) \\ \bar{p}^x(y) \end{Bmatrix} = \begin{Bmatrix} \bar{S}_1^x(y) \\ \bar{S}_2^x(y) \\ \bar{S}_3^x(y) \end{Bmatrix} \qquad (1)$$

where the total source term $\underline{\bar{S}}^x(y)$ consists of the transverse-integrated mass sources and body forces within the node, the transverse-integrated advection terms [which have been included in the source terms in anticipation of the application of the local Green's function method which requires that the equations be *formally linear*], and the effective source vector, $\underline{\bar{Q}}^x(y)$,

$$\begin{Bmatrix} \bar{Q}_1^x(y) \\ \bar{Q}_2^x(y) \\ \bar{Q}_3^x(y) \end{Bmatrix} = \begin{Bmatrix} \nu \dfrac{\partial}{\partial x} u(x,y) \Big|_{-a}^{+a} - p(x,y) \Big|_{-a}^{+a} \\ \nu \dfrac{\partial}{\partial x} v(x,y) \Big|_{-a}^{+a} \\ - u(x,y) \Big|_{-a}^{+a} \end{Bmatrix} \tag{2}$$

A similar transverse integration is done over the y-variable. These two "transverse integrations" result in two coupled sets (coupled through the total source terms) of three coupled ordinary differential equations.

2.2 Local Green's Function Formalism

Each of the two sets of coupled ordinary differential equations are converted to three coupled local integral equations within a node through the use of a Green's tensor (a 3x3 matrix of Green's functions) which satisfies (for the y-dependent, x-integrated quantities)

$$\underline{\underline{A}}^\dagger \, \underline{\underline{G}}^\dagger \, (y|y_o) = \delta(y - y_o) \, I,$$

where $\underline{\underline{A}}^\dagger$ is the adjoint of the matrix operator in Eq. (1). Multiplying Eq. (1) and Eq. (3) by $[G^\dagger \, (y|y_o)]^T$ and $\{\bar{u}^x(y), \bar{v}^x(y), \bar{p}^x(y)\}$ respectively, subtracting, integrating over $(-b_\ell, + b_\ell)$, and applying Gauss' theorem followed by several additional steps (See Chapter 4 of Ref. 5) yields the local integral equations (for the y-dependent, x-integrated variables)

$$\begin{Bmatrix} \bar{u}^x(y_o) \\ \bar{v}^x(y_o) \\ \bar{p}^x(y_o) \end{Bmatrix} = \left\{G_{1i}^\dagger(+b|y_o)\right\} J_{\bar{u}^x}^{y,\,m\to\ell} + \left\{G_{1i}^\dagger(-b|y_o)\right\} J_{\bar{u}^x}^{y,\,k\to\ell}$$

$$+ \left\{G_{2i}^\dagger(+b|y_o)\right\} J_{\bar{v}^x}^{y,\,m\to\ell} + \left\{G_{2i}^\dagger(-b|y_o)\right\} J_{\bar{v}^x}^{y,\,k\to\ell} \tag{4}$$

$$+ \int_{-b}^{+b} dy \Big[\left\{G_{1i}^\dagger(y|y_o)\right\} \bar{S}_1^x(y) + \left\{G_{2i}^\dagger(y|y_o)\right\} \bar{S}_2^x(y)$$

$$+ \left\{G_{3i}^\dagger(y|y_o)\right\} \bar{S}_3^x(y) \Big]$$

where the J-quantities are defined on the (constant-y) surfaces of the computational node as

$$J_{\bar{u}^x}^{m_y \to \ell} = U\,\bar{u}^x(+b) + \nu \frac{d}{dy}\bar{u}^x(y)\bigg|_{+b} \,, \tag{5a}$$

$$J_{\bar{u}^x}^{k_y \to \ell} = U\,\bar{u}^x(-b) - \nu \frac{d}{dy}\bar{u}^x(y)\bigg|_{-b} \,, \tag{5b}$$

$$J_{\bar{v}^x}^{m_y \to \ell} = U\,\bar{v}^x(+b) + \left(\nu \frac{d}{dy}\bar{v}^x(y)\bigg|_{+b} - \bar{p}^x(+b)\right) \,, \tag{5c}$$

$$J_{\bar{v}^x}^{k_y \to \ell} = U\,\bar{v}^x(-b) - \left(\nu \frac{d}{dy}\bar{v}^x(y)\bigg|_{-b} - \bar{p}^x(-b)\right) \,. \tag{5d}$$

Here U is an arbitrary parameter introduced into the formalism. A similar set of equations are developed analogously for the y-integrated, x-dependent quantities. Eq. (4) and its x-dependent analog comprise two sets of local integral equations which are exact expressions for the transverse-integrated velocities and pressures in terms of the J-quantities, which are defined only on the surfaces of the computational node, and a volume integral of the total source terms.

A set of local integral equations for the J-quantities is developed using

$$J_{\bar{g}^z}^{\ell \to i} = 2\,U\bar{g}^z(\pm c) - J_{\bar{g}^z}^{i \to \ell} \,;\; g = u,v;\; z = x,y;\; c = b,a;\; i = k_y, m_y, k_x, m_x, \tag{6}$$

and substituting for the transverse-integrated surface velocities (obtained by evaluating Eq. (4) or its y-integrated, x-dependent counterpart at the node surfaces). The above relation is equivalent to stating that the velocities and stresses are continuous at the nodal interfaces.

The total source terms are evaluated by making zero-order expansions for them and requiring first that the nodal-average velocities and pressures be consistently calculated from the transverse-integrated velocities and pressures, and second that the forces balance within the node. A procedure for higher-order expansions, which has not yet been implemented, is developed in Appendix F of Ref. 5.

2.3 Discrete Equation Formalism

As in the other recently developed local Green's function methods[1-4], the J-quantities are determined first. Since the algebraic equation set for the J-quantities formed by combining the local integral equations with a set of algebraic equations which describe the specified boundary conditions in terms of the J-quantities is nonlinear (due to the advection terms), a Newton-Raphson procedure is used to solve for the J-quantities. The sets of linear algebraic equations in the J-quantities formed during the Newton-Raphson process can be solved using direct inversion, due to the sparsity of the resulting Jacobian, or by an iterative technique in which the J-quantities defined on constant y-surfaces are alternately solved with the J-quantities defined on constant x-surfaces.

After the J-quantities have been determined, the local integral equations for the transverse-integrated velocities and pressure are solved using a local weighted residuals procedure. Independent local (low-order) expansions are made for the transverse-integrated quantities and the expansion coefficients determined. These equations for the expansion coefficients of the transverse-integrated velocities and pressure are independent of each other. Therefore, once the J-quantities are determined it is not necessary to solve for the transverse-integrated velocities in order to solve for the transverse-integrated pressure and vice versa.

3. NUMERICAL EXAMPLES

3.1 Fully-developed Flow

The method was applied to the calculation of fully-developed flow between parallel plates. Because the analytic total source term is a constant for this problem, the method yielded the exact analytical solution, a parabolic distribution.

3.2 Inlet Flow

To obtain an estimate of the computational efficiency of the new method, a flow field was calculated for the inlet section between two parallel plates, using the nodal method and a finite difference code, SOLA[6]. Comparisons of the numerical results were made by averaging the x-directed and y-directed velocities over four equal-area subdomains (quadrants) in the flow field. Reference solutions for these quadrant-average velocities were obtained via a double-Richardson extrapolation of the nodal Green's tensor method results (See Appendix) to eliminate the h^2 and h^4 error terms, since analysis of the SOLA results showed that they were far from exhibiting an asymptotic h or h^2 error behavior. The results of the calculation for Re=10 are summarized in Table 1, where only two quadrant y-directed velocities are shown, since the problem is symmetric about the centerline. As shown in Table 1, the nodal Green's tensor method obtained accurate results on meshes that are twice as coarse as the SOLA meshes [The right quadrant velocity values in Table 1 obtained by the SOLA code are deceptive due to its numerical treatment of the outflow boundary condition. The result of this treatment is that the flow field is forced artifically to have the fully-developed pattern even on relatively coarse meshes. The fact that the SOLA results are not yet fully spatially converged is evidenced by the changing left quadrant velocities.]. The computing times of this nodal method are much shorter than those of the SOLA code, which was run using its steady-state algorithm.

3.3 Modified-Driven Cavity

To further demonstrate the high accuracy of the new method, it was applied to the solution of a modified driven cavity problem. Fig. 1 shows the flow pattern for the cavity region of the problem, Re=1000, as calculated by the new method. The counter-flow pattern shown in the lower right hand corner represents a weak vortex that normally does not appear in the results of finite difference calculations at this Reynolds number unless extremely fine meshes are used. The computational efficiency (computer time for fixed accuracy requirements) of the new method relative to the SOLA code was determined for this problem at Re=1,10, and 100. Fig. 2 shows the CPU time for this problem relative to SOLA for the same average relative accuracy of the quadrant-averaged velocity magnitudes in the cavity region.

4. CONCLUSION

A new nodal method based on the use of local Green's tensors has been developed and applied to the numerical solution of laminar incompressible flow problems. In a series of test problems, the new method was shown to yield very high accuracies even when applied to very large nodes, and thus high computational efficiencies when compared to the finite difference code SOLA[6].

Table 1: Comparison of the Results Obtained for an Inlet Flow Problem for Re=10 with Those Obtained Using the SOLA Code

Mesh	Lower Left Quadrant		Lower Right Quadrant		CPU(s)[b]
	v	%error[a]	v	%error[a]	
SOLA					
12x12	0.884×10^{-3}	20.2	1.215×10^{-4}	14.1	4[c]
16x16	0.948×10^{-3}	14.4	1.204×10^{-4}	14.9	15[c]
24x24	1.006×10^{-3}	9.2	1.217×10^{-4}	13.9	100[c]
32x32	1.034×10^{-3}	6.7	1.227×10^{-4}	13.2	250[c]
LGTM					
4x4	0.911×10^{-3}	17.8	0.313×10^{-4}	77.9	0.6
6x6	1.031×10^{-3}	6.9	0.847×10^{-4}	40.1	2.0
8x8	1.065×10^{-3}	3.9	1.095×10^{-4}	22.6	6.0
12x12	1.089×10^{-3}	1.7	1.272×10^{-4}	10.0	30.0[d]
(EXT)[e]	1.108×10^{-3}		1.414×10^{-4}		

a - relative to extrapolated answer

b - all times for CDC-7600, OPT=1

c - CPU time estimated for specified accuracy of $\varepsilon = 10^{-5}$

d - results obtained from 6x12 half mesh, CPU time is estimate for 12x12 mesh

e - see Appendix

NOTE: All results for specified accuracy, $\varepsilon = 10^{-5}$

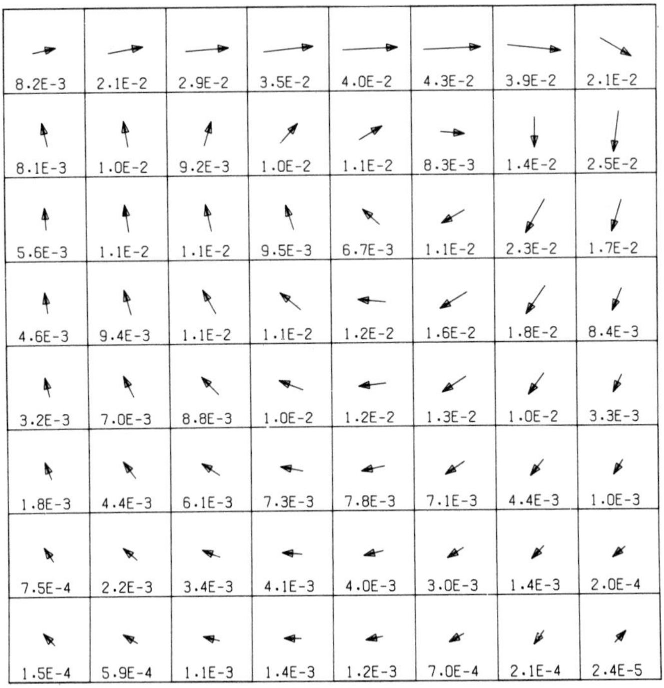

8.2E-3	2.1E-2	2.9E-2	3.5E-2	4.0E-2	4.3E-2	3.9E-2	2.1E-2
8.1E-3	1.0E-2	9.2E-3	1.0E-2	1.1E-2	8.3E-3	1.4E-2	2.5E-2
5.6E-3	1.1E-2	1.1E-2	9.5E-3	6.7E-3	1.1E-2	2.3E-2	1.7E-2
4.6E-3	9.4E-3	1.1E-2	1.1E-2	1.2E-2	1.6E-2	1.8E-2	8.4E-3
3.2E-3	7.0E-3	8.8E-3	1.0E-2	1.2E-2	1.3E-2	1.0E-2	3.3E-3
1.8E-3	4.4E-3	6.1E-3	7.3E-3	7.8E-3	7.1E-3	4.4E-3	1.0E-3
7.5E-4	2.2E-3	3.4E-3	4.1E-3	4.0E-3	3.0E-3	1.4E-3	2.0E-4
1.5E-4	5.9E-4	1.1E-3	1.4E-3	1.2E-3	7.0E-4	2.1E-4	2.4E-5

Fig. 1: Flow map of the cavity region of a modified driven cavity calculated using the local Green's tensor method on an 8 x 8 mesh for Reynolds number, Re = 1,000. Arrows indicate node-averaged velocity magnitudes and directions; numbers indicate node-averaged velocity magnitudes.

NOTE: Counter-flow pattern in lower right corner of cavity is resolved even on this 8 x 8 mesh.

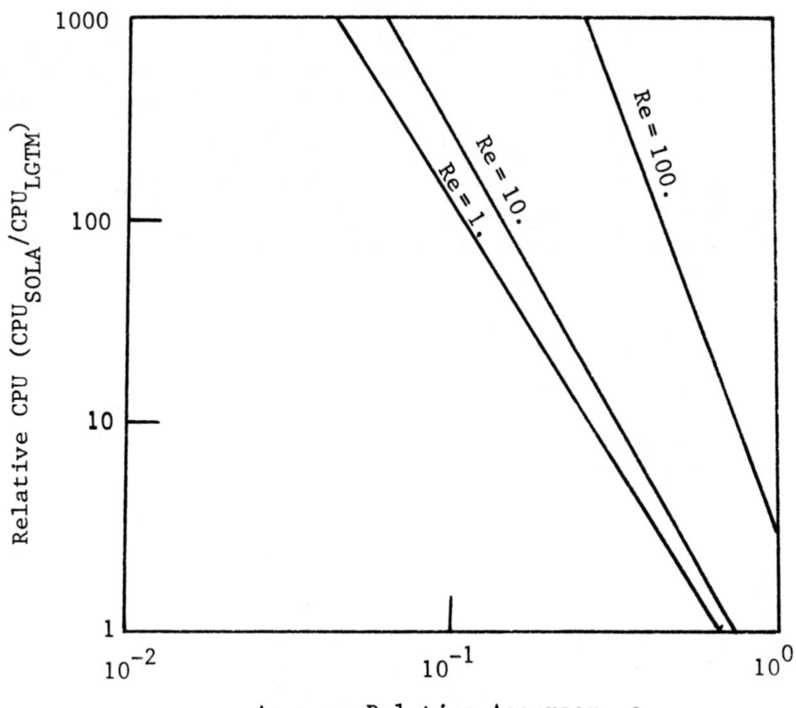

Fig. 2: Relative (cpu) computer time of the local Green's tensor method (LGTM) compared to that of the finite-difference code SOLA for the same average relative accuracy of the quadrant-averaged velocity values in the cavity region of a modified driven cavity for three values of the Reynolds number, Re.

REFERENCES

1. T. J. Burns, "The Partial Current Balance Method: A Local Green's Function Technique for the Numerical Solution of Multidimensional Neutron Diffusion Problems," Ph.D. thesis, Univ. of Illinois, Urbana, IL, 1975. See also, T. J. Burns and J. J. Dorning, "The Partial Current Balance Method: A New Computational Method for the Solution of Multi-dimensional Neutron Diffusion Problems," in Proc.of the Joint NEACRP/CSNI Specialists' Meeting on New Developments in Three-Dimensional Neutron Kinetics and Benchmark Calculations, p. 109, Laboratorium fur Reaktorregelung and Anlagensicherung, Garching (Munich), Germany (1975).

2. R. D. Lawrence, "A Nodal Green's Function Method for Multi-dimensional Neutron Diffusion Calculations," Ph.D. thesis, Univ. of Illinois (1979). See also, R. D. Lawrence and J. J. Dorning, "A Nodal Green's Function Method for Multidimensional Neutron Diffusion Calculations", Nucl. Sci, Eng., 76, 218 (1980).

3. W. C. Horak and J. J. Dorning, "A Local Green's Function Method for the Numerical Solution of Heat Conduction and Fluid Flow Problems, Nucl. Sci, Eng., 64, 192 (1977).

4. W. C. Horak and J. J. Dorning, "A Coarse-Mesh Method for Heat Flow Analysis Based Upon the Use of Locally-Defined Green's Functions", to be published in the Proceedings of the International Conference on Numerical Methods in Thermal Problems, Venice, 1981.

5. W. C. Horak, "Local Green's Function Techniques for the Solution of Heat Conduction and Incompressible Fluid Flow Problems," Ph.D. thesis, Univ. of Illinois (1980).

6. C. W. Hirt, B. D. Nichols, and N. C. Romero, SOLA--A Numerical Solution Algorithm for Transient Fluid Flows, LA-5852 (1975).

APPENDIX--Extrapolated Reference Solutions for the Inlet Flow Problem

For the lower left quadrant, the five decimal-place values, 1.03079, 1.06516, 1.8921, and 0.91105, 1.06516, 1.8921 were double-Richardson extrapolated to eliminate the h^2 and h^4 error terms and obtained the values 1.10815 and 1.10744, which agree to 0.0007, and the former extrapolated value was rounded off to 1.108. Similarly, for the lower right quadrant, 0.31348, 0.84696, 1.09549 and 1.2748 were used to obtain 1.41375 and 1.42131 which agree to 0.007 and the former rounded off to 1.414.

THE NUMERICAL COMPUTATION OF LAMINAR PIPE FLOW

F. J. K. IDERIAH[(i)]

ABSTRACT

The finite-difference EMIT procedure of Ideriah is employed to calculate both the developing and fully-developed regions of a laminar pipe flow. The work thus provides further validation of the procedure. The predictions are compared with both available experimental data and analytical results, and the agreement is very good. The computed results indicate that the length of the developing region is given by $4x/(D.Re) = 0.25$.

1. INTRODUCTION

Since the pioneering work of Hagen[1] and Poiseuille[2], there has been a considerable amount of research work done on laminar flow in circular pipes. Prominent among such works are those of Schiller[3] and Langhaar[4]. The study is of interest in both engineering and medicine : indeed, Poiseuille's research was aimed at understanding the flow of blood through veins[5]. For engineers, the study of flow in pipes is of importance in such areas as water distribution systems, industrial and domestic transmission of hot water, and transportation of crude oil. This class of flow has also formed the basis of the testing of several numerical techniques such as the work of Patankar and Spalding[6].

This paper deals with the numerical computation of both the developing and fully-developed regions of a laminar pipe flow. A finite-difference procedure is employed, the novel feature of which is the EMIT procedure of the author[7] : thus the work provides further application of the procedure. The computed results are compared with both available experimental data and analytical results, and the agreement is very good.

(i) Department of Mechanical Engineering, University of Ibadan, Ibadan, Nigeria.

2. THE GOVERNING EQUATIONS

The basic conservation equations governing laminar flow in a circular pipe, with uniform molecular viscosity, are:

Continuity:

$$\frac{\partial}{\partial x}(r\rho U) + \frac{\partial}{\partial r}(r\rho V) = 0 \qquad (1)$$

Axial momentum:

$$\frac{1}{r}\left\{\frac{\partial}{\partial x}(r\rho U^2) + \frac{\partial}{\partial r}(r\rho UV)\right\} = -\frac{\partial p}{\partial x} + \mu\nabla^2 U \qquad (2)$$

Radial momentum:

$$\frac{1}{r}\left\{\frac{\partial}{\partial x}(r\rho UV) + \frac{\partial}{\partial r}(r\rho V^2)\right\} = -\frac{\partial p}{\partial r} + \mu\left(\nabla^2 V - \frac{V}{r^2}\right) \qquad (3)$$

$$\text{where} \quad \nabla^2 = \frac{\partial^2}{\partial x^2} + \frac{\partial^2}{\partial r^2} + \frac{1}{r}\frac{\partial}{\partial r} \qquad (4)$$

In the foregoing equations, x and r represent the axial and radial co-ordinates of the pipe; U and V are respectively the axial and radial velocity components, and p, ρ and μ the local pressure, density and viscosity.

The boundary conditions are as follows: at the entrance of the pipe the velocities are those of the incoming flow, and no-slip and impermeable boundary conditions require the velocities to vanish at the walls; further, for point P in Fig. 1, the momentum flux through the wall is represented by the wall shear stress τ_w :

$$\tau_w = \left(\frac{\partial U}{\partial r}\right)_{wall} \qquad (5)$$

In addition, at the axis of the pipe,

$$\frac{\partial \phi}{\partial r} = 0 \qquad (6)$$

$$\text{where} \quad \phi = U, V, P \qquad (7)$$

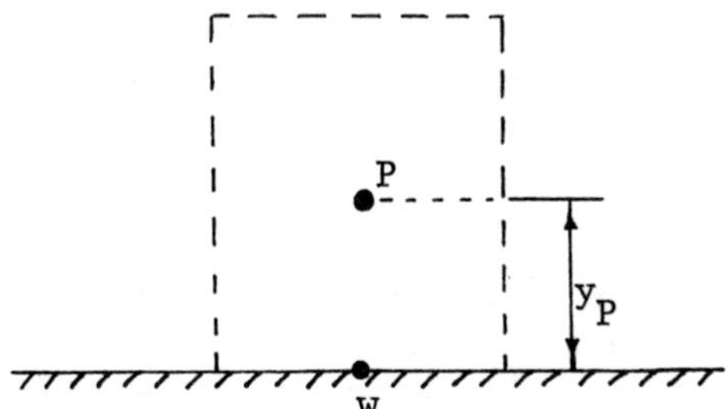

Fig. 1: Representation of near wall region

3. THE SOLUTION PROCEDURE

The foregoing transport equations, coupled with the boundary conditions are solved by a finite-difference technique whose novel feature is the EMIT procedure of the author[7]. Since the details of the procedure have been reported elsewhere[7], the method will be described here only briefly. The flow field is divided into series of grid lines or 'planes', and the procedure is employed to deduce the flow properties as follows:

(i) It solves along only one plane at a time, the elliptic pressure-linked momentum equations by first estimating the velocities from a guessed pressure field, and then correcting both velocities and pressures with a view to satisfying continuity via the solution of a Poisson equation for 'pressure correction'.

(ii) The new pressures thus obtained along a given plane are further re-adjusted so that they are correctly linked to 'upstream' values, based on the requirement that the upstream velocities remain unchanged.

(iii) Steps (i) and (ii) are applied at each plane of the flow domain in an ordered sequence until a complete 'pass' is made through the domain; the procedure is then applied repeatedly (i.e. several 'passes' are made through the flow domain), until a converged solution is obtained.

These constitute the basic features of the EMIT (= Elliptic Marching Integration Technique) procedure.

4. RESULTS AND DISCUSSIONS

4.1 Computational details

In the computations, a non-uniform grid is employed in the radial direction while a uniform grid is used in the axial direction. The overall length of the domain of computation is fixed such that, based on the work of Langhaar[4],

$$\frac{x}{D}\cdot\frac{4}{Re} > 0.4 \qquad (8)$$

where D is the diameter of the pipe, Re (= $\rho U_o D/\mu$) is the Reynolds number of the flow, and U_o is the mean velocity; and, owing to symmetric, the radius of the domain is half the diameter of the pipe. For a typical domain x/D = 18 : in this case, a grid having 21 nodes in the axial direction and 19 in the radial direction gives a grid-independent solution. A uniform velocity ($U = U_o$, and $V = 0$) is prescribed at the entrance of the pipe (x = o); this condition corresponds with the experimental situation with which comparison is made here.

For each pass of the EMIT procedure, the computed values of U and V at each plane are under-relaxed while the pressure corrections are not relaxed. About 50 passes give converged solutions.

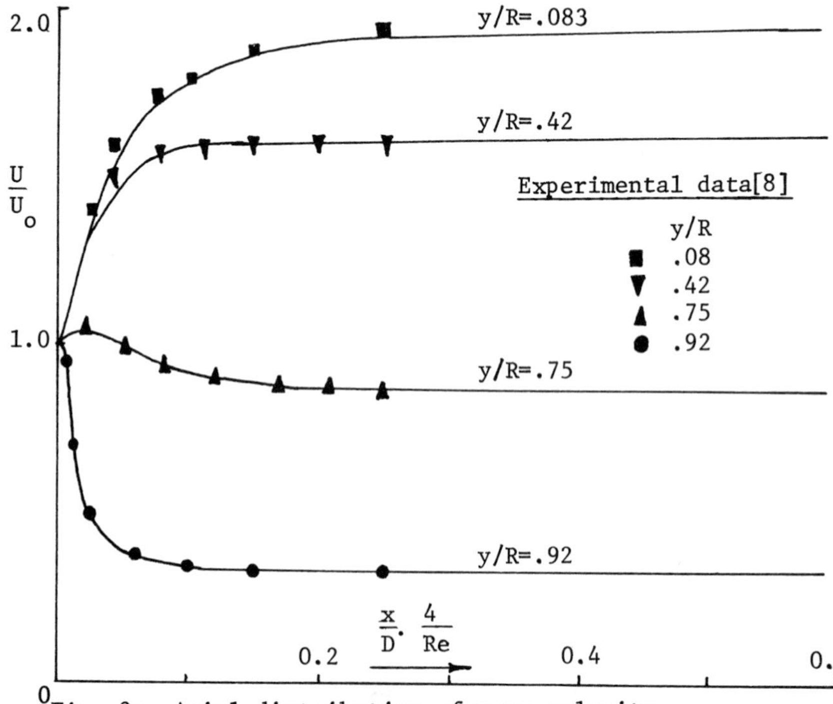

Fig. 2: Axial distribution of mean velocity

4.2 Results of the computations

Fig. 2 shows a comparison with the data of Nikuradse (quoted from Schlichting[8]) of the predicted axial distribution of the streamwise component of velocity for different radii of the pipe, and for both the developing and fully-developed regions of the flow. There y stands for distance from the axis of the pipe, and R is the radius of the pipe. The agreement between the predictions and the data is very good. The computations indicate that the length of the developing region is given by

$$\frac{x}{D}\cdot\frac{4}{Re} = 0.25 \qquad (9)$$

It may be noted that the approximate analysis of Schiller[3] gives a value of 0.115, while that of Langhaar[4] yields a value of 0.228.

Fig. 3 presents the computed radial distribution of the streamwise component of velocity in the fully-developed region

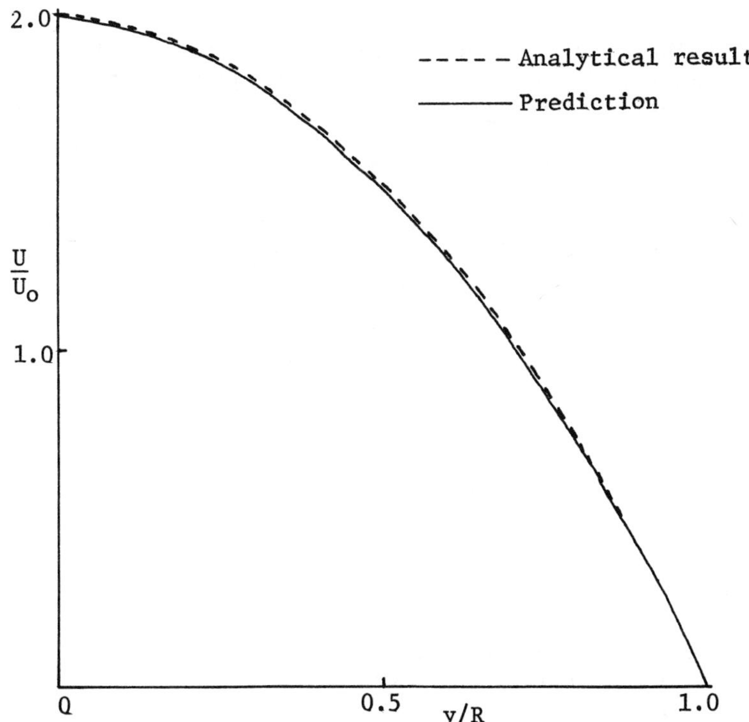

Fig. 3: Radial distribution of mean velocity (fully developed profile)

Also shown there for comparison is the analytical result which is often referred to as the Hagen-Poiseuille parabolic velocity distribution:

$$\frac{U}{U_o} = 2[1 - (\frac{y}{R})^2] \qquad (10)$$

Again, the agreement between the two profiles is very good.

5. CONCLUSION

The EMIT procedure of Ideriah has been employed to compute the developing and fully developed regions of the laminar flow in a pipe, thus providing a further validation of the procedure. The predictions have also been compared with both experimental data and analytical results, and the agreement is very good. The present results have shown that the length of the developing region is given by $4x/(D.Re) = 0.25$.

REFERENCES

1. HAGEN, G. - Uber die Bewegung des Wassers in engen zylindrischen Rohren. Pogg. Ann., Vol. 46, pp. 423-442, 1839.

2. POISEUILLE, J. - Researches experimentelles sur le mouvement des liquides dans les tubes de tres petits diametres. Comptes Rendus, Vol. 11, p. 961 and p. 1041, 1840; Vol. 12, p. 112, 1841; in more detail: Memoires des Savants Etrangers, Vol. 9, 1846.

3. SCHILLER, L. - Untersuchungen uber laminare und turbulente Stromung. ZAMM, Vol. 2, pp. 96-106, 1922; or Phys. Z., Vol. 23, p. 14, 1922.

4. LANGHAAR, H. - Steady flow in the transition length of a straight tube. J. Appl. Mech., Vol. 9, pp. A55-A58, 1942.

5. MASSEY, B.S. - Mechanics of Fluids, Van Nostrand, London, 1970.

6. PATANKAR, S.V. and SPALDING, D.B. - Heat and Mass Transfer in Boundary Layers, Intertext Books, London, 1970.

7. IDERIAH, F.J.K. - A numerical procedure for calculating partly elliptic flows. J. Mech. Engng. Sci., Vol. 21, No. 6, pp. 373-380, 1979.

8. SCHLICHTING, H. - Boundary-Layer Theory, McGraw-Hill, New York, 1968.

THE BOUNDARY ELEMENT METHOD APPLIED TO THE CREEPING MOTION OF A SPHERE

M.B. Bush* and R.I. Tanner**

Dept. of Mechanical Engineering, The University of Sydney

SUMMARY

The drag and flow field associated with a sphere moving at low Reynolds number through a viscous, incompressible fluid are computed using the boundary element method. Results obtained using simple elements show satisfactory agreement with the analytical solution.

1. INTRODUCTION

The boundary integral equation method has received a great deal of attention in the last decade due to its ability to handle complex geometric shapes without the large core requirements and computational effort usually needed for classical finite element techniques. Initial development and implementation of these methods in two and three dimensional elastostatics appears to be due to Rizzo [1] and Cruse [2,3]. The foundation of the method is flexible and has allowed boundary integral formulations to be applied to a wide class of problems including elastodynamics [4], analysis of composites [5] and low Reynolds number fluid dynamics [7].

The advantage of boundary integral equation methods is that discretization and numerical approximation is only necessary at the boundary, and not over the entire field. The solution at any selected interior location is then obtained from the boundary solution without the need for further approximation. This characteristic is particularly important for the class of problems in which the solution on the boundary is of major interest, since it avoids computation of a great deal of unwanted information about the solution in the interior. Such unwanted information would be supplied by a classical finite element solution. Problems falling into

* Research student
** Professor of Mechanical Engineering

this class include those containing a stress raiser or singularity [6], fluid flow with a free surface [7] and internal surface problems. The latter type of problem requires a finite element grid extending into the far (in principle, infinite) field, while the use of boundary elements automatically allows complete satisfaction of the far field conditions [8] without special devices.

In this paper we present a further example of the use of integral equations to solve an interior surface problem. The flow field due to the motion of a sphere through an incompressible viscous fluid at low Reynolds number is obtained using a one-dimensional array of nodes representing the semicircular meridian of the sphere. The flow field and drag so obtained show close agreement with the analytical solution.

2. APPLICATION TO FLUID DYNAMICS

In low Reynolds number fluid dynamics where the effects of inertia are small, the flow can be approximated by the motion of a zero density fluid, giving rise to the linear 'creeping flow' equations. The well known equations of motion and mass conservation for an incompressible viscous fluid can then be written using tensor notation as follows

$$\frac{\partial \sigma_{jk}}{\partial x_k} = 0 \qquad (1)$$

$$\frac{\partial u_k}{\partial x_k} = 0 \qquad (2)$$

where σ_{jk} is the stress tensor, u_k are the velocity components and x_k are the coordinates. Development of a boundary integral equation that provides a solution to (1) and (2) has been reported in [7], and is outlined here for convenience.

The familiar Galerkin procedure can be applied by considering an arbitrary set of fields u_j^*, p^* and σ_{ij}^*, and forming the expression:

$$\int_\Omega \frac{\partial \sigma_{jk}}{\partial x_k} u_j^* \, d\Omega + \int_\Omega \frac{\partial u_k}{\partial x_k} p^* \, d\Omega + \int_{\Gamma_1} (u_j - \hat{u}_j) t_j^* \, d\Gamma = 0 \qquad (3)$$

where Ω and Γ represent the field and boundary respectively, t_j^* is a traction tensor formed from σ_{ij}^* and the unit normal vector components at the surface, and $\hat{u}_j$ is the known velocity on the section of the boundary represented by Γ_1. By using Green's theorem and after some manipulation, (3) can be rewritten as:

$$0 = \int_\Omega \frac{\partial \sigma_{jk}^*}{\partial x_k} u_j \, d\Omega - \int_\Gamma u_j t_j^* \, d\Gamma + \int_\Gamma t_j u_j^* \, d\Gamma \qquad (4)$$

where it is understood that $u_j = \hat{u}_j$ on Γ_1.

If the *-fields are now taken to be the influence in direction j at point Q due to a unit source in direction i at point P and represented for example by $u_{ij}^*(P,Q)$, then following [8] we obtain

$$C_{ij}(P)\, u_j(P) = \int_\Gamma u_{ij}^*(P,Q)\, t_j\,(Q)\, d\Gamma - \int_\Gamma t_{ij}^*(P,Q)\, u_j(Q)\, d\Gamma \qquad (5)$$

where $u_j(Q)$ and $t_j(Q)$ represent the components at point Q and $C_{ij}(P)$ is a tensor of constants depending on the smoothness of the boundary at P. Equation (5) is the starting point for the boundary integral method. The kernel $u_{ij}^*(P,Q)$ is an influence function representing the velocity field due to a unit source and it is well known that this function is equivalent to the corresponding Green's function in linear elasticity; with Poisson's ratio equal to 0.5 (corresponding to the incompressible fluid). As previously indicated, these functions are well known in two and three dimensional coordinates. The kernel $t_{ij}^*(P,Q)$ represents the traction field due to the unit source and can similarly be obtained from the corresponding functions in elasticity.

3. SOLUTION IN CYLINDRICAL COORDINATES

The functions $u_{ij}^*(P,Q)$ and $t_{ij}^*(P,Q)$ have been obtained in axisymmetric coordinates by Kermanidis [9] and Cruse et al [10], and reported in complete form more recently by Mayr et al [11]. The results are achieved by finding the fundamental solution corresponding to a unit load distributed around a ring. The expressions obtained are somewhat more complicated than those in rectangular coordinates but allow the axisymmetric body to be represented for purposes of solution by it's meridian shape alone. The savings in core requirements and data preparation are obvious. The kernel functions are expressed in terms of complete elliptic integral functions of the first and second kind which can be accurately computed using a combination of polynomial approximations [12] and asymptotic expansions [13]. The kernels used for the present work are given in the appendix.

4. BOUNDARY DISCRETIZATION

Solution of equation (5) is achieved by discretizing the surface into elements, each being associated with a given number of nodes. Equation (5) then yields a set of equations for velocity and traction at each node, which can be formed into a matrix equation and solved.

The present application utilises geometrically linear elements of the form shown in figure (1), where the value of the velocity and traction is assumed to be constant along each element. This requires each element to be associated

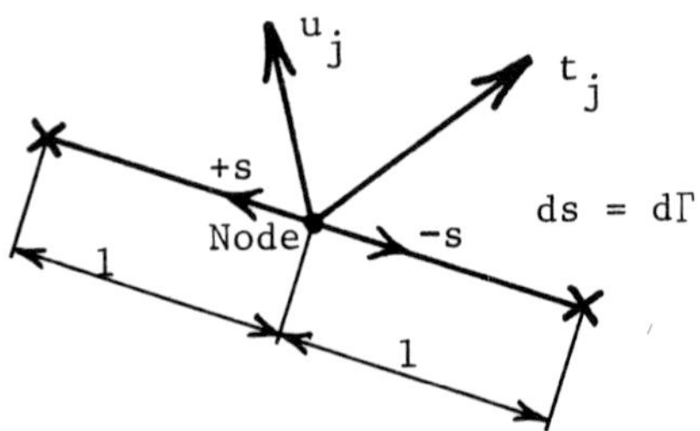

Fig. 1. A simple boundary element

with only one centrally located node, but may require more elements to adequately represent the boundary than if higher order approximations were used. Under such conditions, the tensor $C_{ij}(P)$ in (5) becomes 0.5 δ_{ij} [1, 2, 3, 8, 10], where δ_{ij} is the Kronecker delta function.

Following discretization of the boundary into N elements of this form, equation (5) can be written as:

$$\frac{1}{2} u_i(P) + \sum_{m=1}^{N} u_j(Q) \int_{\Gamma_m} t_{ij}^*(P,Q)\, d\Gamma_m$$

$$= \sum_{m=1}^{N} t_j(Q) \int_{\Gamma_m} u_{ij}^*(P,Q)\, d\Gamma_m. \tag{6}$$

Care must be taken when P and Q both lie on the same element, since the integrands in (6) contain singularities of order $\ell n|s|$ and $1/s$. These have been treated by extracting the singular term and integrating this in exact form. The remaining non-singular portion of the kernel is then integrated using numerical (Gaussian) quadrature.

Equation (6) can be written in matrix/vector form as follows:

$$G\underset{\sim}{u} = H\underset{\sim}{t} \tag{7}$$

where $\underset{\sim}{u}$ and $\underset{\sim}{t}$ are vectors of velocity and traction at the boundary. After specification of the known boundary conditions, equation (7) can be rearranged to yield the linear system of order 2N:

$$A\underset{\approx}{x} = \underset{\sim}{b} \ . \tag{8}$$

This can be solved using standard elimination techniques to yield the unknown quantities $\underset{\sim}{x}$. Once the boundary solution is obtained, equation (5) can be used to compute the interior solution.

5. EXAMPLE - THE MOVING SPHERE

Use of elements over which the function is constant is suitable for application to a sphere since the traction in the direction of motion is constant over the whole surface while the traction perpendicular to the motion is zero. The major error will then be due to the use of geometrically linear elements to approximate a circular boundary.

The particular discretization used is shown in figure (2) where 20 elements approximate the semicircular boundary.

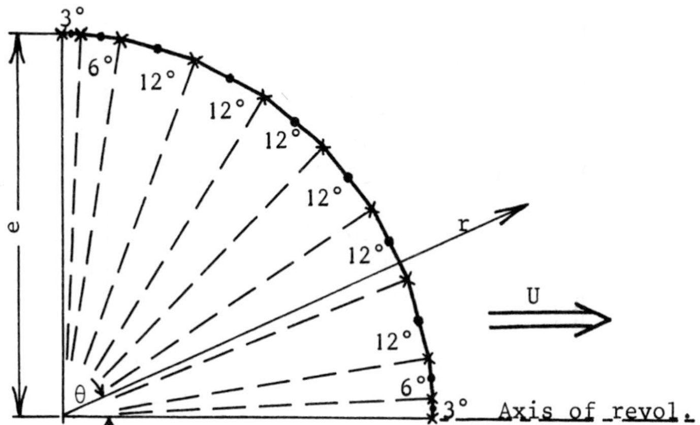

Fig. 2. Representation of sphere

The numerical results were obtained using a CDC Cyber computer in single precision with e = 5, viscosity μ = 1, and U = 10. The presentation is in non-dimensionalised form to provide generality, and compared with analytical solutions obtained from reference [14].

5.1 Results

$$\frac{\text{Drag}}{U\mu e}\ \text{(numerical)} = 5.98\pi,\ \frac{\text{Drag}}{U\mu e}\ \text{(exact)} = 6\pi.$$

θ	0		π/2	
r/e	u/U(num)	u/U(exact)	u/U(num)	u/U(exact)
1.1	.987	.988	.867	.870
1.2	.959	.961	.766	.770
1.3	.924	.926	.688	.691
1.5	.850	.852	.571	.574
2.0	.685	.688	.405	.406
3.0	.480	.481	.258	.259
4.0	.366	.367	.191	.191
5.0	.295	.296	.151	.152
6.0	.247	.248	.126	.126
8.0	.186	.187	.094	.094
10.0	.149	.150	.075	.075

Table 1. Velocity u/U

r/e	C(num)	C(exact)
1.1	1.27	1.24
1.2	1.06	1.04
1.3	.902	.888
1.5	.676	.667
2.0	.379	.375
3.0	.168	.167
4.0	.094	.094
5.0	.060	.060
6.0	.042	.042
8.0	.023	.023
10.0	.015	.015

Table 2. Pressure variation $C = e(p-p_\infty)/U\mu$ on stagnation streamline

5.2 Discussion

The drag has been computed by integrating the traction acting in the direction of motion over the surface. Some averaging of errors would therefore be expected, resulting in the close agreement (~ 0.3% error) observed between numerical and exact values. The velocity field (Table 1) also shows close agreement with the exact solution.

Due to the complicated nature of the axisymmetric kernels, analytical differentiation of equation (5) to obtain stresses at internal points is difficult. For this reason we have chosen to compute the pressure field by utilising the simpler kernels expressed in rectangular coordinates. This is computationally less efficient but more practical and, as shown in Table 2, yields sufficient accuracy. The ratio of central processor time used to obtain the pressure at each selected point to that used to obtain the boundary solution was 1.4.

The accuracy of solution can be varied greatly using various numbers of elements and configurations, and only one possible arrangement is shown here. It is expected that an accurate solution could be obtained using a smaller number of circular elements, yielding a better geometrical boundary approximation.

6. CONCLUSION

The numerical example of a moving sphere presented here demonstrates the unique advantage the boundary element method has when treating external flow problems. The results presented show that sufficient accuracy can be obtained using simple elements and the method can be applied to find the drag and flow field due to the motion of any axisymmetric body or, by superposition, the solution for a body held in a uniform flow.

It is believed that a boundary integral equation solution is possible for the complete Navier-Stokes equations at moderate Reynolds numbers by regarding inertia terms as body force loading and iterating to find the solution. The authors are presently investigating this possibility.

REFERENCES

1. RIZZO, F.J. - An Integral Equation Approach to Boundary Value Problems of Classical Elastostatics. Q. Appl. Math., Vol. 25, No. 1, pp. 83-95, 1967.

2. CRUSE, T.A. - Numerical Solutions in Three Dimensional Elastostatics. Int. J. Solids Struct., Vol. 5, pp. 1295-1274, 1969.

3. CRUSE, T.A. - Application of the Boundary Integral Equation Method to Three Dimensional Stress Analysis. Comp. Struct., Vol. 3, pp. 509-527, 1973.

4. CRUSE, T.A. and RIZZO, F.J. - A Direct Formulation and Numerical Solution of the General Transient Elastodynamic Problem. Int. J. Math. Anal. Applic., Vol. 22, p. 244, 1968.

5. CRUSE, T.A. and SWEDLOW, J.L. - Interactive Program for Analysis and Design Problems in Advanced Composites Technology. AFML-TR-71-268, 1971.

6. BARONE, M.R. and ROBINSON, A.R. - Determination of Elastic Stresses at Notches and Corners by Integral Equations. Int. J. Solids Struct., Vol, 8, pp. 1319-1338, 1972.

7. MILTHORPE, J.F. and TANNER, R.I. - Boundary Element Methods for Free Surface Viscous Flows, 7th Aust. Hyd. and Fluid Mech. Conf., Brisbane, 18-22 Aug., 1980.

8. BREBBIA, C.A. - The Boundary Element Method for Engineers, Billing and Sons, London, 1978.

9. KERMANIDIS, T. - A Numerical Solution for Axially Symmetrical Elasticity Problems. Int. J. Solids Struct., Vol. 11, pp. 493-500, 1975.

10. CRUSE, T.A., SNOW, D.W., WILSON, R.B. - Numerical Solutions in Axisymmetric Elasticity. Comp. Struct., Vol. 7, pp. 445-451, 1977.

11. MAYR, M., DREXLER, W., KUHN, G. - A Semianalytical Boundary Integral Approach for Axisymmetric Elastic Bodies with Arbitrary Boundary Conditions. Int. J. Solids Struct., Vol. 16, pp. 863-871, 1980.

12. HART, J.F. et al - Computer Approximations, J. Wiley and Sons, U.S.A., 1968.

13. ABRAMOWITZ, M. AND STEGUN, I.A. - Handbook of Mathematical Functions, Dover Pub. Inc., U.S.A., 1970.

14. HAPPEL, J. and BRENNER, H. - Low Reynolds Number Hydrodynamics, Prentice-Hall, N.J., 1965.

APPENDIX

In the equations that follow, z is the axial coordinate, r is the radial coordinate, (Z,R) are the coordinates of the ring source, μ is the viscosity, $Q_{-\frac{1}{2}}(\gamma)$ and $Q_{\frac{1}{2}}(\gamma)$ are Legendre's function of the second kind of degree $-\frac{1}{2}$ and $\frac{1}{2}$ respectively, and γ is given by

$$\gamma = \frac{R^2 + r^2 + (Z-z)^2}{2Rr} \; . \tag{A1}$$

The velocity kernels are given by:

$$\lambda_1\, u_{zz}{}^*(P,Q) = \frac{1}{\sqrt{(Rr)}}\, Q_{-\frac{1}{2}}(\gamma) - \frac{(Z-z)^2}{Rr\sqrt{(Rr)}} \frac{dQ_{-\frac{1}{2}}}{d\gamma}$$

$$\lambda_1\, u_{zr}{}^*(P,Q) = \frac{-(Z-z)}{r\sqrt{(Rr)}} \left\{ \frac{1}{2}\, Q_{-\frac{1}{2}}(\gamma) + (\gamma - \frac{r}{R}) \frac{dQ_{-\frac{1}{2}}}{d\gamma} \right\}$$

$$\lambda_1\, u_{rz}{}^*(P,Q) = \frac{(Z-z)}{r\sqrt{(Rr)}} \left\{ \frac{1}{2}\, Q_{\frac{1}{2}}(\gamma) - (\gamma - \frac{r}{R}) \frac{dQ_{\frac{1}{2}}}{d\gamma} \right\} \tag{A2}$$

$$\lambda_1\, u_{rr}{}^*(P,Q) = \frac{1}{\sqrt{(Rr)}}\, Q_{\frac{1}{2}}(\gamma) + \frac{(Z-z)^2}{Rr\sqrt{(Rr)}} \frac{dQ_{\frac{1}{2}}}{d\gamma}$$

where $\lambda_1 = 8\pi^2\mu$.

The kernels $t_{ij}{}^*(P,Q)$ can be obtained using the surface normal vector n_k and the tensor $\sigma_{ijk}{}^*(P,Q)$ (where $\sigma_{ijk}{}^*(P,Q)$ is the stress tensor component σ_{jk} at Q due to a unit source in direction i at P) by forming

$$t_{ij}{}^*(P,Q) = \sigma_{ijk}{}^*(P,Q)\, n_k \tag{A3}$$

where

$$\lambda_2\, \sigma_{zzz}{}^*(P,Q) = \frac{-(Z-z)^3}{(\gamma^2-1)(Rr)^2\sqrt{(Rr)}} \left\{ \frac{1}{4}\, Q_{-\frac{1}{2}}(\gamma) + 2\gamma \frac{dQ_{-\frac{1}{2}}}{d\gamma} \right\}$$

$$\lambda_2\, \sigma_{rzz}{}^*(P,Q) = \frac{(Z-z)^2}{Rr\sqrt{(Rr)}} \left\{ \frac{-3A}{4(\gamma^2-1)}\, Q_{\frac{1}{2}}(\gamma) + \left[\frac{2\gamma A}{(\gamma^2-1)} + \frac{1}{2r} \right] \frac{dQ_{\frac{1}{2}}}{d\gamma} \right\}$$

$$\lambda_2\,\sigma_{zrr}{}^*(P,Q) = \frac{-(Z-z)}{\sqrt{(Rr)}}\left\{\frac{1}{4}\left[\frac{A^2}{(\gamma^2-1)} - \frac{3}{r^2}\right] Q_{-\frac{1}{2}}(\gamma) + \left[\frac{2\gamma A}{(\gamma^2-1)} + \frac{3}{r}\right] A\,\frac{dQ_{-\frac{1}{2}}}{d\gamma}\right\}$$

$$\lambda_2\,\sigma_{rrr}{}^*(P,Q) = \frac{3A(Z-z)^2}{4(\gamma^2-1)Rr\sqrt{(Rr)}}\,Q_{\frac{1}{2}}(\gamma) + \frac{1}{\sqrt{(Rr)}}\left\{A - \frac{(Z-z)^2}{Rr}\left[\frac{2\gamma A}{(\gamma^2-1)} + \frac{3}{2r}\right]\right\}\frac{dQ_{\frac{1}{2}}}{d\gamma} \qquad (A4)$$

$$\lambda_2\,\sigma_{zzr}{}^*(P,Q) = \frac{(Z-z)^2}{2Rr\sqrt{(Rr)}}\left\{\frac{A}{2(\gamma^2-1)}\,Q_{-\frac{1}{2}}(\gamma) + \left[\frac{3}{r} + \frac{4\gamma A}{(\gamma^2-1)}\right]\frac{dQ_{-\frac{1}{2}}}{d\gamma}\right\}$$

$$\lambda_2\,\sigma_{rzr}{}^*(P,Q) = \frac{3(Z-z)}{8\sqrt{(Rr)}}\left\{\frac{1}{(\gamma^2-1)}\left[A^2 - \frac{(Z-z)^2}{(Rr)^2}\right] - \frac{1}{r^2}\right\} Q_{\frac{1}{2}}(\gamma) - \frac{(Z-z)}{2\sqrt{(Rr)}}\left\{\frac{2\gamma}{(\gamma^2-1)}\left[A^2 - \frac{(Z-z)^2}{(Rr)^2}\right] + \frac{2}{Rr} + \frac{2A}{r}\right\}\frac{dQ_{\frac{1}{2}}}{d\gamma}$$

where $\lambda_2 = 4\pi^2$ and $A = \frac{1}{R} - \frac{\gamma}{r}$.

For computational purposes, Legendre's functions can be expressed in terms of complete elliptic integrals of the first and second kinds (represented by $K(\omega)$ and $E(\omega)$ respectively) from reference [13] as follows:

$$Q_{-\frac{1}{2}}(\gamma) = \omega K(\omega)$$

$$Q_{\frac{1}{2}}(\gamma) = \omega\{\gamma K(\omega) - (\gamma+1)E(\omega)\} \qquad (A5)$$

where $\omega = \sqrt{\left(\frac{2}{\gamma + 1}\right)}$.

NUMERICAL SOLUTIONS FOR LAMINAR FLOWS IN DISC-TYPE STIRRERS

Jong H. Wang (I) and Wen-Jei Yang (II)

(I) Research Engineer (II) Professor ME/AM

SUMMARY

A rapidly converging line-relaxation numerical technique is developed to study laminar-flow phenomena in a disc-type stirrer whose vessel walls consist of one cylindrical and two cone-shaped elements. The flow is characterized by a combination of translational and rotational flows. Results are obtained for the distribution of streamline, vorticity, and velocity. The effects of vessel wall geometry, through flow and rotational speed on the flow patterns, are determined. The theoretical prediction of streamlines agrees qualitatively with the results obtained by a flow visualization method.

1. INTRODUCTION

Rotating disc contactors (RDC) are widely used for gas-liquid absorption and liquid-liquid extraction [1,2]. They are more efficient and flexible in operation than the conventional sieve-plate, packed and spray columns. However, axial mixing which tends to decrease the stage efficiency is always a problem in this type of mixing device. Thus, various modifications have been designed to improve the efficiency. Figure 1 shows one of relatively new devices which can be applied to fermentation and oxidation, as well as other complicated processes with or without chemical reaction. Details on the characteristics of the system are

---

I Research Office, Babock and Wilcox Company, Alliance, Ohio 44601 U.S.A.

II Professor of Mechanical Engineering and Applied Mechanics, University of Michigan, Ann Arbor, Michigan 48109 U.S.A.

available in references 3 through 6. In brief, a perforated disc is used as a dispersing means to reduce the size of gas bubbles resulting in a significant increase in the gas-liquid interface for mass transfer at extremely low power consumption. The housing is designed in a conical shape for the purpose of minimizing the occurrence of axial mixing. A numerical study of turbulent flow behavior in disc-type stirrers is available in reference 7. The present work is an extension of the study to the laminar-flow case.

The factors which affect the performance of the apparatus in Fig. 1 include the housing geometry such as the vessel diameter, the disc diameter and angle of the conical wall, the axial liquid flow rate, and the rotating speed of the disc. System performance is measured by the overall mass-transfer coefficient defined based on the mean characteristic velocity of bubbles $\bar{u}_c$ or their residence time $\bar{t}_r$[8,9]. Both $\bar{u}_c$ and $\bar{t}_r$ were empirically determined as functions of the system parameters, but thus far no theoretical model has been developed to predict their relationship. The latter may be achieved by determining $\bar{u}_c$ and $\bar{t}_r$ in relation to the trajectory of bubbles moving through flowing liquid in the appartus. As a first step, the present study investigates laminar flow in a RDC housing a rotating shaft with a disc. A finite-difference numerical method is employed to obtain the solutions for the stream function, vorticity, and velocity components. Theory is compared with flow visualization results.

2. ANALYSIS

Consider an apparatus housing a rotating shaft with a disc, as seen in Fig. 1. A fluid flows into the housing along the shaft downward in the case of liquid or upward in the case of a gas. The effect of gravity on the flow will be neglected. The disc has a radius of R_d and a thickness of 2b. The origin of the cylindrical coordinates is fixed at the center of the disc with z measuring a positive distance in the direction against the flow. The Navier-Stokes equations for steady axisymmetric flow of an incompressible, viscous fluid read

$$u\frac{\partial u}{\partial r} - \frac{v^2}{r} + w\frac{\partial u}{\partial z} = \frac{1}{\rho}\frac{p}{r} + \nu\left(\frac{\partial^2 u}{\partial r^2} + \frac{1}{r}\frac{\partial u}{\partial r} - \frac{u}{r^2} - \frac{\partial^2 u}{\partial z^2}\right) \quad (1)$$

$$u\frac{\partial v}{\partial r} + \frac{uv}{r} + w\frac{\partial v}{\partial z} = \nu\left(\frac{\partial^2 v}{\partial r^2} + \frac{1}{r}\frac{\partial v}{\partial r} - \frac{v}{r^2} + \frac{\partial^2 v}{\partial z^2}\right) \quad (2)$$

$$u\frac{\partial w}{\partial r} + w\frac{\partial w}{\partial z} = -\frac{1}{\rho}\frac{\partial P}{\partial z} + \nu\left(\frac{\partial^2 w}{\partial r^2} + \frac{1}{r}\frac{\partial w}{\partial r} + \frac{\partial^2 w}{\partial z^2}\right) \quad (3)$$

The equation of continuity is

$$\frac{1}{r}\frac{\partial(ru)}{\partial r} + \frac{\partial w}{\partial z} = 0 \tag{4}$$

The appropriate boundary conditions are:

at inlet: $u=v=0$, $w=f(r)$
on stationary walls: $u=v=w=0$
on rotating surfaces: $u=w=0$, $v=rw$ (5)
at exit: $u=v=0$, $\partial w/\partial z=0$.

Baffle plates are installed at the inlet and exit to reduce vortices so that both the radial and tangential velocity components are zero. The actual velocity profile at the exit is unknown. However, it is common to describe it as $\partial w/\partial z=0$ as an approximation.

Two numerical approaches are available to solve the governing equations: the stream function ψ - vorticity ζ method [10, 11] and the velocity-pressure or primitive equations method [11-13]. Roached [11] recommended the former method for solving most flow problems except those with free surfaces or transient compressible flow cases. This is caused by difference in the boundary conditions such that the latter method would take a longer computer time than the former method in achieving an iterative convergence. In the present study dealing with an incompressible flow, the $\psi-\zeta$ method by Gosman, et. al. [10] is employed.

With the use of the vorticity and stream function defined as $\zeta = \partial U/\partial z - \partial w/\partial r$ and $u = -(1/r)(\partial\psi/\partial z)$, $w = (1/r)(\partial\psi/\partial r)$, respectively, equations (1), (2), and (3) may be combined to yield

$$a\left[\frac{\partial}{\partial z}\left(\phi\frac{\partial\psi}{\partial r}\right) - \frac{\partial}{\partial r}\left(\phi\frac{\partial\psi}{\partial z}\right)\right] - \frac{\partial}{\partial z}\left[br\frac{\partial}{\partial z}(c\phi)\right] - \frac{\partial}{\partial r}\left[br\frac{\partial}{\partial r}(c\phi)\right] + rd = 0 \tag{6}$$

Here the quantity ϕ and the coefficients a, b, c, and d are given in Table 1.

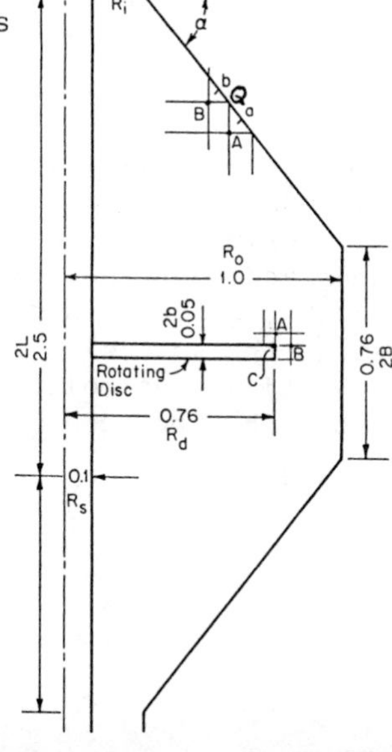

Fig. 1 A schematic of disc-type stirrer and numerical grid (R_o=79.4 mm)

ϕ	a	b	c	d
ψ	0	$1/(\rho r^2)$	1	$-\zeta/r$
ζ/r	r^2	r^2	μ	$-\frac{\partial}{\partial z}(\rho\omega^2)$
$r\omega$	1	μr^2	$1/r^2$	0

Table 1. Coefficients for equation (6)

The boundary conditions (5) are rewritten as follows:

(i) At inlet: Two extreme cases, uniform and fully-developed inlet velocity profiles, are studied. The uniform velocity profile can be described as

$$w=\text{const.},\ v=0,\ \zeta=0,\ \psi=-\rho\omega(R_s^2-r^2)/2 \tag{7}$$

while the fully-developed velocity case corresponds to

$$w = -\frac{1}{4\mu}\frac{\partial P}{\partial z}(R_s^2-R_i^2) + \frac{(n^2-1)R_s^2}{\ell n\ n}\ \ell n\frac{r}{R_s}$$

$$\zeta = -\partial w/\partial r\ ,\qquad \psi = \int_{R_i}^{r} wr\ dr \tag{8}$$

Here, R_s and R_i denote the shaft and outer radii, respectively, and n is R_i/R_s.

(ii) On stationary walls: $v=\psi=0$. Two approaches are adopted: (a) The first-order accuracy vorticity

$$\zeta_P/R = \frac{2(\psi_P-\psi_{NP})}{\rho R^2\Delta^2} \tag{9}$$

is employed for a sloping boundary. The subscript P refers to a nodal point on the wall such as a and b in Fig. 1; NP, the interior point such as A and B which are on the normals from a and b, respectively. Δ denotes the distance between P and NP, while R is the radial location of P. This approach regards the vorticity in the near-wall layer as constant. The vorticity at point Q in Fig. 1 is interpolated between a and b. (b) the second-order accuracy vorticity

$$\zeta_P/R = g_1+g_2\ \zeta_{NP}/(R-\Delta) \tag{10}$$

is used with

$$g_1 = \frac{3(\psi_P-\psi_{NP})}{\rho R^2 \Delta} \quad \text{and } g_2 = -\frac{1}{2}$$

along a constant -z wall; and

$$g_1 = \frac{3(\psi_P-\psi_{NP})}{\rho R^2\Delta^2(1+5S/8)} \text{ and } g_2 = \frac{(1+S)(1/2+3S/8)}{1+5S/8}$$

along a constant -r boundary. S is defined as $k\Delta/R$ where k is a unit vector in the r direction.

(iii) at exit (assuming $\partial w/\partial z=\partial\zeta/\partial z = 0$); v=0. One uses

$$\zeta_N = \zeta_{N-1}, \quad \psi_N = 2\psi_{N-1} - \psi_{N-2} \qquad (11)$$

Here, the subscript N refers to a point at the exit and N-1 corresponds to its upstream adjacent point.

Discontinuous values of vorticity [11, 14] are employed at sharp corners. Take point C in Fig. 1 for example. When a control volume is defined to include points A and C, the vorticity ζ_c at point C is expressed as a function of the stream function ψ_A, ψ_c and vorticity ζ_A. Similarly, if a control volume includes points B and C, then ζ_C is given as a function of ψ_B,ψ_c and ζ_B.

3. RESULTS AND DISCUSSION

The dimension of the stirrer used in the study is given in Fig. 1 in reference to R_o, radius of cylindrical section, which is 7.94 cm.

The validity of the mathematical model and computer program was tested on a well-known back-step flow problem, i.e. flow from a pipe of radius R_1 into another pipe with a larger radius R_2. The flow field was divided into rectangular grids of nonuniform size. Several networks of different grid number were tested to determine the effect of the grid size on the solutions obtained. It was concluded that a 46 x 21 nonuniformly spaced grid network (with 46 grids in the axial direction and 21 grids in the radial direction) was sufficient to assure a good convergence. The predicted variation of the boundary-layer reattachment point measured from the step with the inlet Reynolds number agreed with the results of reference 15.

In the present study one-half of the apparatus was subdivided into nonuniform grid network consisting of 46 meshes in the axial direction by 20 meshes in the radial direction. Numerical procedure was initiated with the assumption of an initial value for each nodal point. Computations were then repeated utilizing the line iteration scheme until the convergence

$$[(\phi^{M}-\phi^{M-1})/\phi^{M-1}_{max}]_{max} \leqq 5x10^{-5}$$

was satisfied at all nodal points. The superscripts M and (M-1) denote two successive iterations, while the subscript "max" refers to a maximum value over the entire field of iterations.

Numerical integrations were performed using an AMDAHL 470/V/7 digital computer. Approximately 0.07 to 9.1 seconds of computer time were consumed in each iteration. The number of iterations required for convergence at each nodal point depends on the system geometry, boundary conditions, axial flow Reynolds Re = $\dot{m}(\mu R_d)$, and Taylor number Ta = $\omega R_d^2/\nu$, where $\dot{m}$ is the mass flow rate and ω is the angular velocity of the disc. In general, approximately 400 to 600 iterations were required. However, if the initial values were assumed based on the results of a lower Reynolds-number flow combined with a proper extrapolation factor accounting for the corresponding flow increase, only 100 to 200 iterations could achieve the convergence.

Numerical results were obtained for three housing geometries of special interest: The angle of conical section $\alpha=\pi/4$ was found to be most efficient for mixing [3, 4], while $\alpha=\pi/2$ was earlier employed in liquid-liquid extraction. $\alpha=0$ corresponds to a flow through a column with a rotating disc. In each housing geometry both the Taylor and Reynolds numbers were varied between zero and the critical values for flow transition. Results were obtained including the distributions of streamlines, vorticity, and tangential, and radial velocities. Only representative results are presented here in the interest of brevity.

Figures 2, 3 and 4 correspond to translational flow, rotational (swirling) flow and their combination with both the Taylor and Reynolds numbers at 500. In each figure a, b, and c illustrate the distributions of streamline, vorticity, and velocity, respectively. It is noted that a defect in the graphic subroutine rather than the grid size has caused several of the streamline and vorticity contours to appear not smooth. Only Fig. 4 has an additional illustration d for the

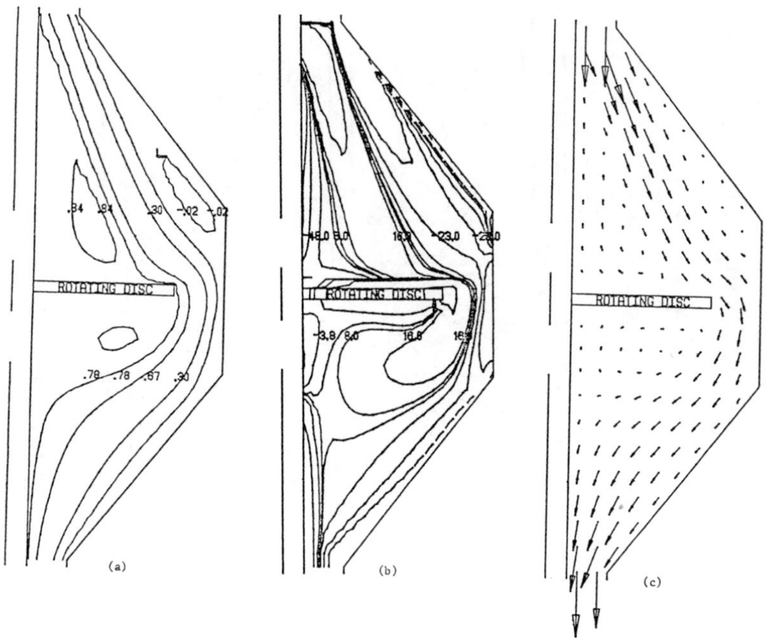

Fig. 2 Translational flow at Re = 500 in RDC with 45-degree cone angle, (a) streamline, (b) vorticity and (c) velocity

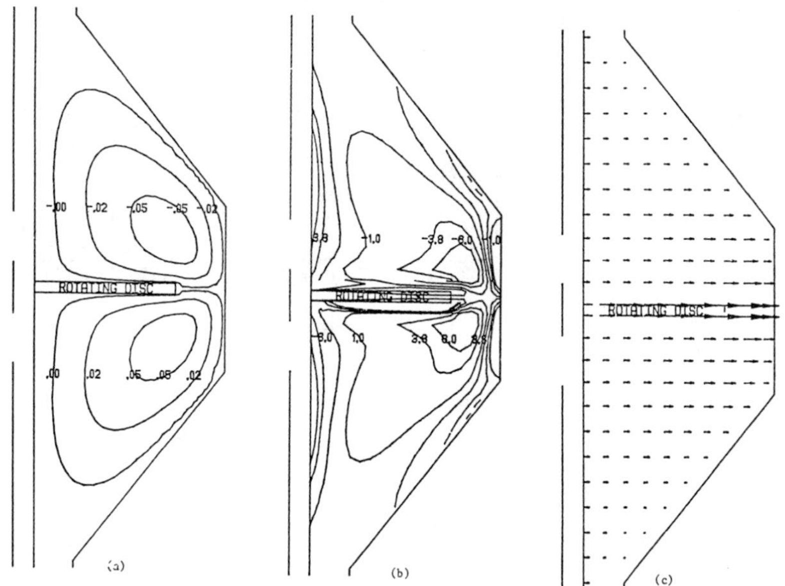

Fig. 3 Pure rotational flow at Ta = 500 in a RDC with 45-degree cone angle, (a) streamline, (b) vorticity, and (c) tangential velocity

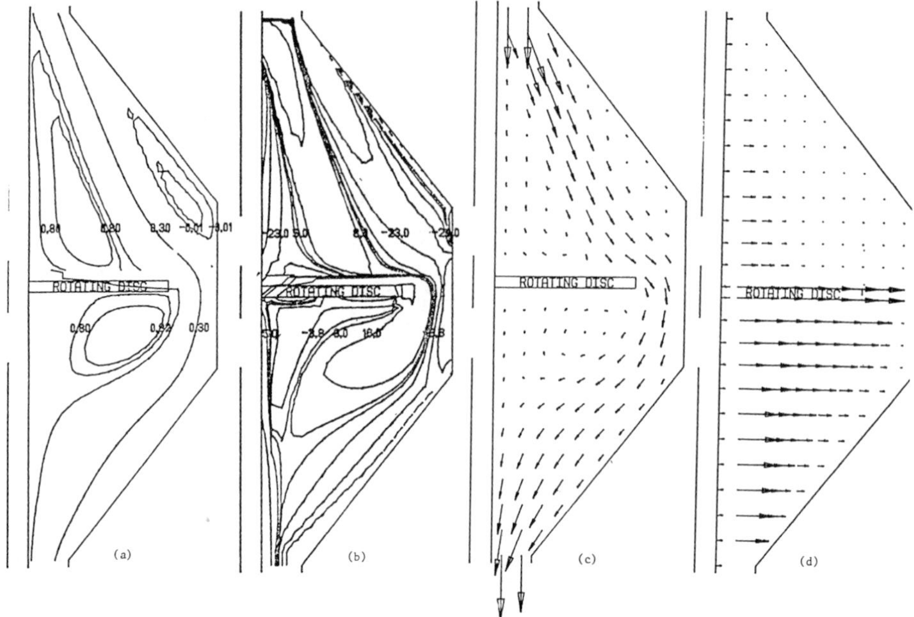

Fig. 4 Combined flow at Re = Ta = 500 in a RDC with 45-degree cone angle, (a) streamline, (b) vorticity, (c) velocity, and (d) tangential velocity

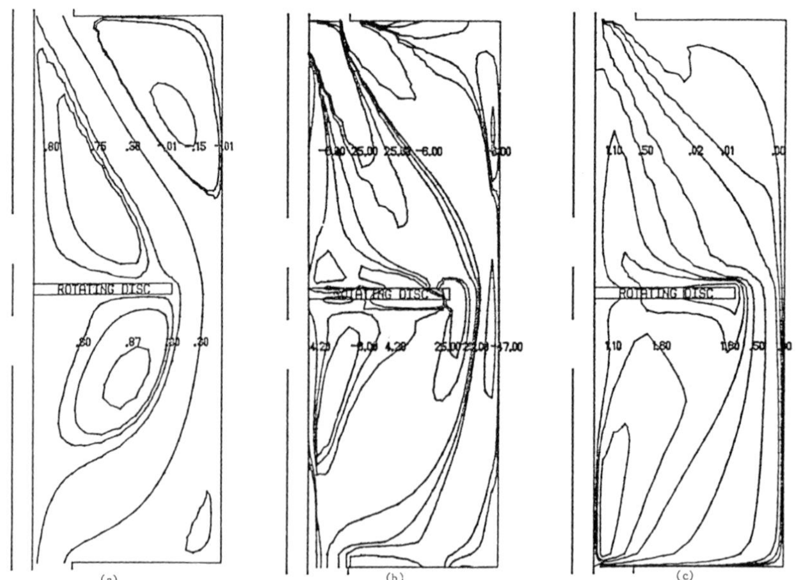

Fig. 5 Combined flow at Re = Ta = 500 in a RDC with 45-degree cone angle, (a) streamline, (b) vorticity, and (c) tangential velocity

tangential velocity component. The numbers in Figs. -a and -b represent the magnitudes of stream function and vorticity, respectively. Figure 2-a demonstrates that the flow separates at the upstream sloping wall and both sides of the stationary disc, resulting in flow circulation in each separation regime. As the axial-flow Reynolds number Re increases, the circulation zone near the sloping wall reduces slightly while those on both sides of the disc continue to expand. Figure 2-b shows the distribution of vorticity whose shape appears similar, but the magnitude of vorticity greatly enhances with an increase in Re. The length of arrows in Figs. 2-c, 3-c, 4-c and -d is a measure of velocity with 1 cm length corresponding to 1.78 cm/sec of the combined velocity or 2.91 cm/sec of tangential velocity. At a low value of Re, the main stream follows the axis toward the stationary disc like an axisymmetrical stagnation flow. As flow is increased, it tends to take the shortest course from the entrance to the exit as expected. The flow consists of radial and axial components, but no tangential velocity exists. In case of pure rotational or swirling flow, both the streamlines and vorticities are symmetrically distributed with respect to the rotating disc, but opposite in signs, as seen in Figs. 3-a and 3-b respectively. The flow consists of radial and tangential components but no axial flow exists.

When duct flow and disc rotation are combined, Figs. 2 and 4 show strikingly similar flow patterns. Figure 4-d illustrates the tangential velocity pattern induced by the rotating disc, which is substantially different from that of pure rotating flow in Fig. 3-c. It is seen in Fig. 4-d that the fluid rotates much faster in the lower half than in the upper half of the flow field where fluid motion is practically limited to the region close to the shaft. The presence of tangential flow enhances the strength of the two vortices near the disc as a result of conversion of the angular momentum in the fluid. Although numerical results were obtained for flows with high values of Re and Ta up to the transition, the flow patterns remain similar to those given in Figs. 2, 3, and 4.

Figure 5 demonstrates the flow patterns for $\alpha=\pi/2$ case, which are different from those for $\alpha=\pi/4$ in Fig. 4. In addition to three vortices previously identified, a fourth vortex is observed at the lower corner of the stationary walls in Fig. 5-a. All four recirculation zones are very extensive, leaving only a narrow street from the inlet to the exit for the main stream. The strength of the two corner vortices is much stronger than that for $\alpha=\pi/4$ in Fig. 4-a. It is seen in Fig. 5-c that main tangential-velocity lines are distributed diagonally in the upper half, but almost in parallel with the vertical stationary wall in the lower half of the vessel. In conclusion, the vessel of $\alpha=\pi/2$ is much less effective for flow mixing than that of $\alpha=\pi/4$ although flow recirculation promotes mixing.

Results of two different inlet flow conditions, uniform and parabolic profiles, are practically identical except in a very narrow region at the inlet. Test runs were conducted to examine the effect of grid size on convergence using finer grid network. No noticeable differences were detected in the results. It was disclosed also that the line-iteration scheme results in faster convergence by a factor of 1.2 to 1.4, depending upon the direction of line marching and the type of flows. In the case of pure rotating flow, a special numerical stability problem was encountered as reported previously [for example, 15-18]. Gosman [15] proposed two methods for remedy: multi-point circulation adjustment and under-relaxation method. The latter was adopted in the present study. At large values of Ta small values of the under-relaxation factors, about 0.5 to 0.6, were used to achieve convergence of the solution. The initial values employed to initiate numerical integrations were decisive in the stability of iterations. The presence of duct flow contributed to aid in the stability. When the ratio of Ta to Re exeeded a certain value depending on the system geometry and the boundary conditions, the solution would not converge. The solutions for the downstream half of the flow field converged much slower than those in the upstream half because the flow in the former region is substantially more complicated than that in the latter region.

4. FLOW VISUALIZATION TESTS

The flow visualization technique was employed to qualitatively confirm the numerical predictions of flow patterns because direct measurements of local velocity were difficult to apply to the present system. Due to the presence of a flow recirculation in the field, ALCOA 1401 aluminum powder was used as a tracer. The RDC consisted of two 45-degree conical sections and one cylindrical section, both made of transparent plastic glass. A plastic disc was mounted on a stainless-steel shaft supported by three thrust ball bearings. The dimensions of the apparatus are indicated in Fig. 1. The shaft was driven by a Zero-Max variable speed a-c motor through a belt-pulley combination. The rotating speed varying from 0 to 1200 RPM was detected by a magnetic pickup and displayed on a digital frequency indicator. Water flowed downward through the apparatus. The aluminum powder (20 g) mixed with citric acid (5 g) and alcol solution (30 cc) was injected into the RDC using a hyperdermic needle which was removed upon the competion of injection. A large vortex is observed one each in the flow fields over and below the disc, as predicted by theory (photographs are not included due to page limitation).

REFERENCES

1. REMAN, G.H. and OLNEY, R.B. - The Rotating-Disc Contractor-a new tool for liquid-liquid extraction. Chemical Engineering Progress, Vol. 51, pp. 141-146, 1966.

2. LADDHA, G.S. and DEGALEESAN, T.F. - Transport Phenomena in Liquid Extraction, McGraw-Hill, New York, 1978.

3. GLAESER, H. BIESECKER, B.O. and BRAUER, H. - Begasung von Flussigkeiten mit Propeller-und Lochscheibenruhren. Verfahrenstechnik, 2, pp. 31-49, 1973.

4. BRAUER, H., SCHMIDT, H. and THIELE, H. - Fluiddynamische Untersuchunger an einem Ruhrkaskaden-Fermenter. Chemie Ingenieur Technik, 16, p. 699, 1974.

5. YANG, W-J. - Mechanism of Power Dissipation in Liquid-Gas Mixing in a Perforated-Disk Type Stirring Cascade. Heat Transfer 1978, Vol. 4, Hemisphere, Washington, D.C., pp. 7-12, 1978.

6. YANG, W-J. - Gas-Liquid Mass Transfer in a Multi-Stage Perforated-Disk type Stirring Cascade. Letters J. Heat Mass Transfer, Vol. 3, pp. 403-422, 1976.

7. WANG, J.H. and YANG W-J. - Turbulent Flows in Disc-Type Stirrers with Cone-Shaped Housing. Numerical Methods for Non-Linear Problems, Vol. 1, Taylor, C., Hinton, E. and Owen, D.R.J., Pineridge, pp. 885-893, 1980.

8. OLNEY, R.B. - Droplet Characteristics in a Counter-Current Contactor. J. AIChE, Vol. 10, pp. 827-835, 1964.

9. KANNAPPAN, R. - Hydrodynamics and Mass Transport in Rotary Disc Contactors, Ph.D. Thesis, University of Madras, 1973.

10. GOSMAN, A.D., PUN, W.M., RÜNCHAL, A.K., SPALDING, D.B. and WOLFSHTEIN, M. - Heat and Mass Transfer in Recirculating Flows, Academic Press, New York, 1969.

11. ROACHE, P.J. - Computational FLuid Dynamics, Hermosa Publishers, New Mexico, 1976.

12. PATANKAR, S.V. and SPALDING, D.B. - A Calculation Procedure for Heat, Mass and Momentum Transfer in Three Dimensional Parabolic Flows. Int. J. Heat Mass Transfer, Vol. 15, pp. 1787-1805, 1972.

13. GOSMAN, A.D. and PUN, W.M. - Lecture Notes for Course entitled, Calculation of Recirculating Flows, Report No. HTS17412, Imperial College of Science and Technology, London, 1973.

14. THOM, A. and APELT, C.J. - Field Computations in Engineering and Physics, C. Van Nostrand Co., 1961.

15. MACAGNO, I.O. and HUNG, T.K. - Computational and Experimental Stability of a Captive Annular Eddy. J. Fluid Mechanics, Vol. 28, No. 7, pp. 43-64, 1967.

16. GOSMAN, A.D. and SPALDING, D.B. - Computation of Laminar Recirculating Flow between Shrouded Rotating Discs, Proc. 2nd Int. Conf. on Numerical Methods in Fluid Dynamics, Springer Verlag, East Berlin, pp. 67-73, 1971.

17. OWN, J.M. and BILIMORIA, E.D. - Heat Transfer in Rotating Cylindrical Cavities. J. Mechanical Engineering Science, Vol. 19, No. 4, pp. 175-187, 1977.

18. LOUGHHEAD, J.N. - Turbulent Recirculating Flow in Shrouded Disc Geometries, A Computer Prediction Procedure, M.Sc. Thesis, Imperial College, University of London, 1971.

NUMERICAL SOLUTION OF THE NAVIER-STOKES EQUATIONS BY MULTI-GRID TECHNIQUES

Tomas Thunell[(1)] and Laszlo Fuchs[(2)]

SUMMARY

The steady state Navier-Stokes equations, in two dimensions are considered. The governing equations are formulated in terms of i) the velocity vector and the pressure, or ii) the stream function, or iii) the stream function and the vorticity. These formulations are solved numerically by using multi-grid methods with different relaxation procedures. The basic principles of the methods are presented and some advantages and disadvantages of each method are discussed. All the methods has been tested for some test problems and a comparison of the results is given.

1 INTRODUCTION

The application of simple iterative techniques to the solution of the discretized steady state Navier-Stokes (N-S) equations result in slowly convergent methods. The rate of convergence is strongly dependent on the mesh-size and the Reynolds number. An improper coupling between the equations or the treatment of the boundary conditions may, further reduce the efficiency of the numerical procedure. Previous experience shows that the multi-grid (MG) technique can be applied to many boundary value problems (i.e. ref. 1-3). The generalization of the MG method to systems of equations is, however, not trivial. A straight forward coupling of several efficient MG solvers does not result,in general, in an efficient numerical procedure for the solution of a system of equations. In the following we discuss some possibilities for coupling among the equations of a system in consistency with the basic ideas of MG techniques.

---

1. Research student, Dep. of Gasdynamics, The Royal Institute of Technology, Stockholm, Sweden.

2. Research Associate, Dep. of Gasdynamics, The Royal Institute of Technology, Stockholm, Sweden.

Different MG methods are applied to some different formulations of the N-S equations. These formulations have different number of dependent variables which in turn has implications on the number of the equations in the system, the treatment of the boundary conditions and the relaxation procedures.Our emphasis is on the efficiency of the numerical schemes and we shall not discuss in detail the question of the accuracy ofthe different approaches.

In the following we describe the basic principles of MG techniques. The relaxation methods for the different formulations of the governing equations are also presented. The results of the numerical methods, for the same test problems, give some information on their efficiency as a function of the mesh size and the Reynolds number. It has also been experienced that the formulation of the boundary conditions (on the vorticity) is an important factor that effects the efficiency of the numerical scheme.

2 SOME PRINCIPLES OF MULTI-GRID METHODS

For simplicity we consider, first, a single linear partial differential equation for φ

$$L\varphi = f, \qquad \text{in a domain } \Omega. \tag{2.1}$$

and

$$\varphi = g, \; g \text{ is given on the boundaries of} \Omega.$$

The operator L is discretized on a uniform grid with a mesh spacing h. The finite-difference approximation to L is denoted by L_h. The system of the algebraic equations to be solved is denoted by:

$$L_h\varphi_h = f_h \tag{2.2}$$

A common method of solving (2.2) is by repeated Gauss-Seidel (G-S) relaxation sweeps, until convergence is obtained. It has often been experienced that the convergence of the method is fast only in the first few iterative steps. This phenomenon can be explained if one considers different Fourier-components of the error. By a mode analysis (see e.g. [2]) it is possible to estimate the magnitude of the factors μ, by which each error component is decreased by a relaxation sweep (μ is called the smoothing rate). G-S relaxations result in a good smoothing rate for those error components which has about the same wave-length as the mesh size (i.e. high frequency error components), while the smoothing rate of slowly varying error components is poor.

The principles of the method can be understood by considering the solution of the correction problem

$$L_h^k \, v = f_h - L_h\varphi_h \tag{2.3}$$

where φ_h is an approximation to the solution φ , and v is an unknown correction such that $v = \varphi - \varphi_h$. L_h^k is an approximation to L.It is given that L_h^k has a smoothing rate μ such that

$$\mu(\vartheta) \begin{cases} < \mu_o < 1, & \vartheta \in \lambda_k \\ \approx 1 \quad , & \vartheta \notin \lambda_k \end{cases}$$

That is, the smoothing rate μ is less than μ_o for a range of frequencies λ_k and it is close to unity for all the other frequencies. For many elliptic operators, the G-S relaxations has similar properties. Fast convergence can be achived if one uses a sequence of m operators L_h^k such that

$$L_h^1 L_h^2 \ldots . L_h^m v = f_h - L_h \varphi_h \qquad (2.4)$$

L_h^k are choosen in such a way that the union of the efficient ranges of the operators covers the whole frequency spectrum. That is, $\cup \lambda_k$ = all frequencies.

One way to construct operators that has different optimal frequency range is by applying the same operator on a sequence of grids. G-S relaxations, for example, are efficient for error components with wave-lengths which are not much longer than the mesh size. A wave-length which is long relative to a fine mesh is shorter relative to coarser meshes. For this reason successive application of G-S relaxation sweeps on a sequence of grids is equivalent, in principle, to (2.4).

A simple MG process can be described in the following way: The residual of the current approximation on the fine mesh (φ_h) is given by

$$R_h = f_h - L_h \varphi_h \qquad (2.5)$$

We solve,approximatly, the correction problem on a coarse grid with spacing H:

$$L_H v_H = I_H^h R_h \qquad (2.6)$$

Here, I_H^h is a projection operator from the fine to the coarse mesh.

The correction v_H is interpolated to a finer mesh, e.g. with a spacing of H/2. Some relaxation sweeps are made on the new correction problem

$$L_{H/2} v_{H/2} = I_{H/2}^h R_h$$

This procedure is repeated until the finest grid is reached. The correction v_h contains errors, but all the error-components have small and relatively the same amplitude. v_h is added to the old approximation, and if the residual is not small enough the procedure can be repeated. This way of working explicitly with the corrections is called the correction scheme (CS).

When the operator L is not linear the coarse grid operator L_H (H>h) should be linearized, which in turn reduces the efficiency of the relaxations. The coarse and the fine grids relaxation operators are not similar. For this and other reasons a different formulation, instead of (2.6), is preferable. We define the coarse grid approximation to be φ_H (= $v_H + I_H^h \varphi_h$). Equation (2.6) is replaced by

$$L_H\varphi_H = f_H \qquad (2.7)$$

where

$$f_H = L_H I_H^h \varphi_h + I_H^h (f_h - L_h\varphi_h)$$

In this way, the form of the operator approximating L is the same on all grids, even if the problem is not linear. It can be shown [4] that $L_H I_H^h \varphi_h - I_H^h L_h \varphi_h$ is an approximation to the differnce between the local truncation errors on the grid with spacing H and the grid with spacing h. When the coarse grid approximation φ_H is used instead of the correction v_H, the method is called the full approximation storage (FAS).

Many more details concerning the choice of the operator I_H^h which transfers the residuals to coarse grids, I_h^H the interpolation operator between coarse and finer grids, the choice of efficient relaxation operators and many other problems are discussed in refences 1,2 and 4.Different versions of MG methods has been applied to second order boundary value problems, linear [3] as well as nonlinear [3],[4] and [5]. The experience has shown that the non-linearity does not effect much the efficiency of the MG method. However, the extension of the methods to systems of equations is not trivial. Some of these difficulties and remedies are described below.

3 THE GOVERNING EQUATIONS

We consider the steady flow of an incompressible viscous fluid in two dimensions, governed by the Navier-Stokes equations. These equations express the conservation of the mass and the two components of the momentum. In these conservation laws the velocity components and the pressure are used as dependant variables :

$$\nabla\cdot\bar{u} = 0 \qquad (3.1.a)$$

$$\nabla^2 u - Re(uu_x + vu_y) - p_x = 0 \qquad (3.1.b)$$

$$\nabla^2 v - Re(uv_x + vv_y) - p_y = 0 \qquad (3.1.c)$$

The system of equations (3.1) is elliptic and it has a solution (u,v,p) if the two velocity components are specified on all boundaries. It should be noted that no boundary conditions should be specified on the pressure.

The number of the equations in the system can be reduced by introducing a stream function ψ such that

$$u = \psi_y \quad \text{and} \quad v = -\psi_x \qquad (3.2)$$

Substituting (3.2) into (3.1) gives

$$\nabla^4\psi + Re[(\psi_x\nabla^2\psi)_y - (\psi_y\nabla^2\psi)_x] = 0 \qquad (3.3)$$

The boundary conditions on ψ and its derivatives normal to the boundaries are found from the boundary conditions on the velocity vector and relation (3.2).

Two second order eqations can be found from (3.3) if the

vorticity ζ is introduced :

$$\zeta = v_x - u_y \qquad (3.4)$$

Then

$$\nabla^2\zeta + Re[(\psi_x\zeta)_y - (\psi_y\zeta)_x] = 0 \qquad (3.5.a)$$

$$\nabla^2\psi = -\zeta \qquad (3.5.b)$$

The boundary condition on the normal derivative of ψ must be replaced by a condition on ζ. The vorticity on the boundaries is a combination of the normal and the tangential derivatives of the components of the velocity vector. The tangential derivative of the velocity on the boundaries is given but the normal derivative can only be found as part of the solution. For this reason the application of the boundary conditions of equations (3.5) is more complicated than for equations (3.1).

4 MG SOLUTION OF THE N-S EQUATIONS

A possible iterative solution of equations (3.1) can be constructed if equations (3.1.b) and (3.1.c) are solved for u and v, respectively. However, the system does not contain a natural equation for the pressure. For this reason, probably, the original system of equations has not been solved by 'standard' iterative techniques.

A formulation which results in an equation 'for the pressure' is obtained by taking the divergence of the momentum equations and using the mass conservation equation. The new system of equations can be written as:

$$\nabla^2 p + (u^2)_{xx} + 2(uv)_{xy} + (v^2)_{yy} = 0 \qquad (4.1.a)$$

$$\nabla^2 u - Re(uu_x + vu_y) - p_x = 0 \qquad (4.1.b)$$

$$\nabla^2 v - Re(uv_x + vv_y) - p_y = 0 \qquad (4.1.c)$$

The system of equations (4.1) requires an additional boundary condition compared with the original system (3.1). A Neumann boundary condition on the pressure can be specified only as part of the solution in such a way that the original equations and the boundary conditions are satisfied. The iterative solution of (4.1) seems to be simpler than that of (3.1), but the boundary conditions for (4.1) are more complex.

Here, we treat only the solution of system (3.1). The method of solution, which is desribed here shortly, is similar to the one suggested by Brandt [4]. The computational domain is coverd by a mesh as in figure 1. The components of the velocity vector are defined in the middle of that side of each computational cell which is normal to the direction of the corresponding velocity component. The pressure is defined in the center of each computational cell. The 'staggered grid' allows a discretization of the equations in a form which corresponds to the conservation laws in each cell.

The finite difference approximations to (3.1) are written as:

$$\Sigma\, \partial_j u_j = 0 \qquad (4.2.a)$$

$$Qu_j - \partial_j p = 0 \qquad (4.2.b)$$

where u_j is the component of the velocity vector in the j-th direction. $Q = \partial_x^2 + \partial_y^2 - \mathrm{Re}\Sigma u_j \overleftarrow{\partial}_j$. ∂_j and $\overleftarrow{\partial}_j$ denote the central difference and the upstream difference approximations to the first derivative in the j-th direction, respectively. $\partial_x^2 + \partial_y^2$ denotes the standard five point approximation to the Laplace operator in the plane.

The boundary conditions on u and v are given. However, the computational boundary of the (staggered) grid do not coincide with the physical boundary (figure 1). Thus, the boundary conditions are applied with an accuracy which is proportional to the mesh size.

The system of algebraic equations (4.2) is solved by a MG technique. The relaxation procedure on each grid has three steps. In the first two steps equations (4.2.b) are solved for u and v, while the pressure is kept constant. In the third step the mass conservation equation is satisfied, pointwise, by changing the velocity vector and the pressure at some points (distributing the correction). The distribution of the correction is done in such a way that the residuals of equations (4.2.b) are not changed. To find the distribution coefficients a characteristic function $\chi_{\bar{x}}$ is defined at the center of each computational cell.

$$\chi_{\bar{x}}(\bar{\xi}) = \begin{cases} 0, & \bar{x} \neq \bar{\xi} \\ 1, & \bar{x} = \bar{\xi} \end{cases}$$

The mass conservation equation is to be satisfied at the cell with the center at $\bar{\xi}$. The j-th component of the velocity vector is changed by Δu_j and the pressure is changed by Δp, such that:

$$\Delta u_j = -\delta \partial_j \chi_{\bar{x}}(\bar{\xi}) \qquad (4.3.a)$$

and

$$\Delta p = \delta Q \chi_{\bar{x}}(\bar{\xi}) \qquad (4.3.b)$$

By substituting (4.3) into (4.2.b) (assuming that $\partial_j Q \simeq Q\partial_j$) one finds that the residuals of (4.2.b) change only slightly.

δ is found by substituting (4.3.a) into (4.2.a) and solving for δ. The correction δ, at each cell is distributed to u_j and p by adding (4.3.a) and (4.3.b), respectively. Some of the distribution coefficients are displayed in figure 1.

It has been shown [4] that the distributed Gauss-Seidel (DGS) relaxation scheme for the mass conservation equation together with point,or line relaxations of the momentum equations has a good smoothing rate for the high frequency error-components. For high Reynolds number flows, the smoothing rate

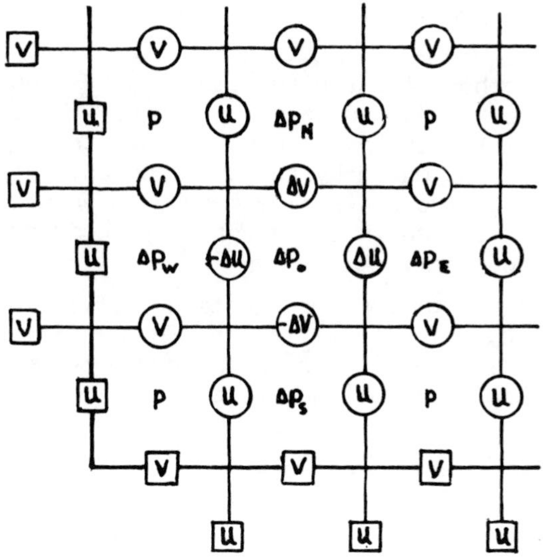

Figure 1: the staggered grid and the distribution of the coorections. When (4.2.a) is satisfied at the cell denoted at its center by Δp_o, the corrections are distributed as shown above. Some of the pressure changes are given by:

$$\Delta p_o = \delta\{-4/h^2-Re[|u|/h+|v|/h]\}$$

$$\Delta p_E = \begin{cases} \delta\{\ 1/h^2+Re\ u/h\} & u<0 \\ \delta\{\ 1/h^2\ \} & u>0 \end{cases}$$

$$\Delta p_W = \begin{cases} \delta\{\ 1/h^2\ \} & u<0 \\ \delta\{\ 1/h^2+Re\ u/h\} & u>0 \end{cases}$$

depends strongly on the direction of the relaxation of the momentum equations. Best results are obtained if the direction of the relaxations coincide with the direction of the flow. In most cases this direction is not known apriori and it is not constant throughout the flow field. A general method, which is efficient in such cases is to relax line-wise back and forth. The smoothing rate of the DGS relaxations is larger than the smoothing rate of the momentum equations for large Reynold numbers. For this reason the number of DGS sweeps should be larger than the number of the relaxation sweeps on the momentum equations, as the Reynolds number increases.

The described relaxation scheme has been used in a FAS multi-grid computer program. The results indicate that the method is efficient even though the relative efficiency is less than for single second order partial differential equations. The results obtained by this, and other methods, are presented in Section 6.

5. MG SOLUTION OF THE VORTICITY AND STREAM FUNCTION FORMULATIONS.

Numerical methods for the solution of the coupled system (3.5) and equation (3.3) has been considered. The Laplace operator is approximated by the five point operator (∇_5^2) and the operator ∇^4 is approximated by the thirteen point operator ($\nabla_{13}^4 = \nabla_5^2 (\nabla_5^2)$). The convective terms, in equations (3.3) and (3.5.a), are written in conservative, second order accurate central difference form.

The boundary condition for equation (3.5.a) (i.e. for ζ) is approximated by a finite difference formulation of (3.5.b) at the boundaries. Several expressions could be derived. Assuming an impermeable wall one can write:

$$\zeta_w = \frac{2}{h^2} (\psi_w - \psi_{w+1}) + \frac{2}{h} (\psi_n)_w + O(h) \qquad (5.1.a)$$

or

$$\zeta_w = \frac{1}{3h^2}\{16\psi_w - 21\psi_{w+1} + 6\psi_{w+2} - \psi_{w+3}\} + \frac{4}{h}(\psi_n)_w + O(h^2) \qquad (5.1.b)$$

where ζ_w is ζ at the wall and $(\psi_n)_w$ is the normal derivative of ψ at the wall. Other formulations are possible [6]. The expression (5.1.a) is most commonly used.

The ∇_{13}^4 difference operator cannot be used at points adjacent to the boundaries. Instead we use at these points $\nabla_5^2\zeta$ ($\zeta = \nabla^2\psi$) where relation (5.1.a) or (5.1.b) is incorporated. In this way the finite difference approximations to (3.3) and (3.5) are related in the same way as the differential equations are. Equation (5.1.a) is of order h for ζ, but it gives an error term of order 1/h when it is used to approximate ∇^4 at points adjacent to the boundaries. Equation (5.1.b) gives a consistent approximation to the ∇^4 operator at these points.

Relaxation of system (3.5) can be done in many ways. A simple straight forward procedure is to relax one equation at a time:

1) update ζ by relaxing equation (3.5.a)
2) update ψ by relaxing (3.5.b) using the latest ζ
3) update the velocities (ψ_x and ψ_y) and the boundary condition (ζ_w).

Steps 1-3 are repeated several times.

In principle, the purpose of the relaxations is not to solve the problem, but to smooth out the error, i.e. eliminate high frequency error components. Due to the coupling between the governing equations, directly and through the boundary conditions, the straight forward relaxation has a poor smoothing rate. Smooth errors in ψ produce, through the updating of the boundary condition, ζ_w, high frequency error components in equation (3.5.a). A simple way to improve the smoothing rate of steps 1-3 has been found to be an incomplete updating of the stream function in step 2. That is, after the relaxation of the whole ψ field, the change in ψ is decreased at each point by a factor α before it is added to the old values of the stream function. The exact value of α is found to be dependant on the degree of coupling between the equations. The largest allowed α can be used when only one relaxation sweep is done each time step 2 is performed. A complete solution of each equation in each step during the iterations is very ineffective.

A more correct method, with respect to the role of the relaxation process in MG procedures, is constructed by relaxing simultaneously equations (3.5). This kind of relaxation takes better care of the coupling between the equations, but the simultaneous relaxations need more computational effort than two uncoupled relaxations of equations (3.5.a) and (3.5.b) However, the coupled relaxations still produce high frequency errors when the boundary conditions (on the vorticity) are computed.

Proper incorporation of the boundary conditions is achieved when no high frequency errors are introduced. An improvement is obtained if the updating of the vorticity boundary condition is incorporated into the relaxation procedure. This more 'strongly' coupled method makes the relaxation more complex in the sense that the computational effort is larger than for steps 1-3 of the 'weakly coupled' method. Our treatment of the equations and the boundary conditions results in a relatively efficient method, but further improvements in the incorporation of the vorticity boundary conditions is needed.

The single equation for the stream function (3.3) is discretized by (central) finite-differences. In this form of discretization, pointwise (or linewise) G-S relaxations are simple, since all the algebraic equations which are solved at each point (or line) are linear. Improved smoothing rate for ∇^4_{13} is obtained by using a weighted Jacobi relaxation procedure [2]. This weighted simultaneous displacement is fast to compute since the Jacobi relaxation sweep is done pointwise. The computational effort is smaller than for the G-S line relaxations where five diagonal matrices must be inverted.

The relaxation schemes described in this section have been incorporated in a full MG procedure. The FAS formulation is more natural in most non-linear cases, even though the CS formulation has been found almost as efficient as the FAS mode. In the FAS mode the right hand side of the coarse grid problem (2.7) is found by using a weighted average transfer of the approximations on the fine grid. This weighted operator (I_H^h) has been found superior to simple injection for equation (3.3) (see also [2]). The later method has been used in the solution of the N-S equations (3.1) and equations (3.5). The interpolations (I_h^H) of the corrections from a coarse grid with spacing H to a fine grid with spacing h, is cubic for equations (3.3) and (3.5). This order of interpolation is necessary since the governing equation (3.3) is of fourth order. In the case of the N-S equations (3.1), the corrections on the coarse grid are interpolated linearly into the fine grid.

6 RESULTS AND DISCUSSION

As a test problem we solved the driven cavity flow for a range of Reynolds numbers up to 200. The efficiency, and in some respect the accuracy, of the different versions of the methods have been compared for Re 1 and 50 on three meshes: 12x12, 24x24 and 48x48 intervals. A second type of problem is more artificial: we define a function ψ and use it to compute the right hand sides of the different equations. These typs of problems have been used to determine the relative accuracy of the different formulations.

The comparison of the efficiency of numerical methods is complex, since the residuals which are commonly used to define accuracy, are not comparable for the different formulations. Another factor is that the initial rate of convergence is strongly dependent on the initial approximation. For this reason we define the asymptotic rate of convergence, which is not sensitive to the initial approximation.

A usual technique to produce good initial approximations is to solve the problem on successively refined meshes until the finest level is reached. If the interpolations are done properly the initial approximations contain (mainly) high frequency error components. After a few relaxation sweeps the error of the approximation is less than the truncation errors. In these cases, the rate of convergence on each grid is better than or equal to the smoothing rate of the relaxation operator for high frequencies. If a more accurate solution of the discrete problem is computed, the convergence rate is slower. (We define the convergence rate as the reduction in the 'error' by a computational effort which is equivalent to one relaxation sweep on the finest grid). The asymptotical convergence rate is obtained when the computations are continued beyond the accuracy due to the truncation errors.

In Table 1 we compare the averaged rate of convergence (θ) and the computational times (T,in seconds) of the methods for the solution of a test problem by equations (3.1) and by (3.3) with (4.1.a) and (4.1.b), Re = 1.

Table 1

	Equation (3.1)		Equation (3.3) with			
			(5.1.b)		(5.1.a)	
Mesh	θ	T	θ	T	θ	T
12x12	0.73	0.33	0.68	0.31	0.82	0.62
24x24	0.77	1.23	0.72	1.28	0.78	2.14
48x48	0.76	4.95	0.68	5.09	0.87	8.82

The test examples, in table 1, have been computed with single precision by the N-S solver, and with double precision by the solver of equation (3.3). It has been experienced that for meshes with fine spacings, the round-off errors with single precision preven the convergence to errors which are needed for the computation of the asymptotical rate of convergence. All the results for equation (3.3), in table 1, have been computed in double precision.

The rate of convergence for the driven cavity flow, for Reynolds numbers 1 and 50 are quite similar to those in table 1. The rate of convergence for the N-S solver is about 0.80, while the solver of (3.3) using relations (5.1.a) or (5.1.b) is about 0.84 or 0.72, respectively.

The solver of (3.5) is slightly less efficient than that of (3.3) for Reynolds numbers under 50, and the efficiency decreases further for larger Re. The required work is increased by as much as 50% while for the later method the increase is about 10%. The FAS and the CS formulations for the solution of the vorticity-stream function equations, has about the same efficiency. This, together with the fact that the efficiency decreases for high Re, indicate that the method in its present form is not optimal. The treatment of the boundary conditions is still not satisfactory. Equation (5.1.a), when it is used in a FAS mode gives slower convergence than when it is used in a CS mode. This is so since the incorporation of (5.1.a) in the approximation to ∇^4, at points adjacent to the boundaries, results in an approximation of order 1/h. Coarse grid FAS equations do not approximate well the problem of the finer grids and therefore the MG procedure is less efficient. Equation (5.1.b), on the other hand, works well both in the FAS and the CS modes. Further improvements in treating the vorticity boundary condition are under study.

The accuracy of the different formulations has also been tested for problems with known solutions. Here we give,shortly, some of these results. Both relations for the boundary conditions (5.1) result in second order accuracy for the stream function and its first derivatives, i.e. the errors are proportional to the square of the mesh size. The N-S solver is first order accurate for the components of the velocity vector, which is expected because of the accuracy in applying the boundary conditions on the staggered grid. The introduction of second order accurate boundary condition into the N-S solver, is involved in difficulties similar to those that are encountered in the solution of (3.5).

Computations of problems which include boundary layers show that (5.1.b) approximates better the stream function than its (first) derivatives, while (5.1.a) gives a better approximation to the derivatives, rather than to the stream function.

We have discussed the relative (asymptotic) efficiency of some numerical methods. In the computation of many flow problems, the iterative process is terminated before the asymptotic convergence is reached. Adequate accuracy may be obtained if the multi-level construction of initial approximations is utilized and the approximation on the finest grid is improved, by some relaxation sweeps (without using a full MG procedure). In such cases, the rate of convergence is better than the asymptotical values which are presented in this work.

The choice of a particular method is natural in some cases. When the primitive variables are to be computed, or when three dimensional problems are to be considered, the N-S solver is most natural. It can be, also, generalized to include compressible flows. For plane or axisymmetrical flows the equation involving only the stream function is the simplest from programing and efficiency point of views.

Our results show that the MG methods are efficient also for systems of equations, but the optimal efficiency has not yet been reached. It is expected that better treatment of the application of the boundary conditions will improve the numerical efficiency.

REFERENCES

1. FUCHS, L. - Finite-Difference Methods for Plane Steady Inviscid Transonic Flows, TRITA-GAD-2, 1977.
2. BRANDT, A. - Multi-Level Adaptive Solutions to Boundary Value Problems, Math. Comp. vol. 31, pp. 333-390, 1977.
3. FUCHS, L. - A Fast Numerical Method for the Solution of Boundary Value Problems, TRITA-GAD-4, 1980.
4. BRANDT, A. and DINAR, N. - Multigrid Solution to Elliptic Flow Problems, Numerical Methods for PDE, Ed. Bramble, J.H. pp. 53-147, 1979.
5. FUCHS, L. - A Newton Multi-Grid Method for the Solution of Non-linear PDE, Proc. of the BAIL I Conference, Ed. Miller, J.J.H., Boole Press, 1980.
6. ROCHE, J.P. - Computational Fluid Dynamics, Hermose Publishers, 1972.

LAMINAR FLOW IN CURVED MOON SHAPED DUCTS

K. Nandakumar, J.H. Masliyah

Department of Chemical Engineering
University of Alberta, Edmonton, Alberta, CANADA

ABSTRACT

Fully developed laminar flow of a Newtonian fluid in a curved duct is considered. The shape of the duct is the region formed by the natural coordinate surfaces in bipolar toroidal coordinate system. This is useful in studying the effects of distortion in the cross-section of circular and semi-circular coiled tubes. The equations of motion in the stream function-vorticity form are solved numerically using a recently proposed multigrid method and the conventional SOR method. The multigrid method has been found to be superior. For certain geometrical shapes dual solutions have been found above a critical Dean number.

1. INTRODUCTION

Laminar flow in curved ducts of various cross-sections has been studied quite extensively. These include circular tubes [1 - 8], elliptical tubes [9], square and rectangular tubes [10 - 12], triangular tubes [13, 14] and semi-circular ducts [15, 16]. Analytical solution for laminar flow in straight moon shaped ducts has been reported in [17]. Out of this long, but incomplete, list, readers' attention is drawn in particular to references [10, 12, 16] because they have a direct bearing on the current work. Cheng *et al.* [10, 12] and Masliyah [16] report on the existence of dual solutions over certain range of flow parameters. Existence of such dual solutions in the presence of slight distortions in the cross-sectional area of curved tubes will be investigated. Small distortions in the cross-section are to be expected when coiled tubes are formed by winding straight tubes of circular and semi-circular cross-section around cylinders.

Distortions in the cross-section will be modelled by choosing desired areas bounded by bipolar coordinates as shown in Figure 1. Hence the equations of motion are developed in a bipolar toroidal coordinate system (ξ_1, ξ_2, ξ_3). The flow situation to be considered is that of a fully developed laminar flow of a Newtonian fluid in a curved tube of zero pitch. The equations of motion are given in the next section in a non-dimensional, stream-function vorticity form.

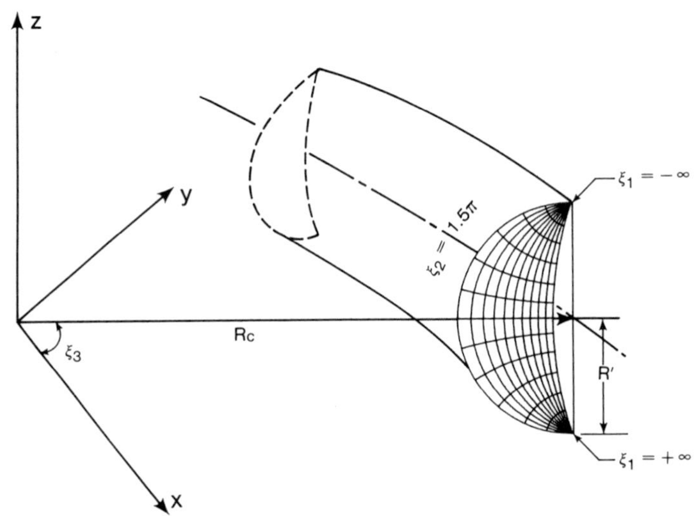

Fig.1 Co-ordinate System

2. GOVERNING EQUATIONS IN BIPOLAR TOROIDAL COORDINATES

The equations of motion are given by

$$\begin{bmatrix} \left\{ f_1 \nabla^2 - C_1 \dfrac{\partial}{\partial \xi_2} - C_2 \dfrac{\partial}{\partial \xi_1} - C_3 \right\} & 0 & C_4 \\ -1 & \left\{ \dfrac{f_1^3}{f_2} \nabla^2 + C_5 \dfrac{\partial}{\partial \xi_2} + C_6 \dfrac{\partial}{\partial \xi_1} \right\} & 0 \\ 0 & 0 & \left\{ f_1 \nabla^2 - C_1 \dfrac{\partial}{\partial \xi_2} - C_2 \dfrac{\partial}{\partial \xi_1} - C_7 \right\} \end{bmatrix} \cdot \begin{bmatrix} \zeta \\ \psi \\ u_3 \end{bmatrix} = \begin{bmatrix} 0 \\ 0 \\ \dfrac{1}{f_2} \dfrac{\partial p}{\partial \xi_3} \end{bmatrix} \qquad (1)$$

where

$$f_1 = \cosh \xi_1 - \cos \xi_2$$

$$f_2 = f_1/[Rc\ f_1 + \sin \xi_2]$$

$$\nabla^2 = \left[\frac{\partial^2}{\partial \xi_1^2} + \frac{\partial^2}{\partial \xi_2^2}\right]$$

$$C_1 = u_2 + (1 - \cos \xi_2 \cosh \xi_1)/f_2$$

$$C_2 = u_1 + \sin \xi_2 \sinh \xi_1/f_2$$

$$C_3 = \frac{f_1}{f_2^2} + u_1 \frac{\sin \xi_2 \sinh \xi_1}{f_1\ f_2} + u_2 \frac{(1 - \cos \xi_2 \cosh \xi_1)}{f_1\ f_2}$$

$$C_4 = \frac{2}{f_2}\left[\sin \xi_2 \sinh \xi_1 \frac{\partial u_3}{\partial \xi_2} - (1 - \cos \xi_2 \cosh \xi_1) \frac{\partial u_3}{\partial \xi_1}\right]$$

$$C_5 = \frac{f_1^2}{f_2^2} (1 - \cos \xi_2 \cosh \xi_1)$$

$$C_6 = \frac{f_1^2}{f_2^2} \sin \xi_2 \sinh \xi_1$$

and

$$C_7 = \frac{f_1}{f_2^2} - u_1 \frac{\sin \xi_2 \sinh \xi_1}{f_1\ f_2} - u_2 \frac{(1 - \cos \xi_2 \cosh \xi_1)}{f_1\ f_2}$$

and the velocity components $\{u_1, u_2, u_3\}$ are in the coordinate directions $\{\xi_1, \xi_2, \xi_3\}$, respectively.

The metric coefficients of the bipolar toroidal coordinate system are:

$$h_1 = h_2 = \cosh \xi_1 - \cos \xi_2$$

and
$$h_3 = \frac{\cosh \xi_1 - \cos \xi_2}{[Rc(\cosh \xi_1 - \cos \xi_2) + \sin \xi_2]}$$

The variables were rendered dimensionless as follows:

$$Rc = \frac{Rc'}{R'} \qquad \underset{\sim}{u} = \frac{\underset{\sim}{u}'}{(\nu/R')} \qquad Q = \frac{R'^2}{\nu\mu}\frac{\partial P'}{\partial \xi_3}$$

$$\psi = \frac{\psi'}{\nu R'} \qquad \zeta = \frac{\zeta'}{(\nu/R'^2)}$$

The prime denotes a dimensional quantity and R' is the radius of the duct measured along $\xi_2 = \pi$, and Rc' is the radius of curvature.

The stream function is defined by

$$u_1 = -\frac{f_1^2}{f_2}\frac{\partial\psi}{\partial\xi_2} \tag{2a}$$

$$u_2 = \frac{f_1^2}{f_2}\frac{\partial\psi}{\partial\xi_1} \tag{2b}$$

and the vorticity is given by

$$\zeta = f_1^2\left[\frac{\partial}{\partial\xi_1}\frac{u_2}{f_1} - \frac{\partial}{\partial\xi_2}\frac{u_1}{f_1}\right] \tag{3}$$

Because of symmetry about the coordinate $\xi_1 = 0$ only the half region $\xi_1 = [0, \infty]$ will be considered. The transformation,

$$\xi_1 = [\exp(\beta) - 1]/2 \tag{4}$$

used to provide more grid points near the coordinate $\xi_1 = 0$.

The friction coefficient C_f defined as

$$C_f = \frac{< \tau_w' >}{\frac{1}{2}\rho < u_3' >^2} \tag{5}$$

is computed from

$$C_f = 2\,De\left(-\frac{\partial p}{\partial\xi_3}\right)\frac{A_x}{A_w}\frac{1}{< u_3 >}\frac{1}{Re} \tag{6}$$

where

$De = 4\frac{V_w}{A_w}$ is the equivalent diameter

A_x is the cross-sectional area

A_w is the wall area/unit ξ_3

V_w is the volume/unit ξ_3

$< u_3 >$ is the mean velocity

and $\quad Re = 2R' < u_3' > / \nu$

C_f can also be calculated by integrating the shear stress distribution along the wall as follows:

$$C_f = 2 < \tau_w > De / < u_3 > / Re \tag{7}$$

where the average wall shear stress is obtained from

$$< \tau_w > = \frac{\oint \tau_{\xi_2\xi_3} \frac{f_2}{f_1^2} d\xi_1}{\oint \frac{f_2}{f_1^2} d\xi_1} \tag{8}$$

where $\tau_{\xi_2\xi_3}$ at the wall is given by,

$$\tau_{\xi_2\xi_3} = f_1 \frac{\partial u_3}{\partial \xi_2} \tag{9}$$

and the integration is performed around the wall. The Reynolds number is given by

$$Re = De < u_3 > \tag{10}$$

and the Dean number is defined as

$$Dn = Re/\sqrt{Rc} \tag{11}$$

2.1 Boundary Conditions

a) Axial velocity:

$$u_3 = 0 \quad \text{on wall} \tag{12}$$

$$\frac{\partial u_3}{\partial \xi_1} = 0 \quad \text{along } \xi_1 = 0 \quad \text{(Symmetry)} \tag{13}$$

b) Stream function:

$$\psi = 0 \quad \text{on all boundaries} \tag{14}$$

c) Vorticity:

$$\zeta = 0 \quad \text{along } \xi_1 = 0 \quad \text{(Symmetry)} \tag{15}$$

$$\zeta = \frac{f_1^3}{f_2} \frac{\partial^2 \psi}{\partial \xi_2^2} \quad \text{along } \xi_2 = \text{constant (two walls)} \tag{16}$$

$$\zeta = \frac{f_1^3}{f_2} \frac{\partial^2 \psi}{\partial \xi_1^2} \quad \text{along } \xi_1 = \text{constant} \tag{17}$$

3. METHOD OF SOLUTION

The governing partial differential equations (1) were discretized using central difference approximations. The resulting set of non-linear algebraic equations could be solve with any standard iterative scheme such as the relaxation method. In this work, the traditional successive over-relaxation (SOR) scheme was used as a reference case against which the recently proposed multigrid method (MGRD) was compared.

The multigrid method was originally proposed by Brandt and co-workers [18 - 20]. The method has been used for generating boundary fitted coordinate systems by Camarero and Younis [21]. The convergence of the multigrid method for specific classes of problems has been studied by various workers [22 - 25]. The reader is referred to the original papers for details of the method. Here only the salient differences in the implementation of the multigrid method will be discussed.

Coupled nonlinear problems of this nature are often solved by treating them as locally linear [4]. In such methods one major cycle of iteration constitutes one sweep for the vorticity and the axial velocity equations followed by 4 or 5 sweeps for the stream function equation. During this major cycle, the nonlinear coefficients C_1 to C_4 and C_7 in equation (1) are frozen at the values of previous major iteration; *i.e.* these coefficients are only updated after each major iteration. As these coefficients depend on the derivative of ψ, the stream function equation is relaxed a few times to smooth out the solution. At low values of Dean number a point over relaxation scheme can be used in conjunction with the above iteration scheme. However, at high values of Dn an under relaxation factor may be needed to keep the iteration stable. This basic iteration scheme will be used in a multigrid setup.

The multigrid method as outlined in [18, 19, 25] is normally implemented in a recursive manner. However, for complex problems, Camarero and Younis [21] found it easier to implement it in a non-recursive manner. The latter scheme is used here. The steps involved are shown schematically in Figure 2. The salient features of the method, as used here, are:

(1) The multigrid method is used as for a linear problem. Several major iterations (20 to 50) are performed at the finest level for the variables $\{\zeta, u_3, \psi\}$ and the residuals are monitored.

(2) Once the convergence rate slows down, a coarse grid correction is applied. The level of coarse grid and the number of major sweeps (normally up to 25) at the coarse grid are specified interactively.

(3) The residuals from the finest grid are injected into the coarse grid and the nonlinear coefficients are frozen at the last major iteration on the finest level.

(4) Since the method is implemented as for a linear problem, only the corrections to the solution $\{\zeta, u_3, \psi\}$ are computed on the coarse grid. This eliminates the need to compute the residuals on the coarse grid.

(5) The corrections are added to the solution at the finest level after a bilinear interpolation.

(6) Steps (1) to (5) are repeated until the residuals are reduced to a desired level.

All the solutions presented in this work were obtained using the multigrid method with two levels and a maximum grid size of 21 x 21. The efficiency of the multigrid method was compared with that of the SOR method for several cases. The four examples presented in Table 1 are typical of the general trend.

Table 1: Comparison of Multigrid and SOR Methods (Rc = 30.0)

- Q	ξ_2 (outer)	ξ_2 (inner)	Method and Grid Size			Type of Solution	CPU(sec) Amdahl 470V/6
			Method	Levels			
800,000	1.05π	1.50π	SOR	-	*	4 vortex	22.3
			MGRD	2	*	4 vortex	8.6
150,000	1.05π	1.50π	SOR	-	*	2 vortex	16.8
			MGRD	2	*	2 vortex	6.3
75,000	0.60π	1.50π	SOR	-	*	2 vortex	24.0
			MGRD	2	*	2 vortex	7.7
10,000	0.60π	1.50π	SOR	-	†	2 vortex	39.6
			MGRD	3	†	2 vortex	17.5

* Grid size Imax = Jmax = 21

† Grid size Imax = Jmax = 41

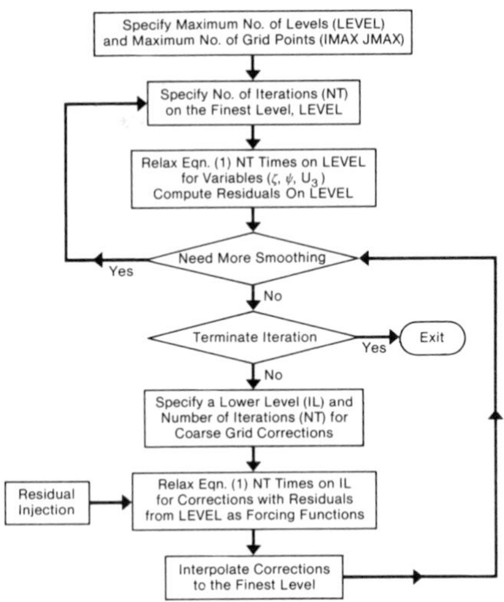

Fig.2 Multigrid Scheme

The multigrid method was at least twice as fast as the SOR method. It was able to predict both the two vortex and four vortex flow patterns. The first case with a pressure drop of Q = - 800,000 (Re = 849.6, Dn = 155.1) required an under relaxation factor of 0.4 for the vorticity and axial velocity equations. As the stream function equation is linear no relaxation factor was used. For the same case, the SOR method remained stable with a relaxation factor of 0.5 for vorticity and axial velocity. For the other cases presented in Table 1, the multigrid method did not require an under relaxation factor, but the SOR method required an under relaxation factor of 0.6.

3.1 Accuracy of Numerical Solution

The bipolar coordinates are such that a semi-infinite region must be spanned in the ξ_1 coordinate direction. But the finite difference method requires the selection of a finite region. Choosing $\beta = 3.0$ ($\xi_1 = 9.5428$ from equation (4)) spans most of the physical space with Z = 0.99986 (as opposed to the exact value 1.0). The region enclosed by the coordinates $\xi_1 \varepsilon [0.0,\ 9.5428]$ and $\xi_2 \varepsilon [0.5\pi,\ 1.5\pi]$ corresponds to a full circle and choosing Rc = 10,000 simulates a straight circular tube. For this case the product C_f Re was 16.07 with a grid size of 21 x 21 which is within 0.45% of the exact value of 16. For a curved circular tube with Rc = 10, Q = - 24,000, the product C_f Re was 26.75 with a 15 x 15 grid and 26.129 with a 21 x 21 grid. The latter is within 0.25% of the value of 26.192 reported in the literature [6]. As a further check to assure accuracy of the numerical results, the friction coefficient, C_f was calculated in two different ways using equations (6) and (7). The difference between the two values was not more than 3% at the highest Reynolds number of 909.9.

4. RESULTS AND DISCUSSION

The question of whether the complete Navier Stokes equation can exhibit dual solution for laminar flow in a torus is still unresolved. It appears to depend to a large extent on the cross-sectional geometry. Existence of dual solution in a square geometry was hinted in [10]. Computational and experimental evidence of a dual solution in semi-circular ducts was presented in [16]. For circular geometry the question is still being argued. Van Dyke [7] presented a series solution for circular tubes and pointed out a discrepancy between his solution and all the earlier numerical computations [2 - 5]. Recently Dennis [8] has presented evidence supporting the accuracy of earlier numerical work. He has suggested that Van Dyke's solution could represent a different solution to the same problem.

In the present work the geometry is chosen in such a way that it could be changed smoothly and continuously from a semi-circular cross-section to a circular cross-section. The shape of the inner wall is held constant at a semi-circle by choosing $\xi_2 = 1.5\pi$. The shape of the outer wall is changed from 1.05π (deformed semi-circular duct) to π (perfect semi-circular duct); 0.6π (deformed circular duct) and 0.5π (perfect circular duct). The objective was to check for dual solutions at sufficiently high Dean number. Dual solutions were generated as follows: for the case of Rc = 30, $\xi_{2,outer} = 1.05\pi$, a series of solutions were generated starting with (-Q) = 25000 and increasing it gradually to (-Q) = 800,000. The primary two vortex solution was obtained for the entire range. Starting with a converged two vortex solution at -Q = 400,000, a four vortex solution was obtained by increasing (-Q) to 800,000 in one step. This is equivalent to providing a large perturbation to the primary flow so that the initial guess is within the region of attraction of the secondary four vortex flow pattern. Once a four vortex flow pattern is established for some Q, a series of runs can be generated by starting with a four vortex solution and changing Q in small steps. A similar procedure can be used with a geometrical parameter, $\xi_{2,outer}$.

The results, for a few selected pressure drops Q, are presented in Table 2. At $\xi_{2,outer} = 1.05\pi$ and Q = - 400,000, only a two vortex solution was obtained. However, on increasing (-Q) to 600,000, both solutions were obtained. Both solutions existed as $\xi_{2,outer}$ was changed to 0.9π, 0.75π and 0.6π.

Table 2: Dual Solutions as $\xi_{2,outer}$ is changed (Rc = 30)

-Q	$\xi_{2,outer}$	← Two vortex →			← Four vortex →		
		Re	Dn	$C_f Re$	Re	Dn	$C_f Re$
400 000	1.05π	514.4	93.92	19.34			
600 000	1.05π	720.8	131.6	20.70	671.4	122.6	22.22
400 000	0.9 π	766.8	140.0	23.00	715.9	130.7	24.63
400 000	0.75π	1013.0	184.9	27.05	990.8	180.9	27.66
200 000	0.6 π	775.6	141.6	26.13	760.2	138.8	26.66

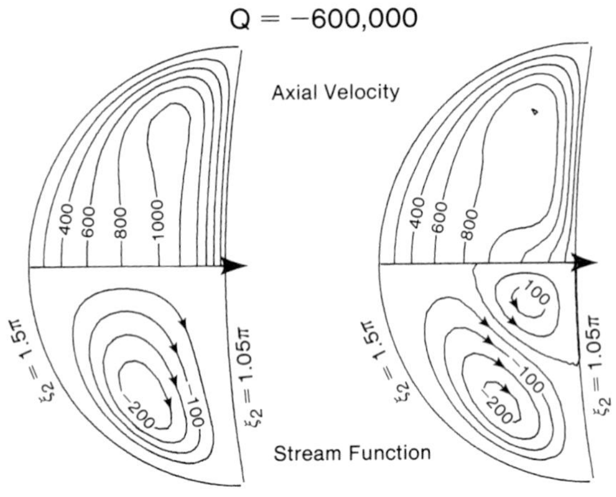

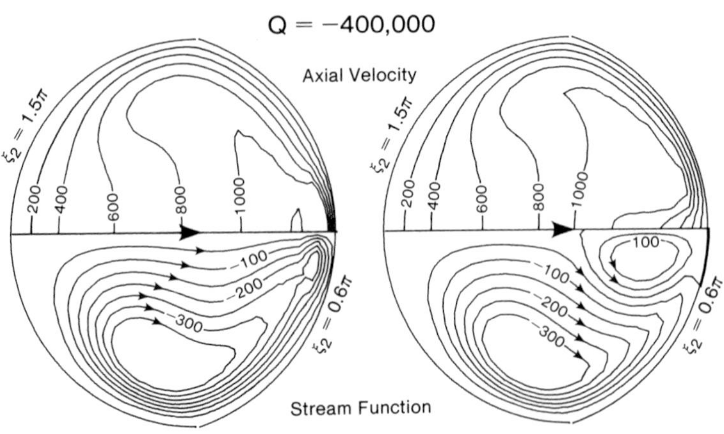

Fig.3 Stream function and axial velocity contours for Rc = 30

For two values of $\xi_{2,outer} = 1.05\pi$ and 0.6π, the contours of stream function and axial velocity are shown in Figure 3 for both types of solution. For $\xi_{2,outer} = 1.05\pi$, the contours are similar to those reported in [16]. For $\xi_{2,outer} = 0.6\pi$, the contours for the two vortex solution is similar to that reported in [4]. The four vortex solution for the same case appears to be new.

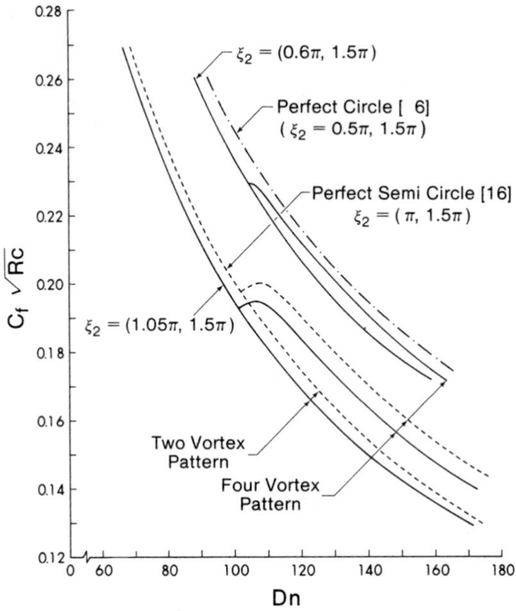

Fig.4 Variation of $C_f \sqrt{Rc}$ with Dean Number

Figure 4 shows the computed values of the friction coefficient plotted as $C_f \sqrt{Rc}$ vs Dn for the two cases $\xi_{2,outer} = 0.6\pi$ and 1.05π. Also shown are the results for a semi-circular duct and a circular duct taken from [6, 16]. The critical Dean number appears to be in the range 100 to 105. It is also interesting to note that the difference between the two solutions for $\xi_{2,outer} = 1.05\pi$ (deformed semi-circular duct) is more pronounced than for $\xi_{2,outer} = 0.6\pi$ (deformed circular duct). This trend indicates then that for a perfect circle, the difference between a two vortex solution and a four vortex one, if one exists, should be much smaller. This is perhaps why it is difficult to obtain the four vortex solution for a full circle. In this study, starting with a four vortex solution of a deformed circle ($\xi_{2,outer} = 0.6\pi$), most often we ended up with a two vortex solution for a full circle. In a few cases, we were able to obtain a four vortex solution for the circle; but it required extreme care in the iterative process using the multigrid method. Hence it appears that dual solutions are possible even for a perfect circle. But the difference between the two solutions appears to be small. The region of attraction of the four vortex solution also appears to be small.

REFERENCES

1. DEAN, W.R., The Stream-line Motion of Fluid in a Curved Pipe, Phil. Mag., Vol. 5, No. 7, pp. 673-695, 1928.

2. McCONALOGUE, D.J., and R.S. SRIVASTAVA, Motion of Fluid in a Curved Tube, Proc. Roy. Soc., Vol. A307, pp. 37-53, 1968.

3. TRUESDELL, L.C., Jr., and R.J. ADLER, Numerical Treatment of Fully Developed Laminar Flow in Helically Coiled Tubes, AIChE J., Vol. 16, p. 1010, 1970.

4. AUSTIN, L.R., and J.D. SEADER, Fully Developed Viscous Flow in Coiled Circular Pipes, AIChE J., Vol. 19, p. 85, 1973.

5. COLLINS, W.M. and S.C.R. DENNIS, The Steady Motion of a Viscous Fluid in Curved Tube, Quart. J. Mech. Appl. Math., Vol. 28, pp. 133-156, 1975.

6. TARBELL, J.M., and M.R. SAMUELS, Momentum and Heat Transfer in Helical Coils, Chem. Eng. J., Vol. 5, p. 117, 1973.

7. VAN DYKE, M., Extended Stokes Series: Laminar Flow Through a Loosely Coiled Pipe, J. Fluid Mech., Vol. 86, pp. 129-145, 1978.

8. DENNIS, S.C.R., Calculation of the Steady Flow through a Curved Tube using a new Finite-Difference Method, J. Fluid Mech., Vol. 99, pp. 449-467, 1980.

9. CUMING, H.G., The Secondary Flow in Curved Pipes, Aeronautical Research Council Reports and Memoranda, #2880, 1952.

10. CHENG, K.C., and M. AKIYAMA, Laminar Forced Convection Heat Transfer in Curved Rectangular Channels, Int. J. Heat Mass Transfer, Vol. 13, pp. 471-490, 1970.

11. JOSEPH, B., *et al.*, Numerical Treatment of Laminar Flow in Helically Coiled Tubes of Square Cross-Section AIChE J., Vol. 21, p. 965, 1975.

12. CHENG, K.C., *et al.*, Fully Developed Laminar Flow in Curved Rectangular Channels, Trans. A.S.M.E.I., J. Fluids Engng., Vol. 98, pp. 41-48, 1976.

13. COLLINS, W.M., and S.C.R. DENNIS, Viscous Eddies Near a 90° and a 45° Corner in Flow Through a Curved Tube of Triangular Cross-section, J. Fluid Mech., Vol. 76, pp. 417-432, 1976.

14. *ibid.*, Steady Flow in a Curved Tube of Triangular Cross-section, Proc. Roy. Soc. A., Vol. 352, pp. 189-211, 1976.

15. MASLIYAH, J.H., and K. NANDAKUMAR, Fully Developed Viscous Flow and Heat Transfer in Curved Semi-Circular Sectors, AIChE J., Vol. 25, pp. 478-487, 1979.

16. MASLIYAH, J.H., On Laminar Flow in Curved Semi-Circular Ducts, J. Fluid Mech., Vol. 99, pp. 469-479, 1980.

17. LONDON, A.L., and R.K. SHAH, Advances in Heat Transfer, Supplement I, Laminar Forced Convection in Ducts, Academic Press, 1978.

18. BRANDT, A., Multi-Level Adaptive Solutions to Boundary-Value Problems, Math. Comp., Vol. 31, pp. 333-390, 1977.

19. BRANDT, A., Multi-Level Adaptive Techniques (MLAT) for Partial Differential Equations: Ideas and Software, Mathematical Software III, Ed. Rice, J.R., Academic Press, 1977.

20. BRANDT, A., J.R. DENDY, and H. RUPPEL, The Multigrid Method for Semi-implicit Hydrodynamics Codes, J. Computational Physics, Vol. 34, pp. 348-370, 1980.

21. CAMARERO, R., and M. YOUNIS, Generation of Body-Fitted Coordinates for Cascade Computations Using Multigrid, Private Communication, 1979.

22. FEDORENKO, R.P., The Speed of Convergence of One Iterative Process, USSR Comp. Math. Math. Phys., Vol. 4, pp. 227-235, 1964.

23. NICOLAIDES, R.A., On Multigrid Convergence in the Indefinite Case, Math. Comp., Vol. 32, pp. 1082-1086, 1978.

24. NICOLAIDES, R.A., On the Observed Rate of Convergence of an Iterative Method Applied to a Model Elliptic Difference Equation, Math. Comp., Vol. 32, pp. 127-133, 1978.

25. HACKBUSCH, W., On the Fast Solutions of Non-linear Elliptic Equations, Numer. Math., Vol. 32, pp. 83-95, 1979.

NUMERICAL SIMULATION OF THREE DIMENSIONAL FLOWS IN DUCT

Laszlo Fuchs(1) and Heshu Zhao(2)

SUMMARY

The numerical solution of incompressible viscous three dimensional flows is considered. The original method of "parabolization" of the governing equations is further developed by relaxing some of the assumptions on the flow field and by introducing more efficient numerical techniques. Two methods for the solution of the parabolized equations are compared with a third multi-grid method for the three-dimensional Navier-Stokes equations. The flow in a straight duct is considered as a test problem. The accuracy of the methods is analysed mainly in the entrance region. It is concluded that the "parabolization" methods do not simulate well the simplified equations or the Navier-Stokes equations.

1 INTRODUCTION

Most flow problems have a three-dimensional character. The governing equations are nonlinear and therefore analytical methods cannot be used in general. Standard numerical methods are not practical for the solution of the three-dimensional Navier-Stokes (N-S) equations since the rate of convergence is too slow and unacceptable computer time and storage are required to obtain an approximation to the solution with a reasonable accuracy.

Here, we shall discuss the computations of flows of incompressible viscous fluids. Various simplifications can be introduced by making different assumptions on the flows.

---

1. Research Associate, Dep. of Gasdynamics, The Royal Institute of Technology, Stockholm

2. Visiting scientist, Dep. of Gasdynamics, The Royal Institute of Technology, Stockholm. Permanently at: Peking Institute of Aeronautics and Astronautics, China.

These assumptions may simplify greatly the governing equations and the corresponding numerical solution procedures. In some cases of high Reynolds number flows through ducts, one may assume that the viscous effects due to the variations of the gradient of the velocity components in the main flow direction are small and can be neglected. This assumption seems to be reasonable since inside the boundary layers the viscous effects are mainly due to the variations in the transversal plane and outside the boundary layers the convective effects dominate. Under this circumstances one can neglect some terms in the N-S equations. The resulting momentum equations become simpler in the sense that they are parabolic with a time-like direction parallel the main flow. Such a simplification has been used by Patankar and Spalding [1] and Briley [2] for the computation of the flow in straight ducts. Ghia and Sokhey [3] Ghia, Ghia and Studerus [4] and Ghia and Goyal [5] have used similar ideas to compute flows in curved ducts. Compressible flow computation is described by Briley and McDonald [7].

In the following we discuss the assumptions that are used in the formulation of the governing equations. Our starting point is the method described by Briley [2]. We simplify the solution procedure by reformulating some of the governing equations. This modified method is called here as Method 1. By modifying an existing N-S two-dimensional solver we have been able to relax some of the assumptions on the governing equations without loosing efficiency or accuracy and obtain a second method (Method 2). These two methods do not differ from each other in the treatment of the primary flow and the mean pressure. A third method (Method 3), which does not include any assumption on the governing equations is also presented shortly.

Our main interest in this work is to evaluate the simplified methods and to determine when such assumptions are valid. All the three methods are applied to the same problem, namely, the flow of incompressible fluids in a straight duct with a square cross section. We analyse the results of Methods 1 and 2 and compare them with that of Method 3. It has been found that the assumptions are not justified uniformly along the duct and that Method 1 results in a oscillatory secondary flow with decreasing amplitude. This oscillation is absent from the solutions obtained by methods 2 and 3.The errors in the results of Methods 1 and 2 are large near the entrance and decrease successively as the fully developed flow is approached. The "parabolization" of the governing equations results in a solution which is not sensitive to the secondary flow and in the first two cross sections the solutions are weakly dependent on the mesh spacing in the main flow direction. The parabolized solutions show also much faster development in the entrance region compared with the non-simplified three-dimensional solution.

2 THE GOVERNING EQUATIONS

The flow of incompressible viscous fluids can be described by the three-dimensional N-S equations

$$\nabla \cdot \bar{u} = 0 \qquad (2.1.a)$$

$$\frac{1}{Re} \nabla^2 \bar{u} - \bar{u} \cdot \nabla \bar{u} - \nabla p = 0 \qquad (2.1.b)$$

inside the domain Ω. $\bar{u} = (u,v,w)$ is the velocity vector with the components u, v and w in the x, y and z directions, respectively. The velocity vector is normalized by a given velocity w_o. A unit length of the coordinates equals some characteristic length D. p is the pressure with a unit equal to ρw_o^2, where ρ is the density of the fluid. The Reynolds number, Re, is defined to be $Re = w_o D \rho/\mu$, where μ is the viscosity coefficient of the fluid.

The boundary conditions for system (2.1) are given by specifying $\bar{u}$ on all boundaries. That is

$$\bar{u} = \bar{g} \qquad \text{on } \partial\Omega \text{ (the boundaries of } \Omega) \qquad (2.2.a)$$

where $\bar{g}$ is a given function.

However, it is required for the existence of the solution that the boundary condition $\bar{g}$ should be compatible with the mass conservation equation (2.1.a). By the divergence theorem

$$\int_\Omega \nabla \cdot \bar{u} \, dv = \int_{\partial\Omega} \bar{u} \cdot d\bar{s}$$

it is required that

$$\int_{\partial\Omega} \bar{g} \cdot d\bar{s} = 0 \qquad (2.2.b)$$

It should be noted that no further boundary conditions can or should be imposed on the system and in particular no conditions can be specified on the pressure.

The system of equations (2.1) is difficult to solve by conventional numerical methods, and only the recent developments of Multi-Grid (MG) techniques has enabled the rapid solution of the full problem. Simplifications can be introduced if the following assumptions are adopted:

1. $\partial^2 q/\partial z^2 = 0$ where q is any of the velocity components.
2. The pressure p can be written as

$$p(x,y,z) = p_m(z) + \hat{p}(x,y;z)$$

These two assumptions may be justified for high Reynolds number flows in ducts since the neglected terms $\partial^2 q/\partial z^2$ are small relative to $(\partial^2/\partial x^2 + \partial^2/\partial y^2)q$ inside the boundary layers and they are small relative to $Re\bar{u} \cdot \nabla q$ outside the boundary layers. It should be pointed out that for the computed test problem, $\partial^2 w/\partial z^2$ is not small relative to $(\partial^2 w/\partial x^2 + \partial^2 w/\partial y^2)$ at the entrance to the duct.

The two assumptions simplify the governing equations

$$u_x + v_y + w_z = 0 \tag{2.3.a}$$

$$ww_z = -uw_x - vw_y + \frac{1}{Re}(w_{xx} + w_{yy}) - p_{m,z} \tag{2.3.b}$$

$$wu_z = -uu_x - vu_y + \frac{1}{Re}(u_{xx} + u_{yy}) - \hat{p}_x \tag{2.3.c}$$

$$wv_z = -uv_x - vv_y + \frac{1}{Re}(v_{xx} + v_{yy}) - \hat{p}_y \tag{2.3.d}$$

Eqs. (2.3.b) - (2.3.d) are parabolic with respect to a time-like direction in the main flow direction z. It is assumed that w is positive everywhere (and therefore separation flows can not be computed without modifications). The solution to system (2.3) can be computed step by step by marching in the positive z direction. The numerical procedure presented by Briley [2] needs further assumptions. Some of these are adopted by the procedure which is called here as Method 1. Method 2 needs no additional assumptions.

3 SOLUTION OF THE PARABOLIZED EQUATIONS

Since equation (2.3.b) is parabolic for the dependent variable w an ADI technique can be used to advance the solution in the z-direction. Assume that $\bar{u}^n$ and p_m^n are given at cross-section n (the plane $z = z_n$) then w^{n+1} is computed in two steps

$$\frac{w^{n+\frac{1}{2}} - w^n}{\Delta z/2} = \frac{1}{Re}(\partial_x^2 w^{n+\frac{1}{2}} + \partial_y^2 w^n) - u^n \partial_x w^{n+\frac{1}{2}} - v^n \partial_y w^n - \frac{p_m^{n+1} - p_m^n}{\Delta z} \tag{3.1.a}$$

$$\frac{w^{n+1} - w^{n+\frac{1}{2}}}{\Delta z/2} = \frac{1}{Re}(\partial_x^2 w^{n+\frac{1}{2}} + \partial_y^2 w^{n+1}) - u^n \partial_x w^{n+\frac{1}{2}} - v^n \partial_y w^{n+1} - \frac{p_m^{n+1} - p_m}{\Delta z} \tag{3.1.b}$$

Where ∂_x, ∂_y, ∂_x^2 and ∂_y^2 are finite difference approximations to first and second derivatives, respectively. Each step of (3.1) needs the inversion of a three-diagonal matrix. This is done by Gaussian elimination. p_m^{n+1} is adjusted by a Newton method in such a way that the integral form of equation (2.3.a) is satisfied. That is, we require that the mass flow through each cross section equals a given constant.

Once w^{n+1} and p_m^{n+1} are known, one can use, as Briley [2], a similar ADI procedure to equations (2.3.c) and (2.3.d) for the secondary flow components. The gradients of $\hat{p}$ are computed from equations (2.3.c) and (2.3.d) at a cross-section just upstream the one that is being computed. The resulting velocity components are denoted by u_p and v_p. When such a procedure is adopted the mass conservation equation is not necessarily satisfied. Briley [2] suggests that corrections u_c and v_c should be added to u_p and v_p, respectively, in such a way that the mass conservation equation is satisfied:

$$(u_c + u_p)_x + (v_c + v_p)_y + w_z = 0 \tag{3.2}$$

Briley introduces a new assumption, namely, that the correction flow field is irrotational.

$$u_c = \Phi_x \qquad\qquad v_c = \Phi_y \tag{3.3}$$

where Φ is the velocity potential.
Eqs. (3.2) and (3.3) result in a Poisson equation for Φ with Neumann boundary conditions. Briley solves also a Poisson equation (by ADI-relaxations) for the pressure $\hat{p}$ using Neumann boundary conditions which are applied on 'a computational boundary'. We note, that normal pressure derivatives at the physical boundaries can be found only when the velocity field is known. In our version of Briley's procedure, Method 1, we solve a Poisson equation for u_c and another one for v_c by using Dirichlet boundary conditions ($u_c=v_c = 0$ on all the boundaries). That is

$$(u_c)_{xx} + (u_c)_{yy} = -[w_{zx}+(u_p)_{xx}+ (v_p)_{xy}] \tag{3.4.a}$$

$$(v_c)_{xx} + (v_c)_{yy} = -[w_{zy}+(v_p)_{xy}+ (v_p)_{yy}] \tag{3.4.b}$$

The solutions of the Poisson problems are obtained very efficiently by the MG computer program which has been developed by Fuchs [6]. The secondary flow is given by

$$u = u_p + u_c \tag{3.5.a}$$

$$v = u_p + v_c \tag{3.5.b}$$

Since only the gradients of $\hat{p}$ are interesting, they are computed, at the current cross-section from eqs. (2.3.c) and (2.3.d) by using the values of the velocity components which have already been computed.

Our second method, Method 2, uses a modified version of a two-dimensional Navier-Stokes Multi-Grid solver [9] to compute u, v and $\hat{p}$. We describe briefly the relaxation procedure for two- and three-dimensional N-S equations in the following. More details on the M-G method in general and Navier-Stokes MG-solvers in particular can be found in references [8-10].

4. A RELAXATION PROCEDURE OF NAVIER-STOKES EQUATIONS

The relaxation procedures for two- and three-dimensional N-S equations are the same in principle. For this reason we shall not distinguish between the two cases when we describe the basic principles of the method. Eqs. (2.1) are discretized on a staggered grid. In this way each velocity components is defined on the middle of each face of a small cube (or rectangle in two dimensions) normal to the coordinate corresponding to the velocity component. The pressure is defined at the middle of the cube. Eq. (2.1.a) is satisfied at the middle of the cube, and each momentum equation is satisfied at points where the corresponding velocity component is defined. All terms except the convective terms in Eqs. (2.1.b) are approximated by upstream finite differences. Because of the structure of the staggered grid, the boundaries of the computational domain do not always coincide with the physical boundaries. This implies that boundary conditions are applied with first order accuracy (in the mesh spacing h). Each momentum equation is solved for the corresponding velocity component by 'freezing' the values of the other velocity components and the pressure. The mass conservation equation is satisfied at each point by simultaneously changing the velocity components in the neighbouring points. At the same time the pressure is changed in some neighbouring points in such a way that the residuals of momentum equations are not changed locally. This ensures that the relaxation of the mass conservation equation does not introduce large changes into the residual of the momentum equations. More exactly, we want to solve the finite difference equations

$$\Sigma \partial_j u_j = 0 \qquad (4.1)$$

$$Q u_j + \partial_j p = 0 \qquad (4.2)$$

where ∂_j is a central finite difference operator in the jth direction, and

$Q = -(\partial_x^2 + \partial_y^2 + \partial_z^2) + \mathrm{Re}\ \Sigma u_j \partial_j$. The momentum equations are relaxed by solving u_j from Eq. (4.2) successively at each point (or for a whole line). Eq. (4.1) is relaxed by a so called distributed relaxation (see [9] and [10]).

We define a characteristic function $X_{\underline{x}}$ at the centre of each computational cell, $\underline{\xi}$:

$$X_{\underline{x}}(\underline{\xi}) = \begin{cases} 0 & \underline{x} \neq \underline{\xi} \\ 1 & \underline{x} = \underline{\xi} \end{cases} \qquad (4.3)$$

u_j and p are changed such that

$$u_j \leftarrow u_j - \delta_o \partial_j X_{\underline{x}} \qquad (4.4)$$

$$p \leftarrow p + \delta_o Q X_{\underline{x}} \qquad (4.5)$$

It is clear that if relations (4.4) and (4.5) are substituted into (4.2) the residuals of the momentum equations are kept almost constant. By substituting (4.4) into (4.1) one can find δ_o for each computational cell. Eq. (4.5) determines in what way should the correction due to the mass conservation equation distributed to the pressure in the neighbouring points.

It has been shown ([9], [10]) that this relaxation procedure can be used efficiently in a multi-grid process. Method 3 uses such a three-dimensional multi-grid solution procedure.

In the case of Method 2 we solve (2.3.b) by the same ADI method as is used by Method 1. The rest of the problem is a two-dimensional one and can be written as

$$u_x + v_y = -w_z \tag{4.6}$$

$$Qu + \hat{p}_x = 0 \tag{4.7}$$

$$Qv + \hat{p}_y = 0 \tag{4.8}$$

where $Q = -(\partial_x^2 + \partial_y^2) + Re(u\partial_x + v\partial_y + w\partial_z)$.
Eqs. (4.6) - (4.8) are solved by a modified two-dimensional multi-grid solver. The modifications are mainly due to the definition of Q, which differ from the original two-dimensional operator in the term $w\partial_z$. One should also observe that the compatibility condition (2.2.b) should be fulfilled by Eq. (4.6). This condition is satisfied by requiring that the mass flow through each cross-section is constant.

5 RESULTS

Following Briley [2],we obtain solutions to the flow in straight ducts by the three methods. At the upstream boundary, the primary flow is taken to be uniform and equal to w_o, while the secondary flow components vanish. The velocity vector vanishes at the side walls of the ducts. The full three dimensional method needs the specification of the velocity vector at a downstream cross-section. Such boundary conditions are not necessary for Methods 1 and 2.

We have solved the duct flow problem for a range of Reynolds numbers between 10 and 1000. Most of the results presented below are, however, for Reynolds number 200. Methods 1 and 2 predict practically the same primary flow profiles which are in good agreement with experiments [11] (see figure 1).

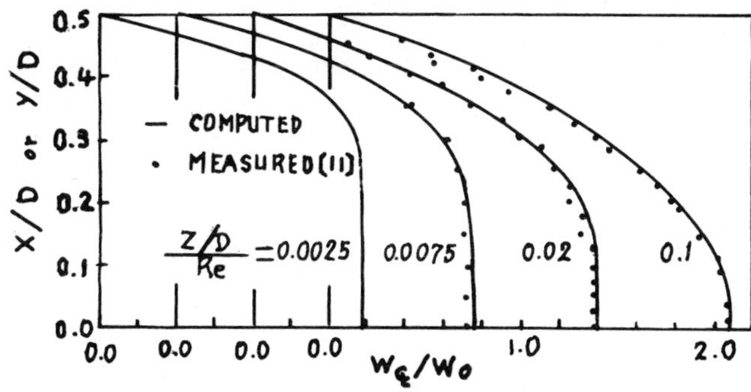

Figure 1

The developement of the primary flow profiles as predicted by Methods 1 and 2 compared with experiments.

The secondary flow which is obtained by Method 1 contains four vortex-pairs which decrease in strength and oscillate in their direction along the axis of the duct. The problem has been recalculated for a series of different "time" steps (Δz). The results show that the wave-length of the oscillations in the secondary vortices is not constant but it is proportional to the step size, Δz. This fact together with the fact that no oscillating vortices are computed by Method 2 (or Method 3), indicate that the oscillations may be attributed to the extra assumption (irrotationality) on the secondary flow corrections u_c and v_c. The mesh refinements show also that the flow not too close to the entrance is well predicted even by relatively large steps Δz. The secondary flow does not effect the primary flow by more than some percents, even if the secondary flow is taken to be identically equal to zero at all the cross sections. For this reason the primary flow as computed by Methods 1 or 2 differ only slightly, and they are in good agreement with experiments at those distances where these are available.

The agreement between experimental and numerical results is expected since the flow becomes more and more two dimensional, and the errors due to the assumptions (section 2) decrease, as the flow approaches its fully developed profile. In the entrance region, the mesh refinements show that the primary flow at the first and the second cross sections is almost independent of the step size Δz. The results are similar for both Method 1 and Method 2. This results indicate that the error in these cross sections is large and it cannot be reduced by reducing the step size. The results, on the other hand, do tend to converge at later cross sections as the step size decreases (see figure 2.b).

In figures 2 the convergence of the gradient of the primary flow at the center of the duct is compared. The results obtained by Method 1 (figure 2.a) have a tendency to slower convergence than those obtained by Method 2 (figure 2.b), as the step size decreases. The later results do not contain oscillations similar to those show in figure 2.a.

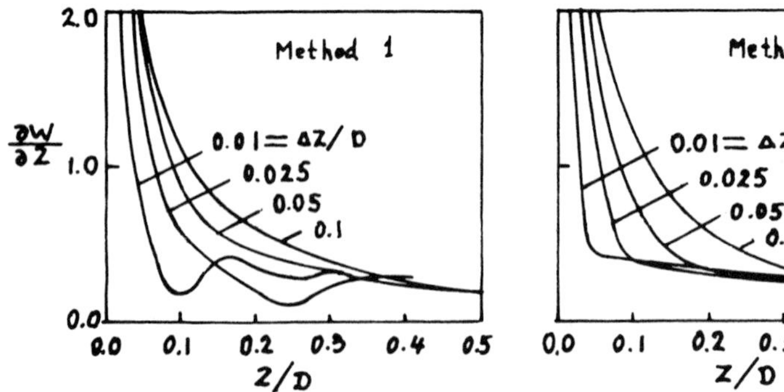

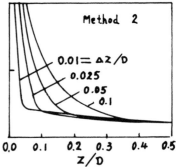

Figure 2.a Figure 2.b

$\partial w/\partial z$ computed at the center of the duct.

Solutions has also been computed by Method 3, using as down-stream boundary conditions the velocity field predicted by Method 1. The general trend of the results is that the parabolic methods show a faster development of the primary flow near the entrance compared to the full three dimensional N-S solver. This effect may be attributed to larger numerical dissipation in Methods 1 and 2.

We define two "error parameters" which are used to compare the accuracy of Methods 1 and 2 in the entrance region, where the validity of the basic assumptions can be questioned:

i) The root mean square (RMS, denoted by $\| \|_2$) of the residual of the mass conservation equation, R_m,

$$R_m = \partial_x u + \partial_y v + \partial_z w \quad (5.1)$$

ii) The maximum, $\| \|_\infty$ and the RMS, $\| \|_2$ values of the "relative error in the pressure" (Δp), which is defined as:

$$\Delta p = \frac{\tilde{p}_z - dp_m/dz}{dp_m/dz} \quad (5.2)$$

where

$$\tilde{p}_z = [\frac{1}{Re}(\partial^2{}_x + \partial^2{}_y + \partial^2{}_z) - (u\partial_x + v\partial_y + w\partial_z)]w$$

Table 1 displays the development of $\| R_m \|_2$, $\| \Delta p \|_\infty$ and $\| \Delta p \|_2$ as function of the distance z, for two values of Δz. These results are computed by Method 1, and they show that the amplitude of the errors decrease with the distance. $\| \Delta p \|_\infty$ has an oscillatory sign with a frequency proportional to the step size (Δz).

Table 1

$\Delta z = 0.025$				$\Delta z = 0.1$			
z	$\| \Delta p \|_\infty$	$\| \Delta p \|_2$	$\| R_m \|_2$	z	$\| \Delta p \|_\infty$	$\| \Delta p \|_2$	$\| R_m \|_2$
0.000	-2.04	0.92	1.15	0.0	-1.04	0.46	0.31
0.025	-1.18	0.51	0.65	0.1	-0.59	0.27	0.16
0.050	-0.74	0.19	0.51	0.2	-0.23	0.10	0.17
0.075	-0.97	0.23	0.38	0.3	-0.30	0.09	0.18
0.100	-1.12	0.34	0.27	0.4	-0.55	0.15	0.15
0.125	-1.13	0.41	0.20	0.5	-0.62	0.19	0.10
0.150	-1.38	0.41	0.17	0.6	-0.50	0.18	0.06
0.175	-0.99	0.30	0.20	0.7	-0.31	0.12	0.06
0.200	-0.30	0.11	0.23	0.8	-0.09	0.04	0.08
0.225	0.47	0.13	0.22	0.9	0.17	0.05	0.08
0.250	0.64	0.21	0.19	1.0	0.30	0.10	0.07
0.275	0.69	0.23	0.14	1.1	0.36	0.12	0.05
0.300	0.63	0.21	0.11	1.2	0.34	0.11	0.04
0.325	0.51	0.17	0.09	1.3	0.27	0.09	0.03
0.350	0.36	0.11	0.08	1.4	0.17	0.06	0.04
0.375	0.20	0.08	0.08	1.5	0.07	0.03	0.04
0.400	-0.19	0.05	0.07	1.6	-0.04	0.01	0.04
0.425	-0.22	0.05	0.05	1.7	-0.08	0.02	0.03
0.450	-0.21	0.06	0.04	1.8	-0.10	0.03	0.02

Table 1 continued

0.475	-0.20	0.06	0.04	1.9	-0.09	0.03	0.01
0.500	-0.15	0.05	0.05	2.0	-0.06	0.03	0.01
0.525	-0.10	0.03	0.06	2.1	-0.04	0.01	0.01
0.550	-0.03	0.01	0.07	2.2	-0.02	0.01	0.02
0.575	0.07	0.02	0.08	2.3	0.03	0.01	0.02
0.600	0.10	0.03	0.08	2.4	0.01	0.01	0.01

Table 2

The errors in "the pressure" as a function of the distance z, and the stepsize Δz. The superscript 1 or 2 on $\|\Delta p\|_2$ denote if the values are obtained from the results computed by Method 1 or 2 respectively.

	$\Delta z=0.025$		$\Delta z=0.05$	$\Delta z=0.1$	
z	$\|\Delta p\|_2^1$	$\|\Delta p\|_2^2$	$\|\Delta p\|_2^2$	$\|\Delta p\|_2^1$	$\|\Delta p\|_2^2$
0.00	0.92	1.01	0.72	0.46	0.51
0.05	0.19	0.26	0.43		
0.10	0.34	0.04	0.17	0.27	0.33
0.15	0.41	0.04	0.05		
0.20	0.12	0.04	0.04	0.10	0.14
0.25	0.21	0.03	0.03		
0.30	0.21	0.03	0.03	0.09	0.04
0.35	0.12	0.02	0.03		
0.40	0.05	0.02	0.02	0.15	0.04
0.45	0.06	0.02	0.02		
0.50	0.05	0.02	0.02	0.19	0.02
0.55	0.01	0.01	0.02		
0.60	0.03	0.01	0.01	0.15	0.02
0.65	0.02	0.01	0.01		
0.70	0.02	0.01	0.01	0.12	0.02

In table 2 Methods 1 and 2 are compared for some step sizes. The error decay with Method 2 is faster than with Method 1 and after about five cross sections the error is within the error which has been prescribed in the computation of the secondary flow components.

6 CONCLUSIONS AND DISCUSSION

Our results show that one has to consider two types of errors: Those errors which are due to the assumptions on the governing equations, and those which are due to the particular numerical technique which is used to solve the (simplified) governing equations. The former errors are eliminated best by using Method 3. The computational time is not much longer than for Methods 1 or 2. A computation of the test problem, by Method 3, on a mesh with 16x16x16 computational cells, takes about 16 sec. The residuals decrease by a factor about 400, and the maximum value of the final residuals is less than 0.01. For comparison, a solution on a 16x16x15 mesh by Method 1 or 2 takes about 8 or 12 sec, respectively. Method 3 requires however a larger computational space.

Method 3 can be improved by using it adaptively. In this way coarse grid solutions determine the region where the gradients of the velocity vector are large. In these region a mesh refinement is made. The resulting mesh will be refined, in the case of the test problem, only in the entrance to the duct. The computational time and memory may be reduced to about the same as required by Methods 1 or 2. Method 2 is preferable over Method 1 since it has better accuracy in solving the simplified governing equation, and it also provides directly the transversal pressure values.

For simple problems, such as flows in straight ducts without separation Method 2 may be used if the entrance region is excluded. For more complicated geometrics, or in general cases, the adaptive form of Method 3 is an efficient and preferable numerical technique.

REFERENCES

1. PATANKAR, S.V. and SPALDING, D.B. - A calculation Procedure for heat, Mass and Momentum Transfer in Three-Dimensional Parabolic Flows, Vol. 15. pp. 1781-1806, 1972.
2. BRILEY; W.B. - Numerical Method for Predicting Three-Dimensional Steady Viscous Flow in Ducts, J. Comp. Physics, Vol. 14, No. 1, pp. 8-24, 1974.
3 GHIA, K.N. and SOKHEY, J.S. - Laminar Incompressible Viscous Flow in Curved Ducts of Regular Cross Sections, J. Fluids Engineering, Vol. 99, pp. 640-648, 1977.
4. GHIA, U., GHIA, K.N. and STUDERUS, C.J. - Three-Dimensional Laminar Incompressible Flow in Straight Polar Ducts, Int. J. Computers and Fluids, Vol. 5, No. 4, pp. 205-218, 1977.
5. GHIA, U., GHIA, K.N. and GOYAL, R.K. - Three-Dimensional Viscous Incompressible Flow in Curved Ducts, AIAA Paper 79-1536, 1979.
6. FUCHS, L. - A Fast Numerical Method for the Solution of Boundary Value Problems, TRITA-GAD-4, April 1980.
7. BRILEY, W.R. and McDONALD, H. - Analysis and Computation of Viscous Subsonic Primary and Secondary Flows, AIAA Paper 79-1453, 1979.
8. BRANDT, A. - Multi-Level Adaptive Solutions to Boundary Value Problems, Math. Comp. Vol. 31, pp. 333-390, 1977
9. THUNELL, T. and FUCHS, L. - Numerical Solution of the Navier-Stokes Equations by Multi-Grid Techniques, To appear 1981.
10. BRANDT, A. and DINAR, N. - Multigrid Solutions to Elliptic Flow Problems, Numerical Methods for PDE, Ed, Bramble J.H. pp. 53-147, 1979.
11. GOLDSTEIN, R.J. and KREID, D.K. - Measurement of Laminar Flow Development in A Square Duct Using A Laser-Doppler Flowmeter, J. Appl. Mech. 813, 1967.

Different Finite Element Formulations for the Navier-Stokes Equations

DHATT, Gouri
Associate Professor

KAMGA FOMO, Bonaventure
Research Associate

Abstract

This study deals with the presentation of three different finite element formulations for solving steady state two-dimensional viscous fluid flows governed by the Navier-Stokes equations. The first consists of employing $\psi-\omega$ variables using an eight nodes quadrilateral twelve degree of freedom element with a new algorithm of implementating the no slip boundary conditions on solid frontiers. The second is a variational formulation called "mixed", using $\psi-\omega$ variables, similar to the Hellinger-Reissner principle used for solving plate bending problems. Finally we present a penalty formulation using a nine nodes quadrilateral eighteen degree of freedom element to obtain rapidly convergent solution for high Reynolds numbers.

1. Introduction

Since last ten years, there is an increasing interest in the application of finite elements to solve two dimensional Navier-Stokes equations. Various formulations are employed using either primitive variables [1-2] (velocity-pressure), the stream function-vorticity variables [1,3-8] ($\psi-\omega$) or the stream function [1,9-10] alone (ψ).

The finite element model using the velocity-pressure variables is of practical interest for engineering problems, but its convergence characteristics are generally limited to flows with low Reynolds numbers due to the importance of the convective terms in the solution process and the difficulty of satisfying the incompressibility constraint. Much effort has been made recently to overcome the problem of the incompressibility by introducing the penalty function in the Galerkin weighted residual method [2,11-12]. Doing so, the pressure is eliminated as independent variable, but can be obtained from the penalty function terms.

The stream function-vorticity formulation satisfies identically the continuity equation through the choice of stream function (ψ), but the problem posed by the no slip condition on solid boundaries is a major difficulty. Many authors handled this in an iterative manner [1,3], transforming the stationary problem to non-stationary one, the stationary case being obtained as the limiting solution in time. The convergence speed of the method is poor and its computational cost is discouraging.

The stream function formulation is employed by few researchers only [1,9,10] due to the difficulty of discretising 4th order partial derivative equations along with complicated boundary conditions.

In this work, we present three different finite element formulations for solving the two-dimensional Navier-Stokes equations: standard ψ-ω, mixed ψ-ω and the penalty function.

2. The standard ψ-ω formulation

2.1 Method proposed In terms of stream function ψ and vorticity ω, the steady state two-dimensional Navier-Stokes equations are:

$$\nabla^2\psi + \omega = 0 \tag{1a}$$

$$\nabla^2\omega - Re(\psi_{,y}\,\omega_{,x} - \psi_{,x}\,\omega_{,y}) = 0 \tag{1b}$$

where Re is the Reynolds number and ∇^2 is the two-dimensional Laplacian operator.

The boundary conditions are:

$$\psi = \bar{\psi} \quad \text{or} \quad \frac{\partial\psi}{\partial n} = \bar{u} \quad \text{on} \quad S \tag{2a}$$

$$\omega = \bar{\omega} \quad \text{or} \quad \frac{\partial\omega}{\partial n} = \bar{w} \quad \text{on} \quad S \tag{2b}$$

where n is the normal directed towards the exterior.

Along solid boundaries, the no slip conditions are:

$$\psi = \bar{\psi} \quad \underline{\underline{\text{and}}} \quad \frac{\partial\psi}{\partial n} = \bar{u} \quad \text{on} \quad S \tag{2c}$$

We notice that two conditions are given for ψ and none for ω. The last condition of (2c) is used to obtain ω values on the solid boundaries.

Instead of the time marching scheme explored extensively in the literature [1,3], the present authors [4,5,7,8] have proposed a direct method of implementing the no-slip condition and have obtained fast convergent solution for very high Reynolds numbers. Both approximations of the condition are given in detail in the reference 4.

This section presents further improvement in the implementation of the second order approximation. The method may be summarised as: ω is defined using the following two conditions on a point (x_o,y_o) of solid boundary:

$$\omega = -\frac{\partial^2\psi}{\partial n^2} \qquad (3)$$

$$\frac{\partial\psi}{\partial n} = \bar{u} \qquad (4)$$

The Taylor series expansion of $\psi(x,y)$ around the point (x_o,y_o) along the direction "n" leads to:

$$\psi(x,y) = \psi(x_o,y_o) + h\frac{\partial\psi}{\partial n}(x_o,y_o) + \frac{h^2}{2}\frac{\partial^2\psi}{\partial n^2}(x_o,y_o) + 0(h^3) \qquad (5)$$

where "h" is the distance of an interior point (x,y) from (x_o,y_o). Using equations (3), (4) and (5), one obtains:

$$\omega(x_o,y_o) + \frac{2}{h^2}[\psi(x,y) - \psi(x_o,y_o)] = \frac{2\bar{u}}{h} \qquad (6)$$

The equation (6) which defines the value of "ω" in terms of $\bar{u}$, $\psi(x,y)$ and $\psi(x_o,y_o)$ is employed to modify the assembled tangent matrix K_T in the following manner: Let the variable "ω" corresponding to the point (x_o,y_o) on the no slip boundary be defined by j^{th} equation in the matrix $[K_T]$. The equation numbers for $\psi(x_o,y_o)$ on the boundary node and $\psi(x,y)$ on the interior point normal to the boundary are represented by "j_1" and "j_2" respectively. For simplicity, we assume that the point (x,y) represents an element node. One may then write the algorithm as follows:

- Calculate the distance "h" between (x,y) and (x_o,y_o).
- Modify the terms of the line j of $[K_T]$ as:

$$K_T(j,j_1) = -K_T(j,j_2) = -\frac{2}{h^2}$$

$$K_T(j,j) = 1$$

$$K_T(j,i) = 0 \quad \text{for all } i \neq j,\ j_1,\ j_2$$

- Replace the jth term of the residue vector by $2\,\frac{\bar{u}}{h}$.

This process is repeated for each "ω" corresponding to all nodes on the solid wall.

2.2 Numerical Example: The recirculating flow in a square cavity This classical problem has been studied by various investigators due to it simple geometry and type of kinematical boundary conditions. Burgraff [13] studied this problem using finite differences for moderate Reynolds numbers. Olson et al. [1,2], Cliffe et al. [2] and Hutton [2] among others applied the finite element method for estimating the flow fields at various Reynolds numbers. They have shown that a finer grid must be employed near the upper corners of the cavity in order to obtain realistic flow patterns for high Reynolds numbers.

The complete boundary of the cavity is subjected to no slip conditions with the upper lid moving from left to right at a unit speed (fig. 1-a). In the figure 1-b, we compare the values of vorticity along the moving lid at Re = 400 obtained by the present investigation (13 × 15, 3 elements, where 3 represents the corner refinement, fig. 1-a) with those obtained by Olson et al., Cliffe et al. and Hutton. In the Table I, we present values of ψ at the primary vortex centre using the same gridwork for Re = 0 to 7000. Another type of refinement has been obtained by confounding mid-nodes of the corner elements to same point, say C, and D and the results are shown in the same table for Re = 0 to 12,000.

TABLE I - Stream Function at the Primary Vortex Centre

Re	Olson	Nallasamy	This study*	This study**
0			.100	.100
100	.1033	.1026	.1044	.1045
400	.1146	.1014	.121	.121
700		.0986	.138	.1378
1,000	.1226	.0977	.1582	.1589
1,500	.1285		.1898	.1915
2,000	.1361	.0951	.2084	.2112
2,500			.2212	.2254
3,000	.1591	.0906	.2322	.2378
4,000	.1911		.2498	.2534
5,000	.232	.0861	.2597	.2780
6,000	.281		.2869	.3094
7,000			.2908	.3121
8,000				.3065
9,000				.2979
10,000		.0873		.2849
11,000				.2606
12,000				.2407

* Refinement by removing the corner element.

** Refinement by confounding nodes of the opening.

These results are compared with those presented in the references 2 and 14. One may observe that the finite difference results of the reference 14 differ widely at high Reynolds numbers since no corner refinement have been employed. Our results start differing from those of Olson et al. (12 × 12 mesh, c^1 triangular element) at Reynolds numbers of 1000 and upwards. It seems that a finer mesh size is necessary to obtain precise values at high Reynolds numbers, and to assess the accuracy of results. In our study we have observed no convergence problem for Reynolds numbers up to 22,000 using a surprisingly low number of iterations with incremental solution strategy coupled with the Newton-Raphson scheme. Two secondary flows at the bottom corners appear at increasing

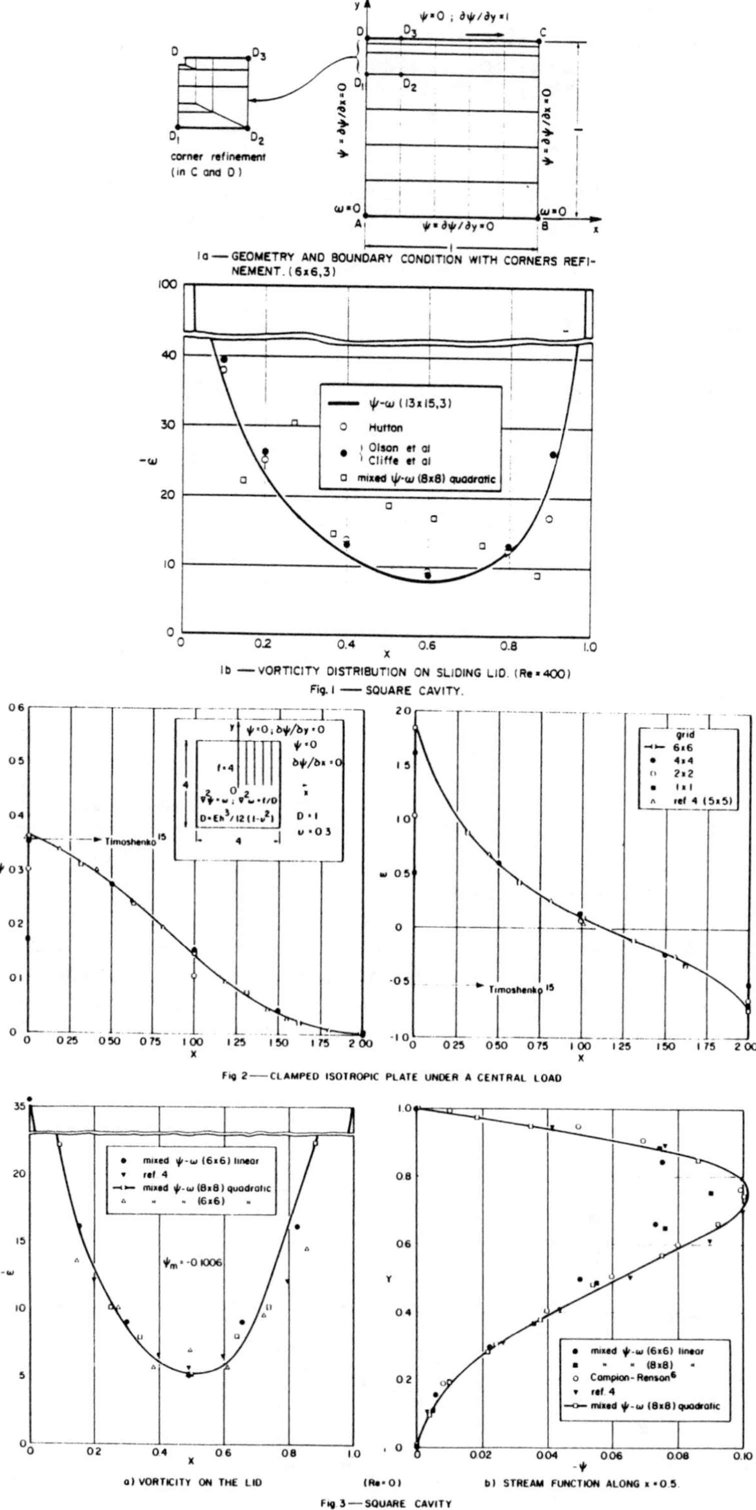

1a — GEOMETRY AND BOUNDARY CONDITION WITH CORNERS REFINEMENT. (6x6,3)

1b — VORTICITY DISTRIBUTION ON SLIDING LID. (Re = 400)

Fig. 1 — SQUARE CAVITY.

Fig 2 — CLAMPED ISOTROPIC PLATE UNDER A CENTRAL LOAD

a) VORTICITY ON THE LID (Re = 0) b) STREAM FUNCTION ALONG x = 0.5

Fig 3 — SQUARE CAVITY

Reynolds numbers. Both regions are relatively small compared with the overall flow fields and ψ,ω values change sign. A third secondary vortex flow appears at the upper left side of the cavity at Re = 1750.

3. The mixed formulation

The application of the Hellinger-Reissner principle to the equations (1a) and (1b) leads to the integral form:

$$W = \int_A \{\delta\omega(\nabla^2\psi + \omega) + \delta\psi[\nabla^2\omega - Re(\psi_{,y}\,\omega_{,x} - \psi_{,x}\,\omega_{,y})]\}\ dA = 0 \quad (7)$$

After integrating by parts terms corresponding to ∇^2, the weak integral form is:

$$W = \int_A [-(\delta\omega_{,x}\psi_{,x} + \delta\omega_{,y}\psi_{,y}) + \delta\omega.\omega - (\delta\psi_{,x}\omega_{,x} + \delta\psi_{,y}\omega_{,y}) - Re\ \delta\omega\ (\psi_{,y}\omega_{,x} - \psi_{,x}\omega_{,y})]\ dA - \int_{C_1} \delta\omega \frac{\partial\psi}{\partial n}\ ds - \int_{C_2} \delta\omega \frac{\partial\omega}{\partial n}\ ds = 0 \quad (8)$$

where C_1 and C_2 are part or boundary where $\partial\psi/\partial n$ and $\partial\omega/\partial n$ are known. The essential boundary conditions are ψ and ω. The cross weighting of contour terms permit to introduce no slip conditions in a direct manner.

Using finite element approximation for $\psi(x,y)$ and $\omega(x,y)$, Eq. (8) may be written in the following form:

$$W = \sum_{\text{element "e"}} W^e = \langle\delta U_n\rangle\ (\{F_n\} - [K]\ \{U_n\}) = \langle\delta U_n\rangle\ \{R\} \quad (9)$$

where $\langle\delta U_n\rangle = \langle\delta\omega_n, \delta\psi_n\rangle$ is the row vector of first variation of $\langle U_n\rangle$. $\{F_n\}$ is the equivalent load vector due to non-homogeneous boundary conditions. $\{R\}$ is the vector called "residue". $[K]$ is the assembled matrix of the system.

Because the matrix $[K]$ depends on $\{U_n\}$, it is necessary to use the Newton-Raphson solution technique to solve the system given by (9). The tangent matrix is then obtained in terms of the discretisation of the first variation of W, which is:

$$\Delta(W) = \int_A [-(\delta\omega_{,x}\Delta\psi_{,x} + \delta\omega_{,y}\Delta\psi_{,y}) + \delta\omega\Delta\omega - (\delta\psi_{,x}\Delta\omega_{,x} + \delta\psi_{,y}\Delta\omega_{,y}) + Re\ \delta\omega(\Delta\psi_{,y}\omega_{,x} - \Delta\psi_{,x}\omega_{,y}) + Re\ \delta\omega(\Delta\omega_{,x}\psi_{,y} - \Delta\omega_{,y}\psi_{,x})]\ dA \quad (10)$$

This leads to

$$\Delta(W) = \langle\delta U_n\rangle\ [K_T]\ \{\Delta U_n\} \quad (11)$$

where the tangent matrix $[K_T]$ can be written symbolically:

$$[K_T] = \left[\begin{array}{c|c} K_{\omega\omega} & K_{\omega\psi} \\ \hline \overline{K_{\psi\omega}} & \overline{K_{\psi\psi}} \end{array}\right] \quad (12)$$

The submatrices $K_{\omega\omega}$ and $K_{\psi\omega}$ are constant whereas $\overline{K_{\psi\omega}}$ and $\overline{K_{\psi\psi}}$ are function of $\{U_n\}$.

Campion et al. [6] also used this method with quadrilateral elements. This study employs triangular elements with 3 nodes and 6 nodes.

The method presented is first tested for the linear case, i.e., for Re = 0. This case corresponds to the plate problem studied analytically by Timoshenko [15]. The problem and results are presented in figure 2. One may remark that the matrix for the linear case is indefinite. The study of cavity flow using this mixed formulation revealed that this method is very sensitive to the orientation of the triangular elements and for increasing Reynolds numbers the convergence is not obtained when elements are ill-oriented. The figure 3-b shows the stream function along the axe x = .5 for Re = 0 in comparison with that obtained in the references 4 and 6. It is clear that quadratic element gives better results. Figures 3-a and 1-b show the vorticity on the lid. The numerical experimentation has shown that linear element deteriorates rapidly with increasing Reynolds numbers. The quadratic element gives accurate result for Re = 0 but for Re = 400 it is necessary to use very fine grid for obtaining acceptable values (fig. 1-b).

Our experience with this formulation indicates that the method is good for low Reynolds numbers. The convergence characteristics is poorer than that of the classical ψ-ω formulation and the accuracy of results suffers too much with the orientation of triangular elements, the mesh size and the increasing Reynolds number. However its advantage resides in the fact that the no slip conditions introduced in a standard manner.

4. The penalty formulation

The penalty formulation uses the u-v variables after elimination of the pressure through the choice of the penalty function. The method was first applied to the full Navier-Stokes equations by Hugues et al. [11], then Bar-Joseph [2] used it for rotating flow in a closed container and Heinrich et al. [12] for convective problems. The main point of the method will be summarised here.

Let us consider the Navier-Stokes equations in terms of u, v, p variables. Instead of satisfying exactly the continuity equation, we try to satisfy the equation:

$$u_{,x} + v_{,y} = - \frac{1}{\lambda} P \tag{13}$$

P is the pressure and λ is called the penalty parameter. Its value depends on the problem, but in general, it must be sufficiently large to satisfy approximately the equation (13). In this way, the integral form after integration by parts is [Eq. (13) is used to eliminate P]:

$$W = \int_A [\delta u_{,x} u_{,x} + \delta u_{,y} u_{,y} + \lambda \, Re \, \delta u_{,x}(u_{,x} + v_{,y}) + Re \, \delta u(uu_{,x} + vu_{,y}) + \delta v_{,x} v_{,x} + \delta v_{,y} v_{,y} + \lambda \, Re \, \delta v_{,y}(u_{,x} + v_{,y}) + Re \, \delta v(uv_{,x} + vv_{,y})] \, dA \quad (14)$$

The first variation of W used to obtain the solution by the Newton-Raphson technique is:

$$\begin{aligned} \Delta(W) = \int_A & [\delta u_{,x} \Delta u_{,x} + \delta u_{,y} \Delta u_{,y} + \lambda \, Re \, \delta u_{,x}(\Delta u_{,x} + \Delta v_{,y}) \\ & + Re \, \delta u(\Delta u u_{,x} + v \Delta u_{,y} + u \Delta u_{,x} + u_{,y} \Delta v) + \delta v_{,x} \Delta v_{,x} + \delta v_{,y} \Delta v_{,y} \\ & + \lambda \, Re \, \delta v_{,y}(\Delta u_{,x} + \Delta v_{,y}) + Re \, \delta v(\Delta u v_{,x} + u \Delta v_{,x} + \Delta v v_{,y} + v \Delta v_{,y}) \, dA \\ & = \langle \delta U_n \;\; \delta V_n \rangle \; [K_T] \begin{Bmatrix} \Delta U_n \\ \Delta V_n \end{Bmatrix} \end{aligned} \quad (15)$$

where

$$K_T = K_1 + \lambda \, Re \, K_2 \quad (16)$$

$$K_1 = \int \left\{ \left[\begin{array}{c|c} N^T_{,x} N_{,x} + N^T_{,y} N_{,y} + Re \, N^T(u_{,x} N + u N_{,x} + v N_{,y}) & Re \, u_{,y} N^T N \\ \hline Re \, v_{,x} N^T N & N^T_{,x} N_{,x} + N^T_{,y} N_{,y} + Re \, N^T(u N_{,y} + v_{,y} N + v N_{,y}) \end{array} \right] \right\} dA$$

$$K_2 = \int \left\{ \left[\begin{array}{c|c} N^T_{,x} N_{,x} & N^T_{,x} N_{,y} \\ \hline N^T_{,x} N_{,y} & N^T_{,y} N_{,y} \end{array} \right] \right\} dA$$

The reduced numerical integration (2 × 2 gaussian points) is used for the penalty matrix K_2 and the matrix K_1.

The cavity flow was studied to assess and to confirm the good characteristics of convergence already shown by Heinrich et al. [2,12], using a nine nodes quadrilateral element. The λ values were selected between 10^4 and 10^{10} with no noticeable change in the results. The flow was studied up to Re = 10,000 without any numerical instability, and two or three iterations only were necessary at each step.

We have reproduced the results of Heinrich et al. [2,12] for 10 × 10 gridwork for Reynolds number of 400 (fig. 4). We show as well the velocity profiles for Re = 1000 and 1500.

Our study has confirmed the fact that the penalty method is powerful, and can be used for high Reynolds. Moreover, 9 nodes element seems to be very precise in predicting complete flows.

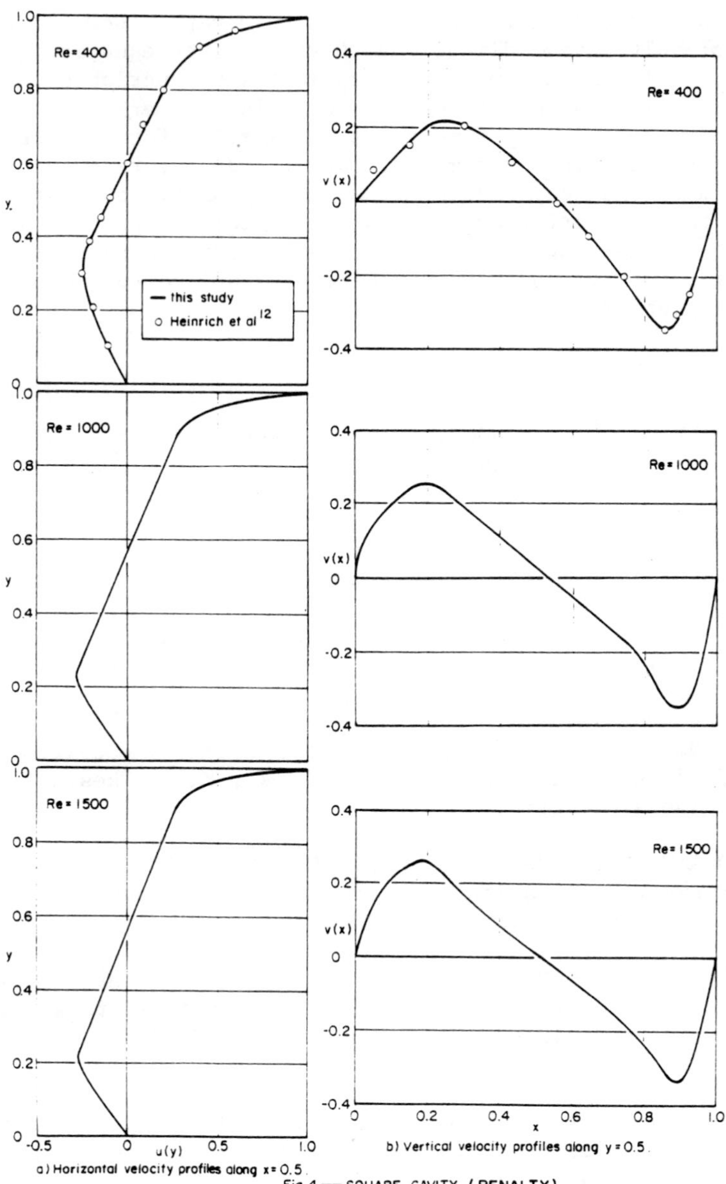

Fig. 4 — SQUARE CAVITY (PENALTY)

5. Concluding remarks

The three methods presented in this paper show different ways of solving two dimensional Navier-Stokes equations. The classical ψ-ω method with its explicit introduction of no slip condition is convenient for high Reynolds numbers but it is necessary to use very fine grids. The mixed formulation is good for low Reynolds numbers. Results deteriorate rapidly with increasing Reynolds numbers. The penalty formulation seems to be superior due to its convergents properties and use of engineering variables u,v for boundary conditions. Moreover this method could be directly applied to 3-dimensional problems.

References

1. TAYLOR, C., and HOOD, P.: "A Numerical Solution of the Navier-Stokes Equations Using the Finite Element Technique", Comp. Fluids, 1, pp. 73-100, 1973.

2. Proceedings of the Third Int. Conf. on F.E.M. in Flow Problems. Banff, Alberta, Canada, 10-13 June, 1980.

3. CHENG, R.T.: "Numerical Solution of the Navier-Stokes Equations by the Finite Element Method", Phys. of Fluid, Vol. 15, No. 12, pp. 2098-2105, 1972.

4. KAMGA FOMO, B.: "Contribution à l'étude du recollement de jets turbulents incompressibles", Ph.D. Thesis, Laval University, 1980.

5. DHATT, G., KAMGA FOMO, B., and BOURQUE, C.: "A ψ-ω Finite Element Formulation for the Navier-Stokes Equations", to appear in I.J.N.M.E., 1980.

6. CAMPION-RENSON, A., and CROCHET, M.J.: "On the Stream Function-Vorticity Finite Element Solutions of Navier-Stokes Equations", I.J.N.M.E., pp. 1809-1818, Vol. 12, 1978.

7. DHATT, G., KAMGA FOMO, B., and BOURQUE, C.: "An Improved (ψ-ω) Finite Element Solution for the Navier-Stokes Equations", paper proposed to I.J.N.M.E., 1980.

8. DHATT, G., and KAMGA FOMO, B.: "Finite Element Solution for Navier-Stokes Equations", 1st Asian Congress of Fluid Mechanics, 8-13 December 1980, Bangalore, India, p. 27.

9. TUAN, S.Y., and OLSON, M.D.: "Review of Computing Methods for Recirculating Flows", J. Comp. Phys., Vol. 29, No. 1, 1-19, 1978.

10. OLSON, M.D.: "Comparison Problem no 1: Recirculating Flow in a Square Cavity", U.B.C., Civil Eng. Struct. Report No. 22, May 1979.

11. HUGHES, T.J.R., TAYLOR, R.L., and LEVY, J.F.: "High Reynolds Number, Steady, Incompressible Flows by a Finite Element Method", F.E. in Fluids, Vol. 3, Wiley, New York, 1978.

12. HEINRICH, J.C., and MARSHALL, R.S.: "Viscous Incompressible Flow by a Penalty Function Finite Element Method", Comp. and Fluids J., Vol. 9, pp. 73-83, 1981.

13. BURGGRAF, O.R.: "Analytical and Numerical Studies of Structure of Steady Separated Flows", J. Fluid Mech., Vol. 24, Part 1, pp. 113-151, 1965.

14. NALLASAMY, M., and KRISHNA PRASAD, K.: "On Cavity Flow at High Reynolds Numbers", J. Fluid Mech., Vol. 79, Part 2, pp. 391-414, 1977.

15. TIMOSHENKO, S., and WOINOWSKI-KRIEGER, S.: "The Theory of Plates and Shells", McGraw-Hill, 2nd ed., 1959.

QUASI THREE-DIMENSIONAL FINITE ELEMENT SOLUTION FOR A TWIN SCREW EXTRUDER

J. Riedler *

Institut für Strömungslehre und Wärmeübertragung, Technische Universität Wien,A-1040 Wien, Austria.

Summary

The finite element method was applied to predict polymer melt flow in twin screw extruders. The temperature distribution was also calculated because the molten polymer has to be kept in a small temperature range. It was possible to develop a quasi three-dimensional solution, where the equations of motions of longitudinal and cross channel flow are not coupled. The dissipative term in the energy equation, however, has to include both the cross sectional flow and the flow in the longitudinal direction. Although we are concerned with creeping flow, the convective terms in the energy equation cannot be omitted because the Péclet number is quite large. The computer program calculates the fluid flow, the temperature, the pressure and a mixing parameter. It is remarkable that under certain conditions the pressure field in the cross section is of the saddle-point type. Besides it was found that leakage gaps are very important for homogenisation and reduction of the peak temperature.

1. Introduction

Nearly all polymeric products pass through a screw extruder at least once during their manufacturing. Although the twin scre extruder is more and more used and the technology of polymer extrusion is well established [2, 3], methods of predicting fluid flow theoretically lag far behind the practical advances. The reason for this is in the geometry and in the complexity of the extrusion

* Assistant Professor

process. A screw extruder has three zones, the feed, melting and metering section. The present paper is only concerned with the metering zone, where the polymer is completely molten and the channels are fully filled.

Methods of predicting fluid flow theoretically in single screw extruders are well established [1, 4]. The solution cannot be transmitted to twin screw extruders because the flow field is entirely different.

In twin screw extruders there are C-shaped chambers between the screws and the barrel. The polymer is conveyed compulsorily in the chambers with leakage flows connecting them. The output rate depends but weakly on the pressure caused by the extrusion head. The channels of a twin screw extruder are not shallow and of a complex shape so that it is necessary to use numerical methods for predicting the fluid flow.

The finite element method is the most suitable one for predicting fluid flow in twin screw extruders, as it will fit arbitrary boundaries very well and the mesh can be adapted according to the problem.

2. Formulation of the Problem

The geometry of the C-shaped chambers (Fig.1) can be simplified because the diameter of the screw is relatively large compared with the width and depth of the cross section. Thus it is possible to consider the fully developed flow in unrolled chambers (Fig.2). The cross section is approximated by a trapezoid with a radial clearance f between the screw and the barrel. The channel width B, the flight width b and the helix angle φ_h are measured at the flight tips. H_c is the hight of the channel and φ_f is the flight angle. The Cartesian coordinate system is fixed relative to the rotating screw. Therefore the barrel surface in Fig. 2 moves with the velocity V relative to the screw, where

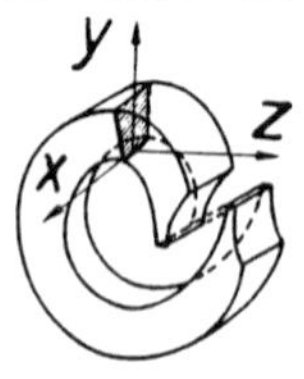

Fig.1: C-shaped chamber [2]

$$V = D\pi n \qquad (1)$$

if the screw with the diameter D rotates at n revolutions per unit time. The relative velocity consists of components parallel and perpendicular to the channel:

$$W = V\cos\varphi_h \qquad U = V\sin\varphi_h \tag{2}$$

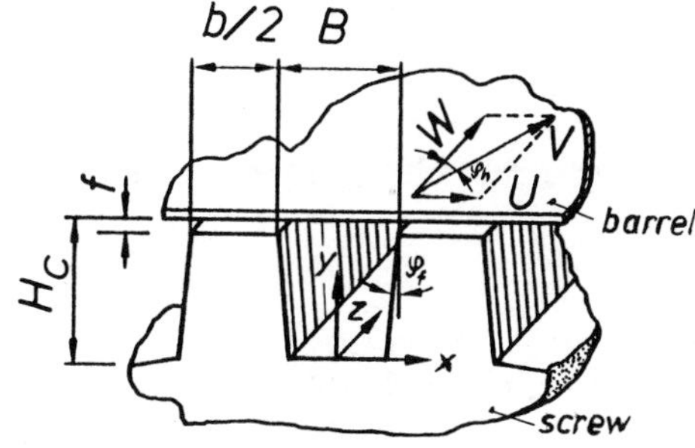

Fig.2: Simplified geometry for the C-shaped chamber

The continuity equation for steady flow becomes

$$\partial u/\partial x + \partial v/\partial y = 0 \tag{3}$$

if we consider the fully developed flow in the unrolled channel and if the density is locally constant. u and v are the velocity components in the cross section.

Inertia forces are negligible in comparison with viscous and pressure forces because the Reynolds number for twin screw extruders is very small. The centrifugal forces will cause the pressure difference

$$\Delta p_f \sim H_c V^2 \varrho 2/D\ . \tag{4}$$

If we compare Δp_f with the characteristic pressure difference

$$\Delta p_c \sim \mu V_Q B/H_c^2 \sim \mu V_Q/H_c\ , \tag{5}$$

where $V_Q = Q_L/f$ is a significant velocity caused by the leakage flow rate Q_L , we recognize that the centrifugal forces are negligible:

$$\Delta p_f/\Delta p_c \sim Re\,(Vf/Q_L)\,2(H_c/D) \ll 1 \tag{6}$$

since $Vf/Q_L \sim \mathcal{O}(1)$, $Re \ll 1$ and $H_c/D \ll 1$. Therefore the equations of motion can be written as

$$\begin{aligned}\partial p/\partial x &= \partial\tau_{xx}/\partial x + \partial\tau_{xy}/\partial y\\ \partial p/\partial y &= \partial\tau_{xy}/\partial x + \partial\tau_{yy}/\partial y\\ \partial p/\partial z &= \partial\tau_{xz}/\partial x + \partial\tau_{yz}/\partial y\ ,\end{aligned} \tag{7}$$

where τ is the shear stress. Notice that $\partial p/\partial z$ is constant for a fully developed flow. Now we have the option of working with a stream function ψ for the cross sectional flow, which satisfies the con tinuity Equation (3) identically. Thus the three-dimensional incompressible creeping Newtonian flow is governed by the biharmonic equation

$$\nabla^4\psi = 0 \tag{8}$$

for the cross sectional flow and the Poisson equation

$$1/\mu \ \partial p/\partial z = \nabla^2 w \tag{9}$$

where μ is the melt viscosity, for the longitudinal velocity w. ∇ is the two-dimensional Nabla-operator, i.e. it contains derivatives with respect to x and y only.

The conservation laws are completed by the energy equation

$$\varrho . Dh/Dt - Dp/Dt = \frac{\partial}{\partial x}(K\cdot\partial T/\partial x) + \frac{\partial}{\partial y}(K\cdot\partial T/\partial y) + \Phi \tag{10}$$

where h is the enthalpy, T is the temperature, K the thermal conductivity of melt and

$$D/Dt = u\cdot\partial/\partial x + v\cdot\partial/\partial y \tag{11}$$

is the substantial derivative. The dissipative term Φ includes both the cross sectional flow and the flow in longitudinal direction:

$$\Phi = \mu[4(\partial^2\psi/\partial x\partial y)^2 + (\partial^2\psi/\partial y^2 - \partial^2\psi/\partial x^2)^2 + (\partial w/\partial x)^2 + (\partial w/\partial y)^2] \tag{12}$$

The enthalpy can be expressed in terms of an equation of state

$$dh = [(1 - T\beta)/\varrho]\ dp + C_p\ dT , \tag{13}$$

where β is the volumetric thermal expansion coefficient and C_p is the specific heat capacity at constant pressure. As Dp/Dt is of the same order as the dissipative term Φ and $T\beta$ is usually very small for polymer melts, we can rewrite Eq.(10) as

$$\varrho C_p\ DT/Dt = K\nabla^2 T + \Phi \tag{14}$$

if ϱ, C_p and K are locally constant. As the Péclet number for polymer melts is quite large the convective term has to be retained in the energy equation (14).

The boundary conditions for the velocity components are based on the assumption that there is no flow normal to a solid wall and there is no slip at the boundaries. Den Otter [5] confirmed the validity of the no slip sondition for a number of polyethylenes. For the cross sectional flow the boundary conditions can be formulated in terms of the stream function ψ and its normal derivative $\partial\psi/\partial n$. As the area of the radial leakage gaps is large compared with the gaps between the intermeshing screws, it is reasonable to assume that there is no leakage in the longitudinal direction. Hence $\partial p/\partial z$ is calculated from

$$\int_\Omega w dA = 0 \quad , \tag{15}$$

where Ω is the area of the cross section. Thermal boundary conditions may be specified in terms of temperatures or temperature derivatives with respect to the direction normal to the boundary. In the middle cross section of a leakage gap between two chambers it is possible to assume Couette flow.

$$x^*=x/H_c,\ y^*=y/H_c,\ f^*=f/H_c,\ B^*=B/H_c,\ b^*=b/H_c$$

$$u^*=u/V,\ v^*=v/V,\ w^*=w/V,\ \psi^*=\psi/VH_c \tag{16}$$

$$T^*=(T-T_B)K/\mu V^2,\ p^*=pH_c/\mu V,\ dp^*/dz^*=(\partial p/\partial z)/\mu V.$$

If we substitute the non-dimensional variables into the Eqs.(8,9,10), we finally get the non-dimensional equations:

$$\nabla^{*4}\psi^* = 0 \tag{17}$$

$$dp^*/dz^* = \nabla^{*2}w^* \tag{18}$$

$$Pe\ (u^*\ \partial T^*/\partial x^* + v^*\ \partial T^*/\partial y^*) = \nabla^{*2}T^* + \Phi^*\ . \tag{19}$$

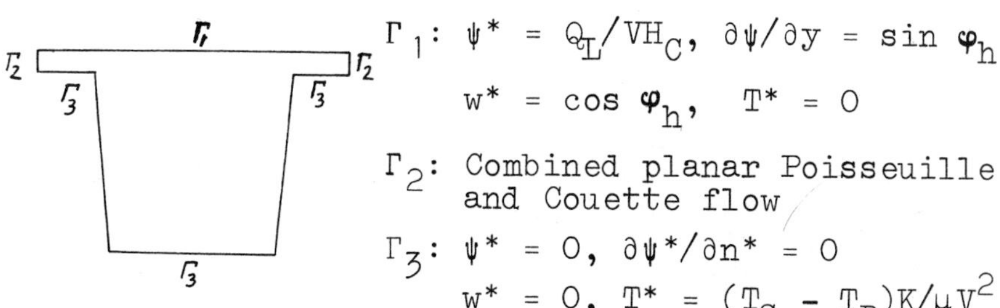

Γ_1
Γ_2
Γ_2
Γ_3
Γ_3
Γ_3
Γ_1: $\psi^* = Q_L/VH_C$, $\partial\psi/\partial y = \sin\varphi_h$
$w^* = \cos\varphi_h$, $T^* = 0$
Γ_2: Combined planar Poisseuille and Couette flow
Γ_3: $\psi^* = 0$, $\partial\psi^*/\partial n^* = 0$
$w^* = 0$, $T^* = (T_S - T_B)K/\mu V^2$

Fig. 3: Non-dimensional boundary conditions for the cross section.

The non-dimensional boundary conditions are shown in Fig. 3.

3. The finite-element scheme

Supposed that the cross section is divided into elements as shown in Fig. 4, we may express the behavior of the unknown functions ψ, w and T within each element as

$$\hat{\psi}^{(e)} = \Sigma L_i \mathbb{P}_i = \lfloor L \rfloor \{\mathbb{P}\}^{(e)} \tag{20}$$

$$\hat{w}^{(e)} = \lfloor M \rfloor \{w\}^{(e)} \tag{21}$$

$$\hat{T}^{(e)} = \lfloor N \rfloor \{T\} \tag{22}$$

by the usual procedure, where $\mathbb{P}_i$, w_i and T_i are nodal values. Our goal here is to derive element equations in terms of the interpolation functions L_i, M_i and N_i. We shall do this by applying either a variational principle or Galerkin's method. With the element equation known, we can evaluate them for the element type of our choice.

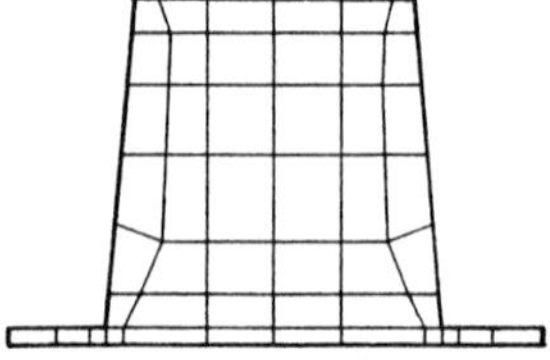

Fig. 4: Finite element mesh for the cross section

3.1 Cross sectional flow

As we shall use only non-dimensional variables we omit the stars.

The biharmonic equation (17) is the Euler equation of the following variational principle[7]:

$$\delta\chi(\psi) = \delta\int_{\Omega} (1/2)\{(\nabla^2\psi)^2 - 4[(\partial^2\psi/\partial x^2)(\partial^2\psi/\partial y^2) - (\partial^2\psi/\partial x\partial y)^2]\}\, dA \tag{23}$$

Thus the approximate solution ψ must be piecewise twice differentiable and as only ψ and its first derivatives are specified at modal points, we don't need the second-order derivatives, we have a C^1 problem and L_i must be chosen to secure at least continuity of $\partial\psi/\partial n$ at element interfaces. χ can be represented as the sum of integrals $\chi(\psi^{(e)})$ over all elements and the minimum condition (23) becomes

$$\delta\chi(\psi) = \partial\chi(\psi)/\partial \mathbb{P}_i = 0 \tag{24}$$

if we substitute Eq.(20) into Eq.(23).

To obtain a compatible triangular element by using only a cubic expansion, the basic element is devided into three subelements as shown in Fig. 5a in which the point O is the centroid of the original triangle. The subtriangle number is identified in the algebraic expression by a superscript.

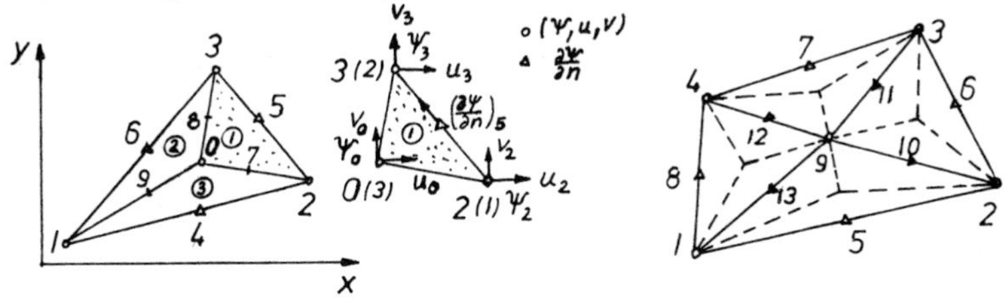

Fig. 5: Elements with C^1 continuity
a) triangular b) quadrilateral

The nodal degrees of freedom to be considered are shown in Fig. 5a. The streamfunction expansion (20) is now assumed over each subtriangle. For the subelement i the vector $\{\mathbb{P}\}^{(i)}$ contains ten components. Notice that node numbering for the degrees of freedom is given in complete element assembly and the set of the ten cubic interpolation polynomials [6] for a triangle are described in normalized triangular coordinates. Thus the vector $\{\mathbb{P}\}$ of all nodal degrees of freedom of the complete triangular element assembly (Fig. 5a)is

$$\{\mathbb{P}\}^T = (\psi_1, u_1, v_1, \psi_2, u_2, v_2, \psi_3, u_3, v_3,$$
$$(\partial\psi/\partial n)_4, (\partial\psi/\partial n)_5, (\partial\psi/\partial n)_6, \psi_0, u_0, v_0) = \tag{25}$$
$$(\{\mathbb{P}_e\}^T \ \{\mathbb{P}_0\}^T).$$

The streamfunction in subtriangle i may be expressed as

$$\psi^{(i)} = (\lfloor L_e^{(i)} \rfloor \ \lfloor L_0^{(i)} \rfloor) \ \{\mathbb{P}\} \tag{26}$$

where $\lfloor L_e^{(i)} \rfloor$ and $\lfloor L_0^{(i)} \rfloor$ represents the interpolation functions for the external and internal degrees of freedom. If we consider two adjacent elements with common interfaces we state that tangential derivatives are uniquely specified by nodal values but normal derivatives differ between the nodes. Thus we cannot represent an internally compatible streamfunction with Eq.(26). Now we introduce the mid-points 7, 8 and 9 at subelement interfaces. By differenciating Eq. (26) continuity of normal outward derivatives is achieved by matching nodal slopes in adjacent subtriangles. From the compatibility requirements we can eliminate the internal degrees of freedom at node 0 and the internally compatible streamfunction for the complete assembly becomes

$$\begin{pmatrix} \hat{\psi}^{(1)} \\ \hat{\psi}^{(2)} \\ \hat{\psi}^{(3)} \end{pmatrix} = \begin{pmatrix} \lfloor \tilde{L}^1 \rfloor \\ \lfloor \tilde{L}^2 \rfloor \\ \lfloor \tilde{L}^3 \rfloor \end{pmatrix} \{ \mathbb{P}_e \} \tag{27}$$

The element matrix of a subtriangle may be calculated from

$$\mathbb{k}_{ij} = \iint [4\tilde{L}_{xyi}\tilde{L}_{xyj} + \tilde{L}_{yyi}\tilde{L}_{yyj} - \tilde{L}_{yyi}\tilde{L}_{xxj} - \tilde{L}_{xxi}\tilde{L}_{yyj} + \tilde{L}_{xxi}\tilde{L}_{xxj}] \, dxdy \quad , \tag{28}$$

where the integration is carried out analytically for polynomials in natural triangular coordinates. For the triangle assembly we have to add the contributions of the three subelements, as they are all expressed in terms of the same set of nodal coordinates. Four such elements may be combined to form a quadrilateral with 23 degrees of freedom (Fig. 5b) of which 16 are associated with the nodes 1 to 8 and the remaining seven degrees of freedom with the nodal points 9 to 13. The seven internal degrees of freedom are eliminated to reduce the order of the algebraic system, so that the final quadrilateral has only 16 degrees of freedom.

The element matrices are written on disk storage. As the complete matrix is subdivided into blocks the boundary conditions are introduced by multiplying them with a large number which is also put into the diagonal of the matrix. The final step is to solve the simultaneous equations for the nodal solution. After that the pressure is recalculated from the Laplace equation, where the boundary terms are calculated from the $\hat{\psi}$ results.

3.2 Second order equations

A finite-element scheme for the energy equation (19) is constructed, because it will also include the Poisson equation for the longitudinal velocity, Eq.(18), and the Laplace equation for the pressure field in the cross section. Applying the method of weighted residuals with Galerkin's criterion to the energy equation (19) we get the system of algebraic equations

$$[\mathbb{k}]\{T\} = \{\mathbb{F}\} \tag{29}$$

where

$$k_{ij} = \iint \{ Pe[u(\partial N_j/\partial x) + v(\partial N_j/\partial y)]N_i + + (\partial N_i/\partial y)(\partial N_j/\partial y)] \, dxdy \tag{30}$$

and

$$\mathbb{F}_i = \iint \Phi N_i dxdy + \int_\Gamma q_w N_i ds \tag{31}$$

As Eq.(30) contains first order derivatives only, we need elements with C^0 continuity. So we can use the isoparametric element with eigth nodes from the Serendipity family [7, 8]. Integrations in Eqs.(30) and (31) are performed numerically by using Gaussian points. The resulting system of equations is solved iteratively by using the Gauss-Seidel method [7].

This finite element program can also be used for non-Newtonian fluids and for temperature dependent viscosity provided that we assume a constant viscosity for each element which differs from one element to another.

4. Numerical results

Results obtained by the finite-element program are presented for Newtonian flow with constant viscosity.

In the cross sectional flow there appears a dividing streamline which separates the circulating flow from the leakage flow. The leakage flow is caused by a moving wall and by a pressure gradient. The streamline separates at the convex corner if the motion is only caused by the moving wall (Fig. 7a). But if the pressure gradient increases in the gaps the main vortex moves deeper into the screw channel (Fig. 7b) until it is divided into two parts.Near the concave corners there is an infinite number of corner vortices as the angle is less than a critical one ($\approx 146°$) [9]. The corner vortices are not shown, however, because they are very small in our example. The computer program for Stokes flow in the cross section was checked by a local analytical solution for the convex corner and good agreement was achived for it [10].

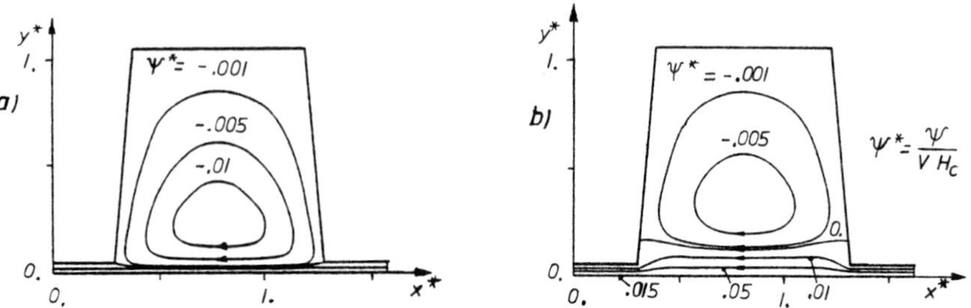

Fig. 7: Streamlines of the cross sectional flow.
H_c=B=0.02m, f*=0.05, φ_f=5°, φ_h=10°,
V =0.125m/s, (Re=0.0048), μ=620Ns/m^2
a) Q_L=-1.085 $10^{-5}m^3$/sm
b) Q_L=-3.75 $10^{-5}m^3$/sm

The pressure is recalculated from the numerical ψ̂ results. If the leakage flow is caused by a small pressure gradient the pressure field for the cross section is of the saddle-point type (Fig.8a). The pressure distribution in an extruder of common type is shown in Fig. 8b.

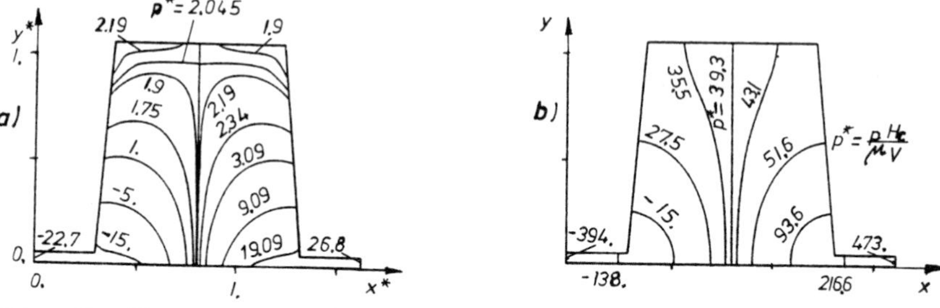

Fig.8: Pressure field in the cross section. Parameters see Figs.7a and 7b.

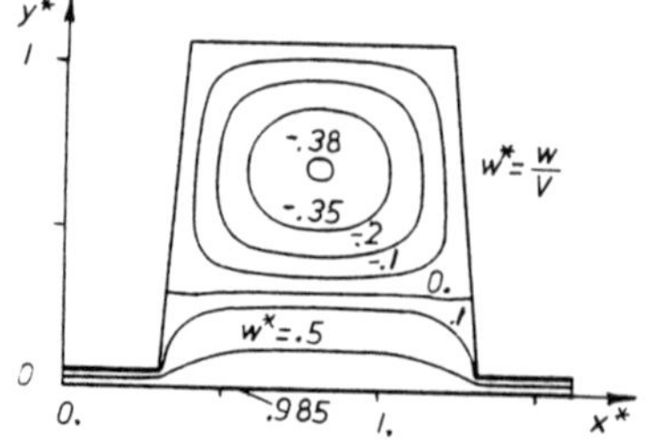

Fig.9: Lines of constant longitudinal velocities. ∂p/∂z= 15.083 bar/m, other parameters see Fig. 7.

The longitudinal velocity w was calculated under the assumption that there is no leakage flow through the end cross sections, cf. Eq.(15). Lines of equal velocity are shown in Fig. 9. This part of the computer program was checked with an analytical solution for the square cavity.

Isothermal boundary conditions were assumed at the screw and the barrel for small and moderate Péclet numbers. For Pe = 0 (Fig.10a) the temperature distribution is symmetric, while at Pe = 34 000 (Fig.10b) the temperature is influenced by the convective terms. The gaps are very important for reducing the peak temperature. If the hight of the clearance is reduced at constant flow rate, the temperature increases rapidly (Fig.11).

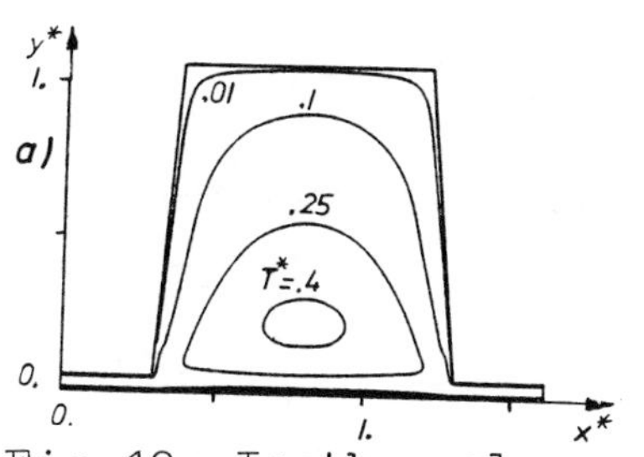

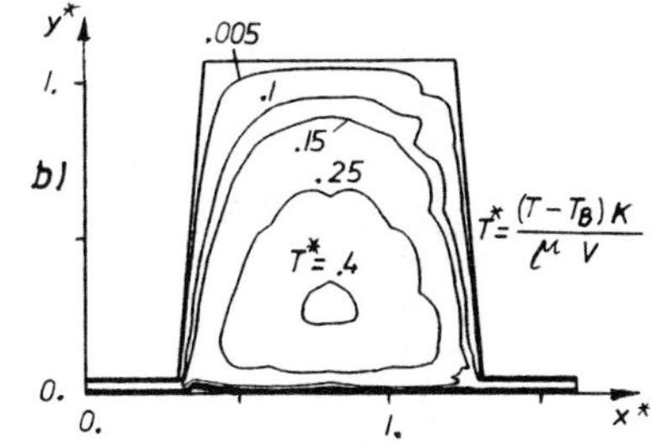

Fig.10: Isothermal curves for ψ from 7b and w from 9

a) Pe = 0 b) Pe = 34 000

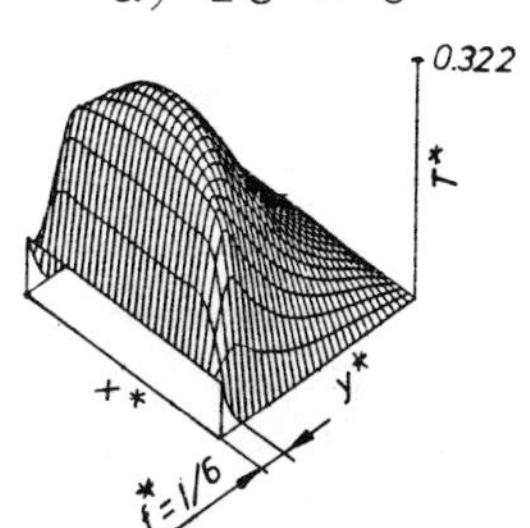

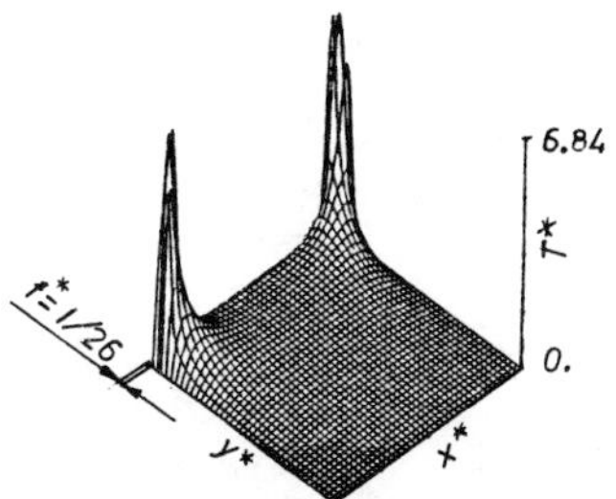

Fig.11: Temperature distribution for constant flow rate and various clearance hights

A global mixing parameter [5] is given by:

$$\bar{M} = (1/\Omega) \iint_\Omega \sqrt{2I_2}\, dA, \quad I_2 = 4(\partial^2\psi/\partial x\partial y)^2 + (\partial^2\psi/\partial y^2 - \partial^2\psi/\partial x^2)^2 + (\partial w/\partial x)^2 + (\partial w/\partial y)^2 , \tag{32}$$

where I_2 is the second invariant of the deformation tensor and Ω is the cross section. $\bar{M}$ is calculated from the results of the finite element method. Fig. 12 shows that the homogeneity is improved by flatter channels, smaller helix angles φ_h and

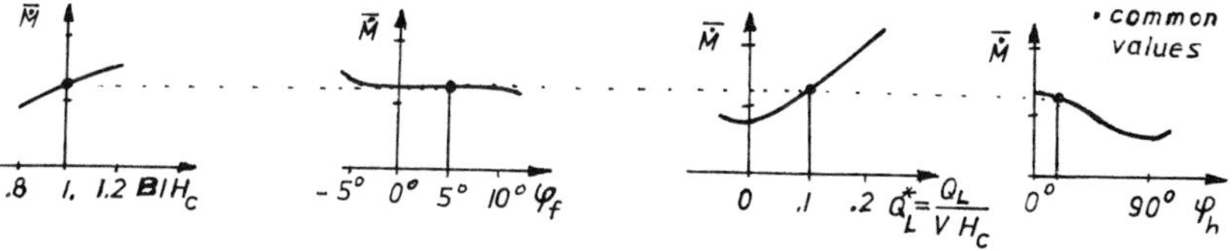

Fig.12: Mixing parameter $\bar{M}$ versus $B/H_c, \varphi_f, \psi^*, \varphi_h$

larger leakage flow rates. The influence of the flight angle φ_f is small.

Acknowledgement - This work is a part of the Ph.D. thesis [10] of the author who wishes to express his deep gratitude to Prof. W. Schneider for initiating this work and contributing essentially to it.

5. References

1. MC KELVEY, J. M. - Polymer Processing, Wiley, New York, 1962.
2. JANSSEN, L. P. - A Phenomenological Study on Twin Screw Extruders, Ph.D.thesis, Delft,1976.
3. KLENK, K. P. - Ein Beitrag zur werkstoffgerechten Verarbeitung von hart-Pulverauf Ein- und Zweischnecken-Extrudern, Plastverarbeiter,21 (1970), 537-544, 642-648, 723-730, 819-821, 881-888; 22(1971), 189-193, 270-275.
4. FENNER, R. T. - Extruder Screw Design,ILIFE, London, 1970.
5. DEN OTTER, J. L. - Some investigation of melt fracture. Reol. Acta 10 (1971), 200-207.
6. CLOUGH, R. W. and FELLIPPA, C. A. - A Refined Quadrilateral Element for Analysis of Plate Bending. Proceedings of the 2nd Conference on Matrix Methods in Structural Mechanics, pp. 399-440. Eds. L. BERKE et al., Wright Patterson Airforce Base Ohio, AFFDL-TR-68-150, 1969.
7. HUEBNER, K. H. - The Finite Element Method for Engineers, Wiley, New York, 1975.
8. KECK, H. - Application of the Finite Element Method to Potential Flow with Free Surface at Critical Conditions, GAMM-Conference on Numerical Methods in Fluid Mechanics, pp.82-89. Eds. HIRSCHEL, E. H. and GELLER, W., DFVLR, Köln, 1975.
9. MOFFATT, H. K. - Viscous and resistive eddies near a sharp corner. J. Fluid Mech. 18(1964), 1-18.
10. RIEDLER, J. - Ein Beitrag zur Berechnung von Strömungen in Doppelschneckenextrudern, Ph.D. thesis, Technische Universität Wien, 1980.

SECTION 2
TURBULENT FLOW

PREDICTION OF TURBULENT FIELDS, INCLUDING FLUCTUATING VELOCITIES CORRELATIONS AND APPROXIMATE SPECTRA,BY MEANS OF A SIMPLIFIED SECOND-ORDER CLOSURE SCHEME : THE ROUND FREE JET AND DEVELOPED PIPE FLOW

G. GOUESBET and A. BERLEMONT

Laboratoire de Thermodynamique - L.A. C.N.R.S. N° 230 -
Faculté des Sciences et Techniques de Rouen
B.P. 67 - 76130 MONT-SAINT-AIGNAN (France)

Abstract :

Mean velocities, turbulent energy, rate of dissipation of turbulent energy and fluctuating velocities correlations are predicted for the round free jet and the pipe flow. Approximate Eulerian spectra are also built up in the pipe flow situation. These predictions are basically obtained from a (k-ε) - model supplemented with algebraic relations deduced from a second-order closure scheme.

The algebraic relations involving cartesian tensors are tensorialized to provide us with relations invariant under any co-ordinates systems transformations, then expressed in the required cylindrical co-ordinates system.

Predictions are compared with experiments and the results are found satisfactory.

Although self-consistent, the present work was a step necessary for the prediction of the behaviour of discrete particles embedded in turbulent flows (see another paper by the same authors, included in the present conferences).

1 - THE (k-ε) - predictions :

The (k-ε) - predictions have been carried out with the aid of the Teach-T computer program [1] in a parabolic version (boundary-layer assumption).

The figure 1 explicits the main notations.

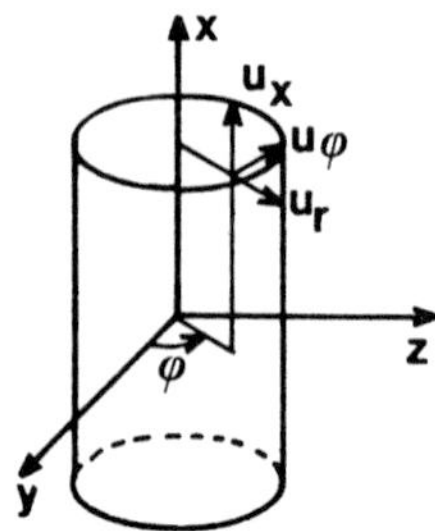

Fig.1- Coordinates system

x is taken as the main flow direction and u_x and u_r are the mean velocities in the x - and r - directions respectively.

For the flows here considered, the governing equations are [1] :

. Continuity equation :

$$\frac{\partial}{\partial x}(r\rho u_x) + \frac{\partial}{\partial r}(r\rho u_r) = 0 \qquad (1)$$

. Momentum equations :

$$\frac{1}{r}\left[\frac{\partial}{\partial x}(r\,\rho\,u_x u_x) + \frac{\partial}{\partial r}(r\rho u_r u_x)\right] = -\frac{\partial p}{\partial x} + \frac{1}{r}\frac{\partial}{\partial r}\left(r\mu_e \frac{\partial u_x}{\partial r}\right) + S_{u_x} \qquad (2)$$

$$\frac{1}{r}\left[\frac{\partial}{\partial x}(r\,\rho\,u_x u_r) + \frac{\partial}{\partial r}(r\rho u_r u_r)\right] = -\frac{\partial p}{\partial r} + \frac{1}{r}\frac{\partial}{\partial r}\left(r\,\mu_e \frac{\partial u_r}{\partial r}\right) - \mu_e \frac{u_r}{r^2} + S_{u_r} \qquad (3)$$

where the source terms S_{u_x} and S_{u_r} read :

$$S_{u_x} = \frac{\partial}{\partial x}\left(\mu_e \frac{\partial u_x}{\partial x}\right) + \frac{1}{r}\frac{\partial}{\partial r}\left(r\,\mu_e \frac{\partial u_r}{\partial r}\right) \qquad (4)$$

$$S_{u_r} = \frac{\partial}{\partial x}\left(\mu_e \frac{\partial u_x}{\partial r}\right) + \frac{1}{r}\frac{\partial}{\partial r}\left(r\,\mu_e \frac{\partial u_r}{\partial r}\right) - \mu_e \frac{u_r}{r^2} \qquad (5)$$

. Transport equation for the turbulence energy k :

$$\frac{1}{r}\left[\frac{\partial}{\partial x}(r\,\rho\,u_x k) + \frac{\partial}{\partial r}(r\,\rho u_r k)\right] = \frac{1}{r}\frac{\partial}{\partial r}\left(r\frac{\mu_e}{\sigma_k}\frac{\partial k}{\partial r}\right) + G\ C_D\rho\,\varepsilon \qquad (6)$$

where ε is the rate of dissipation of k and G is the production of k given by :

$$G = \mu_T\left\{2\left[\left(\frac{\partial u_x}{\partial x}\right)^2 + \left(\frac{\partial u_r}{\partial r}\right)^2 + \left(\frac{u_r}{r}\right)^2\right] + \left(\frac{\partial u_x}{\partial r} + \frac{\partial u_r}{\partial x}\right)^2\right\} - \frac{2}{3}\mu_T\left[\frac{1}{r}\frac{\partial}{\partial r}(r\,u_r) + \frac{\partial u_x}{\partial x}\right]^2 \qquad (7)$$

where the turbulent dynamic viscosity μ_T is :

$$\mu_T = C_\mu\,\rho\,k^2/\varepsilon \qquad (8)$$

. Transport equation for the rate of dissipation ε :

$$\frac{1}{r}\left[\frac{\partial}{\partial x}(r\rho u_x\varepsilon) + \frac{\partial}{\partial r}(r\rho u_r\varepsilon)\right] = \frac{1}{r}\frac{\partial}{\partial r}\left(r\frac{\mu_e}{\sigma_\varepsilon}\frac{\sigma_\varepsilon}{\partial r}\right) + C_1\frac{\varepsilon}{k}G - C_2\rho\frac{\varepsilon^2}{k} \qquad (9)$$

ρ is the fluid density and p the mean pressure. The

effective dynamic viscosity μ_e is equal to the molecular dynamic viscosity μ plus the turbulent dynamic viscosity μ_T. The closure constants are C_1, C_2, C_D, C_μ, σ_k and σ_ε.

Boundary conditions must be included according to the specific problem under study. In the present boundary layers situations, these conditions involve the specification of the relevant fields in an upstream section, plus transverse conditions. For the round free jet, the transverse condition is the nullity of velocity gradients on the axis and far away from it. For the (confined) pipe flow, wall functions are used.

Some examples of predictions are now given. More details, results and discussions,are available from the references [2], [3] and [4].

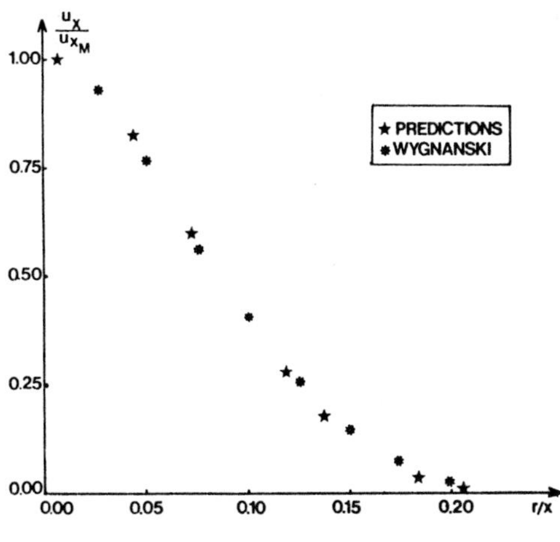

Fig.2 - Velocities

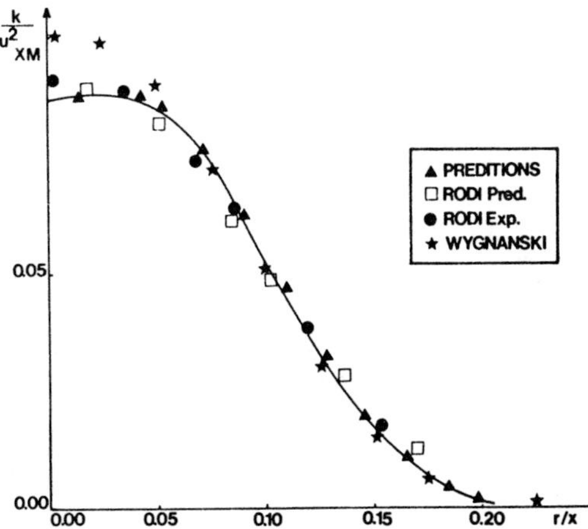

Fig.3- Turbulence energy

The figures 2 and 3 concern the round free jet. The upstream inlet profiles correspond to the nozzle outlet (laminar fully developed pipe flow). A "Finite Difference Equations" grid diverging with the flow has been used. The constants are

$C_1 = 1.44$

$C_D = 1.0$

$\sigma_k = 1.0$ and

$\sigma_\varepsilon = 1.3$

Following Launder and Spalding [5] and Rodi [6], the constants C_2 and C_μ are replaced by :

$$C_2=1.92-0.0667\ f \quad (10)$$

$$C_\mu=0.09-0.04\ f \quad (11)$$

where f is given by :

$$f=\left[\frac{y}{2\Delta u}\left(\frac{\partial u_{xM}}{\partial x}-\left|\frac{\partial u_{xM}}{\partial x}\right|\right)\right]^{0.2} \tag{12}$$

where y is the jet half-width and Δu the change of u_x in y and u_{xM} is the value of u_x on the axis.

The figure 2 shows the profile u_x/u_{xM} versus r/x in the similarity region of the jet. The comparison with experimental results from Wygnanski and Fiedler [7] is quite satisfactory.

The figure 3 shows the profile $k/u_{x}{}^{2}{}_{M}$ versus r/x in the similarity region of the jet. The comparison with predictions and experiments from Rodi [6] is quite satisfactory. Comparison with experiments from Wygnanski and Fiedler [7] exhibits a "slight" disagreement near the center of the jet.

For the pipe flow, f is zero and C_1 is taken equal to 1.56. The grid is a rectangular one with 2 8 nodes in the r-direction between the axis and the wall. The radius r_o of the pipe is 0.123m and the velocity at the inlet of the pipe where a flat profile is assumed is equal to u_{xo} =30.5m/s. The fluid is air at ambient temperature.

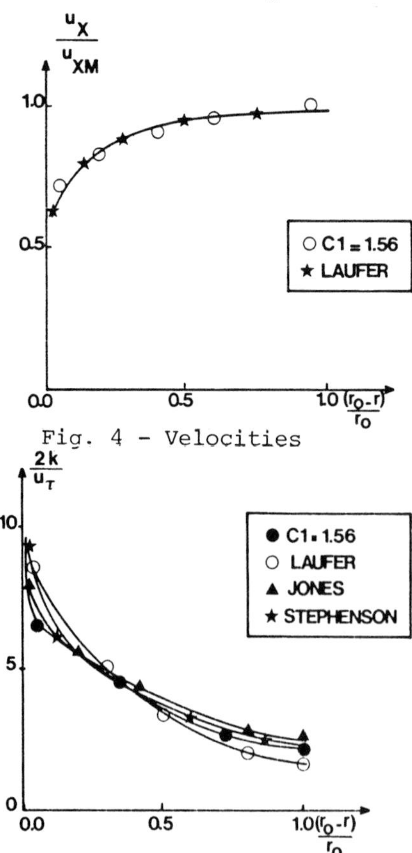

Fig. 4 - Velocities

Fig.5 - Turbulence energy

The figure 4 shows the predicted profile of u_x/u_{xM} versus $(r_o-r)/r_o$ and its comparison with Laufer's experiments [8] for the same Reynolds number ($\sim 5.10^5$).The results are quite satisfactory.

The figure 5 shows the predicted profile of $2k/u_\tau$, where u_τ is the friction velocity, versus $(r_o-r)/r_o$, and its comparison with Jones' predictions [9], Stephenson's predictions [10] and Laufer's experiments [8]. The agreement between all predictions is very satisfactory but the comparison with experiments is not so good near the wall and the axis of the pipe. The discrepancy is attributed to the model itself .

Results for the developing flows, for ε and for G,are not given here because of the lack of room. Nevertheless, results for ε and G are implicitly included in the next section.

2 - FLUCTUATING VELOCITIES CORRELATIONS :

The fluctuating velocities correlations can be computed from algebraic relations deduced from a second-order closure scheme. According to Rodi [11], these algebraic relations read, in cartesian coordinates systems :

$$\overline{u'_i u'_j} = k\left[\frac{2}{3}\delta_{ij} + \frac{1}{(P-\varepsilon + c'_1\varepsilon)}\left[(1-\gamma_1)\left(P_{ij} - \frac{2}{3}\delta_{ij}P\right) - \gamma_2 k\left(\frac{\partial u_i}{\partial x_j} + \frac{\partial u_j}{\partial x_i}\right) - \gamma_3\left(D_{ij} - \frac{2}{3}\delta_{ij}P\right)\right]\right] \quad (13)$$

$$P_{ij} = -\left[\overline{u'_j u'_k}\frac{\partial u_i}{\partial x_k} + \overline{u'_i u'_k}\frac{\partial u_j}{\partial x_k}\right] \quad (14)$$

$$D_{ij} = -\left[\overline{u'_j u'_k}\frac{\partial u_k}{\partial x_i} + \overline{u'_i u'_k}\frac{\partial u_k}{\partial x_j}\right] \quad (15)$$

where u_i, u'_i are the mean and fluctuating velocities respectively, in the i th direction, δ_{ij} is the cartesian Kronecker tensor, and c'_1, γ_1, γ_2, γ_3 are new constants. Furthermore the production term P is G/ρ.

The set (13)-(15) involves cartesian tensors. Its tensorialized form (invariant under any coordinates systems transformations) is, according to the general tensor calculus (see [12], [13] for instance) :

$$\overline{u'_i u'_j} = k\left[\frac{2}{3}g_{ij} + \frac{1}{(P-\varepsilon + c'_1\varepsilon)}\left[(1-\gamma_1)\left(P_{ij} - \frac{2}{3}g_{ij}P\right) - \gamma_2 k\left(\frac{Du_i}{Dx^j} + \frac{Du_j}{Dx^i}\right) - \gamma_3\left(D_{ij} - \frac{2}{3}g_{ij}P\right)\right]\right] \quad (16)$$

$$P_{ij} = -g^{\alpha k}\left[\overline{u'_j u'_\alpha}\frac{Du_i}{Dx^k} + \overline{u'_i u'_\alpha}\frac{Du_j}{Dx^k}\right] \quad (17)$$

$$D_{ij} = -g^{\alpha k}\left[\overline{u'_j u'_\alpha}\frac{Du_k}{Dx^i} + \overline{u'_i u'_\alpha}\frac{Du_k}{Dx^j}\right] \quad (18)$$

where g^{ik} and g_{ik} are the contravariant and covariant metric tensors respectively and Du_i/Dx^k is the covariant derivative:

$$\frac{Du_i}{Dx^k} = \frac{\partial u_i}{\partial x^k} - \Gamma^{\alpha}_{ki}\, u_\alpha \quad (19)$$

where the Christoffel symbol Γ^j_{km} is :

$$\Gamma^j_{km} = \frac{1}{2} g^{ji}\left(\frac{\partial g_{mi}}{\partial x^k} + \frac{\partial g_{ik}}{\partial x^m} - \frac{\partial g_{km}}{\partial x^i}\right) \quad (20)$$

Note that the subscripts appearing in the coordinates became superscripts to point out the contravariant character of an infinitesimal variation dx^k.

The set (16)-(18) is valid whatever the coordinates system. We shall use it to find the specific relations valid in the (x, r, ϕ)- system of the figure 1. Only steps in calculations and results will be given here. More details are available from [14] and [4]. The steps are as following :

1°) In the (x, r, ϕ)-system, we have :

$$x^1 = x \qquad x^2 = r \qquad x^3 = \phi \tag{21}$$

and the metric tensors are :

$$g_{km} = \begin{bmatrix} 1 & o & o \\ o & 1 & o \\ o & o & r^2 \end{bmatrix} , \quad g^{km} = \begin{bmatrix} 1 & o & o \\ o & 1 & o \\ o & o & 1/r^2 \end{bmatrix} \tag{22}$$

2°) From the relation (20), the non-zero components of the Christoffel symbol are found to be :

$$\Gamma^2_{33} = - r \quad \text{and} \quad \Gamma^3_{23} = \Gamma^3_{32} = 1/r \tag{23}$$

3°) Express the covariant derivatives with the aid of (23) and (19)

4°) Substitute in the set (16)-(18), the above expressions for the metric tensors and covariant derivatives. At this stage, the set is expressed in the (x, r, ϕ)-system. Yet,the work is not achieved since, in the formalism we have obtained, the modulus $|V|$ of whatever a vector V_k must be computed with the aid of the contravariant metric tensor according to the relation :

$$|V| = (g^{ik} V_k V_i)^{1/2} \tag{24}$$

Thus the components of the vector u_k (for instance) are not the components u_x, u_r, u_ϕ shown on the figure 1 but are linked to them through the relation (24). To made clear the point, let us consider the third component of u_k, namely u_3, and consider it as a vector $u_{(3)}$ whose components $u_{(3)k}$ are :

$$u_{(3)k} = u_3 \, \delta_{3k} \tag{25}$$

According to (24) the modulus of $u_{(3)}$ is :

$$\begin{aligned} |u_{(3)}| &= (g^{ik} u_{(3)k} u_{(3)i})^{1/2} \\ &= (g^{ik} u_3 \delta_{3k} u_3 \delta_{3i})^{1/2} \\ &= (g^{33} u_3 u_3)^{1/2} = \frac{u_3}{r} \end{aligned} \tag{26}$$

But $|u_{(3)}|$ is equal to u_ϕ. Thus u_3 must be replaced by ru_ϕ.

The foregoing discussion explains the next step :

5°) In the set of relations which have been obtained at the step numbered 4°), change u_1, u_2, u'_1, u'_2 to u_x, u_r, u'_x, u'_r and u_z, u'_3 to ru_ϕ, ru'_ϕ respectively. For similar reasons, change V_{11}, V_{22}, V_{12}, V_{13}, V_{23} and V_{33} (where the letter V stands for P or D) to V_{xx}, V_{rr}, V_{xr}, $rV_{x\phi}$, $rV_{r\phi}$ and $r^2V_{\phi\phi}$ respectively.

6°) We are here concerned with axisymmetric flows. Thus the set obtained at the fifth step simplifies. Due to

$$u_\phi = o \quad \text{and} \quad \frac{\partial}{\partial\phi} = o, \quad \text{we have :}$$

$$P_{x\phi} = P_{r\phi} = D_{x\phi} = D_{r\phi} = o \tag{27}$$

$$\overline{u'_x\, u'_\phi} = \overline{u'_r\, u'_\phi} = o \tag{28}$$

and :

$$\overline{u'^2_x} = k\left[\frac{2}{3} + \frac{1}{P-\varepsilon+c'_1\varepsilon}\left[(1-\gamma_1)\ (P_{xx} - \frac{2}{3}P) - 2\gamma_2 k\ \frac{\partial u_x}{\partial x} - \gamma_3\ (D_{xx} - \frac{2}{3}P)\right]\right] \tag{29}$$

$$\overline{u'^2_r} = k\left[\frac{2}{3} + \frac{1}{P-\varepsilon+c'_1\varepsilon}\left[(1-\gamma_1)(P_{rr} - \frac{2}{3}P) - 2\gamma_2 k\ \frac{\partial u_r}{\partial r} - \gamma_3(D_{rr} - \frac{2}{3}P)\right]\right] \tag{30}$$

$$\overline{u'^2_\phi} = k\left[\frac{2}{3} + \frac{1}{P-\varepsilon+c'_1\varepsilon}\left[(1-\gamma_1)(P_{\phi\phi} - \frac{2}{3}P) - 2\gamma_2 k\ \frac{u_r}{r} - \gamma_3(D_{\phi\phi} - \frac{2}{3}P)\right]\right] \tag{31}$$

$$\overline{u'_x\, u'_r} = \frac{k}{P-\varepsilon+c'_1\varepsilon}\left[(1-\gamma_1)P_{xr} - \gamma_2 k\ (\frac{\partial u_x}{\partial r} + \frac{\partial u_r}{\partial x}) - \gamma_3 D_{xr}\right] \tag{32}$$

$$P_{xx} = -2\left[\overline{u'^2_x}\ \frac{\partial u_x}{\partial x} + \overline{u'_x\, u'_r}\ \frac{\partial u_x}{\partial r}\right] \tag{33}$$

$$D_{xx} = -2\left[\overline{u'^2_x}\ \frac{\partial u_x}{\partial x} + \overline{u'_x\, u'_r}\ \frac{\partial u_r}{\partial x}\right] \tag{34}$$

$$P_{rr} = -2\left[\overline{u'_x\, u'_r}\ \frac{\partial u_r}{\partial x} + \overline{u'^2_r}\ \frac{\partial u_r}{\partial r}\right] \tag{35}$$

$$D_{rr} = -2\left[\overline{u'_x\, u'_r}\ \frac{\partial u_x}{\partial r} + \overline{u'^2_r}\ \frac{\partial u_r}{\partial r}\right] \tag{36}$$

$$P_{xr} = -\left[\overline{u'^2_x}\ \frac{\partial u_r}{\partial x} + \overline{u'_x\, u'_r}\ (\frac{\partial u_x}{\partial x} + \frac{\partial u_r}{\partial r}) + \overline{u'^2_r}\ \frac{\partial u_x}{\partial r}\right] \tag{37}$$

$$D_{xr} = -\left[\overline{u'^2_x}\ \frac{\partial u_x}{\partial r} + \overline{u'_x\, u'_r}\ (\frac{\partial u_x}{\partial x} + \frac{\partial ur}{\partial r}) + \overline{u'^2_r}\ \frac{\partial u_r}{\partial x}\right] \tag{38}$$

$$P_{\phi\phi} = D_{\phi\phi} = -2\ \overline{u'^2_\phi}\ .\ \frac{u_r}{r} \tag{39}$$

The set (29) - (39) is basically a set of four independent equations ((29) to (32)) with four unknown terms

($\overline{u'^2_x}$, $\overline{u'^2_r}$, $\overline{u'^2_\phi}$ and $\overline{u'_x u'_r}$) since the other quantities appearing in the relations are known from the (k − ε) - computations or are constants to adjust. Although it could be classically solved by the Kramer's method, we used another solving procedure. The correlations are first estimated from a gradient transport hypothesis with the aid of the (k − ε) - results :

$$\overline{u'_i u'_j} = - \frac{\mu_t}{\rho} \left(\frac{Du_i}{Dx^j} + \frac{Du_j}{Dx^i} \right) + \frac{2}{3} g_{ij} k \qquad (40)$$

These estimates are substituted in the r.h. sides of the set (29)-(39) to produce new estimates of the correlations corresponding to the l.h. sides of the relations (29)-(32). These new estimates are again substituted in the r.h.s., and so on, until convergence of the process. The required number of iterations is usually small, say, three or four.

For the round free jet, the Fig.6 shows the variances predictions in the similarity region which are compared with Rodi's experiments [6]. The constants are $c'_1 = 2.5$, $\gamma_1 = 0.76$, $\gamma_2 = 0.17$, $\gamma_3 = 0.20$.

The agreement is very satisfactory.

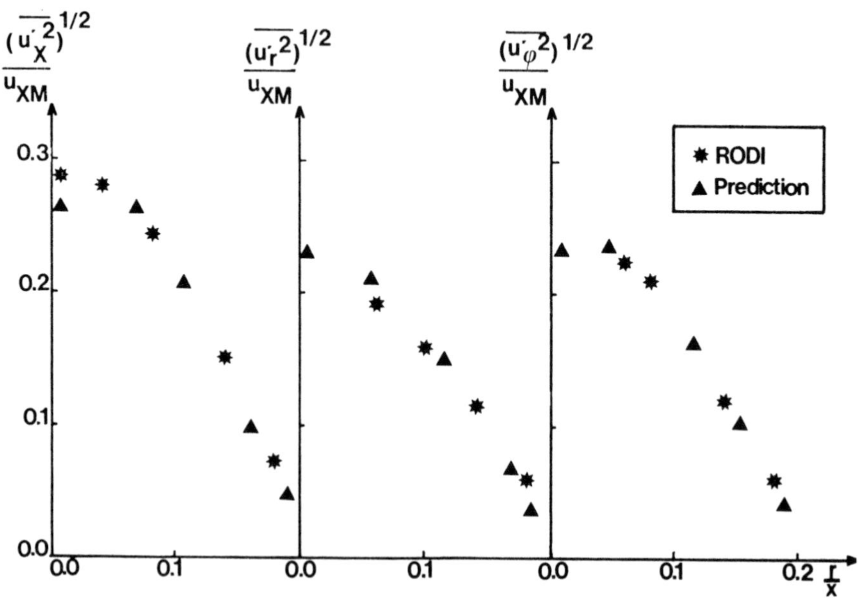

Fig. 6 - Variances

The figure 7 shows the tangential correlations predictions when using the foregoing formalism ($P \neq \varepsilon$) and also when adding the assumption that the production term P is equal to the dissipation term ($P = \varepsilon$ in relation (32)). Experiments (and predictions !) are more difficult for tangential than normal correlations ; so, only the envelope of experimental data from Rodi [6] and Wygnanski and Fiedler [7] is drawn. The comparisons are considered as satisfactory.

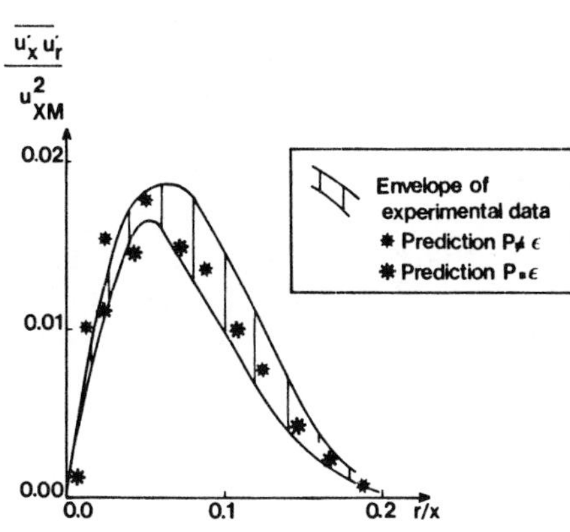

Fig.7-Tangential correlation

For the fully developed pipe flow, variances predictions are shown on the figure 8 and compared with Laufer's experiments [8]. The constants are c'_1=1.80, γ_1=0.76, γ_2=0.18, γ_3= 0.20. Results are satisfactory except again near the wall and the flow axis, a fact which is not surprising anyway (look at the figure 5).

The figure 9 shows the covariance predictions with and without the assumption P = ε. When P is taken different than ε (as predicted by the (k-ε)-model), no adjustment of c'_1

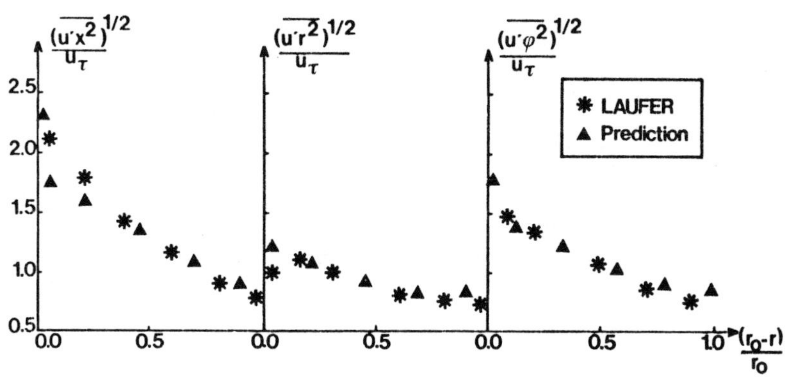

Fig. 8 - Variances

enabled us to obtain quite satisfactory results. But, when P is put equal to ε in the relation (32), then the comparison with Laufer's experiments becomes impressive. It is not clear to us whether this fact contains a deep meaning or is only a mere "coincidence".

3 - Eulerian spectra :

From now on, only the fully developed pipe flow will be discussed. Our aim is now to predict approximate longitudinal

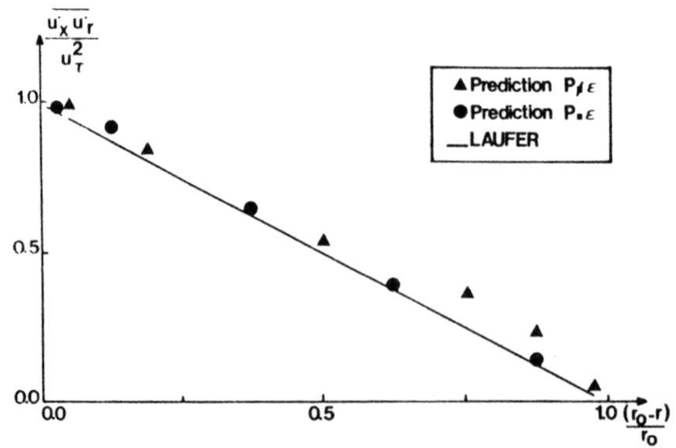

Fig. 9- Tangential correlation

and radial monodimensional spectral densities $E_{fE,x}(k_x)$ and $E_{fE,r}(k_r)$ respectively. Let us only discuss $E_{fE,x}(k_x)$, although the discussion would be exactly similar for the radial density. We have :

$$\overline{u'^2_x} = \int_0^\infty E_{fE,x}(k_x)\, d\, k_x \tag{41}$$

The approximate spectra is built in two parts, separated by a critical wave-number $k_{c,x}$. For $k_{c,x} \leqslant k_x \leqslant \infty$, we write:

$$E_{fE,x}(k_x) = \frac{18}{55}\,\alpha\,\varepsilon^{2/3}\,k_x^{-5/3} \tag{42}$$

where α is taken equal to 1.7 according to Pao [15]

For $o \leqslant k_x \leqslant k_{c,x}$, we write :

$$E_{fE,x}(k_x) = \frac{18}{55}\,\alpha\,\varepsilon^{2/3}\,k_{c,x}^{-5/3} \tag{43}$$

that means the spectral density is taken as a constant in this range of wave-numbers.

The critical wave-number is determined by demanding that the approximate spectra comply with the relation (41). Thus after some straightforward calculations :

$$k_{c,x} = \left(\frac{11}{9\alpha}\ \overline{u'^2_x}\right)^{-3/2} . \varepsilon \tag{44}$$

where ε is known from section I and $\overline{u'^2_x}$ from section II.

The figure 10 compares the predicted spectrum and Laufer's experiments at the center of the pipe [8]. This comparison is considered as satisfactory. More results are available from [4]

Conclusion :

This paper was basically aimed at the predictions of approximate eulerian spectral densities. Preliminary steps required to use a (k-ε)-model supplemented with algebraic relations deduced from a second-order closure scheme.

Although self-consistent, this paper is actually a part of a more general work, namely the prediction of the behaviour of discrete particles embedded in turbulent flows. This more general work is discussed in the present conferences [16].

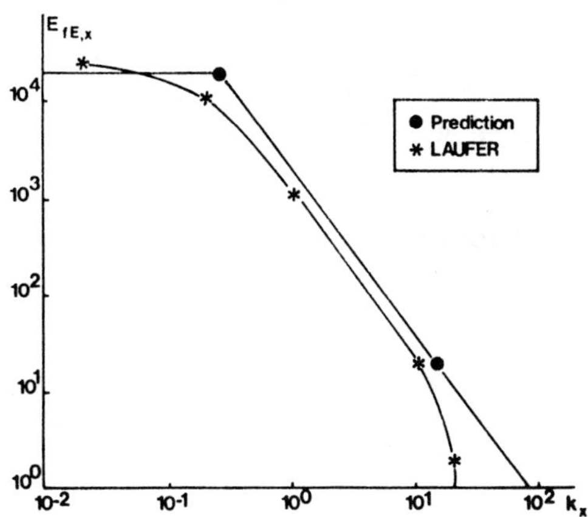

Fig. 10 - Spectrum

REFERENCES :

[1] GOSMAN A.D., IDERIAH P.J.K., Teach-T. Dept Mech. Eng. Imperial College - London - 1976

[2] A. BERLEMONT, G. GOUESBET, Prédiction du jet rond libre turbulent à l'aide d'un modèle en (k-ε) : résultats et discussion. Rapport C.F.R. Juillet 1979

[3] A. BERLEMONT, G. GOUESBET, Prédiction d'un écoulement turbulent établi dans un tuyau à l'aide d'un modèle en (k-ε). Internal Report TTI/BG/79/09/07

[4] A. BERLEMONT, Modélisation et prédiction du comportement de particules discrètes dans des écoulements turbulents. Thèse de 3ème cycle. Université de Rouen, 1981.

[5] LAUNDER B.E., SPALDING D.B., Comp. Meth. Appl. Mech. and Eng., 3, 269, 1974

[6] RODI W., Ph. D. Thesis, Mech. Eng. Dept, Imperial College, London, 1972

[7] WYGNANSKI I. and FIEDLER H., J. of Fluid Mechanics, 38,3, 577, 1969

[8] LAUFER J., Nat. Ad. Com. Aero., Report 1/74, 1954

[9] JONES W.P. and LAUNDER B.E., Int. Journal Heat Mass Trans., 15, 301, 1972

[10] STEPHENSON P.L., Int. Journal Heat Mass Trans., 19, 413, 1976

[11] RODI W., Turbulence models for environmental problems. Prediction methods for turbulent flows. Von Karman Institute for Fluid Dynamics, Lecture Series 1979-2, January 15-19, 1979

[12] LUMLEY J.L. Stochastic tools in turbulence, Academic Press, New York, 1970

[13] WINOGRADSKI J., Les méthodes tensorielles de la physique Tome 1 : calcul tensoriel dans un continuum amorphe, Masson, 1979. Tome 2 : Calcul tensoriel dans un continuum structuré : en préparation

[14] A. BERLEMONT and G. GOUESBET, Prediction of turbulence characteristics, including fluctuating velocities variances, by use of a (k-ε)-model and second-order closure relations, with the aid of general tensor calculus : the fully developed pipe flow, and the round free jet. Internal Report TTI/BG/80/01/II

[15] PAO Y.H., Phys. Fluids, 8, 6, 1965

[16] A. BERLEMONT and G. GOUESBET, Prediction of the behaviour of a cloud of discrete particles released in a fully developed turbulent pipe flow, using a non-discrete dispersive approach. Present conferences.

NUMERICAL SOLUTION OF ELECTROMAGNETICALLY DRIVEN LIQUID METALS

P.M.Bauer, F.Hoislbauer, H.Bartosch, Ch.Jaquemar

VOEST-ALPINE AG, Linz, Austria

SUMMARY

Our research is directed toward the computation of the velocity field of liquid metal in a duct with a rectangular cross-section in the presence of an external electromagnetic field. In this paper all computations are done for an isothermal flow. The velocity and force distribution is presented within the liquid metal.

1. INTRODUCTION

The aim of this paper is to show a method of calculating electric and magnetic fields and the fluid flow motion due to an electromagnetic force. First an additional special notation will be introduced. All vector quantities are indicated by an arrow above the symbol and are 2 or 3 dimensional. A tensor quantity is indicated by 2 arrows above the symbol. A subscript o indicates an initial value.

$\vec{E}$ electric field strength $[Vm^{-1}]$

$\vec{J}$ electric current density $[Am^{-2}]$

σ_E electric conductivity $[\Omega^{-1}m^{-1}]$

$\vec{H}$ magnetic field strength $[Am^{-1}]$

$\vec{B}$ magnetic flux density $[Vsecm^{-2}]$

μ_M magnetic permeability $[VsecA^{-1}m^{-1}]$

mmf magneto-motoric force $[A-t]$

φ scalar potential $[A-t]$

U potential difference $[A-t]$

$\vec{F}$ electromagnetic force $[Nm^{-3}]$

$\vec{x}$	cartesian coordinate [m]
$\vec{\nabla}$	gradient operator
$\vec{v}$	velocity of medium [$msec^{-1}$]

An electric conducting liquid metal can be driven by a body force $\vec{F} = \vec{J} \times \vec{B}$. There are several ways of realisation. Our approach is a constant current $\vec{J}$ and a constant magnetic flux $\vec{B}$ in the following way.

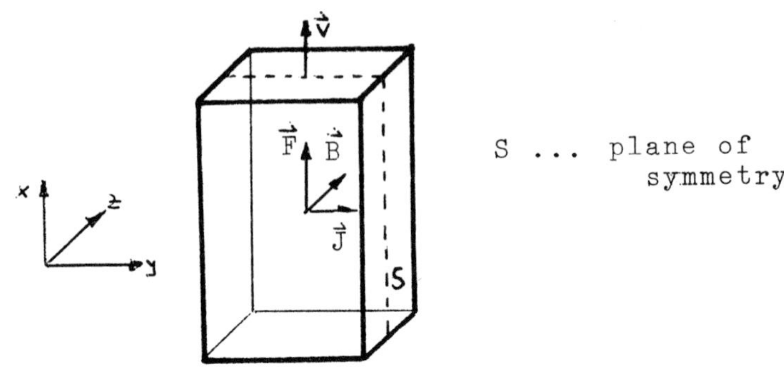

Fig. 1.1

The computations involve the calculation of the fields and the body force $\vec{F}$ including mutual induction due to the induced velocity $\vec{v}$. This velocity is computed via turbulent Navier-Stokes-Equations using the k-ϵ-model. In section 2 the basic equations and assumptions are given.The field computations are done in 3 dimensions (scattering effects in the magnetic case). The Navier-Stokes-Equations are solved only in 2 dimensions (in the plane of symmetry, see Fig. 1.1). Section 3 gives a little comment on the solution procedure and section 4 presents the results.

2. BASIC EQUATIONS

In this chapter a model,which describes the flow phenomena in an electromagnetic field, is developed. The following assumptions are made:

(1) steady state current through the fluid metal
(2) steady state magnetic field
(3) slow initial velocity of the fluid metal through the electromagnetic field (no relativistic effects, one order less than computed velocity)
(4) the solid duct has the same electromagnetic properties as the fluid
(5) the solid duct is fixed in space (zero velocity)
(6) isothermal flow

2.1 Current through an electric conducting medium

The classical problem of electrostatic field theory starts with the set of Maxwell Equation [1].

$$\vec{\nabla} \times \vec{E} = \vec{0} \qquad (2.1a)$$

$$\vec{\nabla} \vec{J} = 0 \qquad (2.1b)$$

$$\vec{J} = \sigma_E \vec{E} \qquad (2.1c)$$

In the classical scalar potential approach, the system (2.1) is reduced by the introduction of a so-called electric potential φ_E, which defines $\vec{E}$ by

$$\vec{E} = - \vec{\nabla} \varphi_E \qquad (2.2)$$

By substituting $\vec{J}$ from (2.1c) into (2.1b) and on eliminating $\vec{E}$ from (2.1b) the governing equation for the scalar potential becomes

$$\vec{\nabla}(\sigma_E \vec{\nabla} \varphi_E) = 0 \qquad (2.3)$$

As σ_E is constant, equation (2.3) can be reduced to

$$\vec{\nabla}(\vec{\nabla} \varphi_E) = 0 \qquad (2.4)$$

and equation (2.1a) is automatically satisfied. It should also be mentioned that this electric current will generate a magnetic field (Biot-Savart's Law)

$$\vec{H}_J = \int \frac{\vec{J} \times (\vec{x} - \vec{x}')}{|\vec{x} - \vec{x}'|^3} \, d^3x' \qquad (2.5)$$

2.2 Magnetic field

The classical problem of magnetostatic field theory starts again with a set of Maxwell Equations [1].

$$\vec{\nabla} \times \vec{H} = \vec{J} \qquad (2.6a)$$

$$\vec{\nabla} \vec{B} = 0 \qquad (2.6b)$$

$$\vec{B} = \mu_M \vec{H} \qquad (2.6c)$$

In the classical vector potential approach, the system (2.6) is reduced by introduction of a so-called magnetic vector potential $\vec{A}$, which defines $\vec{B}$ by its curl as

$$\vec{B} = \vec{\nabla} \times \vec{A} \qquad (2.7)$$

Equation (2.6b) is automatically satisfied and by using (2.7) and (2.6c) for equation (2.6a) the governing equation for the vector potential yields

$$\vec{\nabla} \times \frac{1}{\mu_M} (\vec{\nabla} \times \vec{A}) = \vec{J} \qquad (2.8)$$

In three dimensions any numerical formulation must be solved for all three components of $\vec{A}$ simultaneously. There are several approaches to overcome these difficulties, e.g. to split $\vec{H}$ into two parts [2].

$$\vec{H} = \vec{H}_c + \vec{H}_M \tag{2.9}$$

with

$$\vec{\nabla} \times \vec{H} = \vec{\nabla} \times \vec{H}_c = \vec{J} \tag{2.1o}$$

that leads to

$$\vec{\nabla} \times \vec{H}_M = \vec{0} \tag{2.11}$$

Again it is possible to introduce a scalar potential called magnetic potential φ_M defining $\vec{H}_M$ by its gradient,

$$\vec{H}_M = -\vec{\nabla}\varphi_M \tag{2.12}$$

This results in a single governing equation (Poisson

$$\vec{\nabla}\,(\mu_M(\vec{\nabla}\varphi_M)) + \vec{\nabla}\,(\mu_M\vec{H}_c) = 0 \tag{2.13}$$

$\vec{H}_c$ is the magnetic field strength due to the electric current $\vec{J}$ through the coil and is computed via Biot-Savart's Law (2.5).

Our approach is a little bit different and described in more detail elsewhere [6]. If the interest in the magnetic field is not in the vicinity of the coil but in the field distribution in a greater distance of the source, it is admissible to replace in first order the coil by a permanent magnet with the same physical effects (magnetomotoric force = mmf).

In this case system (2.6) is reduced to

$$\vec{\nabla} \times \vec{H} = \vec{0} \tag{2.14a}$$

$$\vec{\nabla}\,\vec{B} = 0 \tag{2.14b}$$

$$\vec{B} = \mu_M\vec{H} \tag{2.14c}$$

Now system (2.14) is equivalent to system (2.1) and following the same approach one will find the governing equation for the magnetostatic problem

$$\vec{\nabla}\,(\mu_M(\vec{\nabla}\varphi_M)) = 0 \tag{2.15}$$

To guarantee the same physical effects as in the case of a coil, the following condition must be fulfilled

$$U_M^c + U_M^o = mmf \tag{2.16}$$

Equation (2.16) means that the sum of the potential difference in the coil and outside the coil must be equal to the given Amperé-turns [A-t] of the coil.

The superscript c,o indicates coil and outside resp.. To solve the problem (2.15) under the condition (2.16) a double iterative solution procedure will be needed:

(1) in general μ_M depends on $|\vec{H}|$ and therefore on φ_M,

$$\mu_M = \mu_M(|\vec{H}|) = \mu_M((H_x^2 + H_y^2 + H_z^2)^{1/2}) \quad (2.17)$$

(2) in equation (2.16) only the r.h.s. is known.

The calculation is started with an assumed U_M^o (= boundary condition), then equ. (2.15) is solved and U_M^c computed. U_M^o is corrected until (2.16) is fulfilled.
It should be mentioned that a standard heat conduction program can be used for the electric but also for the magnetic case.

2.3 Fluid Flow

The turbulent motion of the fluid medium is given by the time-averaged equations for momentum and continuity (Navier-Stokes-Equations) with an additional body force term due to the electromagnetic force $\vec{F}$,

$$\vec{F} = \vec{J} \times \vec{B} \quad (2.18)$$

and by the well-known k-$\mathcal{E}$-model [7] .

$$\vec{\nabla}\vec{v} = 0 \quad (2.19a)$$

$$\rho(\vec{v}.\vec{\nabla})\vec{v} = -\vec{\nabla}p - \vec{\nabla}\vec{\vec{\tau}} + \vec{F} \quad (2.19b)$$

+ transport-equations for k,$\mathcal{E}$

2.4 Correction of $\vec{J}$ and $\vec{B}$

In the case of an external magnetic field and of a moving electric conducting flowing medium, the initial electric field strength E_o is altered by the Lorentz-force

$$\vec{E} = \vec{E}_o + (\vec{v} \times \vec{B}) \quad (2.2o)$$

$\vec{B}$ is changed by the magnetic field (2.5),

$$\vec{B} = \vec{B}_o + \mu_M\vec{H}_J \quad (2.21)$$

Obiously this will result in a new body force $\vec{F}$ and therefore a new computation of the velocity-distribution is necessary. This iterative procedure is stopped when the differences in $\vec{F}$ and $\vec{v}$ in successive iterations become negligible.

2.5 Useful relations

a) The magnetic Reynolds number

Following other authors [3] , the so-called magnetic Reynolds number Re_M is introduced defined by

$$Re_M = \frac{Lv_o}{\eta_M} \tag{2.22}$$

where

L characteristic length of the problem
v_o characteristic velocity of the medium
$\eta_M = (\sigma_E \mu_M)^{-1}$ magnetic diffusivity

In experiments using liquid metals Re_M is in the order of 10^{-2} to 1.

b) The relationship between velocity and force

If the inertial force dominates (i.e. at high Reynolds number) then as a first approximation the first two terms of the r.h.s. of equation (2.19b) may be neglected. The order of magnitude of the remaining terms are:

$$\varrho(\vec{v}.\vec{\nabla})\vec{v} \sim v_c^2/L \tag{2.23a}$$

$$\vec{J} \times \vec{B} \sim J_c B_c = F_c \tag{2.23b}$$

where

J_c, B_c, F_c, v_c, L are characteristic quantities of the system.
From equation (2.23a) and (2.23b) we can write

$$v_c \sim (J_c B_c)^{1/2} = (F_c)^{1/2} \tag{2.24}$$

which indicates that the velocity is proportional to the squareroot of the force.

3. SOLUTION PROCEDURE

The first step will be the computation of the electric and magnetic scalar potentials. This leads to an initial current density $\vec{J}_o$ and an initial magnetic flux density $\vec{B}_o$.
Then the Navier-Stokes-Equations are solved in the presence of $\vec{F}_o = \vec{J}_o x \vec{B}_o$. The results of this computation will be (among others) the velocity distribution within the fluid metal.
The next step will be the update for the initial electric current density $\vec{J}_o$ and the initial magnetic

flux density $\vec{B}_o$ via equation (2.2o + 2.1c), (2.24) resp.. The calculation of the velocity is repeated with a new electromagnetic body force density $\vec{F}$. This iterative solution process is continued until the difference in velocity and body force is negligible.
It should be noted that the magnitude of the correction term $\mu_M \vec{H}_J$ of $\vec{B}$ in equation (2.21) is one order less than the correction term ($\vec{v}$ x $\vec{B}$) of $\vec{E}$ in equation (2.2o). Therefore the update of $\vec{B}$ is neglected.

4. COMPUTED RESULTS

All computational details of the calculation of the initial current density $\vec{J}_o$ and initial flux density $\vec{B}_o$ is given elsewhere [6]. It should be noticed that the computation of the electric and magnetic fields were done in 3 dimensions (because of scattering effects in the magnetic case).

4.1 Grid and Boundary conditions for fluid flow equations

The turbulent motion in the liquid metal was restricted to be in 2 dimensions (in the symmetry-plane of Fig. 1.1). The domain of the solution of the turbulent Navier-Stokes Equations including the electromagnetic body force term is of rectangular shape.

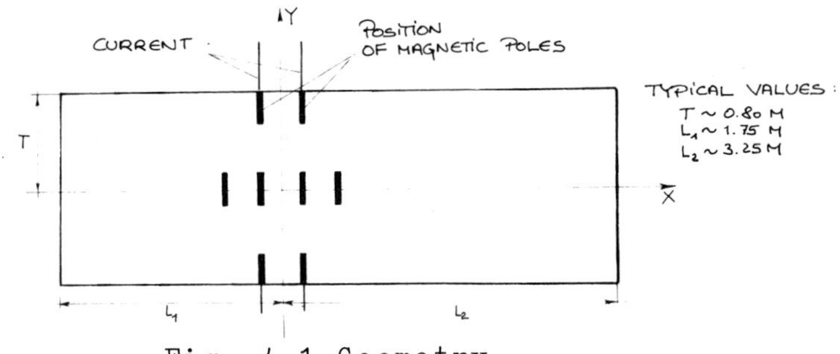

Fig. 4.1 Geometry

Because of symmetry reasons the solution domain can be restricted to ($0 \leq y \leq T$, $-L_1 \leq x \leq L_2$).
The boundary conditions are as follow:

(1) $y = 0$, $-L_1 \leq x \leq L_2$: $v = 0$

(2) $y = T$, $-L_1 \leq x \leq L_2$: $u = v = 0$

(3) $x = -L_1$, $0 \leq y \leq T$: $u = \text{const}$, $v = 0$

(4) $x = L_2$, $0 \leq y \leq T$: $\partial u / \partial n = 0$

u, v are the components of the velocity $\vec{v}$ in x, y-

direction resp..
L_2 was chosen so large to guarantee that B.C. (4) is realistic. For the B.C. of k,ε see [7]. The initial guess of k,ε were taken according to [6]. A 36 x 9 grid was taken for the computation of the fluid flow motion. The grid was nearly uniform in y-direction and graded in x-direction with a fine grid near the position of the poles and the current input and a coarse grid elsewhere.

4.2 Distribution of the initial body force $\vec{F}_o$ and the resulting velocity $\vec{v}_o$

The results of the calculated initial body force $\vec{F}_o$ is given in [6]. A comparison of some values of $\vec{F}_o$ with the updated force $\vec{F}$ are given in Tab. 4.1 (see below). The distribution of the velocity $\vec{v}_o$ due to $\vec{F}_o$ is of the following shape.

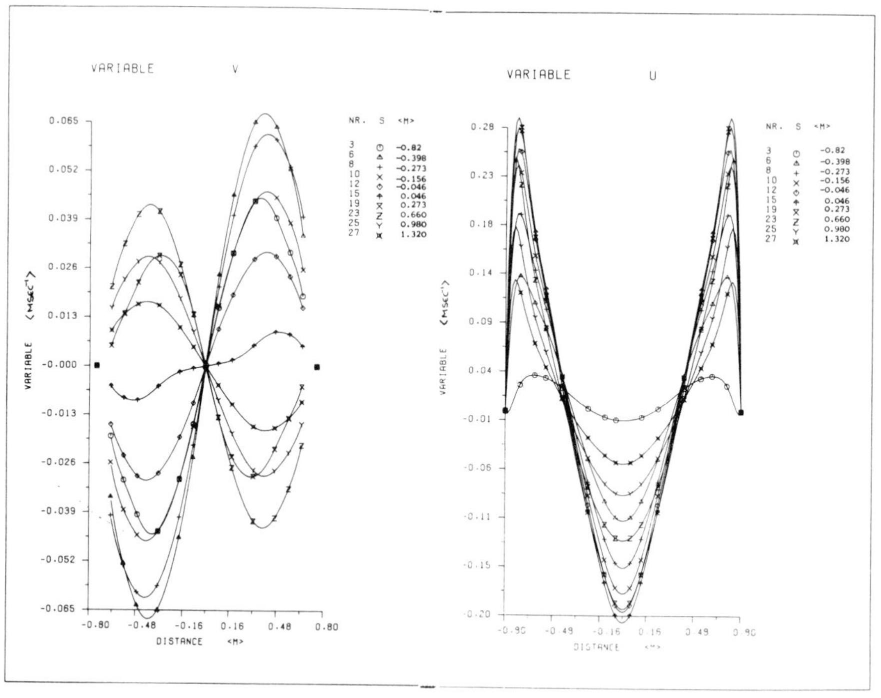

Fig. 4.2a
v_o-distribution

Fig. 4.2b
u_o-distribution

Fig.4.3 shows the position of some I = const. grid-lines.

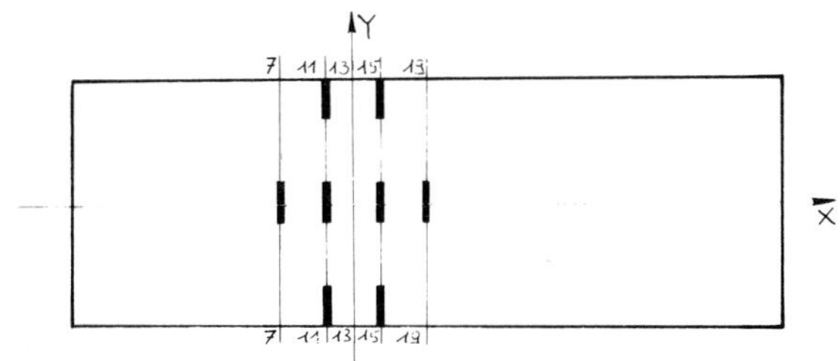

Fig. 4.3
Position of I = const. grid-lines

One can immediate conclude that the motion within the fluid metal will be of the following form.

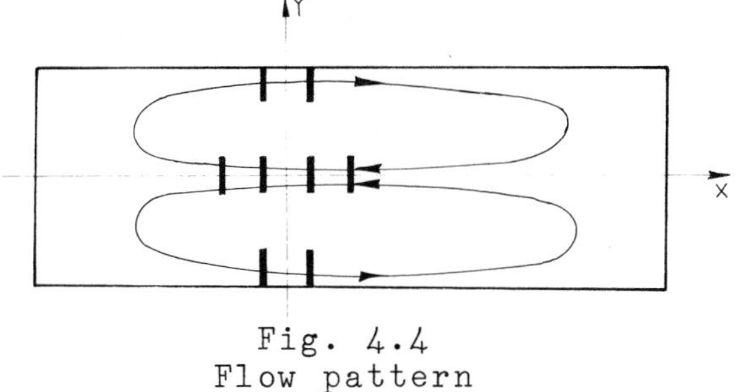

Fig. 4.4
Flow pattern

4.3 Solution of the Navier-Stokes equation with the updated body force $\vec{F}$

After 5 iterations (correction of $\vec{J}$ and $\vec{B}$ by equation (2.2o + 2.1), (2.21) resp., computation of (2.19)) the body force and the velocity has converged satisfactorily to the following solution (see Fig. 4.6, 4.7).
The next figure 4.5 shows the maximum of the velocity u, u_{max}, vs the number of iterations.

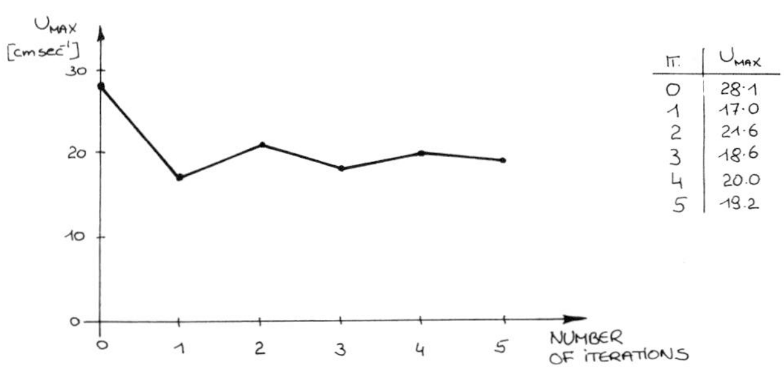

IT.	U_{MAX}
0	28·1
1	17·0
2	21·6
3	18·6
4	20.0
5	19.2

Fig. 4.5
u_{max} vs number of iterations

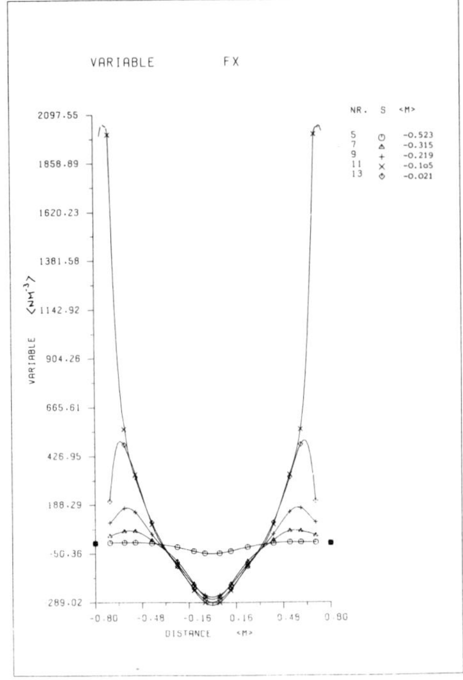

Fig. 4.6
x-component of $\vec{F}$

		$\lvert F_x^{max} \rvert$ [Nm^{-3}]	$\lvert F_y^{max} \rvert$ [Nm^{-3}]
surface	initial	2570	927
	updated	2097	890
centerline	initial	726	32
	updated	290	46

Tab. 4.1

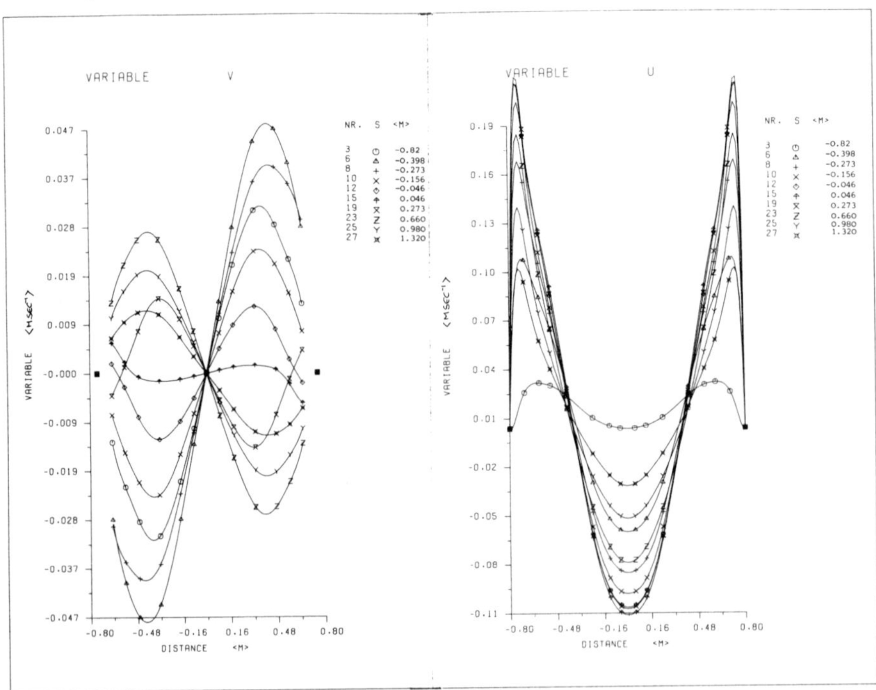

Fig.4.7a v-distribution Fig.4.7b u-distribution

4.4 Verification of equation (2.24); the magnetic Reynolds-number

(1) Several runs were done for varying the body force $\vec{F}$. Fig. 4.8 shows the maximum velocity as function of a relative force.

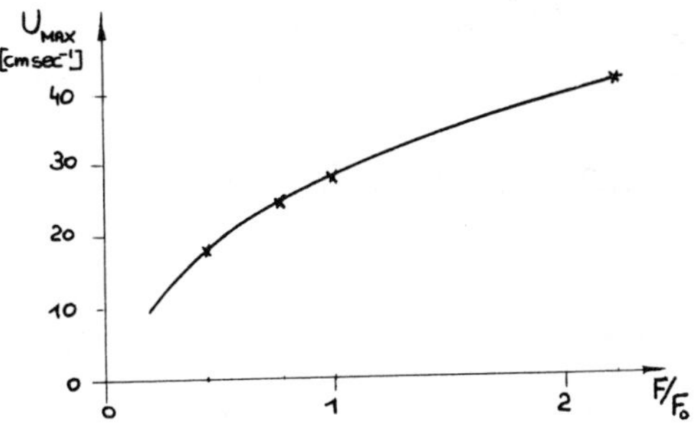

F/F_o	U_{max} comp.	U_{max} theor.
0.45	18.7	18.7
0.78	23.9	24.8
1.00	28.1	28.1
2.23	41.1	41.9

Fig. 4.8
velocity vs. force

(2) Taking the maximum velocity of the solution (o.2 $msec^{-1}$) one will find $Re_M \sim 10^{-1}$ in good agreement with experimental data.

4.5 Some facts on the computer code and running time

The turbulent flow computations were done with an extended version of the FD code TUFC1 of HTFS [7], [6] on an IBM 3o31. The typical CPU-time for a run was about 2o min, the number of iterations about 1ooo.

5. CONCLUSION

Considering the main physical effects of electromagnetically driven flow problems, the computations led to rather encouraging results. The next aim will be the inclusion of non-isothermal effects in the calculation [4]. Other phenomena, usually one order less than the main ones, will be taken into consideration step by step.

REFERENCES

1. STUMPF, H., SCHULER, W. - Elektrodynamik, Vieweg, Braunschweig, 1973.

2. ZIENKIEWICZ, O.C., LYNESS, J., OWEN, D.R.J. - Three-dimensional magnetic field determination using a scalar potential - a finite element solution, IEEE Transactions on magnetics, Vol, MAG-13, No. 5, pp. 1649 - 1656, Sept. 1977.

3. JEFFREY, A. - Magnetohydrodynamics, Oliver and Boyd Ltd., Edingurgh, 1966.

4. BAUER, P.M., HOISLBAUER, F., BARTOSCH, H., JAQUEMAR, C. - Electromagnetic and thermal driven liquid metals in a duct, submitted to International Conference on Numerical Methods for Coupled Problems (Sept. 7th - 14 th, 1981, Swansea).

5. SPALDING, D.B., LAUNDER, B.E. - The numerical computation of turbulent flows, Computer Methods in Applied Mechanics and Engineering, Vol. 3, pp. 269 - 289, 1974.

6. BAUER, P.M. - Berechnung turbulenter Strömungen in elektromagnetischen Feldern, VA-Internal Report, 1981.

7. SYKES, J. - HTFS User guide for TUFC1, AERE-Harwell, Sept. 1978.

ON THE FINITE ELEMENT SIMULATION OF INCOMPRESSIBLE TURBULENT FLOW IN GENERAL TWO-DIMENSIONAL GEOMETRIES

A. G. Hutton and R. M. Smith(I)

The difficulties which distinguish the finite element analysis of turbulent from that of laminar flow in general geometries are examined. Various turbulent wall modelling techniques which have been devised and tested by the authors in high Reynolds number flows are reviewed. Subsequently, a new wall model is introduced which is designed to accomodate both low and high Reynolds number flows as well as regions in the vicinity of zero wall shear stress. It is shown that this element can produce a stable solution for the situation where laminar pipe flow suddenly encounters a turbulent viscosity field. Finally, a finite element discretisation of the k-ε turbulence model is described which does not rely on large and complicated element matrices. The performance of this model in simulating turbulent pipe flow is examined.

1. INTRODUCTION

The finite element (FE) method has now met with considerable success in the simulation of laminar flow and related processes, e.g. [1]. However, in extending such models (and, indeed, any numerical technique) to deal with turbulent flow in general geometries two characteristic difficulties are encountered. Firstly, for high Reynolds numbers, the flow in the vicinity of a wall is likely to exhibit regions of steep velocity gradient necessitating special modelling procedures. In general configurations this problem is further compounded since such a region may not extend over the whole wall (e.g. the wall shear stress, and hence the velocity gradient, reduces to zero at a point of reattachment). Secondly the turbulent stresses are not, in general, locally determined and thus complex transport equations for turbulence quantities governing the state of stress must be introduced.

(I) Central Electricity Generating Board, Berkeley Nuclear Laboratories, Berkeley, Gloucestershire, U.K.

The severity of these problems is perhaps reflected in the fact that a comparatively small number of FE turbulent solutions have so far appeared in the literature and, with few exceptions, [2], [3], these have only dealt with simple idealised situations. The aim of the present paper is to review the nature of these problems and report the progress being made by the present authors in resolving them. The work of other researchers in this area will be only briefly mentioned in order to contain the length of the paper within acceptable limits. Various turbulence models will be introduced with little justification or discussion of their limitations, these being available elsewhere (see [4]). Finally, all variables will be rendered dimensionless with respect to a characteristic length L and velocity V (resulting in the appearance of the Reynolds number $Re \equiv LV/\nu$) and summation convention will be used throughout.

2. BASIC EQUATIONS

The dimensionless equations governing the plane or axi-symmetric mean flow of an incompressible, turbulent fluid are,

$$u_m \frac{\partial u_n}{\partial x_m} + \frac{\partial p}{\partial x_n} - \frac{1}{x_2^\alpha} \frac{\partial}{\partial x_m} (x_2^\alpha \Gamma_{mn}) + 2\alpha \frac{\mu_e}{Re} \frac{u_2}{x_2^2} \delta n2 = 0 \qquad n=1,2, \underline{x} \varepsilon \Omega \tag{1}$$

$$\frac{1}{x_2^\alpha} \frac{\partial}{\partial x_m} (x_2^\alpha u_m) = 0 \qquad \underline{x} \varepsilon \Omega \tag{2}$$

where p is the mean pressure and $\underline{u}$ is the mean velocity (the usual overbar notation has been omitted for convenience). The velocity components u_1, u_2 are in the coordinate directions x_1, x_2 which, for axi-symmetric flows, are the axial and radial directions respectively. The quantity Γ_{mn} is given by

$$\Gamma_{mn} = \frac{\mu_e}{Re} \left(\frac{\partial u_m}{\partial x_n} + \frac{\partial u_n}{\partial x_m} \right) \tag{3}$$

where , if μ_T is the turbulent viscosity, then $\mu_e = 1 + \mu_T$. That is the Boussinesq analogy is assumed.

These equations describe plane flows on setting $\alpha=0$, and axi-symmetric flows on setting $\alpha=1$. They must be solved subject to the following general set of boundary conditions,

$$\underline{u} = \underline{\hat{u}} \qquad \underline{x} \varepsilon \partial\Omega_1 \text{ [walls, inlet]} \tag{4}$$

$$T_1 = u_2 = 0 \qquad \underline{x} \varepsilon \partial\Omega_2 \text{ [symmetry line, fully developed outlet]} \tag{5}$$

$$T_1 = T_2 = 0 \qquad \underline{x} \varepsilon \partial\Omega_3 \text{ [general outlet]} \tag{6}$$

where $\partial\Omega_i$, $1 \le i \le 3$ is a non-intersecting partition of the boundary $\partial\Omega$ and, if $\underline{n}$ is the outward pointing unit normal to $\partial\Omega$,

$$T_\ell = \left[-p\, \delta_{\ell m} + \frac{\mu_e}{Re} \left(\frac{\partial u_\ell}{\partial x_m} + \frac{\partial u_m}{\partial x_\ell} \right) \right] n_m \tag{7}$$

As indicated above, most types of physical boundaries encountered in practice can be handled by these conditions. Equations (1) - (7) are very similar to those describing laminar flow, the only difference being the variable viscosity μ_e (which, of course, has yet to be specified to close the problem). Consequently they should be amenable to those FE methods of discretisation which have proved successful in laminar flow analysis. The formulation adopted here (summarised in Appendix 7.1) is based upon the Galerkin method of approximation [5] with the continuity constraint (eq. 2) being incorporated by the penalty-augmented Lagrangian multiplier (PALM) method [6]. The authors have found that, for quite moderate values of the penalty parameter π (e.g. $1\sim100$), the PALM method, used on standard elements ($C^{(o)}$ quadratic velocity/ linear pressure), can produce a marked improvement in continuity satisfaction over the more usual formulation (viz. π=0).

As was mentioned above, it remains to specify the effective viscosity field, μ_e, in order to complete the formulation. For many simple uni-directional flows (e.g. boundary layers, channel flows etc.), μ_e is satisfactorily represented by the Van Driest model,

$$\mu_e = \begin{cases} \{1 + \left[1 + 4\kappa^2 y^{+2}\left(1-\exp(-y^+/A^+)\right)^2\right]^{\frac{1}{2}}\}/2 & y\leq\lambda\delta/\kappa \quad 8(i) \\ \lambda\delta^+ & y>\lambda\delta/\kappa \quad 8(ii) \end{cases}$$

where κ and A^+ are universal, experimentally determined constants (0.419 and 26 respectively), δ is the boundary layer thickness, and λ is an adjustable constant. The coordinate y^+ is related to y, the normal distance from the wall into the flow, by the equation,

$$y^+ = \mathrm{Re}|\tau_s|^{\frac{1}{2}}y \tag{9}$$

where τ_s is the dimensionless wall shear stress. For more complicated flows, in which the transport of turbulence quantities from surrounding points in the flow strongly influence the local structure (e.g. separated flows with regions of recirculation) such simple algebraic specifications of μ_e are inadequate. Perhaps the simplest (certainly the most popular) turbulence model which suffices for such situations is the so called k-ε model [4]. Under this description, the turbulent viscosity μ_T is determined by the relation,

$$\mu_T = \mathrm{Re}\ C_\mu k^2/\varepsilon \tag{10}$$

where the turbulence kinetic energy, k, and its rate of dissipation, ε, are governed by transport equations of the form,

$$u_m \frac{\partial k}{\partial x_m} - \frac{1}{\mathrm{Re}}\frac{1}{x_2^\alpha}\frac{\partial}{\partial x_m}\left(x_2^\alpha[1 + \mu_T/\sigma_k]\frac{\partial k}{\partial x_m}\right)$$
$$= \frac{\mu_T}{\mathrm{Re}}\left(\left[\frac{\partial u_n}{\partial x_m} + \frac{\partial u_m}{\partial x_n}\right]\frac{\partial u_n}{\partial x_m} + 2\alpha\frac{u_2^2}{x_2^2}\right) - \varepsilon \tag{11}$$

$$u_m \frac{\partial \varepsilon}{\partial x_m} - \frac{1}{Re} \frac{1}{x_2^\alpha} \frac{\partial}{\partial x_m} \left(x_2^\alpha [1 + \mu_T/\sigma_\varepsilon] \frac{\partial \varepsilon}{\partial x_m} \right)$$

$$= C_\mu \, C_{\varepsilon 1} \, k \left(\left[\frac{\partial u_n}{\partial x_m} + \frac{\partial u_m}{\partial x_n} \right] \frac{\partial u_n}{\partial x_m} + 2\alpha \frac{u_2^2}{x_2^2} \right) - C_{\varepsilon 2} \frac{\varepsilon^2}{k} \qquad (12)$$

The parameters C_μ, $C_{\varepsilon 1}$, $C_{\varepsilon 2}$, σ_k, σ_ε appearing in these equations must be determined by experiment and here, the "recommended" values 0.09, 1.44, 1.92, 1.0, 1.3 respectively will be adopted [4]. Suitable boundary conditions on k and ε must also be specified. There is no difficulty at inlet boundaries (k,ε specified) or symmetry and outlet boundaries (normal derivatives of k and ε vanish), but wall boundaries present certain problems since equation (10) is not valid when the turbulence Reynolds number is low. This difficulty is further discussed in the next section, whilst discussion on the implementation of the k-ε model in the finite element method is reserved to Section 4.

3. SPECIAL WALL TREATMENT

One of the numerical difficulties which distinguishes the turbulent formulation from its laminar counterpart (i.e. μ_e=1) arises out of the behaviour of μ_e close to walls. Equations (8), together with eq. (9) show that for large $Re|\tau_s|^{\frac{1}{2}}$, μ_e varies extremely rapidly with y in the close vicinity of the wall. For example, in the case of fully developed pipe flow with Re = 5×10^5 (based on pipe diameter) μ_e, which is unity at the wall, climbs to a value of 10^2 at about 0.015 diameters from the wall. If now the x_1-axis is chosen to be parallel to the wall, in the direction of flow (i.e. y is along or in the opposite direction to x_2), then the local shear stress close to the wall is $\tau \approx \frac{\mu_e}{Re} \frac{\partial u}{\partial y}$. If τ does not vary steeply with y, then this rapid variation in μ_e must be accomodated by a rapid variation in $\partial u_1/\partial y$ and hence u_1. In order to capture this variation with conventional elements an excessive and unacceptable degree of mesh refinement would be necessary. Consequently, for "large $Re|\tau_s|^{\frac{1}{2}}$" uni-directional boundary layers, a special, economical wall treatment is essential. The problem is further compounded in more general geometries since $Re|\tau_s|^{\frac{1}{2}}$ might vary quite considerably over a given wall. For example, $Re|\tau_s|^{\frac{1}{2}}$ vanishes at a flow reattachment point even though it may be very large further along the wall. It follows that a generally applicable wall treatment must be able to naturally accomodate a range of local wall conditions.

Before examining FE procedures for dealing with these problems it is appropriate to briefly review the nature of turbulent flow in the vicinity of a wall. If it is assumed that a) $|\tau|/|\tau_s|$ = 1-2y (as is the case for fully developed duct flow) and b) μ_e is given by the following asymptotic forms of eq. 8(i) close to the wall,

$$\mu_e = \begin{cases} 1 & y^+ \lesssim 11 \\ \kappa y^+ & y^+ \gtrsim 30 \end{cases} \qquad \begin{matrix} 13(i) \\ 13(ii) \end{matrix}$$

then, it is not difficult to show [9],

$$u_1^+ = y^+(1-y) \quad y^+ \lesssim 11 \qquad 14(i)$$

$$u_1^+ = 1/\kappa \, \ln y^+ + C - 2y/\kappa \quad y^+ \gtrsim 30 \qquad 14(ii)$$

where u_1^+ denotes $u_1/|\tau_s|^{\frac{1}{2}}$ and κ and C are constants (typically 0.419 and 5.45 respectively). For small y, eqs. 14 differ insignificantly from the more usual "law of the wall" (i.e. omitting the non-universal terms). However, eqs. (14) are to be preferred for duct flow analysis in order to preserve numerical consistency when using turbulence model, eqs. 13. The behaviour of u_2 can now be derived by invoking continuity. Hutton [7] has shown that, provided u_1^+ is a function of y^+ only, then u_2^+ $(\equiv u_2/|\tau_s|^{\frac{1}{2}})$ is given by,

$$u_2^+ = \pm\, Q\, y^+ \, u_1^+/Re \qquad (15)$$

where Q denotes $d|\tau_s|^{-\frac{1}{2}}/dx$ and the plus sign is assumed unless x_2 is in the opposite direction to y. If the derivation is reworked with eqs. (14) (i.e. u_1^+ is a function of y and y^+), additional terms appear in the right hand side of eq. (15). However, these are of higher order in y and can be omitted for sufficiently small y-values. It remains to establish the behaviour of k and ε. This will only be considered at points for which $y^+>30$ since the role of ε in eq. (10) is valid only for high turbulence Reynolds numbers. Furthermore, it will be assumed that transport effects are negligible close to the wall. Under these conditions it can be shown that $k=|\tau|/C_\mu^{\frac{1}{2}}$, [5] so that if τ varies linearly with y,

$$k = |\tau_s|(1-2y)/C_\mu^{\frac{1}{2}}; \quad \varepsilon = C_\mu k^2/\mu_e \qquad (16)$$

3.1 Procedures for Large $Re|\tau_s|^{\frac{1}{2}}$

The very different asymptotic forms of both the velocity (eqs. (14)) and μ_e (eqs. (13)) at small and large y^+ render it very difficult to devise a special wall element which can adequately span both types of behaviour. Although Taylor et al., [8] have experimented with the construction of such an element, perhaps the simplest course of action is to displace the element edge to a position $y=\Delta h$ in the flow such that $\Delta h^+ \geq 30$. Incidentally this also removes the difficulty of having to model low turbulence Reynolds number regions when using the k-ε closure. If $Re|\tau_s|^{\frac{1}{2}}$ is large, Δh can be extremely small so that the mesh boundary remains sensitive to the wall's geometrical details (e.g. in pipe flow at $Re=5\times10^5$ based on pipe diameter, Δh can be as small as 0.00148). Only the behaviour of eq. (14(ii)) and its corresponding u_2 expression need now be considered within the mesh although, of course, mesh boundary or "slip" conditions must be devised. These must match the numerical solution to the flow conditions enforced by eq. (14(ii)) and eq. (15) (it is assumed that Δh is sufficiently small for the previously mentioned higher order terms in eq. (15) to be omitted). As demonstrated by Hutton [9] this can be conveniently arranged if the normal derivatives of velocity are included as nodal parameters at each mesh-edge node as well as the usual velocity values. If $\partial u_{i,1}$, $\partial u_{i,2}$ represent these normal derivatives evaluated at the ith boundary node then eq. (14(ii)) implies that, at the corresponding wall location,

$$|\tau_s|^{\frac{1}{2}} = \kappa\Delta h_i \ \partial u_{i,1}/(1-2\Delta h_i) \tag{17}$$

Thus, once again using eq. 14(ii), $u_{i,1}$ must satisfy the "slip" condition

$$u_{i,1} = \kappa\Delta h_i \ \partial u_{i,1}\left[\frac{1}{\kappa}\ \ell n\ (Re|\tau_s|^{\frac{1}{2}}\Delta h_i) + C - 2\Delta h_i/\kappa\right]/(1-2\Delta h_i) \tag{18}$$

where $|\tau_s|^{\frac{1}{2}}$ is given by eq. (17). A "slip" condition for u_2 now emerges on differentiating eq. (15) with respect to y^+ and using the result to eliminate Q from eq. (15). This yields,

$$u_{i,1}\ u_{i,2} = \Delta h_i\ (u_{i,1}\ \partial u_{i,2} - u_{i,2}\ \partial u_{i,1}) \tag{19}$$

Equations (17)-(19) replace the usual wall boundary conditions (viz. $\underline{u}$=0) and eqs. (16) can be invoked to provide boundary conditions on k and ε. In principle, these conditions can be used in conjunction with any element-type provided it supports normal derivative parameters, $\partial \underline{u}_i$, at the boundary nodes. Figure 1 depicts three such elements. The element designated "type 3" has been used extensively by the authors in laminar flow analysis [10], [11]. The associated velocity variation is quadratic parallel to the wall and cubic normal to the wall. The second element, designated "linlog", was introduced by Hutton [9] and supports a quadratic velocity variation parallel to the wall whereas the normal variation at any stream-wise station is of the form $a\ell ny + by + c$. In both cases pressure is defined only at the corner nodes (i.e. is bilinear over the element) whereas μ_e varies quadratically in the fashion of the eight-

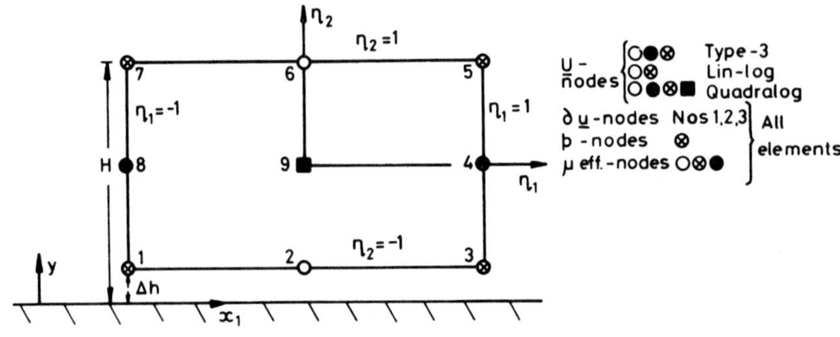

FIGURE 1 Schematic of various wall elements

noded "serendipity" element [12]. This differs slightly from the linlog description given in [9] which used a bilinear μ_e variation. It should be noted that both elements are compatible with an interior structure of what will be designated type 2 elements (i.e. eight-noded "serendipity" variations for u_1, u_2 and μ_e). Table 1 compares the performance of both elements in

			Turbulent				Laminar	
		x_2	INLET	OUTLET			INLET	OUTLET
				Lin-Log	Type-3	Quadralog		Quadralog
Type-3 El.	u_1-Values	0.0	1.2042	1.2042	1.2629	1.2042	2.0	2.0
		0.0625	1.198	1.198	1.2566	1.198	1.9688	1.9687
		0.125	1.1793	1.1793	1.2373	1.1793	1.1875	1.1875
Type 2-Els.		0.1875	1.1482	1.1482	1.2038	1.1482	1.7188	1.7187
		0.250	1.1047	1.1047	1.1545	1.1047	1.5	1.5
		0.3125	1.0487	1.0487	1.0846	1.0487	1.2188	1.2187
		0.375	0.98027	0.98027	0.9896	0.98027	0.875	0.875
		0.40638	0.94122	0.94122	0.93084	0.94122	0.67884	0.67884
Wall El		0.43776	0.89904	0.89905	0.86387	0.89904	0.46693	0.46693
		0.46733	0.84101	0.84101	0.77828	0.84102	0.25283	0.25283
		0.4969	0.61352	0.61356	0.41523	0.61353	0.024745	0.024747
	∂u_1		31.749	31.75	22.716	31.749	7.9504	7.9511

TABLE 1. Fully developed pipe flow, Re=388000

fully developed pipe flow at a Reynolds number of 388000. A simple algebraic specification of μ_e was used (namely eqs. (8) with eq. (8(i)) replaced by eq. (13(ii)) since this permits an analytic solution of the momentum equations. Assuming Prandtl's law of friction, this solution satisfies overall continuity with $\lambda=0.05216$ [9]. It was imposed at the nodes across the inlet and the flow was allowed to develop for ten diameters to outlet where eq. (5) was imposed. As can be seen, with Δh set to 0.0031 (corresponding to $\Delta h^+=50$) the linlog element performs remarkably well returning almost exactly the analytical solution at outlet. The type-3 element on the other hand is less than satisfactory (Table 1). Indeed, it was found that Δh^+ had to be in excess of 300 in order to limit errors to one per cent (at which point Δh was 43% of the width of the element). A special wall element containing the essential near-wall behaviour would therefore seem desirable for general use. The linlog element has also been used to analyse developing flow in a pipe entrance [9]. It was found that the solution exhibited spatial oscillations in the axial direction which were (incorrectly) attributed to an internal inconsistency in the numerical model. It has since been established that these oscillations reflect errors in continuity satisfaction and, indeed, a smooth solution can be quite simply obtained by adopting the PALM method with $\pi=10$.

3.2 General Wall Model

The wall region behaviour (eqs. (14)) is valid over a layer, the physical width of which, Y, depends upon the depth of penetration of far-field influences and the y^+-width of which depends on the value of $Re|\tau_s|^{\frac{1}{2}}$. Thus it is perfectly possible that, at certain wall locations, Y^+ maybe too small for the logarithmic regime to become established (e.g. in the vicinity of reattachment). At such locations the variation in velocity with y is likely to be less extreme and should be adequately modelled by the polynomial variations used in laminar analysis (e.g. quadratic). On the other hand, at other points on the same wall a strong, large $Re|\tau_s|^{\frac{1}{2}}$, boundary layer could be established (e.g. downstream of a reattachment point) necessitating the above described special wall treatment. What is clearly required is an element (together with slip conditions) which can adjust naturally to both types of behaviour. The nine-noded "quadralog" element (fig. 1) was designed for this purpose. As can be seen from Appendix 7.2 (which lists the velocity-basis functions in terms of local iso-parametric co-ordinates), the velocity components vary quadratically parallel to the wall whereas the normal variation is of the form $a\ell ny + by^2 + cy + d$. The "slip" boundary conditions now required must be more general than eqs. (18) and (19) which evidently only cope with $\Delta h^+ \geq 30$. Consider the case $\Delta h^+ \leq 11$. An appropriate u_1-condition can be developed from eqs. (14(i)) and (13(i)) (the latter implying $\tau = 1/Re\ \partial u/\partial y$) giving,

$$u_{i,1} = \tau\ Re\Delta h_i \frac{(1-\Delta h_i)}{(1-2\Delta h_i)} = \partial u_{i,1}\ \Delta h_i \frac{(1-\Delta h_i)}{(1-2\Delta h_i)} \tag{20}$$

Equation (19) is retained for u_2. A generally applicable u_1-condition can now be constructed from the asymptotic expressions eqs. (18) and (20) as follows. If $\Delta h^+ \leq Y_L^+$, eq. (20) is applied, whereas if $\Delta h^+ \geq Y_T^+$ eq. (18) is applied. If $Y_L^+ < \Delta h^+ < Y_T^+$ then a weighted mean of these two equations is used, the weighting varying cubically with $\partial u_{i,1}\ \Delta h_i/(1-2\Delta h_i)$. The nominal values of Y_L^+ and Y_T^+ are of course 11 and 30, but these can be slightly adjusted for numerical stability.

Table 1 demonstrates the performance of the quadralog element in conjunction with this slip treatment in analysing both fully developed turbulent flow ($\Delta h^+ = 50$) and fully developed laminar flow (same Δh) at Re=388000. As can be seen, it is able to handle both extremes with great accuracy. In order to test the general stability of the scheme over a range of wall conditions, the same mesh (with Δh=0.004) was stretched to thirty diameters in the axial direction and then used to solve a pseudo-transition from laminar to turbulent flow. That is to say, the fully developed laminar profile was imposed at inlet, μ_e was set to its laminar value of unity for $x_1 \leq 10$ whereas, for $x_1 > 10$ μ_e was set equal to the fully developed turbulent pipe flow values previously described. The Reynolds number was once again set to 388000, Y_L^+ and Y_T^+ were given the values 7.5 and 75 respectively and the penalty parameter, π, was set to 100. The resulting streamwise development of the u_1-velocity component is given in figure 2. As can be seen, the solution is remarkably smooth and stable except for small-amplitude oscillations upstream of $x_1 = 10$. These oscillations

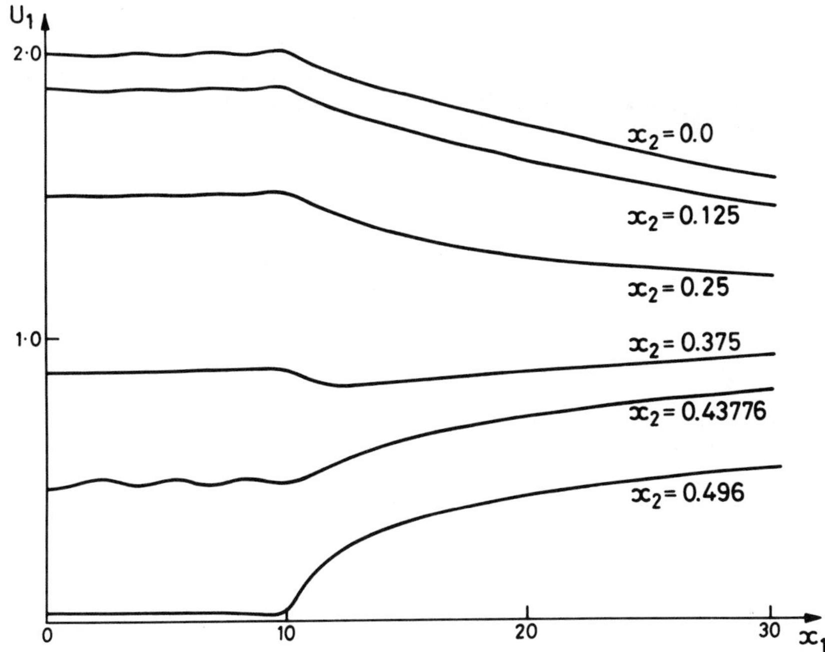

FIGURE 2. Laminar-to-turbulent development of u_1-velocity. Re-388000.

are perhaps to be expected in view of the sharp streamwise gradient at x_1=10 and the fact that "upwind weighting" was not employed. Indeed it was found that they could be easily smoothed by enhancing the streamwise diffusion of momentum in the region $x_1 \leq 10$ (formally equivalent to introducing upwind weighting).

4. FE TREATMENT OF THE k-ε TRANSPORT EQUATIONS

The problems associated with the FE discretisation of equations (10)-(12) have been examined by Hutton [7]. There is no real difficulty with the convective and diffusive terms (i.e. the left-hand sides) of equations (11) and (12), since they are very similar to the corresponding terms in the momentum equations. However, the source and sink terms on the right-hand sides are extremely complex and pose special difficulties. If the Galerkin method is applied in the usual way (as described for the momentum equations in Appendix 7.1), with $\underline{u}$, μ_T, k and ε all discretised using a single set of basis functions, then the source terms yield an exceptionally complicated matrix quantity which is very expensive to calculate, store and manipulate. Furthermore, in the resulting FE discrete form of the sink term, $C_{\varepsilon 2}\ \varepsilon^2/k$, the integrals are <u>not</u> independent of the nodal values of k, and cannot therefore be pre-calculated before numerical solution of the complete equation system. Both these problems can, of course, be overcome by dispensing with pre-calculated elemental matrices and evaluating the integrals in the source terms every iteration of the numerical procedure (e.g. refs. [3] and [13]).

Alternatively, by introducing new variables to represent source/sink terms, one can seek a simpler discretisation which can be generated by standard pre-calculated elemental matrices. This is the approach adopted here. Firstly, the right-hand sides of equations (11) and (12) are written as $S_k - D_k$ and $S_\varepsilon - D_\varepsilon$ respectively, where

$$S_k = \mu_T\, S/Re \quad , \quad D_k = \varepsilon$$
$$S_\varepsilon = C_\mu\, C_{\varepsilon 1}\, kS \quad , \quad D_\varepsilon = C_{\varepsilon 2}\, \varepsilon^2/k \tag{21}$$

and

$$S = \left(\frac{\partial u_n}{\partial x_m} + \frac{\partial u_m}{\partial x_n}\right)\frac{\partial u_n}{\partial x_m} + 2\alpha\left(\frac{u_2}{x_2}\right)^2 \tag{22}$$

The discretisation strategy is then as follows. The k and ε fields are both discretised in the same way as the velocity field, whilst the FE approximations to μ_T and to the above source/sink quantities are all formed from linear combinations of piece-wise quadratic basis functions generated by eight-noded serendipity elements. The usual Galerkin method is then applied to equations (11) and (12) (with right-hand sides written as above) and to eq. (22), whilst the algebraic expressions (10) and (21) are required to be satisfied only at the nodes. Such a procedure involves matrices (all of which can be pre-calculated) which are no more complicated or numerous than those required to calculate a general (varying-viscosity) laminar flow.

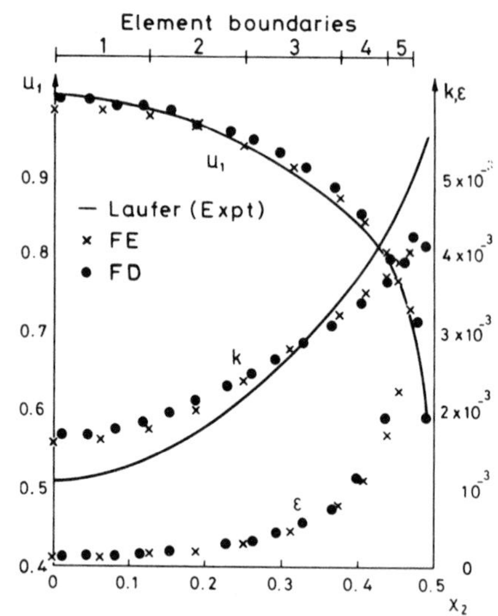

FIGURE 3. Fully developed pipe flow results compared with experiment.

Examples of FE pipe-flow solutions using the above discretisation were obtained from a simultaneous solution of the k-ε and momentum/continuity equations using Newton-Raphson iteration. In both cases, the grids consisted of five elements across the stream, namely type-3 elements at the centre line and at the wall ($\Delta h_+ \gtrsim 400$) and three interior type-2 elements. Figure 3 shows comparisons of the FE results for fully developed flow with both experiment [14] and finite difference results (generated on a uniform 30 x 30 mesh). The Reynolds number, based on experimental centre-line velocity and pipe diameter is 5 x 10^5. Agreement between the numerical results is good, though neither predicts the experimental k-levels very accurately using the "recommended" model constants. Very little change is made to the FE predictions by increasing the number of elements across the pipe from 5 to 15, demonstrating that the result is grid-converged and also indicating that

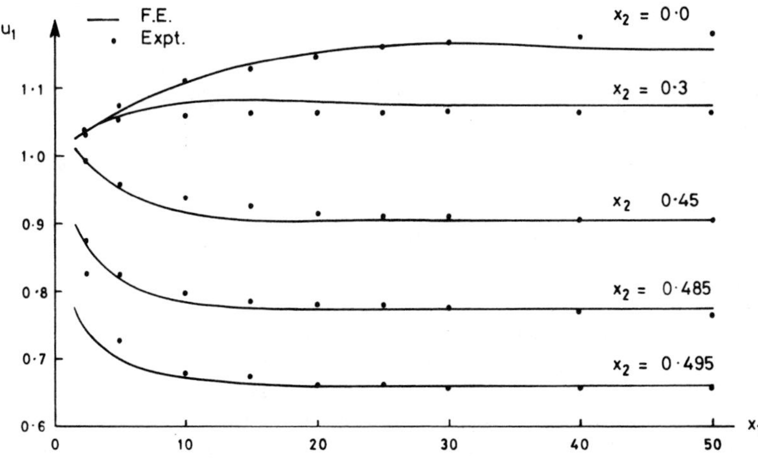

FIGURE 4. Developing Pipe Flow Results compared with experiment

the novel approximations made in the discretisation procedure do not limit the accuracy of the method. Figure 4 shows a comparison between FE predictions and the experimental data (Re = $3x10^5$ based on bulk velocity) of Richman and Azad [15] for developing flow in a pipe entrance. As can be seen, agreement is reasonably good, though it should be noted that the inlet levels of k could have an appreciable effect on the predictions. No turbulence measurements were cited in [15], and an inlet value k = 0.005 was used, as recommended in [16].

5. CONCLUDING REMARKS

A general approach to the FE treatment of turbulent flow has been proposed. When used with conventional polynomial elements in the near-wall region, considerable attention to mesh design is required at high Reynolds numbers, detracting from its practical usefulness. On the other hand, the use of specially designed elements (e.g. linlog, quadralog) render the scheme comparatively robust. The quadralog element appears especially promising as a basis for a general wall model due to its demonstrated capacity to accomodate a range of local conditions.

The simplified discretisation of the k-ε model introduced in this paper is capable of producing accurate solutions on fairly coarse meshes. However, an inconveniently good initial guess has been found necessary to achieve convergence of the Newton-Raphson method. Further work is in progress to obviate this difficulty and to evaluate the model in more complex geometries.

6. ACKNOWLEDGEMENT

This paper is published by permission of the Central Electricity Generating Board.

7. APPENDICES

7.1 The Galerkin-PALM Finite Element Approximation

Construct function spaces $S_u(\Omega)$, $S_p(\Omega)$ such that,

$$S_u(\Omega) = \{\underline{\tilde{v}}_i ; \tilde{v}_n = \sum_{i=1}^{M} \tilde{v}_{i,n} W_i(\underline{x}) + \sum_{i=1}^{N} \partial\tilde{v}_{i,n} \bar{W}_i(\underline{x}),\ \tilde{v}_i = 0 \text{ on } \partial\Omega_1, \ \tilde{v}_{i2} = 0 \text{ on } \partial\Omega_2\}$$

$$S_p(\Omega) = \{\tilde{q} ; \tilde{q} = \sum_{s=1}^{M} \tilde{q}_s X_s(\underline{x})\}$$

where $W_i(\underline{x})$, $\bar{W}_i(\underline{x})$, $X_s(\underline{x})$ are $C^{(o)}$ continuous, FE generated basis functions, each X_s being bilinear on quadrilateral elements, each W_i, $\bar{W}_i$, being of higher order on the same elements. In the usual way, the coefficients $\underline{\tilde{v}}_i$, $\tilde{q}_s$ represent the nodal values of $\tilde{v}$ and $\tilde{q}$ (s indexing corner nodes only) and the coefficients $\partial\underline{\tilde{v}}_i$ represent normal derivatives of $\underline{v}$ at certain boundary nodes. Finite element approximations $\underline{\tilde{u}}$, $\tilde{p}$ to the velocity and pressure respectively are now sought in the form,

$$\tilde{u}_n = \sum_{i=1}^{M} \tilde{u}_{i,n} W_i(\underline{x}) + \sum_{i=1}^{N} \partial\tilde{u}_{i,n} \bar{W}_i(\underline{x}) \tag{A1}$$

$$\tilde{p} = \sum_{s=1}^{M} \tilde{p}_s X_s(\underline{x}) \tag{A2}$$

where $\tilde{u}_{i,2} = 0$ at nodes on $\partial\Omega_2$ and $\underline{\tilde{u}} = \underline{\hat{u}}$ at nodes on $\partial\Omega_1$ (i.e. the essential boundary conditions are satisfied). The Galerkin-PALM equations can now be written

$$\int_\Omega \tilde{u}_m \frac{\partial\tilde{u}_n}{\partial x_m} \tilde{v}_n \, d\underline{x} + \int_\Omega \frac{\mu_e}{Re}\left(\frac{\partial\tilde{u}_m}{\partial x_n} + \frac{\partial\tilde{u}_n}{\partial x_m}\right)\frac{\partial\tilde{v}_n}{\partial x_m} d\underline{x} + \frac{2\alpha}{Re}\delta_{n2}\int_\Omega \mu_e \frac{\tilde{u}_2\tilde{v}_n}{x_2^2} d\underline{x}$$

$$- \int_\Omega \tilde{p} \operatorname{div} \underline{\tilde{v}} \, d\underline{x} + \pi \int_\Omega \operatorname{div} \underline{\tilde{u}} \operatorname{div} \underline{\tilde{v}} \, d\underline{x} = 0 \text{ for all } \underline{v} \varepsilon S_u(\Omega) \tag{A3}$$

$$\int_\Omega \operatorname{div}\underline{\tilde{u}}\, q \, d\underline{x} = 0 \text{ for all } q \varepsilon S_p(\Omega) \tag{A4}$$

which, on invoking the definitions of $S_u(\Omega)$ and $S_p(\Omega)$ and inserting eqs. (A1) and (A2) for $\underline{\tilde{u}}$ and $\tilde{p}$, are seen to reduce to a non-linear algebraic system for the nodal variables. The penalty term involving the parameter π is included in eq. (A3) in order to reduce continuity errors when using $C^{(o)}$ - continuous pressure spaces. By comparing the last two terms of eq. (A3) it can be seen that the introduction of this term can be interpreted as a broadening of the pressure representation to the following $C^{(-1)}$ form,

$$\tilde{p} = \sum_{s=1}^{M} \tilde{p}_s X_s(\underline{x}) - \pi \operatorname{div} \underline{\tilde{u}}$$

7.2 Quadralog Wall-element Velocity Basis Functions

With reference to figure 1, the velocity field within a quadralog element can be written,

$$\tilde{u}_n = \sum_{i=1}^{9} \tilde{u}_{i,n} W_i(\underline{\eta}) + \sum_{i=1}^{3} \partial\tilde{u}_{i,n} \bar{W}_i(\underline{\eta})$$

where the local basis functions, $W_i(\underline{\eta})$, $\bar{W}_i(\underline{\eta})$ are constructed in terms of local isoparametric co-ordinates η_1, η_2 as follows. If the functions $f_i(\eta_2)$, $\bar{f}_i(\eta_2)$ are defined as

$$\Gamma f_i(\eta_2) = \begin{cases} \{\ell n[(\gamma+1)/2][\eta_2^2+2\eta_2-3] + \dfrac{(1-\gamma)}{2\gamma}[\eta_2^2-\eta_2] \\ \quad + 3\ell n[\gamma+(1+\eta_2)(1-\gamma)/2]\} \quad 1\le i\le 3 \\ \{\ell n\gamma[3-2\eta_2-\eta_2^2] + \dfrac{(1-\gamma)}{\gamma}[1-\eta_2^2]-4\ell n[\gamma+(1+\eta_2)(1-\gamma)/2]\} \\ \quad i=4,8,9 \\ \{\ell n[2/1+\gamma]+\ell n[2\gamma/1+\gamma][\eta_2^2+2\eta_2] + \dfrac{(1-\gamma)}{2\gamma}[\eta_2^2+\eta_2] \\ \quad + \ell n[\gamma+(1+\eta_2)(1-\gamma)/2]\} \quad 5\le i\le 7 \end{cases}$$

$$\Gamma\bar{f}_i(\eta_2) = \frac{H(1-\gamma)}{2}\{2\ell n[(1+\gamma)/2][\eta_2^2-1] + \ell n\,\gamma[\eta_2-\eta_2^2] + 2\ell n[\gamma+(1+\eta_2)(1-\gamma)/2]\} \quad 1\le i\le 3$$

where $\gamma=\Delta h/H$ and $\Gamma = \{3\ \ell n\gamma + 4\ell n\ [2/1+\gamma] + (1-\gamma)/\gamma\}$ and, if the functions $Z_i(\eta_1)$ are defined as,

$$Z_i(\eta_1) = [(1+\eta_{i,1}\ \eta_1)\eta_{i,1}\ \eta_1/2 + (1-\eta_{i,1}^2)(1-\eta_1^2)] \quad 1\le i\le 9$$

where $\eta_{i,1}$ is the value of η_1 at the i'th node, then,

$$W_i(\underline{\eta}) = Z_i(\eta_1)\ f_i(\eta_2) \quad 1\le i\le 9$$

$$\bar{W}_i(\underline{\eta}) = Z_i(\eta_1)\ \bar{f}_i(\eta_2) \quad 1\le i\le 3$$

8. REFERENCES

1. NORRIE, D.H. (ed) - Finite Elements in Flow Problems, Proc. of 3rd International Conference, Banff, Canada, 1980.

2. BAKER, A.J. - Finite element analysis of turbulent flows, Numerical Methods in Laminar and Turbulent Flow Ed. Taylor, C., Morgan, K. and Brebbia, C.A., Pentech Press, 1978.

3. TAYLOR, C., THOMAS, C.E. and MORGAN, K.-Confined Turbulent Flow utilising the Finite Element Method, Annual Winter Meeting, A.S.M.E., AMD 34, 1979.

4. RODI, W. - Turbulence Models and their Application in Hydraulics, IAHR State-of-the Art Paper, Delft, Netherlands, 1980.

5. HUTTON, A.G. - A Survey of the Theory and Application of the Finite Element Method in the Analysis of Viscous Incompressible Newton Flow, CEGB Report RD/B/N3049, 1974.

6. BERNARD, M. and GRANDOTTO, M. - Problems Mathematiques et

algorithmes Numeriques pour la Resolution des Equations de Navier-Stokes, Note Technique DRE/STRE/LMTA 79/245, Centre D'Etudes Nucleaires de Cadarache, 1979.

7. HUTTON, A.G. - Finite Element Analysis of Turbulent, Incompressible Flow Bounded by Smooth Walls, Preprints 2nd International Conference on 'Finite Elements in Flow Problems', Rapallo Italy, pp527-538, 1976.

8. TAYLOR, C., HUGHES, T.G. and MORGAN, K. - A Numerical Analysis of Turbulent Flow in Pipes, Int.J.Computers and Fluids, 5, 191-204, 1977.

9. HUTTON, A.G. - Progress in the development of a finite element wall model for high Reynolds number turbulent flow, Appl Math. Modelling 3, 322-326, 1979.

10. HUTTON, A.G. and SMITH, R.M. - The prediction of Laminar Flow over a Downstream-Facing Step by the Finite Element Method, CEGB Report RD/B/N3660, 1979.

11. HUTTON, A.G. - Finite Element Boundary Techniques for Improved Performance in Computing Navier-Stokes and related Heat Transfer Problems, CEGB Report RD/B/N4651, 1979.

12. ZIENKIEWICZ, O.C. - The Finite Element Method - 3rd Ed., McGraw Hill, London, 1977.

13. TAYLOR, C., HUGHES, T.G. and MORGAN, K., A Finite Element Model of One and Two Equation Models of Turbulent Flow, art. in Recent Advances in Numerical Methods in Fluids 1, Pineridge Press, Swansea, 1980.

14. LAUFER, J. - The Structure of Turbulence in Fully Developed Pipe Flow NACA Report 1174, 1954.

15. RICHMAN, J.W. and AZAD, R.S., - Developing Turbulent Flow in Smooth Pipes, Appl. Sci. Res. 28, 419-441, 1973.

16. PUN, W.M. and SPALDING, D.B. - A General Computer Program for Two-Dimensional Elliptic Flows, Imperial College Heat Transfer Report HTS/76/2, 1977.

THE PREDICTION OF LOCAL MEAN FLOW CHARACTERISTICS FOR FULLY DEVELOPED TURBULENT FLOW IN STRAIGHT NON-CIRCULAR PASSAGES.

C.W. Rapley

Department of Mechanical Engineering, Sunderland Polytechnic, U.K.

SUMMARY

A finite-difference method has been developed to predict the main features of the mean flow in fully-developed turbulent flow in passages of arbitrary non-circular cross-section. The transport effects of secondary flow have been simulated in a simple way which eliminates solution of the cross-plane momentum and continuity equations. The outcome of these simplifications are discussed in a comparison of the predictions for three different passage shapes with experiment and with predictions from other methods.

1. INTRODUCTION

The design and development of compact heat exchangers, including those in nuclear reactor cores, depends as much on knowledge of the local mean flow characteristics as on the overall flow and heat transfer. Fully developed turbulent flow in straight non-circular passages is considerably more complex than in circular tubes due to the presence of turbulence-driven secondary flow in the passage cross-plane. These flows cause the main flow to spiral through the passage and although they are relatively weak compared with the main flow, they have a significant influence on the local mean-flow distributions of interest, chiefly the wall shear stress and axial velocity.

Until recently the main source of information on local mean flow has been experiment. However, with the almost overwhelming variety of passage shapes that could be of interest, it appears more effective to supplement experiment with a calculation method that can provide useful local data with, for example, a much wider range of geometry and flow conditions than could be contemplated with experiment. Many efforts have been made to develop such calculation methods for particular geometries, mainly using finite-differences and for the square duct, and the axial flow passages in rod-bundles. A brief overview of these methods can be found in Gosman and Rapley [1] from which it is evident that such methods may include the simultaneous solution of from five to eleven coupled non-linear partial differential equations (pde's). The cross-plane momentum and continuity and the axial momentum equations account for four of these pde's, the remainder being required for calculation of the Reynolds stresses appearing in the momentum equations. The square duct procedures of Naot et al [2] and Reece [3] are examples where each required Reynolds stress was calculated from its partial differential transport equation. However, in many of the most recent of these finite difference methods [1][4][5][6][7][8][9] the number of pde's for the stresses have been kept to a minimum by using an algebraic stress transport model(ASTM) first derived for square duct calculations [4][10]. This was obtained by systematic simplification of the Reynolds stress transport equations which included the neglect of transport by convection and diffusion. The resulting equations express the Reynolds stresses as functions of axial velocity gradients, the turbulence kinetic energy k and its dissipation rate ε.

In the procedures developed for calculation of flow in a square duct [4], rod bundles [5][6] and an equilateral triangular duct [7], k was obtained from its modelled partial differential transport equation [9] and ε from the widely used formula obtained from dimensional analysis [11]

$$\varepsilon = C_D k^{3/2}/\ell \tag{1}$$

where C_D is an empirical constant and ℓ a turbulence length scale which was prescribed either from experimental measurement [5][6] or by the Buleev [12] geometric formula [4][7].

In the ASTM based method developed by Gosman and Rapley [1] for arbitrary shaped passages, both k and ε were obtained from modelled partial differential transport equations [11], thus entailing the solution of a total of six pde's. These solutions were obtained by finite-differences on an orthogonal curvilinear mesh fitted into the passage cross-section. The present work seeks to establish how much this latter type of procedure can be simplified by eliminating some of the pde's and producing a calculation method still capable of predicting the main features, if not the detail, of local axial velocity and wall shear stress in non-circular passages of arbitrary cross-section. It is recognised that the effects of secondary flow must be included if the main features are to be predicted. From previous experiments and calculations [1][2][3][4][5][6][7][8], secondary flow tends to make wall shear stress and axial velocity more uniform than it would otherwise be and a particular feature of the latter is the bulging of axial velocity contours into the corners of the duct.

The present paper reports some of the predictions obtained from a method in which the cross-plane momentum and continuity equations are eliminated leaving the simultaneous solution of only three pde's (k, ε and axial velocity).

2. THE MATHEMATICAL PROBLEM

2.1 Axial momentum

The axial momentum equation can be written firstly in Cartesian co-ordinate form for clarity, for time averaged fully developed turbulent flow in straight passages as:

$$\partial(\rho uw)/\partial x + \partial(\rho vw)/\partial y = -\partial p/\partial z + \partial(\mu\partial w/\partial x)/\partial x + \partial(\mu\partial w/\partial y)/\partial y - \partial(\rho\overline{u'w'})/\partial x - \partial(\rho\overline{v'w'})/\partial y \quad (2)$$

where u and v are the cross-plane secondary velocity and w the axial velocity components.

The ASTM of Launder and Ying [4] was further analysed by Gessner and Emery [10] to yield the following relations for the Reynolds stresses in equation (2):

$$\rho\overline{u'w'} = -C_4\rho(k^2/\varepsilon)\partial w/\partial x \quad (3)$$
$$\rho\overline{v'w'} = -C_4\rho(k^2/\varepsilon)\partial w/\partial y \quad (4)$$

Taking $C_4\rho k^2/\varepsilon$ = turbulent viscosity = μ_t (5)

equation (2) can be written in terms of effective viscosity $\mu_{eff} = \mu_t + \mu$ as

$$\partial(\rho uw)/\partial x + \partial(\rho vw)/\partial y = -\partial p/\partial x + \partial(\mu_{eff}\partial w/\partial x)/\partial x + \partial(\mu_{eff}\partial w/\partial y)/\partial y \quad (6)$$

Solution of this equation requires values of k and ε for μ_t and of secondary velocity components u and v. The latter would normally require solution of the cross-plane momentum and continuity equations, with the attendant problems of re-circulating flow. However, in a wall bounded region the transport effects contained in the two terms on the ℓ.h.s. of equation (6) can be simplified. Study of secondary velocity measurements in square ducts [4][13], rectangular ducts [14][15] and an equilateral triangular duct [7] and of calculations which show the convective effects of these flows [1], show that in the wall region, secondary flow normal to the wall is negligible and the region is dominated by flow parallel to the wall. If a wall is assumed parallel to the x direction, then in the ℓ.h.s. of equation (6) the first-term dominates the second which can be neglected.

In the present work a gross simplification is made by replacing the convection transport effect of the first term by diffusion transport i.e. let

$$\partial(\rho uw)/\partial x = -\partial(B\mu_{eff}\partial w/\partial x)/\partial x \quad (7)$$

where B is an empirical constant to be determined. It must be emphasised that there is no theoretical justification for equation (7) since the two transport mechanisms are physically quite different and the constant B cannot be derived from theoretical considerations. The simplification is made here since it provides a significant transport effect parallel to the wall and is expedient in eliminating the difficulties and expense of calculating secondary velocities. Equation (6) now becomes

$$- \partial p/\partial z + \partial(1+B)\mu_{eff}\partial w/\partial x)/\partial x + \partial(\mu_{eff}\partial w/\partial y)/\partial y = 0 \quad (8)$$

giving, in effect, anisotropic effective viscosities with an anisotropy of (1+B).

Calculation methods neglecting secondary flow and assuming anisotropic turbulent viscosities have of course been used before for passage flows, mainly in rod-bundle calculations (e.g. [16][17][18]). It is also of interest to note that Trupp and Aly [6] found that secondary flow and anisotropic viscosities had similar effects on calculated wall shear stress in rod bundles. It was also apparent that an assumed uniform anisotropy gave results comparable to the more complex distributions implied by the measurements of Rehme [19]. Thus, as a first step in the present work, a constant and uniform value of B will be assumed to apply throughout the flow fields and universal to all passage geometries calculated. From a parametric study of a range of geometries the value B = 2.5 was adopted.

2.2 Turbulence equations

The modelled transport equations used to calculate k and ε are the now accepted forms appropriate to high Reynolds number flows in which viscous effects are deemed negligible [11][20]. For fully developed flow in straight passages they are written in Cartesian form as

$$\partial(\rho uk)/\partial x + \partial(\rho vk)/\partial y = \partial((\mu_t/\sigma_k)\partial k/\partial x)/\partial x + \partial(\mu_t/\sigma_k)\partial k/\partial y)/\partial y + P - \rho\varepsilon = 0 \quad (9)$$

$$\partial(\rho u\varepsilon)/\partial x + \partial(\rho v\varepsilon)/\partial y = \partial(\mu_t/\sigma_\varepsilon)\partial\varepsilon/\partial x)/\partial x + \partial((\mu_t/\sigma_\varepsilon)\partial\varepsilon/\partial y)/\partial y + \varepsilon(C_{\varepsilon 1}P - C_{\varepsilon 2}\rho\varepsilon)/k \quad (10)$$

where σ_k and σ_ε are the turbulent Prandtl (Schmidt) numbers for k and ε respectively, $C_{\varepsilon 1}$ and $C_{\varepsilon 2}$ are constants and P the production rate of k which, neglecting secondary velocity gradients, is given by

$$P = - \rho\overline{u'w'}\ \partial w/\partial x - \rho\overline{v'w'}\ \partial w/\partial y \quad (11)$$

The Reynolds stresses appearing in equation (11) are calculated via equations (3), (4) and (5).

The same simplifications are made to the convection transport terms on the ℓ.h.s. of equations (9) and (10) as made in the axial momentum equation yielding

$$\partial((1+B)(\mu_t/\sigma_k)\partial k/\partial x)/\partial x + \partial((\mu_t/\sigma_k)\partial k/\partial y)/\partial y + P - \rho\varepsilon = 0 \quad (12)$$

$$\partial((1+B)(\mu_t/\sigma_\varepsilon)\ \partial\varepsilon/\partial x)/\partial x + \partial((\mu_t/\sigma_\varepsilon)\ \partial\varepsilon/\partial y)/\partial y + \varepsilon(C_{\varepsilon 1}P - C_{\varepsilon 2}\rho\varepsilon)/\ k \quad (13)$$

2.3 The equations in orthogonal curvilinear co-ordinates

The pde's to be solved have been reduced to diffusive transport equations (8) (12) and (13) which can be written in the relatively simple common form:

$$\partial((1+B)D_\phi\partial\phi/\partial x)/\partial x + \partial(D_\phi\partial\phi/\partial y)\partial y + C_\phi \ = 0 \quad (14)$$

where ϕ stands for w, k or ε and D_ϕ is the diffusion coefficient associated with each ϕ. C_ϕ represents the source and any other terms not contained in the other components.

The Cartesian equation (14) can be transformed into general orthogonal curvilinear co-ordinate form (see for example Pope [21]) giving an equation for the form

$$\partial((h_2/h_1)(1+B)D_\phi\partial\phi/\partial\zeta_1)/\partial\zeta_1 + \partial((h_1/h_2)D_\phi\partial\phi/\partial\zeta_2)/\partial\zeta_2 + C_\phi \ = 0 \quad (15)$$

where h_1 and h_2 are the metric coefficients in the cross-plane curvilinear co-ordinate directions ζ_1 and ζ_2 respectively. Table 1 below summarises D_ϕ and C_ϕ appropriate to each ϕ with axial velocity now appearing as u_3 in the straight (h_3 = 1) axial co-ordinate direction ζ_3

ϕ	D_ϕ	C_ϕ
u_3	μ_{eff}	$- h_1h_2\partial p/\partial\zeta_3$
k	μ_t/σ_k	$h_1h_2(P - \rho\varepsilon)$
ε	μ_t/σ	$h_1h_2(C_{\varepsilon 1}P - C_{\varepsilon 2}\rho\varepsilon)/k$

Table 1 Coefficients in the general transport equation (15)

In table 1, the production rate P of turbulence kinetic energy is now written as

$$P = - \rho\overline{u_1'u_3'}\ \partial u_3/h_1\partial\zeta_1 - \rho\overline{u_2'u_3'}\partial u_3/h_2\partial\zeta_2 \quad (16)$$

and the turbulent stresses as

$$\left.\begin{aligned}\rho\overline{u_1'u_3'} &= - \mu_t\partial u_3/h_1\partial\zeta_1\\ \rho\overline{u_2'u_3'} &= - \mu_t\partial u_3/h_2\partial\zeta_2\end{aligned}\right] \quad (17)$$

where μ_t is given by equation (5)

3 THE NUMERICAL SOLUTION METHOD

The transport equation (15) is solved by finite differences on a mesh of orthogonally intersecting grid lines in three dimensions. In the cross-plane of the passage the grid lines are curvilinear and intersect orthogonally with each other and with the passage boundaries. The grid nodes are at the points of intersection, through which pass the straight axial lines which complete the three dimensional mesh. A portion of a typical cross-plane mesh is shown in figure 1 which also shows the contiguous control volumes or 'cells' surrounding each grid node. The boundaries of these cells, shown with broken lines, are formed in the cross-plane by a mesh of lines representing axial planes, mid-way between the main grid lines, and in the axial plane by a pair of cross-sectional planes. A typical grid node is denoted by P, its nearest neighbours by N,S,E,W. These letters are used as suffices to denote the value of a variable at that location.

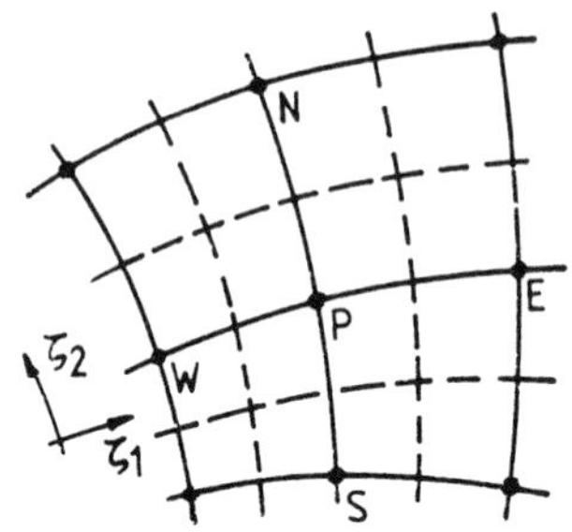

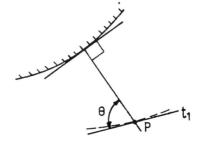

Figure 1 The grid control volumes

Figure 2 Orientation of grid line tangents

The finite-difference equivalent of equation (15) is obtained for each interior location by integrating each term over the control volume, using central differencing with C_ϕ linearised and assumed uniform over the control volume. This yields a finite-difference equation of the standard form

$$\phi_P = A_N\phi_N + A_S\phi_S + A_E\phi_E + A_W\phi_W + e \tag{18}$$

in which e is a component of the linearised source and the diffusion ('A') coefficients contain an appropriate proportion [16] of anisotropy according to the orientation θ of the grid line tangent t_1 (see figure 2) at that location.

Equation (18) was solved using a line-by-line alternating direction method based on the tridiagonal matrix (Thomas) algorithm (see for example [22]), for each ϕ in the sequence u_3, k, ε, updating the turbulent viscosity after each sequence.

Wall functions were used to match the interior flow with the wall conditions thus avoiding the large number of grid nodes that would otherwise be needed in this region of high gradients. These functions were applied to the cells next to the wall which was assumed to be a region of constant shear in local equilibrium. This led to conventional wall functions in which the local wall shear stress τ_s was calculated from the well known velocity log-law

$$u_3 = u_3^* \ln(E S^+)/\kappa \tag{19}$$

in which u_3^* is the local friction velocity and $S^+=u_3^*S/\nu$ with S the distance to the wall.

Turbulence kinetic energy in a near wall control volume was calculated as in the interior with diffusion to the wall set to zero and production and dissipation of k in the source calculated using τ_s.

The near-wall value of ε was obtained by neglecting transport of ε by convection and diffusion and assuming a length scale that varied linearly with S [11] so that

$$\varepsilon = u_3^{*3}/\kappa S \qquad (20)$$

This value was imposed directly at the near-wall node i.e. the finite-difference equation was not used there.

The numerical solution was assumed converged when the sum of the absolute axial momentum sources over the field was less than 0.001 of the axial momentum flux. Reduction of the residual sources below this gave negligible changes to the solution. Discretisation errors were minimised in the present work by using the meshes and differencing methods selected in [1] after a comprehensive series of accuracy tests which included grid refinement, comparisons between solutions with different orthogonal meshes, extensive comparisons of laminar flow calculations with analytical solutions and symmetry tests with turbulent flow. Details of these tests can be found in [24].

The values of the empirical constants used were mainly taken from previous work [1] and were assumed universal to all passage geometries. They were: $C_4 = 0.085$, $\sigma_k = 1.0$, $\sigma_\varepsilon = 1.22$, $E = 9.02$, $\kappa = 0.42$.

4 PREDICTIONS

4.1 Square duct

This duct provides a useful test case since extensive measurements are available and many previous authors have applied their duct flow calculation methods to this case. A 12 x 12 cartesian mesh was used in a duct quadrant with a variable spacing grid to give more nodes in the near wall region.

The present predictions are compared with experiment and previous calculations in figure 3 which also shows the predictions with no allowance for secondary flow (i.e. isotropic turbulent viscosities). The present and previous predictions all show axial velocity contours to bulge well into the corner and the wall shear stress to be more uniform than with no secondary flow effect. These main effects thus appear to

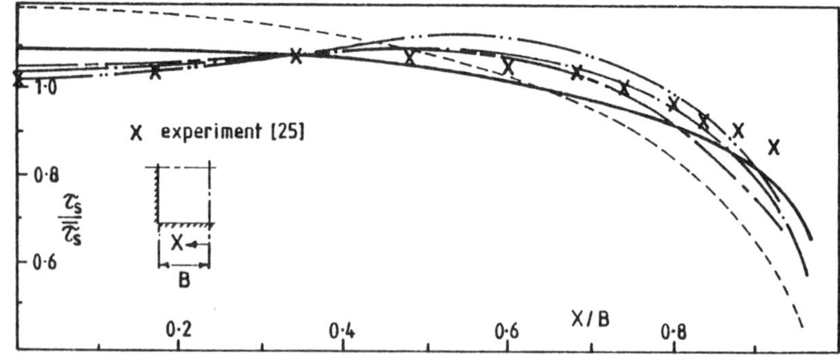

(a) Wall shear stress, $R_e = 3.4 \times 10^4$

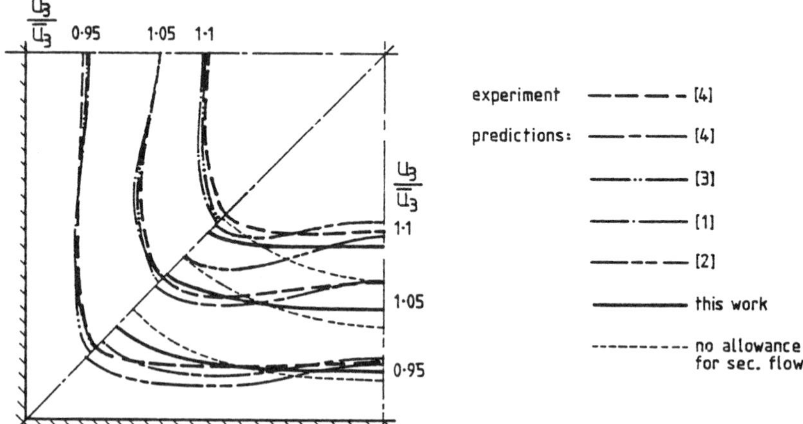

(b) Axial velocity, $R_e = 2.15 \times 10^5$

Figure 3 Square duct

have been simulated with the present method, although not quite as well in detail as with the other more complex calculation methods.

4.2 Rod bundle channel

With its important application in nuclear reactor cores, this passage shape has also received much attention from both experimenters and numerical analysts. A 16 x 10 orthogonal curvilinear mesh similar to that shown in figure 4 was used for calculations in symmetry sub-channels of an equilateral triangular infinite array with rod pitch/diameter (P/D) ratios of 1.1 and 1.123. A simple method based on the numerical solution of Laplace equations was used to generate the grid. Details are given in Rapley [24].

Figure 5 compares the present and other predictions with experiment showing all predictions of wall shear stress to be in good agreement with experiment. The effects of secondary

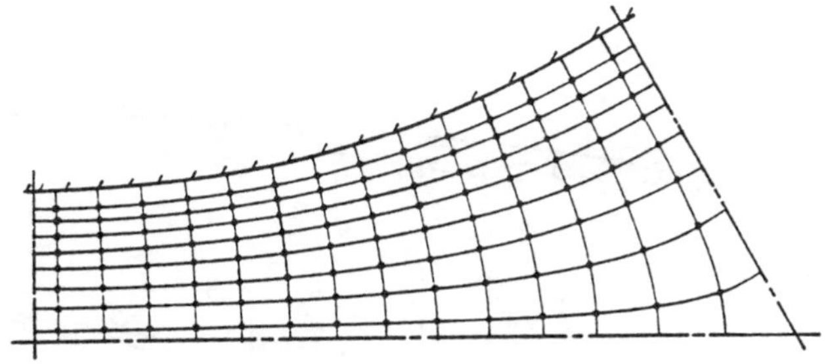

Figure 4. Orthogonal grid for rod bundle calculations

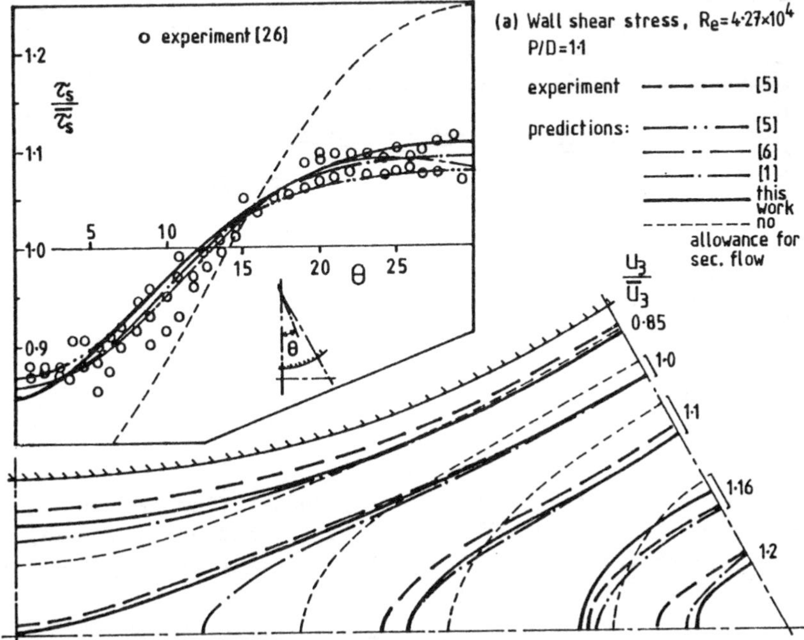

Figure 5. Rod bundle channel

flow in making the wall shear stress profile more uniform is well simulated by the present method. The calculated axial velocity contours from the present method are also in good agreement with experiment, simulating well the effect of secondary flow in making the contours bulge into the gap region. It is of interest to note that a typical CPU time for the above solutions including generation of the grid, was only 7.6 mins on an ICL 2950 digital computer.

4.3 Rectangular duct

A 22 x 12 Cartesian mesh was used for calculations in a quadrant of a rectangular duct of 3:1 aspect ratio. The present calculations are compared with the experiments of Leutheusser [25] and calculations from Gosman and Rapley [1]

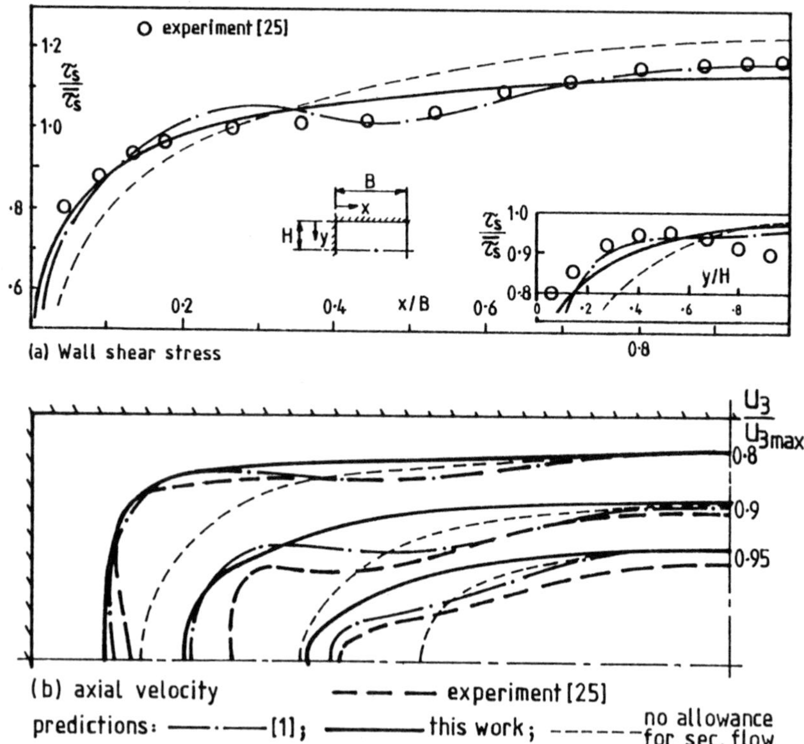

Figure 6. Rectangular duct

in figure 6. The main effects of secondary flow are once again seen to be simulated by the present method, although, as in the square duct case, the detail is better predicted by the Gosman and Rapley [1] method.

4.4 Friction factor characteristics

The present predictions are seen in figure 7 to be in reasonable agreement with experiment and as good if not better than most other predictions shown.

5 CONCLUSIONS

The foregoing comparisons have shown the present method to be capable of simulating the main effects of secondary flow, producing acceptable mean flow predictions in fully developed turbine flow. In some cases, the detail has not been predicted as well as with other more complex procedures, but this is accepted as a not too severe penalty for the large simplifications made in the calculation method. These simplifications have the benefit however of eliminating the stability and convergence problems inherent in the more complex methods where the recirculating secondary flow is calculated, and reducing the required computer storage and CPU time. Compared

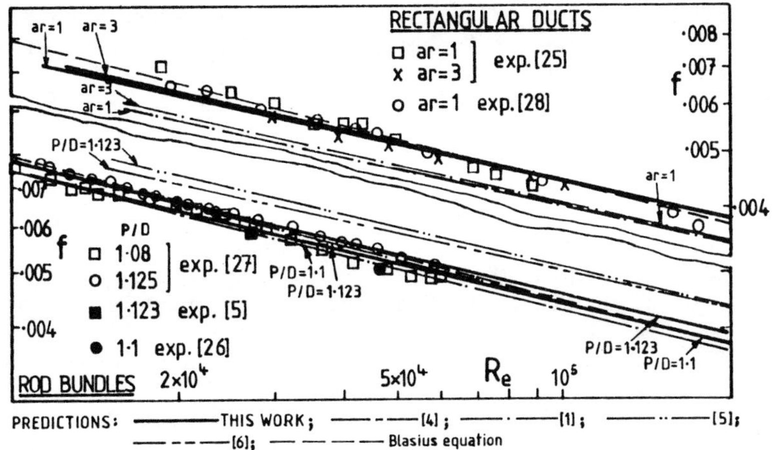

Figure 7. Friction factor characteristics

with the 6 pde method of Gosman and Rapley [1], a calculation with the present method required only two thirds of the storage and one third of the CPU time.

6 REFERENCES

1. GOSMAN, A.D. and RAPLEY, C.W. - Fully-Developed Flow in Passages of Arbitrary Cross-Section. Recent Advances in Numerical Methods in Fluids, Ed. Taylor, C. and Morgan, K, Pineridge Press, 1980.
2. NAOT, D., SHAVIT, A. and WOLFSHTEIN, M. - Numerical Calculation of Reynolds Stresses in a Square Duct with Secondary Flow. Wärme-und-Stoffubertragung, Vol.7,p151,1974.
3. REECE,G.J. - A Generalised Reynolds Stress Model of Turbulence. Ph.D. Thesis, Uni.of London,1977
4. LAUNDER,B.E. and YING, W.M. - Prediction of Flow and Heat Transfer in Ducts of Square Section. Proc.I.Mech.E., Vol.187, p455,1973.
5. CARAJILESCOV, P. and TODREAS, N.E. - Experimental and Analytical Study of Axial Turbulent Flows in an Interior Sub-channel of a Bare-rod bundle. J.Heat Trans., Trans ASME, Vol.98, p262, 1976.
6. TRUPP, A.C. and ALY, A.M.M. - Predicted Secondary Flows in Triangular Array Rod Bundles. J.Fluids Eng., Trans ASME vol 101, p354, 1979.
7. ALY, A.M.M., TRUPP, A.C. and GERRARD, A.D. - Measurement and Prediction of Fully Developed Flow in an Equilateral Triangular Duct. J.Fluid Mech., Vol 85, p57, 1978.
8. NETI, S. and EICHHORN, R. - Computations of Developing Turbulent Flow in a Square Duct. ASME Symp. Turbulent Boundary Layers, Niagara Falls, p179, 1979.
9. GESSNER, F.B. and EMERY, A.F. - The Numerical Prediction of Developing Turbulent Flow in Rectangular Ducts,2nd Symp. Turbulent Shear Flows, Imperial Coll.,U.K.p17.1, 1979

10.GESSNER, F.B., and EMERY, A.F. - A Reynolds Stress Model for Turbulent Corner Flows, Part 1. J.Fluids Eng., Trans ASME 76-F-ED, 1976.
11.LAUNDER, B.E. and SPALDING, D.B. - Mathematical Models of Turbulence, Academic Press, 1972.
12.BULEEV, N.I. - Theoretical Model of the Mechanism of Turbulent Exchange in Fluid Flows. AERE Trans. 957, 1963.
13.BRUNDRETT, E. and BAINES, W.D.-The Production and Diffusion of Vorticity in Duct Flow. J.Fluid Mech.,Vol 19,p375,1964.
14.GESSNER, F.B. and JONES, J.B. - On Some Aspects of Fully-Developed Turbulent Flow in Rectangular Channels. J.Fluid Mech., Vol 23, p689, 1975.
15.TRACY, H.J. - Turbulent Flow in a Three-dimensional Channel. Proc. ASCE(HY6), Vol 91, p9, 1965.
16.MEYDER, R. - Turbulent Velocity and Temperature Distributions in the Central Sub-Channel of Rod Bundles. Nuclear Eng. & Design, Vol 35, p181, 1975.
17.NIJSING, R. and EIFLER, W. - Temperature field in liquid-metal-cooled rod assemblies. Prog.in Heat Mass Trans., Vol 7, p117, 1973.
18.VONKA, V. and BOONSTRA, B.H. - Calculated Heat Transfer Development in Bundles. Nuclear Eng.& Design, Vol 31,p337,1974
19.REHME, K. - Anisotropic Eddy Viscosities in the Turbulent Flow Through a Rod Bundle. Symp. Turbulent Shear Flows,Penn. State Uni., p8.41, 1977.
20.REYNOLDS, W.C. and CEBECI, T. - Calculation of Turbulent Flows. Turbulence, Ed.Bradshaw, Springer Verlag, 1978
21.POPE, S.B. - The Calculation of Turbulent Recirculating Flows in General Orthogonal Co-ordinates. J.Comp.Physics, Vol 26, p197, 1978.
22.FORSYTHE, G.E. and WASOW, R.W. - Finite-difference Methods for Partial Differential Equations, J.Wiley, 1960.
23.GOSMAN, A.D. and RAPLEY, C.W. - A Prediction method for fully Developed Flow Through Non-circular Passages. Paper No.IC/FS/78,Int.Conf.Numerical Methods Lam.Turb.Flow, Swansea, U.K. 1978.
24.RAPLEY, C.W. - Fluid and Heat Flow in Tubes of Arbitrary Cross-section. Ph.D. Thesis, Uni.of London, 1980.
25.LEUTHEUSSER, H.J. - Turbulent Flow in Rectangular Ducts. Proc. ASCE, Vol 89, p 1, 1963.
26.SUBBOTIN, V.I. et al - Velocity Field of Turbulent Fluid Flow in a Longitudinal Streamline of Clusters of Rods. AEC-tr-7189, 1971
27.REHME, K. - Pressure-drop Performance of Rod Bundles in Hexagonal Arrangements. Int.J.Heat Mass Trans. Vol 15, p2499, 1972.
28.HARTNETT, J.P., KOH, J.C. and McCOMAS, S.T. - A Comparison of Predicted and Measured Friction Factors for Turbulent Flow Through Rectangular Ducts. Trans.ASME,Vol 84,p82,1962.

A Refined PUMPIN (Pressure Update by Multiple Path INtegration) Method for Updating Pressures in the Numerical Solution of the Incompressible Fluid Flow Equations.

Z. MAZHAR and G. D. RAITHBY

Department of Mechanical Engineering
University of Waterloo
Waterloo, Ontario, CANADA

SUMMARY

This paper reports a modification of the PUMPIN method [1] of updating pressures, in which Binomial weights are employed. The motivation for this choice is discussed and, through two test problems, the revised method is shown to greatly enhance stability and promote faster convergence. A few remarks concerning the choice of a suitable reference point are made. A generalization of the new method to three-dimensions is also presented.

1. GOVERNING EQUATIONS

The equations of motion (mass, x-momentum and z-momentum respectively) for an incompressible fluid are

$$\frac{\partial u}{\partial x} + \frac{\partial w}{\partial z} = 0, \tag{1}$$

$$\frac{\partial(\rho u)}{\partial t} + \frac{\partial(\rho uu)}{\partial x} + \frac{\partial(\rho uw)}{\partial z} = \frac{\partial}{\partial x}[\mu_e \frac{\partial u}{\partial x}] + \frac{\partial}{\partial z}[\mu_e \frac{\partial u}{\partial z}] - \frac{\partial p}{\partial x} + \dot{q}^u \tag{2}$$

and

$$\frac{\partial(\rho w)}{\partial t} + \frac{\partial(\rho wu)}{\partial x} + \frac{\partial(\rho ww)}{\partial z} = \frac{\partial}{\partial x}[\mu_e \frac{\partial w}{\partial x}] + \frac{\partial}{\partial z}[\mu_e \frac{\partial w}{\partial z}] - \frac{\partial p}{\partial z} + \dot{q}^w, \tag{3}$$

in which laminar and turbulent momentum transfer have been both modelled through the effective viscosity coefficient, μ_e. The term $\dot{q}$ accounts for any other momentum sources not explicitly written in the equations. Possible couplings between these equations and temperature and concentration are not the central issue here so these equations have been omitted. The primary interest is the steady-state solution of Eqs. (1)-(3) subject to specified boundary conditions.

In this set of equations there is no explicit equation for pressure. Instead, the correct pressure field must be found which, when used in the momentum equations, yields velocities which satisfy the velocity constraint expressing mass conservation. Because the pressure is buried in the equations, without an equation of its own, special solution procedures are required and the name "velocity-pressure coupling problem" has been adopted.

2. DISCRETIZATION OF THE GOVERNING EQUATIONS

Fig. 1 shows a section of the rectangular, non-uniform, staggered grid that was used, while Fig. 2 shows the typical storage locations for u, w and p. Using a 5-point, upstream-

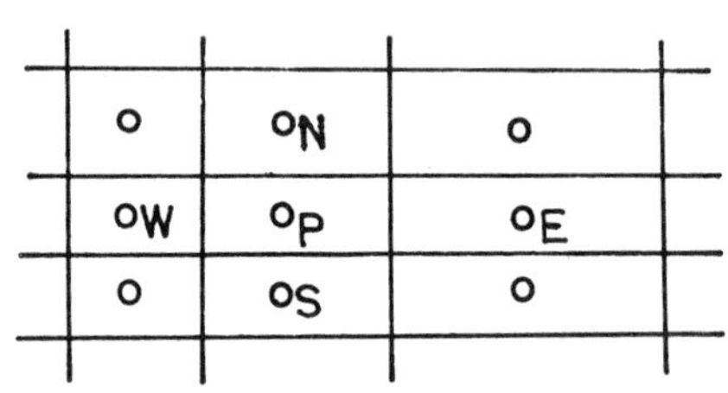

Figure 1. Section of the solution region.

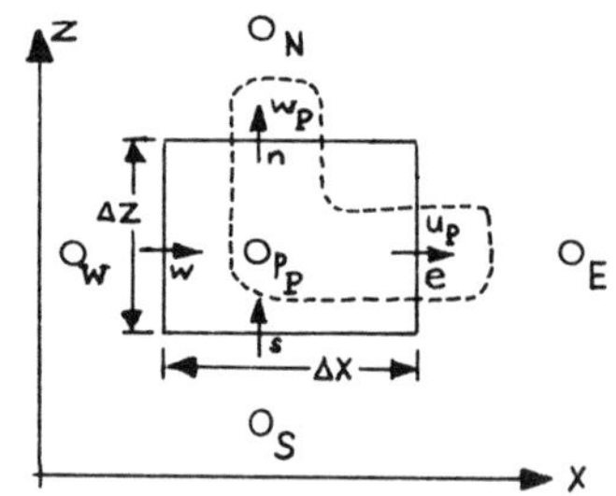

Figure 2. Typical storage configuration.

weighting scheme [2] and the E-factor formulation [1], Eqs. (1)-(3) can be discretized to give equations of the form

$$(u_e - u_w)\,\Delta Z + (w_n - w_s)\,\Delta X = 0, \tag{4}$$

$$A^u_P u_P = A^u_e u_E + A^u_w u_W + A^u_n u_N + A^u_s u_S + b^u_P + \Delta Z\,(p_P - p_E) \tag{5}$$

and

$$A^w_P w_P = A^w_e w_E + A^w_w w_W + A^w_n w_N + A^w_s w_S + b^w_P + \Delta X\,(p_P - p_N) \tag{6}$$

respectively, in which

$$A^u_P = (A^u_e + A^u_w + A^u_n + A^u_s)(1+E)/E, \tag{7}$$

$$A^w_P = (A^w_e + A^w_w + A^w_n + A^w_s)(1+E)/E, \tag{8}$$

$$b^u_P = \rho u^t_P / E + \dot{q}^u_P\,, \tag{9}$$

and

$$b^w_P = \rho w^t_P / E + \dot{q}^w_P\,, \tag{10}$$

where the superscript "t" denotes previous time level. Similar equations can be obtained for each grid point P in the solution domain.

3. SEGREGATED SOLUTION PROCEDURES

Various procedures have been proposed in the literature for the solution of Eqs. (4)-(6) [1,3,4]. All of these procedures work in a guess-and-correct fashion as follows:

a) Guess a preliminary pressure fielp p*.
b) Solve for u and w from Eqs. (5) and (6), respectively.
c) Correct this velocity field such that mass is conserved (i.e., Eq. (4) is satisfied).
d) Update the pressure field.
e) Repeat steps (b)-(d) until convergence.

Steps (a)-(b) are straightforward. Step (c) involves the utilization of a "pressure-correction" p' variable in order to be able to update the velocity field obtained in step (b) such that mass is conserved. For this purpose, the Patankar-Spalding-p' method [3] is utilized in this study. The details of this method are omitted here; however it should be noted that the method requires the solution of a "Poisson-like" equation and this, at times, may be expensive.

The pressure field can also be updated in step (d) by the solution of a "Poisson-like" equation for p (see, for example, PUP (Pressure Update by Patankar's scheme) [4] and PULS (Pressure Update by Least Squares) [1]). This step is the main focus of the present study.

The desire to be able to update the pressure field without the need of any "expensive" Poisson solving lead Raithby and Schneider [1] to the so-called PUMPIN method. The original method is described in the next Section in detail.

4. THE PUMPIN METHOD

Once the velocity field has been updated in step (c) so that mass is conserved, the main objective is to find that pressure field which, along with the corrected velocity field (hereafter called u^c and w^c), will "best" satisfy the two discrete momentum equations. At this point the old A^u and A^w coefficients can be used. However, experience of the authors have shown that updating the A^u and A^w coefficients using the u^c and w^c field prior to pressure-update promotes faster convergence. Therefore these coefficients were updated before pressure-update.

Inserting u^c and w^c velocities into Eqs. (5)-(6) and rearranging gives

$$p_P = p_E + F_P^u \tag{11}$$

and

$$p_P = p_N + F_P^w \tag{12}$$

where

$$F_P^u = [A_P^{u^c} u_P^c - (A_e^{u^c} u_E^c + A_w^{u^c} u_W^c + A_n^{u^c} u_N^c + A_s^{u^c} u_S^c + b_P^u)] / \Delta Z \tag{13}$$

and

$$F_P^w = [A_P^{w^c} w_P^c - (A_e^{w^c} w_E^c + A_w^{w^c} w_W^c + A_n^{w^c} w_N^c + A_s^{w^c} w_S^c + b_P^w)]/\ \Delta X \quad . \tag{14}$$

The idea here is to update the pressure at the point P once the pressures at the east and north sides of P are known, using some weighted average of the p values from Eqs. (11)-(12). For this purpose, a reference point must be chosen "at an appropriate location" in the solution domain where the pressure is "known". Usually, any value can be chosen for the pressure at the reference point (e.g., zero) since only the pressure gradients are of interest here.

The choice of the reference point was not discussed originally [1] but it will be illustrated later that this choice may play a crucial role in the stability and convergence of the method.

5. TEST PROBLEMS

Two test problems are considered. The first problem is described schematically in Fig. 3, where the boundary conditions and the general flow pattern are shown. The problem is two-dimensional and the flow is laminar. Water enters the square tank at one corner, deflected by the opposing corner and flows out of the diagonally opposite corner. A large recirculation zone is developed within the tank.

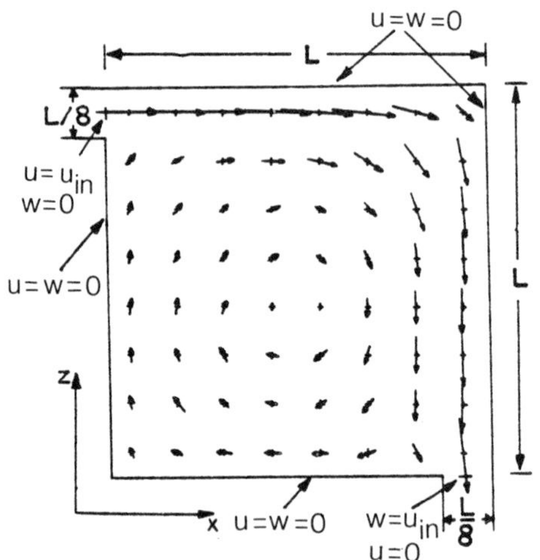

Figure 3. Geometry for the square tank problem. [L=0.04[m], u_{in}=0.1[m/s]].

Fig. 4 describes the second problem. In this case, the vertical length of the tank is four times the horizontal length. Relatively high velocities are imposed at the inlet and outlet, as shown.

Hereafter the first problem will be called "the square tank problem" and the second problem will be called "the rectangular tank problem".

A 10x10 mesh (i.e., 64 interior pressure points) was used for each problem. Very precise numerical solutions for both problems for this grid were obtained and stored for later reference.

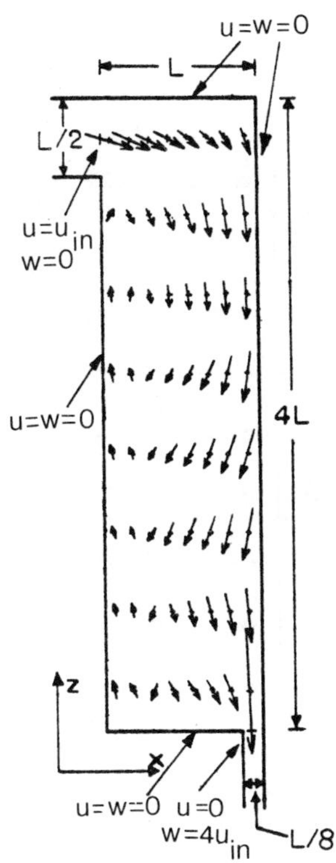

Figure 4. Geometry for the rectangular tank problem. [L=0.04[m], u_{in}=0.4[m/s]].

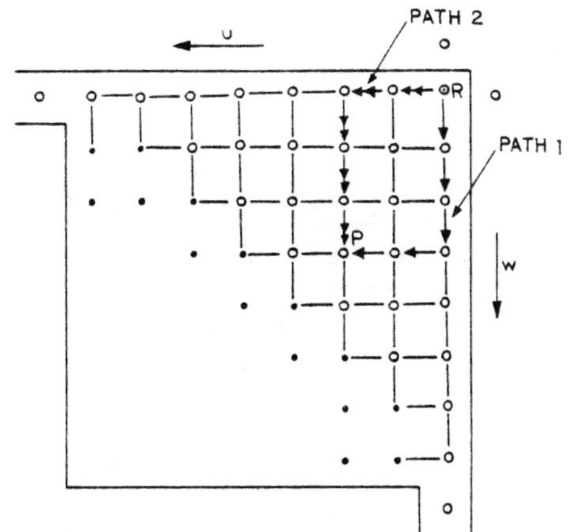

Figure 5. Grid for the square tank problem.

6. ORIGINAL APPLICATION OF THE PUMPIN METHOD

Fig. 5 shows the grid used for the solution of the square tank problem. Suppose that the pressure has been specified at the interior grid point denoted by R. Further suppose that the pressure at the point P is sought. The original PUMPIN method uses Eqs. (11)-(12) to step from grid point to grid point between R and P along the two paths shown. The pressure at P is then computed to be the average value obtained from the two paths. This procedure is repeated for each interior pressure point. The pressure field thus computed satisfies the momentum equations in some "average" sense.

7. THE NEW PUMPIN METHOD

It is desirable [1] to obtain an average p_P in Fig. 5 by integrating over <u>all</u> possible paths between R and P. A better approximation to this ideal than the two-path approximation used originally, will now be described. In this, the integration directions are restricted to those of the arrows shown in Fig. 6. The number shown beside each point in Fig. 6 indicates the number of all possible paths between R and that point.(I) Following these paths individually for each point is costly,

(I) These numbers form the so-called Tartaglia's Rectangle (Pascal's or the Binomial Triangle if a complete triangle).

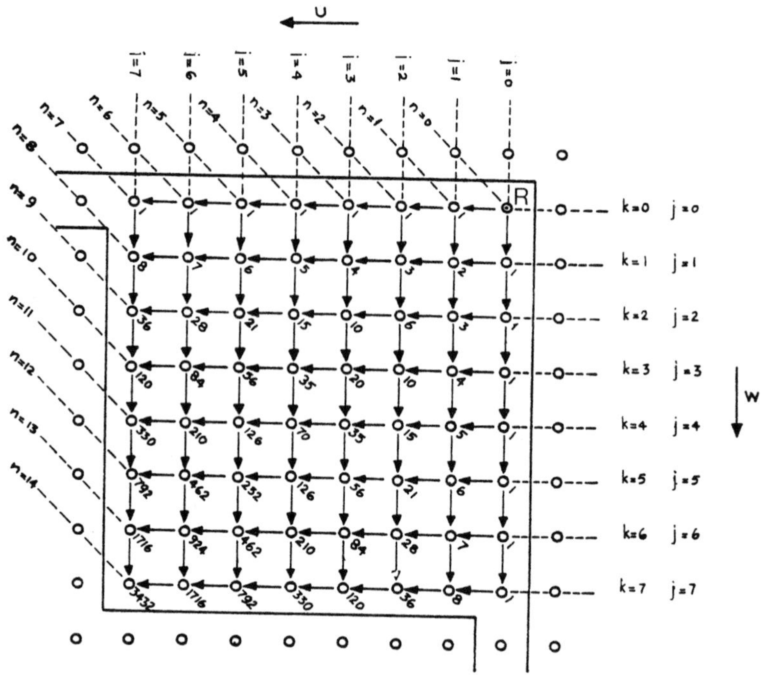

Figure 6. Path directions and number of paths.

even for the 10x10 grid considered here. The cost will be prohibitive for example, for a 30x30 or larger grid which is typical of most physical applications.

The technique developed here avoids the necessity of actually following all the possible paths. Consider a point (n,k) in Fig. 6. There are $\binom{n}{k}$ (I) possible paths to the point (n,k). However, to reach the point (n,k), the point (n-1,k-1) must be visited $\binom{n-1}{k-1}$ times and the point (n-1,k) $\binom{n-1}{k}$ times.(II) Weighting the pressure at the point (n,k) equally for each possible path results in

$$p(n,k)=\left\{\binom{n-1}{k}[p(n-1,k)+F^{u}_{n,k}] + \binom{n-1}{k-1}[p(n-1,k-1)+F^{w}_{n,k}]\right\}/\binom{n}{k} \quad (15)$$

On the other hand, it can be verified that

$$\binom{n-1}{k}/\binom{n}{k} = 1 - k/n \quad (16)$$

and

$$\binom{n-1}{k-1}/\binom{n}{k} = k/n \quad (17)$$

(I) $\binom{n}{k} = \frac{n!}{(n-k)!k!}$

(II) Note that $\binom{n-1}{k-1} + \binom{n-1}{k} = \binom{n}{k}$.

Therefore, Eq. (15) becomes

$$p(n,k) = (1 - \frac{k}{n})[p(n-1,k)+F^{u}_{n,k}] + \frac{k}{n}[p(n-1,k-1)+F^{w}_{n,k}] , \quad (18)$$

which holds for every point (n,k).

It can be seen from Eq. (18) that p(n,k) is just a simple (Binomial-weighted) average of the pressures that are calculated by applying Eqs. (11)-(12) between (n,k) and the two "upstream" pressure points.

Along the two lines of pressure points which have 1 written beside them in Fig. 6, the pressures can be determined immediately. The pressures at the remaining interior points can than be determined through Eq. (18).

Moreover, the formulation can be simplified further by considering the "rectangular" (i,j) notation as shown in Fig. 6. It can be seen from the Figure that

$$k = j \quad (19)$$

and

$$n = i + j, \quad (20)$$

so that

$$k/n = j/(i+j). \quad (21)$$

Hence Eq. (18) can be written as

$$p(i,j) = (1 - \omega_{ij})[p(i-1,j)+F^{u}_{i,j}] + \omega_{ij}[p(i,j-1)+F^{w}_{i,j}] , \quad (22)$$

where

$$\omega_{ij} = j/(i+j) . \quad (23)$$

8. STABILITY AND CONVERGENCE CHARACTERISTICS

The two problems described earlier were solved using both PUMPIN methods with various values of E. (It should be noted that the E-factor serves as an effective relaxation parameter for the overall iterations). Initially, the velocity and pressure fields were set to zero everywhere except the velocities at the inlet and at the outlet. A maximum of 20 iterations were performed in all cases. The quantity reported is

$$\epsilon_p = \max_{ij} [p_{ij} - p^{exact}_{ij}] / \frac{1}{2}\rho u^2_{in} \quad (24)$$

where the numerator is the maximum error in the pressure field for any given iteration and the denominator is the dynamic pressure head of the entering fluid. The reference point for pressure was chosen to be the point at the north-east corner.

Fig. 7 compares the OLD and NEW PUMPIN methods applied to the square tank problem. It can be seen that for this problem both methods converge for a small E value of 4. However, although the OLD PUMPIN method seems to converge a little faster, irregularities in the convergence pattern **can** be observed.

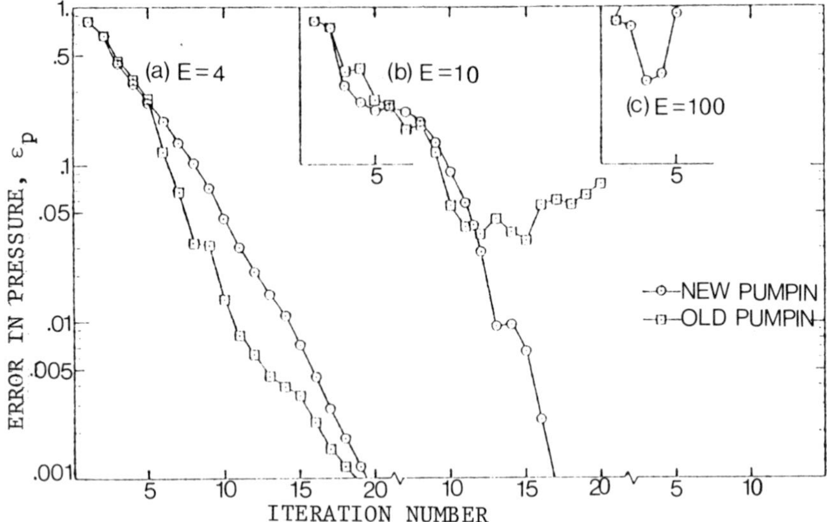

Figure 7. Results for the square tank problem.

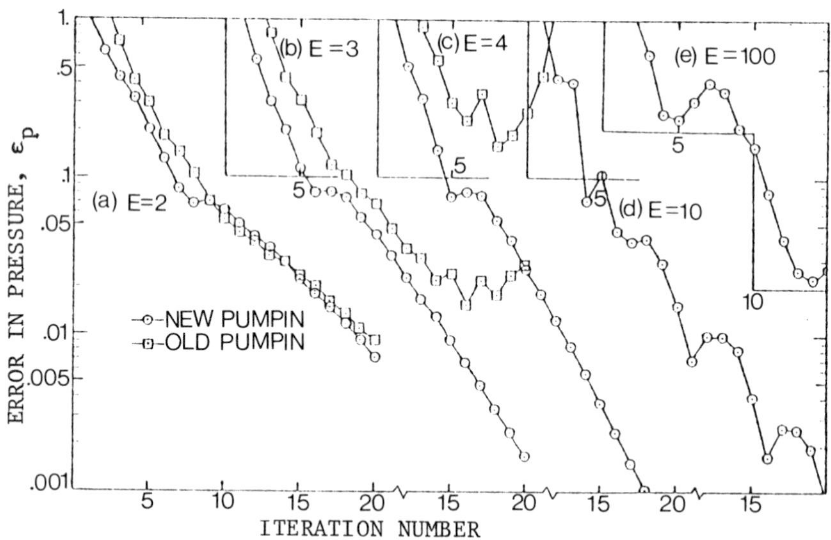

Figure 8. Results for the rectangular tank problem.

For a moderate E=10, the OLD PUMPIN method diverges, whereas the NEW PUMPIN method converges to ϵ_p=0.001 in 17 iteratios, although some irregularities are produced. For a "large" E=100, both methods diverge, but the divergence of the OLD PUMPIN method is faster.

Fig. 8 compares the performance of the OLD and NEW PUMPIN methods for the rectangular tank problem. For a "very small" E=2, both methods converge. The convergence is slow however, and an ε_p=0.01 could only be reached in 20 iterations. For E=3, convergence to ε_p=0.01 cannot be reached with the OLD PUMPIN method. The NEW PUMPIN method converges to ε_p=0.01 in 15 iterations and to ε_p=0.001 in about 22 iterations. For E=4 the OLD PUMPIN method diverges immediately while the NEW PUMPIN method is still stable and strongly convergent. Convergence to ε_p=0.001 can be reached in 18 iterations. No convergence could be reached for higher E values with the OLD PUMPIN method, whereas it can be seen from Fig. 8(d,e) that the NEW PUMPIN method can produce fast convergence with E=10, although oscillations may occur. For E=100, oscillations become more intense, but no divergence is encountered.

The main observation here is that the NEW PUMPIN method promotes faster convergence and has stronger stability characteristics over a considerably wider range of E values compared with the OLD PUMPIN method.

9. CHOICE OF THE REFERENCE POINT

At this point, the freedom of choice for the reference point, as mentioned earlier, lead the authors to a series of numerical experiments to determine whether the choice of some other reference point will make any difference in the convergence characteristics. First, reference pointswere studied at the remaining three corners. Tests were conducted for both problems, both methods and for various values of E. These tests, using OLD PUMPIN, showed that the best location was the north-east corner, but the other choices led to only marginally worse results. A marginal improvement was observed for the square tank problem with the NEW PUMPIN method when the reference point was placed at the south-east (outlet) corner. For the rectangular tank problem both methods performed better with the reference point at the north-east corner.

Secondly, the reference point was placed at the "center" of the solution domain. Fig. 9 shows results for the square tank problem when the NEW PUMPIN method is utilized. It can be seen that the oscillatory behaviour is suppressed for E=10 and the performance is better for E=100 when compared with other reference points. Figs. 10 and 11 show the performance of the OLD and NEW PUMPIN methods respectively, for the rectangular tank problem. Both methods performed surprisingly better for this problem. It can be seen from Fig. 10 that the OLD PUMPIN results were substantially improved. Even an E=100 produced very good convergence characteristics (ε_p=0.001 in 18 iterations) when compared with the cases of other reference points, in which no convergence was reached unless E was <u>very</u> small (i.e., E=2).

The results for the NEW PUMPIN method were more surprising. Fig. 11 reveals that very tight convergence can be reached with

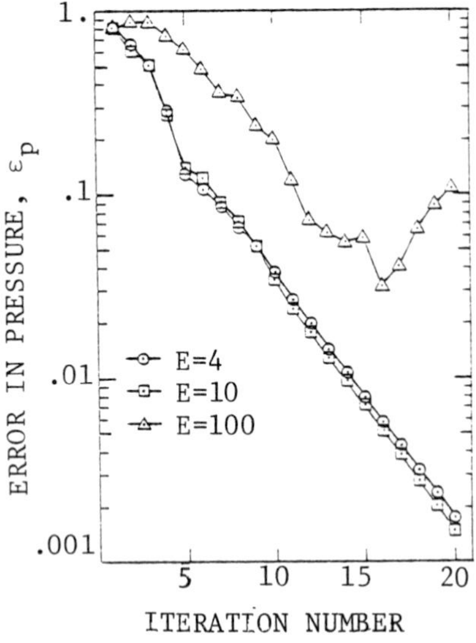

Figure 9. Results for the square tank problem (NEW PUMPIN, reference at center).

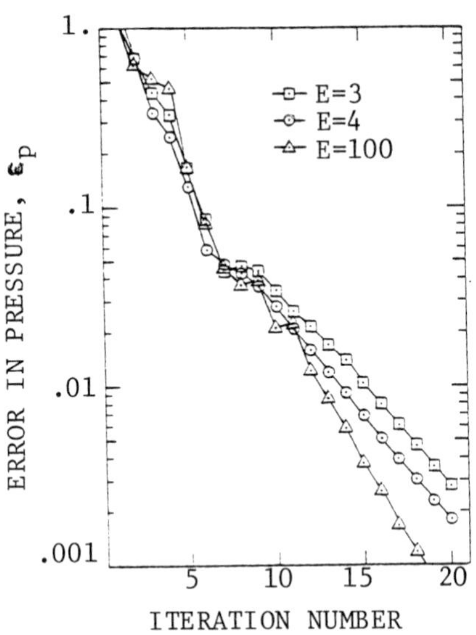

Figure 10. Results for the rectangular tank problem (OLD PUMPIN, reference at center)

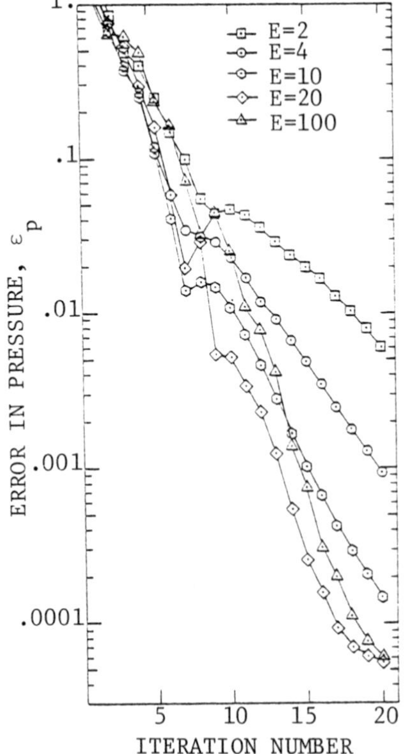

Figure 11. Results for the rectangular tank problem. (NEW PUMPIN, reference at center).

virtually any E value. With E=20 at least one order of magnitude difference can be observed (at the end of the 20th iteration) between the "best" solution obtainable with other choices of the reference point.

Before discussing this effect, it is observed that the square tank problem has one large, circular-shaped recirculation zone which fills the entire solution domain. Therefore small velocity and pressure changes are expected almost everywhere in the region, over two consecutive iterations. In the rectangular tank problem, however, the recirculation zone is slender and confined to the left half portion of the solution domain. In this problem, therefore, higher velocity and hence pressure changes are expected to occur in this half region.

By choosing the reference point at the center of the solution domain, the entire solution domain is effectively subdivided into four subregions where the pressure effects have been isolated by the separate application of the PUMPIN method to these four regions. (see Fig. 12). It should be further noted that recirculation is present in all four subregions for the square tank problem, whereas in the rectangular tank problem the recirculation is confined to the two left subregions. Therefore, the improvement achieved for the latter problem can be attributed to the isolation of the recirculation region from the rest of the flow.

10. GENERALIZATION OF THE NEW PUMPIN METHOD TO THREE DIMENSIONS

As it was mentioned before, it is most desirable that a method developed for two-dimensional problems be readily applicable to three-dimensional problems as well. The method developed here can in fact be easily generalized to three-dimensions. Using the notation shown in Fig. 13 the pressure is found from

$$p(i,j,k) = \frac{i}{i+j+k}[p(i-1,j,k)+F^u_{i,j,k}] + \frac{j}{i+j+k}[p(i,j-1,k)+F^v_{i,j,k}] + \frac{k}{i+j+k}[p(i,j,k-1)+F^w_{i,j,k}] \quad . \tag{25}$$

It should be noted that Eq. (25) reduces to Eq. (22) when terms corresponding to v are dropped (in which case i and j in Eq. (22) correspond to i and k in Eq. (25), respectively).

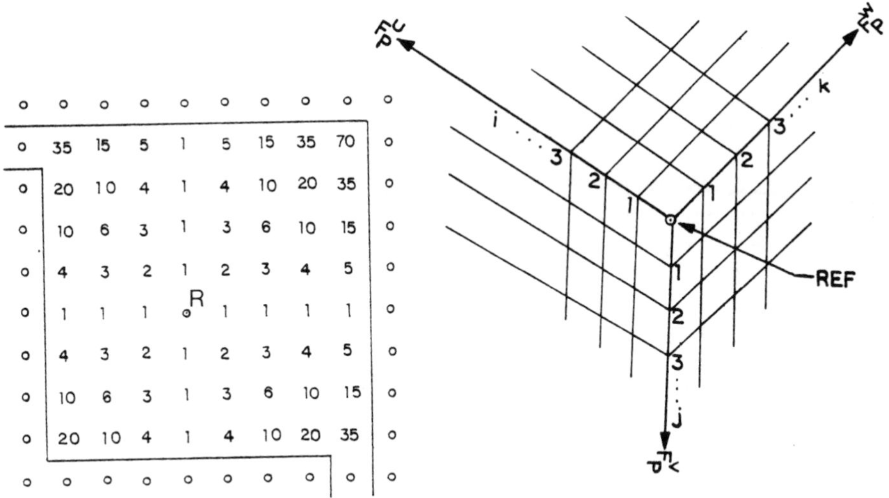

Figure 12. Application of NEW PUMPIN; reference at center.

Figure 13. Three-dimensional domain.

11. CONCLUDING REMARKS

A new method has been developed for updating the pressure field within the segregated solution of incompressible fluid flow problems through an efficient application of the PUMPIN method. This method is applied to two test problems and it is illustrated that it produces faster convergence and it is more stable than the original counterpart, although the cost remains the same. It is furthermore illustrated that considerable freedom is achieved with the new method for the choice of the important E-factor. A three-dimensional formulation is also presented for possible use in a three-dimensional flow problem.

The nature of the PUMPIN method necessitates the choice of a reference point. Although general guidelines for this choice could not be drawn for any physical situation, two important facts should be remembered: a) Choose a reference point which possibly isolates any recirculation region from the rest of the flow, and b) Locate the reference point at a minimal distance from the rest of the pressure points than any other reference point. It should also be noted here that more than one reference point can be chosen, if necessary, in which case only one of the reference points should be regarded as the "actual" reference point. At the moment, this issue remains as an art rather than a science however, and deserves further study.

REFERENCES

[1] RAITHBY,G.D., and SCHNEIDER, G.E., Numerical Solution of Incompressible Fluid Flow: Treatment of the Velocity-Pressure Coupling. Num. Heat Transfer, Vol. 2, pp. 417-440, 1979.

[2] RAITHBY, G.D., and TORRANCE, K·E·, Upstream-Weighted Differencing Schemes and their Application to Elliptic Problems Involving Fluid Flow. Computers and Fluids, Vol. 2, pp. 191-206, 1974.

[3] PATANKAR, S·V·, and SPALDING, D.B., A calculation Procedure for Heat, Mass and Momentum Transfer in Three-Dimensional Parabolic Flows. Int. J. Heat Mass Transfer, Vol. 15, pp. 1787-1806, 1972.

[4] PATANKAR, S.V., Numerical Heat Transfer and Fluid Flow, Hemisphere Publishing, 1980.

PREDICTION OF DEVELOPING TURBULENT FLOW BY THE FINITE ELEMENT METHOD

W. Slagter and H.A. Roodbergen

Netherlands Energy Research Foundation (ECN)
3, Westerduinweg, 1755 LE PETTEN, The Netherlands

ABSTRACT

The finite element method is applied to solve two-dimensional incompressible turbulent flow problems using the mixing length model of turbulence. The velocity components are approximated over each element by quadratic interpolation functions, while linear interpolation is used for the pressure. The solution of the resulting set of non linear algebraic equations is evaluated by employing a suitable interative technique. The applicability of the method is demonstrated by numerical examples of developing flow in a parallel plate duct in comparison with experimental data.

INTRODUCTION

The purpose of this paper is to present the finite element method as a viable numerical technique for analysing turbulent flow problems. Initially the method was applied to fully established flow conditions only, using several models of turbulence.
In most cases the technique has, in general, certain advantages over other numerical methods. These advantages, that include the ease with which irregular geometries and non-uniform meshes can be handled and the imposition of natural boundary conditions, have been demonstrated in fully developed turbulent flow analyses.

The present work deals with the application of the finite element method for developing turbulent flow in a parallel plate duct. This requires the solution of the time-averaged Navier-Stokes equations for steady state conditions.
The contribution of turbulence to the transport of momentum is introduced by the concept of eddy viscosity. The values of the eddy viscosity coëfficient are calculated according to Prandtl's mixing length theory.

Since primitive dependent variables are used and no special elements are applied to the solid boundaries, the imposition of natural or forced boundary conditions becomes particularly straightforward.
The solution of the resulting system of non linear equations is evaluated by employing a suitable iterative technique.
The results are compared with those obtained experimentally by means of the Laser Doppler Anemometry. These measurements are carried out in a duct with rectangular cross section using water as fluid.

2. THEORETICAL FORMULATION

2.1. Governing equations

The Navier-Stokes equations for a steady state two-dimensional flow of an incompressible fluid, ignoring body forces, can be written in the form

$$\rho u_j \frac{\partial u_i}{\partial x_j} = \frac{\partial}{\partial x_j} (\sigma_{ij}) \qquad i,j = 1,2 \qquad (1)$$

together with the continuity equation,

$$\frac{\partial u_j}{\partial x_j} = 0 \qquad j = 1,2 \qquad (2)$$

Here ρ is the fluid density, u_i and σ_{ij} are the components of the velocity vector and the stress tensor, respectively. Equations (1) and (2) are written in Cartesian tensor notation with summation implied over repeated subscripts. For turbulent flows of a Newtonian viscous fluid the stress tensor is given by

$$\sigma_{ij} = -p\delta_{ij} + \tau_{ij} - \rho\overline{u_i'u_j'} \qquad (3)$$

where p is the averaged pressure, δ_{ij} is the Kronecker delta, and the Stokes stress tensor τ_{ij} is defined as

$$\tau_{ij} = \rho\nu \left(\frac{\partial u_i}{\partial x_j} + \frac{\partial u_j}{\partial x_i}\right) \qquad (4)$$

where ν is the actual kinematic fluid viscosity.
Introducing the concept of eddy viscosity the Reynolds stresses $\overline{u_i'u_j'}$ can be written as

$$-\overline{\rho u_i'u_j'} = \mu_t \left(\frac{\partial u_i}{\partial x_j} + \frac{\partial u_j}{\partial x_i}\right) \qquad (5)$$

where μ_t is the so-called turbulent dynamic viscosity.

Substitution of the defining equations (3), (4) and (5) into Eq. (1) yields the conservation equation

$$\rho u_j \frac{\partial u_i}{\partial x_j} = - \frac{\partial p}{\partial x_i} + \frac{\partial}{\partial x_j} \left\{ (\mu+\mu_t) \left(\frac{\partial u_i}{\partial x_j} + \frac{\partial u_j}{\partial x_i} \right) \right\} \tag{6}$$

By a suitable manipulation of the above equation it can be shown [1] that Eq. (6) can be rewritten as

$$\rho u_j \frac{\partial u_i}{\partial x_j} = - \frac{\partial p}{\partial x_i} + \frac{\partial}{\partial x_j} \left(\mu_e \frac{\partial u_i}{\partial x_j} \right) + \frac{\partial u_j}{\partial x_i} \frac{\partial \mu_t}{\partial x_j} \tag{7}$$

where the effective viscosity is related to the fluid viscosity μ via the relationship

$$\mu_e = \mu + \mu_t \tag{8}$$

Equation (7), together with the continuity Eq. (2), describes the turbulent flow of an incompressible fluid using the eddy viscosity concept.
To solve the equations (2) and (7) over a bounded domain Ω with boundary Γ, two types of boundary conditions are considered.

$$u_i = u_i^o \text{ on } \Gamma_o \tag{9}$$

where u^o is a prescribed velocity on the part Γ_o of the boundary Γ. The second type of condition considered here is that of prescribed velocity gradients

$$\mu_e \frac{\partial u_i}{\partial x_j} n_j = \alpha_i \text{ on } \Gamma_\alpha = \Gamma - \Gamma_o \tag{10}$$

where n_j are the components of the unit outward normal to the boundary. In this work we will only consider the case when conditions of the type (10) are homogeneous, i.e. $\alpha_i = 0$.

2.2. Mixing length viscosity model

Following the Prandtl mixing length hypothesis [2] which indicates that the turbulent dynamic viscosity can be written in the form

$$\mu_t = \rho \ell^2 \, |J| \tag{11}$$

where

$$J^2 = \tfrac{1}{2} \left(\frac{\partial u_k}{\partial x_j} + \frac{\partial u_j}{\partial x_k} \right) \left(\frac{\partial u_k}{\partial x_j} + \frac{\partial u_j}{\partial x_k} \right)$$

and ℓ, called the mixing length, is a function of position which has to be specified.

For the problem under consideration i.e., developing flow in a parallel plate channel, the mean flow velocities transverse to the main direction of flow are relatively small and consequently the mixing length model of Eq. (11) may be reduced to

$$\mu_t = \rho \ell^2 \left|\frac{\partial u}{\partial y}\right| \tag{12}$$

Here u is the velocity in the axial stream direction and y denotes the coordinate perpendicular to the wall. The transverse velocity is indicated by v, while x is the axial coordinate.
The following empirical relationship [3] is employed for ℓ

$$\ell = y_o \left\{0.14 - 0.08\,(1 - y/y_o)^2 - 0.06\,(1 - y/y_o)^4\right\} \tag{13}$$

where y_o is half the channel width.

Since the solution domain considered includes the flow in the immediate vicinity of the wall, the mixing length has to be corrected for the viscous sublayer and transition layer. For this reason the expression of the mixing length Eq.(13) is multiplied by Van Driest's [4] formula.

$$(1 - \exp\,(-y^+/C)) \tag{14}$$

where C is a constant and y^+ is the dimensionless wall distance defined by

$$y^+ = \frac{y}{\nu}\sqrt{\frac{\tau_s}{\rho}} \tag{15}$$

where τ_s is the local wall shear stress.
Other quantities of interest for turbulent flows are the local shear stress velocity

$$u^* = \sqrt{\frac{\tau_s}{\rho}} \tag{16}$$

and the dimensionless velocity

$$u^+ = u/u^* \tag{17}$$

3. FINITE ELEMENT FORMULATION

3.1. Finite element method

The velocity-pressure (primitive variable) formulation is adopted for solving the problem under consideration. The flow domain to be analysed is discretised by the use of triangular elements and the mixed interpolation technique devised by Hood and Taylor [5] is adopted.
This means that flow region, Ω, is divided into triangular

elements with quadratic interpolation function for the velocity u_i and linear variations in pressure p. The element interpolation functions are expressed by

$$u_i = \langle M\rangle \{u_i\} \text{ and } p = \langle N\rangle \{p\} \tag{18}$$

where $\langle M\rangle$ and $\langle N\rangle$ are the row vectors of the interpolation functions, while $\{u_i\}$, $\{p\}$ are column vectors of the velocity and pressure at nodal points.

Since the finite element method acts on an integral formulation, the method of weighted residuals is applied to reformulate the governing differential equations.

Applying the Galerkin principle to select the weighting functions, the momentum equation (7) is weighted by the velocity interpolation function $\langle M\rangle$ and the continuity equation (2) is weighted by the pressure interpolation function $\langle N\rangle$. The resulting system of finite element equations can be written as

$$[A]\{\beta\} = \{B\} \tag{19}$$

where the non linear terms are incorporated in the matrix [A] and the column vector $\{\beta\}$ is made up of the column vectors of unknown nodal values

$$\{\beta\} = \begin{Bmatrix} u \\ v \\ p \end{Bmatrix} \tag{20}$$

The right-hand side vector {B} contains the prescribed velocity gradients at the surface resulting from the application of Green's theorem to the integrals which contain second order derivatives.

3.2. Solution procedure

Since the equations to be solved are non linear, their solution requires necessarily some form of iteration. The successive-substitution scheme as well as the Newton-Raphson method are employed to cope with the non linearity of Eq. (19). For the current problem it has been found that the iteration process can be started from zero initial values of u, v and p. The resulting "laminar solution" is then used to evaluate the turbulent viscosity, wall shear stress and a basis for the next iteration is available. In each iterative stage, the linearized set of equations is solved by the frontal solution technique [6]. The iteration process is repeated until the relative difference between two successive iterations becomes less than a specified limit, ε, in the Euclidean norm (L_2) of the solution vector.

4. COMPUTED RESULTS AND DISCUSSION

4.1. Turbulent entrance region

The solution technique outlined in the previous sections has been applied to solve the problem of developing turbulent flow in the entrance region of a parallel plate duct. For this problem the following boundary conditions apply.

Upstream

$$\left.\begin{array}{l} u = \overline{u} \\ v = 0 \end{array}\right\} \quad \begin{array}{l} \text{on } x = 0 \\ \text{for } 0 < y \leq y_o \end{array}$$

where u is the mean velocity of the flow and y_o is half the distance between the plates.

Downstream

$$\left.\begin{array}{l} \frac{\partial u}{\partial x} = 0 \\ v = 0 \end{array}\right\} \quad \begin{array}{l} \text{on } x = 140\ y_o \\ \text{for } 0 < y \leq y_o \end{array}$$

At the wall

$$\left.\begin{array}{l} u = 0 \\ v = 0 \end{array}\right\} \quad \begin{array}{l} \text{on } y = 0 \\ \text{for } x \geq 0 \end{array}$$

At the symmetry line

$$\left.\begin{array}{l} \frac{\partial u}{\partial y} = 0 \\ v = 0 \end{array}\right\} \quad \begin{array}{l} \text{on } y = y_o \\ \text{for } x \geq 0 \end{array}$$

The downstream section has been placed at $x = 140\ y_o$ since the flow considered can be assumed to be fully developed for this entrance length.
The pressure is defined at the downstream boundary. The Reynolds number is defined as

$$Re = \frac{4 y_o \overline{u}}{\nu} \tag{21}$$

where $\overline{u}$ is the upstream mean velocity.

The computations are carried out for half a channel width of $y_o = 0.0125$ m with water as fluid. The fluid properties are assumed to be

$$\mu = 10^{-3}\ \text{kg/msec and } \rho = 10^{3}\ \text{kg/m}^3$$

The upstream mean velocity is chosen to be $\overline{u} = 0.20$ m/sec, corresponding to a Reynolds number $Re = 10^4$.

The finite element mesh used for this analysis is shown in fig. 1 and contains 294 triangular elements with the solution domain extended right up to the wall.
To illustrate the mesh and the computed results in a proper way, both coordinate axes are plotted on a logarithmic scale. The first mesh line next to the wall is positioned at y^+ of about 2.

4.2. Results and comparison

Converged results were produced by the finite element method using the mesh distribution of fig. 1. A plot of the numerical results showing the velocity vectors is given in fig. 2. The velocity components are normalized with the maximum value in their corresponding direction.
Figure 2 indicates that the fluid moves from the wall region into the core of the flow. The relative axial velocities versus the axial coordinate are shown in fig. 3 for some relevant transverse positions y/y_o.
A comparison is made between present numerical results and the experimental results obtained by Vonka [7].

A further comparison between numerical results of the axial velocity distribution in the downstream section and the universal velocity profiles is shown in fig. 4.
The agreement is striking and consequently the computed velocity profile is at the fully developed state. The resulting solution of the velocities permits an accurate evaluation of the wall shear stress τ_s as shown in fig. 5.

To achieve a relative accuracy in the Euclidean norm of $\varepsilon < 10^{-6}$ from either of the two iteration techniques, the Newton-Raphson method requires eight iterations, while fourteen are necessary for the successive substitution scheme. The latter procedure only uses a relaxation technique with a relaxation factor of 0.5. The computations are carried out on a CDC Cyber-175 requiring 200 CP and 50 IO seconds for the Newton-Raphson method, while the successive-substitution scheme consumes 335 CP and 80 IO seconds.

4.3. Conclusions

The present paper has demonstrated that the finite element method, using a simple turbulent viscosity model, can be successfully employed in the analysis of developing turbulent flows including the boundary layer right up to the wall.
The present results are encouraging and the method will be applied to compute the turbulent flow distribution in a parallel plate channel with a flow obstruction at the centre of the channel.

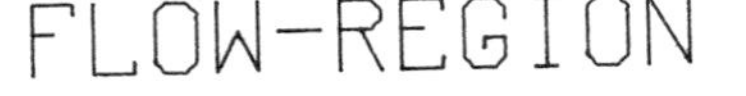

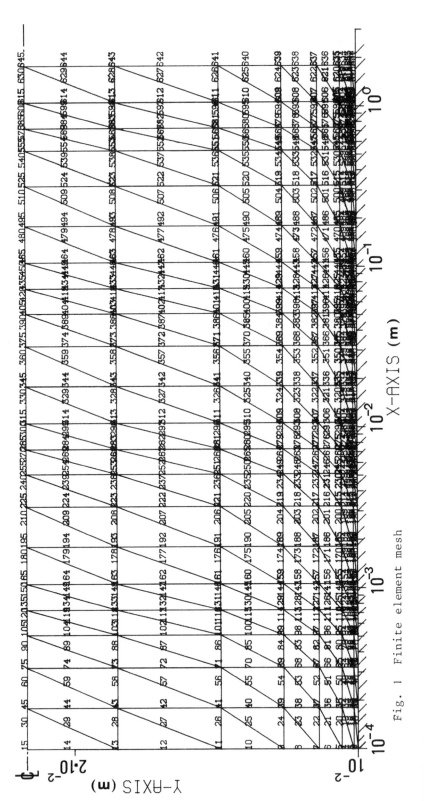

Fig. 1 Finite element mesh

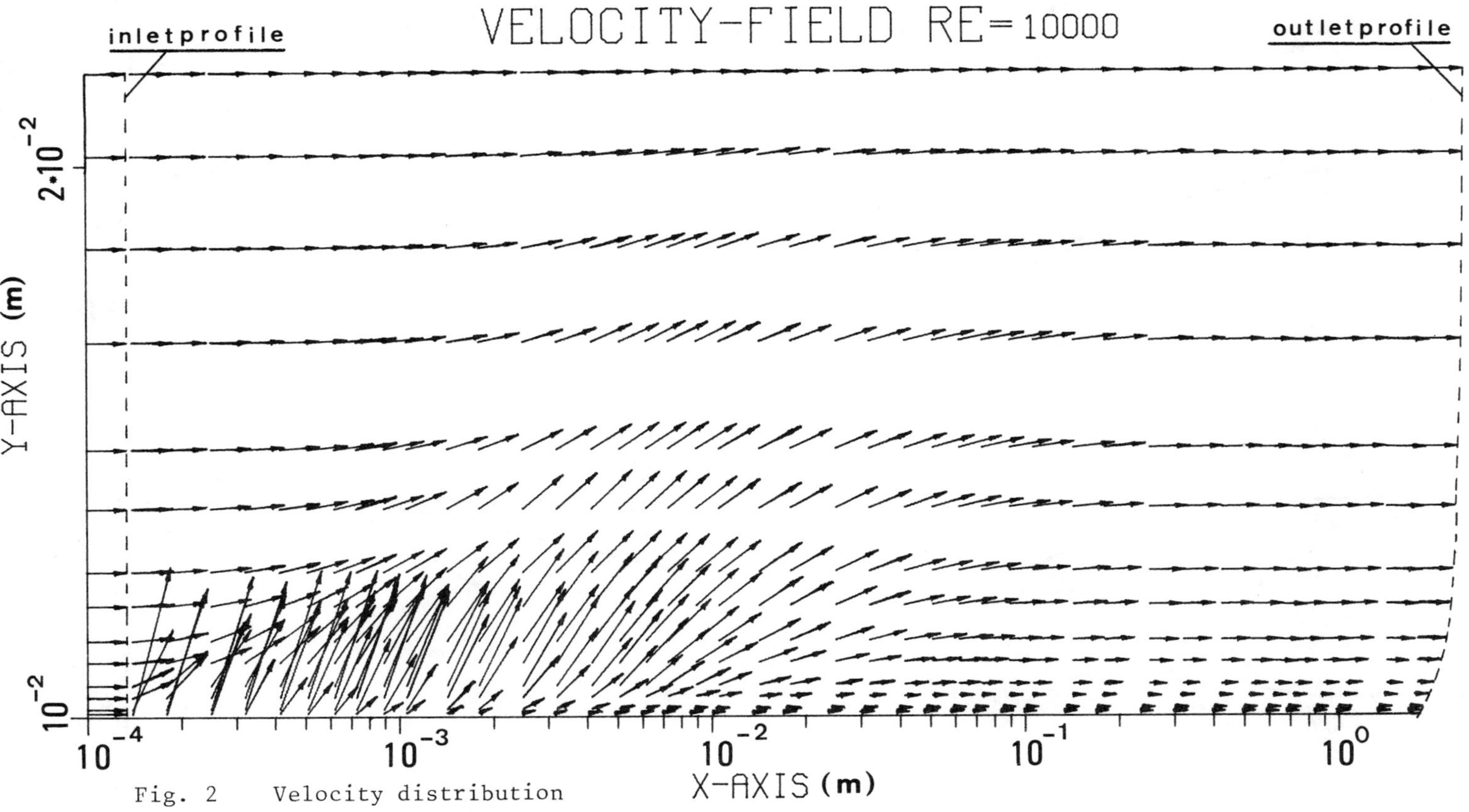

Fig. 2 Velocity distribution

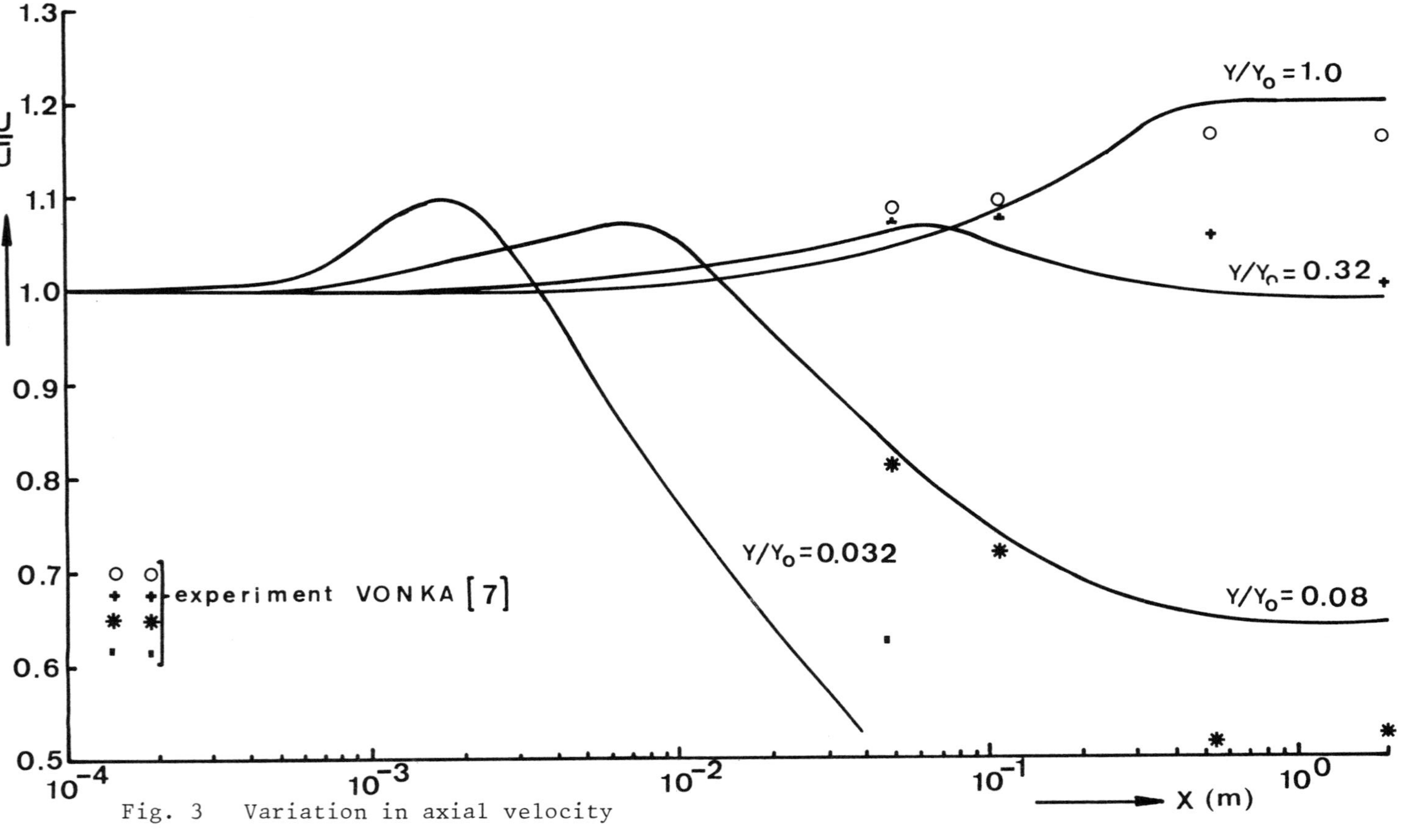

Fig. 3 Variation in axial velocity

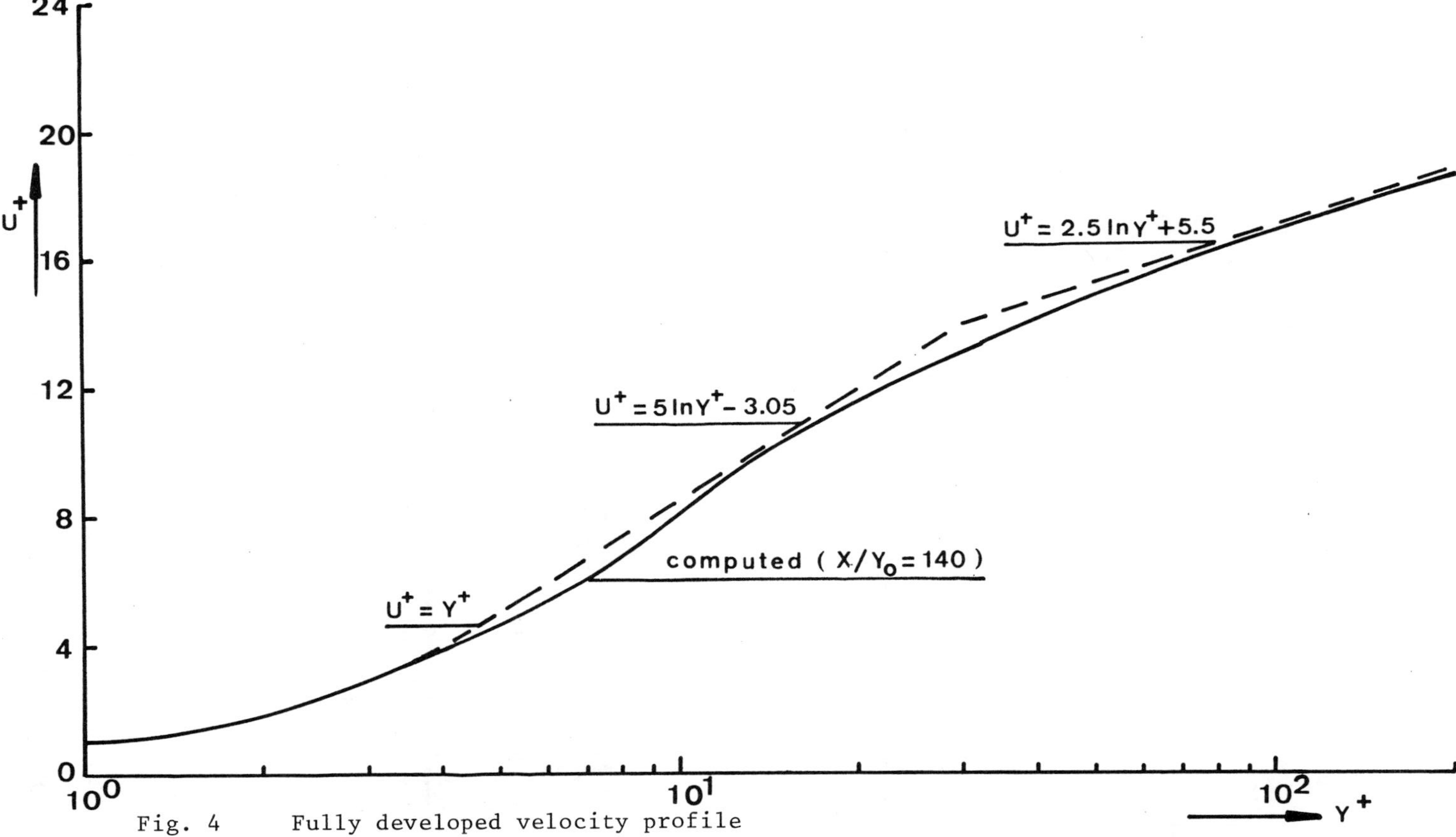

Fig. 4 Fully developed velocity profile

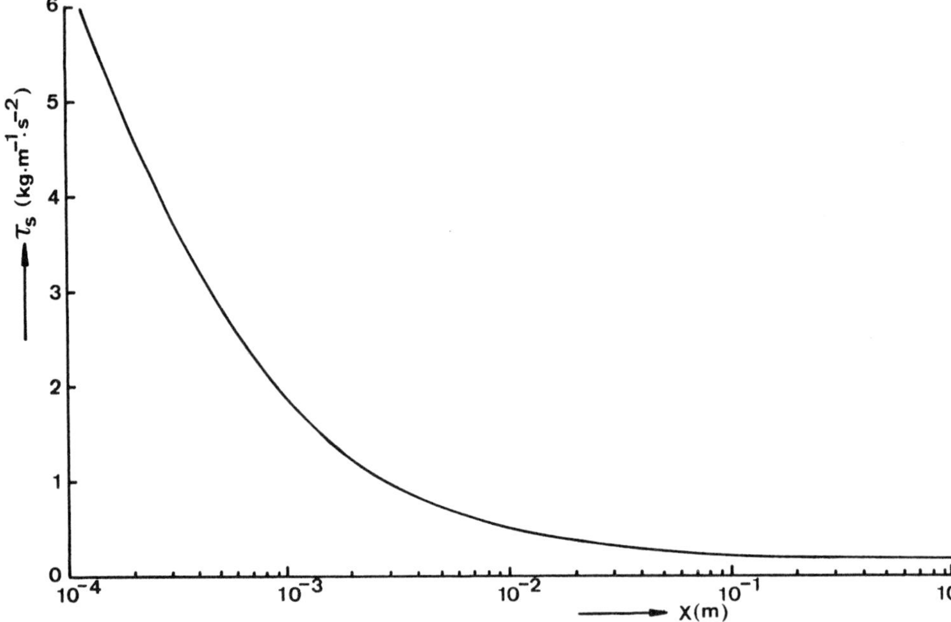

Fig. 5 Wall shear stress distribution

REFERENCES

[1] HINZE, J.O. - Turbulence, an introduction to its mechanism and theory, McGraw-Hill, New York, 1975.

[2] PRANDTL, L. - Bericht über Untersuchungen zur Ausgebildeten Turbulenz, ZAMM, 5, 136, 1925.

[3] NIKURADSE, J. - Gesetzmässigkeit der turbulenten Strömung in glatten Rohren. Forsch. Arb. Ing. Wes., Heft 356, 1932.

[4] Van DRIEST, E.R. - On Turbulent Flow near a Wall, J. Aero. Sci., vol. 23, 1007, 1956.

[5] HOOD, P. and TAYLOR, C. - Navier-Stokes equations using mixed interpolation, Proc. Int. Symposium on Finite Elements in Flow Problems, Ed. Oden, J.T. et al, UAH Press, 1974.

[6] HOOD, P. - Frontal Solution Program for Unsymmetric Matrices, Int. J. Num. Meth. Eng. 10, 379, 1976.

[7] VONKA, V. - Private communication.

THE NUMERICAL SIMULATION OF FLOW AROUND SQUARES

R. W. Davis and E. F. Moore(I)

This paper presents a numerical simulation of two-dimensional unsteady flow around squares in infinite domains with uniform upstream velocity profiles. The Reynolds number regime investigated is from 100 to 2800, with computed Strouhal numbers in this range being compared with those obtained from a wind tunnel test. Variations in the behavior of lift and drag with Reynolds number are discussed. Passive marker particles are used to visualize the onset and subsequent development of vortex shedding at a Reynolds number of 1000. The finite difference scheme employed in this simulation utilizes third-order accurate upwind differencing for convection and a Leith-type of temporal differencing. Variations in convective differencing near the corners of the square and at the outflow boundary of the mesh are described.

1. INTRODUCTION

The problems associated with bluff-body flows have recently become of great concern because of the significance of these flows to the energy conservation effort [1,2,3]. Drag reduction of road vehicles and reduction of convective heat loss from buildings are examples of areas in which bluff-body flows are important. In spite of their importance, however, these flows are rather poorly understood. Most of the research effort to date has been concentrated on flow around simple circular or rectangular cylinders, with the former receiving most of the attention. Some experimental studies of flow around rectangular cylinders have been conducted in high Reynolds number turbulent flow (e.g., [4]). An early two-dimensional computer simulation of flow around rectangles at low Reynolds

(I) Fluid Engineering Division, FM 105, National Bureau of Standards, Washington, D. C. 20234, U. S. A.

numbers was flawed by the use of central differencing for convection at large cell Reynolds numbers [5]. At that time (1963) the resulting spatial oscillations ahead of the rectangle were not understood. Since then, these spatial oscillations have been the subject of numerous studies [6]. First-order upwind differencing has been the most common way of eliminating them. Unfortunately the dependence of a flow on Reynolds number may well be masked by the large numerical diffusion associated with this type of differencing.

In the present paper, two-dimensional unsteady flow around a square in an infinite domain is studied numerically. The undisturbed upstream velocity profile is uniform, and Reynolds numbers range from 100 to 2800. Drag and Strouhal number are calculated as functions of Reynolds number, with the predicted Strouhal number behavior comparing well with the results of a wind tunnel test for Reynolds numbers below 1000. The use of variously shaped passive marker particles provides exceptional visualization of the vortices both as they form and as they travel downstream from the square.

The simulation described here was initially attempted with a weighted mixture of central and upwind differencing for convective modeling. Changes in weighting, however, produced large changes in Strouhal number, probably due to variations in numerical diffusion. Also, heavy weighting of central differencing produced spatial oscillations ahead of the square. The numerical scheme finally selected for this study is based upon the two-dimensional steady QUICK and one-dimensional unsteady QUICKEST methods which have been recently proposed [7,8]. This numerical scheme will be described in the next section.

2. NUMERICAL METHOD

The two-dimensional Navier-Stokes and continuity equations for an incompressible fluid are

$$\frac{\partial \underset{\sim}{q}}{\partial t} + (\underset{\sim}{q} \cdot \nabla)\underset{\sim}{q} = - \nabla p + \nu \nabla^2 \underset{\sim}{q} \tag{1}$$

$$\nabla \cdot \underset{\sim}{q} = 0 \ , \tag{2}$$

where $\underset{\sim}{q} = (u,v)$ is the velocity vector in a cartesian reference frame, p is the ratio of pressure to constant density, ν is kinematic viscosity, and t is time. A solution of these equations is required for flow around a square in an infinite domain. Reynolds number is defined as $Re = Ub/\nu$, where U is the constant approach velocity, and b is the length of the square's side. Henceforth, all lengths are nondimensionalized with respect to U, time with respect to b/U, and p with respect to U^2. Strouhal number, S, is simply the nondimensional shedding frequency $fb/U = f$.

Equations (1) and (2) are finite differenced and solved on a variably-spaced staggered mesh in which pressures are defined at cell centers and normal velocities at the center of cell faces. The finite differencing of the convective terms in Eq. (1) is performed using third-order upwind differencing in the manner of the two-dimensional steady QUICK scheme [7]. This method performs interpolations by means of quadratic fits through upstream weighted groups of grid points. In any given coordinate direction, it is generally a five-point scheme. The temporal differencing of Eq. (1) is based upon the explicit Leith-type of differencing utilized by the one-dimensional unsteady QUICKEST method [8]. This type of scheme can be derived by expanding the unknown in a Taylor series in time and then, using the original partial differential equation, replacing time derivatives with space derivatives. Convective differencing is once again third-order accurate and the Courant number, $u\Delta t/\Delta x$, must be less than unity. Finally, the QUICKEST scheme has been shown to perform well in the modeling of moving free shear layers at Reynolds numbers up to a few thousand [9].

Following the solution of Eq. (1), the solution of the continuity equation, Eq. (2), is accomplished by adjusting the pressure and velocity fields interatively by successive overrelaxation [10]. This adjustment process proceeds until the sum of the absolute values of the mass residuals over all mesh cells is less than 0.2 percent of the mass flow entering the computational domain.

A uniform flow profile is specified at the inlet to the computational domain a distance of 4.5 upstream of the front face of the square. The freestream velocity is specified at distances of 6 above and below the centerline of the square. The distance from the rear face of the square to the exit from the computational domain is either 9.5 or 14.5 depending on the particular mesh being employed. In order to avoid the possibility of upstream error propagation, it is important to allow the unsteady wake to leave the mesh with minimal interference. Unfortunately, nothing is known a priori about the exit flow except that fluid leaves the mesh everywhere. Therefore, first-order upwind differencing is used at the exit with the diffusion terms in Eq. (1) being ignored and the pressure gradient being assumed constant. Adjustments are made to ensure that the mass flow leaving the computational domain is equal to that entering. Around the square itself, no-slip boundary conditions are employed. Quadratic extrapolations are used to obtain any data required by the numerical scheme at grid points interior to the square. Because of the singularities at the corners of the square, the corner regions must be treated specially. This is accomplished at the two front corners by using two-dimensional linear interpolations among the nearest nonzero nodal values of the unknown velocities in order to obtain the convective fluxing across the cell half-faces nearest each corner. At the rear corners, only minor variations of the basic

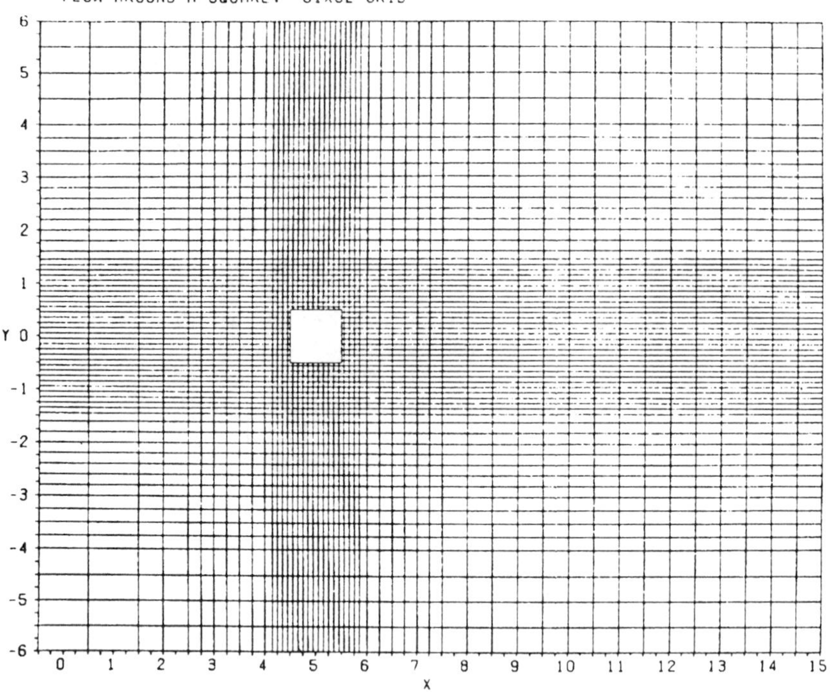

Fig. 1. Nonuniform Mesh

numerical scheme are required. Use of these procedures at the corners results in slightly better comparisons of the computed Strouhal numbers with those obtained in the experiment to be described later.

Three nonuniform computational meshes were employed in this study: 41 x 40, 51 x 62, and 61 x 74, with the second one of these illustrated in Fig. 1. Initial conditions for a computation were either a uniform flow everywhere or the results of a previous calculation, often at a different Reynolds number. The time step was 0.05 for the first two meshes and 0.025 for the finest mesh, which was employed sparingly. Computation times on the NBS UNIVAC 1108 ranged up to 24 hours when starting from a uniform flow with the 61 x 74 mesh. Generally, computations took 6 to 9 hours to obtain constant Strouhal number vortex shedding when starting from the end of a previous calculation at a different Reynolds number. No perturbations were required to initiate vortex shedding.

3. NUMERICAL RESULTS

Strouhal numbers and lift and drag coefficients have been obtained for the flow around a square utilizing the numerical method discussed previously. The calculated variations of

Strouhal number with Reynolds number for the three meshes are compared with the results of a wind tunnel test in Fig. 2. The test data was obtained from a hot-wire placed behind a 3.175 mm square steel rod spanning a 0.94 m wide low velocity wind tunnel with freestream turbulence intensity less than 0.05 percent. The differences between the two test runs give a crude indication of experimental scatter. As seen in Fig. 2, the computed and experimental results appear to be in reasonable agreement for Re $\lesssim$ 1000.

The computed values of average drag coefficient for the three meshes are plotted against Reynolds number in Fig. 3. The differences among the three grids are within seven percent for Re < 1000. The computed average lift coefficients were $O(10^{-2})$ or less.

Figure 4 illustrates the initiation and subsequent development of the vortex-shedding phenomenon at Re = 1000 using the 51 x 62 grid starting from a uniform flow everywhere. Flow visualization is accomplished by the injection of variously-shaped passive marker particles ahead of the square. A given shape of particle is injected at every other time step along its particular streakline. At the end of every time step, a particle is moved a distance $|q|\ \Delta t$, where q is the average velocity of the particle over the time step Δt. The particles become entrained in the vortices behind the body and are shed with them. Thus the evolution of these large coherent structures as they break away from the square and move downstream is clearly visualized. The sequence in Fig. 4 shows the bottom recirculation zone moving back and rotating counterclockwise. As it does so, it separates the top recirculation zone into two parts. Because of the large amount of fluid in the recirculation zones, the initial vortices are very large. Finally, the vortex shedding settles down into a pattern illustrated by Figs. 4e and 4f. The vortex patterns shown here repeat themselves every two shedding cycles, i.e., with a frequency S/2. This is consistent with the plot of lift coefficient versus time in Fig. 5a, which shows the presence of a subharmonic in the lift oscillation. The drag coefficient in Fig. 5b is even more irregular. The irregularities in the lift and drag fluctuations appear to be a function of Reynolds number, as both these fluctuating quantities are simple sine waves at Re = 250. In fact, at Re = 250, the vortices themselves are very regularly shaped, as seen in Fig. 6. Once steady-state vortex shedding has been reached at this Reynolds number, the basic pattern in Fig. 6 remains unchanging. Finally, Fig. 7 shows the results of a computation at Re = 1000 with an extended downstream region. The variety of vortices seen here is striking and is an indication of the applicability of marker particles to the study of large coherent structures.

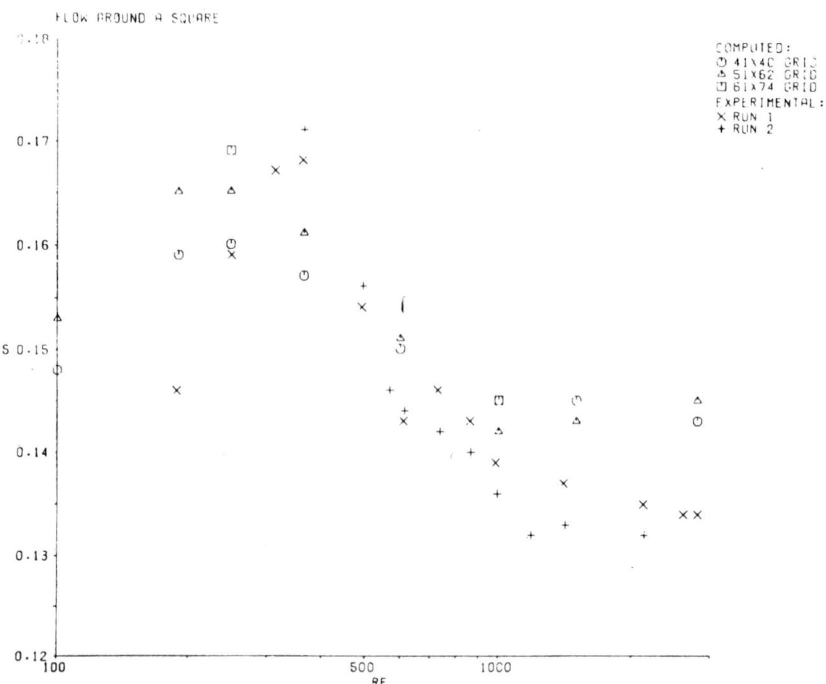

Fig. 2. Numerical-Experimental Strouhal Number Comparison

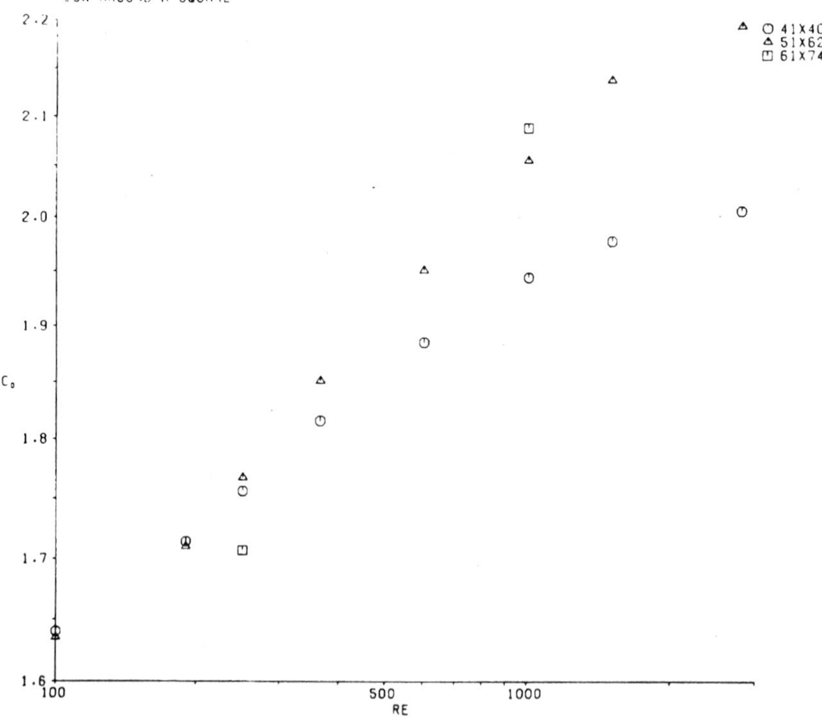

Fig. 3. Computed Average Drag Coefficients (Log-Log Plot)

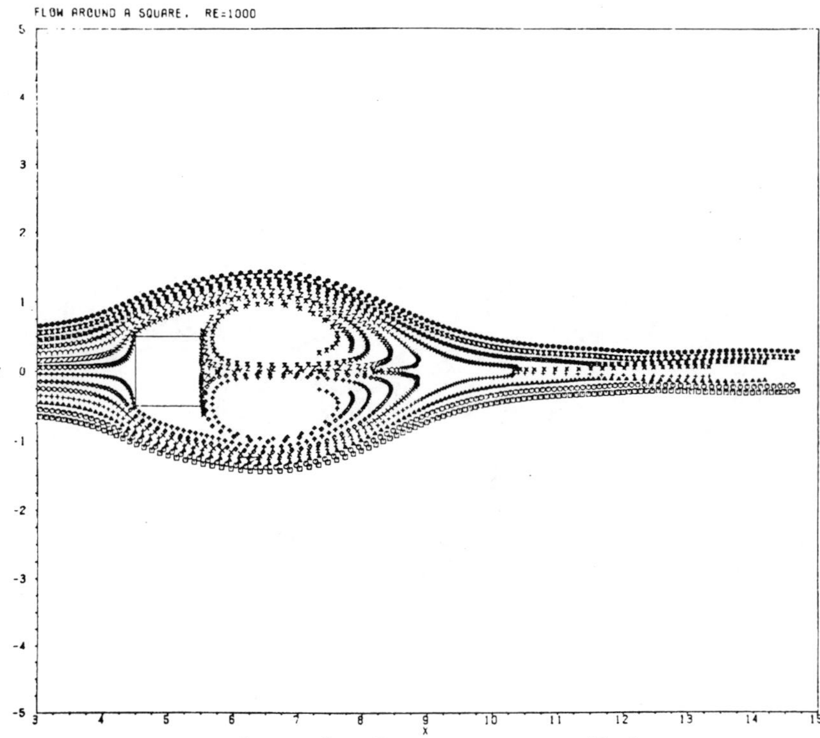

Fig. 4a. Flow Pattern at t = 11.5

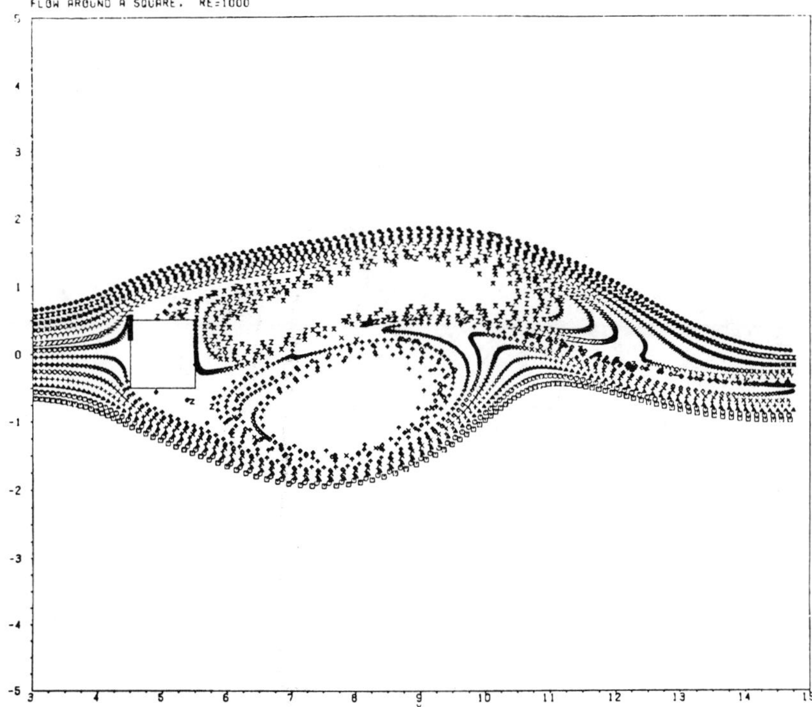

Fig. 4b. Flow Pattern at t = 34.5

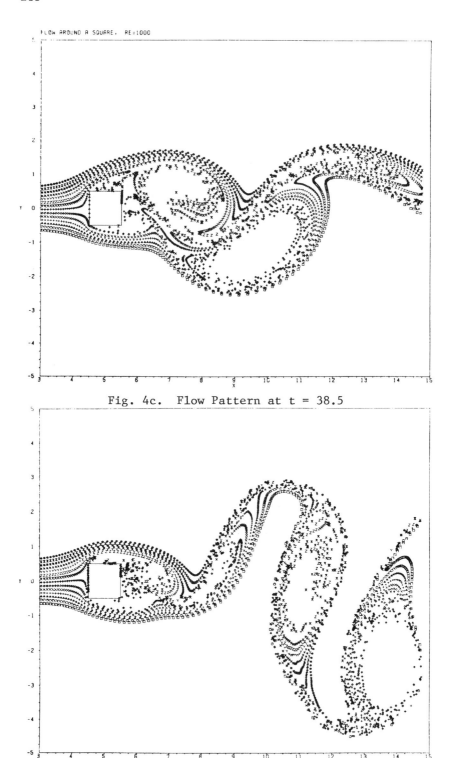

FLOW AROUND A SQUARE, RE=1000

Fig. 4c. Flow Pattern at t = 38.5

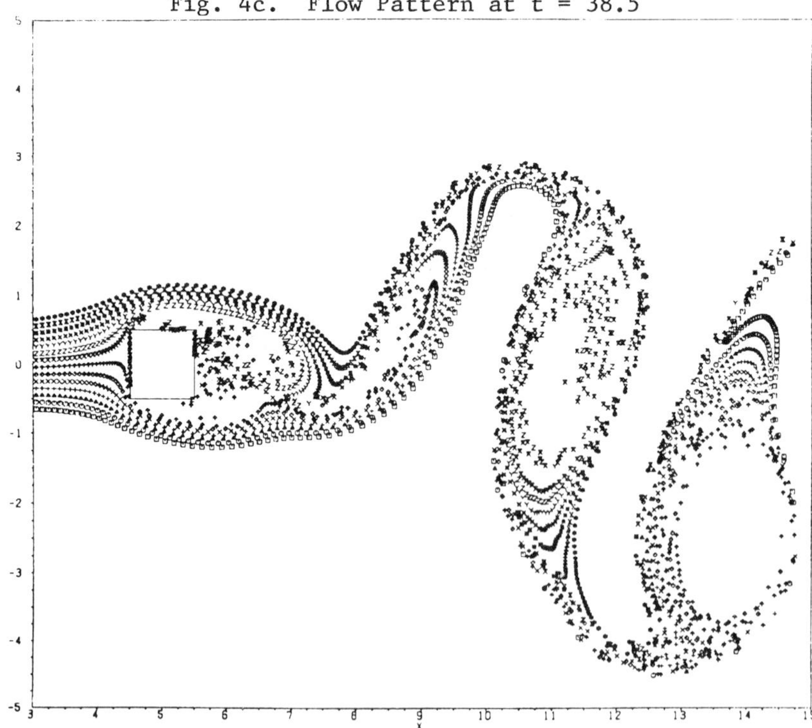

Fig. 4d. Flow Pattern at t = 43.5

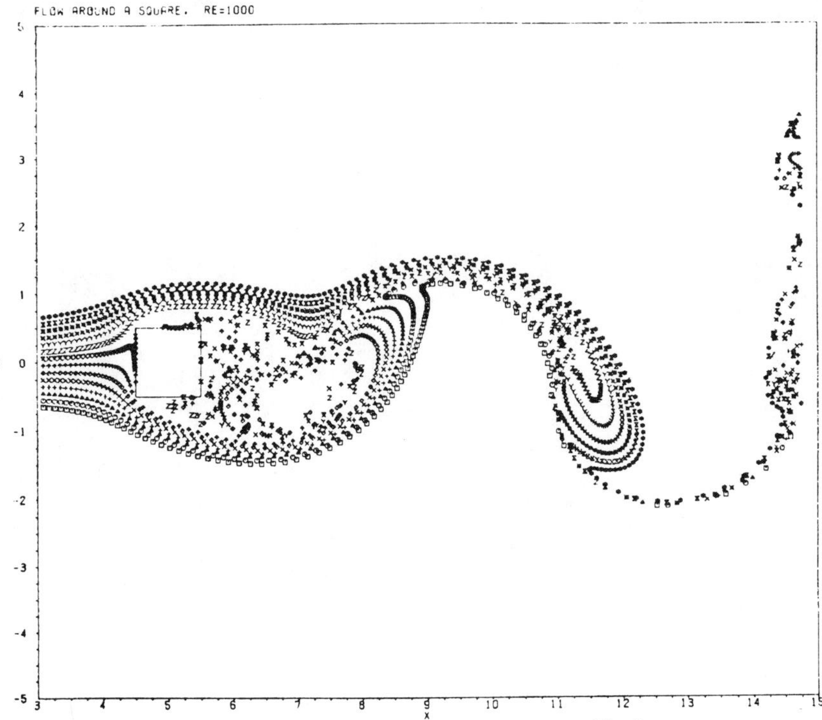

Fig. 4e. Flow Pattern at t = 63.3

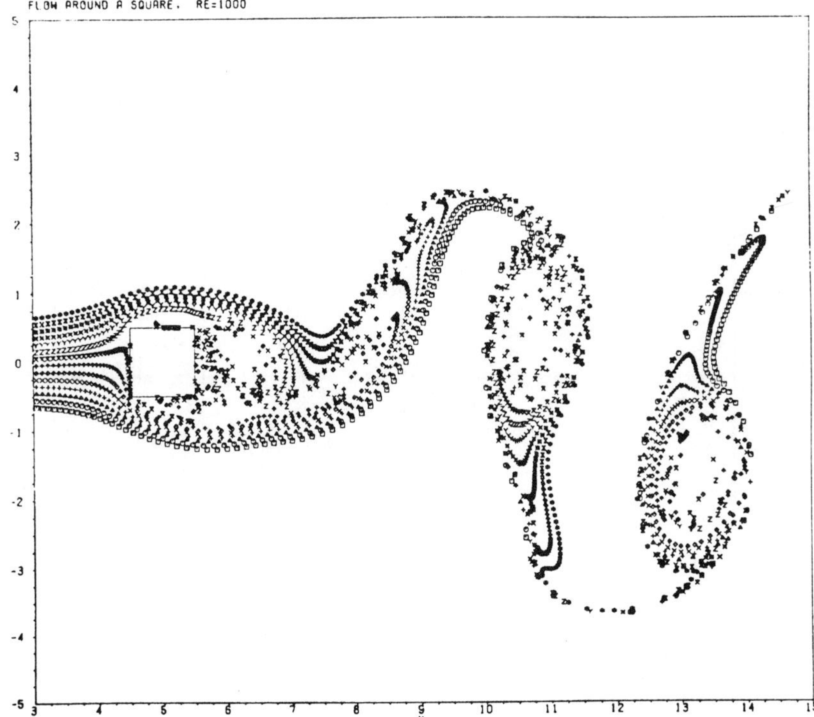

Fig. 4f. Flow Pattern at t = 70.7

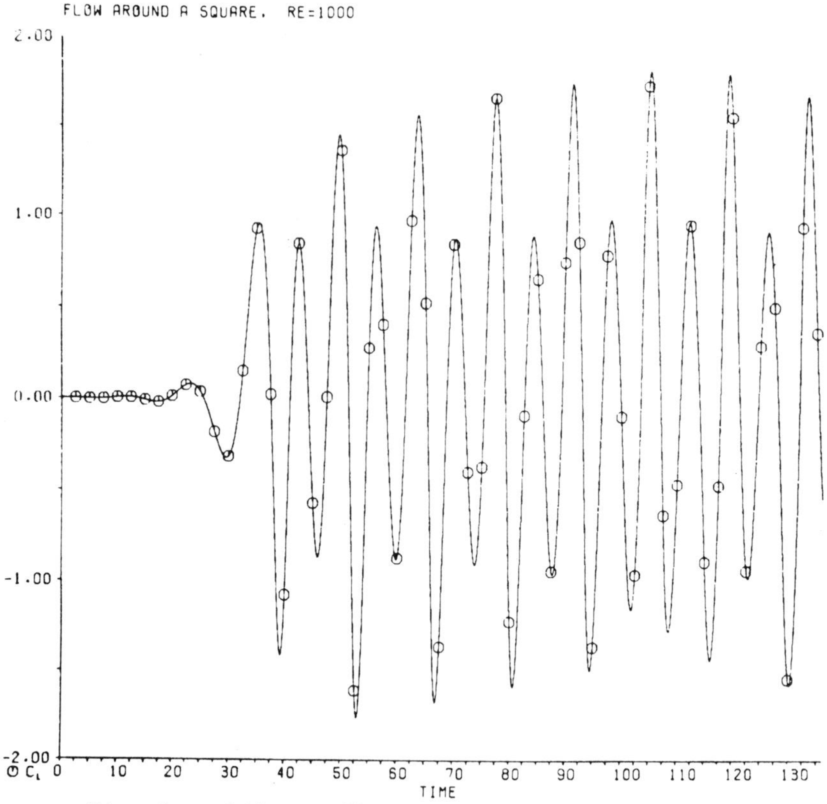

FLOW AROUND A SQUARE. RE=1000

TIME

Fig. 5a. Lift Coefficient; Symbols Every 50 Δt

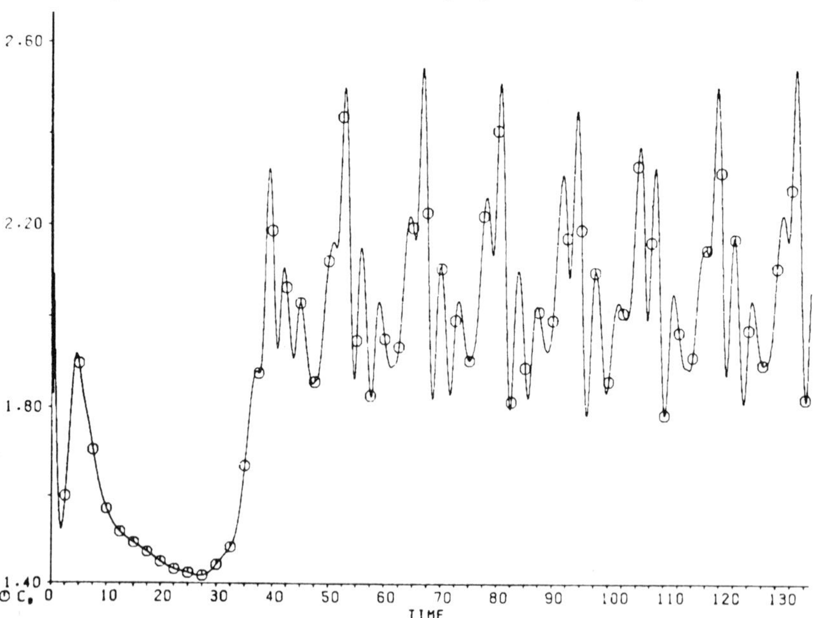

Fig. 5b. Drag Coefficient; Symbols Every 50 Δt

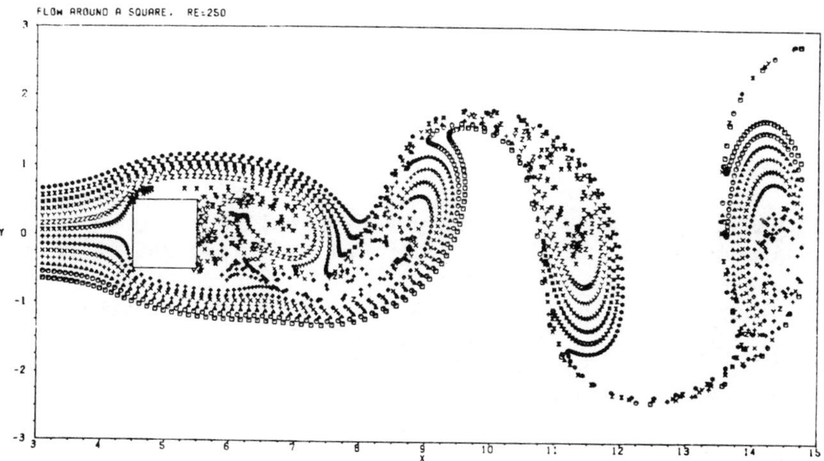

Fig. 6. Flow Pattern for Re = 250

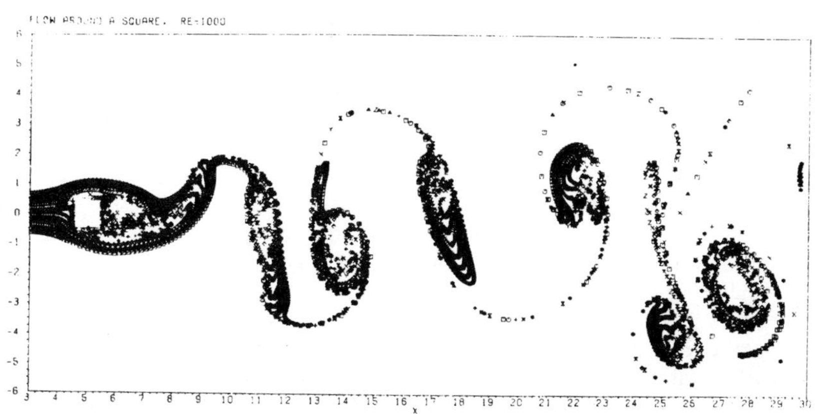

Fig. 7. Extended Downstream Region for Re = 1000

4. CONCLUSIONS

It has been demonstrated that a reasonably accurate simulation of vortex shedding from a square at Reynolds numbers up to about 1000 can be obtained by means of an extension of the QUICKEST method to two dimensions. The use of a higher-order numerical scheme such as this greatly reduces the problems associated with first- and second-order schemes. Offsetting this advantage, though, is the increased complexity in applying boundary conditions. The great value of marker particles in visualizing this type of unsteady flow has also been demonstrated.

It is possible to extend the simulation presented here to more complicated situations. These include nonzero angles of attack, nonuniform upstream velocity profiles, and various rectangular body shapes. A future paper will discuss these extensions to the basic problem as well as providing more detail on the numerics than was possible here.

5. REFERENCES

1. BEARMAN, P. W. - Bluff Body Flows Applicable to Vehicle Aerodynamics. J. Fluid Eng., Vol. 102, pp. 265-274, 1980.
2. BEARMAN, P. W. and GRAHAM, J. M. R. - Vortex Shedding From Bluff Bodies in Oscillatory Flow: A Report on Euromech 119. J. Fluid Mech., Vol. 99, pp. 225-245, 1980.
3. SIMIU, E. and SCANLAN, R. H. - Wind Effects on Structures, John Wiley, New York, 1978.
4. VICKERY, B. J. - Fluctuating Lift and Drag on a Long Cylinder of Square Cross-Section in a Smooth and in a Turbulent Stream. J. Fluid Mech., Vol. 25, pp. 481-494, 1966.
5. FROMM, J. E. and HARLOW, F. H. - Numerical Solution of the Problem of Vortex Street Development. Phys. Fluids, Vol. 6, pp. 975-982, 1963.
6. ROACHE, P. J. - Computational Fluid Dynamics, Hermosa, Albuquerque, 1976.
7. LEONARD, B. P., LESCHZINER, M. A. and MCGUIRK, J. - Third-Order Finite-Difference Method for Steady Two-Dimensional Convection, Numerical Methods in Laminar and Turbulent Flow, Ed. Taylor, C., Morgan, K. and Brebbia, C. A., John Wiley, New York, 1978.
8. LEONARD, B. P. - A Stable and Accurate Convective Modelling Procedure Based on Quadratic Upstream Interpolation. Comp. Meth. Appl. Mech. and Eng., Vol. 19, pp. 59-98, 1979.
9. BAUM, H. R., CIMENT, M., DAVIS, R. W. and MOORE, E. F. - Numerical Solutions for a Moving Shear Layer in a Swirling Axisymmetric Flow, Proceedings of Seventh International Conference on Numerical Methods in Fluid Dynamics, Springer-Verlag, New York, 1980.
10. HIRT, C. W., NICHOLS, B. D. and ROMERO, N. C. - SOLA - A Numerical Solution Algorithm for Transient Fluid Flows, Los Alamos Scientific Laboratory Report LA-5852, 1975.

TURBULENT FLOW AND TURBULENT MASS TRANSFER IN PARTLY CLOSED PIPES AND CHANNELS

O. Kvernvold, A.K. Rastogi, and T. Søntvedt

Det norske VERITAS, Research Division, Norway

ABSTRACT

The flow and mass transfer phenomena in a two-dimensional channel with a regulating valve are studied numerically applying the k-ε model of turbulence. The results for the mean velocity distribution and the length of the regions with recirculating flow are compared with LDV measurements of the velocity distribution in the regulating valve region of a pipe with corresponding dimensions.

The results indicate that the two-dimensional model gives a qualitative correct picture of the flow characteristics in a partly closed pipe. Quantities such as mass transfer and pressure loss given from the numerical solution have not been compared with measurements due to lack of experimental results.

NOMENCLATURE

C	- concentration
C_w	- wall value of concentration
C^+	- normalized concentration
C_μ, C_1, C_2	- empirical constants in the turbulent model
G	- production of turbulent kinetic energy
$J_{c,w}$	- diffusion of concentration to rigid walls
k	- turbulent kinetic energy
p	- pressure
Re	- Reynolds number
S	- source term in governing equations

Sc	- Schmidt number
Sc_T	- turbulent Schmidt number
Sh	- Sherwood number
u,v	- velocity components
u^+	- normalized velocity component
u_τ	- friction velocity
x,y	- coordinates
y^+	- normalized coordinate

Greek symbols

ρ	- density
ε	- dissipation rate of turbulent kinetic energy
ϕ	- general dependent variable
μ_e	- effective viscosity
μ_T	- turbulent viscosity
σ_ϕ	- effective Schmidt number
τ_w	- wall shear stress
$\varkappa$	- von Karman's constant
δ_v	- laminar sublayer

1. INTRODUCTION

Turbulent flow in pipes and channels containing obstructions of different kinds - valves, welded joints and fences - are of major importance in many practical flow situations. The obstructions give rise to a highly turbulent flow with recirculating regions in which augmented mass transfer of aggressive substances (CO_2 or O_2) to the surface, separation of water from emulsions of oil and brine and fatigue of passive surface films may lead to severe erosion/corrosion attacks. Damages on both onshore and offshore production/transportation equipment indicate that areas with joints, valves and in bends are most prone to erosion/corrosion attacks. A detailed survey of the flow characteristics matched with erosion/corrosion monitoring is therefore expected to lead to an understanding of the influence of the flow on the erosion/corrosion damages and thereby to give some ideas for improving the construction of the equipment.

In a recent paper |1| the turbulent flow field in a regulating valve has been investigated experimentally by LDV-equipment. Flow characteristics such as mean velocity profiles, length of the region with recirculating and wall shear stress have been measured in the vertical plane through the axis |Fig. 1| of the pipe for both one and two phase flows.

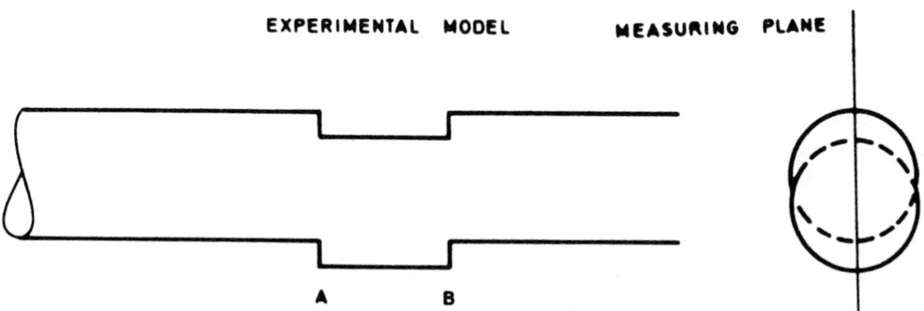

Fig. 1 Experimental model of regulating valve - cross-sectional area at A and B

In the present paper a numerical analysis based on the high Reynolds number version of the k-ε model of turbulence |2|, is performed for flow in a channel with a two-dimensional valve |Fig. 2|. The mean velocity distribution and the length of the regions with recirculation are determined. Similar computations of flow in channels with fences have been performed in Refs. |3, 4|, and the results show satisfactory agreement with experiments.

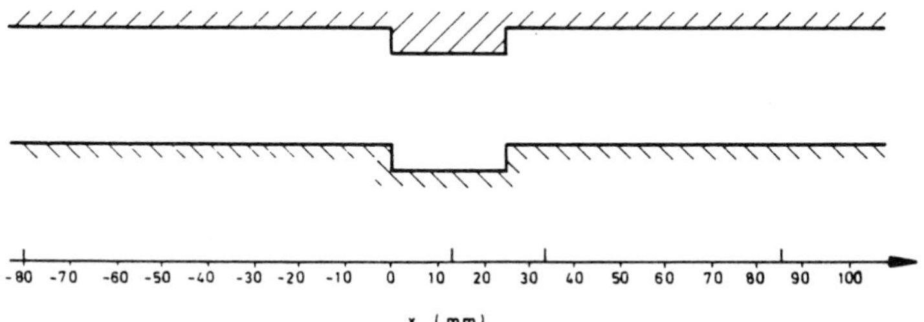

Fig. 2 Two-dimensional theoretical model

The dimensions of the two-dimensional model is choosen in accordance with the dimensions of the experimental model reported in |1|. The width of the channel corresponds to the pipe diameter, the height of the obstruction corresponds to the height of the pipe obstruction in the measuring plane |Fig. 1|, and the length of the obstruction is equal in both cases. This choice of dimensions gives a relative closure of 26% for the experimental model and 20% for the theoretical one. It is the purpose of the present work to examine to what extent the two-dimensional theoretical model may quantify the flow characteristics - mean velocity distribution, pressure distribution and length of the regions with recirculation - for the semi-closed valve reported in |1|. In the present analysis also the mass transfer of inert species to the wall under diffusion controlled conditions is calculated. Numerical calculations |5, 6| have shown to provide the heat and/or mass transfer coefficients reasonably well in turbulent recirculating flow. For the present model no mass transfer measurements have been performed. However, such recordings are planned in near future.

2. THE GOVERNING EQUATIONS

The turbulent flow fields is determined by the time-averaged Navier-Stokes equations, including the Boussinesq turbulent-viscosity concept. The inert species concentration is found from a time-averaged diffusion equation. The governing equations may be written in the following general form |2|:

$$\frac{\partial}{\partial x}(\rho u\phi) + \frac{\partial}{\partial y}(\rho v\phi) = S_\phi + \frac{\partial}{\partial x}\left(\frac{\mu_e}{\sigma_\phi}\frac{\partial\phi}{\partial x}\right) + \frac{\partial}{\partial y}\left(\frac{\mu_e}{\sigma_\phi}\frac{\partial\phi}{\partial y}\right) \qquad (2.1)$$

where:

$$\phi = u,\ S_\phi = -\frac{\partial p}{\partial x} + \frac{\partial}{\partial x}\left(\mu_e\frac{\partial u}{\partial x}\right) + \frac{\partial}{\partial y}\left(\mu_e\frac{\partial v}{\partial x}\right),\ \sigma_\phi = 1 \qquad (2.1a)$$

for the x-momentum equation

$$\phi = v,\ S_\phi = -\frac{\partial p}{\partial y} + \frac{\partial}{\partial x}\left(\mu_e\frac{\partial u}{\partial y}\right) + \frac{\partial}{\partial y}\left(\mu_e\frac{\partial v}{\partial y}\right),\ \sigma_\phi = 1 \qquad (2.1b)$$

for the y-momentum equation,

$$\phi = 1,\ S_\phi = 0 \qquad (2.1c)$$

for the continuity equation,

$$\phi = c,\ S_\phi = 0 \text{ and } \sigma_\phi = 0.9 = Sc_T;\ \frac{\mu_e}{\sigma_\phi} = \frac{\mu}{Sc} + \frac{\mu_T}{Sc_T} \qquad (2.1d)$$

for the transport equation for concentration.

The effective turbulent viscosity is defined by the k-ε model of turbulence |2|:

$$\mu_e = \mu + C_\mu \rho k^2/\varepsilon;\ C_\mu = 0.09 \qquad (2.2)$$

k and ε are obtained by solving the partial differential equations which represent their conservation and have the same form as eq. (2.1), where

$$\phi = k,\ S_\phi = G - \rho\varepsilon \text{ and } \sigma_\phi = 1 \qquad (2.1e)$$

for the turbulent kinetic energy equation

$$\phi = \varepsilon,\ S_\phi = C_1 \frac{\varepsilon}{k} G - C_2 \rho \frac{\varepsilon^2}{k},\ \sigma_\phi = 1,3,\ C_1 = 1.44 \text{ and}$$

$$C_2 = 1.92 \qquad (2.1f)$$

for the turbulent dissipation equation.

G represents the production of turbulent kinetic energy by the mean motion and is given by:

$$G = \mu_T \left| 2 \left(\frac{\partial u}{\partial x}\right)^2 + 2\left(\frac{\partial v}{\partial y}\right)^2 + \left(\frac{\partial v}{\partial x} + \frac{\partial u}{\partial y}\right)^2 \right| \text{ with } \mu_T = C_\mu \rho\, k^2/\varepsilon$$

Boundary conditions have to be specified at rigid walls and the inflow and outflow faces of the calculation domain. At the wall nearest grid-point the velocity parallel to the wall, u_p, is related to the wall shear stress by the logarithmic law of the wall giving

$$u^+ = \frac{U_p}{U_\tau} = \frac{1}{\varkappa} \ln|Ey^+| = \frac{1}{\varkappa} \ln\left|E \frac{y}{\mu/\sqrt{\tau_w \rho}}\right| \qquad (2.3)$$

The wall shear stress in the above expression is obtained by assuming balance between production and dissipation of turbulent energy at the wall nearest grid point, giving:

$$\tau_w = C_\mu^{1/2} \rho k \qquad (2.4)$$

The velocity component normal to rigid walls is taken to be zero.

The value of species concentration at the wall nearest grid point is determined by assuming a logarithmic concentration profile

$$c^+ = (C_p - C_w) \sqrt{\rho \tau_w}/(- J_{c,w}) = \frac{\mu}{\sigma_c} \left|\frac{1}{\varkappa} \ln Ey^+ + F(Sc, Sc_T)\right| \qquad (2.5)$$

where F is a function of laminar and turbulent Schmidt number given by |2|:

$$F(Sc, Sc_T) = 9.24 \cdot ((Sc/Sc_T)^{3/4} - 1) \cdot (1 + 0.28 \exp$$

$$(-0.007 \cdot Sc/Sc_T)) \qquad (2.6)$$

Fully developed flow conditions were applied at the upstream boundary, and $\frac{\partial\phi}{\partial x} = 0$ boundary conditions were used at the downstream boundary.

The value of the turbulent kinetic energy at the wall nearest grid line is obtained by integrating the equation for k over the wall nearest control volume. The diffusion of k to the wall is neglected. The turbulent shear stress is assumed to be constant over the control volume and equal to the wall shear stress. Similarly k is assumed to be constant and equal to the value at the wall nearest grid point.

3. RESULTS AND COMPARISONS

The numerical calculations give results for the mean velocity distribution, the pressure distribution and the mass transfer to the wall. However, at the moment only the mean velocity distribution may be compared with experimental results obtained in Ref. |1|, where the variation of the longitudinal velocity component along the walls and velocity profiles in the obstruction half of the valve were measured with LDV.

In Fig. 3 is displayed both the theoretical and experimental results for the longitudinal velocity distribution at a distance of 1 mm from the walls. The measurements are performed in a vertical plane through the pipe axis for Re = 26000. Comparison between experimental and numerical results show acceptable agreement, except for some underestimation of the velocity in regions with recirculation. Some of this discrepancy is probably due to three-dimensional effects in the experimental model, but there is also a general trend of the present calculation procedure to underestimate the flow velocity in regions with recirculation |3|. Some of the discrepancy may, however, also be due to experimental difficulties in measuring exactly 1 mm from the wall. In the wall region where the velocity gradient is large only a small change in the distance from the wall will give large variation in the velocity.

The regions with recirculation are of interest, because these areas are prone to corrosion attacks due to augmented mass transfer to the wall. For the present model the length of the recirculating region is found to be about three obstruction heights which is with an accuracy within 5% of the measured value. Both measurements and theory give that the length of the recirculating region is independent of the Reynolds number (for the same geometrical configuration). This is, however, a general and well-known result for turbulent flow past obstacles |3, 4|.

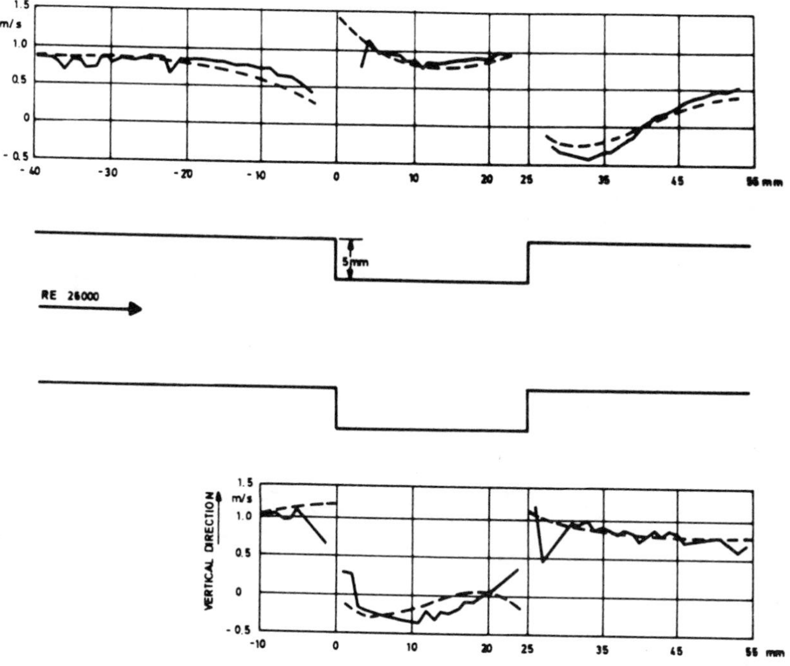

Fig. 3 Velocity distribution at a distance of 1 mm from the walls. Re = 26000
—— Experimental results
--- Theoretical results

Measurements of velocity profiles in the obstruction half of the pipe are displayed in Fig. 4 together with results of the numerical computation. The velocities were measured in a vertical plane through the pipe axis for Re = 15000. The agreement between the theoretical and experimental results is observed to be very good, but also in this case the calculation routine underestimates the velocity in regions with recirculation. To give a more detailed picture of the flow pattern, the stream lines for this case are given in Fig. 5. Especially to be noticed is the streampattern in the cavity. Due to the obstruction the main stream is forced to enter the cavity. This will deform the region with recirculating flow in the cavity and give a tendency to create two separate zones with recirculation.

The pressure loss in the valve region is shown in Fig. 6 together with the pressure loss for fully turbulent pipe flow. The Reynolds number for this case is Re = 15000. No measurements of the pressure distribution in the valve region have been performed and therefore comparison between experimental and numerical results are not possible. However, the numerical results show, as expected, a large pressure drop across the valve compared with the pressure drop in fully developed pipe flow.

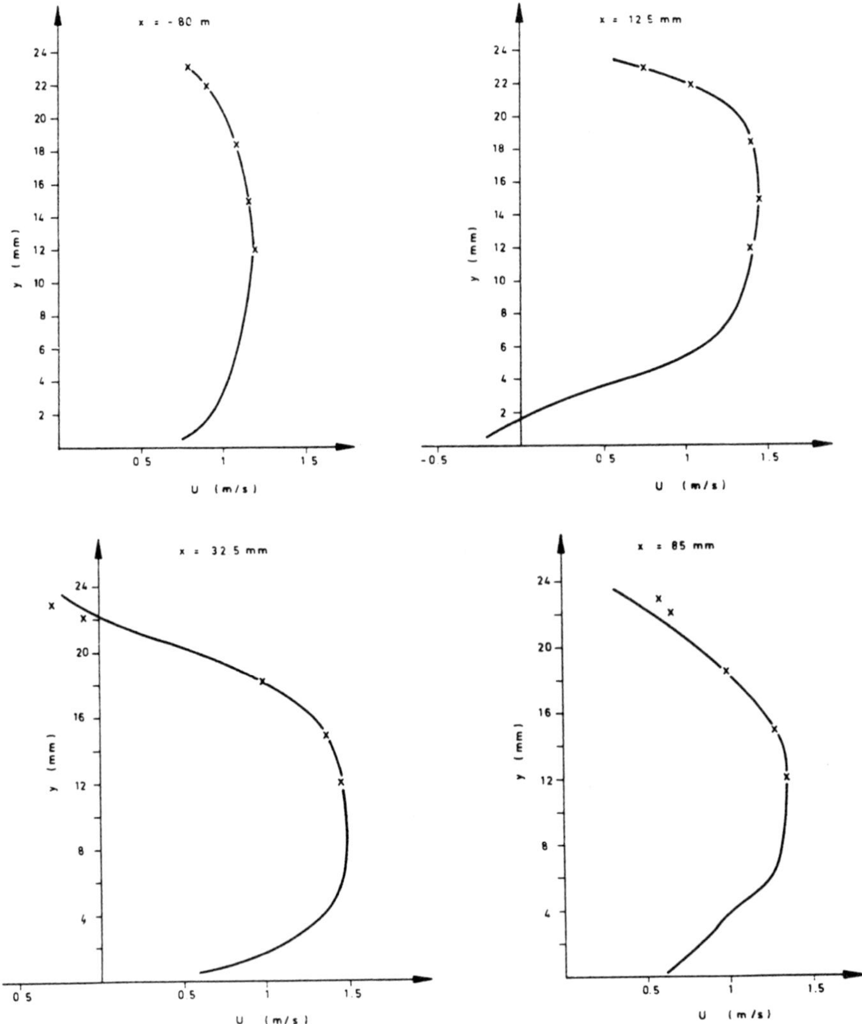

Fig. 4 Velocity profiles for Re = 15000. Position relative to the valve is indicated at Fig. 2
x Experimental results
— Theoretical results

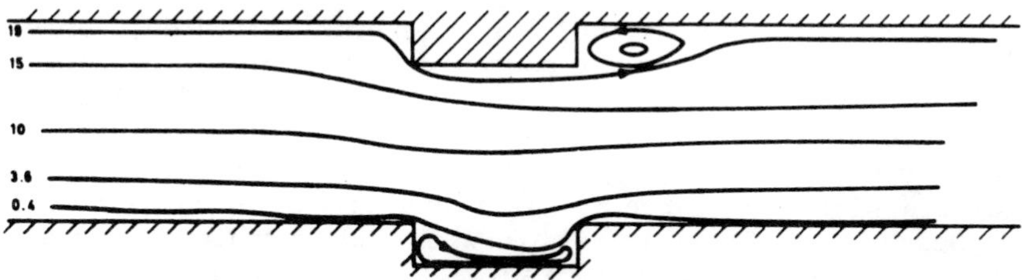

Fig. 5 Streamlines for Re = 15000

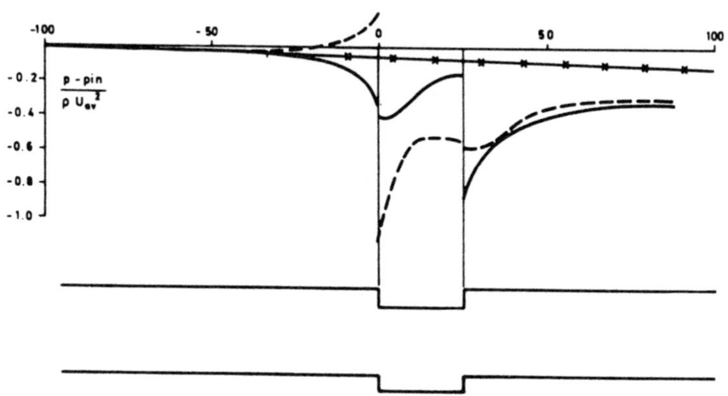

Fig. 6 Pressure losses
---- along the wall at the obstruction side of the channel
—— along the wall at the cavity side of the channel
—x— fully developed pipe flow

Fig. 7 and Fig. 8 show the mass transfer to the surface - given by the Sherwood number - with zero wall concentration for Re = 15000 and Re = 26000, respectively. Schmidt number is 1670 which corresponds to a 0.01 molar solution of Cu So_4 in 1 molar NaCl-solution. The mass transfer increases strongly in the valve region due to the high production of turbulence. In the obstruction half of the channel the maximum mass transfer is found at the front of the obstruction and near the reattachment point for the flow. This is in accordance with results obtained for flow over a rectangular obstruction in a pipe |5|. On the cavity side of the channel the maximum mass transfer occurs at the downstream edge of the cavity. It is worth noticing that the maximum mass transfer on the cavity side of the channel is larger than in the recirculating region. The augmented mass transfer, the increased turbulence level and regions with recirculating flow make the valve region more prone to corrosion/ erosion attack than is a straight tube. In many practical situations, however, the formation of corrosion products on the surface will decrease the mass transfer (compared to the mass transfer with zero wall concentration) and therefore act as a protective layer. In regions with large fluctuations in wall shear stress and pressure these passive films may be removed with corresponding increase in mass transfer. Large fluctuations in wall shear stress and pressure are found in regions with high turbulent intensity as for example in a valve region. This combination of removal of passive corrosion films and corrosion is termed erosion-corrosion and is one of the most dangerous destruction mechanisms in oil/gas production and transportation systems.

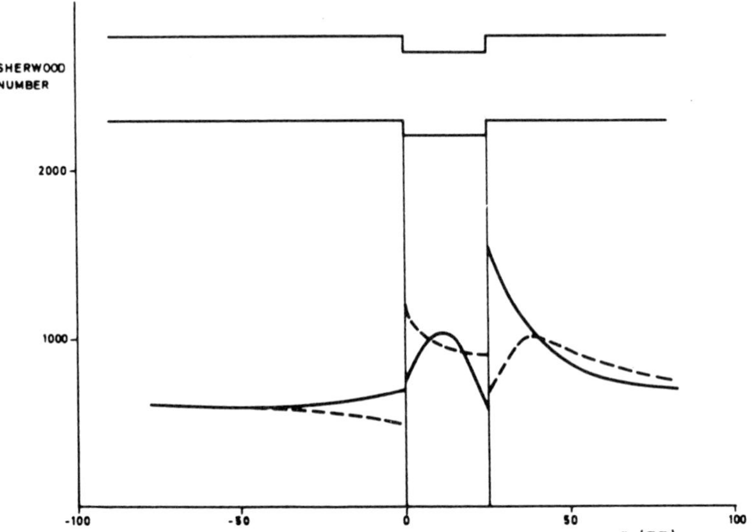

Fig. 7 Mass transfer to the wall given by the Sherwood number. Re = 15000
--- the obstruction side of the channel
—— the cavity side of the channel

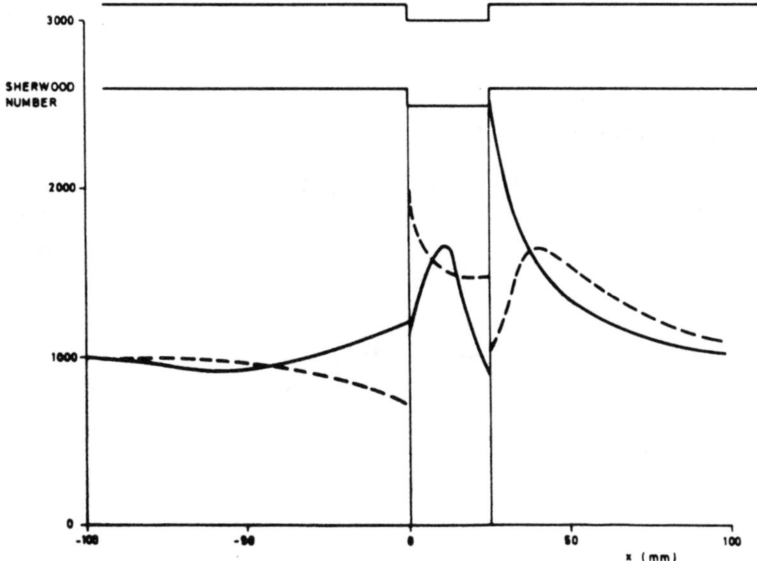

Fig. 8 Mass transfer to the wall given by the Sherwood number. Re = 26000
--- the obstruction side of the channel
—— the cavity side of the channel

4. CONCLUSIONS

The above results indicate that the present two-dimensional calculation procedure may be applied to give a qualitative correct picture of the flow conditions in the regulating valve at small closure reported in |1|. The lengths of the recirculating region are very well determined. So are also the mean velocity profiles. The main defect is the underestimation of the flow velocity in the recirculating regions. Quantities such as mass transfer and pressure drop given from the numerical calculations have not been compared with experiments due to lack of experimental results. The numerical procedure should also be compared with experiments with other closures of the valve.

5. REFERENCES

1. VINDØY, V., OLSEN, S., SØNTVEDT, T. - Application of LDV-System for Measurement of One and Two-phase Flow in a Regulating Valve. VERITAS Report No. 80-0360 (1980).

2. LAUNDER, B.E. and SPALDING, D.B. - Mathematical Models of Turbulence, Academic Press, London and New York, (1973).

3. DURST, F. and RASTOGI, A.K. - Theoretical and Experimental Investigations of Turbulent Flows with Separation, Turbulent Shear Flows I, Springer Verlag (1979).

4. DURST, F. and RASTOGI, A.K. - Turbulent Flows over Two-Dimensional Fences, Turbulent Shear Flow II, Springer Verlag, pp. 218-232, (1980).

5. RASTOGI, A.K., KVERNVOLD, O., SØNTVEDT, T. - Flow and Mass Transfer in a Pertubated Turbulent Pipe Flow, ASME Paper 80-HTD-13, ASME Winter Annual Meeting, Chicago, November 16-21, pp. 91-96, 1980.

6. CHIENG, C.C. and LAUNDER, B.E. - On the Calculation of Turbulent Heat Transport Downstream from an Abrupt Pipe Expansion, TF/10/79, Dept. Mech. Eng., University of California, Davis, USA, (1979).

7. GASMAN, A.D. and PUN, W.M. - Lecture Notes for Course Entitled - Calculation of Recirculating Flow, HTS/74/2, Heat Transfer Section, Imperial College, London (1973).

IMPROVEMENTS OF THE OPTIMAL CONTROL METHOD TO SOLVE TURBULENT FLOWS

M. BUFFAT ; D. JEANDEL ; J.F. BRISON

Laboratoire de Mécanique des Fluides, Ecole Centrale de Lyon
69130 Ecully, France

SUMMARY

We investigate in this paper suitable modifications of the optimal control method to allow its application to the numerical solution of turbulent flows. They are

a) choice of a more appropriate linear operator ;

b) introduction of a Petrov-Galerkin formulation with optimal test functions.

1. INTRODUCTION

The recent development of the performant Optimal Control Method applied to the solution of the Navier-Stokes equations, suggests to extend these techniques to turbulent flow predictions. Making use of the classical statistical analysis, the calculation of mean quantities needs to solve a set of partial differential equations of Navier-Stokes type. Two main difficulties appear, namely the non-linear convective term and the additional non-linear term depending on the chosen turbulence model.

The least square formulation is now extensively used at moderate Reynolds number to solve incompressible laminar viscous flow problems [1]. However at high Reynolds numbers, the Galerkin approximation of the non-linear convective acceleration induces inaccuracies and spatial oscillations in the solution. Furthermore, the speed of convergence decreases as the Reynolds number is increased.

We propose here suitable modifications of the standard control technique to overcome these difficulties. A detailed presentation of this approach is given for the model equation of Burgers ; a new least square formulation is developed (proper choice of the linear operator and of the cost function) and an appropriate test space is incorporated in the discrete formulation. The resulting method is applied to the two-dimensional Navier-Stokes equations. Considering the non-linear turbulent source term, a one-dimensional physical problem is examined. The procedure will undoubtedly help in the solution of two-dimensional turbulent flows.

2. PHYSICAL CONCEPTS AND MODEL EQUATIONS

The equations governing the turbulent incompressible flow field follow from the Navier-Stokes equations by the usual decomposition of the velocity field into mean and fluctuating components $u_i = \bar{u}_i + u_i'$. We have then

$$\frac{\partial \bar{u}_i}{\partial t} - \nu \frac{\partial^2 \bar{u}_i}{\partial x_j^2} = - \frac{1}{\rho} \frac{\partial \bar{p}}{\partial x} - \bar{u}_j \frac{\partial u_i}{\partial x_j} - \frac{\partial}{\partial x_j} \overline{u'_i u'_j} \tag{1}$$

$$\frac{\partial \bar{u}_i}{\partial x_i} = 0$$

where $\overline{u'_i u'_j}$ is the Reynolds stress tensor.

At high turbulent Reynolds number, the two non-linear terms in the right hand side of (1) are simultaneously dominant. In order to point out the influence of each term, two model equations are proposed.

a) The Burgers equation $\frac{\partial u}{\partial t} - \nu \frac{\partial^2 u}{\partial x^2} = - u \frac{\partial u}{\partial x}$ (2)

underlines the effect of the non-linear convective term;

b) The simplified equation for the mean turbulent velocity field between two parallel planes

$$\frac{\partial u}{\partial t} - \nu \frac{\partial^2 u}{\partial x^2} = K - \frac{\partial \tau}{\partial x} \tag{3}$$

takes into account the additional turbulent terms; the shear stress τ will be firstly expressed in terms of the mean velocity gradient

$$\tau = - \nu_T \frac{\partial u}{\partial x}$$

A suitable model for the local turbulent viscosity ν_T will be analysed further.

3. FIRST MODEL EQUATION : BURGERS EQUATION

We consider the model problem in one dimension

$$\frac{\partial u}{\partial t} - \nu \frac{\partial^2 u}{\partial x^2} = - u \frac{\partial u}{\partial x} \quad \text{in } \Omega = [0,1] \tag{4}$$

with the boundary and initial conditions

$u(x,t) = u_B$ on boundary Γ and $u(x,0) = u_\circ(x)$.

We adopt an implicit time discretization. A first order approximation gives

$$\frac{u(t + \Delta t) - u(t)}{\Delta t} - \nu \frac{\partial^2 u(t + \Delta t)}{\partial x^2} = - \gamma u(t) \frac{\partial u(t + \Delta t)}{\partial x}$$
$$- (1 - \gamma) u(t + \Delta t) \frac{\partial u(t + \Delta t)}{\partial x} \qquad (5)$$

where $\gamma \in [0,1]$ is a relaxation parameter.

A choice of γ different from zero allows the linearization of a part of the convective term. Such a decomposition will be of great importance for the convergence of the method.

The initial problem may then be written in the form

$$\begin{aligned} &Lu = B(u) \text{ in } \Omega \\ &\text{with } Lu = au - \nu\Delta u + \gamma(u_t.\nabla)u \qquad a = \frac{1}{\Delta t} \\ &\qquad Bu = -(1 - \gamma)u.\nabla u + au_t \end{aligned} \qquad (6)$$

3.1. Optimal control and least square formulation

Let Ω be a bounded region in R and consider the problem (P_1) $Lu = B(u)$ in Ω and $u/_\gamma = u_o$.
L is a linear continuous, coercive, elliptic operator from $H^1(\Omega)$ to $H^{-1}(\Omega)$ (dual space of H^1_o).
We denote by $H^1(\Omega)$ the Hilbert space of all functions whose derivatives up to order one have finite norm over Ω.
We also require the following subspace of H^1

$$H_o^{1}(\Omega) = \{v \in H^1(\Omega)/v = 0 \text{ on } \gamma\} \text{ with a scalar product } ((.,.))$$

We note $(u,v) = \int_\Omega uv d\Omega$ the scalar product in $L_2(\Omega)$.
The problem (P_1) is equivalent to the following minimization problem

(P_2)

find λ^* optimal $\in H^1_o$ such that
$J(\lambda^*) = \operatorname{Min} J(\lambda)$

with $J(\lambda) = \frac{1}{2} ||u\lambda - \lambda||^2$ the cost functional to be defined in 3.3 and $u\lambda$ solution of the state equation

$Lu\lambda = B(\lambda)$

The solution λ^* of (P_2) is characterized by an optimality system, which cannot be solved in practice. So we use an iterative gradient method to compute λ^*. Thus, the gradient $G\lambda \in H^1_o(\Omega)$ of the cost function must be calculated.

It verifies

$$\forall\ \Phi \in H_o^1\ ((G\lambda,\Phi)) = -\ ((\lambda - u\lambda,\Phi)) + (B'(\lambda)\Phi,p\lambda)$$

with $p\lambda \in H_o^1$ solution of the adjoint equation

$$\forall\ \Phi \in H_o^1\ (L\Phi,p\lambda) = -\ ((\lambda - u\lambda,\Phi))$$

The continuous iterative algorithm may be written
step 0 $\quad \lambda_o \in H^1$ given
step m $\quad \lambda_m \in H^1$ given

- compute the state u_m by
  $Lu_m = B(\lambda_m) \qquad u_m/\Gamma = u_o$
- compute the cost
  $J(\lambda_m) = \frac{1}{2}\ ||u_m - \lambda_m||^2$
- compute the adjoint state $\quad p_m \in H_o^1$
  $(L\Phi,p_m) = -\ ((\lambda_m - u_m,\Phi)) \qquad \forall\ \Phi \in H_o^1$
- compute the gradient G_m of $J(\lambda_m)$
  $((G_m,\Phi)) = ((\lambda_m - u_m,\Phi)) + (B'(\lambda_m)\Phi,p_m) \quad \forall\Phi \in H_o^1$
- compute the descent parameter
  $\rho^* = \text{Min } J(\lambda_m - \rho G_m) \quad \rho \in R^+$
- then the new control
  $\lambda_{m+1} = \lambda_m - \rho^*\ G_m$
- m = m + 1 GOTO STEP m.

.2. Finite element discrete formulation

Suppose the domain Ω is discretized with a finite element mesh; a set $\{N_j\}$ of trial functions spans a subspace S^h of $H^1(\Omega)$ and a set $\{\psi_j\}$ of test functions spans a subspace T_o^h of $H_o^1\ (\Omega)$.

The finite element approximations $\tilde{v}_m$, $\tilde{\lambda}_m$, $\tilde{G}_m$ of v_m, λ_m, G_m are expanded in basis functions $\{N_j\}$ and the approximation $\tilde{p}_m$ of the adjoint state is taken in T_o^h.

The discrete algorithm may be written
step 0 $\quad \tilde{\lambda}_o \in S^h$ given
step m $\quad \tilde{\lambda}_m \in S^h$ given

- solve the state equation : find $\tilde{u}_m \in S^h$ by
  $(L\ \tilde{u}_m^i\ N_i,\psi_j) = (B(\tilde{\lambda}_m^i\ N_i),\psi_j) \qquad \forall\ \psi_j \in T_o^h$ i,j from 1 to n
- compute the cost $J(\tilde{\lambda}_m) = \frac{1}{2}||(\tilde{u}_m^i - \tilde{\lambda}_m)\ N_i||^2$

- compute the adjoint state : Find $\tilde{p}_m \in T_o^h$ by
  $(LN_i, \tilde{p}_m^j \psi_j) = -(((\tilde{\lambda}_m^j - \tilde{u}_m^j) N_j, N_i)) \;\; \forall N_i \in S^h \;\; i,j = 1,n$
- compute the gradient $\tilde{G}_m \in S^h$ by
  $((\tilde{G}_m^i N_i, N_j)) = (((\tilde{\lambda}_m^i - \tilde{u}_m^i) N_i, N_j)) + (B'(\tilde{\lambda}_m) N_j, \tilde{p}_m^i \psi_i)$
  $\forall N_i \in S^h$
- compute the descent parameter
  $\rho^* = \text{Min} \;\; J(\tilde{\lambda}_m - \rho \tilde{G}_m)$
- then the new control $\tilde{\lambda}_{m+1} \in S^h$ is given by
  $\tilde{\lambda}^i_{m+1} = \tilde{\lambda}^i_m - \rho^* \tilde{G}^i_m \qquad i = 1,n$

A Petrov-Galerkin (generalised Galerkin) method is proposed to solve the state equation in a weighted formulation. The gradient initially written

$$J'(\lambda)\delta\lambda = \lim_{\rho \to o} \frac{(J(\lambda + \rho\delta\lambda) - J(\lambda))}{\rho} = ((\lambda - u\lambda, \delta\lambda)) - ((\lambda - u_\lambda, \delta u))$$

may be turned in the convenient discrete form

$$((G\lambda, \delta\lambda)) = ((\lambda - u\lambda, \delta\lambda)) + (B'(\lambda)\delta\lambda, p\lambda)$$

using the weighted formulation of the differentiated state equation

$$(L\delta u, \Phi) = (B'(\lambda)\delta\lambda, \Phi) \qquad \forall \; \Phi \in T_o^h$$

and $p\lambda$ solution of : $p\lambda \in T^h$

$$(L\delta u, p\lambda) = -((\lambda - u\lambda, \delta u)) \qquad \forall \; \delta u \in S^h$$

At each iteration, the iterative procedure requests the solution of six linear systems (state equation, adjoint equation, gradient equation and three state equations for the computation of ρ^*). A lower-upper factorisation is used to split the unsymmetric state matrix.

3.3. Choice of the norm and choice of the test space

For Burgers equation, the convergence of the optimal control method depends on the cost function

$$J(\lambda) = \frac{1}{2} \, ||u_\lambda - \lambda||^2$$

In the case of a symmetric operator L, a convenient choice as found in [1] is

$$J(\lambda) = \frac{1}{2} (L(u_\lambda - \lambda), u_\lambda - \lambda)$$

while for unsymmetric L given by

$$Lu = au - \nu \Delta u + \gamma u_t . \nabla u$$

a good choice seems to be

$$J(\lambda) = \frac{1}{2} || u_\lambda - \lambda ||^2$$

with $||u||^2 = a(u,u) + \nu(\nabla u, \nabla u) + \gamma(u_t u, u_t u)$.

Let us now investigate the choice of the test function and by way of illustration, we first consider the steady problem (2) (with $a = 0$ and $\gamma = 0$)

$$- \nu \Delta u = - u . \nabla u \quad \text{in } [0,1] \tag{7}$$

with $u(0) = u_o$ and $u(1) = u_1$.

We seek a piecewise linear approximation to u in the form

$$u(x) = \sum_{j=1}^{n} u^j N_j(x)$$

Using the definitions introduced in the last section, the optimal control problem may be written

find $\lambda^* \in S^h$ such that
$J(\lambda^*) = \text{Min } J(\lambda) \quad \lambda \in S^h$

$$\text{with } J(\lambda) = \frac{1}{2} \int_0^1 ((\lambda^i - u_\lambda{}^i) \frac{dN_i}{dx})^2 \, dx$$

and u_λ solution of the state equation

$$\nu \, u_\lambda^i (\frac{dN_i}{dx}, \frac{d\psi_j}{dx}) = - (\lambda^i N_i \frac{d\lambda^i N_i}{dx}, \psi_j) \quad \forall \psi_j \in T_o^h$$

A simple choice of the trial function ψ_j is obviously $\psi_j = N_j$ which give the well-known Galerkin approximation. But, at high Reynolds numbers $Re = 1/\nu$, we observe spurious oscillations in the approximate solution (fig. 1), a phenomenon common to finite difference methods when central differencing of the convective term is used. By studying the linear diffusive - convective equation in one dimension, Heinrich et al. [2] have derived test functions which eliminate the oscillations. They are

$$\psi_i = N_i - \alpha N_i(1 - N_i) \text{ for } x \in [x_{i-1}, x_i]$$
$$\psi_i = N_i + \alpha N_i(1 - N_i) \text{ for } x \in [x_i, x_{i+1}]$$

with α given by $\alpha = \dfrac{1}{\tanh Pe} - \dfrac{1}{Pe}$ and Pe a local Peclet number [3].

Using these functions for problem (7), leads to excessive dissipation and lost of accuracy, as the Reynolds is increased (see table 1). To deduce a convenient form of the test functions for Burgers equation, we first study the discretized problem (7) using three nodes

TABLE I MIDDLE NODAL VALUES FOR GALERKIN (U_G), PETROV GALERKIN APPROXIMATION [WITH HEINRICH (U_{UH}) AND PRESENT TEST FUNCTIONS (U_{UP})] AND TRUE SOLUTION (u).

$n = 3$ $H = 0.5$ $u(0) = 0$ $u(1) = 1$

Re	1	10	100	1 000	10 000
U_G	0.44	0.058	-0.71	-0.93	-1.
U_{UH}	0.44	0.24	0.27	0.27	0.27
U_{VP}	0.44	0.18	0.03	$3.9\ 10^{-3}$	$3.9\ 10^{-4}$
u	0.44	0.20	0.03	$3.1\ 10^{-3}$	$3.1\ 10^{-4}$

TABLE II NODAL VALUES OBTAINED WITH UNSTEADY SCHEME AT Re = 100 000 and $T \to \infty$

$n = 11$ $H = 0.1$ $u(0) = 0$ $u(1) = 1$ $\gamma = 1/3$

NODE J	6	7	8	9	10
U_{UP}	$0.42\ 10^{-4}$	$0.60\ 10^{-4}$	$0.94\ 10^{-4}$	$0.19\ 10^{-3}$	1.
u	$0.43\ 10^{-4}$	$0.61\ 10^{-4}$	$0.96\ 10^{-4}$	$0.19\ 10^{-3}$	1.

Tables 1 and 2 - Nodal values for Burgers equation.

We seek test functions of the same form as Heinrich [2] $\Psi_j = N_j \pm \alpha N_j(1 - N_j)$ (but with α to be determined). Indeed, these functions have the advantage to weight unsymmetrically the convective term only.

The Petrov-Galerkin procedure gives the assembled equation for the middle node

$$\frac{\nu}{h}\left[2u_2 - u_1 - u_3\right] = -(u_2 - u_1)\left(u_2\left(\frac{1}{3} + \frac{\alpha}{4}\right) + u_1\left(\frac{1}{6} + \frac{\alpha}{4}\right)\right) - (u_3 - u_2)\left(u_2\left(\frac{1}{3} - \frac{\alpha}{4}\right) + u_3\left(\frac{1}{6} - \frac{\alpha}{4}\right)\right) \quad (8)$$

For $\alpha = 0$ (Galerkin method) the unique solution u_2 of (8) may be written

$$u_2 = (u_3 + u_1)\frac{\left(\frac{6}{\beta} - 1\right)}{\left(\frac{12}{\beta} + 1\right)} \qquad \text{with } \beta = \frac{H(u_3 - u_1)}{\nu}$$

We note that for β greater than 6 the approximate solution at the middle node is negative and as Re $\to \infty$,

$u_2 \to -(u_3 + u_1)$ (on the contrary $u_2 \to -\infty$ in the linear case).

For $\alpha \neq 0$ (Petrov-Galerkin procedure) the physical solution of (7) is

$$u_2 = \frac{(u_3 - u_1)}{\alpha}\left(\frac{2}{\beta} + \frac{1}{6}\right)\left(-1 \mp \sqrt{1 + \frac{2\alpha}{|2/\beta + 1/6|^2}\,\frac{\left[(u_3^2 + u_1^2)\frac{\alpha}{4} - (u_3^2 - u_1^2)\left(\frac{1}{6} - \frac{1}{\beta}\right)\right]}{(u_3 - u_1)^2}}\right)$$

and

$$\lim_{\beta\to\infty} u_2 = \frac{(u_3 - u_1)}{6\alpha}\left(-1 \mp \sqrt{1 + 12\alpha\frac{\left[(u_3^2 + u_1^2)\frac{3\alpha}{2} - \frac{(u_3^2 - u_1^2)}{6}\right]}{(u_3 - u_1)^2}}\right)$$

With the given boundary conditions, the physical solution

$u_2 \to u_1$ or $u_2 \to u_3$ implies $\alpha = 1 \mp 1/3$.

For finite values of β a crude choice for α seems to be

$$\alpha = \left(1 \mp \frac{1}{3}\right)\left(1 - \frac{2}{\beta}\right) \quad \text{for } \beta \geqslant 2 \tag{9}$$

$$\alpha = 0 \qquad\qquad \text{for } \beta < 2$$

The nodal values u_2 for different Reynolds number are given in table 1.

Using n nodes, solutions of (7) can be obtained by the above technique and comparisons with some other weighting procedures are made at moderate and high Re in figure 1 and tables 1, 2.

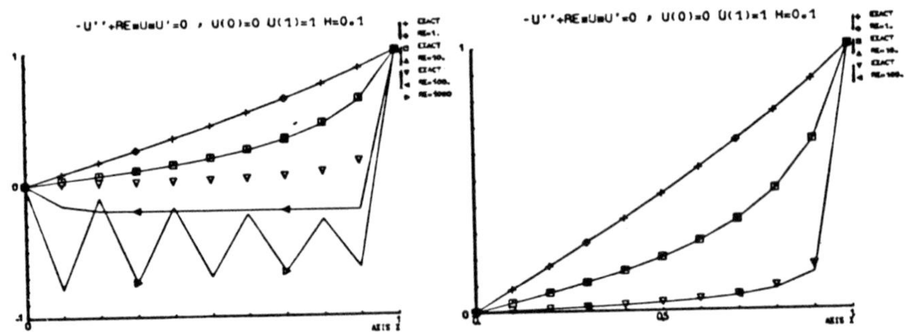

Fig. 1 - Solution of Burgers equation.
1a) Galerkin method - 1b) Present method.

In figure 2 are pointed out the effects of γ on the convergence of the iterative optimal control process. The decrease of the cost function is given for the successive boundary conditions $u(o) = 0$, $u(1) = 1$ and $u(0) = 1$, $u(1) = 0$.

In the first situation (fig. 2b), the effect of γ on the rate of convergence is not determinant. Obviously as the velocity u (and u_t) falls to zero, the contribution of the linear part of the convective term is not dominant. On the contrary for $u(0) = 1$ and $u(1) = 0$ the behaviour of convergence of the algorithm with $\gamma = 0$ appears to be very poor, and a good decrease of the cost function is obtained for $\gamma = 0.2$ (fig. 2a).

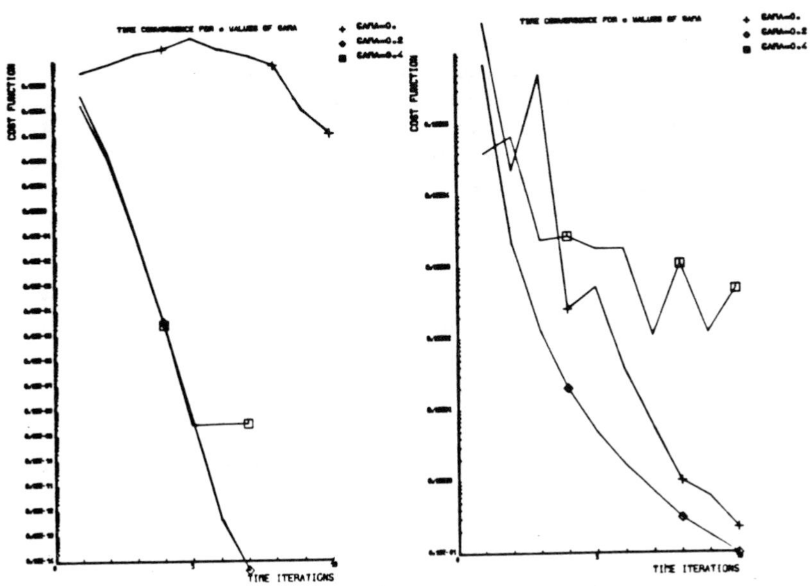

Fig. 2 - Time convergence to the steady state.
$\nu = 1.E\text{-}04$ $\quad H = 0.1$
2a) $u(o) = 1$ $\quad u(1) = 0$
2b) $u(o) = 0$ $\quad u(1) = 1$

4. TWODIMENSIONAL APPLICATION

We describe briefly the extension of our modified optimal control method to the Navier-Stokes equations.

Let consider the general Navier-Stokes problem

$$\frac{\partial u}{\partial t} - \nu \, \Delta \, u = - \frac{1}{\rho} \nabla p - u.\nabla u \text{ in } \Omega \qquad (10)$$

$\nabla.u = 0$ with $u|_\Gamma = u_B$ and $u(o,x) = u_\circ(x)$

using the definition introduced in the last section and after time discretization the resulting equations are treated with a least square formulation, ref. [1], setting the problem as an optimal control problem solved with F.E.M.

Find $\lambda^* \in V$ such that $J(\lambda^*) = \text{Min } J(\lambda)$

with $J(\lambda) = \frac{1}{2}||u_\lambda - \lambda||^2$

and $||u||^2 = a \int_\Omega u^2 d\Omega + \nu \int_\Omega \nabla u^2 d\Omega + \gamma \int_\Omega (u_t . u)^2 d\Omega$

where u_λ is solution of the state equation

$$au - \nu\Delta u + \gamma(u_t.\nabla)u = -\frac{1}{\rho}\nabla p - (1-\gamma)(\lambda.\nabla)\lambda$$
$$\nabla.u = 0$$

A triangular mesh is used and a piecewise linear approximation of the velocity is searched. The weighting functions are those developped by Huyakorn [3], with the modified value of parameter α as given by (9). By way of illustration two examples are presented here.

Example 1 - We study here the entrance flow between two parallel plates and solve the two-dimensional stationary problem (10) with Neumann conditions downstream. We used a mesh of 384 three-noded triangular elements. Figure 3 shows results for Reynolds number equal to 500.

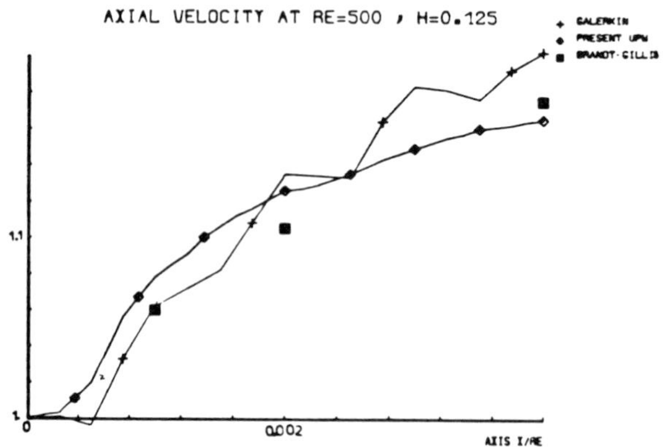

Fig. 3 - Flow entrance between parallel plates.

Example 2 - This second example involves steady separated flow around a cylinder. The mesh has 1 000 three-noded triangular elements. In figure 4 we show the center-line velocity at Reynolds number equal to 67.

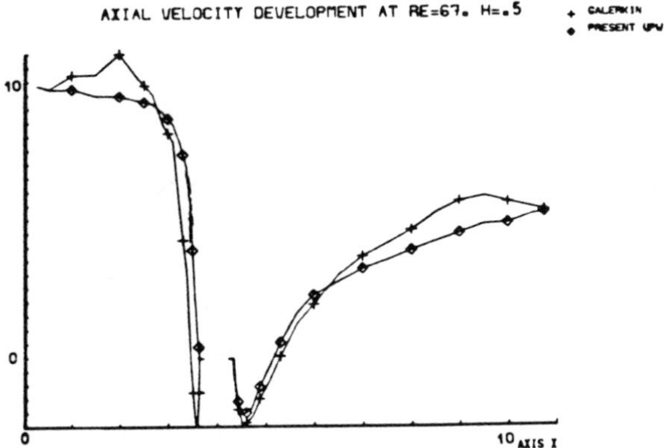

Fig. 4 - Flow around a cylinder.

In this two cases we observe the stabilization of the solution, especially in front of the cylinder, where the Galerkin solution oscillates.

5. TURBULENCE MODEL EQUATION

We consider the turbulent fluid in motion between two parallel planes. The mean motion $\bar{u}$ is described by the equation

$$\frac{\partial \bar{u}}{\partial t} - \frac{\partial^2 \bar{u}}{\partial y^2} = - \frac{1}{\rho} \left(\frac{\partial p}{\partial x}\right) - \frac{\partial}{\partial y} \overline{u'_1 u'_2} \tag{11}$$

where $- \frac{1}{\rho}\left(\frac{\partial p}{\partial x}\right) = K$ is given constant.

To close equation (11), we consider the simple viscosity model

$$\overline{u'_1 u'_2} = - \nu_T \left(\frac{\partial \bar{u}}{\partial y}\right)$$

in which the turbulent viscosity ν_T is prescribed as function of two principal arguments namely : the distance y from the wall and the skin friction velocity

$$u_f = \sqrt{\nu \left(\frac{\partial \bar{u}}{\partial y}\right)_{y=0}}$$

$$\nu_T = \nu K y^+ \left(1 - \exp\left(\frac{-y^+}{A}\right)\right)^2 \qquad y^+ \leqslant y_o^+ \text{ with } y^+ = \frac{y u_f}{\nu}$$

$$\nu_T = \nu_T (y_o^+) \qquad y^+ > y_o^+$$

We must notice that the local non linear term depends mainly on the wall velocity gradient $\left(\frac{\partial u}{\partial y}\right)_{y=0}$. The standard optimal control technique gives good results as shown in figure (5).

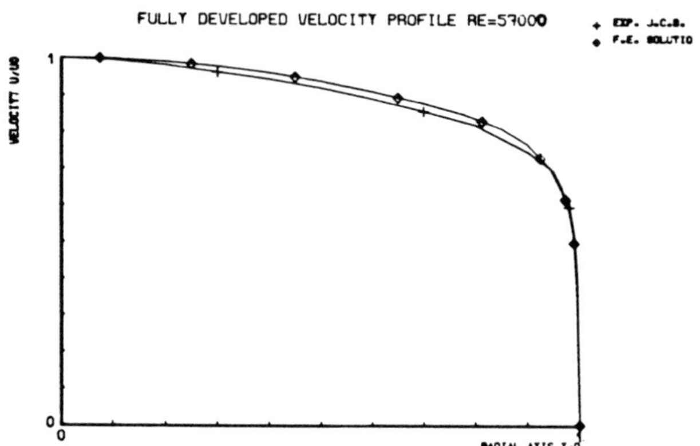

Fig. 5 - Turbulent flow between two parallel plates (fully developed flow).

6. CONCLUSION

From the above study is appears that the optimal control formulation remains available for solving turbulent flows at high Reynolds number. Concerning the convective term the rate of convergence is greatly improved by way of a convenient decomposition (new linear operator and new cost function). A Petrov Galerkin method (modified test functions) has been proposed to eliminate the inaccuracies and oscillations obtained with a Galerkin method. At the present time, the numerical extension of the adapted optimal control method to higher dimensions is currently underway.

7. REFERENCES

1. GLOWINSKI, R; MANTEL, B.; PERIAUX, J. and PIRONNEAU, D. - H^{-1} least squares method for the Navier-Stokes equations. 1st. Int. Conf. on Numerical Methods in Laminar and Turbulent Flow, Ed TAYLOR C., MORGAN K., and BREBIA C.A., Pentech Press, SWANSEA, 1978.

2. HEINRICH, J.C.; HUYAKORN, P.S.; MITCHELL, A.R. and ZIENKIEWICZ, O.C. - An upwind finite element scheme for two dimension al convective transport equations. Int. J. Num. Mech. Engng. 11, pp. 131-143, 1977.

3. HUYAKORN, P.S. - Solutions of steady-state convective transport equation using an upwind finite element scheme. App. Num. Modelling, 1, pp. 187-195, 1977.

LAMINAR TO TURBULENT FLOW STUDY BY MEANS OF F.E.M.

P. MELE[(i)], M. MORGANTI[(ii)], A. DI CARLO[(i)], A. TATONE[(iii)]

SUMMARY

This work aims at the development of a versatile numerical procedure to be used in a combined experimental/numerical analysis of the stability of plane parallel flows. To this purpose, a finite element approximation of the Orr-Sommerfeld equation is obtained using the Galerkin method with Hermite cubics as trial/test functions (Sections 2 and 3). The ensuing well-conditioned eigenvalue problem is solved by means of a QR algorithm. Plane Poiseuille and Couette flows are considered as test cases (Section 4): a number of results are presented and compared with those previously obtained with different techniques (finite differences and Galerkin with orthogonal functions). When extended to unbanded domains (Section 5), the present method shows promise for the analysis of boundary layer flows.

(i) Associate Professor, University of Rome, Italy.
(ii) Assistant Professor, University of Rome, Italy.
(iii) Associate Professor, University of L'Aquila, Italy.

1. INTRODUCTION

The phenomenon of the loss of stability of laminar flow and the consequent transition to turbulence has continued to attract much attention, after O. Reynolds performed his classical experiments on the instability of Poiseuille flow in a pipe towards the end of the last century. A large number of researches, both theoretical and experimental, has since been undertaken on this subject, and numerous significant advances have been made. Rayleigh and, after him, Tollmien, Schlichting and Lin contributed fundamental results on the stability of parallel flows under small disturbances; afterwards, the linear stability theory was further developed and applied to boundary-layer flows.

In the last two decades, different numerical methods have been implemented on high-speed computers, and applied to the analysis of a certain number of standard problems, such as the triggering of instability of Couette and Poiseuille flows.

In this paper a finite element approximation of the (one-dimensional) Orr-Sommerfeld equation is obtained, making use of standard Hermite cubic interpolating functions (plus a special infinite element, when needed). The proposed method is applied to the analysis of plane Couette and Poiseuille flows, for which numerical results obtained with different methods are already available, so that a comparison is allowed.

The final aim of this work is to obtain a versatile numerical procedure for linear stability analysis of arbitrary velocity profiles, prescribed - analytically or numerically - on bounded or unbounded domains, in order to apply it to experimental velocity profiles of locally unstable flows with defined dominant fluctuations close to the walls.

2. BASIC EQUATIONS

A two-dimensional shear flow in the x,y plane will be considered as the basic flow, with velocity u parallel to the x-axis and constant along each line y = constant. The decomposition of an arbitrary (divergence-free) two-dimensional disturbance ([1]) into Fourier components, allows the disturbance stream function ψ to be assumed to have the form

$$\psi(x,y) \equiv \varphi(y) \exp [i\alpha (x - ct)] \qquad (1)$$

where α is real and $c = c_R + ic_I$ is complex: α represents the

---

([1]) Squire theorem assures that the limitation to two-dimensional disturbances is not unduly restrictive.

wave number, c_R the phase velocity, and c_I is a measure of damping ($c_I > 0$ corresponds to an amplification of the disturbance).

If quadratic terms in φ are neglected, the (incompressible) Navier-Stokes equations reduce to the complex Orr-Sommerfeld equation

$$L\varphi = 0 \tag{2}$$

The formal part of the linear fourth-order differential operator L is given by

$$L = (D^2 + \alpha^2)^2 + i\alpha\,Re\,[\,(u(y)-c)\,(D^2 + \alpha^2) + d^2u/dy^2] \tag{3}$$

where

$$D \equiv -\,i\,\frac{d}{dy} \tag{4}$$

Re is the Reynolds number, and u represents the basic velocity field. The domain of L is the set of sufficiently smooth functions φ defined over the bounded (i) interval $I \equiv [y_a, y_b]$, vanishing together with their first derivatives at the walls $y = y_a$, $y = y_b$:

$$\left.\begin{aligned} \varphi &= 0 \\ D\varphi &= 0 \end{aligned}\right\} \text{ on } \partial I \tag{5}$$

Boundary conditions impose in fact that both disturbance velocity components u', v' vanish at the walls, while eq. (1) implies

$$\begin{aligned} u' &\equiv \partial\psi/\partial y = D\varphi \;\; i\,\exp[\,i\alpha(x - ct)] \\ v' &\equiv \partial\psi/\partial x = -\,\varphi \;\; i\alpha\,\exp\,[\,i\alpha(x - ct)] \end{aligned} \tag{6}$$

A weak form of eq. (2) is obtained by imposing that

$$\int_I (L\,\varphi)\,\bar{w}\;dy = 0 \tag{7}$$

for any (complex-valued) function w such that its second derivative is square integrable on I, while w itself and its first derivative vanish on ∂I. In short, using standard notations:

$$<L\,\varphi, w> = 0 \qquad \forall w \in H^2_o(I) \tag{8}$$

A weak solution can be looked for within the same function space $H^2_o(I)$. In fact, the formal differential operator (3) may be split into

$$L = D^4 + L_2 \tag{9}$$

where

$$L_2 \equiv \alpha^2(2D^2 + \alpha^2) + i\alpha Re\,\left[(u-c)(D^2 + a^2) + \frac{d^2u}{dy^2}\right] \tag{10}$$

Integration by parts yields

(i) Unbounded intervals will be considered in Section 5.

$$\langle L\varphi, w\rangle = \langle D^4\varphi, w\rangle + \langle L_2\varphi, w\rangle$$
$$= \langle D^2\varphi, D^2 w\rangle + \langle L_2\varphi, w\rangle \tag{11}$$

Hence, integral (7) makes sense provided that both φ and w belong to $\overset{o}{H}{}^2(I)$.

A finite dimensional subspace $\Phi_N \subset \overset{o}{H}{}^2(I)$ is deviced through the choice of N (real-valued) linearly independent shape functions $\varphi_1, \varphi_2, \ldots, \varphi_N \in \overset{o}{H}{}^2(I)$:

$$\Phi_N = \{f | f = \sum_{k=1}^{N} z_k \varphi_k,\ z_k \in \mathbb{C},\ \varphi_k \in \overset{o}{H}{}^2(I)\} \tag{12}$$

If both φ and w are constrained to belong to Φ_N

$$\varphi = \sum_{j=1}^{N} z_j \varphi_j, \qquad z_j \in \mathbb{C} \tag{13^1}$$

$$w = \sum_{j=1}^{N} y_j \varphi_j, \qquad y_i \in \mathbb{C} \tag{13^2}$$

the following Galerkin equation is obtained

$$\sum_{j=1}^{N} \langle L\varphi_j, \varphi_i\rangle z_j = 0, \qquad i = 1,2, \ldots, N \tag{14}$$

which in matrix form is written as

$$\underline{L}\ \underline{z} = \underline{0} \tag{15}$$

with

$$L_{ij} = \langle L\varphi_j, \varphi_i\rangle \tag{16}$$

In order to make apparent the role played by the parameter c, L may be split into

$$L = B - c\,C \tag{17}$$

where

$$B \equiv (D^2 + \alpha^2)^2 + i\alpha Re\,[u\ (D^2 + \alpha^2) + d^2u/dy^2]$$
$$C \equiv i\alpha Re\ (D^2 + \alpha^2) \tag{18}$$

Correspondingly, eq. (15) may be restated as the algebraic eigenvalue problem

$$\underline{B}\ \underline{z} = c\ \underline{C}\ \underline{z} \tag{19}$$

with

$$B_{ij} = \langle B\varphi_j, \varphi_i\rangle$$
$$C_{ij} = \langle C\varphi_j, \varphi_i\rangle \tag{20}$$

Matrix $\underline{C}$ is nonsingular. In fact

$$\underline{C} = i\alpha Re\ \underline{G} \tag{21}$$

$$G_{ij} \equiv < D\varphi_j, D\varphi_i > + \alpha^2 < \varphi_j, \varphi_i > \tag{22}$$

Eq. (22) shows that matrix $\underline{G}$ is real, symmetric and positive definite; hence, it has an inverse, and

$$\underline{C}^{-1} = - i\ (\alpha Re)^{-1}\ \underline{G}^{-1} \tag{23}$$

In conclusion, the eigenvalue problem (19) may be reduced to the standard form

$$\underline{A}\ \underline{x} = c\ \underline{x} \tag{24}$$

with

$$\underline{A} = \underline{B}\ \underline{C}^{-1} \tag{25^1}$$

$$\underline{x} = \underline{C}\ \underline{z} \tag{25^2}$$

An eigenvalue c of matrix $\underline{A}$ with positive imaginary part

$$c_I \equiv I(c) > 0 \tag{26}$$

corresponds to a disturbance growing in size. If all the eigenvalues have negative imaginary part, the basic flow is stable.

3. HERMITE CUBICS

A mesh is defined on $I \equiv [y_a, y_b]$ by the finite set of reals (internal mesh points)

$$\{y_1, y_2, \ldots, y_n\} \tag{27}$$

such that

$$y_a \equiv y_0 < y_1 < y_2 < \ldots < y_n < y_{n+1} \equiv y_b \tag{28}$$

Any internal mesh point y_i (i = 1, 2, ..., n) belongs to two sub-intervals (elements)

$$I_i^- \equiv [y_{i-1}, y_i] \qquad I_i^+ \equiv [y_i, y_{i+1}] \tag{29}$$

and is associated with the following two shape functions on I (Hermite cubics):

$$\varphi_{2i-1}(y) = \begin{cases} h_1(1-\xi^-) & \text{in } I_i^- \\ h_1(\xi^+) & \text{in } I_i^+ \\ 0 & \text{elsewhere} \end{cases} \tag{30^1}$$

$$\varphi_{2i}(y) = \begin{cases} -\ell^- h_2(1-\xi^-) & \text{in } I_i^- \\ \ell^+ h_2(\xi^+) & \text{in } I_i^+ \\ 0 & \text{elsewhere} \end{cases} \tag{30^2}$$

with

$$\ell^- \equiv y_i - y_{i-1} \qquad \ell^+ \equiv y_{i+1} - y_i \tag{31^1}$$

$$\xi^- \equiv (y - y_{i-1})/\ell^- \qquad \xi^+ \equiv (y - y_i)/\ell^+ \tag{31^2}$$

and

$$h_1(\xi) \equiv 1 - 3\xi^2 + 2\xi^3 \tag{32^1}$$

$$h_2(\xi) \equiv \xi - 2\xi^2 + \xi^3 \tag{32^2}$$

It is straightforward to check that the set of functions

$$\{\varphi_k \mid k = 1,\ 2,\ \ldots,\ N \equiv 2\,n\} \tag{33}$$

is linearly independent, and that it spans a subspace Φ_N of $H_o^2(I)$.

Moreover, the basis (33) is computationally convenient: $\underline{L}$ is a block tri-diagonal matrix (each block being 2x2), whose entries are trivially integrated; $\underline{G}$ is well-conditioned for any reasonable mesh, and its triangular decomposition is numerically stable. It is worth noting that also the basic velocity profile u may usefully - though not necessarily - be represented trough Hermite cubics:

$$u(y) = h_1(\xi)u(y_i)+h_1(1-\xi)u(y_{i+1})+\ell h_2(\xi)u_y(y_i)-\ell h_2(1-\xi)u_y(y_{i+1}) \tag{34}$$

in $[y_i,\ y_{i+1}]$ $\quad (i = 0,\ 1,\ 2,\ \ldots,\ n)$

where $u_y \equiv du/dy$ and ℓ, ξ are defined as ℓ^+, ξ^+ in (31).

4. NUMERICAL RESULTS

Plane Couette and Poiseuille flows were chosen as test cases, and analysed for a standard set of values of α and Re, previously considered by different authors. For each problem, the complete set of eigenvalues and the eigenvector describing the fastest growing disturbance were computed using a QR algorithm and an inverse iteration procedure [1,2]. All cases were run in single precision arithmetic on a UNIVAC 1100/22. The computer time needed for assembling and solving a problem turned out to be roughly proportional to N^3.

4.1 Poiseuille flow

The basic velocity profile is given by

$$u(y) \equiv 1-y^2 \qquad \text{in } I \equiv [-1,\ +1] \tag{35}$$

The Reynolds number based on channel half-width and centre-stream velocity equals the inverse of the kinematic viscosity. The values

$$\alpha = 1, \qquad Re = 10^4 \tag{36}$$

are considered. Table 1 lists the eigenvalues with the largest positive imaginary part obtained with different non-uniform meshes - finer near the walls - in the range from n = 15 to n = 43.

n	eigenvalue
15	0.2371315 + i 0.0064739
21	0.2371629 + i 0.0039411
*21	0.2371626 + i 0.0039414
23	0.2372525 + i 0.0038414
25	0.2373214 + i 0.0037951
27	0.2373724 + i 0.0037685
29	0.2374104 + i 0.0037569
33	0.2374575 + i 0.0037491
37	0.2374853 + i 0.0037404
*43	0.2374854 + i 0.0037399

Table 1: Poiseuille flow (α = 1, Re = 10^4), present method; eigenvalues corresponding to the fastest growing disturbance obtained with different meshes; (*) only (n+1)/2 mesh points actually considered, accounting for symmetry.

Typically, the ratio between minimum and maximum element size within the same mesh is 1/5. Because the corresponding eigenfunction is even, all the results of Table 1 could have been obtained on the smaller computational domain [-1,0], with the single symmetry condition $D\varphi = 0$ prevailing at y = 0. This has actually been done for the two starred entries; since in Table 1 n always refers to the entire mesh defined in [-1, +1], N = 2(n-1)/2+1 = n when the symmetry is exploited (n being obviously odd).

Table 2 is a collection of results from the literature, to be compared with those presented in Table 1. The earliest numerical results are due to Thomas [3], who in 1953 used a five-point difference scheme on a uniform mesh: values (a) and (b) were obtained with 50 and 100 mesh points, respectively.

Author		eigenvalue
Thomas, 1953	(a)	0.2375006 + i 0.0035925
	(b)	0.2375243 + i 0.0037312
Grosch & Salwen, 1968		0.237413 + i 0.003681
Gary & Helgason, 1970	(a)	0.2373074 + i 0.0037562
	(b)	0.2375296 + i 0.0037424
Orszag, 1971		0.2375265 + i 0.0037397

Table 2: Poiseuille flow (α = 1, Re = 10^4), different methods.

More sophisticated difference techniques were used by Gary and Helgason [4] on a uniform mesh defined directly in the physical domain (value (a)) or after a suitable coordinate stretching was performed (value (b)). Symmetry was accounted for in both cases, and 43 computational mesh points adopted. A sharply different class of methods (Galerkin with orthogonal functions) was employed about the same years by Grosch and Salwen [5], and by Orszag [6]. The value cited in Table 2 was obtained by Grosh and Salwen using 50 even eigenfunctions of the self-adjoint operator $(D^2 + \alpha^2)^2$, with boundary conditions (5). Expansions in terms of Chebyshev polynomials were adopted by Orszag, whose results are seemingly the most accurate ever obtained for Poiseuille flow; the eigenvalues computed using more than 25 even-degree Chebyshev polynomials (from 26 to 50) agree with the value cited in Table 2 to within one part in 10^8.

Although an approximation to the Orr-Sommerfeld equation in terms of Hermite cubics has severe theoretical handicaps with respect to Chebyshev (or Legendre) expansions [6,7], present results compare quite well with those by Orszag. Notice also that the results in Table 1 were computed with a round-off error $\simeq 10^{-8}$, while the CDC 6600 used by Orszag had a far smaller round-off ($\simeq 10^{-14}$). This is to be stressed, because an increase of the round-off error from 10^{-14} to 10^{-8} spoils the accuracy of the Chebyshev method, as is demonstrated in Ref. [6].

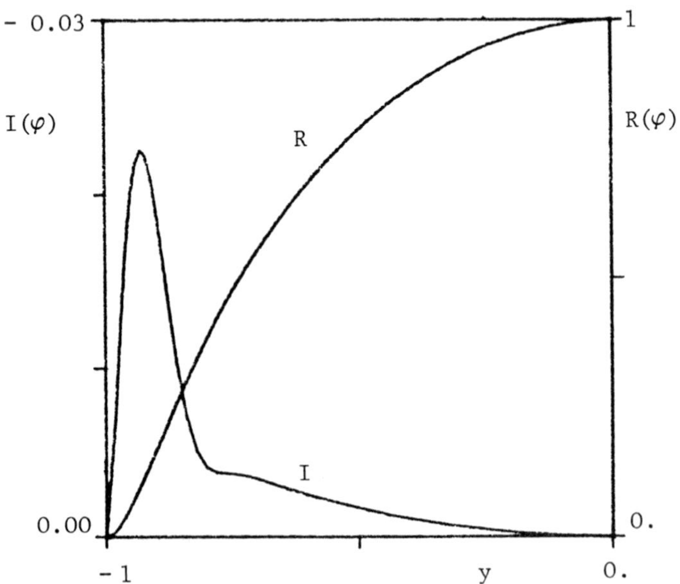

Fig. 1: Poiseuille flow (α=1, Re=10^4); real (R) and imaginary (I) parts of the eigenfunction corresponding to the fastest growing disturbance.

The graphs of the real and imaginary parts of the "most instable" mode are depicted in Fig. 1. Similar - and, in fact, nearly identical results - were presented in Refs. [3] and [5].

4.2 Couette flow

The basic velocity profile is given by

$$u(y) \equiv y \qquad \text{in} \quad I \equiv [-1, +1] \tag{37}$$

and is universally considered stable for any value of α and Re [8, 9]. Combining asymptotic analysis and numerical calculations Davey [9] established that, for sufficiently large values of Re (with reasonable accuracy for Re $\geqslant$ 200) and α not too small

$$Re^{1/2}c_I = f(Re^{-1/2}\alpha) \tag{38}$$

c_I being the largest imaginary part of the corresponding eigenvalues.

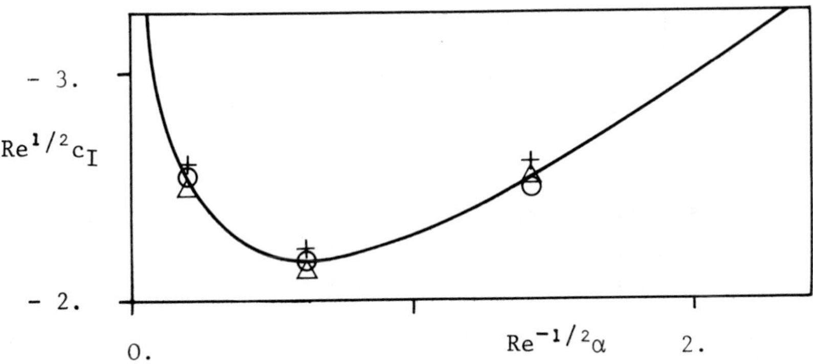

Fig. 2: Couette flow, (——) Davey, Re $\geqslant$ 200; present method: (O) Re = 200, (+) Re = 500, (△) Re = 800.

The graph of function (38) is reproduced from Ref. [9] in Fig. 2, where the results presently obtained for Re = 200, 500 and 800 are also plotted. According to Davey, function (38) attains its maximum at $Re^{-1/2}\alpha = 0.6321$: the corresponding values predicted by Davey and by the present method are compared in Table 3.

Author	Davey	present method		
Re	$\geqslant 200$	200	500	800
$Re^{-1/2} c_I$	- 2.1772	- 2.1828	- 2.2397	- 2.1473

Table 3: Couette flow ($\alpha = 0.6321\ Re^{1/2}$).

On the other hand, Gallagher and Mercer [8] determined that, when Re is fixed and α is sufficiently small ($\alpha < 75\ Re^{-1}$), all eigenvalues are purely imaginary. This finding is essentially in agreement with the results of the present analysis: for Re = 25 and $\alpha = 1$ all computed eigenvalues have real parts satisfying

$$|c_R| < 0.63 \cdot 10^{-4} \tag{39}$$

5. UNBOUNDED DOMAIN

As stated in the Introduction, the present study was undertaken aiming essentially at developing a combined numerical and experimental analysis of the stability of boundary layer flows, such as the wall jets analysed by Scibilia and Durox [10, 11]. This leads to considering an unbounded domain, say the interval $I \equiv [0, +\infty[$. Typically, basic velocity profiles are considered such that

$$u(y) = u_\infty \qquad \forall y \in [y_\infty, +\infty[\tag{40}$$

where u_∞, y_∞ are finite real constants.

The eigenfunction φ is subjected to the same boundary conditions (5), where now it is intended that

$$\lim_{y \to +\infty} \varphi(y) = \lim_{y \to +\infty} D\varphi(y) = 0 \tag{41}$$

The behaviour at infinity of φ, asymptotically valid for $Re \to \infty$, is easily established by solving the reduced Orr-Sommerfeld equation

$$(D^2 + \alpha^2)\varphi = 0 \qquad \text{in }]y_\infty, +\infty[\tag{42}$$

The set of solutions of eq.(42) satisfying the condition at infinity (41) is the one-parameter family of functions

$$\varphi(y) = \varphi(y_\infty) \exp[-\alpha(y - y_\infty)] \tag{43}$$

A simpleminded extension of the discretization procedure described in Section 3 is therefore obtained by defining the set of internal mesh points $\{y_1, y_2, \ldots, y_n\}$ such that

$$0 \equiv y_o < y_1 < y_2 < \ldots < y_n < y_{n+1} \equiv y_\infty < y_{n+2} \equiv +\infty \qquad (44)$$

All internal mesh points are treated as prescribed in Section 3 (see eqs. (30)); one further d.o.f. is attached to the mesh point y_{n+1}, associated with the shape function

$$\varphi_{2n+1}(y) = \begin{cases} h_1(1-\xi^-) + \alpha\ell^- h_2(1-\xi^-) & \text{in } I^-_{n+1} \\ \exp[-\alpha(y - y_\infty)] & \text{in } I^+_{n+1} \\ 0 & \text{elsewhere} \end{cases} \qquad (45)$$

which represents a "bubble" function over the infinite element $[y_n, +\infty[$ (refer to eqs. (29), (31) and (32) for the meaning of symbols).

On the basis of the results presented in Section 4 and of the reasonable treatment of the behaviour of eigenfunctions at infinity, it is deemed that the present numerical model should perform better than those devised for the analysis of the stability of a wall jet by Scibilia and Durox. In Ref. [10], these authors use as basis functions the set

$$\varphi_k(y) = y^2 e^{-ky} \qquad (46)$$

which generates an extremely ill-conditioned algebraic problem: when computing in double precision arithmetic with an IBM 360/168, the round-off blurs all information as soon as N = 10. In Ref. [11], after discarding Laguerre polynomials for similar reasons, Durox and Scibilia opt for

$$\varphi_k(y) = [\sin(k\pi y/y_\infty)]^2 \qquad (47)$$

integrating over the finite interval $[0, y_\infty]$. While not pathologically affected by the round-off errors, the results presented in Ref. [11] exhibit an undue dependence on the artificial choice of y_∞, seemingly in consequence of the tricky treatment of boundary conditions.

6. CONCLUDING REMARKS

It is felt that the present method, while simple in concept and application, should yield satisfactory results on a wide range of problems. Hopefully, numerical evidence supporting this claim will be given in forthcoming papers. In the writers' opinion, a crucial improvement lies in the development of rational, computer-generated coordinate transformations along the lines sketched in Ref. [12], coupled with the present finite element technique

REFERENCES

1. WILKINSON, J.H. and REINSCH, C.-Handbook for Automatic Computation, Vol. 2, Linear Algebra, Part.2, Springer-Verlag, New-York, 1971.

2. SMITH, B.T. et al.-Matrix Eigensystem Routines - EISPACK Guide, Lecture Notes in Computer Science 6, Springer-Verlag, Berlin, 1976.

3. THOMAS, L.H. - The Stability of Plane Poiseuille Flow. Phys. Rev., Vol. 91, pp. 780-783, 1953.

4. GARY, J. and HELGASON, R. - A Matrix Method for Ordinary Differential Eigenvalue Problems. J. Comp. Phys., Vol. 5, pp. 169-187, 1970.

5. GROSH, C.E. and SALWEN, H. - The Stability of Steady and Time-dependent Plane Poiseuille Flow. J. Fluid. Mech. Vol. 34, pp. 177-205, 1968.

6. ORSZAG, S.A. - Accurate Solution of the Orr-Sommerfeld Stability Equation. J. Fluid. Mech., Vol. 50, pp. 689-703, 1971.

7. ORSZAG, S.A. - Galerkin Approximations to Flows within Slabs, Spheres, and Cylinders. Phys. Rev. Lett., Vol. 26, pp. 1100-1103, 1971.

8. GALLAGHER, A.P. and MERCER, A.McD. - On the Behaviour of Small Disturbances in Plane Couette Flow. J. Fluid. Mech., Vol. 13, pp. 91-100, 1962.

9. DAVEY, A. - On the Stability of Plane Couette Flow to Infinitesimal Disturbances. J. Fluid. Mech., Vol. 57, pp. 369-380, 1973.

10. DUROX, D. and SCIBILIA, M.F. - Stabilité d'un Jet Pariétal Plan. Laboratoire d'Aérothermique - C.N.R.S., rapport 79-1, 1979.

11. SCIBILIA, M.F. and DUROX, D. - Stability of a Wall Jet. Private Communication.

12. DI CARLO, A., PIVA, R. and GUJ, G. - On Coordinate Transformation and Curvilinear Cell Models for Large Reynolds Number Flows, in Boundary and Interior Layers - Computational and Asymptotic Methods, Edited by J.J.M. Miller, pp. 246-250, Boole Press, Dublin, 1980.

PREDICTION OF THE BEHAVIOUR OF A CLOUD OF DISCRETE PARTICLES RELEASED IN A FULLY DEVELOPED TURBULENT PIPE FLOW, USING A NON-DISCRETE DISPERSIVE APPROACH

A. BERLEMONT and G. GOUESBET

Laboratoire de Thermodynamique - L.A. C.N.R.S. N° 230
Faculté des Sciences et Techniques de Rouen
B.P. 67 76130 MONT-SAINT-AIGNAN (France)

Abstract :

A non-discrete dispersive approach for the prediction of the behaviour of discrete particles embedded in turbulent flows is described. In this approach,a cloud of monodispersed particles is basically considered as a continuous field for which a transport equation is written and solved.

This transport equation contains a space-and time-dependent tensor of dispersion whose components can be computed from the characteristics of the turbulence and the properties of the particles.

The basic computer program corresponding to the above model is described and applied to the case of a cloud of monodispersed, non-vaporizing, discrete particles released on the axis of a fully developed turbulent pipe flow.

I - THE NON-DISCRETE DISPERSIVE APPROACH (N D D A)

The N D D A has been suggested previously in a 1 D-[1] then in a 3 D-version [2]. Assumptions are pointed out in the reference [3] and more extensively discussed in the reference [4]. This approach can be considered as a synthesis between the 3 D-diffusion of fluid particles described by Batchelor [5] and the 1 D-dispersion of discrete particles studied by Tchen [6] and described in a more familiar form by Hinze [7]. It is different of the tracking approach where the trajectory of particles is computed with the aid of the fundamental law of dynamics (see the references [8], [9], [10] and [11] among others). In the N D D A, a cloud of monodispersed, non-vaporizing (spherical) particles is considered as a continuous field of mean number-density n. The word dispersion is used,rather than the word diffusion, to point out the fact that the spreading phenomenon is mostly dependent on particles properties.

When the brownian dispersion is neglected, the transport equation for n reads :

$$\frac{\partial n}{\partial t} + u_{f,i} \frac{\partial n}{\partial x_i} = \frac{\partial}{\partial x_i} \varepsilon_{p,ij} \frac{\partial n}{\partial x_j} \quad (1)$$

where the Einstein rule of summation applies. The subscripts f and p refer to "fluid" and "particles" respectively, $u_{f,i}$ is the ith component of the mean fluid velocity in the cartesian coordinates system x_i. Mean fluid velocity $u_{f,i}$ and mean particle velocity $u_{p,i}$ are assumed equal in the present stage of development (an assumption to be removed in the near future). t is the time. $\varepsilon_{p,ij}$ is the tensor of particles dispersion.

The inhomogeneity of the turbulence appears explicitly through the fact that $\varepsilon_{p,ij}$ is spatially dependent, thus preventing us to write it outside the operator $\partial/\partial x_i$. Yet we shall assume that $\varepsilon_{p,ij}$ can be computed in the framework of homogeneous turbulence theories (local homogeneous hypothesis).Thus, this tensor reads [2], [3], [4] :

$$\varepsilon_{p,ij} = \int_0^\infty E_{pL,(ij)}(\omega) \frac{\sin \omega t}{\omega} d\omega \quad (2)$$

where ω is the angular frequency of an harmonic component into which random motion can be resolved, and $E_{pL(ij)}$ is the symmetric part of a Lagrangian (subscript L) spectral tensor $E_{pL,ij}$ characterizing the particles velocities fluctuations. This symmetric tensor is given by [2], [4] :

$$E_{pL,(ij)}(\omega) = \eta^2(\omega) E_{fL,(ij)}(\omega) \quad (3)$$

where $E_{fL,(ij)}(\omega)$ is the symmetric part of the tensor $E_{fL,ij}(\omega)$ defined through a cosine Fourier transform :

$$S_{L,ij}(\tau) = \int_0^\infty E_{fL,ij}(\omega) \cos \omega\tau \, d\omega \quad (4)$$

The tensor of correlation $S_{L,ij}(\tau)$ expresses the correlation $\overline{u'_{fL,i}(t_o)\, u'_{fL,j}(t_o + \tau)}$ between the i th component of the fluctuating velocity of a given fluid particle at the time t_o and the j th component of the fluctuating velocity of the same fluid particle at the time $(t_o + \tau)$. The reference time t_o does not matter since we only consider here a steady turbulence.

$\eta(\omega)$ is the so-called Tchen's amplitude ratio [6] which involves the properties of particles. It can be readily expressed through simple algebraic relations [7], [1], [2], [3], [4]. These very classical relations are not repeated here because of the lack of room. Nevertheless, it is worthwhile to point out that the amplitude ratio $\eta(\omega)$ can be expressed as a function of two adimensional numbers [12], namely the ratio $s = \rho_p/\rho$ of the particle and fluid densities respectively and the Stokes number $N_S = [\nu / \omega d^2]^{1/2}$ where ν is the fluid kinematic viscosity and d the diameter of particles.

This formalism can be (in principle) easily generalized to the case of polydispersed, vaporizing or coalescing particles [2] but our ambition will be here much more limited.

II - THE STRUCTURE OF THE CODE DISCO - 1

A basic computer program has been built in order to make computations in the framework of the N D D A. This program, the so-called DISCO - 1 (DISCO standing for DISpersion COmputing), involves some simplifications of the previous analysis and will be here applied to the case of a cloud of monodispersed, non-vaporizing spherical droplets of water released on the axis of a fully developed turbulent pipe flow of air at ambient temperature. The complete listing is available from [4]. The structure of this code is now described.

Part I : turbulence

This part is aimed at the prediction of the turbulence characteristics included in the relation (1), namely the mean velocities $u_{f,i}$ and the spectral tensor $E_{fL,(ij)}(\omega)$ required to compute the tensor of dispersion $\varepsilon_{p,ij}$. It contains four sections.

Section I-1 : (k-ε)-model

This section predicts the mean velocities $u_{f,i}$, the turbulence energy k, its rate of dissipation ε, its production G (or P) due to the interaction between the velocities fluctuations and the mean velocities, with the aid of a (k-ε)-model. See the reference [13] in the present conferences.

Section I-2 : fluctuating velocities correlations

This section predicts the fluctuating velocities correlations $\overline{u'^2_x}$, $\overline{u'^2_r}$, $\overline{u'^2_\phi}$ and $\overline{u'_x u'_r}$ with the aid of algebraic relations deduced from a second-order closure scheme. Again, see [13].

Section I-3 : Eulerian spectra

The code DISCO-1 does not take into account for the non-diagonal terms of the tensor of dispersion, although they should be reintroduced in a more refined version.

The particles are released on the axis of an axisymmetric turbulent flow. Thus, the problem of particles dispersion is also axisymmetric and there is no angular dispersion.

The axisymmetric flow is fully developed in a pipe : $u_{f,\phi} = u_{f,r} = o$ (see figure 1 of the reference [13] ; the only change in notation is the appeareance of the subscript f).

Thus, in the (x, r, ϕ)-system, the transport equation reads :

$$\frac{\partial n}{\partial t} + u_{f,x} \frac{\partial n}{\partial x} = \frac{\partial}{\partial x} \varepsilon_{p,xx} \frac{\partial n}{\partial x} + \frac{1}{r} \frac{\partial}{\partial r} r\varepsilon_{p,rr} \frac{\partial n}{\partial r} \tag{5}$$

The tensor of dispersion reduced to two coefficients of dispersion which will be noted from now on $\varepsilon_{p,x}$ and $\varepsilon_{p,r}$.

From the relations (2) and (3), we have :

$$\varepsilon_{p,x} = \int_0^\infty \eta^2 (\omega)\ E_{fL,(xx)} (\omega)\ \frac{\sin \omega t}{\omega}\ d\omega \tag{6}$$

$$\varepsilon_{p,x} = \int_0^\infty \eta^2 (\omega)\ E_{fL,(xx)}(\omega)\ \frac{\sin \omega t}{\omega}\ d\omega \tag{7}$$

where $E_{fL,(xx)}$ will be from now on noted $E_{fL,x}$. From the relation (4) :

$$\overline{u'^2_{fL,x}} = \int_0^\infty E_{fL,x}(\omega)\ d\omega \tag{8}$$

Similar relations hold with the subscript r instead of x.

In order to compute the monodimensional Lagrangian spectra $E_{fL,x}(\omega)$ and $E_{fL,r}(\omega)$, this section I-3 carries out a first step, namely the computations of monodimensional Eulerian spectra $E_{fE,x}(k_x)$ and $E_{fE,r}(k_r)$. These computations have been discussed in the reference [13].

Section I-4 : Lagrangian spectra

The aim of this section is to operate a transformation from the Eulerian densities $E_{fE,i}(k_i)$, where i stands either for x or r, to the Lagrangian densities $E_{fL,i}(\omega_i)$ (A subscript i will be sometimes used to specify ω, for convenience). This transformation involves a change in the argument (from a wave-number k_i to an angular frequency ω_i) plus a change in the frame of reference (from an Eulerian frame to a Lagrangian one). This is indeed a very difficult problem and there is apparently no rigorous solution available at the present time. Thus, we have chosen a very simple rule to achieve this transformation :

$$E_{fL,i} (\omega_i) = C_{EL}\ E_{fE,i}\ (k_i)\ /(\overline{u'^2_{(i)}})^{1/2} \tag{9}$$

where a subscript between parentheses means that the Einstein rule of summation does not apply.

The relation between ω_i and k_i is :

$$\omega_i = k_i\ (\overline{u'^2_{(i)}})^{1/2} \tag{10}$$

The physical basis for this transformation is given in the reference [4] although it is mainly justified by its consequences discussed some distance below. Nevertheless, the authors are not completely satisfied with it and a new effort is planned on this point to build the next version of the code.

The constant C_{EL} has been adjusted to 0.5 (see section III of this paper)

Part II : Particles

The part II is concerned with the behaviour of the discrete particles in the above predicted turbulent field. It is assumed that the suspension is diluted enough so that the turbulence is not affected by the presence of the particles.

This part contains two sections.

Section II-1 : coefficients of dispersion

This section computes the coefficients of dispersion $\varepsilon_{p,x}$ and $\varepsilon_{p,r}$. Results are discussed in the section III of this paper.

Section II-2 : solving the transport equation

This section solves the transport equation (5). Details and results will be discussed in the section IV of this paper.

III - THE COEFFICIENTS OF DISPERSION

The time t appearing in the relation (7) is taken as x/u_{xM} where x is the distance between the release location of the cloud and its present location, and u_{xM} is the mean flow velocity on the axis.

Results will be discussed under the form of ratios ε_i defined by :

$$\varepsilon_i = \varepsilon_{p,i}\ (t)\ /\ \varepsilon_{p,i}\ (t \rightarrow \infty) \tag{11}$$

where $\varepsilon_{p,i}\ (t \rightarrow \infty)$ is the coefficient of dispersion of discrete particles for an infinite time of dispersion. This asymptotic value is equal to the asymptotic value of the coefficient of diffusion $\varepsilon_{f,i}\ (t \rightarrow \infty)$ and is given by (examine the relation (7)) :

$$\varepsilon_{p,i}\ (t \rightarrow \infty) = \varepsilon_{f,i}\ (t \rightarrow \infty) = \frac{\pi}{2}\, E_{fL,i}(o) \tag{12}$$

Thus, from (12) and (7) :

$$\varepsilon_i = \frac{2}{\pi\, E_{fL,i}(o)} \int_o^\infty \eta^2(\omega)\ E_{fL,i}(\omega)\ \frac{\sin \omega t}{\omega}\ d\,\omega \tag{13}$$

From the Eulerian spectral densities modelling [13] and the Eulerian → Lagrangian transformation previously described, we have :

(i) if $\omega \leqslant \omega_{c,i}$: $E_{fL,i}(\omega_i) = E_{fL,i}(\omega_{c,i}) = E_{fL,i}(0)$ (14)

(ii) if $\omega \geqslant \omega_{c,i}$: $E_{fL,i}(\omega_i) = E_{fL,i}(0) \cdot (\omega/\omega_{c,i})^{-5/3}$ (15)

where the critical angular frequency $\omega_{c,i}$ corresponds to the critical wave-number $k_{c,i}$ through the relation (10).

Thus :

$$\varepsilon_i = \frac{2}{\pi} \left\{ \int_0^{\omega_{c,i}} \eta^2 \frac{\sin \omega t}{\omega} d\omega + \int_{\omega_{c,i}}^{\infty} \eta^2 \left(\frac{\omega}{\omega_{c,i}}\right)^{-5/3} \frac{\sin \omega t}{\omega} d\omega \right\} \quad (16)$$

This ratio evolves from 0 to 1 when the time t evolves from 0 to ∞. Generally speaking, it depends on the dispersion time t, on the diameter d of particles (for a given particle material in a given fluid) and on the turbulence through the single parameter $\omega_{c,i}$.

Integrals have been computed by means of the classical Simpson's method [14]. We observed that, in the relevant ranges of ω and d :

$$\eta^2 (\omega,d) = g (\omega d^2) \quad (17)$$

The figures 1 to 3 show the ratios ε_i for three radial locations, and for different dispersion times, versus diameters. In agreement with the model, ε_i decreases with the diameter and increases with the time. For r = 0.25cm, ε_x and ε_r are found equal at any time. This fact illustrates the isotropic character of the turbulence near the axis. For r=3.77cm and 8.79cm, ε_x and ε_r are different except when the time t is high enough for both ratios having reached (or nearly reached) their asymptotic values.

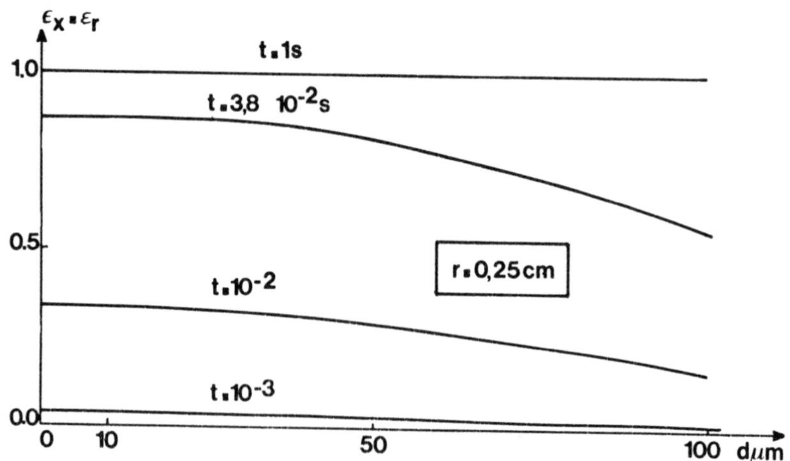

Fig. 1 - Reduced dispersion coefficients

ε_x or ε_r

$t = 1s$ ($\varepsilon_x = \varepsilon_r$)

1.0

r = 3,77cm

0.5

$t = 10^{-2}s, \varepsilon_x$

$t = 10^{-2}s, \varepsilon_r$

$t = 10^{-3}s$ ε_r ε_x

0.0

0 10 50 100 dμm

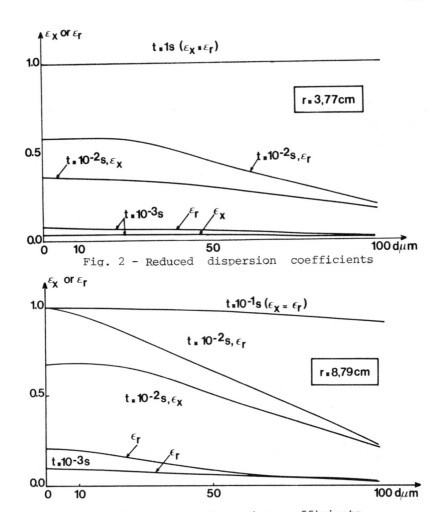

Fig. 2 - Reduced dispersion coefficients

ε_x or ε_r

1.0

$t = 10^{-1}s$ ($\varepsilon_x = \varepsilon_r$)

$t = 10^{-2}s, \varepsilon_r$

r = 8,79cm

0.5

$t = 10^{-2}s, \varepsilon_x$

ε_r

ε_r

$t = 10^{-3}s$

0.0

0 10 50 100 dμm

Fig. 3 - Reduced dispersion coefficients

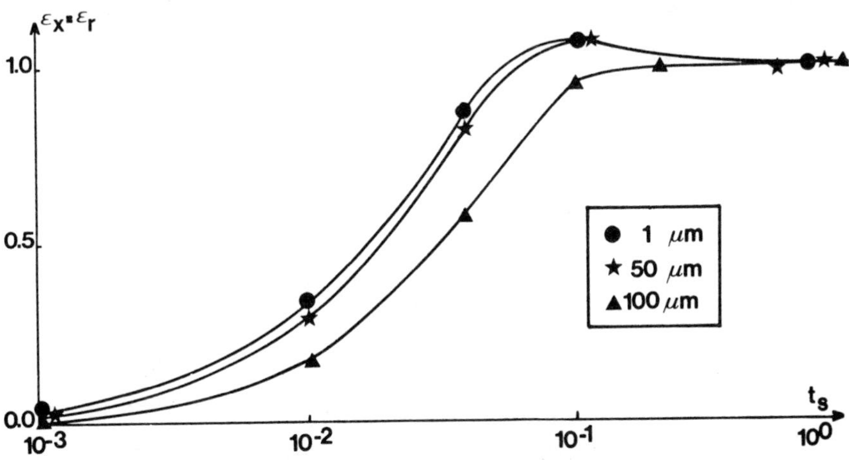

Fig. 4 -Reduced dispersion coefficients

The figure 4 shows $\varepsilon_x = \varepsilon_r$ at $r = 0.25$cm versus the dispersion time, for three different diameters. Starting from 0 at $t = 0$, the ratios increase to reach their asymptotic values 1 when t is high enough. Nevertheless, this increase is not monotone since the ratios become higher than 1 before decreasing asymptotically to 1. This behaviour is certainly not correct and should certainly be considered as a discrepancy to remove.

The increase of ε_i can be characterized by a relaxation time t_r (d). At the present stage of development, we are not able to give a completely relevant definition of it. So, it is fair to define it as the time for which ε_i=0.5. In agreement with the model, t_c(d) increases with the diameter. Furthermore, we found $t_c(0) = t_c(1) \sim 2.10^{-2}$s (see figure 4). This time $t_c(0)$ is a relaxation time for the evolving of the coefficients of diffusion of fluid particles. So, it should have the same order of magnitude as the Lagrangian time scale τ_L [4,7,15] defined by :

$$\tau_L = \varepsilon_{f,r} \ (t \to \infty) \ / \ \overline{u'^2_{fL,r}} \tag{18}$$

To independently estimate τ_L, we shall use for $\varepsilon_{f,r}$ $(t \to \infty)$ a value taken from Flint et al [16] (see also further discussion) and estimate $\overline{u'^2_{fL,r}}$ by $\overline{u'^2_r}$ predicted in the reference [13].This last assumption ($\overline{u'^2_{fL,r}} = \overline{u'^2_r}$) is rigorous for incompressible, homogeneous, turbulence [17]. Then, it is found $\tau_L \sim 1.10^{-2}$s which has indeed the same order of magnitude as t_c (0), as expected.

For small dispersion times, the increase of the ratios with respect to the time should be linear according to the limiting law :

$$\varepsilon_i = \{ \frac{2}{\pi} \cdot \frac{1}{E_{fL,i}(0)} \int_0^\infty \eta^2 \ (\omega) \ E_{fL,i} \ (\omega) \ d\omega \}.t \tag{19}$$

Comparisons have been made between results given by (19) and (16) and the agreement has been found very satisfactory [4].

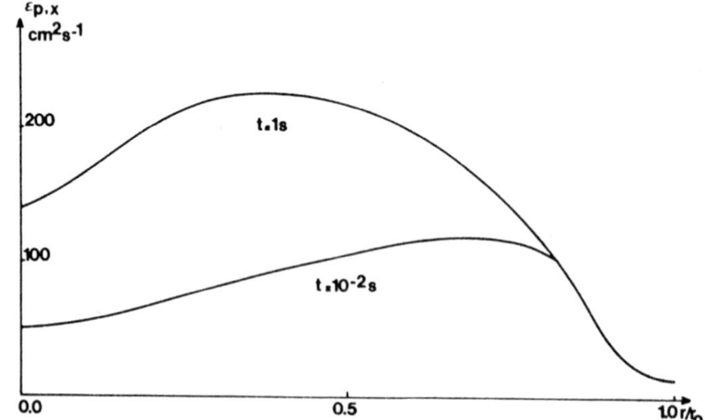

Fig. 5 - Dispersion coefficient

The figure 5 shows $\varepsilon_{p,x}$ for d=5 μm as a function of r at two different dispersion times. In agreement with the model, we again observe that $\varepsilon_{p,x}$ increases with time. We also observe that the maximal value of $\varepsilon_{p,x}$ is located at different locations for different times. This fact should be commented. At small times, we have from (13), (9), (10) and (41) in [13]:

$$\varepsilon_{f,x}\ (t \to 0) = \{ \int_0^\infty E_{fL,x}\ (\omega_x)\ d\,\omega_x \}.\ t$$

$$= \{ \int_0^\infty \frac{C_{EL}}{(\overline{u_x'^2})^{1/2}} E_{fE,x}(k_x)\ d\,k_x \overline{(u_x'^2)}^{1/2} \}.t = C_{EL}\ \overline{u_x'^2}\ .t \qquad (20)$$

Thus the maximal value of $\varepsilon_{p,x}(t \to 0)$ tends when $d \to 0$ to have the same location as the maximal value of $\overline{u_x'^2}$, that means to be located near the wall. On the other hand, at large diffusion times, the maximal value of $\varepsilon_{f,x}$ corresponds to the maximal value of $E_{fL,x}$ (0) according to the relation (12).

According to Flint et al [16], $\varepsilon_{f,r}$ $(t \to \infty)$ is given near the axis by :

$$\varepsilon_{f,r}\ (t \to \infty) = 0.00082\ u_{xM}\ (2r_0) \qquad (21)$$

corresponding to a Peclet number u_{xM} $(2r_0)$ $/\varepsilon_{f,r}$ equal to $Pe \sim 1200$. In the present case u_{xM} = 36 m/s and r_0 =0.123m. Thus we should have $\varepsilon_{f,r}$ $(t \to \infty) \sim 75 cm^2 s^{-1}$. This value has been used to adjust the constant C_{EL}. With C_{EL} = 0.5 and incorporating Laufer's experimental spectrum $E_{fE,x}$ (k_x) in our computations, we also find $\varepsilon_{f,r}$ $(t \to \infty) \sim 75 cm^2.s^{-1}$. But, with a modelled spectrum, we have $\varepsilon_{f,r}$ $(t \to \infty) \sim 140\ cm^2.s^{-1}$, that means a value basically two times too high. This discrepancy points out that the Eulerian spectra predictions, particularly for k_x=o, should be improved.

Nevertheless, the value $\varepsilon_{f,r}$ $(t \to \infty)$ $\sim 140 cm^2.s^{-1}$ gives $Pe \sim 630$. According to a review by Gohar et al [18], the Peclet number Pe is a constant at high Reynolds numbers. But experimental data basically spreads in the range 600/1000. Taking into account for this experimental spreading, our predictions can be considered as satisfactory.

A study of the dispersion of droplets of water in a turbulent flow of air has been carried in Rouen. Some results are available from [18] and more results will be soon published [19]. The dispersion takes place in a turbulent wake produced by the encounter of two parallel fully developed channel flows. Thus a precise comparison between our predictions and experiments cannot be made since turbulent situations were different. Such a precise comparison is planned to be done in the near future. The main result is that both our predictions and Gohar's experiments conclude that the dispersion decreases when the particles diameters increase, an important conclusion since others authors claimed a different result. See, for instance, the reference [20].

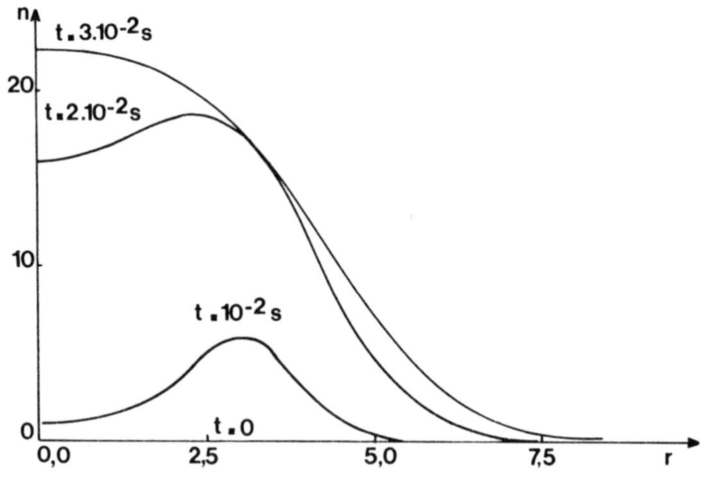

Fig. 6 - Tangential dispersion

IV - EVOLVING OF THE CLOUD

The equation (5) has then been solved by means of a semi-implicit method with alternate directions [14]. Details and extensive results are available from the reference [4]. We shall here limit our discussion to only one figure.

When t = 0, the cloud of particles (d = 5μm) is cylindrical. Its length is 3.8cm and its diameter is 6.6cm. By convention, n is equal to 100 in the cloud (and 0 outside).

The figure 6 shows the radial dispersion 3.8cm behind the median plane. At t = 0, there is no particles in this part of the flow. At $t = 10^{-2}$s, we observe that the maximal value of n is obtained for r = 3.3cm, due to a complex coupling between the transport by dispersion and by differential convection (particles which are far from the axis are going slower than on the axis). The dispersion (random motion) will progressively wash out in this plane this structure mainly attributed to the differential convection (ordered motion). At $t=3.10^{-2}$s, the profile of n in this plane looks like the one observed at the same time in the median plane.

Basically, the net result of the coupling between convection and dispersion should be to produce an apparent dispersion higher in the x-direction than in the r-direction. This enhancement of apparent dispersion in shear flows is a well-known phenomenon in atmospheric layers [15]

CONCLUSION :

A non-discrete dispersive approach for modelling the behaviour of discrete particles embedded in turbulent flows has been presented.

A computer program has been built on to achieve the predictions corresponding to the NDDA. It has been applied to the case of a cloud of monodispersed spherical particles of water released on the axis of a fully developed turbulent pipe flow of air.

The results lead us to the conclusion that the NDDA should be further developed in the future since this approach appears to be very promising.

REFERENCES :

[1] GOUESBET G., BERLEMONT A., C.R. Acad. Sc., t.288, 961, 1979.

[2] GOUESBET G., BERLEMONT A., A three-dimensional dispersion model for the behaviour of vaporizing particles in a turbulent fluid. Proceedings of the Sixth Biennial Symposium on turbulence. University of Missouri-Rolla, Oct. 8-10, 1979.

[3] GOUESBET G., BERLEMONT A., Assumptions and problems connected with the dispersive approach of two-phase flows modelling. Internal Report TTI/GB/79/04/11.

[4] BERLEMONT A., Modélisation et prédiction du comportement de particules discrètes dans des écoulements turbulents. Thèse de 3ème Cycle, Université de Rouen, 1981.

[5] BATCHELOR G.K., Aust. Journal Sci. Res., 2, 436,1949.

[6] TCHEN C.M., Mean value and correlation problems connected with the motion of small particles suspended in a turbulent fluid. Ph. D. dissertation, Martinus Nijhoff. The Hague, 1947.

[7] HINZE J.O., Turbulence, Mc Graw Hill Book Company, Inc., 1959.

[8] DOMINGOS J.J.D., ROZIZ L.C.F., Dynamics of an evaporating or burning droplet. European Symposium on Combustion, Sheffield, 1973.

9 JUREWICZ J.T.,STOCK D.E., A numerical model for turbulent diffusion in gas-particle flows. ASME publication (Received at ASME Headquarters : august 16, 1976 - Contributed by the FED of the ASME for presentation at the Winter Annual Meeting, New York).

10 LOCKWOOD F.C., SALOOJA A.P. and S.A. SYED
Some exploratory calculations of the flow and combustion in a ciment kiln. Department of Mechanical Engineering. Imperial College of Science and Technology, March 1978.

11 BOYSON F. and SWITHENBANK J.,
Spray evaporation in recirculating flow, 17th International Symposium on Combustion, Leeds, 1978.

12 HJELMFELT A.T. and MOCKROS L.F.
Appl. Sci. Res., 16, 149, 1966.

13 GOUESBET G. and BERLEMONT A.
I. Prediction of turbulent fields, including fluctuating velocities correlations and approximate spectra, by means of a simplified second-order closure scheme : the round free jet and developed pipe flow. Present conferences.

14 CARNAHAN B., LUTHER H.A., WILKES J.O.
Applied numerical methods, John Wiley and Sons, Inc., 1969.

15 CSANADY G.T., Turbulent diffusion in the environment. Reidel Publishing Company, 1973.

16 FLINT D.L., KADA H., HANRATTY T.J., AIChE Journal, 6, 2, 325, 1960.

17 TENNEKES H. LUMLEY J.L.,
A first course in turbulence, M I T-Press, Cambridge, Mass, 1972.

18 GOHAR M., ALLANO D., LEDOUX M.,
Rapport A.T.P., 3273, Août 1980.

19 GOHAR M.,
Thèse de 3ème Cycle, Université de Rouen. To be Published.

20 LILLY G. P.,
Ind. Eng. Chem. Fundam. 12, 3, 268, 1973.

Prediction of Turbulent Mixing In Confined Co-axial Reacting Jets

M.M.M. Abou-Ellail[(I)] and T.W. Abou-Arab[(II)]
Mechanical Engr. Dept., Cairo Univ.,Cairo, Egypt

SUMMARY

Theoretical studies of fluid dynamic characteristics of gaseous flames are carried out. This study considers a co-axial diffusion flame existing in a circular air duct.

The present work describes a numerical solution procedure, embodied in a developed computer program applicable to parabolic reacting and non-reacting flows. It shares some common features with the numerical methods used for solving nonlinear, strongly coupled differential equations. These methods are mainly based upon the solution of the finite difference form of momentum, mass and energy governing equations with the assistance of digital computers. The solution of the above mentioned equations requires physical models for turbulence, chemistry, and also for the thermodynamic properties.

The calculations are carried out for different velocity ratios of fuel and air. The calculated results for both reacting and nonreacting coaxial jets show an increase in the mean flow velocity and turbulence intensity resulting from combustion. The developed computation procedure is applied to methane and hydrogen flames to assess its validity.

(I) Associate Prof. of Heat Engines: Present address: Information and computer services depart.,Qater G.P.Co. P.O. Box 3212-Doha-Qater.

(II) Dr.-Ing. Present address: University of California, Irvine School of Engineering, Mechanical Engineering Department, Irvine, CA 92717, U.S.A.

INTRODUCTION

Combustion chamber analysis embrace the subject area of: fluid mechanics, convective heat and mass transfer, turbulence, thermodynamics, chemical kinetics and thermal radiation. Although many modern combustion devices now in operation have reached a high level of performance, they still are considered far from ideal. It is, then, small wonder that combustion devices design has developed as more of an 'art' than a science, guided by faith and a myriad of empirical, and largely algebraic, correlations each of which has been eked out of result for a very restricted range of experimental conditions. Recent developments of optical measuring instruments (e.g. Laser velocimetry & Raman scattering) have made measurements of velocity, temperature, and major combustion species in a combustor feasible in the future.

The availability of very large, high speed digital computer has encouraged efforts to simulate various types of combustors. The ultimate goal is a general prediction procedure enabling the performance of a proposed combustor to be accurately determined in the absence of a costly experimental program. Its attainment demands the adoption by the designer of a more fundamental approach. At the fundamental level the probelm is two-fold. Firstly, it is necessary to be able to solve, with acceptable economy and accuracy, the partial diffrential equations governing the mass, momentum, and energy transfer. Secondly, good physical models must be available for the turbulence, chemistry, and also for the thermodynamic properties.

The solution of the governing equations is a purely mathematical problem which, with the assistance of the computer and modern finite difference techniques, is a tractable one. The need to construct physical models arises because many of the physical and chemical processes which occur in a combustion chamber are far too complex to be handled at a completely fundamental level (e.g. the time scales of the smallest turbulence eddies [1]). These numerical solutions are widely used in many practical applications [2,3].

This present work is another attempt to investigate numerically reacting and non-reacting parabolic flows, utilizing a developed numerical procedure [4,5] that solves the continuum equations of conservation of mass, momentum and energy, equipped with appropriate models of turbulence and reaction kinetics. The developed numerical scheme consists of discretizing the continuum equations and solving the finite difference analoges after linearization iteratively. The two-equation turbulence model, used to represent the turbulent characteristics, has been appraised in detail by Gosman et. al [6].

In the present investigation, the validity of the developed numerical procedure is assessed by comparing the numerical results with the experimental ones which were produced recently by Wang et al [7] on a similar system.

FORMULATION OF THE PROBLEM

As it is usually the case in turbulent flow problems the instantaneous value of any dependent variable ϕ is expressed in terms of a time-averaged component $\bar{\phi}$ and a fluctuating one ϕ', i.e. $\phi = \bar{\phi} + \phi'$. However the fluctuating component of density is neglected i.e. $\rho = \bar{\rho}$.

The conservation equations can thus be written in term of these components in compact tensor notations as follows:

Mass continuity
$$\frac{\partial}{\partial x_i}(\rho \bar{u}_i) = 0 \tag{1}$$

Momentum equations
$$\frac{D\rho\bar{u}_i}{Dt} = \frac{\partial}{\partial x_j}[\mu(\frac{\partial \bar{u}_i}{\partial x_j} + \frac{\partial \bar{u}_j}{\partial x_i} - \frac{2}{3}\frac{\partial \bar{u}_k}{\partial x_k}\delta ij)$$
$$- \bar{p}\partial ij] - \frac{\partial}{\partial x_j}(\rho\overline{u_i'u_j'}) \tag{2}$$

Energy equation
$$\frac{D\rho\bar{\tilde{h}}}{Dt} = \frac{\partial}{\partial x_i}[\Gamma_h \frac{\partial \bar{\tilde{h}}}{\partial x_i}] - \frac{\partial}{\partial x_i}[\overline{(\rho u_i')\tilde{h}'}]$$
$$+ S \tag{3}$$

The equation of conservation of a chemical species, ℓ, is
$$\frac{D\rho\bar{m}_\ell}{Dt} = \frac{\partial}{\partial x_i}[\Gamma_\ell \frac{\partial \bar{m}_\ell}{\partial x_i}] - \frac{\partial}{\partial x_i}[\overline{(\rho u_i')m_\ell'}]$$
$$+ R_\ell \tag{4}$$

In the above equations t is time, u_i is velocity in the direction of coordinate x_i, ρ is density, p is pressure, μ is the laminar dynamic viscosity and the operator δ_{ij} (Kronecker Delta) is unity for i = j and zero when i ≠ j. In equation (3) $\tilde{h}$ represents stagnation enthalpy, Γ_h is equal to the fluid thermal conductivity divided by its constant pressure specific heat and S is a source term. In equation (4) m_ℓ stands for the mass fraction of ℓ, Γ_ℓ is the mass diffusion coefficient for this specie, and R_ℓ is its chemical rate of production.

TURBULENCE AND COMBUSTION MODELS

In order to solve the above set of equations further relations should be provided for the fluctuating component correltations. The present work made use of the two-equation k-ε turbulence model [6]. This model utilizes an eddy viscosity concept, and expresses the Reynolds stresses ($\overline{\rho u'_i u'_j}$) in terms of mean velocity gradients and the eddy viscosity in the following manner

$$- \rho\overline{u'_i u'_j} = \mu_T[(\frac{\partial \bar{u}_j}{\partial x_i} + \frac{\partial \bar{u}_i}{\partial x_j}) - \frac{2}{3}\frac{\partial \bar{u}_k}{\partial x_k}\delta ij] - \frac{2}{3}\rho k\delta_{ij} \quad (5)$$

where μ_T is a turbulent viscosity which may be related to the kinetic energy of turbulence (k) and its dissipation rate (ε) through the following equation

$$\mu_T = C_\mu \rho k^2/\varepsilon \quad (6)$$

where C_μ is a constant of the model. k and ε are to be determined from the solution of their transport equations which are respectively:

$$\frac{D\rho k}{Dt} = \frac{\partial}{\partial x_i}(\frac{\mu_T}{\sigma_k}\frac{\partial k}{\partial x_i}) + \mu_T \frac{\partial \bar{u}_i}{\partial x_j}(\frac{\partial \bar{u}_i}{\partial x_j} + \frac{\partial \bar{u}_j}{\partial x_i}) - C_\mu \rho^2 \frac{k^2}{\mu_T} \quad (7)$$

and

$$\frac{D\rho \varepsilon}{Dt} = \frac{\partial}{\partial x_i}(\frac{\mu_T}{\sigma_\varepsilon}\frac{\partial \varepsilon}{\partial x_i}) - C_1 C_\mu \rho k \frac{\partial \bar{u}_i}{\partial x_j}(\frac{\partial \bar{u}_i}{\partial x_j} + \frac{\partial \bar{u}_j}{\partial x_i}) - C_2\rho\frac{\varepsilon^2}{k} \quad (8)$$

As for the other quantities, enthalpy and mass fraction of species, turbulent transport correlations are determined from the Bossinesq approximation

$$\overline{(\rho u'_i)\phi'} = \Gamma_\phi \frac{\partial \bar{\phi}}{\partial x_i} \quad (9)$$

where $\Gamma_\phi = \frac{\mu_T}{\sigma_\phi}$ (10)

where σ_ϕ is a turbulent Prandtl/Schmidt number of order unity.

Regarding the combustion model global kinetics are assumed. This supposes that the reaction between the fuel and oxidant takes place in one-step, irreversible reaction, i.e. fuel + oxidant → products. As a consequence of the global kinetic model only one equation is required to describe the conservation of various species in the system.

This simple supposition allows the heat release rate to be well evaluated, but, it is not essential. The further conventional assumption is made, which is entirely justifiable, that the turbulent exchange coefficients are all equal. A consequence of these two assumptions is that the quantity of

$$\phi = sm_{fu} - m_{ox} \tag{11}$$

where s is the stoichiometric oxygen requirement by mass, and m_{fu} and m_{ox} are the fuel and oxidant mass fractions, becomes a passive scalar. It is convenient to work with a dimensionless ϕ, defined by

$$f = (\phi - \phi_o)/(\phi_1 - \phi_o) \tag{12}$$

where the subscripts 1 and o designate the fuel- and oxidant-bearing streams. The time averaged value of the balance equation for the mixture fraction is expressed by

$$\frac{\partial \rho \bar{u}_i f}{\partial x_i} = \frac{\partial}{\partial x_i} (\Gamma_f \frac{\partial \bar{f}}{\partial x_i}) - \frac{\partial}{\partial x_i} \overline{(\rho u_i')f'} \tag{13}$$

In addition, the variation of specific heats of various individual species with temperature is considered by the calculation.

The constants in the above mentioned equations are given in Table 1 and were taken from [6] and [8].

FINITE DIFFERENCE EQUATIONS

The solution of the above set of conservation equations is difficult even for nonreacting laminar flows. It can only be approached by numerical methods. The present method of solution entails subdividing the circular air duct (solution domain) into a number of finite volumes or 'cells' each of which encloses an imaginary grid node, at which all scaler variables ($\bar{\bar{h}}$, p, k, ε, and $\bar{f}$) are stored, while the velocity components $\bar{u}_i$ are chosen to lie on the cell boundaries where they are needed for mass flux computations. The set of equations to be integrated may be compactly represented in the following appropriate form:

$$\frac{\partial}{\partial x_i}(\rho \bar{u}_i \bar{\phi} - \Gamma_\phi \frac{\partial \bar{\phi}}{\partial x_i}) = S\phi \tag{14}$$

where $\bar{\phi}$ is any of the variables mentioned above, Γ_ϕ is the corresponding effective exchange coefficient and S_ϕ is a source term. This equation may formally be integrated, over a typical cell having volume V_p, to yield

$$\Sigma_b \, [\int_{A_b} (\rho \bar{u}_i \bar{\phi} - \Gamma_\phi \frac{\partial \bar{\phi}}{\partial x_i} \, dA_b] = \int_{V_p} S_\phi dV \tag{15}$$

where the summation is over all boundaries of the cell, A_b is the area of the cell boundary in question, and $\bar{u}_i$ and x_i now represent the velocity and coordinate normal to the boundary.

The integrals on the left-hand side, each of which represents the total transport across each of the cell boundaries by convection and diffusion. The integral on the right-hand side is expressed in linearized form as $S_u + S_p\bar{\phi}_p$, where the S's are deduced from the differential equation in question. The resultant difference equation has the following form[2]

$$a_p\bar{\phi}_p = \sum_n a_n \bar{\phi}_n + S_u \tag{16}$$

where a_n are influence coefficients, $a_p = \sum a_n - S_p$ and $\sum_n$ denotes summation over neighbouring nodes of a typical grid node p. The a_n coefficients are calculated using hybrid differences in order that they are always positive or zero[2].

Finally, the continuity equation is integrated to give

$$\sum_b \dot{m}_b = \sum_b \rho_b \bar{u}_{i,b} A_b = 0 \tag{17}$$

where $\dot{m}_b$ is mass flow across each cell boundary.

SOLUTION PROCEDURE

A preliminary set of velocities $\bar{u}_i^*$ is obtained from the solution of equation (16) for an estimated pressure field p^*. Solution of equation (16) is obtained using a Gauss-elimination line-by-line alternating-direction algorithm.

The computed velocities $\bar{u}_i^*$ will not, in general, satisfy the local continuity equation (17) but will produce a net mass source at each grid note. p^* and $\bar{u}_i^*$ are then corrected so as to reduce the mass source at each grid node to zero in the following manner:

$$p = p^* + p' \tag{18}$$

$$\bar{u}_i = \bar{u}_i^* + D_n\,(p'_p - p'_n) \tag{19}$$

where p is pressure correction and D_n $(= \partial\bar{u}_i/\partial(p_p - p_n))$ is evaluated from the relevant momentum equation. Equation (19) is substituted into the continuity equation (17) to give

$$a_p\, p'_p = \sum_n a_n\, p'_n + S_o \tag{20}$$

where S_o is the local continuity imbalance based upon $\bar{u}_i^*$. Equation (20) is of the same form as equation (16) and is solved to yield p'; thereupon p and $\bar{u}_i$ are computed at each node from equations (18) - (19).

The calculation of the remaining dependent variables, as well as the updating of the thermodynamic properties and exchange coefficients, are imbedded into the above sequence.

Since the difference equations are non-linear, the above calculation steps are repeated few times with due care to avoid numerical instabilities. The iteration is terminated when the current solution satisfies the difference equations to within 1% or less of a global measure such as the total mass flow rate or inlet momentum, depending on the variable in question.

RESULTS AND DISCUSSION

The present developed numerical procedure is applied to the prediction of flow and energy transferred in confined non-reacting and reacting jets. The calculation presented in this section are compared with the available experimental data.

Figures 1 and 2 show the predicted radial profiles and the corresponding data by Wang et al [7] of the mean axial velocity obtained at $z/D = 2$ and two hydrogen-to-air speed ratio (S·R) conditions with burning. At this station, the axial velocity is independent of burning because only a minimum amount of reaction has taken place (see Fig. 3) in the flow with burning. The velocity profiles of fuel jet at the nozzle exit were step functions with respect to the velocity profiles of the surrounding air flow, as expected. The axial velocity of air flow is assumed flat around the fuel jet and equal to 14.5m/s. The fuel-to-air speed ratios, S·R, were determined as the ratio of the centerline velocities of fuel jet (165 and 55m/s) and the axial air velocity and are referred to by the nominal values of 11 and 4. The agreement between the experimental and predicted results is fairly good.

The predicted results for the radial distributions of the mean axial velocity and the r.m.s. of its fluctuations for both reacting and non-reacting flow conditions are depicted in Fig. 4.

Figure 5 shows the centerline mean axial flow velocity distribution for both reacting and non-reacting flows vs z/D at S·R=4. The axial turbulence intensity distribution along the centerline of burner is also depicted in Fig. 6. For the fuel-air jet without combustion, both the mean axial velocity and the turbulence intensity distributions were similar to those obtained in a free jet flow. The presence of combustion (see Fig. 7) lengthened the potential core of the fuel jet and moved the location of maximum turbulence intensity in the upstream direction.

Table 1: Constants in Eqs. (6) to (14)

C_μ	C_1	C_2	σ_ε	σ_k	σ_f
0.09	1.44	1.92	1.22	0.7	1.0

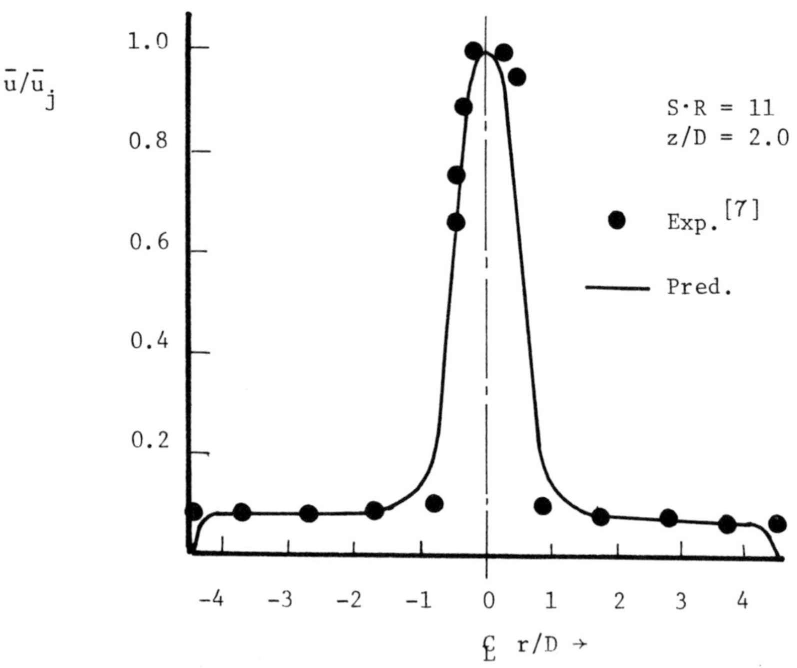

Fig. 1: Radial profile of mean axial velocity near the fuel jet outlet (H_2 - Air Flame).

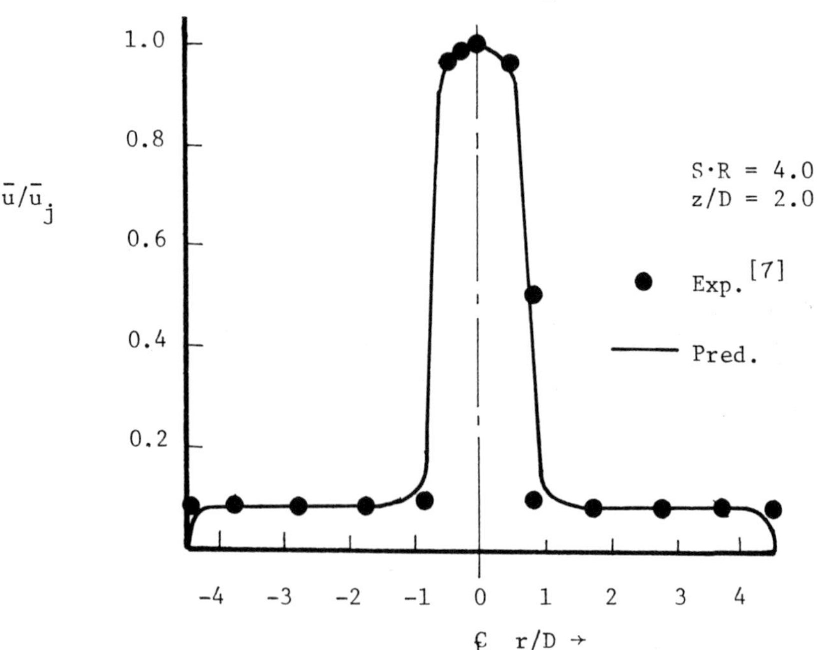

Fig. 2: Radial profile of mean axial velocity near the fuel jet outlet (H_2 - Air Flame).

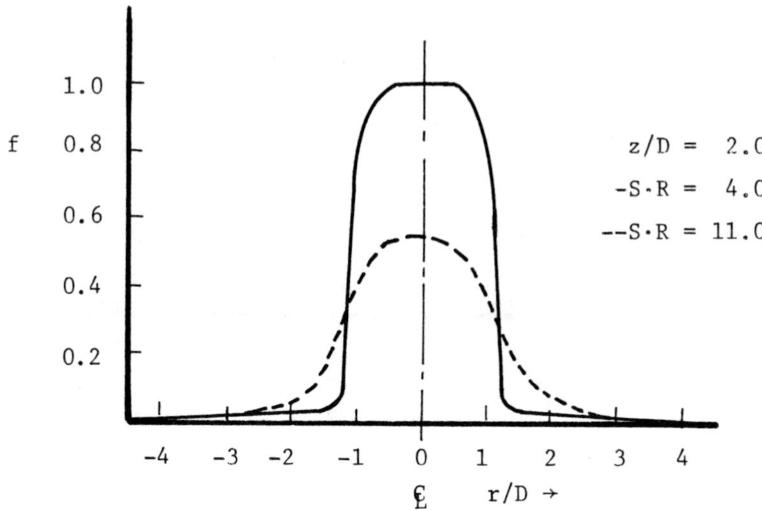

Fig. 3: Radial distribution of fuel mass fraction near the fuel jet (H_2 - Air Flame).

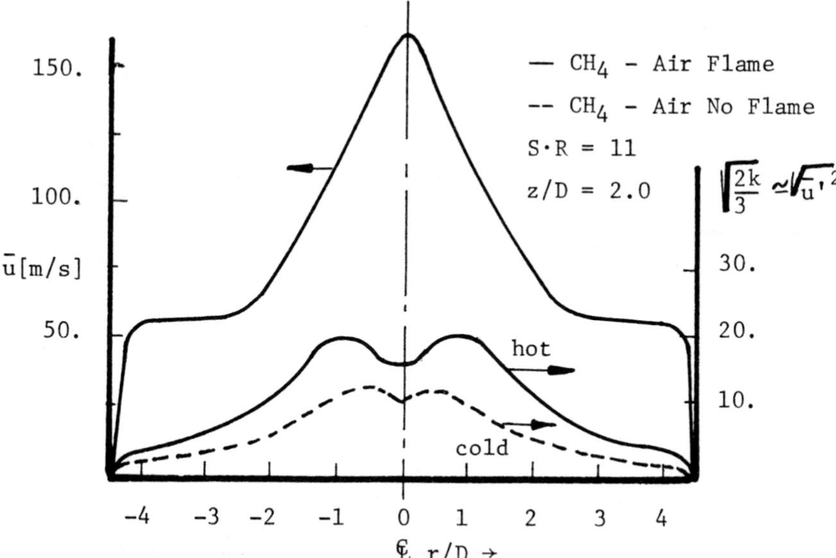

Fig. 4: Radial distribution of mean axial velocity and normal turbulent shear stress component.

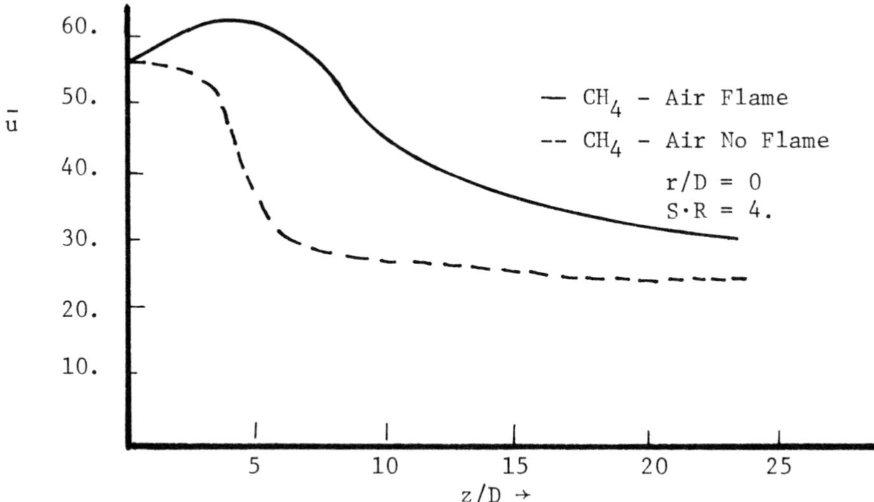

Fig. 5: Centerline mean axial velocity distribution

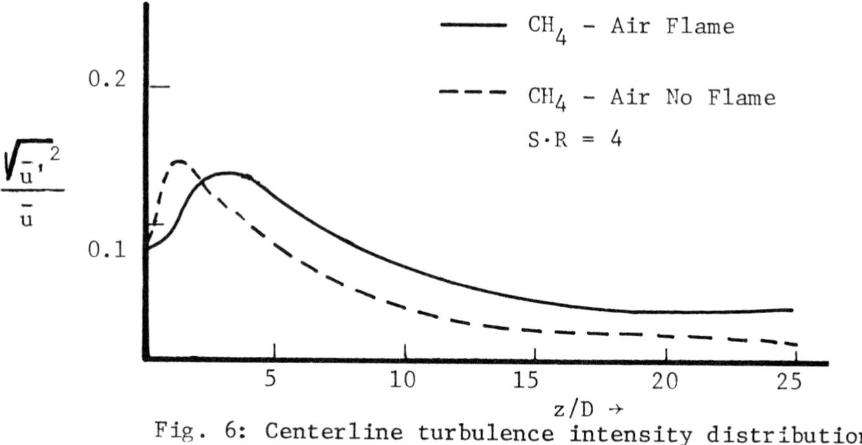

Fig. 6: Centerline turbulence intensity distribution

1.0
0.8
f
0.6
0.4
0.2

CH_4 - Air Flame
S·R = 4

5 10 15 20 25
z/D →

Fig. 7: Axial distribution of centerline fuel mass fraction

CONCLUDING REMARKS

The present developed numerical procedure was shown to solve the non-linear partial differentail equations governing the mass, momentum, and energy in an iterative manner. The calculated results were compared with the corresponding experimental data and the level of agreement obtained support the use of the present numerical procedure for other practicle applications. Near the fuel jet exit, similar distributions of both mean axial velocity and the turbulence intensity were obtained for the flows with and without reaction, except that z/D scale was shifted upstream from those without burning. However, the values of the turbulence intensity and the axial velocity for the reacting flows are increased due to the presence of combustion. The increase in mean axial velocity at further downstream stations $z/D \geq 30$ with burning, can be attributed to the decrease in density resulting from the combustion.

REFERENCES

1. LAUNDER, B.E. and SPALDING, D.B. - Mathematical Models of Turbulence. Academic Press, London and New York, 1972.
2. ABOU-ELLAIL, M.M.M.,et al. - Description and Validation of a Three Dimensional Procedure for Combustion Chamber Flows. AIAA Journal of Energy, Vol. 2, No. 2, pp. 71-80, March-April 1978.
3. RICHTER, W. Mathematische Modelle Technisher Flammen Grundlagen und Anwendungen für achssymmetrische Systeme Dissertation, Stuttgart University, West Germany, 1978.
4. ABOU-ARAB, T.W. and ABOU-ELLAIL, M.M.M. - Calcuations of flow and heat transfer in pipe with a Reynolds stress model of turbulence. Numerical Methods for Non-Linear Problems, Taylor-Hinton-Owen (Eds), Pineridge Press Ltd. Swansea, England 1980.
5. ABOU-ELLAIL, M.M.M. and ABOU-ARAB, T.W. - A Computer Model for Two Dimensional Steady and Nonsteady reacting Flows, (To be published).
6. GOSMAN, A.D. et. al. - The Calculation of Two-Dimensional Turbulent Recirculating Flows, Turbulent Shear Flows I, Ed. Durst, Springer Verlag, 1979.
7. WANG, J.C.F. and GERHOLDT, B.W. Measurements on Turbulent Hydrogen Flames in a Circular Air Duct, Paper 77-48 at the AIAA, 15th Aerospace Science Meeting, Los Angeles, CA, Jan. 24-26, 1977.
8. ABOU-ARAB, T.W. - Zur Modellierung der Turbulenz in eingeschlossenen drallfreien und verdrallten Strömungen. Dissertation, Stuttgart University, West Germany, 1978.

ON THE TRANSITION TO TURBULENCE

J.K. PLATTEN, J.M. LUIJKX and P. FLANDROY

UNIVERSITY OF MONS, BELGIUM

0. SUMMARY.

In the context of the transition to turbulence, some numerical experiments are performed, in the case of plane Poiseuille flow, in order to try to understand the first stages of this transition. Nonlinear instability is described, and, at others wavenumbers, periodic flows. These results are connected to fundamental theoretical work.

1. INTRODUCTION.

From the physicist point of view, the "nature of turbulence" is still an open question, and therefore a better understanding of the earlier steps of the transition from laminar to turbulent flows, should contribute in a better knowledge of turbulence itself. Questions that should be resolved are for example :

- does turbulence appear suddenly, or gradually after a finite (or infinite) number of "bifurcations", each new bifurcation adding a new frequency (generally incommensurate) to the existing flow field ?
- does turbulence appear below the critical point determined in the frame of a linear hydrodynamic stability theory ?
- does a "numerical experiment" throw some light on the nature of this transition, and, to some extent, what is the influence of the numerical method adopted, on the information that we get, by solving numerically the Navier-Stokes equations.

From the point of view of Landau [1], turbulence appears as the effect of successive instabilities, of a sequence of bifurcations. Each new instability introduces into the flow field a new cyclical component and the relative phase Φ_n of the cyclical components with incommensurate periods ω_n is undetermined. Therefore the behaviour of the function

$$\sum_n A_n \cdot \cos (\omega_n t + \Phi_n) \cdot V_n (x, y, z) \tag{1}$$

which is used to represent the flow field, becomes more and

more complicated and confused, when the number of components increases. Thus in the Landau picture, turbulence happens gradually and there is no well defined onset of turbulence. Even when the Reynolds number is very large, the flow appears to be a superposition of periodic modes with generally incommensurate frequencies and is therefore quasiperiodic. This seems not to be a very realistic picture. Indeed turbulent flows are nonperiodic (or chaotic) rather than quasiperiodic. The qualitative difference between a nonperiodic and a quasiperiodic behaviour should be emphasized : the former is very sensitive to initial conditions. In a quasiperiodic regime, two flows with nearly identical initial conditions, will remain for all time almost identical. On the contrary in a nonperiodic regime, the two flows will, for large time, evolve completely differently, no matter how small was the difference between the two initial conditions : a nonperiodic flow is unstable. Therefore we have an alternative to the experimentally unverified picture of Landau : a deterministic flow (a flow whose future may be predicted given the initial conditions, as e.g. described by the Navier-Stokes equations) can appear chaotic rather than quasiperiodic ; this is a consequence of the nonlinearity of the system ; the equations can have an exceedingly complex solution, that does not look as if it was the result of a deterministic process. This was discussed for the first time in the famous paper by Lorentz [2] "Deterministic nonperiodic flow". Finite models of this type of behaviour are usually discussed in the study of the transition to turbulence in thermal convection (the Rayleigh-Bénard problem). Another view of the transition to turbulence has been given by Ruelle and Takens [3]. Instead of an infinite sequence of bifurcations, as in the Landau picture, after the third or fourth bifurcation, there is a bifurcation into a "nonperiodic attractor" : the orbits in phase space will be attracted to more complex subsets of the configuration space, which they call "strange attractors". Ruelle and Takens propose that these strange attractors are characteristics of turbulent flows. In this picture, there is thus an abrupt transition at a well defined value of the Reynolds number (or the Rayleigh number in thermal convection), from quasiperiodic to turbulent behaviour. Sometimes a flow which is stable with respect to small perturbations may be unstable when the initial amplitude of these perturbations exceeds some threshold. Thus instability may appear below the critical point determined in the frame of a linear hydrodynamic stability theory, and the laminar flow cannot remain the stable solution : the initially large disturbances grow with time and the flow evolves towards a completely different configuration, that we also call "turbulence".

Thus there are many routes to turbulence, and different type of flows will become turbulent following different routes. Even in the very simple case of thermal convection (convection in a box heated from below, the so-called Rayleigh-Bénard convection), the routes to turbulence are not the same for fluids enclosed in "small size" containers or in "large" containers !

Regarding the transition in Poiseuille flow there is little numerical work that was performed. This paper is an attempt is this direction. There is also a fundamental difference between pipe Poiseuille flow, and plane Poiseuille flow. Indeed for plane Poiseuille flow there is a critical Reynolds number ($Re^{crit} = 5772$) at which infinitesimal disturbances must grow. On the contrary in pipe Poiseuille flow a linear hydrodynamic stability theory predicts that the flow is always stable ($Re^{crit} = \infty$) and therefore a "Hopf bifurcation" (bifurcation into a periodic regime when $\varepsilon = (Re - Re^{crit})/Re^{crit}$ is small) is not applicable and the transition is always subcritical. This paper is only related to plane Poiseuille flow. Before describing some numerical results that we have obtained, we should like to summarize fundamental theoretical results, in order to have a better understanding of the numerical results.

It has been proven that steady bifurcation solutions which bifurcate supercritically ($Re > Re^{crit}$) are stable and that those which bifurcate subcritically ($Re < Re^{crit}$) are unstable. For time periodic solutions, Joseph and Sattinger [4] have proven that if a fluid system admits a steady state when a stability parameter, say the Reynolds number Re, is small, if this steady state loses its stability when Re is increased beyond a critical value Re^{crit}, and if "a pair of complex conjugate eigenvalues associated with the linearized stability problem for the steady motion crosses the imaginary axis as Re crosses Re^{crit}", then

(i) there exists an unique bifurcating time periodic motion which may be constructed as a Taylor series in a small parameter ε

(ii) this time periodic motion is stable when ε is small and $Re > Re^{crit}$

(iii) this time periodic motion is unstable when ε is small and $Re < Re^{crit}$.

For plane Poiseuille flow, the surface on which periodic solutions to the Navier-Stokes equations exists in the space (Re, α, ε^2) has been constructed by Chen et al. [5], and is sketched on Fig. 1. Here α is the wavenumber of a Fourier component, as introduced by the linear stability theory. The dotted lines indicate an unstable time periodic solution, a full line, a stable one. We see at $\alpha < \alpha_o \simeq 0.9$, that a supercritical stable time periodic solution exists.

2. SOME RESULTS OF NUMERICAL EXPERIMENTS.

The equation to be integrated is :

$$\frac{\partial}{\partial t}(\nabla^2\psi) - \frac{\partial\psi}{\partial y}\cdot\frac{\partial}{\partial x}(\nabla^2\psi) + \frac{\partial\psi}{\partial x}\cdot\frac{\partial}{\partial y}(\nabla^2\psi) = \frac{1}{Re}\cdot\nabla^4\psi \qquad (2)$$

which is an alternative form of the Navier-Stokes equations in the two-dimensional space $x - y$, ψ being the stream function. There is a basic flow in the x direction, $\bar{U}_x = (1 - y^2)$, (limited by two horizontal boundaries at $y = \pm 1$) to which corresponds a basic stream function $\bar{\psi} = (1/3)\, y^3 - y$. We superpose a disturbance $\psi'(x, y, t)$, and by introducing

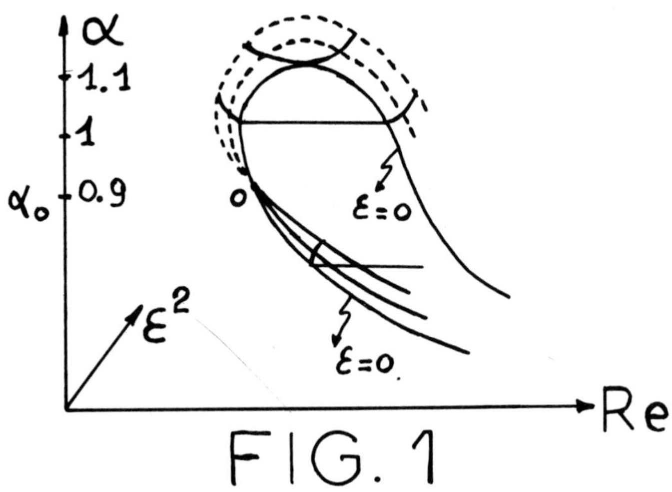

FIG. 1

$$\psi(x,y,t) = \bar{\psi}(y) + \psi'(x, y, t) \tag{3}$$

into Eq. (2), we get the time evolution for ψ', associated to the following "no slip" boundary conditions at $y = \pm 1$

$$\frac{\partial \psi'}{\partial x} = \frac{\partial \psi'}{\partial y} = 0 \tag{4}$$

The problem is numerically resolved by a very classical Galerkin type technique. Thus we write

$$\psi' = \sum_{n=-\infty}^{+\infty} \hat{\psi}_n \; (y,\, t) \exp\,(in\alpha x) \tag{5}$$

$$\hat{\psi}_n \; (y,t) = \sum_{i=1}^{N} a_{n,i} \; (t) \; f_{n,i} \; (y) \tag{6}$$

It should be emphasized here that the trial functions $f_{n,i}(y)$ are not the same for every x-harmonic : indeed $\hat{\psi}_1$, $\hat{\psi}_3$, $\hat{\psi}_5$ ... are even functions of y ; $\hat{\psi}_o$, $\hat{\psi}_2$, $\hat{\psi}_4$ ... are odd in y. The trial functions that we use are Tchebychev polynomials, that we multiply by an appropriate "weighting function" in order to satisfy the boundary conditions. By inserting Eqs (3) (5) (6) into Eq. (2), we get of course the time evolution of the coefficients $a_{n,i}$. These equations are then integrated numerically, given some initial conditions $\{a_{n,i} \; (0)\}$, using a method given by Gear [6] , and based on Adams formula, of variable order. All numerical details may be found elsewhere [7] .

For numerical evidence, the series (5) must be truncated. We have used the following harmonics : $n = 0, \pm 1, \pm 2$. Also in Eq (6) we put $N = 20$. We have thus to integrate a nonlinear system of 100 equations.

In a first step, we have tried to recover the results of the linear theory, by imposing very small initial perturbations such as $\{a_{n,i} \; (0) \sim 10^{-6}\}$. Thus nonlinear terms are negligible. We recover in this way at $\alpha = 1$ the well known result from the Orr-Sommerfeld equation $Re^{crit} = 5814$ (1).

In a next step, we try to know if subcritical instabilies exist. Experimentally, this should be the case : laminar flow in rectangular ducts can only be maintained if $Re \lesssim 1000$ or 2000, unless a drastic control is imposed in order to reduce the level of "residual turbulence" in the inlet region (in order to have only small perturbations "compatible" with the linear theory). We have indeed found such instabilities at $Re < 5814$. These results were already published by our team in 1980 [7] and we shall therefore be as brief as possible. In our numerical study the initial condition was

$$a_{n,i}(0) = 0 \;\forall_{n,i} \text{ except } a^R_{1,1} = K_A \tag{7}$$

where $a^R_{n,i}$ is the real part of $a_{n,i}$ (indeed the coefficients $a_{n,i}$ may be complex). Thus K_A is a measure of the initial amplitude of a particular perturbation. From the knowledge of the 100 coefficients $a_{n,i}(t)$, we can compute E(t), the kinetic energy associated with the perturbed flow field. For $K_A \to 0$, $Re^{crit} \to 5814$; for $K_A = 0.1$, we find that E(t) decreases with time (not necessarily monotonically) for $Re \lesssim 2500$. On the contrary for $Re \gtrsim 3000$, E(t) increases indicating an instability (see Fig. 2 ; time is reduced, see [7] ; time steps are typically $\sim 10^{-3}$ or 10^{-2}). Let us recall that at

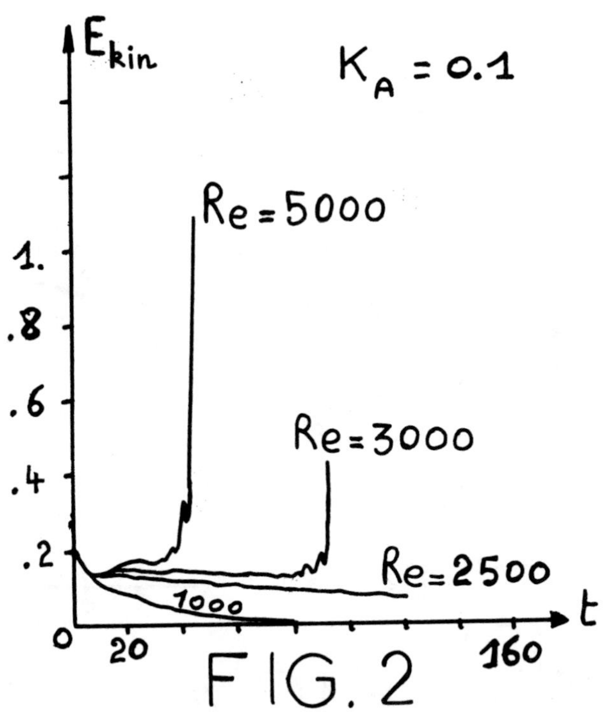

FIG. 2

-----

(1) The value $Re^{crit} = 5772$ is obtained for $\alpha = \alpha^{crit} = 1.0205$

Re = 5000, the flow was stable when $K_A \to 0$. Thus instability may be found in subcritical conditions. If K_A is still increased (K_A = 0.25), instability is found even at Re = 2500 and Re = 2000, a situation which was stable at K_A = 0.1 ; however the flow is still stable at Re = 1000. All the numerical results are summarized on Fig. 3. They are compatible with the

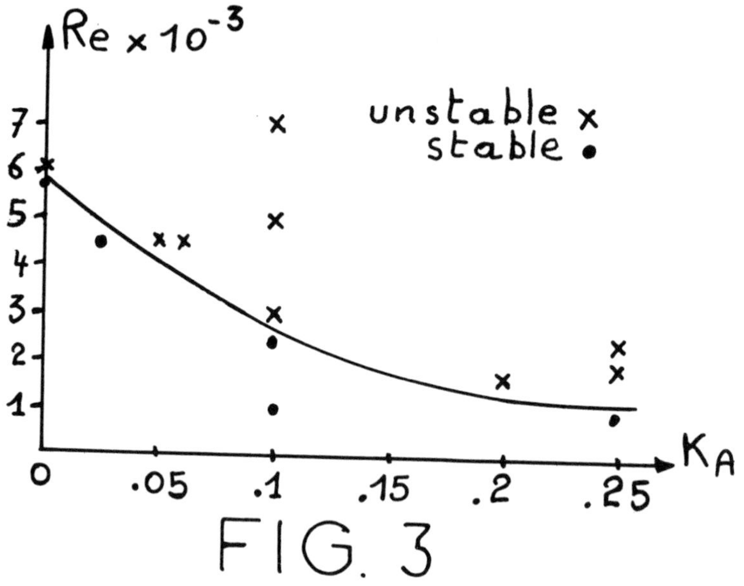

FIG. 3

existence of an unstable time periodic regime at Re < Re^{crit}. (see Fig. 1 and also the bifurcation diagram of Fig. 4.a).

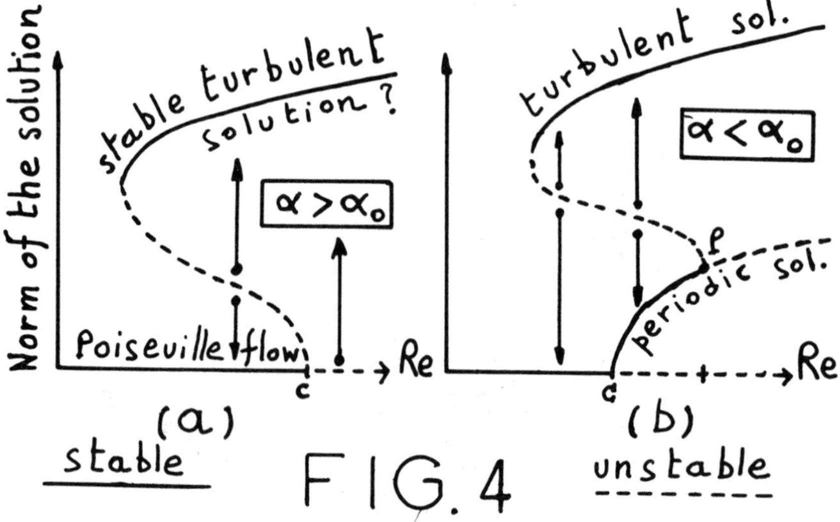

FIG. 4

Let us recall that experimentally $1000 \lesssim Re^{crit} \lesssim 2000$. The question to know towards what kind of solution the system is attracted beyond instability (labelled "stable turbulent solution" on Fig. 4.a) remains till now unanswered. This is the main point in the study of the transition towards turbulence in plane Poiseuille flow, and seems to be a very difficult task. We hope that numerical experiments could help to resolve, at least partially, this question, but very likely much more sophisticated methods (than the very simple Galerkin technique) should be used.

The next point to be examined, is the possible existence of a stable time periodic regime at $\alpha < \alpha_o$ and $Re > Re^{crit}(\alpha)$ (see Fig. 1 and also Fig. 4.b). An important difference with the study of subcritical instabilities, is that now we have to determine the new state reached by the system after instability and therefore, the final integration time t_f is much larger (typically of the order of $t_f \sim 2000$ or 3000, instead of $t_f \sim 200$). With the present Galerkin technique, the computational costs become prohibitive ; we had to reduce the number of trial functions in y, namely N = 6 instead of N = 20 (see Eq. 6). We have thus now to integrate a system of 30 differential equations (instead of 100) and this becomes practicable even up to $t_f \sim 3000$. The price we have to pay is of course a loss of accuracy on the critical point : we have found $\alpha^{crit} \simeq 1.3$ (instead of 1.02) and $Re^{crit} \simeq 3520$ (instead of 5772), i.e. 30 % error on α^{crit} and 40 % on Re^{crit}. Anyway the order of magnitude is preserved. Now, $\alpha_o \sim 0.9 \times \alpha^{crit} \sim 1.1$. Therefore the system is still integrated with $\alpha = 1$ (if $\alpha = 1$, $Re^{crit} = 4087$). Integration was performed with Re = 4187 , Re = 4287 and Re = 4587. A stable time periodic solution was found of period T_1 (see e.g. Fig. 5 where we show, in arbitrary units, the time behaviour of one of the coefficients, namely $a^R_{1,1}$; each coefficient, and of course the kinetic energy exhibit the same behaviour). The amplitude of this time periodic motion, and its frequency was determined for various Reynolds numbers. But the main point is that this time periodic solution is stable if $\varepsilon = (Re - Re^{crit})/Re^{crit}$ is small. Indeed at the "final" integration time (say $t_f \simeq 3500$, see Fig. 5) we multiply all the 30 coefficients $a_{n,i}$ by a constant factor K_B, and close to unity, e.g. $K_B = 1.01$ (we perturb slightly the time periodic motion). The kinetic energy return to its "initial" (at $t \simeq 3500$) value by oscillations (of period T_2) of decreasing amplitude (see Fig. 6). The stability analysis of the T_1-periodic motion, introduces a second component of period T_2, but at small $\varepsilon = (Re - Re^{crit})/Re^{crit}$, this second component is not excited. However it is clear that if the Reynolds number is still increased (say beyond point P, Fig. 4.b), the amplitude of this second T_2-periodic component will no longer decrease with time (see Fig. 7). Thus the time periodic solution becomes "linearly" unstable at point P. Even if the T_1-periodic motion was linearly stable for $Re^{crit} < Re < Re_P$, this does not mean "stability for all kind of perturbations". Indeed if we multiply all the coefficients $a_{n,i}$ at $t_f \simeq 3500$ (see

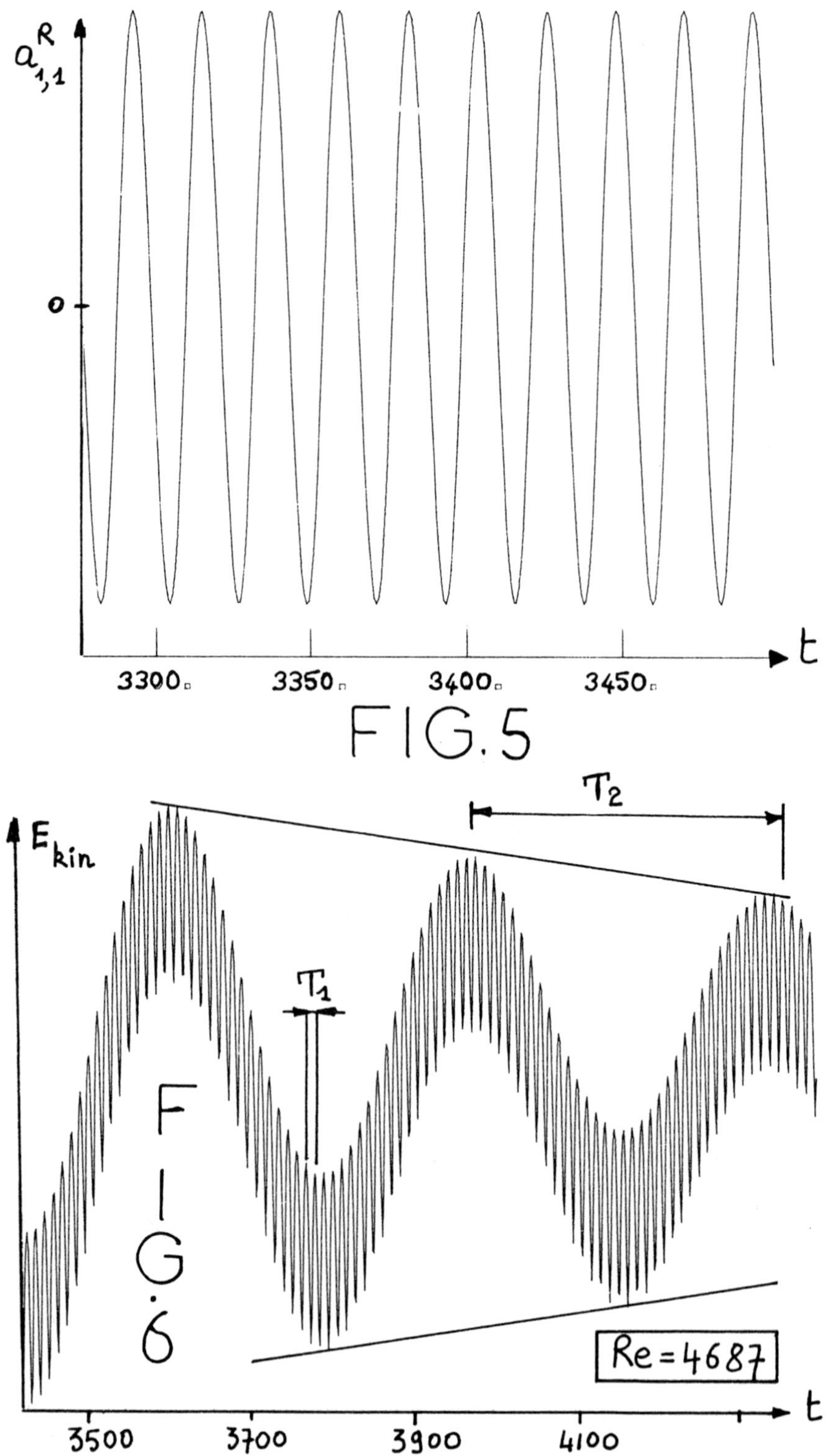

$a^R_{1,1}$
0
t
3300.
3350.
3400.
3450.
FIG.5
T_2
E_{kin}
T_1
FIG.6
Re=4687
t
3500
3700
3900
4100

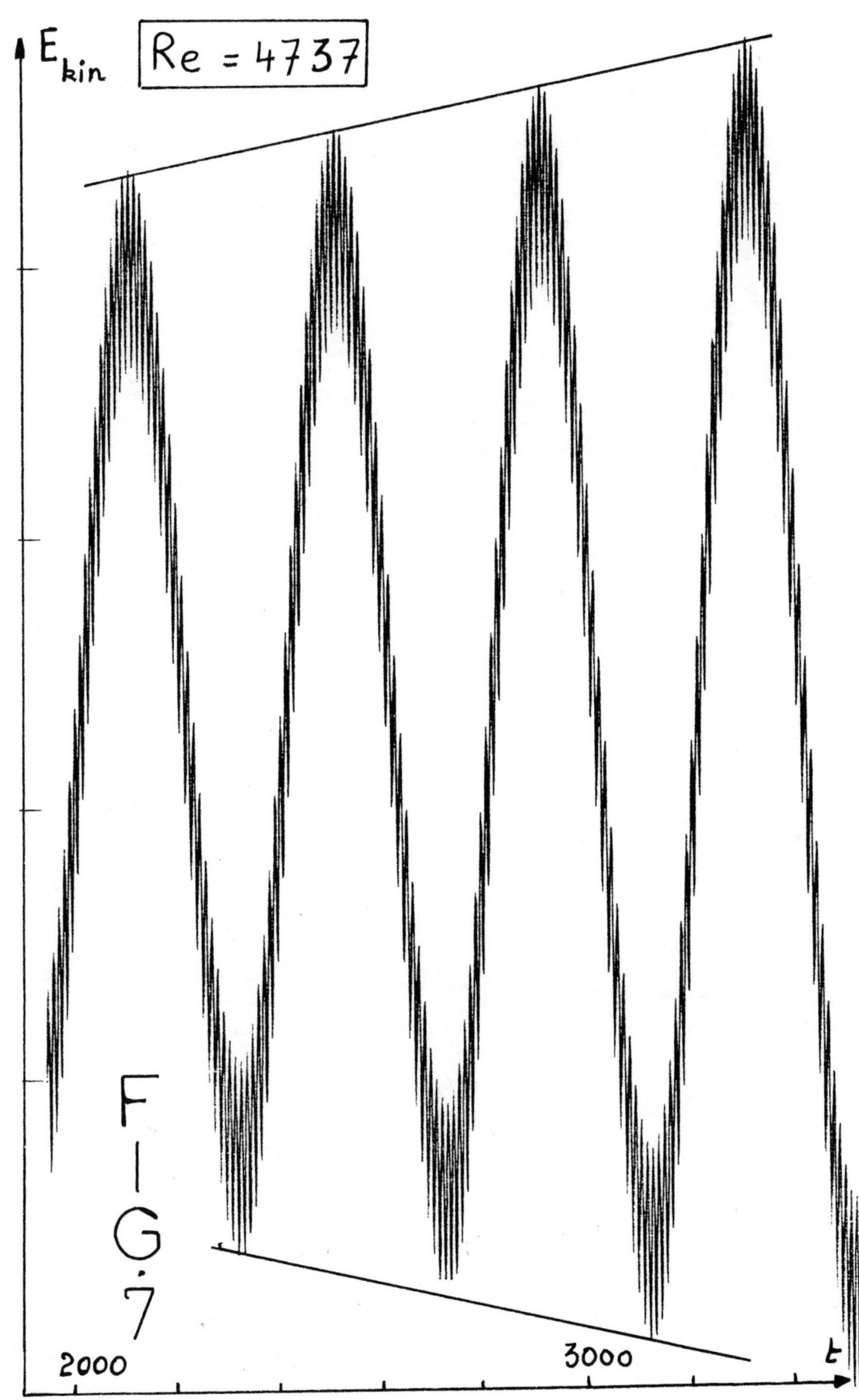

E_{kin}
Re = 4737
2000
3000
t
FIG. 7

Fig. 5) by $K_B = 2$ (instead of $K_B = 1.01$), introducing thus a very large perturbation, the system does not return to the T_1-periodic motion; on the contrary the kinetic energy increases with time. Thus at Re = 4737 a second component of period T_2 is introduced in the flow field. The question to know if there is a gap in Reynolds numbers in which this second component may be stabilized (implying the existence of a quasiperiodic motion) is not yet resolved. However for Re >> Re^P, this quasiperiodic motion is in turn unstable : after several cycles of period T_2 (a situation comparable to Fig. 7) there is a sudden "catastrophic" growth of the kinetic energy. The unanswered question is to know the reason of this catastrophic growth. It might be that the system is attracted towards "something else" (also called "turbulence" on Fig. 4.b), or on the contrary that the catastrophic growth of the kinetic energy is a "numerical artefact", due to the fact that in our numerical scheme, there is no quadratic invariant, such as energy (or perhaps enstrophy, a vocable for mean square vorticity) in the limit of zero viscosity (or $Re \to \infty$), but this is conjecture.

4. CONCLUSIONS.

The preliminary, and very incomplete, numerical results presented in this paper, are all in agreement with fundamental theoretical work on the behaviour of the Navier-Stokes equations, and, in some sense, go further than what is permitted by the analytical study (e.g. numerical analysis is not restricted to small values of $(Re - Re^{crit})/Re^{crit}$). All our results are summarized on the two "possible" bifurcation diagrams of Fig. 4.

(i) For $\alpha \simeq \alpha^{crit} > \alpha_o$, our results are consistent with the existence of an unstable limit cycle : at a given Reynolds number, the perturbed laminar flow returns to laminar flow if the initial perturbation is small enough, or evolves towards a solution of "large norm" (this is indicated by arrows). The critical amplitude of the initial perturbation, needed for instability increases when Re decreases, and goes to zero when $Re \to Re^{crit}$.

(ii) For $\alpha < \alpha_o \simeq 0.9\,\alpha^{crit}$, there is a stable limit cycle for $0 < \varepsilon = (Re - Re^{crit})/Re^{crit} < (Re^P - Re^{crit})/Re^{crit}$.

Moreover, when $Re \to Re^{crit}$, the frequency coïncides with the one predicted by the linear theory (with the eigenvalues deduced from the Orr-Sommerfeld equation). The new features, obtained by numerical analysis, but absent in the analytical study, is the nonlinear instability of this limit cycle for $Re^{crit} < Re < Re^P$, and its linear instability for $Re > Re^P$, introducing a second frequency. Finally, it is also verified that laminar plane Poiseuille flow may be unstable even when $Re < Re^{crit}$ (all these features are indicated by arrows in Fig. 4.b), but, in contradistinction with Fig. 4.a, the amplitude needed for instability does not go to zero when $Re \to Re^{crit}$.

Finally the main question raised by this paper and addressed to the participants of this conference, is the following : is it worth to continue such numerical experiments, and in case of positive answer, how may we reduce computational costs, in order to get really very important and new physical information on the transition mechanism to turbulence, i.e. what kind of numerical methods suitable for this problem should be used. It is indeed clear that a simple Galerkin technique has serious practical limitations, as e.g.

(i) no possibility to increase the number of x-harmonics, or the number of y-trial functions in order to study the convergence of the numerical process

(ii) no possibility to run the program for very large times (say $t_f \simeq 10^4$ with time steps of 10^{-3} or 10^{-2}).

(iii) no possibility to run the program for large Reynolds numbers (practically $Re > 1.2\ Re^{crit}$) due to the limitation of a small number of harmonics.

(iiii) no possibility to study three-dimensional disturbances, etc...

REFERENCES.

[1] L. LANDAU, e.g. C.R. Acad. Sci. URSS, 44, 311 (1944) see also Mécanique des Fluides, MIR, Moscou (1959).

[2] E.N. LORENZ, J. Atm. Sci., 20, 448 (1963).

[3] D. RUELLE & F. TAKENS, Commun. Math. Phys., 20, 167 (1971).

[4] D.D. JOSEPH & D.H. SATTINGER, Arch. Rat. Mech. Anal., 45, 79 (1972).

[5] T.S. CHEN & D.D. JOSEPH, J. Fluid Mech., 58(2), 337 (1973).

[6] C.W. GEAR, Numerical Initial Value Problem in Ordinary Differential Equations, Prantice Hall, 1971.

[7] P. FLANDROY and J.K. PLATTEN, J. Mécanique (in press), 19(4), p. ?, 1980.

Mr J.M. LUIJKX and Dr P. FLANDROY are indebted to the I.R.S.I.A. (Institut pour l'Encouragement de la Recherche Scientifique dans l'Industrie et l'Agriculture) Brussels, for a grant.

COMPUTER MODELLING OF TURBULENT RECIRCULATING FLOWS IN ENGINEERING APPLICATIONS

E.E.KHALIL & H.M.W.ASSAF
Faculty of Engineering,
Cairo University, Egypt

SUMMARY

This work is an attempt to verify that the calculation procedures of the type discussed here, are indispensible and valuable means of providing a parametric investigation for swirling flows. Two flows are discussed namely the flow in a rotary garbage classifier and the flow in vortex tubes. The governing conservation equation of mass and momentum are solved, in finite difference form, simultaneously and iteratively. A two-equation turbulence model is employed to represent the turbulent characteristics. The conclusion that can be drawn from this investigation, is that such procedures can readily yield results, at reasonable costs, which exhibit the effects of varying the operating and geometrical conditions. Hence, it can be safely said that such procedures are most helpful tool to the designer.

I.INTRODUCTION

The present investigation is another attempt in the series of investigations in the field of swirling flows, whose wide applicability of such flows has long tempted investigators. A numerical procedure, which solves the governing conservation equations of mass and momentum in terms of velocities and pressures, is employed, [1]. The elliptic nature of the flow together with the non-linearity of the equations, make it necessary to solve them simultaneously and in an iterative manner. This procedure is embodied in the computer program of references [2 & 3], whose validity was assessed in a wide range of turbulent flows. Khalil et al [4], Gosman et al [3] and Cebeci et al [5], all reported reasonable agreement between the measured and predicted flow patterns and turbulence characteristics for separating boundary layer flows, recirculating flows and wake flows.

This work is not intended primarily as an assessment of the calculation procedure, by way of comparison with experiments, but is intended to demonstrate that such a procedure is indeed an invaluable tool in providing a parametric investigation of each of the flows considered.

The two-equation turbulence model was used to represent the turbulence characteristics, this model has been appraised in detail ,[3]. In the present investigation the two cases of a rotary air classifier and the flow in vortex tubes are discussed. Results are represented for different operating conditions in each case. Three different values of the tangential momentum of the inlet air along the conical surface, and two different sets of classifier arrangements, are the different flow conditions which were investigated in the case of the rotary classifier. In the vortex tubes, results were calculated for two different length to diameter ratios, and for two mass ratios. The results were only compared to experimental data, whenever these were available; e.g. for the vortex tube flow, the calculations were compared to the measurements of Bruun [6].

The paper describes briefly the governing equations, turbulence models and solution procedure in one section. The results are presented and discussed in another, and the paper ends with a summary of conclusions.

II. EQUATIONS and SOLUTION PROCEDURE

The governing differential conservation equations of mass and three components of momentum, which are all non-linear, can be expressed, for the steady state and time averaging, as follows:

$$\frac{\partial}{\partial x_j} \quad \bar{\rho}\,\bar{u}_j \quad = 0 \tag{1}$$

and

$$\frac{\partial}{\partial x_j}\left(\bar{\rho}\,\bar{u}_j\bar{u}_i + \bar{\rho}\,\overline{u'_j u'_i}\right) = -\frac{\partial \bar{P}}{\partial x_i} + \mu\,\frac{\partial}{\partial x_j}\left(\frac{\partial \bar{u}_i}{\partial x_j} + \frac{\partial \bar{u}_j}{\partial x_i}\right) \tag{2}$$

In deriving these equations, all triple correlations and density fluctuations correlations were neglected. The only terms that contain the turbulence characteristics are the Reynolds stresses terms $\bar{\rho}\overline{u'_j u'_i}$.

II.1 Turbulence Model

In the present work, the two-equation turbulence model [7] is employed. This model employs an eddy viscosity concept, and expresses the Reynolds stresses in terms of the mean velocity gradient and the eddy viscosity. The effect of the rotation invariant and the rate of strain are incorporated for strongly swirling flows, Khalil [8], as follows,

$$-\bar{\rho}\,\overline{u'_i\, u'_j} = \frac{2}{3}\, k\, \delta_{ij} + A_{ij}(\, a_{ij}\,,\, b_{ij}\,)k \tag{3}$$

where a_{ij} and b_{ij} are tensors representing the rate of strain and the rotation invariant respectively. The tensor A_{ij} is a finite tensor polynomial in terms of all the linearly independent second order tensors, and coefficients of the polynomial depend on the values of k and e. The values of k and e, the kinetic energy of turbulence and its dissipation rate are calculated from their corresponding transport equations [7].

II.2 Calculation Procedure

The calculation procedure, used here within, employs the SIMPLE algorithm embodied in the TEACH-T code of Gosman et al [2], and modified for swirling flows by Gosman et al [3] and Khalil [8]. The steady state, time averaged conservation equations governing the transport of mass, three components of momentum, kinetic energy of turbulence and its dissipation rate, can be expressed for axisymmetric flows, as;

$$\frac{\partial}{\partial x}\,\bar{\rho}\,\bar{u}\bar{\emptyset} + \frac{1}{r}\frac{\partial}{\partial r}\, r\,\bar{\rho}\bar{v}\bar{\emptyset} = \frac{\partial}{\partial x}\Gamma_{\emptyset}\frac{\partial\bar{\emptyset}}{\partial x} + \frac{1}{r}\frac{\partial}{\partial r}\Gamma_{\emptyset} r\frac{\partial\bar{\emptyset}}{\partial r} + \bar{S}_{\emptyset} \tag{4}$$

Where $\bar{\emptyset}$ is the general dependent variable representing $\bar{u}, \bar{v}, \bar{w}$, k and e. The continuity equation is realized by equating $\bar{\emptyset}$ to unity and $\bar{S}_{\emptyset}$ to zero. The values of the turbulent exchange coefficients $\Gamma_{\emptyset}$ equals the effective viscosity for $\bar{u}, \bar{v}, \bar{w}$ and k. For e, the turbulent exchange coefficient is 1.22 times the effective viscosity. $\bar{S}_{\emptyset}$ denotes the source/sink of the generation or destruction of the entity in question. The expressions of $\bar{S}_{\emptyset}$ are similar to those of ref. [3].

Equations of the form of equation 4 are both non-linear and elliptic, therefore they must be solved simultaneously and iteratively. The finite difference form of the above equations is solved at the nodes of the intersection of the orthogonal grid lines superimposed on the solution domain. Economic consideration and available computers restrict the grid size.

III.RESULTS and DISCUSSION

III.1 Flow in a Rotary Air Classifier

The local flow properties in a conical rotary air classifier,of the type used to sort out garbage containing light and heavy constituents,are predicted. Compressed air is admitted tangentially along the conical wall through various nozzles. The conical drum rotates along its axis;this rotation together with the compressed air separate the heavy and light constituents. Due to the adverse pressure created by the compressed air,air is induced through the small end. Different operating conditions are discussed:

a. Three values of the air jet velocities corresponding to three values of the air flow rate issuing from 25 air nozzles:

i. w_j = 10 m/s corresponding to $\dot{m}$=0.153m^3/s,case1
ii. w_j =17.8m/s corresponding to $\dot{m}$=0.275m^3/s,case2
iii. w_j = 40 m/s corresponding to $\dot{m}$=0.611m^3/s,case3

where w_j is the tangential velocity of the jet and $\dot{m}$ is the volume flow rate. The dimensions of the classifier were: large diameter=2.286m,small diameter= 1.08m and length=4.74m,the rotation speed= 20 r.p.m.

b. Two classifier arrangements using 10 air nozzles:

i. large diameter=1.125m,small diameter=0.823m and length = 1.703 m (case 4)
ii. large diameter=1.125m,small diameter=0.823m and length =1.703m,with a small cylinder blocking the recirculation zone at the large diameter side and along the classifier axis, (case 5).

The prevailing boundary conditions are:

1. Along the centreline, $\partial \phi / \partial r = 0$
2. At the inlet section: floating conditions were assumed.
3. At the exit plane : $\partial u / \partial x)_{x=L} = 0$
4. Along the inclined wall: the influence of the rotation on the flow field was represented in the source term of the w equation along the wall as;

   $S_{w1} = N.y_w/60$ where N is the rotation speed and y_w is the radius at a given location along wall. The injection of high speed air jets,necessitates the addition of source terms at the wall as; $S_{w2} = w_j$ and the total source $S_w = S_{w1} + S_{w2}$
   Along the wall,the radial and axial velocities were assumed zero,and the shear stress was obtained from the wall functions,references[7 and 8].

In figure 1 the calculated flow patterns,for cases 1,2 and 3 are shown. A high velocity region existed near the wall and a wall jet behaviour was observed. A central recirculation zone,created by the rotation and the adverse pressure field,existed at the centre of the flow and extended up to the large end. This recirculation zone is responsible for the classification of the garbage. The velocity profile of the air induced through the small end started as fairly uniform,but then developed into a high wall jet as a result of the rotation and the effect of the compressed air. It is clear from the figure that increasing the jet velocity,i.e. the tangential mass flow rate,increased the size of the recirculation zone,thus ensuring better and quicker classification of the charge. The calculated velocity profiles at various axial locations are shown on the same figure; regions of forward and backward flows are identified clearly.

The tangential and radial velocity profiles, for case 2,are displayed in figure 2 as an example of the profiles for the different cases. The tangential velocity exhibited a peak value in the vicinity of the wall,but decreased to zero at the centreline.

A parametric study for the effects of the rotational speed,cone angle and geometry of the exit section,was performed. The effect of the rotational speed was found to be small relative to the effect of increasing the jet velocity. This is because relatively small rotation speeds were used due to practical considerations. The cone angle is important in determining the flow pattern and separation,however its value,in this application,was limited to small angle (5 → 7 degrees) to prevent separation at the wall region.

For cases 4 and 5 , the half angle was 5 and the compressed air volume is 0.0275m^3/s,while the induced air volume is 0.45 m^3/s. Figure 3 shows the axial velocity profiles for the two cases at various axial locations. Relatively small swirling motion was identified; the velocity profiles were influenced by the swirl and adverse pressure gradients at locations greater than 0.4 of the classifier length,compared with 0.1 for cases 1 to 3 . The small solid cylinder ,placed along the axis,decreases the size of the recirculation zone,but causes small changes in the radial and tangential velocity profiles of the order of 5%.

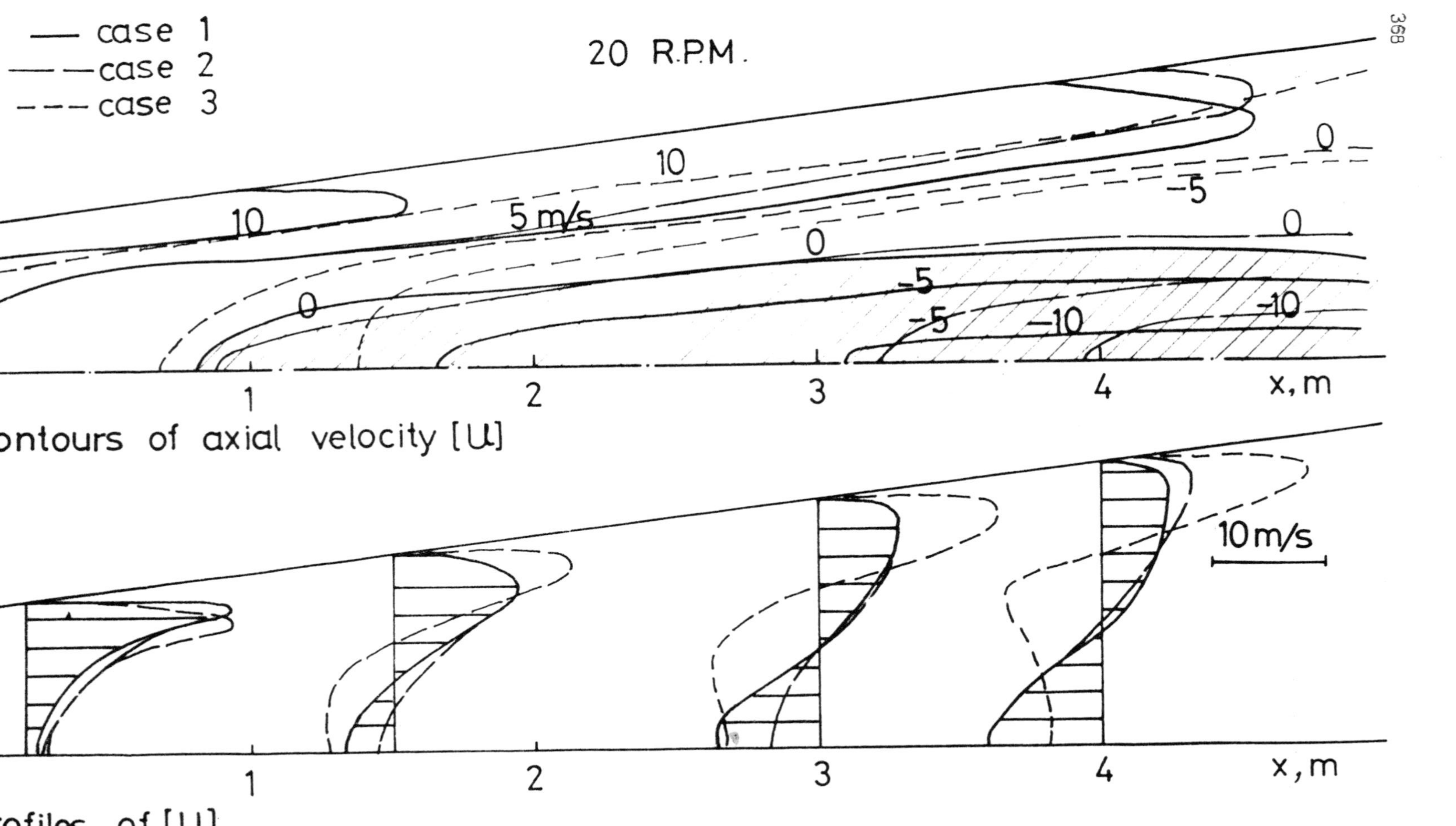
case 1
case 2
case 3
20 R.P.M.
10
10
5 m/s
0
0
0
-5
-5
-5
-10
-10
0
x, m
1
2
3
4
Contours of axial velocity [U]
10 m/s
x, m
1
2
3
4
Profiles of [U]

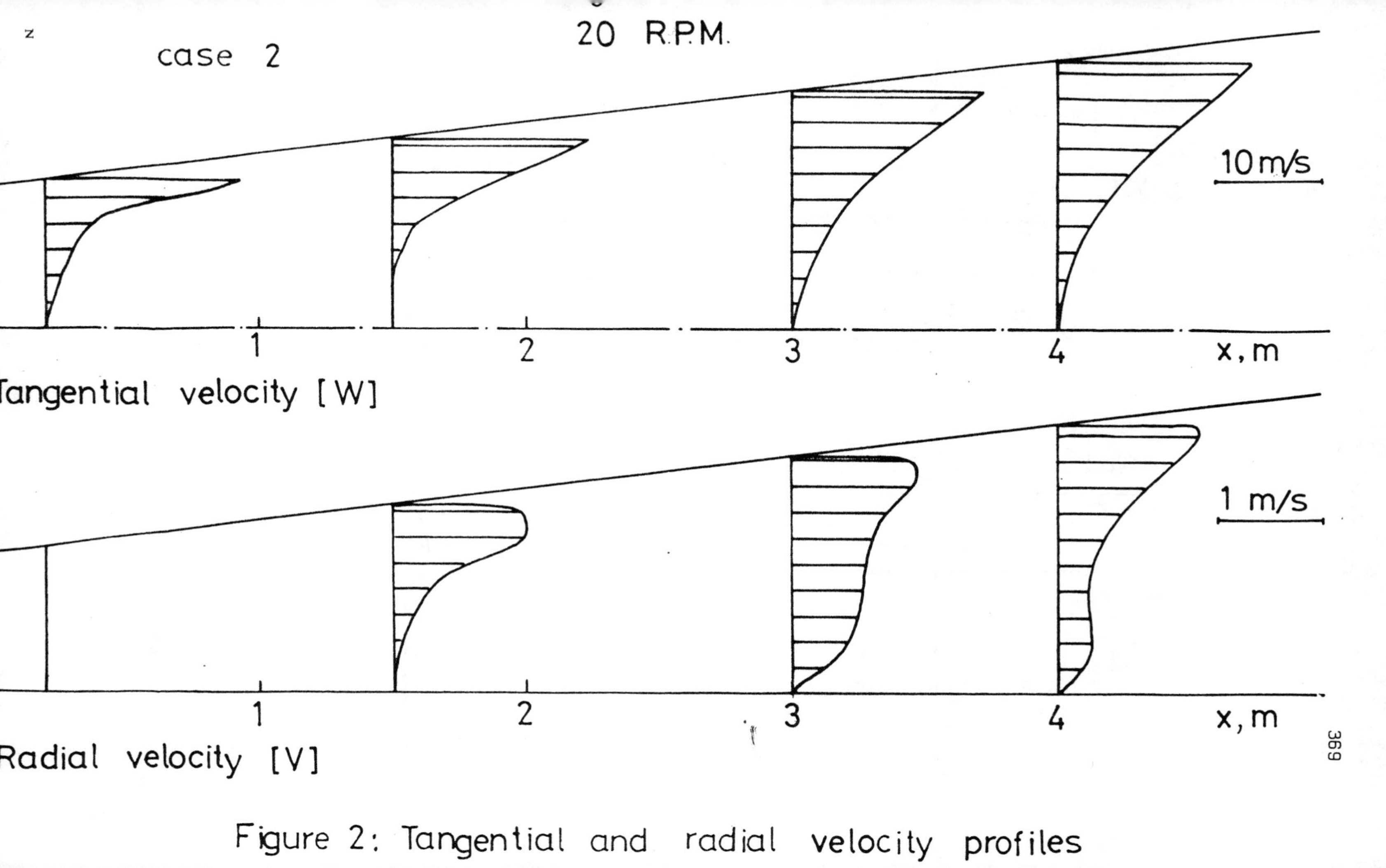

Figure 2: Tangential and radial velocity profiles

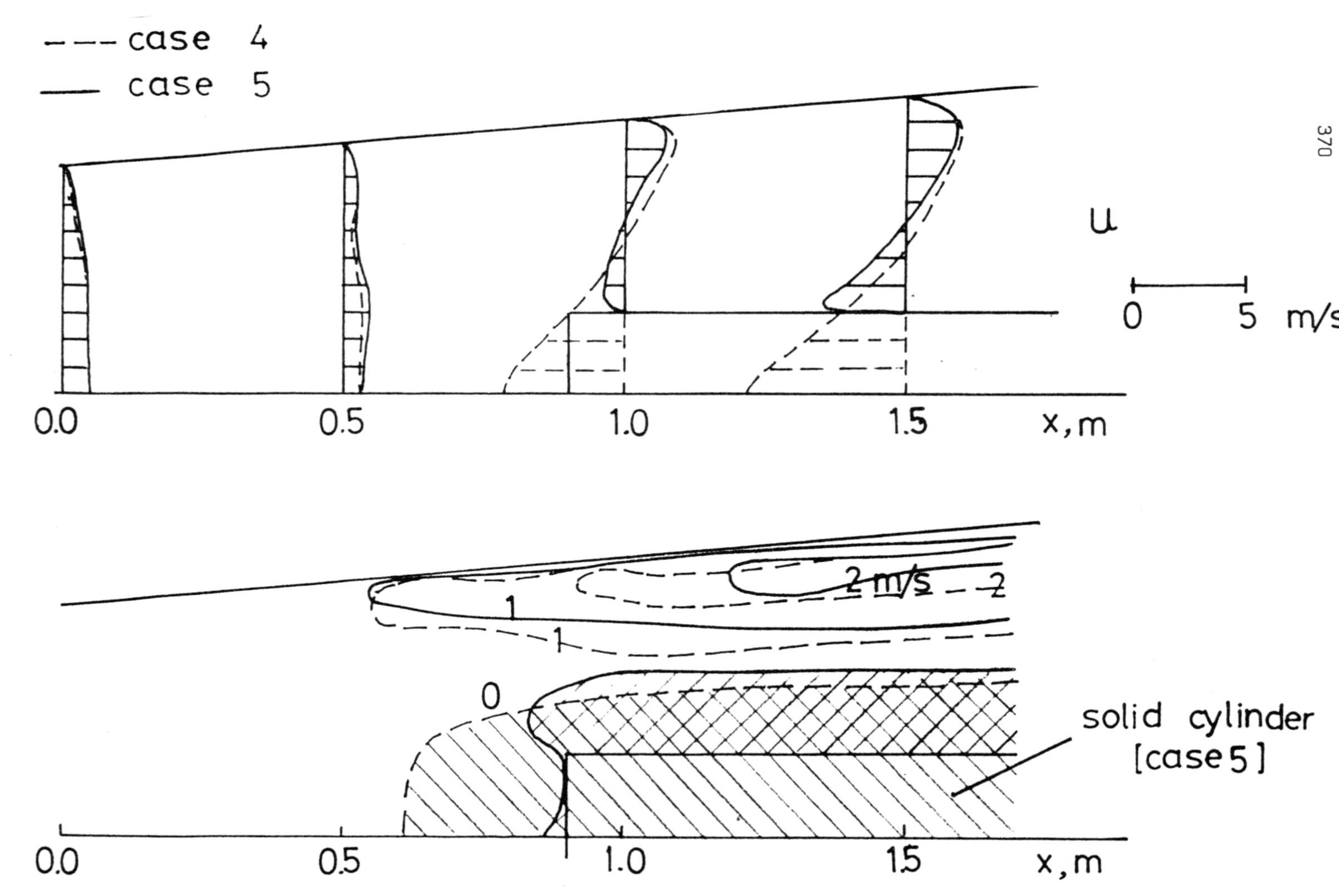

---- case 4
— case 5
u
0 5 m/s
0.0
0.5
1.0
1.5
x, m
2 m/s
2
1
1
0
solid cylinder
[case 5]
0.0
0.5
1.0
1.5
x, m

III.2 Flow in a Vortex Tube

The compressed air,admitted tangentially at one end,is separated by the vortex action into a hot stream and a cold one. The cold stream is discharged through a restriction at the centre of the tube and adjacent to the inlet nozzles,figure 4. The hot air is discharged through the hot valve located at the other end. The temperature difference between the hot and the cold streams is 25 °C,which promotes the potential of the vortex tube as a simple cooling device. The operating conditions were:

a. Two length to diameter ratios :L/D=5.5 and 18.
b. Two ratios of mass flow rates of hot and cold streams : $\dot{m}_h / \dot{m}_c = 3.34$ and 1.0.

The following boundary conditions were employed:

1. Along the centreline,$\partial \emptyset / \partial r$) = 0
2. At the exit plane: the mass flow ratio was specified and $\partial u / \partial x = 0$.
3. Along the tube wall, zero velocities were presumed, and the wall functions of references [5,7 and 8] were used to express the shear stresses.
4. Inlet nozzles : The tangential nozzles were replaced in the calculations by an annular slit yielding the same momentum and flow rate,in order to preserve the axisymmetry of the flow.

The calculated flow pattern,for L/D=5.5 and $\dot{m}_h/\dot{m}_c = 3.34$, is shown in figure 4. The flow regimes were identified clearly. A well defined hot wall jet region was observed in the vicinity of the wall,which was discharged through the hot valve end. A large central recirculation zone was also observed;this reversed cold flow was discharged through the cold orifice. The strength of the recirculation zone was considerable as the cold mass flow is just under one third of the hot flow. Figure 5 exhibits the predicted profiles of kinetic energy of turbulence. Regions of high turbulence intensity existed in the vicinity of steep velocity gradients,due to the production of kinetic energy from the shear work. Earlier work, [9], compared the obtained calculated results to the experimental data of Bruun [6].

The predicted velocity profiles and turbulence characteristics for L/D=18 and $\dot{m}_h/\dot{m}_c=1$ are shown in figure 6. A much smaller central recirculation zone was observed. It is clear that the cooling effect was hindered by the large L/D and by the relatively small $\dot{m}_h/\dot{m}_c$.

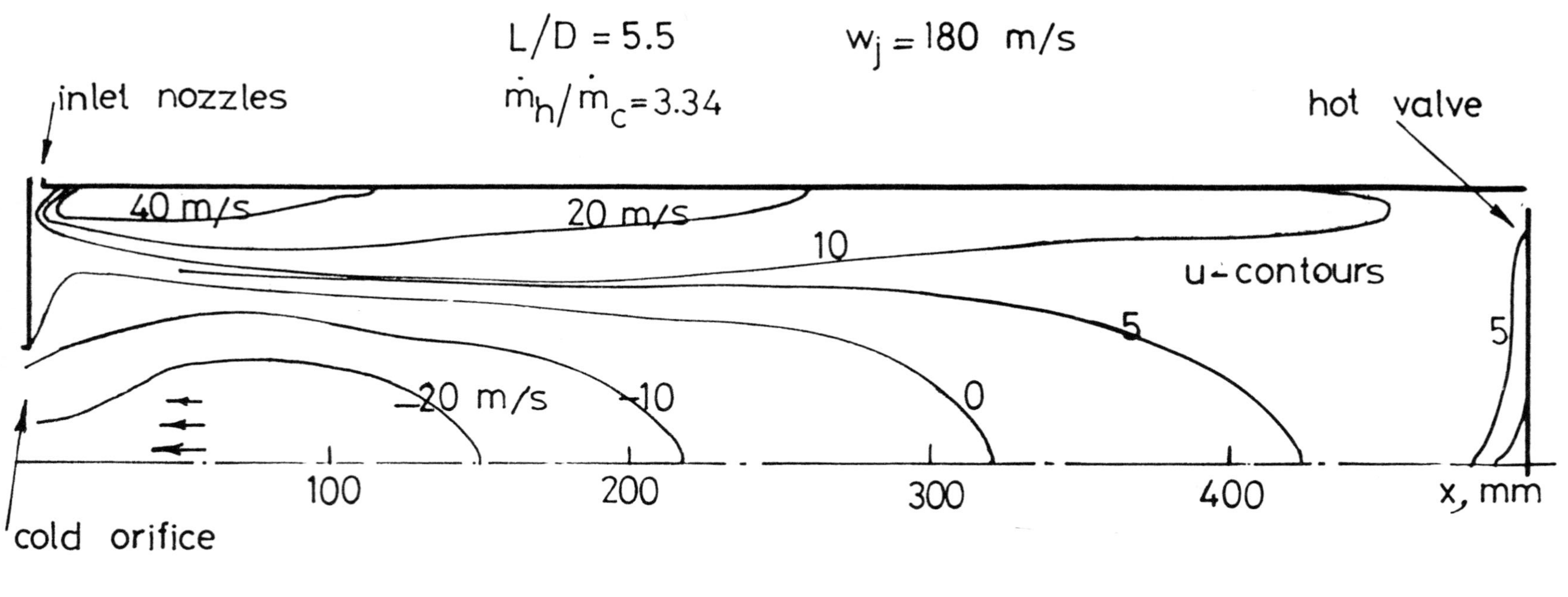

Figure 4 : Flow pattern in vortex tube

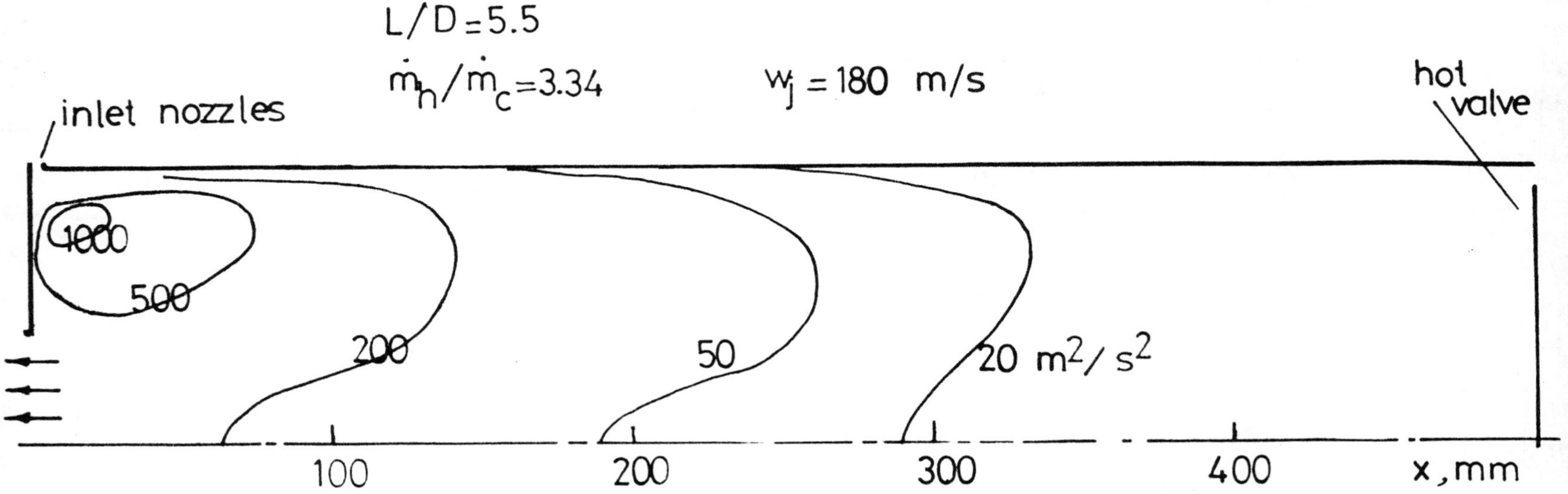

Figure 5 : Contours of constant 'k'

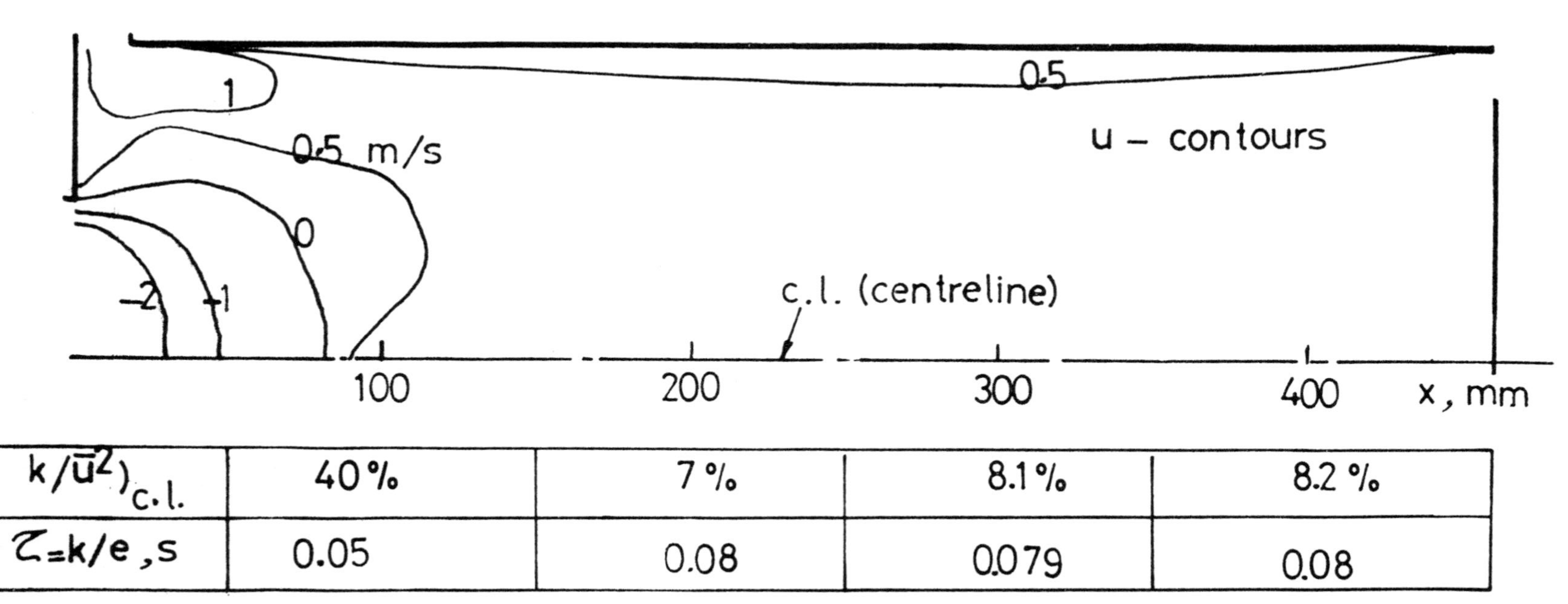

$k/\bar{u}^2)_{c.l.}$	40%	7%	8.1%	8.2%
$\tau = k/e$, s	0.05	0.08	0.079	0.08

Figure 6 : Long vortex tube characteristics

IV.CONCLUDING REMARKS

This work intended to demonstrate that it was both feasible and economical to obtain calculations of the flow pattern and separation characteristics in practical turbulent flow systems. The numerical procedure was shown to solve the non-linear partial differential equations governing the transport of mass and momentum. Each flow was investigated under various parameters. Further development and modifications of the existing turbulence models and numerical techniques are needed to improve the physical representation of the flow characteristics.

REFERENCES:

1. PATANKAR,S.V.and SPALDING,D.B.,A Computer Code for Three Dimensional Flow in Furnaces,Proc.14th Symp. (Int.) on Combustion,pp605-614,1973.
2. GOSMAN,A.D.and PUN,W.M.,Calculation of Recirculating Flows,Mech.Eng.Dept.Report HTS/74/2,Imperial College,1974.
3. GOSMAN,A.D.,KHALIL,E.E.and WHITELAW,J.H.,The Calculation of 2-Dimensional Turbulent Recirculating Flows,Turbulent Shear Flows I,Edt.Durst et al,Springer-Verlag,1979.
4. KHALIL,E.E.,SPALDING,D.B.and WHITELAW,J.H.,The Calculation of Local Flow Properties in Two-Dimensional Furnaces,Int.J.H.Mass Transfer,Vol.18,pp775-791, 1975.
5. CEBECI,T.,KHALIL,E.E.and WHITELAW,J.H.,Calculation of Separated Boundary Layer Flows,AIAA Journal,Vol.17, pp 1291-1292,1979.
6. BRUUN,H.H.,Experimental Investigation of the Energy Separation in Vortex Tubes, J.Mech.Engng.Sci.,Vol. 11,No.6,pp 567-582,1969.
7. LAUNDER,B.E.and SPALDING,D.B.,The Numerical Computations of Turbulent Flows,Comput.Methods Appl.Mech. Eng,Vol.3,pp 269-289,1974.
8. KHALIL,E.E.,Numerical Computations of Turbulent Flow Structure in a Cyclone Chamber,ASME paper number 79-HT-31,1979.
9. KHALIL,E.E.and ASAAD,S.,Flow Regime in Vortex Tubes, Proc.3rd Int.Conf.for Mech.Power Engng,Cairo, Sept.1980.

ON THE CALCULATION OF THE LAMINAR AND TURBULENT FLOW BETWEEN PARALLEL DISKS

A. Pollard

Assistant Professor, Department of Mechanical Engineering, University of Calgary, Calgary, Alberta, Canada, T2N 1N4

ABSTRACT

Flow between parallel disks is used as a simplified case of a flow splitting tee-junction. This reduces a three-dimensional flow to one that is two-dimensional. For extremely fine finite-difference grid distributions, the laminar flow between parallel disks is accurately calculated. For turbulent flows, using k-ε, poor agreement between experiments and the calculations is noted. It is concluded that the standard k-ε model is incapable of predicting this flow.

1.0 INTRODUCTION

The author, for some time, has been concerned with calculating the uniformly dividing three dimensional turbulent flow in tee-junctions (Pollard and Spalding [1] [2], Pollard [3]); but, because of computer storage limitations, predictions that are in good agreement with experimental data have yet to be obtained, even though every effort has been made to simplify both the flow situation and the numerical procedure thereby permitting the maximum possible finite-difference grid density to be used. It would appear that a two-dimensional flow situation that exhibits the same features as the flow in a tee-junction is needed so that grid independent results can be achieved, thereby permitting a turbulence model to be more fairly tested.

It is believed that radial flow between parallel stationary disks fills this need as Figure 1 indicates. It is true that this flow situation exhibits no swirl (provided the inlet flow does not possess this attribute), but it does exhibit an additional feature, namely the provision for the transition from turbulent to laminar flow, which is commonly referred to as "relaminarization".

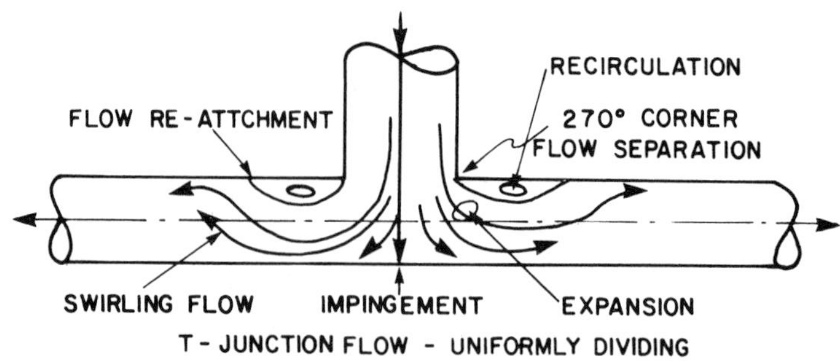

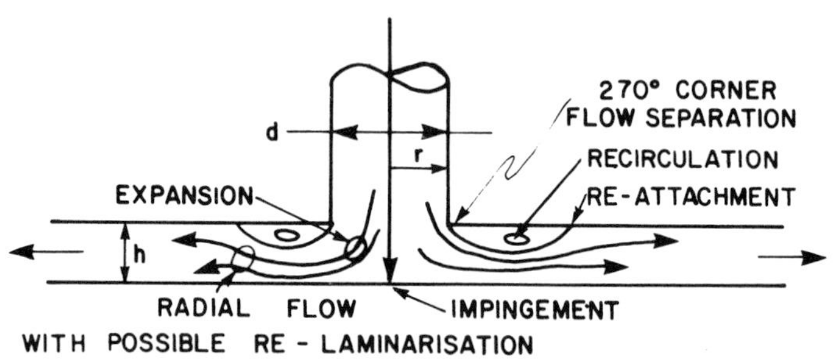

Figure 1: Comparison of Flow Features

The main purpose of this paper is to determine whether the k-ε model used to predict the three-dimensional uniformly dividing turbulent flow in a tee-junction (FLIRT, Pollard [4]), can predict successfully the turbulent radial flow between parallel disks.

The computation of the flow between parallel disks would tend to exhibit the same deficiencies as those encountered when computing the uniformly dividing flow in a tee-junction. However, in two-dimensional confined recirculating flow, when grid independence has been stated to be assured, the currently favoured k-ε model of turbulence (Launder and Spalding [5]) has performed reasonably well. Examples include Durst and Rostogi [6], Gosman et al [7] and Elgobashi et al [8]. However, there is evidence that for flows undergoing strong curvature, the k-ε model requires a modifying term, and that unless extremely fine grid distributions are used, the hybrid difference formulation (Spalding [9]) will not reflect these modifications to the turbulence model; an excellent review of this area for unconfined recirculating flows is given by Leschziner and Rodi [10]. These authors employed the false-diffusion absent or reduced schemes of Raithby [11] and Leonard [12], coupled with a streamline curvature and normal stress modifications to the standard form of the k-ε model. Their results show clearly that the schemes of [11] [12], when coupled with the aforementioned turbulence model modifications, yield very good agreement with experimental data; the hybrid/standard k-ε formulation faired poorly.

The apparent inconsistency between the use of hybrid differencing/k-ε model when applied to confined and unconfined turbulent flows is disturbing, yet encouraging from the point of view of the confined flow situation considered here.

A Brief Overview of the Literature

A brief review of the literature relevant to the laminar and turbulent radial flow between parallel disks is now provided; a more detailed account can be found in [13].

In the open literature there appears to be but two studies that are directly applicable to the range of flows to be considered in this paper. These are the studies of McGinn [14] and Moller [15]; these papers have been isolated because they provide, among other things, wall pressure data; and, since these flows are pressure driven, it seems judicious to use these data for comparison purposes.

McGinn [14] provides an analytical formulation for the fully developed radial pressure distribution between parallel disks containing completely laminar flow. This result is essentially only valid for radii greater than approximately three times the inlet hole radius. The analytical results were confirmed experimentally.

Moller [15] provided an analytical expression for the radial pressure distribution for both purely laminar and purely turbulent flows; in general, good agreement was obtained bebetween his analytical expressions for pressure and his experimental data. The data provided by Moller does not include the inlet conditions; reference is made to the fact that the inlet pipe, which guides the fluid through the surface of one of the disks, is of a length that provides fully developed pipe flow that exhibits "little or no swirl".

2.0 MATHEMATICAL FORMULATION

2.1 The Equations

The equations describing the steady, axisymmetric radial flow between parallel disks are given below, noting that the effective viscosities (μ_e), which reflect turbulent stresses, can be replaced by the molecular viscosity (μ) when the flow is laminar. The equations, in their most general form, are:

$$\text{div}\,(\rho\bar{V}\phi) = \text{div}\,\frac{\mu_e}{\sigma_\phi}\,\text{grad}\,\phi + S_\phi \qquad (1)$$

where ρ is the density, $\bar{V}$ the velocity vector, ϕ is any scalar variable, σ_ϕ the Prandtl number, S_ϕ the source term. These equations represent also the k-ε model of turbulence; details of which can be found in [5].

The boundary conditions for the laminar and turbulent flow situations will be given in detail later.

2.2 Solution Procedure

The equations (and their boundary conditions) are solved using a two-dimensional axisymmetric version of the FLIRT code [4]. This code is based on the SIMPLE algorithm of Patankar and Spalding [16]; it also uses the hybrid differencing scheme [9]. The reader who is unfamiliar with these techniques is referred to, for example, [17].

3.0 COMPUTATIONAL DETAILS

3.1 Geometry

The general geometrical arrangement is shown in Figure 2.

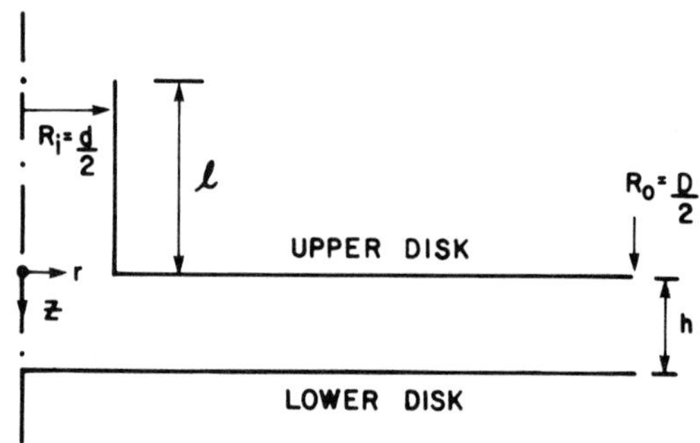

Figure 2: Parallel Disk-Geometry Used

The geometry used for the turbulent flow models the experimental situation of [15]. Since a closed form solution for the radial pressure distribution is provided by [14] for the laminar flow case, the geometry used for this case is the same as that specified for the turbulent flow. The details of the non-dimensional geometric parameters used are: $R_o/R_i = 6.0$, $h/2R_i = 0.267$ and $\ell/2R_i = 0.5$ or 0.0; the latter parameter indicates that the inlet pipe is one-half diameter in length or this pipe is replaced by a hole in the upper grid.

3.2 Finite-Difference Grid

The geometry is overlaid with a family of orthogonal intersecting grid lines; the density and non-uniformity of these grid lines can be altered so as to obtain results that are grid independent. An example of a grid independence test is shown in Figure 3. In this figure the centre-plane velocity, turbulence energy and a modified pressure coefficient are plotted as functions of radial distance from the centre-line of the two disks. It is seen that it is the turbulence kinetic energy that dominates the grid selection procedure: the 74 x 22 (r-z) uniformly distributed grid gives grid independent results but the differences in the results obtained using this grid and the 59 x 14 non-uniformly distributed grid is seen to be very minor. The 59 x 14 grid was used for all the calculations because the computer time using this grid is about 1/4 of that for the 74 x 22 grid. The 59 x 14 grid was used for the laminar flow results.

The false diffusion for the turbulent flow situation can be easily estimated [17]. For the 74 x 22 grid, the ratio (maximum) of false to real diffusion is estimated to be of order 1; for the 59 x 14 grid, when cognisance is given to the range of turbulent inlet magnitudes used, this ratio is estimated to vary between 0.1 and 10.

It is appropriate to note here that the 270° corner region requires special attention, particularly for the velocity equations. There are two points to note: the first is that this region is crucial to the accuracy of the predictions downstream of the corner; this requires that for laminar flow, the grid spacing in this region be of the same order of magnitude as the Stokes radius ($r_s \equiv \mu_e/\rho\bar{U}$, $\bar{U}$ = local velocity vector in corner region) as pointed out by Blowers [19] and Castro [20]. For turbulent flow the ratio between grid spacing and r_s can be somewhat larger than unity [20]. For the turbulent flow case, and using the 59 x 14 grid, the ratio of grid spacing to Stokes radius is noted to be of order 5. For the 74 x 22 grid, this ratio is of order 50. The difference here is due to the non-uniform clumping of the grid in the corner region for the 59 x 14 grid. The second point concerns the modifications to the shear stress terms: the use of the standard staggered grid arrangement results in the velocity cell bordering the corner region being bounded only by part of the disk or inlet pipe; the coefficients of the discretized equations account for this arrangement. Details of this special attention can be found in [4] [6].

3.3 Boundary Conditions

For the two flow situations considered here, the following table indicates the boundary conditions that are employed.

Note that in Table 1, WF ≡ Wall Functions [5], FDF Fully Developed Profile, and the double entry under the "V" and "W" are for laminar/turbulent flows respectively. The r^* term indicates that the inlet values of k and ε are modified

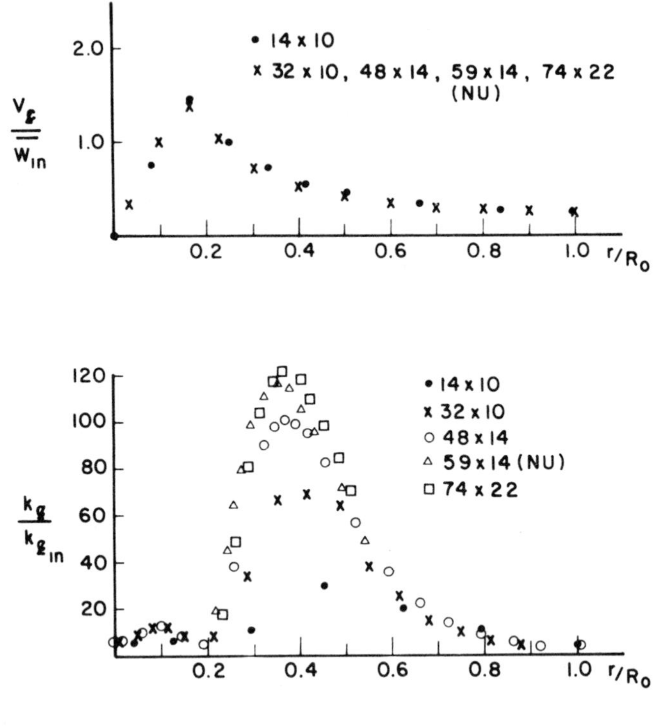

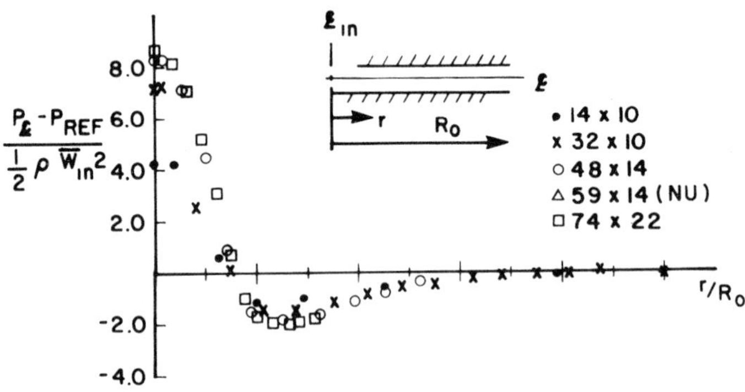

Figure 3: Grid Independence Test

to reflect the non-linear radial variation of these quantities found in fully developed pipe flow. C_k and C_ε are constants that are used to vary the intensity of k and ε at the inlet. In addition to the inlet conditions for k-ε, the data of Laufer (see [18]) were used.

3.4 Convergence and Computer Costs

The SIMPLE algorithm is a semi-implicit iterative

Boundary	V	W	k	ε
Inlet	0	FDP	$C_k r^* W_{IN}^{-2}$	$C_\mu k^{3/2} r^* / C_\varepsilon r$
Walls	0/WF	0/WF	WF	WF
Symmetry Axis	0	$\frac{\partial w}{\partial r} = 0$	$\frac{\partial k}{\partial r} = 0$	$\frac{\partial \varepsilon}{\partial r} = 0$
Exit	$\frac{\partial v}{\partial z} = 0$	$\frac{\partial w}{\partial z} = 0$	$\frac{\partial k}{\partial z} = 0$	$\frac{\partial \varepsilon}{\partial z} = 0$

Table 1: Boundary Conditions

procedure that starts from initial guesses of pressure, velocity, etc. and converges to the solution of the equations and their specified boundary conditions. Convergence was deemed complete when the residuals of each equation summed over the entire computational domain and suitably normalised by the inlet flux of each quantity reached a pre-determined level; this level of convergence was specified to be 0.001 and required, for the laminar/turbulent flow situations, 200 and 400 iterations respectively. These iterations required about 30 and 230 minutes of computer time. (Note that the 70 x 22 turbulent flow results required ∿ 500 iterations and about 12 hours of computer time!) The computer used was a Honeywell series 6000; this machine is about 40 times slower than a CDC 7600.

4.0 PRESENTATION AND DISCUSSION OF RESULTS

4.1 Laminar Flow

The laminar flow results are presented and discussed first since they are uncomplicated by the presence of turbulence; the numerical calculations of the radial gauge pressure distribution can be found in Figure 4. Figure 4 also contains the theoretical curve provided by McGinn [14]. This figure shows for two inlet pipe Reynolds numbers, the gauge pressures along the upper and lower disks. The horizontal axis (r/R_o) indicates the location of the inlet duct radius R_i and $3R_i$; this latter figure is noted in [14] as that minimum radius for which the theoretically derived pressure equation can be expected to be valid. It is seen from the figure that for Re_d = 120, the calculations are seen to depart from the theoretical curve at about $r = 3R_i$.

The results shown in Figure 4 imply that for Re_d = 6, no fluid recirculation takes place while for the larger value of Re_d, such fluid motion is present; this is evidenced by the inflection point in the calculations for pressure.

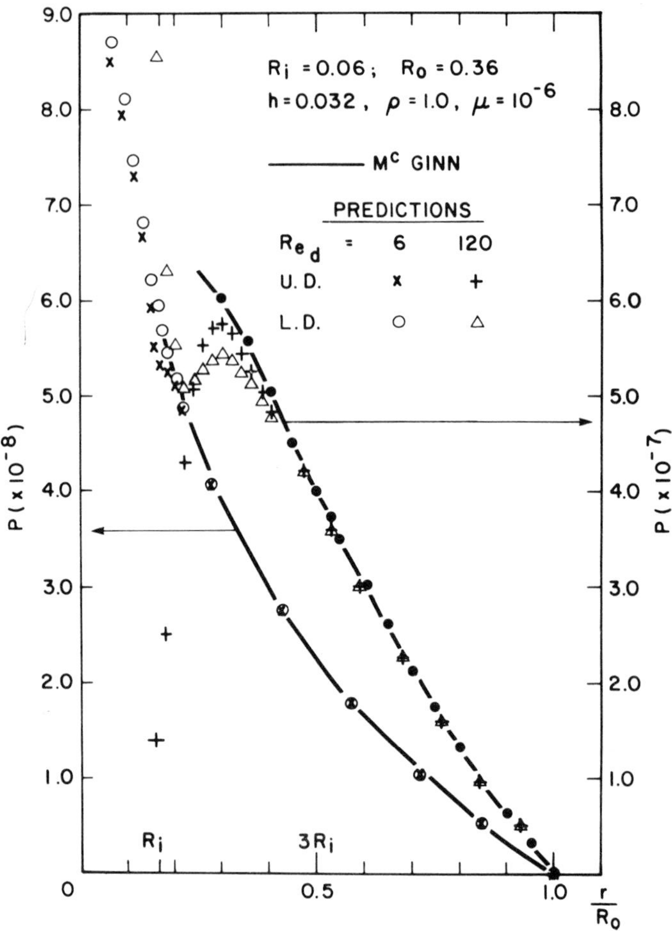

Figure 4: Laminar Flow

4.2 Turbulent Flow

The results of the calculations for turbulent flow are shown in Figure 5; here a modified pressure coefficient is plotted against the non-dimensional radius from the centre of the upper disk to its outer radius. The inlet Reynolds number is 1.73×10^5. The curve in the figure represents a mean line through the data of [15], and the various symbols indicate the results of the predictions. The circled symbols denote the points of re-attachment obtained using different turbulent inlet conditions.

Different turbulence inlet conditions were used when it became evident that Laufer's data produced very poor agreement, as noted in Figure 5. The inlet conditions were then changed to those given by the entry in Table 1 and a systematic

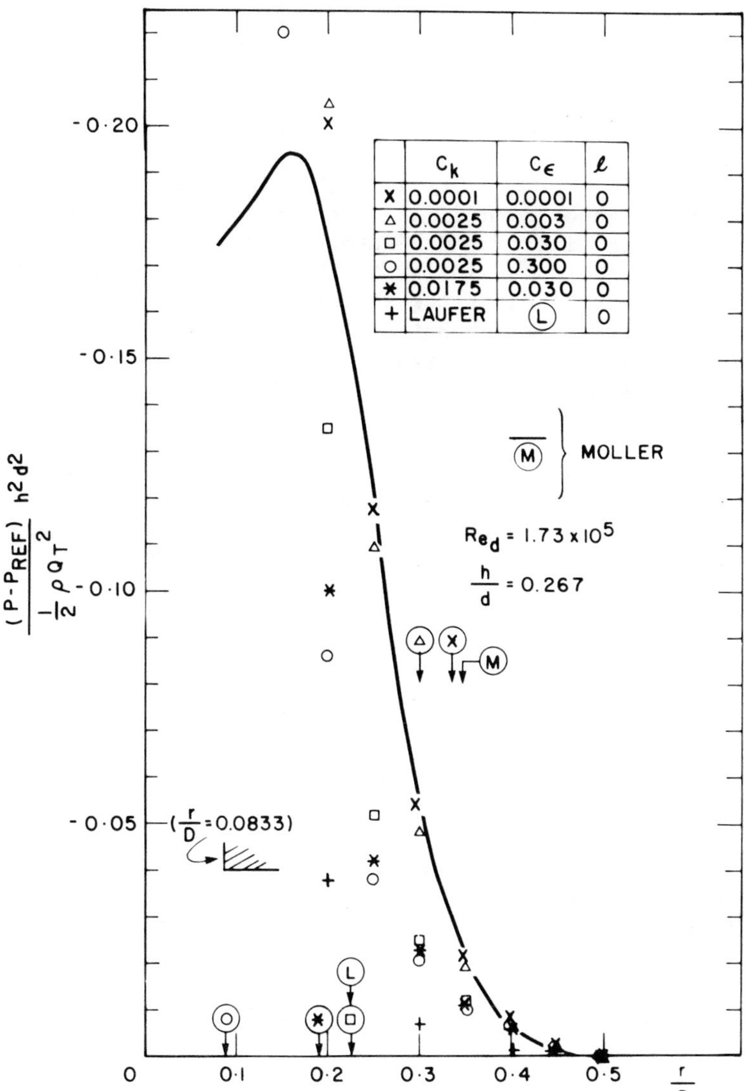

Figure 5: Turbulent Flow

variation of C_k and C_ε was performed; a sample of the results obtained are shown in the figure.

At first glance the results look quite promising; particularly those that correspond to low turbulence levels at the inlet. However, it is not shown, due to space limitations on the graph, that for $r/D \lesssim 0.20$, the inflection in the data curve of [15] is not reflected by the predictions: these predictions increase with decreasing radius such that at $r/D = 0.1$, the pressure coefficient reaches the value of 0.53 for that case denoted by the open circle (O). However, contrary to the laminar flow situation, fluid recirculation is present next to the junction of the inlet pipe and upper disk.

The reasons for the discrepancies shown in Figure 5 are not clear; in an attempt to alter the predictions, the hole in the upper disk was replaced by a pipe of length $\ell = 0.5d$. Using initially a grid of 14 x 58 that included the inlet pipe, and using the inlet conditions that agreed best with the experimental data (i.e.: those denoted by (x) in Figure 5), results were obtained that are similar to those results obtained when Laufer's data are used at the entrance to the hole in the upper disk. It would appear then that by specifying an almost zero turbulence level flow gives the best results; the use of the inlet pipe allows the turbulence to develop before entering the space between the disks.

The implication of the last statement and the entries in Figure 5 is that the turbulence model has little effect upon the calculations particularly in the vicinity of the inlet to the parallel disks.

In [10] it is noted that "curvature strongly influences turbulent transport in shear layers". In the present flow situation, there are two areas of curvature: flow about the 270° corner region and the recirculating zone of fluid next to this corner. Since the standard form of the k-ε model does not account for strain effects (i.e. anisotropy between normal stresses) due to curvature (see, for example, [21]), it is probably not surprising that the calculations are in poor agreement with the experimental data, even though in these regions the grid Peclet number is of order 1. Whether the curvature modification term introduced in [10] would have a beneficial effect upon the present results is difficult to say since, in [10], false diffusional effects are still apparent; thus the poor results obtained using hybrid differencing may still be due to false diffusional problems.

5.0 CONCLUSIONS

The examination of the laminar and turbulent flow between parallel disks allows the following conclusions to be drawn:

- laminar flow between parallel disks can be accurately predicted using hybrid differencing.

- turbulent flow cannot be accurately predicted using hybrid differencing and the standard k-ε model of turbulence even when great attention is paid to eliminating numerical diffusion.

ACKNOWLEDGEMENTS

Financial support provided by NSERC of Canada is

gratefully acknowledged. This paper is based on U. of C. Mech. Engg. Report #179.

REFERENCES

1. Pollard, A., Spalding, D.B., "The Prediction of the Three-Dimensional Turbulent Flow Field in a Flow-Splitting Tee-Junction", Computer Methods in Applied Mechanics and Engineering, Vol. 13, pp. 293-306, 1978.
2. Pollard, A., Spalding, D.B., "Turbulent Flow and Heat Transfer in a Tee-Junction", ASME, Paper No. 79-WA/HT-47, ASME, New York, 1979.
3. Pollard, A., "Computer Modelling of Flow in Tee-Junctions", submitted for publication, 1980.
4. Pollard, A., "Flow in Tee-Junctions", Ph.D. Thesis, University of London, 1978.
5. Launder, B.E., Spalding, D.B., "The Numerical Computation of Turbulent Flows", Comp. Meth. App. Mech. Engg., Vol. 3, pp. 269, 1974.
6. Durst, F., Rastogi, A.K., "Theoretical and Experimental Investigations of Turbulent Flow with Separation", Proc. 1st Symp. Turbulent Shear Flows, Penn. State (1977), Springer-Verlag, New York, 1979.
7. Gosman, A.D., Khalil, E.E., Whitelaw, J.H., "The Calculation of Two-Dimensional Turbulent Recirculating Flows", Proc. 1st Symp. Turbulent Shear Flows, Penn. State (1977), Springer-Verlag, New York, 1979.
8. Elgobashi, S.E., Samuelson, G.S., Wierer, J.E., LaRue, J.C., "Prediction and Measurement of Mass, Heat and Momentum Transport in a Non-Reacting Turbulent Flow of a Jet in an Opposing Stream", Proceedings, 2nd Symposium Turbulent Shear Flows, Imperial College, 1979.
9. Spalding, D.B., "A Novel Finite-Difference Formulation for Differential Expressions Involving Both First and Second Derivatives", Int. Jn. Num. Methods Engg., Vol. 4, pp. 551-559, 1972.
10. Leschziner, M.A., Rodi, W., "Calculation of Annular and Twin Parallel Jets Using Various Discretization Schemes and Turbulence Model Variants", SFB 80/T/159, February 1979, University of Karlsruhe.
11. Raithby, G.D., "Skew-Upwind Differencing Schemes for Problems Involving Fluid Flow", Computer Methods in Applied Mechanics and Engineering, Vol. 9, pp. 153-164, 1976.
12. Leonard, B.P., "A Stable and Accurate Convective Modelling Procedure Based on Quadratic Upstream Interpolation", Computer Methods in Applied Mechanics and Engineering, Vol. 19, pp. 59-98, 1979.
13. Pollard, A., Siu, A.L., "A Literature Survey on the Steady Incompressible Radial Source and Radial Sink Flows Between Fixed Parallel Disks", University of Calgary, Dept. Mech. Engg., Report 168, 1980.

14. McGinn, J.H., "Observations on the Radial Flow of Water Between Fixed Parallel Plates", Appl. Sci. Res., Section A, Vol. 4, 1955.
15. Moller, P.S., "Radial Flow Without Swirl Between Parallel Disks", Aero, Quart., Vol. 14, pp. 163-186, 1963.
16. Patankar, S.V. and Spalding, D.B., "A Calculation Procedure for Heat, Mass and Momentum Transfer in Three-Dimensional Parabolic Flows", Int. Jn. Heat Mass Transfer, Vol. 15, pp. 1787-1806, 1972.
17. Patankar, S.V., "Numerical Heat Transfer and Fluid Flow", Hemisphere Publishing, Washington, D.C., 1980.
18. Hinze, J.O., "Turbuliner", McGraw Hill, New York, pp. 728-730, 1975.
19. Blowers, R.M., "The Calculation by Finite Differences of Steady Two-Dimensional Laminar Flow in a T-Junction", Ph.D. Thesis, University of Surrey, 1973.
20. Castro, I.P., "Numerical Difficulties in the Calculation of Complex Turbulent Flows", Proc. 1st Turbulent Shear Flow, Es. Durst, Launder, Schmidt and Whitelaw, Springer-Verlag, New York, 1979.
21. Rodi, W., "Influence of Buoyancy and Rotation on Equations for the Turbulent Length Scale", Proceedings, 2nd Symposium Turbulent Shear Flows, Imperial College, London, pp. 1037-1042, 1979.

ON THE NEUMANN PROBLEM FOR THE PRESSURE IN A NAVIER-STOKES MODEL

B.J. Alfrink

Project Engineer, Mathematics Branch, Delft Hydraulics Laboratory, The Netherlands.
Temporarily at Electricité de France, Laboratoire National d'Hydraulique, France.

ABSTRACT

The present paper treats the numerical solution of two-dimensional viscous incompressible fluid flow. As the transient pressure solution is required, the differential system is formulated in terms of primitive variables. The equation of motion is split into a divergence-free and a curl-free part. The pressure is solved using the Marker-and-cell (MAC) algorithm. The Neumann problem for the pressure must satisfy Gauss'theorem. Care has to be taken to fulfil this compatibility condition in discrete form. It will be shown that by using a regular grid this is not realized in general. Particular attention will be paid to the treatment of boundary conditions on a space-staggered grid.

1. INTRODUCTION

The study presented here, originates from a research project dealing with the numerical computation of two- or three-dimensional, steady or unsteady, turbulent flow in complicated geometries (Alfrink [1]). The pressure is assumed to be non-hydrostatic thus yielding a Navier-Stokes model. Turbulence closure will be obtained by means of a two-equation ($k - \varepsilon$) model. For the present the numerical model is based on a classical finite difference method. In order to attack arbitrary geometries, general boundary-fitted curvilinear coordinates are used.

In several areas of application within the working field of the Delft Hydraulics Laboratory the transient pressure field is needed. So a primitive variables formulation has been chosen. This choice seems particulary suitable when considering free surface or three-dimensional flow.

The present paper describes the numerical method chosen. Particular attention will be paid to the Neumann problem for the pressure.

For ease of notation the equations will be written only in a two-dimensional rectilinear coordinate system. Also a constant viscosity will be assumed here. However, the results are valid for the general case as defined above.

2. GOVERNING EQUATIONS AND BOUNDARY CONDITIONS

Using the variable a for the constant viscosity and defining the pressure ϕ as

$$\phi = \frac{p}{\rho}$$

the differential system for two-dimensional viscous incompressible fluid flow can be written as

$$\frac{\partial \underline{u}}{\partial t} + \nabla.(\underline{u}\underline{u}) + \nabla\Phi - a\,\nabla^2\,\underline{u} = \underline{f} \tag{1}$$

$$\nabla.\underline{u} = 0, \tag{2}$$

where $\underline{u} = (u\ w)^t$ is the two-dimensional velocity-vector and f is an arbitrary forcing function. Boundary conditions are primarily formulated in terms of velocities, such as

$$\underline{u} = \underline{\alpha} \tag{3}$$

$$\underline{u}.\underline{\nu} = \alpha \tag{4a}$$

$$\frac{\partial}{\partial \nu}(\underline{u}.\underline{\tau}) = \beta, \tag{4b}$$

and

$$\frac{\partial}{\partial \nu}(\underline{u}.\underline{\nu}) = \alpha \tag{5a}$$

$$\underline{u}.\underline{\tau} = \beta, \tag{5b}$$

where $\underline{\nu}$ and $\underline{\tau}$ are the unit normal and tangential vectors at the boundary. Taking $\alpha = \beta = 0$ yields for Equation (4) the symmetry or free-slip condition and for Equation (5) a "free" outflow condition. Equation (3) is called the no-slip condition when $\underline{u}.\underline{\nu} = 0$.

By taking the divergence of Equation (1) a Poisson equation for the pressure can be derived. A boundary condition of the Neumann-type can be found by taking the dot product of Equation (1) and $\underline{\nu}$, resulting for the no-slip condition in

$$\frac{\partial \phi}{\partial \nu} = a\,\frac{\partial^2}{\partial \nu^2}(\underline{u}.\underline{\nu}) + \underline{f}.\underline{\nu}, \tag{6}$$

for the symmetry-condition in

$$\frac{\partial \phi}{\partial \nu} = \underline{f}.\underline{\nu}, \tag{7}$$

and for the free outflow condition in

$$\frac{\partial \phi}{\partial \nu} = a\nabla^2(\underline{u}.\underline{\nu}) + \underline{f}.\underline{\nu} - \frac{\partial}{\partial t}(\underline{u}.\underline{\nu}). \tag{8}$$

So it can be seen that in general the Neumann condition will be inhomogeneous.

3. NUMERICAL METHOD

According to the idea of Chorin [2] the equation of motion is split into a divergence-free and a curl-free part. The pressure is solved using the algorithm introduced by Harlow and Welch [3] for the Marker-and-cell (MAC) method. If explicit finite differences are used, this yields the following scheme:

$$\frac{\underline{u}^{aux} - \underline{u}^n}{\Delta t} + \nabla.(\underline{u}\underline{u})^n - a\nabla^2\underline{u}^n = \underline{f}^n \tag{9}$$

$$\nabla^2\Phi^{n+1} = \frac{\nabla.\underline{u}^{aux}}{\Delta t} \tag{10}$$

$$\frac{\underline{u}^{n+1} - \underline{u}^{aux}}{\Delta t} + \nabla\Phi^{n+1} = 0. \tag{11}$$

First of all, the equation of motion excluding the pressure gradient is solved, yielding an auxiliary velocity field, which contains the correct vorticity but has non-zero divergence. The auxiliary velocity is then modified in such a way as to bring the divergence to zero while preserving the vorticity. Using Equation (11) the boundary condition for the pressure can be written here as

$$\frac{\partial \Phi^{n+1}}{\partial \nu} = \frac{1}{\Delta t}(\underline{u}^{aux} - \underline{u}^{n+1}).\underline{\nu}. \tag{12}$$

The scheme proposed has been used amongst others by Fortin et al [4].

4. NEUMANN PROBLEM FOR THE PRESSURE

4.1 Compatibility condition

If the inhomogeneous Neumann problem, expressed by Equations (10) and (12) has a solution, the right-hand members must satisfy the following compatibility condition, obtained by applying Gauss' theorem:

$$\int_\Omega \frac{\nabla.\underline{u}^{aux}}{\Delta t}\, d\Omega = \int_\Gamma \frac{1}{\Delta t}(\underline{u}^{aux} - \underline{u}^{n+1}).\underline{\nu}\, d\Gamma, \tag{13}$$

where Ω denotes the considered two-dimensional region enclosed by the boundary Γ. In case there is conservation of mass, i.e.

$$\int_\Gamma \underline{u}^{n+1}.\underline{\nu}\, d\Gamma = 0, \tag{14}$$

Equation (13) reduces to

$$\int_\Omega \nabla.\underline{u}^{aux}\, d\Omega = \int_\Gamma \underline{u}^{aux}.\underline{\nu}\, d\Gamma. \tag{15}$$

Care has to be taken to satisfy the discrete analogue of Equation (13). Consider for simplicity a rectangular region, which has been discretised uniformly on a regular grid with grid spacing h (Figure 1).

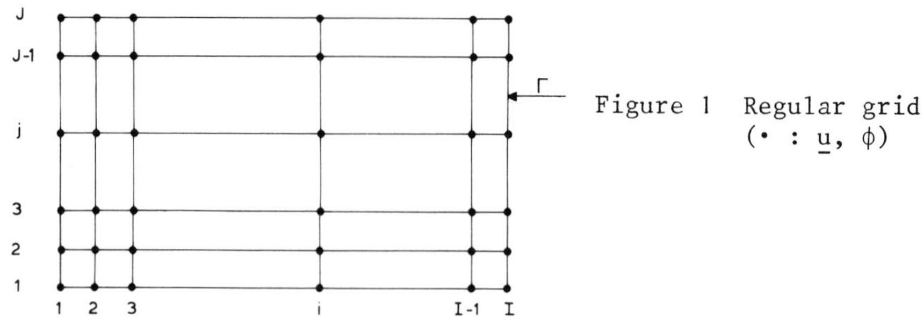

Figure 1 Regular grid ($\cdot$: $\underline{u}$, ϕ)

The simplest discretisation of the Neumann problem consists of the application of Equation (10) for the interior region and of Equation (12) for the boundary in the following way:

$(2 \leqslant i \leqslant I - 1;\ 2 \leqslant j \leqslant J - 1)$:

$$\frac{1}{h^2}(\phi^{n+1}_{i+1,j} + \phi^{n+1}_{i-1,j} + \phi^{n+1}_{i,j+1} + \phi^{n+1}_{i,j-1} - 4\phi^{n+1}_{i,j}) =$$

$$= \frac{1}{2h\Delta t}(u^{aux}_{i+1,j} - u^{aux}_{i-1,j} + w^{aux}_{i,j+1} - w^{aux}_{i,j-1}) \tag{16}$$

$(2 \leqslant i \leqslant I - 1)$:

$$\frac{1}{h}(\phi^{n+1}_{i,2} - \phi^{n+1}_{i,1}) = \frac{1}{\Delta t}(w^{aux}_{i,1} - w^{n+1}_{i,1}) \tag{17a}$$

$$\frac{1}{h}(\phi^{n+1}_{i,J-1} - \phi^{n+1}_{i,J}) = \frac{1}{\Delta t}(w^{n+1}_{i,J} - w^{aux}_{i,J}) \tag{17b}$$

$(2 \leqslant j \leqslant J - 1)$:

$$\frac{1}{h}(\phi^{n+1}_{2,j} - \phi^{n+1}_{1,j}) = \frac{1}{\Delta t}(u^{aux}_{1,j} - u^{n+1}_{1,j}) \tag{17c}$$

$$\frac{1}{h}(\phi^{n+1}_{I-1,j} - \phi^{n+1}_{I,j}) = \frac{-1}{\Delta t}(u^{aux}_{I,j} - u^{n+1}_{I,j}) \tag{17d}$$

Note that the corner points can be neglected. When multiplying Equation (16) by h^2 and Equation (17) by h, and adding for all specified nodal points, the left-hand members cancel. This can be interpreted as the divergence theorem in discrete form. However, the assembled right-hand member will in general be non-zero, yielding a disturbed compatibility condition:

$$\text{RHS} = \frac{h^2}{2h\Delta t}\{\sum_{i=2}^{I-1}\sum_{j=2}^{J-1}(u^{aux}_{i+1,j} - u^{aux}_{i-1,j} + w^{aux}_{i,j+1} - w^{aux}_{i,j-1})\} +$$

$$+ \frac{h}{\Delta t}\{\sum_{i=2}^{I-1}(w^{aux}_{i,1} - w^{aux}_{i,J}) + \sum_{j=2}^{J-1}(u^{aux}_{1,j} - u^{aux}_{I,j})\} =$$

$$= \frac{h^2}{2h\Delta t}\{\sum_{i=2}^{I-1}\left[-(w_{i,1}^{aux} + w_{i,2}^{aux}) + (w_{i,J}^{aux} + w_{i,J-1}^{aux})\right] +$$

$$+ \sum_{j=2}^{J-1}\left[-(u_{1,j}^{aux} + u_{2,j}^{aux}) + (u_{I,j}^{aux} + u_{I-1,j}^{aux})\right]\} +$$

$$+ \frac{h}{\Delta t}\{\sum_{i-2}^{I-1}(w_{i,1}^{aux} - w_{i,J}^{aux}) + \sum_{j=2}^{J-1}(u_{1,j}^{aux} - u_{I,j}^{aux})\} \neq 0, \qquad (18)$$

where it has been assumed that there is conservation of mass, i.e. Equation (14) holds.

Only by means of a particular choice of boundary conditions for the auxiliary velocity field, the expression (18) can be made equal to zero. Taking for example

$$w_{i,1}^{aux} = w_{i,2}^{aux} \qquad (19a)$$

$$(2 \leqslant i \leqslant I - 1)$$

$$w_{i,J}^{aux} = w_{i,J-1}^{aux} \qquad (19b)$$

$$u_{1,j}^{aux} = u_{2,j}^{aux} \qquad (19c)$$

$$(2 \leqslant j \leqslant J - 1)$$

$$u_{I,j}^{aux} = u_{I-1,j}^{aux} \qquad (19d)$$

it can be seen in Equation (18) that compatibility is achieved on a regular grid. However, as the Neumann condition for the pressure is directly related to the boundary condition for the auxiliary velocity field, now the pressure is actually solved with a physically unrealistic boundary condition (compare with Equation (6)). Considering the elliptic character of the pressure equation, its solution has to be distrusted.

In the analysis given above, the Poisson equation for the pressure is not applied at the boundary. This could be seen as a possible cause of the unsatisfied compatibility condition. Therefore, the Neumann problem has been examined too, using virtual pressure points (see Figure 2). Due to limited space, only a verbal description will be given.

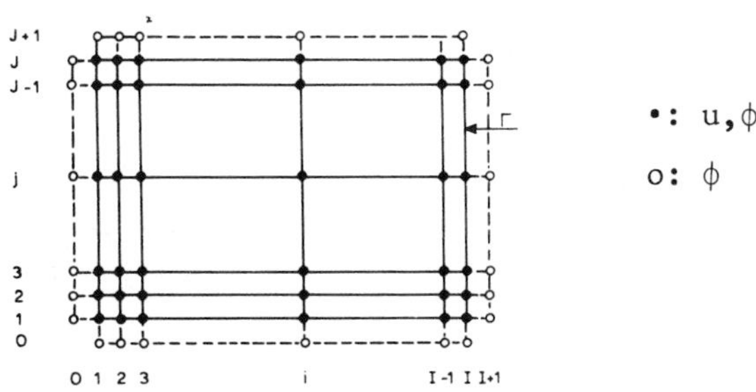

Figure 2 Regular grid with virtual pressure points

Now, the Poisson equation can be applied for the entire physical region, including the corners ($1 \leq i \leq I$; $1 \leq j \leq J$). The pressure variables defined at the virtual points are eliminated, using boundary condition (12). Adding the equations with weights 4 : 2 : 1 for the interior, boundary and corner points respectively, then yields a left-hand member equal to zero, but the right-hand member will again not vanish in general. Taking the right-hand member equal to zero, gives an integral relation for both the normal and tangential auxiliary velocity components on the boundary.
A systematic approach yields, after some computational effort, the following relations for the normal components:

$$w^{aux}_{i,1} = 2\, w^{aux}_{i,2} - w^{aux}_{i,3} - 4\, w^{n+1}_{i,1} \qquad (20a)$$

$$(2 \leq i \leq I - 1)$$

$$w^{aux}_{i,J} = 2\, w^{aux}_{i,J-1} - w^{aux}_{i,J-2} - 4\, w^{n+1}_{i,J} \qquad (20b)$$

$$(2 \leq j \leq J - 1)$$

$$u^{aux}_{I,j} = 2\, u^{aux}_{2,j} - u^{aux}_{3,j} - 4 u^{n+1}_{1,j} \qquad (20c)$$

$$u^{aux}_{I,j} = 2\, u^{aux}_{I-1,j} - u^{aux}_{I-2,j} - 4 u^{n+1}_{I,j}. \qquad (20d)$$

It will be clear that again the use of a regular grid implies the use of a boundary condition for the pressure that cannot be interpreted physically (Equation 6).

Ghia et al [5] have examined the problem of satisfying the compatibility condition when using a regular grid too. They have tried to "solve" the problem by modifying the right hand side of the Poisson equation uniformly at every grid point by an amount E/Ω where E is given as

$$E = \int_{\Omega} \frac{\nabla . \underline{u}^{aux}}{\Delta t}\, d\Omega - \int_{\Gamma} \frac{1}{\Delta t} (\underline{u}^{aux} - \underline{u}^{n+1}) . \nu \, d\Gamma. \qquad (21)$$

In principle, this ad-hoc approach can give rise to false results for the pressure, and therefore cannot be accepted here. Alternative discretisations, described in [5] seem to modify the physical nature of the Neumann problem too. So, these have not been considered here further.

Next, consider a space - staggered grid where the velocity -and pressure- variables are defined at different nodal points as used, for example, by Fortin et al [4] (Figure 3).

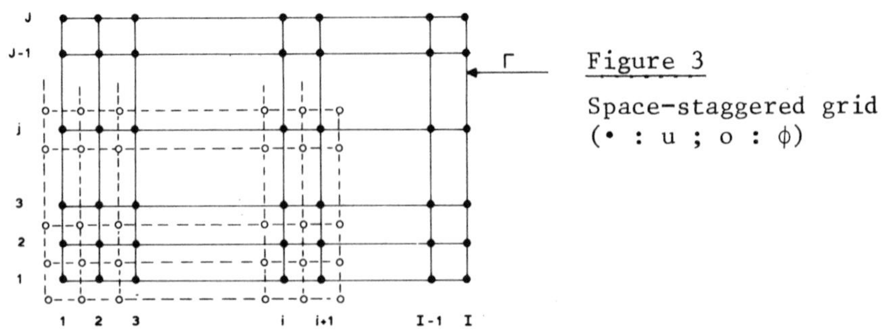

Figure 3

Space-staggered grid
(• : u ; o : ϕ)

As in the analysis for the regular grid of Figure 1, the Poisson equation for the pressure is applied only for the interior points:

$(1 \leq i \leq I - 1;\ 1 \leq j \leq J - 1)$:

$$\frac{1}{h^2}(\phi^{n+1}_{i+\frac{3}{2},j+\frac{1}{2}} + \phi^{n+1}_{i-\frac{1}{2},j+\frac{1}{2}} + \phi^{n+1}_{i+\frac{1}{2},j+\frac{3}{2}} + \phi^{n+1}_{i+\frac{1}{2},j-\frac{1}{2}} - 4\phi^{n+1}_{i+\frac{1}{2}.j+\frac{1}{2}}) =$$

$$= \frac{1}{2h\Delta t}(u^{aux}_{i+1,j} - u^{aux}_{i,j} + u^{aux}_{i+1,j+1} - u^{aux}_{i,j+1} +$$

$$+ w^{aux}_{i,j+1} - w^{aux}_{i,j} + w^{aux}_{i+1,j+1} - w^{aux}_{i+1,j}) \tag{22}$$

Boundary condition (12) can now be discretised centrally, thus preserving the second-order accuracy of the total numerical scheme:

$(1 \leq i \leq I - 1)$:

$$\frac{1}{h}(\phi^{n+1}_{i+\frac{1}{2},\frac{3}{2}} - \phi^{n+1}_{i+\frac{1}{2},\frac{1}{2}}) = \frac{1}{2\Delta t}(w^{aux}_{i+1,1} + w^{aux}_{i,1} - w^{n+1}_{i+1,1} - w^{n+1}_{i,1}) \tag{23a}$$

$$\frac{1}{h}(\phi^{n+1}_{i+\frac{1}{2},J-\frac{1}{2}} - \phi^{n+1}_{i+\frac{1}{2},J+\frac{1}{2}}) = \frac{1}{2\Delta t}(w^{n+1}_{i+1,J} + w^{n+1}_{i,J} - w^{aux}_{i+1,J} - w^{aux}_{i,J}) \tag{23b}$$

$(1 \leq j \leq J - 1)$:

$$\frac{1}{h}(\phi^{n+1}_{\frac{3}{2},j+\frac{1}{2}} - \phi^{n+1}_{\frac{1}{2},j+\frac{1}{2}}) = \frac{1}{2\Delta t}(u^{aux}_{1,j+1} + u^{aux}_{1,j} - u^{n+1}_{1,j+1} - u^{n+1}_{1,j}) \tag{23c}$$

$$\frac{1}{h}(\phi^{n+1}_{I-\frac{1}{2},j+\frac{1}{2}} - \phi^{n+1}_{I+\frac{1}{2},j+\frac{1}{2}}) = \frac{1}{2\Delta t}(u^{n+1}_{I,j+1} + u^{n+1}_{I,j} - u^{aux}_{I,j+1} - u^{aux}_{I,j}) \tag{23d}$$

Multiplying Equation (22) by h^2 and Equation (23) by h, and adding for all specified nodal points, gives again a vanishing left-hand member. Due to the averaging on the space-staggered grid, now the right-hand member is also equal to zero. So here, the compatibility condition is satisfied exactly, without imposing specific boundary conditions for the auxiliary velocity field. Now, it is possible to apply the physically realistic boundary condition for the pressure.

4.2 Boundary conditions in case of a space-staggered grid

Suppose the velocities to be known on the boundary Γ. The Neumann condition for the pressure (12) has been expressed in the components of $\underline{u}^{n+1}$ and $\underline{u}^{aux}$ normal to the boundary. So $\underline{u}^{aux}\cdot\underline{\nu}$ has to determined to be able to impose the Neumann condition. Fortin et al. [4] propose to choose

$$\underline{u}^{aux}\cdot\underline{\nu} = \underline{u}^{n+1}\cdot\underline{\nu} \quad \text{on } \Gamma, \tag{24}$$

resulting in a homogeneous Neumann condition for the pressure:

$$\frac{\partial\phi^{n+1}}{\partial\nu} = 0 \tag{25}$$

This homogeneous condition generally does not make physical sense as shown in Paragraph 2. Moreover, the computed solution is not independent of the applied boundary condition for the pressure, contrary to what is stated in [4]. Fortin et al. explain this remarkable proposal by writing the difference equations for the one-dimensional analogue of the Neumann problem. Indeed, the one-dimensional case is illustrative for explaining the idea, but it lacks one important feature of the two-dimensional case: a tangential direction. Therefore, here a full analysis is given of the difference solution of the two-dimensional Neumann problem as described by Equations (10) and (12). Essential assumptions in this are:

i The auxiliary velocity field has been computed with an explicit scheme.

ii A space-staggered grid is used; choose, for example, the grid as depicted in Figure 3.

iii The boundaries are parallel to the coordinate lines.

Consider the rectangular area of Figure 3 and choose the line $j=1$ as the boundary that will be analysed. Equations (22) and (23a) yield respectively

$$\Phi^{n+1}_{i+\frac{3}{2},\frac{3}{2}} + \Phi^{n+1}_{i-\frac{1}{2},\frac{3}{2}} + \Phi^{n+1}_{i+\frac{1}{2},\frac{5}{2}} + \Phi^{n+1}_{i+\frac{1}{2},\frac{1}{2}} - 4\,\Phi^{n+1}_{i+\frac{1}{2},\frac{3}{2}} = \tag{26}$$

$$= \frac{h}{2\Delta t}(u^{aux}_{i+1,1} - u^{aux}_{i,1} + u^{aux}_{i+1,2} - u^{aux}_{i,2} + w^{aux}_{i,2} - w^{aux}_{i,1} + w^{aux}_{i+1,2} - w^{aux}_{i+1,1})$$

$$\Phi^{n+1}_{i+\frac{1}{2},\frac{3}{2}} - \Phi^{n+1}_{i+\frac{1}{2},\frac{1}{2}} = \frac{h}{2\Delta t}(w^{aux}_{i+1,1} + w^{aux}_{i,1} - w^{n+1}_{i+1,1} - w^{n+1}_{i,1}). \tag{27}$$

After the addition of Equations (26) and (27) it can be seen that the variables $\Phi^{n+1}_{i+\frac{1}{2},\frac{1}{2}}$, $w^{aux}_{i,1}$ and $w^{aux}_{i+1,1}$ have disappeared. This means that the computation for the pressure at the discrete time interval (n,n+1) is independent of the specific choice for the auxiliary velocity normal to the boundary at the discrete time interval considered. A similar analysis can be given for the corner points [1].

However, due to the use of a space-staggered grid, it is necessary to specify the auxiliary velocity tangential to the boundary too. Using Equation (11) the following boundary condition can be derived:

$$\underline{u}^{aux}\cdot\underline{\tau} = \underline{u}^{n+1}\cdot\underline{\tau} + \Delta t\,(\frac{\partial\Phi}{\partial\tau})^{n+1}. \tag{28}$$

Due to the occurrence of Φ^{n+1} in the right-hand side, substitution of Equation (28) in the Poisson equation for the pressure yields a coupled system of equations. For computational reasons this is not desired. Instead, Fortin et al. [4] propose to use the expression

$$\underline{u}^{aux}\cdot\underline{\tau} = \underline{u}^{n+1}\cdot\underline{\tau} + \Delta t(\frac{\partial\Phi}{\partial\tau})^{n}. \tag{29}$$

Equation (29) seems to be the most reasonable choice for the auxiliary velocity tangential to the boundary.

It is important to note that the computed result for the tangential component $\underline{u}^{aux}\cdot\underline{\tau}$ on the boundary at the time interval (n,n+1) is partially determined by the Neumann condition for the pressure at the previous time interval (n-1,n). So it can be concluded that even under the limiting assumptions i - iii given above, the calculated pressure field is dependent of its imposed Neumann condition.

Summarizing, it can be stated that use of a homogeneous Neumann condition can give rise to false numerical results. The alternative is to impose the real boundary condition for the pressure as described in paragraph 2.

5. RESULTS AND DISCUSSION

In this paragraph some experiments will be described to illustrate the analysis of the compatibility condition. For this purpose, it is not necessary to choose realistic flow problems, so treated are the classical driven cavity and the developed flow in a channel.

Firstly, two results of the driven cavity problem are given, using a regular grid. In case no specific measures are taken, the compatibility condition is not satisfied, yielding a strongly disturbed solution in the neighbourhood of the nodal point, where the pressure is prescribed (see Figure 4a). Taking the boundary conditions for the auxiliary velocity as given by Equation (19), the problem can be made compatible, resulting, however, in a false boundary condition for the pressure. The numerical result of Figure 4b seems to be wrong in the upper corner regions.

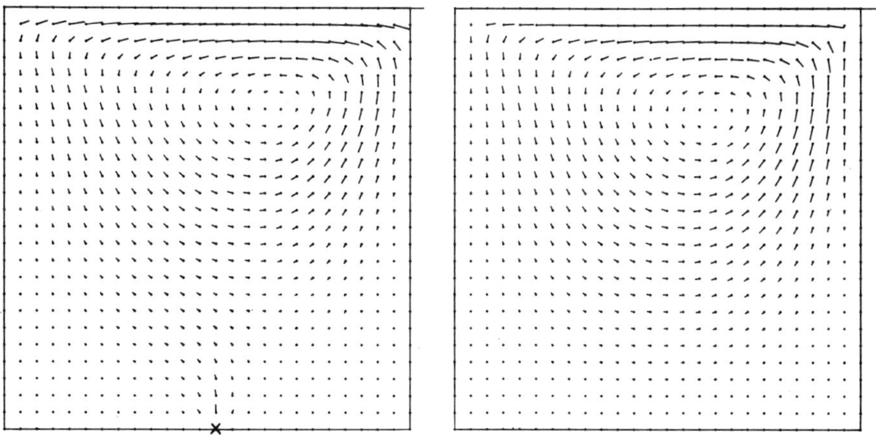

Figure 4 Driven cavity flow on a regular grid ($Re = 100; h = \frac{1}{25}$), compatibility condition not satisfied (left) or satisfied (right).

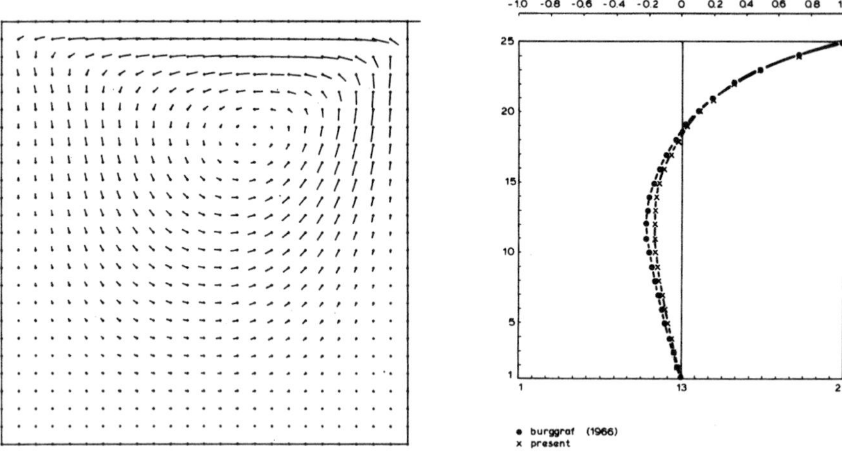

Figure 5 Driven cavity flow (Re = 100; $h=\frac{1}{24}$) on a space-staggered grid, flow pattern (left), centre line velocity profile (right).

Using the space-staggered grid of Figure 3, the compatibility condition is satisfied "naturally". Figure 5 shows the result for boundary conditions (24) and (29), which can be seen to agree reasonably well with corresponding results of Burggraf [6].

In engineering practice often flow problems occur, where no reliable information concerning the velocity field is available, especially at outflow. These problems can be treated well by prescibing the velocity profile at inflow and using the free outflow condition, i.e. Equation (5) with $\alpha = \beta = 0$, at outflow. However, starting the calculation from initial rest, yields, in general, again a completely disturbed numerical result, even when a space-staggered grid is used. Figure 6a shows this result for laminar channel flow. The imposed velocity profile is parabolic; at the centre line a symmetry condition has been applied.

The remarkable result of Figure 6a can be explained by regarding the combination of boundary and initial conditions. Due to this combination there is no conservation of mass at the initial stage of the time-stepping procedure, yielding again a disturbed compatibility condition. Because the pressure and velocity calculations are coupled, the numerical process cannot correct the initially disturbed pressure field. However, choosing a slightly different initial condition, for example, a velocity distribution as shown in Figure 6b, the compatibility condition is satisfied again, yielding quite rapidly the parabolic velocity solution (see Figure 6c). It appeared to be important to use a second order approximation for the symmetry condition. A more detailed description of these and other experiments can be found in [1].

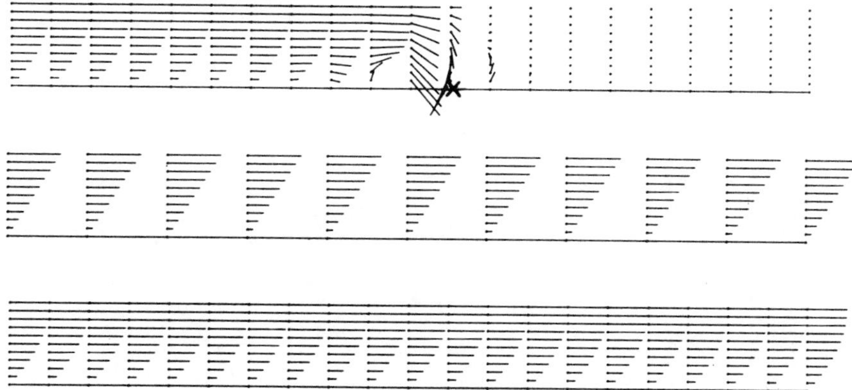

Figure 6 Developed channel flow (Re = 10). Result of initial condition violating global continuity (above), alternative initial condition (middle), converged solution (below).

At the moment of writing this contribution no clear example is available, showing significant differences, when either the homogeneous or the real Neumann condition is used. However, especially when boundaries are not parallel to grid lines or when regions of important acceleration near the boundary do occur it is expected to be important to use the physically realistic boundary condition for the pressure. Therefore, in future more general flow problems will be investigated extensively.

ACKNOWLEDGEMENT

The author wishes to thank prof. dr. C.B. Vreugdenhil for his valuable compents and suggestions, which were found to be most helpful.

REFERENCES

1. ALFRINK, B.J. - Numerical solution of the Navier-Stokes equations using curvilinear finite differences, Delft Hydraulics Laboratory, Research Report S 384-I, 1980.
2. CHORIN, A.J. - Numerical solution of the Navier-Stokes equations, J. Computational Phys. 2, 1967, 12-26.
3. HARLOW, F.H., WELCH, J.E. - Numerical calculation of time dependent viscous incompressible flow of fluid with free surfaces, Phys. Fluids 8, 1965, 2182-2189.
4. FORTIN, M., PEYRET, R., TEMAN, R. - Résolution numérique des équations de Navier-Stokes pour un fluide incompressible, Journal de Mécanique 10, 1971, 357-390.
5. GHIA, K.N. HANKEY Jr., W.L., HODGE, J.K. - Study of incompressible Navier-Stokes equations in primitive variables using implicit numerical technique, AIAA Paper 77-648, 1977.
6. BURGGRAF, O.R. - Analytical and numerical studies of the structure of steady separated flows, J. Fluid Mech. 24, 1966, 113-151.

A NUMERICAL STUDY OF TURBULENT, CONFINED, SWIRLING JETS

J.I. Ramos*

Assistant Professor

ABSTRACT

The flow field in a model combustor composed of two confined, coaxial, swirling jets under noncombusting conditions has been numerically investigated by means of the k/ε model of turbulence. Mean flow results are obtained for different flow conditions to determine the effect of the inner and outer swirl on the recirculation zone (which is used for flame stabilization under combustion conditions). The results show that a recirculation zone occurs at the centerline and is in the form of a one-celled toroidal vortex. The diameter of this recirculation zone first decreases and then increases when the magnitude of the outer swirl is first decreased from counter-swirl to zero, and then increased to give co-swirl flow conditions.

1. INTRODUCTION

Swirling flows have been commonly used for a number of years for the stabilization of high-intensity combustion processes [1]. It is well known that under appropriate conditions, swirl can be used to induce recirculation and stabilize combustion. Because of their significance, swirl effects in combustion have been studied for some time [1,2]. However, a fundamental understanding of swirl and its role in combustion eludes us and there is a clear need for continued study of swirling flows and their application to combustion devices.

In swirling flows the term vortex breakdown is used to refer to a number of different flow phenomena observed in non-reacting flows [3]. These phenomena may be characterized by high levels of swirl, one or more stagnation points along the vortex core, and erratic behavior of the vortex core. Our main

*Department of Mechanical Engineering, Carnegie-Mellon University, Pittsburgh, PA 15213 (USA).

concern here is with the so-called axisymmetric form of vortex breakdown [4]. The other forms of vortex breakdown are generally observed at lower swirl levels [3]. Vortex breakdown has been observed in both compressible and incompressible flows, and in free and confined flows [3]. It is believed that vortex breakdown is the agent for recirculation in combusting flows where swirl is used to induce recirculation. Unfortunately, a completely satisfactory theory for vortex breakdown in the absence of reaction is not available. In the past, there has been considerable interest in vortex breakdown; and there are a number of theories [3,4], none of which are completely satisfactory. The most valuable theory seems to be that of the critical state as developed by Benjamin [5-7] and extended by Leibovich and co-workers [8-10]. The critical state theory views vortex breakdown as occurring in a transition region between supercritical and subcritical flow states. Criticality is defined in terms of the ability of axisymmetric waves to propagate upstream. If upstream propagation is not possible, the flow is supercritical. According to Leibovich, the erratic behavior ascribed to the vortex breakdown is due to waves trapped at the critical state [10]. Extension of this inviscid flow theory allows for the prediction of the position and general shape of the axisymmetric form of vortex breakdown. On the basis of numerical calculations [11] and Leibovich's results, it seems reasonable to assume that the transition from a supercritical state to a subcritical state is the result of flow divergence and viscous diffusion of vorticity. This paper presents the results of numerical calculations of turbulent swirling flows in the presence of vortex breakdown for a simplified combustor and represents the first step in building a picture of primary zone combustion.

Prediction of the gross flow features and the time mean values of turbulent, confined, swirling flow requires the solution of the governing Reynolds equations subject to known boundary conditions. The major problem is that of finding a model for the Reynolds stresses to close the conservation equations. The closure model employed here was developed by Jones & Launder [12]. The Reynolds stresses are expressed in terms of a scalar viscosity which depends on the turbulent kinetic energy, k, and the dissipation rate of turbulent energy, ε. Differential equations for both k and ε are solved along with the other governing equations. Good performance of the model has been observed for confined, swirling flows [13, 14]. Some investigators claim to have observed an anisotropic turbulent viscosity in swirling flows [15,16]. For these calculations it was assumed that the turbulent viscosity may be expressed as a scalar. Calculations for free swirling [17] flows indicate that a scalar viscosity is adequate. Other authors have also studied turbulent, confined, swirling flows. Habib & Whitelaw [18] studied the influence of different swirl numbers on the flow field of a model combustor. Their experimental results were compared with a numerical model. This comparison showed that for weak swirl flow conditions, the model was unable to predict the minimum velocity at the centerline. Kubo & Gouldin [19] solved the conservation equations in terms of the vorticity and streamfunction, and found that a recirculation zone exists for both co- and counter-swirl flow conditions. Similar results were found by Ramos [13,14]. In this paper, isothermal flow calculations are presented for the flow configuration shown in Fig. 1. The axial and swirl velocities of both the inner and outer flows can be independently varied.

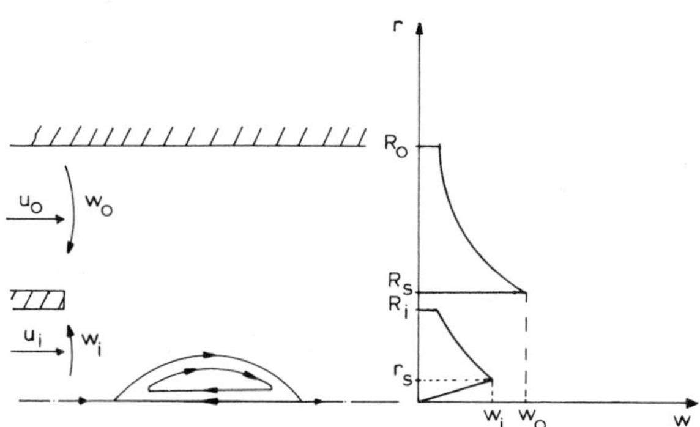

Fig. 1. Geometrical arrangement and swirl velocity profiles at the inlet.

2. THE GOVERNING EQUATIONS

The governing equations of motion for turbulent, axisymmetric, incompressible flows can be written as,

$$\frac{\partial}{\partial t}(\rho u) + \frac{1}{r}\frac{\partial}{\partial r}[r(\rho vu - \mu_e \frac{\partial u}{\partial r})] + \frac{\partial}{\partial z}[\rho uu - \mu_e \frac{\partial u}{\partial z}] = -\frac{\partial p}{\partial z} + \frac{1}{r}\frac{\partial}{\partial r}[r\mu_e \frac{\partial v}{\partial z}] + \frac{\partial}{\partial z}[\mu_e \frac{\partial u}{\partial z} - \frac{2}{3}\rho k] \tag{1}$$

$$\frac{\partial}{\partial t}(\rho v) + \frac{1}{r}\frac{\partial}{\partial r}[r(\rho vv - \mu_e \frac{\partial v}{\partial r})] + \frac{\partial}{\partial z}[\rho uv - \mu_e \frac{\partial v}{\partial z}] = -\frac{\partial p}{\partial r} + \rho \frac{w^2}{r} + \frac{1}{r}\frac{\partial}{\partial r}[r(\mu_e \frac{\partial v}{\partial r} - \frac{2}{3}\rho k)] - 2\mu_e \frac{v}{r^2} + \frac{2}{3}\frac{\rho k}{r} \tag{2}$$

$$\frac{\partial}{\partial t}(\rho w) + \frac{1}{r}\frac{\partial}{\partial r}[r(\rho vw - \mu_e \frac{\partial w}{\partial r})] + \frac{\partial}{\partial z}[\rho uw - \mu_e \frac{\partial w}{\partial z}] = -\rho\frac{vw}{r} - \frac{w}{r^2}\frac{\partial}{\partial r}(r\mu_e) \tag{3}$$

for the z (axial), r (radial), and θ (tangential) directions, together with the continuity equation,

$$\frac{1}{r}\frac{\partial}{\partial r}(r\rho v) + \frac{\partial}{\partial z}(\rho u) = 0 \tag{4}$$

The stress terms in these equations are the sum of laminar and turbulent stresses. Introducing a scalar turbulent viscosity,

$$\mu_T = C_\mu \rho \frac{k^2}{\varepsilon} \tag{5}$$

the effective viscosity can be written as,

$$\mu_e = \mu + \mu_T$$

Two additional differential equations are required to account for the transport of k and ε. For high Reynolds number flows, these equations may be reduced to simpler forms [12],

$$\frac{\partial}{\partial t}(\rho k) + \frac{1}{r}\frac{\partial}{\partial r}[r(\rho v k - \mu_e \frac{\partial k}{\partial r})] + \frac{\partial}{\partial z}[\rho u k - \mu_e \frac{\partial k}{\partial z}] = P - \rho\varepsilon \tag{6}$$

$$\frac{\partial}{\partial t}(\rho\varepsilon) + \frac{1}{r}\frac{\partial}{\partial r}[r(\rho v - \frac{\mu_e}{\sigma_\varepsilon}\frac{\partial \varepsilon}{\partial r})] + \frac{\partial}{\partial z}(\rho u \varepsilon - \frac{\mu_e}{\sigma_\varepsilon}\frac{\partial \varepsilon}{\partial z}) = C_1 \frac{\varepsilon}{k} P - C_2 \rho \frac{\varepsilon^2}{k} \tag{7}$$

P is the turbulent kinetic energy production term and is defined as,

$$P = \mu_e[2(\frac{\partial u}{\partial z})^2 + 2(\frac{\partial v}{\partial r})^2 + 2(\frac{v}{r})^2 + (\frac{\partial w}{\partial r} - \frac{w}{r})^2 + (\frac{\partial w}{\partial z})^2 + (\frac{\partial u}{\partial r} + \frac{\partial v}{\partial z})^2] \tag{8}$$

Values of the empirical constants were taken from the work of Launder & Spalding [20] on free shear flows. In the present problem the shear layers between the two jets and around the recirculation zone are the dominant sources of turbulence. $C_\mu = 0.09$, $C_1 = 1.44$, $C_2 = 1.92$, $\sigma_\varepsilon = 1.2$.

3. THE NUMERICAL PROCEDURE

The governing equations are elliptic, and for each equation boundary conditions must be given for all boundaries. The finite difference technique employed is an adaptation of the method developed by Ramos, et al. [21]. The following is a summary of the technique; a detailed description can be found elsewhere [21]. The conservation equations of the axial (1), radial (2) and tangential (3) momentum components, the turbulence kinetic energy (6) and its dissipation rate (7) can be written in the general form:

$$\frac{\partial}{\partial t}(\rho\phi) + \frac{1}{r}\frac{\partial}{\partial r}[r(\rho v\phi - \Gamma_\phi \frac{\partial \phi}{\partial r})] + \frac{\partial}{\partial z}[\rho u\phi - \Gamma_\phi \frac{\partial \phi}{\partial z}] = S_\phi \tag{9}$$

where ϕ, Γ_ϕ and S_ϕ stand for any transportable quantity (e.g. k), diffusivity and source term.

These equations are converted to finite difference form using a backward difference scheme. This results in a fully implicit numerical scheme. To avoid convective instability a hybrid difference scheme, which employs upwind/central differences, is used.

3.1 Initial and Boundary Conditions

At the inlet plane, axial and tangential velocity profiles are specified. The axial velocity profile is assumed uniform for both the inner jet (u_i) and outer jet ($u_o = \alpha u_i$). The swirl velocity profile for the inner jet is assumed to be that of a solid body rotation ($w = w_i r/r_s$) up to $r_s = 0.96$ cm. This solid body rotation is matched to a free vortex structure ($w = w_i r_s/r$). The outer flow is assumed to have a swirl velocity profile of a free vortex ($w = w_o R_o/r$); where w_i and w_o are the swirl velocity values at r_s and $R_o = 14.5$ cm, respectively (Fig. 1). The radius and thickness of the inner pipe are $R_i = 3.14$ cm and $R_s - R_i = 1.04$ cm. In a certain sense the boundary conditions are arbitrary, specially those for the inner flow. Actual conditions will depend on the particular apparatus used to generate swirl. It is believed that the swirl number and the average velocity are the important features of the inlet velocity profiles.

The value of k at the inlet was taken from Schlichting [22],

$$k = (0.035\ u_i)^2 [2 + 8(\frac{r}{R_i})^2] \qquad 0 \leq r \leq R_i \qquad (10)$$

$$k = (0.035\ u_o)^2 [2 + 4(\frac{r}{R_s})^2 + r(\frac{r}{R_o})^2] \qquad R_i \leq r \leq R_o \qquad (11)$$

while ε was calculated from equation (5) after μ_T is estimated by mixing length theory. For the present calculations the following values of ε were used,

$$\varepsilon = k^{3/2}/(0.005\ R_i) \qquad 0 \leq r \leq R_i \qquad (12)$$

$$\varepsilon = k^{3/2}/(0.005\ R_o) \qquad R_i \leq r \leq R_o \qquad (13)$$

At the wall boundaries, the normal velocity was set to zero, while the law of the wall was used for the parallel velocity to the wall. The gradient of turbulent kinetic energy was set to zero, and the dissipation was calculated assuming an equilibrium boundary layer.

At the centerline the gradients of axial velocity, turbulence kinetic energy and its dissipation rate are zero, while the radial and swirl velocities are zero. At the combustor exit, the flow is assumed to be fully developed, i.e., the gradients of all transportable quantities are set to zero. The selection of the proper combustor length that allows a fully developed flow was the subject of numerous tests, from which it was concluded that the minimum length to achieve fully developed flow was 8 times R_o. Initially the flow field was guessed, and the calculations were performed until the results were time-independent. Although the numerical model was initially developed to analyze the transient response of swirl combustors, only steady state results are presented here.

4. PRESENTATION OF RESULTS

Figure 2 shows the normalized mean axial velocity profiles at different axial locations along the pipe for the following inlet velocities: u_i = 20.30 m/s, u_o = 13.53 m/s, w_i = 19.95 m/s and w_o = -6.90 m/s, i.e. counter-swirl. The velocity has been normalized by the inner jet axial velocity at the inlet. At z/R_i = 0.110 the inner jet is lifted by centrifugal forces. There is also a minimum velocity at r/R_i = 1.0 due to the

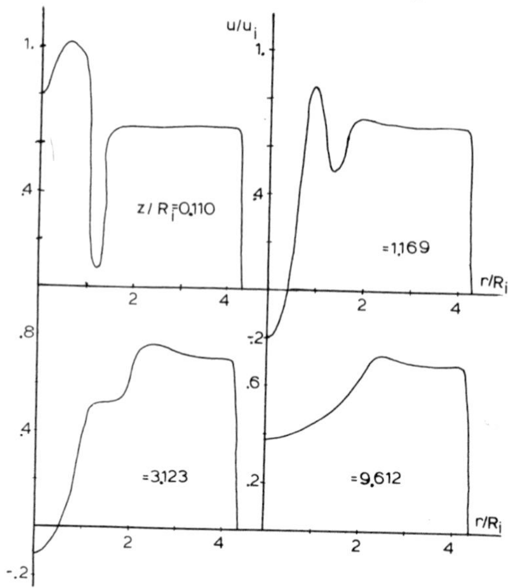

Fig. 2. Axial velocity profiles.

finiteness of the inner pipe. At z/R_i = 1.169, the axial velocity at the centerline is negative, i.e. a recirculation zone has appeared. The maximum and minimum velocities have been dissipated, however the outer flow preserves its structure longer. At z/R_i = 3.123, there is still a recirculation zone at the centerline, whose diameter is greater than at z/R_i = 1.169. This recirculation zone has disappeared at z/R_i = 9.612, where the flow velocity is positive everywhere. Figures 3 and 4 show the normalized swirl velocity profiles. At z/R_i = 0.110 the solid body rotation and the free vortex structure of the inner jet are apparent. There is a very steep swirl velocity gradient at r/R_i = 1.0 followed by the free vortex structure of the outer flow. As we will see in later figures, this inter-jet shear layer gives rise to very

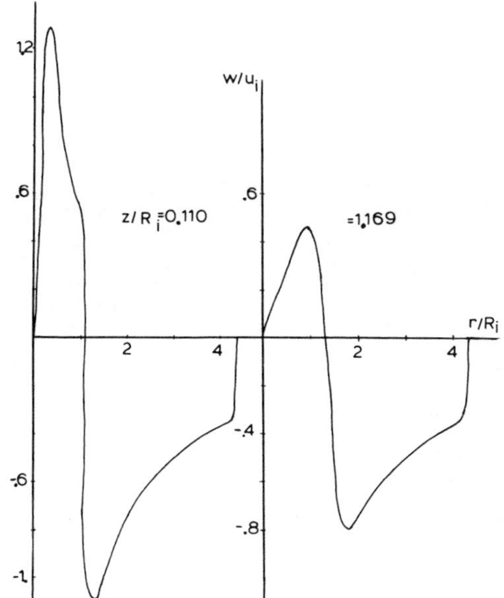

Fig. 3. Tangential velocity profiles.

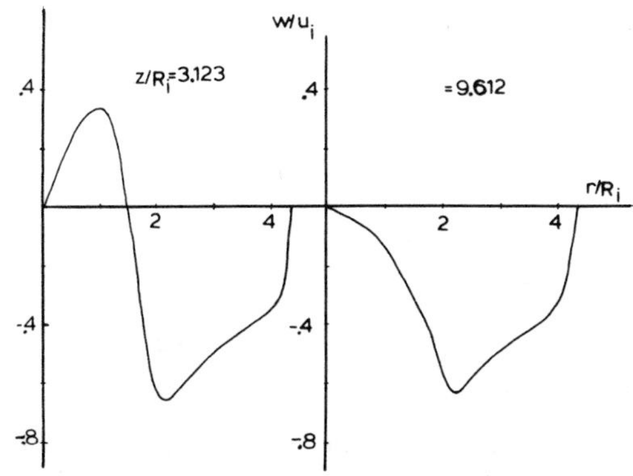

Fig. 4. Tangential velocity profiles.

high levels of turbulence kinetic energy and dissipation. At $z/R_i = 1.169$, the inner jet velocity profile is about 2.2 times smaller than at $z/R_i = 0.110$, while the outer flow still preserves its structure. This extremely rapid damping of the inner jet velocity seems due to the large levels of dissipation at the inter-jet shear layer. At $z/R_i = 3.123$ the inner jet profiles are smaller than before. Notice that the main damping occurs from $z/R_i = 0.110$ to 1.169, where maximum dissipation of turbulence energy is taking place. At $z/R_i = 9.612$ the inner flow velocity is negative while the outer flow still preserves its free vortex structure. Figure 5 shows the turbulent intensity profiles at different axial locations. This turbulence intensity has been defined as $i = \sqrt{2k/3}\,/u_i$. At $z/R_i = 0.255$, there are two peaks of turbulent intensity. The first appears at $r/R_i = 1.0$, i.e. at the inter-jet shear layer;

and the other at the recirculation zone, i.e. at the centerline. The magnitude of the latter is higher than the magnitude of the former. The axial location $z/R_i = 0.255$ almost corresponds to the free stagnation point of the recirculation zone. At $z/R_i = 1.049$, the peak at the inter-jet shear layer still persists, while the peak at the recirculation zone no longer appears at the centerline but at the outer edge of the recirculation zone. Both peaks are damped further downstream along the combustor because of diffusion and dissipation. Notice that there is another turbulent intensity peak at the outer pipe; this peak is due to the very steep velocity gradient there. Figure 6 shows the velocity at the centerline ($u_{℄}$) along the combustor. This velocity rapidly decreases

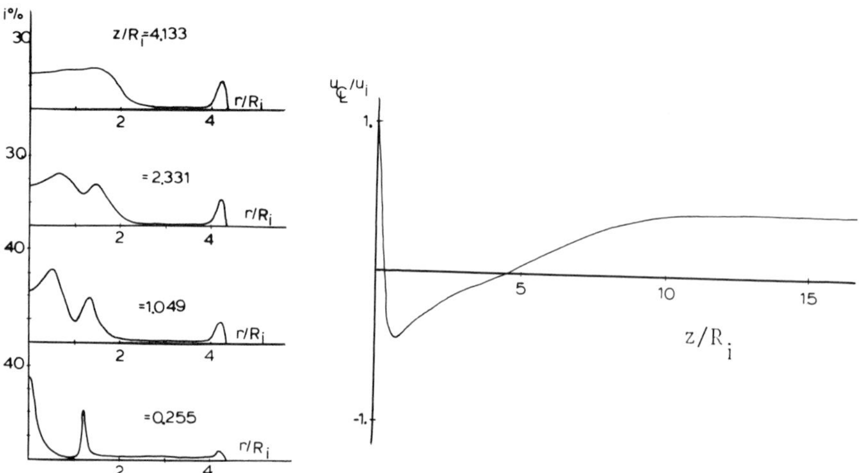

Fig. 5. Turbulent intensity profiles.

Fig. 6. Centerline velocity.

and becomes negative close to the inlet port. The minimum centerline velocity occurs near $z/R_i \simeq 1.0$ and the maximum value of the recirculation zone velocity is close to $0.5\ u_i$. At $z/R_i = 10$ the flow is almost fully developed and the velocity is necessarily positive.

Similar results have been found for zero-outer swirl and co-swirl conditions [14]. For zero outer swirl, the calculations show that the diameter of the recirculation zone is smaller than for counter-swirl. Also, the tangential velocity peak is more slowly damped. This can easily be understood, because the velocity gradients at the inter-jet shear layer are smaller than for counter-swirl. The results of a parametric study conducted to investigate the influences of swirl in the recirculation zone are shown in Figures 7, 8 and 9.

Figure 7 shows that the diameter ($2r_r$) of the recirculation zone is independent of the Reynolds number; while its length (ℓ_r) slightly increases at first and then remains constant. Both the velocity (u_r) and recirculating mass (ψ_r) are monotonically increasing functions of the Reynolds number. The recirculating mass has been nondimensionalized with the mass flow rate of the inner jet. The Reynolds number has been obtained using the outer pipe diameter and the average velocity at the inlet.

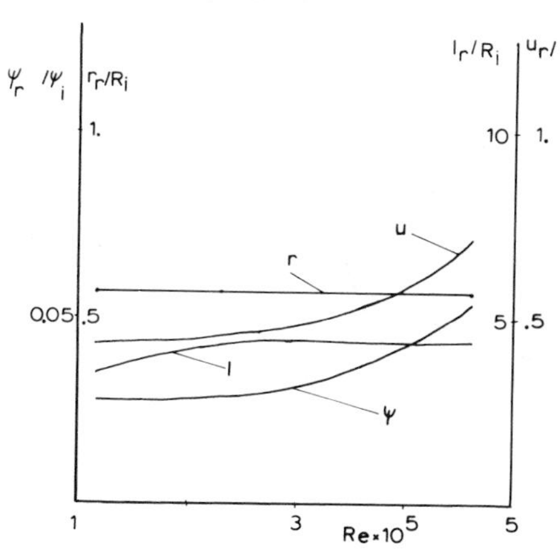

Fig. 7. Recirculation zone parameters vs. Reynolds number.

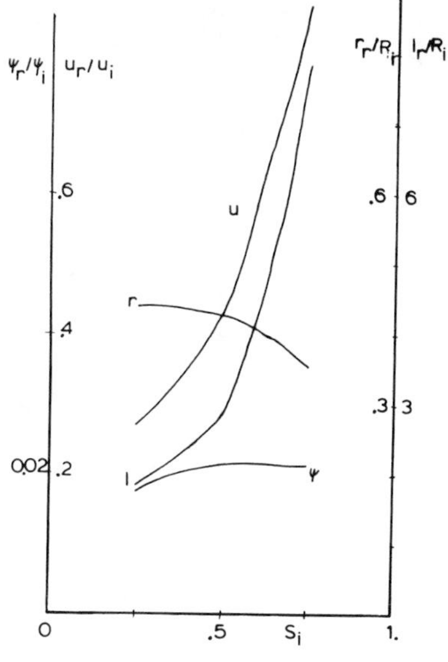

Fig. 8. Recirculation zone parameters vs. inner swirl number.

Figure 8 shows that the diameter of the recirculation zone is a monotonically decreasing function of the inner swirl number $S_i(=0.5\ w_i/u_i)$, while its length and velocity are monotonically increasing functions.

Figure 9 shows a most interesting behavior. In this figure negative values of $S_o(=w_o/u_o)$ correspond to counter-swirl, and positive values to co-swirl. As the magnitude of the outer swirl is first decreased from counter-swirl to zero and then increased to give co-swirl flow conditions, the diameter, length, velocity and mass of the recirculation zone first decrease and then increase. The maximum recirculating mass occurs for counter-swirl and is about 10% of the inner jet mass flow rate. This finding is not in agreement with the experimental results of Vu and Gouldin [23] who showed that the recirculation zone only occurs for counter-swirl; while its diameter and velocity are monotonically decreasing functions of the outer-swirl number. Possible reasons for this discrepancy may be associated with the streamline curvature of the flow and the inappropriateness of the presumed isotropic viscosity. Rodi [24] and Gibson [25] have shown that C_μ is a decreasing function of the Richardson number. The streamline curvature of the flow is presently under investigation.

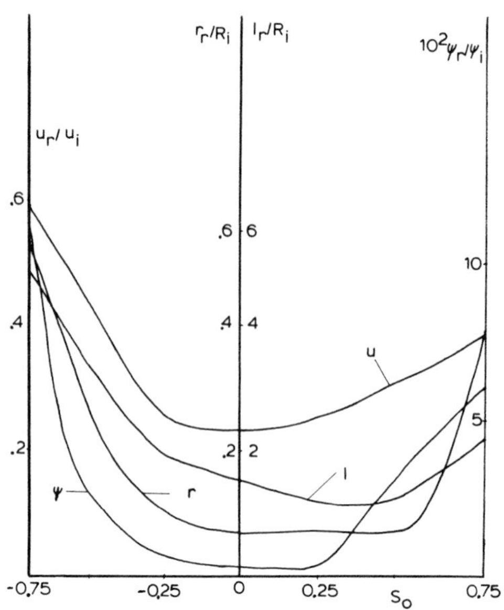

Fig. 9. Recirculation zone parameters vs. outer swirl number.

5. CONCLUSIONS

a) Co- and counter-swirl create a recirculation zone at the centerline. This recirculation zone is in the form of a one-celled toroidal vortex.

b) The diameter, velocity, mass and length of the recirculation zone first decrease and then increase, when the outer swirl number is first decreased from counter-swirl to zero and then increased to maximum co-swirl flow conditions.

c) The diameter of the recirculation zone is a monotonically decreasing function of the inner swirl number.

d) Further work is required to assess the importance of streamline curvature and anisotropy in the numerical simulation of swirling flows.

ACKNOWLEDGMENT

This research has been supported by the Ford Motor Company Research Fund.

REFERENCES

1. SYRED, N. and BEER, J.M. - Combustion in Swirling Flows: A Review. Combustion and Flame, Vol. 23, pp. 143-201, 1974.

2. NOVICK, A.S., MILES, G.A. and LILLEY, D.G. - Numerical Simulation of Combustor Flow Fields. Paper No. 78-949, presented at the AIAA/SAE 14th Joint Propulsion Conference, Las Vegas, 1978.

3. HALL, M.G. - Vortex Breakdown. Annual Review of Fluid Mechanics, Vol. 4, pp. 195-218, 1972.

4. LEIBOVICH, S. - The Structure of Vortex Breakdown. Annual Review of Fluid Mechanics, Vol. 10, pp. 221-246, 1978.

5. BENJAMIN, T.B. - Theory of Vortex Breakdown Phenomenon. J. Fluid Mech., Vol. 14, pp. 593-629, 1962.

6. BENJAMIN, T.B. - Some Developments in the Theory of Vortex Breakdown. J. Fluid Mech., Vol. 28, pp. 65-84, 1967.

7. BENJAMIN, T.B. - A Unified Theory of Conjugate Flows. Phil. Trans. Roy. Soc., Series A, Vol. 269, pp. 587-647, 1971.

8. LEIBOVICH, S. - Weakly Nonlinear Waves in Rotating Fluids. J. Fluid Mech., Vol. 42, pp. 803-822, 1970.

9. LEIBOVICH, S. and RANDALL, J.D. - Amplification and Decay of Long Nonlinear Waves. J. Fluid Mech., Vol. 53, pp. 481-493, 1973.

10. RANDALL, J.D. and LEIBOVICH, S. - The Critical State: A Trapped Wave Model of Vortex Breakdown. J. Fluid Mech., Vol. 58, pp. 495-515, 1973.

11. GRABOWSKI, W.J. and BERGER, S.A. - Solutions of the Navier-Stokes Equations for Vortex Breakdown. J. Fluid Mech., Vol. 75, pp. 525-544, 1975.

12. JONES, W.P. and LAUNDER, B.E. - The Prediction of Laminanization with a Two Equation Model of Turbulence. Int. J. Heat and Mass Transfer, Vol. 15, pp. 301-314, 1972.

13. RAMOS, J.I. - A Numerical Study of Incompressible, Turbulent, Confined, Swirling Flows. Part I: The Vortex Sheet Layer. Report CO/80/4, Carnegie-Mellon University, Pittsburgh, PA, 1980.

14. RAMOS, J.I. - A Numerical Study of Incompressible, Turbulent, Confined, Swirling Flows. Part II: The Effect of the Inner Pipe Thickness. Report CO/80/5, Carnegie-Mellon University, Pittsburgh, PA, 1980.

15. LILLEY, D.G. and CHIGIER, N.A. - Nonisotropic Turbulent Stress Distribution in Swirling Flows from Mean Value Distributions. Int. J. Heat and Mass Transfer, Vol. 14, pp. 573-585, 1971.

16. LILLEY, D.G. - Combustor Swirl Flow Modeling. AIAA J., Vol. 13, pp. 419-420, 1975.

17. KUBO, I. - A Numerical Study of Confined Turbulent Swirling Flows. Ph.D. Thesis, Cornell University, Ithaca, NY, 1974.

18. HABIB, M.A. and WHITELAW, J.H. - Velocity Characteristics of Confined Coaxial Jets with and without Swirl. ASME Trans., J. Fluids Eng., Vol. 102, pp. 47-53, 1980.

19. KUBO, I. and GOULDIN, F.C. - Numerical Calculations of Turbulent Swirling Flow. ASME Trans., J. Fluids Eng., Vol. 97, pp. 310-315, 1975.

20. LAUNDER, B.E. and SPALDING, D.B. - Mathematical Models of Turbulence, Academic Press, New York, 1972.

21. RAMOS, J.I., HUMPHREY, J.A.C. and SIRIGNANO, W.A. - Numerical Prediction of Axisymmetric, Laminar and Turbulent Flows in Motored, Reciprocating Internal Combustion Engines. SAE Trans., Vol. 88, pp. 1217-1242, 1979.

22. SCHLICHTING, H. - Boundary Layer Theory. McGraw Hill Co., New York, 1968.

23. VU, B.T. and GOULDIN, F.C. - Flow Measurements in a Model Swirl Combustor. Paper No. 80-0076 presented at the AIAA 18th Aerospace Sciences Meeting, Pasadena, CA, 1980.

24. RODI, W. - A New Algebraic Relation for Calculating the Reynolds Stresses. Z. Ang. Math., Vol. 56, p. 219, 1976.

25. GIBSON, M.M. - An Algebraic Stress and Heat-Flux Model for Turbulent Shear Flow with Streamline Curvature. Int. J. Heat Mass Transfer, Vol. 21, p. 1609, 1978.

THE PREDICTION OF THE DRAG ON STRUCTURAL BEAMS

I Grant[(i)], F H Barnes[(ii)] and T H Loon[(i)]

SUMMARY

The form drag of a two-dimensional bluff body may be obtained from the pressure distribution around a section of the body. However, in practice it is often difficult and expensive to measure the pressure distribution. Consequently, semi-empirical potential flow models have been used to predict the pressure distribution over the front face of the body. The attraction of these models is that for sharp edged bluff bodies with clearly defined separation points they require only a single experimental measurement, that of the base pressure, for the entire pressure distribution around the body to be calculated.

In the present study three theoretical models have been used, the hodograph, the wake-source and the vortex lattice models, to calculate the drag of two bluff bodies. The bodies had I and L cross-sectional shapes and were chosen for their practical importance in structures. The theoretical results are compared with each other and also with experimental values obtained from wind tunnel studies.

1. INTRODUCTION

Bluff bodies with large aspect ratios are features which occur in industrial structures such as partially clad buildings and lattice towers. When significant forces are anticipated, information on the drag and vortex shedding properties of the individual elements is required for the design of these structures. This paper gives a brief description of three theoretical methods for predicting the drag of two-dimensional bluff bodies and of their application

(i) Department of Offshore Engineering, Heriot-Watt University, Edinburgh

(ii) Department of Physics, University of Edinburgh, Edinburgh

to two particular bodies which are frequently used as structural elements.

At large Reynolds numbers the flow around a bluff body divides into two main regions, the wake and the remaining part of the flow which is largely irrotational. The properties of the wake are difficult to describe theoretically since it is highly rotational and often contains a vortex street. Many experimental investigations of the vortex shedding from simple cylindrical shapes have been reported [1,2], and data on the drag and vortex shedding properties of structural elements is also available [3,4]. Outside of the wake the flow can usually be assumed to be irrotational. If this assumption is reasonable then the pressure distribution on the front part of the body between the boundary layer separation points will be, to a good approximation, equal to the pressure distribution appropriate to the flow of an inviscid fluid over the body. Furthermore, if the bluff body has sharp edges, such as a flat plate held normal to the flow, then the separation points will be fixed at the edges.

This paper describes how the pressure distribution over the front face of two structural elements may be calculated using three potential flow methods, viz the hodograph, the wake-source and the vortex-lattice methods. The drag, C_D, can then be derived as the pressure over the entire rear of each element can be assumed equal to the base pressure, C_{pb}. The flow configurations which are examined are those in which the I- and L-beams are held symmetrically to the air flow, as shown in Figure 1.

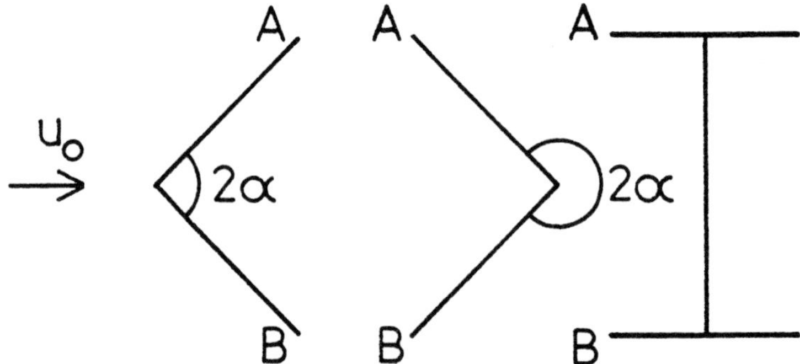

Figure 1: Flow configurations of the L- and I-beams

It has been assumed that the flows separate at the sharp edges A and B, and also that the flow velocity at the separation points can be derived from the base pressure using Bernoulli's equation.

2. THE HODOGRAPH METHOD

The hodograph method was developed in the 19th century to solve flows with free streamlines and it has been described by Lamb [5] amongst others. The essence of the method is to introduce a new complex variable, which can be either the complex velocity ν or, more usually, the logarithm of the inverse of the complex velocity, Ω. ν and Ω are given by

$$\nu = dw/dz = qe^{-i\theta} \quad \text{and}$$

$$\Omega = \ln dz/dw = \ln q^{-1} + i\theta$$

respectively, where $z = x + iy$, $w = \phi + i\psi$ and q and θ are the magnitude and direction of the velocity vector. In the ν-, or hodograph, plane the boundaries of the L- and I-beams become radial lines, and streamlines of constant velocity become circular arcs. In the Ω-plane the imaginary part of Ω is constant along the straight boundaries, and the real part of Ω is constant on streamlines of constant velocity.

The flow problem can be solved if a mapping between the ν-plane, or Ω-plane, and the w-plane can be found. Then z as a function of w can be obtained by integration.

In each of the flows which are considered the streamline which passes through the stagnation point is given the value $\psi = 0$ in the w-plane. At the stagnation point the streamline divides into two to follow the front face of the body, from which it separates at the edges as two free streamlines. In the w-plane therefore, each boundary is represented by two lines coincident with the real axis, and because the flows are symmetrical the lines are of equal length. The required pressure distribution meant that knowledge of the velocity distribution on the boundaries only was needed. This simplified the problem therefore, as all the necessary integrations were performed along the real axis in the w-plane.

2.1 The I-Beam

For the particular configuration being considered the rear legs of the I-beam may be ignored, so that the problem becomes the flow over a cup with vertical sides, as in Figure 2.

This problem was first considered by Love [6] who made the standard hodograph assumption that the velocity on the free streamline was constant and equal to the free stream velocity u_o. This assumption implies that C_{pb} is zero, which is not appropriate for bluff bodies in air flow where C_{pb} usually takes a large negative value. Love derived a solution for C_D

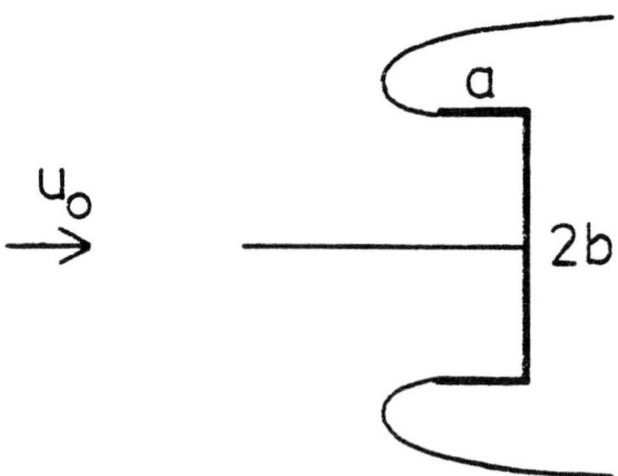

Figure 2: The flow over a cup

in terms of the ratio of the length of the side to the length of the bottom of the cup, a/2b, when a << 2b. A complete solution for a/b and C_D has been provided by Birkhoff and Zarantonello [7] in terms of elliptic functions. Integration of the full expressions in Love's solution give, of course, the same results as those of Birkhoff and Zarantonello. Figure 3 shows the pressure distribution over the front face for a/2b = 0.33.

The hodograph method was not modified to allow for a separation velocity other than u_o. It was felt that for the case a/2b = 0.33, for which there was an experimental value of C_{pb}, the pressure distribution on the bottom of the cup would not be significantly altered.

2.2 The L-Beam

Roshko [8] modified the hodograph method by allowing the separation velocity to be determined by an experimental value of C_{pb}. The velocity on the free streamlines was assumed to be constant from the separation points to the downstream position where they became parallel to the free stream. From the latter point the velocity decreased to u_o at infinity. In the hodograph plane this introduced a notch in the circular arc which represented the free streamlines.

The L-beam in the two configurations being considered can be treated as a symmetrical wedge with a nose angle of 2α. 2α then takes the values 90° or 270°. Roshko derived a mapping between the ν-plane and the w-plane via a number of intermediate planes, and found that z was given by

$$z(w_1) = (\pm i)^n\left(\frac{k^2+1}{2k}\right)^n \int_o^{w_1}\left(\left(\frac{1}{w} - \frac{1}{a_1{}^2}\right)^{\frac{1}{2}} + \left(\frac{1}{w} - 1\right)^{\frac{1}{2}}\right)^n dw \quad , \qquad (1)$$

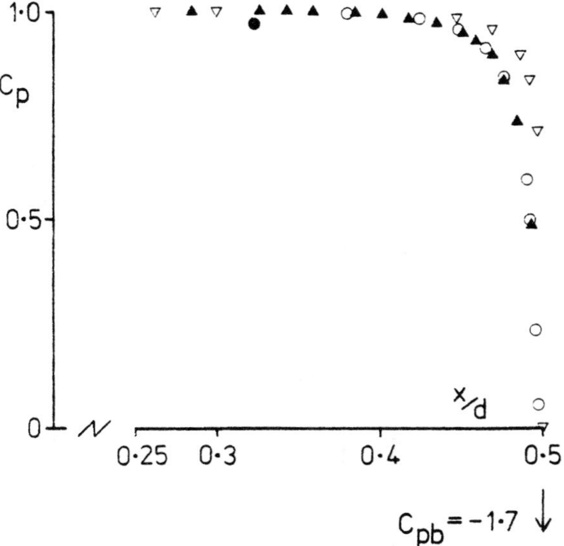

Figure 3: I-beam. ▽, hodograph method; ○, wake-source method; ▲, vortex-lattice method; ●, experiment. x is the distance from centre of body and d is the length of front face of the body.

where $n = 2\alpha/\pi$, $a_1 = (k^2+1)/(k^2-1)$ and k is dependent on the experimental value C_{pb}, $C_{pb} = 1-k^{2n}$.

The integration of equation (1) was performed on the Edinburgh Regional Computing Centre ICL 2980 computer for the two cases $n = \frac{1}{2}$ and $n = 3/2$. The NAG routine D01AGF was used which utilised an internal subdivision strategy based on the Clenshaw-Curtis quadrature method. The singularity at $\text{Re}(w) = \phi = 0$ was removed from the numerical procedure by using the analytical approximations as $\phi_m \to 0$,

$$\int_0^{\phi_m} z'(\phi)\,d\phi \approx \left(\frac{4\sqrt{2}}{3}\right)\phi_m^{3/4}, \quad n = \tfrac{1}{2}$$

$$\approx 4(\sqrt{2})^3\phi_m^{1/4}, \quad n = 3/2$$

As $\phi_m \to 0$ the sum of the analytical and numerical elements of the integration became constant and a satisfactory approximation to the complete integral was obtained. The pressure distributions for the two cases, $n = \frac{1}{2}$ and $3/2$, are plotted in Figures 4 and 5.

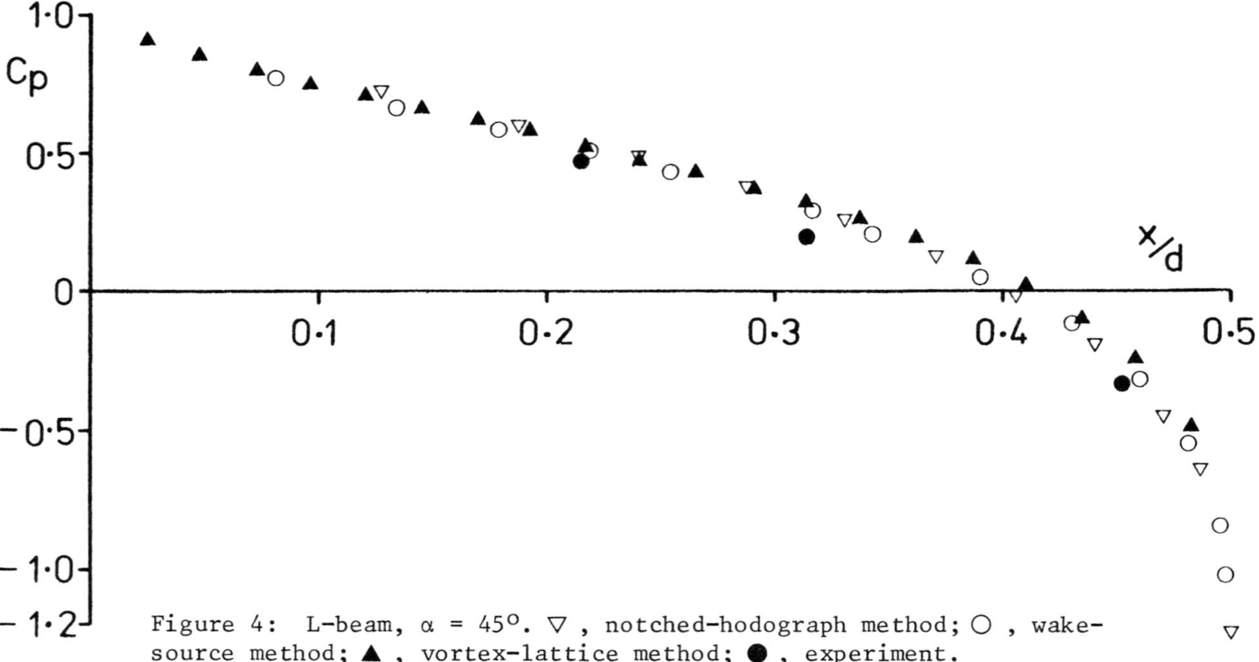

Figure 4: L-beam, $\alpha = 45^{o}$. ▽ , notched-hodograph method; ○ , wake-source method; ▲ , vortex-lattice method; ● , experiment.

3. THE WAKE-SOURCE METHOD

The wake-source method of Parkinson and Jandali [9] models the potential flow external to a bluff body and its wake. It requires an experimental knowledge of the separation positions and also of the separation velocity. In the method a slit in the physical z-plane, corresponding to the front face of the body between the separation points, is transformed to a slit on the real axis in the s-plane. In the s-plane sources are placed on the boundary of the slit to produce stagnation points at the positions corresponding to the separation points in the z-plane. The positions and strengths of the sources are decided by the positions of the separation points and an experimental value of C_{pb}. The flow in the s-plane is known once the positions and strengths of the sources has been found. The pressure distribution on the front face of the body and the shape of, and the pressure distribution along, the free streamlines in the z-plane can then be calculated.

3.1 The L-Beam

In both flows, in which $\alpha = 45^o$ or 135°, a Schwarz-Christoffel transformation was used to map the slit in the z-plane onto the slit in the s-plane.

For $\alpha = 45^o$, which was one of the flows considered by Parkinson and Jandali, the transformation

$$dz/ds \quad = \quad H(s+1)^{-1/4}s(s-1)^{-3/4} \qquad (2)$$

was used, where H is a constant specifying the change of scale and the rotation introduced by the transformation. The upper half of the z-plane only was mapped as the flow is symmetrical. Integration of equation (2) gave z as a function of s, and the x-coordinate of the points on the surface of the wedge was given by Re(z) for values of s, $-1 \leqslant s \leqslant 0$. H was determined by the value of x when s = 0.

The sources in the s-plane were placed at $(\varepsilon,0)$, and so the complex potential in the s-plane was given by

$$w(s) \quad = \quad v_o s + \frac{Q}{\pi}\ln(s-\varepsilon) \; , \qquad (3)$$

where v_o was the free-stream velocity and Q was the source strength. The complex velocity was therefore given by

$$\nu(s) \quad = \quad v_o + \frac{Q}{\pi(s-\varepsilon)} \; . \qquad (4)$$

From equation (2) it can be seen that as $s \to \infty$ so $dz/ds \to H$.

Now

$$\nu(z)dz/ds = \nu(s) \quad ,$$

where $\nu(z)$ and $\nu(s)$ are the complex velocities at corresponding points in the z- and s-planes. It followed that

$$v_o = Hu_o \quad . \tag{5}$$

The velocity in the s-plane was made zero at the origin and so, using equations (4) and (5),

$$Q = \pi v_o \varepsilon = \pi H u_o \varepsilon \quad . \tag{6}$$

From equations (2), (4) and (6) it can be shown that

$$\nu(z) = (s+1)^{1/4}(s-1)^{3/4}(s-\varepsilon)^{-1} \tag{7}$$

At the separation point, s = 0, the velocity is given by

$$|\nu(z)| = u_o K = u_o(1 - C_{pb})^{\frac{1}{2}} = u_o/\varepsilon \quad . \tag{8}$$

The pressure coefficient on the front surface is therefore given by

$$C_p = 1 - \frac{|\nu(z)|^2}{u_o^2} = 1 - \left|\frac{(s+1)^{1/4}(s-1)^{3/4}}{(s-1/K)}\right|^2 , \quad -1 \leqslant s \leqslant 0 \ . \tag{9}$$

The x-coordinate of points on the surface of the wedge and C_p were computed for values of s in the range $-1 \leqslant s \leqslant 0$ for a given C_{pb}. C_{pb} was put equal to -1.23 which was the experimental value found by the present authors.

For $\alpha = 135^o$, the transformation

$$dz/ds = H(s+1)^{-3/4}s(s-1)^{-1/4} \tag{10}$$

was used. In order to broaden the scope of the method a numerical integration scheme was used to obtain Re(z). A singularity existed at s = -1, and was an obvious difficulty in the integration. A transformation was therefore necessary in the vicinity of the singularity, and it was obtained by noting that the leading behaviour of dz/ds as $s \to -1$ was $(s+1)^{-3/4}$. A new variable t was introduced, where

$$t = (s+1)^{1/4} \quad , \tag{11}$$

and the function that was integrated close to the singularity was

$$dz/dt = dz/ds.ds/dt = 4H(t^4-1)(t^4-2)^{-1/4} \quad . \tag{12}$$

In this case the pressure coefficient was given by

$$Cp = 1 - \left|\frac{(s+1)^{3/4}(s-1)^{1/4}}{(s-1/K)}\right|^2, \ -1 \leqslant s \leqslant 0 \tag{13}$$

Values of Cp were calculated for $C_{pb} = -1.13$.

The pressure distributions for $\alpha = 45^o$ and 135° are shown in Figures 4 and 5.

3.2 The I-Beam

In Section 2.1 the flow over an I-beam was assumed to be identical to the flow over a cup, as shown in Figure 2. In applying the wake-source method to the flow over an I-beam, the base of the body was allowed to extend to infinity, as shown in Figure 6. This was permissible as the method does not model the flow within the wake and it can be assumed that the afterbody has no effect on the front surface pressure distribution. Again the upper half of the body was mapped into the s-plane, using the transformation

$$dz/ds = Hs(s+c)^{-\frac{1}{2}}(s+1)^{-\frac{1}{2}} \quad , \tag{14}$$

where c was determined by the ratio a/b. Equation (14) gives

$$z = (s+1)^{\frac{1}{2}}(s+c)^{\frac{1}{2}} - (c+1)(\ln((s+c)^{\frac{1}{2}} + (s+1)^{\frac{1}{2}})) + z_o \ , \tag{15}$$

where z_o is a constant which can be found by evaluating equation (15) when s = -1.

The ratio a/b was determined by the value of c. By inspection it was found that c = 0.83 gave a/b = 0.66, which corresponded to the dimensions of the model used in the wind tunnel experiments.

The pressure coefficient was given by

$$C_p = 1 - \left|\frac{(s+c)^{\frac{1}{2}}(s+1)^{\frac{1}{2}}}{s-\varepsilon}\right|^2 \quad , \tag{16}$$

where $\varepsilon = \sqrt{c}/K$. In choosing a value for the base pressure parameter, K, some caution was required. Inspection of the experimental pressure distribution showed that C_{pb} was greater than the pressure coefficient on the outside lip of the body just downstream of the separation point. The latter coefficient, $C_p = -1.7$ was therefore used to calculate K. The pressure distribution is shown in Figure 3.

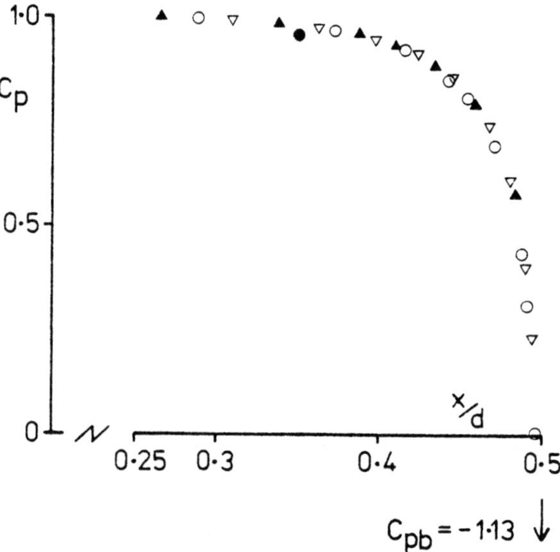

Figure 5: L-beam, α = 135°. ▽ , notched-hodograph method; ○, wake-source method; ▲ , vortex-lattice method; ● , experiment.

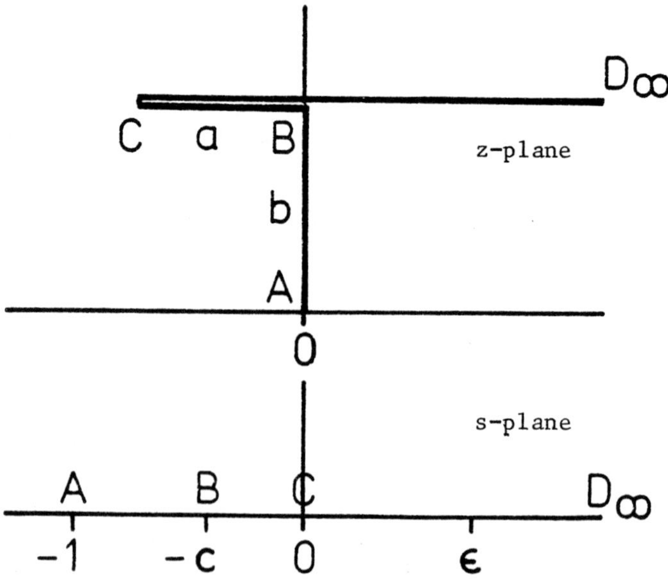

Figure 6: The mapping from the z-plane to the s-plane for the I-beam.

4. THE VORTEX-LATTICE METHOD

An examination of the Parkinson and Jandali wake-source method enabled Bearman and Fackrell [10] to apply the vortex-lattice method to the case of bluff-body flow. Bearman and Fackrell noted that the wake-source method implied that the front face of a bluff body between the separation points could be replaced by a sheet of vorticity and a pair of sources. The distribution of the vorticity and the strength and positions of the sources had to be consistent with the condition of zero normal velocity at the surface of the body, the positions of the separation points and also the separation velocity. Bearman and Fackrell proposed that the vortex sheet could be replaced by a set of discrete vortices distributed along the front face of the body. Bearman and Fackrell showed that the vortex-lattice method gave excellent agreement with the wake-source method, and that it was possible to extend it to two-dimensional bodies of arbitary shape and also axisymmetric bluff bodies.

Bearman and Fackrell's procedure has been followed for the three flows which are examined in this paper. A set of (N+1) control points was placed on the front surface of the body. The control points were spaced an equal distance, 2h, apart and the first and last points were h/2 from the separation points of the body. N vortices were then placed on the surface, each half way between adjacent control points. Two pairs of sources were used, as suggested by Bearman and Fackrell, and these were placed at adjacent control points near each end of the surface.

A set of (N+1) equations was obtained by setting the normal velocity to the surface at the control points equal to zero. Two further equations were obtained by making the tangential velocity at the separation points consistent with C_{pb}. Finally the condition that the net circulation about the body was zero was satisfied by setting the total vorticity equal to zero. There were therefore (N+4) unknowns, the strengths of N vortices and four sources, and (N+4) equations. The equations were solved simultaneously using a matrix inversion scheme. The correct solution, as pointed out by Bearman and Fackrell, was the one in which the strengths of all four sources was positive. The positions of the sources were altered until this was achieved. Once the values of the strengths of the vortices and sources had been calculated, the tangential velocity at each of the vortex positions was readily found. The pressure distribution along the front surface of the body was then calculated using Bernoulli's equation. The pressure distributions for the three cases considered are shown in Figures 3, 4 and 5.

5. EXPERIMENTAL RESULTS

The experimental values for the pressure coefficients were obtained by the authors, who used the closed-circuit wind tunnel in the Department of Physics, University of Edinburgh. The bluff-body models were made in the Civil Engineering workshops at Heriot-Watt University. A full description of the experiment and the results has been given in [3].

6. CONCLUSION

Three potential flow methods for the modelling of the flow over a bluff body have been discussed. All three methods make similar assumptions regarding the physical properties of the flow. The pressure distributions given by the notched-hodograph, wake-source and vortex-lattice methods all showed good agreement between themselves, and also with the limited number of experimental results which were presented for the three flows which were considered. There was, however, a difference between the methods in the difficulty of applying them to a particular flow. Both the hodograph and wake-source methods are limited to bluff bodies with shapes for which transformations could be derived. The vortex-lattice method however can be applied to bluff bodies of arbitary shape.

REFERENCES

1. ROSHKO A - On the drag and shedding frequency of two-dimensional bluff bodies. NACA Tech Note N°3169, 1954.
2. BEARMAN P W - On vortex street wakes. J Fluid Mech, Vol 28, Part 4, pp625-641, 1967.
3. GRANT I and BARNES F H - The vortex shedding and drag associated with structural angles. Proc 4th Colloquium on Industrial Aerodynamics, Part 2, pp45-57, Aachen 1980.
4. TASK COMMITTEE ON WIND FORCES. Struct Division ASCE, Trans ASCE, Vol 126, pp1124-1197, 1961.
5. LAMB H - Hydrodynamics. Cambridge University Press, Cambridge, 1952.
6. LOVE A E H - On the theory of discontinuous fluid motions in two dimensions. Proc Camb Phil Soc, Vol 7, pp175-201, 1891.
7. BIRKHOFF G and ZARANTONELLO E H - Jets, Wakes and Vortices. Academic Press, New York, 1957.
8. ROSHKO A - A new hodograph for free streamline theory. NACA Tech Note N°3168, 1954.
9. PARKINSON G V and JANDALI T - A wake source model for bluff body potential flow. J Fluid Mech, Vol 40, Part 3, pp577-594, 1970.
10. BEARMAN P W and FACKRELL J E - Calculation of two-dimensional and axisymmetric bluff-body potential flow. J Fluid Mech, Vol 72, Part 2, pp229-241, 1975.

NUMERICAL PREDICTION OF CONFINED VORTEX FLOWS

F. BOYSAN and J. SWITHENBANK

Department of Chemical Engineering and Fuel Technology, University of Sheffield, Sheffield, U.K.

Abstract

The problem of confined turbulent vortex flows in cyclone chambers is formulated in terms of the equations of conservation of mass and momentum and solved on a digital computer using finite differences. It is shown that the usual two-equation model of turbulence which entails the solution of two transport equations for the kinetic energy of the fluctuating motion and its dissipation rate, fails to reproduce many important features of confined vortex flows. An attempt is made to account for the anisotropic nature of turbulent and a new approach is employed to model the pressure-strain terms of the transport equations of the Reynolds stresses. Algebraic relations between the correlations of the velocity fluctuations and the mean velocity gradients derived from the transport equations are utilized in the conservation of momentum equations. Results display good agreement between the predicted and experimental profiles of axial and tangential velocity components in the cyclone chamber.

1. INTRODUCTION

Cyclone dust separators, fluidic vortex amplifiers and swirl combustors are only a few of the many engineering applications of the cyclone chamber. The vortex motion in this device is created by the tangential introduction of the fluid into a cylindirical chamber with a single axial exit in one end plate. Numerous configurations are possible depending on the number and position of the tangential entry ports, the location of the axial exit and the length to diameter ratio of the cylindirical chamber. Despite their simple geometry the aerodynamics of these chambers are extremely complicated and are characterized by several annular zones of forward and reverse flow, entrainment of the fluid from the outside and the high degree of preservation of swirl.

In the past there have been several experimental studies of the vortex structure in the cyclone chamber. Baluev and Troyankin [1] have reported measurements of all three components of velocity in 23 designs with the aid of a calibrated five channel probe. Hot-wire anemometer measurements of not only the mean but also the fluctuating velocity components have been made by Ustimenko and Bukhman [2]. Observations of the flow pattern in a cyclone chamber using smoke have been conducted by Smith [3], who also measured the axial and tangential time averaged velocity components using a cobra probe to determine direction and a special pilot-static to measure the magnitude. Unfortunately, vortex flows are extremely sensitive to disturbances created by probes [3,4,5] and hence these measurements are perhaps of questionable value. Non-intrusive optical diagnostic techniques have recently started to become available and at the time of writing of the paper LDA measurements in a vortex tube have been reported by Escudier et al [6].

The theoretical analysis of confined vortex flows on the other hand is extremely difficult and little progress in this area has so far been achieved. Apart from the fact that the governing equations are non-linear and strongly coupled, the structure of turbulence is non-homogeneous and anisotropic. For example, the dependence of the turbulence structure on the shape of the angular momentum profile was demonstrated both by Eskinazi and Yeh [7] and by Margolis and Lumley [8]. Despite these difficulties, most of the existing turbulence models assume that the turbulence is isotropic and although more complex Reynolds stress models are being developed, these have not so far led to successful predictions of strongly swirling flows.

The object of the present paper is to provide numerical predictions of turbulent vortex flow in cyclone chambers using a simplified version of Erdogan's [9] algebraic Reynolds stress closure.

2. MATHEMATICAL FORMULATION

The equations required for the description of flow patterns in cyclone chambers express the fluid flow balance of mass and momentum and are given here in cylindirical co-ordinates (z,r,θ), best suited to the geometry of the problem. In view of the experimental evidence that the flow in multiple entry cyclones loses its three dimensional character at a short distance from the inlet and becomes axially symmetric, all the derivatives with respect to the tangential direction have been neglected.

Continuity

$$\frac{\partial u}{\partial z} + \frac{1}{r}\frac{\partial}{\partial r}(rv) = 0 \tag{1}$$

z-momentum

$$u\frac{\partial u}{\partial z} + v\frac{\partial u}{\partial r} = -\frac{1}{\rho}\frac{\partial p}{\partial z} - \frac{\partial}{\partial z}\overline{u'^2} - \frac{1}{r}\frac{\partial}{\partial r}\overline{ru'v'} \tag{2}$$

r-momentum

$$v\frac{\partial v}{\partial r} + u\frac{\partial v}{\partial z} - \frac{w^2}{r} = -\frac{1}{\rho}\frac{\partial p}{\partial r} - \frac{\partial}{\partial z}\overline{u'v'} - \frac{1}{r}\frac{\partial}{\partial r}r\overline{v'^2} + \frac{\overline{w'^2}}{r} \tag{3}$$

θ-momentum

$$u\frac{\partial w}{\partial z} + v\frac{\partial w}{\partial r} + \frac{vw}{r} = \frac{\partial}{\partial z}\overline{u'w'} - \frac{1}{r}\frac{\partial}{\partial r}\overline{rv'w'} - \frac{\overline{v'w'}}{r} \tag{4}$$

The above equations can be solved only when the terms in angled brackets, i.e. the Reynolds stresses, are specified. A widely used method of relating these terms to known or calculable quantities is to stipulate that the generalized Boussinesq hypothesis applies. This hypothesis can be expressed in tensor form as

$$\overline{u_i' u_j'} = \frac{2}{3}k\delta_{ij} - \nu_T\left(\frac{\partial u_i}{\partial x_j} + \frac{\partial u_j}{\partial x_i}\right) \tag{5}$$

where ν_T is the turbulence kinematic viscosity, k is the turbulence energy and δ_{ij} is the Kronecker delta tensor. The distribution of ν_T is provided by the turbulence model employed.

Many models of turbulence exist, some involve the calculation of the turbulence viscosity from a prescribed length scale, whilst others require the solution of one or more partial differential equations. The two-equation k-ε model, which is of moderate complexity, has been extensively tested and proven to be adequate over a wide range of applications. This model entails the solution of two transport equations for the turbulence characteristics, namely the local energy of the fluctuating motion k and its dissipation rate ε. These equations can be written as [10],

$$u\frac{\partial k}{\partial z} + v\frac{\partial k}{\partial r} = \frac{1}{r}\frac{\partial}{\partial r}\left(r\frac{\nu_T}{\sigma_k}\frac{\partial k}{\partial r}\right) + \frac{\partial}{\partial z}\left(\frac{\nu_T}{\sigma_k}\frac{\partial k}{\partial z}\right) + P - C_D\varepsilon \tag{6}$$

$$u\frac{\partial \varepsilon}{\partial z} + v\frac{\partial \varepsilon}{\partial r} = \frac{1}{r}\frac{\partial}{\partial r}\left(r\frac{\nu_T}{\sigma_\varepsilon}\frac{\partial \varepsilon}{\partial r}\right) + \frac{\partial}{\partial z}\left(\frac{\nu_T}{\sigma_\varepsilon}\frac{\partial \varepsilon}{\partial z}\right) + (C_1P - C_2\varepsilon)\frac{\varepsilon}{k} \tag{7}$$

where, P stands for the rate of generation of turbulence energy. Solution of the above equations allows the local turbulence and effective viscosities to be evaluated from

$$\nu_T = C_\mu k^2/\varepsilon \quad \text{and} \quad \nu_e = \nu + \nu_T \tag{(8),(9)}$$

respectively. The coefficients C_1, C_2, C_D, C_μ, σ_k and σ_ε are constants which are assigned the following values, $C_1 = 1.44$, $C_2 = 1.92$, $C_D = 1.0$, $C_\mu = 0.09$, $\sigma_k = 1.0$ and $\sigma_\varepsilon = 1.3$ as recommended by Launder, Priddin and Sharma [11].

Despite the success with which the k-ε model allowed the prediction of numerous flows, its ability to reproduce the strong influence of anisotropy in swirling flows has been found wanting. It was suggested in the past the effects of rotation can be accounted for by making either C_1 or C_2 a function of the 'swirl' Richardson number [11,12] or by employing a different effective viscosity, somewhat lower than that calculated from the k-ε model, in the θ-momentum equation [13]. In an earlier effort by the present authors it was shown that both these modifications indeed lead to improved predictions [14]. It was concluded however that the Richardson number approach was restricted to situations where the swirl was weak and that no universal rule for the selection of the swirl effective viscosity existed.

In the light of the above mentioned theoretical developments and recent experimental results [15] which challenge the assumption of isotropy on which the k-ε model rests, it is necessary to adopt a more fundamental approach to the modelling of turbulence as applied to swirling flows.

The exact differential equations for the transport of Reynolds stresses can be derived directly from the Navier-Stokes equations which can be written compactly as,

$$T_{ij} = P_{ij} - \frac{2}{3}\,\varepsilon\,\delta_{ij} + \psi_{ij} \tag{10}$$

where, T_{ij} stands for the transport terms, P_{ij} is the production, ε is the dissipation and ψ_{ij} is the pressure-strain term. These equations can be solved numerically along with those for the conservation of mass and momentum provided that the pressure-strain term is appropriately modelled. Since, the solution of six additional partial differential equations increases the computational effort required considerably, it is desirable to reduce these into algebraic form wherever possible. Employing the model proposed by Erdogan [9] for the pressure-strain term, which offers maximum flexibility and replacing the transport terms by the approximation which is due to Rodi [16],

$$T_{ij} = \frac{\overline{u_i' u_j'}}{k}\,(P-\varepsilon) \tag{11}$$

the following algebraic relations are obtained for the six Reynolds stresses,

$$\overline{u_i'u_j'} = \alpha \frac{k}{P}\left(P_{ij} - \frac{2}{3}\delta_{ij}P\right) + \beta\frac{k}{P}\left(D_{ij} - \frac{2}{3}\delta_{ij}P\right) + \gamma\frac{k^2}{P}\left(\frac{\partial u_i}{\partial x_j} + \frac{\partial u_j}{\partial x_i}\right) \tag{12}$$

$$P_{ij} = -\left[\overline{u_i'u_k'}\frac{\partial u_j}{\partial x_k} + \overline{u_j'u_k'}\frac{\partial u_i}{\partial x_k}\right]$$

$$D_{ij} = -\left[\overline{u_i'u_k'}\frac{\partial u_k}{\partial x_j} + \overline{u_j'u_k'}\frac{\partial u_k}{\partial x_i}\right]$$

$$P = -\overline{u_i'u_j'}\frac{\partial u_i}{\partial x_k}$$

where, α, β and γ are functions of P/ε.

It is assumed in the present study that the only significant terms in these equations are those which contain $\partial w/\partial r$ and w/r, since the tangential component of velocity is many times larger than the other two at all the points in the cyclone chamber. This assumption leads to simple explicit expressions for the Reynolds stresses such as for example

$$\overline{u'v'} = -\frac{\left[\frac{2}{3}(1-\alpha-\beta)(\alpha+\beta) - \gamma\right]\frac{\varepsilon}{P}}{1 - \frac{k^2}{P^2}\left(\alpha\frac{\partial w}{\partial r} - \beta\frac{w}{r}\right)\left(\beta\frac{\partial w}{\partial r} - \alpha\frac{w}{r}\right)}\frac{k^2}{\varepsilon}\frac{\partial u}{\partial r} \tag{13}$$

and similar ones for the others. It is interesting to note that the first group of terms on the right hand side of the above equation corresponds to the constant C_μ of the k-ε model and that swirl tends to reduce its magnitude. The functions α, β and γ take the forms given below:

$$\alpha = \frac{0.2115\,(P/\varepsilon)^3}{0.175 + (P/\varepsilon)^3},\quad \beta = \frac{-0.1764\,(P/\varepsilon)^3}{2.92 + (P/\varepsilon)^3},\quad \gamma = \frac{-0.135(P/\varepsilon)^3}{0.927+(P/\varepsilon)^3} \tag{14}$$

when tuned in accordance with the data of Harris, Graham and Corrsin [17] and Champagne, Harris and Corrsin [18] on nearly homogeneous plane shear flow.

3. SOLUTION PROCEDURE

The insertion of the expressions for the correlations of the fluctuating velocities presented in the preceding section into the equations of conservation of momentum results in a set of coupled elliptic partial differential equations which contain second order derivatives with respect to both z and r directions. The numerical solution procedure therefore needs two dimensional storage and requires iteration. Before an attempt is made to solve these equations, they must first be reduced into their finite difference analogues. This is achieved by integration over a computational cell. The resulting algebraic equations which relate the value of a dependent variable at a grid point P to those at the four neighbouring points, can be presented in the following form:

$$\left(\sum_{i=N,S,E,W} A_i - S_P\right) \phi_P = \sum_{i=N,S,E,W} A_i \phi_i + S_U \tag{15}$$

where the A's are the coefficients which contain contributions from the convective and diffusive fluxes and S_U and S_P are the components of the linearized source term. The set of simultaneous algebraic equations (15) are solved by the SIMPLE algorithm of Patankar and Spalding [19]. This aglorithm involves the solution of the momentum equations for a guessed pressure distribution to give a first estimate of the velocity field. Corrections to both the pressure and the velocity components are made after solving an equation for pressure correction derived from the continuity equation. These corrections are such that the resulting velocity field satisfies mass conservation.

4. BOUNDARY CONDITIONS

Because of the elliptic nature of the equations, a complete description of the problem considered necessitates the specification of the boundary conditions at all the boundaries of the domain of integration.

In the near wall regions, the well known wall functions are matched with the algebraic equations (15) to preclude fine grid calculations in this region. The usual practice is to cut the link between the boundary and near-wall points by setting the appropriate coefficient to zero, and to insert the wall influence by way of the linearized source terms. The specific wall functions employed in the present study are

$$\tau_s = \frac{u_p}{y_p \chi^{-1}} \frac{\mu y_p^+}{\ln (E y_p^+)} \tag{16}$$

$$y_p^+ = \rho (K_p C_\mu^{\frac{1}{2}})^{\frac{1}{2}} \frac{y_p}{\mu} \tag{17}$$

$$\varepsilon_p = (C_\mu^{\frac{1}{2}} K_p)^{3/2} / \chi y_p \tag{18}$$

where, τ_s is the shear stress at the wall, u_p, K_p and ε_p are the velocity, turbulence energy and dissipation rate at the near wall node respectively, y_p is the distance from the wall, χ and E are constants.

The boundary conditions at the axis of symmetry are of the zero normal gradient type for all the variables except the radial and tangential components of velocity which are themselves zero there.

Although the conditions at the inlet are specified once and for all and do not need updating during the course of the solution procedure, those at the exit are not known beforehand. An analytical difficulty associated with confined vortex flows is their extreme sensitivity to downstream boundary conditions [20]. In order to minimize the effects of the downstream conditions on the flow inside the cyclone chamber, the domain of integration is extended further downstream of the chamber exit. The boundary conditions imposed at this end are of the gradient type for all the dependent variables.

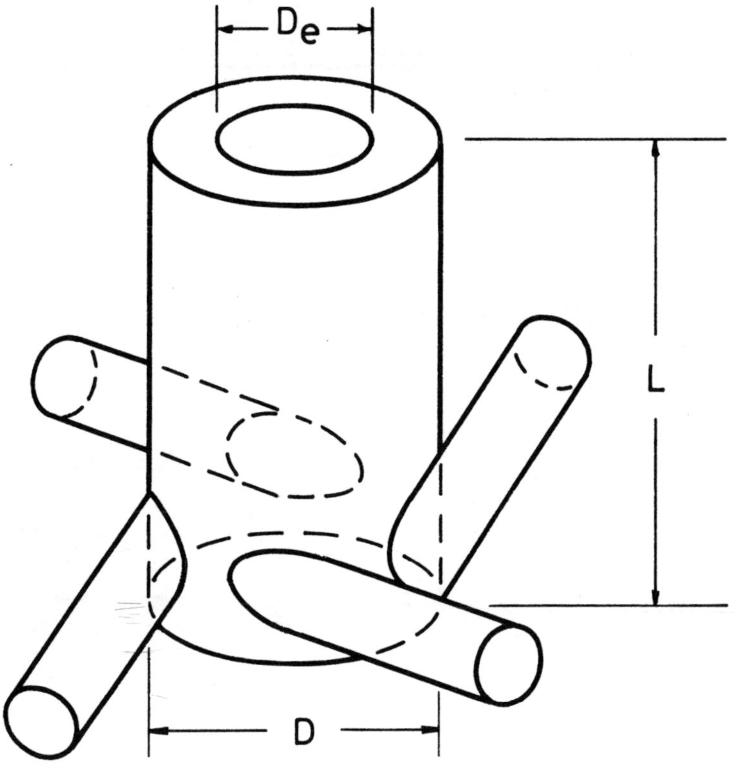

Fig. 1. Multiple Entry Cyclone Chamber

5. RESULTS

The calculations were performed for a cyclone chamber of diameter D = 0.25 m, length L = 0.386 m and exit throat diameter D_e = 0.1 m. The schematic diagram of the cyclone is shown in Fig. 1. The fluid was assumed to be fed into the

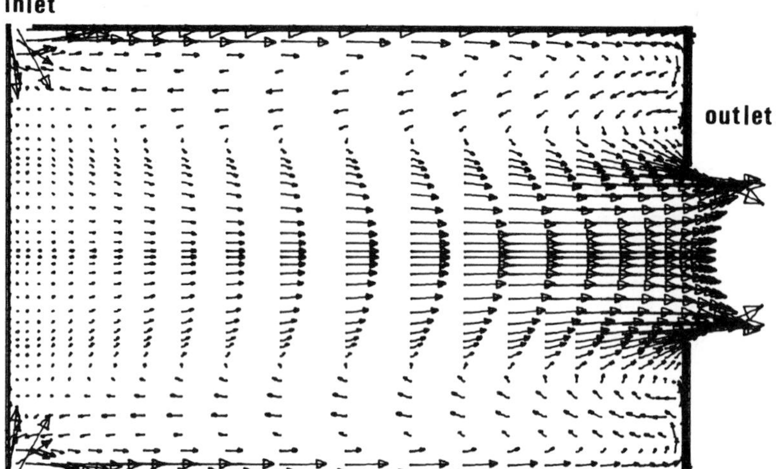

Fig. 2 Distribution of the Velocity Vectors in the Cyclone Chamber Showing both Direction and Relative Magnitude. (k-ε model predictions)

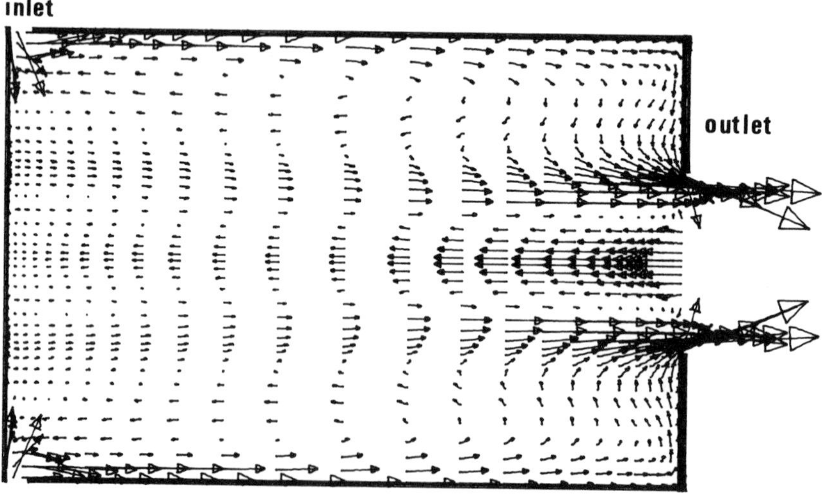

Fig. 3 Distribution of the Velocity Vectors in the Cyclone Chamber Showing both Direction and Relative Magnitude. (Algebraic Reynolds Stress Model Predictions)

chamber through a circumferential slot, the width of which was adjusted so that the flow rate and the inlet velocities matched the experimental conditions reported by Ustimenko and Bukhman [2]. The ratio of tangential to radial velocity components at the inlet was of the order of 6. In order to preserve the stability of the solution procedure and to obtain convergence for such high degrees of swirl,it was found necessary to reduce the under-relaxation parameters of all the three velocity components to 0.25. The number of iterations required was also in direct proportion with the degree of swirl. The calculations were performed on a 40 x 21 non-uniform grid in the z and r directions respectively.

The distribution of velocity vectors in the cyclone chamber predicted using the k-ε and the algebraic stress models are presented in Figs. 2 and 3. These show that the incoming flow follows the cylindirical walls until the downstream base plate is reached from which the fluid rebounds and forms a reverse flowing stream. The fluid leaves the chamber after a final change of direction. A toroidal vortex exists between the uppermost forward and reverse streams which extends in both cases from the upstream base plate to about half the chamber length. Although there is some similarity between the two distributions especially in the vicinity of the cylindirical wall, experimentally observed entrainment of the outside fluid is predicted by the algebraic model alone.

The comparison between the measured (2) and calculated axial velocity profiles are displayed in Fig. 4. It is interesting to note that the predicted curves show good agreement with the experimental data in the region $0.5\ r_o < r < r_o$ at all the axial locations. In the neighbourhood of the axis, however, the k-ε model predictions deviate considerably from the measurements, while tha algebraic model leads to generally acceptable agreement.

Fig. 5 shows both the measured and predicted tangential velocity profiles at four axial stations in the cyclone. The agreement between the algebraic Reynolds stress model predictions and the measurements are indeed very good. Not only the general shape of the profiles but also the degree of preservation of swirl velocity throughout the chamber could be reproduced with this model. There is a slight descrepancy between the predictions and the experiments however in the neighbourhood of the axis of symmetry which probably arises from the omission of all the velocity gradients except $\partial w/\partial r$ and w/r in the final phase of the modelling. It is evident that the k-ε model fails to reproduce even the qualitative features of the measured profiles.

Finally, Fig. 6 displays the profiles of turbulence kinetic energy in the cyclone both calculated and measured. Although the agreement between the predictions and measurements

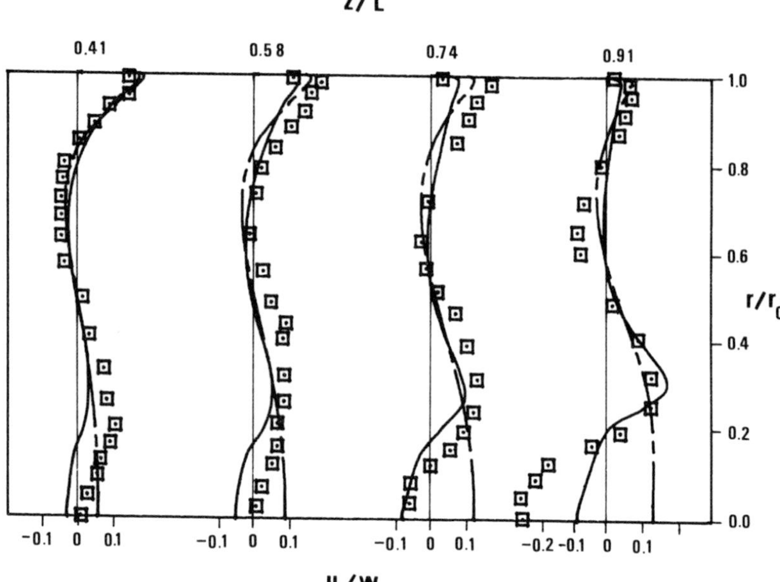

Fig. 4 Comparison of predicted and measured axial velocity profiles at four axial stations. ⊡ experiments, —— algebraic stress model, —-— k-ε model.

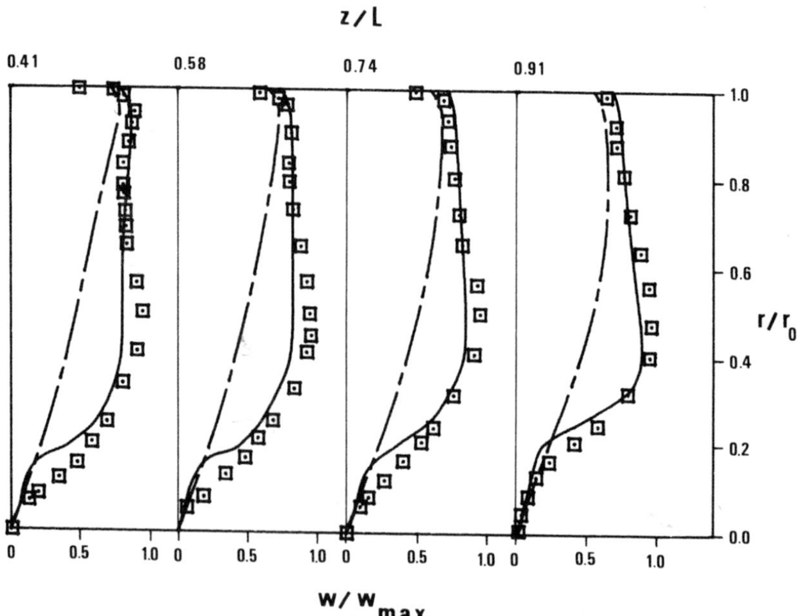

Fig. 5 Comparison of predicted and measured tangential velocity profiles. For legend see Fig. 4.

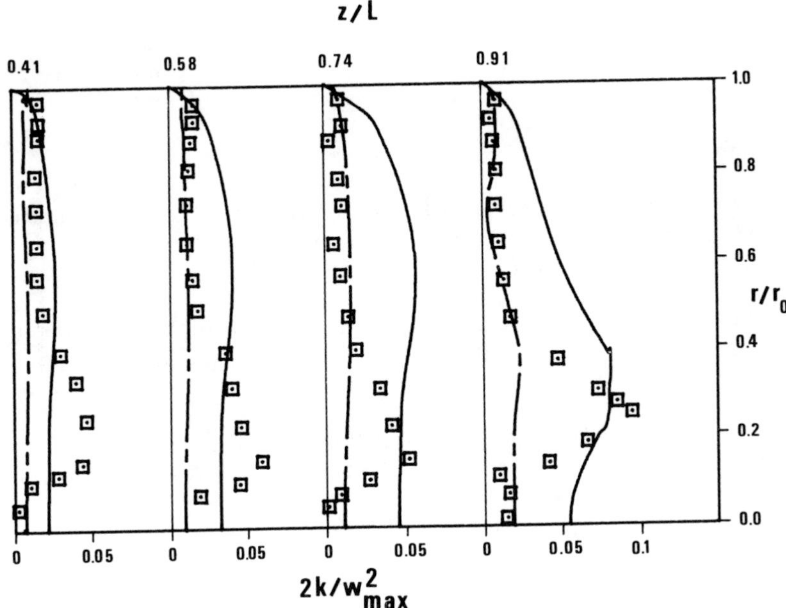

Fig. 6 Comparison of predicted and measured tubulence kinetic energy profiles. For legend see Fig. 4.

are not as good as in the case of the axial and tangential velocity components, at least the predicted levels of turbulence are not too different from the experiments. It is difficult to judge with regard to the shapes of the profiles whether the source of the error is in the experiments or turbulence modelling.

6. CONCLUDING REMARKS

The present study has demonstrated that it is possible to approach the problem of confined turbulent vortex flows with an algebraic Reynolds stress closure based on Erdogan's [9] modelling of the pressure-strain term and Rodi's [16] approximation for the transport terms. The comparison between the predictions and what experimental data is available in the literature have been encouraging and further work in this area is justified.

On the other hand, since most of the measurements in cyclone chambers have been made by the insertion of probes into the flow, it is difficult to judge whether the source of the present discrepancies lies in the experiments or in the turbulence modelling. At this stage, therefore, one must conclude that detailed measurements of confined vortex flows by means of non-intrusive optical techniques are required urgently for further validation of the turbulence model.

7. ACKNOWLEDGEMENTS

This work was supported by the U.K. Science Research Council and USAF under contract AFOSR 80-0174. Thanks are also due to Professor E. Erdogan for valuable discussions.

8. REFERENCES

1. BALUEV, E.D. and TROYANKIN, V., 'Study of the Aerodynamic Structure of Gas Flow in a Cyclone Chamber', Teploenergetika 14 (1), pp. 63-65, 1967.

2. USTIMENKO, B.P. and BUKHMAN, M.A., 'Turbulent Flow Structure in a Cyclone Chamber', Teploenergetika 15 (2), pp. 64-67, 1968.

3. SMITH, J.L., 'An Experimental Study of the Vortex in the Cyclone Separator', ASME J. Basic Eng., December 1962, pp. 602-608.

4. HOLMAN, J.P. and MOORE, G.D., 'An Experimental Study of Vortex Chamber Flow', ASME J. Basic Eng., 83, p. 632, 1961.

5. SWITHENBANK, J., Flow Visualization in Cyclone Chambers - Film, 1975.

6. ESCUDIER, M.P., BORNSTEIN, J. and ZEHNDER, N., 'Observations and LDA Measurements of Confined Turbulent Vortex Flow', J. Fluid Mech. 98, Part 1, pp. 49-63, 1980.

7. ESKINAZI, S. and YEH, M., 'An Investigation of Fully Developed Turbulent Flows in a Curved Channel', J. Aerospace Sci., Vol. 23, No. 1, January, 1956.

8. MARGOLIS, D.P. and LUMLEY, J.L., 'Curved Turbulent Mixing Layer', Physics of Fluids, Vol. 8, No. 10, October 1965.

9. ERDOGAN, M.E., 'On the Calculation of Reynolds Stresses', University of Sheffield, Department of Chemical Engineering and Fuel Technology, Report no. HIC 347.

10. LAUNDER, B.E. and SPALDING, D.B., 'Mathematical Models of Turbulence', Academic Press, London, 1972.

11. LAUNDER, B.E., PRIDDIN, C.H. and SHARMA, B.I., 'The Calculation of Turbulent Boundary Layers on Spinning and Curved Surfaces', J. Fluids Eng., March 1977, p. 231.

12. RODI, W., 'Influence of Bouyancy and Rotation on Equations for the Turbulent Length Scale', 2nd Symposium on Turbulent Flows, London, 1979.

13. LILLEY, D.G., 'Prediction of Inert Turbulent Swirl Flows', AIAA Journal, 11, No. 7, p. 955, 1973.

14. BOYSAN, F. and SWITHENBANK, J., 'Numerical Prediction of Confined Turbulent Vortex Flows', University of Sheffield, Department of Chemical Engineering and Fuel Technology, Report No. HIC 360, 1980.

15. PRATTE, B.D. and KEFFER, J.F., 'The Swirling Turbulent Jet', J. Basic Eng., 95, 1973.

16. RODI, W., 'A New Algebraic Relation for Calculating the Reynolds Stresses', ZAMM, 56, p. 219, 1976.

17. HARRIS, V.G., GRAHAM, J.A. and Corrsin, S., 'Further Experiments in Nearly Homogeneous Shear Flow', J. Fluid Mech., Vol. 81, p. 657, 1977.

18. CHAMPAGNE, F.H., HARRIS, V.G. and CORRSIN, S., 'Experiments on Nearly Homogeneous Turbulent Shear Flow', J. Fluid Mech, Vol. 41, p. 81, 1970.

19. PATANKAR, S.V. and SPALDING, D.B., 'A Calculation Procedure for Heat Mass and Momentum Transfer in Three Dimensional Parabolic Flows', Int. J. Heat Mass Transfer, 15, 1972.

20. RAZGAITIS, R. and HOLMANN, J.P., 'A Survey of Heat Transfer in Confined Swirl Flows', presented at the Summer Seminar of the International Centre for Heat and Mass Transfer, Dubrovnik, Yugoslavia, August 1975.

FINITE-DIFFERENCE SOLUTION FOR AXIALLY DEVELOPING FLOW IN AN ANNULUS

Vijay K. Garg(i)

Based on the boundary-layer approach, an implicit finite-difference method is developed for determining the axially developing laminar flow in a concentric annulus. The scheme is universally stable and is computationally about two orders of magnitude faster than the Sparrow & Lin's method [1]. It is also simpler and more economical than the method of Shah & Farnia [2]. Results for pressure drop and velocity distribution are given for a wide range of annular diameter ratio (k) varying from 0.001 to 0.8. Presently calculated pressure drop variation with axial distance matches very well with that given by Sparrow & Lin and also with that given by Shah & Farnia. In the near entry region the present velocity profiles differ from those of Sparrow & Lin by at most 7% and the difference decreases as the flow develops downstream.

1. INTRODUCTION

The developing flow in a duct is of interest not only because of its direct engineering application but also for accurate predictions of viscosity using capillary viscometers. However, while several solutions are available for the developing flow in circular tubes and parallel-plate channels, only three, somewhat unsatisfactory, solutions are available for the flow in annular ducts. These solutions are by Sparrow & Lin [1], by Heaton et al. [3], and by Shah & Farnia [2]. The first two studies obtained continuous solutions for the velocity distribution by linearizing the inertia terms in the momentum

---

(i)Professor, Department of Mechanical Engineering, Indian Institute of Technology Kanpur, India.

equation. Langhaar [4] has, however, pointed out that linearization is valid only in the core and fully-developed regions but not in the boundary-layer region of the developing flow. Besides Sparrow & Lin's method involves summation of a series consisting of Bessel functions of both first and second kind and of order 0 and 1 whose evaluation requires considerable computational time. It was found that the computation time taken in determining the velocity field at any cross-section by the Sparrow and Lin method is about two orders of magnitude larger than that taken by the present method. Like the method of Shah & Farnia, the present method employs a boundary-layer approach but unlike their use of Patankar & Spalding's rather involved method [5], we use a very simple and economical finite-difference method first suggested by Bodoia & Osterle for the channel flow [6]. Moreover, Shah & Farnia concentrate more on comparing their results for the developing flow in pipes and channels with those of earlier investigators. In fact, they do not provide any velocity profiles for the developing flow in annular ducts. Instead, they report only some friction factor values for k = 0.005 and 0.1.

2. PRESENT METHOD

It is well known that the boundary-layer model neglects the axial transport of vorticity and considers the pressure as a function of axial distance only. As shown by Hornbeck et al. [7], this model is good for Reynolds numbers greater than 200 for the developing flow in ducts. The boundary-layer equations for the coordinate system shown in Fig. 1 are

$$u\frac{\partial u}{\partial z} + v\frac{\partial u}{\partial r} = -\frac{1}{\rho}\frac{dp}{dz} + \nu\left(\frac{\partial^2 u}{\partial r^2} + \frac{1}{r}\frac{\partial u}{\partial r}\right) ,$$

and (1)

$$r\frac{\partial u}{\partial z} + \frac{\partial (vr)}{\partial r} = 0 .$$

The boundary conditions are

$$u(r,o) = u_0 , \quad u(r_1, z) = u(r_2, z) = 0 ,$$
$$v(r_1, z) = v(r_2, z) = 0 , \quad p(o) = p_0 , \qquad (2)$$

where r_1 and r_2 are the inner and outer radii of the annulus, and p_0 and u_0 are the pressure and axial velocity distribution at the inlet section. For the present study, u_0 is taken to be uniform. As shown by Wang & Longwell [8], this assumption is justified for Reynolds numbers greater than 600.

Equations (1) and (2) are non-dimensionalized by use of the following dimensionless variables

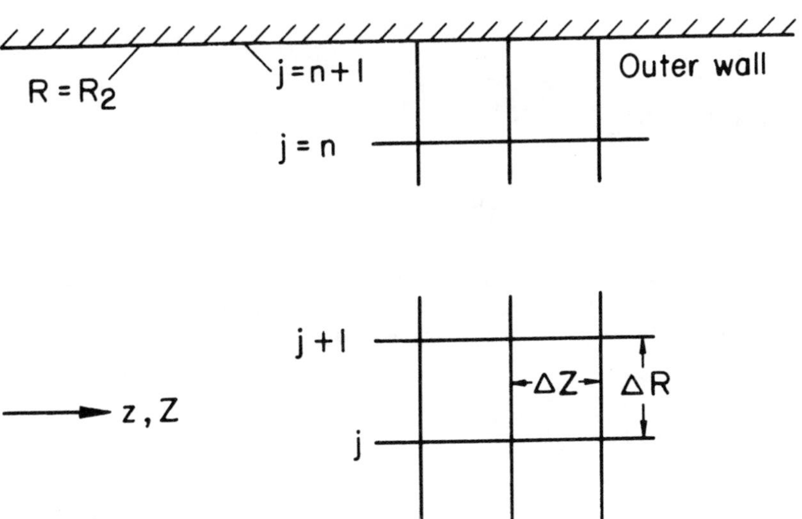

Fig.1 Co-ordinate system and grid

$$R = \frac{2r}{r_2 - r_1}, \quad Z = \frac{2z}{(r_2 - r_1)} \cdot \frac{1}{Re}, \quad Re = \frac{u_a(r_2 - r_1)}{2\nu}$$

$$U = \frac{u}{u_a}, \quad V = \frac{vRe}{u_a}, \quad P = \frac{p - p_0}{\rho u_a^2}, \tag{3}$$

where u_a is the average velocity of flow through the annulus. For the network of grid points shown in Fig. 1, the finite-difference form of these equations at any grid point (i, j) is [6]

$$U_{i+1,j-1}\left[\frac{1}{2R_j} - \frac{1}{\Delta R} - \frac{V_{i,j}}{2}\right]\frac{1}{\Delta R} + U_{i+1,j}\left[\frac{U_{i,j}}{\Delta Z} + \frac{2}{(\Delta R)^2}\right]$$

$$+ U_{i+1,j+1}\left[\frac{V_{i,j}}{2} - \frac{1}{2R_j} - \frac{1}{\Delta R}\right]\frac{1}{\Delta R} + \frac{P_{i+1}}{\Delta Z} = \frac{P_i + U_{i,j}^2}{\Delta Z} , \qquad (4)$$

and

$$R_j V_{j+1} - R_j V_j = \frac{\Delta R}{2\Delta Z}\left[R_{j+1}(U_{i+1,j+1} - U_{i,j+1}) + R_j(U_{i+1,j} - U_{i,j})\right] , \qquad (5)$$

for $j = 1(1)n$, where n is the number of mesh points along the radial direction between the inner and outer walls of the annulus. Equation (5) when summed up for $j = 1(1)n$ gives

$$\sum_{j=1}^{n} R_j U_{i,j} = \sum_{j=1}^{n} R_j U_{i+1,j} \quad . \qquad (6)$$

Equations (4) for $j = 1(1)n$ along with equation (6) give a set of (n+1) simultaneous equations for the (n+1) unknowns ($U_{i+1,j}$ at n mesh points and P_{i+1}) at section (i+1) in terms of known values of $U_{i,j}$ and P_i at section i. The (n+1) X (n+1) matrix that results is of tridiagonal type except for the row corresponding to (6). The matrix is, therefore, partitioned to take advantage of tridiagonality. An improved version of Gauss elimination [9] is used to solve the equations. Having obtained U's at section (i+1), the radial component of velocity at these mesh points is determined from (5). This marching technique enables determination of the velocity and pressure field in the entire flow.

Since discretization of the cross-section with $U_{o,j} = 1$ for all $j = 1(1)n$ at the inlet section leads to a reduction in the volumetric flow rate owing to no slip boundary conditions at the annulus walls, $U_{o,j}$ was taken as

$$U_{o,j} = (n+1)/n \quad \text{for all} \quad j = 1(1)n. \qquad (7)$$

After some numerical experimentation, $\Delta R = 0.005$ and $\Delta Z = 10^{-5}$, 10^{-4}, $5X10^{-4}$ and 10^{-3} respectively for $0 < Z \leqslant 0.01$, $0.01 < Z \leqslant 0.1$, $0.1 < Z \leqslant 0.2$, and $0.2 < Z \leqslant 0.8$ were selected.

3. RESULTS

3.1 Pressure Drop Distribution

Fig. 2 shows the variation of the pressure drop P(Z) with Z for different values of $k = r_1/r_2$. On the log-log plot, this variation is almost linear and changes little with the diameter ratio specially in the near inlet region. In fact, even at Z = 0.2, the pressure drop changes by only 3% as k

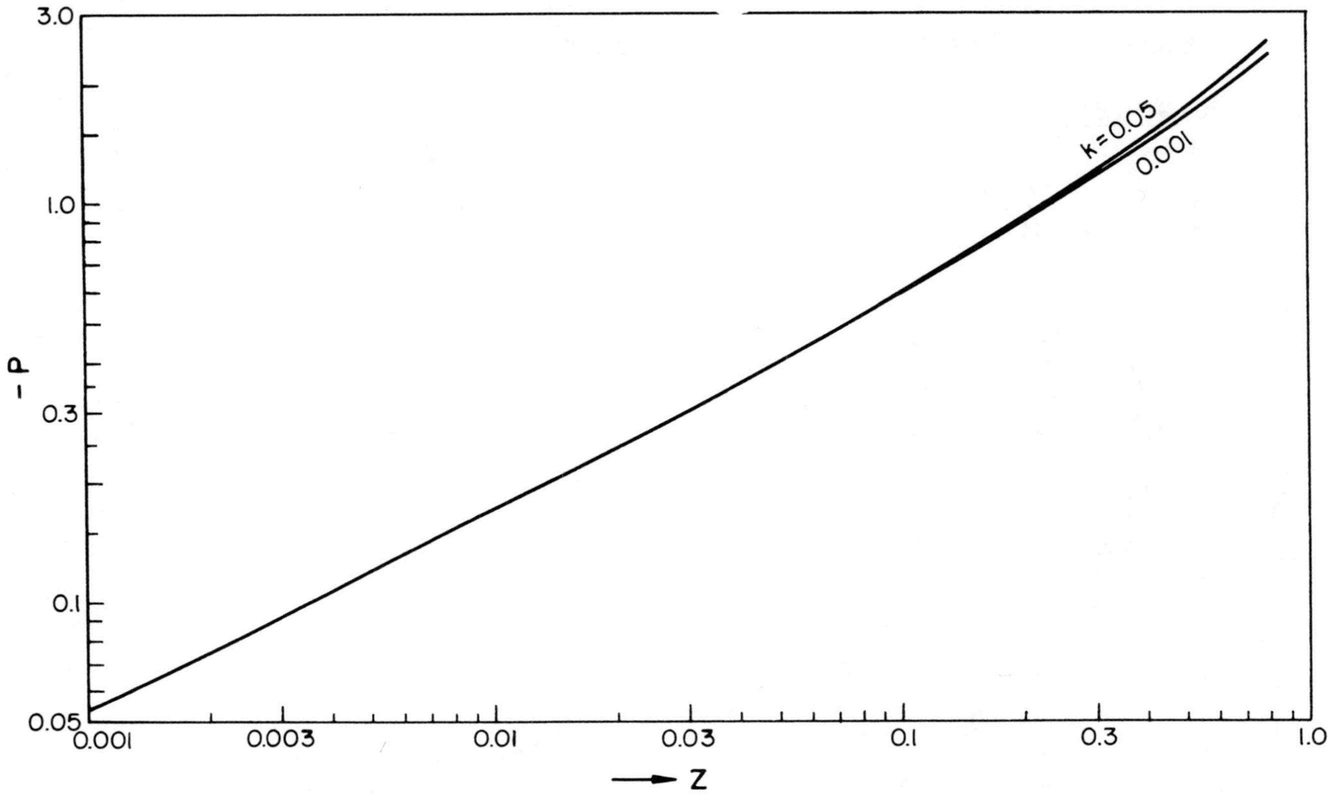

Fig.2 Pressure drop variation with Z.

changes from 0.001 to 0.8. Thus for $Z < 0.2$, the curve shown in Fig. 2, though plotted for $k = 0.001$, holds good for any k. The present results match very well with those given by Sparrow & Lin [1] and Shah & Farnia [2].

3.2 Velocity Distribution

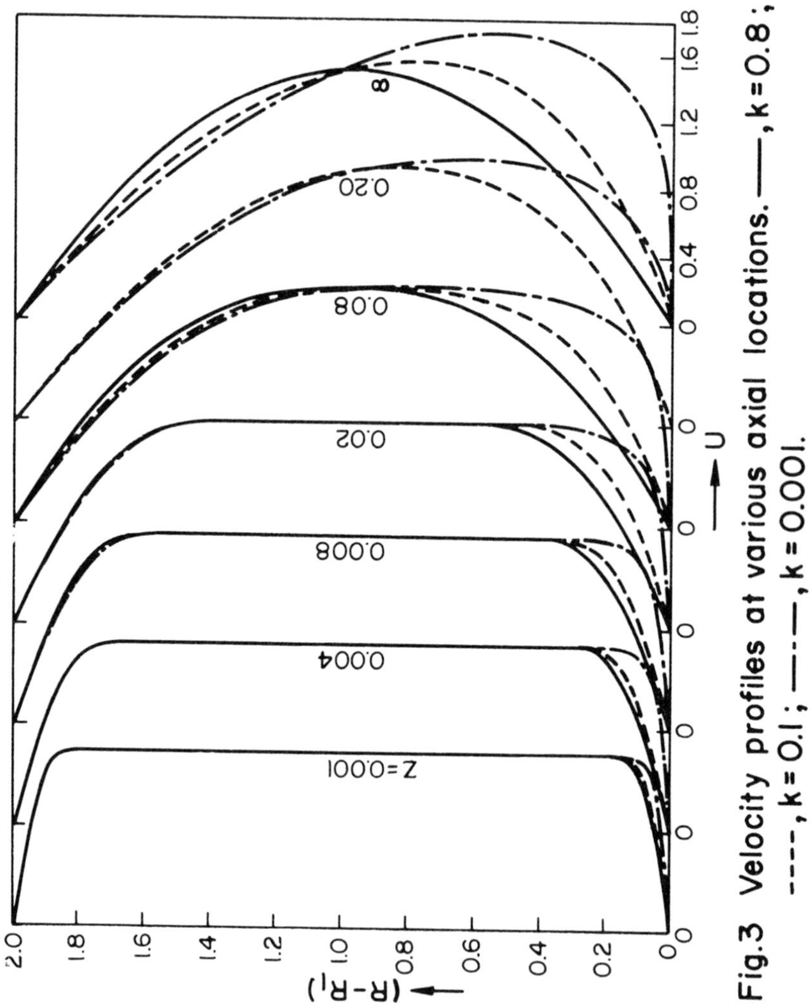

Fig.3 Velocity profiles at various axial locations. ——, k=0.8; ----, k=0.1; —·—, k=0.001.

Fig. 3 shows the variation of axial component of velocity U across the cross-section at several axial locations for k = 0.001, 0.1 and 0.8. Figs. 4, 5 and 6 show the variation of U with Z for k = 0.01, 0.1 and 0.4 at nine radial locations $b = (r-r_1)/(r_2-r_1) = 0.1$ to 0.9 along the cross-section. The latter also display Sparrow & Lin's results for comparison. It is observed that in the near-entry region the two velocity pro-

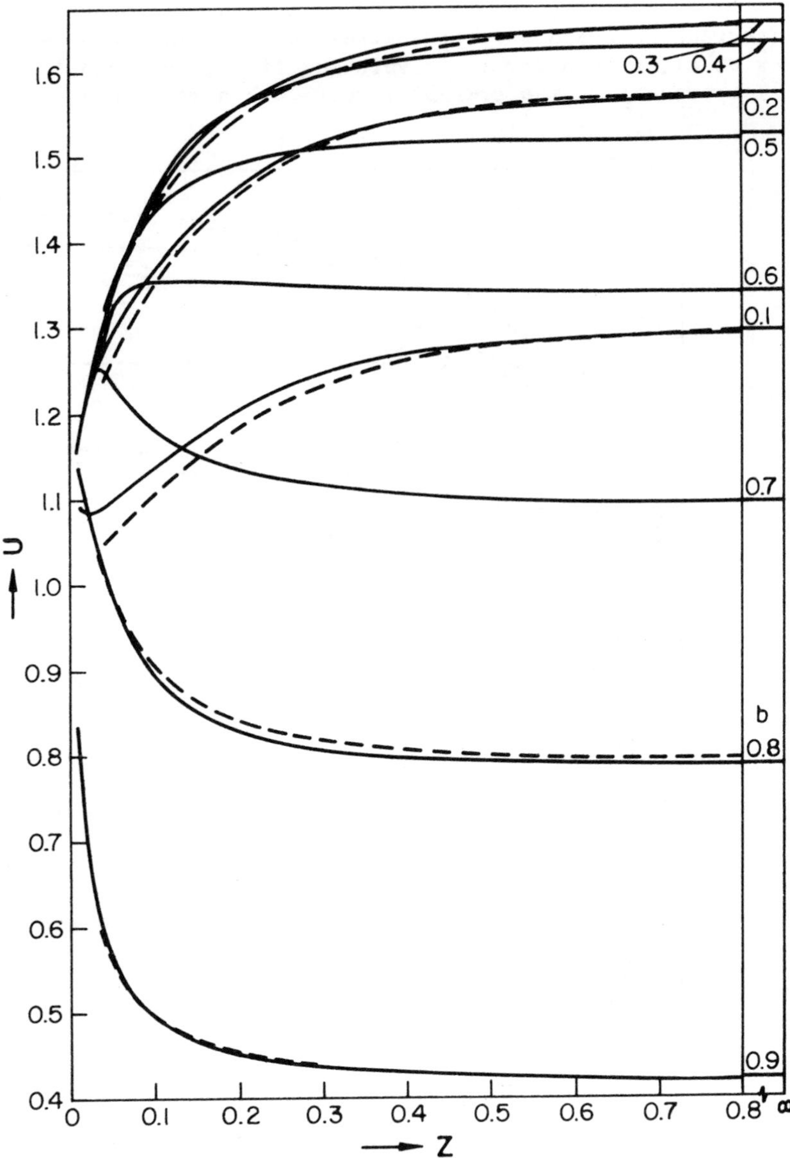

Fig.4 Axial velocity at various radial positions for k=0.01. ——, present work; ----, Sparrow and Lin's work

files differ by as much as 7% but the difference decreases as Z increases. Due to lack of experimental data, it is impossible to predict which of the two methods gives a more accurate velocity field description. However, since the finite-difference technique is known to yield better results [10] than the Sparrow et al. method for the developing flow in a channel [11], it is expected that the present results are more accurate.

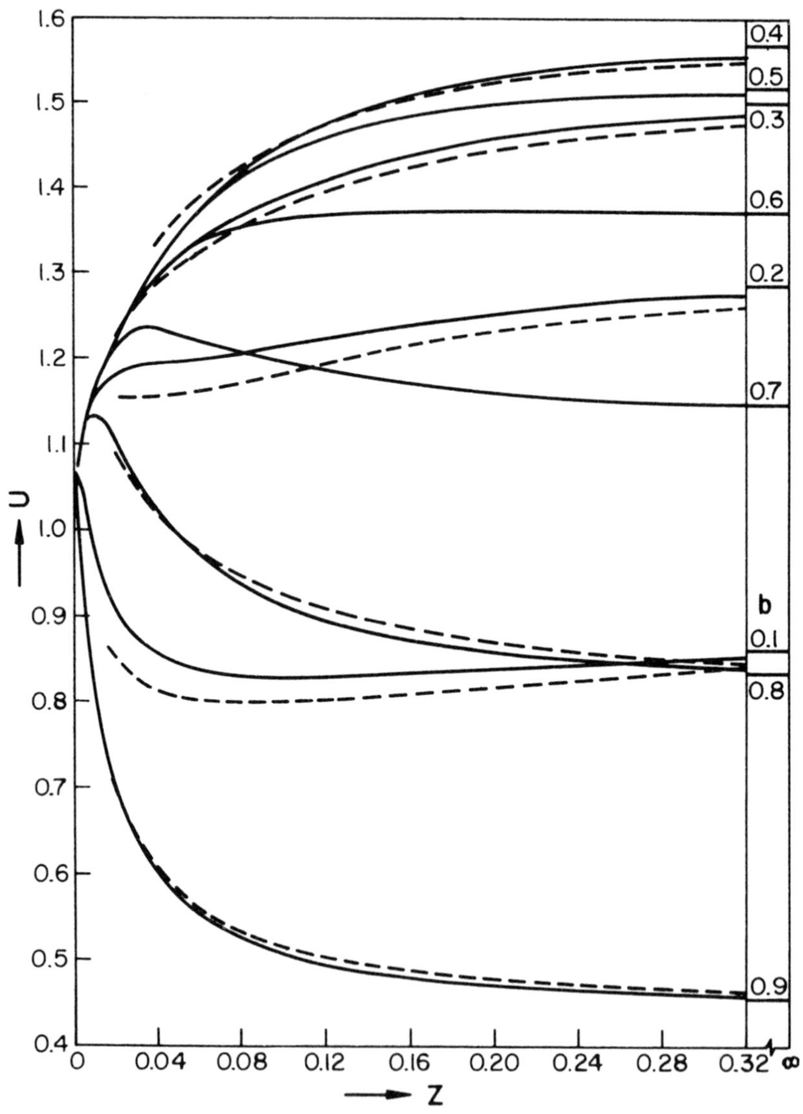

Fig.5 Axial velocity at various radial positions for k=0.1. ——, present work; ----, Sparrow and Lin's work.

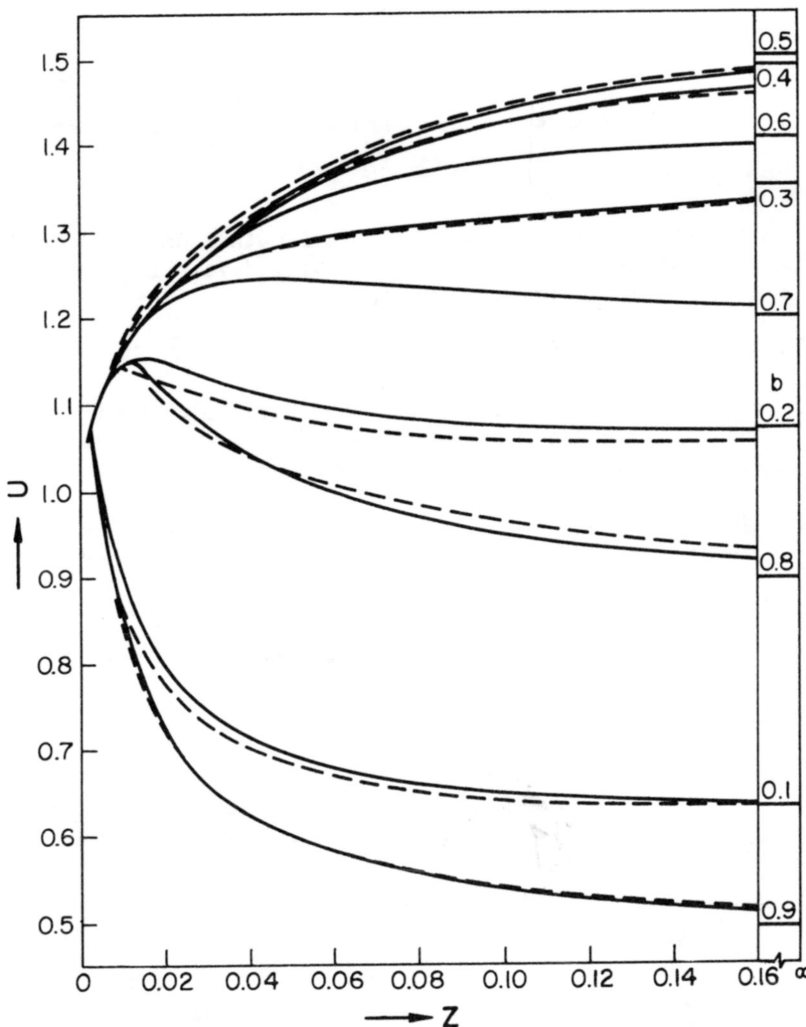

Fig.6 Axial velocity at various radial positions for k = 0.4. ——, present work; ----, Sparrow and Lin's work

These figures show that the velocity profile is asymmetric about the mid-point between the inner and outer walls, and as the diameter ratio decreases, the asymmetry increases such that the point of maximum velocity shifts closer to the inner wall. From Fig. 3 it is also clear that change in k affects the velocity profile more near the inner wall. Also, for small k, the boundary layer at the outer wall is thicker than that at the inner wall. Fig. 6 for k = 0.4 shows that U decreases with Z for b = 0.1, 0.2, 0.8 and 0.9. However, as k decreases, the U vs. Z curves (Figs. 4 and 5) for b = 0.8 and 0.9 continue to drop while those for $b < 0.5$ tend to shift upwards. This implies that for smaller k, the velocity profile rises sharply from its zero value at the inner wall, remains relatively flat over a substantial portion of the duct cross-section and decreases relatively gradually to a zero value at the outer wall.

4. CONCLUSIONS

An extremely simple and economical method has been developed for determining the hydrodynamically developing laminar flow field in a concentric annulus.

REFERENCES

1. SPARROW, E.M. and LIN, S.H. - The Developing Laminar Flow and Pressure Drop in the Entrance Region of Annular Ducts. J. Basic Engg., pp. 827-834, 1964.

2. SHAH, V.L. and FARNIA, K. - Flow in the Entrance of Annular Tubes. Compt. & Fluids, Vol. 2, pp. 285-294, 1974.

3. HEATON, H.S., REYNOLDS, W.C. and KAYS, W.M. - Heat Transfer in Annular Passages. Int. J. Heat Mass Trans., Vol. 7, pp. 763-781, 1964.

4. LANGHAAR, H.L. - Steady Flow in Transition Length of a Straight Tube. J. App. Mech., Vol. 9, pp. 55-58, 1942.

5. PATANKAR, S.V. and SPALDING, D.B. - Heat and Mass Transfer in Boundary Layer, Intertext Books, London, 2nd Edn. 1970.

6. BODOIA, J.R. and OSTERLE, J.F. - Finite Difference Analysis of Plane Poiseuille and Couette Flow Developments. App. Sci. Res., Vol. A10, pp. 265-276, 1961.

7. HORNBECK, R.W., ROULEAU, W.T. and OSTERLE, J.F. - Effects of Radial Momentum Flux on Flow in the Entrance of a Porous Tube. J. App. Mech., Vol. 32, pp. 195-197, 1965.

8. WANG, Y.L. and LONGWELL, P.A. - Laminar Flow in the Inlet Section of Parallel Plates. A.I.Ch.E. J., Vol. 10, pp. 323-329, 1964.

9. HORNBECK, R.W. - Numerical Methods, Quantum Pub., New York, 1975.

10. SHAH, R.K. - A Correlation for Laminar Hydrodynamic Entry Length Solutions for Circular and Noncircular Ducts. J. Fluids Engg., Vol. 100, pp. 177-179, 1978.

11. SPARROW, E.M., LIN, S.H. and LUNDGREN, T.S. - Flow Development in the Hydrodynamic Entrance Region of Tubes and Ducts. Phys. Fluids, Vol. 7, pp. 338-347, 1964.

F.E.M. AND THE TWO EQUATION MODEL OF TURBULENCE

C. TAYLOR[*], C.E. THOMAS[**] and K. MORGAN[***]

DEPARTMENT OF CIVIL ENGINEERING, UNIVERSITY COLLEGE, SINGLETON PARK, SWANSEA SA2 8PP, U.K.

SUMMARY

A two equation model of turbulence is employed where the transport of turbulence kinetic energy and dissipation rates are depicted using transport type equations, i.e. the two equation model of turbulence. The results obtained are compared with other models and experimental results for flow over a backward facing step. Generally, the model was found to be underpredictive with regard to the reattachment length although the velocity distribution and turbulence energy could be predicted quite accurately.

INTRODUCTION

During recent years the F.E.M. has been employed quite extensively in predicting both laminar [1-5] and turbulent flow [6-13]. The technique now compliments other methods for solving problems where the flow is governed by the generalised Navier Stokes equations and has become a useful addition to the scientists' repertoire of methods for solving such problems.

The present paper is an investigation of the application of the F.E.M. to predict turbulent flow over a downstream facing step [13]. The fluid motion is assumed to be governed by the Navier Stokes equations, the equations of continuity, a turbulence transport equation and an equation depicting local dissipation [14]. The F.E.M. is used to

---

[*] Reader, [**] Research Assistant, [**] Lecturer.

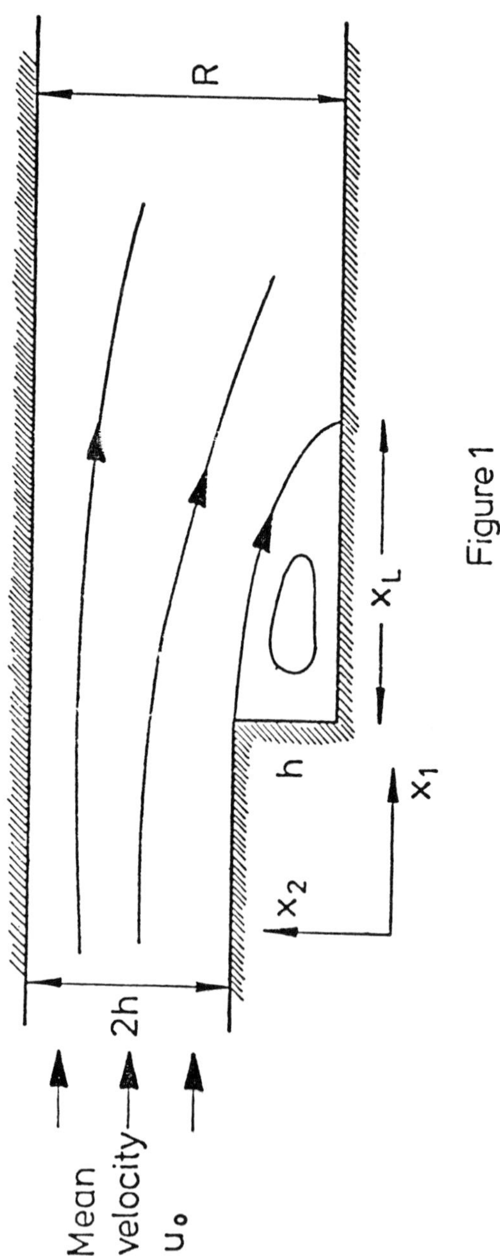

Figure 1

effect a spatial discretisation and variation in the primitive variables and the resulting discrete form of the governing equations are then solved, iteratively, in order to ascertain the spatial variation in the pertinent variables. The resulting solutions obtained are compared with known experimental [15] and numerical results [16].

THEORETICAL FORMULATION

A schematic representation of the region of interest is shown on Figure 1. The flow is assumed to be two dimensional, steady and both the laminar viscosity and density are constant.

A set of differential equations which are commonly used to depict the flow under the prescribed conditions are,

$$\rho\, u_j \frac{\partial u_i}{\partial x_j} = -\frac{\partial p}{\partial x_i} + \frac{\partial}{\partial x_j}\left[\mu_e\left(\frac{\partial u_i}{\partial x_j} + \frac{\partial u_j}{\partial x_i}\right)\right] \qquad (1)$$

and

$$\rho\, \frac{\partial u_i}{\partial x_i} = 0 \qquad (2)$$

in which u_i denotes the velocity vector in the ith coordinate direction, p the local pressure, ρ the density and μ_e the effective viscosity. All variables are time averaged and the effective viscosity is written

$$\mu_e = \mu + \mu_t \qquad (3)$$

where μ is the laminar viscosity and μ_t the turbulent viscosity whose value varies pointwise throughout the flow domain.

The magnitude of the turbulent viscosity can be defined [17,18] in terms of the turbulence kinetic energy, k, and ℓ, the turbulent length scale, and a constant C_μ,

$$\mu_t = C_\mu\, \rho\, k^{\frac{1}{2}} \ell \qquad (4)$$

The present text is concerned with the evaluation of both k and ℓ from two further transport equations. These are,

$$\rho u_j \frac{\partial k}{\partial x_j} = \frac{\partial}{\partial x_k}\left[\left(\mu + \frac{\mu_t}{\sigma_k}\right)\frac{\partial k}{\partial x_k}\right] + \mu_t \frac{\partial u_i}{\partial x_j}\left(\frac{\partial u_i}{\partial x_j} + \frac{\partial u_j}{\partial x_i}\right)$$

$$- C_D\, \rho k^{3/2}/\ell \qquad (5)$$

for the turbulence kinetic energy and,

$$\rho u_j \frac{\partial \varepsilon}{\partial x_j} = \frac{\partial}{\partial x_k}\left[\left(\mu + \frac{\mu_t}{\sigma_\varepsilon}\right)\frac{\partial \varepsilon}{\partial x_k}\right] + C_1 \rho k \frac{\partial u_i}{\partial x_j}\left(\frac{\partial u_j}{\partial x_i} + \frac{\partial u_i}{\partial x_j}\right)$$

$$- C_2 \rho \frac{\varepsilon^2}{k} \qquad (6)$$

where, σ_k, σ_ε, C_D, D_1 and C_2 are usually considered to be constant [14]. A comprehensive investigation regarding the evaluation of the above constants has been presented [19] in which the values are given as,

$C_\mu = 0.22$, $C_D = 0.092$, $\sigma_k = 1.00$, $\sigma_\varepsilon = 1.3$, $C_1 = 1.45$ and $C_2 = 0.18$.

BOUNDARY CONDITIONS

The following boundary conditions were imposed,

$$\left.\begin{array}{l} u_1 \text{ - specified} \\ u_2 = 0 \\ \ell \text{ - specified} \\ \varepsilon \text{ - specified} \end{array}\right\} \text{upstream}$$

The specified values of ℓ and ε are those corresponding to fully developed flow in a channel equal in width to that upstream of the step.

At the downstream end of the domain,

$$\left.\begin{array}{l} \dfrac{\partial u_1}{\partial x_1} = 0 \\ \dfrac{\partial}{\partial x_1} = 0 \\ \dfrac{\partial k}{\partial x_1} = 0 \end{array}\right\} 0 < x_2 < R$$

ℓ - fully developed values from the one equation model

$\varepsilon = \dfrac{k^{3/2}}{\ell}$, $p = 0$

WALL

If the walls are smooth, rigid, impermeable and the no-slip condition is valid, then all variables assume a zero value at the wall. However, the variation in such quantities quite close to the wall render the imposition of zero values impractical unless a very fine mesh discretisation is used near the wall. Indeed, such refinement could lead to excessive core and c.p.u. requirements and is usually discarded.

A generally accepted technique which obviates the necessity to follow rapid near wall variations is to terminate the mesh at some distance away from the wall, usually, and utilise the normal laws depicting the variation in shear velocity in the near wall region [20],

$$\left.\begin{aligned} u_i^* &= \lambda^* && 0 \le \lambda^* \le 5 \\ u_i^* &= (-3.05 + 5.0 \log \lambda^*)\left(\frac{\tau_w}{|\tau_w|}\right) && 5 \le \lambda^* \le 30 \\ u_i^* &= (-5.5 + 2.5 \log \lambda^*)\left(\frac{\tau_w}{|\tau_w|}\right) && \lambda^* > 3 \end{aligned}\right\} \qquad (7)$$

in which

$$u_i^* = u_i/\sqrt{|\tau_w/\rho|} \; ; \quad \lambda^* = \left(\frac{\lambda}{\mu}\right) \sqrt{|\tau_w/\rho|} \qquad (8)$$

$i = 1$ for boundaries parallel to u_1 axis

$i = 2$ for boundaries parallel to u_2 axis

λ = distance measured normal to wall

The shear stress at the limit of the near wall region is assumed to be identical with that at the wall, $\tau_w = \mu \dfrac{\partial u_i}{\partial \lambda}$. Once the gradient in velocity and associated shear stress can be evaluated then the near wall value of k can be calculated from

$$k = \frac{|\tau_w|}{C_D^{\frac{1}{2}} \rho} \qquad (9)$$

where the absolute magnitude for τ_w is included since k must always be positive. This is derived from the usual assumption that the variation in static pressure normal to a wall, can be ignored and derivatives of the pertinent variables parallel to the wall are small compared to those normal to the wall. Using these assumptions becomes a particular solution of the generalised equation depicting transport of 'k', provided that the location under consideration is within the fully turbulent region.

The near wall values of ε may now be found if the length scale is defined. Following the procedure adopted for the one equation model and $\ell = \lambda$, the values of ε can be defined since

$$\varepsilon = \frac{k^{3/2}}{\ell} \tag{10}$$

Since the discretised domain terminates at some small distance away from the wall the above conditions will also apply at the upstream and downstream extremities of the domain.

METHOD OF SOLUTION

Quadrilateral isoparametric elements are used and the now standard approach to equation formulation into matrix form is adopted [13]. The resulting matrix equation can be written in a generalised form,

$$\underset{\sim}{H} \cdot \underset{\sim}{\beta} = \underset{\sim}{f} \tag{11}$$

in which the matrix $\underset{\sim}{H}$ is non-symmetric. Details of the coefficients of $\underset{\sim}{H}$ and composition of $\underset{\sim}{f}$ are presented in [13]. The only additional equation to be incorporated is (6) which can either be included in $\underset{\sim}{H}$ in exactly the same manner as (5) leading to a single matrix where,

$$\beta^k = \begin{Bmatrix} u_1^k \\ p^k \\ u_2^k \\ k^k \\ \varepsilon^k \end{Bmatrix} \qquad \text{for a corner node}$$

and

$$\beta^k = \begin{Bmatrix} u_1^k \\ u_2^k \\ k^k \\ \varepsilon^k \end{Bmatrix} \qquad \text{for a mid-side node}$$

ITERATIVE TECHNIQUE

The solution technique adopted involved isolating the ε equation and solving an $\underset{\sim}{H}$ matrix for the remaining variables and then calculating ε separately, resulting in the following iteration scheme.

i) Set all initial values to zero within the flow domain and assume that the effective viscosity corresponds to the molecular viscosity,

ii) Solve for u_i, p and k for a fixed distribution of ℓ (3 iterations on 1st entry to problem),

iii) Estimate near wall and inlet boundary conditions for ε using $\varepsilon = C_\mu \dfrac{k^{2/3}}{\ell}$

iv) Solve the ε equation using fixed values of u_i, p and k,

v) Test on convergence of ε; if not converged go to (iii)

vi) Update ℓ using $\ell = C_\mu \dfrac{k^{2/3}}{\varepsilon}$

vii) Update μ_t using the new k and ℓ values,

viii) Using equation (8), re-estimate the wall shear stress and therefore the boundary conditions on u_i and k,

ix) Repeat from (ii) until convergence criteria satisfied.

It was found, during the application of the above iterative technique that (v) would initiate several iterations for ε for set values of the remaining variables. This would sometimes cause divergence since the distribution is highly sensitive to variations in either the upstream or downstream boundary conditions. For better stability and distribution was obtained when fully developed values on u_i, k and p were imposed upstream and a corresponding $\dfrac{\partial \varepsilon}{x_1} = 0$ as opposed to $\varepsilon = C_\mu \dfrac{k^{3/2}}{\ell}$.

NUMERICAL CALCULATIONS

The spatial discretisation used for the present calculations is shown on Figure 2.

During numerical calculations the initial values were taken as zero throughout the domain although special care was exercised when imposing boundary conditions. The upstream boundary conditions were evaluated by conducting an analysis of flow in a channel equal in width to that upstream of the step. The one equation model boundary conditions were imposed upstream and the flow allowed to develop to a fully developed profile downstream. The mesh used and boundary conditions imposed are shown on Figure 3. The fully developed values of velocity and k were then employed as upstream boundary conditions for the step problem. Gradients of all variables, except pressure, were taken as zero on the downstream face, again implying fully developed conditions. It was found, however, during trial calculations, that the imposition of boundary conditions on ε on other than fully developed values on the upstream boundary did not change the results to any marked degree. The values of this variable seemed to develop to essentially the same values at the step irrespective of small changes in ε on the upstream boundary.

The problem analysed corresponds to a Reynolds number of 3025 and the coefficients,

$C_\mu = 0.22$, $\sigma_k = 1.00$, $C_1 = 1.45$, $C_2 = 0.18$, $\sigma_\varepsilon = 1.3$ and
$C_D = 0.092$

which correspond to those taken by Atkins [19]. These are compared on Figure 4 which also includes the experimental results of Denham [15].

CONCLUSIONS

A marked difference in the recirculation length as calculated 4.5h and measured 6h was observed although the overall velocity distribution seemed within reasonable agreement. This is less accurate than that obtained using the one equation model [12].

The correllation between the 'k' values calculated using the F.E.M. gave a better approximation than the finite difference method to the measured values.

It has been shown that although the F.E.M. seems a viable technique for the solution of the governing equations, more effort is required to rationalise on the application of accepted turbulence techniques to solve such problems.

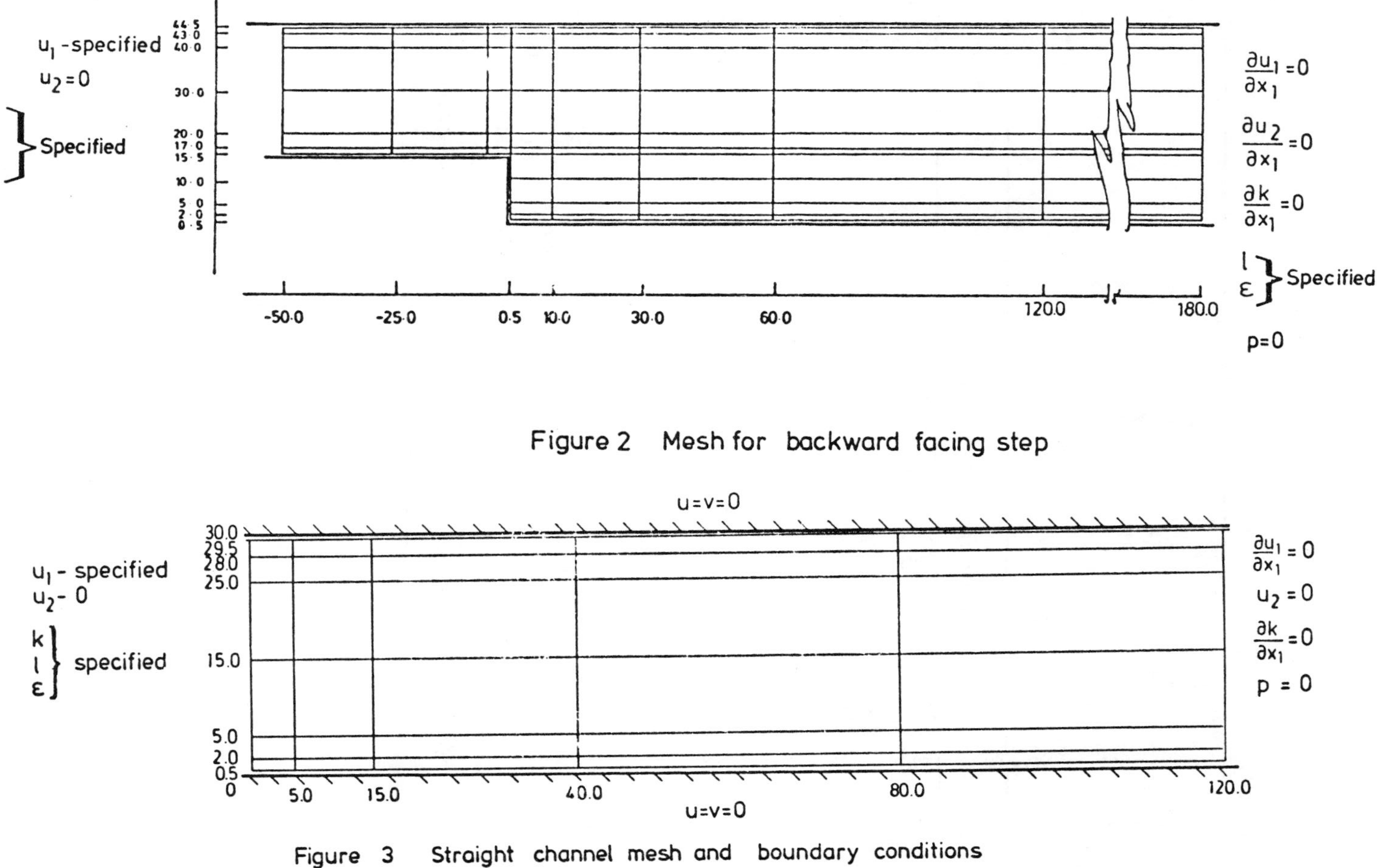

Figure 2 Mesh for backward facing step

Figure 3 Straight channel mesh and boundary conditions

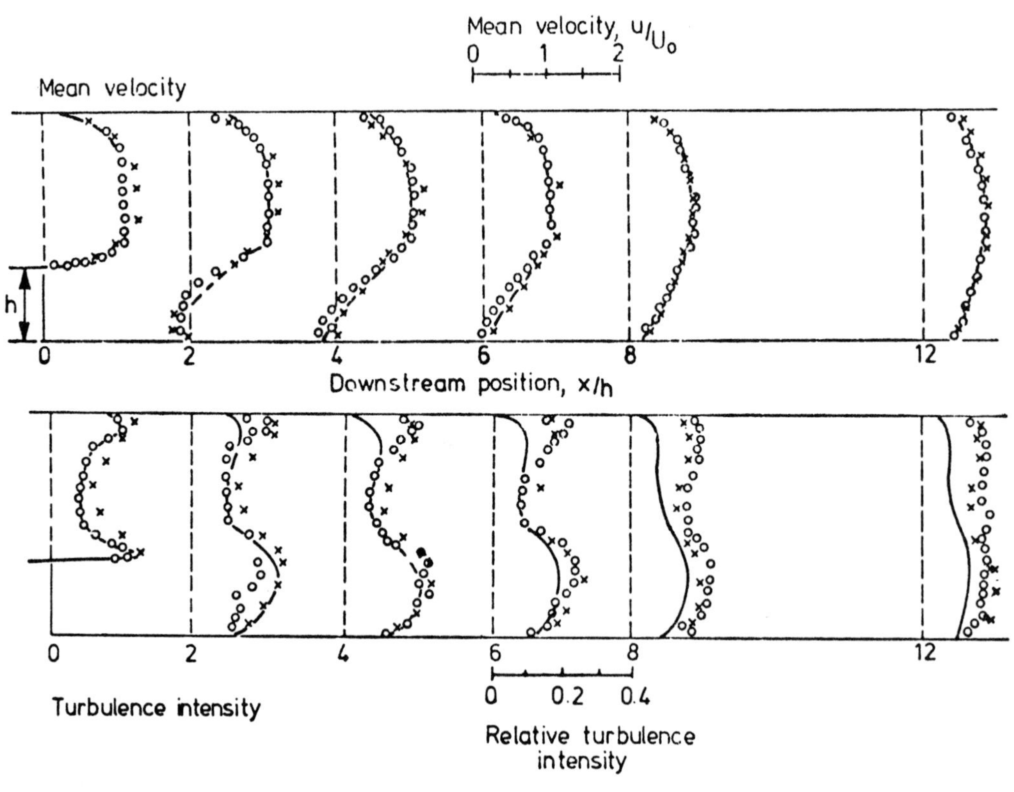

Figure 4 Velocity and turbulence intensity plots for Re=3025 two equation model

REFERENCES

1. J.T. ODEN, Finite Elements of Non-Linear Continua, McGraw-Hill, 1972.

2. C. TAYLOR and P. HOOD, 'A numerical solution of the Navier-Stokes equations using the Finite Element Technique', Int. Journ. Comp. and Fluids, Vol.1, No.1, pp. 73-100, 1973.

3. C. TAYLOR and P. HOOD, 'Navier-Stokes equations using mixed interpolation', Proc. Int. Conf. on F.E.M. in Flow Problems, Swansea, pp. 121-132, 1974.

4. T.J.R. HUGHES, R.L. TAYLOR and J.F. LEVY, 'A finite element method for incompressible viscous flows', Proc. 2nd Int. Conf. on F.E.M. in Flow Problems, Rapallo, (Italy), pp. 1-16, 1976.

5. R.M. SMITH, 'A study of laminar flow entrance sections using the F.E.M.', C.E.G.B. Report No. RD/B/M3513, 1975.

6. R. GERRARD, 'Finite element solution of flow in non-circular conduits', Proc. A.S.C.E., Journ. HYD. DIV., 100, HY3, pp. 425-441, 1974.

7. C. TAYLOR, T.G. HUGHES and K. MORGAN, 'A numerical analysis of turbulent flow in pipes', Int. Journ. Comp. and Fluids, 5, pp. 191-204, 1977.

8. K. MORGAN, T.G. HUGHES and C. TAYLOR, 'A numerical model of turbulent shear flow behind a prolate spheroid', Applied Math. Modelling, 2, pp. 271-274, 1978.

9. A.J. BAKER, 'Finite element analysis of turbulent flows' Proc. 1st Int. Conf. Num. Meth. in Turbulent Flows, Swansea, pp. 203-229, 1978.

10. C. TAYLOR, T.G. HUGHES and K. MORGAN, 'Finite element solution of one-equation models of turbulent flow', Journ. Comp. Phys., 29, 2, pp. 163-172, 1978.

11. K. MORGAN, T.G. HUGHES and C. TAYLOR, 'The analysis of turbulent free shear flows by the F.E.M.', Comp. Meth. in Appl. Mech. and Eng., 19, pp. 117-125, 1979.

12. C.E. THOMAS, I. MORGAN and C. TAYLOR, 'Finite element analysis of flow over a backward facing step', Comp. and Fluids, to be published, 1981.

13. C. TAYLOR, T.G. HUGHES and K. MORGAN, 'A finite element model of one and two equation models of turbulent flow' Recent Advances in Num. Meth. in Fluids, 1, Pineridge Press, pp. 311-334, 1980.

14. B.E. LAUNDER and D.B. SPALDING, Mathematical Models of Turbulence, Academic Press, London/New York, 1972.

15. M.K. DENHAM, P. BRIARD and M.A. PATRICK, 'A directionally sensitive laser anemometer for velocity measurements in highly turbulent flow', Journ. of Physic E: Scientific Instruments, 8, pp. 681-683, 1975.

16. D.J. ATKINS, S.J. MASKELL and M.A. PATRICK, 'Numerical prediction of separated flows', Int. J. Num. Meth. in Engng. 15, 1, pp. 129-144, 1980.

17. L. PRANDTL, 'Uber ein neues Formelsystem fur die ausgebildete Turbulenz', Nachr. Akad. der Wissenschaft in Gottingen, 1945.

18. A.N. KOLMOGOROV, 'Equations of turbulent motion of an incompressible fluid', FSV. Akad. Nauk. SSSR Ser. Phys. VI, 1-2, pp. 56-68, 1942.

19. D.J. ATKINS, 'Numerical studies of separated flows', Ph.D. Thesis, Exeter University, 1974.

20. J.T. DAVIES, 'Turbulence Phenomena', Academic Press, N.Y. 1972.

SECTION 3
BOUNDARY LAYER ANALYSIS

REGULAR BOUNDARY INTEGRAL EQUATIONS FOR FLUID FLOW

C. Patterson and M.A. Sheikh

Dept. of Mechanical Engineering, University of Sheffield

1 INTRODUCTION

The Boundary Element Method is now well known as a general numerical technique for the solution of field problems In contrast with Finite Element Method, freedoms need only be defined on the boundary of the domain of the problem. Once these are determined, the solution within the domain is obtained using appropriate surface integrals of the boundary solution.

Central to the method is the generation of Boundary Integral Equations which properly state the problem to be solved in terms of unknown field functions on the boundary only. These equations are usually obtained [1], using the fundamental solution of the given problem with the singular point located on the boundary. (The equations for the solution at the interior are obtained similarly, by locating the singular point within the domain of the problem). There ensues an infinite system of singular surface integral equations, one for each boundary point. The system is discretized by defining boundary elements, after the manner of finite elements, and the resulting finite system of singular integrals is evaluated, thereby giving a system of linear algebraic equations.

Two discomforting features are apparant in this conventional approach. Firstly, accurate evaluation of singular integrals requires special and careful treatment in the neighbourhood of the singular point. Secondly, the class of problems for which the method is well defined, may be unduly restrictive because of divergence of the integrals.

In this paper it will be shown that Regular Boundary Integral Equations can quite readily be derived which also properly state the given problem. These are obtained by the simple device of moving the singularity of the fundamental solution outside the domain of the **problem**. The resulting system of equations tolerates higher order singularities in the solution than previously and requires no special attention to a singular integrand.

2 THEORY

Consider a Potential Function ϕ over the domain Ω, which satisfies the governing (Laplace's) equation,

$$\nabla^2\phi = 0 \quad \text{in } \Omega \tag{1}$$

The boundary conditions for the problem are (Fig.1).

(a) $\phi = \bar{\phi}$ on Γ_1

(b) $\dfrac{\partial\phi}{\partial n} = \overline{\dfrac{\partial\phi}{\partial n}}$ on Γ_2

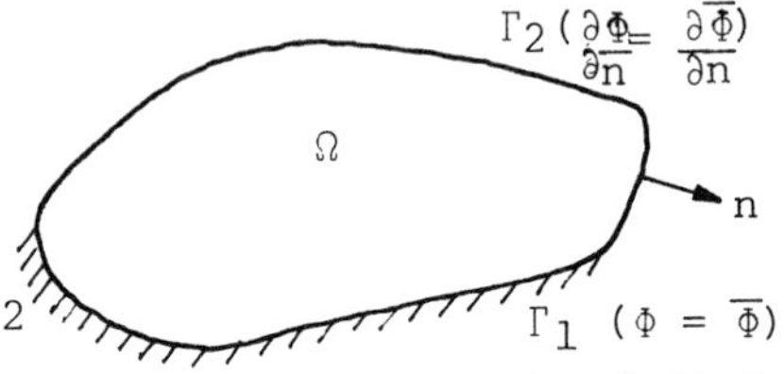

$\Gamma_2\left(\frac{\partial\Phi}{\partial n} = \frac{\partial\bar{\Phi}}{\partial n}\right)$ Ω n $\Gamma = \Gamma_1 + \Gamma_2$ Γ_1 $(\Phi = \bar{\Phi})$

Figure 1. Prob. definition

and the total boundary is given as $\Gamma = \Gamma_1 + \Gamma_2$

Introducing function ϕ^* for numerical solution, we can write,

$$\int_\Omega (\nabla^2\phi)\phi^* d\Omega = \int_{\Gamma_2}\left(\frac{\partial\phi}{\partial n} - \overline{\frac{\partial\phi}{\partial n}}\right)\phi^* d\Gamma - \int_{\Gamma_1}(\phi - \bar{\phi})\frac{\partial\phi^*}{\partial n}d\Gamma \tag{2}$$

Integrating by parts twice the Laplacian term **of** eqn.(2) we get:

$$\int_\Omega (\nabla^2\phi^*)\phi d\Omega = -\int_{\Gamma_2}\overline{\frac{\partial\phi}{\partial n}}\phi^* d\Gamma - \int_{\Gamma_1}\frac{\partial\phi}{\partial n}\phi^* d\Gamma + \int_{\Gamma_2}\phi\frac{\partial\phi^*}{\partial n}d\Gamma + \int_{\Gamma_1}\bar{\phi}\frac{\partial\phi^*}{\partial n}d\Gamma \tag{3}$$

Assuming a concentrated source at a point 'i', the governing equation is

$$\nabla^2\phi^* + \Delta^i = 0 \tag{4}$$

where Δ^i is the Dirac delta function for which,

$$\int_\Omega \phi\,(\nabla^2\phi^* + \Delta^i)d\Omega = \int_\Omega \phi\nabla^2\phi^* d\Omega + \phi^i \tag{5}$$

If eqn. (4) is to be satisfied by the fundamental solution, then

$$\int_\Omega \phi\,(\nabla^2\phi^*)d\Omega = -\phi^i \tag{6}$$

hence eqn.(3) becomes

$$\phi^i + \int_{\Gamma_1}\bar{\phi}\frac{\partial\phi^*}{\partial n}d\Gamma + \int_{\Gamma_2}\phi\frac{\partial\phi^*}{\partial n}d\Gamma = \int_{\Gamma_1}\frac{\partial\phi}{\partial n}\phi^* d\Gamma + \int_{\Gamma_2}\overline{\frac{\partial\phi}{\partial n}}\phi^* d\Gamma \tag{7}$$

If $\bar{\phi}$ and $\overline{\frac{\partial\phi}{\partial n}}$ values are known for respective parts of the boundary, we can write, in general, eqn. (7), for the total boundary, as,

$$c^i\phi^i + \int_\Gamma \phi\frac{\partial\phi^*}{\partial n}d\Gamma = \int_\Gamma \frac{\partial\phi}{\partial n}\phi^* d\Gamma \tag{8}$$

in which ϕ^* is the fundamental solution and c^i is the unknown coefficient.

If the said point 'i' is located outside the domain of the given problem, this coefficient c^i equals zero and eqn. (8) becomes

$$\int_\Gamma \phi \frac{\partial \phi^*}{\partial n}\, d\Gamma = \int_\Gamma \frac{\partial \phi}{\partial n}\, \phi^*\, d\Gamma \tag{9}$$

Eqn. (9) forms the basis of 'Regular Boundary Integrals', thereby giving a system of equations; one for each singular point corresponding to the boundary node under the boundary element discretization and located outside the domain of the problem.

2.1 Constant boundary elements

The boundary is discretized into N elements (say) and the values of potential and its derivative are assumed to be constant on each element, and equal to the value at the mid-node of the element. The singular point corresponding to this mid-node 'i' is located at an arbitrary distance (taken as L/2 in our case study, where L is the length of the element) from 'i' and along the positive normal. (fig.2).

Eqn. (9) for this singular point corresponding to the boundary node 'i' becomes in discretized form,

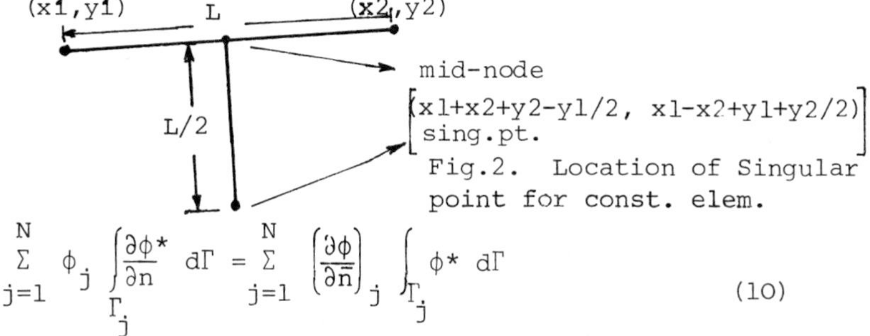

Fig.2. Location of Singular point for const. elem.

$$\sum_{j=1}^{N} \phi_j \int_{\Gamma_j} \frac{\partial \phi^*}{\partial n}\, d\Gamma = \sum_{j=1}^{N} \left(\frac{\partial \phi}{\partial n}\right)_j \int_{\Gamma_j} \phi^*\, d\Gamma \tag{10}$$

The integrals in eqn. (10) can be evaluated numerically for all segments over the boundary including the one containing the node 'i'. N such equations are obtained and solved for the boundary unknowns. Once the solution over the whole boundary is obtained, interior solution can be generated using eqn. (8) in which case c^i equals unity.

2.2 Linear elements

The variation of ϕ and $\frac{\partial \phi}{\partial \eta}$ is assumed to be linear within each element. For N elements, eqn. (9) can be written as,

$$\sum_{j=1}^{N} \int_{\Gamma_j} \phi \frac{\partial \phi^*}{\partial n}\, d\Gamma = \sum_{j=1}^{N} \int_{\Gamma_j} \frac{\partial \phi}{\partial n}\, \phi^*\, d\Gamma \tag{11}$$

The values of ϕ and $\frac{\partial \phi}{\partial \eta}$ at any point of an element can be written in terms of their nodal values and interpolation functions F_1 and F_2 as:

$$\phi\ (\xi) = F_1\ \phi_1 + F_2\ \phi_2$$

$$\frac{\partial\phi}{\partial n}\ (\xi) = F_1\ \left(\frac{\partial\phi}{\partial n}\right)_1 + F_2\ \left(\frac{\partial\phi}{\partial n}\right)_2 \qquad (12)$$

where ξ is the dimensionless co-ordinate $\xi = x/L/2$ (Fig.3) and

$$F_1 = 1/2(1-\xi) \qquad F_2 = 1/2(1+\xi) \qquad (13)$$

The singular point corresponding to a node 'i' is located at any arbitrary distance (taken as L/2 in the present study) from 'i' and along the positive normal. (Fig.4)

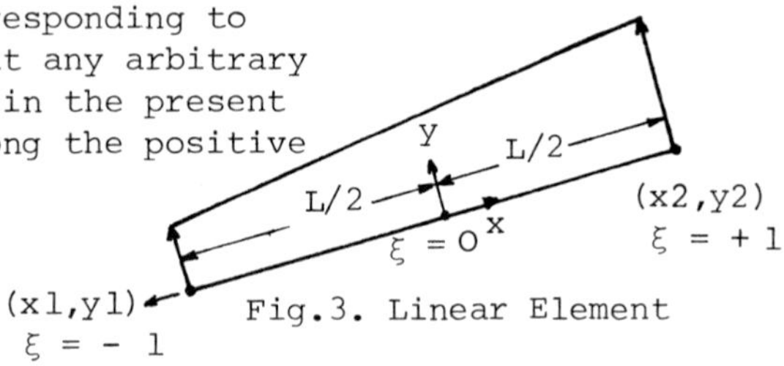

Fig.3. Linear Element

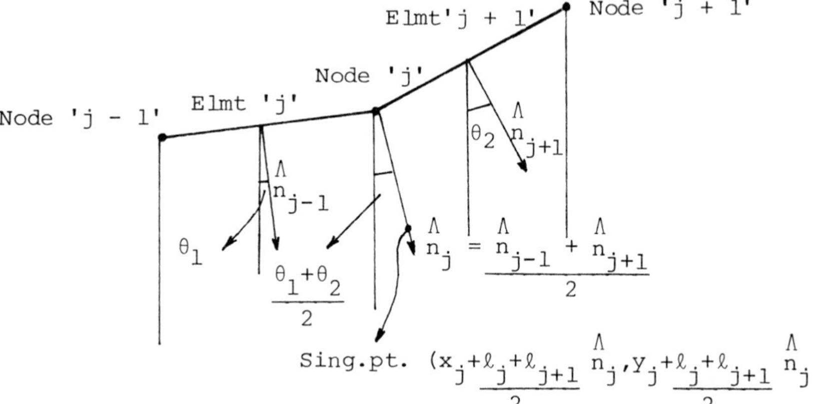

Fig.4. Location of sing.pt. for Linear Elmt.

Eqn.(11) for any singular point corresponding to the boundary node 'i', can now be written as:

$$\sum_{j=1}^{N} \int_{\Gamma_j} [F_1\ F_2]\ \frac{\partial\phi^*}{\partial \hat{n}} d\Gamma \begin{Bmatrix}\phi_1\\ \phi_2\end{Bmatrix} = \sum_{j=1}^{N}\ [F_1\ F_2]\phi^* d\Gamma \begin{Bmatrix}\partial\phi_1/\partial n\\ \partial\phi_2/\partial n\end{Bmatrix} \qquad (14)$$

A system of N 'Regular Boundary Integral Equations' can now be obtained and solved as for constant elements.

3 APPLICATIONS

Three 2-dimensional irrotational fluid flow test problems are analysed using constant and linear elements for both conventional and Regular Boundary Integral Methods and a critical comparison of the results is made.

3.1 Flow past a circular obstacle in a channel (Fig.5a)

This is a typical fluid flow problem solved using the Finite Element Method with 55 nodes and 80 elements [2] and 72 nodes and 110 elements [3]

The governing equation is:

$$\frac{\partial^2\psi}{\partial x^2} + \frac{\partial^2\psi}{\partial y^2} = 0 \qquad (15)$$

where ψ is the stream function.

Only a quarter of the domain is considered because of symmetry. Fig.5b shows the prescribed boundary conditions where $\frac{\partial\psi}{\partial\eta}$ are the velocities along the boundary. The problem is solved using 32 constant elements and 37 linear elements with two corner nodes near to each other. The boundary solutions for 'conventional' and'regular' boundary integral methods are shown in Fig.6a and Fig.6b for constant and linear elements respectively. Streamline ψ = 1, inside the domain is also plotted.

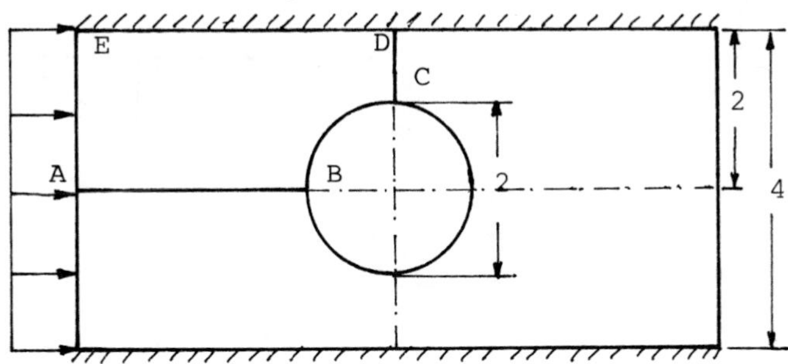

Fig. 5a. Flow past a circular obstacle in a channel

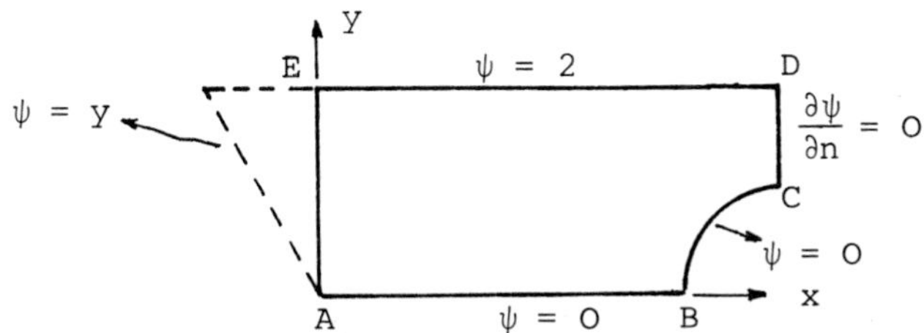

Fig.5b. Boundary Conditions

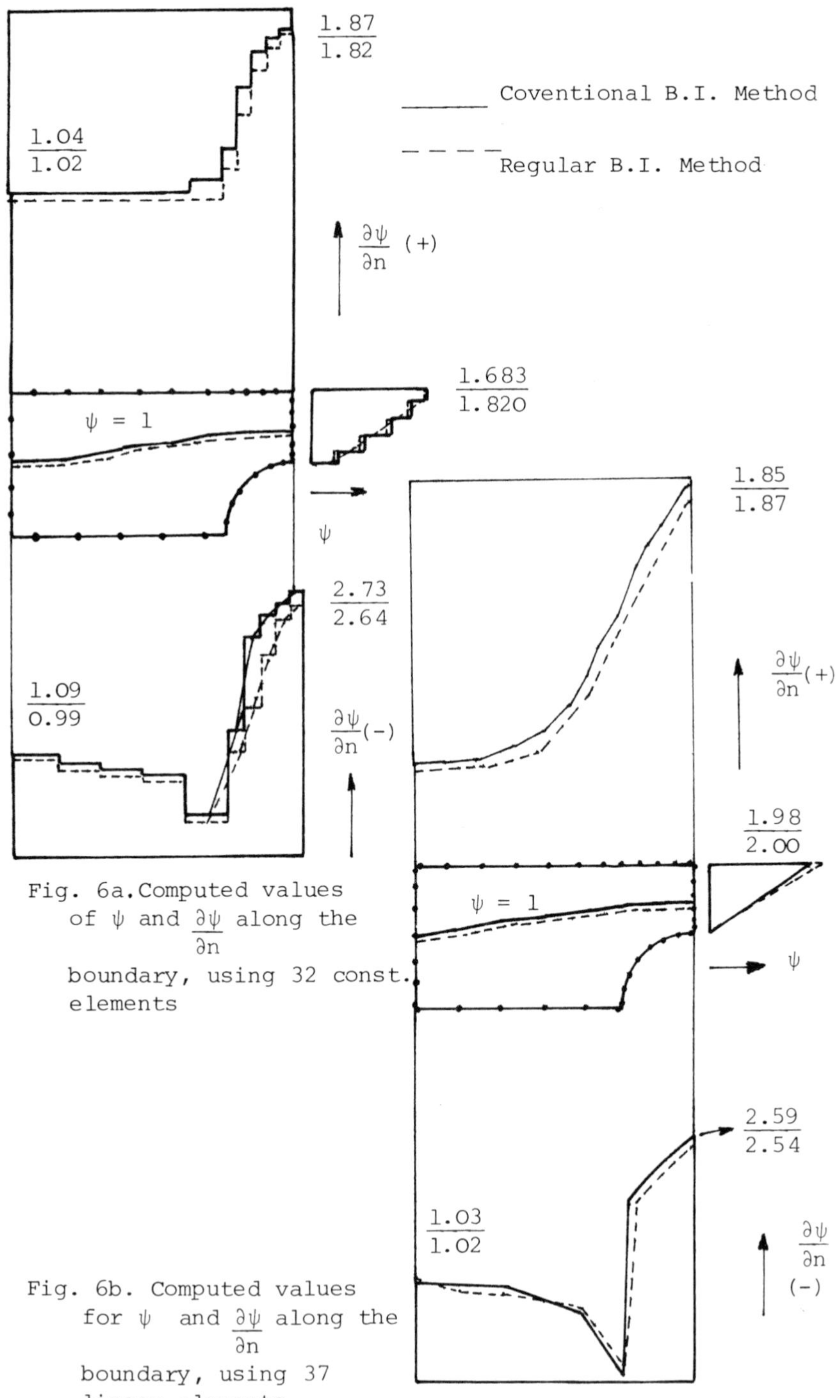

Fig. 6a. Computed values of ψ and $\frac{\partial \psi}{\partial n}$ along the boundary, using 32 const. elements

Fig. 6b. Computed values for ψ and $\frac{\partial \psi}{\partial n}$ along the boundary, using 37 linear elements.

3.2. Flow past an aerofoil

A symmetric aerofoil (NACA 0018) in uniform flow is shown in fig.7. Because of symmetry, only half of the aerofoil need be considered.

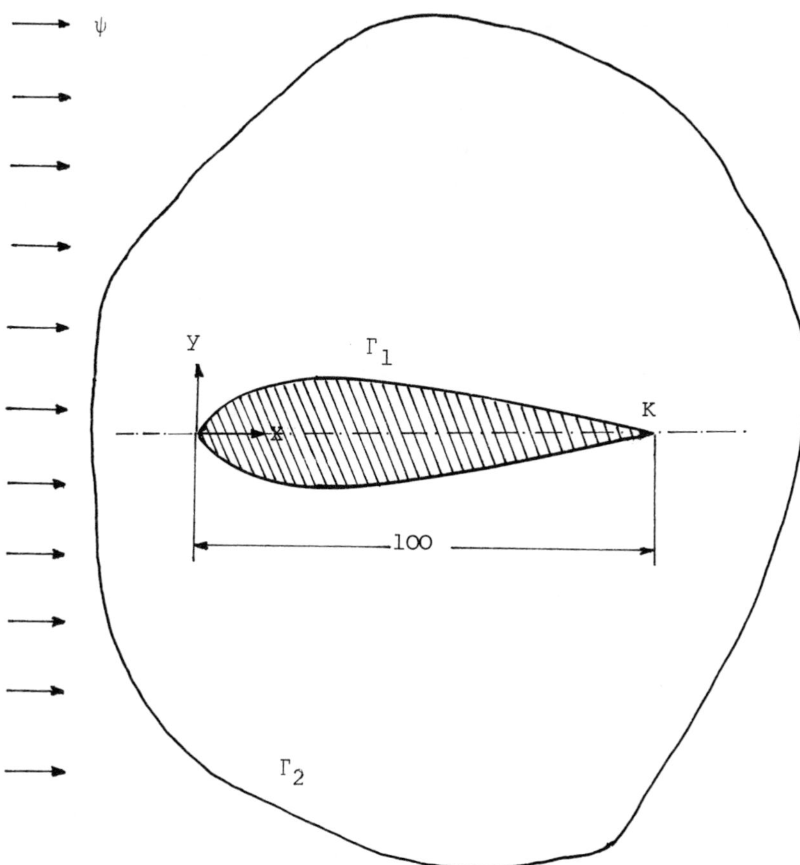

Fig. 7. NACA (0018) Aerofoil

Let Γ_2, the outer boundary, be taken remote from the aerofoil, Γ_1. If the stream function is ψ the velocity components (u,v) are $\left(\frac{\partial\psi}{\partial y}, -\frac{\partial\psi}{\partial n}\right)$. For uniform horizontal flow on Γ_2 (u,v) = (U,o). Thus if $\psi = Uy$ on Γ_2 and on the symmetry line (y=o) and $\psi = o$ on Γ_1, the remote flow is uniform and point k is a stagnation point (Kutta condition [4]).

The results for the tangential velocity U_t obtained using linear elements for conventional and regular boundary integrals are shown in Table 1., along with the analytical solution by NACA

		u_t/U	
x	Analytical	Conventional B.I. Method	Regular B.I.M.
0.0	0.000	0.000	0.000
1.25	0.926	0.932	0.932
2.50	1.103	1.096	1.101
5.00	1.228	1.217	1.219
7.50	1.267	1.262	1.254
10.00	1.276	1.271	1.267
15.00	1.278	1.273	1.279
20.00	1.275	1.269	1.272
25.00	1.262	1.252	1.258
30.00	1.247	1.241	1.240
40.00	1.205	1.199	1.209
50.00	1.157	1.161	1.159
60.00	1.116	1.121	1.119
70.00	1.074	1.091	1.081
80.00	1.025	1.052	1.034
90.00	0.966	0.987	0.979
95.00	0.914	0.951	0.925
100.00	0.000	0.000	0.000

Table 1. Comparison of tangential velocities for NACA aerofoil

3.3 Flow past a plate in a channel

Let the flow in the channel be at normal incidence to the flat plate and represented by stream function ψ (Fig.8a). Only a quarter of the domain is considered for analysis due to symmetry. Boundary conditions are shown in Fig.8b. An infinite speed will be acquired by the stream at point O, the edge of the plate [5], giving rise to a singularity in the mathematical solution.

Results obtained for both conventional and regular Boundary Integrals, using 26 constant elements and 32 linear elements are shown in Figures 9a. and 9b. respectively. Streamlines for $\psi = 1, 2$ are also plotted for both cases to make a comparison for the interior solution.

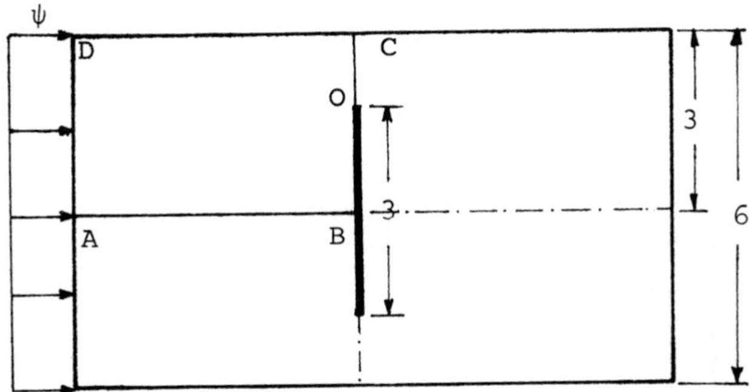

Fig.8a. Flow past a plate in a channel

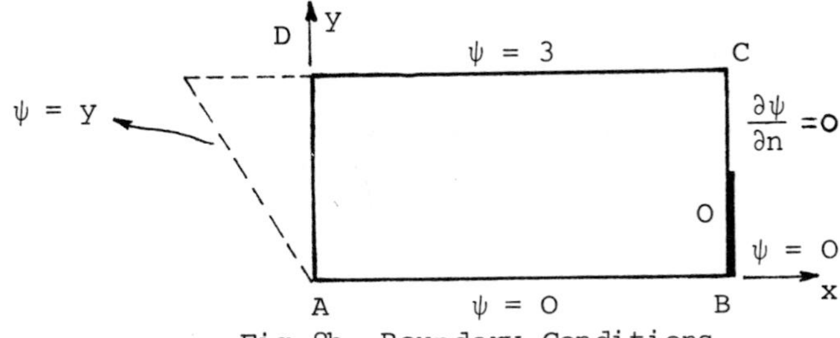

Fig.8b. Boundary Conditions

DISCUSSION AND CONCLUSIONS

In this paper a Regular Boundary Integral Method has been presented. Here the singularity of the fundamental solution is taken outside the domain of the problem giving an infinite system of regular integral equations. On discretizing this system, in the usual manner, the resulting kernels are everywhere regular over the boundary so that no special attention to singular integrends is required. The class of problems to which the usual method may be applied is limited by the requirement that the singular boundary integrals do not diverge, the corresponding limitation of the regular method is substantially weaker because the kernels are everywhere regular on the boundary.

The case study results presented here, for harmonic fluid flow problems, show that the conventional and regular methods have similar convergence characteristics. On the boundary, variations of around 2% were found, using constant and linear elements, but the computed interior results were in closer agreement. The benchmark solution of the NACA aerofoil is in good agreement with both methods, but favours the regular method slightly. The solution is singular in the case of flow past a plate, nevertheless both methods perform well in the neighbourhood of the singularity, the regular method again being marginally better.

A systematic study was undertaken of

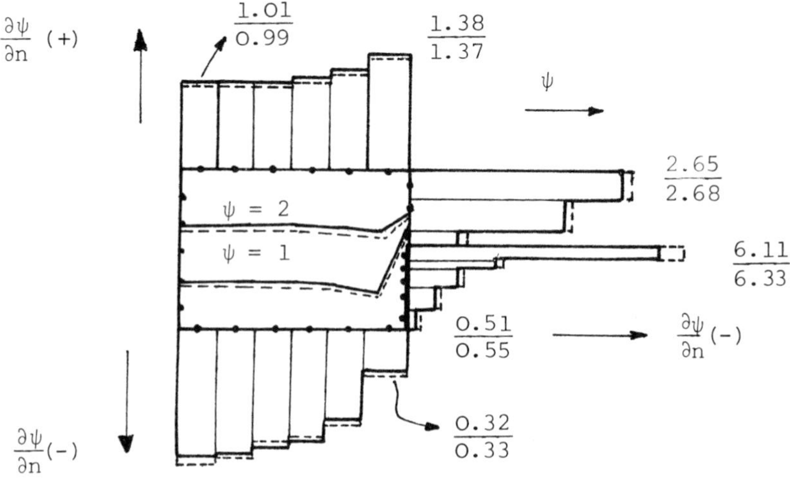

Fig. 9a. Computed values of ψ and $\frac{\partial\psi}{\partial n}$, along the boundary, using 26 constant elements.

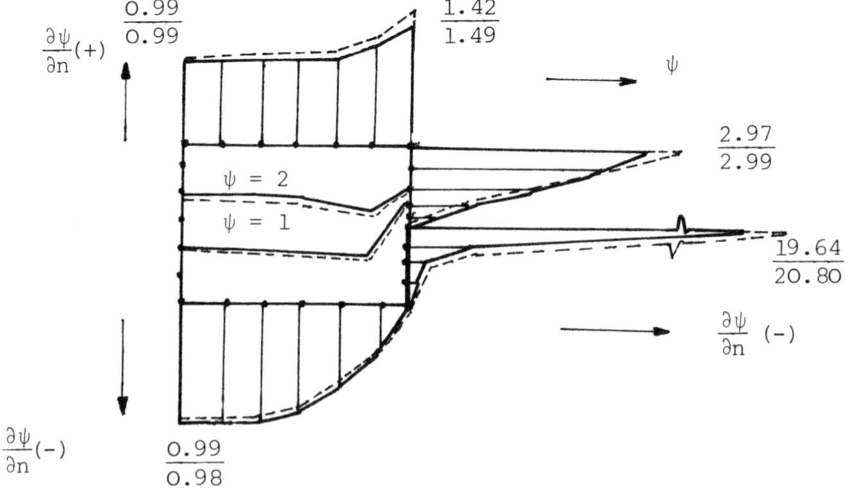

Fig. 9b. Computed values of ψ and $\frac{\partial\psi}{\partial n}$ along the boundary, using 32 linear elements.

the best location of the singular point of the fundamental solution. As expected the quality of solution deteriorated as the singularity was moved away from the domain of the problem. Deterioration was also observed as the singularity was brought close to the boundary. Best results were obtained with the singularity at one half the element length away from the element along the outward normal to a node.

Clearly, since no special attention is needed to handle singular kernels, and integrends are more slowly varying, the integrals in the regular method may be accurately determined at less computational cost than in the usual method.

In conclusion : a regular boundary integral method has been presented and applied to harmonic fluid flow problems; in applications close agreement with established results and with the conventional boundary element method has been observed; marginally better results were obtained using the regular, rather than the conventional, method at lower computational cost.

REFERENCES

(1) BREBBIA, C.A. and WROBEL, L.C. Steady and Unsteady Potential Problems using the Boundary Element Method, Recent Advances in Numerical Methods in Fluids, Ed. Taylor, C. and Morgan, K., Pineridge Press, 1980.

(2) SEGERLIND, L.J. Applied Finite Element Analysis, John Wiley & Sons, New York, 1976.

(3) MARTIN, H.C. Finite Element Analysis of Fluid Flows, Matrix Methods in Structural Mechanics, AFFDL TR 68-150, Patterson Air Force Base, 1969.

(4) DeVRIES, G. and NORRIE, D.H. The Application of Finite Element Technique to Potential Flow Problems, J., of Applied Mechanics, Vol.38, pp.778-802, 1971.

(5) MILNE THOMPSON, L.M. Theoretical Hydrodynamics, Macmillan, London, 1968.

ACCURATE SOLUTIONS FOR LAMINAR AND TURBULENT BOUNDARY LAYERS AT VERY LARGE PRESSURE GRADIENT PARAMETERS

A. Postan*, O. Pade*, D. Anshelovitz* and M. Wolfshtein**

SUMMARY

Numerical solutions are obtained for non-similar laminar and turbulent compressible and incompressible boundary layers at large favourable pressure gradients. The Lees-Dorodnytzin transformation is used, together with mesh stretching and an OCI fourth order scheme in the lateral direction. An apparent three level second order scheme is applied in the longitudinal direction. Iterations are used to ensure full implicitness of the non linear marching, and the self-similar laminar solution serves as an initial guess. The method allows good accuracy at a relatively small number of mesh points, despite the steepness of the velocity profile near the wall. Agreement with previous solutions and experimental data is good.

---

* Ministry of Defence, Scientific Department, P.O. Box 2250, Haifa, Israel.

** Technion, Israel Institute of Technology, Haifa, Israel.

1. INTRODUCTION

This work is concerned with compressible and incompressible flows at large favourable pressure gradients. We use the Lees-Dorodnytzin [1,2,3] parameter $\beta=(2\xi/M_\infty)(dM_\infty/d\xi)$ as a measure of the pressure gradient, rather than the Stewartson [4] definition for β. The Stewartson definition appears to be inapplicable for compressible laminar flows with $\beta > 2$, as his similarity transformation cannot be transformed back to the physical plane in such cases.

For example, Christian et al.[5] obtained self-similar solutions in the range of $2<\beta\leq 20$ but the momentum thickness could not be calculated by their method for these values of β, because it was not possible to calculate the distance from the wall, y. To overcome these problems we used a modified Lees-Dorodnytzin transformation [1,2,3]

$$\xi = \xi_0 + \int_0^x \rho_\infty \mu_\infty u_\infty dx$$

$$\eta = \frac{1}{\delta(x)} \int_0^x \frac{\rho}{\rho_\infty} dy \tag{1.1}$$

$$\delta(x) = \sqrt{2\xi}/\rho_\infty u_\infty$$

where the constant ξ_0 is defined by $\xi_0 = 0.5\beta\ M_{\infty,0}/(dM_\infty/d\xi)_0$ and $M_{\infty,0}$ is the edge Mach number at ξ_0. It should be noted that in the commonly used form of the Lees-Dorodnytzin transformation $\xi_0 = 0$.

The transformation (1.1) is used to evaluate incompressible and compressible laminar and turbulent, boundary layer flows at arbitrarily large values of β. Two main problems are encountered in such cases: First, the velocity and temperature profiles in the transverse direction are very steep and second, the longitudinal variations are very rapid. These characteristics cause great difficulties in solving the boundary layer equations by numerical methods, and especially so when high accuracy is required. As a result, a very fine mesh must be used in both the longitudinal and the transverse direction.

Efficient numerical algorithms for the solution of incompressible and compressible laminar and turbulent boundary layer equations, are available [6,7,8] e.g. the method of Patankar and Spalding [6,7], the method of Cebeci and Smith [8]. The latter approach is based on the Keller's box scheme [8,9,10]. These methods are based on finite difference techniques of second order accuracy in the transverse direction and first or second order accuracy in the longitudinal direction. Although these methods are very efficient, their application to problems with large pressure gradient requires very fine meshes, increasing the computation time as well as the price of a computer run, and causing accumulation of round-off errors.

In this work we developed an accurate method of fourth order in the transverse direction and of second order in the longitudinal direction. The fourth order method is based on the three points OCI schemes [11,12,13,14,15]. Further improvement in accuracy is achieved by mesh stretching in the transverse direction near the solid walls, where the velocity gradient becomes very steep for large β. The method is fully implicit and therefore iterations are required at each step. To increase the efficiency the initial guess for the laminar cases is taken as the self similar solution of the local β.

It turns out that because of the large acceleration (at large pressure gradient) the similar solution at the new station is a better initial approximation to the accurate solution than the numerical solution at the previous station.

The turbulent viscosity is described by the mixing length model. We did not attempt to improve the turbulence model, as our main objective was to improve the numerical methods of solution.

The mathematical problem is formulated in section 2. The numerical method is outlined in section 3. Comparison with previous solutions, new results and discussion appear in section 4.

2. MATHEMATICAL FORMULATION

2.1 The governing equations

The continuity equation for a two dimensional steady flow is:

$$\frac{\partial}{\partial x}(\rho u) + \frac{\partial}{\partial y}(\rho v) = 0 \tag{2.1}$$

the momentum equation (in the x direction) is

$$\frac{\partial}{\partial x}(\rho uu) + \frac{\partial}{\partial y}(\rho vu) = -\frac{\partial p}{\partial x} + \frac{\partial}{\partial y}[(\mu_e \frac{\partial u}{\partial y})] \tag{2.2}$$

and the energy equation is

$$\rho u \frac{\partial H}{\partial x} + \rho v \frac{\partial H}{\partial y} = \frac{\partial}{\partial y}[\mu_e(1-\frac{1}{P_r}) u\frac{\partial u}{\partial y} + \frac{\mu_e}{P_r}\frac{\partial H}{\partial y}] \tag{2.3}$$

Transform to the (ξ,η) coordinates (see eq.(1.1)) yields the momentum equation

$$(c\hat{\varepsilon}_m f'')' + ff'' + \beta(g-f'^2) = 2\xi(f'\frac{\partial f'}{\partial \xi} - f''\frac{\partial f}{\partial \xi}) \tag{2.4}$$

and the energy equation

$$(c\hat{\varepsilon}_h g')' + fg' = 2\xi(f'\frac{\partial g}{\partial \xi} - g'\frac{\partial f}{\partial \xi}) \tag{2.5}$$

where we assumed $P_r = 1$ and $P_{r_t} = 0.9$ (P_r and P_{r_t} are the laminar and turbulent Prandtl numbers respectively) although these numbers may assume other values (especially P_{r_t} which may be variable).

The new non-dimensional quantities are

$f' = u/u_\infty$

$g = H/H_\infty$

$c = \rho\mu/\rho_\infty\mu_\infty$

$\beta = (2\xi/M_\infty)(dM_\infty/d\xi)$

Equations (2.4) and (2.5) form a set of two coupled non-linear equations. We solve them for f' and g, using a marching finite-difference scheme, as explained below. The normalized stream function, f, is obtained by quadrature of f'.

2.2 The turbulence model

The mixing length model used here is based on the formulation of Cebeci and Smith [8] who use

$$\hat{\varepsilon}_m = 1 + \frac{\rho \nu_T}{\mu} = \frac{\mu_e}{\mu} \tag{2.6}$$

where $\nu_T = \varepsilon_{in}$ in the inner sublayer and $\nu_T = \varepsilon_{out}$ in the outer range. ε_{in} and ε_{out} are calculated by the following equations:

$$\varepsilon_{in} = L^2 \left|\frac{du}{dy}\right| \gamma_{tr}. \tag{2.7}$$

$$\varepsilon_{out} = \bar{\alpha} u_\infty |\delta_k^*| \gamma_{tr}. \tag{2.8}$$

where

$L = 0.41y(1-e^{-y/A})$ is the modified Van Driest mixing length,

$$A = 26 \frac{\mu}{\rho N_c} \left(\frac{\tau_s}{\rho}\right)^{-\frac{1}{2}} , \tag{2.9}$$

$$\tau_s = \mu_s (du/dy) , \tag{2.10}$$

$$N_c = [1-11.8 \frac{\mu_s}{\mu_\infty} \left(\frac{\rho_\infty}{\rho_s}\right)^2 p^+]^{\frac{1}{2}} , \tag{2.11}$$

$$p^+ = \frac{\nu_\infty u_\infty}{u_\tau^2} \left(\frac{du_\infty}{dx}\right) , \quad u_\tau = \left(\frac{\tau_s}{\rho_s}\right)^{\frac{1}{2}} ,$$

$$\gamma_{tr} = [1-e^{-(G/u_\infty)(x-x_{trans.})^2}] , \tag{2.12}$$

$$G = \frac{3}{G^2} \frac{u_\infty^2}{\nu^2} Re_{x_{trans.}}^{-1.34} ,$$

$$C = 60 + 4.86 M_\infty^{1.92} , \tag{2.13}$$

x_{trans} and $Re_{x_{trans}} = \rho_\infty (u_\infty x_{trans.})/\mu_\infty$ are determined by assuming $Re_{\theta_{trans.}} = 450$

where

$$\theta = \int_0^\infty (\rho u/\rho_\infty u_\infty)(1-u/u_\infty)\, dy.$$

The other empirical functions are:

$$\bar{\alpha} = 0.02604/(1 + \Pi) \tag{2.14}$$

$$\Pi = 0.55 [1-e^{-0.243\sqrt{z_o}} - 0.298 z_o] \tag{2.15}$$

and

$$z_o = (Re_{\theta_k}/425) - 1 \qquad (2.16)$$

The kinematic thicknesses θ_k and δ_k*are given by:

$$\theta_k = \int_0^\infty \frac{u}{u_\infty}\left(1 - \frac{u}{u_\infty}\right) dy \qquad (2.17)$$

$$\delta_k^* = \int_0^\infty \left(1 - \frac{u}{u_\infty}\right) dy \qquad (2.18)$$

and

$$Re_{\theta_k} = \frac{u_\infty \theta_k \rho_s}{\mu_s} \qquad (2.19)$$

$\hat{\varepsilon}_h$ which appears in the transformed energy equation (2.5) is similar to $\hat{\varepsilon}_m$ and is given by

$$\hat{\varepsilon}_h = 1 + \rho \frac{\nu_T}{Pr_t \cdot \mu} \quad (Pr_t = 0.9) \qquad (2.20)$$

2.3 Boundary conditions

The variables f' and g are governed by second order parabolic equations. The boundary conditions for these variables are

at	$\eta = 0$	$f' = 0$,	$g = T_w/T_{o,\infty}$
at	$\eta \to \infty$	$f' = 1$	$g = 1$

For the fourth order method higher derivatives are required on the boundaries. On the outer boundary we assumed

$$f''_\infty = g'_\infty = 0$$

On the surface these derivatives were computed numerically, as outlined below.

2.4 Coordinate transformations

If the wall temperature is lower than the free stream temperature, the transverse coordinate transformation causes the mesh to become coarser in the vicinity of the wall. This happens as $\rho \propto T^{-1}$ and by (1.1):

$$d\eta = \frac{1}{\delta(x)} \frac{\rho}{\rho_e} dy \ . \qquad (2.23)$$

This situation raises an additional difficulty in solving equations (2.4) and (2.5) by a finite difference scheme, especially with large pressure gradient, as in such cases the velocity and the temperature profiles become very steep.

To overcome this difficulty we used an additional transformation, viz.

$$\zeta = \frac{\eta}{A + \eta} . \qquad (2.24)$$

The constant A is so chosen as to ensure that a given number of mesh points m falls within a region $0 \leq \eta \leq \eta_{in}$, where the steepest gradients are expected. Then

$$A = \frac{N - m}{m} \quad \eta_{in}$$

where N is the total number of mesh points in the lateral direction.

Simple algebra shows that the stretching is limited by the number of mesh points as follows

$$\frac{\eta_N - \eta_{N-1}}{\eta_1} = \frac{N}{2} (N - 1)$$

with $\eta_N = \eta_\infty$ and N is the number of mesh points.

3. THE NUMERICAL METHOD

The solution is obtained by a second order accurate marching procedure, in the x direction, and a fourth order procedure for the transverse coordinate. These two procedures are described below.

3.1 Axial direction

A second order expression for the first derivative in the x direction is evaluated by a 3-level formula. For a continuous first derivative this is

$$\left(\frac{\partial \Psi}{\partial x}\right)_{x=x_n} = - \left(\frac{\partial \Psi}{\partial x}\right)_{x=x_{n-1}} + 2 \frac{\Psi_n - \Psi_{n-1}}{x_n - x_{n-1}} + O(h_n{}^2) \qquad (3.1)$$

where $h_n = x_n - x_{n-1}$.

3.2 Lateral direction

The lateral derivatives of a function v are evaluated by a fourth order method due to Kreiss [11]

$$\frac{v_{i+1} - v_{i-1}}{2h} = \frac{v'_{i+1} + 4v'_i + v'_{i-1}}{6} \qquad (3.2)$$

$$\frac{v_{i+1} - 2v_i + v_{i-1}}{h^2} = \frac{v''_{i+1} + 10v''_i + v''_{i-1}}{12} \tag{3.3}$$

Both the energy and momentum equations have the same formal description

$$(c\hat{\varepsilon}v')' + (f+2\xi \frac{\partial f}{\partial \xi})v' + J\beta(g-f'v) = 2f'\xi \frac{\partial v}{\partial \xi} \tag{3.4}$$

where J=1 and v = f' for the momentum equation and J=0 and v=g for the energy equation. For the momentum equation $\hat{\varepsilon}$ is replaced by $\hat{\varepsilon}_m$ and for the energy equation, by $\hat{\varepsilon}_h$.

By discretization we obtain

$$c\hat{\varepsilon}v_j'' + [c_j'\hat{\varepsilon}_j + c_j\hat{\varepsilon}_j' + f_j + 2\xi(\frac{\partial f}{\partial \xi})_j]v_j' - f_j'(J\beta + 2\xi/\Delta\xi)v_j$$

$$= - J\beta g_j - 2\xi f_j'[(\frac{\partial v}{\partial \xi})_{j-1} + 2 \frac{v_{j-1}}{\Delta\xi}] \tag{3.5}$$

By substitution of (3.5) in (3.3) we get a system of two equations in v and v'. This is a block-tri-diagonal system which can be easily solved by standard algorithms. As equation (3.5) represents two non-linearly coupled equations, the tri-diagonal solver is applied iteratively until convergence is obtained.

3.3 Boundary conditions

The values of u and T on the wall and at the edge of the boundary layer are given. For the procedure described above f_s'', f_∞'', g_s', and g_∞' are required. From eq. (2.21) and (2.22) we have

$$f_\infty'' = g_\infty' = 0$$

The values of f_s'' and g_s' are obtained from eq. (3.5) to fourth order accuracy. The expressions are given below.

$$f_s'' = [16\, f_1' - f_2' + 6h^2 \frac{\beta g_s}{\alpha_s} + \frac{4h^3}{3} \frac{\beta}{\alpha_s} (g_s' - 2 \frac{\alpha_s' g_s}{\alpha_s}] \frac{1}{D} + O(h^4)$$

$$g_s' = [16g_1 - g_2 - 15g_s - \frac{8h^3}{3\alpha_s} f_s'' \xi g_\xi] \frac{1}{D} + O(h^4)$$

$$D = 14h - 6h^2 \frac{\alpha_s'}{\alpha_s} + \frac{4}{3} \frac{h^3}{\alpha_s} (\frac{2\alpha_s'}{\alpha_s} - \alpha_s'')$$

$$\alpha = c\, \nu_e$$

The functions β, ξ, have already been defined, and h is the increment in ζ.

3.4 Description of the algorithm

Equation (3.5) represents a system of two coupled non-linear equations, which is solved iteratively. The two equations are first iterated to resolve the linear coupling between them. Then all the non-linear coefficients are updated, and the linear iterations are renewed. The inner iterations are stopped when the condition $||v_n - v_{n-1}||_\infty < \sigma_1$ is satisfied, where $\sigma_1 \approx 10^{-7}$. The outer iteration is stopped when the residue of eq. (3.5) is smaller than σ_2 where $\sigma_2 \approx 10^{-3}$.

3.5 Initial values for the iterative process

At low pressure gradient, when moving downstream from X^k to X^{k+1} we calculate an initial guess of the coefficients using the value of the functions g and f at X^k. However, at the presence of large pressure gradient, it turns out that the coefficients and the values of $\frac{\partial f}{\partial \xi}$ and $\frac{\partial f'}{\partial \xi}$ are better approximated by the self-similar laminar solution for the local β, obtained from tabulated solutions.

3.6 Numerical integrations

All the integrals for evaluation of parameters as θ, δ^* are computed by the Simpson's rule. To compute it we use the following method: Since f' and f" are known, we apply (3.2) and (3.3) to compute f_1 and f_2 and then by (3.2) compute f_i for $i \geq 3$ (i denotes here the points along the lateral direction, $i = 0$ denotes the point at the wall).

4. RESULTS AND CONCLUSIONS

The procedure described above was applied to a number of problems, in order to check the accuracy and scope of the method. Fig. 1a and 1b show Richardson extrapolations for f_s'' and $f'(\eta=0.8)$. The slopes are proportioned to h^4 as expected. A comparison of the results for flat plate laminar boundary layer with those of ref.[5] is given in fig. 2. The agreement is very good. Fig. 3 shows a comparison with the experimental data of Lewis [16] for compressible turbulent boundary layers. The agreement is good. Figs.4 and 5 illustrate high pressure gradient runs for laminar and turbulent flows. It is seen that the method can produce reliable results, at moderate size grids.

REFERENCES

1. DORRANCE, W.H. - <u>Viscous Hypersonic Flow</u>, McGraw-Hill, New York, 1962.

2. SCHLICHTING, H. - <u>Boundary Layer Theory</u>, McGraw-Hill, New York, 1978.

3. PADE, O., POSTAN, A. and ANSHELOVITZ, D. - Revision of Common Transformations of Boundary Layer Equations, Submitted for publication

4. STEWARTSON, K. - <u>The Theory of Laminar Boundary Layers in Compressible Flows</u>, Oxford University Press, 1965.

5. CHRISTIAN, J.W., HANKY, W.L. and PETTY, J.S. - ARL 70-0023 AD 705581, 1970.

6. CRAWFORD, M.E. and KAYS, W.M. - Nasa CR-2742, 1972.

7. PATANKAR, S.V. and SPALDING, D.B. - <u>Heat and Mass Transfer in Boundary Layers</u>, Intertext Books, London, 1971.

8. CEBECI, T. and SMITH, A.M.O. - <u>Analysis of Turbulent Boundary Layers</u>, Academic Press, New York, 1974.

9. KELLER, H.B. and CEBECI, T. - Accurate Numerical Methods for Boundary Layer Flows - I. Two-Dimensional Laminar Flows, Lecture Notes in Physics, Vol.8, Proceedings of the Second International Conf. on Numerical Methods in Fluid Dynamics, p. 92, Springer-Verlag, New York, 1971.

10. KELLER, H.B. and CEBECI, T. - Accurate Numerical Methods for Boundary Layer Flows - II. Two Dimensional Turbulent Flows, AIAA J., Vol.10, p.1193, 1972.

11. KREISS, H.O. - Methods for the approximate Solution of Time Dependant Problems, GARP Report No.13, Geneva, 1975.

12. KRAUSE, E., HIRSCHEL, E.H., and KORDULLA, W. - Fourth Order "Mehrstellen"-Integration for Three-Dimensional Turbulent Boundary Layers, Computers and Fluids, Vol.4, pp.77-92, 1976.

13. HIRSH, R.S. - Higher Order Accurate Difference Solutions of Fluid Mechanics Problems by a Compact Differencing Technique, J. of Computational Physics, Vol.19, pp.90-109, 1979.

14. ADAM, Y. - Highly Accurate Compact Implicit Methods and Boundary Conditions, J. of Computational Physics, Vol.24, pp.10-22, 1977.

15. CIMENT, M., LEVENTHAL, S.H., and WEINBERG, B.C. - The Operator Compact Implicit Method for Parabolic Equations, J. of Computational Physics, Vol.28, pp.135-166, 1978.

16. LEWIS, J.E., GRAN, R.L. and KUBOTA, T. - An experiment on the adiabatic compressible turbulent boundary layer in adverse and favourable pressure gradients, J. Fluid Mech., Vol. 51, pp.657-672, 1972.

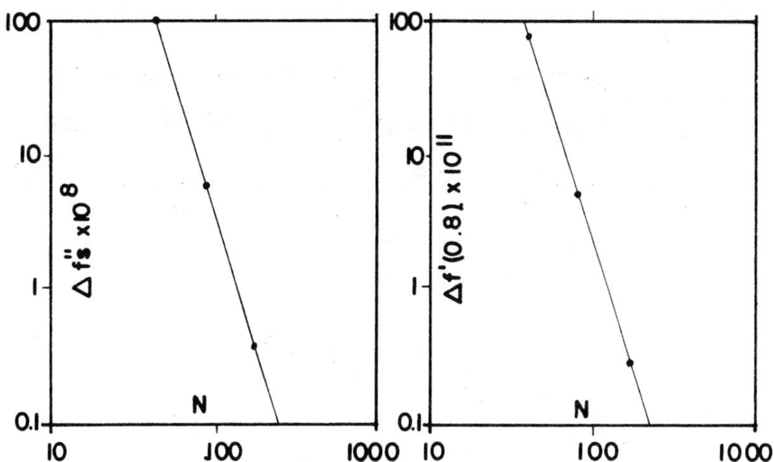

Fig.1: Richardson estrapolation in the y direction for (a) f''_s and (b) f' (0.8); laminar similar solution, β=0, M_∞=0.1, no stretching; (Δ is the difference between the exact and computed values).

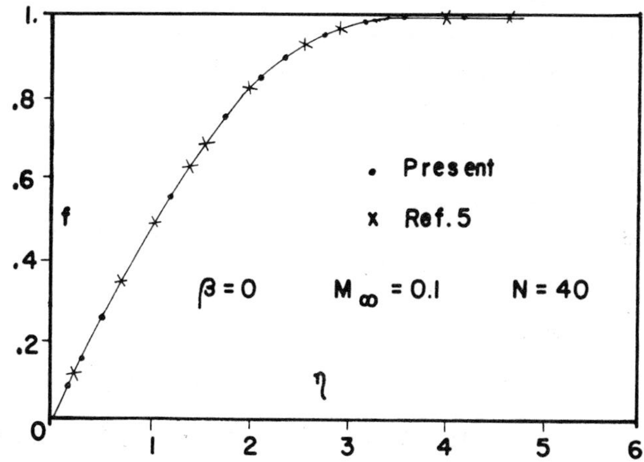

Fig.2: Laminar flat plate velocity profiles; similar solution.

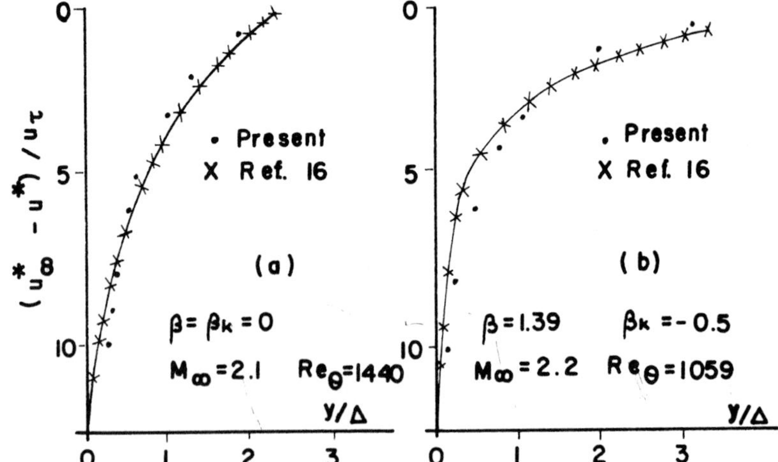

Fig.3: Turbulent velocity profiles; adiabatic flow; u^*, β_k and Δ are as in ref. 16

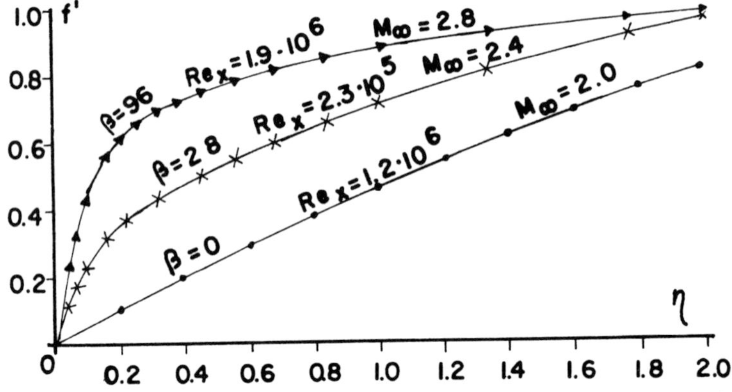

Fig.4: Laminar velocity profiles, g_s=0.5, A=0.427, N=20.

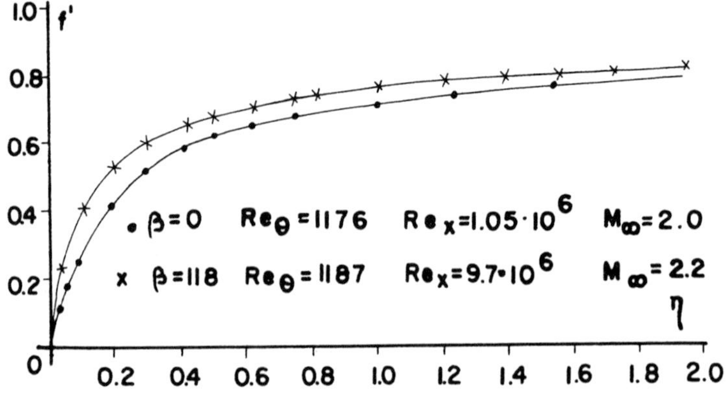

Fig.5: Turbulent velocity profiles, g_s=0.5, A=0.724, N=40.

THE NUMERICAL SIMULATION OF THE TURBULENT BOUNDARY LAYERS AT A ROUGH, AIR-WATER INTERFACE

Stephen P. Klotz[i]

Robert L. Street[ii]

SUMMARY

Coupled gas-liquid boundary-layer problems have direct engineering relevance to the design and operation of industrial process equipment and are important in meteorological and geophysical applications requiring knowledge about the transport processes at the air-ocean interface. The boundary-layer equations of heat, mass, and momentum in turbulent flow above and below an air-water interface roughened by wind-waves are solved numerically using the efficient and versatile Keller Box scheme and compared with experimental data acquired in a laboratory research facility.

1. INTRODUCTION

Gas-liquid boundary-layer flows with heat and mass transfer comprise an important class of problems for which there is a dearth of experimental data or theoretical analysis. Generally speaking, previous investigations have focused on either the gaseous or liquid regimes separately and have disregarded the possible interactions of one fluid on the other by treating boundary conditions at the common interface as known or capable of being specified. Indeed, the work of Lock [1], Kotake [2], and Schröppel and Thiele [3], each of whom considered only laminar systems, and the work of Street [4], who considered only the turbulent transport of heat and water-vapor across an air-water interface, represent the available literature on coupled boundary-layer flows.

---

[i]Research Assistant, Department of Civil Engineering, Stanford University, U.S.A.

[ii]Professor, Department of Civil Engineering, Stanford University, U.S.A.

In another paper in this volume we considered the coupled problem in the context of laminar flow at a smooth, phase-changing interface. In this paper we extend the analysis to the turbulent case and present solutions to the transport problem at an air-water interface with surface waves. Solutions are obtained numerically using the efficient Box scheme of Keller and Cebeci [5] and compared with experimental data acquired in the Stanford Wind, Water-Wave Research Facility (SWWWRF).

2. BASIC EQUATIONS

We consider the forced-convection of a binary mixture of dry air and water vapor over water (Figure 1). The turbulent flow in each fluid regime is assumed to be two-dimensional and steady in the mean and to have constant-properties. Because the air consists of a two-phase mixture we treat enthalpy, rather than temperature, and vapor concentration as conserved properties. A free-stream pressure gradient in the air may exist but we neglect any mean current far beneath the interface. Wind-waves on the water surface will be regarded as interfacial roughness which must be specified by, say, the mean wave height of all frequency components of the wave field as a function of fetch (downstream distance) in the boundary-layer flow.

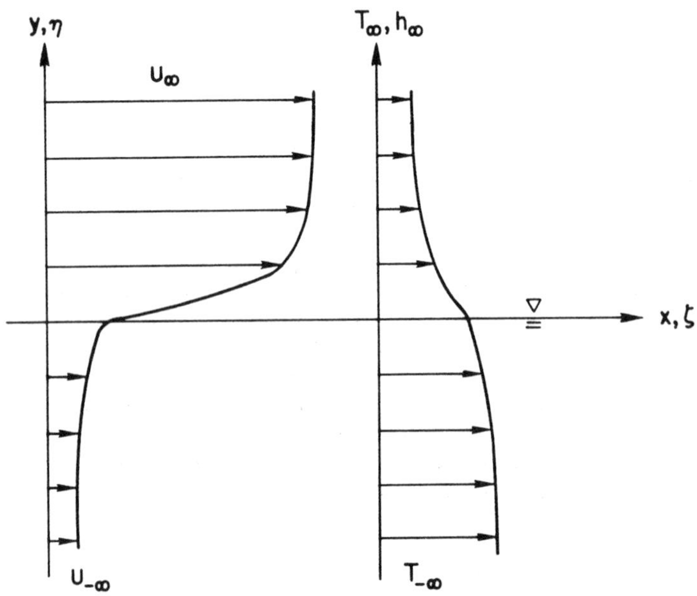

Figure 1. The coupled gas-liquid problem.

The differential equations governing the mean flow which we must solve are directly analogous to those of [6] with the inclusion of the appropriate turbulence terms. In stream function coordinates these equations are (subscripts 1 and 2 refer, respectively, to the air and aqueous regimes):

(air:momentum)

$$\frac{\partial}{\partial \eta_1}[(1 + \varepsilon^+_{M_1}) \frac{\partial^2 f_1}{\partial \eta_1^2}] + \frac{m+1}{2} f_1 \frac{\partial^2 f_1}{\partial \eta_1^2} + m[1 - (\frac{\partial f_1}{\partial \eta_1})^2]$$

$$= \zeta(\frac{\partial f_1}{\partial \eta_1} \frac{\partial^2 f_1}{\partial \zeta \partial \eta_1} - \frac{\partial f_1}{\partial \zeta} \frac{\partial^2 f_1}{\partial \eta_1^2}) \tag{1a}$$

(air:enthalpy)

$$\frac{1}{Pr_1} \frac{\partial}{\partial \eta_1}[(1 + Pr_1 \varepsilon^+_{G_1}) \frac{\partial g_1}{\partial \eta_1}] + \frac{m+1}{2} f_1 \frac{\partial g_1}{\partial \eta_1} = \zeta[\frac{\partial f_1}{\partial \eta_1} \frac{\partial g_1}{\partial \zeta} - \frac{\partial f_1}{\partial \zeta} \frac{\partial g_1}{\partial \eta_1}] \tag{1b}$$

(air:water vapor)

$$\frac{1}{Sc} \frac{\partial}{\partial \eta_1}[(1 + Sc\ \varepsilon^+_H) \frac{\partial h}{\partial \eta_1}] + \frac{m+1}{2} f_1 \frac{\partial h_1}{\partial \eta_1} = \zeta[\frac{\partial f_1}{\partial \eta} \frac{\partial h}{\partial \zeta} - \frac{\partial f_1}{\partial \zeta} \frac{\partial h}{\partial \eta_1}] \tag{1c}$$

(water:momentum)

$$\frac{\partial}{\partial \eta_2}[(1 + \varepsilon^+_{M_2}) \frac{\partial^2 f_2}{\partial \eta_2^2}] + \frac{m+1}{2} f_2 \frac{\partial^2 f_2}{\partial \eta_2^2} + m(\frac{\partial f_2}{\partial \eta_2})^2$$

$$= \zeta(\frac{\partial f_2}{\partial \eta_2} \frac{\partial^2 f_2}{\partial \zeta \partial \eta_2} - \frac{\partial f_2}{\partial \zeta} \frac{\partial^2 f}{\partial \eta_2^2}) \tag{1d}$$

(water:enthalpy)

$$\frac{1}{Pr_2} \frac{\partial}{\partial \eta_2}[(1 + Pr_2 \varepsilon^+_{G_2}) \frac{\partial g_1}{\partial \eta_2}] + \frac{m+1}{2} f_2 \frac{\partial g_2}{\partial \eta_2}$$

$$= \zeta\left(\frac{\partial f_2}{\partial \eta_2}\frac{\partial g_2}{\partial \zeta} - \frac{\partial f_2}{\partial \zeta}\frac{\partial g_2}{\partial \eta_2}\right) \tag{1e}$$

In the above equations f_1 and f_2 are the non-dimensional stream functions, g_1 and g_2 are enthalpies above and below the interface normalized by i_∞, and h is the mass fraction of water vapor in the air. Furthermore $m = (\zeta/U_\infty)(dU_\infty/d\zeta)$ is the pressure-gradient parameter of the free stream and ε_M^+, ε_G^+, and ε_H^+ are, respectively, the eddy viscosity and eddy diffusivities of heat and vapor concentration.

The interfacial and far-field conditions are

$$\left.\frac{\partial f_1}{\partial \eta_1}\right|_o = \left.\frac{\partial f_2}{\partial \eta_2}\right|_o \quad \text{(continuity of velocity)} \tag{2a}$$

$$\left.(1 + \varepsilon^+_{M_{1o}})\frac{\partial^2 f_1}{\partial \eta_1^2}\right|_o = \left.\Lambda(1 + \varepsilon^+_{M_{2o}})\frac{\partial^2 f_2}{\partial \eta_2^2}\right|_o$$

(stress continuity, $\rho_1 u_{*1}^2 = \rho_2 u_{*2}^2$) (2b)

$$\Lambda \equiv \frac{\rho_2}{\rho_1}\sqrt{\frac{\nu_2}{\nu_1}}$$

$$g_{1o} - \left(1 + (d_1 - 1)h_o\right)\left(d_2 g_{2o} - d_3\right) = 0$$

(continuity of temperature) (2c)

$$d_1 = c_{p_A}/c_{p_V}, \quad d_2 = c_{p_A}/c_W, \quad d_3 = (L/i_\infty)d_2$$

$$(m + 1)f_{1o} + 2\zeta\left.\frac{\partial f_1}{\partial \zeta}\right|_o = \Lambda\left[(m + 1)f_{2o} + 2\zeta\left.\frac{\partial f_2}{\partial \zeta}\right|_o\right]$$

(conservation of mass) (2d)

$$f_{1o} = \frac{2}{m + 1}\left[\frac{\partial h/\partial \eta_1|_o}{Sc(1 - h_o)} - \zeta\left.\frac{\partial f_1}{\partial \zeta}\right|_o\right]$$

(conservation of mass) (2e)

$$\dot{M}\frac{L}{i_\infty} - \frac{1}{Pr_1}\frac{\partial g_1}{\partial \eta_1}\bigg|_o - \frac{\Lambda}{Pr_2}\frac{\partial g_2}{\partial \eta_2}\bigg|_o = 0$$

(conservation of energy) (2f)

$$\dot{M} = -\frac{1}{Sc(1 - h_o)}\frac{\partial h}{\partial \eta_1}\bigg|_o$$

$$h_o = F(T_o) , \quad T_o = \frac{i_\infty g_{1o}}{c_{p_G}\left[1 + (d_1 - 1)h_o\right]} \tag{2g}$$

$$\frac{\partial f_1}{\partial \eta_1} \to 1 , \quad g_1 \to 1 , \quad h \to h_\infty \quad \text{as} \quad \eta_1 \to \infty \tag{2h}$$

$$\frac{\partial f_2}{\partial \eta_2} \to 0 , \quad g_2 \to \frac{i_{-\infty}}{i_\infty} \quad \text{as} \quad \eta_2 \to -\infty$$

Equation (2b) assumes stress continuity across the interface with no momentum transfer to wave growth (see, e.g. [9] for a discussion of this point). Equation (2g) relates the concentration of water vapor at the interface to the saturation temperature there; we use an empirical expression for $F(T_o)$.

In deriving (1)-(2) we do not address ourselves to the questions of radiative heat transfer, which is generally of secondary importance in many applications (Coantic [7]), or spray formation, which only occurs at high windspeeds ($U_\infty \gtrsim$ 13 m/s in the SWWWRF) when large-scale wave breaking occurs. When surface waves are present, the interfacial coupling is applied at the mean water level.

Equations (1)-(2) comprise the complete problem to be solved and our solution method employs the well-known Box scheme which was applied in [6] to laminar boundary-layers with a density interface. Here our only additional requirement is to parameterize the eddy viscosity and diffusivities in each fluid regime.

3. THE TURBULENCE MODEL

We have extended the roughness model of Street [4], which strictly applies to the diffusive transport of heat and specie across an air-water interface (and which hereafter will be referred to as the YKSM model because it is based on the earlier work of Yaglom and Kader [8] and Street and Miller [9]), to the momentum problem. The extension recognizes the fundamental difference between the transport of momentum and a passive scalar such as heat or specie near a rough surface;

the mechanism of the former is characterized by form drag and small-scale separation around roughness elements whereas the latter must be molecularly dominated in sublayers near the surface (see, e.g. Cebeci and Chang [10] and Ligrani et al. [11]). Consequently, we postulate the existence of a non-zero eddy viscosity in the immediate vicinity of the interface and assume a cubic variation of ε_M^+ with distance away from the roughened surface. Specifically, in wall coordinates with $y_+ = yu_*/\nu$ we require $\varepsilon_M^+ \sim (y_+ + \Delta_+)^3$ when $y_+ \sim O(1)$. Here Δ_+ is a shift parameter which must be specified empirically. The cubic estimate is then patched to the law-of-the-wall form $\varepsilon_M^+ \sim y_+ + \Delta_+$ at an appropriate height above (and below) the interface.

In order to retain the original formulation of the YKSM model one can demonstrate that $\Delta_+ \sim O(\sqrt{k_+})$. We assume the roughness Reynolds number k_+ is based on the mean wave height and that all waves in the wind-wave spectrum contribute to the roughness (a reasonable assumption for laboratory-generated wind-waves in the frequency range 2-10 Hz). In our model,

$$\Delta_{1_+} = a\sqrt{k'_{1_+} - b} \tag{3a}$$

$$\Delta_{2_+} = a\sqrt{k_{2_+} - b} \tag{3b}$$

where $a \simeq 0.3$ and $b \simeq 30$ are constants and

$$k'_{1_+} = k_{1_+} \exp(\kappa c_o/u_{*_1}) \tag{3c}$$

Here c_o/u_{*_1} is the wave age based on the celerity c_o of the dominant wave in the wind-wave spectrum. The exponential factor in (3c) represents a correction (typically O(1)) to account for the mobility of the interface and is suggested by a simple argument of Kitaigorodskii [12]; the correction cannot be applied below the interface because the aqueous regime is characterized by a different stage of wave development (with $c_o/u_{*_2} >> 1$).

4. SOME SOLUTIONS

We present now turbulent boundary-layer computations using the eddy viscosity model outlined in Section 3 and compare the results with data acquired by Chambers et al. [13] and by Mangarella et al. [14] in the SWWWRF. That data consists of profiles of mean air data, rms wave height, and dominant wave frequency at different fetches in the facility with only a bulk water temperature far below the interface. We have no other data in the aqueous regime with which to compare our simulations.

Figure 2 compares computed momentum thicknesses above the interface with measured values at two different windspeeds ($U_\infty \simeq 6.5$, 11 m/s) for which $k_{1_+} \gtrsim 70$ (corresponding approximately to fully rough conditions at the interface). The computations commence with input profiles obtained from the rough surface theory outlined in Section 3 (see, e.g. Street [4] for a description of this procedure) matching the data at the initial station; these profiles are analytic and readily generated when u_*, T_*, and h_* are known. Agreement between the computed and measured momentum thicknesses is generally good.

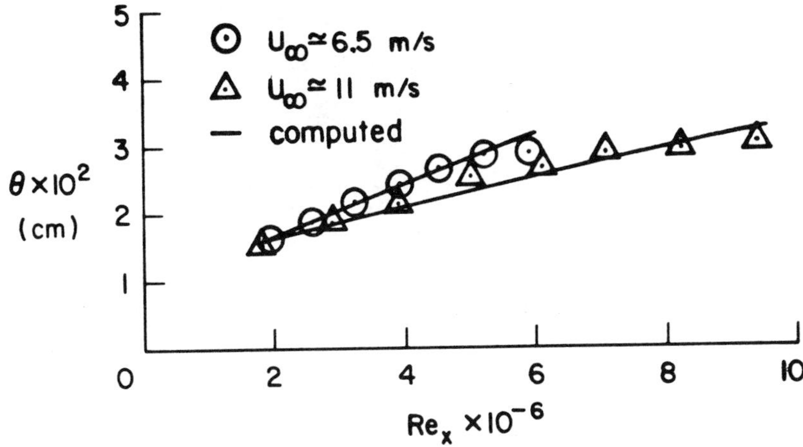

Figure 2. Momentum thickness, θ versus Re_x. a) $U_\infty \simeq 6.5$ m/s, $T_\infty \simeq 18.3°C$, $T_{-\infty} \simeq 33.3°C$, $h_\infty \simeq 0.007$; b) $U_\infty \simeq 11$ m/s, $T_\infty \simeq 20.2°C$, $T_{-\infty} \simeq 31.7°C$, $h_\infty \simeq 0.008$.

Figure 3 contains a normalized profile of mean velocity for one of these windspeeds at the final downstream station where measurements are available. Agreement between the experimental and predicted profiles is good. Skin-friction coefficients and Stanton numbers at both windspeeds are compared in Figure 4. Some discrepancy exists between measured and computed values of c_f, but St calculations agree well with the data.

5. SUMMARY

Comparison of calculated boundary-layer parameters and mean profiles with measurements acquired in a wind, water-wave research facility confirm the applicability of the numerical method for simulating the effects of wind-waves on the transport processes at an air-water interface. Furthermore, these computations point to the ability of the method to treat the coupled, turbulent gas-liquid boundary-layer problem.

ACKNOWLEDGMENT

This material is based upon work supported by the National Science Foundation, U.S.A., under Grant No. CME-7901176 from the Heat Transfer Program.

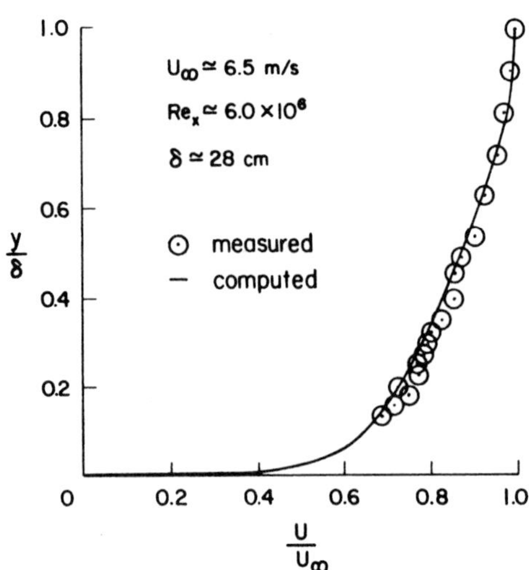

Figure 3. Computed and measured mean velocity profiles ($U_\infty \simeq$ 6.5 m/s, $Re_x \simeq 6 \times 10^6$).

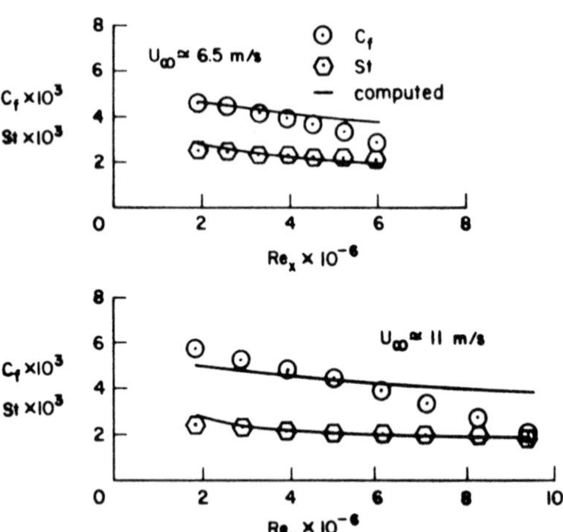

Figure 4. c_f and St versus Re_x.

REFERENCES

1. LOCK, R.C. - The Velocity Distribution in the Laminar Boundary Layer Between Parallel Streams. Quart. J. Mech. and App. Math. Vol. 4, No. 1, pp. 42-63, 1951.
2. KOTAKE, S. - Heat Transfer and Skin Friction of a Phase-Changing Interface of Gas-Liquid Laminar Flows. Int. J. Heat..Mass Transfer. Vol. 16, pp. 2165-2176, 1973.
3. SCHROPPEL, J., and THIELE, F. - Numerical Method for the Calculation of Binary Gas Mixture Condensation in Boundary Layer Flow, Proc. First Int. Conf. on Numerical Methods in Laminar and Turbulent Flow, John Wiley & Sons, 1978.

4. STREET, R. L. - Turbulent Heat and Mass Transfers Across a Rough Air-Water Interface: A Simple Theory. Int. J. Heat Mass Transfer. Vol. 22, pp. 885-899, 1979.
5. KELLER, H.B., and CEBECI, T. - Accurate Numerical Methods for Boundary Layer Flows. I. Two-Dimensional Laminar Flows, in Lecture Notes in Physics, 8, Proc. Second Int. Conf. on Numerical Methods in Fluid Dynamics, Springer-Verlag, Berlin and New York, pp. 92-100, 1971.
6. KLOTZ, S. P., and STREET, R. L. - On the Simulation of the Coupled Laminar Boundary Layer at a Smooth, Phase-Changing Gas-Liquid Interface, Proc. Second Int. Conf. on Numerical Methods in Laminar and Turbulent Flow, Venice, Italy, 1981.
7. COANTIC, M. F. - Coupled Energy Transfer and Transformation Mechanisms Across the Ocean-Atmosphere Interface. Sixth Int. Heat Transfer Conf., Toronto, pp. 73-87, 1978.
8. YAGLOM, A. M., and KADER, B. A. - Heat and Mass Transfer Between a Rough Wall and Turbulent Flow at High Reynolds and Peclet Numbers. J. Fluid Mech. Vol. 62, No. 3, pp. 601-623, 1974.
9. STREET, R. L., and MILLER, A. W. - Determination of the Aqueous Sublayer Thicknesses at an Air-Water Interface. J. Phys. Ocean. Vol. 7, pp. 110-117, 1977.
10. CEBECI, T., and CHANG, K. C. - Calculation of Incompressible Rough-Wall Boundary-Layer Flows. AIAA J. Vol. 16, No. 7, pp. 730-735, 1978.
11. LIGRANI, P. M., KAYS, W. M., and MOFFAT, R. J. - The Thermal and Hydrodynamic Behavior of Thick, Rough-Wall, Turbulent Boundary Layers. Rpt. No. HMT-29, Dept. Mech. Eng., Stanford University, 1979.
12. KITAIGORODSKII, S. A. - The Physics of Air-Sea Interaction. Israel Program for Scientific Translations, Jerusalem (TT 72-50062, U.S. Department of Commerce, Springfield, Virginia), 1973.
13. CHAMBERS, A. J., MANGARELLA, P. A., STREET, R. L., and HSU, E. Y. - An Experimental Investigation of Transfer of Momentum at an Air-Water Interface. TR 133, Dept. Civil Eng., Stanford University, 1970.
14. MANGARELLA, P. A., CHAMBERS, A. J., STREET, R. L., and HSU, E. Y. - Energy and Mass Transfer Through an Air-Water Interface. TR 134, Dept. Civil Eng., Stanford University, 1971.

APPENDIX - Nomenclature

a,b constants

c_f skin-friction coefficient

c_L specific heat of water

c_{p_G}, c_{p_V} constant-pressure specific heat of dry air and water-vapor, respectively

c_o celerity of the dominant wave in the wind-wave spectrum f dimensionless stream function

g dimensionless enthalpy, i/i_∞

h mass fraction of vapor in the binary gas mixture

h_* friction vapor concentration

i specific enthalpy

k_+ roughness Reynolds number based on mean wave height

L latent heat of vaporization

m pressure-gradient parameter

$\dot{M}$ surface mass transfer rate

Pr Prandtl number

Re_x Reynolds number based on streamwise distance

R_v molecular gas-constant of water-vapor

Sc Schmidt number

St Stanton number

T temperature

T_B boiling temperature of the liquid at the ambient pressure of the gas-liquid system.

T_* friction temperature

u_* friction velocity

U horizontal velocity

$U_\infty(x)$ free-stream velocity distribution

x,y horizontal and vertical coordinates, respectively

y_+ yu_*/ν

Greek Symbols

δ boundary-layer thickness

Δ_+ shift parameter

ε_G^+, ε_H^+ eddy diffusivity of heat and mass, respectively

ε_M^+ eddy viscosity

κ von Karman constant

Λ density-viscosity ratio, $\frac{\rho_2}{\rho_1}\sqrt{\frac{\nu_2}{\nu_1}}$

θ momentum thickness

ρ density

ν kinematic viscosity

ζ,η Falkner-Skan coordinates

Subscripts

1,2 air, water

∞, $-\infty$ evaluated in the free stream and bulk liquid, respectively

o evaluated at the interface

A FINITE ELEMENT-DIFFERENTIAL METHOD FOR A CLASS OF COMPRESSIBLE LAMINAR BOUNDARY-LAYER FLOWS

Chen-Chi Hsu and Antonios Liakopoulos
Department of Engineering Sciences
University of Florida, Gainesville, FL 32611

ABSTRACT

A numerical method developed and tested successfully upon a class of incompressible flows has been investigated for its application to the solution of compressible flows. In this method the transformed solution profile at a streamwise station is represented by a classical cubic spline, and the reduced initial value problem is integrated numerically by a predictor-corrector method. The numerical results obtained have shown that the method is indeed very simple and efficient and can provide highly accurate results for the class of compressible boundary-layer flows considered.

INTRODUCTION

For a complex flow problem an accurate solution to the entire boundary-layer flow region can be crucial to the viscous-inviscid interaction analysis and, hence, to an accurate prediction of drag and lift forces, e.g. [1]. It is therefore important to have a general numerical technique which is simple and yet capable of solving accurately and efficiently the complicated set of boundary-layer equations.

There have been numerous approximate methods developed for boundary-layer flow problems. In recognizing the advantages as well as the difficulties of numerical techniques available, a finite element-differential method has been developed and tested upon a class of steady two-dimensional incompressible laminar forced convection boundary-layer flows [2]. In this method the transformed flow region is divided into a very limited number of unequal strips parallel to the fixed boundary, and at a streamwise station the classical cubic spline is employed to approximate the solution profiles. The governing partial differential equations are then reduced to a system of first order nonlinear ordinary differential equations by a subdomain collocation method, in which the resulting definite integrals are evaluated with Gauss-Legendre quadrature formulas. The reduced initial value problem is finally solved by a predictor-corrector method having the integration step automatically controlled by an assigned accuracy requirement. The numerical experiments conducted for flows past a circular cylinder of constant temperature showed that the method is indeed very simple and efficient; it can give extremely accurate wall shear stress distribution and heat transfer coefficient distribution for the entire boundary layer region. In fact the numerical integration can be carried out directly with fairly large steps to within one integration step upstream of the separation point.

It seems that the finite element-differential method developed is a rather promising solution technique for more complex flow problems. The objective of this study is, therefore, to further investigate the application as well as the accuracy and efficiency of the approach to compressible laminar boundary-layer flow problems.

GOVERNING EQUATIONS

For the class of steady, two-dimensional, compressible, laminar, ideal gas flows past a submerged body, neglecting the buoyancy force but taking into account the temperature dependence of the viscosity and the thermal conductivity, the governing boundary-layer equations are

$$\frac{\partial(\rho u)}{\partial x} + \frac{\partial(\rho v)}{\partial y} = 0, \tag{1}$$

$$\rho\left(u\frac{\partial u}{\partial x} + v\frac{\partial u}{\partial y}\right) = -\frac{dP}{dx} + \frac{\partial}{\partial y}\left(\mu\frac{\partial u}{\partial y}\right), \tag{2}$$

$$\rho C_p\left(u\frac{\partial T}{\partial x} + v\frac{\partial T}{\partial y}\right) = u\frac{dP}{dx} + \frac{\partial}{\partial y}\left(K\frac{\partial T}{\partial y}\right) + \mu\left(\frac{\partial u}{\partial y}\right)^2, \tag{3}$$

$$P = \rho RT. \tag{4}$$

The associated boundary conditions considered in this study are

$$\text{at } y = 0: \quad u = 0, \ v = 0, \quad T = T_W(x), \tag{5}$$

$$\text{at } y \to \infty: \quad u \to Ue(x), \quad T \to Te(x). \tag{6}$$

It is well known that the system of equations, (1) - (6), can be reduced to the form almost the same as the governing equations for incompressible flows by the Illingworth-Stewartson transformation if one assumes that the viscosity is linearly proportional to the temperature and that the Prandtl number is constant. Furthermore, from the numerical technique point of view it is advantageous to carry out a number of additional transformations [3]. For instance, the reduced problem can be transformed so that the Reynolds number does not appear in the equation and that the boundary conditions are the same for all conceivable problems of the class. Moreover, the application of a numerical method to a boundary-layer flow requires that the conditions at infinity be imposed at a sufficiently large finite distance H from the body; a proper choice of the distance H is crucial to the efficiency of the numerical scheme as well as to the accuracy of the solution. Since the growth rate of boundary-layer thickness is not known beforehand, it seems advantageous to introduce a transformation of Falkner-Skan type so that the transformed boundary-layer thickness may remain about the same order of magnitude throughout the entire flow region. Also, the application of von Mises transformation is of particular advantages to the solution technique; it provides additional end conditions for the spline approximation and at the same time it reduces the number of dependent variables by one. If one carries out all these transformations, in addition to the Illingworth-Stewartson transformation, one is led to solve the following initial boundary value problem

$$\frac{\partial w}{\partial \xi} = \frac{w^{\frac{1}{2}}}{4}\left[\frac{1}{\eta^2}\frac{\partial^2 w}{\partial \eta^2} - \frac{1}{\eta^3}\frac{\partial w}{\partial \eta}\right] + \frac{\eta}{4}\frac{\partial w}{\partial \eta} + 2\left[(1-w) + \theta(1-\phi^{\frac{1}{2}})\right]\frac{d\ell n U}{d\xi}, \tag{7}$$

$$\frac{\partial \phi}{\partial \xi} = \frac{w^{\frac{1}{2}}}{4P_R}\left[\frac{1}{\eta^2}\frac{\partial^2 \phi}{\partial \eta^2} - \frac{1}{\eta^3}\frac{\partial \phi}{\partial \eta} + \frac{1}{2\eta^2}\frac{\partial \ell n(w/\phi)}{\partial \eta}\frac{\partial \phi}{\partial \eta}\right] + \frac{\eta}{4}\frac{\partial \phi}{\partial \eta}$$

$$+ 2(\phi^{\frac{1}{2}}-\phi)\frac{d\ell n\theta}{d\xi} - \frac{\alpha(\xi)}{2\theta}\left(\frac{\phi}{w}\right)^{\frac{1}{2}}\left[\frac{w}{\eta^2}\frac{\partial^2 w}{\partial \eta^2} + \frac{1}{2\eta^2}\left(\frac{\partial w}{\partial \eta}\right)^2 - \frac{w}{\eta^3}\frac{\partial w}{\partial \eta}\right], \tag{8}$$

$$\alpha(\xi) \equiv \frac{P_R - 1}{P_R} \cdot \frac{\frac{1}{2}(\gamma-1)M^2(\xi)}{1 + \frac{1}{2}(\gamma-1)M^2(\xi)}, \tag{9}$$

$$w(\xi,0) = 0, \quad \phi(\xi,0)=0, \; w(\xi,H) = 1, \quad \phi(\xi,H) = 1, \tag{10}$$

$$w(0,\eta) = w_o(\eta), \quad \phi(0,\eta) = \phi_o(\eta), \tag{11}$$

where $w(\xi,\eta)$ and $\phi(\xi,\eta)$ are, respectively, the square of the transformed dimensionless velocity and temperature, while $U(\xi)$, $\theta(\xi)$ and $M(\xi)$ are related to the given boundary conditions and reference parameters. The additional end conditions resulted from the physical boundary conditions and von Mises transformation are

$$\frac{\partial w}{\partial \eta} = \frac{\partial \phi}{\partial \eta} = 0 \quad \text{at} \quad \eta = 0 \quad \text{and at} \quad \eta = H. \tag{12}$$

Note that the conditions at infinity have been imposed at a sufficiently large finite distance $\eta = H$.

SOLUTION METHOD

In the proposed method the transformed solution profiles $w(\xi,\eta)$ and $\phi(\xi,\eta)$ at a streamwise station ξ are approximated by the classical cubic spline. Suppose that the interval $0 \le \eta \le H$ is divided into n elements with nodes at $0 = \eta_o < \eta_1 < \ldots < \eta_n = H$ and the element size $h_i = \eta_i - \eta_{i-1}$ for $i = 1,\ldots, n$. Also, denote the unknown functional value at ith mode as

$$w_i(\xi) = w(\xi, \eta_i), \qquad \phi_i(\xi) = \phi(\xi, \eta_i) . \tag{13}$$

Let $\bar{w}_i(\xi,z)$ and $\bar{\phi}_i(\xi,z)$ be the cubic polynomials approximating $w(\xi,\eta)$ and $\phi(\xi,\eta)$ in the ith element, where the local coordinate $z = \eta - \eta_{i-1}$. Then the continuity of the function, its slope and curvature at the node and the application of the end conditions (12) give [2]

$$\bar{w}_i(\xi,z) = \sum_{j=0}^{n} a_{ij}(z)w_j(\xi) , \quad \bar{\phi}_i(\xi,z) = \sum_{j=0}^{n} a_{ij}(z)\phi_j(\xi). \tag{14}$$

for $i = 1, \ldots, n$, where $a_{ij}(z)$ are known polynomials of degree three in z. One notes that the cubic polynomial approximating a function in the ith element depends linearly upon all the nodal values of the function.

The classical cubic spline approximating a function is the sum of the element cubics derived. Therefore, the assumed solution for $w(\xi,\eta)$ and $\phi(\xi,\eta)$ can be written as

$$w(\xi,\eta) = \sum_{i=1}^{n} \delta_i \bar{w}_i(\xi,z) = \sum_{j=0}^{n} N_j(\eta)w_j(\xi), \tag{15}$$

$$\phi(\xi,\eta) = \sum_{i=1}^{n} \delta_i \bar{\phi}_i(\xi, z) = \sum_{j=0}^{n} N_j(\eta)\phi_j(\xi), \tag{16}$$

where $N_j(\eta)$ are the known interpolation functions defined as

$$N_j(\eta) = \sum_{i=1}^{n} \delta_i a_{ij}(z), \qquad \eta = z + \eta_{i-1}, \tag{17}$$

and $\delta_i = 1$ when $\eta_{i-1} \le \eta \le \eta_i$, otherwise $\delta_i = 0$. The assumed solutions, (15) and (16), and the boundary conditions (10) show that there are only (n-1) unknowns in $w_j(\xi)$ and in $\phi_j(\xi)$. For the application of a subdomain collocation method we have chosen the following set of weight functions for both equations (7) and (8).

$$\gamma_k(\eta) = \begin{cases} 1, & \text{when } \eta_{k-1} \le \eta \le \eta_{k+1} \\ 0, & \text{otherwise} \end{cases} , \tag{18}$$

for k = 1, 2, ..., (n-1). Accordingly, we obtain the following system of equations:

$$\int_0^{h_k} \frac{\partial \bar{w}_k}{\partial \xi} dz + \int_0^{h_{k+1}} \frac{\partial \bar{w}_{k+1}}{\partial \xi} dz = \int_0^{h_k} R_w \, dz + \int_0^{h_{k+1}} R_w \, dz , \quad (19)$$

$$\int_0^{h_k} \frac{\partial \bar{\phi}_k}{\partial \xi} dz + \int_0^{h_{k+1}} \frac{\partial \bar{\phi}_{k+1}}{\partial \xi} dz = \int_0^{h_k} R_\phi \, dz + \int_0^{h_{k+1}} R_\phi \, dz , \quad (20)$$

for k=1, ..., (n-1). Here R_w and R_ϕ represent the right hand side of (7) and (8), respectively. Since the cubic polynomials $\bar{w}_k(\xi,z)$ and $\bar{\phi}_k(\xi,z)$ depend linearly upon the unknown nodal functions $w_j(\xi)$ and $\phi_j(\xi)$, respectively, equations (19) and (20) can be written in the matrix form

$$[Q] \{\frac{d\tilde{w}}{d\xi}\} = \{r_1(\xi)\} , \quad [Q] \{\frac{d\tilde{\phi}}{d\xi}\} = \{r_2(\xi)\} , \quad (21)$$

where [Q] is an (n-1) by (n-1) non-singular constant matrix, and the kth element of $\{r_1\}$ and $\{r_2\}$ is given by the two integrals on the right hand side of (19) and (20), respectively. The undetermined (n-1) dimensional vectors $\{\tilde{w}\}$ and $\{\tilde{\phi}\}$ are

$$\{\tilde{w}\}^T = [w_1(\xi), \ldots, w_{n-1}(\xi)], \quad \{\tilde{\phi}\}^T = [\phi_1(\xi), \ldots, \phi_{n-1}(\xi)]. \quad (22)$$

In this study the reduced initial value problem, (21) and (11), is solved with Hamming's modified predictor-corrector method which is of fourth order accuracy; consequently, the initiator employed is a fourth order Runge-Kutta scheme. The integration step size taken is automatically controlled by an assigned accuracy requirement of 10^{-6} and, hence, no iteration is made in the predictor-corrector phase of the integration. Because of the complex expression for R_w and R_ϕ, explicit relations in terms of the unknown functions $w_j(\xi)$ and $\phi_j(\xi)$ cannot be obtained for elements of $\{r_1(\xi)\}$ and $\{r_2(\xi)\}$. Therefore the integrals on the right hand side of (19) and (20) have been approximated effectively by a Gauss-Legendre quadrature formula, e.g.[4]. The computer used in this study is Amdahl 470 V/6-II, and all the numerical results are obtained with double precision computations.

SIMILAR SOLUTIONS

For the compressible boundary-layer flow considered, there exist classes of similarity solution if one assumes that the Prandtl number $P_R = 1$. A rather important class of similar solutions is governed by

$$f''' + ff'' = \beta[(f')^2 - 1 - S] \;, \tag{23}$$

$$S'' + fS' = 0 \;, \tag{24}$$

$$f(0) = f'(0) = 0, \qquad S(0) = S_w = \text{const.} \;, \tag{25}$$

$$f'(\zeta \to \infty) \to 1, \qquad S(\zeta \to \infty) \to 0 \;, \tag{26}$$

in which $f(\zeta)$ is a transformed stream function and the enthalpy function $S(\zeta)$ is defined as

$$S = \frac{C_p T + \frac{1}{2}u^2}{C_p T_o} - 1 \;. \tag{27}$$

Accurate solutions for (23)-(26) have been obtained by Cohen and Reshotko for different values of β and S_w [5]. It is known that the last term of (23) is responsible for the phenomenon of velocity overshoot. Therefore, the similarity solution can be a real test for the accuracy of the proposed finite element-differential method.

For the similarity solution considered, the right hand side of (7) and (8) does not depend upon the variable ξ since $\theta(\xi)$ becomes a constant related to S_w and

$$\alpha(\xi) = 0 \;, \qquad \frac{d\ell nU}{d\xi} = \frac{1}{2}\beta \;. \tag{28}$$

For the numerical experiment we have chosen $S_w = 1.0$ which implies that $\theta(\xi) = 1.0$ and have considered the following cases:

$$\beta = 0.0, \quad H = 4.0 \;, \quad h_i = |0.1,\ 0.2,\ 0.3,\ 0.4,\ 6{*}0.5| \;,$$

$$\beta = 0.3, \quad H = 3.4 \;, \quad h_i = |0.2,\ 0.3,\ 0.4,\ 5{*}0.5| \;,$$

$$\beta = 1.0, \quad H = 3.0 \;, \quad h_i = |0.2,\ 0.3,\ 3{*}0.5,\ 1.0| \;,$$

$$\beta = 1.5, \quad H = 3.5 \;, \quad h_i = |0.2,\ 0.3,\ 6{*}0.5| \;.$$

A rather arbitrary profile satisfying the boundary conditions (10) and the additional end conditions (12) has been chosen for the initial conditions $w_o(\eta)$ and $\phi_o(\eta)$ to start the integration of the reduced initial value problem. The computed results and Cohen & Reshotko's results are given in the table, on the following page, for comparison. It clearly indicates that the proposed finite element-differential method can give highly accurate results for the compressible flow problems considered. Additional numerical experiments have been carried for different values of H and different element size, h_i, distributions. The results show that the increase in H does not affect the accuracy of the solution if

β	Computed results f''_w/S'_w	Cohen & Reshotko's f''_w/S'_w	Percentage difference %
0.0	0.4679 - 0.4679	0.4696 - 0.4696	0.36 0.36
0.3	0.9842 - 0.5444	0.9829 - 0.5457	0.13 - 0.24
1.0	1.7350 - 0.6160	1.7368 - 0.6154	- 0.10 0.10
1.5	2.1370 - 0.6411	2.1402 - 0.6425	- 0.15 - 0.22

the largest element used is of 0.5 in size; however, the choice of larger element size in the outer region does deteriorate the accuracy of the solution. For instance, the results of the case

$$\beta = 1.0, \quad H = 6.0, \quad h_i = |0.2,\ 0.3,\ 0.5,\ 0.8,\ 1.2,\ 2{*}1.5|$$

gives $f''_w = 1.7302$ and $S'_w = 0.6621$ which are off by -0.38% and 7.59%, respectively.

FLAT PLATE PROBLEM

The similarity solution also exists for a compressible laminar flow past a flat plate of constant temperature, since $U(\xi)$, $\theta(\xi)$ and $\alpha(\xi)$ appeared in the governing equations become constant. In fact the transformed momentum equation (7) is now independent of the enthalpy function, while the transformed temperature equation (8) depends not only upon w but also upon the Prandtl number and free stream Mach number. If we assume that $T_w = 2To$, then we have

$$S_w = 1.0, \quad \theta(\xi) = 1.0, \quad \alpha(\xi) = \frac{P_R - 1}{P_R} \cdot \frac{(\gamma-1)M^2_\infty}{2 + (\gamma-1)M^2_\infty}, \qquad (29)$$

in which $P_R = 0.72$ and $\gamma = 1.4$ have been used in the computation. It is of interest to find out the effect of the Prandtl number and Mach number upon the heat transfer coefficient which is linearly related to S'_w .

For the numerical result we have chosen the finite element model

$$H = 4.0, \quad h_i = |0.2,\ 0.3,\ 7{*}0.5| \ ,$$

and a discretized Blasius solution for $w_o(\eta)$ and $\phi_o(\eta)$. The computed S'_w for different Mach numbers is given in the table on the following page.

M_∞	10^{-6}	1.0	2.0	3.0	4.0	5.0	1000.
$-S'_w$	.4183	.4279	.4441	.4558	.4629	.4672	.4772

The result also gives $f''_w = 0.4731$ which is 0.75% greater than the accepted value of 0.4696. It seems to indicate that the accuracy of the computed S'_w is within 1%. These results show that the assumption of the Prandtl number $P_R = 1.0$ for low Mach number flows can overpredict the heat transfer coefficient by as much as 10%.

This work was sponsored by the AFOSR, AFSC, USAF under Grant No. AFOSR-80-0033. The interest of Dr. C. L. Keller, the program manager,in this research is much appreciated.

REFERENCES

1. ROM, J. - Flows with Strong Interaction between the Viscous and Inviscid Regions. Proc. International Symposium on Modern Developments in Fluid Dynamics, Ed. Rom, J., SIAM, 1977.

2. HSU, C.C. - A Finite Element-Differential Method for a Class of Boundary-Layer Flows. Proc. 3rd International Conference on Finite Elements in Flow Problems, Ed. Norrie, D. H., 1980.

3. HSU, C.C. - The Use of Splines for the Solution of the Boundary-Layer Equations. Tech. Rep. AFFDL-TR-75-158. Air Force Wright Aeronautical Laboratories, WPAFB, Feb. 1976.

4. KRYLOV, V.I. - Approximate Calculations of Integrals, MacMillian, 1962.

5. COHEN, C.B. and RESHOTKO, E. - Similar Solutions for the Compressible Laminar Boundary Layer with Heat Transfer and Pressure Gradient. NACA Tech. Rep. 1293, 1956.

ON THE OLP PREDICTION OF THE UNSTABLE MODES OF THE FLAT PLATE TURBULENT BOUNDARY LAYER(i)

Aspi Rustom Wadia(ii) and Fred R. Payne
Aerospace Engineering
The University of Texas at Arlington
Arlington, Texas 76019, U.S.A.

SUMMARY

An energy extremum principle is used to predict the eigenmodes that correspond to the turbulent large eddy structure of the fully developed, two dimensional, flat plate turbulent boundary layer. Lumley's stability equations are solved with a piecewise continuous, linear eddy viscosity model and a mean velocity profile based on the experiments of Schultz-Grunow. The numerical results are compared with the experimental observations of both Grant and Lemmerman. The presence of the "burst and sweep" phenomena is identified in this numerical investigation.

1. INTRODUCTION AND OVERVIEW

1.1 Statement of the Problem

Can Lumley's [1] extractive scheme (PODT) be applied successfully to hot wire velocity covariance data of Grant [2] and Tritton [3] taken in the flat plate turbulent boundary layer? The major uncertainty resided in the lack of completely definitive data; a reason for such lack, in addition to costs, was that the data were taken prior to the development of sophisticated analysis methods.

The major goal of the project [8] undertaken at the University of Texas since 1975 was to extract the "Lumley eddies"

(i) Partially supported by a NASA/AMES Grant NSG 2077, Dr. M. W. Rubesin, Technical Monitor.

(ii) Presently with the Garrett Turbine Engine Company, #503-4X, 111 S. 34th Street, P.O. Box 5217, Phoenix, AZ 85010, U.S.A.

from these empiric data and interpret their structure for compatibility with well-established intuitive and semi-empiric concepts developed over the past several decades by workers in the field of turbulence.

The present study involves the development of the OLP (Orr-Lumley-Payne) stability analysis to the mean profile of Schultz-Grunow [4] in a fully three dimensional analysis. Homogeneity assumption in the downstream (similarity coordinates) and cross-stream directions allowed the reduction of the PDE to two coupled linear ODE's. Green's functions converted the ODE's to coupled integral equations [8,9] which were solved for the "dominant" (largest eigenvalue) neutral mode, corresponding to the lowest turbulent Reynolds number.

1.2 Why Study the Large Eddy Structure?

Ever since Richardson [5] was attributed the ditty about "eddy cascades" it has become even more apparent that the largest scales of the turbulent motion are primarily responsible for the predominant transfer of mass, momentum, heat and contaminants in turbulent flow. Such transfer rates are typically two or more orders of magnitude greater than molecular transports.

Hence, improved insight into the largest scales is crucial to the understanding and modeling, for quantitative calculations, of these most diverse phenomena. A presumed "fall out" of PODT (and OLP, Lumley's predictive method) could be the determination of the range of length scales which cannot be modeled crudely (i.e., efficiently in the calculational sense) and those scales which can, with good precision, be so modeled.

1.3 Brief History of PODT and OLP Methods

John Lumley sought (1964-1966 while the second author was his student at Penn State) a rational definition of the large scale structure of turbulent shear flows. For some 4-5 decades investigators had intuitively discussed, based upon various experimental observations and a "guessing game" type of analysis (i.e.--assume an eddy shape and compare its predictions with experiment), the "big eddies." Townsend [6] and Grant [2] interpreted their data sets differently and generated contradictory large scale model structures in the 2-D wake.

This contradiction sharpened Lumley's interest and he formulated the Proper Orthogonal Decomposition Theorem (PODT) as an extension of the Harmonic Orthogonal Decomposition Theorem of Loeve [7]. Lumley's PODT maximized, in a Hilbert space, the mean square of the inner product of the stochastic velocity vector with a deterministic candidate vector function of space and time. This results in the classic integral eigenvalue problem [1]:

$$\int_{\Omega} R_{ij}(\cdot\, , \cdot') \, \phi_j(\cdot') \, d(\cdot') = \lambda\phi_i(\cdot) \quad (1)$$

where R_{ij} is the two point velocity co-variance, the dot and dot-prime denote the 3-vector of position and scalar time of two different space-time points and Ω indicates the integration is over all $(\cdot') = (\underline{x}', t)$.

Lumley [1] showed that for homogeneous (or stationary)coordinates, Eq. (1) reduces to HODT (Loeve [8]) for which the eigen solutions in those coordinates are simply the harmonic functions and λ has a dense distribution in spectral space (wave number and/or frequency); i.e. homogeneous spectra are continuous, not discrete.

However, in the coordinates for which statistical homogeneity does not prevail, the eigenvalues have a discrete spectrum. Likewise, each orthogonal (deterministic) $\phi_i \to \phi_i^{(n)}$; $n = 1,2, \ldots$.

Payne applied the PODT to Grant's data [2] and confirmed much of Grant's semi-intuitive structure but also obtained some details not given in Grant's paper. Payne's result (1966) showed a single structure (Grant had to assume two in orthogonal planes) and verified Lumley's [1] conjecture that the $\phi_i^{(n)}$ were indeed the "big eddies".

Lumley later formulated what the authors have named OLP (Orr-Lumley-Payne). Orr's energy method [10] of flow stability analysis has the merit (over, say Orr-Sommerfeld) of not requiring any assumptions about the disturbance form or its amplitude which had been linearized in most analyses. The later assumption of linearization totally precludes meaningful turbulent profile stability analysis.

OLP includes Orr's original global extremum principle for the disturbance kinetic energy and adjoins Lumley's [11] maximization principle for the eddy viscosity. This resulted in Lumley's eigenvalue problem:

$$S_{ij}\, u_j = \phi_{,i} + [\nu(u_{i,j} + u_{j,i})]_{,j} \quad ; \quad u_{i,i} = 0 \quad (2)$$

where S_{ij} is the mean rate-of-strain tensor, u_j the neutral eigen function (velocity), ν the sum of the molecular and eddy viscosities and ϕ is essentially a Lagrange multiplier which satisfies the constraint of imcomressibility and is pressure-like. If Townsend [6] is correct, presumably the eigen solutions of Eq. (2) will relate to his big eddies which are in energy equilibrium with the mean profile. Townsend assumed that the large-scale-structure will be dominated by those motions which most efficiently extract energy from the mean profile and, in turn, and lose energy at about equal rates to smaller motions. Hence, in a sense, Lumley [11] formalizes Townsend's maximal principle [6].

The first successful application of Lumley's stability analysis was to the 2-D wake of a circular cylinder [12]. The simple wake geometry permitted use of a constant eddy viscosity based upon Grant's data. Comparisons with PODT results were favorable [12].

Further studies of the PODT and the OLP were made by Lemmerman and Payne [13-15]. Lemmerman [13] presented the full 7 x 7 wave number grid data for the first eigenmode, i.e., largest eigenvalue which, ala Lumley, he interpreted as the mean square kinetic energy per unit mass. Two spectral peaks were noted which seem to correspond with the "wall eddies" of relatively small scale compared to the largest scales which probably are associated with those eddies doing most of the transport and/or "bursting" and subsequent "sweep". Higher order modes were also obtained by Lemmerman and they were found to change the eddy structure very little, merely serving to "round-out" or smooth the gradients. One is reminded of Bakewell's [17] one dimensional result that only the first eigenvalue was required for the wall layer; his other four calculated eigenvalues were negative. This is non-physical due to their interpretation as the energy content of the appropriate mode. Payne [14] outlined the PODT-SAS and extended the two stage eddy viscosity model and indicated the dual, complimentary aspects of OLP and PODT. Conjectures were made on the application of the PODT-SAS (Structural Analysis System, which emphasizes the structural aspects of PODT as opposed to the direct dynamics of OLP) to grid generation for "sub-grid" modeling and possible generation of proto-type families of fundamental modes for various geometries virtually independent of Reynolds number and, probably, influenced primarily by pressure gradient and its derivatives. Later Payne [12] concentrated upon the first fully three dimensional stability analysis of turbulent flows via the OLP and comparisons with fully 3-D extractions from 2-D wake experiments.

Hong [16] applied a limited, i.e., 2-D, OLP stability analysis to the mean profile of Schultz-Grunow. Qualitative comparisons with Lemmerman's [13] PODT eddies were favorable. With the encouraging results of Payne's [12] fully 3-D stability analysis for the wake of a circular cylinder and Hong's [18] analysis for the flat plate boundary-layer the next logical step was to extend Hong's OLP analysis into three dimensions which is summarized here with further details given in [9]

2. DERIVATION OF THE GOVERNING EQUATIONS

The general Lumley equations are:

$$S_{ij}\, u_j = \phi_{,i} + \frac{\partial}{\partial x_j}[\nu(u_{i,j} + u_{j,i})] \;'\; u_{i,i} = 0 \qquad (3)$$

where S_{ij} = mean rate-of-strain, ν = total effective viscosity, u = perturbation velocity and ϕ is essentially a Lagrange multiplier which satisfies the constraint of incompressibility or a pressure, depending upon point of view. The mean rate of strain is (where U is the mean velocity):

$$S_{ij} = \frac{1}{2}[U_{i,j} + U_{j,i}] \qquad (4)$$

Assuming two dimensional mean flow (U_1=U(y)) and a viscosity model which is a function of y alone, the Lumley equations, on elimination of ϕ by cross differentiation become:

$$\frac{1}{2}\frac{dU}{dy}\frac{\partial v}{\partial z} = 2\nu\nabla^2(\omega_y) + \frac{d\nu}{dy}\frac{\partial \omega_y}{\partial y} \qquad (5)$$

$$\frac{1}{2}\frac{dU}{dy}\frac{\partial u}{\partial z} = -2\nu\nabla^2(\omega_x) - \frac{\partial \omega_x}{\partial y}\frac{d\nu}{dy} - \frac{d\nu}{dy}\nabla^2 w - 2\frac{d^2\nu}{dy^2}\gamma_{yz} \qquad (6)$$

where ω_x, ω_y and ω_z are the angular velocity components in the x,y, and z directions respectively and γ_{yz} is the rate-of-strain velocity (coordinate system is shown in Figure 1).

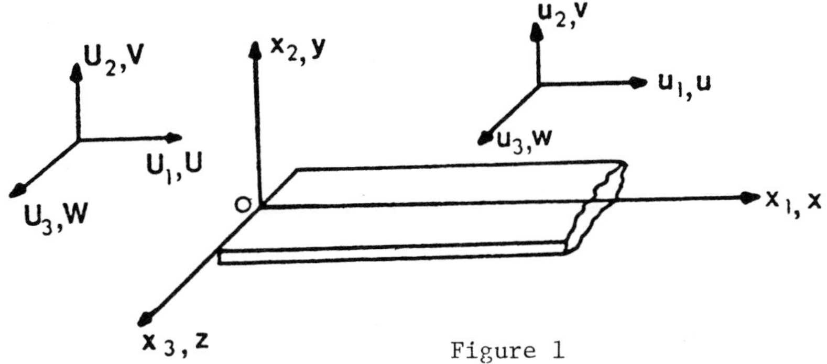

Figure 1

Assuming homogeneity in the x and z directions and using 2-D Fourier transforms defined as

$$\underline{u}(\underline{x}) = (2\pi)^{-2} \iint_{-\infty}^{\infty} e^{i\underline{k}\cdot\underline{x}}\, \underline{\psi}(y,\underline{k})\, d\underline{k}$$

where, $\underline{k}$ = wave number pair (k_1,0,k_3), $\underline{x}$ = (x,y,z), leads (via conservation of mass) to

$$\psi_3 = \frac{iD^+\psi_2 - k_1^+\psi_1}{k_3^+} \qquad (7)$$

$$[k^{+2}(\nu^+\nabla^{+2} + \frac{d\nu^+}{dy^+} D^+)]\psi_1 = [\frac{1}{2}\frac{dU^+}{dy^+} k_3^{+2} + ik_1^+\nu^+\nabla^{+2}D^+ + ik_1^+ \frac{d\nu^+}{dy^+} D^{+2}]\psi_2 \quad (8)$$

$$[\nu^+\nabla^{+2}(k_3^{+2} - D^{+2}) + \frac{d\nu^+}{dy^+}\{k_3^{+2} + k^{+2} - 2D^{+2})D^+\} - \frac{d^2\nu^+}{dy^{+2}}(k_3^{+2} + D^{+2})\psi_2$$

$$= [\frac{1}{2} k_3^{+2} \frac{dU^+}{dy^+} + ik_1^+(\nu^+\nabla^{+2} + \frac{d\nu^+}{dy^+} D^+)D^+ + ik_1^+ \frac{d\nu^+}{dy^+}\nabla^{+2} + ik_1^+ \frac{d^2\nu^+}{dy^{+2}} D^+]\psi_1 \quad (9)$$

where the superscript "+" denotes the nondimenstional parameters:

$$L = \frac{\nu_o}{v^*},\ v^* = \sqrt{\frac{\tau\ wall}{\rho}},\ U^+ = \frac{U}{v^*},\ \underline{x}^+ = \underline{x}/L\ ,\ k^+ = \underline{k}L$$

$$\nu^+ = \frac{\nu}{\nu_o},\ D^+ = \frac{d}{dy^+},\ \nabla^{+2} = D^{+2} - k^{+2},\ v^* = \text{friction velocity}$$

ν_o = molecular viscosity, τ = shear stress, and ρ = density.

Cross differentiating partly to untangle* the $\psi_1 - \psi_2$ couplings leads to

$$-\nu^+\nabla^{+4}{}_2 = \frac{1}{2} U^{+'} k^{+2}\psi_1 + \frac{1}{2} ik_1^+ D^+(U^{+'}\psi_2) + 2\frac{d\nu^+}{dy^+}\nabla^{+2}D^+\psi_2 + k^{+2}\frac{d^2\nu^+}{dy^{+2}}(1 + \frac{1}{k_3^{+2}} D^{+2})\psi_2 \quad (10)$$

$$-\nu^+\nabla^{+4}\psi_1 = \frac{1}{2} ik_1^+ D^+(U^{+'}\psi_1) + \frac{1}{2}(k_3^{+2} - D^{+2})(U^{+'}\psi_2) + \frac{d\nu^+}{dy^+}[\nabla^{+2}(3D^+\psi_1 - ik_1^+\psi_2)] + \frac{d^2\nu^+}{dy^{+2}}[(3D^{+2} - k^{+2})\psi_1 + 2ik_1^+(\frac{1}{k_3^{+2}} D^{+2} - 1)D^+\psi_2] + \frac{d^3\nu^+}{dy^{+3}}[D^+\psi_1 + ik_1^+(1 + \frac{1}{k_3^{+2}} D^{+2})\psi_2] \quad (11)$$

where, $U^{+'} = \frac{dU^+}{dy^+}$

Equations (10) and (11) are the most general system of ODE's which can be used to predict the eigen-modes of the flat-plate boundary-layer. Equations (10) and (11) reduce to Payne's [12] equations for a constant eddy viscostiy model which was verified empirically for the central 90% of the wake. Because the non-dimensionalization has been done with the appropriate length and velocity scales, the total viscosity cofficient ν^+, becomes a reciprocal turbulent Reynolds number, $\nu^+ \to (R_T)^{-1}$ for the case of constant eddy viscosity. Henceforth, the superscript "+" will be dropped for brevity.

---

*Simplify

3. MEAN VELOCITY PROFILE AND THE EDDY VISCOSITY MODEL

The mean velocity profile and the eddy viscosity model based on the experiments of Schultz-Grunow [4] and curve fitted by Hong [16] used herein are:

$$U = \begin{cases} R(V^{**})^2 y & 0 \leqslant y \leqslant 0.0024 \\ 0.09669 \,\ell n\, y + 0.93161 & 0.0024 \leqslant y \leqslant 0.20 \\ 0.3232 \sin\left(\frac{\pi}{2} y\right) + 0.6768 \quad ; & 0.20 \leqslant y \leqslant 1.0 \\ 1.0 \quad ; & y > 1.0 \end{cases} \tag{12}$$

where U is non-dimensionalized with the free stream velocity, R is the Reynolds number based on the boundary layer thickness, δ; V^* is the wall friction velocity, V^{**} is the ratio of the wall friction velocity to the free stream velocity and y has been normalized using δ.

The total effective viscosity, ν, is modeled as:

$$\nu = \begin{cases} 1.0 & 0 \leqslant y \leqslant 0.0006623 \\ 0.37\, R(V^{**}) y & 0.0006623 \leqslant y \leqslant 0.147 \\ 0.016\, U_o \delta^*/\nu_o \quad ; & 0.147 \leqslant y < 1.0 \\ 1.0 & y \geqslant 1.0 \end{cases} \tag{13}$$

where, the total effective viscosity, ν, is normalized with the molecular viscosity, ν_o. The numerical values of the parameters used in Eq. (12) and (13) are: $U_o = 1940$ cm/sec; $x_1 =$ 530 cm; $V^{**} = 0.0358$; $\delta = 8.4$cm; $\delta^* = 1.022$ cm and $\nu_o = 0.142963$ cm^2/sec.

Based on the above non-dimensional parameters and the piece-wise linear eddy viscosity model, the Lumley stability equations (10) and (11) in the inner, overlap and outer layers reduce to [5,10]:

$$\nabla^4\psi_1 = -\frac{1}{2}\, ik_1 D(U'\psi_1) - \frac{1}{2}(k_3^2 - D^2)(U'\psi_2) = F_{11}; \quad 0 \leqslant y \leqslant 0.0006623,\ y \geqslant 1 \tag{14a}$$

$$\nabla^4\psi_2 = -\frac{1}{2}\, k^2 U'\psi_1 - \frac{ik_1}{2} D(U'\psi_2) = F_{21}; \quad 0 \leqslant y \leqslant 0.0006623,\ y \geqslant 1 \tag{14b}$$

$$\nabla^4\psi_1 = -\frac{1}{2}\frac{ik_1}{\nu} D(U'\psi_1) - \frac{1}{2\nu}(k_3^2 - D^2)(U'\psi_2) - \frac{A}{\nu}\nabla^2(3D\psi_1 - ik_1\psi_2) = F_{12}; \quad 0.0006623 \leqslant y \leqslant 0.147 \tag{14c}$$

$$\nabla^4\psi_2 = -\frac{1}{2}\frac{k^2}{\nu} U'\psi_1 - \frac{ik_1}{2\nu} D(U'\psi_2) - \frac{2}{\nu} A\nabla^2\psi_2 = F_{22}; \quad 0.0006623 \leqslant y \leqslant 0.147 \tag{14D}$$

where, $A = 0.37\, RV^{**}$

$$\nabla^4\psi_1 = -\frac{1}{2}\frac{ik_1}{\nu} D(U'\psi_1) - \frac{1}{2\nu}(k_3^2-D^2)(U'\psi_2) = F_{13} \;;\; 0.147 \leqslant y < 1.0 \tag{14e}$$

$$\nabla^4\psi_2 = -\frac{1}{2}\frac{ik^2}{\nu} U'\psi_1 - \frac{ik_1}{2\nu} D(U'\psi_2) = F_{23};\quad 0.147 \leqslant y < 1.0 \tag{14f}$$

respectively.

4. BOUNDARY CONDITIONS

The boundary conditions along the wall (no-slip) and far from the plate are

$$\psi_1(0) = \psi_2(0) = \psi_1(\infty) = \psi_2(\infty) = 0 \tag{15a}$$

where the condition at $y\to\infty$ is justified by the finite total energy of the perturbations. Continuity yields:

$$D\psi_2(0) - D\psi_2(\infty) = 0 \tag{15b}$$

The shear stress and vorticity equations give:

$$D\psi_1(\infty) = 0 \tag{15c}$$

Assuming the value of the eddy viscosity to be dominant with respect to the molecular viscosity results in:

$$D^2\psi_1(0) = 0 \tag{15d}$$

5. METHOD OF SOLUTION

The governing differential equations (14) are converted in to Fredholm integral equations [9,16]:

$$\psi_1(y^*,\underline{k}) = \int_o^{a_1} F_{11}G_1(y,y^*)dy + \int_{a_1}^{a_2} F_{12}G_1(y,y^*)dy + \int_{a_2}^{a_3} F_{13}G_1(y,y^*)dy$$
$$+ \int_{a3}^{\infty} F_{11}G_1(y,y^*)dy \tag{16a}$$

$$\psi_2(y^*,\underline{k}) = \int_o^{a_1} F_{21}G_2(y,y^*)dy + \int_{a_1}^{a_2} F_{22}G_2(y,y^*)dy + \int_{a_2}^{a_3} F_{23}G_2(y,y^*)dy$$
$$+ \int_{a_3}^{\infty} F_{21}G_2(y,y^*)dy \tag{16b}$$

where the Green's function G_1, G_2 are given in [9].

Numerical differentiation of the $G_i \in C^2$ cannot be avoided.

Iterative solution of eqs. (16) ala Lumley [1] consists of any "reasonable" initial guess for ψ_i, and rewriting Eq. (16) as a matrix enginvalue problem:

$$K_{ij}\psi_i = \lambda\psi_i \tag{17}$$

Lumley has shown uniform convergence in y*:

$$\lim_{n\to\infty} \frac{\psi_i^{(n+1)}}{\psi_i^{(n)}} \to \lambda_1 \quad \text{, largest eignevalue} \tag{18}$$

where, $K_{ij} = \sum_{n=1}^{\infty} \lambda_n \psi_i^{(n)}(y^*)\psi_j^*(y)$; $\lambda_1 \geqslant \lambda_2 \geqslant \lambda_3 \ldots \geqslant 0$ and the ψ_i are orthonormal. K_{ij} Hermitian ensures all eigenvalues $\geqslant 0$. Since the scheme will fail or convergence slowly if $\psi_i^{(o)}$ are orthogonal to the true eigenfunction, ψ_1. Some care is necessary in the initial approximation.

The wave number grid was spaced logarithmically $\in[0.5,3.2]$. The upper limit of the itegral was chosen by numerical experiments which suggested changes of less than 5% for beyond 5 . The convergence criterion was 10^{-4} (relative).

6. VELOCITY RECONSTITUTION

The velocity field can be written as

$$u_i(\underline{x}) = -\frac{2}{\pi}\int_0^\infty\int_0^\infty \sqrt{\lambda^{(1)}}\,[\mathrm{Re}(\psi_i)\cos k_1x_1 + \mathrm{Im}(\psi_i)\sin k_1x_1]\cos k_3x_3\,d\underline{k};\quad i=1,2 \tag{19a}$$

$$u_3(\underline{x}) = -\frac{2}{\pi}\int_0^\infty\int_0^\infty \sqrt{\lambda^{(1)}}\,[\mathrm{Re}(\psi_i)\sin k_1x_1 + \mathrm{Im}(\psi_i)\cos k_1x_1]\sin k_3x_3\,d\underline{k} \tag{19b}$$

which are calculated for $0 \leqslant x_1 \leqslant 2.0$ and $0 \leqslant x_3 \leqslant 2.0$, equivalent a "snapshot" of the dominant mode. Fourier transforms destroy all phase information. Phase relationships for any $\underline{k}$-pair are correct; how the solutions over $\underline{k}$-space are relatively aligned so that constructive (destructive) interactions are preserved is unclear. The eigenvalue was assumed proportionate to the mean square energy of each mode (Lemmerman [13]).

7. RESULTS

Calculations on a 7x7 grid yielded Figs. 2-3. The eigenvalues are tabulated in Table I. Table II shows location of the maximum ψ_i, [10].

The results agree qualitatively with those obtained by PODT extraction [13]. It is likely that the predicted modes are instabilities of the mean profile alone, whereas the "real" large scale structure undoubtedly is modified by its own instabilities and self-interactions.

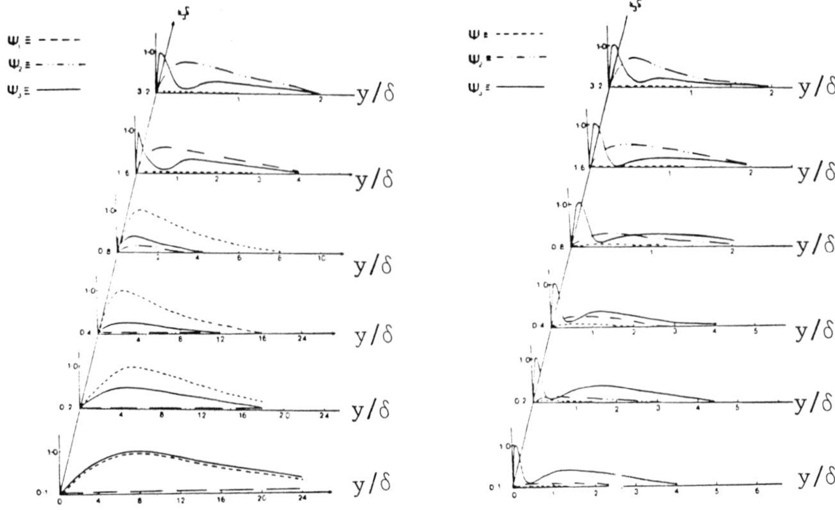

Fig. 2. Eigenfunction Amplitude for $k_1\delta=0.1$

Fig. 3. Eigenfunction Amplitude for $k_1\delta= 1.6$

k_3 \ k_1	0.05	0.10	0.20	0.40	0.80	1.60	3.20
0.05	0.64	0.63	0.59	0.52	0.43	0.39	0.30
0.10	0.63	0.62	0.59	0.53	0.43	0.43	0.31
0.20	0.59	0.59	0.56	0.51	0.43	0.43	0.31
0.40	0.52	0.52	0.51	0.47	0.43	0.42	0.31
0.80	0.43	0.43	0.43	0.43	0.42	0.40	0.31
1.60	0.40	0.40	0.40	0.40	0.39	0.36	0.31
3.20	0.30	0.30	0.30	0.30	0.31	0.29	0.26

Table I. Largest Eigenvalues for Each Wave Number Pair

k_3 \ k_1	0.05	0.1	0.2	0.4	0.8	1.6	3.2
0.05	14.10	8.95	4.85	2.50	1.30	0.70	0.40
0.1	8.95	7.05	4.50	2.40	1.20	0.70	0.40
0.2	4.85	4.50	3.50	2.20	1.20	0.70	0.40
0.4	2.50	2.40	2.20	1.80	1.10	0.60	0.30
0.8	1.30	1.20	1.20	1.10	0.90	0.60	0.30
1.6	0.70	0.70	0.70	0.60	0.60	0.50	0.30
3.2	0.40	0.40	0.40	0.30	0.30	0.30	0.30

Table II. The Location $((y/\delta)_{max}$ of the Maximum Amplitude

8. BRIEF DISCUSSION OF THE NUMERICAL RESULTS

The ψ_1,ψ_3 amplitudes show a strong relationship with k_1/k_3 ratio; for $k_1,k_3 < 1.0$ results agree well with Ref.[13]. These results, however, do not agree with OLP predictions for large wave number pairs (smaller scales).

This are surprising qualitative agreements with Hong's [16] analysis for some scale ranges due to the 2-D field of the dominant mode, i.e., in some regions the "big eddy" is mostly x-y motion, x-z in others, etc. Rarely are all three components of comparable magnitueds at the same scale. This may explain the so-called "tendency toward isotropy" observed far downstream of a grid. Lumley's PODT solution [1] for isotropic turbulence contained pure harmonics interlocked in a "3-D chainmail." Generally $Re_{critical}$ increased with wave number (suggesting that large scales are more unstable and energetic than small scales) but there are two "peaks" corresponding with Lemmerman's. Magnitudes display at least two "families" of eddies, roughly corresponding to the "wall eddies" (small scale) and "big eddies" (probably "bursting", "sweep", etc.)

The neutral modes [20] may dominate fully developed turbulence; conversely, lack of such a dominant family may imply either post-instability regions of transition and/or "developing turbulence": the basic OLP assumption is global instantaneous, stationary energy in the disturbance mode (turbulent "eddy" if the mean profile is turbulent), hence, the only way to change either eigenvalue phase or amplitude to vary the mean flow. Payne [12] observed an abrupt change in PODT mode shape at $\underline{k}$-values for which the first and second eigenvalues became equal. He interpreted [12] this as a change from "symmetric" to "anti-symmetric" modes. Therefore, one should suspect two (or more) eigenvalues of nearly the same magnitude if the calculated functional form of the eigenmodes exhibits "rough" rather than "smooth" behaviour across the wave number grid. The modes in Fig. 2 showed smooth behaviour. Hence, the conclusion is that one eddy family is dominant.

9. CONCLUSIONS

The two OLP analysis, Hong's [16] 2-D and the current 3-D, of the flat-plate flow agreed rather well qualitatively; i.e., the 3-D neutral modes generally were nearly planar but in different orthogonal planes across wave number space. Herein may lie the driver toward isotropy noted far downstream of a grid. A caution is in order here due to OLP results being incapable, for now, of rational inverse transformation back to laboratory coordinates; this could modify the last statement but such modifications should be minor rather than dramatic.

Doubling the value of the eigenvalue generally corresponds to a decrease in the neutral mode wave length by a factor of about 60 which is consistent with the concept that increasing Reynolds number excites many new modes. Only the first mode has been calculated but one could deflate the matrix and calculate higher modes. Comparison of the predicted neutral mode with Lemmerman's eddies is good over most of the gird. At extreme scales results are less satisfactory due to either breakdown of homogeneity or small scales where probe size is inadequate. OLP results should be valid for the small scales (to extend empiric PODT) but probably not for the largest scales.

10. PROGNOSES

PODT-SAS extracts the energy and 3-D structure for the dominant eddies from empiric data and is complimentary to conditional sampling, a filtered "slice" of the complete structure. PODT determines the largest scale which can be modeled via an eddy viscosity approach. OLP is predictive in nature, involves a seventh-order, coupled, linear PDE system, and provides the eigenmodes spectral extent for global energy equilibrium.

OLP depends upon two extrema: (a) Orr's of stationary global disturbance kinetic energy and (b) Lumley's of maximal eddy viscosity. The equations are determined by energy forms. The mean velocity and eddy viscosity must be known (functional form only for ν_e, not amplitude). The eigenmodes are in global energy equilibrium with both the mean flow (Orr) and the "smaller" eddies of the turbulent motion (ala Townsend) and are those modes which drain maximum energy from the mean flow (Townsend, Lumley) if the mean flow is in a state of energetic equilibirum or "full developed". The eigenvalues are stability parameters. Intuitively, one expects the OLP-eddies to resemble, at least, the PODT-eddies; however, each new flow geometry should be verified for this expectation.

11. REFERENCES

1. LUMLEY, J.L., Dok. Akad. Nauk., Moscow, USSR, 1966.
2. GRANT, H.L., JFM, 4, pp. 149-190, 1958.
3. TRITTON, D.J., JFM, 28, pp. 439-462, 1967.
4. SCHULTZ-GRUNOW, F., NACA-TM-986, 1940.
5. RICHARDSON, See Tennekes, H. & Lumley, J.L., A First Course in Turbulence, MIT Press, 1972.
6. TOWNSEND, A.A., The Structure of Turbulent Shear Flow, Cambridge Press, 1956.
7. LOEVE, M., Probability Theory, V. Nostrand Co., 1960.
8. PAYNE, F.R., Final Report, NASA/AMES Grant NSG-2077, 1980.
9. WADIA, A.R., Ph.D. Dissertation, UTA Arlington, TX, 1979.
10. ORR, W.M., Proc. Royal Irish Acad.,Vol.28,A, pp.122, 1907.
11. LUMLEY, J.L., Internal Memo, ORL/Penn State Univ., 1967.
12. PAYNE, F.R., Predicted Large Eddy Structure of a Turbulent Wake, Report to USN/ONR, NONR656(33), 1968.
13. LEMMERMAN, L.A. & PAYNE, F.R., AIAA Paper No. 77-717, 1977.
14. PAYNE, F.R., Intl.Conf. on Non-Linear Eqs. in Abstract Spaces, Academic Press, pp. 417-437; 1978.
15. PAYNE, F.R., NASA CP-2032, pp. 260-266, 1977.
16. HONG, S.K., MSAE Thesis, UT-Arlington, TX, 1978.
17. BAKEWELL, H.P., Ph.D. Thesis, Penn State Univ., 1966.
18. WADIA, A.R., J. of Comp & Appl. Math, Vol. 6, No. 2, 1980.
19. WADIA, A.R., Intl. J. of Comp. Math., Vol. 8, No. 2, 1980.
20. PAYNE, F.R., NonLinear Phenomena in Mathematical Sciences, Academic Press, June 18-20, 1980.

ON THE SIMULATION OF THE COUPLED LAMINAR BOUNDARY LAYERS AT A SMOOTH, PHASE-CHANGING GAS-LIQUID INTERFACE

Stephen P. Klotz[i]

Robert L. Street[ii]

SUMMARY

There are a number of industrial and physical processes which depend on the laws of heat and mass transfer with phase conversion. In particular, engineering problems such as two-phase flow, film cooling, condensation, gas adsorption, and drying require knowledge of the simultaneous heat and mass transfer in gas-liquid flows with a phase-changing interface between the fluids. The coupled, laminar boundary-layer equations of heat, mass, and momentum at a smooth interface are solved numerically for flows with and without pressure gradients using the Keller Box method. For self-similar flows, Runge-Kutta integration provides a basis with which to compare calculations obtained by the more general finite-difference scheme.

1. INTRODUCTION

Many industrial and physical processes rely on a phase-changing gas-liquid interface for the transport of heat and specie from one fluid regime to the other. Engineering applications include drying, film cooling, and condensation and occur in such process equipment as cooler condensers, gas adsorption and distillation columns, dehumidifiers, evaporators, and catalytic reactors. Gas-liquid boundary-layer flows are also relevant to problems involving ablation or combustion and to the physical processes governing the exchange of heat and water vapor between the oceans and atmosphere.

---

[i]Research Assistant, Department of Civil Engineering, Stanford University, U.S.A.

[ii]Professor, Department of Civil Engineering, Stanford University, U.S.A.

In this paper we consider the coupled boundary-layer problem in the context of laminar flow at a smooth, phase-changing interface. Lock [1] analyzed the coupled momentum problem without heat or mass transfer and provided several self-similar solutions which were obtained numerically. More recently, Kotake [2] included the effects of heat and mass transport in the self-similar problem. He pointed out that all parameters of interest, such as the surface velocity, shear stress, interfacial temperature, and surface mass transfer rate, should be obtained implicitly from solution of the governing differential equations in each fluid subject to the appropriate boundary conditions at the common interface.

We solve the laminar problem using the Box scheme of Keller and Cebeci [3] and obtain solutions for both similar and non-similar boundary-layer flows. This scheme, which is efficient, versatile, and well documented, is extended here to treat the coupled problem. Self-similar solutions can also be obtained numerically using standard techniques for integrating ordinary differential equations, thereby providing a test of accuracy for the more general finite-difference method.

2. BASIC EQUATIONS

We consider, as depicted in Figure 1, the problem of a constant-property binary mixture of perfect gases flowing over a single component liquid. Across the common interface between them the transport of liquid vapor can occur by condensation or evaporation. The flow field is assumed to be two-dimensional, steady, and laminar in both fluid regimes and the interface is assumed to remain horizontal. We treat enthalpy and specie concentration as conserved properties in the heat and mass transfer problems. A free-stream pressure gradient in the gaseous regime and a mean current far beneath the interface in the liquid may exist.

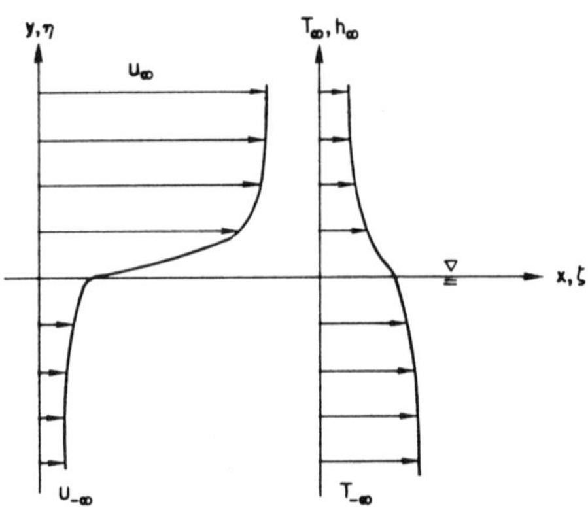

Figure 1. The coupled gas-liquid problem.

The equations we wish to solve are analogous to, but much more general than, those considered by Kotake [2]. Let ψ be the stream function in physical (x,y) coordinates and f a non-dimensional stream function in the transformed coordinate system (ζ,η). Using the classical Falkner-Skan transformation (subscripts 1 and 2 refer, respectively, to the gas and liquid regimes),

$$\psi_1(x,y) = \sqrt{U_\infty \nu_1 x}\; f_1(\zeta,\eta_1)$$

$$\psi_2(x,y) = \sqrt{U_\infty \nu_2 x}\; f_2(\zeta,\eta_2)$$

$$\eta_1 = \sqrt{\frac{U_\infty}{\nu_1 x}}\;, \quad \eta_2 = \sqrt{\frac{U_\infty}{\nu_2 x}}\;, \quad x = \zeta$$

and the definition of the stream function, the differential equations governing this flow are given by

(gas:momentum)

$$\frac{\partial^3 f_1}{\partial \eta_1^3} + \frac{m+1}{2} f_1 \frac{\partial^2 f_1}{\partial \eta_1^2} + m\left[1 - \left(\frac{\partial f_1}{\partial \eta_1}\right)^2\right] = \zeta\left(\frac{\partial f_1}{\partial \eta_1}\frac{\partial^2 f_1}{\partial \zeta \partial \eta_1} - \frac{\partial f_1}{\partial \zeta}\frac{\partial^2 f_1}{\partial \eta_1^2}\right) \tag{1a}$$

(gas:enthalpy)

$$\frac{1}{Pr_1}\frac{\partial^2 g_1}{\partial \eta_1^2} + \frac{m+1}{2} f_1 \frac{\partial g_1}{\partial \eta_1} = \zeta\left(\frac{\partial f_1}{\partial \eta_1}\frac{\partial g_1}{\partial \zeta} - \frac{\partial f_1}{\partial \zeta}\frac{\partial g_1}{\partial \eta_1}\right) \tag{1b}$$

(gas:mass concentration)

$$\frac{1}{Sc}\frac{\partial^2 h_2}{\partial \eta_1^2} + \frac{m+1}{2} f_1 \frac{\partial h_1}{\partial \eta_1} = \zeta\left(\frac{\partial f_1}{\partial \eta_1}\frac{\partial h}{\partial \zeta} - \frac{\partial f_1}{\partial \zeta}\frac{\partial h}{\partial \eta_1}\right) \tag{1c}$$

(liquid:momentum)

$$\frac{\partial^3 f_2}{\partial \eta_2^3} + \frac{m+1}{2} f_2 \frac{\partial^2 f_2}{\partial \eta_2^2} + m\left[\frac{\rho_1}{\rho_2} - \left(\frac{\partial f_2}{\partial \eta_2}\right)^2\right] = \zeta\left(\frac{\partial f_2}{\partial \eta_2}\frac{\partial^2 f_2}{\partial \zeta \partial \eta_2} - \frac{\partial f_2}{\partial \zeta}\frac{\partial^2 f_2}{\partial \eta_2^2}\right) \tag{1d}$$

(liquid:enthalpy)

$$\frac{1}{Pr_2}\frac{\partial^2 g_2}{\partial \eta_2^2} + \frac{m+1}{2} f_2 \frac{\partial g_2}{\partial \eta_2} = \zeta\left(\frac{\partial f_2}{\partial \eta_2}\frac{\partial g_2}{\partial \zeta} - \frac{\partial f_2}{\partial \zeta}\frac{\partial g_2}{\partial \eta_2}\right) \tag{1e}$$

Here g_1 and g_2 are non-dimensional enthalpies defined by $g_1 = i_1/i_\infty$ and $g_2 = i_2/i_\infty$, h is the mass concentration of vapor in the binary mixture, and $m = (\zeta/U_\infty)(dU_\infty/d\zeta)$ is the pressure-gradient parameter relating the streamwise pressure gradient to the inviscid external-velocity distribution far above the interface.

The interfacial and far-field conditions in stream function coordinates are

$$\left.\frac{\partial f_1}{\partial \eta_1}\right|_o = \left.\frac{\partial f_2}{\partial \eta_2}\right|_o \tag{2a}$$

$$\left.\frac{\partial^2 f_1}{\partial \eta_1^2}\right|_o = \Lambda \left.\frac{\partial^2 f_2}{\partial \eta_2^2}\right|_o , \quad \Lambda = \frac{\rho_2}{\rho_1}\sqrt{\frac{\nu_2}{\nu_1}} \tag{2b}$$

$$g_{1o} - \left(1 + (d_1 - 1)h_o\right)\left(d_2 g_{2o} - d_3\right) = 0 \tag{2c}$$

$$d_1 = c_{p_G}/c_p , \quad d_2 = c_{p_G}/c_L , \quad d_3 = (L/i_\infty)d_2$$

$$(m + 1)f_{1o} + 2\zeta \left.\frac{\partial f_1}{\partial \zeta}\right|_o = \Lambda\left[(m + 1)f_{2o} + 2\zeta \left.\frac{\partial f_2}{\partial \zeta}\right|_o\right] \tag{2d}$$

$$f_{1o} = \frac{2}{m + 1}\left[\frac{\partial h/\partial \eta_1|_o}{Sc(1 - h_o)} - \zeta \left.\frac{\partial f_1}{\partial \zeta}\right|_o\right] \tag{2e}$$

$$\dot{M}\frac{L}{i_\infty} - \frac{1}{Pr_1}\left.\frac{\partial g_1}{\partial \eta_1}\right|_o - \frac{\Lambda}{Pr_2}\left.\frac{\partial g_2}{\partial \eta_2}\right|_o = 0 \tag{2f}$$

$$\dot{M} = -\frac{1}{Sc(1 - h_o)}\left.\frac{\partial h}{\partial \eta_1}\right|_o$$

$$h_o = \exp\left[\frac{L}{R_v}\left(\frac{1}{T_B} - \frac{1}{T_o}\right)\right] , \quad T_o = \frac{i_\infty g_{1o}}{c_{p_G}\left(1 + (d_1 - 1)h_o\right)} \tag{2g}$$

$$\frac{\partial f_1}{\partial \eta_1} \to 1 , \quad g_1 \to 1 , \quad h \to h_\infty \quad \text{as} \quad \eta_1 \to \infty \tag{2h}$$

$$\frac{\partial f_2}{\partial \eta_2} \to \frac{U_{-\infty}}{U_\infty} \equiv \lambda , \quad g_2 \to \frac{i_{-\infty}}{i_\infty} \quad \text{as} \quad \eta_2 \to -\infty \tag{2i}$$

Equations (2a) and (2b) are continuity relations for the interfacial velocity and stress, respectively, and (2c) follows from continuity of temperature across the interface. Equations (2d) and (2e) are expressions for the conservation of mass, (2f) is an equation for the conservation of energy, and (2g) relates the concentration of vapor in the saturated mixture at the interface to the temperature there.

In the derivation of equations (1)-(2) we have considered only dissipationless fluids at low velocity with negligible buoyancy and have neglected the diffusive transport of heat by any vapor concentration gradient in the gas. The additional terms corresponding to this transport can be included easily in the analysis, but they do not materially affect the results and conclusions we present here. We have also assumed that i_∞, $i_{-\infty}$, and h_∞ are constants and have used the following equations relating temperature and enthalpy (with a datum of zero enthalpy in the gas at 0°C):

$$i_1 = c_{p_G}\left(1 + (d_1 - 1)h\right)T_1$$

$$i_2 = c_L T_2 - L$$

3. SOLUTION METHOD

Equations (1)-(2) constitute the mathematical formulation of the coupled boundary-layer problem in transformed coordinates. This set of equations is parabolic and solutions must be obtained in a marching scheme commencing with input profiles at an initial streamwise station and proceeding downstream in the direction of the flow. To solve the problem (1)-(2) we employ the Box scheme of Keller and Cebeci [3] which has been developed for flows of this type and used to solve a variety of boundary-layer problems. We first rewrite the equations as a system of non-linear first-order equations before applying the centered difference approximations inherent in the method. The resulting system of non-linear difference equations are second-order accurate, fully implicit in the cross-stream direction η, and must be solved iteratively at each streamwise station. Details of this process are described elsewhere [4,5,6] and are omitted here. A typical non-uniform net for the coupled problem at which the difference approximations are applied is contained in Figure 2. The solution to all the difference equations is known at ζ^{n-1} and sought at the station represented by ζ^n. The interfacial coupling requires that the interface in this net consist of two arbitrarily close, horizontal grid lines, one in the gaseous regime and the other in the liquid.

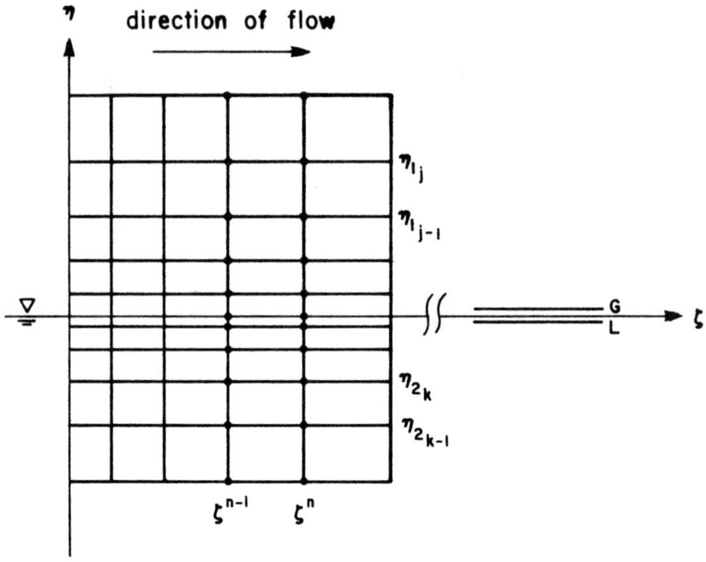

Figure 2. A typical net with arbitrary spacing. The interfacial coupling is performed along two superimposed horizontal grid lines, one in each fluid regime.

The algorithm for solving the gas-liquid problem closely parallels that employed for boundary layers at solid surfaces (see, e.g. [4]). Since we consider constant-property fluids, the only coupling between the momentum and the heat and mass transfer problems occurs in the interfacial condition (2e). Consequently, it is convenient to linearize the difference equations for the momentum problem and solve them separately from the enthalpy and specie-concentration equations using an efficient block tri-diagonal factorization scheme. Since the system of linearized equations is implicit, the scheme is unconditionally stable. Once a momentum solution has been generated, the difference equations for the heat and mass transfer problem become linear and can be solved directly with their own factorization schemes.

Consider for a moment only the momentum problem. Adopting the notation of Keller [6], let $\underset{\sim}{\delta}_{1_j}$ be a vector of the Newton iterates at the point (n,j) above the interface (see Figure 2). Since the Box scheme requires us to rewrite (1)-(2) as a system of first-order equations, the momentum equation (1a) becomes a system of three first-order, nonlinear equations. Hence, at any streamwise station we can write the iterates for the p+1 iteration as

$$\underset{\sim}{\delta}_{1_j} \equiv \begin{bmatrix} \delta f_{1_j} \\ \delta u_{1_j} \\ \delta v_{1_j} \end{bmatrix} = \begin{bmatrix} f_{1_j}^{n,p+1} \\ u_{1_j}^{n,p+1} \\ v_{1_j}^{n,p+1} \end{bmatrix} - \begin{bmatrix} f_{1_j}^{n,p} \\ u_{1_j}^{n,p} \\ v_{1_j}^{n,p} \end{bmatrix}$$

where $u_1 = \partial f_1/\partial\eta_1$ and $v_1 = \partial u_1/\partial\eta_1$. The resulting linearized system of equations can then be written in the form

$$B_{1_j}^{n,p}\underset{\sim}{\delta}_{1_{j-1}}^{n,p+1} + A_{1_j}^{n,p}\underset{\sim}{\delta}_{1_j}^{n,p+1} + C_{1_j}^{n,p}\underset{\sim}{\delta}_{1_{j+1}}^{n,p+1} = \underset{\sim}{r}_{1_j}^{n,p} \ , \qquad j = 0,\ldots,J$$

Similarly, below the interface we obtain the system

$$C_{2_k}^{n,p}\underset{\sim}{\delta}_{2_{k-1}}^{n,p+1} + A_{2_k}^{n,p}\underset{\sim}{\delta}_{2_k}^{n,p+1} + B_{2_k}^{n,p}\underset{\sim}{\delta}_{2_{k+1}}^{n,p+1} = \underset{\sim}{r}_{2_k}^{n,p} \ , \qquad k = -K,\ldots,0$$

Here J and K index the net points above and below the interface, respectively, at which the far-field boundary conditions are applied. The B_2 and C_2 are 3×3 matrices which preserve the special structure of the B_1 and C_1 matrices which appear in the difference equations for flows over solid surfaces (see, e.g. [5]).

$$B_{1_j}^{n,p} = \begin{bmatrix} x & x & x \\ x & x & x \\ 0 & 0 & 0 \end{bmatrix} \quad 1 \leqslant j \leqslant J \ ; \quad C_{1_j}^{n,p} = \begin{bmatrix} 0 & 0 & 0 \\ 0 & 0 & 0 \\ x & x & x \end{bmatrix} \quad 0 \leqslant j \leqslant J-1$$

$$B_{2_k}^{n,p} = \begin{bmatrix} x & x & x \\ x & x & x \\ 0 & 0 & 0 \end{bmatrix} \quad -K \leqslant k \leqslant -1 \ ; \quad C_{2_k}^{n,p} = \begin{bmatrix} 0 & 0 & 0 \\ 0 & 0 & 0 \\ x & x & x \end{bmatrix} \quad -K+1 \leqslant k \leqslant 0$$

$$B_{1_o}^{n,p} = \begin{bmatrix} 0 & 0 & 0 \\ 0 & 0 & x \\ 0 & 0 & 0 \end{bmatrix} \ , \quad B_{2_o}^{n,p} = \begin{bmatrix} 0 & 0 & 0 \\ 0 & x & 0 \\ 0 & 0 & 0 \end{bmatrix}$$

In block tri-diagonal form the system of difference equations becomes

$$\underset{\sim}{P}\ \underset{\sim}{\delta} = \underset{\sim}{r} \tag{3a}$$

where

$$
\underset{\sim}{P} = \left[\begin{array}{cccccccccc}
A_{2_{-K}} & B_{2_{-K}} & & & & & & & & \\
C_{2_{-K+1}} & A_{2_{-K+1}} & B_{2_{-K+1}} & & & & & & & \\
 & \cdot & \cdot & \cdot & & & & & & \\
 & & C_{2_{-1}} & A_{2_{-1}} & B_{2_{-1}} & & & & & \\
 & & & C_{2_o} & A_{2_o} & B_{2_o} & & & & \\
 & & & & B_{1_o} & A_{1_o} & C_{1_o} & & & \\
 & & & & & B_{1_1} & A_{1_1} & C_{1_1} & & \\
 & & & & & & \cdot & \cdot & \cdot & \\
 & & & & & & & B_{1_{J-1}} & A_{1_{J-1}} & C_{1_{J-1}} \\
 & & & & & & & & B_{1_J} & A_{1_J}
\end{array}\right] \qquad (3b)
$$

interface coupling matrices

and $\underset{\sim}{r}$ is a vector containing terms computed at the last (upstream) station and terms calculated from the previous iteration at the current station.

Similar analysis of the heat- and mass-transfer problem leads to a system of equations of the form

$$B_{H_{1_j}}^{n,p} \underset{\sim}{\Delta}_{H_{1_{j-1}}}^{n,p+1} + A_{H_{1_j}}^{n,p} \underset{\sim}{\Delta}_{H_{1_j}}^{n,p+1} + C_{H_{1_j}}^{n,p} \underset{\sim}{\Delta}_{H_{1_{j+1}}}^{n,p+1} = \underset{\sim}{r}_{H_{1_j}}^{n,p} \quad ; \quad j = 0,\ldots,J \qquad (4a)$$

$$B_{M_j}^{n,p} \underset{\sim}{\Delta}_{M_{j-1}}^{n,p+1} + A_{M_j}^{n,p} \underset{\sim}{\Delta}_{M_j}^{n,p+1} + C_{M_j}^{n,p} \underset{\sim}{\Delta}_{M_{j+1}}^{n,p+1} = \underset{\sim}{r}_{M_j}^{n,p} \quad ; \quad j = 0,\ldots,J \qquad (4b)$$

in the gas and of the form

$$B_{H_{2_k}}^{n,p} \underset{\sim}{\Delta}_{H_{2_{k-1}}}^{n,p+1} + A_{H_{2_k}}^{n,p} \underset{\sim}{\Delta}_{H_{2_k}}^{n,p+1} + C_{H_{2_k}}^{n,p} \underset{\sim}{\Delta}_{H_{2_{k+1}}}^{n,p+1} = \underset{\sim}{r}_{H_{2_j}}^{n,p} \quad ; \quad k = -K,\ldots,0 \qquad (4c)$$

in the liquid. Here, however, the A, B, and C matrices are 2×2 and the $\underset{\sim}{\Delta}$ vectors contain the unknowns to be computed. Of course the system (4a) for the enthalpy problem in the gaseous regime is coupled to the system (4c) through the interfacial conditions and the associated $\underset{\sim}{P}_H$ matrix in tri-diagonal form is analogous to the $\underset{\sim}{P}$ matrix in (3b) for the momentum problem.

As is evident in (3b), the interfacial boundary conditions are incorporated directly into the block tri-diagonal matrices for the momentum and enthalpy problems. Consequently, the simplicity and efficiency inherent in the Keller Box method is maintained, and since the difference equations at each streamwise station are fully implicit, quadratic convergence characteristic of Newton's method is preserved.

5. SOME SOLUTIONS

Solutions to the boundary-layer problem (1)-(2) exist for the following restricted classes of flows:

$$\text{(a)} \quad m = 0 \;, \quad 0 \leq \lambda < 1 \;; \text{ and}$$

$$\text{(b)} \quad m \neq 0 \;, \quad \lambda = \sqrt{\rho_1/\rho_2}$$

Self-similar solutions to the laminar problem exist if all dependent variables are independent of the streamwise coordinate ζ and if the pressure-gradient parameter m is a constant; the boundary-layer equations (1)-(2) then reduce to a set of coupled, non-linear ordinary differential equations which can be solved using standard techniques. Cebeci and Bradshaw [5] describe a shooting method using Runge-Kutta integration for obtaining solutions to the coupled momentum problem first solved by Lock [1]. We have used that procedure to treat the self-similar problem with heat and mass transfer, and to provide a test of accuracy for the finite-difference method.

					4th-order Runge Kutta				Finite Difference			
T_∞ (°C)	T_∞ (°C)	h_∞	m	λ	f_o'	f_o''	$\dot{M}$	T (°C)	f_o'	f_o''	$\dot{M}$	T (°C)
100	80	0	0	0	.2512	.2311	.1503	68.25	.02512	.2310	.1502	68.26
100	80	0	0	.1	.1087	.2076	.2028	73.22	.1087	.2074	.2027	73.22
100	80	0	0	.2	.2059	.1960	.2306	74.42	.2059	.1959	.2304	74.42
100	80	0	-.05	$\sqrt{\rho_1/\rho_2}$	.03943	.0970	.1532	71.20	.03942	.09681	.1530	71.21
100	80	0	.1	$\sqrt{\rho_1/\rho_2}$	.05057	.3747	.1936	70.59	.05057	.3746	.1934	70.59
100	80	0	.2	$\sqrt{\rho_1/\rho_2}$	.05250	.4893	.2083	70.41	.05250	.4891	.2081	70.41
10	20	0	0	0	.02313	.3257	.008696	19.03	.02311	.3253	.008684	19.03
10	20	.01	0	0	.02328	.3288	.004235	19.45	.02326	.3284	.004229	19.45

Table 1. Comparison of Computed Interfacial Parameters for Some Self-Similar Flows

Table 1 contains interfacial values of the velocity $f_o' = df/d\eta_1|_o$, shear stress $f_o'' = d^2f/d\eta^2|_o$, temperature T_o, and mass transfer rate $\dot{M}$, computed by the two methods for the air-water system under different bulk fluid conditions. All properties are evaluated at the bulk fluid temperatures, and one case can be directly compared to the results of Kotake [2]. The fourth-order Runge-Kutta method, which automatically adjusts step-size to control errors, is probably the more accurate of the two schemes because truncation errors are smaller (assuming other arithmetic errors are of comparable magnitude). Differences in the values computed by the two methods are generally limited to one or two digits in the last significant figure. The Runge-Kutta procedure suffers, however, from the disadvantages of reduced computational efficiency and the necessity of specifying initial estimates of the surface parameters. In some cases the specification of these must be inspired. Figures 3 and 4 contain mean profiles above and below the interface computed for one case using each method.

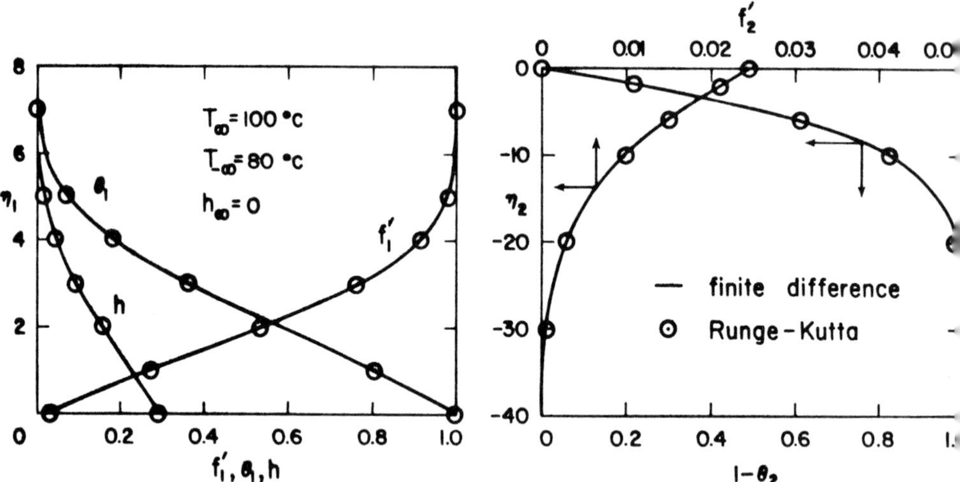

Figures 3 and 4. Mean profiles above and below the interface for the air-water system with T_∞ = 100°C, $T_{-\infty}$ = 80°C, h_∞ = 0, m = 0, and λ = 0. Here $\theta_1 = (T_\infty - T)/(T_\infty - T_0)$ and $\theta_2 = (T_{-\infty} - T)/(T_{-\infty} - T_0)$ are normalized temperature profiles, f_1' and f_2' are velocity profiles, and h is the concentration profile.

The effect of a mean current and pressure gradient on the interfacial quantities is also evident in Table 1. Increasing either the pressure gradient parameter m, or the mean current λ, results in increasing the surface velocity, shear stress, and mass transfer rate. One interesting case illustrates the additional effect of a small change in free-stream concentration on the surface parameters and, in particular, the surface temperature.

In contrast to the numerical calculations of Kotake [2][iii] our results clearly indicate that when the surface mass

[iii] We believe the differences between our work and that of Kotake for self-similar gas-liquid flows are related to computational errors in calculation of interfacial temperatures. As a consequence, his mean profiles which were illustrated graphically and which are strongly affected by the surface mass transfer rate, appear to be incorrect. In fact, we programmed his equations, which are correct, and integrated

transfer rate (which is primarily controlled by the bulk-liquid temperature and the free-stream vapor-concentration, and not by the temperature of the external flow) at an evaporating surface is sufficiently high, the interfacial temperature may be markedly depressed below the fluid temperatures far from the interface. This is not a surprising result when one recognizes that evaporating fluid must absorb its latent heat when it undergoes phase change.

Finally, we present the results of one non-similar boundary-layer calculation using the finite-difference method, and chosing a linearly accelerating flow for which $U_\infty = 1 + \zeta$, $T_{-\infty} = 80°C$, $T_\infty = 100°C$, and $\lambda = \sqrt{\rho_1/\rho_2}$. The variations of the interfacial velocity, shear stress, and the surface mass transfer rate, $\dot{M}$ with downstream distance are illustrated in Figure 5. The calculations were started from the self-similar solution at $\zeta = 0$ and were arbitrarily terminated at $\zeta = 0.55$ ($Re_\zeta \simeq 3 \times 10^4$).

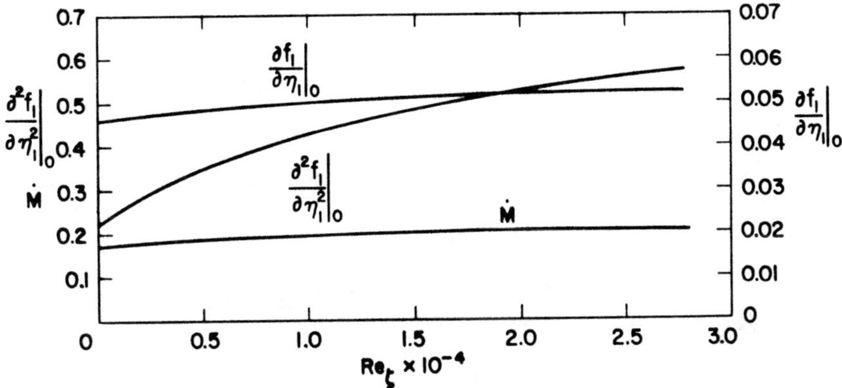

Figure 5. Streamwise variation of surface velocity, shear stress, and mass-transfer rate for a linearly accelerating flow ($U_\infty = 1 + \zeta$, $T_\infty = 100°C$, $T_{-\infty} = 80°C$, $h_\infty = 0$, $\lambda = \sqrt{\rho_1/\rho_2}$).

The Box scheme is also capable of solving non-similar flows near separation and we have encountered no particular difficulty in obtaining solutions to gas-liquid boundary-layer flows in an adverse pressure gradient. We do not report those results here, however.

---

them numerically using Runge-Kutta techniques. We obtained results agreeing with the methods outlined here.

6. CONCLUSIONS

This work extends that of Kotake [2] and considers a broader class of coupled boundary-layer problems involving laminar, gas-liquid flows. In Sections 3 and 4, we demonstrated the applicability of the efficient Keller Box method for solving external flows with a density, phase-changing interface and presented some solutions. Comparisons with numerical results obtained independently for self-similar flows using Runge-Kutta integration indicate that the coupling has been correctly accomplished in the finite-difference scheme. The results of computations of gas-liquid flows with turbulence are presented in another paper in this volume.

ACKNOWLEDGMENT

This material is based upon work supported by the National Science Foundation, U.S.A., under Grant No. CME-7901176 from the Heat Transfer Program.

REFERENCES

1. LOCK, R.C. - The Velocity Distribution in the Laminar Boundary Layer Between Parallel Streams. Quart. J. Mech. and App. Math. Vol. 4, No. 1, pp. 42-63, 1951.
2. KOTAKE, S. - Heat Transfer and Skin Friction of a Phase-Changing Interface of Gas-Liquid Laminar Flows. Int. J. Heat Mass Transfer. Vol. 16, pp. 2165-2176, 1973.
3. KELLER, H.B., and CEBECI, T. - Accurate Numerical Methods for Boundary Layer Flows. I. Two-Dimensional Laminar Flows, in Lecture Notes in Physics, 8, Proc. Second Int. Conf. on Numerical Methods in Fluid Dynamics, Springer-Verlag, Berlin and New York, pp. 92-100, 1971.
4. CEBECI, T., and SMITH, A.M.O. - Analysis of Turbulent Boundary Layers, Academic Press, New York, 1974.
5. CEBECI, T., and BRADSHAW, P. - Momentum Transfer in Boundary Layers, Hemisphere Publishing Corporation, Northampton and London, 1977.
6. KELLER, H.B. - Numerical methods in boundary-layer theory. Ann. Rev. Fluid Mech., pp. 417-433, 1978.

APPENDIX - Nomenclature

A, B, C	matrices of coefficients of the unknowns in the finite-difference equation of the Box scheme
c_L	specific heat of liquid
c_{p_G}	constant-pressure specific heat of dry gas
c_{p_V}	constant-pressure specific heat of vapor
d_1	c_{p_G}/c_{p_V}
d_2	c_{p_G}/c_L

d_3	$(L/i_\infty)d_2$
f	dimensionless stream function
g	dimensionless enthalpy, i/i_∞
h	mass fraction of vapor in the binary gas mixture
i	specific enthalpy
L	latent heat of vaporization
m	pressure-gradient parameter
$\dot{M}$	surface mass transfer rate
$\underset{\sim}{P}$, $\underset{\sim}{P}_H$	block tri-diagonal matrices
Pr	Prandtl number
$\underset{\sim}{r}$	vector of inhomogeneous terms in the difference equations of the Box scheme
Re_ζ	Reynolds number based on streamwise distance
R_v	molecular gas-constant of liquid-vapor
Sc	Schmidt number
T	temperature
T_B	boiling temperature of the liquid at the ambient pressure of the gas-liquid system.
u	$\partial f/\partial\eta$
$U_\infty(x)$	free-stream velocity distribution
v	$\partial^2 f/\partial\eta^2 \equiv \partial u/\partial\eta$
x,y	horizontal and vertical coordinates, respectively

<u>Greek Symbols</u>

$\underset{\sim}{\delta}$, $\underset{\sim}{\Delta}$	vectors of the unknowns in the finite-difference equations of the Box scheme
δf, δu, δv	iterates in the momentum problem
λ	$U_{-\infty}/U_\infty$
Λ	density-viscosity ratio, $\frac{\rho_2}{\rho_1}\sqrt{\frac{\nu_2}{\nu_1}}$
θ	dimensionless temperature
ρ	density
ν	kinematic viscosity
ζ,η	Falkner-Skan coordinates
ψ	stream function in physical coordinates

Subscripts

1,2	gas, liquid
∞	evaluated in the free stream
$-\infty$	evaluated in the bulk liquid
o	evaluated at the interface
j	index for horizontal grid lines in the finite-difference net above the interface ($0 \leq j \leq J$)
k	index for horizontal grid lines in the finite-difference net below the interface ($-K \leq k \leq 0$)

Superscripts

n	index of vertical grid lines in the finite-difference net
p	iteration index

NUMERICAL PREDICTION OF TURBULENT BOUNDARY LAYER DEVELOPMENT ON A TWO-DIMENSIONAL CURVED WALL

V. GANESAN and B.H.L. GOWDA

Asst. Professors, I.I.T., Madras, INDIA.

ABSTRACT :

This paper is concerned with a numerical prediction and experimental comparison of the development of turbulent boundary layer under adverse pressure gradient on a convex wall. For the prediction a finite difference procedure with marching integration technique has been used. Experiments have been carried out in a low speed wind tunnel. From the comparison of results it may be infered that the numerical method employed in this investigation could be advantageously used for the prediction of flows on mildly curved walls with mild adverse presssure gradiants. However, when the pressure gradients become strong one may have to adopt a partially parabolic procedure.

INTRODUCTION :

The flow over a curved wall has great practical relevence in many branches of engineering. The development of boundary layer on such walls is of practical interest in many engineering applications like flow over aerofoil, flow in curved diffusers and channels. Different configurations are possible both in axi-symmetric and two dimensional geometries. To make a systematic and careful analysis of different geometries by means of experimental study will be costly and time consuming. Hence, it will be of great use and help if numerical methods could be developed for making analysis of flows in systems described above. However, experimental results are a must for comparing the predicted results.

A theoretical method would become acceptable when it gives plausible results and becomes well established if the predicted results compare favourably with the experimental results.

Here is an attempt in this direction to investigate some of the problems described above.

Previous Investigations :

The boundary layer development on curved surfaces is influenced to a very large extent by the curvature. The influence of convex curvature on the mean turbulent characteristics of a boundary layer has been investigated by many researchers [1-7]. However, all the results are on experimental side. Theoretical investigations are very scanty. Moreover, when there is an adverse pressure gradient the problem becomes quite complex. Investigations of flows on systems where both curvature and pressure gradient are present are almost nil except the one by So and Mellor [8]. Hence it is worthwhile to analyse such flows theoretically since it can offer more insights.

Present Contribution :

The main aim of the present paper is to use a finite difference procedure to predict the flow and boundary layer development on a two dimensional convex wall. The axial velocity distribution at various downstream stations have been predicted. From the predicted velocity profiles the boundary layer characteristics have been computed. Experiments have been conducted in a wind tunnel and measurements are made using a pitot tube. From the measured velocity profiles various boundary layer characteristics have been calculated. The predicted results have been compared with the measured ones.

Geometry Considered:

The geometry of the system under consideration is given in Fig. 1. The local slopes at various axial locations in the downstream direction are also given in the figure. The inner wall has a radius of curvature 1160 mm. while the outer wall is straight.

Experimental set-up and measurements :

The measurements were carried out in a low speed, straight through, suction type wind tunnel. For the measurement on the curved surface the geometry shown in Fig. 1 was fabricated out of teakwood. Pitot tube was traversed at various locations shown in Fig.1. The wall static pressures at the various stations were measured by surface static taps.

All measurements were made at a tunnel velocity of 20 m/sec, measured at the entrance to the test section. The variation in the tunnel velocity was within 1 %. .

The Governing Differential Equation :

In the prediction procedure, first the governing differential equation in the x-y coordinate system is transferred to Patankar - Spalding coordinate (x-ω) system using appropriate transformations [9]. In the x-ω coordinate system there is no net transfer of mass in the lateral direction. The equation to be solved for the problem under consideration is for the u momentum. This can be written in x-ω coordinides as

$$\frac{\partial u}{\partial x} + (a+b\,\omega)\,\frac{\partial u}{\partial \omega} = \frac{\partial}{\partial \omega}\left(c\,\frac{\partial u}{\partial \omega}\right) + d \qquad (1)$$

where

$$a = -\frac{1}{\varphi_E - \varphi_I}\,\frac{d\,\varphi_I}{dx}$$

$$b = -\frac{1}{\varphi_E - \varphi_I}\,\frac{d}{dx}\,(\varphi_E - \varphi_I)$$

$$c = r^2 \rho\, u\, \mu_{eff} \,/\, (\varphi_E - \varphi_I)^2$$

$$d = \frac{1}{u}\left(F_x - \frac{dp}{dx}\right)$$

Where u is the axial velocity, x is the axial distance, F_x is the body force; p is the pressure φ is the stream function, ρ is the density, μ_{eff} is the effective viscosity, ω is the non dimensional stream function. I and E denote internal and external boundaries respectively.

Turbulence Model :

In the case of turbulent flows the effective viscosity, μ_{eff}, is a combination of both laminar and turbulent components.

$\mu_{eff} = \mu_l + \mu_t$ where μ_l and μ_t are the laminar and turbulent viscosities respectively. For the calculation of μ_t the turbulence model selected here is a simple one, viz., Prandtl mixing length hypothesis given by the equation

$$\rho l_m^2 \left(\frac{\partial u}{\partial y} \right)$$

where l_m is the mixing length.

Solution Procedure :

Only the salient points of the solution procedure are enumerated here. The complete details may be found in [9].

(i) the procedure is capable of solving the partial differential equations of the form given in equation (1).

(ii) the grid is made up of lines of constant x, intersecting lines of constant nondimensional stream function (ω).

(iii) the finite difference equations are formed from the differential ones by integration over control volumes.

(iv) Implicit forms of finite difference equations are solved by tridiagonal matrix algorithm.

Results and Discussion :

Figure.2 shows the predicted and measured velocity profiles in the boundary layer at various downstream locations. It is seen that in the first three stations (a,b,c) the agreement is very good while at station d the agreement is only satisfactory. At stations e and f the agreement is poor. It is seen that the present prediction method which is mainly used for parabolic flows is doing extreamely well up to station c where local slopes (Fig.1) are less than 5° and adverse pressure gradient effects are mild. Beyond this point the effect of adverse pressure gradient increases and the flow separates (station g in fig.1). So between stations e and g, the flow is partially parabolic, pressure effects are transmitted upstream which is also evident from the measurements at stations e and f. At station g there is onset of separation identified by the tuft probe. As this has not been taken into account in the prediction proceduce the agreement is poor at stations e and f.

Fig. 3 shows the variation of maximum velocity at downstream locations. Similar trends as discussed above is seen.

Fig. 4 shows the variation of predicted and measured boundary layer thickness along the convex wall. It is seen that the agreement is good.

Fig. 5 shows the variation of displacement, momentum and energy thickness at various axial location. It is seen that upto about 200 mm from the upstream the agreement is quite satisfactory. Beyond this point, the agreement is poor.

Figs. 6 and 7 show the variation of skin friction coefficient and the shape factor respectively. The agreement between experiment and prediction is similar to what is explained in the above paragraph.

On the whole the overall agreement between the prediction and measurement is satisfactory.

From the present study it may be infered that the numerical method used in this investigation could be advantegeously employed for the prediction of flows even when adverse pressure gradients are present. However, for better agreement a partially parabolic procedure with a two equation model of turbulence should be tried.

REFERENCES

1. Eskinazi, S. and Yeh, H., An investigation on fully developed flow in a curved channel, J. of Aero, Sciences, 23. (1956), p.23.

2. Patel, V.C., The effects of curvature on the turbulent boundary layer, ARC - 30427, Great Britain, (1968 a)

3. So, R.M.C., and Mellor, G.L. Experiments on convex curvature effects in turbulent boundary layers, J. Fluid Mechanics, 60 (1973), p. 43.

4. Ellis, L.B., and Joubert, P.N., Turbulent shear flow in a curved duct, J. Fluid Mechanics, 62 (1974), p.65.

5. Meroney, R.N., and Bradshaw, P., Turbulent boundary layer growth over a longitudinally curved surface, AIAA, 13. No.11, (1975), p.1448.

6. Ramapriyan, B.R. and Shivaprasad, B.G., Mean flow measurements in turbulent boundary layers along mildly curved surfaces, AIAA, 15, No.2 (1977), p. 189.

7. Ramapriyan, B.R. and Shivaprasad, B.G., The Structure of turbulent boundary layer along mildly curved surfaces, J. Fluid Mechanics, 85 (1978) p. 273.

8. So, R.M.C. and Mellor, G.L., An experimental imvestigation of turbulent boundary layers along curved surfaces, NASA CR - 1940, (1972)

9. Spalding D.B., 'Genmix', A general computer program for two dimensional parabolic phenomena, Imperial College Report HTS/75/17, London.

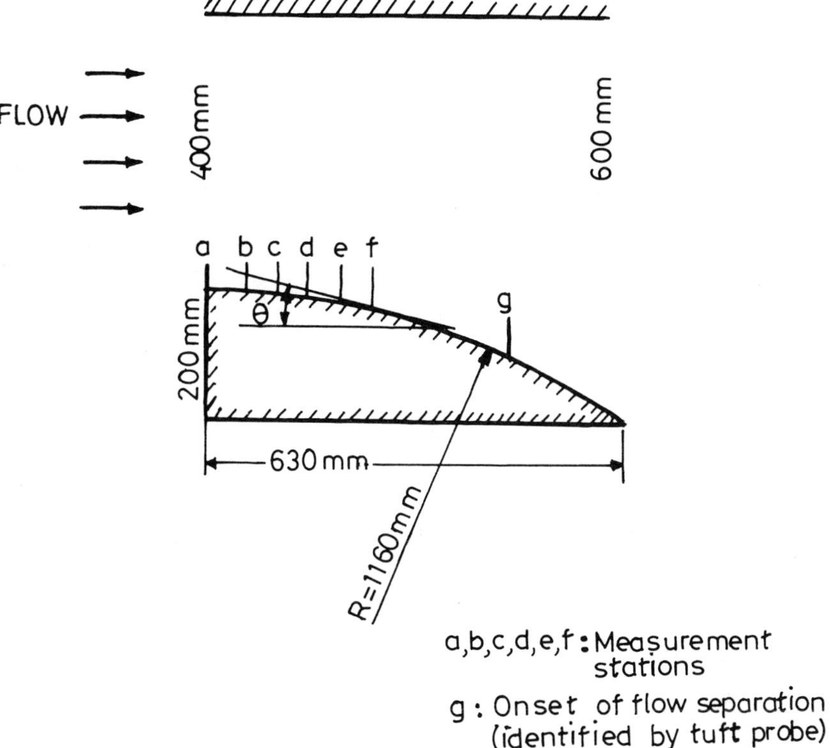

STATION	LOCAL SLOPE θ
a	0°
b	2.5°
c	4.5°
d	7.5°
e	11.5°
f	14.5°

FIG.1. GEOMETRY

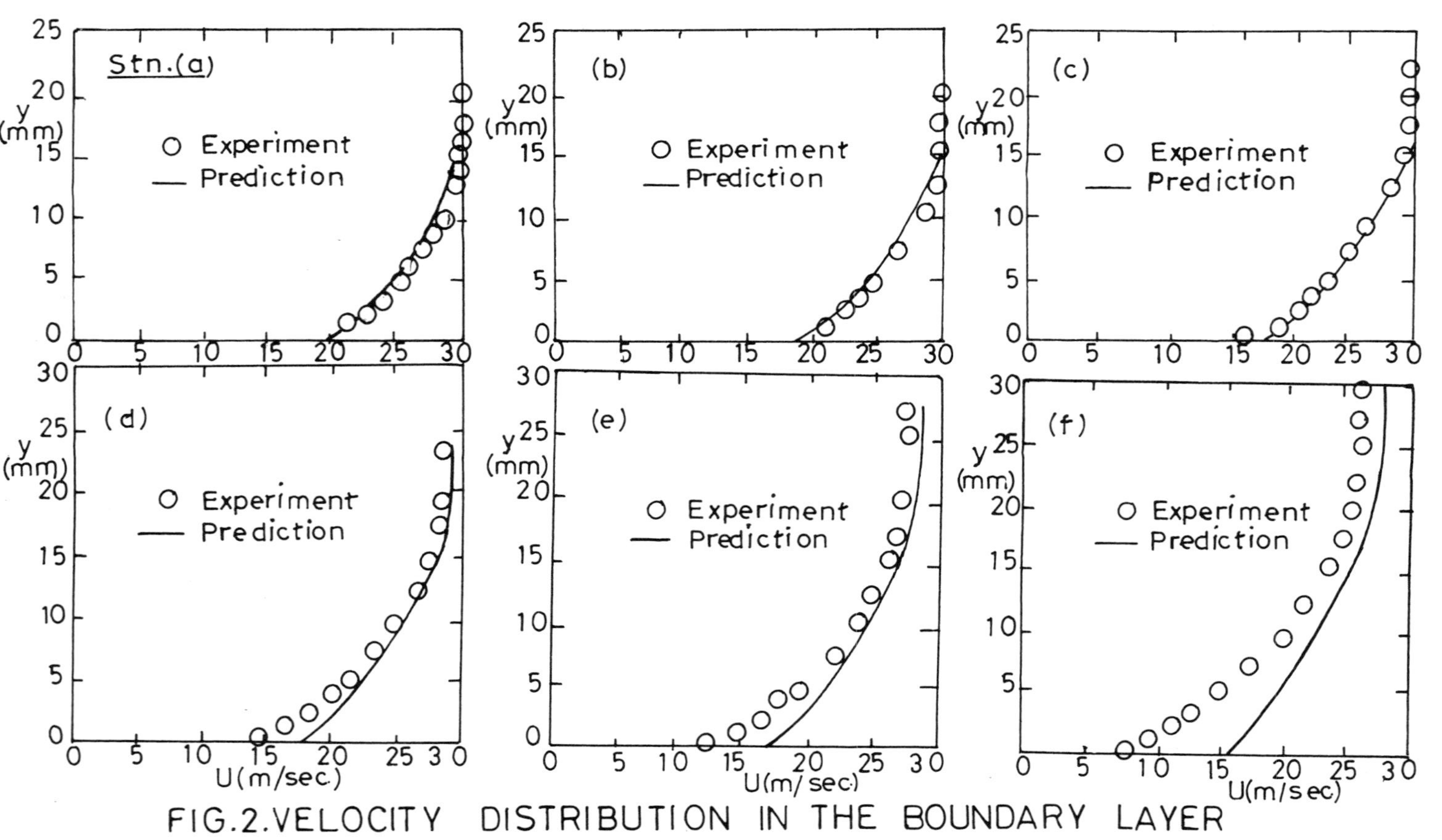

FIG.2.VELOCITY DISTRIBUTION IN THE BOUNDARY LAYER

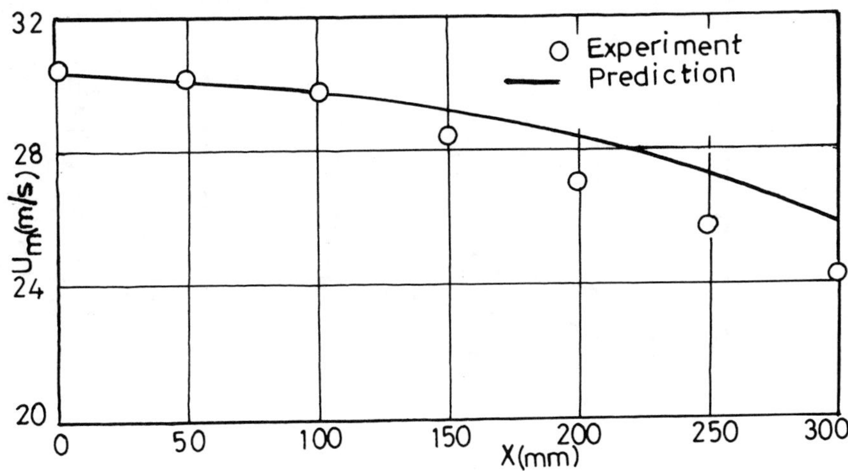

FIG.3.VARIATION OF MAXIMUM VELOCITIES

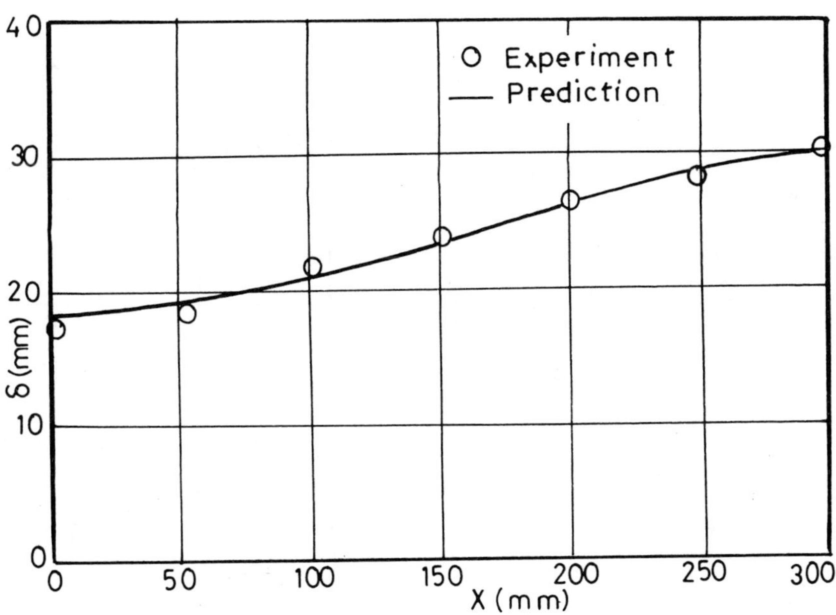

FIG.4.BOUNDARY LAYER THICKNESS

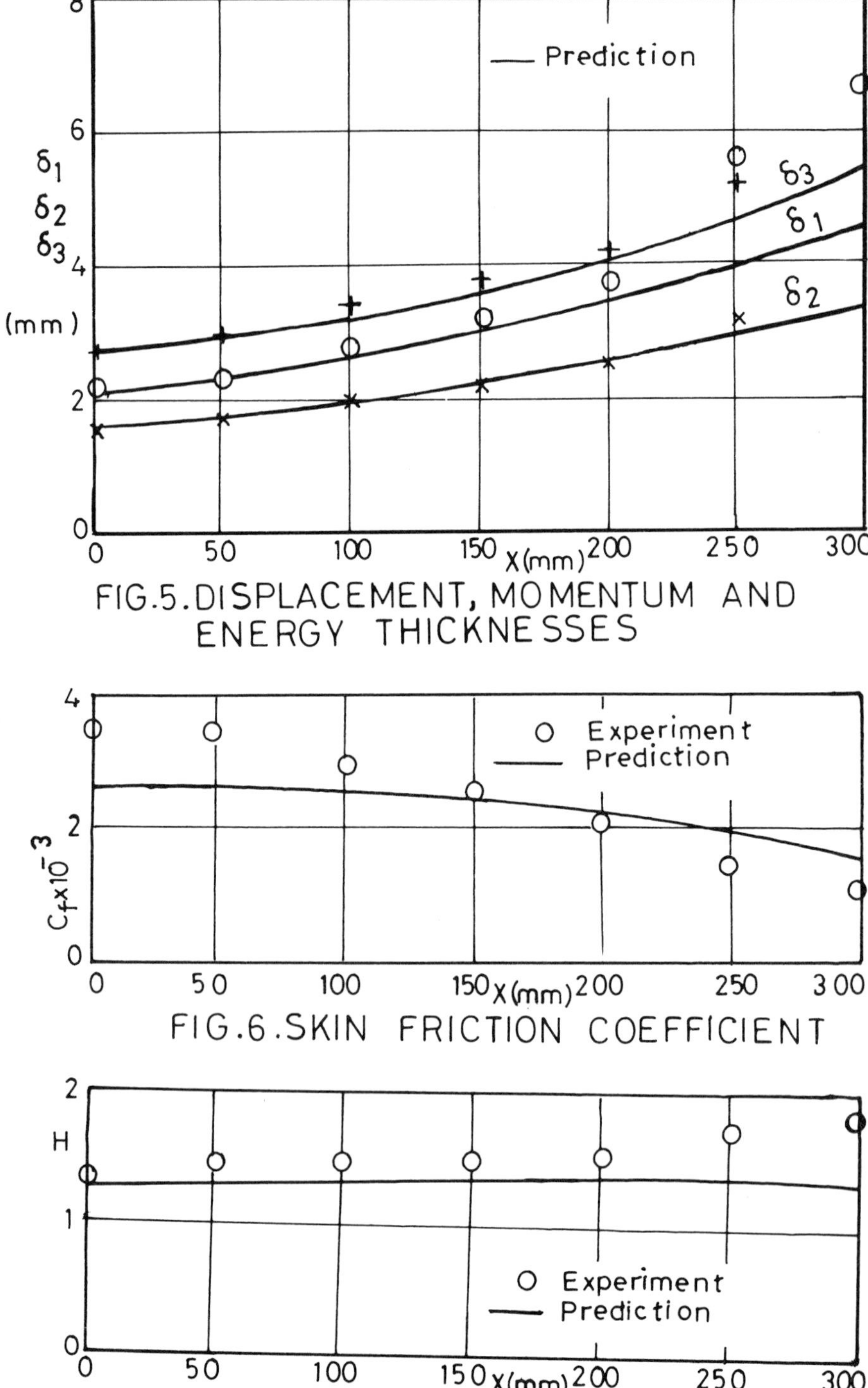

FIG.5. DISPLACEMENT, MOMENTUM AND ENERGY THICKNESSES

FIG.6. SKIN FRICTION COEFFICIENT

FIG.7. SHAPE FACTOR

SECTION 4
FLOW WITH HEAT TRANSFER

A STABLE, ACCURATE, ECONOMICAL, AND COMPREHENDIBLE ALGORITHM FOR THE NAVIER-STOKES AND SCALAR TRANSPORT EQUATIONS

B.P. Leonard*

1. SUMMARY

The author's method of Quadratic Upstream Interpolation for Convective Kinematics (QUICK) [1] is applied to a streamfunction-vorticity formulation of the Navier-Stokes equations and scalar transport in steady 2D flows. High-convection stability is achieved by using a third-order accurate upstream bias in the convective terms. Truncation error is equivalent to introducing a spatial fourth derivative term [2], which is a great improvement over the destabilizing (wiggly) third-derivative term of second-order central differencing or the artificially diffusive second-derivative term of first-order methods [3]. Although there are more operations per grid point than with lower-order methods, the high accuracy allows the use of very practical grid sizes so that overall computer usage is reduced while accuracy is increased dramatically. The QUICK 2D transport algorithm is based on a physically motivated control-volume formulation using locally quadratic interpolation functions for the transported variable. It is equivalent to flux formulations of second-order methods [4] with the addition of a stabilizing, upstream-biassed normal curvature term in modelling the convected control-volume face value. For brevity, results are confined here to the benchmark problem [5] of driven laminar flow in a square cavity.

2. TWO-DIMENSIONAL QUICK ALGORITHM

In two (or three) dimensions, the *exact* control-volume integral formulation for the convection and diffusion of a scalar in the presence of source terms can be written

---

*Associate Professor of Engineering Science
City University of New York
College of Staten Island

$$\iiint_V (\phi^{n+1} - \phi^n)\, dV$$

$$= \int_0^{\Delta t} \left[-\oiint (\underset{\sim}{f}_C + \underset{\sim}{f}_D)\cdot d\underset{\sim}{A} + \iiint_V S\, dV \right] d\tau \quad (1)$$

where $\underset{\sim}{f}_C$ and $\underset{\sim}{f}_D$ are convective and diffusive influx terms.

The two-dimensional QUICK algorithm is based on a local quadratic interpolation ϕ-surface for estimating both $\underset{\sim}{f}_C$ and $\underset{\sim}{f}_D$ on each control-volume face individually. The grid distribution is chosen to favour locally upstream nodes, thus guaranteeing convective stability. Figure 1 shows the appropriate grid points for the left face, when the convecting velocity components are both positive as shown.

Using the local (ξ,η) coordinate system shown, a six-point quadratic interpolation surface has the form

$$\phi = C_1 + C_2\xi + C_3\xi^2 + C_4\eta + C_5\eta^2 + C_6\xi\eta \quad (2)$$

where of course the C's are evaluated in terms of the six node values outlined in Figure 1.

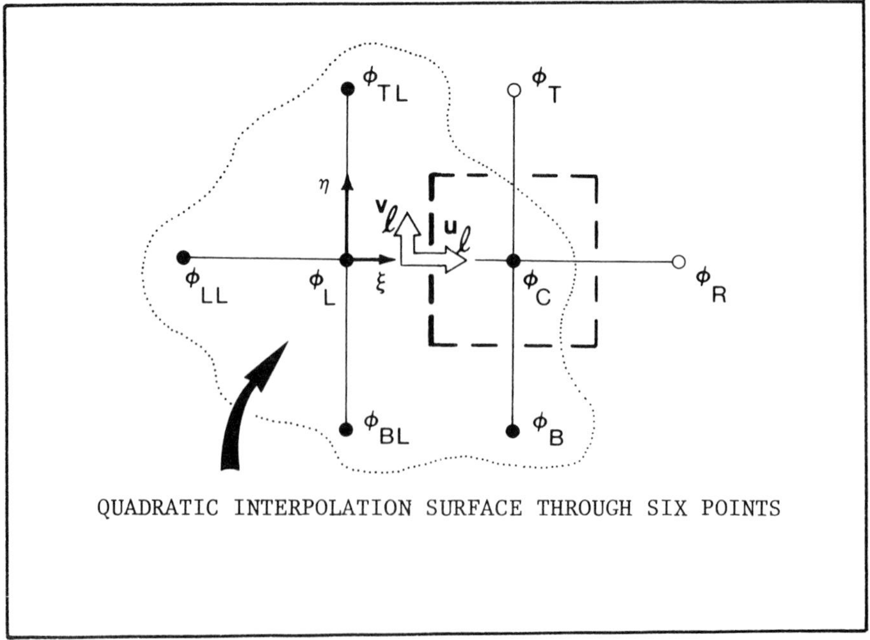

Figure 1. Quadratic interpolation surface for left CV face.

2.1 Average face value

In two dimensions the average face value being convected across the left face is defined, quite generally, as

$$\phi_\ell = \frac{1}{\Delta y}\int_{-\Delta y/2}^{\Delta y/2} \phi(\Delta x/2,\eta)\, d\eta \tag{3}$$

Substituting (2) into (3) results in the basic QUICK 2D formula:

$$\boxed{\phi_\ell = \phi_{LIN} - \frac{1}{8}\, \text{CURVN} + \frac{1}{24}\, \text{CURVT}} \tag{4}$$

where the first term is the two-point linear interpolation, equivalent to second-order central differencing,

$$\phi_{LIN} = \frac{1}{2}(\phi_L + \phi_C) \tag{5}$$

CURVN is the all-important stabilizing upstream-weighted second difference normal to the control-volume face, in this case,

$$\text{CURVN} = \phi_C - 2\phi_L + \phi_{LL} \qquad (u_\ell > 0) \tag{6}$$

and the last term in (4) represents a small upstream-weighted transverse curvature effect, in this case,

$$\text{CURVT} = \phi_{TL} - 2\phi_L + \phi_{BL} \qquad (u_\ell > 0) \tag{7}$$

Note that (for steady flow) the sign of v_ℓ does not affect Equation (4); nor does the value of ϕ_B (when $u_\ell > 0$). However, if u_ℓ is negative, upstream weighting results in

$$\text{CURVN} = \phi_R - 2\phi_C + \phi_L \qquad (u_\ell < 0) \tag{8}$$

and

$$\text{CURVT} = \phi_T - 2\phi_C + \phi_B \qquad (u_\ell < 0) \tag{9}$$

2.2 Average normal gradient

The corresponding normal gradients appearing in the diffusive terms are estimated in a consistent manner. In general, the average normal gradient at the left face is given by

$$\left(\frac{\partial\phi}{\partial x}\right)_\ell = \frac{1}{\Delta y}\int_{-\Delta y/2}^{\Delta y/2} \frac{\partial\phi}{\partial x}(\Delta x/2,\eta)\, d\eta \tag{10}$$

and substitution of (2) into (10) results in

$$\left(\frac{\partial\phi}{\partial x}\right)_{\ell} = \frac{\phi_C - \phi_L}{\Delta x} \tag{11}$$

which of course is identical to the second-order central-difference formula [6].

For two-dimensional quasi-steady flow, an update algorithm based on Equation (1) can be written

$$\phi^{NEW}(i,j) = \phi^{OLD}(i,j) + FLUXL(i,j) - FLUXR(i,j)$$
$$+ FLUXB(i,j) - FLUXT(i,j) + \bar{S}(i,j) \tag{12}$$

using an obvious notation. However, not all these terms need be computed, since

$$FLUXR(i,j) = FLUXL(i+1,j) \tag{13}$$

and

$$FLUXT(i,j) = FLUXB(i,j+1) \tag{14}$$

At the left face, define the local x-component Courant number

$$CXL(i,j) = u_{\ell}(i,j)\Delta t/\Delta x \tag{15}$$

where u_{ℓ} is the average convecting velocity normal to the left face as shown in Figure 1. The corresponding diffusion parameter is given by

$$DXL(i,j) = \Gamma_{\ell}(i,j)\Delta t/\Delta x^2 \tag{16}$$

where Γ_{ℓ} is the effective diffusion coefficient at the left face. Similar formulas are used for the bottom face.

The convective and diffusive fluxes through the left face of each (i,j) cell are computed explicitly from current values as sketched in the following, using a left-to-right (increasing i) sweep:

$$\text{Set:} \quad GRADL = GRADR \tag{17}$$

$$\text{Compute:} \quad GRADR = \phi(i+1,j) - \phi(i,j) \tag{18}$$

$$\text{Set:} \quad CURVL = CURVR \tag{19}$$

Compute: $\text{CURVR} = \text{GRADR} - \text{GRADL}$

$$- \frac{1}{3}\,[\phi(i,j+1) - 2\phi(i,j) + \phi(i,j-1)] \qquad (20)$$

Compute and store:

$$\begin{aligned} \text{FLUXL}(i,j) = {} & \frac{1}{2}\,\text{CXL}(i,j)\cdot[\phi(i,j) + \phi(i-1,j)] \\ & - \frac{1}{16}\Big\{[\text{CXL}(i,j) - |\text{CXL}(i,j)|]\cdot\text{CURVR} \\ & \qquad + [\text{CXL}(i,j) + |\text{CXL}(i,j)|]\cdot\text{CURVL}\Big\} \\ & - \text{DXL}(i,j)\cdot\text{GRADL} \end{aligned} \qquad (21)$$

The i sweep is then repeated for each j value to generate the complete FLUXL(i,j) array. The bottom flux array, FLUXB(i,j), is set up in a similar fashion. Then the explicit update formula can be written very simply as

$$\boxed{\begin{aligned} \text{Set:} \quad \phi(i,j) = {} & \phi(i,j) + \text{FLUXL}(i,j) - \text{FLUXL}(i+1,j) \\ & + \text{FLUXB}(i,j) - \text{FLUXB}(i,j+1) + \bar{S}(i,j) \end{aligned}} \qquad (22)$$

3. QUICK 2D NAVIER-STOKES CODE

3.1 Vorticity-streamfunction formulation

The two-dimensional Navier-Stokes equations can be written, in nondimensionalized form, as [5]

$$\frac{\partial v_x}{\partial t} + \nabla\cdot(\underset{\sim}{v}\, v_x) = -\frac{\partial p}{\partial x} + \frac{1}{Re}\nabla^2 v_x \qquad (23)$$

$$\frac{\partial v_y}{\partial t} + \nabla\cdot(\underset{\sim}{v}\, v_y) = -\frac{\partial p}{\partial y} + \frac{1}{Re}\nabla^2 v_y \qquad (24)$$

with the incompressibility constraint

$$\nabla\cdot\underset{\sim}{v} = 0 \qquad (25)$$

Although it is possible to devise so-called 'primitive-variable' methods to solve these equations directly, the QUICK 2D method is more suited to the vorticity-streamfunction formulation. Specifically the incompressibility constraint can be automatically satisfied by introducing a streamfunction ψ such that

$$\underset{\sim}{v} = \nabla\times(\psi\hat{\underset{\sim}{k}}) \tag{26}$$

where $\hat{\underset{\sim}{k}}$ is the unit vector normal to the plane of the flow. The vorticity is defined as

$$\zeta = (\nabla \times \underset{\sim}{v})_k \tag{27}$$

thus, ψ satisfies the (scalar) Poisson equation

$$\nabla^2\psi = -\zeta \tag{28}$$

The vorticity transport equation is readily found by cross-differentiating Equations (23) and (24), thus eliminating the pressure, to give

$$\frac{\partial\zeta}{\partial t} + \nabla\cdot(\underset{\sim}{v}\,\zeta) = \frac{1}{Re}\nabla^2\zeta \tag{29}$$

In a numerical scheme, a typical time-step (or iteration loop) consists of solving (28) for ψ using 'old' values for ζ, then finding $\underset{\sim}{v}$ from (26), and finally solving (29) for ζ.

3.2 Control-volume equations

To develop the control-volume form, Equations (28) and (29) must be integrated spatially over the control-volume cell, taken here to be a uniform square, $\Delta x = \Delta y = h$. Using the 'ℓrtb' notation, this gives for the control-volume Poisson equation:

$$\left(\frac{\partial\psi}{\partial x}\right)_r - \left(\frac{\partial\psi}{\partial x}\right)_\ell + \left(\frac{\partial\psi}{\partial y}\right)_t - \left(\frac{\partial\psi}{\partial y}\right)_b$$

$$= -\frac{1}{\Delta x\Delta y}\iint \zeta\,dxdy = -\bar{\zeta} \tag{30}$$

The normal gradient on the left face consistent with *quadratic* interpolation is

$$\left(\frac{\partial\psi}{\partial x}\right)_\ell = \frac{\psi_C - \psi_L}{\Delta x} \tag{31}$$

and similarly for the other gradient terms. Consistent modelling of the average vorticity simply generates the arithmetic mean

$$\bar{\zeta} = \frac{1}{4}\sum\zeta_{corners} \tag{32}$$

Thus the control-volume discretization for the streamfunction Poisson equation becomes

$$\psi_L + \psi_B + \psi_R + \psi_T - 4\psi_C = -\frac{h^2}{4}\sum\zeta_{corners} \tag{33}$$

There are several convenient and accurate methods for solving this form of the discrete Poisson equation [6], so this need not be discussed further here.

The control-volume form of (29) using QUICK modelling for face values and gradients results in an equation identical to (22) (with $\bar{S} = 0$ and ζ replacing ϕ, of course). The Courant numbers appearing in the flux terms are based on average normal velocity components computed by differencing adjacent streamfunction values. On the left face of the vorticity control-volume cell, for example,

$$CXL = (\psi_{TL} - \psi_{BL})\Delta t/(\Delta x \Delta y) \tag{34}$$

as shown in Figure 2.

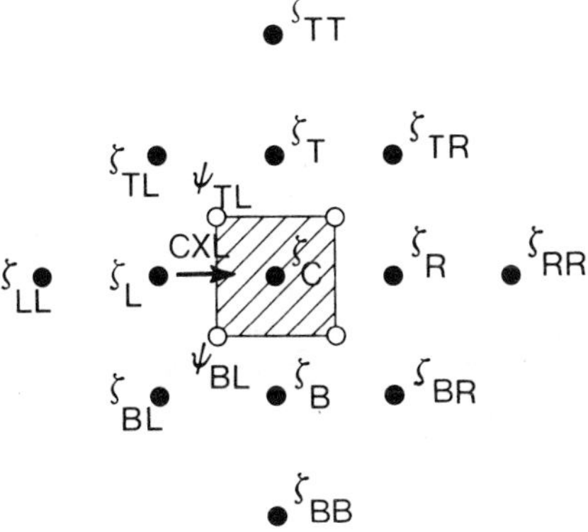

Figure 2. Vorticity control-volume cell and relevant nodes.

3.3 Boundary conditions

The physical boundary conditions at solid boundaries are (i) the impermeable-wall condition:

$$\psi_b = \text{const} \tag{35}$$

and (ii) the no-slip condition:

$$\left(\frac{\partial\psi}{\partial n}\right)_b = U \tag{36}$$

where n is a normal coordinate and U a prescribed velocity. At a free surface it is possible to specify an applied shear stress, as in modelling a wind-driven lake, for example [7]. This can be interpreted as specifying the boundary vorticity:

$$\zeta_b = \text{const} \tag{37}$$

There are, of course, other possibilities; e.g. at inflow boundaries, the transverse streamfunction gradient may be given in terms of a specified velocity profile. In any case, *two* independent conditions are needed on a closed boundary. In the vorticity-streamfunction formulation, boundary condition implementation may be straight-forward (as in the case of an applied shear stress where (35) is used for the ψ equation and (37) for ζ), or more difficult (as in the case of a solid boundary where two conditions are specified on ψ, but none on ζ). In this latter case, only one ψ condition can be used on the Poisson equation (otherwise it would be over-specified), then the boundary vorticity behaviour has to be specified in such a way that it is consistent with the other ψ condition.

In the QUICK 2D formulation, boundaries are most conveniently situated at streamfunction nodes. Then solid-wall conditions can be implemented as follows:

(i) for each boundary streamfunction node, compute the value of ψ at an adjacent external node (in the normal direction) by *cubic* extrapolation using ψ values at two adjacent internal nodes, the previously *computed* ψ value at the boundary, and the specified normal gradient condition.

(ii) choose two external ζ values at the appropriate diagonally adjacent ζ-nodes so that, in combination with the two corresponding internal ζ-node values, the *average* vorticity is adjusted so that the subsequent ψ computation will generate the correct (specified) boundary ψ value.

In this way, both streamfunction conditions are satisfied to a high degree of accuracy.

Boundary vorticity information gets into the flow domain *via* diffusion (both physically and algorithmically) from the boundary, thus it is critical to correctly model the boundary vorticity normal gradient, $(\zeta_{ext} - \zeta_{int})/h$; and since ζ_{int} is known (from its most recently computed value), it can be seen that the consistent modelling of ζ_{ext} is fundamental to a successful simulation. In this respect, it is essential to use cubic extrapolation for the boundary streamfunction -- thus allowing a finite ζ-gradient (proportional to the third normal derivative of ψ). Models which assume only quadratic behaviour [8] are inherently inconsistent, implying $\partial\zeta/\partial n = 0$, although the computed vorticity clearly does not (and cannot) have this behaviour.

4. LAMINAR FLOW IN A SQUARE CAVITY

Figure 3 shows the results of using the QUICK 2D code for computing two-dimensional steady laminar flow in a square cavity with a specified lid velocity and no-slip conditions on the remaining walls. In this case, the Reynolds number, based on lid velocity and the length of one side, is set at Re = 100. The computation uses a 13 × 13 grid. The streamline and velocity-profile results confirm those most generally accepted at this Reynolds number [5], although a possible second recirculation cell (at the lower left corner, as reported by some investigators using second-order methods) was apparently too weak and small to be resolved on this grid.

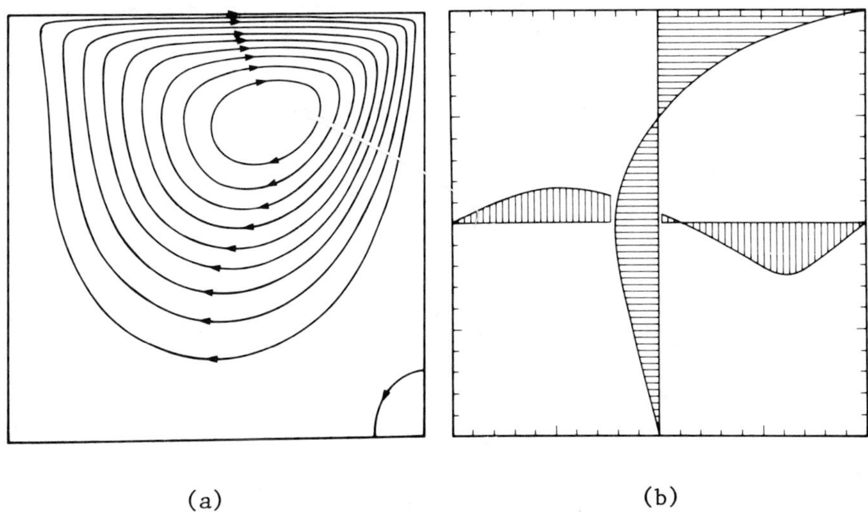

(a) (b)

Figure 3. Re = 100: (a) Streamlines, in increments of 0.01; $|\psi|_{max} = 0.103$. (b) Centerline velocity profiles.

At Re = 1000, Figure 4 shows that the main vortex has become stronger and moved toward the geometric centre, approximating solid-body rotation in that region. Two extremely weak lower recirculation cells are seen, and another is just noticeable at the top left corner. The latter grows stronger with increasing Re. Note the velocity kink in the upright profile representing the convection (without much diffusion) of the low-momentum fluid from the top-left wall region. This effect also grows stronger at higher Re values.

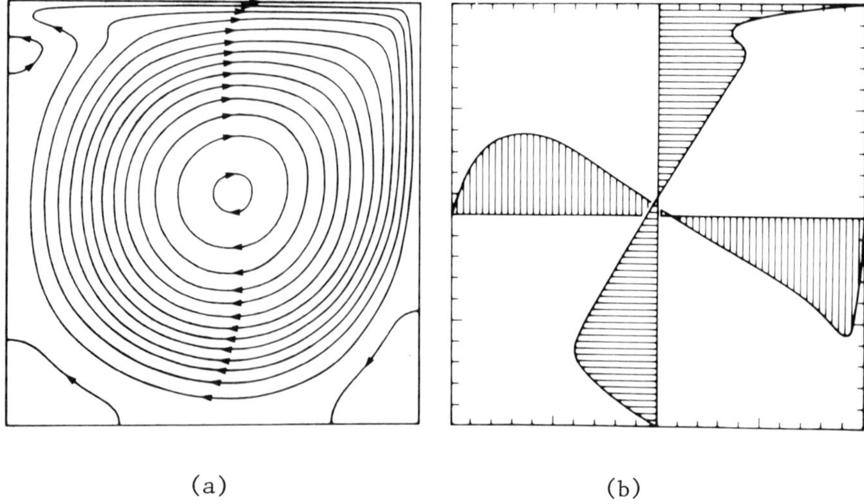

(a) (b)

Figure 4. Re = 1000: (a) Streamlines, in increments of 0.01. $|\psi|_{max} = 0.131$. (b) Centerline velocity profiles.

5. HEAT TRANSFER IN A DRIVEN CAVITY

Figure 5 shows QUICK 2D results for convective-diffusive heat transfer in a driven cavity flow, with a prescribed solenoidal velocity field, shown by the dashed streamlines in (a), and a constant effective thermal diffusion (conduction) coefficient. Thermal boundary conditions prescribe T = 1 at the right boundary, T = 0 at the left, and adiabatic conditions ($\partial T/\partial y = 0$) at top and bottom. The Péclet number is based on the lid velocity, side length, and effective diffusion coeffient.

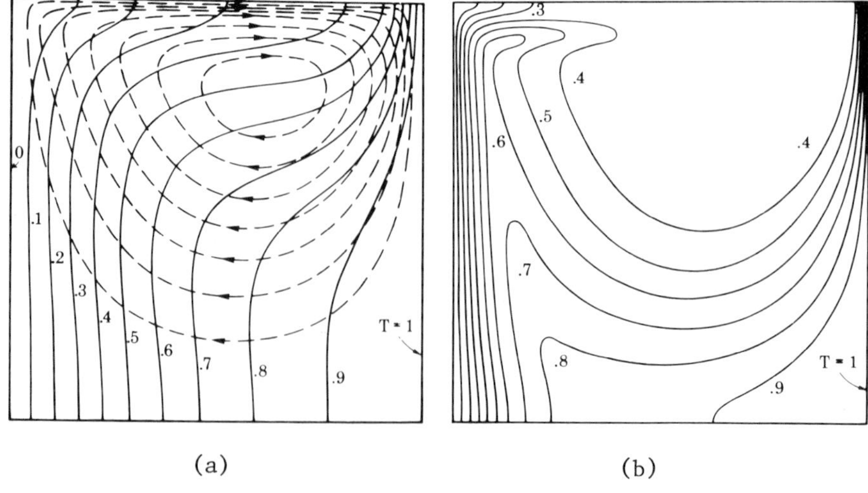

(a) (b)

Figure 5. (a) Streamlines and isotherms, Pe = 50.
(b) Isotherms, Pe = 1000.

REFERENCES

1. LEONARD, B.P. -- A Stable and Accurate Convective Modelling Procedure Based on Quadratic Upstream Interpolation. *Computer Methods in Applied Mechanics and Engineering* 19, 59, 1979.

2. LEONARD, B.P. -- A Survey of Finite Differences with Upwinding for Numerical Modelling of the Incompressible Convective Diffusion Equation, *Recent Advances in Numerical Methods in Fluids* 2, Ed. C. Taylor, Pentech Press, 1981.

3. ROACHE, P.J. -- *Computational Fluid Dynamics*, Hermosa Publishers, 1976.

4. GOSMAN, A.D. & W.M. PUN -- Calculation of Recirculating Flows, *Lecture Notes*, Department of Mechanical Engineering, Imperial College, London, Report Number HB/74/2, 1974.

5. TUANN, S.-Y. & M.D. OLSEN -- Review of Computing Methods for Recirculating Flows. *Journal of Computational Physics* 29, 1, 1978.

6. CARNAHAN, B., H.A. LUTHER & J.O. WILKES -- *Applied Numerical Methods*, John Wiley & Sons, Inc., 1969.

7. MAY, R.L. -- A Numerical Solution of the Navier-Stokes Equation in a Rectangular Basin, *Ph.D. Thesis*, Department of Mathematics, University of Adelaide, 1978.

8. THOM, A. & C.J. APELT -- *Field Computations in Engineering Physics*, C. Van Nostrand Co., Ltd., 1961.

THERMAL HYDRAULIC CALCULATIONS OF WIRE-WRAPPED BUNDLES USING A FINITE-ELEMENT METHOD. THESEE CODE

Ph. ROUZAUD (I) - J. CHINARDET (II)
B. GAY (III) - R. VERBIEST (IV)

This paper presents the physical and mathematical models used in the THESEE code now under development by the CEA/CEN Cadarache.

The objective of this code is to predict the fine three-dimensional temperature field in the sodium in a wire-wrapped rod bundle.

Numerical results of THESEE are compared with measurements obtained by Belgonucleaire in 1976 in a sodium-cooled seven-rod bundle.

(I) DRNR/SEDC Centre d'Etudes Nucléaires de Cadarache
13115 - St Paul lez Durance
(II) CISI/Cadarache 13115 - St Paul lez Durance
(III) Ecole Centrale Lyon - 36 route de Dardilly-69130 ECULLY
(IV) Belgonucléaire - 25rue du champ de Mars -
B- 1050 BRUXELLES (Belgique)

INTRODUCTION

Liquid Metal Fast Breeder Reactors (LMFBR) currently have hexagonal assemblies of wire-wrapped fuel rods. The wires are all wrapped around the rods in a unidirectional helix and, under nominal conditions, pumps create an upward-forced-convection sodium flow around the rods.

In recent years, numerous attempts have been made to develop computing programs to calculate heat transfers in the subassemblies.

These codes can be classified into three categories according to their approximations :

1) The porous medium approximation : the assembly is considered as a single channel partially obstructed by the bundle. This approach is often used for safety computations.

2) The subchannel approach : the assembly is divided into subchannels and simplified physical models are used to compute mean velocity and temperature values in each subchannel. This is the most widely used model.

3) The local or fine approach : each subchannel is subdivided into fine meshes (up to a thousand points) and the general Navier-Stokes equations are solved. As far as we know this analysis is devoted to fine flow computations such as fully developed turbulent flows through bundles without wires

The code THESEE (Thermo Hydraulique des Ecoulements Sodium En Elements finis) is aimed to allow the prediction of fine three-dimensional temperature fields in wire-wrapped rod bundles. The approach used is intermediate between the subchannel approach and the local approach, in the sense that a subchannel contains only a few mesh-points.

The first part of this paper details the modelisation of the problem. Solution algorithms and the finite element discretization used are briefly presented in the second part. The last part covers the description of the Belgonucleaire experiment and the comparison with THESEE results.

1) FORMULATION OF EQUATIONS

The governing equations written in cartesian variables (x, y, z) with primary flow in the z-direction are, with classical notations :

$$\begin{cases} -(U.\nabla)\, U + \nabla\,(\nu\nabla U) = \dfrac{1}{\rho}\,\nabla p & (1) \\ \nabla\,(\rho U) = 0 & (2) \\ -(U.\nabla)T + \nabla(\alpha\nabla T) = 0 & (3) \end{cases}$$

In these equations :

$-\nabla = (\frac{\partial}{\partial x}, \frac{\partial}{\partial y}, \frac{\partial}{\partial z})$

$-\alpha$: denotes the thermal diffusivity.

Boundary conditions will be detailed later.

It is assumed that :

H1 : Since the flow is predominantly axial (forced convection) streamwise viscous diffusion and streamwise thermal conduction can be dropped. All second derivative with respect to z are thus discarded and the system becomes parabolic. [1] [2]

H2 : The pressure gradient in the z-direction is defined by a mean pressure pm (z) [1], [2].

This pressure is computed to maintain the mass flow rate.

H3 : Wire spacers can be treated by slipping conditions : at some points, the velocity direction is specified.

H4 : At each level (z) velocities are independent of the local temperature field.
Only the mean density variation is considered in the code.

H5 : Because of the geometrical complexity a simplified turbulence model is used. Computation of the axial viscosity ν_{ax} will be discussed below. The transverse viscosity ν_{tr} is assumed to be constant and thermal diffusivity α is assumed to be proportional to ν_{ax}.

With these assumptions the governing equations (1), (2), (3) become :

$$\begin{cases} -u\frac{\partial w}{\partial x} - v\frac{\partial w}{\partial y} - w\frac{\partial w}{\partial z} + \frac{\partial}{\partial x}(\nu_{ax}\frac{\partial w}{\partial x}) + \frac{\partial}{\partial y}(\nu_{ax}\frac{\partial w}{\partial y}) = \frac{1}{\rho}\frac{d}{dz}(p_m) & (4) \\ \int_\Omega \rho\, w\, dx\, dy = \dot{m} & (5) \end{cases}$$

$$\begin{cases} -u\frac{\partial u}{\partial x} - v\frac{\partial u}{\partial y} - w\frac{\partial u}{\partial z} + \frac{\partial}{\partial x}(\nu_{tr}\frac{\partial u}{\partial x}) + \frac{\partial}{\partial y}(\nu_{tr}\frac{\partial u}{\partial y}) = \frac{1}{\rho}\frac{\partial p}{\partial x} & (6) \\ -u\frac{\partial v}{\partial x} - v\frac{\partial v}{\partial y} - w\frac{\partial v}{\partial z} + \frac{\partial}{\partial x}(\nu_{tr}\frac{\partial v}{\partial x}) + \frac{\partial}{\partial y}(\nu_{tr}\frac{\partial v}{\partial y}) = \frac{1}{\rho}\frac{\partial p}{\partial y} & (7) \\ \frac{\partial}{\partial x}(\rho u) + \frac{\partial}{\partial y}(\rho v) + \frac{\partial}{\partial z}(\rho w) = 0 & (8) \end{cases}$$

$$\left\{ -u\frac{\partial T}{\partial x} - v\frac{\partial T}{\partial y} - w\frac{\partial T}{\partial z} + \frac{\partial}{\partial x}(\alpha\frac{\partial T}{\partial x}) + \frac{\partial}{\partial y}(\alpha\frac{\partial T}{\partial y}) = 0 \right. \quad (9)$$

2.) NUMERICAL METHOD

Since the system to be solved is parabolic, stepwise integration can be used in the axial oz direction.

The adopted axial step size (Δz) is, up to now, equal to one sixth of the wire pitch.

For any level ($z + \Delta z$) :

** The solutions of the system at level (z) (i.e. velocities u^-, v^-, w^-, mean pressure p_m^-, temperature T^-) are assumed to be known.

For the first level it is necessary to specify these quantities at the assembly inlet.

**Iterations are necessary since the system is non linear (terms $(U.\nabla)\,U$).

At iteration (i) convective terms are written :

$$(U\cdot\nabla)U = (\tilde{U}^{(i)}\nabla)\,U^{(i)}$$

$$\text{Where : } \tilde{U}^{(i)} = \omega\tilde{U}^{(i-1)} + (1-\omega)\,U^{(i-1)}$$

$$0 \leqslant \omega < 1 \qquad \tilde{U}^{(0)} = U^{(0)} = U^-$$

and $U^{(i-1)}$ is the solution of the previous iteration.

**A finite difference discretization is used in the z-direction. So every term $\frac{\partial\phi}{\partial z}$ where ϕ is a dummy symbol, becomes :

$$\frac{\partial\phi}{\partial z} = \frac{\phi\,(z+\Delta z) - \phi(z)}{\Delta z} = \frac{\phi^{(i)} - \phi^-}{\Delta z}$$

**A finite element discretization is used for transverse variables x, y.

The domain is divided into subchannels. These subchannels are discretized using isoparametric Lagrangian elements : nine-node quadrilaterals or six-node triangles, (see Fig.F1.).

The same quadratic approximation is used for velocities and temperatures.

The operator defined by :

$$\left\{ -u^{(i)}\frac{\partial\cdot}{\partial x} - v^{(i)}\frac{\partial\cdot}{\partial y} - w^{(i)}\frac{\cdot}{\Delta z} + \frac{\partial}{\partial x}\nu_\xi\frac{\partial\cdot}{\partial x} + \frac{\partial}{\partial y}\nu_\xi\frac{\partial\cdot}{\partial y} \right. \qquad (10)$$

where ξ is either (ax) or (tr), will be designated $\mathcal{L}_\xi^{(i)}$

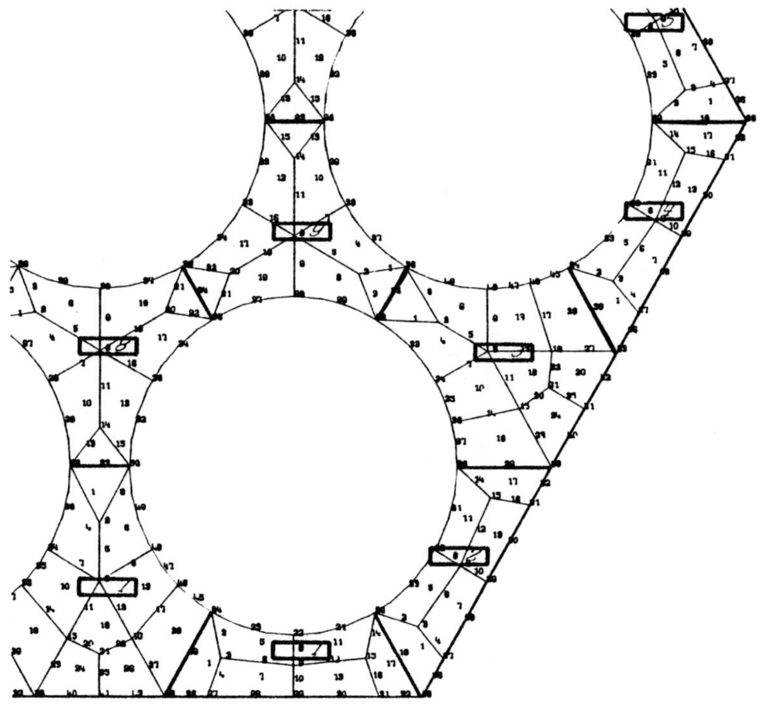

FIGURE F1 : PART OF A SEVEN-ROD MESH-GRID

** Subchannels are assimilated with pipes having the same hydraulic diameter, and a Nikuradse eddy diffusivity model is used to compute the axial viscosity ν_{ax} [3].

On the basis of the pipe results, ν_{tr} is assumed to be equal to one-half of the mean value of ν_{ax}, the thermal diffusivity model used is the one suggested by CHUANG [4].

** The system is solved as follows :

S1 : After linearization and discretization equations (4) and (5) become :

$$\begin{cases} \mathcal{L}_{ax}^{(i)} \, w^{(i)} = \dfrac{-\, p_m^{(i)} - p_m^{-}}{\rho \Delta z} - \dfrac{\tilde{w}^{(i)} \quad w^{-}}{\Delta z} & (11) \\ \displaystyle\int_{\Omega} \rho \, w^{(i)} \, dxdy = \dot{m} & (12) \end{cases}$$

$w^{(i)}$ is computed from equation (11) and

$p_m^{(i)}$ is determined (without iterations since the constraint (12) is linear) to satisfy the axial mass flow relation (12).

Boundary conditions are no slip condition everywhere on the walls (rods and wrapper tube).

S2 : In the same way, equations (6), (7), (8) become :

$$\left\{\begin{array}{ll} \mathcal{L}_{tr}^{(i)} u^{(i)} = -\frac{1}{\rho}\frac{\partial p}{\partial x} - \frac{\tilde{w}^{(i)}\bar{u}}{\Delta z} & (13) \\ \mathcal{L}_{tr}^{(i)} v^{(i)} = -\frac{1}{\rho}\frac{\partial p}{\partial y} - \frac{w^{(i)}\bar{v}}{\Delta z} & (14) \\ \frac{\partial}{\partial x}(\rho u^{(i)}) + \frac{\partial}{\partial y}(\rho v^{(i)}) = -\frac{\rho w^{(i)} - \rho \bar{w}}{\Delta z} & (15) \end{array}\right.$$

This system is similar to a two-dimensional time-dependent Navier-Stokes system with a modified continuity equation :

$$\left\{ \frac{\partial}{\partial x}(\rho u^{(i)}) + \frac{\partial}{\partial y}(\rho v^{(i)}) = g^{(i)}(x, y) = -\frac{\rho w^{(i)} - \rho \bar{w}}{\Delta z} \right. \quad (16)$$

A penalty formulation of equation (16) is used [5], [6].

Briefly, this method assumes that the pressure can be written :

$$\left\{ p = -\lambda \, \mathrm{div}\, (\rho U) \right. \quad (17)$$

where λ is large enough (greater than 10^3), then uses this expression (17) to eliminate pressure from the momentum equations (13), (14).

The same no-slip conditions as for axial calculations are used, except at each point concerned with the wire at the level $(z + \Delta z)$.

At these points, since $w^{(i)}$ is known, it is easy to compute transverse velocities (uf, vf) such that the total velocity (uf, vf, $w^{(i)}$) has the same direction as the wire. These velocities (uf, vf) are used as boundary conditions for transverse computations.

Now it is possible to solve (13) and (14), and to obtain $u^{(i)}$ and $v^{(i)}$.

These two steps (S1) and (S2) are repeated until the convergence (I) on the velocity field is obtained.

(I) The convergence criterion is :

$$\varepsilon_\phi = \frac{\sum_k (\phi_k^{(i)} - \tilde{\phi}_k^{(i)})^2}{\sum (\phi_k^{(i)})^2} \qquad \text{where } \phi \text{ represents } u, v \text{ or } w$$

Let $U = \text{limit } (U^{(i)})$

S3 : After discretization equation (9) becomes :

$$\left\{ -u \frac{\partial T}{\partial x} - v\frac{\partial T}{\partial y} - w\frac{\partial T}{\partial z} + \frac{\partial}{\partial x} (\alpha\frac{\partial T}{\partial x}) + \frac{\partial}{\partial y} (\alpha \frac{\partial T}{\partial y}) = 0 \right. \qquad (18)$$

The boundary conditions can be written :

$$\frac{\partial T}{\partial n} = a (T - T_o) + b$$

Various values of a and b can be used to describe adiabatic walls or specified temperatures and heat fluxes.

Using the velocity field U, which is known at this step, it is then possible to compute the temperature field T at level $(z + \Delta z)$.

It should be noted that, to our knowledge, no theoretical proof exists of the convergence of such an algorithm using finite elements. The difficulty lies in the particular form of continuity equation (16).

3.) NUMERICAL RESULTS

Tests of mathematical assumptions have been carried out in a three-rod bundle.

From these tests it was concluded that :

- a value of $\omega = 0.5$ gives a good convergence : 5 or 6 iterations are enough to reach an accurancy of 10^{-4}.

- there is no significant difference between computations using a value of λ greater than 10^4.

- computing time at one level (6 levels are necessary for a wire pitch) is almost 15 sec and 260 k are needed (IBM 370/168), these quantities are nearly linear (I) with respect to the number of subchannels (30 sec, 350 k for a 7-rod bundle ; 120 sec, 700 k for a 37-rod bundle).

The 7-rod bundle tests performed by BELGONUCLEAIRE in its facilities located at the University of Brussels were used as a first check of the physical models used in THESEE.

(I) In order to reduce the storage a substructure method has been implemented : every subchannel appears as a substructure.

3.1.) Short description of the 7-rod-bundle tests

The test bundle consists of seven 6 mm O.D. heated rods wrapped with 1.1. mm O.D. wire in a righthand helix, the axial wire winding pitch being 90 mm. This bundle is hanging free within a hexagonal wrapper tube, 21 mm across inside flats, 1 mm thick and 968 mm long, the outer surface of which is covered by thermal insulation. The seven rods are electrically heated, one being instrumented. The heated zone is 200 mm long, beginning 400 mm downstream from the bundle inlet. Wall temperatures are measured in eight equally spaced horizontal planes by means of thermocouples, the hot junctions of which are spotwelded on the wrapper tube outer wall, at mid-face. The main levels A, B, C are indicated on figure F2.

Spacer wire temperatures are measured at level B by 1.1mm O.D. thermocouples wrapped on 5 of the rods instead of standard spacers, as indicated on figure F2.

Rod cladding temperatures are measured by thermocouples embedded in the central heated rod cladding, two thermocouples being placed at each of the A, B and C levels, such that one rod thermocouple at level B is overlayed by the wire spacer of this rod (see figure F2). An experimental calibration has been performed in order to transform cladding thermocouple readings into rod surface temperature measurements.

Differential temperature measurements are made relative to the bundle inlet temperature with an accurancy better than $\pm$ 2.5 K.

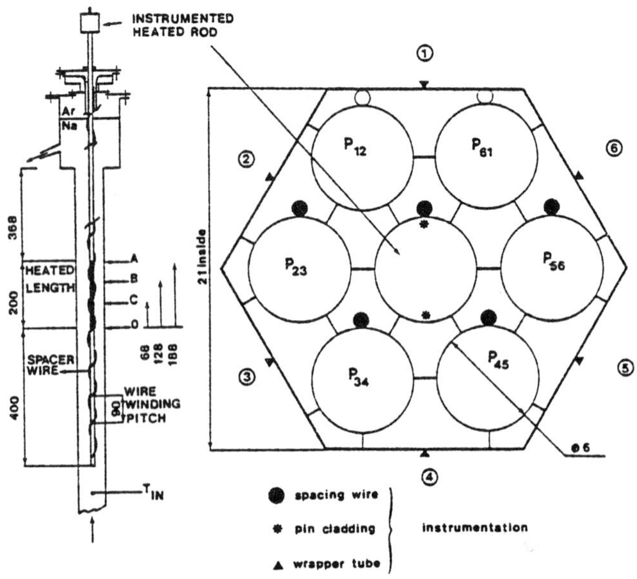

FIGURE F2 :TEST SECTION AND THERMOCOUPLES AT LEVEL B

The loop flow rate is determined by an electromagnetic flow meter within 3 % and the heated rod power is measured with 1 % accurancy.

The experimental programme included uniform power distribution conditions as well as plane power gradients and partial heating configurations.

The following ranges of experimental conditions were investigated : - sodium inlet temperature... 130 to 450° C
-flow rate .... 0.8 to 3.2 m^3/h
- linear power............. 80 to 325 W/cm
thus covering Reynolds numbers from 1×10^4 to 4.5×10^4

3.2.) Comparison of THESEE computed results with experimental data

Two tests were recalculated with the THESEE code. The results are compared with experimental data, they correspond to the following experimental conditions :

	Test No. 1	Test No. 2
Sodium inlet temperature	300°C	301°C
Bundle flow rate	1.65 m^3/h	1.0 m^3/h
Total power	34500 W	7230 W
Number of heated rods	7:uniform power	1 : pos. R 34
Reynolds number	22700	13760

Figures (F3) and (F4) show a typical pattern of transverse flow (F3) and axial flow (F4) through the assembly.

As expected the modelization of the wire seems to provide a correct description of transverse flows :

- the maximum of peripheral swirl flow is reached on the opposite side from the wire,

- smaller flows occur through the bundle from one wrapper side to the opposite side.

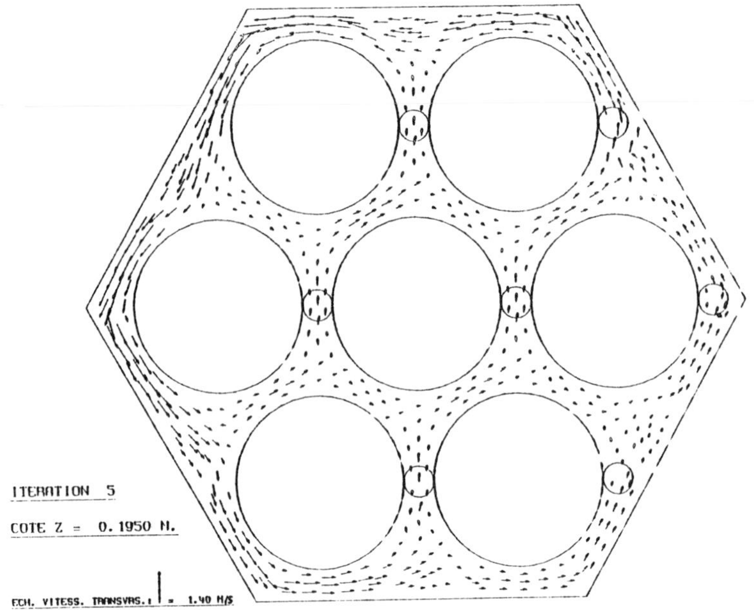

FIGURE F3- TRANSVERSE FLOW AT LEVEL 0.195 m.

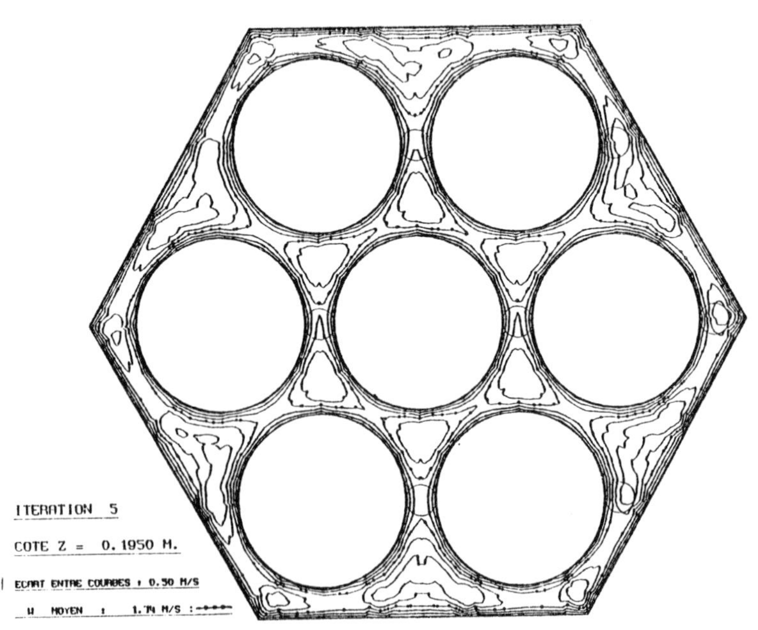

FIGURE F4 - AXIAL FLOW AT LEVEL 0.195 m.

Test N$_o$ 1 showed good agreement between computed and measured values for the subchannel sodium temperatures and the wrapper tube side temperatures. Nevertheless, in this case a good agreement was also obtained with subchannels codes such as DISTAV /7/.

Figure (F5) plots, in the case of test N_o 2, the computed and measured temperatures of the two sides nearest the heating rod versus the axial coordinate z. DISTAV results are also plotted : (D)

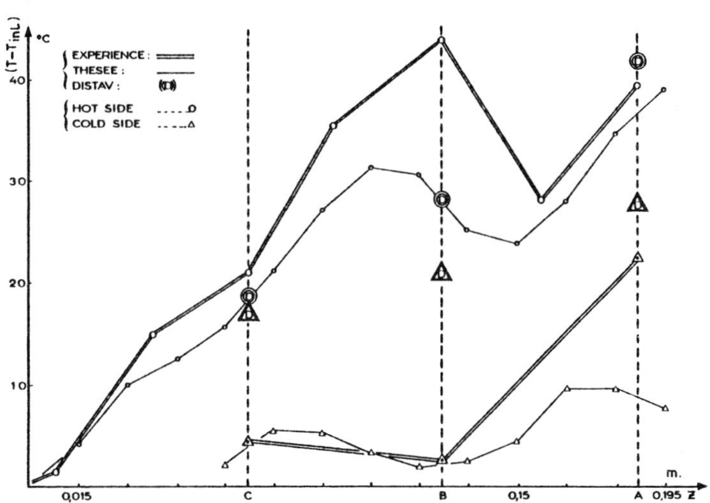

FIGURE F5 - TEMPERATURES ON SIDES 3 (Δ) and 4 (o)

The subchannel sodium temperatures were also measured at level B (figure F6)

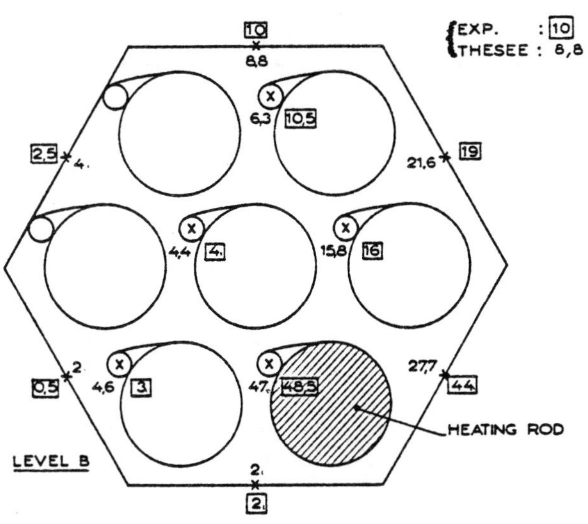

FIGURE F6 - TEMPERATURE AT LEVEL B

The greatest discrepancy between DISTAV and the experimental findings involves the temperature of the cold side (A)(see figure F5). Discrepancies between THESEE computations and experiments are within measurement tolerance limits. Some significant deviations remain however (hot side at level B, cold side at level A (cf F5)), but they are probably attributable to the fact that, up to now, THESEE calculations did not take clearances into account.

The next stage in developing the THESEE code will be to allow for the description and computation of distorted geometries and clearances.

CONCLUSION

A computing method of fine velocity field and heat transfer in a wire-wrapped rod bundle without geometrical disturbances has been presented.

This method requires some simplifying assumptions :

- the 3-D flow can be treated using the 2 1/2 D approximation (parabolization of the governing equations)
- turbulent effects are weak compared to wire effects, so that some crude eddy viscosity models can be used.

From the tests carried out using THESEE it can be concluded that :

- a penalty algorithm jointly used with a finite element method seems to be a promising way of treating the complex geometry of a bundle.
- transverse flows seem to be correctly described and wire effects are well approximated by slipping conditions.
- heat transfer descriptions are improved in comparison with subchannel codes, especially in the case of a single heating rod.

Further tests are still necessary in order to confirm these models :

- computations of larger bundles (e.g.the 19-rod bundle tested by LAFAY /̅8̅/.
- assessment of the influence of clearances.

Following this work, the scope of the THESEE code will be extended to cover more realistic situations including distorted geometries, wrapper tube clearances and heat conduction in the rods.

REFERENCES

[1] W.R. BRILEY "Numerical Method for Predicting Three-Dimensional..." Journal of Comp. Physics, 14, 8-28,(1974)

[2] S.V. PATANKAR, D. B. SPALDING Int. J. Heat Mass Transfer, 15, 1787, (1972)

[3] B.E. LAUNDER et al., "Mathematical Models of Turbulence, Academic Press, (1972)

[4] M.C. CHUANG et al. "Three-Dimensional Thermal-Hydraulic N.S.E., 64, 244-257 (1977)

[5] M. BERCOVIER, M . ENGELMAN, "A finite Element for Numerical Solution of Viscous Incompressible Flows", J. of Comp. Physics, 30, 181-201, (1979)

[6] T. J. R. HUGHES et al. " Finite Element Analysis of Incompressible Viscous Flows by the Penalty Function Formulation"
j. of comp. Physics 30, 1-60, (1979)

[7] D. LETEINTURIER et al. "Code de calcul DISTAV..."
Sixth Int. Heat Transfer Conference Toronto 1978

[8] J. LAFAY et al. "Influence of Helical Wire-Wrap Spacer System in a Water 19-Rod Bundle"
1975 National Heat Transfer Conference, San Francisco

ON THE MODELLING OF TURBULENT HEAT AND MASS TRANSFER FOR THE COMPUTATION OF BUOYANCY AFFECTED FLOWS

Pierre-Louis VIOLLET

EDF - Laboratoire National d'Hydraulique - 6 quai Watier
78400 Chatou - France

Abstract : The $k - \varepsilon$ eddy viscosity turbulence model is applied to simple test cases of buoyant flows. Vertical as horizontal stable flows are nearly well represented by the computation, and in unstable flows the mixing is underpredicted. The general agreement is good enough for allowing application to thermal-fluid engineering problems.

1 - INTRODUCTION

For the multi-dimensional practical computation of buoyancy affected flows, such as thermal discharges in the environment or thermal flows in reactor vessels, it is necessary to use a turbulence modelling. This modelling must be suitable for mean flow and temperature prediction, but is wished to remain as simple as possible, and not to be too heavy to handle.

The "$k - \varepsilon$" eddy viscosity model has been widely used for the computation of isothermal flows, and it seems important to know how this model deals with thermal effects. The goal of this paper is to present simple test cases in highly buoyant situations, and an example of practical application to a more complex flow.

2 - THE TURBULENCE MODEL

2 1 - Background

The simplest way of turbulence modelling is the use of algebraic expressions of the eddy viscosity and eddy diffusivity coefficients ν_T and K_T. This kind of modelling may suit for simple flows as jets or boundary layers, but is not accurate enough for complex recirculating flows with buoyancy effects. On the other hand, the writing modelling and solving of a transport equation for each of the unknown turbulent stresses and fluxes $\overline{u'_i u'_j}$ and $\overline{u'_i T'}$ is still now too heavy for practical applications. As a consequence, the choice of the "k - ε" model appears as a good compromise.

Defining k the turbulent kinetic energy per unit of mass, and ε its dissipation rate :

$$k = \frac{1}{2} \overline{u'_i u'_i} \qquad (1)$$

$$\varepsilon = \nu \overline{\frac{\partial u'_i}{\partial x_j} \frac{\partial u'_i}{\partial x_j}} \qquad \text{(summation over i and j)} \quad (2)$$

the eddy viscosity and eddy diffusivity coefficients are given, from dimensional analysis, by the Prandtl-Kolmogorov expression :

$$\sigma_T K_T = \nu_T = C\mu \frac{k^2}{\varepsilon} \qquad (3)$$

where σ_T (turbulent Prandtl number) and C_μ have to be prescribed.

The transport equations which theoretically allow to determine k and ε , are written from the Navier-Stokes equation, and can be solved after modelling of the unknown terms.

As shown in the state of the art review by Rodi [1] , the resulting model gives satisfactory results in most of isothermal situations, though moderate deficiencies appear in some cases.

Improvements of this model (see [1]), in which $\overline{u'_i u'_j}$ and $\overline{u'_i T'}$ are given by algebraic expressions from k and the local energy balances, without using ν_T and K_T, have been used for the computation of stably stratified simple flows, subjected to parabolic numerical modelling (Launder [2] , Hossain and Rodi [3]) ; practical applications to complex recirculating flows of this improved model never seem to have been attempted because of the lack of universality in the writing of the algebraic expressions, and of the difficulties tied to the explicit numerical treatment of the turbulent stresses and fluxes in the Navier-Stokes and energy equations

2.2 - Present model

The simplest version of the "k - ε" model is used in this study. The turbulent Prandtl number σ_T (see eq. 3) is taken as a constant : it has been shown that in thermally unstable situation σ_T should remain in the range (0.5, 1), while its increasing in stable situation should become important only when both K_T and ν_T have fallen to very low values. Thus it can be expected that this assumption does not affect too much the flow and temperature field.

The modelled form of the k equation is,as usual :

$$\frac{\partial k}{\partial t} + u_i \frac{\partial k}{\partial x_i} = \frac{\partial}{\partial x_i} \left(\frac{\nu_T}{\sigma_k} \frac{\partial k}{\partial x_i}\right) + \mathbb{P} + \mathbb{G} - \varepsilon \tag{4}$$

with :

$$\mathbb{P} = \nu_T \frac{\partial u_i}{\partial x_j} \left(\frac{\partial u_i}{\partial x_j} + \frac{\partial u_j}{\partial x_i}\right) \tag{5}$$

(energy production by mean velocity field)

$$\mathbb{G} = - K_T \beta g \frac{\partial T}{\partial z} \tag{6}$$

(gravitational work)

An important problem arises in the modelling of the ε equation : among the unknown source terms which are usually either neglected or modelled by assuming proportionality with production $\mathbb{P}$ and dissipation ε (Launder [4]), is a term involving the fluctuating temperature field. The question is if this term is to be neglected, or modelled by assuming proportionality to $\mathbb{G}$. It has been found that this term should be neglected in horizontal stable flows (Launder [2]), as shown figure 2.3 about the first of the present test cases. For universality of the modelling, it is assumed in the present study that this term is negligible in all high Reynolds number circumstances, the ε equation being written, as a consequence :

$$\frac{\partial \varepsilon}{\partial t} + u_i \frac{\partial \varepsilon}{\partial x_i} = \frac{\partial}{\partial x_i} \left(\frac{\nu_T}{\sigma_\varepsilon} \frac{\partial \varepsilon}{\partial x_i}\right) + \frac{\varepsilon}{k} \left(C_{\varepsilon 1} \mathbb{P} - C_{\varepsilon 2} \varepsilon\right) \tag{7}$$

The values of the constants are, from Launder and Spalding [5]:

C_μ	σ_k	σ_ε	$C_{\varepsilon 1}$	$C_{\varepsilon 2}$
0.09	1.	1.3	1.44	1.92

The turbulent Prandtl number is taken equal to 1, except for the test reported in section 5 (buoyant jet), where a value $\sigma_T = 0.67$ has been choosen.

2.3 - Modelling of wall interaction

The above model being valuable only in high Reynolds number regions of the flow, and in another way because of the cost and difficulty of grid refining to the thickness of the viscous sublayer close to a wall, a wall interaction must be submitted to a particular modelling.

The first computing mesh is assumed to be inside the logarithmic layer where local equilibrium ($\mathbb{P} = \varepsilon$) can be assumed, and buoyancy effects are small.

Defining n and S the directions normal and tangential to the wall, u_* the friction velocity ($= \sqrt{\tau_s / \rho}$), and u_o the tangential velocity computed at the previous time step (or iteration), at the second mesh point assumed to be distant of δn to the wall, the following set of boundary conditions is used :

$$\begin{cases} k = \dfrac{u_*^2}{\sqrt{C\mu}} \quad , \quad \varepsilon = 10 \dfrac{u_*^3}{\alpha \delta n} \\ \vec{V}.\vec{n} = 0 \quad , \quad \dfrac{\partial}{\partial n} (\vec{V}.\vec{S}) = \dfrac{u_*^2}{\nu + \nu_T} \end{cases} \tag{8}$$

where α is the Karman constant (0.41), and u_* is computed from:

$$u_o = \frac{u_*}{\alpha} (\text{Log} \frac{u_* . \delta n}{\nu} + 5) \tag{9}$$

For a free surface, normal derivatives of $\vec{V}.\vec{S}$, k, ε are set equal to zero.

3 - NUMERICAL METHODS USED IN PRESENT TESTS

The turbulence model described above can be of course associated with any numerical method for two or three dimensional computations. All numerical results presented in this paper are obtained from 2 D finite difference transient computations, with an algorithm using the velocity components and stream function as auxiliary variables (Viollet, Benque, Bisch [6]). The treatment of convective terms involves the method of characteristics, which allows to minimise numerical diffusion. The equations which are solved are, after use of the Boussinesq approximation :

$$\frac{\partial u_j}{\partial x_j} = 0 \tag{10}$$

$$\frac{\partial u_i}{\partial t} + u_j \frac{\partial u_i}{\partial x_j} = - \frac{1}{\rho_o} \frac{\partial P}{\partial x_i} + \frac{\partial}{\partial x_j} (\nu + \nu_T) \frac{\partial u_i}{\partial x_j} + g_i \beta (T-T_o) \tag{11}$$

$$\frac{\partial T}{\partial t} + u_j \frac{\partial T}{\partial x_j} = \frac{\partial}{\partial x_j} (K + K_T) \frac{\partial T}{\partial x_j} \tag{12}$$

with i = 1, 2, coupled with equations 3 to 7.

A rectilinear regular grid is used (computer program SBIRE-T).

It is to be noted that in highly stable flows, the coupling between the turbulence model, energy equation 12 and momentum equation 11 may become unstable. The method used in the present study consists in iterating inside a time step in order to tend to an implicit treatment of the non-linearities of the convection and turbulence terms. The computing time is of course increased, but, as a low number of such iterations appears to be necessary for stabilising the computation (5 or less), this method is found to be much more efficient than reducing the time step.

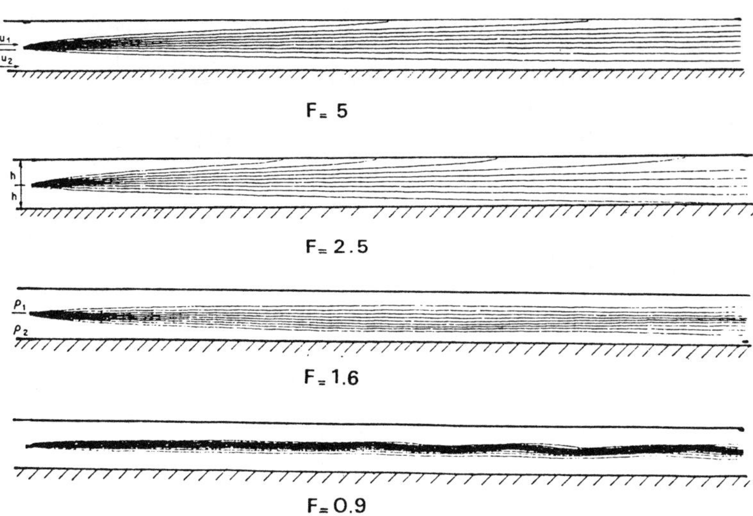

COMPUTED ISOTHERMAL LINES
IN A STABLE SHEAR FLOW FOR INCREASING
(from up to down) DENSITY EFFECT

Figure : 1

4 - HORIZONTAL STABLE OR UNSTABLE SHEAR LAYER

The experimental study is made in a water flume 0.4 m wide, in which a system of two layers of different velocities and temperatures is created ; the height of each layer is h = 0.1 m, and the velocity of the cold layer is twice the velocity of the hot layer (see Viollet [7]).

The comparison with the computation is done for four values of the reduced Froude number :

$$F = \frac{|u_2 - u_1|}{\sqrt{g\,\beta\,h\,|T_2 - T_1|}}$$ with a vertical mesh size $\delta z = \frac{h}{10}$.

The isothermal lines computed in the stable case (the upper layer is heated) are drawn on figure 1 ; figure 2 shows, for the same cross-section x/h = 30, the vertical profiles of velocity, temperature, and turbulent kinetic energy (which could not be measured), as well as the influence of a temperature term in the ε equation upon a temperature profile. The

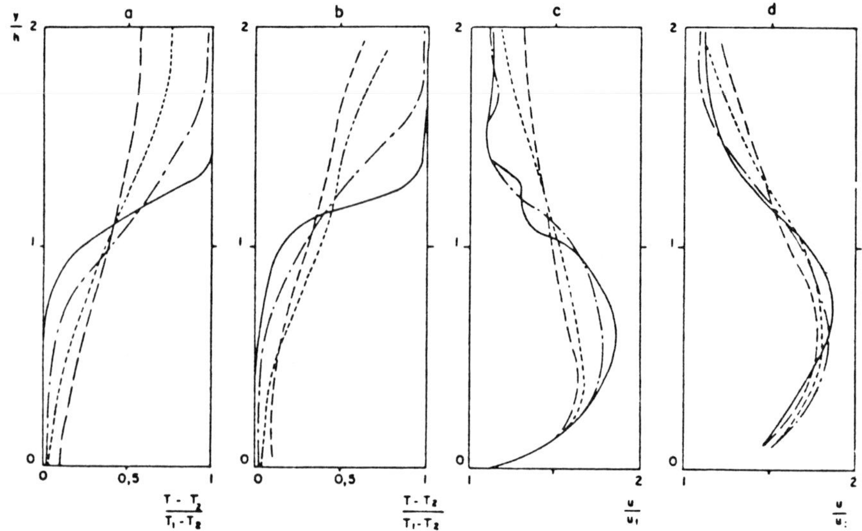

Figure 2.1 : - Mean temperature and velocity profiles computed and mesured at $\frac{x}{h} = 30$ _ a. computed temperature prof. b_ measured temperature prof. c_ computed velocity prof. d_ measured velocity prof.

———	F = 0,9
— —·—	F = 1,6
- - - - -	F = 2,5
— — —	F = 5

$\frac{u_2}{u_1} = 2$

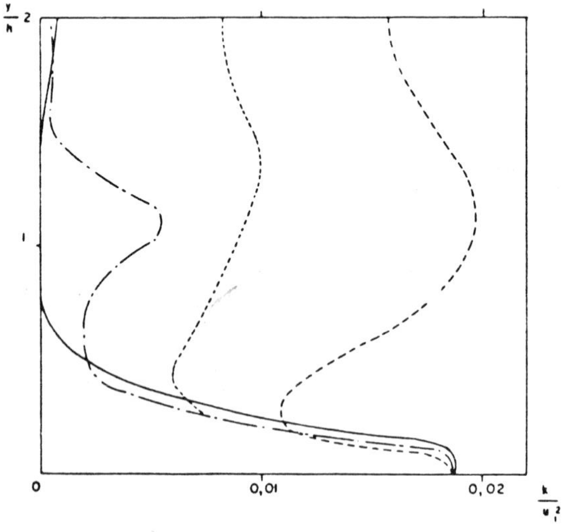

Figure 2.2 : - Computed profiles of turbulent kinetic energy at x/h = 30 (same legend as figure 2.1)

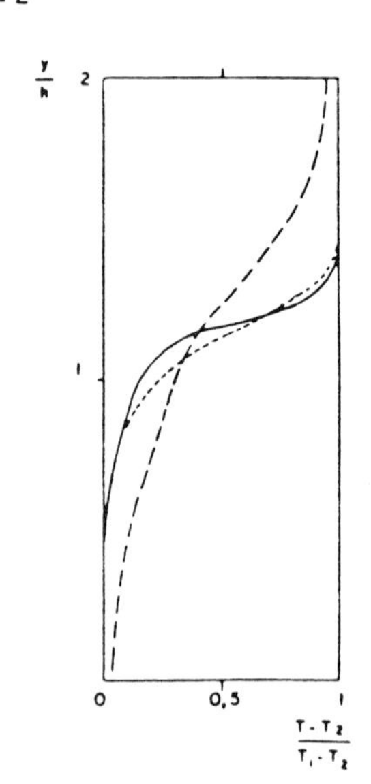

Figure 2.3 : - Influence of temperature term in the ϵ - equation, on the temperature profile at x/h = 30, for F = 0,9

- - - - - no temperature term
— — — with $\frac{\varepsilon}{k} C_{\varepsilon 1} G$
——— experimental result

Figure 2 - Stable horizontal shear layer _ Vertical profiles at x / h = 30

comparison between experiment and computation appears to be fairly good. It can be seen from fig. 1 and 2.2 that in the extreme case F = 0.9 the turbulent energy falls to nearly zero in the stratified layer, and that the mixing is completely inhibited.

Figure 3 shows the comparison between measured and computed velocity and temperature profiles, at x/h = 10, in the unstable case (the lower layer is heated). In that case, the increase of the turbulent mixing due to gravity effects appears both from computation and experiment, but is under-estimated by the computation.

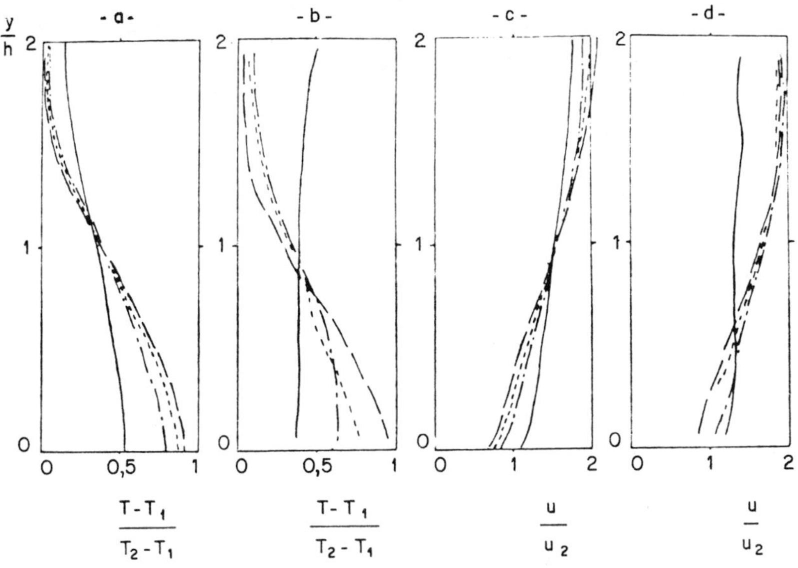

- Figure 3 : Unstable horizontal shear layer. Mean temperature and velocity profiles computed and measured at x/h = 10.

————	F = 0,9
—·—	F = 1,6
-------	F = 2,5
— — —	F = 5

$\frac{U_2}{U_1} = 2$

5 - VERTICAL BUOYANT JET

Preliminary tets have shown that the rate of spreading of the axisymetric isothermal jet is nearly well represented by the computation (the rate of spreading is over predicted by less than 10 %).

When such an upward vertical jet is heated, experimental results (Viollet [8]) show that the centerline velocity w_c increases from its exit value w_o, which induces quicker mixing than in the isothermal case. A free surface existing in the

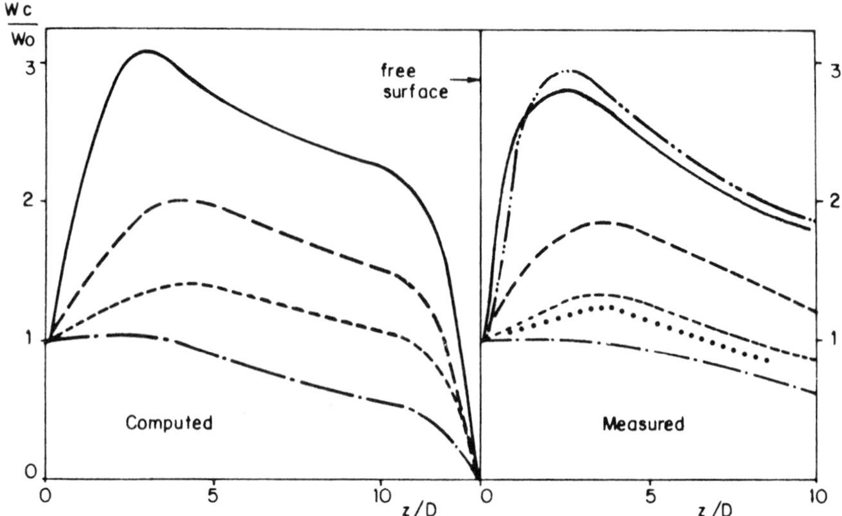

– Figure 4 : Computed and measured vertical velocity along the axis of the vertical buoyant jet

F = 0,57 F = 2,4
F = 0,65 F = 3,4
F = 1,4 F = 9,3

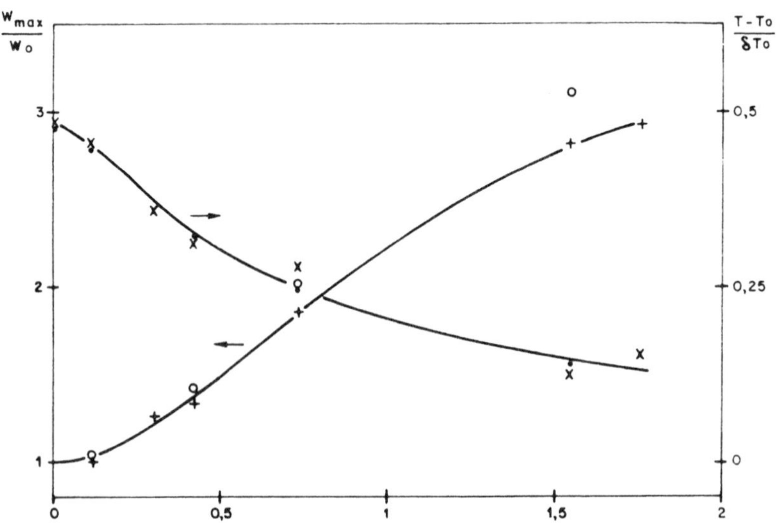

– Figure 5 : Influence of the reduced Froude number upon
(left): the maximum vertical velocity + measured , o computed
(right): the temperature excess at z/D = 10 x measured , • computed

experiments at z/D = 13 above the exit nozzle (D being the diameter of the nozzle), is represented in the computation, whose mesh sizes are $\delta r/D = 0.125$ and $\delta z/D = 0.295$.

Figure 4 shows the centerline velocity computed and measured for different values of the reduced Froude number F (based upon the exit velocity w_0, diameter D, and density difference). Figure 5 shows the influence of F upon the maximum value of the vertical velocity, and upon the non-dimensional centerline temperature difference at z/D = 10.

From these results, it seems that the buoyancy effects in the vertical jet are nearly well represented in the computation.

6 - THERMAL-FLUID BEHAVIOUR OF POOL-TYPE LIQUID METAL REACTOR HOT PLENUM

In a liquid-metal fast breeder nuclear reactor, similar to the Super-Phenix prototype, the fluid from the core circulates in a pool-type free surface hot plenum, and leaves it through the heat exchangers windows. Experimental studies as well as three-dimensional computations (Esposito, Taillifet, Viollet [9]) have shown that three-dimensional effects have moderate influence, so that a two-dimensional computation with properly modelled variable width along the azimutal third dimension is possible. The turbulent jet from the core, after being deflected by the "cover" (designed as BCC on fig. 6), is known from experimental data and taken as boundary conditions of the computation. The flow through the heat exchangers windows is represented using a sink term in the continuity equation.

Figure 6 shows the steady isothermal flow for a flat-redan plenum for which transient scale modelling results are available. On figure 7 are plotted the transient non-dimensional temperatures at three different locations, versus time, resulting from a quick shut-down of the reactor (the total temperature decrease at point S is 160°C). It is found, both from computation and experiment, that a stable stratification appears close to the free surface, and is wiped away by the flow a few minutes later. The velocity field and isothermal lines computed at t = 4 mn, before the stratification being destroyed, are plotted fig. 8. Figure 9 shows through vertical profiles of k, how the turbulence has been damped by thermal effects, especialy in the sharp gradient region. Similar computations in more complicated geometries (oblique redan) are presented in [6] and [9].

7 - CONCLUSION

Simple tests on buoyancy influenced horizontal and vertical shear flows suggest that the two-equation "k- ε" eddy viscosity turbulence model can be used for the prediction of thermal

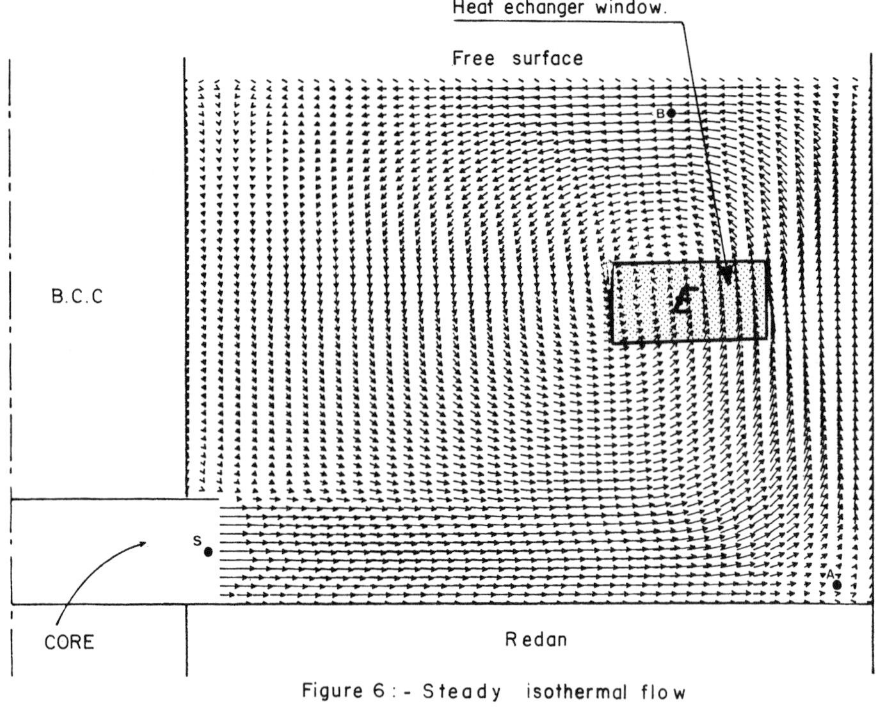

Figure 6 :- Steady isothermal flow

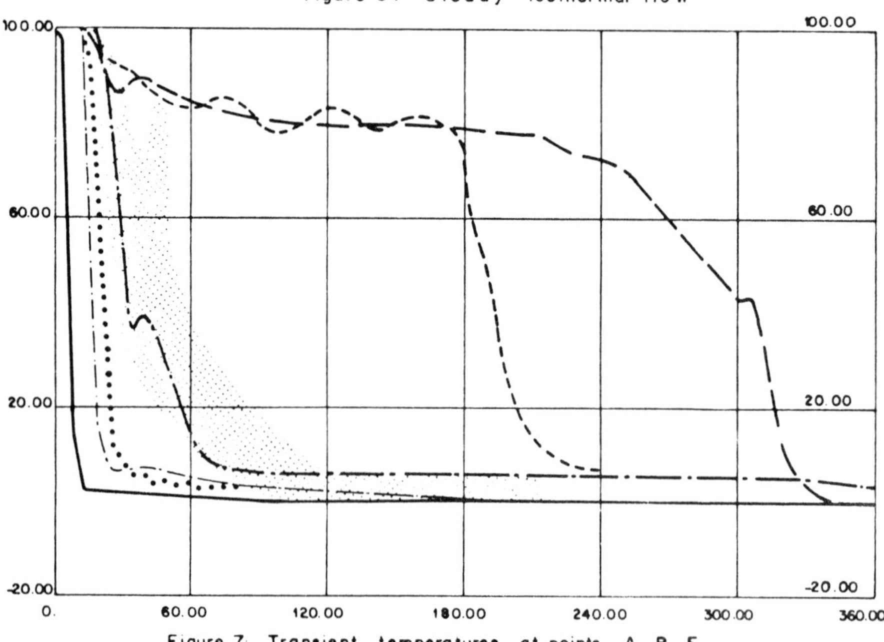

Figure 7 _Transient temperatures at points A , B , E .

S (boundary condition)
A Computed .... A Measured
B Computed ------- B Measured
E Computed E Measured

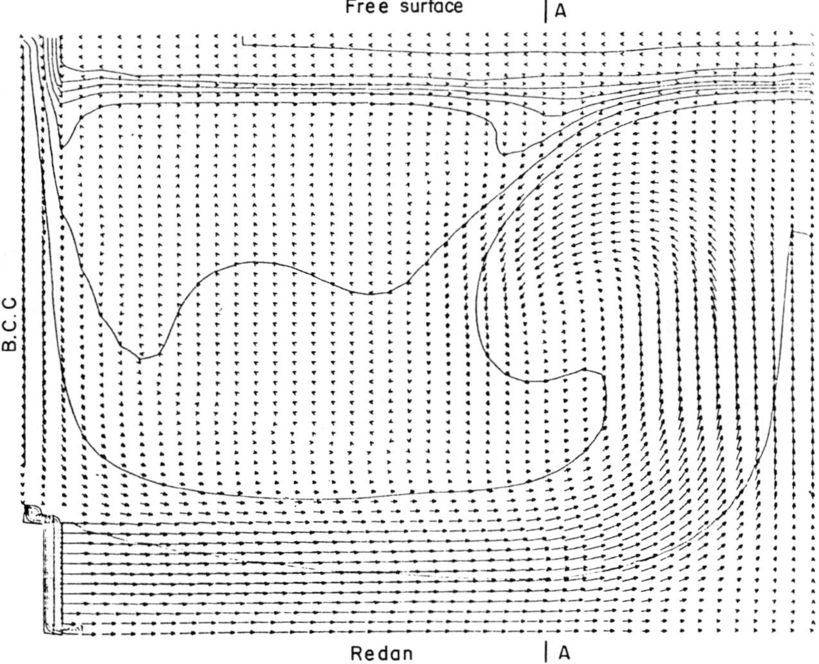

- Figure 8 : Velocity field and isothermal lines computed at t = 240s

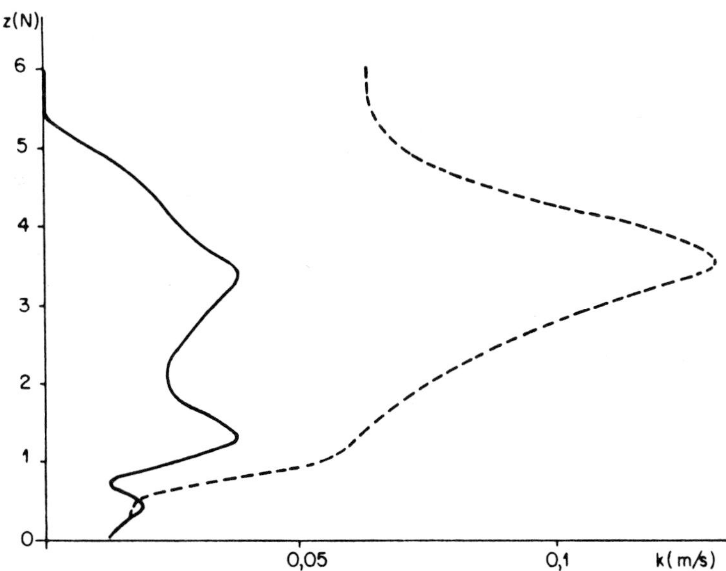

- Figure 9 : Vertical profiles of turbulent kinetic energy in section A (see fig. 8)

----- t = 0 (isothermal steady-state)

——— t = 240 s

fluid interaction. The computation of highly stably stratified flows with sharp density gradients is found to be possible, but in unstable situations the mixing seems to be underpredicted. An improvent of the model, which is wished to allow better predictions of the mixing increase in thermally unstable flows, might be obtained through the use of algebraic expressions for the C_μ constant and for the turbulent Prandtl number.

REFERENCES

1 - W. RODI : "Turbulence modelling : a state of the art review". pub. by IAHR, Delft, 1980.

2 - B.E. LAUNDER : "On the effects of a gravitational field on the turbulent transport of heat and momentum" J. Fluid Mech. 67-3, 1975.

3 - M.S. HOSSAIN - W. RODI : "Mathematical modelling of vertical mixing in stratified channel flow". 2nd int. symp. on stratified flows. Trondheim - June 1980.

4 - B.E. LAUNDER : "Turbulence transport models for numerical computation of complex turbulent flows". Von Karman Inst. for Fluid Dyn. Lecture series 1980-3.

5 - B.E. LAUNDER - D.B. SPALDING : "The numerical computation of turbulent flows. Computer meth. in app. mech. and eng. - 3-1974.

6 - P.L. VIOLLET - J.P. BENQUE - A.M. BISCH : "Plane modelling of unsteady non-isothermal incompressible flows". ANS - ENS int. topical meeting on Advances in Math. Methods for Nuclear Eng. Problems.Munich - April 1981.

7 - P.L. VIOLLET : "Turbulent mixing in a two-layer stratified shear flow" - 2nd int. symp. on stratified flows. Trondheim - June 1980.

8 - P.L. VIOLLET : "Etude de jets dans des courants traversiers et dans des milieux stratifiés" - Thèse - Univ. Paris 6 - February 1977.

9 - P. ESPOSITO - D. TAILLIFET - P.L. VIOLLET : "Thermohydraulic study of large plena in fast breeder reactors" - SMIRT 6 Conf.-Paris - August 1981.

DEVELOPING COMPRESSIBLE FLOW BETWEEN PARALLEL HEATED PLATES

J.A. REIZES*

SUMMARY

An extrapolation method is presented for obtaining the downstream boundary conditions in problems of flow between parallel plates in which fully developed flow is not attained. The compressible stream function-vorticity formulation is used. The results for channel lengths as short as two channel heights and for Reynolds numbers between 10 and 10^3 are compared for consistency since no previous solution is available. The results show that the proposed technique leads to satisfactory results except where separation due to heat transfer occurs and the channel length is such that re-attachment would occur further downstream. Results for isothermal compressible flow are presented and discussed.

1 INTRODUCTION

A number of solutions of the incompressible developing flow between parallel plates have appeared in the literature [2,3,5,8,10 and 13]. The main difficulty in this type of problem is the treatment of the boundary conditions at the outlet from the region of solution. The usual assumption [2,3,5,8,10 and 13] has been to take a sufficiently long region for fully developed flow to occur at the outlet. As a result a large number of mesh points are required most of which lie in a region in which little change occurs. The use of variable mesh size has allieviated this disadvantage to some extent [10,13]. Reed and Oberkampf [11] have proposed an approach in which the boundary conditions for a modified vorticity equation (the curl of the vorticity equation was taken) were extrapolated from internal values. Although their method permits the calculation of a flow field which is not fully developed at the outlet, the three methods of extrapolation lead to significant differences in the solution, so that, it is not clear which extrapolation should be used in

----

* University of N.S.W., Kensington, Australia, 2033.

general. Further, if the plates were heated the energy equation would have to be included, with the result that outlet boundary conditions for the temperature would also have to be established.

In the case of subsonic compressible flow between parallel plates in which viscous effects are included, fully developed flow never occurs,since the gas is accelerated along the passage until choking occurs. The energy equation has to be solved as well as the momentum equation,even in the case of "isothermal" flow in which the plates are at the same temperature as the incoming gas. In fact, even in this simple case there will be heat transfer from the plates to the gas because, as the fluid is accelerated the gas temperature drops below that of the plates.

Walter and Larsen [13] when solving for the developing boundary layer along a flat plate used the upstream boundary condition as the downstream boundary conditions and obtained good results. They state [13] that the downstream boundary conditions have no influence on the solution because of the strongly parabolic nature of the problem. However, when solving for the flow between parallel plates a fully developed profile was assumed by them as the outlet boundary condition [13]. Following Walter and Larsen it appears possible to treat the downstream boundary conditions rather "arbitrarily" and still obtain good solutions. A method for extrapolating the outlet boundary conditions from internal values is proposed for the geometry shown in Figure 1. A discussion is presented of solutions for regions of various channel lengths, including

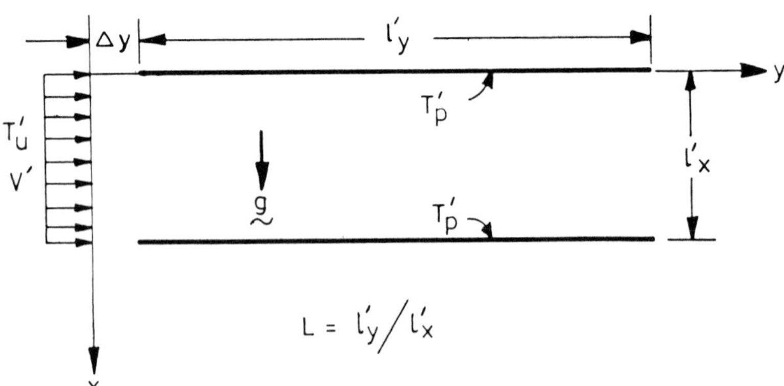

Figure 1. Definition sketch.

lengths which are much shorter than those required to achieve fully developed flow at the outlet. These results are compared with those obtained by other authors. Some results for isothermal compressible flow solutions are also given.

2 MATHEMATICAL FORMULATION

The problem to be studied is illustrated in Figure 1. It is assumed that the fluid is air which can be treated as a perfect gas. The parallel plates, which are of infinite extent into the paper, are at a uniform temperature, T_p', and are impermeable and non-slip. The fluid is flowing at a uniform velocity, V', upstream of the passage. The passage is described by the aspect ratio, L, defined as the ratio of the length of the passage, ℓ_y', to the height of the passage, ℓ_x'.

Since non-dimensional parameters are often used to generalise the scope of experimental and numerical results a non-dimensional approach is used here, although Leonardi and Reizes [6] have shown that when dealing with flows with variable properties many of the advantages of non-dimensionalisation are lost. Because comparisions with results obtained by other workers are to be made, the non-dimensional approach has been retained. It has been found convenient to use a Reynolds number-Froude number non-dimensionalisation rather than the conventional [10] Reynolds number-Rayleigh number approach, since the Rayleigh number has to be artificially introduced into the equations [6]. In fact, the Rayleigh number can always be calculated from

$$R_A = P_R Re^2 \varepsilon\eta / F_R \qquad (1)$$

in which, $R_A = P_R \beta_r' g' \ell_x'^3 (T_p - T_u)/\nu_r^2$, $Re = \ell_x' V'/\nu_r'$, F_R is the Froude number, $V'^2/g'\ell_x'$, ε is the non-dimensional temperature difference $(T_p' - T_u')/T_u'$, in which all the temperatures are absolute, $\eta = \beta_r' T_r'$, $P_R = C_{pr}' \mu_r'/K_r'$, ℓ_x' is the height of the passage, the subscripts p, u, and r refer to the plates, upstream of the plates and to the reference conditions respectively, the prime is used to denote a dimensional variable and non-dimensional variables appear without primes.

The reference temperature, T_r', is taken as the temperature of the incoming gas, T_u', with the result that the coefficients of viscosity, μ_r', thermal conductivity, K_r', and thermal expansion, β_r', are all defined at that temperature. Air only will be discussed in this paper. Since air can be treated as a perfect gas in the range of temperatures and pressures considered, it may be easily shown that $\eta=1$.

The equations to be solved are written in terms of the compressible stream function, ξ', defined by

$$\rho' u' = \partial\xi'/\partial y' \quad \text{and} \quad \rho' v' = -\partial\xi'/\partial x' \qquad (2)$$

and the vorticity, $\zeta'(=(\partial v'/\partial x') - \partial u'/\partial y')$. The non-dimensional form of the vorticity transport equation may be written as,

$$\mathrm{Re}\left[\frac{\partial\rho u\zeta}{\partial x} + \frac{\partial\rho v\zeta}{\partial y}\right] = \left(\frac{\partial^2\mu\zeta}{\partial x^2} + \frac{\partial^2\mu\zeta}{\partial y^2}\right) - \frac{\mathrm{Re}}{F_r}\frac{\partial\rho}{\partial y}$$
$$+ \mathrm{Re}\left[\left(u\frac{\partial u}{\partial x} + v\frac{\partial v}{\partial x}\right)\frac{\partial\rho}{\partial y} - \left(u\frac{\partial u}{\partial y} + v\frac{\partial v}{\partial y}\right)\frac{\partial\rho}{\partial x}\right]$$
$$+ \quad 2\left[\frac{\partial y}{\partial y}\frac{\partial^2\mu}{\partial x^2} + \frac{\partial^2\mu}{\partial x\partial y}\left(\frac{\partial v}{\partial y} - \frac{\partial u}{\partial x}\right) - \frac{\partial v}{\partial x}\frac{\partial^2\mu}{\partial y^2}\right]$$
$$+ \quad 2\left[\frac{\partial\mu}{\partial x}\left(\frac{\partial^2 u}{\partial x\partial y} + \frac{\partial^2 v}{\partial y^2}\right) - \frac{\partial\mu}{\partial y}\left(\frac{\partial^2 u}{\partial x^2} + \frac{\partial^2 v}{\partial x\partial y}\right)\right] \quad (3)$$

in which x' and y' have been non-dimensionalised with ℓ_x', and u' and v' with V', ρ' with ρ_r', ζ' with V'/ℓ_x' and μ' with μ_r'. The compressible stream function is related to the vorticity [6] by

$$\frac{\partial^2\xi}{\partial x^2} + \frac{\partial^2\xi}{\partial y^2} = -\rho\zeta + \frac{1}{\rho}\left[\frac{\partial\xi}{\partial x}\frac{\partial\rho}{\partial x} + \frac{\partial\xi}{\partial y}\frac{\partial\rho}{\partial y}\right] \quad (4)$$

in which ξ' has been non-dimensionalised with $\rho_r'\ell_x'V'$. The temperature can be calculated from

$$\mathrm{ReP_R}\left(\frac{\partial\rho\theta u}{\partial x} + \frac{\partial\rho\theta v}{\partial y}\right) = K\left(\frac{\partial^2\theta}{\partial x^2} + \frac{\partial^2\theta}{\partial y^2}\right) + \frac{\partial K}{\partial x}\frac{\partial\theta}{\partial x} + \frac{\partial K}{\partial y}\frac{\partial\theta}{\partial y}$$
$$+ \frac{\mathrm{ReP_N P_R}}{\varepsilon}\left(u\frac{\partial p}{\partial x} + v\frac{\partial p}{\partial y}\right) + F_R G_N P_R \Phi \quad (5)$$

in which θ is the non-dimensional temperature, $(T'-T_u')/(T_p'-T_u')$, T' is the local absolute temperature, $P_N=p_r'/(\rho_r'C_p'T_r')=(\gamma-1)/\gamma$ for a perfect gas , γ is the ratio of specific heat capacities C_p'/C_v', $G_N=g'\ell_x'/(C_p'T_r')$ and Φ is the dissipation function, K' has been non-dimensionalised with K_r', p' with p_r' which is the pressure at some convenient point in the solution region, $\rho_r'=p_r/R'T_r'$ and R' is the specific gas constant. If a unique solution is sought for the pressure the divergence of the Navier-Stokes equations must be taken [6] leading to,

$$\frac{\partial^2 p}{\partial x^2} + \frac{\partial^2 p}{\partial\ ^2} = G_N P_N \hat{\underset{\sim}{g}}\cdot\nabla\rho - \frac{F_R G_N}{P_N}\nabla\cdot(\rho\underset{\sim}{V}\cdot\nabla\underset{\sim}{V})$$
$$+ \frac{F_R G_N}{\mathrm{ReP_N}}\Big((\nabla\times\underset{\sim}{\zeta})\cdot\nabla(\lambda+\mu) + \nabla^2\lambda(\nabla\cdot\underset{\sim}{V}) + \nabla(\lambda+\mu)\cdot\nabla(\nabla\cdot\underset{\sim}{V})$$
$$+ (\lambda+2\mu)\nabla\cdot(\nabla^2\underset{\sim}{V}) + \nabla^2\underset{\sim}{V}\cdot\nabla(\lambda+2\mu) + \nabla\cdot(\underset{\sim}{V}\cdot\nabla\nabla\mu)$$
$$+ \nabla\big((\nabla\cdot\underset{\sim}{V})\cdot(\nabla\mu)\big) - v\cdot\nabla(\nabla^2\mu)\Big) \quad (6)$$

in which λ is the non-dimensional second coefficient of viscosity.

The pressure equation has been left in vector form because it is too long in the expanded form which may be found in reference [6].

The equation of state for a perfect gas may be written in non-dimensional form as,

$$\rho = p/(\varepsilon\theta + 1) \qquad (7)$$

The viscosity and thermal conductivity coefficients may be determined from the Sutherland Equation [4,6] which becomes

$$K = \mu = s(\varepsilon\theta + 1)^{1.5}/(\varepsilon\theta + s) \qquad (8)$$

in which $s=1+1.47T_b'/T_r'$ and T_b' is the absolute temperature at the boiling point.
Since in general λ has very little effect [6], it has been set to zero in all the solutions obtained in this paper.

For "isothermal" flow, in which it has been assumed that the incoming gas and the plates are at the same temperature, θ and ε have to be re-defined as $\theta=(T'-T_u')/T_u'$ and $\varepsilon=1$. With this change all the equations may be used for isothermal flow.

The conventional equations have been used for incompressible flow, see for example [10].

It remains only to determine the boundary conditions in order to fully specify the problem.

3 BOUNDARY CONDITIONS

The necessary boundary conditions on each of the boundaries (Figure 1) are discussed separately.

3.1 Solid Boundaries

The solid boundaries at x=0 and x=1 are isothermal, impermeable and non slip walls. That is $\theta=1$ if $T_p'>T_u'$ or $\theta=0$ if $T_p'=T_u'$. Also u=v=0 on both boundaries so that ξ is a constant but different value on each of the boundaries. Since,

$$\xi'_{x'=0} - \xi'_{x'=\ell_x'} = \dot{m}', \qquad (9)$$

it follows that

$$\xi_{x'=0} - \xi_{x=1} = 1 \qquad (10)$$

so that $\xi_{x=1}$ was arbitrarily set to zero and $\xi_{x=0}$ set to one. The vorticity at the solid boundaries can be calculated from [6]

$$\zeta_s = -(1/\rho)(\partial^2\xi/\partial n^2) \qquad (11)$$

in which the subscript s refers to the boundaries at x=0 and x=1.
The boundary conditions for the pressure can be obtained from

the Navier-Stokes Equations which become on x=0 or x=1

$$\frac{ReP_N}{F_R G_N}\frac{\partial p}{\partial x} = \frac{Re\rho}{F_R} + \frac{\partial u}{\partial x}\left(\frac{\partial \lambda}{\partial x} + \frac{2\partial \mu}{\partial x}\right) + (\lambda+2\mu)\left(\frac{\partial^2 u}{\partial x^2} + \frac{\partial^2 v}{\partial x \partial y}\right)$$

$$+ \frac{\partial \mu \partial v}{\partial y \partial x} - \frac{\mu \partial \zeta}{\partial y} \qquad (12)$$

3.2 Inlet

Reed and Oberkampf [11] mention that the solution of the developing velocity profile is strongly affected by the assumptions made at the inlet. They assumed a uniform velocity at the inlet (as do many others eg [10,13]), but this is hardly possible since the plates are assumed to be non-slip. Numerically the mass flow rate would change with the mesh size used so that this does not appear to be an appropriate assumption. To avoid this difficulty the plates are started at a distance Δy downstream of the inlet, as may be seen in Figure 1. Thus at the inlet, y = 0, it is assumed that,

$$v = 1 \quad \text{and} \quad \zeta = 0. \qquad (13)$$

The temperature of the fluid at inlet was assumed uniform at $\theta=0$. Since the temperature is uniform at inlet and the flow field is uniform upstream of y = 0 the Navier-Stokes equations reduce to

$$\partial p/\partial x = \rho(G_N/P_N) \qquad (14)$$

Further, the pressure change across the inlet is small and the density may be taken to be a constant. Because the boundary conditions on the pressure are of the Neumann type, the value of the pressure has to be specified at one point. The "corner" pressure was specified at the origin of the coordinate system. This pressure was taken as the reference pressure although this need not be the case.

3.3 Outlet

It is possible at each iteration to directly determine the boundary conditions for the pressure at the outlet by an equation similar to Equation (12). However, the boundary conditions of the other variables cannot be established in this way. For a developing boundary layer Walter and Larsen [13] suggest that the downstream boundary conditions are not important since somewhere downstream in the field, in their case one mesh interval upstream of the downstream boundary, the solution is reasonably accurate. If it is argued that it is really the upstream boundary conditions which control the development of the flow field, then it would appear possible to calculate the values of the variable at the outlet by extrapolation from internal values.

For large aspect ratio, L, in incompressible flow, provision was made for the usual boundary conditions, namely,

$$(\partial\zeta/\partial y) = (\partial\xi/\partial y) = (\partial\theta/\partial y) = 0 \tag{15}$$

so that a comparison between the two approaches could be made.

The problem has now been fully specified, a numerical scheme is required for a solution.

4 METHOD OF SOLUTION

The solution procedure was the same as that used by Leonardi and Reizes [6]. The finite difference approximations (FDA) to equations (4) and (6) were solved using a combined Fourier Analysis-Fast Fourier Transform method. Equations (3) and (5) were modified into false transient equations [7] and the resulting FDAs were solved using the Samarskii - Andreyev ADI scheme. Central differences were used throughout except when equations (15) were invoked when a mirror image approach was employed.

The values of the vorticity on the solid walls were obtained by an integration of equation (11) to yield [6] on say the x=0 boundary,

$$\zeta_{1,j} = -\left\{\frac{12}{\Delta x^2}\,\xi_{2,j} + \zeta_{2,j}\,(\rho_{1,j} + \rho_{2,j})\right\}\Big/\left\{3\rho_{1,j} + \rho_{2,j}\right\} \tag{16}$$

A similar relation can be derived for the x=1 boundary.

For the outlet boundary conditions it is proposed to use a five point formula developed from the Gregory-Newton extrapolation equation viz,

$$\Lambda_{i,n} = \Lambda_{i,n-5} - 5\Lambda_{i,n-4} + 10\Lambda_{i,n-3} - 10\Lambda_{i,n-2} + 5\Lambda_{i,n-1} + 0(h^5) \tag{17}$$

in which n is the index of the last mesh point in the y direction and Λ is one of the variables ζ, ξ or θ.
Other formulations (three, four, six or more points) are possible, but it was felt that the five points used were sufficiently far upstream to yield good boundary values. In fact reasonable results have been obtained with a three point formulation for $\Lambda_{i,n}$, but Equation (17) was employed to generate the results discussed in this paper.

5 RESULTS

As a first step incompressible isothermal flow was studied to determine whether equation (17) could be used to evaluate the down stream values of the vorticity, stream function and the temperature. Solutions obtained using Equation (17) were compared with results generated using Equation (15) as the downstream boundary condition. A 33×65 non-uniform mesh was employed for an aspect ratio L=50 at Reynolds numbers of 10, 100 and 1000. Surprisingly the results did not depend on which equation was used, since the solutions were identical at Re = 10 and Re = 100. Also, these results agree well with previously obtained solutions, as may be seen in Figure 2. In the case of Re=1000 the solution in which Equation (15) was employed yielded results which differed by 5-10% from those calculated using Equation (17). This is to be expected since the development length for Re=1000 is of the same order as the solution length and Equation (15) can only be used if the length of the solution region is very much longer than the development length. For Re=10 and 100 good agreement was also obtained when other aspects of the flow were compared with the results of previous writers.

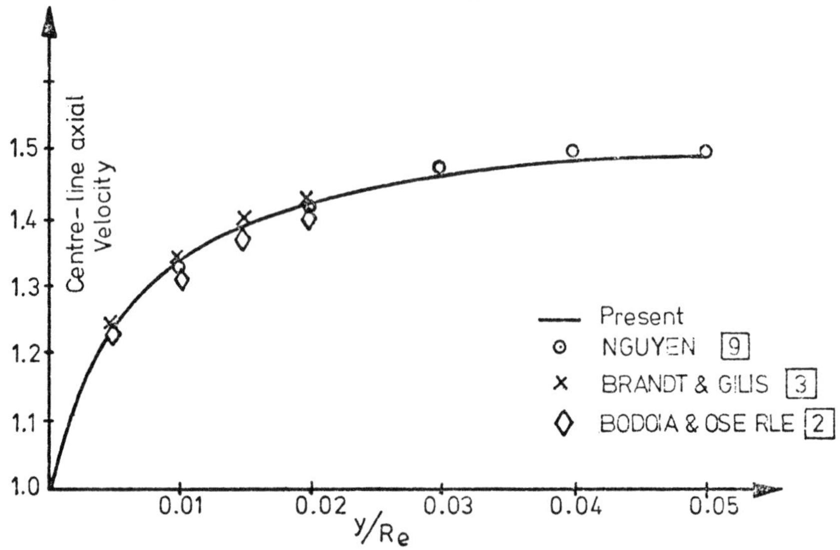

Figure 2. Development of the Centre-Line Axial Velocity in Incompressible Flow at Re = 100.

As may be seen in Figure 3, near the inlet the axial velocity profiles in both compressible and incompressible flow have a maximum which does not occur at the centre of the passage. This solution is similar to the results of Brandt and Gillis [3] and Morihara and Cheng [8], but differs from those of Nguyen [9], Reed and Oberkampf [11] and Walter and Larsen [13] who show the maximum axial velocity as occurring on the centre line of the passage. Abarbanel et al [1] confirm that the "bulges" exist as part of the analytical solution of the problem and are not the result of numerical errors or the discontinuity in velocity at the leading edge of the plate. In fact, it is unlikely that the vorticity could be zero at the inlet, as Nguyen [9] and Nguyen et al [10] have shown that the flow is affected a long distance upstream of the entrance. A better possibility may be to assume that the axial vorticity gradient is zero at the entrance. When the inlet boundary condition $\zeta=0$ was altered to $\partial\zeta/\partial y=0$ (as was assumed by Nguyen [9]) the "bulges" in the velocity profile disappeared. Thus, as Reed and Oberkampf [11] state, the inlet boundary conditions affect the solution significantly and may be more important than the downstream boundary conditions. This, of course, is the justification for the use of Equation (17) as the downstream boundary condition.

Equation (15) could not be used when solutions for L=2 or 4 were sought at any of the Reynolds numbers studied (10, 100 and 1000) so that comparisons with solutions for L=50 have to be made. The results for L=4 were generated with a 33×65 non-uniform mesh whereas the solutions for L=2 were obtained with a 33×65 square mesh. In all cases, including L=50, the distance between the first mesh point and the leading edge of the plates was kept constant at a value $\Delta y=1/32$ (see Figure 1). No differences were found between the results for L=50, L=4 and L=2 which could not be attributed to a change in mesh size. For example the centre-line velocity for Re=100 at y=2 was 1.414 for L=50, 1.408 for L=4 and 1.397 for L=2. Similar results were obtained for all the variables at any position in the solution region for all values of Reynolds number. Thus Equation (17) can be used as the outlet boundary condition in incompressible isothermal developing flow and is much simpler and easier to implement than the method proposed by Reed and Oberkampf [11].

No comparison with other solutions could be made in the case of compressible flow, since none were found in the literature. The only relevant test which could be performed was to solve the same problems (i.e. the same Reynolds number, Froude number and ε respectively) for various aspect ratios. Three values of Reynolds number 10, 100 and 1000 were used with $F_R=2.81\times10^5$, in isothermal flow for aspect ratios L=2, 4 and 50. L=100 was also run for Re=100 with a non-uniform 17×129 mesh. The meshes for the other aspect ratios were the same as those employed for incompressible flow. Δy was kept constant at a value of 1/32 as before.

Remarkable agreement, similar to that for incompressible flow, was also obtained in these cases. Again, choosing the centre line velocity at y=2 for Re=100 the results were 1.420 for L=100, 1.431 for L=50, 1.435 for L=4 and 1.413 for L=2; a difference which is easily explained as being due to the different mesh sizes used. Similar good agreement was obtained for all the other variables, including the temperature, at all points in the solution region. This good agreement means that at a particular Reynolds number the same flow field has been generated for all the aspect ratios studied. Therefore Equation (17) is a satisfactory formulation for the outlet boundary conditions in compressible through flow.

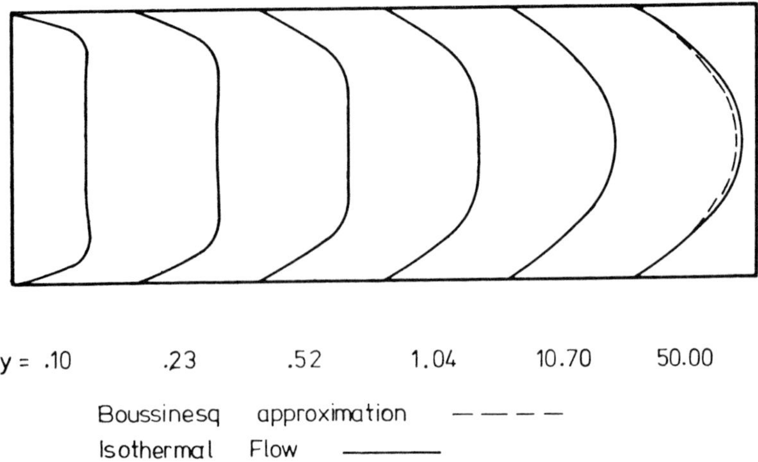

Figure 3. Development of Axial Velocity Profiles in Compressible and Incompressible Flow Re=100

This conclusion was also supported by the results for both compressible and incompressible flow with heating which yielded excellent agreement between the three aspect ratios studied. The only exceptions were in cases in which separation, due to heating, occurred. As expected, if L is chosen such that the flow is separated at the outlet from the solution region, the solution cannot be converged. Results for aspect ratios approximately twice the distance from the inlet to the re-attachment point were in very close agreement with solutions in which much longer aspect ratios were employed. Thus, it appears that Equation (17) can be used to calculate the outlet boundary values of the vorticity, the compressible stream function and the temperature provided it is applied sufficiently far from a separated region.

The remainder of the paper is devoted to a discussion of some results for compressible isothermal flow. A comparison of the development of the axial velocity profiles for both compressible and incompressible flows at Re=100 is presented in Figure 3. There is no discernible difference between the profiles near the inlet, but at y=50, the centre-line velocity for incompressible flow was 1.498 (a value of 1.5 is obtained if a five point difference formula is adopted) whereas it was 1.598 in the case of compressible flow. In fact the centre-line velocity increased to 1.669 at y=100.

The development of the centre-line axial velocity for Re=10, 100 and 1000 is given in Figure 4. The differences between compressible and incompressible cases is particularly noticeable for Re=10. For incompressible flow, fully developed flow occurs at y≈0.6 and the centre-line velocity remains essentially at 1.5 for y>0.6, whereas, in the case of compressible flow the velocity increases to 1.769 at y=50. This increase in velocity is accompanied by a drop in temperature, θ, which on the centre-line at y=50 is -0.0021 for Re=1000, -0.0040 for Re=100 and -0.0013 for Re=10. Of course if the channel were longer the temperature would continue to drop. This is in agreement with a temperature of -0.0045 calculated on the centre-line at y=100 for Re=100. The temperature drop is accompanied by heat transfer from the upper and lower boundaries. A bigger temperature drop is therefore to be expected in the case of adiabatic flow.

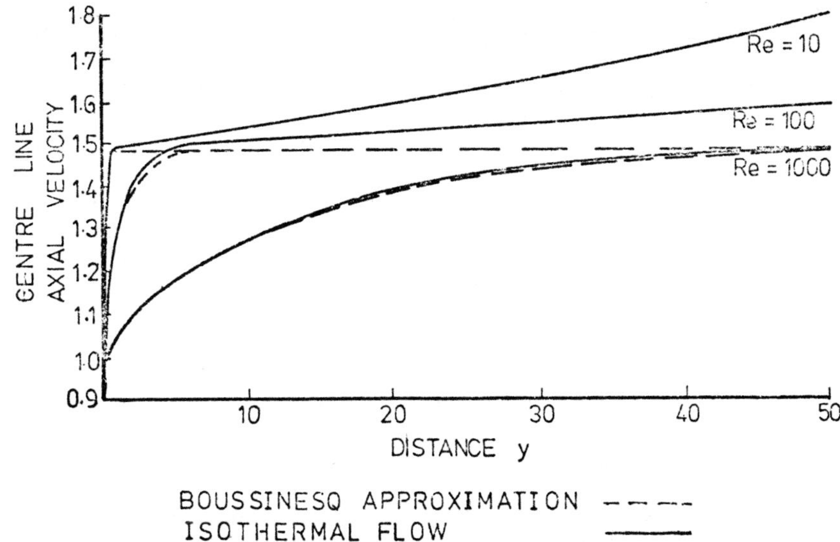

Figure 4. Development of the Centre-Line Axial Velocity in Compressible and Incompressible Flows.

In fact the development of the temperature profile is quite interesting. For example, for Re=100 at x=1/32 θ drops from θ=0 at y=0 to θ=-0.00032 at y=0.06 due to the high transverse velocities. Then θ rises due to heat transfer from the walls to θ=-0.00014 at y≈0.6. For y>0.6 θ falls again because the heat transfer from the walls is now less than that conducted to the faster moving gas nearer the centre, until at y=50 θ=0.00043. The v velocity has a similar history. For example, v at x=1/16 drops from v=1 at y=0 to v=0.353 at y=6.1 and then increases to v=0.393 at y=100. The changes in velocity are accompanied by changes in the value of the wall vorticity which decreases from 54.4 at the leading edge to 6.1 at y=6.1 and then increases to 6.38 at y=50 and 6.87 at y=100. All this is in contrast to the case of steady incompressible flow, which, once fully developed flow has been attained, has a parabolic velocity profile with a centre-line velocity equal to 1.5 and a wall vorticity ζ_s=6 at any position further downstream.

6 REFERENCES

1. ABARBANEL, S., BENNETT, S., BRANDT, A. and GILLIS, J. Velocity Profiles for Flow at Low Reynolds numbers, Trans. ASME J. of Fluids Engg., Vol. 92, p2, 1970.
2. BODOIA, J.R. and OSERLE, J.F. - Finite Difference Analysis of Plane Poiseuille and Couette Flow Development. Applied Scientific Research, Sec. A, Vol. 10 pp.265-276, 1961.
3. BRANDT, A. and GILLIS, J. - Mangetohydrodynamic Flow in the Inlet Section of a Straight Channel. Phys. of Fluids, Vol. 9, No. 4, 1964.
4. BRETSZNAJDER, S. - Prediction of Transport and other Physical Properties of Fluids, Vol 2, Pergamon Press Ltd., Oxford, 1971.
5. HAN, L.S. - Hydrodynamic Entrance Lengths for Incompressible Laminar Flow in Rectangular Ducts, Trans. ASME, J. Appl. Mechs., Vol. 82, p403, 1960.
6. LEONARDI, E. and REIZES, J.A. - Convective Flows in Closed Cavities with Variable Fluid Properties, Chap. 18 in Numerical Methods in Heat Transfer, Ed. Lewis, R.W., Morgan, K. and Zienkiewicz, O.C., John Wiley and Sons, 1981.
7. MALLINSON, G.D. and de VAHL DAVIS, G. - The Method of the False Transient for the Solution of Coupled Elliptic Equations, J. Comp. Phys. Vol. 12, pp.435-461, 1973.
8. MORIHARA, H. and CHENG, R.T. - Numerical Solution of the Viscous Flow in the Entrance Region of Parallel Plates, J. Comp. Phys., Vol. 12, pp.550-572, 1973.
9. NUGYEN, T.V. - Parallel Plate Heat Exchangers, Ph.D. Thesis, University of N.S.W., Sydney, Australia, 1980.
10. NGUYEN, T.V. MACLAINE-CROSS, I.L. and de VAHL DAVIS, G. - The Effect of Free Convection on the Entry Flow Between Parallel Plates, Chap. 15 in Numerical Methods in Heat Transfer, Ed. Lewis, R.W., Morgan, K. and Zienkiewicz, O.C., John Wiley and Sons, 1981.
11. REED, D.B. and OBERKAMPF, W.L. - A New Formulation for Computational Fluid Dynamics, Trans. ASME J. of Fluids Engg., Vol. 101, pp.453-460, 1979.
12. SAMARSKII, A.A. and ANDREYEV, V.B. - On a High Accuracy Difference Scheme for an Elliptic Equation with Several Space Variables, U.S.S.R. Comp. Math. and Math. Phys., Vol. 3, pp.1373-1382, 1963.
13. WALTER, K.T. and LARSEN, P.S. - The FON Method for the Steady Two-Dimensional Navier-Stokes Equations, Danish Center for App. Maths. and Mechs., The Tech. Uni. of Denmark, Report No. 176, 1980.

SPATIAL RESOLUTION REQUIREMENTS FOR NUMERICAL SIMULATION OF INTERNALLY HEATED FLUID LAYERS

Günther Grötzbach

Kernforschungszentrum Karlsruhe
Institut für Reaktorentwicklung
7500 Karlsruhe, Fed. Rep. Germany

ABSTRACT

Direct numerical simulations are performed for turbulent convection in an infinite horizontal fluid layer. At a Rayleigh number of 4 x 10^6 grids with 16^3 to 64^2 x 32 nodes are used to test criteria for the spatial resolution capabilities of grids. It is seen that Nusselt numbers and other statistical quantities are insensitive to the number of nodes, whereas energy spectra and flow patterns are rather critical results. These demand about 32^3 nodes in minimum for sufficient accuracy. This value is also predicted by calculating the coefficient of a subgrid scale heat flux model.

1 INTRODUCTION

For the detailed numerical description of turbulent heat transport in volumetrically heated fluid layers direct numerical simulation and statistical turbulence models are preferably used. The statistical models as in /_1,2_/ are based on time averaged conservation equations. Therefore, models have to be introduced which must represent the total information about turbulence. The direct methods are based on the complete, non-steady, three-dimensional conservation equations. Only when coarse grids are used, subgrid scale models are necessary for the unresolved turbulence elements. In general the direct method is not fully implemented for this type of flow: The papers /_2-5_/, e.g., are based on two-dimensional basic equations. As compared to statistical models higher universality and real 'prediction' capabilities would be the advantages of direct methods using the complete conservation equations /¯6_/.

In this paper the computer code TURBIT-3 /¯7_/ is used for direct numerical simulation of turbulent convection. This code is based on a finite difference form of the complete conservation equations for mass, momentum and heat. To justify the neglection of subgrid scale models, the most important spatial resolution requirements for grids are investigated by numerical simulations with different grids at a fixed Rayleigh number of $Ra = g\beta\dot{Q}D^5/(\nu\alpha K) = 4 \times 10^6$, with $\dot{Q}$ = volumetric heat source, D = channel height, α = thermal diffusivity. A theoretical way to test grids in advance of applications is also applied.

2 NUMERICAL SIMULATION MODEL

The simulation model TURBIT-3 is based on the complete, three-dimensional, non-stationary conservation equations for mass, momentum and heat. Constant material properties and the validity of the Boussinesq approximation are assumed. Cartesian coordinates are used with x_1 and x_2 horizontal and x_3 directed upwards.

To get a finite difference scheme these equations are integrated formally over the mesh volume $V = \Delta x_1 \, \Delta x_2 \, \Delta x_3$. Application of the Gaussian theorem to the volume average of partial derivatives directly gives a finite difference form $\delta_i {}^i\overline{y}$ for surface average values ${}^i\overline{y}$, where y is any variable and i is the direction normal to the mesh cell surface. A staggered grid is used to get exact mass conservation, but this does not eliminate the need for some linear averages $\overline{y}^j$ to approximate variables between two nodes. As deduced in /¯8_/, this results in the following explicit finite difference scheme, in which the space averaging bars have been omitted, the summation convention is used for repeated lower indices, and the superscript n refers to the time step $t^n = n \, \Delta t$:

$$(\tilde{u}_i^{n+1} - u_i^{n-1})/(2\,\Delta t) = -\,\delta_j(\overline{u_j}^i\,\overline{u_i}^j)^n + \delta_j(\nu\,\delta_j u_i)^{n-1} - g\beta\delta_{i3}(T_{ref} - T)^n \tag{1a}$$

$$\frac{1}{\rho}\,\delta_i\,\delta_i\,p^n = \delta_i\,\tilde{u}_i^{n+1}/(2\,\Delta t) \tag{1b}$$

$$i = 1,2,3$$

$$u_i^{n+1} = \tilde{u}_i^{n+1} - 2\,\Delta t/\rho\;\delta_i\,p^n \tag{1c}$$

$$(T^{n+1} - T^{n-1})/(2\Delta t) = -\delta_j(u_j\overline{T}^j)^n + \delta_j(\alpha\delta_j T)^{n-1} + \dot{Q}/(\rho C_p) \tag{1d}$$

As boundary conditions for an infinite plane channel periodicity is assumed in both horizontal directions with periodicity lengths X_1 and X_2. In the vertical direction the velocities are set to zero at the walls, and the wall shear stresses and wall heat fluxes are approximated by linear finite differences using the velocities and temperatures in the wall adjacent grid cells.

Equations (1) already contain turbulence assumptions. The unknown averaged convective terms are split in a large scale part ${}^j\overline{y}$ and in a subgrid scale part y',

$$ {}^j\overline{u_i u_j} = {}^j\overline{u_i}\,{}^j\overline{u_j} + {}^j\overline{u_i' u_j'} \tag{2a} $$

$$ {}^j\overline{u_j T} = {}^j\overline{u_j}\,{}^j\overline{T} + {}^j\overline{u_j' T'} \tag{2b} $$

and the subgrid scale terms are neglected. To justify this neglection it has to be investigated by use of calculated Nusselt numbers, energy spectra and vortex patterns, e.g., whether the grid widths chosen are small enough to resolve the smallest relevant turbulence elements.

3 SPATIAL RESOLUTION REQUIREMENTS

The most important requirements for spatial resolution of grids used in direct numerical simulations are prescribed by the main features of laminar and turbulent convection:

1. The flow patterns in channels with large horizontal extensions consist of large scale vortex systems with maximum wavelengths of typically some channel heights /_9,10,11_/. The periodicity lengths X_i should be chosen large enough to capture these large scale structures. Comprehensive data on the maximum wavelengths observed are not known. Thus, X_i must be determined by numerical tests.
2. In addition, the convection patterns contain very small scale structures. The horizontal extension of the observed cold blobs falling from the upper wall to the channel center, e.g., is about one order of magnitude smaller than the channel height /_10_/. Again, the grid widths necessary to resolve these small scale structures have to be determined by numerical tests.
3. Most statistical flow data in channel flows show strong space dependence near the walls, like the steep temperature gradients in the thermal boundary layer δ, and the sharp maximum values of velocity and temperature fluctuations, and the turbulent heat flux in the transition layer to the isothermal turbulent core. A recommendation for the vertical grid width distribution can be deduced from experimentally deduced temperature profiles /_9_/ and from numerical experience /_12_/: It seems sufficient to have three to four nodes within the thermal boundary layer,

and about the same number in the transition layer $\delta < x_3 < 2.5\ \delta$. The layer thickness $\delta/D = 1/Nu$ can be evaluated from known Nusselt number correlations [9,10,13] in advance of simulations.

In case of limited computing time and storage capabilities, criterion 2 may be replaced by use of subgrid scale models, and criterion 3 by use of universal profiles to approximate the wall conditions. This is the approach chosen for the simulation of high Reynolds number forced convection [7,8].

4 NUMERICAL SEARCH FOR REQUIRED GRID WIDTHS

4.1 Initial Conditions

Four simulations with different grids have been performed for a Prandtl number of Pr = 6 and for a Rayleigh number of Ra = 4 x 10^6 (Table 1). The horizontal node numbers N_1 and N_2 and the respective grid widths Δx_1 and Δx_2 have been varied together to get fixed periodicity lengths $X_i = N_i\ \Delta x_i = 2.8$. This value has been used for simulations of Bénard convection [12,14]. The vertical grid widths near the lower and upper wall, Δx_{3w1} and Δx_{3w2}, agree for all cases with criterion 3.

Case	$N_1=N_2$	N_3	$\Delta x_1=\Delta x_2$	Δx_{3w1}	Δx_{3w2}	Δx_{3max}	t_{max}	NT	CPU-time min/IBM
5	16	16	.175	.044	.027	.158	143.1	4120	127.5/168
7	32	16	.0875	.044	.027	.158	168.9	7880	542.7/3033
10	32	32	.0875	.0325	.020	.0325	123.6	9600	1358.1/3033
11	64	32	.04375	.0325	.020	.0325	40.2	4040	2293.4/3033

Table 1: Case specifications and time intervals. t_{max} = time interval simulated, NT = number of time steps.

The three dimensional velocity fields at time t=0 have been set to zero. The initial temperature fields were generated to have an isothermal mean temperature (average over horizontal planes) in the turbulent core and a linear decrease within the thermal boundary layers to the constant wall temperatures $T_{w1} = T_{w2}$. To this trapezoidal profile random fluctuations have been superposed with a maximum amplitude of 25 % of the time mean temperature difference $\Delta T_o = \langle T_{max} - T_w \rangle$. This type of initial conditions reduces the computing times necessary to approach fully developed flow and makes the computed statistical data independent of initial conditions [12].

An approximate universal presentation of numerical results is forced by normalizing Eq. (1) with a length D, a time D^2/ν, a velocity $\sqrt{g\beta\Delta T_0 D}$, and a temperature ΔT_0. The latter is calculated using a dependence between the Damköhler and Nusselt numbers: $Da = \dot{Q}\, D^2/(K\, \Delta T_0) = Nu_1 + Nu_2$. Application of the correlations in /13/ results in $Da(t=0) = 15.72$.

4.2 Examination of Numerical Results

Equations (1) are integrated in time until steady-state conditions, in a statistical sense, are established for a period suitable for evaluation. The respective problem times and numbers of time steps are indicated in Table 1. To get reasonable statistical data from the non-steady numerical results, averages <y> are formed over horizontal planes and over 15 to 60 arbitrary time steps.

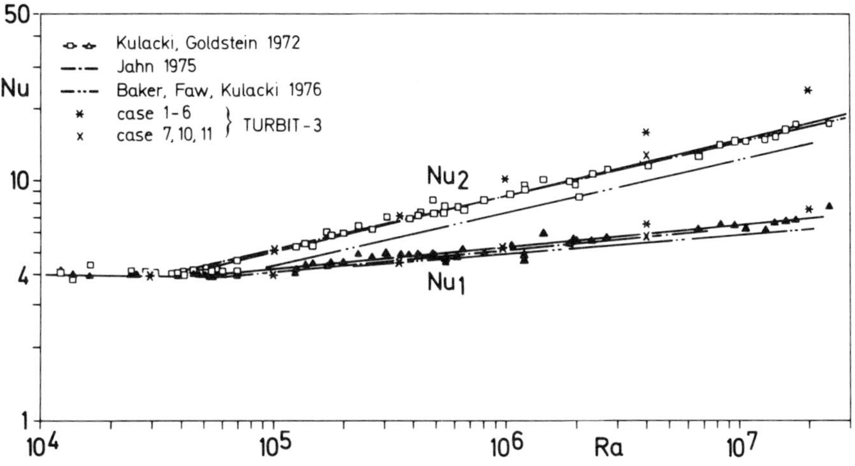

Fig. 1: Rayleigh dependence of the Nusselt numbers

The Nusselt numbers calculated for both walls are compared to experimental data in Fig. 1. The figure contains further results for different Rayleigh numbers but equal grids. The numerical results for small Ra agree with experimental data. With increasing Ra the experimental data are overestimated. Only cases 7, 10 and 11 are within the scatterband of experimental data. The decrease in Nu_1 from case 5 to 7 is obtained for constant vertical grid widths changing the horizontal ones (Table 1). Thus the region near the wall (criterion 3) seems to be resolved adequately by grid 5, but not the turbulent core.

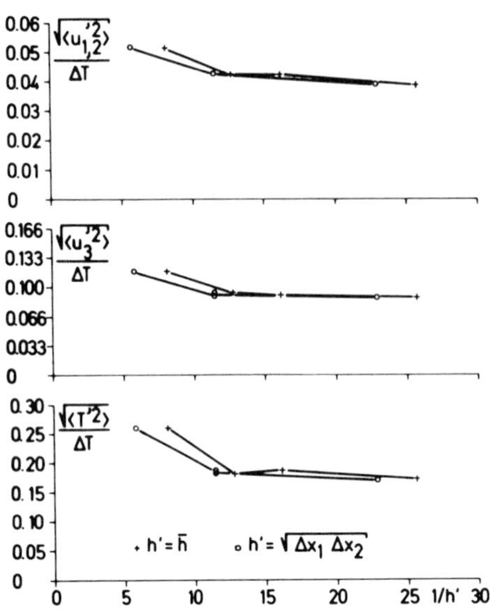

The maximum values of the vertial root mean spuare velocity and temperature fluctuation profiles are shown in Fig. 2 as a function of the reciprocal mean grid width $\overline{h} = (\Delta x_1 \Delta x_2 \overline{\Delta x_3})^{1/3}$, with $\overline{\Delta x_3} = 1/N_3$. The results are normalized by $\Delta T_{calculated}$ to eliminate its deviation from ΔT_0. As expected, the local statistical data on turbulent fluctuations are more sensitive to the grid widths than the Nusselt numbers. The slight decrease of the rms values from grid 7 and 10 to 11 is within the statistical uncertainty of $\pm$ 5 % of these results.

Fig. 2: Dependence of the maximum rms values on the mean grid width

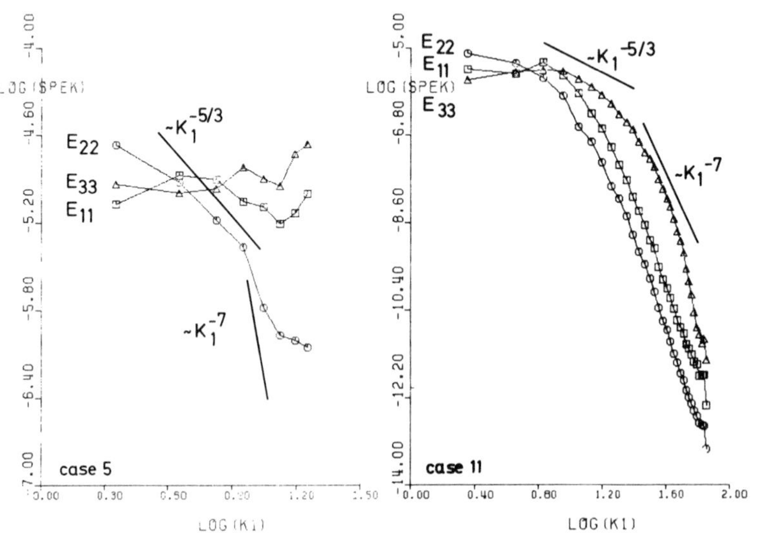

Fig. 3: Energy spectra for the velocity fluctuations ${}^i\overline{u}_i' = {}^i\overline{u}_i - \langle {}^i\overline{u}_i \rangle$, calculated from space dependent results at $x_3 \approx 0.89$

Most sensitive indicators for insufficient spatial resolution are calculated energy spectra of velocity fluctuations (Fig. 3). Two spectra of case 5 are completely non-physical, because the energy does not decrease at high wave numbers $K1 = \pi/(m \Delta x_1)$, with $1 \leq m \leq N_3/2$. Comparison to case 11 shows that grid 5 neglects about 1 % of energy of the horizontal fluctuations, and less than 10 % of the most important vertical fluctuations. Of course, these fractions would be much larger in terms of vorticity spectra, and so grid 5 is the only one which definitely calls for a subgrid scale model to represent the damping caused by the small vortices containing this energy.

Fig. 4: Contour line plots of temperature fields for case 5 at t=122. The contour line increment Δ is .0625 for the vertical section, and .025 for the horizontal section. Origins of sections are denoted by additional arrows.

A 1 % limit has also been found with direct simulations of Bénard convection /_14_/.

Contour line plots of instantaneous temperature fields are given in Figs. 4 to 6 and are compared to experimental results /_10_/. In the vertical sections the cold blobs falling from the narrow cold upper thermal boundary layer into the nearly isothermal hot core of the channel are recorded appropriately by grid 11 (Fig. 5) and are in qualitative agreement with the interference pictures from /_10_/, but are insufficiently recorded by grid 5 (Fig. 4).

The reason why the horizontal grid widths have the largest influence on statistical data becomes evident from the horizontal sections through the upper half of the channel: Very narrow spoke-pattern like structures are formed in

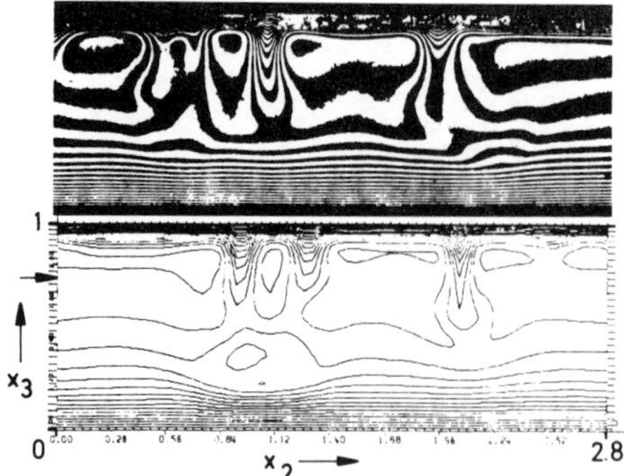

experiment /_10_7 and also in high resolution numerical studies like in case 11 (Fig. 6). Such structures are totally absent in the horizontal section for case 5 (Fig. 4). Thus, the spatial resolution of case 5 is far from being sufficient.

Fig. 5: Vertical temperature field for Ra = 3.7×10^6 /_10_7 and below for case 11 at t = 31. Δ = .0625. The arrow denotes the origin of Fig. 6.

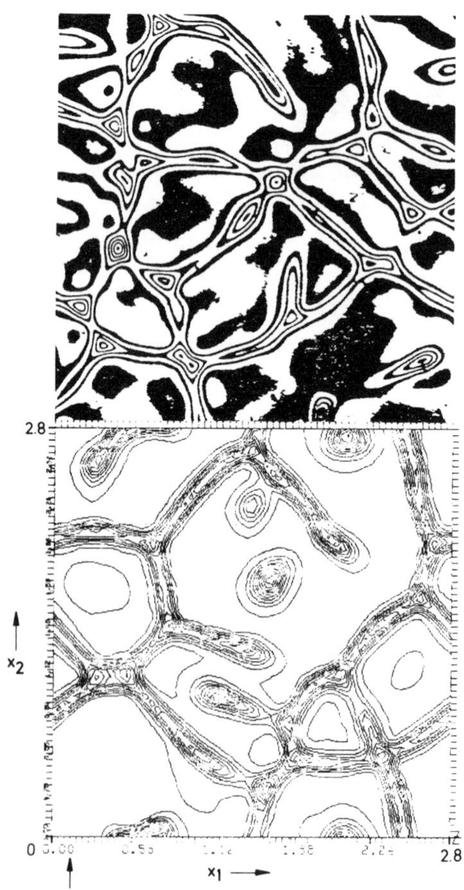

Fig. 6: Horizontal temperature fields deduced experimentally /_10_7 and below for case 11. Δ = .0625. The arrow denotes the origin of Fig. 5

5 THEORETICAL INVESTIGATION OF THE REQUIRED GRID WIDTHS

The subgrid scale heat flux model implemented in the TURBIT-code has successfully been used to judge on the spatial resolution capabilities of grids for Bénard convection [14] and for forced liquid metal flows [15]. Therefore, this model is applied here to the internally heated fluid flow problem.

The subgrid scale heat flux is proportional to a coefficient C_{T2} [15]. Assuming local isotropy for subgrid scale turbulence, this coefficient can be calculated on the basis of the Kolmogorov and Batchelor energy spectra for turbulent velocity and temperature fluctuations. For the constants in the spectra α = 1.5 and β = 1.3 are used. This procedure gives:

$$C_{T2} = \frac{1 - \beta\, f_1(\Delta x_i)(Re_o Pr)^{-1}\, V^{-4/9}\, \langle\varepsilon\rangle}{\beta\, \alpha^{1/2}\, f_2(\Delta x_i)} \qquad (3)$$

The functions $f_j(\Delta x_i)$, which are of order one and which depend only on grid parameters, are irrelevant here. The second term of the numerator only is important for small mesh volumes or for small Reynolds numbers $Re_o = u_o D/\nu$, where $u_o = \sqrt{g\beta\, \Delta T_o\, D}$. For these limits an approximation for the dissipation $\langle\varepsilon\rangle$ of turbulence energy has to be introduced to determine C_{T2}.

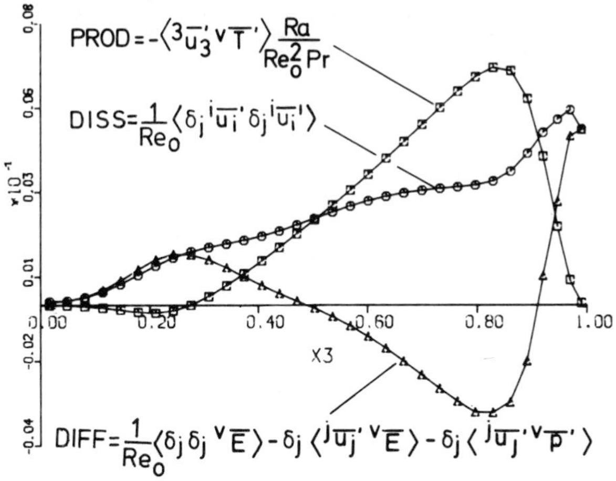

Fig. 7: Vertical profiles of terms of the conservation equation for the resolved turbulence energy ${}^v\overline{E} = \langle {}^i\overline{u}_i'^2 \rangle / 2$

The dissipation, calculated from the results of case 11 (Fig. 7), is nonzero in the conduction controlled lower area, where in accordance with estimations in [16] a small negative production is predicted. The nonzero dissipation is caused by the diffusion term which removes energy from the upper half to the lower half of the channel. For a rough estimate, the resulting smooth dissipation profile is approximated by a linear increase between the wall values $\langle\varepsilon_{W1}\rangle = 0$ and $\langle\varepsilon_{W2}\rangle = (Nu_2 - Nu_1)\, Ra/(Re_o^3\, Pr^2)$.

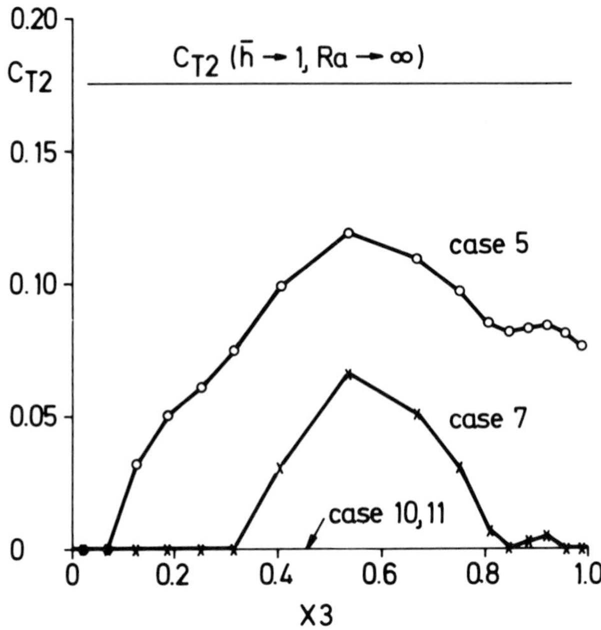

Fig. 8: Vertical profiles for the calculated coefficient C_{T2}

This approximation together with the complete theory indicated in Eq. (3) gives the results for C_{T2} shown in Fig. 8. The largest values appear for case 5; smaller ones for case 7; for case 10 and 11 the calculated values are zero. As the subgrid scale heat flux is proportional to C_{T2}, a subgrid scale model is necessary for cases 5 and 7, whereas practically all small scale turbulence elements are resolved in cases 10 and 11.

6 CONCLUSIONS

The Nusselt numbers calculated are slightly too high. This may indicate insufficient resolution of large vortices caused by too small periodicity lenghts, because for the similar Bénard convection an increase in turbulent heat flux for decreasing wavelenghts is known /‾12_/. Nusselt numbers and other statistical data are not very sensitive to insufficient resolution of small vortices. Sensitive results are energy spectra and contour line plots. These are completely nonphysical in case 5, although all computed statistical data deviate by not more than 25 % from those of the other cases. The sharp requirements to the horizontal grid widths are due to the existence of flow structures with very small horizontal extensions.

Compared to Bénard simulation requirements /12,14/, finer grids have to be chosen here despite lower degrees of turbulence, and other quantities are the real indicators for insufficient resolution. Thus, the indicators and the required grid widths depend strongly on the type of flow. For all flow types cited in this paper the statistical data usually considered are insensitive to insufficient subgrid scale modelling or insufficient spatial resolution. Thus, coarse grid simulations have to be carefully checked. The theory to calculate the coefficient of a subgrid scale heat flux model turned out to be a useful tool to predetermine grids with sufficient spatial resolution.

7 ACKNOWLEDGEMENTS

The author wishes to appreciate the lively discussions and many useful comments given to him by Dr. Ch. Homann and Dr. U. Schumann, Institut für Reaktorentwicklung, during the preparation of this paper. The careful typing by Mrs. H.Jansky is also kindly acknowledged.

8 REFERENCES

1 MAYINGER, F., JAHN, M., REINECKE, H.H., STEINBERNER, U. - Untersuchung thermohydraulischer Vorgänge sowie Wärmeaustausch in der Kernschmelze. Abschlußbericht BMFT-RS 48/1, Teil I, 1975

2 BIASI, L., CASTELLANO, L., HOLTBECKER, H. - Molten pool theoretical studies. PAHR-Information Exchange, ANL-78-10, Argonne, November 2-4, 1977

3 PECKOVER, R.S., HUTCHINSON, I.H. - Convective rolls driven by internal heat sources. Physics of Fluids, Vol. 17, pp. 1369-1371, 1974

4 REINECKE, H.H. - Numerische Berechnung der thermohydraulischen Vorgänge in einer Kernschmelze, Diss., Tech. University Hannover, 1974

5 EMARA, A.A., KULACKI, F.A. - Studies of heat source driven natural convection: A numerical investigation. RF 3746, Ohio State University

6 SAFFMAN, P.G. - Problems and progress in the theory of turbulence. Structure and Mechanisms of Turbulence II, Ed. H. Fiedler, Lecture Notes in Physics, Vol. 76, Springer-Verlag Berlin, 1978

7 GRÖTZBACH, G. - Numerical investigation of radial mixing capabilities in strongly buoyancy-influenced vertical, turbulent channel flows. Nuclear Engineering and Design, Vol. 54, pp. 49-66, 1979

8 SCHUMANN, U., GRÖTZBACH, G., KLEISER, L. - Direct numerical simulation of turbulence. Prediction Methods for Turbulent Flows, Ed. W. Kollmann, Hemisphere Publ. Corp., pp. 123-258, 1980

9 KULACKI, F.A., GOLDSTEIN, R.J. - Thermal convection in a horizontal fluid layer with uniform volumetric energy sources. J. Fluid Mech., Vol. 55, pp. 271-287, 1972

10 JAHN, M. - Holographische Untersuchung der freien Konvektion in einer Kernschmelze. Diss., Techn. University Hannover, 1975

11 TVEITEREID, M. - Thermal convection in a horizontal fluid layer with internal heat sources. Int. J. Heat Mass Transfer, Vol. 21, pp. 335-339, 1978

12 GRÖTZBACH, G. - Direct numerical simulation of laminar and turbulent Bénard convection. To appear in J. Fluid Mech.

13 BAKER Jr., L., FAW, R.E., KULACKI, F.A. - Post-accident heat removal I: Heat transfer within an internally heated nonboiling liquid layer. Nucl. Science Engng., Vol. 61, pp. 222-230, 1976

14 GRÖTZBACH, G. - Über das räumliche Auflösungsvermögen numerischer Simulationen von turbulenter Bénard-Konvektion. KfK 2981 B, Kernforschungszentrum Karlsruhe, 1980

15 GRÖTZBACH, G. - Numerical simulation of turbulent temperature fluctuations in liquid metals. To appear in Int. J. Heat Mass Transfer, 1981

16 KULACKI, F.A., GOLDSTEIN, R.J. - Eddy heat transport in thermal convection with volumetric energy sources. Fifth Int. Heat Transfer Conf., Tokyo, Vol. 3, pp. 64-68, 1974.

HEAT AND MASS TRANSFER IN TUBULAR REACTORS: A FINITE ELEMENT APPROACH

Stefano DEL GIUDICE ([1]) e Antonio TROTTA

Istituto di Impianti Chimici, Istituto di Fisica Tecnica ([1]) dell'Università di Padova, via Marzolo 9, 35100 PADOVA (Italy).

ABSTRACT

A finite element method is applied to the study of the temperature and composition distributions and heat transfer characteristics for a reacting fluid flowing through a tube. Results are succesfully compared with those obtained by finite difference methods.
The method described may be used in the basic design or analysis of laminar flow reactor.

INTRODUCTION

The development of efficient numerical integration procedures and the speed of modern computers have enabled the extensive use of two dimensional models for the simulation of tubular reactors, both empty or packed with a catalyst.
This paper is concerned with the application of the finite element method (FEM) to these problems traditionally solved by finite difference methods.
In a previous paper ([1]) the authors applied the FEM to the study of plug flow catalytic reactors. In the simpler plug flow model the velocity, the thermal conductivity and the diffusivity are assumed to be constant in the radial direction. This assumption is quite acceptable when the reactant is flowing in the turbulent regime, however, significant error can be introduced when it is applied to systems in the laminar regime.

Radial gradients of the axial velocity cause each cylindrical shell of reacting fluid to have a different residence time in the reactor. This results in a radial composition gradient which, in turn, tends to be diminished by molecular diffusion and, under some conditions, by free convection. Moreover, owing to the temperature gradient, there is a variation in the reaction rate constant across the tube. Thus reaction, flow and diffusional effects are related.

BASIC EQUATIONS

The basic equations are derived from the complete transport equations ([2]) by elimination of terms in accordance with the physical aspects of the system. If the reaction is assumed to be irreversible and first order with respect to the single reactant, the mass and energy balances are as follows

$$v' Le \partial C'/\partial z' = (1/r')\partial/\partial r'(r'\partial C'/\partial r') - \chi^2 C' exp\{-\gamma/(T'+1)\} \quad (1)$$

$$v' \partial T'/\partial z' = (1/r')\partial/\partial r'(r'\partial T'/\partial r') + \chi^2 \beta C' exp\{-\gamma/(T'+1)\} \quad (2)$$

The following dimensionless variables are used:
$z' = \frac{1}{2} z\, \lambda/(R^2 G\, c_p)$, $r' = r/R$, $C' = C/Co$, $T' = (T-To)/To$,
$v' = \frac{1}{2} v/v_{av}$.

The parameters are the following dimensionless assemblies of the properties of the system:
$Le = \lambda/(\rho \mathcal{D} c_p)$, $\chi = R(Mc_p Pk_o/\lambda)^{\frac{1}{2}}$, $\beta = (-\Delta H)\, Co/(M\, c_p\, To)$,
$\gamma = E/(R_g\, To)$

The initial and boundary conditions for the case of constant wall temperature ($T_w = To$) are

$$C' = 1,\ T' = 0 \quad at \quad z' = 0 \quad (3)$$

$$T' = 0,\ \partial C'/\partial r' = 0 \quad at \quad r' = 1 \quad (4)$$

Derived quantities

Bulk values The bulk properties are defined as the temperature and composition which result when the fluid at a given axial position is mixed at constant enthalpy and no further reaction occurs:

$$C'_b = \int_0^1 C' \, v' \, r' \, dr' / \int_0^1 v' \, r' \, dr' \tag{5}$$

$$T'_b = \int_0^1 T' \, v' \, r' \, dr' / \int_0^1 v' \, r' \, dr' \tag{6}$$

Nusselt numbers The local Nusselt number may be written as

$$Nu_{loc} = - 2(\partial T'/\partial r')_{r'=1}/T'_b \tag{7}$$

The average Nusselt number is based on the heat flux integrated along the length of the reactor and the arithmetic mean temperature difference

$$Nu_{av} = \{ -T'_b + \beta(1 - C'_b)\} / (z' \, T'_b) \tag{8}$$

where T'_b and C'_b are evaluated at z'.
It is worth noting here that the FE formulations lead naturally to the heat flux at each point of the boundary where conditions (4) applies ([3]), whereas the finite difference formulation cannot be so readily applied ([4,5]).

FINITE ELEMENTS DISCRETISATION

Let the unknown functions $C'$and T' be approximated by the expansions $C' = \underset{\sim}{N}\, \underset{\sim}{C}$, $T' = \underset{\sim}{N}\, \underset{\sim}{T}'$, where $\underset{\sim}{N}$ are the shape functions. Application of the Galerkin method gives rise to a system of differential equations which may be written in matrix form as

$$\underset{\sim}{H}\, \underset{\sim}{\phi} + \underset{\sim}{P}\, \partial\underset{\sim}{\phi}/\partial z' + \underset{\sim}{F} = 0 \tag{9}$$

where the vectors $\underset{\sim}{\phi}$, $\underset{\sim}{F}$ are defined by

$$\underset{\sim}{\phi} = [\underset{\sim}{C}', \underset{\sim}{T}']^{\mathbf{T}} \;;\quad \underset{\sim}{F} = [\, \underset{\sim}{S}, - \beta\, \underset{\sim}{S}]^{\mathbf{T}}$$

and $\underset{\sim}{H}$, $\underset{\sim}{P}$ are $2m \times 2m$ matrices:

$$\underset{\sim}{H} = \begin{bmatrix} \underset{\sim}{K} & 0 \\ 0 & \underset{\sim}{K} \end{bmatrix}; \quad \underset{\sim}{P} = \begin{bmatrix} Le\, \underset{\sim}{M} & 0 \\ 0 & \underset{\sim}{M} \end{bmatrix}$$

Typical matrix elements are

$$k_{ij} = \Sigma \int_{\Delta r'^{e}} \partial N_i/\partial r' \; \partial N_j/\partial r' \; r' \, dr' \qquad (i,j=1,\ldots,m)$$

$$m_{ij} = \Sigma \int_{\Delta r'^{e}} v' \, N_i \, N_j \, r' \, dr'$$

$$s_i = \phi^2 \, \Sigma \int_{\Delta r'^{e}} C' \, exp\{-\gamma/(T'+1)\} \, N_i \, r' \, dr'$$

where $\Delta r'^{e}$ are the lengths of the line elements and the summation are taken over the contribution of each element. The numerical solution of equation (9) is accomplished by the use of a three level algorithm ([6]):

$$\underset{\sim}{\phi}^{n+1} = \{\underset{\sim}{H}/3+\underset{\sim}{P}/(2\Delta z')\}^{-1}\{-\underset{\sim}{H}/3(\underset{\sim}{\phi}^{n-1}+\underset{\sim}{\phi}^{n})+\underset{\sim}{P}/(2\Delta z')\underset{\sim}{\phi}^{n-1}-\underset{\sim}{F}^{n}\}$$

Once concentrations and temperature at location z' have been calculated, bulk values and Nusselt numbers may be evaluated by means of equations (5-8).

RESULTS

As a test of the validity and correctness of the formulation, a laminar flow situation in a empty reactor was analyzed. The values of the parameters were chosen to represent a more or less typical gaseous reaction system. Figure 1 and 2 present the axial temperature and concentration profiles. In figure 3 point and average Nusselt numbersare shown for the same operating conditions.

Results obtained demonstrate excellent agreement with the finite difference models of Rothenberg and Smith ([4]) and Andersen and Coull ([5]). The solution for the case of zero heat reaction ($\beta = 0$) can be compared with the results of Cleland and Wilhelm ([7]) which they considered exact. At plotting accuracy no significant difference can be detected between the two solutions.

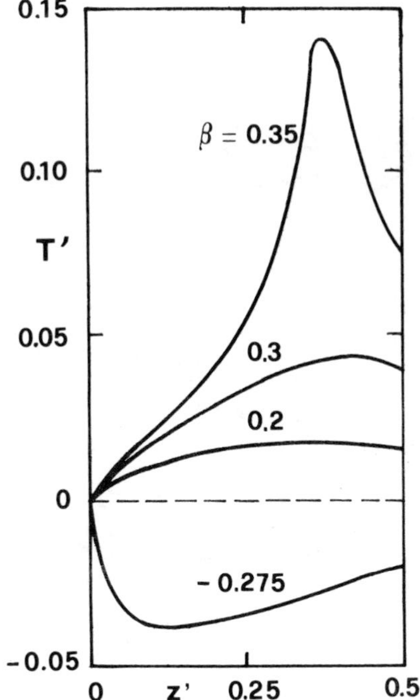

Fig.1 - Axial temperature profiles for exothermic (Le=1, $\chi^2=10^{13}$, γ = 30.7) and endothermic (γ = 28.4) reactions for a laminar flow reactor with radial diffusion.

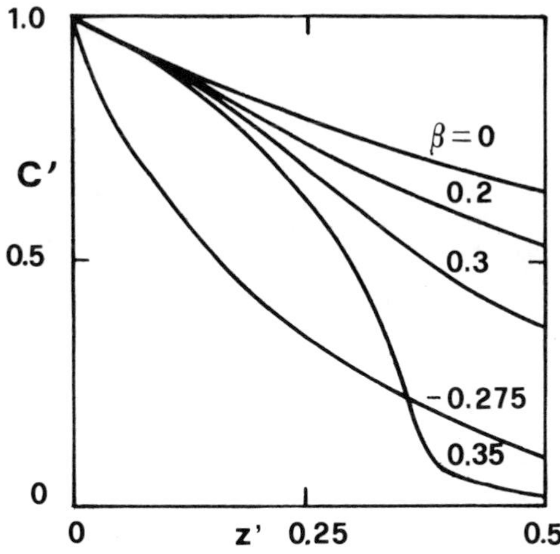

Fig.2 - Axial reactant concentration profiles for exothermic and endothermic reactions (same parameters as in Fig.1).

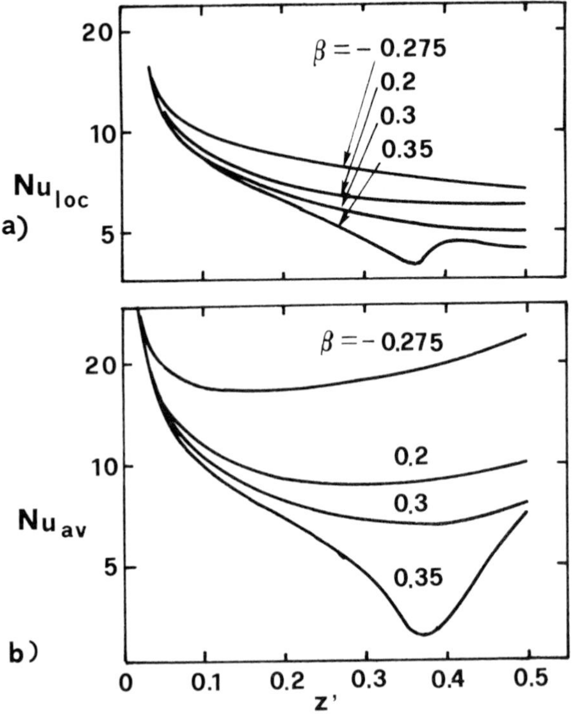

Fig.3 - Point (a) and average (b) Nusselt numbers for exothermic and endothermic reactions (Same parameters as in Fig.1).

SPECIAL APPLICATION: FIXED BED CATALYTIC REACTOR

In recent years, there has been a growing interest in the role of hot spots and temperature excursion in affecting the malfunctioning of catalytic packed bed reactors. It has been generally recognized that hot spot formation is associated with flow maldistribution which may be aggrvated with spatially non uniform temperatures. In the reactor tubes, packed with catalyst pellets, the mean local porosity is not constant. At a given location along the length of the tube the porosity and velocity are functions of the radius. This induces a radial variation of the effective thermal conductivity. At the wall, the void fraction approaches one; the velocity changes drastically and the resistance to the heat transfer is altered. This increased resistance can be modelled by a heat transfer coefficient.

That being stated, the mathematical model consists of the following continuity and energy equations

$$v(r)\,\partial C_i/\partial z = (1/r)\,\partial/\partial r(r\,\mathcal{D}_{er}\,\partial C_i/\partial r) + \rho_b(r)\,\mathcal{R}_i \quad (i = 1,\dots,n) \tag{10}$$

$$\rho_f v(r) c_p\,\partial T/\partial z = (1/r)\,\partial/\partial r(r\,\lambda_{er}\,\partial T/\partial r) + \rho_b(r)\,\Sigma_i(-\Delta H_i)\,\mathcal{R}_i \tag{11}$$

$$C_i = Co_i,\ T = To \quad at \quad z = 0 \tag{12}$$

$$\lambda_{er}\ \partial T/\partial r = h_w[\,T_w - T(R,z)\,]$$

$$\partial C_i/\partial r = 0 \quad at \quad r=R$$

This model served as a basis for the simulation of a practical case: the catalytic oxidation of ortoxylene to phthalic anhydride in a multitubular reactor. The reaction scheme, the rate equations and operating conditions were those proposed by Froment ([8,9]). The velocity profile used was the experimental one of Scwartz and Smith ([10]), taken under isothermal conditions and assuming the velocity profile would remain similar in the non isothermal case. The porosity profile is assumed on the basis of the experimental data reported in literature ([11]). The static contribution to effective conductivity was calculated by means of the Zehner-Schlunder equation ([12]). The dynamic contribution by means of the De Wasch and Froment equation ([13]). Since the bulk catalyst density ρ_b depends upon the void fraction, point values were also taken for the former. No point values are required for $\mathcal{D}_{er}$. The influence of $\mathcal{D}_{er}$ is much smaller than that of λ_{er}.

Element equations for problems of this type can be derived by using the method of weighted residuals with Galerkin's criterion. The matrix equations obtained this way have the same form as those given in the previous section; however, in this case, the coefficient matrix $\underset{\sim}{H}$ is function of ϕ.

Figure 4 compares the isotherm contour map for the model with porosity and velocity profile with the corresponding map calculated by means of the model with uniform velocity and porosity profiles. It is wotrh noting that the former model

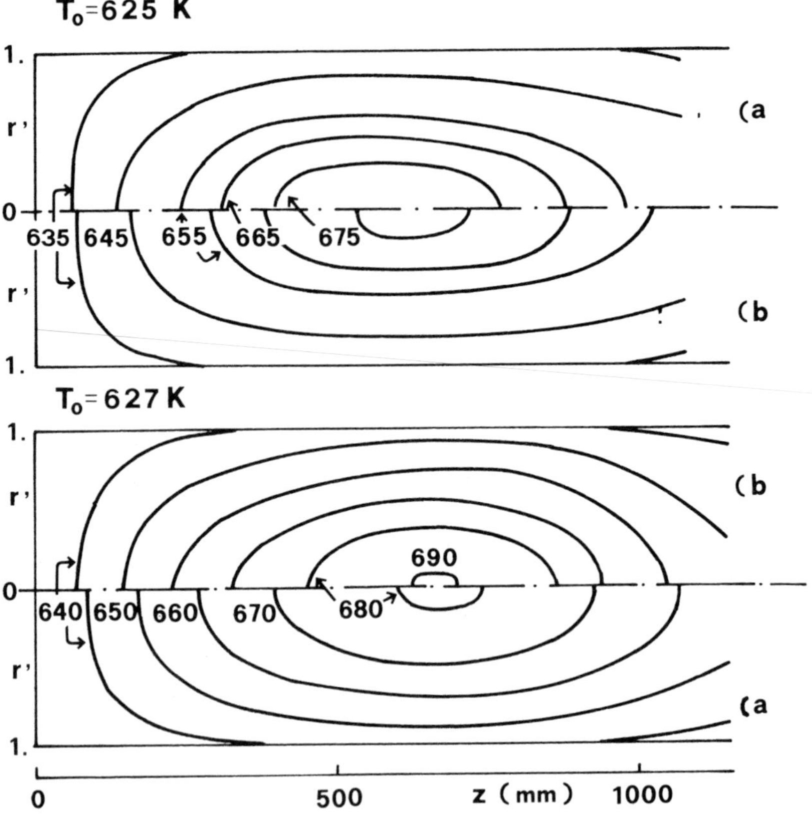

Fig.4 - Catalyzed air oxidation of ortoxylene into phthalic anhydride. Temperature maps in the reactor: (a) model with uniform velocity and porosity profiles; (b) model with velocity and porosity profiles.

is more conservative in the prediction of critical situations. The results obtained can be compared qualitatively with the temperature profiles published by Lerou and Froment ([9]). Their logitudinal profiles are similar in shape to those of Fig. 4. A quantitative comparison is not possible because of the different values of the porosity and velocity profiles.

CONCLUSION

The FEM is an efficient numerical method for solving the equations governing tubular reactors with radial gradients. This model served as a basis for the simulation of a reaction scheme of a relatively complex nature and representative of a commercial hydrocarbon oxidation.

ACKNOWLEDGEMENT

The authors aknowledge the financial support of the "Ministero della Pubblica Istruzione".

NOMENCLATURE

C	concentration	Nu	Nusselt number
c_p	specific heat	P	total pressure
$\mathscr{D}$	diffusion coefficient	$\mathscr{R}$	reaction rate
G	mass velocity	R_g	ideal gas constant
ΔH	heat of reaction	R	tube radius
h	heat tranfer coefficient	r	radial coordinate
Le	Lewis number	T	temperature
M	molecular weight	v	axial velocity
m	number of nodes	z	axial coordinate
n	number of components		

Greek letters

β	heat of reaction group	ρ_b	catalyst bulk density
γ	activation energy group	ρ_f	fluid density
λ	thermal conductivity	χ	frequency factor group

Subscript and superscripts

b	bulk mean value	o	inlet of reactor
er	radial effective	w	evaluated at wall
f	fluid	$'$	dimensionless variable

REFERENCES

(1) A. TROTTA, S. DEL GIUDICE, Internal report, Istituto di Impianti Chimici, Università di Padova, 9, 1979.

(2) R.B. BIRD, W.E. STEWART and E.N. LIGHTFOOT, *Transport Phenomena,* Wiley, New York (1960).

(3) S. DEL GIUDICE and A. TROTTA, *Chem. Engng Sci.* 33, 697, 1978.

(4) R.J. ROTHENBERG and J.M. SMITH, *A.I.Ch.E.J.*, 12,213, 1966.

(5) T.S. ANDERSEN and J. COULL, *A.I.Ch.E.J.*, 16, 542, 1970.

(6) M. LEES, *Maths. Comp.*, 20, 516, 1966.

(7) F.A. CLELAND and R.H. WILHELM, *A.I.Ch.E.J.*, 2, 489, 1956.

(8) G.F. FROMENT, *Ind. Engng Chem.*, 19, 59, 1967.

(9) J.J. LEROU and G.F. FROMENT, *Chem. Engng Sci.*, 32, 567, 1977.

(10) C.E. SCHWARTZ and J.M. SMITH, *Ind. Engng Chem.*, 45, 1209, 1953.

(11) M.C. THADANI and F.N. PEEBLES, *I.& E. Ch. Proc. Des. Dev.*, 5, 265, 1966.

(12) P. ZEHNER and E.U. SCHLUNDER, *Chemie Ing. Techn.*, 42,933, 1970, 44, 1303, 1972, 45, 272, 1973.

(13) A.P. DE WASCH and G.F. FROMENT, *Chem. Engng Sci.*, 25, 567, 1972.

A NUMERICAL SOLUTION OF RIBBON THICKNESS FORMATION DURING MELTSPINNING.

L. Katgerman*

SUMMARY

A model of the meltspinning process (rapid solidification by an impinging jet of molten metal on a cold moving substrate) is described, in which the equations of heat and fluid flow are coupled by the moving solid-liquid phase boundary. The finite difference calculation is executed in terms of dimensionless quantities for Nusselt numbers where Newtonian or ideal cooling models can no longer be applied ($0.01 < Nu < 25$). The thermal and momentum boundary layers which determine the final thickness of the meltspun ribbon are calculated. It is found that the momentum boundary layer is substantially decreased as a result of the moving solid-liquid interface.

1. INTRODUCTION.

The meltspinning process is one of the rapid quenching techniques for solidification of metals and alloys at high cooling rates in the range of $10^5 - 10^7$ K/s. Rapidly quenched metals have shown important improvements in structure and properties [1-3].

In the meltspinning process a jet of liquid metal impinges on the surface of a cold rotating wheel, where it solidifies to produce a continuous ribbon or sheet. A metal ribbon made this way usually exhibits dimensional variations which can give non-uniform properties in the case of crystalline rapidly solidified metals. In further processing of rapidly solidified ribbons such dimensional variations are often undesirable.

* Research Associate; Laboratory for Metallurgy,
Delft University of Technology,
2628 AL DELFT, The Netherlands.

The development of a mathematical model can be used to determine quantitatively the effect of meltspinning process variables on ribbon thickness formation. To calculate the thickness of the ribbon the thermal and velocity profile must be determined. In all analyses sofar the interaction of heat flow and fluid flow has been neglected [4-6], steady state conditions were assumed [7] or only the two asymptotic cases of heat flow (Newtonian and ideal) have been considered [8]. The purpose of this work is to calculate the ribbon thickness formation taking into account the interaction of the solidifying boundary with the velocity profile for all cases of heat flow and non-steady state conditions.

2. FORMULATION.

Consider the liquid puddle during meltspinning to be a semi-infinite body of liquid metal bounded by a flat chill surface (Fig. 1). Initially the liquid metal and the chill surface are at rest. At time t = 0 the chill surface is set in motion in the positive x-direction with a velocity V_o; heat is extracted from the liquid metal by the substrate and the liquid metal starts to solidify.

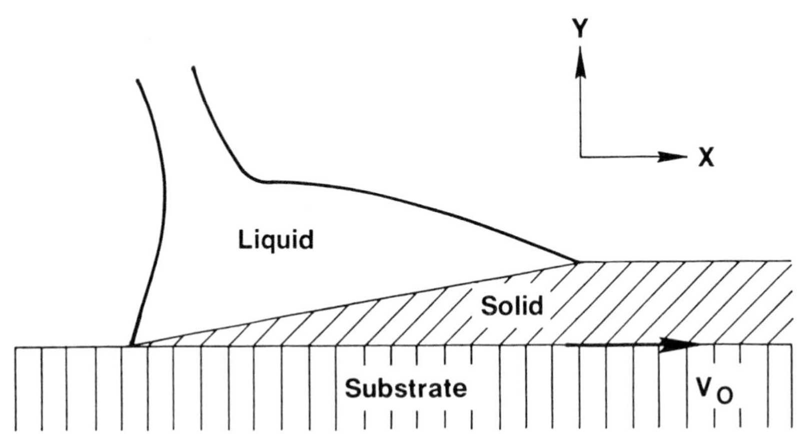

Fig. 1. Schematic representation of the melt puddle.

From the physical description it follows that fluid flow and heat flow are interrelated.
The further assumptions inherent in the formulation of the governing equations are the following:

1. The thermal and physical properties of liquid and solid metal are constant.
2. the heat flow is geometrically in one dimension (y-direction) and can be characterized by a convective heat-transfer coefficient α.
3. the liquid metal is initially at the melting point T_M and there is no thermal gradient in the liquid metal.
4. the solid-liquid interface is plane.
5. the velocity components in y and z direction are zero and consequently by application of the equation of continuity the velocity component in x-direction is a function of y and t alone ($u = u(y,t)$).
6. there is no gravity force or pressure gradient in the x-direction and the fluid flow is laminar.

For a Cartesian coordinate system the equation of motion relative to the moving solidification front takes the following form

$$\frac{\partial u}{\partial t} = \frac{\partial^2 u}{\partial y'^2} + R(t) \cdot \frac{\partial u}{\partial y'} \qquad (1)$$

in which $y' = y - S(t)$ is the distance from the solid-liquid interface, $S(t)$ is the thickness of the solidified crust and $R(t) = dS/dt$ is the solidification rate.
Boundary conditions are:

$t = 0$	$u = 0$	for all y'	(2)
$y' = 0$	$u = V_o$	for all $t > 0$	(3)
$y' = \infty$	$u = 0$	for all $t > 0$	(4)

The second boundary condition (Eq.(3)) expresses that the solidified metal is at surface velocity V_o and that slip between solidified metal and moving substrate is neglected. The interface velocity is determined by the cooling conditions. An exact solution for the solidification rate from the unidirectional heat flow equation can only be obtained for the ideal case ($\alpha = \infty$, perfect thermal contact) or for the Newtonian case (constant heat flux $q = \alpha\,(T_M - T_o)$). It is convenient to introduce the following dimensionless variables and parameters:
(see Nomenclature for additional notation).

$$\tau = \frac{\alpha^2 a}{K_s^2}\, t; \quad \xi = \frac{\alpha y'}{K_s}; \quad \xi_T = \frac{\alpha S}{K_s} \qquad (5)$$

$$\text{Stef} = {}^{L}f/C_p\,(T_M - T_o);$$

The dimensionless solidification rate R^* in the ideal case then takes the form [9]:

$$R^{*} = d\xi_T/d\tau = \gamma\tau^{-\frac{1}{2}} \tag{6}$$

where γ is implicitely given by

$$\gamma \exp(\gamma^2) \operatorname{erf}(\gamma) = \mathrm{Stef}/\sqrt{\pi} \tag{7}$$

for Newtonian cooling R^{*} is constant and given by R^{*} = Stef. In the case of intermediate cooling numerical methods have to be used [10,11] and from these calculations it was found [10] that for $\xi_T > 30$ and $\xi_T < 0.015$ the analytical expressions derived above can be applied to ideal or Newtonian cooling respectively.

It was shown by Jones [12] that second approximations of plane-front solidification with a finite heat-transfer coefficient gives solidification rate and solidified thickness to an accurancy of within a few percent relative to the exact solution.

In this paper we will use the approximation obtained by Megerlin [13], giving for the solidification rate

$$R^{*} = M/(1 + \xi_T) \tag{9}$$

in which M is a function of Stef and ξ_T (or τ).Eq. (9) can be integrated analytically to give the solidified thickness and to give the explicite expression of the solidification rate as a function of τ.

The result in the case of Al (Stef = 1.74) is given in Fig. 2. It can be seen that the approximate analytical solutions converges to the asymptotic solutions for ideal and Newtonian cooling.

When Eq. (1) is written in terms of dimensionless parameters we get

$$\frac{\partial u}{\partial \tau} = P_R \frac{\partial^2 u}{\partial \xi^2} + R^{*}(\tau) \cdot \frac{\partial u}{\partial \xi} \tag{10}$$

The total ribbon thickness ξ_R at time τ when it leaves the puddle is calculated as

$\xi_R = \xi_T + \xi_M$, where ξ_T is the solidified thickness and ξ_M is the momentum boundary layer thickness taken as that distance for which u has dropped to a value of 0.01 V_o.

For the case of Newtonian and ideal cooling analytical solutions of Eq. (10) can be obtained and the momentum boundary layer is readily calculated [8].

Eq. (10) for intermediate cooling was solved by the finite difference technique using a four-point explicite difference scheme with forward time difference. The stability criterion for the finite difference equation remains the same as in the case without first order term [14] and requires:

$$P_R \cdot {}^{\Delta\tau}/(\Delta\xi)^2 < 0.5$$

where $\Delta\tau$ and $\Delta\xi$ are the increments of τ and ξ.

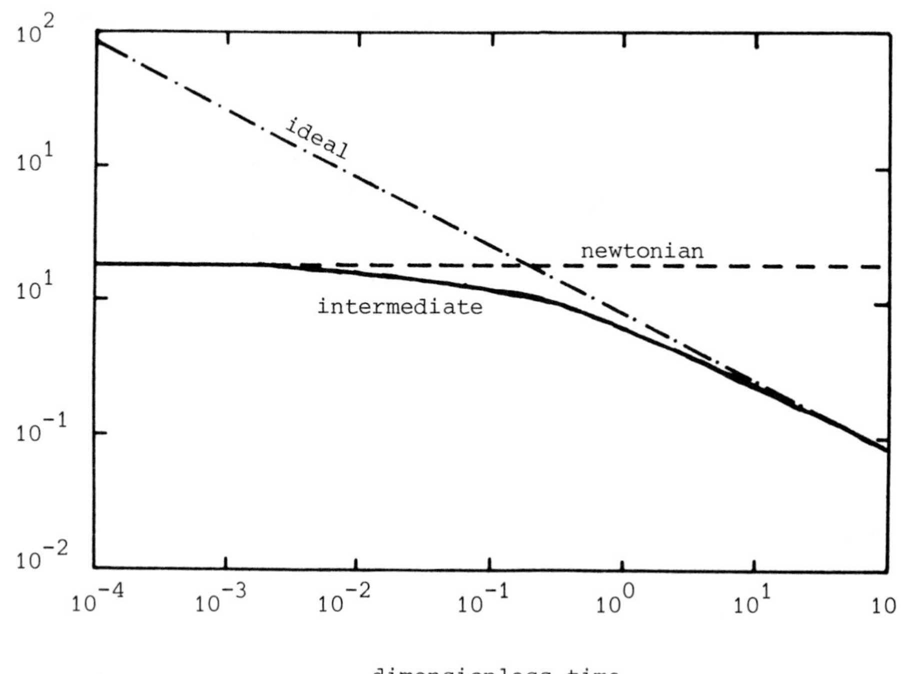

Fig. 2. Dimensionless solidification rate as a function of the dimensionless time for Newtonian, intermediate and ideal cooling.

At each time step the velocity profile was calculated and the momentum boundary-layer thickness ξ_M was determined using a 5-point Lagrange interpolation schema in combination with a regula-falsi procedure. The computer programme was first checked with the analytical solution for Newtonian cooling conditions from ref. [8]. The agreement was highly satisfactory and the relative error in the calculated shear boundary layer was less than 1%.
The calculations were carried out using the IBM 370/158 digital computer of the Delft University of Technology.

3. COMPUTED RESULTS AND DISCUSSION;

The procedure described in the previous section was used to obtain velocity profiles as a function of ξ and τ and the momentum boundary layer thickness as a function of τ.
The computations have been carried out for the case of aluminium. The values of the parameters used in the computations are listed in Table 1.

Table 1. Values of physical parameters of Al used in computation.

$\mu = 2.5\ 10^{-3}\ \mathrm{Ns/m^2}$	$L_f = 3.93\ 10^5\ \mathrm{J/kg}$
$\rho = 2.7\ 10^{-3}\ \mathrm{kg/m}$	$T_M = 660\ ^\circ C$
$c_p = 1.08\ 10^{-3}\ \mathrm{J/kg.K}$	$T_o = 25\ ^\circ C$
$K_s = 100\ \mathrm{W/m.K}$	$Stef = 1.74$
$\alpha = 10^5\ \mathrm{W/m^2.K}$	$P_R = 2.69\ 10^{-2}$

The results of the boundary layer calculation are shown in Fig. 3. It can be seen that the momentum boundary layer thickness is substantially decreased as a result of the moving solidification front.
This effect is less drastic than in the case of strictly Newtonian cooling but is still large with respect to the non-interaction boundary layer calculation*.
For a good control of ribbon thickness formation during meltspinning it is important to know whether the process is thermal transport controlled or momentum transport controlled. Therefore the thermal boundary layer thickness (solidified crust) and momentum boundary layer should be compared. In Fig. 4 these results have been summarized. For Newtonian cooling conditions the thickness formation is momentum transport controlled for $\tau < 3.5\ 10^{-2}$. The results of the intermediate cooling computations shown that this region is extended to $\tau < 4.5\ 10^{-2}$.
In the case of non-interaction the momentum transport is dominant for $\tau < 1.5\ 10^{-1}$ which has lead tot the conclusion that the meltspinning process is momentum transport controlled [6]. The present calculations clearly demonstrate that the thermal transport plays a more significant role during ribbon thickness formation than may be concluded from non-interaction calculations.

* The non-interaction velocity profile can be expressed as $u(y,t) = V_o\ \mathrm{erfc}\ (y/2\sqrt{\nu t})$. [15].

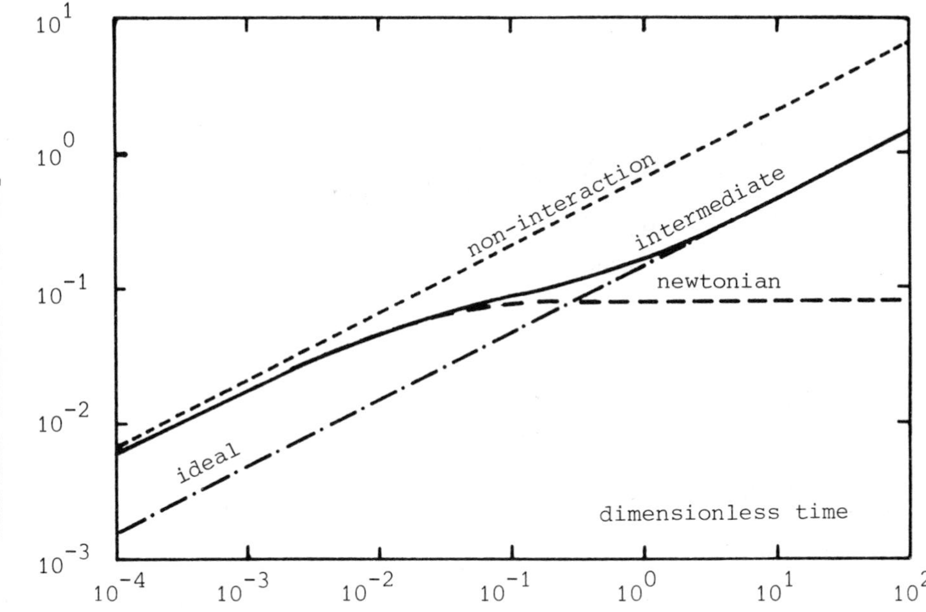

Fig. 3. Dimensionless momentum boundary layer thickness as a function of dimensionless time for different conditions of cooling.

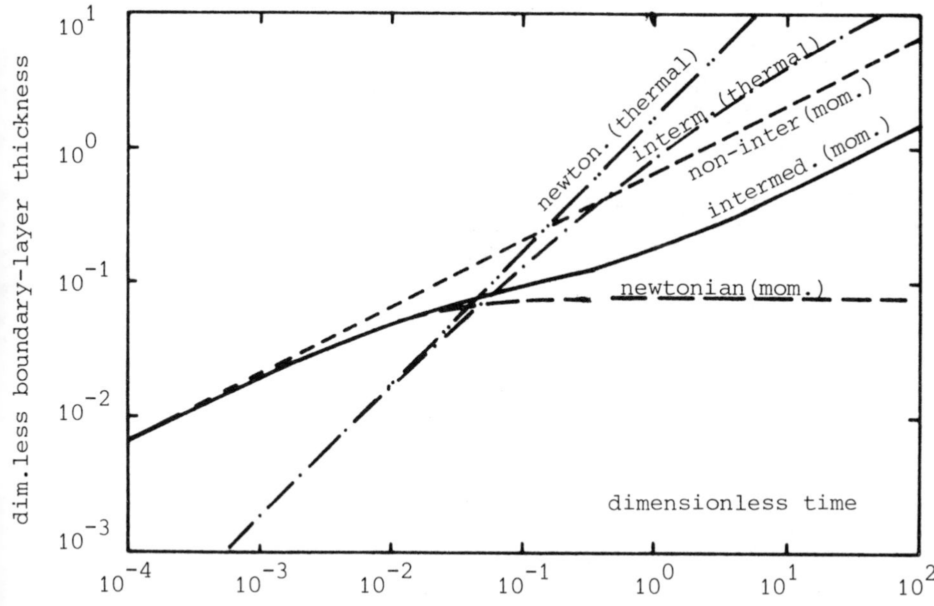

Fig. 4. Thermal and momentum boundary layer thickness for Newtonian and intermediate cooling as a function of dimensionless time.

ADDITIONAL NOMENCLATURE

$a = K_s/\rho c_p$ thermal diffusivity (m^2/s)
L_f latent heat of fusion (J/kg)
R solid-liquid interface velocity (m/s)
S thickness solidified (m)
$Stef = L_f/c_p(T_M-T_o)$ dimensionless Stefan-number for solidification
T_M melting temperature
T_o substrate temperature

α convective heat-transfer coefficient (W/m^2k)
$\tau = \alpha^2 at/K_s^2$ dimensionless time
$\xi = \alpha y/K_s$ dimensionless distance

ACKNOWLEDGEMENTS

The author is very grateful to Mrs. Lia van Aken for accurate typing of the manuscript and to the Industrial Aluminium Research Foundation (BIRA), The Netherlands, for partial financial support.

REFERENCES

1. M. LEBO, N.J. GRANT - Structure and Properties of a Splat Cooled 2024 Aluminium Alloys, Met. Trans. 5, 1547 (1974).

2. H. JONES - Developments in Aluminium Alloys by Solidification at High Cooling Rates, Aluminium 54, 274 (1978).

3. N.J. GRANT - Applications of metastable microcrystalline alloys, Rapidly Quenched Metals III Vol. 2, 172, Ed. B. Cantor, The Metals Society, London (1978).

4. S. KAVESH - Principles of Fabrication, Metallic Glasses, American Society for Metals, Metals Park OH. (1978).

5. H. HILLMAN, H.R. HILTZINGER - On the formation of amorphous ribbon by the meltspin technique. p 22, see reference [3].

6. J.H. VINCENT, H.A. DAVIES - High Speed Continuous Casting of Strip by Chill-Block Meltspinning, Paper 78 to be published in Solidification Technology in the Foundry and Cast House, Warrick, The Metals Society.

7. T.R. ANTHONY, H.E. CLINE - On the Uniformity of Amorphous Metal Ribbon Formed by a Cylindrical Jet Impinging on a Flat Moving Substrate, J. Appl. Phys. 49 829 (1978).

8. L. KATGERMAN - Theoretical Analysis of Ribbon Thickness Formation during Meltspinning, Scripta Met. 14, 861 (1980).

9. H.S. CARSLAW, J.C. JAEGER - Conduction of Heat in Solids, Oxford University Press, Oxford (1959).

10. R.C. RUHL - Cooling Rates in Splat Cooling, Mat. Sci. Eng. 1. 313 (1967).

11. Moving Boundary Problems Ed. D.G. Wilson, A.D. Solomon, P.T. Boggs, Academic Press New York (1978).

12. H. JONES - A Comparison of Approximate Analytical Solutions of Freezing from a Plane Chill, J.I.M. 97, 38 (1969).

13. F. MEGERLIN - Geometrisch Eindimensionable Warmeleitung Beim Schmelzen und Erstarren, Diss. T.H. Aachen, Germany (1965).

14. R.D. RICHTMYER, K.W. MORTON - Difference Methods for Initial Value Problems, Interscience New York (1967).

15. R.D. BIRD, W.E. STEWART, E.N. LIGHTFOOT - Transport Phenomena, John Wiley, New York (1960).

NUMERICAL "TURBULENT TRANSITION" LOCUS FOR CONTINUOUS CASTING TYPE FLOWS

Lee A. Bertram

Sandia National Laboratories, Albuquerque, NM 87185

ABSTRACT

Torrance and Rockett [1] numerically examined free convection in a closed cylindrical chamber heated by an axisymmetric disc on its floor. All other surfaces were held isothermal at a lower temperature. The solution for constant properties became time dependent (periodic) at $G_R \simeq 4 \times 10^{10}$. Related experiments [2] underwent turbulent transition at $G_R = 1.2 \times 10^9$. Prandtl number was 0.7.

The present work reports results of a similar numerical study of the parameters at which the computed motions in a pool of liquid metal ($P_R = 0.0435$) begin to display time dependence. The metal pool represents the liquid atop a cylindrical ingot being formed in a continuous casting process (vacuum consumable arc remelting) in which large electrical currents are present. Thus the dimensionless parameters of the problem include a Peclet number P_E and an Alfven number A_o as well as P_R and G_R.

The numerical scheme employed was a close relative of that in [1]; however, its use on that problem does not produce periodic solutions at large G_R (up to 4×10^{12}) even though solutions at low G_R agree. For the casting flows, with

$$0 \leq P_E \leq 5.0 \text{ and } 10^{-3} \leq A_o \leq 10^{10} \quad ,$$

all periodic solutions of significant magnitude can be accounted for by boundary condition dynamics or can be removed by changing the computational mesh. Consequently, it is conjectured that numerical steady laminar solutions to both of these convection-enclosure problems exist at quite large G_R far beyond the level at which turbulent transition occurs in the physical system.

*This work supported by the U.S. Department of Energy.

x

1. INTRODUCTION

The numerical-experimental correlation reported by Torrance et al. [1,2] in their combined study of turbulent transition in axisymmetric thermal convection in a cylinder (Fig. 1) suggested that physical meaning might be assigned to similar time-dependent numerical solutions. The immediate question this raises is "How wide is the class of physical problems and numerical techniques for which such correlation can be expected?"

The sensitivity of the correlation to numerical technique is examined to a limited degree in Section 2 where the ZAP code, which uses an algorithm very similar to that in Ref. [1] is applied to the same problem. On similar (coarse) meshes, ZAP produces steady solutions three orders of magnitude further than those computed in [1], reaching $G_R = 4 \times 10^{12}$.

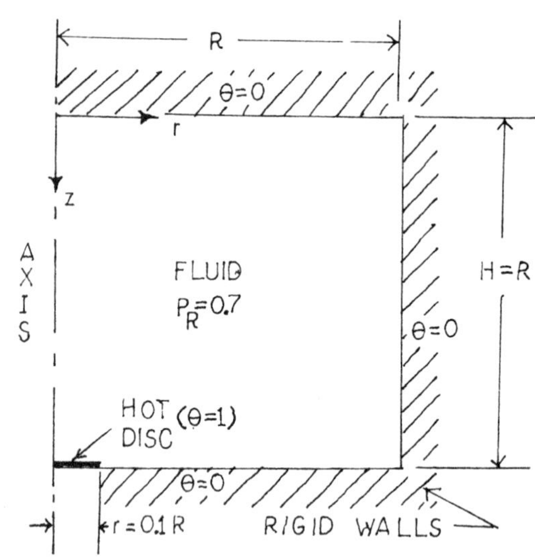

Fig. 1 Cylinder with heated disc on floor. θ denotes dimensionless temperature.

A second class of problems, arising from a study of the vacuum consumable arc remelt casting process [3,4] is examined in Section 3. For these solutions, a case is considered to have displayed a periodic solution only if it continues to show time dependence on an altered mesh. This proves to be a stringent requirement; the only periodic solutions surviving this sifting are those associated with a nonlinear boundary condition.

Although significant questions of convergence remain for the solutions in both problem classes, the present results suggest that the restriction to axisymmetric flow stabilizes the numerical solutions far beyond the experimentally observed [2] turbulent transition at $G_R = 1.2 \times 10^9$. Such results are becoming familiar in the numerical fluid mechanics literature; e.g., several supercritical steady convective numerical solutions [5,6] are known, Fornberg [7] has computed the flow over a circular cylinder to Re = 300 when symmetric 2D flow is assumed; and Orszag and Kells [8] report that plane Couette flow displays no neutrally stable 2D disturbances up to Re = 5000.

The effort needed to resolve these questions by high resolution calculations would be better spent on a benchmark problem of more general interest. In this regard, the two dimensional convection in a square enclosure as put forward by I. P. Jones [9] would seem to be the leading candidate. Like the flow over a two-dimensional cylinder, it could provide a standard for testing of numerical methods on realistic problem of its class, thereby justifying the effort of computing a nonphysical solution. On the other hand, the casting flows will receive such an extended treatment in the near future in order to sharpen convergence criteria. These issues are touched on in Section 4.

2. CYLINDER WITH HOT SPOT

The original computations on this problem [1] were done with a streamfunction-vorticity form of the Navier-Stokes-Boussinesq equations. A uniform mesh of 51 X 21 points was used. Boundary conditions included a second-order Taylor series expansion for boundary vorticity in terms of the newly updated streamfunction. The explicit time integration used fully upwind differenced advection terms and centered space differences on diffusion terms. Time step was 0.95 times the neutrally stable time step. The solution for each G_R was obtained by starting from $\underset{\sim}{u} \equiv \underset{\sim}{0}$, $\theta \equiv 0$ initial conditions. Convergence to steady flow was determined by requiring global energy outflow to balance global inflow.

The driving force in this problem is the heated disc, so that use of a 21-point radial mesh (which allows only three gridpoints on the disc) raises the question of numerical resolution. Examination of the axial heat flux on the floor of the enclosure (Fig. 2) clearly indicates that the resolution is not sufficient, since q_z undergoes O(1) changes on each of the first three gridpoints. This means that all results here will be coarse mesh approximations rather than fully converged solutions since all used the 21-point radial mesh.

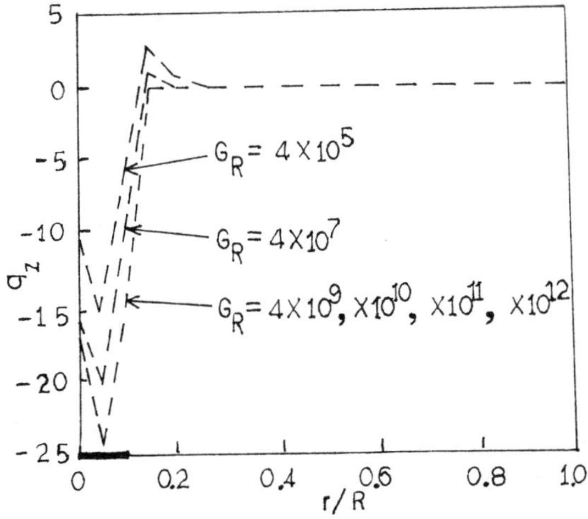

Fig. 2. Heat flux over floor and heated disc on 21-point radial mesh.

The final feature of the original computation noted here is the use of centered differencing of the streamfunction to obtain velocities which were then averaged to obtain the advection velocities which appear in the Lagrangian derivative. For example, on a uniform mesh at point $z_i = (i-1)\Delta z$, $r_j = (j-1)\Delta r$, the axial advection velocity was given by

$$U_{i,j} = \frac{1}{2r_j \Delta r} \left[\frac{\psi_{ij+1} + \psi_{i+1,j+1}}{2} - \frac{\psi_{i,j-1} + \psi_{i+1,j-1}}{2} \right] \quad (1)$$

The ZAP code differs from the above in that a slightly coarser 41 x 21 mesh was used, and a new G_R value was reached by simple continuation from the previous solution rather than starting from static initial conditions. The sequence of G_R values was 4×10^5, 4×10^7, 4×10^9, 4×10^{10}, 4×10^{11}, and 4×10^{12}.

In addition, the "kinematically consistent" differencing introduced by Parmentier and Torrance [10] was used to obtain velocities. For the velocity component in (1), the corresponding expression would be [I]

$$U_{i,j} = \frac{1}{2r_j \Delta r} \left[\frac{j-3/4}{j-1/2} \frac{\psi_{i,j+1} + \psi_{i+1,j+1}}{2} - \frac{j-1}{2(j-1/2)(j-3/2)} \frac{\psi_{ij} + \psi_{i+1j}}{2} - \frac{j-5/4}{j-3/2} \frac{\psi_{i,j-1} + \psi_{i+1,j-1}}{2} \right] \quad (2)$$

[I] Since the ZAP code used strained coordinates, (2) is given only for comparison purposes. The notion of kinematic consistency is simply to define the discrete velocity components so that Stokes' Theorem holds for the rectangular control volume centered on the gridpoint. It can be shown that the formulation given in [10] reduces to interpolating the streamfunction quadratically in the radial direction, linearly in the axial direction. This could be summarized by stating that "The streamfunction should be represented by an interpolation polynomial of at least the degree sufficient to represent a constant normal velocity along the edges of the computational cell." Thus $u = \partial\psi/r\partial r$ requires $\psi = ar^2+b$ for for $r_{j-1/2} < r < r_{j+1/2}$ in order that $U_{i,j} = 2a$ can be piecewise constant

By comparison of (1) and (2) it is clear that the two expressions are the same at large j values, but differ significantly near the axis (j = 1). The computational molecules are the same six-point array, but only the four corners appear in (1).

The resulting solutions are virtually indistinguishable from those reported in [1] at small G_R. Even at large G_R the flow is extremely close to their Fig. 7 for $G_R = 4 \times 10^{10}$; see Fig. 3. Their value for $|\psi|$ max (366) differs only 10 percent from the ZAP value (405). The fact that ZAP predicts more intense flows can be attributed to the difference between Eqs. (1) and (2) near the axis where the hot plume rises. The different axial velocities in this region must be accompanied by different buoyancy fluxes; hence, different total circulations.

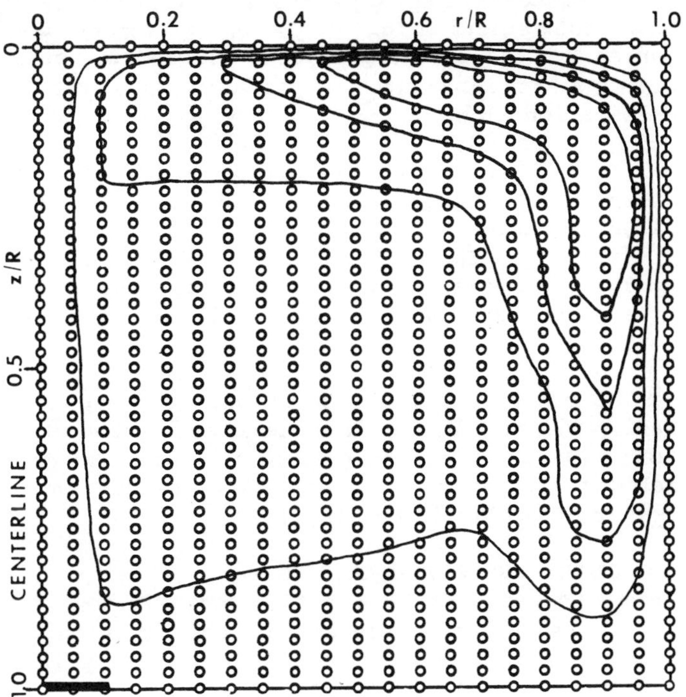

Fig. 3. Streamlines for cylinder with hot spot. $G_R > 4 \times 10^9$; contours at 0.2 $|\psi|$ max.

As stated in the abstract, the ZAP solutions are steady up to $G_R = 4 \times 10^{12}$, the largest value computed. For these solutions,

$$|\psi|_{max} = 0.002 G_R^{1/2} \; ; \; G_R \gtrsim 4 \times 10^9 \qquad (3)$$

serves as a curve fit for the circulation strength which changes while the streamlines remain essentially the same as in Fig. 3. Further, the isotherms have virtually ceased changing at these large G_R values as well. These scalings, namely,

$$\underset{\sim}{u}(r,z;G_R = G_R)\ \underset{\sim}{u}_o(r,z) \text{ and } \theta(r,z;G_R) = \theta_o(r,z) \tag{4}$$

are, of course, just the required forms for an asymptotic solution in which diffusion is negligible and inertial forces balance buoyancy:

$$\begin{aligned} u_o \ \nabla\zeta_o &= P_R^2\ \partial\theta_o/\partial r \\ \underset{\sim}{u}_o\ \nabla\zeta_o &= 0 \end{aligned} \tag{5}$$

The "core flow" governed by (4), (5) (which appears to have the solution $\underset{\sim}{u}_o = \underset{\sim}{0}$, θ_o = const) would be accompanied by thin boundary layers. These are spread over several mesh points in Figure 3, probably as a result of numerical dissipation alone.

In the experiments reported in [2], the boundary layers underwent turbulent transition at $G_R = 1.2 \times 10^9$, only slightly above the G_R at which they first become defined ($\sim 4 \times 10^8$). Therefore, it seems likely that a proper analysis of transition by numerical means would require a treatment like that of Fasel [11] for the Blasius boundary layer.

As noted in the Introduction, such a calculation is more easily justified for the benchmark convection problem of Jones [9] than for the present problem. To date, such calculations are apparently not available for any G_R at the low P_R values of liquid metals [12].

3. VACUUM ARC REMELT CASTING

The solutions discussed in this section all satisfy the boundary conditions indicated schematically in Figure 4. The parameter ψ_1 is the dimensionless flow rate into the ingot, given by $U_I/2(\kappa_o/R) = P_E/2$ in terms of the Peclet number defined by casting speed U_I. G_R is based on the superheat-to-immobilization temperature difference, 176.9K. The body force is given by a magnetostatic solution of Maxwell's equations [4], and Alfven number A_o is defined as the ratio of diffusion speed (κ_o/R) to Alfven wave speed $(\mu_o/\rho_o\ I_m/2\pi R)$ where μ_o is the permeability of free space and I_m is the current passing through the pool.

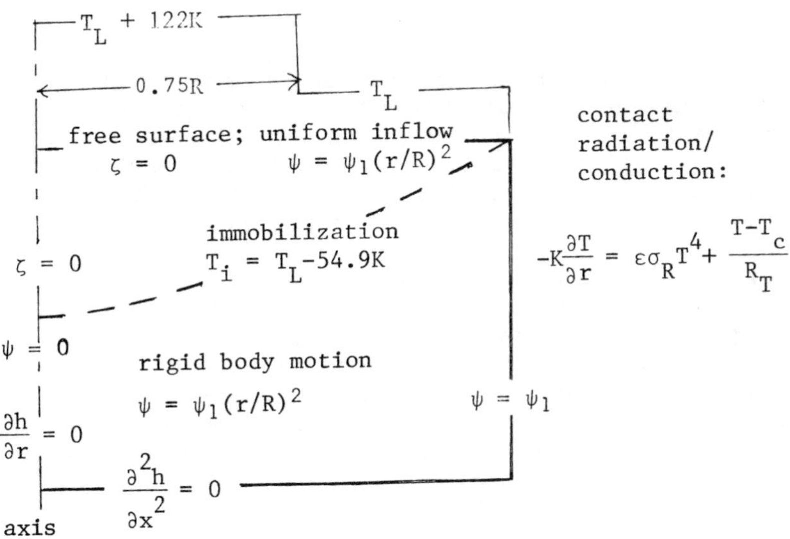

Figure 4. Boundary conditions for casting calculations.

The variable resistance boundary condition on the ingot lateral surface includes parameters ε (ingot surface emissivity), σ_R (Stefan-Boltzmann radiation constant), T_c (coolant temperature), and thermal resistance function R_T. The latter is a nonincreasing function of wall temperature T given by

$$R_T = \begin{cases} R_o\ ; & T > T_s \\ R_o \dfrac{200K}{T - T_s + 200K}\ ; & T_s - 199.98K \leq T \leq T_s \\ R_o x10^4\ ; & T < T_s - 199.98K \end{cases} \quad (6)$$

with R_o = 1 W/cm^2-K.

The calculation proceeds in the same sequence described above; however, since the immobilization locus (T_i isotherm) is an unknown free boundary, it is determined at the end of each time step by scanning the mesh of new temperatures, T^{n+1}. When the locus has been fixed, rigid body motion is prescribed in the solid by setting $\psi = \psi_1(r/R)^2$. With ψ now defined everywhere, boundary vorticity can be obtained by finite differencing ψ^{n+1} along the immobilization locus to yield ζ^{n+1}.

A major difference in behavior near the boundary is caused by the use of strongly temperature dependent viscosity near immobilization [3]. Typically, kinematic viscosity varies by a factor of 250 between liquidus and solidus temperatures, so changing the immobilization gridpoint can result in an order of magnitude change in the perceived viscosity.

The two convergence questions--achievement of steady flow and full numerical resolution--arise here also. Steady flow is presumed when dimensionless time has reached one or two conductive diffusion times (0.25-0.50 dimensionless time units). This is conservative for the cylinder with hot spot problem and must certainly be conservative here when $P_E > 1$.

Direct confirmation of numerical convergence has not been obtained by high resolution computations. The alternative of using strained coordinates to locally refine the mesh only where needed by the structure of the solution has been implemented. Calculations to date have been carried out on uniform meshes of 27 x 20, 35 x 20, and 40 x 20. Because the 27 x 20 mesh shows some effect of the outflow boundary condition on the whole solution,it is not useful except at very low G_R and P_E. Strained meshes of 35 x 20 and 40 x 20 have also been used.

The general features of the resulting solutions are present in Figure 5, where a solution with nearly equal Lorentz and buoyancy forces is given. Unlike the previous problem, sharp boundary layers are not formed here until G_R becomes very large. This is due to two features: (1) the presence of latent heat release (with its attendant lower temperature gradients in the mushy zone) and (2) the increase of viscosity near the immobilization locus (which smears the viscous boundary layer). This means that most of the cases of interest here with G_R between 10^8 and 10^{10}, will not need to deal with boundary layer behavior.

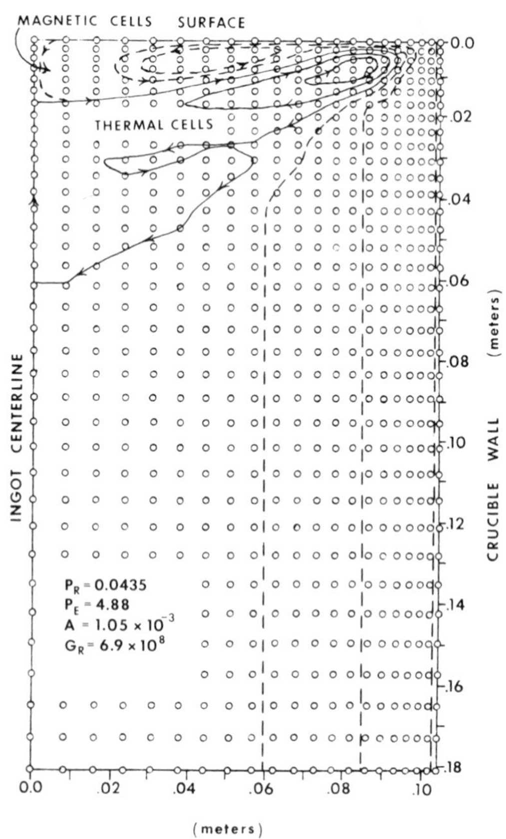

Figure 5. Calculated flow streamlines in ingot with comparable Lorentz and buoyancy forces. Temperature has fourth digit sinusoidal oscillation.

The absence of thin boundary layers reduces the resolution problem to providing a sufficiently fine mesh in the zone of large gradients in Figure 5; namely, near the mensiscus contact zone at the top of the ingot. This is also the only zone in which (6) gives significant conductive heat transfer at the wall. Such resolution can be provided to some degree by straining of the mesh (open circles).

Perhaps the most striking feature of the solutions of this class is their persistent time dependences. These can be placed into the following three categories.

The first appears to be due to the competition between nearly equal buoyancy and Lorentz forces. It consists of a periodic flow in which vortices of opposite signs are sequentially produced beneath the step in the surface temperature function, migrate to the axis, and are annihilated by the following vortex. When

$$\frac{1}{\left(P_R A_o\right)^2} \lesssim G_R \lesssim \frac{7}{\left(P_R A_0\right)^2} \tag{7}$$

holds, the buoyancy and Lorentz sources in the vorticity equation are comparable, and such motions appear in the 0.05R x 0.05R square uniform mesh solutions. If the same problem is then run with a strained mesh, this time dependence vanishes, raising the question of whether or not the motion is physically realistic. Since the coarsest part of the strained mesh is in the critical near-axis zone, it may be that the increased numerical diffusivity is sufficient to damp out such motions, so the straining may serve to decrease accuracy rather than enhance it. Only refined uniform mesh calculations will allow firm conclusions to be reached on these questions.

The second type of oscillation is associated with the contact boundary condition, Eq. (6). A complicated interaction results when the meniscus contact point temperature oscillates about the immobilization temperature, introducing large changes by remelting, freezing, remelting, etc. This also dramatically affects the interior flow as well since the viscosity appearing at this point changes dramatically during each such cycle. This causes large changes in the calculational timestep, which in turn results in large changes in the interior flow's Courant and diffusion numbers. Since the numerical method's phase and diffusive errors are functions of these latter two parameters, these errors undergo large excursions also. For this oscillation, then, the driving force is physical but the effects on the solution are both physical and numerical. This oscillation is always present on uniform mesh problems by $G_R = 8 \times 10^9$ and by 6.5×10^{10} on strained meshes.

The final form is sinusoidal oscillations, varying from barely detectable (5th digit) to finite amplitude. These seem to be always present except at very small G_R and very large (gasdynamic) A_o. However, the requirement that they persist on the strained mesh before being considered physically plausible removes all such motions. Here, an amplitude smaller than the tolerance of the SSOR computation for streamfunction (for these cases, about 1.4×10^{-4}) is ignored. Of course, once G_R has become large enough to produce flows of the second type, these sinusoids are swamped, so their behavior at G_R above the 8×10^9 threshold has not been investigated. Since it seems to be most closely related to the continuously-evolving type of solution reported in [1,2], this is the most likely mode to be associated with transition if it is physically realistic.

4. DISCUSSION

The main task above was to determine the onset of numerical unsteadiness in the ZAP code solutions to the two model problems and dependence of the transition value of G_R on the other parameters of the problems.

The results of section 2 indicate that the seemingly minor change from the algorithm used in [1] is sufficient to make the ZAP code solution steady well above the G_R at which that solution became periodic. Examination of the solutions indicates that several issues remain open.

First, the correct core flow solution to (4), (5) is not known. If there is a steady axisymmetric asymptotic solution, it can take several forms. On the fixed uniform mesh, it may be either the $\underset{\sim}{u}_o = \underset{\sim}{0}$, $\theta_o = a + bz$ form reported in [1] or the $\underset{\sim}{u}_o = \underset{\sim}{0}$, $\theta_o = \text{const}$ form suggested by ZAP. Other possibilities include $\zeta_o = \text{const}$, $\theta_o = \text{const}$ and also the "flywheel" flow of Jones, Moore, and Weiss [5] in which $\theta_o = \text{const}$ and ζ_o is constant on streamlines. Choice cannot be made among these possibilities on the basis of the calculations performed to date. The effect of a fully kinematically consistent treatment with first order vorticity boundary conditions has also not been investigated.

Second, settling these numerical issues leaves the problems of relating the numerical solutions to the physical system. As noted by the authors in [2], the working fluid (air) was no longer a Boussinesq fluid at the transition G_R. Further, the actual mode of transition observed was related to the deceleration of the plume, a feature not present in either of the numerical solutions.

Third, the general issue of whether or not two dimensional numerical solutions can be correlated with actual transition in convective flows remains unresolved. The most studied problem is Benard convection, for which Krishnamurti [13,14,15] has provided a global experimental investigation, including also the low P_R values appropriate to the casting problem.

Two careful conservative (but not upwind differenced) finite difference numerical solutions of the Benard problem done with P_R order unity are of interest. Deardorff [6] used Arakawa's method to obtain steady motions to R_A = 395 $R_{A,crit}$ in Cartesian coordinates. He also checked this result experimentally, obtaining steady motion at 300 $R_{A,crit}$ in a Hele Shaw type cell, indicating that no unstable two dimensional modes exist at this R_A. The second calculation by Jones et al. [5] reached 100 $R_{A,crit}$ in cylindrical coordinates with steady flows of the convective roll type. Since the cylinder with hot spot also yields this type of solution, a steady solution at large supercritical G_R would not be very surprising.

Thus it seems that the lack of computed two dimensional unsteady Benard solutions reflects the actual behavior of the system. On the other hand, when such solutions are expected to be present, as in the transition problem for the flow in a vertical slot, calculations have succeeded in capturing them. Gill and Davey [16] showed that these solutions can bifurcate at either a steady secondary flow or a traveling wave secondary flow, both two dimensional. The numerical solution of Mallinson and de Vahl Davis [17] reflected both behaviors at the analytically derived G_R values for P_R = 1000. Agreement with the experiments of Elder [18] is excellent.

Turning to the casting problem, the first difficulty is judging the effects of very small P_R. Krishnamurti's straight line fit to the Willis and Deardorff values of $R_{A,t}$ (turbulent transition value) as a function of P_R is roughly

$$R_{A,t} = 1.59 \times 10^4 \, P_R^{1.36} \qquad (8)$$

for the range $0.5 < P_R < 20$. Since the only data point below $P_R = 0.5$ is that for mercury, the shape of the curve at low P_R can only be conjectured; presumably it is monotone decreasing to $R_{A,crit}$ as $P_R \to 0$. The precise nature of that dependence is still under theoretical discussion. In any event, the implication is that a low P_R fluid will undergo unsteady motion very soon after the onset of convection. This argues for a very early transition in the casting problem. The absence of two dimensional unsteadiness in the numerical solutions, if it is valid, argues for transition by three dimensional modes. Some sort of indirect experimental evidence will have to be sought for transition in alloy problems with their smeared out boundary layers, and refined numerical solutions will have to be obtained to reliably fix the numerical transition locus.

5. ACKNOWLEDGEMENT

The author is extremely grateful to Drs. Frank Zanner and Alex Treadway of Sandia National Laboratories and Professor Ken Torrance of Cornell for useful discussion in the course of this work.

6. BIBLIOGRAPHY

1. TORRANCE, K. E. and ROCKETT, J. A. - Numerical Study of Natural Convection in an Enclosure with Localized Heating from Below. J. Fluid Mech., 36, 1, 33-54. 1969.

2. TORRANCE, K. E., ORLOFF, L., and ROCKETT, J. A. - Experiments on Natural Convection in Enclosures with Localized Heating from Below. J. Fluid Mech., 36, 1, 21-31. 1969.

3. ZANNER, F. J. and BERTRAM, L. A. - Computational and Experimental Analysis of a U-6w/oNb Vacuum Consumable Arc Remelt Ingot. SAND80-1156, Sandia National Laboratories, Albuquerque, New Mexico, 1980.

4. BERTRAM, L. A. - Coupled Conduction, Convection and Nonequilibrium Freezing Effects in Alloy Casting. Num. Meth. Therm. Probs. 1981.

5. JONES, C. A., MOORE, D. R., and WEISS, N. O. - Axisymmetric Convection in a Cylinder. J. Fluid Mech. 73, 2, 353-388. 1976.

6. DEARDORFF, J. W. - A Numerical Study of Two-Dimensional Parallel Plate Convection. J. Atmos. Sci. 21, 419-438. 1964.

7. FORNBERG, B. - A Numerical Study of Steady Viscous Flow Past a Circular Cylinder. J. Fluid Mech. 98, 4, 819-855. 1980.

8. ORSZAG, S. A. and KELLS, L. C. - Transition to Turbulence in Plane Poiseuille and Plane Couette Flow. J. Fluid Mech. 96, 1, 159-205. 1980.

9. JONES, I. P. - Natural Convection in an Enclosed Cavity: A Comparison Problem. J. Fluid Mech. 95, 4, inside back cover. 1979.

10. PARMENTIER, E. M. and TORRANCE, K. E. - Kinematially Consistent Velocity Fields for Hydrodynamic Calculations in Curvilinear Coordinates. J. Comp. Phys., 19, 4, 404-417. 1975.

11. FASEL, H. - Investigation of the Stability of Boundary Layers by a Finite-Difference Model of the Navier-Stokes Equations. J. Fluid Mech. 78, 2, 355-383. 1976.

12. BETTS, P. L., HASLAM, J. C., and LIDDER, J. S. - Comparisons of Four Computer Programs for Two-Dimensional Convection in Closed Cells. Num. Meth. in Therm. Probs., R. W. Lewis and K. Morgan, eds., 243-252, Pine- ridge Press, Swansea, 1979.

13. KRISHNAMURTI, R. - On the Transition to Turbulent Convection, Part I. J. Fluid. Mech. 42, 2, 295-307. 1970.

14. Ibid., 309-321.

15. KRISHNAMURTI, R. - Some Further Studies on the Transition to Turbulent Convection. J. Fluid Mech. 60, 2, 285-303. 1973.

16. GILL, A. E. and DAVEY, A. - Instabilities of a Buoyancy-Driven System. J. Fluid Mech. 35, 4, 775-795. 1969.

17. DE VAHL DAVIS, G. and MALLINSON, G. D. - A Note on Natural Convection in a Vertical Slot. J. Fluid Mech. 72, 1, 87-93. 1975.

18. ELDER, J. W. - Laminar Free Convection in a Vertical Slot. J. Fluid Mech. 23, 1, 77-98. 1965.

NUMERICAL CALCULATIONS OF TURBULENT HEAT TRANSFER DOWNSTREAM OF A REARWARD-FACING STEP

A.M. Gooray*, C.B. Watkins**, and W. Aung***

Howard University, Washington, D.C. 20059

SUMMARY

Momentum and temperature fields are numerically calculated for turbulent recirculating flow behind a rearward-facing step at low to moderate Reynolds number. The computations described were performed using the standard k-ε, two-equation, model of turbulence in an extended version of the TEACH-T computer code. For a constant step height, results are obtained for three different step-height-based Reynolds numbers. The results compare favorably with previously-published experimental measurements. The study indicates that results can be obtained of sufficient accuracy to evaluate heat transfer coefficients for engineering purposes. The results are sensitive to the approach profiles assumed. The proper specification of the turbulent kinetic energy is especially critical for computations simulating a flow with a tripped approach boundary layer. Further improvements in the results can be anticipated by refinements in the turbulence modeling to account for low Reynolds number effects, especially in the near wall region.

1. INTRODUCTION

The prediction of wall convective heat transfer characteristics for recirculating flows form a very important aspect of engineering heat transfer research. Such flow, consisting of separation and reattachment, may be accidentally induced in some systems; and in others, it may be deliberately introduced so as to augment the wall convective heat transfer rates.

---

*Research Assistant

**Professor and Chairman, Department of Mechanical Engrg.

***Adjunct Professor

Among the various types of separated flows, those due to abrupt changes in body geometry are of great interest. The sudden change in geometry over which the flow occurs results in highly turbulent recirculation just downstream of the step. Experimental studies have shown that in this recirculating region the heat transfer coefficient is many times larger than that in the attached region. This increase in heat transfer, in principle, can be visualized as a result of the increase in the streamwise turbulent kinetic energy in the mixing layer (region remote from the wall). Recently, Fraser and Siddig [1]and Moss and Baker [2] measured the turbulent kinetic energy distributions in recirculating flow fields. They showed that the streamwise turbulent intensity ($\overline{u'^2}/U_o^2$) is small within the buffer zone and then increases quite sharply to a maximum in the mixing layer, before falling off sharply again to a low value in the main flow. Furthermore, the peak value of the kinetic energy increases initially with downstream distance from the sudden change in geometry, reaching a maximum about the reattachment point, and then decreasing after reattachment. Beyond reattachment where the flow tends toward a reestablished boundary layer (or fully-developed flow in a pipe or channel) the turbulent kinetic energy takes on the smooth boundary layer profile. The high values of the turbulent, kinetic energy in the mixing layer particularly around reattachment, is associated with spreading of the layer or an increase in turbulent diffusion coefficients in the main turbulent flow. This then results in a "shrinking" of the near-wall viscous sub-layer, thus allowing greater passage of heat by molecular diffusion.

Hence, to aid in the understanding of this complex heat transfer phenomenon the authors have undertaken the present research. It is the intention here to present examples of turbulent predictions of flow-field patterns and local heat transfer coefficients for the two-dimensional rearward-facing step by employing the two-equation (k-ε) model. The strong dependence on inlet turbulent kinetic energy is demonstrated as well as the importance of near-wall modeling.

The qualitative nature of separated flows have been known for a long time, as discussed by Prandtl and Tietjens [3].The most common geometries studied are sudden expansions in ducts and pipes, airfoils at large angles of attack, forward-facing or rearward-facing steps, blockages and blunt bodies. Separated heat transfer flows, for most of the above geometries, have been reviewed previously by Fletcher, et al. [4] and by Aung and Watkins [5].

In other relevant work, experimental measurements were obtained by several investigators. Very few, however, included heat transfer effects. In this category, are the data of Abbot and Kline [6], Bradshaw and Wong [7], Narayanan, et al. [8], Denham and Patrick [9], Etheridge and Kemp [10], Symth [11] and Kuehn [12]. Recently, measurements were also taken by Kim, et al. [13], Armaly, et al. [14] and Moss and Baker [2]. All of the above measurements indicated that the reattachment length lies between 5.0 and 7.5 step heights downstream. For recirculating heat transfer flows, Aung and Goldstein [15] obtained what is apparently the only existing low-speed data for the rearward-facing step. Work involving other related geometries includes the data of Seki, et al. [16], Seki, et al. [17], Koram and Sparrow [18], Smyth [19], and Zemanick and Dougall [20]. All of these measurements confirm the peak in the heat transfer coefficient near the reattachment point, the peak valve being many times larger that the fully developed value.

Very few studies involving mathematical modeling of turbulent recirculating heat transfer have been carried out. The most significant work was recently reported by Chieng and Launder [21]. They present very detailed calculations of recirculating heat transfer in a sudden pipe expansion using the k-ε model, for both high and low turbulent Reynolds number flows. Their results agree fairly well with the measurements of Zemanick and Dougall [20]. The scope of success obtained by Chieng and Launder [21] has served as partial motivation for the present research.

2. MATHEMATICAL AND PHYSICAL MODEL

2.1 Basic Equations for the Two Equation Model of Turbulence

Predictions of turbulent flow for the two-dimensional step were performed numerically by solving the governing elliptic differential equations. Following the practice of several studies, for example, Jones and Launder [22], the turbulent viscosity is obtained by the k-ε viscosity model. In this model the turbulent viscosity is uniquely determined by the local values of density,turbulent kinetic energy, k, and a turbulent dissipation length scale, ε. Description of this model is given elsewhere [22,23]. The geometry considered is shown in Figure 1.

The governing equations are as follows:

x - momentum

$$\rho u \frac{\partial u}{\partial x} + \rho v \frac{\partial u}{\partial y} = -\frac{\partial p}{\partial x} + \frac{\partial}{\partial x}\left(\mu \frac{\partial u}{\partial x}\right) + \frac{\partial}{\partial y}\left(\mu \frac{\partial u}{\partial y}\right) - \frac{\partial}{\partial x_j}(\overline{\rho u_1' u_j'}) \quad (1)$$

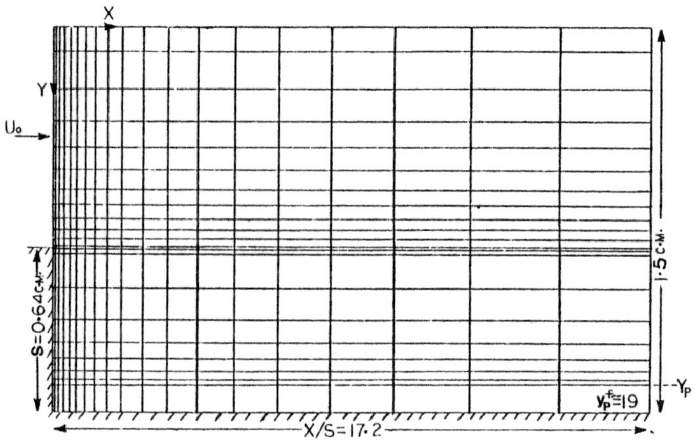

Fig. 1 Typical Grid Distribution

y - momentum

$$\rho u \frac{\partial v}{\partial x} + \rho v \frac{\partial v}{\partial y} = -\frac{\partial p}{\partial y} + \frac{\partial}{\partial x}\left(\mu \frac{\partial v}{\partial x}\right) + \frac{\partial}{\partial y}\left(\mu \frac{\partial v}{\partial y}\right)$$

$$- \frac{\partial}{\partial x_j} (\overline{\rho u_2' u_j'}) \tag{2}$$

energy

$$\rho u \frac{\partial T}{\partial x} + \rho v \frac{\partial T}{\partial y} = \frac{\partial}{\partial x}\left(\Gamma_{eff} \frac{\partial T}{\partial x}\right) + \frac{\partial}{\partial y}\left(\Gamma_{eff} \frac{\partial T}{\partial y}\right) \tag{3}$$

turbulent kinetic energy

$$\frac{\partial}{\partial x}\left(\rho uk\right) + \frac{\partial}{\partial y}\left(\rho vk\right) = \frac{\partial}{\partial x}\left(\frac{\mu_t}{\sigma_k} \frac{\partial k}{\partial x}\right) + \frac{\partial}{\partial y}\left(\frac{\mu_t}{\sigma_k} \frac{\partial k}{\partial y}\right)$$

$$+ \mu_t G - \rho\varepsilon \tag{4}$$

turbulent diffusion

$$\frac{\partial}{\partial x}\left(\rho u\varepsilon\right) + \frac{\partial}{\partial y}\left(\rho v\varepsilon\right) = \frac{\partial}{\partial x}\left(\frac{\mu_t}{\sigma_\varepsilon} \frac{\partial \varepsilon}{\partial x}\right) + \frac{\partial}{\partial y}\left(\frac{\mu_t}{\sigma_\varepsilon} \frac{\partial \varepsilon}{\partial y}\right)$$

$$+ \frac{\varepsilon}{k} C_{\varepsilon 1} \mu_t G - C_{\varepsilon 2} \rho \frac{\varepsilon^2}{k} \tag{5}$$

In the above equations the overbars have been omitted for the mean flow fluid-dynamic variables and the following definitions apply:

$$\rho\,\overline{u_i'u_j'} = -\mu_t\left(\frac{\partial u_i}{\partial x_j}+\frac{\partial u_j}{\partial x_i}\right),\quad \mu_t = C_\mu\,\rho k^2/\varepsilon$$

$$\Gamma_{eff}=\left(\frac{\mu}{P_R}+\frac{\mu_t}{P_{R_T}}\right)$$

$$G=2\left[\left(\frac{\partial u}{\partial x}\right)^2+\left(\frac{\partial v}{\partial y}\right)^2\right]+\left(\frac{\partial u}{\partial y}+\frac{\partial v}{\partial x}\right)^2$$

C_μ, $C_{\varepsilon 1}$, $C_{\varepsilon 2}$, σ_k and σ_ε are constants given in Ref. 22.

2.2 Wall Functions

For solution of the above equations, levels of near-wall kinetic energy and dissipation rates are needed as well as the usual temperature and velocity boundary conditions. This is obtained by the "wall-function" formulation. The relations given below and used in the present work are those proposed by Launder and Spalding [24].

(i) Velocity: The wall shear stress

$$\tau_w=\rho_w u_p C_\mu^{1/4} k_p^{1/2} \frac{\kappa}{\ln(Ey_p+)},\quad \kappa=.42,\ E=9.8 \qquad (6)$$

where subscripts "w" and "p" refer to values at the wall and a point p close to the wall. The point y_p is established by the value of y_p+.

$$y_p+=\frac{y_p}{\nu}\left(C_\mu^{1/2}k_p\right)^{1/2} \qquad (7)$$

For our calculations $y_p+ \simeq 19$ for most cases.

(ii) Temperature: The heat flux,

$$\dot{q}''_w=(T_w-T_p)\rho C_p C_\mu^{1/4} k_p^{1/2}/P_{R_T}\left[U^+ + \frac{1}{P_{R_T}}P(P_R)\right] \qquad (8)$$

The function $P(P_R)$ has been modelled by Jayatilleke [25]

$$\frac{P(P_R)}{P_{R_T}}=9.24\left[\left(\frac{P_R}{P_{R_T}}\right)^{3/4}-1\right]\left[1+0.28\,e^{-0.007\,P_R/P_{R_T}}\right] \qquad (9)$$

(iii) Turbulent kinetic energy:

$$k_p = \rho c_\mu^{3/4} k^{\frac{1}{2}} \frac{1}{\kappa} \ln(Ey_p^+) \quad (10)$$

(iv) Dissipation rate of turbulent energy

$$\int_0^{y_p} \varepsilon dy = C_\mu \frac{k_p^{3/2}}{\kappa} \ln\left[\frac{Ey_p C_\mu^{\frac{1}{4}} k_p^{\frac{1}{2}}}{\nu}\right] \quad (11)$$

The above relations describe wall functions for the horizontal walls, the obvious modifications are made for the vertical walls. A variable turbulent Prandl number, P_{RT}, in the near-wall region is used in the present calculations [26].

2.3 Approximations for Approach Conditions

Experience gained from preliminary computations indicated that, for a fixed step height and Reynolds number, the recirculating flow is quite sensitive to the assumptions made regarding approach profiles. The only available experimental data for validating calculations for the present geometry in the present Reynolds-number regime is the previously-mentioned data of Aung and Goldstein [15]. Unfortunately, their work does not include complete measurements of the upstream (approach) profiles since the focus was on heat-transfer coefficients. They do however, include measurements of the approach temperature boundary-layer profile. Their experiment was performed utilizing a tripped turbulent approach boundary layer with a fairly short,but not entirely insignificant,thermal entry length. To approximate the inlet profiles for u,k, and ε, turbulent boundary layer data from another experiment was correlated to the specified thermal profile. The data used for this purpose were the velocity and temperature fits of Abbot and Kline [6] and the turbulent kinetic energy fit of Hinze [27] based on the data of Klebanoff [28].

3. NUMERICAL ANALYSIS

Calculations were performed using a version of the TEACH-T computer code [29],modified to include heat transfer. This code solves the two-dimensional Navier-Stokes and energy equations in primitive variables, and it has the capability of solving other coupled differential equations,such as in this case, those for k and ε. The governing equations are integrated over the entire flow region and then a line iteration is performed. Details of the code are given in [29].

Figure 1 illustrates a typical grid distribution used in the calculations. This grid was devised to locate as many grid lines as possible near the walls. The results were found to be not particularly sensitive to the position of the near wall grid point, i.e. calculated heat transfer coefficients remain fairly constant for $38 < y_p^+ < 19$.

For a 20x22 non-uniform grid, such as the one shown in Figure 1, convergence was obtained after approximately 190 iterations. Central processor times on an IBM 370/158 averaged about six minutes.

4. RESULTS AND DISCUSSION

Initial test calculations were made for the sudden pipe expansion case of Chieng and Launder [21], comparing the test results with their computed results, and with the experimental results of Zemanick and Dougall [20]. As shown in Figure 2, this comparison proved favorable. The difference between the results of the present test calculations and the Chieng and Launder results can be attributed principally to differences in the wall functions assumed. However, efforts were made to incorporate into the present calculations the "improved" functions as described by Chieng and Launder [21]. For still unresolved reasons, the modified wall functions did not yield reasonable solutions.

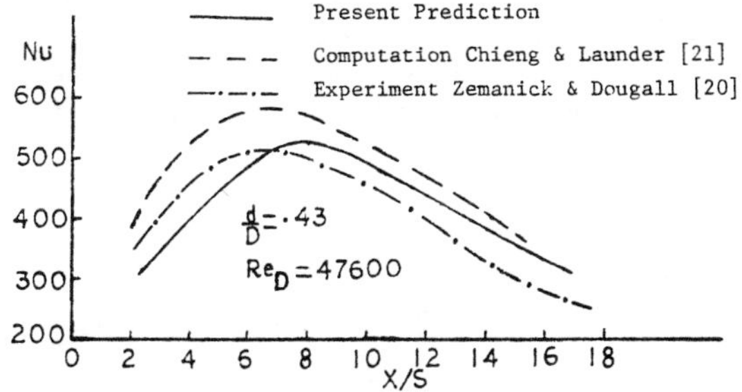

Fig. 2 Nusselt number distribution for sudden expansion in a pipe

To investigate the rearward-facing step, calculations were made for the three Reynolds-number cases of Aung and Goldstein [15] and the results compared with their experiments. In attempting to employ the Hinze [27] fit for the approach profile, it was discovered that the level of turbulent kinetic energy in the innermost of the three regions of the fit led to physically unreasonable results for the recirculating flow. To eliminate this difficulty it was necessary to increase the level of turbulent kinetic energy in this region approximately three-fold to a value of $0.3U_o^2$. This higher level of turbulent energy is not completely arbitrary since it is consistent with the value used in the pipe-expansion calculations of Chieng and Launder [21]. Moreover, the presence of the trip in the experiments would seem likely to produce turbulence levels persisting downstream of the trip higher than those of the corresponding Klebanoff data for the same Reynolds number.

The results of the calculations are summarized in Table 1.

U_o	$Re=U_oS/\nu$	$(Nu/Nu_s)_{max}$	X_r/S	Y_p^+
4.55	1728.4	1.24	4.3	19.6
3.15	1329.5	1.26	4.3	17.8
1.65	626.78	1.36	4.3	14.5

Table 1. Variation of maximum Nusselt number with Reynolds number

The reattachment distances of approximately 4.3 step heights downstream of the step given in the table are quite close to the distances of 4.5 step heights observed by Aung and Goldstein [15].

A comparison of the streamwise variation of the computed heat transfer coefficients with experiment is given in figures 3 through 5. The Nusselt numbers are normalized with the approach Nusselt number, Nu_s, before the step. The agreement is good considering the limited accuracy of the measurements and the inherent uncertainties in turbulent flow calculations.

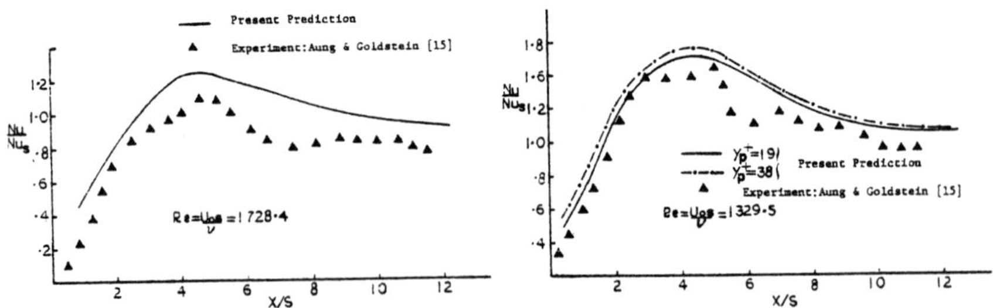

Fig. 3 Streamwise variation of Nusselt number downstream of step

Fig. 4 Streamwise variation of Nusselt number downstream of step

Only limited improvement in the accuracy of the results can be obtained by optimizing the position of the near-wall grid line. The effect of two such choices for the location of the near-wall grid are shown in Figure 4. The indicated y_p^+ is the approximate average value in the recirculation zone. The constraints on the choice of the near wall grid

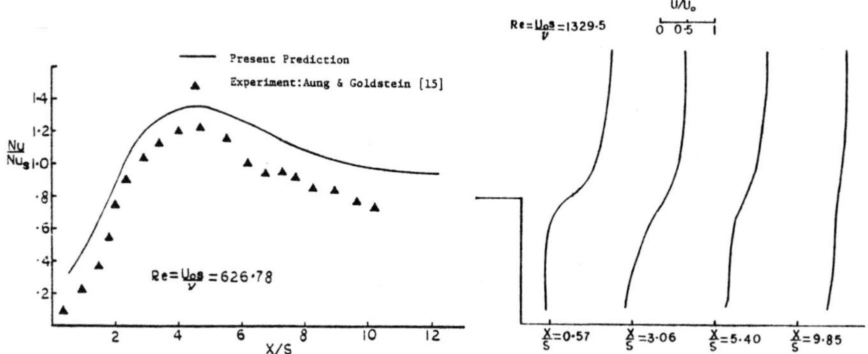

Fig. 5 Streamwise variation of Nusselt number downstream of step

Fig. 6 Streamwise velocity profiles downstream of step

point are that it must lie outside the linear sublayer and yet, be close enough to the wall to provide accurate resolution of the flow in the near-wall region.

Since the Reynolds numbers of the present flow calculations for the step are comparatively low, in an attempt to further improve the results, the low Reynolds number form of the two-equation model as presented by Chieng and Launder [21] was employed as well as the traditional model described in Section 2. The heat transfer results obtained using the low Reynolds number model were an order of magnitude too small. Such a model, if it had been successful, should have yielded an additional improvement by permitting increased resolution of the flow in the near-wall region downstream of the step. This is especially important when a local extremum in the velocity profile occurs near the wall as is the tendency in the present cases.

Increasing near-wall resolution was also attempted by modifying the code to patch calculations using empirical inner-law turbulent viscosity models to the two-equation model further from the wall. Patching a laminar sublayer calculation to the two-equation model was also attempted. Neither of these approaches yielded acceptable results.

Typical computed velocity profiles are displayed in Figure 6. Figure 7 is a comparison of computed temperature profiles for one Reynolds number with the corresponding data of Aung and Goldstein and is typical of the agreement of results at the other two Reynolds numbers as well.

Figures 8 and 9 depict the computed behavior of the turbulent kinetic energy. Figure 8 shows the profiles at selected streamwise locations and Figure 9 is a plot of the stream-

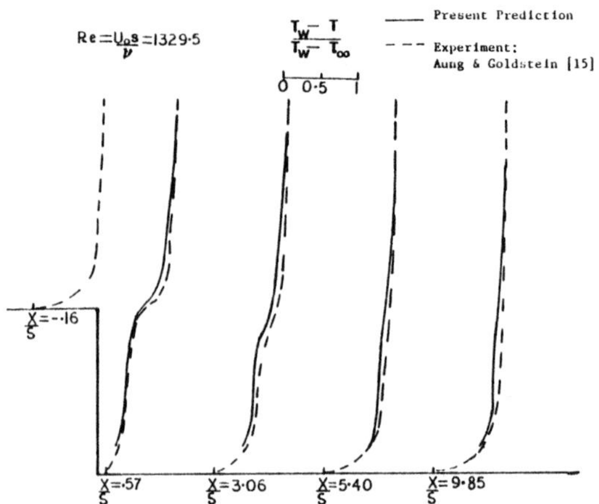

Fig. 7 Temperature distribution downstream of step

wise variation of the maximum value. The trends are qualitatively as described in the experimental work summarized in Section 1, particularly the work of Fraser and Siddig [1] and Moss and Baker [2].

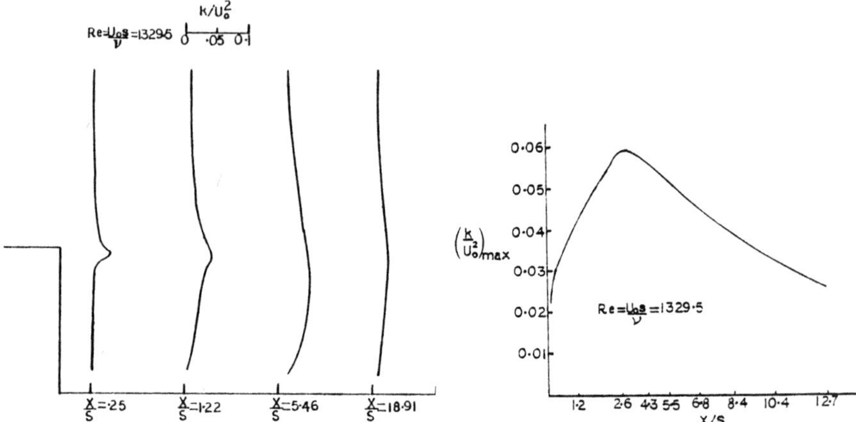

Fig. 8 Turbulent kinetic energy downstream of step

Fig. 9 Peak values of turbulent kinetic energy downstream of step

5. CONCLUSIONS

The present results indicate that, using the usual two-equation model of turbulence, reasonably accurate predictions can be made of the heat-transfer coefficients for turbulent recirculating flow downstream of a step. The accuracy of these predictions relies on proper modeling of the approach profiles. In the case of computations involving simulation of a tripped approach boundary layer, enhancement of the approach boundary layer kinetic energy appears to be necessary. It is likely that further improvements can be anticipated as more reliable methods are developed for including low Reynolds number effects.

This work was supported, in part, by the Office of Naval Research.

6. REFERENCES

1. FRASER, S.M. and SIDDIG, M.H. - Turbulent Flow Over a Normal Wall. J. of Mech. Engrg. Sc., Vol. 22, No. 4, 1980.

2. MOSS, W.D. and BAKER, S. - Re-circulating Flows Associated with Two-Dimensional Steps. The Aeronautical Quarterly, Vol. 31, Part 3, 1980.

3. PRANDTL, L. and TIETJENS, O. - Fundamentals of Hydro and Aero Mechanics, McGraw-Hill, New York, 1934.

4. FLETCHER, L.S. et al. - Heat Transfer in Separated and Reattaching Flows: An Annotated Review. Israel J. of Technology, Vol. 12, 1974.

5. AUNG, W. and WATKINS, C.B. - Heat Transfer Mechanisms in Separated Forced Convection, Proceedings of the NATO Institute on Turbulent Forced Convection in Channels and Rod Bundles, Hemisphere Publications Inc., Washington, D.C., 1979.

6. ABBOTT, D.E. and KLINE, S.J. - Experimental Investigation of Subsonic Turbulent Flow over Single and Double Backward Facing Steps. J. Basic Engrg. Trans. ASME, D83, pp. 317-325, 1962.

7. BRADSHAW, P. and WONG, F.Y.F. - The Reattachment and Relaxation of a Turbulent Shear Layer. J. Fluid Mech., 52, part 1, pp. 113-135, 1972.

8. NARAYANAN, B. et al. - Similarities in Pressure Distribution in Separated Flow Behind Backward - Facing Steps. The Aeronautical Quarterly, Vol. 25, part 4, pp. 305-312, 1974.

9. DENHAM, M.R. and PATRICK, M.A. - Laminar Flow over a Downstream - Facing Step in a Two-dimensional Flow Channel. Trans. Instn. Chem. Engrs. 52, pp. 361-367, 1974.

10. ETHERIDGE, D.W. and KEMP, P.H. - Measurements of Turbulent Flow Downstream of a Rearward-Facing Step. J. Fluid Mech., 86, part 3, pp. 545-566, 1978.

11. SMYTH, R. - Turbulent Flow over a Plane Symmetric Sudden Expansion. J. of Fluids Eng. Trans. ASME, 100, pp. 348-353, 1979.

12. KUEHN, D.M. - Effects of Adverse Pressure Gradient on the Incompressible Reattaching Flow over a Rearward-Facing Step. AIAA Journal, 18, No. 3, pp. 343-344,1980.

13. KIM, J. et al. - Investigation of a Reattaching Turbulent Shear Layer: Flow over a Backward-Facing Step. J. Fluids Eng. Trans. ASME, Vol. 102, pp. 302-308, 1980.

14. ARMALY, B.F. et al. - Measurements and Predictions of Flow Downstream of a Two-Dimensional Single Backward Facing Step, Sonderforschungsbereich 80-172, Universitat Karlsruhe, July 1980.

15. AUNG, W. and GOLDSTEIN, R.J. - Heat Transfer in Turbulent Separated Flow Downstream of a Rearward-Facing Step. Israel J. of Tech., 10, No. 1-2, pp. 35-41,1972.

16. SEKI, N. et al. - Turbulent Fluctuations and Heat Transfer for Separated Flow Associated with a Double Step at Entrance to an Enlarged Flat Duct. J. Heat Transfer Trans. ASME, 98, pp. 588-593, 1976.

17. SEKI, N. et al. - Effect of Stall Length on Heat Transfer in Reattached Region Behind a Double Step and Entrance to an Enlarged Flat Duct. Int. J. Heat and Mass Transfer, 19, pp. 700-702, 1976.

18. KORAM, K.K. and SPARROW, E.M. - Turbulent Heat Transfer Downstream of an Unsymmetric Blockage in a Tube. J. Heat Transfer Trans. ASME, 100, pp. 588-594, 1978.

19. SMYTH, R. - Turbulent Heat Transfer Measurements in Axisymmetric External Separated and Reattached Flows. Letters in Heat and Mass Transfer, 6, pp. 405-412, 1979.

20. ZEMANICK, P.P. and DOUGALL, R.S. - Local Heat Transfer Downstream of Abrupt Circular Channel Expansion. J. Heat Transfer Trans. ASME, 92, p. 53, 1970.

21. CHIENG, C.C. and LAUNDER, B.E. - On the Calculation of Turbulent Heat Transport Downstream from an Abrupt Pipe Expansion. Numerical Heat Transfer, Vol. 3, pp. 189-207, 1980.

22. JONES, W.P. and LAUNDER, B.E. - The Prediction of Laminarization with a 2 - Equation Model of Turbulence. Int. J. Heat Mass Transfer, Vol. 15, p. 301, 1972.

23. LAUNDER, B.E. and SPALDING, D.B. - Mathematical Models of Turbulence. Academic Press, London, 1972.

24. LAUNDER, B.E. and SPALDING, D.B. - The Numerical Computation of Turbulent Flows. Comput. Methods Appl. Mech. Eng., Vol. 3, pp. 269-289, 1974.

25. JAYATILLEKE, C.L.V. - The Influence of Prandtl number and Surface Roughness on the Resistance of the Laminar Sub-Layer to Momentum and Heat Transfer. Prog. Heat Mass Transf., Vol. 1, p. 193, 1969.

26. ANTONIA, R.A. - Behavior of the Turbulent Prandtl Number Near the Wall. Int. J. Heat Mass Transfer, Vol. 23, pp. 906-908, 1980.

27. HINZE, J.O. - Turbulence, McGraw-Hill, New York, 1975.

28. KLEBANOFF, P.S. - Characteristics of Turbulence in a Boundary Layer with Zero Pressure Gradient. NACA TN, 3178, 1954.

29. GOSMAN, A.D. - The TEACH-T Computer Program - Structure, Flow, Heat and Mass Transfer in Turbulent Recirculating Flows - Prediction and Measurements, McGill University, Canada, 1976.

LOW REYNOLDS NUMBER HEAT TRANSFER IN CYLINDRICAL PACKED BEDS

M. S. Khader

Associate Professor
Mechanical Engineering Dept.
Cairo University
Giza, Egypt

J. S. Goodling

Professor
Mechanical Engineering Dept.
Auburn University,
Alabama 3684o, USA

ABSTRACT

In the present paper, the problem of low Reynolds number heat transfer in cylindrical backed beds is investigated numerically. The packed bed is considered to be packed with spherical particles. A homogeneous model for heat transfer is adopted throughout the analysis to describe the thermal behavior of the bed.

Allowance for the spatial variation of void fraction (porosity) and its influence on both the thermal properties and the velocity distribution is made. The governing differential equation of the heat transfer in the entrance region of the packed bed is non-dimensionalized and then solved numerically using a marching finite difference scheme. Temperature distributions in the radial direction are given for different axial locations and for different values of Reynolds numbers. The constant wall temperature local Nusselt numbers, are plotted for variety of flow and geometrical parameters of the bed.

NOMENCLATURE

A_{ij}	coefficient in the implicit finite difference equation
B_{ij}	coefficient in the implicit finite difference equation
C_{ij}	coefficient in the implicit finite difference equation
C_f	fluid specific heat
C	experimental constant in Eq. (6)
d	particle diameter
D	bed diameter
EXP ( )	exponential function

K_f	fluid thermal conductivity
K_s	solid thermal conductivity
$\overline{K}_e$	non-dimensional effective thermal conductivity
L	bed height
M	number of grid points in the axial direction
N	number of grid points in radial direction
Nu	Nusselt number, hR/K_f
$\Delta P/\Delta L$	pressure gradient
Re	Reynolds number, $\frac{u_o d}{\nu}$
r	radial coordinate
T	temperature
T_o	initial temperature
T_w	wall temperature
u	velocity
u_o	mean velocity
$\overline{u}$	non-dimensional velocity u/u_o
x	non-dimensional radial coordinate
z	axial coordinate
$\overline{z}$	non-dimensional axial coordinates, z/L
α	ratio of K_s/K_f
Δ	grid spacing in radial direction
δ	grid spacing in axial direction
ρ_f	fluid density
ε	void fraction
ε_o	bulk void fraction
θ	non-dimensional temperature
ν	kinematic viscosity (μ/ρ)
i,j	subscript defining node i,j

INTRODUCTION

Heat transfer in packed beds is an important phenomenon of study because of the large number of practical applications found in chemical, food processing and petroleum industries. Packed beds are commonly used in catalytic converters, ion exchangers and fluidized heat exchangers.

Despite the considerable amount of experimental and theoretical efforts to understand transport processes, in packed beds, their behavior is far from known. This fact is due to the complexity of the flow within the bed and the lack of knowledge of bed bulk properties [1].

Heat transfer with stagnent or moving fluid in beds presents itself in many bed designs. Again this subject is still open for further investigation and understanding. Heat transfer in packed beds with flowing fluid has been the subject of many investigators [2]. The complexity of the mechanisms contributing to the heat flow deters from finding any general model for the transfer of heat in such beds. The continued proliferation of alternate models and experiments attests to these difficulties. A recent review of the heat transfer in packed beds [1] outlines the major efforts in this field.

In packed beds of particles, experiments [3,4] have shown that structural arrangements of the particles are not uniform. For cylindrical packed beds, this structural arrangement is expressed by a non-dimensional quantity called void fraction or porosity. These experiments indicate that the void fraction is not uniform in the radial direction of the bed. This non-uniformity is caused by the existence of the container wall. Reference [4] established the term "particle boundary layer" to describe the distribution of the void fraction near the bed wall, Figure 1. This phenomenon in turn leads to a non-uniformity in the distribution of a fluid velocity within a bed. Accordingly, a region of high velocity exists where high void fraction exists and vice versa. In addition to this non-uniform velocity distribution, the variation of the void fraction influences the thermal properties of the bed. Since the effective thermal conductivity of the bed is a direct function of the void fraction, a spatial distribution of thermal conductivity also results [5].

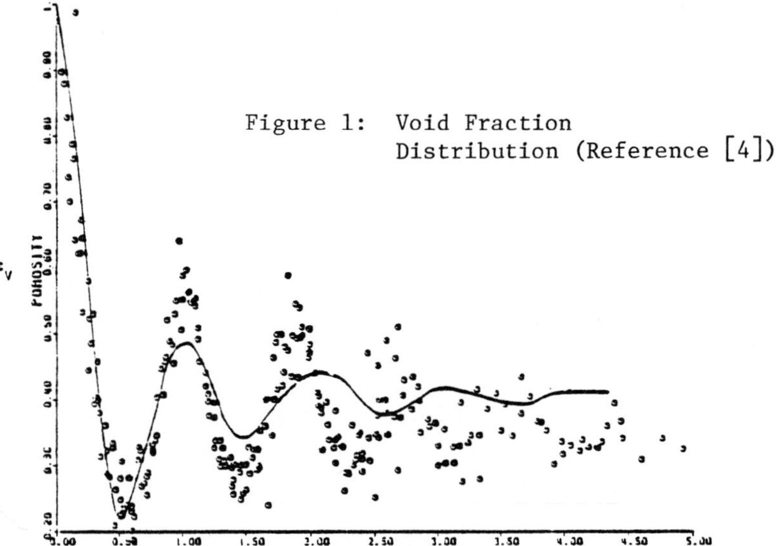

Figure 1: Void Fraction Distribution (Reference [4])

In this paper, an attempt is made to consider these spatial variations of the porosity and thermal conductivity for heat transfer in packed beds under the condition of low Reynolds number, i.e., laminar flow. A model to predict the velocity distribution across the bed is formulated using the experimental void fraction distribution and the resulting spatial thermal conductivity of the bed.

ANALYSIS

Consider a cylindrical bed packed with particles of the same diameter as shown in Figure 2. The fluid enters the bed at Z = 0 with a constant temperature, T_o. Under the steady state assumption, the energy equation in cylindrical coordinates for axisymmetric flow is [6]

$$\rho_f C_f u \frac{\partial T}{\partial Z} = \frac{1}{1} \frac{\partial}{\partial r} \left[K_e r \frac{\partial T}{\partial r} \right] \tag{1}$$

The above equation is based on the assumption that the bed acts like a homogenous continuum. The effective thermal conductivity, K_e and the velocity u are given functions of the radial coordinate r.

According to the physical configuration of the present problem, the above energy equation is subject to the following initial and boundary conditions:

$$\begin{aligned} T &= T_o \quad \text{at} \quad Z = 0 \\ T &= T_w \quad \text{at} \quad r = R \\ \frac{\partial T}{\partial r} &= 0 \quad \text{at} \quad r = 0 \end{aligned} \tag{2}$$

The following dimensionless variables and parameters are introduced.

$$\theta = \frac{T - T_w}{T_o - T_w}, \quad \bar{Z} = Z/L, \quad x = r/R$$

$$\bar{u} = u/u_o, \quad \bar{K}_e = K_e/K_f, \quad R_e = \frac{u_o d}{\nu}$$

The energy equation in its non-dimensionalized form, becomes

$$\bar{u} \frac{\partial \theta}{\partial \bar{Z}} = \frac{1}{PrRe} \left(\frac{L}{R}\right) \left(\frac{d}{R}\right) \frac{1}{x} \frac{\partial}{\partial x} \left[x \bar{K}_e \frac{\partial \theta}{\partial x} \right] \tag{3}$$

Inspection of equation (3) reveals that the temperature in this case takes on the functional relationship

$$\theta = \theta(PrRe, \frac{L}{R}, \frac{d}{R}, x, \bar{Z})$$

However, a reduction in the number of the problem parameters may be achieved if a new axial coordinate is defined as

$$Y = \left(\frac{L}{R}\right) \left(\frac{d}{R}\right) \bar{Z}/PrRe$$

Subsequently, equation (3) reduces to

$$\bar{u} \frac{\partial \theta}{\partial Y} = \frac{1}{x} \frac{\partial}{\partial x} \left[x \bar{K}_e \frac{\partial \theta}{\partial x} \right] \tag{4}$$

with initial and boundary conditions

$$\begin{aligned} \theta &= 1 \quad \text{at} \quad Y = 0 \\ \theta &= 0 \quad \text{at} \quad x = 1 \\ \frac{\partial \theta}{\partial x} &= 0 \quad \text{at} \quad x = 0 \end{aligned} \tag{5}$$

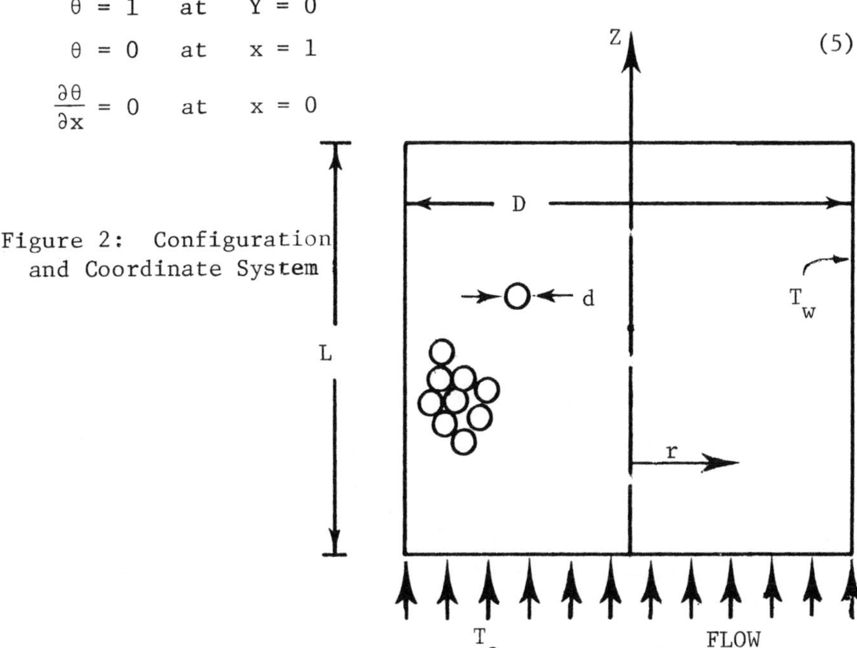

Figure 2: Configuration and Coordinate System

Velocity Distribution

As has been mentioned before for a cylindrical packed bed with spherical particles, the void fraction is a function of the radial position, Figure 1. The distribution shown in the figure indicates that larger void fraction values exist near the wall region than in the central bed region. This type of distribution causes in turn high fluid velocities near the wall and lower velocities at the central part of the bed. In the present analysis, a simplified velocity distribution is formulated accounting for such spatial distribution of the void fraction. Ergun's [7] formula for the pressure gradient in packed material for low flow Reynolds numbers is given as

$$u_o = \left(\frac{\Delta P}{\Delta L}\right) C \frac{d^2}{\mu} \frac{\varepsilon_o^3}{(1-\varepsilon_o)^2} \tag{6}$$

for the average velocity u_o and bed average void fraction ε_o. If one assumes that the flow in a packed bed consists of many channels in parallel with infinitesimal area, the fluid velocity for each area is

$$\frac{u}{u_o} = \frac{(\varepsilon/\varepsilon_o)^3}{\left[\frac{1-\varepsilon}{1-\varepsilon_o}\right]^2} \qquad 1<\varepsilon\leqslant\varepsilon_o \tag{7}$$

with $\frac{u}{u_o} = 0$ at the wall.

In the above equation ε is the local void fraction which is a function of the radial position. The present authors [4] have developed an experimental correlation for the void fraction as a function of the radial position for cylindrical packed beds with spherical particles of diameter d as

$$\varepsilon = \varepsilon_o + (1-\varepsilon_o)\text{EXP}\left[-1.8\frac{D}{d}x\right]^{(1-\varepsilon_o)}\cos\left[2\pi\frac{D}{d}x\right] \tag{8}$$

Upon substituting the above relation into equation (7), the velocity distribution as a function of the non-dimensional radial coordinate is obtained.

Effective Thermal Conductivity

Since the energy equation (1) has been written based on the assumption of a continuum bed, a continuum or effective bed thermal conductivity must be used throughout the analysis as well. In the present investigation a simplified model for the effective thermal conductivity will be adapted based on a model developed by Zehner [8]. The present version of the Zehner model neglects the radiation heat transfer within the bed (i.e., low level of temperatures) and the effect of low pressure. The effective thermal conductivity as a function of the local void fraction is written as

$$\overline{K}_e = K_1(1-\varepsilon)^{\frac{1}{2}} + \left[1-(1-\varepsilon)^{\frac{1}{2}}\right] \tag{9}$$

with

$$K_1 = \frac{2\alpha^2}{(\alpha-1)^2}\left[\text{LN}(\alpha) + \frac{1-\alpha}{\alpha}\right]$$

With the use of equation (8) for ε, the thermal conductivity of the bed becomes a function x, the radial coordinate. Experimental work [9] for the heat transfer from packed beds has indicated that at the wall the thermal conductivity is not equal to the fluid thermal conductivity, as indicated by (9).

This is due to the fact that solid particles next to the wall are also contributing to the heat transfer from the bed to the wall. Accordingly, at the wall region ($x = 1 - d/2R$), an average value of $\overline{K}_e$ is used instead of $\overline{K}_e = 1$. This average value is obtained by using (9) with porosity value obtained at $x = 1 - d/2R$.

SOLUTION

Substitution of equations (7), (8) and (9) into the energy equation (4) completes the formulation of the heat transfer model for the present problem. Unfortunately, such equation does not lend itself to any analytical method of solution. Therefore a numerical solution technique is applied in the present analysis. For equation (4), a finite difference marching technique is used to obtain the solution. The thermal field region is defined as the rectangular strip bounded by $x = 0$, $Y = 0$, $x = 1$ and Y. Due to the symmetry of the field about $x = 0$, a solution obtained for one half of the region may be used to describe the entire thermal field.

To construct the finite difference equations, the region of interest is divided into a grid of mesh lines parallel to the x and Y coordinates. The spacing between the lines in x direction is uniform while the step size in Y direction is varied. The general form of the finite difference equation developed is

$$C_i\theta_{i-i,j+1} + B_i\theta_{i,j+1} + A_i\theta_{i+1,j+1} = D_i \tag{10}$$

where

$$i = 2 \rightarrow (N-1) \text{ and } j = 2 \rightarrow M$$

$$C_i = \left[\frac{\overline{K}_e}{4x\Delta} + \frac{1}{4\Delta}\left(\frac{\partial \overline{K}_e}{\partial x}\right) - \frac{\overline{K}_e}{2\Delta^2}\right]_i$$

$$B_i = \left[\overline{u}/\delta + \overline{K}_e/\Delta^2\right]_i$$

$$A_i = \left[-\frac{\overline{K}_e}{4x\Delta} - \frac{1}{4\Delta}\left(\frac{\partial \overline{K}_e}{\partial x}\right) + \overline{K}_e/2\Delta^2\right]_i$$

$$D_i = \left[\theta_{i+1,j}\quad \frac{\overline{K}_e}{4x\Delta} + \frac{1}{4\Delta}\left(\frac{\partial \overline{K}_e}{\partial x}\right) + \frac{\overline{K}_e}{2\Delta^2}\right]_i$$

$$+ \theta_{i,j}\left[\overline{u}/\delta - \overline{K}_e/\Delta^2\right]_i$$

$$+ \theta_{i-1,j} \left[- \frac{\bar{K}_e}{4x\Delta} - \frac{1}{4\Delta} \left(\frac{\partial \bar{K}_e}{\partial x}\right) + \frac{\bar{K}_e}{2\Delta^2} \right]_i$$

The above equation is valid only for any interior point. For the boundary points 1 and N+1 the following equations are developed.

$$B_1 \theta_{1,j+1} + A_1 \theta_{2,j+1} = D_1 \tag{11}$$

$$B_1 = \left[\bar{u}/\delta + 2\, \bar{K}_e/\Delta^2 \right]_1, \quad A_1 = \left[- 2\, \bar{K}_e/\Delta^2 \right]_1 ;$$

$$D_1 = \theta_{1,j} \left[\bar{u}/\delta - \frac{2\bar{K}_e}{\Delta^2} \right]_1 + \theta_{2,j} \left[2\, \bar{K}_e/\Delta^2 \right]_1$$

and

$$C_N \theta_{N-1,j+1} + B_N \theta_{N,j+1} = D_N \tag{12}$$

$$C_N = \left[\frac{\bar{K}_e}{4x\Delta} + \frac{1}{4\Delta} \left(\frac{\partial \bar{K}_e}{\partial x}\right) - \frac{\bar{K}_e}{2\Delta^2} \right]_N$$

$$B_N = \left[\bar{u}/\delta + \bar{K}_e/\Delta^2 \right]_N$$

$$D_N = \theta_{N,j} \left[\bar{u}/\delta - K_e/\Delta^2 \right]_N + \theta_{N-1,j} \left[- \frac{\bar{K}_e}{4x\Delta} - \frac{1}{4\Delta} \left(\frac{\partial \bar{K}_e}{\partial x}\right) + \bar{K}_e/2\Delta^2 \right]_N$$

These equations are solved for the N nodal points using a modified Gaussian elemination method. The procedure is repeated M times as the solution advances in the Y direction.

After the temperature distribution is evaluated, the bulk temperature can be determined as

$$\theta_b = \int_0^1 2\, x\, \bar{u}\, \theta\, dx \tag{13}$$

For the present case of a constant wall temperature, the local Nusselt number NU(Y) can be written as

$$Nu = - \frac{\left(\frac{\partial \theta}{\partial x}\right)_{x=1}}{\theta_b} \tag{14}$$

RESULTS AND DISCUSSION

To conserve space, only results for particular cases are illustrated. However, the present results are of an indicative nature for the capability of the present analysis. In addition, no attempt is made to compare the present theortical predictions with experimental results due to the lack of well defined parameters associated with the available data. Subsequently, general physical trends of the theortical predication are compared with the available experimental results when possible.

Figure 3 shows the present theortical model prediction of the velocity distribution for a laminar flow through a packed bed of D/d = 9.09 in accordance with equation (7). As one notices, the velocity near the bed wall reaches a higher value than the rest of the bed. This result is due to the porosity changes across the bed. This type of velocity distribution has been observed experimentally by [10] and [11].

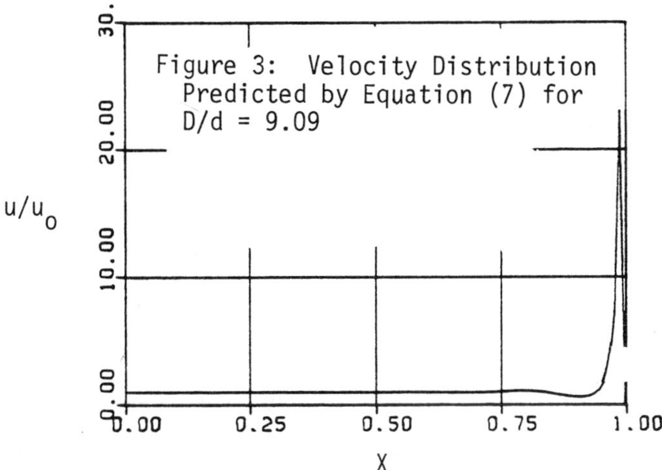

Figure 3: Velocity Distribution Predicted by Equation (7) for D/d = 9.09

Figure 4 represents a typical radial temperature distribution at different Y values, for D/d = 9.09. At the inlet of the bed, the temperature distribution behaves in a boundary layer fashion, while towards the outlet, i.e., at high values of Y, the temperature distribution developes in the same manner as the case of channel flow. These characteristics of temperature distribution coincide with the experimental data of Reference [10]. Also, as Y tends to a large value, the temperature of fluid will reach the wall temperature. This result of equality of temperature could be achieved practically by increasing the value of [(L/R) (d/R)/PrRe].

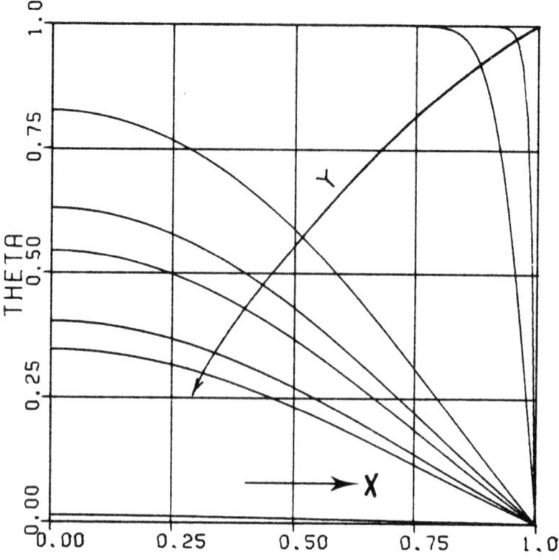

Figure 4: Radial Temperature Distributions at Different Axial Locations

The local values of the wall heat transfer represented by the values of Nusselt number are shown in Figure 5 and 6. The Nusselt number is based on the local bulk temperature as given by equation (14). For low values of Y the local Nusselt number is high as the difference between the fluid temperature and the wall temperature is large. As the flow progresses through the packed bed its temperature increases and the wall heat transfer decreases until it reaches a constant value.

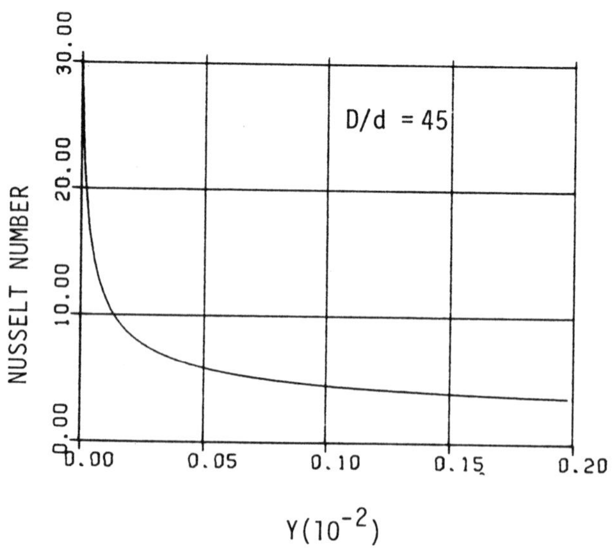

Figure 5: Local Nusselt Number as a function of the Axial Coordinate

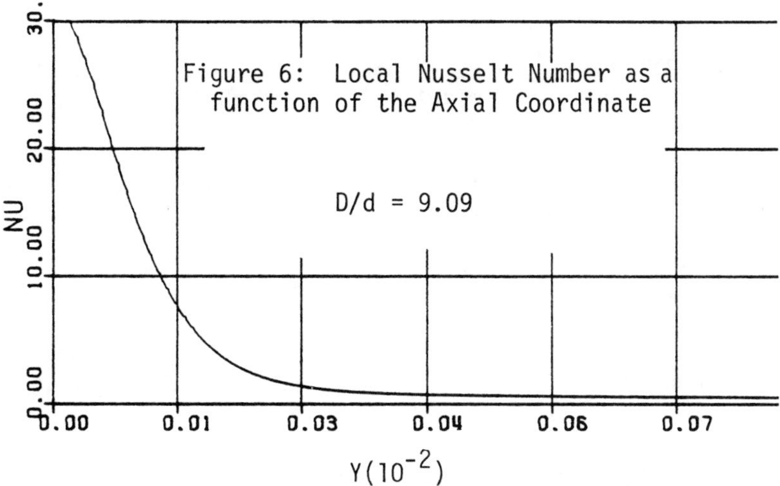

Figure 6: Local Nusselt Number as a function of the Axial Coordinate

Finally, the analysis shown here is made general for any value of (L/R)/PrRe by the use of a non-dimensional axial coordinate Y. In other words, the results obtained for a fixed (d/R) may be used for a variety of values of (L/R)/PrRe.

CONCLUSION

The problem of heat transfer in packed cylindrical bed under laminar flow condition has been numerically analyzed. Allowance for the spatial variation of volume fraction and its influence on both thermal properties and velocity distribution is included. The numerical results are made general for any value of (L/R)/PrRe by choosing a proper non-dimensional axial coordinate.

REFERENCES

1. SCHLUNDER, E.U. (1978), "Transport Phenomena in Packed Bed Reactors", ACS Symposium Series 72 ACHs, p. 110-161.

2. LI, C., FINLAYSON, B.A. (1977), "Heat Transfer in Packed Beds - A Reevaluation", Chemical Engineering Science, 32, p. 1055-1066.

3. BENENATI, R.F., BROSILOW, C.B. (1962), "Void Fraction Distribution in Beds of Spheres", AICHE Journal, 8, 3, p. 359-361.

4. GOODLING, J.S., VACHON, R.I., KHADER, M.S., "Thermal Conductivity of Heterogeneous Mixtures", Final Report, NSF-ENG. 77-08393, 1979.

5. KHADER, M.S., GOODLING, J.S., VACHON, R.I., "Effective Thermal Conductivity of Granular Materials in Cylindrical

Beds", 16th Int. Thermal Conductivity Conference, Chicago, IL, USA, 1979.

6. GINIELINSKI, V. (1980), "Wärme - and Stoffübertragung in Festbetten", Chemical - Ing. - Tech., 52, 3, S, p. 228-236.

7. ERGUN, S. (1952), "Fluid Flow through Packed Columns", Chemical Engineering Prog., 48, p. 89-94.

8. ZEHNER, P. (1972), Doktor-Ingenieurs Dissertation, University Karlsruhe (TH), p. 18.

9. KWONG, S.S., SMITH, J.M. (1956), "Radial Heat Transfer in Packed Beds", Industrial and Engineering Chemistry, 48, 5, p. 894-903.

10. SCHWARTZ, C.E., SMITH, J.M. (1953), "Flow Distribution in Packed Beds", Industrial and Engineering Chemistry, 45, 6, p. 1209-1218.

11. SCHERTZ, W.W., BISCHOFF, K.B. (1969), "Thermal and Material Transport in Nonisothermal Packed Beds", AICHE Journal, 15, 4, p. 597-604.

MOMENTUM AND HEAT TRANSFER CHARACTERISTICS IN TURBULENT FLOW

T.DROZD /I/

SUMMARY

The shearing stress at the wall τ_w spreads to the stream due to the viscous ν and turbulent ν_T viscosity and velocity gradient $d\varphi/d\eta$ in the normal to the wall direction and similarly the heat flux at the wall q_w spreads to the stream due to the molecular α and eddy diffusivity α_T for heat and temperature gradient $dT/d\eta$ in the normal to the wall direction. For numerical treating the problem of evaluating of the velocity and temperature fields in the stream it is necessary to know $\nu_T(\eta)$ and $\alpha_T(\eta)$.
There is presented a coupled empirical thermoanemometric and numerical data processing method of evaluating $\nu_T(\eta)$ and $\alpha_T(\eta)$ in the arbitrary flow problem. On the base of the solution of the energy equation the local and instantaneous heat flux on the wall may be calculated and, in some special cases, compared with the empirical data, presented by the criterial formulae for the Nusselt number. Also another characteristics of the turbulence based on the double auto- and spatial correlation $\varrho_{w_i'w_j'}$ of the velocity fluctuations, the spectral analysis $S_{w_i'w_j'} = \overline{v'^2(f)}$ with the Fourier transforms F, F^{-1} and amplitude probability analysis $p(\frac{x}{\sigma})$ $P(\frac{x}{\sigma})$ are mentioned, as all they needs very special method of the instantaneous experimental data storage and processing [1,2] and are important in the various closure schemes of the conservation equations.

1. INTRODUCTION

The mathematical model of the turbulent flow and the turbulent heat transfer needs for closing some additional dependences. In the averaged basic equations system some unknown terms $\overline{w_i'w_j'}$ and $\overline{w_i't'}$ appear which are to be evaluated

/I/ Assist.prof.dr eng.,Aviation Institute,
Al.Krakowska 110/114, 02-256 Warszawa, Poland

on the experimental way. Even the most sophisticated statistical concept of the turbulence appeal to the experiment for evaluating some constants. The simplest way of the closure of the system of the conservation equations is to express the turbulent viscosity $\nu_T(\eta)$ by the spatial derivative of the velocity profile φ'_η . But this assumption may be taken into account only for the fully developed velocity profile on the enough long distance in the straight duct with the constant cross section.

The further progress was made when the anemometry measurements were used to independent evaluating of the velocity $\varphi(\eta)$ and the temperature profile $T(\eta)$ the eddy diffusivity for momentum /the turbulent viscosity/ $\nu_T(\eta)$ and for heat $a_T(\eta)$ /3/. Then the local turbulent Prandtl number $Pr_T = \nu_T/a_T$ can be evaluated. If the hot wire anemometry is used it is necessary to separate the velocity components in the coordinate K_3 system $w_i^2 = (\overline{w}_i + w_i')^2$ from the measured effective velocities $v_i^2 = (\overline{v}_i + v_i')^2$ by the hot wire situated in some chosen direction defined by two angles φ and β . And this is the first basic step for further empirical analysis of the turbulence.

2. THE COMPONENTS OF THE REYNOLDS STRESS TENSOR AND AVERAGED VELOCITY

The effective velocity for the wire situated as on the Fig.1

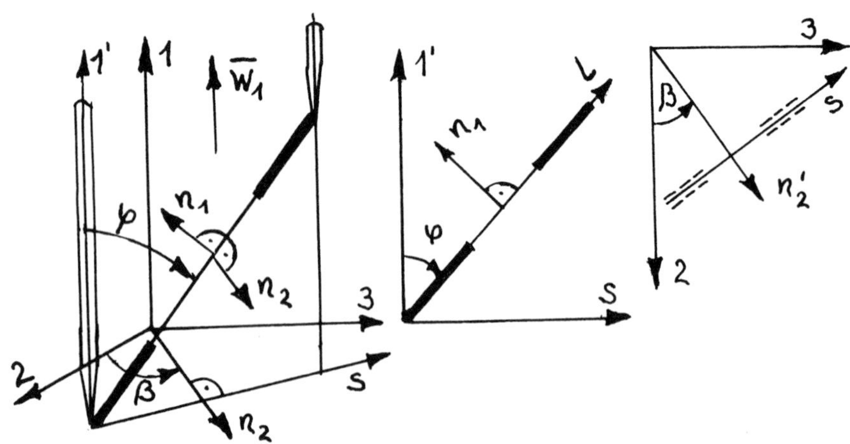

Fig.1. The spatial situation of the hot wire l in K_3

and can be expressed, according to the definition

$$v_{ef}^2 = w_{n1}^2 + w_{n2}'^2 + k^2 w_l^2 = [(\overline{w}_1 + w_1')\sin\varphi - w_s'\cos\varphi]^2 + w_{n2}'^2 + k^2[(\overline{w}_1 + w_1')\cos\varphi + w_3'\sin\varphi]^2 \qquad /1/$$

where

$$W_3' = w_3' \cos\beta - w_2' \sin\beta \, , \qquad w_{n2}' = w_3' \sin\beta + w_2' \cos\beta \qquad /2/$$

and $k = \vartheta_{\parallel}/\vartheta_{\perp}$ is a coefficient, reducing heat transfer sensitivity of a wire to the parallel flow from the perpendicular one and is approximately equal 0.2. The real value of k can be easy evaluated from the angle characteristic of the wire using the linearized signal of the averaged velocity. After introducing trigonometrical dependences between squared functions and double angle functions and after rearranging, the equation /1/ takes the form

$$\begin{aligned} \vartheta_{ef}^2 = {} & \left(\overline{W}_1^2 + 2\overline{W}_1 w_1' + w_1'^2\right)\tfrac{1}{2}\left[(1+k^2) - (1-k^2)\cos 2\varphi\right] + \\ & + \tfrac{1}{2}\left(w_2'^2 + w_3'^2\right)\left\{1 + \tfrac{1}{2}\left[(1+k^2) + (1-k^2)\cos 2\varphi\right]\right\} + \\ & + \tfrac{1}{2}\left(w_2'^2 - w_3'^2\right)\left\{1 - \tfrac{1}{2}\left[(1+k^2) + (1-k^2)\cos 2\varphi\right]\right\}\cos 2\beta + \\ & + \left(\overline{W}_1 + w_1'\right) w_2' (1-k^2)\sin 2\varphi \sin\beta + \qquad /3/ \\ & - \left(\overline{W}_1 + w_1'\right) w_3' (1-k^2)\sin 2\varphi \cos\beta + \\ & + w_2' w_3' \left\{1 - \tfrac{1}{2}\left[(1+k^2) + (1-k^2)\cos 2\varphi\right]\right\}\sin 2\beta \end{aligned}$$

The terms with $\overline{W}_1$ can be suppresed by the electronics using the AC /alternating current/ input circuit. After averaging all terms $\overline{W_1 w_i'} = 0$ as $\overline{w_i'} = 0$ For $\varphi = \pi/4$ on the base of /3/ the squared effective velocity fluctuations in the dependence on β can be expressed as

$$\begin{aligned} \overline{\vartheta'^2(\beta)} = {} & \left\{\frac{1+k^2}{2}\overline{w_1'^2} + \frac{3+k^2}{4}\left(\overline{w_2'^2} + \overline{w_3'^2}\right)\right\} + (1-k^2)\overline{w_1' w_2'}\sin\beta + \\ & - (1-k^2)\overline{w_1' w_3'}\cos\beta + \tfrac{1}{2}(1-k^2)\overline{w_2' w_3'}\sin 2\beta + \frac{1-k^2}{4}\left(\overline{w_2'^2} - \overline{w_3'^2}\right)\cos 2\beta \qquad /4/ \end{aligned}$$

In the equation /4/ there are 6 unknown values $\overline{w_i' w_j'}\, ; i,j = 1,2,3$ and only 4 lineary independent trigonometric functions of β and one unknown term in parentheses, independent on β. For every set of β the set of $\overline{w_i' w_j'}$ must remain unchanged. So it is possible to choose only 5 different values of β and to evaluate 4 coefficients at the trigonometric functions and one constant in parentheses. But on this way it is impossible to evaluate all $\overline{w_i'^2}$ as for them there are only 2 equations: one from the term in parentheses with three components $\overline{w_i'^2}$ and the second from the coefficient in the last term with two components $\overline{w_i'^2}$. So it is necessary to establish on the base /3/ the similar equation to /4/ but for another chosen φ e.g. $\varphi = \pi/2$

After introducing the notation $\left[\overline{\vartheta'^2(\beta)}\right]_{\varphi=\pi/2} = \overline{u'^2(\beta)}$

$$\overline{u'^2(\beta)} = \left\{ \overline{w_1'^2} + \frac{1}{2}(1+k^2)(\overline{w_2'^2}+\overline{w_3'^2}) \right\} + (1-k^2)\overline{w_2'w_3'}\sin 2\beta + \frac{1}{2}(1-k^2)(\overline{w_2'^2}-\overline{w_3'^2})\cos 2\beta \qquad /5/$$

The solution of the equation system /4/ and /5/ is

$$\begin{aligned}
\overline{w_1'^2} &= \frac{1}{(2+k^2)(1-k^2)}\left[-(1+k^2)(\overline{v_A'^2}+\overline{v_B'^2}) + (2+k^2)\overline{u_H'^2} + \overline{u_L'^2}\right] = \\
&= \frac{1}{(2+k^2)(1-k^2)}\left[-(1+k^2)(\overline{v_C'^2}+\overline{v_D'^2}) + \overline{u_H'^2} + (2+k^2)\overline{u_L'^2}\right] \\
\overline{w_2'^2} &= \frac{1}{(2+k^2)(1-k^2)}\left[(\overline{v_A'^2}+\overline{v_B'^2}) - (1+k^2)\overline{u_L'^2}\right] = \\
&= \frac{1}{(2+k^2)(1-k^2)}\left[(\overline{v_C'^2}+\overline{v_D'^2}) + \overline{u_H'^2} - (2+k^2)\overline{u_L'^2}\right] \\
\overline{w_3'^2} &= \frac{1}{(2+k^2)(1-k^2)}\left[(\overline{v_A'^2}+\overline{v_B'^2}) - (2+k^2)\overline{u_H'^2} + \overline{u_L'^2}\right] = \\
&= \frac{1}{(2+k^2)(1-k^2)}\left[(\overline{v_C'^2}+\overline{v_D'^2}) - (1+k^2)\overline{u_H'^2}\right]
\end{aligned} \qquad /6/$$

where

$$v_A' = v'(0), \quad v_B' = v'(\pi), \quad v_C' = v'\left(\tfrac{3}{2}\pi\right), \quad v_D' = v'\left(\tfrac{\pi}{2}\right), \quad v_E' = v'\left(\tfrac{\pi}{4}\right)$$

$$u_H' = u'(0), \quad u_L' = u'\left(\tfrac{\pi}{2}\right), \quad u_K' = u'\left(\tfrac{\pi}{4}\right), \quad u_M' = u'\left(\tfrac{3}{4}\pi\right)$$

$$\begin{aligned}
\overline{w_1'w_2'} &= \frac{1}{2(1-k^2)}(\overline{v_D'^2}-\overline{v_C'^2}) \\
\overline{w_1'w_3'} &= \frac{1}{2(1-k^2)}(\overline{v_B'^2}-\overline{v_A'^2}) \\
\overline{w_2'w_3'} &= \frac{1}{2(1-k^2)}(\overline{u_K'^2}-\overline{u_M'^2}) = \frac{1}{2(1-k^2)}\left[-(\overline{v_A'^2}+\overline{v_B'^2}) + (\overline{v_C'^2}+\overline{v_D'^2}) + \right. \\
&\left. + 2(\overline{u_K'^2}-\overline{u_L'^2})\right] = \frac{1}{1-k^2}\left[2\overline{v_E'^2} + \frac{\sqrt{2}-1}{2}(\overline{v_B'^2}+\overline{v_C'^2}) - \frac{\sqrt{2}+1}{2}(\overline{v_A'^2}+\overline{v_D'^2})\right]
\end{aligned} \qquad /7/$$

On the base of the solution analysis some identicies can be deduced

$$\overline{v_A'^2} + \overline{v_B'^2} + \overline{u_L'^2} \equiv \overline{v_C'^2} + \overline{v_D'^2} + \overline{u_H'^2} \qquad /8/$$

$$\overline{u_M'^2} + \overline{u_K'^2} \equiv \overline{u_H'^2} + \overline{u_L'^2} \qquad /9/$$

and for the unmeasurable value $\overline{v_1'^2}$ on the direct way, because of the disturbances in the wake of the probe prongs for the heated wire, situated in parallel to the axis 1 and mean flow direction $\overline{w_1}$, the formula takes the form

$$\overline{v_1'^2} \equiv \overline{v_A'^2} + \overline{v_B'^2} - \overline{u'^2_H} \equiv \overline{v_C'^2} + \overline{v_D'^2} - \overline{u'^2_L} \qquad /10/$$

The energy of the turbulent fluctuations with the reference to the unitary mass is expressed as

$$E = \frac{1}{2}\left(\overline{w_1'^2} + \overline{w_2'^2} + \overline{w_3'^2}\right) = \frac{1}{2(2+k^2)}\left(\overline{v_1'^2} + \overline{v_2'^2} + \overline{v_3'^2}\right) = \qquad /11/$$

$$= \frac{1}{2(2+k^2)}\left(\overline{v_A'^2} + \overline{v_B'^2} + \overline{u'^2_L}\right) = \frac{1}{2(2+k^2)}\left(\overline{v_C'^2} + \overline{v_D'^2} + \overline{u'^2_H}\right)$$

The formulae /6/ and /7/ define all 9 tensor components $\overline{w_i'w_j'}$; $i,j = 1,2,3$ as $\overline{w_i'w_j'} = \overline{w_j'w_i'}$. The formulae /6/ can be used for evaluating 3 components of the averaged velocity fluctuations $\sqrt{\overline{w_i'^2}}$ or instantaneous velocity fluctuations $\sqrt{w_i'^2}$ if v'^2 are also instantaneous /before averaging/. The formulae for averaged velocity $\overline{w_i}$ take this same form as /6/ if instead v'^2 the v^2 are measured using DC /direct current/ input circuit and averaging output circuits.
But as the hot wire sensor is insensitive to the change of the flow direction on the opposite one there is an undeterminateness of the velocity vector $\vec{w} = \vec{\overline{w}} + \vec{w}'$ because of evaluating only the absolute values of the components, therefore

$$\vec{\overline{w}} = \pm\sqrt{\overline{w_i}^2}\,\vec{i} \pm \sqrt{\overline{w_2}^2}\,\vec{j} \pm \sqrt{\overline{w_3}^2}\,\vec{k} \qquad /12/$$

$$\vec{\overline{w'}} = \pm\sqrt{\overline{w_1'^2}}\,\vec{i} \pm \sqrt{\overline{w_2'^2}}\,\vec{j} \pm \sqrt{\overline{w_3'^2}}\,\vec{k} \qquad /13/$$

Each vector $\vec{\overline{w}}$ and $\vec{\overline{w'}}$ can take 2^3 spatial situations so the vector $\vec{w}$ can take $2^3 \cdot 2^3 = 64$ values and directions at 6 constant absolute values of the components $\sqrt{\overline{w_i}^2}$ and $\sqrt{\overline{w_i'^2}}$ Only the absolute values of every vector $\vec{\overline{w}}$ and $\vec{\overline{w'}}$ are known exactly and are equals 2E acc.to /11/ for w_i and w_i' . For evaluating the signs at components of the averaged velocity vector $\vec{\overline{w}}$ another measuring methods /e.g. lasser-doppler anemometry, Pitot-tube must be used. Mostly in turbulence investigations it is enough to know the squared values $w_i'^2$ and $\overline{w_i'^2}$ or $\sqrt{w_i'^2}$ and $\sqrt{\overline{w_i'^2}}$
For evaluating the work of the viscous shear stresses ε_{IV} and the viscous dissipation by the turbulent motion ε_V /4/, per unit of mass and of time

$$\varepsilon_{IV} = \nu \frac{\partial}{\partial x_i}\overline{w_j'\left(\frac{\partial w_i'}{\partial x_j} + \frac{\partial w_j'}{\partial x_i}\right)} \; , \quad \varepsilon_V = -\nu\overline{\left(\frac{\partial w_i'}{\partial x_j} + \frac{\partial w_i'}{\partial x_i}\right)\frac{\partial w_j'}{\partial x_i}} \qquad /14/$$

it is necessary to know the spatial derivatives of the velocity fluctuations.

3. EVALUATING OF THE TURBULENT VISCOSITY

The turbulent viscosity ν is defined by the formula

$$M_{Tij} = 1 + \frac{\nu_{Tij}}{\nu} = 1 - \frac{\overline{w_i' w_j'}}{\nu\, 2D_{ij}} \qquad /15/$$

where D_{ij} are components of the rate of the deformation tensor. In rectangular cartesian coordinates

$$2D_{xy} = 2D_{yx} = \frac{\partial \bar{w}_x}{\partial y} + \frac{\partial \bar{w}_y}{\partial x}, \quad 2D_{xz} = 2D_{zx} = \frac{\partial \bar{w}_x}{\partial z} + \frac{\partial \bar{w}_z}{\partial x}, \quad 2D_{yz} = 2D_{zy} = \frac{\partial \bar{w}_y}{\partial z} + \frac{\partial \bar{w}_z}{\partial y} \qquad /16/$$

or in cylindrical coordinates

$$2D_{r\varphi} = 2D_{\varphi r} = r\frac{\partial}{\partial r}\left(\frac{\bar{w}_\varphi}{r}\right) + \frac{1}{r}\frac{\partial \bar{w}_r}{\partial \varphi}, \quad 2D_{rz} = 2D_{zr} = \frac{\partial \bar{w}_r}{\partial z} + \frac{\partial \bar{w}_z}{\partial r}, \quad 2D_{\varphi z} = 2D_{z\varphi} = \frac{\partial \bar{w}_\varphi}{\partial z} + \frac{1}{r}\frac{\partial \bar{w}_z}{\partial \varphi} \qquad /17/$$

a/ For the flow in the vicinity of the flat plate /Fig.2/

$$\nu_T = -\frac{\overline{w_r' w_z'}}{\frac{d\bar{w}_z}{dr}} = \frac{\overline{w_3' w_1'}}{\frac{d\bar{w}_1}{dx_3}} = \frac{\Delta x_3}{2(1-K^2)} \frac{\overline{v_B'^2} - \overline{v_A'^2}}{\bar{v}_{L1} - \bar{v}_{L2}} = \frac{\left(\overline{w_3' w_1'}\right)_{max}}{\frac{\bar{w}_{1\,max}}{\varphi_c}\frac{\varrho_0}{r_0}} \frac{\mathcal{W}_w(\eta)}{\frac{d\varphi}{d\eta}} \qquad /18/$$

where

$$\varphi \equiv u^+ = \frac{\bar{w}_1}{v^*} = \varphi_m \omega = \varphi_c \omega_c, \quad \omega_c = \frac{\bar{w}_1}{\bar{w}_{1c}}, \quad \omega = \frac{\bar{w}_1}{\bar{w}_{1m}}$$

$$\eta \equiv y^+ = \frac{x_{3c} v^*}{\nu}, \quad v^* = \sqrt{\frac{\tau_0}{\varrho_g}}, \quad \tau_0 = \frac{1}{2} f \varrho_g \bar{w}_1^2, \quad f = \frac{\Delta p}{4\frac{l}{d}\varrho_g \frac{\bar{w}_1^2}{2}}$$

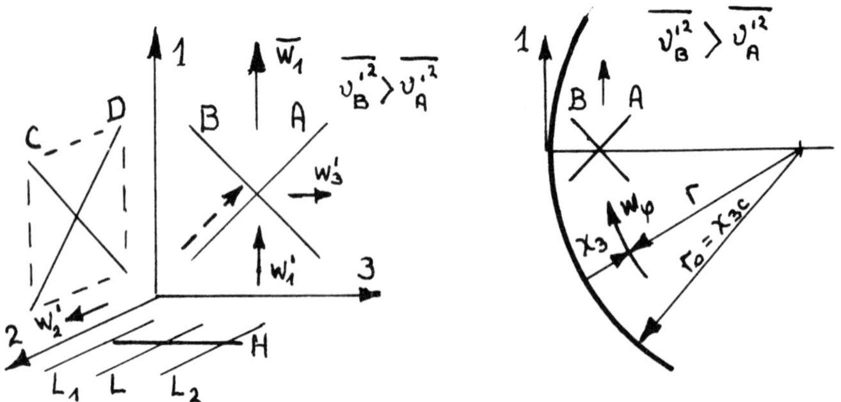

Fig. 2,3. The scheme of the situation of the hot wires sensors of the thermoanemometric probes

b/ For the rotating flow, /Fig.3/

$$\nu_T = -\frac{\overline{w_r' w_\varphi'}}{r\frac{d}{dr}\left(\frac{\overline{w_\varphi}}{r}\right)} = \frac{\overline{w_3' w_1'}}{(x_{3c}-x_3)\dfrac{d\left[\frac{\overline{w_1}}{x_{3c}-x_3}\right]}{dx_3}} \qquad /19/$$

c/ For the flow in the rectangular channel with cross section 80x100 mm and $\overline{w}_1 = 10 \div 40$ m/s, $Re = 0.5\cdot 10^5 \div 2\cdot 10^5$ the results of the performed measurements have been approximated by the formulae

$$\varphi = 21.45 - 31\,\eta^{-0.47}, \quad \frac{d\varphi}{d\eta} = 15\,\eta^{-1.47}; \quad 5 \leqslant \eta \leqslant 300 \qquad /20/$$

$$\mathcal{W}_w = \frac{\overline{w_3' w_1'}}{(w_3' w_1')_{max}} = \begin{cases} 0.02\,\eta, & 5 < \eta < 50 \\ 1, & 50 < \eta < 80 \\ 80\,\eta^{-1}, & 80 < \eta < 800 \end{cases} \qquad /21/$$

$$(\overline{w_3' w_1'})_{max} \cong (0.065 \div 0.04)\,\overline{w_1}$$

$$\frac{\nu_T}{\nu} = \begin{cases} 0.01886\,\eta^2, & 5 < \eta < 50 \\ 0.1338\,\eta^{1.5}, & 50 < \eta < 80 \\ 10.7070\,\eta^{0.5}, & 80 < \eta < 800 \end{cases} \qquad /21'/$$

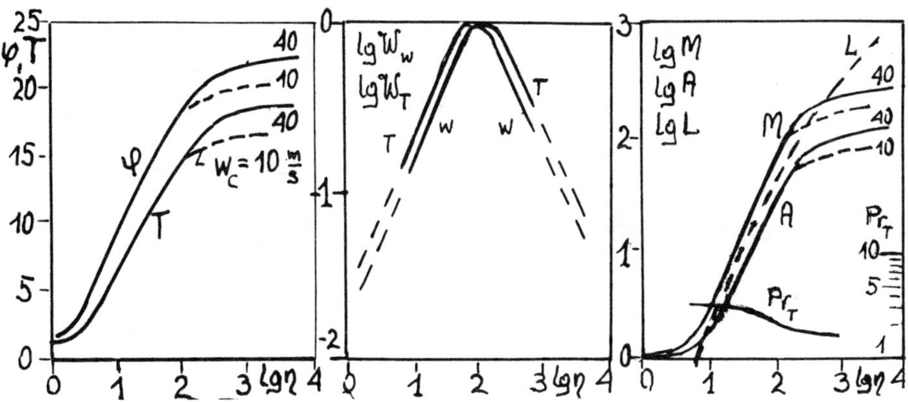

Fig. 4,5,6. The results of measured characteristic values for the turbulent flow; dimensionless velocity φ and temperature T, $\mathcal{W}_w$, $\mathcal{W}_t$, M, A, the Prandtl mixing lenght $L^2 = \frac{M-1}{\varphi'_\eta}$ where $L = \frac{l v^*}{\nu}$

4. EVALUATING OF THE EDDY DIFFUSIVITY FOR HEAT

The eddy diffusivity for heat a_T is defined by the formula

$$A_{Ti} = 1 + \frac{a_{Ti}}{a} = 1 - \frac{\overline{w_i' t'}}{a\,grad_i\,\overline{t}} \qquad /22/$$

where

$$\mathrm{grad}_x t = \frac{\partial \bar{t}}{\partial x}, \quad \mathrm{grad}_y \bar{t} = \frac{\partial \bar{t}}{\partial y}, \quad \mathrm{grad}_z \bar{t} = \frac{\partial \bar{t}}{\partial z} \qquad /23/$$

or

$$\mathrm{grad}_r \bar{t} = \frac{\partial \bar{t}}{\partial r}, \quad \mathrm{grad}_\varphi \bar{t} = \frac{1}{r}\frac{\partial \bar{t}}{\partial \varphi}, \quad \mathrm{grad}_z t = \frac{\partial \bar{t}}{\partial z} \qquad /24/$$

For the flow near the flat wall

$$a_T = -\frac{\overline{w_r' t'}}{\frac{d\bar{t}}{dr}} = \frac{\overline{w_3' t'}}{\frac{d\bar{t}}{dx_3}} = \frac{(\overline{w_3' t'})_{max}}{\frac{t_w - t_c}{T_c}\,\frac{\varrho_0}{r_0}}\;\frac{\mathcal{W}_T(\eta)}{\frac{dT}{d\eta}} \qquad /25/$$

$$\overline{w_3'^2} = \frac{1}{(2+k^2)(1-k^2)}\left[\overline{v_c'^2} + \overline{v_D'^2} - (1+k^2)\,\overline{u_H'^2}\right] \qquad /26/$$

$$\overline{w_3' t'} = \sqrt{\overline{w_3'^2 t'^2}} = \frac{1}{\sqrt{(2+k^2)(1-k^2)}}\sqrt{\overline{v_c'^2 t'^2} + \overline{v_D'^2 t'^2} - (1+k^2)\,\overline{u_H'^2 t'^2}} \qquad /27/$$

$$\frac{d\bar{t}}{dx_3} = \frac{\bar{t}_{L1} - \bar{t}_{L2}}{\Delta x_3}$$

$$T \equiv t^+ = R(\xi)\cdot T_c = \frac{t_w - \bar{t}}{t_w - \bar{t}_c}\;\frac{\varrho_g \overline{w_{1m}} \sqrt{\frac{f}{2}}\, c_p}{\alpha_c} = (t_w - \bar{t})\,\frac{\varrho_g v^* c_p}{q_w}$$

$$T_c \equiv t_c^+ = \frac{t_m^+}{R(m)} = \frac{1}{Sp\,R(m)} = \frac{1}{Sp_c} = \frac{\sqrt{\frac{f}{2}}}{St\,R(m)} = \frac{2\varrho_0 Pr}{Nu}\,\frac{1}{R(m)} = \frac{2}{\frac{Nu_c}{\varrho_0 Pr}} = \frac{\lambda v^*}{\alpha_c a}$$

$$(\overline{w_3' t'})_{max} \cong 0.7\sqrt{\overline{w_3'^2}_{max}}\sqrt{\overline{t'^2}_{max}} = (0.6 \div 1.8)\,\frac{f}{2}\,v^* T_c\,\Delta t_{max}$$

For the flow in the rectangular channel as in section 2 and over that with heated one vertical wall on the distance 600 mm up to the 400°C the following approximating formulae have been obtained

$$T = 18.08 - 46.7\,\eta^{0.6}, \quad \frac{dT}{d\eta} = 23\,\eta^{-1.6}; \quad 7 < \eta < 300 \qquad /28/$$

$$\mathcal{W}_T = \frac{\overline{w_3' t'}}{(\overline{w_3' t'})_{max}} = \begin{cases} 0.025\,\eta, & 5 < \eta < 40 \\ 1, & 40 < \eta < 120 \\ 120\,\eta^{-1}, & 120 < \eta < 800 \end{cases} \qquad /29/$$

$$\frac{a_T}{a} = \begin{cases} 0.0694\ \eta^2, & 5 < \eta < 40 \\ 0.0304\ \eta^{1.5}, & 40 < \eta < 120 \\ 4.0809\ \eta^{0.5}, & 120 < \eta < 800 \end{cases} \qquad /30/$$

5. EVALUATING OF THE TURBULENT PRANDTL NUMBER

The general definition for the turbulent Prandtl number is

$$Pr_{Ti(j)} = \frac{\nu_{Tij}}{a_{Ti}} = \frac{\dfrac{\overline{w_i' w_j'}}{2D_{ij}}}{\dfrac{\overline{w_i' t'}}{grad_i \bar{t}}} \qquad /31/$$

For the flow near the flat plate

$$Pr_{T31} = Pr_{T13} = \frac{\nu_{T31}}{a_{T3}} \quad Pr_{T23} = Pr_{T32} = \frac{\nu_{T23}}{a_{T2}}, \quad Pr_{T12} = Pr_{T21} = \frac{\nu_{T12}}{a_{T1}} \qquad /32/$$

$$Pr_{T31} = \frac{\nu_{T31}}{a_{T3}} = \frac{\overline{w_3' w_1'}}{\overline{w_3' t'}} \frac{d\bar{t}}{d\bar{w}_1} = C \frac{\mathcal{W}_w(\eta)}{\mathcal{W}_T(\eta)} \frac{dT(\eta)}{d\varphi(\eta)} \qquad /33/$$

where

$$C = \frac{(\overline{w_3' w_1'})_{max}}{(\overline{w_3' t'})_{max}} \frac{\varphi_c}{T_c} \frac{t_w - \bar{t}_c}{\bar{w}_{1max}} \qquad /34/$$

After introducing some simplifying assumptions for the shear flow

$$\frac{\tau}{\tau_w} = M\varphi_\eta' = \xi \quad \text{from where} \quad \frac{\nu_T}{\nu} = \frac{\xi}{\varphi_\eta'} - 1 \qquad /35/$$

$$\frac{q}{q_w} = \frac{1}{Pr} A T_\eta' = \xi \quad \text{from where} \quad \frac{a_T}{a} = Pr \frac{\xi}{T_\eta'} - 1 \qquad /36/$$

and after introducing to the definition of Pr_T

$$Pr_T = \frac{\nu_T}{a_T} = \frac{\dfrac{\xi}{\varphi_\eta'} - 1}{Pr\left(\dfrac{\xi}{T_\eta'} - \dfrac{1}{Pr}\right)} \ ; \ (\varphi_\eta')_{\eta=0} = 1 \ , \ (T_\eta')_{\eta=0} = Pr \qquad /37/$$

In the vicinity of the wall Pr_T is approximately ≈ 1. But this assumption is valid only for the stabilized flow in the straight duct with constant cross section at the fully developed profile, far from the disturbances caused by the inlet and outlet and with constant heat flux on the wall. If Pr_T is known the $A(\eta)$ can be based only on the measurements of

$$A(\eta) = 1 + \frac{a_T}{a} = 1 + Pr\frac{1}{Pr_T}\frac{\nu_T}{\nu} = 1 + Pr\frac{1}{Pr_T}\left[M(\eta) - 1\right] = 1 + Pr\frac{1}{Pr_T}L^2\varphi'_\eta \qquad /38/$$

6. THE ENERGY EQUATION

In general, the temperature field in the turbulent flow is described by the energy equation

$$\frac{d\bar{t}}{d\tau} = div\left[a A_{Ti}\, grad_i \bar{t}\right] + q_1(\tau) \qquad /39/$$

and in developed form in cylindrical coordinates

$$\frac{\partial \bar{t}}{\partial \tau} + \bar{w}_r\frac{\partial \bar{t}}{\partial r} + \bar{w}_\varphi \frac{1}{r}\frac{\partial \bar{t}}{\partial \varphi} + \bar{w}_z\frac{\partial \bar{t}}{\partial z} = \frac{1}{r}\frac{\partial}{\partial r}\left[a A_{Tr} r\frac{\partial \bar{t}}{\partial r}\right] + \qquad /40/$$

$$+ \frac{1}{r^2}\frac{\partial}{\partial \varphi}\left[a A_{T\varphi}\frac{\partial \bar{t}}{\partial \varphi}\right] + \frac{\partial}{\partial z}\left[a A_{Tz}\frac{\partial \bar{t}}{\partial z}\right] + \frac{a}{\lambda}\left[\frac{\partial \varphi}{\partial \tau} + \rho\hat{Q}(\tau)\right]$$

As usually $\frac{\partial \bar{t}}{\partial \varphi} = 0$, $\bar{\vartheta}_r = 0$ this equation reduces to

$$\frac{\partial \bar{t}}{\partial \tau} + \bar{w}_z(r)\frac{\partial \bar{t}}{\partial z} = \frac{1}{r}\frac{\partial}{\partial r}\left[a A_{Tr}(r)\frac{\partial \bar{t}}{\partial r}\right] + \frac{\partial}{\partial z}\left[a A_{Tz}(z)\frac{\partial \bar{t}}{\partial z}\right] + Q\, q_1(\tau) \qquad /41/$$

For fully developed turbulent flow the term with A_{Tz} may be omitted. The last term presents instantaneous heat source inside the flow and Q is a constant. The temperature $t(r,z,\tau)$ needs boundary conditions for $\tau = 0$ and $\tau > 0$, for the initial cross section at z = 0 and on the wall at $r = r_0$,where the temperature $t(r_0, z, \tau)$ or the gradient of the temperature $(\partial t/\partial r)_{r=r_0}$ may be given. The solution of the equation /40/ presents the three dimensional problem. After introducing some simplifying assumptions this problem can be reduced to two-dimensional problem or even to the simplest one-dimensional equation /36/. The solutions of the equation /41/ have been obtained using various analytical and numerical method for various boundary conditions of the first and second kind in two and three dimensional space [5]. The first step to the solution of the equation /41/ is to establish on the base of the empirical data or the semiempirical formulae the functions $\bar{w}_z(r) = \bar{w}_c \cdot \omega_c(\xi)$, $A_{Tr}(r) = 1 + a_T/a$ acc.to /25/ and on similar way $A_{Tz}(z) = 1 + a_{Tr}/a$

7. EVALUATING OF THE NUSSELT NUMBER

On the base of the solution of the energy equation /40/ it is possible to evaluate the gradient of the temperature on the wall $\left(\partial\bar{t}/\partial r\right)_{r=r_0}$ in the flow and the mean local differen-ce of the temperature in the cross section $\Delta\bar{t}_m = t_w - \bar{t}_m$

$$Nu_m = \frac{2\left(\frac{\partial\bar{t}}{\partial\xi}\right)_{\xi=1}}{\Delta t_m} = -2\left|\frac{\partial\left[\frac{t_w-\bar{t}}{t_w-\bar{t}_m}\right]}{\partial\left[\frac{y}{r_0}\right]}\right|_{y=0} = \frac{\alpha_m d_e}{\lambda} = \frac{1}{R_m} Nu_c = \frac{q_w d}{(t_w-\bar{t}_m)\lambda} \qquad /42/$$

It should be noticed that for the varying conditions with regard to z or τ, $t_w-\bar{t}_m$ may become zero but at this same spatial-time position heat flux on the wall $q_w \neq 0$ so $Nu_m = \pm\infty$. The formula /12/ therefore is useful for the constant or lineary with regard on z changing boundary con-ditions of the first or second kind.

The assymptotical value of the Nusselt number for above mentioned conditions with some simplifying assumptions is

$$Nu_c = \frac{q_w}{t_w-\bar{t}_c}\frac{d}{\lambda} = \frac{\alpha_c d}{\lambda} = \frac{2}{\int_0^1 \frac{\xi d\xi}{A(\xi)}} = \frac{2}{\frac{1}{\varrho_0}\int_0^{\varrho_0}(1-\eta/\varrho_0)\frac{1}{A(\eta)}d\eta} = \frac{2\varrho_0 Pr}{T_c} \qquad /43/$$

$$T(\eta) = Pr\int_0^{\eta}(1-\eta/\varrho_0)\frac{1}{A(\eta)}d\eta \qquad /44/$$

$$T_c = \left[T(\eta)\right]_{\eta=\varrho_0} = \frac{\sqrt{\frac{f}{2}}}{St\, R(m)} = \frac{1}{Sp\, R(m)} = \frac{1}{Sp_c} = \frac{1}{St\,\varphi_m R(m)}$$

$$\varphi(\eta) = \int_0^{\eta}\frac{(1-\eta/\varrho_0)d\eta}{M(\eta)}, \qquad \varphi_c = \left[\varphi(\eta)\right]_{\eta=\varrho_0} \qquad /45/$$

The assymptotical unitary profile of the temperature

$$R(\xi) = \frac{T[\eta(\xi)]}{T_c} = 1 - \frac{\int_0^{\xi}\frac{\xi' d\xi'}{A(\xi')}}{\int_0^1\frac{\xi d\xi}{A(\xi)}} \qquad /46/$$

$$R(m) = 2\int_0^1 \xi\,\omega(\xi)R(\xi)d\xi, \quad T_m = 2\int_0^1 \xi\,\omega(\xi)T(\xi)d\xi = \frac{\sqrt{\frac{f}{2}}}{St} = \frac{1}{St\varphi_m} = \frac{1}{Sp} \qquad /47/$$

Comparing the calculated Nu on the base of $A(\eta)$ or $M(\eta)$ with the empirical Nu_{emp} evaluated from e.g. formula

$$Nu_{emp} = 0.024\, Re^{0.8} Pr^{0.4} \quad for \quad 10^4 < Re < 10^6, \; 0.7 < Pr < 7 \qquad /48/$$

the equivalent turbulent Pr_T can be evaluated.
For investigated flow on this way the formula

$$Nu = \frac{0.0067\ Re^{0.89}}{1 - 1590\ Re^{-0.69}} \qquad /49/$$

was established and $Pr_{Teq} = 2.3$ evaluated.

8. INVESTIGATIONS OF THE TURBULENT STRUCTURE OF THE STREAM

The most important region in the stream because of the transport phenomena is the region in the vicinity of the wall and this region is of the range about 10 times of the thickness of the boundary layer displacement δ^*. In the investigated turbulent air flow in the channel 100x100 mm at the mean velocity 10÷40 m/s, the displacement thickness $\delta^* = 0.77 \div 0.67$ mm. The maximum of the averaged effective fluctuations $\sqrt{\overline{v_3'^2}} = 1.2 \div 4.5$ m/s appeared in the distance from the wall $x_3 = 0.4 \div 0.6$ mm. The maximum of the $\overline{v'^2(f)}$ on the base of the spectral analysis in the range $f = 10 \div 10^5$ Hz appeared at 300 Hz. The temperature fluctuations $(\sqrt{\overline{t'^2}})_{max} = 6.8 \div 2.6$ K appeared on the distance $x_3 = 0.5 \div 0.8$ mm at $\Delta t_{max} = 116$ K between the wall and the fluid. The size of the microedies in the inertial range of spectrum, evaluated as for isotropic turbulence on the base of the energy dissipation, was $\eta = 0.06 \div 0.046$mm and microscale of time $\tau_E = 0.12 \div 0.06$ ms. The lateral microscale evaluated from autocorrelation function $\lambda_g = 0.78 \div 1.56$ mm.

On the base of the amplitude analysis probability of the velocity fluctuation it was revealed that the density distribution and the distribution function are only coarsely similar to those based on the Gauss's error function. The squeness factor S in the vicinity of the wall reaches the values 0.3 and -0.4 and then asymptotically 0 instead on whole range of x_3 the constant value equal zero and the flatness factor F approaches maximum value 0.5 and then asymptotically 0.118 instead of 3 for whole range of x_3 as for Gaussian distribution. It should be mentioned that in fact these measurements was based on the effective velocity $v_2'^2 = w_1'^2 + k^2 w_2'^2 + w_3'^2$ instead of on one component eg. $w_3'^2$ what may cause some error. Between correlation function $R_{w_i'w_j'}(\tau)$ and the spectral density $S_{w_i'w_j'}(f)$ the Fourier direct and inverse transform is applied. For even normalized functions $[R_{w_i'^2}(\tau)]_N$ and $[S_{w_i'^2}(f)]_N$ it can be noticed that if $[R_{w_i'^2}(\tau)]_N$ is more flat its Fourier transform $[S_{w_i'^2}(f)]_N$ is more sharp and if $[R_{w_i'^2}(\tau)]_N$ takes also the negative values, the $[S_{w_i'^2}(\tau)]_N$ has the maximum shifted in the direction of the greater values of $\omega = 2\pi f$.

CONCLUSIONS

1. The first step to the evaluating momentum and heat transfer characteristic in turbulent flow is to evaluate the

Reynolds stress tensor components $\tilde{R}$, turbulent viscosity ν_T and eddy diffusivity for heat a_T.
2. After establishing the function $A = 1 + a_T/a$, or, with simplifying assumption that Pr_T is known, only the function $M = 1 + \nu_T/\nu$, the energy equation describing the temperature field in the turbulent flow can be solved using various analytical and numerical method and on the base of the solution the heat flux on the wall can be evaluated.
3. Additional informations on the turbulent structure of the flow can be supplied by the correlation functions $R_{w_i'w_j'}(\tau)$ spectral analysis $S_{w_i'w_j'}(f)$ and Fourier transforms of them, amplitude probability analysis for evaluating the density distribution $p(\frac{x}{\sigma})$ and the distribution function $P(\frac{x}{\sigma})$, the squeness S and the flatness F factors.

REFERENCES

1. OPPENHEIM,A.V.,SHAFER,R.W. - Digital Signal Processing, Prentice-Hall Inc.,Englewood Cliffs,New Jersey,1975.

2. LAWRANCE,L.,GOLD,B. - Theory and Application of Digital Signal Processing, Prentice-Hall Inc.,Englewood Cliffs, New Jersey, 1975.

3. DISA Information, No 1 ÷ 25, DISA Elektronik A/S, Skovlunde,1965 ÷ 1980.

4. HINZE,J.O. - Turbulence, McGraw-Hill, USA, 1975.

5. DROZD,T. - Auxiliary termoanemometry measurements for evaluation of local heat transfer characteristics on walls of arbitrary shapes, pp. 1-15, paper presented on conference and in additional copy to Proceedings of the First International Conference on Numerical Methods in Thermal Problems, ed. Lewis,R.W.,Morgan,K. Pineridge Press,Swansea G.B.,July 2nd-6th 1979.

SECTION 5

FREE SURFACE FLOW

A FINITE ELEMENT METHOD FOR THE SHALLOW WATER EQUATIONS

G. LABADIE - J.P. BENQUE - B. LATTEUX

Laboratoire National d'Hydraulique - E.D.F.
6, quai Watier, Chatou, France

SUMMARY

This paper presents a new finite element method for the solution of the shallow water equations. This method is suitable when dealing with complex boundaries by virtue of the finite element domain approximation and when a good description of boundary layers is needed - the latter occurs with flows strongly influenced by advection and turbulent viscosity phenomena such as tidal currents in a harbour.

Many other methods can be found in the literature for the shallow water equations but they generally imply to handle wide banded non-symmetric systems. An interesting feature of the algorithm described below is that all the linear systems to be solved are symmetric and involve only one scalar unknown function. The non-linear advection terms are computed explicitely with a characteristics method which provide a natural way of upwinding.

I. GOVERNING EQUATIONS AND TIME-DISCRETIZATION

The system to solve consists of the mass conservation equation :

$$(1) \qquad \frac{\partial h}{\partial t} + \nabla . \underset{\sim}{Q} = 0$$

and the two momentum equations :

(2) $$\frac{\partial Qi}{\partial t} + \nabla . \, Qi \, \underset{\sim}{u} + gh \frac{\partial h}{\partial x_i} - K\Delta Qi = fi \quad \text{for } i = 1,2$$

where :

$\underset{\sim}{u}$ denotes the velocity vector (components ui)

h denotes the depth of water

$\underset{\sim}{Q} = \underset{\sim}{u}h$ denotes the flow rate per unit length(components Qi)

fi includes terms coming from the Coriolis and friction forces and from the variation of the bottom depth.

The unknown functions $\underset{\sim}{u}$, $\underset{\sim}{Q}$, h are defined on an open set Ω of R^2 and for a time interval $[0,T]$. The boundary conditions can be of several types. We will assume for simplicity that the two components of the flow rate $\underset{\sim}{Q}$ are given on the boundary Γ of Ω.

Let Dt be the time step and $\underset{\sim}{Q}^n$, $\underset{\sim}{u}^n$, h^n denote the approximations of $\underset{\sim}{Q}$, $\underset{\sim}{u}$, h at time nDt. Then the chosen time-discretization is as follows :

(3) $$\frac{h^{n+1} - h^n}{Dt} + \nabla . \underset{\sim}{Q}^{n+1} = 0$$

(4) $$\frac{Qi^{n+1} - Qi^n}{Dt} + \nabla . Qi^n \, \underset{\sim}{u}^n + gh_m \frac{\partial h^{n+1}}{\partial xi}$$

$$- K\Delta Qi^{n+1} = fi - g\zeta^n \frac{\partial h^n}{\partial xi}$$

(h_m being the average of h in time and $\zeta^n = h^n - h_m$)

The advection terms are explicit in time and so, can be computed separately. Introducing an auxiliary variable $\hat{\underset{\sim}{Q}}$, we can split system (3) (4) into a set of two subsystems :

(5) $$\frac{\hat{Q}i - Qi^n}{DT} + \nabla . (Qi^n \, \underset{\sim}{u}^n) = 0 \quad \text{for } i = 1, 2$$

$$(6)\quad \begin{cases} \dfrac{Qi^{n+1} - \hat{Q}i}{Dt} + gh_m \dfrac{\partial h^{n+1}}{\partial xi} - K\Delta Qi^{n+1} \\ = Si \ (= fi - g\zeta^n \dfrac{\partial h^n}{\partial xi}) \\ \dfrac{h^{n+1} - h^n}{Dt} + \nabla . \underset{\sim}{Q}^{n+1} = 0 \end{cases}$$

Systems (5) and (6) are solved in two successive steps :

. 1st step : Calculation of $\hat{\underset{\sim}{Q}}$ from system (5) which is the discrete form of the conservative advection equation :

$$(7)\qquad \frac{\partial Qi}{\partial t} + \nabla . Qi \underset{\sim}{u} = 0$$

This step is described in section 2 below.

. 2nd step : Calculation of $\underset{\sim}{Q}^{n+1}$ and h^{n+1} from system (6), which is a discretization of a propagation system with a viscosity term.

This "diffusion-propagation" step is described in section 3.

II - ADVECTION STEP

2.1. Approach :

We propose a method of characteristics to solve this step. This method has already been used in a simpler way for the Navier-Stokes equations [1].

Let us first consider the more general problem :

$$(8)\qquad \frac{\partial f}{\partial t} + \underset{\sim}{u} . \nabla f = G$$

where $\underset{\sim}{u}$ and G are known functions of space and time variables $\underset{\sim}{x}$ and t.

Solving (8) between time $t^n = nDt$ and time $t^{n+1} = (n+1)\ Dt$ requires an initial condition for f at time t^n and a boundary condition on the part Γ^- of the boundary Γ where the flow enters (i.e. where $\underset{\sim}{u}.\underset{\sim}{n} < 0$, if $\underset{\sim}{n}$ is the unit outward normal vector). Let ϕ be the given value of f on Γ^-.

Along the characteristic line (or path line) defined by :

$$\frac{d\underset{\sim}{x}}{dt} = \underset{\sim}{u}$$

equation (8) can be written as :

$$\frac{df}{dt} = G$$

We can then derive the following algorithm for the discrete calculation of f^{n+1} on a finite element mesh :

For any node Mj :

. Compute the characteristic line Cj leading to Mj at time t^{n+1}, i.e. solve backwards in time from t^{n+1} to t^n :

(9) $$\frac{d\underset{\sim}{x}}{dt} = \underset{\sim}{u} \quad \text{with} \quad \underset{\sim}{x}\,(t^{n+1}) = Mj$$

. Compute $fj^n = f^n(Pj)$ where Pj is the foot of the path line, i.e. the point $\underset{\sim}{x}(t^n)$.

. Solve between t^n and t^{n+1} :

(10) $$\frac{df}{dt} = G \text{ along the path line}$$

with initial condition : $f(t^n) = fj^n$.

The result at time t^{n+1} of the integration of (10) is the expected value for f^{n+1} (Mj).

Equations (9) and (10) are ordinary differential equations. Their discrete solution can be carried out by a Runge-Kutta method.

All the evaluations of $\underset{\sim}{u}$, f^n and G use the finite element approximation defined for these functions.

The boundary condition is taken into account when, between t^n and t^{n+1}, the path line crosses the boundary. In such a case, the integration of (9) stop at time τ ($t^n \leq \tau < t^{n+1}$) and gives the intersection M_Γ

between the characteristic and the boundary. M_Γ obviously belongs to Γ^-. Hence the two last steps of the algorithm can be modified as follows :

. Compute $\phi(\tau, M_\Gamma)$, which is the boundary condition for f at time τ and point M_Γ.

. Solve equation (10) between τ and t^{n+1}, with initial condition $f(\tau) = \phi(\tau, M_\Gamma)$.

2.2 Application to the advection step :

In this step the system to solve is an explicit discretization of :

$$\frac{\partial Qi}{\partial t} + \underset{\sim}{u}^n . \nabla Qi = Gi \quad \text{for } i = 1, 2$$

where $Gi = - Qi^n \nabla . \underset{\sim}{u}^n$

The algorithm described in 2.1 can be applied, replacing successively f by the two components Qi of $\underset{\sim}{Q}$ and Gi by the known functions $- Qi^n \nabla . \underset{\sim}{u}^n$.

This explicit scheme avoids the resolution of non symmetric systems. Although explicit, it is unconditionnaly stable. So the time step is not limited by the C.F.L. criterion.

III - DIFFUSION AND PROPAGATION STEP

In order to get convenient notations, we shall omit in this section the superscript n + 1 and set :

$$\alpha = \frac{1}{Dt}, \quad \underset{\sim}{T} = \frac{\underset{\sim}{Q}}{Dt} + \underset{\sim}{S}^n \quad \text{and} \quad R = \frac{h^n}{Dt}$$

We shall assume that h_m (time-average of h) is a constant although it is not a requirement for the method.

The boundary-value problem to solve in this step is :

$$(11) \qquad \alpha \underset{\sim}{Q} - K \Delta \underset{\sim}{Q} + g h_m \nabla h = \underset{\sim}{T} \text{ in } \Omega$$

(12) $\alpha h + \nabla . \underset{\sim}{Q} \qquad = R \text{ in } \Omega$

(13) $\underset{\sim}{Q}|_{\Gamma} \qquad = \underset{\sim}{\phi} \text{ on } \Gamma$

where $\underset{\sim}{\phi}$ is a given boundary data.

3.1. Outline of the method

The discretization of equations (11), (12), (13) lead to a big linear system : if h and $\underset{\sim}{Q}$ have the same F.E.M. approximation and if N denotes the number of nodes in the triangulation, the system is of order 3N.

In order to reduce this size, we notice that, applying the divergence operator to (11) and plugging (12) into it, we can set up an equation where h is the only unknown :

(14) $\alpha^2 h - (K\alpha + gh_m)\Delta h = W = \alpha R - K\Delta R - \nabla . \underset{\sim}{T}$

Provided that we can deduce proper boundary values for h, the solution procedure is then straight forward :

. find h solution of (14),

. solve (11) (13) where $gh_m\nabla h$ is now a known function i.e. solve a set of two Dirichlet problems on the two components Qi of $\underset{\sim}{Q}$.

The problem then amounts to finding the trace of h on Γ. The procedure used for this purpose in an extension of the technique proposed by Glowinski and al in [2] for the Stokes problem. It is described in the next section.

3.2. Calculation procedure of $h|_{\Gamma}$

Let us consider the following set of Dirichlet problems :

(15a) $\begin{cases} \alpha^2 h_\lambda - (K\alpha + gh_m)\Delta h_\lambda = 0 \\ h_\lambda|_\Gamma = \lambda \end{cases}$

(15b) $\begin{cases} \alpha \underset{\sim}{Q}_\lambda - K\Delta \underset{\sim}{Q}_\lambda = -gh_m\nabla h_\lambda \\ \underset{\sim}{Q}_\lambda|_\Gamma = 0 \end{cases}$

(15c) $\begin{cases} \alpha^2\psi\lambda - (K\alpha + gh_m)\Delta\psi\lambda = \nabla . \underset{\sim}{Q}\lambda + \alpha h_\lambda \\ \psi\lambda|_\Gamma = 0 \end{cases}$

and let A be the operator defined by :

$$A\lambda = -\frac{\partial\psi\lambda}{\partial n}|_\Gamma$$

Note that A only depends on the time step and on some physical constants K, h_m, g.

Let us introduce another set of Dirichlet problems :

(16a) $\begin{cases} \alpha^2 h_o - (K\alpha + gh_m)\Delta h_o = W \\ h_o|_\Gamma = 0 \end{cases}$

(16b) $\begin{cases} \alpha \underset{\sim}{Q}_o - K\Delta \underset{\sim}{Q}_o = - gh_m \nabla h_o + \underset{\sim}{T} \\ \underset{\sim}{Q}_o|_\Gamma = \underset{\sim}{\emptyset} \end{cases}$

(16c) $\begin{cases} \alpha^2\psi_o - (K\alpha + gh_m)\Delta\psi_o = \nabla . \underset{\sim}{Q}_o + \alpha h_o - R \\ \psi_o|_\Gamma = 0 \end{cases}$

and let $B = \frac{\partial\psi_o}{\partial n}$

B obviously depends on time (through the right-hand sides W, $\underset{\sim}{T}$ and R) and on the boundary conditions $\underset{\sim}{\emptyset}$.

The algorithm rely on the two following propositions :

. <u>Proposition 1</u> : A is an isomorphism from $H^{-1/2}(\Gamma)$ onto $H^{1/2}(\Gamma)$ and the bilinear form a defined by :

$$a(i, j) = \int_\Gamma A\lambda i . \lambda j \, ds$$

is continuous, symmetric and positive definite.

. <u>Proposition 2</u> : The function λ of $H^{-1/2}(\Gamma)$ such that :

(17) $A\lambda = B$

is the trace of h, solution of system (11), (12), (13).

The proof for proposition 1 is similar to that given in [3] and [4] for the Stokes problem. Proposition 2 is established in [5].

From those two propositions, we can deduce for the discrete case the following method :

. Construct the matrix A_h approximating the operator A. This implies to solve (15a), (15b) and (15c) for every basis function λi defined on the boundary Γ (i.e. a number of times equal to the number of nodes on the boundary).

. Calculate the discrete right-hand side B_h of equation (17) by solving (16a), (16b) and (16c).

. Solve equation (17), the result of which is the trace of h on the boundary.

As a matter of fact, a variational formulation is given for all the problems (15) - (16) and (17) before discretization.

Further details can be found in [5].

3.3. Some computational aspects

Building the matrix A_h is of course time-consuming but it has to be done only once since A_h does not depend on time. It is a full matrix but its order is not too big (equal to the number of boundary nodes). In addition to A_h two other matrices have to be stored : one for the problems on h and and one for the problem on the components of $\underset{\sim}{Q}$. They are symmetric positive definite and sparse. An incomplete Choleski conjugate gradient method, well adapted to the solution of linear systems involving such matrices has been used. It enables an in-core computation with a fairly high number of nodes (see [4]).

3.4. Choice of the finite element

Unlike the Stokes problem where the pressure and velocity must have different discretizations, we can adopt for h and $\underset{\sim}{Q}$ the same degrees of freedom.

A conforming element, quadratic for both variables $\underset{\sim}{Q}$ and h was used. It was compared in a few preliminary tests to the Taylor-Hood element (quadratic for $\underset{\sim}{Q}$ and linear for h) and showed a slightly better accuracy without any loss in the performance of the code.

3.5. Other boundary conditions

The method described above can be straightforwardly extended to boundary conditions such as :

$$\begin{cases} \tilde{Q}\cdot\tilde{n} = \phi & \text{on } \Gamma \\ \dfrac{\partial \tilde{Q}}{\partial n} \cdot \tilde{\tau} = \pi & \text{on } \Gamma \end{cases}$$

where $\tilde{\tau}$ is the unit vector tangent to the boundary and ϕ, π are some known boundary data. This allows a modelling of boundary layers if necessary. However the equations on the components Q_i are coupled by such conditions and the size of the systems increases heavily.

IV - NUMERICAL RESULTS

The method was tested on the well-known example of Massachusetts Bay for which many measurements and computation results are available (see for instance |6|). In this first experiment, emphasis has been laid on the general behaviour of the method rather than on the comparison with actual data. Thus neither the friction and Coriolis forces, nor the varying bottom topography were simulated. The domain was approximated by a rectangle (fig. 1) divided into 192 triangular elements with 429 nodes for a piecewise quadratic approximation. Both components of the flux were taken to be zero on the land boundaries. On the ocean limit, the inflow $\tilde{Q}_b$ was assumed to be normal to the boundary and uniform, with a time variation of the form : $Q_b(t) = Q_0 \sin \frac{2\pi t}{T}$, where Q_0 is a constant and T is the tidal period. A time step of 360 s was chosen, so that $Dt = 2 \frac{DX}{c}$ (where DX is the mesh size and $c = \sqrt{gh_m} = 19$ m/s is the celerity), leading to a Courant number of 2 for propagation.

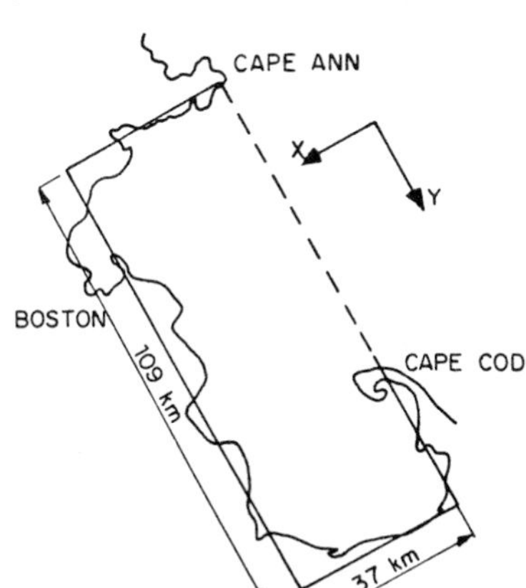

Figure 1
Domain approximation of Massachusetts Bay

Figure 2 below shows the velocity field when the inflow is maximum. On figure 3 are plotted the surface contour lines at different times of the tide. It can be noticed that the difference of surface elevation in the domain is maximum at high and low tide as expected. However this difference is lower than this actually observed, probably due to the absence of bed friction in the computation.

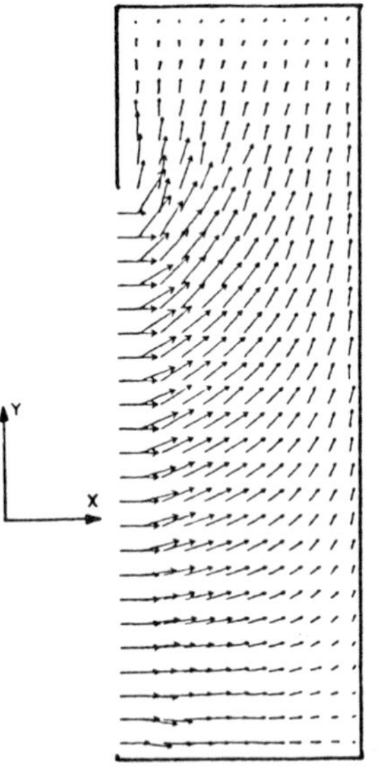

Figure 2
Velocity field
after 11 000 s
(0.25 tidal cycles)

A simple verification of the mass conservation can be made : integrating the continuity equation in the domain yields :

$$S \frac{d\bar{h}}{dt} = lQ_o \sin \frac{2\pi}{T} t$$

where S is the area of the domain, $\bar{h}$ the spatial average of h and l the length of the ocean boundary. It follows that :

$$\bar{h} = h_o + \frac{T}{2\pi} \frac{lQo}{S} \left(1 - \cos \frac{2\pi}{T} t\right)$$

where $A = \frac{T}{\pi} \frac{lQo}{S}$ is the amplitude of the tide. Figure 4 shows the history of the computed average surface elevation compared to the above law.

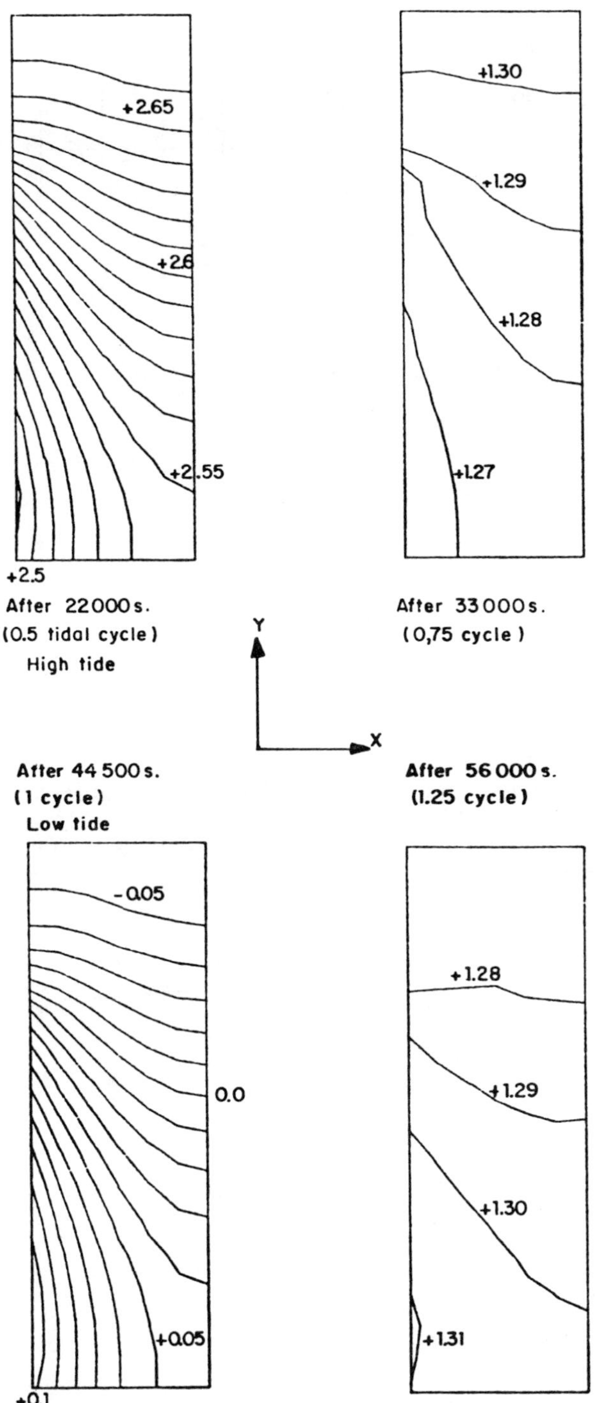

Fig 3 _ Surface contour lines _ The elevations are given above M.L.W.

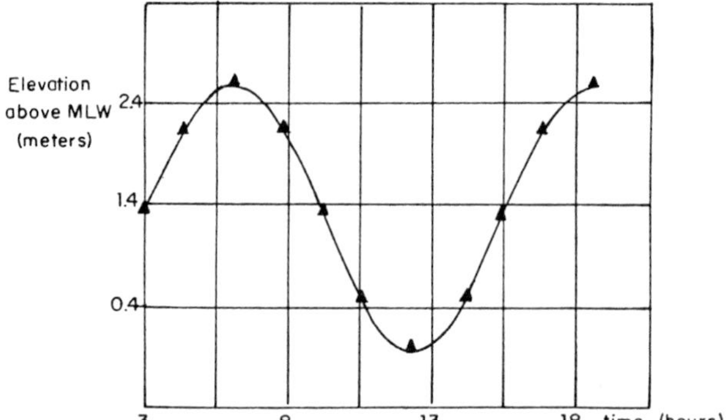

Figure 4 - Average surface elevation

▲ $\bar{h} = h_0 + \frac{A}{2}(1 - \cos \frac{2\pi}{T} t)$

〜 computed values of $\bar{h}$

REFERENCES

1. J.P. BENQUE, G. LABADIE, B. IBLER : A finite element method for Navier-Stokes equations - First International Conference on Numerical Methods for Non Linear Equations - Swansea U.K., Sept. 2nd-5th 1980.

2 M.O. BRISTEAU, R. GLOWINSKI, J. PERIAUX, P. PERRIER, O. PIRONNEAU and G. POIRIER : Application of optimal control and finite element methods to the calculation of transonic flows and incompressible viscous flows - Rapport de recherche 78-294, IRIA Laboria 78150 Le Chesnay France.

3 R. GLOWINSKI, O. PIRONNEAU : On a mixed finite element approximation of the Stokes problem (II). Solution of the approximate problem (to appear).

4 B. IBLER : Resolution des équations de Navier Stokes par une méthode d'éléments finis - Thèse de 3ème cycle - Université de Paris-Sud, January 1981.

5 G. LABADIE : Résolution des équations de Saint-Venant par une méthode d'éléments finis - Rapport E.D.F. (to appear).

6 J.J. CONNOR, C.A. BREBBIA : Finite element techniques for fluid flow - Newnes-Butterworth, 1976.

THE FINITE ELEMENT METHOD APPLIED TO TURBULENT OPEN CHANNEL FLOW

By: E.P.Querner [1]

ABSTRACT

The velocity distribution for steady flow in open channels with arbitrary geometry, is simulated by means of the finite element method. In the general equation of motion, the eddy viscosity accounts for the effect of turbulence on the mean flow. Secondary flows, which occur in open channels, are neglected in this analysis.

The boundary of the finite element model is situated at a prescribed distance from the wall outside the region where rapid variation of velocity occurs. Here the velocity distribution normal to the wall is determined by the well-known equation for two-dimensional flow.

To illustrate various features of the application, the mathematical model is applied to different channel cross-sections with known velocity distribution and boundary shear. The results of this analysis compare well with those obtained from field measurements.

1. INTRODUCTION

Numerous authors have attempted to give equations for a continuous universal velocity profile expression reaching all the way from the wall out into the turbulent flow region. These equations are mostly applicable to two-dimensional flow. They are less suitable when side walls have a pronounced effect on the velocity distribution, as in this study.

The prediction of the velocity distribution in a flowing stream requires knowledge of turbulence in every point of

[1] Hydraulic Engineer, DHV Consulting Engineers, Amersfoort, The Netherlands.

the section and of the boundary shear stress along the wetted perimeter. The effect of turbulence can be modelled by using the eddy viscosity concept. The average shear stress can be readily computed from the hydraulic radius and energy slope. The actual shear stress at any point on the boundary cannot be determined successfully by any known theoretical method, thus requiring an approximate solution. Because of the complexity of turbulent flow a numerical solution would be the only practical method, provided that the unknown parameters can be modelled by means of empirical relations. A numerical solution based on the variational principle for laminar flow in rectangular and square ducts has been reported by Delleur and Sooky [1]. Gerard [2] used the finite element method to analyse turbulent flow and secondary currents in non-circular conduits.

The modern finite element method, which is formulated in terms of a variational approach, is an extension of the classical Ritz or Galerkin method. Complex configurations can be assembled from relatively simple element shapes. The velocity and boundary shear stress distributions are modelled for an open channel with irregular boundaries. The model allows for non-uniform roughness along the wetted perimeter, if necessary. The secondary flows have been neglected in the analysis, as they are relatively small as compared with the primary flow.

2. THE EQUATION OF MOTION FOR TURBULENT FLOW

The method of variational techniques uses certain broad minimum principles and can be formulated in the following manner: the motion of an incompressible fluid satisfying the continuity equation is such that its functional is minimized. This principle was first stated by Helmholtz (see Lamb Ref. 6). He provided the governing function for steady flow. For steady turbulent flow this function can be written as:

$$\Phi_m = \int_A \mu_T \left[\left(\frac{\partial w}{\partial x}\right)^2 + \left(\frac{\partial w}{\partial y}\right)^2 \right] dA - 2 \int_A g\ S\ w\ dA \tag{1}$$

in which Φ_m is the excess of dissipation and μ_T denotes the turbulent kinematic viscosity (eddy viscosity). The first term in equation (1) can be interpreted as the total kinetic energy of the moving fluid, while the second term is related to the amount of work done by an impulsive pressure starting the motion from rest. Delleur and Sooky [1] showed that the minimized form of equation (1) is identical to the Navier-Stokes equation, with its inertia term neglected.

3. THE FINITE ELEMENT SOLUTION

The basic concept of the finite element method is that

the media can be considered as an assemblage of individual elements, which can be two-dimensional or three-dimensional. The elements are usually chosen to be either quadrilateral, or triangular in shape. They can be arranged in a variety of ways to represent exceedingly complex shapes.

Initially each element is considered independently, and the equations for each element are assembled. Provided that continuity of behaviour between adjacent elements can be assured, the solution of the equations should converge upon the correct numerical solution as the finite element mesh is refined.

A quadratic displacement (velocity) function is used to describe the steep velocity gradient near the wall. If this were not done, very small elements would be required when a linear displacement function would be chosen and this would not ensure a continuous variation of the velocity across element boundaries. The variation of velocity within the domain of the element can be represented by the polynomial:

$$w = \alpha_1 + \alpha_2 x + \alpha_3 y + \alpha_4 x^2 + \alpha_5 xy + \alpha_6 y^2 \qquad (2)$$

with the α parameters known as the generalized co-ordinates, which will specify or fix the magnitude of the prescribed distribution of the velocity. With this quadratic displacement function and a one degree of freedom per node, a 6 node triangular element is required, as shown in Figure 1.

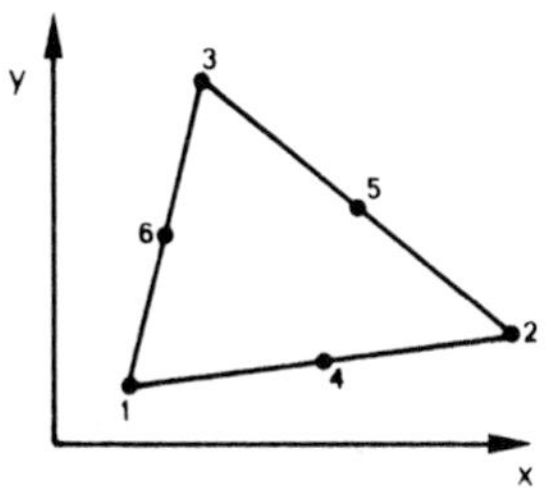

Figure 1 - Quadratic two-dimensional model.

Equation (2) can be written in matrix notation and evaluated for each of the six nodes, giving the expressions:

$$w = [A]\,\{\alpha_i\} \qquad (i = 1 \text{ to } 6) \qquad (3)$$

$$\{w\} = [B]\,\{\alpha_i\} \qquad (4)$$

The velocity components for the x and y direction can be written as:

$$\begin{bmatrix} \dfrac{\partial w}{\partial x} \\ \\ \dfrac{\partial w}{\partial y} \end{bmatrix} = [C]\,\{\alpha_i\} \qquad (5)$$

Substitute equations (3-5) in equation (1) and rearranged it becomes:

$$\Phi_m = \{w_i\}^T [K_e] \{w_i\} - 2\{F_e\}^T \{w_i\} \quad (6)$$

The matrix $[K_e]$ is the fluid element "stiffness". It expresses the relation between the nodal velocities and the associated forces. It is a function of the geometry and turbulence intensity (eddy viscosity) within the element. A linear variation of the eddy viscosity distribution is taken because it also increases rapidly in the region of a boundary.
The vector $\{F_e\}$ contains the external forces acting on the element.

For each element the functional defined by equation (6) must be minimized with respect to each of the undetermined parameters. They are represented by the values of the velocity at the nodes for a typical element. The influence of other elements on the general minimum equation allows an identical approach, and to obtain the general equations it is only necessary to add the contributions of all elements. Consequently,

$$\frac{\partial \Phi}{\partial w_i} = \sum_{m=1}^{N} \frac{\partial \Phi_m}{\partial w_i} \quad (7)$$

where N is the number of elements.

The change in the total value of the functional due to a variation in the value of the velocity at a node is given by the sum of the changes in the value of the elemental functional of the elements connected to this node. The final system of simultaneous equations is obtained by substituting equation (6) in equation (7). The result can be written as:

$$\frac{\partial \Phi}{\partial w} = [K_t] \{w\} - \{F_t\} = 0 \quad (8)$$

in which $[K_t]$ is the assembled fluid element stiffness matrix and $\{F_t\}$ the assembled load vector. The above expression establishes an equilibrium condition at each node.

Once the final matrix is assembled the boundary conditions must be applied so as to permit a solution of the simultaneous equations. The water surface and lines of symmetry are not constrained in any way and no special boundary conditions need to be specified along them.
The Crout reduction procedure was used to solve the simultaneous equations.

4. TURBULENT FLOW

There are two approaches to turbulent flow analysis: a) a statistical theory of turbulent correlation functions and b) a semi-empirical analysis of turbulent mean quantities. The mean-flow analysis can be used to predict the gross properties of a turbulent shear flow, such as mean velocity profiles. The semi-empirical relationships, which have been derived for turbulent flow, mostly use the eddy viscosity to account for the effect of turbulence. Various authors have shown the validity of the eddy viscosity concept and therefore it has been used in this solution technique.

4.1 Eddy Viscosity

The turbulent shear stress can be written as:

$$\tau = \mu_T \frac{\partial w}{\partial z} \qquad (9)$$

in which μ_T is the kinematic eddy viscosity and can be written as:

$$\mu_T = \rho\, \ell_p^2 \frac{\partial w}{\partial z} \qquad (10)$$

where the distance ℓ_p was defined by Prandtl as the mixing length.
Using the Prandtl mixing length hypothesis, Van Driest [13] derived an expression that reflects the effect of the wall in suppressing turbulent transport:

$$\mu_T = \nu \left[1 + \{1 + 4K^2 y^{+2} (1 - e^{-y^+/A^+})^2\}^{\frac{1}{2}}\right] /2 \qquad (11)$$

where:

$$y^+ = \frac{y\, u_*}{\nu} \qquad (12)$$

A^+ is a constant characterizing the thickness of the wall layer flow and K is Von Kármán's constant. For depths greater than 0.25 d, the eddy viscosity is assumed to be constant and has the value given at y = 0.25 d, where d is the total waterdepth at the point in the section considered. Equation (11) has been shown to be applicable to both a smooth and a rough wall.

Schlinger [11] gives an equation for the eddy viscosity expressed in terms of the Fanning friction factor and which is applied here to a rough wall. It can be written as:

$$\mu_T = K\, u_* \left(\frac{d - y}{d}\right) y \qquad (13)$$

Based on experimental results Schlinger limited the validity of the expression to the region where y is less than 0.7 d. For a rough wall, the eddy viscosity distribution, using equation (11) and equation (13) independently, correlated closely, resulting in a negligeable difference between the velocity distributions.

4.2 Flow near a wall

Turbulent flow near a solid wall is given by the well-known logarithmic velocity distribution:

$$\frac{w}{u_*} = 5.75 \log \frac{y}{k_s} + 8.5 \tag{14}$$

in which k_s is the boundary roughness height. The semi-empirical relation, valid for a rough wall, is often called the Von Kármán - Prandtl equation. Strictly it should apply only to the constant stress layer relatively close to the wall, but measurements have shown that it is valid for the entire boundary layer.

The above equation is used for a specified distance from the boundary, to match the wall conditions with the interior flow domain, where the finite element solution is applied. The distance should be taken so, that it will incorporate the area where rapid variation of the velocity occurs, but will be still close enough, for the flow to be considered two-dimensional. It also avoids a large number of elements in this region which would otherwise be required.

4.3 Boundary shear stress

The boundary shear stress in open channels is generally not uniformly distributed over the wetted perimeter. This is caused by the existence of a free surface, where the shear stress is negligeably small, and the cross-sectional shape. The average shear stress in terms of channel slope and hydraulic radius is given by:

$$\bar{\tau}_s = \rho \, g \, R \, S \tag{15}$$

with the shear velocity defined as:

$$u_* = \sqrt{\frac{\tau_s}{\rho}} \tag{16}$$

The boundary shear stress depends on the velocity, turbulence, and secondary currents. The complexity of this problem has been simplified by using the d'Arcy-Weisbach equation. Using this equation the shear velocity can be expressed as:

$$u_* = \bar{V} \sqrt{\frac{f}{8}} \tag{17}$$

in which $\bar{V}$ is the laterally averaged velocity and f is a boundary drag coefficient (Fanning friction factor). It can be determined from either Manning's or Chézy's equation:

$$\sqrt{\frac{f}{8}} = \frac{\sqrt{g}}{C} = \frac{n\sqrt{g}}{d^{1/6}} \tag{18}$$

in which C is Chézy's coefficient and n is Manning's coefficient. Substituting equation (18) in (17) it becomes:

$$u_* = \bar{V} \frac{\sqrt{g}}{C} \tag{19}$$

Chezy's equation is preferred, because of its scientific value. It is directly correlated to the boundary roughness height by the Colebrook-White equation:

$$C = 18 \log \frac{12\ R}{k_s} \tag{20}$$

After the shear velocity has been calculated along the wetted perimeter, it must be made equal to the average shear velocity specified for the cross-section (Eq. 15).

5. SOLUTION PROCEDURE

An iteration process is required, because the velocity distribution, eddy viscosity and the boundary shear are interrelated. The boundary shear velocity for all the boundary node points is assumed to be equal to the average shear velocity (Eq. 15 and 16) for the first iteration. The velocity is then calculated at the finite element boundary (Eq. 14) and used as the boundary condition for the calculation of the flow over the whole cross-section by the finite element method. The boundary shear is calculated (Eq. 19) for this velocity distribution and then modified to obtain the average value prescribed. These modified shear values are used in the next iteration.

6. EXAMPLES

To illustrate the method, the mathematical model is applied to different channel cross-sections with known velocity distribution.

First, for rectangular and trapezoidal sections the average velocity, boundary shear stress and the distance of the element grid from the wall were analysed. Table 1 gives the average velocity calculated by means of the finite

Table 1 - Summary of results for trapezoidal and rectangular sections.

Run	Section	Maximum Water depth (m)	Bedslope	Average Velocity (m/s)	
				Chézy Equation	Finite Element Method
3	Trapezoidal	0,55	1/400	1,51	1,48
4	"	0,55	1/200	2,14	2,10
5	"	0,55	1/100	3,02	2,96
6	"	1,05	1/400	2,30	2,31
7	"	1,05	1/200	3,25	3,26
8	"	1,05	1/100	4,60	4,61
9	"	1,55	1/400	2,95	2,98
10	"	1,55	1/200	4,17	4,21
11	"	1,55	1/100	5,90	5,95
12	"	2,05	1/400	3,52	3,54
13	"	2,05	1/200	4,98	5,01
14	"	2,05	1/100	7,04	7,08
15	Rectangular	1,0	1/400	2,20	2,18
16	"	1,0	1/200	3,11	3,08
17	"	1,0	1/100	4,40	4,36
18	"	1,5	1/400	2,93	2,92
19	"	1,5	1/200	4,13	4,13
20	"	1,5	1/100	5,85	5,84
21	"	2,0	1/400	3,41	3,44
22	"	2,0	1/200	4,82	4,86
23	"	2,0	1/100	6,82	6,87

element method and the Chézy equation. The average velocities agreed to better than 2% and for most sections better than 1%.

With the distance of the finite element grid from the wall being greater than the boundary roughness height (k_s), the velocity distribution as calculated by the finite method agreed satisfactorily with the actual values found by experiments.
The boundary shear stress for runs 7 and 16 was compared with the shear stress distribution given by Lane [7]. The results (Figure 2) show good correlation for the trapezoidal section and the invert of the rectangular section. In the corner of the rectangular section the difference can be attributed to secondary flows and the approximation technique used to calculate the shear stress.

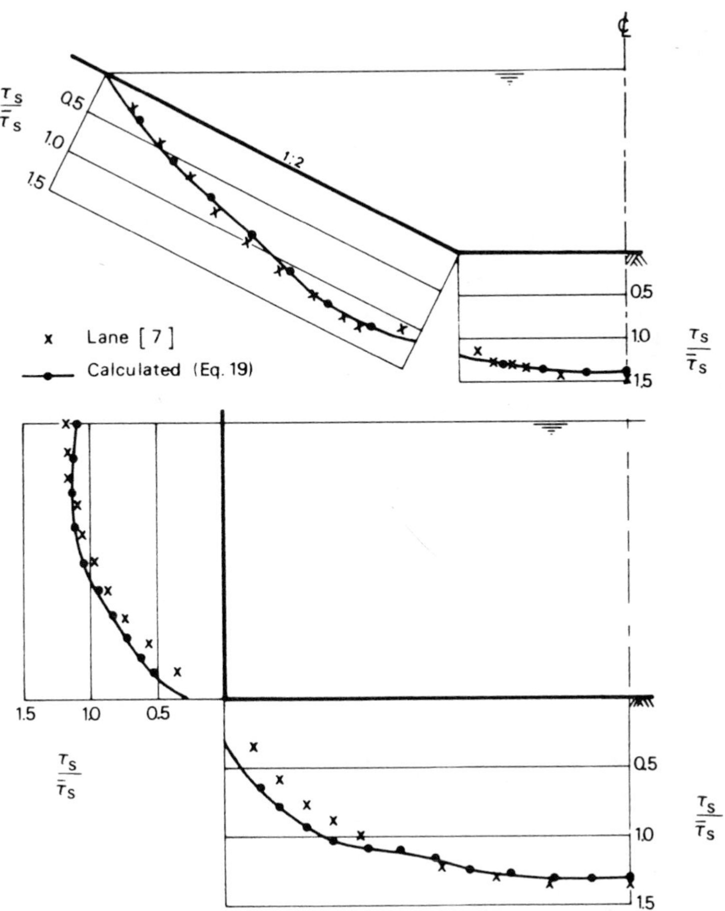

Figure 2 - Distribution of boundary shear stress.

For a rectangular and a parabolic section the measured and the calculated velocity distributions are shown in Figure 3. The presence of secondary flow causes the actual maximum velocity to occur below the water surface at mid-channel. The present method ignores the secondary flow, which results in the maximum velocity occurring at the water surface. With a small width to depth ratio, as in the case of the rectangular section, the difference in position of maximum velocity is appreciable. A difference in the magnitude of the velocity is also noticeable in the shallow region of the parabolic section. In other regions a reasonable correlation of results can be observed.

For a compound section, consisting of a concrete-lined low flow channel with grass overbanks, the velocity distribution from a hydraulic model was available [5]. With the

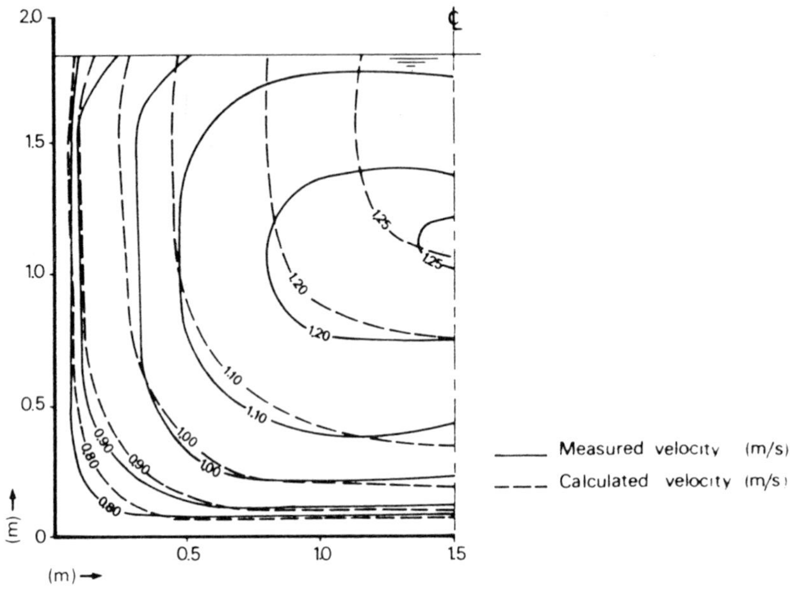

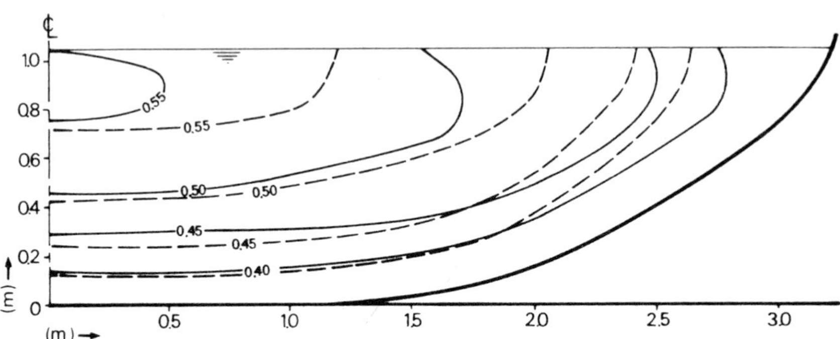

Figure 3 - Velocity distribution in rectangular and parabolic sections

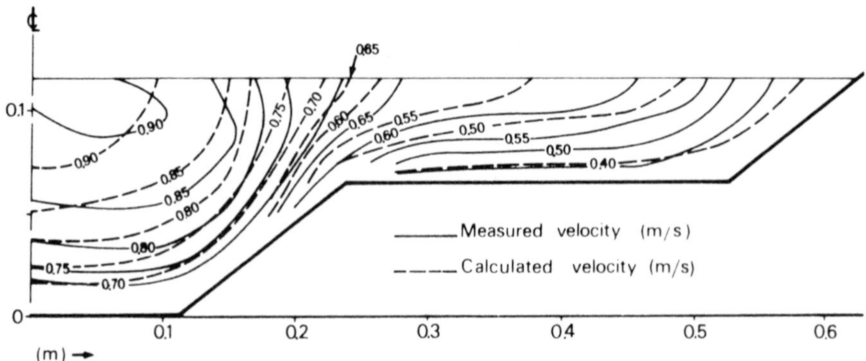

Figure 4 - Velocity distribution in composite section

finite element method the flow was modelled using the computed boundary roughness for the main channel and flood plain (see Figure 4).
The calculated velocity distribution near the boundary of the main channel is approximately 5% lower than the measured velocity. For the flood plain this becomes approximately 14%. The effect of linear-momentum flux into the overbank section resulting from eddies generated on the intersection of the two surfaces, results in a propulsive force on the overbank section. The present method partly takes these forces into account, but the effect of secondary flows contributes largely to the difference in velocity on the flood plain. In general a satisfactory agreement of the actual velocities for engineering practice can be obtained.

7. CONCLUSION

As shown, the finite element method is suitable for predicting the velocity distribution in open channels under steady condition. It can be assumed that the secondary flow largely contributes to the difference in measured and calculated velocities. Consequently, if secondary flow can be included in the method this imperfection will be overcome.

The method for estimating the boundary shear stress is subject to discussion. For sections with regular boundary conditions, such as a trapezoidal section, the method is suitable, but for complex boundary shapes the method is likely to lead to small errors.

8. ACKNOWLEDGEMENTS

The work reported herein was carried out while the author was a post-graduate student at the University of Cape Town, South Africa. The writer wishes to express his sincere gratitude to Professor W.S. Doyle and the staff of Ninham Shand Incorporated for their encouragement and assistance.

References

1. DELLEUR; J.W. and SOOKY, A.A. - Variational Methods in Fluid Dynamics, Proc. ASCE, Vol. 87, No. EM6, pp. 57-77, 1961.

2. GERARD, R. - Finite Element Solution for Flow in Non-circular Conduits, Proc. ASCE, Vol. 100, No. HY3, pp. 425-441, 1974.

3. HINZE, J.O. - Turbulence, McGraw-Hill, New York, 1959.

4. HOPKINS, D. - The Calculation of the Roughness for a 2,5 m Diameter Old Concrete Pipe (in Afrikaans), Trans. of the South African Inst. Civ. Eng., Vol. 20, No. 5, pp. 103-105, 1978.

5. HUISMAN, I.L. - Stream flow Analysis of the Canalised Elsieskraal River (in Afrikaans), Thesis submitted to the University of Stellenbosch in partial fulfilment of the requirements for the Bachelor of Science in Civil Engineering, 1978.

6. LAMB, H.C. - Hydrodynamics, 6th Edition, Cambridge University Press, Cambridge, 1932.

7. LANE, E.W. - Design of Stable Channels, Trans. Amer. Civ. Eng., Vol. 120, pp. 1234-1279, 1955.

8. MEEK, J.L. - Matrix Structural Analysis, McGraw-Hill, New York, 1971.

9. QUERNER, E.P. - Calculation of Velocity Distribution in Open Channels using the Finite Element Method, Thesis submitted to the University of Cape Town in partial fulfilment of the requirements for the Diploma in Engineering, 1978.

10. QUERNER, E.P. and DOYLE, W.S. - The Velocity Distribution in Open Channels, An Analysis by the Finite Element Method (Submitted to Trans. of the South African Inst. Civ. Eng., 1980).

11. SCHLINGER, W.G. et al - Temperature Gradients in Turbulent Gas Streams; Non-uniform Flow, Industr. and Eng. Chemistry, Vol. 45, pp. 662-664, 1953.

12. TAYLOR, C.A., HUGHES, T.G. and MORGAN, K. - A Numerical Analysis of Turbulent Flow in Pipes, Computers and Fluids, Vol. 5, pp. 191-203, 1977.

13. VAN DRIEST, E.R. - On Turbulent Flow near a Wall, J. Aero. Science, Vol. 23, pp. 1007-1011, 1956.

A NUMERICAL AND EXPERIMENTAL STUDY OF FREE SURFACE WAVE FLOW OVER A HALF-CYLINDER

E. W. Miner,[1] O. M. Griffin,[1] S. E. Ramberg,[1] and M. J. Fritts[2]

ABSTRACT

A research program is underway at NRL to develop new methods for the numerical calculation of surface wave effects on marine structures. Sufficient progress has been made with one particular computer code, SPLISH, to demonstrate the attractiveness of numerical calculations for problems such as the wave flow over a submerged obstacle or structure. A brief description is given in this paper of the numerical method and of some recent computational results. The present results demonstrate that realistic time-varying local flow fields, pressures and forces on and near structures such as a half-cylinder on the ocean floor can be determined from numerical calculations for certain conditions. Comparisons between the numerical results from SPLISH and recent linear and fifth-order solutions for wave flow over a bottom seated half-cylinder show good agreement as do comparisons between the numerical results and experiments in a wave channel.

INTRODUCTION

The effects of surface waves on marine structures are of great practical interest, but until quite recently the complexities of the problem have limited most theoretical calculations to idealized flows and structural geometries. In the absence of flow separation the complexity arises primarily from the presence of the free surface, which easily dominates the flow. This is particularly true if wave-breaking, surging or slamming and wave/current interactions are involved. The development of means to calculate the effects of these complex flows poses intractable problems not only analytically, but also numerically. Only the latest generation of computers has made practical calculations even remotely feasible.

While there are many approaches that might be taken for the numerical calculation of free-surface waves, e.g. the finite-element work of Bai and Yeung (1974) and Wellford (1978), and the marker-and-cell codes of Nichols and Hirt (1977), there are distinct advantages in using a technique which naturally tracks the wave flow field and which employs a grid by which the free-surface and structural boundaries can be readily defined. The non-orthogonal curvilinear coordinate systems

[1] Applied Mechanics Branch, Naval Research Laboratory, Washington, D.C. 20375, USA
[2] Laboratory for Computational Physics, Naval Research Laboratory, Washington, D.C. 20375, USA

developed by Shanks and Thompson (1977) can accommodate irregular boundaries without interpolation but they require transformation calculations at each time step.

A research program is underway at NRL to develop means for determining the effects of surface waves on submerged and partially submerged structures in the ocean. One aspect of this research program has centered upon calculations using the computer code SPLISH which was developed at NRL by Fritts and Boris (1975, 1977b). SPLISH couples the advantages of a Lagrangian approach to a triangular grid which can be restructured easily in a conservative, reversible manner to meet the needs of calculations of transient phenomena.

This paper presents a brief description of the numerical method upon which SPLISH is based. The numerical results obtained with the code are compared with linear and fifth-order wave theory results and the numerical and theoretical results are compared with experimental data. For one range of conditions the agreement is good between the present numerical results and both theory and experiment. However, the results obtained for another set of conditions characterized by large wave reflection from a submerged obstacle indicate the need for alternate boundary conditions in the code.

THE NUMERICAL METHOD

The governing equations are solved by a finite-difference method which was developed especially for use on a triangular grid which moves with the flow. The finite-difference method, the Lagrangian approach, and the triangular grid are discussed in depth by Fritts, Miner and Griffin (1980).

A Lagrangian grid will distort in any non-trivial flow field, and as grid distortion becomes severe the calculation quickly loses accuracy. However a triangular grid can be manipulated locally in several ways to extend realistic calculations of transient flows. Each interior grid line defines a quadrilateral diagonal. The opposite diagonal can be chosen whenever vertices move in the flow to positions which favor that connection. Such a reconnection involves just the four vertices describing the quadrilateral all of which remain stationary. No fluid moves relative to the quadrilateral, eliminating one form of numerical dissipation. Vertices can also be added or deleted to preserve the desired resolution by local algorithms which involve only those vertices in the vicinity of the grid anomaly. Major advantages of this technique are that the algorithms can be conservative, they permit a minimum of numerical dissipation and yet they require very little computer time since most of the grid remains unaltered.

The current version of the SPLISH code employs centered implicit integration to advance the physical variables in time and has been shown to be second-order accurate. Vorticity is conserved identically and the pressure is advanced to conserve the divergence about each vertex. These new pressures are then used to advance the velocities and update the grid positions.

The use of periodic boundary conditions restricts the calculational domain to an integral number of wavelengths, so that the domain represents one element of an array of identical domains. For many conditions of wave flow this yields a quite realistic situation and there is no concern for numerical damping at radiative boundaries. This restriction on the problem has in fact permitted considerable progress to be made in the numerical calculation of free-surface, progressive waves, even for cases with obstacles such as bottom-seated half-cylinders in the flow. The present findings show that when wave reflection from an obstacle is small, the periodicity of domains does not prevent good agreement between numerical results, experimental data and theoretical results.

The calculation is started by a sinusoidal pressure pulse on the free surface which initiates a standing wave. At the quarter period of the standing wave ($t = T/4$ sec) a second pulse (phase shifted a quarter wave length) is applied to form the progressive wave. This initialization procedure may not be the most appropriate, especially when an obstacle is in the flow field, but it is simple, easy to apply, and exhibits good numerical stability for standing waves (Fritts and Boris, 1977a) and for progressive waves in a straight channel. Also, it works well with the periodic boundary conditions and introduces minimal grid distortion and variation in the triangle areas.

NUMERICAL RESULTS AND COMPARISON WITH THEORY

Results from calculations for one principal set of conditions are presented and a comparison is made with results obtained from other models based on linear and fifth order wave theories. Although the numerical formulation is not limited to linear waves, most existing models are. Thus, the calculations were performed for small amplitude wave conditions to permit meaningful comparisons.

The triangular grid for the SPLISH code is illustrated in Figure 1 for a case of wave flow (from left to right) in a uniform depth channel. Some triangle reconnections have taken place as may be seen by the departures from the initial regular grid. Also, under the periodic boundary conditions, some triangles have been moved from the right side of the domain to the left side in order to keep the domain compact. For wave flow over a bottom seated half-cylinder, an initial automatic grid adjustment makes room for the half-cylinder by shifting upward the triangle vertices above it.

Calculations with SPLISH for uniform depth cases have given good agreement with classical wave theory for the wave period. Previous calculations with SPLISH (Fritts and Boris, 1977) had given wave period values for a standing wave which agreed very well with theory. In fact, the period converged to the theoretical value as the grid step size was reduced.

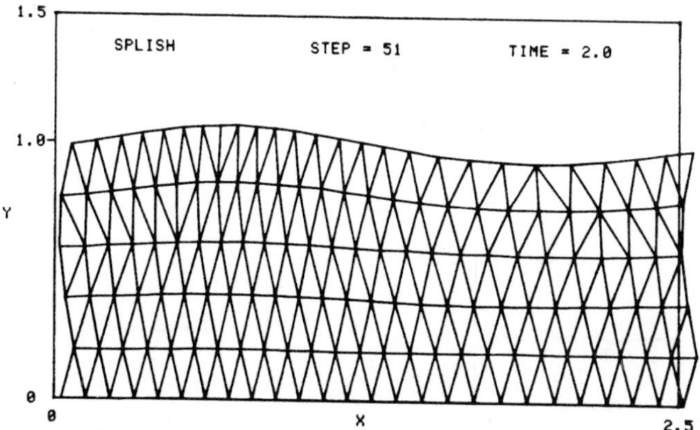

Figure 1 — An example of a SPLISH-generated computational grid

Low Reflection Wave Flow over a Half-Cylinder

The configuration which we consider is the wave flow over a bottom seated half-cylinder, which might correspond to a half buried pipe or a storage tank on the ocean floor. However, with periodic boundary conditions the numerical simulation actually corresponds to an array of pipes which are spaced one or more wavelengths apart. Fortunately, there are combinations of wavelength, depth, and cylinder radius for which only a very small amount of the incident wave is reflected by the half-cylinder. Conditions were chosen so that with the assumption of linear theory, the reflection coefficient $R = 0.03$ (Naftzger and Chakrabarti, 1979). With a reflected wave having only 3% of the amplitude of the incident wave (and thus less than 1% of the energy), neighboring half-cylinders should have negligible effect on the flow. A series of calculations was made for these low reflection conditions.

Figure 2 shows the free surface contour and the resultant bottom pressure P_b at one instant in the passage of a wave (from left to right) over a half-cylinder (radius $a = 0.5$ m) in a channel (depth $d = 1$ m) for a wave length $\lambda = 2.5$ m. The finite amplitude of the wave ($H = 0.044$ m, from the undisturbed free surface) causes only a very small deviation from a sinusoidal surface shape in spite of the presence of the bottom seated half-cylinder. Note that the pressure scale increases downward. Since the calculation was started from still water, the pressures at step 1 correspond to the hydrostatic pressure on the channel bottom and on the surface of the half-cylinder. At step 25 the wave gives an increased pressure on the left side of the half-cylinder and a reduced pressure on the right side. This is responsible for a net force (at this instant) from left to right. At other steps in the calculation, the pressure fluctuation accurately follows the passage of peaks and troughs of the wave as will be shown.

There is also a depth dependence of the pressure fluctuation. As the crest and the trough of the wave pass, the pressure fluctuation on top of the half cylinder (at a depth of 0.5 m) is approximately twice that on the channel bottom (at a depth of 1.0 m). At intermediate depths (i.e. along the sides of the cylinder) the magnitude of the pressure fluctuation is between that on the channel bottom and that on the top of the cylinder.

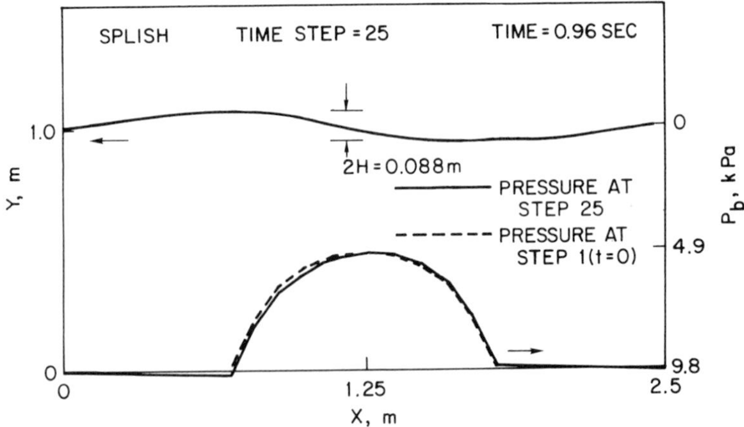

Figure 2 — The surface contour and the bottom pressure from a numerical calculation of the wave flow over a bottom seated half-cylinder

In Figures 3 and 4 the fluctuation in P_b, i.e. ΔP, is normalized by its maximum value (ΔP_{REF}) which occurs at the top of the cylinder as the wave crest passes by. Figure 3 shows the pressure distribution around the half-cylinder at times $t = 1.28$ and $t = 1.92$ sec which are the instants when the wave crest and trough pass over the half-cylinder. Figure 4 shows the pressure distributions around the half-cylinder at $t = 0.96$ sec and at $t = 1.64$ sec which correspond to times when the wave crest is first to the left of the cylinder and then to the right of the cylinder. In these two figures the values of ΔP from the SPLISH numerical calculations are shown by the open symbols. The solid and dashed curves shown in these figure are the values of ΔP obtained for these conditions using the approach of Chakrabarti and Naftzger (1974) which was based on Stokes' fifth-order wave theory and the assumption that the effect of the free-surface on the reflected wave potential could be neglected. For the present low-reflection conditions (with a reflection coefficient of 3%) such an assumption is reasonable. For these conditions the SPLISH results are in good agreement with the fifth-order theory results.

The time history for the pressure (in kiloPascals or kN/m^2) at a vertex on the top of the half-cylinder is shown in Figure 5. The solid curve shows the SPLISH data and the open symbols are values obtained using a full fifth-order solution obtained by Chakrabarti and Naftzger (1974) for wave flow over a bottom seated half-cylinder. The pressure here represents the actual gauge pressure that might be measured. The wave amplitude was assumed to be $H = 0.038$ m for the fifth-order calculations, and the agreement between the numerical calculations and the fifth-order calculation is quite good. With the finite grid size employed here the period given by the numerical results is about 4% greater than that from the theoretical dispersion relation, so that the theoretical and numerical results shift slightly relative to each other with time. From fifth-order theory $T = 1.269$ sec and from linear theory $T = 1.274$, while SPLISH gives here a value of $T = 1.30$ sec based on the intervals between maxima or between minima in the pressure-time history in Figure 5.

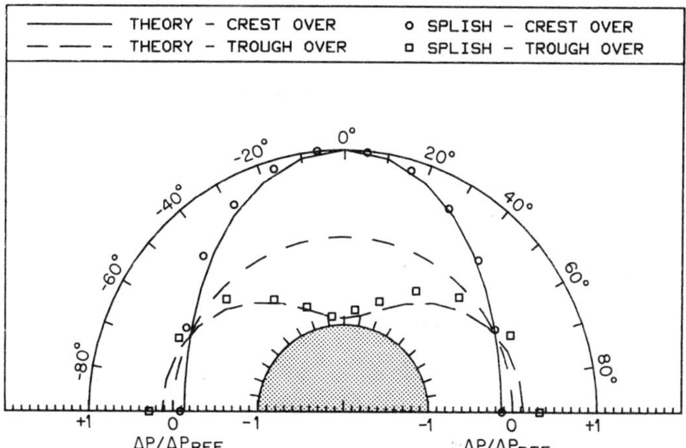

Figure 3 — A comparison of the distribution of pressure fluctuations around the half-cylinder from fifth-order wave theory and from a SPLISH calculation, for the wave crest and trough passage (wavelength $\lambda = 2.5$ m, depth $d = 1$ m, radius $a = 0.5$ m)

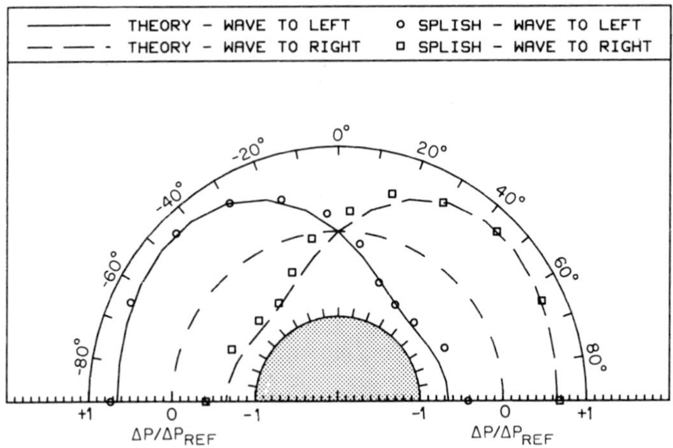

Figure 4 — A comparison of the distribution of pressure fluctuations around the half-cylinder from fifth-order wave theory and from a SPLISH calculation, for the wave crest to the left and the right of the half-cylinder (wavelength $\lambda = 2.5$ m, depth $d = 1$ m, radius $a = 0.5$ m)

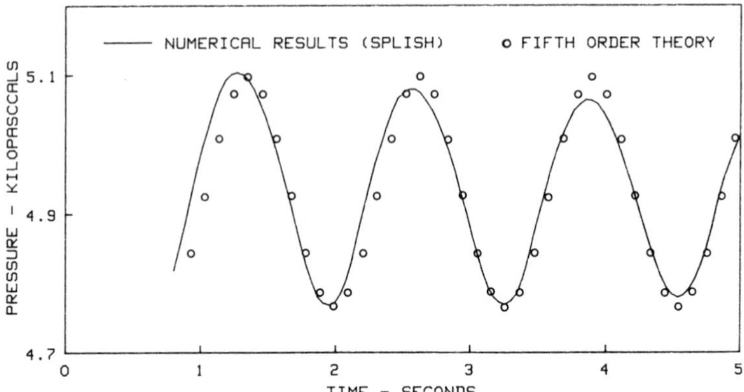

Figure 5 — A comparison of the pressure time history at the top of a bottom seated half-cylinder from a SPLISH calculation and from fifth-order wave theory (wavelength $\lambda = 2.5$ m, depth $d = 1$ m, radius $a = 0.5$ m, wave amplitude $H = 0.038$ m)

The forces and pressures on the half-cylinder are the principal results of the numerical calculations, but at times the necessary information about the quality of the numerical solution is not given by the pressure and force data. Plots of the Lagrangian particle paths have proven to be a very useful diagnostic aid for determining the quality of the solution and for examining the physical mechanism in some wave flow situations. Figure 6 shows the Lagrangian paths of three surface vertices, in Figure 6a the vertex initially at $x = 0$ m, in 6b the vertex at $x = 0.6$ m, and in 6c the vertex at $x = 1.2$ m. The particle paths begin at step 11 after the standing wave is fully formed and the second pressure pulse has been applied. The presence of the half-cylinder in the flow seems to affect the path of the surface particle above it only slightly. A larger effect would be seen for a larger wave amplitude or for vertices nearer to the cylinder. The particle paths in Figure 6 also exhi-

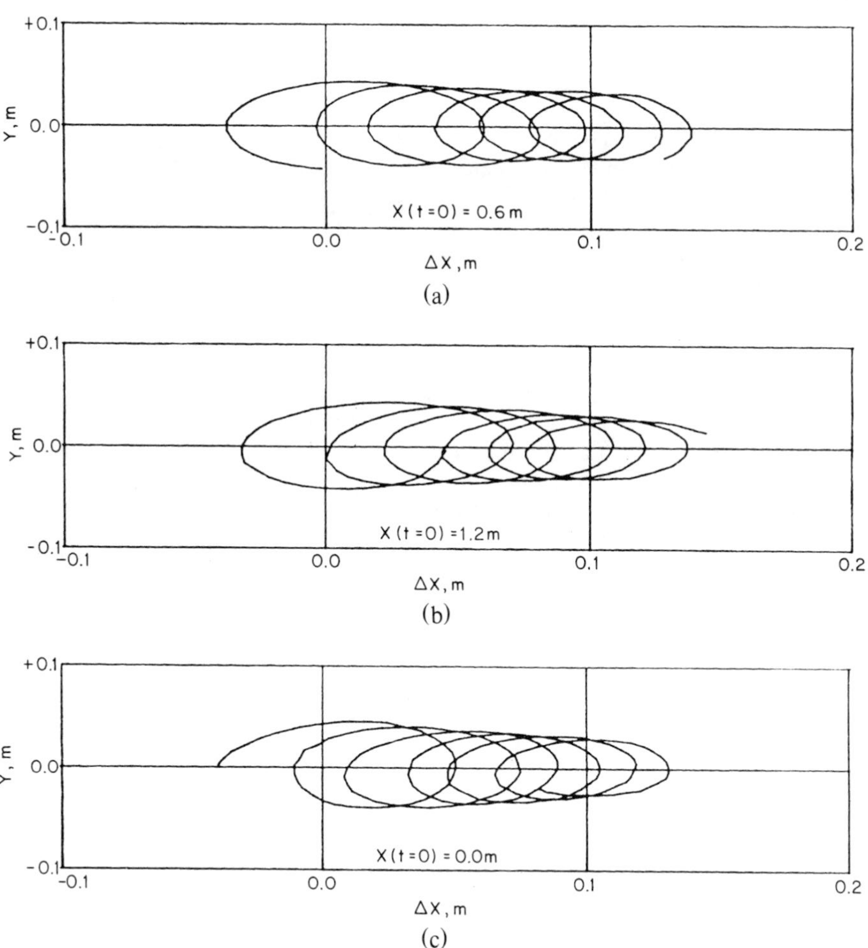

Figure 6 — Particle paths on the free surface from a numerical calculation of wave flow over a bottom seated half-cylinder (wavelength $\lambda = 2.5$ m, depth $d = 1$ m, radius $a = 0.5$ m)

bit a slight decrease in amplitude with time. This decrease in wave amplitude (and thus wave energy) is not due to viscous effects in the governing equations, but it is caused by an effective dissipation of energy due to the incomplete convergence of the successive-over-relaxation (SOR) method used for the solution of the Poisson equation for the pressure.

Even though there is some loss of energy in the numerical solution (due to incomplete convergence of the SOR method mentioned above), the Lagrangian particle paths clearly show that, for these low wave reflection conditions, SPLISH is generating a stable, well-behaved solution for the surface wave flow over the half-cylinder. The particle paths seem to indicate that the wave reflection from the half-cylinder is sufficiently low that it neither adversely affects the quality of the solution nor significantly changes the physical situation which is being modeled numerically. We consider in the next section a case in which the the amount of wave reflection is significant.

High Reflection Wave Flow over a Half-Cylinder

We also have performed numerical calculations with SPLISH for a case with significant wave reflection. For this case the depth $d = 1.0$ m, the radius $a = 0.667$ m and the wavelength $\lambda = 8.4$ m. From the results of Naftzger and Chakrabarti (1979) the wave reflection coefficient $R = 0.4$ and the reflected wave has 16% of the energy of the incident wave.

Surprisingly, some of the results from the SPLISH calculation show little adverse effect from the wave reflection. For example, the time history of the pressure at a point on top of the half-cylinder, shown in Figure 7, agrees quite well with fifth order wave theory calculations (Chakrabarti and Naftzger, 1974). However, other numerical results for these conditions (not shown here) agree rather poorly with the theoretical and experimental results. This disagreement appears to arise from two factors. The first is due to a mismatch in timing for this case of the second pressure pulse initializing the calculation. The second is that boundary conditions other than the periodic ones used are needed for cases with significant wave reflection. The periodic boundary conditions allow the reflected wave to reenter the calculation instead of radiating outward from the obstacle. For high wave reflection, the disturbance to the flow field is significant and the intended physical situation is not properly modeled numerically. In fact it is somewhat puzzling that the pressure-time history results shown in Figure 7 agree as well as they do.

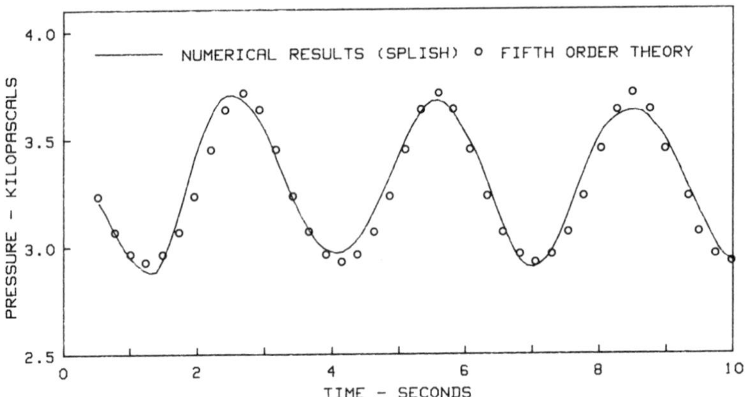

Figure 7 — A comparison of the pressure-time history at the top of a bottom seated half-cylinder from a SPLISH calculation and fifth-order wave theory calculation for high reflection wave flow (wave length $\lambda = 8.4$ m, depth $d = 1$ m, radius $a = 0.667$ m wave amplitude $H = 0.038$ m)

In the above discussion we have only compared the results obtained with SPLISH with results from classical linear and fifth-order wave theories. Even though most of these comparisons showed good to very good agreement, it is also appropriate to compare the calculations with experimental data. This is done in the next section.

COMPARISONS WITH EXPERIMENTAL DATA

Several wave flow experiments were performed in the NRL wave channel with a bottom seated half-cylinder so that actual pressure measurements could be compared both to the numerical results and to the models developed by Chakrabarti and Naftzger (1974, 1979). A 1.07 m (3-1/2 ft) diameter cylinder, which spanned the entire width of the channel, was placed about one-half of the channel's length from the mechanical wavemaker. At the other end of the wave channel a sloping, porous beach with a rubberized horsehair blanket served to absorb nearly all of the incident wave energy. Nineteen equally-spaced ($\Delta\theta = 10°$) pressure taps were located around the circumference of the half-cylinder at its midsection. The individual taps were connected to a differential pressure transducer by a rotary pressure switch. The wave height along the channel was obtained from several traversing capacitance-type wave gauges. Calibrations were performed on all sensors before and after each test series to insure that the overall accuracy of the measurement system remained well within ±5 percent. The pressure and waveheight signals were digitized and processed by means of a Hewlett-Packard 5420A Digital Signal Analyzer. The experimental systems and methods are discussed further in a related NRL report (Miner, Griffin, Ramberg, and Fritts, 1980).

The cylinder reflection coefficient R, defined as the ratio of reflected to incident wave amplitudes, was obtained for a range of wavelengths and water depths. Reflection coefficients were obtained using a modified form of the method, introduced by Ursell, Dean and Yu (1959), based upon modulations of the waveheight along the channel. The results of these measurements are plotted in Figure 8 against the wavenumber $ka = 2\pi a/\lambda$ and for several values of the relative water depth d/a. Also shown in the figure (as solid lines) are theoretical values based on linear theory which were calculated by Naftzger and Chakrabarti (1979). There is general agreement with the observed cylinder reflections.

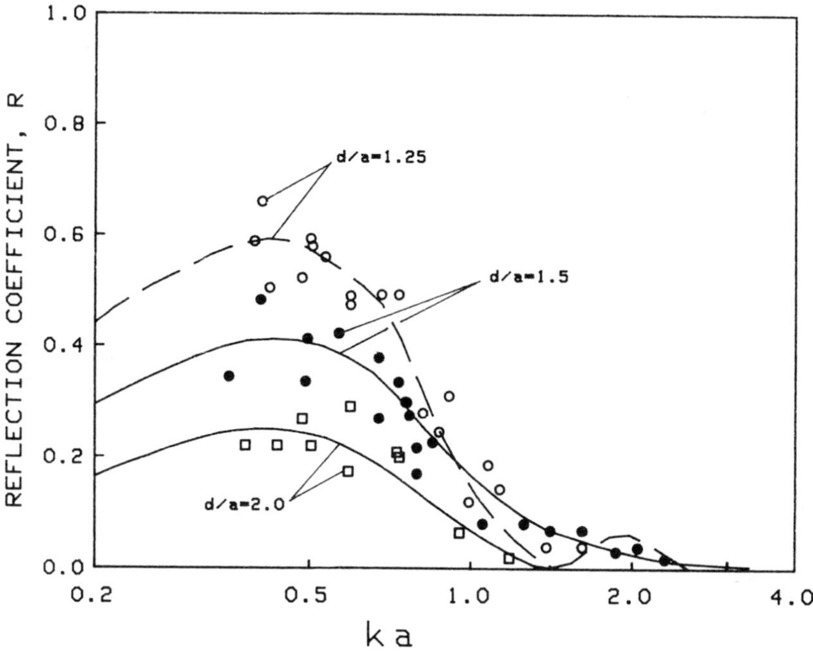

Figure 8. Experimental (•) and theoretical (—) values for the wave reflection coefficient of a bottom-seated half cylinder. Theoretical values were calculated using linear wave theory by Naftzger and Chakrabarti (1979).

It should be noted, however, that some care was taken to avoid finite amplitude effects through the use of small wave steepness ratios ($H/\lambda \lesssim 0.05$). The wave steepness values during the $d/a = 1.25$ tests were reduced further, typically to less than 0.02, in order to avoid second harmonic wave generation at the cylinder. This nonlinear effect seemed to be associated with the finite waveheight being a significant fraction of the finite water depth at the top of the cylinder. The second harmonic wave could be seen as the formation of a secondary crest when a trough was over the cylinder.

A rule of thumb emerged from this observation and from several wave gauge spectral records wherein second harmonic amplification did not occur for waveheights less than about 1/7 to 1/10 of the water depth over the cylinder. This criterion for the validity of the analytical method is considerably more stringent than the one originally proposed by Naftzger and Chakrabarti. The onset and form of this nonlinearity will be the subject of a further investigation in the near future.

From the range of water depths and wavelengths shown in Figure 8, two cases were selected for detailed pressure studies and for comparisons with the theoretical and numerical results. The first is a relatively low reflection case ($d/a = 2.0$ and $ka = 1.25$), so that $R \lesssim 0.05$. This represents a situation where good agreement between the three sets of results was expected. The second case is a relatively high reflection condition ($d/a = 1.5$, $ka = 0.5$, $R \approx 0.4$) which was selected as a significant test for SPLISH with regard to the use of periodic boundary conditions in the code.

Low Reflection Case

A comparison between the experimental results and a linear approximation to the fifth order model (Chakrabarti and Naftzger, 1974) was obtained. This approximation was selected for the comparison because of its simplicity and relative accuracy. The theoretical and experimental results agreed well except near the intersection of the half-cylinder sides and the wave channel floor. These differences could be a consequence of a small gap which existed between the bottom of the half-cylinder and the wave channel.

Since the numerical results are Lagrangian whereas the theoretical and experimental results are Eulerian, the simplest format for comparing all three is the pressure distribution around the cylinder at selected times in the wave cycle. Figures 9 and 10 present several such comparisons for the experimental conditions cited above. The three sets of results compare well except near the bottom of the cylinder. Although the discrepancy appears to be small in such a plot the effect on a measured horizontal component of the wave force can be significant.

High Reflection Case

Similar results and comparisons for a relatively high reflection case ($d/a = 1.5$, $ka = 0.5$, $R \approx 0.4$) are shown in Figures 11 and 12. As before, the pressures are normalized by the magnitude at the top of the cylinder. In this case, however, the maximum pressure fluctuation occurs on the upstream side of the cylinder where both the incident and reflected waves are present. The experimental pressures shown in the figures were obtained from the data after a correction for the additional standing waves in the laboratory channel. This correction, often large, neglected secondary reflections and was based on linear wave theory which

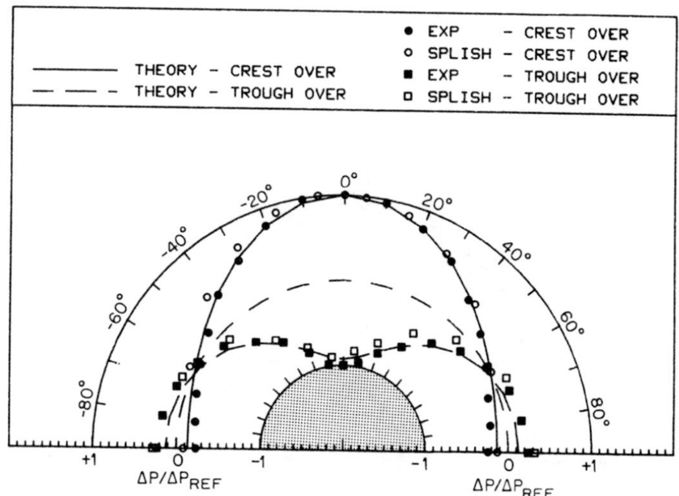

Figure 9. A comparison of theoretical (Chakrabarti and Naftzger, 1974), numerical, and experimental results for selected instantaneous pressure distributions on a bottom-seated half-cylinder in waves, for wave crest and trough directly over the cylinder (low reflection wave flow, $d/a = 2.0$, $ka = 1.25$, $R \lesssim 0.05$)

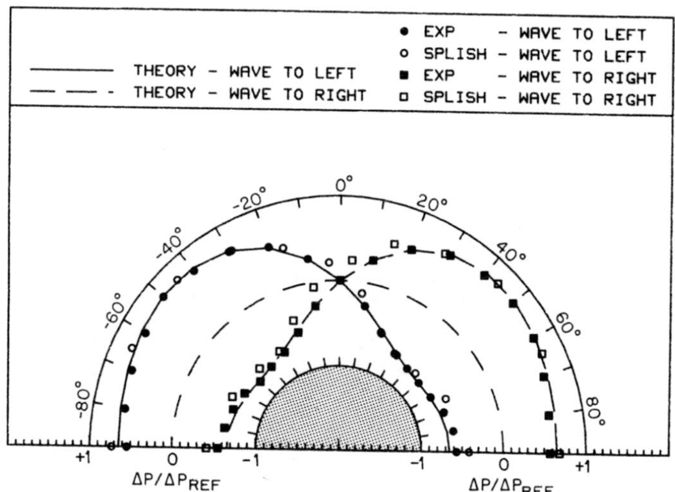

Figure 10. A comparison of theoretical (Chakrabarti and Naftzger, 1974), numerical, and experimental results for selected instantaneous pressure distributions on a bottom-seated half-cylinder in waves, for wave crest to the left and the right of the half-cylinder (low reflection wave flow, $d/a = 2.0$, $ka = 1.25$, $R \lesssim 0.05$)

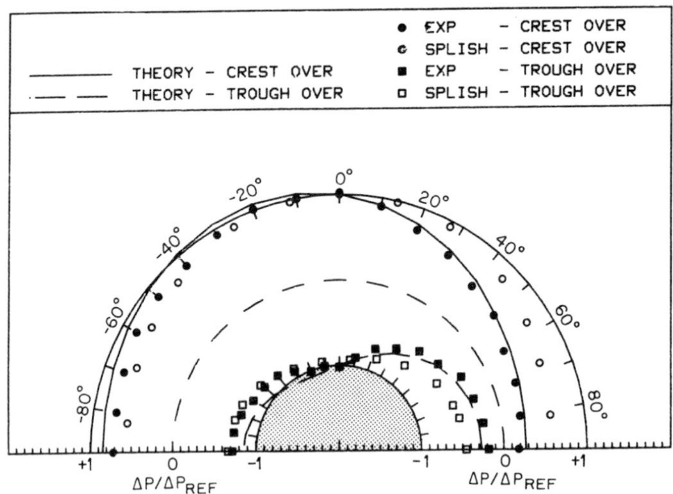

Figure 11. A comparison of theoretical (Naftzger and Chakrabarti, 1979), numerical, and experimental results for selected instantaneous pressure distributions on a bottom-seated half-cylinder in waves, for wave crest and trough directly over the cylinder (high reflection wave flow, $d/a = 1.5$, $ka = 0.5$, $R \approx 0.4$)

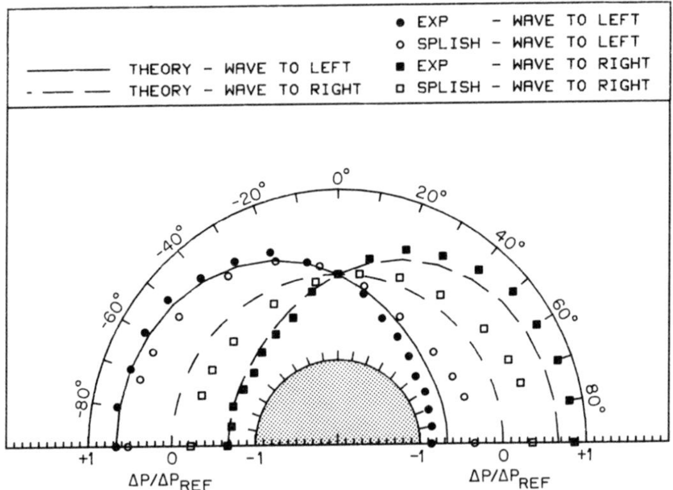

Figure 12. A comparison of theoretical (Naftzger and Chakrabarti, 1979), numerical, and experimental results for selected instantaneous pressure distributions on a bottom-seated half-cylinder in waves, for wave crest to the left and the right of the half-cylinder ($d/a = 1.5$, $ka = 0.5$, $R \approx 0.4$)

may be responsible for all or part of the discrepancies between results. The computed results shown for the method developed by Naftzger and Chakrabarti were provided by them (R.A. Naftzger, private communication, 1980).

In spite of the difficulties implied above, the agreement between the linear theory and the experiments is relatively good. The previously mentioned restriction on wave amplitude ought to be reiterated. The SPLISH results show the effect of the periodic boundary conditions and perhaps also of the improper wave initialization. There is less satisfactory agreement between the numerical calculations and both the experimental data and the theoretical results.

SUMMARY AND CONCLUSIONS

A finite-difference numerical method for solving the governing equations of motion for inviscid, irrotational flow with a free surface using a Lagrangian triangular grid has been used and shown to yield reasonable results. Calculations for progressive surface wave flows have given results for the wave period, the drift velocity and the surface particle movements which are in good agreement with results obtained from classical wave theories.

Calculations for the passage of waves over a submerged obstacle are encouraging and show some promise of providing practical results over a relatively wide range of wave conditions. These calculations demonstrate the adaptability which the triangular grid provides. The advantages of the Lagrangian formulation are shown in that the grid conforms to the fluid area and that no interpolation is needed to locate the free surface or the surface of the submerged obstacle.

Two cases of flow over a submerged, half-cylindrical obstacle have been considered. The results for the low wave reflection case indicate that, even with the periodic boundary conditions, a code such as SPLISH can be employed with reasonable confidence to calculate the flow of waves over obstacles (thus wave-structure interactions). However, for cases with significant wave reflection from an obstacle, continuative or radiative boundary conditions are necessary, since the periodic boundary conditions lead to the simulation of a flow quite different from that desired. Also, alternative techniques for initiating the traveling surface wave seem to be required.

While additional work on non-periodic boundary conditions, initial conditions and gridding of the flow field clearly is necessary, the present numerical method is a promising computational tool for predicting the velocities and pressure fields produced by nonlinear, finite-amplitude regular wave motions. It is hoped that future developments will extend the capabilities of codes such as SPLISH to transient phenomena such as wave breaking and wave slamming and that significant reductions in computing times and costs can be achieved.

ACKNOWLEDGMENTS

This paper has been prepared as part of the research program of the Naval Research Laboratory. The authors wish to thank S.K. Chakrabarti and R.A. Naftzger of the Chicago Bridge and Iron Company for providing detailed computational results for several of the test cases discussed in this report.

REFERENCES

1. Bai, K. J. and R. W. Yeung (1974). "Numerical Solutions to Free-Surface Flow Problems," *Proceedings of the Tenth Symposium on Naval Hydrodynamics,* MIT, Cambridge, Mass., 609-647.

2. Boris, J. P., K. L. Hain, and M. J. Fritts (1975), "Free Surface Hydrodynamics Using a Lagrangian Triangular Grid," *Proceedings of the First Int'l Conference on Numerical Ship Hydrodynamics,* NBS, Gaithersburg, MD., pp. 683-716.

3. Chakrabarti, S. K. and R. A. Naftzger (1974), "Nonlinear Wave Forces on Half Cylinder and Hemisphere," *Proceedings of the ASCE, Journal of the Waterways, Harbors and Coastal Engineering Division,* Vol. 100, 189-204.

4. Fritts, M. J. and J. P. Boris (1977a), "Solution of Transient Problems in Free-Surface Hydrodynamics," NRL Memorandum Report 3446.

5. Fritts, M. J. and J. P. Boris (1977b), "Transient Free Surface Hydrodynamics," *Proceedings of the Second International Conference on Numerical Ship Hydrodynamics,* Berkeley, CA., pp. 319-328.

6. Fritts, M. J., E. W. Miner and O. M. Griffin (1980), "Numerical Calculation of Wave-Structure Interactions," *Computer Methods in Fluids,* Pentech Press; London, pp. 1-25.

7. Miner, E. W., M. J. Fritts and O. M. Griffin (1978), "A finite-difference method for calculating free surface waves," *Proceedings of the First Int'l Conference on Numerical Methods in Laminar and Turbulence Flow,* Swansea, U. K., pp. 597-608.

8. Miner, E. W., O. M. Griffin, S. E. Ramberg and M. J. Fritts (1980), "Numerical Calculation of Surface Wave Effects on Marine Structures," Naval Research Laboratory Memorandum Report 4395.

9. Naftzger, R. A. and S. K. Chakrabarti (1979), "Scattering of Waves by Two-Dimensional Obstacles," *Journal of Ship Research,* Vol. 23, 32-42.

10. Nichols, B. D. and C. W. Hirt (1977), "Nonlinear Hydrodynamic Forces on Floating Bodies," *Proceedings of the Second Int'l Conference on Numerical Ship Hydrodynamics,* Berkeley, CA. pp. 382-394.

11. Shanks, S. P. and J. F. Thompson (1977), "Numerical Solutions of the Navier-Stokes Equations for 2D Hydrofoils in or below a Free Surface," *Proceedings of the Second Int'l Conference on Numerical Ship Hydrodynamics,* Berkeley, CA., pp. 202-220.

12. Ursell, F., R. G. Dean and Y. S. Yu (1959), "Forced small-amplitude water waves: a comparison of theory and experiment," *Journal of Fluid Mechanics,* Vol. 7, 33-52.

13. Wellford, L. C., Jr. (1978), "Calculation of Free Surface Hydrodynamics Problems Using a Finite Element Method with a Hybrid Lagrange Line," *Proceedings of the First Int'l Conference on Numerical Methods in Laminar and Turbulent Flow,* Swansea, U. K., pp. 995-1006.

STRATIFIED FLOW MODELS FOR LAKE ERIE

D.C.L. Lam

National Water Research Institute, Canada Centre for Inland Waters, Burlington, Ontario, Canada L7R 4A6

ABSTRACT

Two types of thermal stratification models for Lake Erie are discussed. The one-dimensional (z) model simulates the seasonal changes in the temperature profile in a vertical water column. The two-dimensional (x-z) model computes the temperature distribution and flow patterns in a vertical plane along the length of the lake. The wind mixing, thermal buoyancy, earth's rotational effects, internal wave oscillations and bottom turbulence have been parameterized and used in the formulation of the eddy diffusivities and viscosities. The computed results and some preliminary applications on water quality problems are discussed with reference to the observational data collected in the lake environment.

1 INTRODUCTION

Lake Erie is the shallowest lake in the Great Lakes system in North America. It can be divided into three basins: Western, Central, and Eastern, according to the mean depths (10 m, 25 m and 60 m, respectively). During the summer period, thermal stratification strongly influences the circulations and the distribution of water quality parameters in the lake. Because of the topographical differences, the magnitude and direction of the transports across these basins are sensitive to the changes in the heat input and environmental turbulence.

As part of the environmental simulation program at the National Water Research Institute (NWRI), we have developed a hierachy of models ranging from the mass balance box models to the time-dependent three-dimensional models [1,2]. One of the main objectives of the program is to examine the long term predictive capability of these models by verifying them against the observational data collected at NWRI over the past decade. For the simple mass balance models, the physical and biochemical processes are often parameterized by lumped coefficients. As the complexity of models increases, more details of the processes can be incorporated. However, there will be an increase in computational time and storage associated with the increase in model complexity. In terms of the objective of verifying the models against long term data, the cost of running the complex ones such as the fully three-dimensional models can be prohibitive and the computed information can be copious. In addition to the efficiency problem, there is also the aspect of treating data collected from the lake environment. In contrast to the laboratory situations, environmental experiments are more subjected to weather elements and limited instrumental resources (e.g., ship times). As a result, there are inevitably temporal and spatial gaps in the data so that the fine grid details in the model results are difficult to be verified.

Because of these restrictions in environmental modelling, emphasis has been placed on models with intermediate complexity. Specifically, it has been found that one-dimensional and two-dimensional models, which retain the essential features of the physical processes, are quite efficient to run, particularly with long term data. In this paper, we give examples of the one- and two-dimensional models for the stratified flows in Lake Erie and some of their applications.

2 ONE-DIMENSIONAL (z) MODEL

A conventional one-dimensional thermocline model describing the vertical temperature structure in a water column is generally given by the following equation:

$$\frac{\partial T}{\partial t} = \frac{1}{A}\frac{\partial}{\partial z}\left(AK_v\frac{\partial T}{\partial z}\right) + \frac{q}{\rho C_p} \qquad (1)$$

where T is temperature; t is time; z is depth, A is cross-sectional area; K_v is vertical turbulent thermal diffusivity; q is the source term; ρ is density and C_p is specific heat. The bottom boundary condition is $\partial T/\partial z = 0$.

There are a number of models [3-6]using the turbulent diffusivity concept to bring the heat conservation equation into closure form. The main difference in these models is the definition of K_v which can be functions of wind mixing, Coriolis force, buoyance effects, internal-wave frequencies and turbulent boundary layer thicknesses. For example, in Mellor and Durbin [6], the diffusivity is parameterized through the balance of kinetic energy and energy dissipation. In Simons [5], the two-unknown, two-equation system is condensed into a one-equation form by considering the geophysical dynamics of Ekman [7].

The following formulation, however, has been tested with Great Lakes data and has produced satisfactory results. Briefly, the vertical turbulent thermal diffusity is defined over three layers:

Epilimnion: $K_v = K_o(1 + \sigma R_i)^{-1} - \gamma\partial\rho/\partial z$ (2a)

Mesolimnion: $K_v = \alpha K_{TC}(N_{TC}^2 / N^2)$ (2b)

Hypolimnion: $K_v = K_B - \gamma\partial\rho/\partial z$ (2c)

where K_o is the air-water diffusion parameter; K_B is the bottom diffusion parameter; σ, γ, α are model coefficients; ρ is water density; R_i is the Richardson number; N is the Brunt-Väisälä frequency; TC denotes the value at the thermocline where the maximum temperature gradient occurs. Mathematically, matching of the three regions of formulations given in Equation (2) is required to ensure the continuity in the diffusivity. For example, in the middle layer (mesolimnion), the diffusivity decreases to its smallest value at the thermocline and then increases to a constant value as the lower layer (hypolimnion) is reached. It is further assumed that the bottom turbulent diffusivity, K_B, is a constant, unless free convection of the Bernard type occurs. The lateral transports in and out of the basin are assumed to be approximately equal in magnitude and hence cancel each other.

There are two approaches of defining the Richardson number, Ri, which essentially compares the thermal buoyancy effect to the current shear mixing effect:

$$Ri = \frac{-g \frac{\partial\rho}{\partial z}}{\rho\left(\frac{\partial u}{\partial z}\right)^2} \quad (3)$$

The first approach is to assume certain semi-empirical velocity profiles, such as the Ekman layer [5] or the von Karman's logarithmic profile [3,4], and eliminates the need of solving for the momentum equation. The second approach is to solve the velocity gradient from the momentum equation directly. We have adopted the first approach with the von Karman profile in the one-dimensional model, but used the second approach in the two-dimensional case. By comparing to the 1967 to 1978 observed lake temperature data [8], it is found that by choosing the values $\sigma = 0.03$, $\gamma = 10^4$ and $\alpha = 0.2$, the model gives satisfactory agreement for all three basins. The bottom turbulent diffusivity, K_B, however, seems to depend at least on the depth of the hypolimnion. For example, in the Central Basin a value of $K_B = 2K_{TC}$ is used, whereas in the Eastern Basin $K_B = 5K_{TC}$ appears to be more appropriate. Both finite difference and finite element methods were used to solve Equation (1) with the

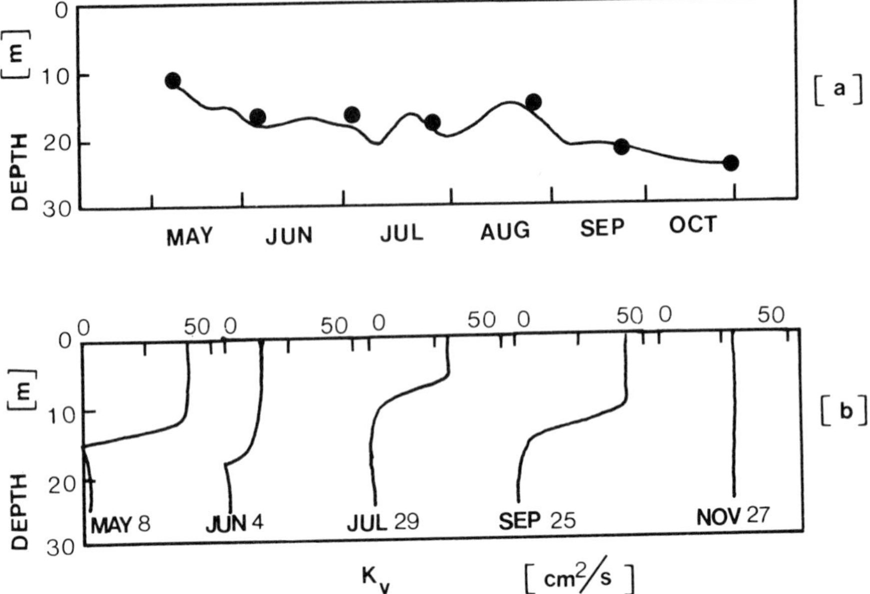

Figure 1. 1-D (z) model results for Central Basin, L. Erie, 1970. [a] Computed (——) and observed (•) thermocline positions; [b] computed vertical eddy diffusivities K_V.

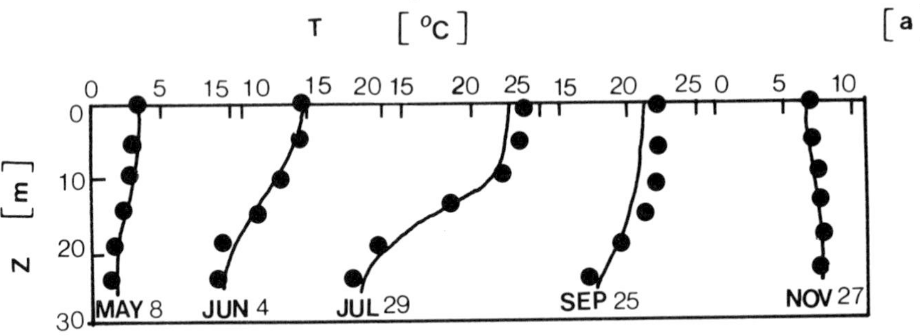

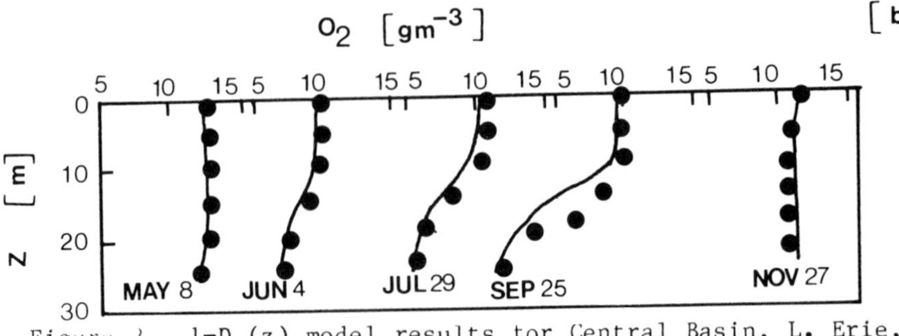

Figure 2. 1-D (z) model results for Central Basin, L. Erie, 1970. [a] Computed (——) and observed (•) temperatures; [b] computed (——) and observed (•) oxygen concentration.

non-linear coefficient K_z given by Equation (2). There were slight differences in the results of the two numerical methods and the differences have been reported in an earlier paper [9].

The present paper reports some further development and application of the model. Figure 1(a) shows the computed and observed positions of the thermocline in the vertical column for the Central Basin of Lake Erie in 1970. The epilimnetic layer depths are affected mainly by wind impulses whereas the hypolimnetic layer depths are influenced by bottom turbulent responses. Figure 1(b) shows the computed eddy viscosity for the Central Basin by using a semi-empirical formula $K_o = 4.5 \times 10^{-3} W$, where W is the scalar wind speed in cm/s. Figure 2(a) shows the computed and observed temperature profiles of the Central Basin in 1970. Because of the shallowness of the Basin (25 m depth), the hypolimnion layer is rather thin (3 to 4 m) as shown in Figures 1(b) and 2(a). In terms of this physical effect on the water quality, the oxygen in the bottom layer may easily be depleted by the water and sediment oxygen demands. Indeed, in 1970 an extensive anoxic area in the basin was recorded [10].

The one-dimensional mass conservation equation for oxygen can be written as:

$$\frac{\partial O_2}{\partial t} = \frac{1}{A}\frac{\partial}{\partial z} AK_v \frac{\partial O_2}{\partial z} + (R + P) - S \qquad (4)$$

where O_2 is the oxygen concentration; R is the reaeration of oxygen from air; P is the oxygen produced during photosynthesis by algae; S is a general source or sink term. As a first approximation, the vertical turbulent eddy diffusivity, K_v, derived from Equation (2) can be used in Equation (4). A further simplification can be made by assuming that the oxygen in the epilimnion is always saturated so that the oxygen concentration is proportional to the computed temperature. Since the term S is dominated by the oxygen demand rates in the sediment and the water, the values derived from a budget calculation [10] can be used. Figure 2(b) shows the computed and observed oxygen profiles of the Central Basin in 1970. Towards the end (September-October) of the stratification period, anoxic levels are indicated near the lake bottom in both the computed and observed results. A more detailed modelling approach is currently carried out to relate the oxygen processes to the nutrient-plankton relationships and phosphorus dynamics.

3 TWO-DIMENSIONAL (x-z) MODEL

The one-dimensional (z) model describes the vertical variations in the temperature, averaged in both x and y directions. To incorporate the variations in the x-direction, the variables will be averaged in the y direction and hence represent more or less the changes in temperature in an x-z plane along the centre line midway between the north and south shores, as in the case of Lake Erie. One of the main objectives of developing an x-z model is to study the possibility of oxygen replenishment of the Central Basin through the hypolimnetic transports from the deeper and oxygen-richer Eastern Basin.

The governing equations of the two-dimensional model are the continuity equation, the momentum equation and the thermodynamic energy equation. Hydrostatic flow is assumed and hence the equation of motion in the vertical reduces to a pressure distribution condition. Non-linear advection terms in the other equations of motion can also be neglected because their effects have been shown to be small in models of this type with small Rossby numbers [11]. The turbulent friction and heat mixing are parameterized by eddy viscosities and diffusivities. Variations in the y directions are only taken at the x-z planes at the centre line and the two shores, as discussed in detail in [11]. Thus, although the equations are three-dimensional, the variables are considered known on the shores while the unknowns are confined on the central x-z plane governed by the following equations of continuity, motions and energy, respectively (see [11]):

$$\frac{\partial u}{\partial x} + \frac{\partial v}{\partial y} + \frac{\partial w}{\partial z} = 0 \tag{5a}$$

$$\frac{\partial u}{\partial t} = \frac{\partial}{\partial x} A_h \frac{\partial u}{\partial x} + \frac{\partial}{\partial y} A_h \frac{\partial u}{\partial y} + \frac{\partial}{\partial z} A_v \frac{\partial u}{\partial z} + fv - \frac{\partial}{\partial x}(P+Q) \tag{5b}$$

$$\frac{\partial v}{\partial t} = \frac{\partial}{\partial x} A_h \frac{\partial v}{\partial x} + \frac{\partial}{\partial y} A_h \frac{\partial v}{\partial y} + \frac{\partial}{\partial z} A_v \frac{\partial v}{\partial z} - fu - \frac{\partial}{\partial y}(P+Q) \tag{5c}$$

$$\frac{\partial T}{\partial t} = -\frac{\partial uT}{\partial x} - \frac{\partial vT}{\partial y} + \frac{\partial}{\partial x} K_h \frac{\partial T}{\partial x} + \frac{\partial}{\partial y} K_h \frac{\partial T}{\partial y} - \frac{\partial wT}{\partial z} + \frac{\partial}{\partial z} K_v \frac{\partial T}{\partial z} \tag{5d}$$

where u, v and w are the velocities along the three coordinates x, y and z, respectively; A_h and A_v are the horizontal and vertical eddy viscosities; K_h and K_v are the horizontal and vertical thermal eddy diffusitives; f is the Coriolis parameter; P and Q represent the barotropic and baroclinic pressure components as defined below:

$$P = g\eta + P_s/\rho; \quad Q = \int_{-z}^{\eta} \frac{\Delta\rho}{\rho}\,dz; \quad \frac{\Delta\rho}{\rho} = \beta(4-T)^2 \tag{6}$$

where η denotes the free surface elevation; g is the gravitational acceleration; P_s is atmospheric pressure; β is the thermal expansion coefficient.

The boundary conditions at surface (s) and bottom (b) are:

$$\rho A_v \frac{\partial u}{\partial z}\Big|_s = \tau_{sx} \quad \rho A_v \frac{\partial v}{\partial z}\Big|_s = \tau_{sy} \quad \rho C_p K_v \frac{\partial T}{\partial z}\Big|_s = q \tag{7a}$$

$$\rho A_v \frac{\partial u}{\partial z}\Big|_b = \tau_{bx} \quad \rho A_v \frac{\partial v}{\partial z}\Big|_s = \tau_{by} \quad \frac{\partial T}{\partial z}\Big|_b = 0 \tag{7b}$$

where τ_s is the wind stress and τ_b is the bottom stress formulated as a quadric function of the bottom velocities.

The finite-difference staggered grid scheme with the rigid lid assumption is described in [11]. In this paper, an implicit scheme has been applied to the vertical diffusion term and a flux-corrected transport scheme has been used to remove some of the oscillation caused by the small eddy diffusivities near the thermocline [9]. The topographical variations of Lake Erie is also taken into account in the model using uniform rectangular grids. An alternative method is to use the finite-element method which can approximate the irregular domain more closely with triangular elements of different sizes. Some tests have been performed on the following simplified version of Equation (5) by fully assuming no variations in the y direction:

$$\frac{\partial u}{\partial x} + \frac{\partial w}{\partial z} = 0 \tag{8a}$$

$$\frac{\partial u}{\partial t} = \frac{\partial}{\partial x} A_h \frac{\partial u}{\partial x} + \frac{\partial}{\partial z} A_v \frac{\partial u}{\partial z} - \frac{\partial}{\partial x} (P + Q) \tag{8b}$$

$$\frac{\partial T}{\partial t} = - \frac{\partial uT}{\partial x} + \frac{\partial}{\partial x} K_h \frac{\partial T}{\partial x} - \frac{\partial wT}{\partial z} + \frac{\partial}{\partial z} K_v \frac{\partial T}{\partial z} \tag{8c}$$

Briefly, two types of basis functions [12] are used: the quadratic basis function $[\phi]$ is used for u and w, and the linear basis function $[\psi]$ is used for T. The Galerkin method is then applied to the Equation (8) resulting in:

$$\int_e \hat{\phi} \left(\frac{\partial \phi}{\partial x} \hat{u} + \frac{\partial \phi}{\partial z} \hat{w}\right) dxdz = 0 \tag{9a}$$

$$\int_e \phi^T \left(\phi \frac{d\hat{u}}{dt} + A_h \frac{\partial \hat{\phi}}{\partial x} \frac{\partial \phi}{\partial x} \hat{u} + (\psi \hat{A}_v) \frac{\partial \hat{\phi}}{\partial z} \frac{\partial \phi}{\partial z} \hat{u} + \frac{\partial \psi}{\partial x} \hat{Q}\right) dxdz$$

$$- \int_s \hat{\phi}\phi \ \hat{\tau}_{sx} dx + \int_b \hat{\phi}\phi \ \hat{\tau}_{b1} dl = 0 \tag{9b}$$

$$\int_e \psi^T \left(\frac{d\hat{T}}{dt} + \frac{\partial (\phi\hat{u})(\psi\hat{T})}{\partial x} + K_h \frac{\partial \hat{\psi}}{\partial x} \frac{\partial \psi}{\partial x} \hat{T} + \frac{\partial (\phi\hat{w})(\psi\hat{T})}{\partial z}\right.$$

$$\left. + (\psi \hat{K}_v) \frac{\partial \hat{\psi}}{\partial z} \frac{\partial \psi}{\partial z} \hat{T}\right) dxdz - \int_s \hat{\psi}\left(\psi \frac{\hat{q}}{C_p \rho}\right) dx = 0 \tag{9c}$$

where $\hat{\phi}$ is the tranpose of ϕ etc; e indicates the domain of the elements; s and b indicate surface and bottom boundaries; and the surface integrals are obtained from integration by parts. A predictor-corrector time integration method [9] is used to advance the solution from the one time level to the next. Notice that whereas the baroclinic pressure component, Q, can be evaluated in the predictor-corrector scheme, the barotropic component P is evaluated as a correction term based on the kinematic boundary condition:

$$\frac{\partial \eta}{\partial t} + \frac{\partial}{\partial x}\left\{\int_{-D}^{\eta} u dz\right\} = 0 \qquad (10)$$

where D is the mean depth. Because of the rigid lid approximation, Equation (10) indicates that the total lateral (x) transports across any vertical (z) section should be equal to the hydraulic inflow and hence the outflow. Since the pressure should distribute evenly along this vertical section, the evaluation of the pressure term P consists of simply correcting each of the velocities along the section by a uniform amount such that the total lateral transport obeys Equation (10). With this pressure correction technique and the mixed basis functions approach, preliminary results show rather stable and smooth solution. Further tests are being carried out to compare the finite difference and the finite element methods.

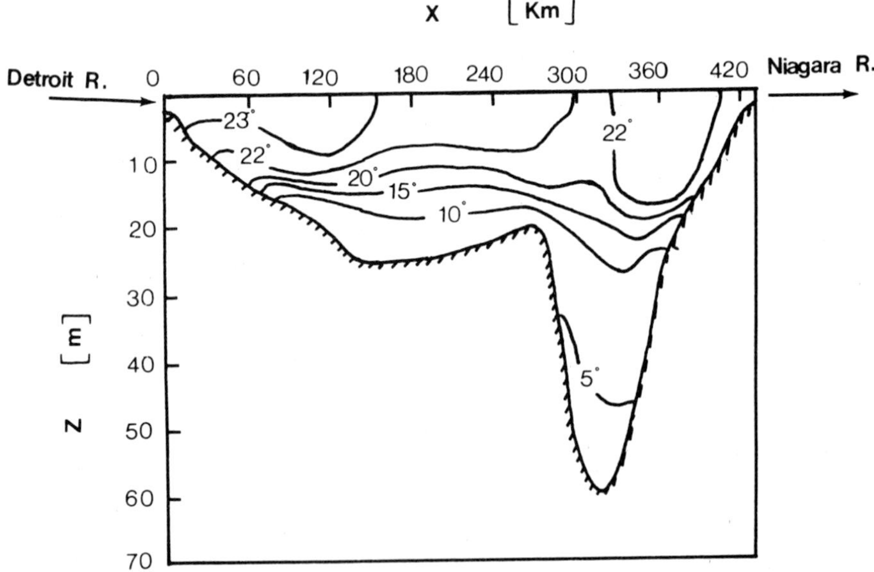

Figure 3. Observed L. Erie temperature data (Aug. 18, 1978) used in 2-D (x-z) model.

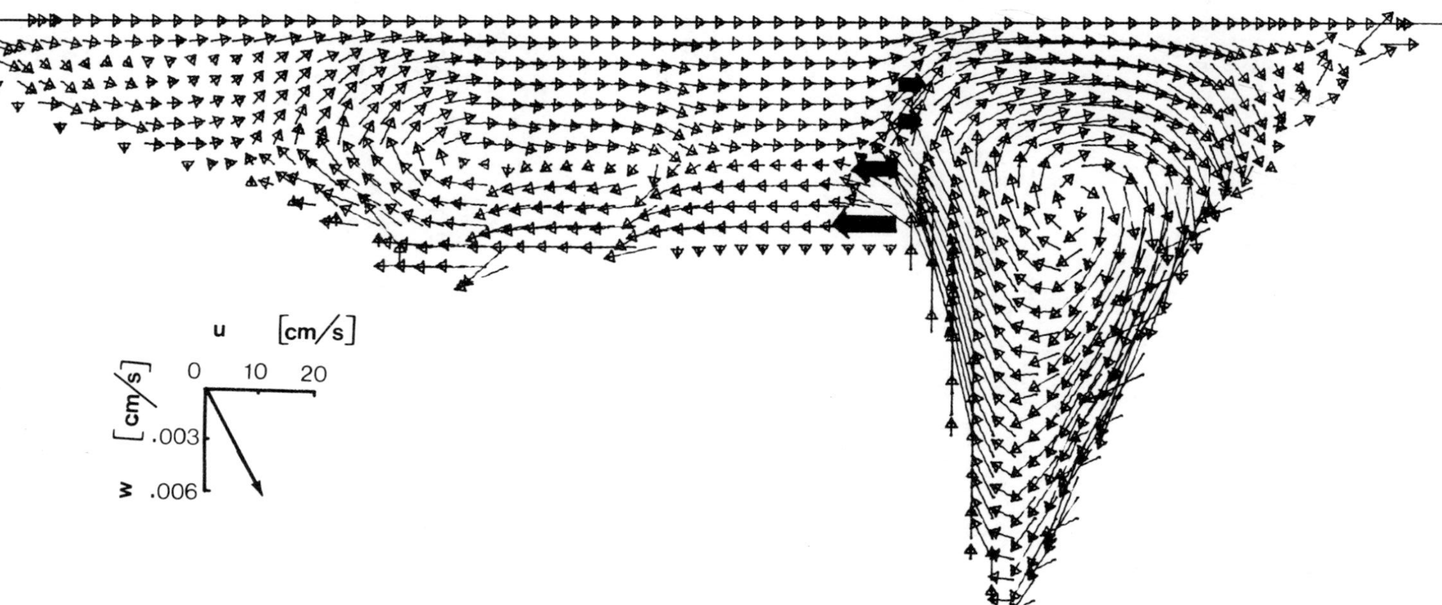

Figure 4. 2-D (x-z) model results for L. Erie (Aug. 18-20,1978). Thin arrows are computed current vectors (u and w) ; thick arrows are observed u components.

Figure 3 shows an observed map of the temperature distribution for Lake Erie in 1978. Using this as the initial temperature distribution and other climatological inputs, we use the eddy diffusivities as defined in Equation (2) and solve for the velocities from a cold start (u = 0 and w = 0). Figure 4 shows the computed velocities from the finite difference model after three days and the observed measurement at the section between Central and Eastern Basin. In both computed and observed results, there is a dominant eastward current in the upper layer but a westward return flow in the hypolimnion.

4 CONCLUSION

The results of a one-dimensional (z) and a two-dimensional (x-z) model are compared to observed data. In the one-dimensional case, long term simulation studies indicate that only small changes in the semi-empirical parameters are needed to obtain satisfactory results. In the two-dimensional case, further tests are required to understand the predictive capability of the model.

ACKNOWLEDGEMENTS

T.J. Simons provided the two-dimensional finite-difference code; W.M. Schertzer, A.S. Fraser and F. Chiocchio provided the observed data. The author also thanks Mrs. J. Hodson for programming support and Mrs. D.C. Crabtree for typing the manuscript.

REFERENCES

1. SIMONS, T.J. and LAM, D.C.L. - Some Limitations of Water Quality Medels for Large Lakes: A Case Study of Lake Ontario. Water Resources Res., Vol. 16, No. 1, pp. 105-116, 1980.

2. LAM, D.C.L. and SIMONS, T.J. - Numerical Computations of Advective and Diffusive Transports of chloride in Lake Erie, 1970. J. Fish. Res. Bd. Canada, Vol. 33, No. 3, pp. 537-549, 1976.

3. KRAUS, E.B. and TURNER, J.S. - A One-Dimensional Model of the Seasonal Thermocline, Part II. Tellus, Vol. 19, pp. 98-105, 1967.

4. WALTERS, R.Z., CAREY, G.F. and WINTER, D.F. - Temperature Computation for Temperate Lakes. Appl. Math. Modelling, Vol. 2, pp. 41-48, 1978.

5. SIMONS, T.J. - Verification of Seasonal Stratification Models. Departmental Report, Institute for Meteorology and Oceanography, University of Utrecht, The Netherlands, 12 p., 1980.

6. MELLOR, G.L. and DURBIN, P.A. - The Structure and Dynamics of the Ocean Surface Mixed Layer. J. Phys. Ocean., Vol. 5, pp. 718-728, 1975.

7. EKMAN, V.W. - On the Influence of the Earth's Rotation on Ocean Currents. Ark. Mat. Astr. Fys., Vol. 2, No. 11, 52 p., 1905.

8. LAM, D.C.L. and SCHERTZER, W.M. - Modelling the Interaction of Climate and Aquatic Regimes of Large Lakes. Proc. Canadian Climate/Water Workshop, University of Alberta, Feb. 28-29, 1980. A.E.S., Downsview, Ontario, Canada. (In Press.)

9. LAM, D.C.L. - Finite element Simulation of Anisotropic Turbulent Diffusion in Lakes. Proc. 3rd Int. Conf. on Finite Elements in Water Resources, Ed. Wang, S.Y. et al., University of Mississippi, University, Miss., U.S.A., pp. 3.20-3.35, 1980.

10. BURNS, N.M. - Oxygen Depletion in the Central and Eastern Basins of Lake Erie, 1970. J. Fish. Res. Bd. Canada, Vol. 33, No. 3, pp. 512-519, 1976.

11. HOLLAN, E. and SIMONS, T.J. - Wind Induced Changes of Temperature and Currents in Lake Constance. Arch. Met. Geophys. Bioclimatology, Series A, Vol. 27, No. 3-4, pp. 333-373, Springer-Verlag, N.Y., 1978.

12. HOOD, P. and TAYLOR, C. - Navier - Stokes Equations Using Mixed Interpolation, Finite Element Methods in Flow Problems, Ed. Oden, J.T., UAH Press, Huntsville, pp. 121-132, 1974.

NUMERICAL MODELS FOR THE COMPUTATION OF RESIDUAL CURRENTS

Niek Praagman

Mathematical Department
Technical University Delft
Delft, The Netherlands

SUMMARY

Numerical models for the computation of residual currents are considered. A new method is proposed. Some preliminary results are shown. Advantages and flaws of the methods are given.

1. INTRODUCTION

As a consequence of the enormous increase of the pollution of water in rivers, estuaries and coastal seas during the last decennia, the interest in mathematical models that treat, although still in a very simplified form, the behaviour of the pollutants, has augmented.
These mathematical models, often called waterquality models, are quite complicated, since various processes of hydrodynamical, sedimental, biological and chemical type play an important role with respect to the pollution.

Water quality models are based on a system of partial differential equations that represent conservation of mass and momentum ([1]):

$$(1)\quad \frac{\partial}{\partial t}(H\bar{c}_i) + \frac{\partial}{\partial x}(H\overline{uc}_i) + \frac{\partial}{\partial y}(H\overline{vc}_i) = \bar{Q}_i + \bar{I}_i + \bar{D}_i\ ,\qquad i = 1(1)N$$

$$(2)\quad \frac{\partial}{\partial t}(H) + \frac{\partial}{\partial x}(H\bar{u}) + \frac{\partial}{\partial y}(H\bar{v}) = 0$$

$$(3)\quad \frac{\partial}{\partial t}(H\bar{u}) + \frac{\partial}{\partial x}(H\bar{u}^2) + \frac{\partial}{\partial y}(H\overline{uv}) - fH\bar{v} = - gH\frac{\partial\zeta}{\partial x} - \frac{g\bar{u}\sqrt{\bar{u}^2+\bar{v}^2}}{C^2} + D_x + F_x$$

$$(4)\quad \frac{\partial}{\partial t}(H\bar{v}) + \frac{\partial}{\partial x}(H\bar{u}\bar{v}) + \frac{\partial}{\partial y}(H\bar{v}^2) + fH\bar{u} =$$

$$-gH\frac{\partial \zeta}{\partial y} - g\,\frac{\bar{v}\sqrt{\bar{u}^2+\bar{v}^2}}{C^2} + D_y + F_y$$

with

$$\bar{c}_i = \frac{1}{H}\int_{-h}^{\zeta} c_i(x,y,z,t)\,dz\ ,$$

the (vertically averaged) concentration of the considered quantity i at point (x,y) and time t.

$\zeta(x,y,t)$: water elevation at location (x,y) and time t.
$h(x,y)$: depth at location (x,y).
$H(x,y,t) = h(x,y) + \zeta(x,y,t)$, i.e. the total depth at location (x,y) at time t.
$\bar{u},\bar{v}$: the (vertically) averaged velocity components.
$\bar{Q}_i$: term representing sources and sinks for quantity i.
$\bar{I}_i$: term in which biological, chemical and sedimental interactions can be taken into account.
$\bar{D}_i$: term representing the dispersive processes.
f : parameter of Coriolis.
C : parameter of Chézy.
g : acceleration of gravity.
D_x,D_y : terms representing dispersion.
F_x,F_y : terms to take into account external forces, for instance windstress.

Remark 1: In the derivation of the above system two assumptions have been made. Firstly it has been assumed that depth-mean variables can be used, or in other words that no significant stratification is present.
Secondly it has been supposed that a moving time average can be applied to the equations in order to separate the turbulent fluctuations from the tidal oscillations [7].

Remark 2: From now on, the bar sign, used to indicate depth mean variables, will be omitted for simplicity reasons.

Remark 3: Accompanied with appropriate boundary conditions and initial values equations (1), (2), (3) and (4) constitute the mathematical model that can be used for waterquality computations.

In the present study we confine ourselves to the so-called long term water quality models. In these long term models the time scales of the processes considered, vary from several days up to several months. Considering the speed of, even large, modern computers, it is not attainable to compute tidal currents and related processes, for more than only a few tidal cycles, i.e. two or three days, with the above given system of equations. In order to overcome this difficulty use is made of time averaging techniques.

As will be shown in chapter 2, the time averaged equations (1) do contain time averaged velocity components. Usually these components are referred to as residual currents. Several residual quantities are in use and in chapter 2 definitions used in this paper are given.

The main goal of the present study is the computation of the "residual currents". In chapters 3 and 4 two methods that are most commonly used are treated and some numerical results are given. In addition, in chapter 5 an alternative method, that is indeed a special combination of the methods treated in chapters 3 and 4, is presented, accompanied by some numerical results. In chapter 6 conclusions and recommendations are given.

2. RESIDUAL CURRENTS: A BACKGROUND.

In order to get a mathematical model that is manageable for long term applications system (1) is integrated using a moving time integration. It is assumed that the principal component of the tide is the semi-diurnal tide, usually called the M_2-tidal component. Corresponding with the M_2-tide, a constant T is defined, that equals the length of one M_2-period (i.e. approximately 12 hours and 25 minutes).
In addition some other quantities are defined:

(5) $U = H\,u$

and

(6) $V = H\,v$

Corresponding to the variables

$u,\ v,\ c_i,\ U,\ V,\ \zeta$ and H

are the variables

$u^0,\ v^0,\ c_i^0,\ U^0,\ V^0,\ \zeta^0$ and H^0

defined in the following way,

(7) $$u^0(x,y,t) = \frac{1}{2T}\int_{t-T}^{t+T} u(x,y,\tau)\,d\tau$$

Differences of two related variables are denoted by the superscript 1, for instance:

(8) $$u^1(x,y,t) = u(x,y,t) - u^0(x,y,t)$$

Integration of system (1) leads to:

(9) $$\frac{1}{2T}\int_{t-T}^{t+T}\frac{\partial}{\partial\tau}(Hc_i)\,d\tau + \frac{1}{2T}\int_{t-T}^{t+T}\frac{\partial}{\partial x}(Uc_i)\,d\tau +$$

$$+\frac{1}{2T}\int_{t-T}^{t+T}\frac{\partial}{\partial y}(Vc_i)\,d\tau = \frac{1}{2T}\int_{t-T}^{t+T}(Q_i+I_i+D_i)\,d\tau \qquad i=1(1)N.$$

In many applications a periodic tidal regime is assumed with period 2T and as a consequence the quantities

u^0, v^0, U^0, V^0, ζ^0 and H^0

are time independent.

Utilizing the above definitions and assumptions system (9) transforms into

$$(10)\quad \frac{\partial}{\partial t}(H^0c_i^0) + \frac{\partial}{\partial x}(U^0c_i^0) + \frac{\partial}{\partial y}(V^0c_i^0) = \text{R.H.S.}$$

Remark: The right hand side terms (R.H.S.) are of great importance for the long-term water quality model but are of no interest for the sequel of the present study; hence no detailed information is given. (See [6]).

With use of equation (2), equation (10) can be simplified to:

$$(11)\quad H^0\frac{\partial c_i^0}{\partial t} + U^0\frac{\partial c_i^0}{\partial x} + V^0\frac{\partial c_i^0}{\partial y} = \text{R.H.S.}$$

Usually the quantities U^0, V^0 and H^0 are called residuals since the tidal components U^1, V^1 and H^1 have been "subtracted" from the original quantities U, V and H.

For the numerical solution in time of equations (11) the quantities

$$(12)\quad \frac{U^0}{H^0} \text{ and } \frac{V^0}{H^0} \text{ have to be computed.}$$

Since, in general, $\zeta^0 \ll h$ often the quantities

$$(13)\quad \frac{U^0}{h} \text{ and } \frac{V^0}{h} \text{ are computed as a first approximation of (12)}$$

3. THE COMPUTATION OF RESIDUALS I.

The residual quantity U^0 is defined by

$$(14)\quad U^0 = \frac{1}{2T}\int_{t-T}^{t+T}(\zeta+h)U\,d\tau$$

An approximation $\tilde{U}^0(x,y)$ of $U^0(x,y)$ can be obtained using the composite trapezoid rule:

$$(15)\quad \tilde{U}^0(x,y) = \frac{\Delta t}{2T}\left(\sum_{i=1}^{N} U(x,y,t_0+i\Delta t)\right) \quad \text{with } N\Delta t = 2T.$$

For many coastal seas and estuaries hydrodynamical models exist that are based on equations (2), (3) and (4) of chapter 2. Utilizing these models approximations of the values

$$U(x,y,t_0+i\Delta t)$$

can be obtained and hence, using (15), approximations of the residual quantities.

In fig. 1 the result of such a computation is shown for the Southern part of the North Sea. (For a description of the underlying hydrodynamical model see [5]).

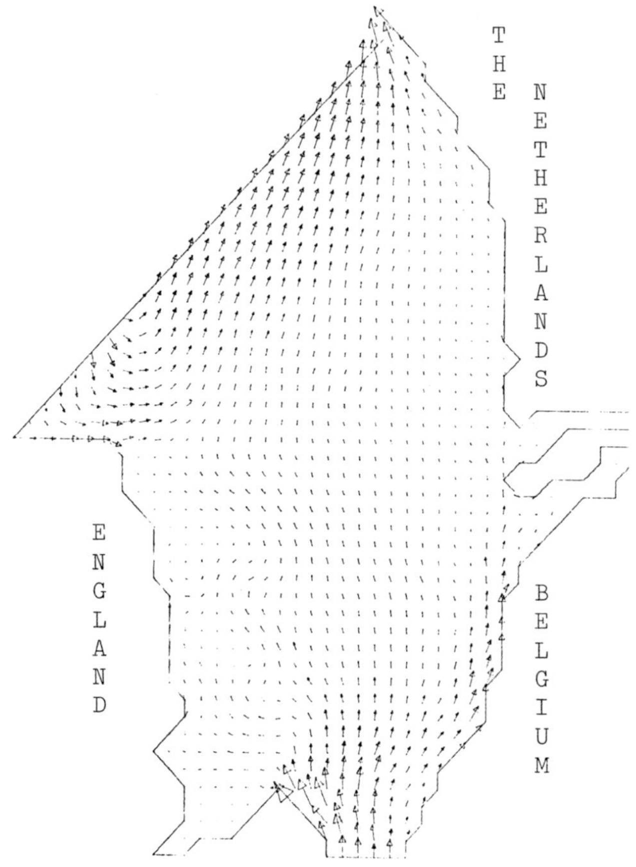

fig. 1 Residuals $(\frac{\tilde{U}^0}{h},\frac{\tilde{V}^0}{h})$, computed by the method of chapter 3.

A serious imperfection of the above method may be the following. In most coastal seas the two components U^0 and U^1 differ considerable in magnitude. In the North Sea, for instance, a typical value of U^1 may be 20-50 times as large as a typical U^0 value. When U is approximated using a hydrodynamical model an error of about 2% is very acceptible. However in the case that such an error is systematic an error of 10-100% can be expected in the numerical approximation of U^0.

4. COMPUTATION OF RESIDUALS II.

Unsatisfied with the method of chapter 3 an alternative method has been formulated by Nihoul and Ronday. ([4]) Application of a moving time average to equations (2), (3) and (4) leads to:

$$(16)\quad \frac{\partial U^0}{\partial x} + \frac{\partial V^0}{\partial y} = 0$$

$$(17)\quad -fV^0 + gh\,\frac{\partial \zeta^0}{\partial x} = -\tau_x - g\left(\frac{U\sqrt{U^2+V^2}}{H^2C^2}\right)^0$$

$$(18)\quad +fU^0 + gh\,\frac{\partial \zeta^0}{\partial y} = -\tau_y - g\left(\frac{V\sqrt{U^2+V^2}}{H^2C^2}\right)^0$$

with

$$(19)\quad \tau_x = \frac{\partial}{\partial x}\left(\left(\frac{U^2}{H}\right)^0\right) + \frac{\partial}{\partial y}\left(\left(\frac{UV}{H}\right)^0\right) + g\,\frac{\partial}{\partial x}\left(\left(\tfrac{1}{2}\zeta^2\right)^0\right)$$

$$(20)\quad \tau_y = \frac{\partial}{\partial x}\left(\left(\frac{UV}{H}\right)^0\right) + \frac{\partial}{\partial y}\left(\left(\frac{V^2}{H}\right)^0\right) + g\,\frac{\partial}{\partial y}\left(\left(\tfrac{1}{2}\zeta^2\right)^0\right)$$

In the following, in a first approximation, the "bottom-stress" terms are linearized, (see [3]), with respect to the residuals U^0 and V^0:

$$(21)\quad g\left(\frac{U\sqrt{U^2+V^2}}{H^2C^2}\right)^0 \approx g\left(\frac{\sqrt{U^2+V^2}}{H^2C^2}\right)^0 U^0 = KU^0$$

and

$$(22)\quad g\left(\frac{V\sqrt{U^2+V^2}}{H^2C^2}\right)^0 \approx g\left(\frac{\sqrt{U^2+V^2}}{H^2C^2}\right)^0 V^0 = KV^0$$

In order to satisfy equation (16) a stream function ψ is introduced, such that

$$(23)\quad U^0 = -\frac{\partial \psi}{\partial y} \quad \text{and} \quad V^0 = \frac{\partial \psi}{\partial x}$$

Using (21), (22) and (23) equations (16) and (17) can be transformed into:

$$(24)\quad K\left(\frac{\partial^2 \psi}{\partial x^2} + \frac{\partial^2 \psi}{\partial y^2}\right) - \frac{K}{h}\left(\frac{\partial h}{\partial x}\frac{\partial \psi}{\partial x} + \frac{\partial h}{\partial y}\frac{\partial \psi}{\partial y}\right) +$$

$$+ \left(\frac{\partial K}{\partial x}\frac{\partial \psi}{\partial x} + \frac{\partial K}{\partial y}\frac{\partial \psi}{\partial y}\right) + \frac{f}{h}\left(\frac{\partial h}{\partial x}\frac{\partial \psi}{\partial y} - \frac{\partial h}{\partial y}\frac{\partial \psi}{\partial x}\right) =$$

$$= h\frac{\partial}{\partial y}\left(\frac{\tau_x}{h}\right) - h\frac{\partial}{\partial x}\left(\frac{\tau_y}{h}\right)$$

Since (24) is an elliptic equation boundary conditions have to be added in order to obtain a well-posed problem. However to get accurate boundary values is very difficult. For results of computations see [2] and [4]. In these publications no indication has been given how sensitive the results are with respect to inaccuracies in the boundary values. In [6] results are given that show this strong dependence.

5. COMPUTATION OF RESIDUALS III.

In the method of chapter 4 approximations have been used, that are reasonable but not necessary to obtain a, numerically, solvable problem.
Consider again equations (17) and (18).
This is a system of two partial differential equations in the three unknowns

ζ^0, U^0 and V^0.
Approximations and a stream function are not needed in the following approach.
Firstly the quantities

$$\zeta^0, \left(\frac{U^2}{H}\right)^0, \left(\frac{UV}{H}\right)^0, (\zeta^2)^0, \left(\frac{U\sqrt{U^2+V^2}}{H^2C^2}\right)^0$$

are computed utilizing a hydrodynamical model. Secondly equations (17) and (18) are used to compute the residuals U^0 and V^0 by either a finite difference or a finite element method. The advantage over the method of chapter 4 is the fact that no approximations have to be made and moreover the computational simplicity of the present method.
It seems also to be an advantage that for instance wind influences or extra residual elevations due to meteorological forcing can be taken into account in a first approximation, without recalculating the tidal stresses, due to the linearity of the model. However it has to be investigated how sensitive the tidal stresses and residual elevations are for altering circumstances ([6]).

In fig. 2 a plot of numerical results, obtained using model 3, is given.
Although globally the results given in fig. 1 and fig. 2 are the same the results near to the boundary of model 3 are questionable which may be due to the fact that derivatives of the residual elevation ζ^0 had to be calculated. However the results do not support the statements of Nihoul and Ronday [4] concerning the influence of errors of tidal computations.

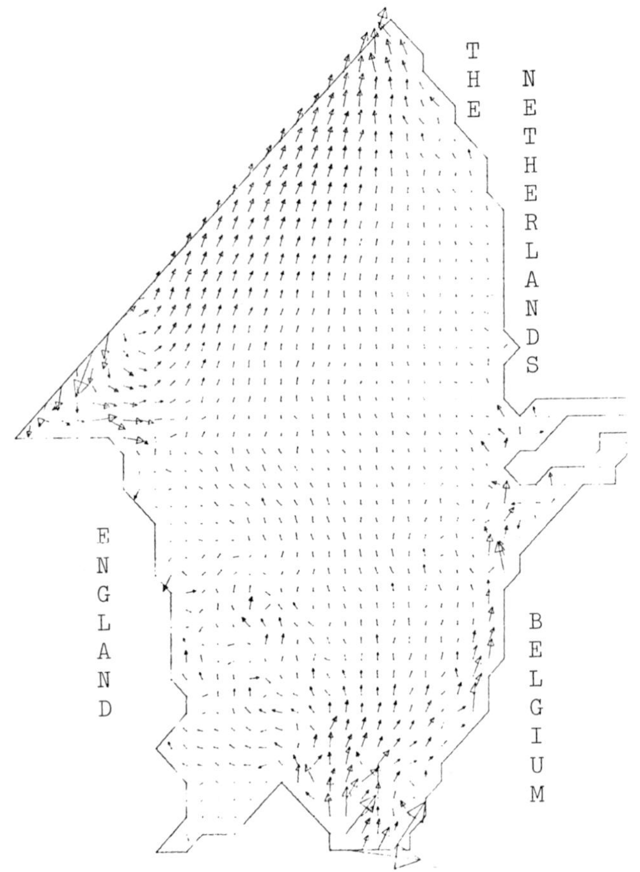

fig. 2 Residuals $(\frac{\tilde{U}^0}{h},\frac{\tilde{V}^0}{h})$, computed by the method of chapter 5.

6. CONCLUSIONS

Mathematical models for the computation of residual currents have been considered. In addition to two already well-known methods (type 1 and type 2) a third, more or less intermediate, method (type 3) has been proposed. Although this method seems to be preferable over a type 2 method (no approximations and computational simplicity) the advantage over a type 1 method is dubious since it is not yet clear, whether the superposition of residuals that are due to several causes is allowed.

REFERENCES:

1. ALFRINK, B.J. - Een vereenvoudigd lange termijn waterkwaliteitsmodel voor de noordzee. WAKWON-report, Rijkswaterstaat & T.H. Delft, 1977.

2. DIJKSTRA, T. - Numerical computations of residual currents in the southern part of the North Sea. DIV-Report, no. 1980627, 1980.

3. GROEN, P. and GROVES, G.W. - Surges, The Sea, ed. N.M. Hill, Interscience Publishers, New York, 1962.

4. NIHOUL, J.C.J. and RONDAY, F.C. - The influence of the "tidal stress" on the residual circulation, Tellus, 27, pp. 484-489, 1975.

5. PRAAGMAN, N. - Numerical solution of the shallow water equations by a finite element method, Ph.D., T.H. Delft, 1979.

6. PRAAGMAN, N. - The computation of residual currents, NA report, T.H. Delft, 1981

7. VREUGDENHIL, C.B. - Computation of gravity currents in estuaries, Ph.D., T.H. Delft, 1970.

Current and Density Structure in Shelf Waters due to Fresh Water Discharge: A Numerical Study

Timothy W. Kao

Professor, Department of Civil Engineering
The Catholic University of America
Washington, D.C. 20064 U.S.A.

ABSTRACT

The shelf water from Cape Cod to Cape Hatteras is a shallow sea which receives substantial freshwater input from the rivers along the East Coast of North America, especially during Spring freshening. Hydrographic observations have shown that there is substantial salinity and density contrast across the shelf, and little along-shelf variability in salinity. In terms of the currents, it is known that there is a mean along-shelf flow towards the southwest.

In this paper, we present the results of a two-dimensional numerical model of the structure of the shelf water due to a line source of fresh water input from the shore. The full time-dependent diffusion and Navier-Stokes equations for a constant Coriolis parameter are solved by a finite difference scheme. A stretched co-ordinate is used in the horizontal direction. The governing equations are then appropriately scaled. The structures of the density field and the along shelf and across-shelf velocity fields are found. The structure of the benthic boundary layer is also resolved. The results are interpreted physically.

1. INTRODUCTION

The shelf water from Cape Cod to Cape Hatteras, the Middle Atlantic Shelf (MAS), is a shallow sea which receives substantial freshwater input from the rivers along the east coast of North America. Hydrographic observations have shown, Bigelow [1], Bigelow and Sears [2], that there is substantial salinity and density contrast across the shelf; vertical stratification (except during early winter); and, aside from the near shore regions in the vicinity of the river mouths, no substantial longshore variation in salinity and density in the mid cross-shelf region. There is also an advance of the isohalines of lower salinity seawards as the seasons progress from Spring onwards. In terms of the currents, in spite of their variability, due largely to wind effects, it is also known from drifter studies, Bumpus [3], as well as moored current meter studies, Beardsley, Boicourt and Hansen [4], that there is a mean along-shelf flow towards the southwest.

Stommel and Leetma [5] gave a steady-state two-dimensional model of the shelf circulation driven by freshwater discharge and wind using the linear momentum equations coupled with the non-linear diffusion equation for the density field. No detailed solutions were obtained although estimates of the characteristic scales of shoreward salt penetration were given. Using the same model, Pietrafesa and Janowitz [6] recently obtained similarity solution for a prescribed buoyancy flux and wind stress at the surface.

In this paper we give a time-dependent model incorporating the full non-linear advection terms in the momentum equations. Solutions are obtained by a finite difference scheme. A more comprehensive paper, discussing the detailed aspects of the solution and their significance in the physical oceanography of the Middle Atlantic Shelf water is in preparation and will appear elsewhere.

2. THE GOVERNING EQUATIONS AND BOUNDARY CONDITIONS

We let (x^*, y^*, z^*) be the co-ordinates in the cross-shelf, along-shelf and vertical directions respectively, (u^*, v^*, w^*) be the corresponding velocity components and t^* be the time. The flow is assumed to be independent of y^*.

The initial value problem corresponds to starting an inflow discharge of Q_e per unit length along the y^*-direction at $x^* = 0$ with density ρ_e into an ambient fluid of constant density ρ_o and depth d. We donote $\rho_o - \rho_e$ by $(\Delta\rho)_e$ and $(\rho - \rho_o)/\rho_o$ by γ, the density anomaly. Then $\gamma_e = -(\Delta\rho)_e/\rho_o$. We let $g' = g(\Delta\rho)_e/\rho_o$. Using d as a reference length, $U = Q_e/d$ as a reference velocity d^2/Q_e as a reference time, we denote the dimensionless variables by dropping the asterisk.

We define ζ, ψ to be the dimensionless y-component of vorticity and stream function respectively. The governing equations for an incompressible, viscous, diffusive, Boussinesq fluid in non-dimensional form are

$$\frac{\partial \gamma}{\partial t} + \frac{\partial}{\partial x}(u\gamma) + \frac{\partial}{\partial z}(w\gamma) = \frac{1}{ScRe} \nabla^2 \gamma \tag{1}$$

$$\frac{\partial \zeta}{\partial t} + \frac{\partial}{\partial x}(u\zeta) + \frac{\partial}{\partial z}(w\zeta) - \frac{1}{Ro}\frac{\partial v}{\partial z} = \frac{1}{F^2}\frac{\partial \gamma}{\partial x} + \frac{1}{Re} \nabla^2 \zeta \tag{2}$$

$$\nabla^2 \psi = \zeta \tag{3}$$

$$\frac{\partial v}{\partial t} + \frac{\partial}{\partial x}(uv) + \frac{\partial}{\partial z}(wv) + \frac{1}{Ro} u = \frac{1}{Re} \nabla^2 v \tag{4}$$

where $Re = Ud/\nu$, $F = U/(gd)^{\frac{1}{2}}$, $Sc = \nu/K$ and $Ro = U/(fd)$ with ν, K, and f denoting the kinematic viscosity, the diffusivity, and the Coriolis parameter respectively. These equations have been given by Kao, Pao and Park [7] where some exploratory results were presented.

The boundary conditions are as follows:

(a) at the free surface, there is no shear, so that $\zeta = 0$, and no mass transfer so that $\partial\gamma/\partial z = 0$.
(b) at the solid boundaries, there is no slip, $(u, v, w) = 0$, and no mass transfer.
(c) at the open side (or offshore) where $x = \infty$, the vorticity and density remain undisturbed.
(d) at the inlet the inflow velocity is uniform.

3. NUMERICAL SOLUTION AND RESULTS

The governing system is solved by a finite difference scheme. The central difference in space and the forward difference in time are used except for the non-linear terms, for which the three-point non-central differencing method [8] is adopted. A stretched coordinate $\bar{x} = 1 - \exp(-ax)$ where a is an appropriate constant is also used following [9]. The difference equations are standard and are similar to those outlined in [10]. Truncation errors in the present case when Re is 0(10) are small, and the physics are faithfully modelled. In the present case we assume the diffusivity and viscosity to be eddy diffusivity and viscosity of the same magnitude so that Sc = 1. It is clear that with the three parameters Re, F, and Ro in the governing equations the range of possible solutions is impossibly large. It is however conceivable that when proper scaling is adopted the parametric regime may be considerably narrowed. This has been done by Kao [11] for the almost inviscid case and applied to the Gulf Stream. In the present case, a remarkably sharp scaling can be found so that under the new scaling the governing system of equations depend on only one parameter, a dimensionless depth $\tilde{d}$, through the boundary condition.

We introduce a time scale $T = 1/f$ and a velocity scale $U_d = (g'Q_e)^{1/3}$. The following are then independent length scales of the problem: $L_o = U_d/f$, $h_o = (Q_e^2/g')^{1/3}$, $h_\nu = (\nu/f)^{\frac{1}{2}}$ and d. These form three new dimensionless parameters of the problem: $R_\nu = L_o/h_\nu$, $E = (h_\nu/h_o)^2$ and $\tilde{d} = d/h_\nu$. For a typical shelf problem with freshwater inflow R_ν is $0(10^3)$ and E of $0(10^2)$ for $\nu \sim 0(1)$ in cgs units. The variables are now to be scaled as follows:

$$(\tilde{x}, \tilde{y}, \tilde{z}) = (x^*/L_o, y^*/L_o, z^*/h_\nu) \tag{5}$$

$$\tau = t^*f \tag{6}$$

$$(\tilde{u}, \tilde{v}, \tilde{w}) = (u^*, v^*, R_\nu w^*)/U_d \tag{7}$$

$$\tilde{\rho} = E^{\frac{1}{2}}(\gamma/\gamma_e) \tag{8}$$

The governing equations then become fully scaled, i.e. with no parameter appearing, if terms of $0(R_\nu^{-2})$ are dropped. These terms are associated with the horizontal diffusion. We thus need to find only one solution with R_ν sufficiently large for any fixed $\tilde{d}$.

Some results for $\tilde{d} = 11$ which corresponds to a shelf depth of 0(10) meters are shown herein for $\tau = 90$ when truly steady conditions are reached.

Briefly the results show a frontal region of the freshening event progressing steadily in the off-shore direction at a slow speed, of about 0.5 cm/s for a typical averaged freshwater discharge of 100 cm^3/s /cm of the Middle Atlantic Shelf (MAS) of North America. This is rather representative of the Spring to Summer progression of the isohalines as observed by Bigelow and Sears [2]. The frontal region is defined by a surface half-jet in the along-shelf component of velocity $\tilde{v}$. The half-jet exhibits strong horizontal and vertical shear. Behind the frontal region the flow is steady. The balance of forces is linear in the cross-shelf direction where the pressure gradient force induced by the buoyant fluid is balanced by friction and the Coriolis force due to the along-shelf velocity. The balance of forces in the along-shelf direction is non-linear in that the non-linear momentum advection terms play an equally important role as the other two forces present, i.e. the frictional force and the Coriolis force induced by the gravitational circulation in the cross-shelf direction.

The distribution of isopycnics from the calculations at 14 inertial periods after the initiation of the freshening event, is shown in Fig. 1 and is typical of the shelf density and, more particularly, salinity structure in the MAS. The leading isopycnic, representing $0.05(\Delta\rho)_e$, defines the front of the freshening water and propagates slowly but steadily across the shelf. The along-shelf velocities $\tilde{v}$ at the surface and at mid-depth are shown in Fig. 2 as functions of the cross-shelf distance $(\tilde{x} - \tilde{x}_f)$ where $\tilde{x}_f$ is the surface location of the leading isopycnic. The vertical profiles of the cross-shelf velocity $\tilde{u}$ is shown in Fig. 3 at two cross-shelf locations. The profile at a location behind the frontal region indicates a strong gravitational component with onshore flow in the lower part of the water column while at a location well ahead of the leading isopycnic, the flow follows linear Ekman dynamics with the Ekman spiral at the bottom developed in response to the offshore pressure gradient associated with a net discharge across the shelf, of Q_e per unit length. The typical balance of the y-component vorticity and momentum at a position behind the frontal region is shown in Fig. 4. At this position steady-state condition has been achieved. It is seen that the above stated linear dynamics for the cross-shelf direction as manifested in the y-vorticity is rather precisely satisfied. It is also seen that the non-linear inertial term is important in the balance of forces in the along-shelf direction. This last statement is a most significant finding of the proposed model, and shows the need to include the non-linear terms in the along-shelf momentum balance.

4. ACKNOWLEDGMENTS

This work is supported by the National Science Foundation under Grants ENG 77-01496 and OCE 79-25061. The computations were done while the author was on an IPA assignment at Goddard Space Flight Center, NASA. The hospitality of members of the Laboratory of Atmospheric Sciences is greatly appreciated. The author wishes to thank Dr. James Mueller specifically for his support and interest. Mr. Fuh-Shing Pan rendered much service during the preparation of this paper.

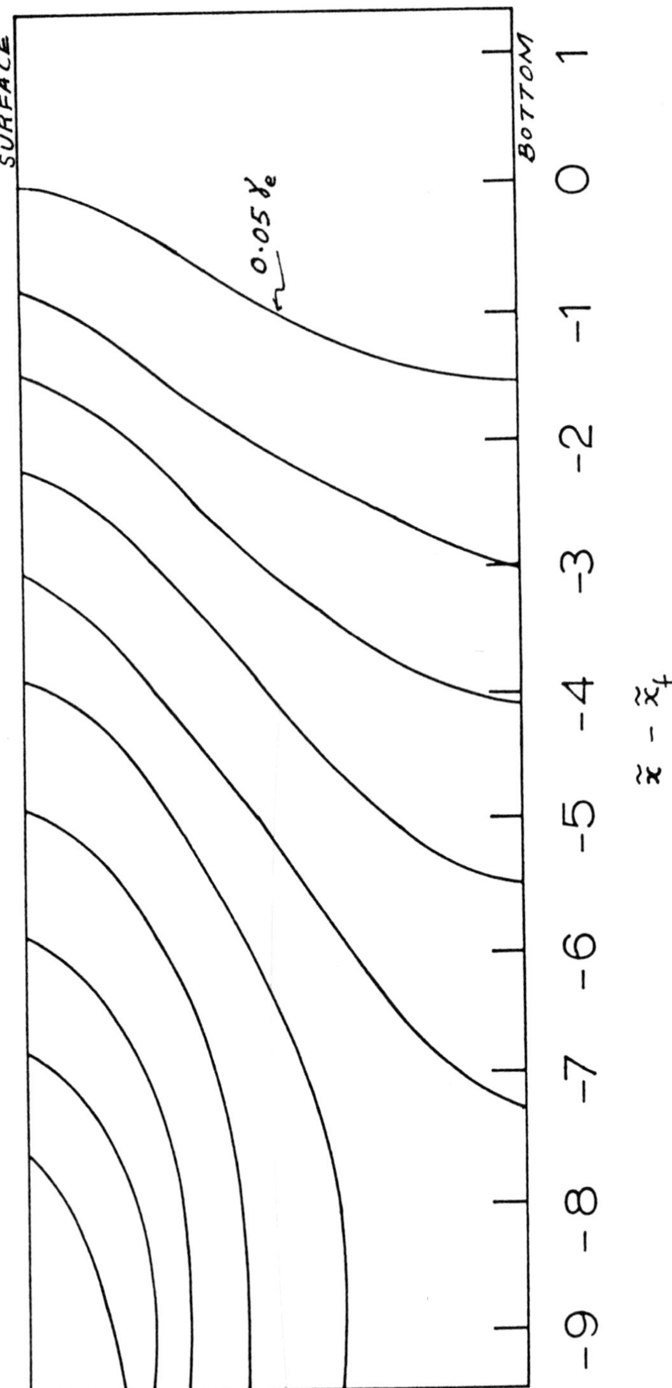

Figure 1. The distribution of isopycnics across the shelf from the model.

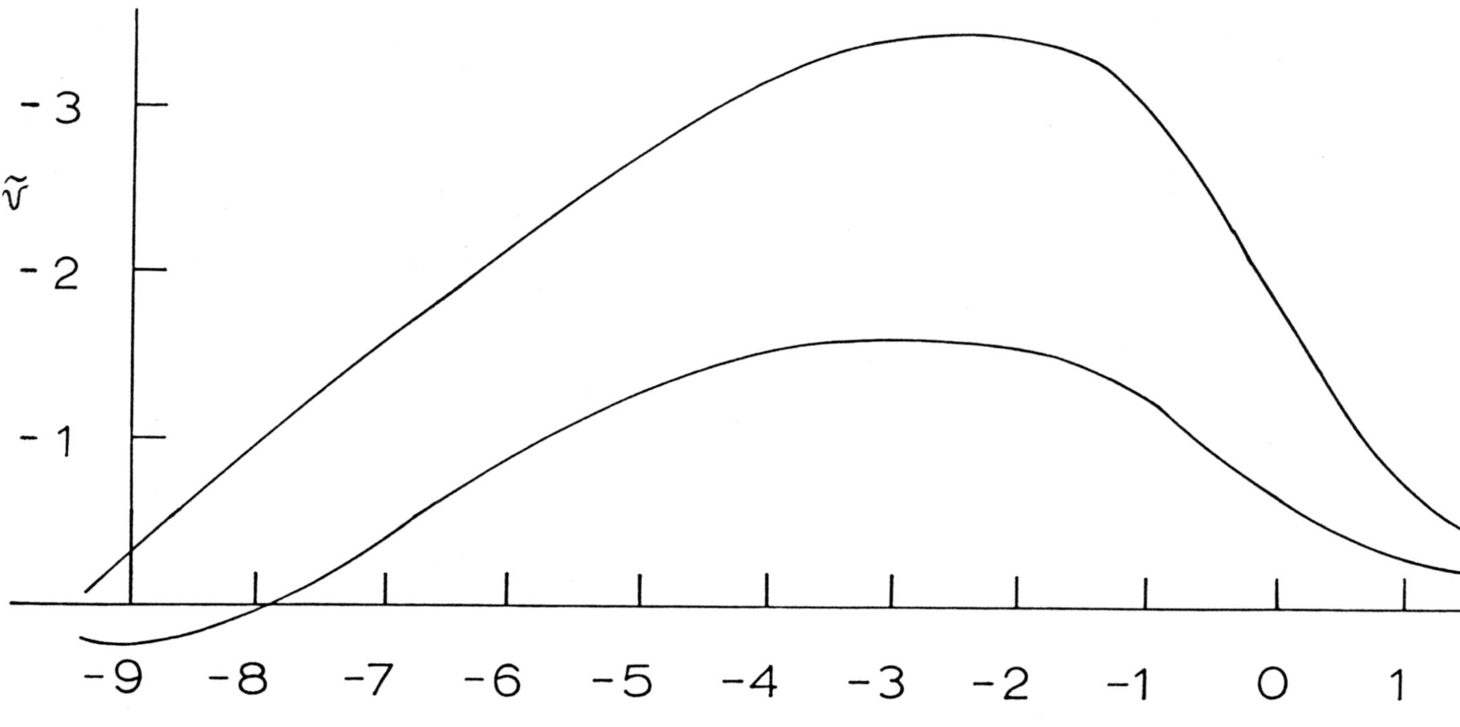

Figure 2. The along-shelf velocity v at the surface at mid-depth.

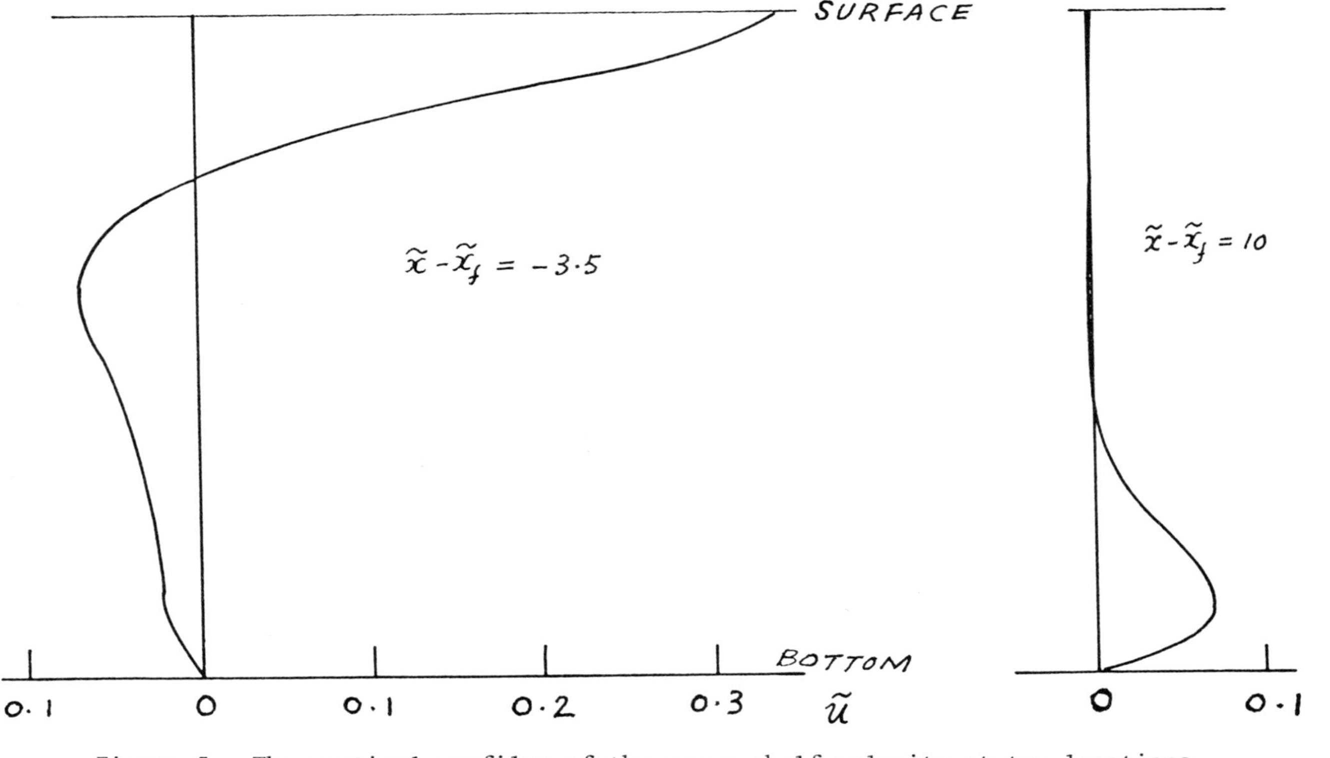

Figure 3. The vertical profiles of the cross-shelf velocity at two locations.

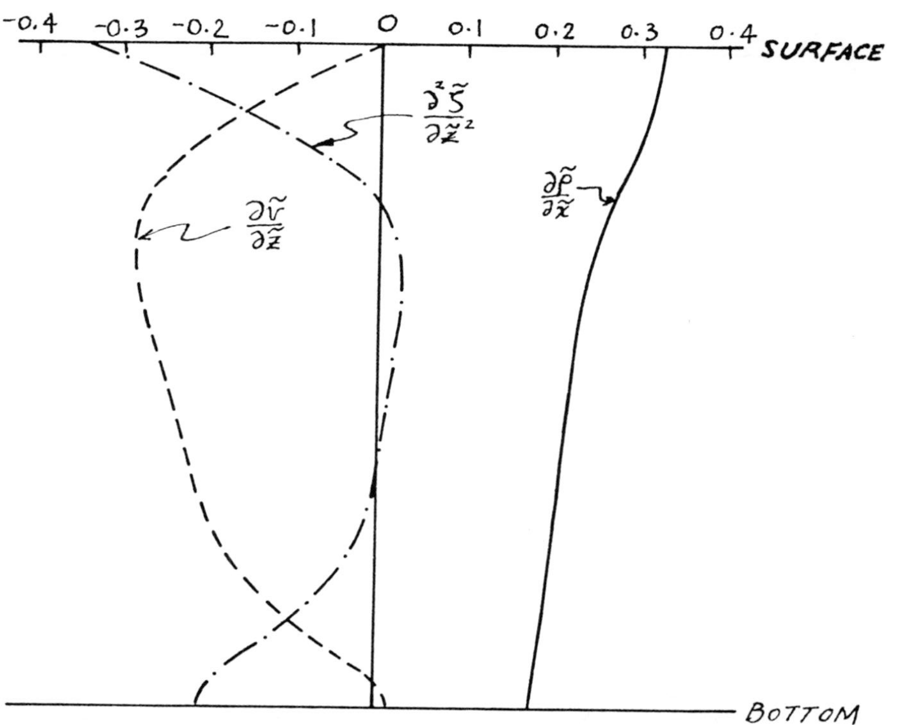

Figure 4(a). Balance of y-component of vorticity rates at $\tilde{x} - \tilde{x}_f = -6$.

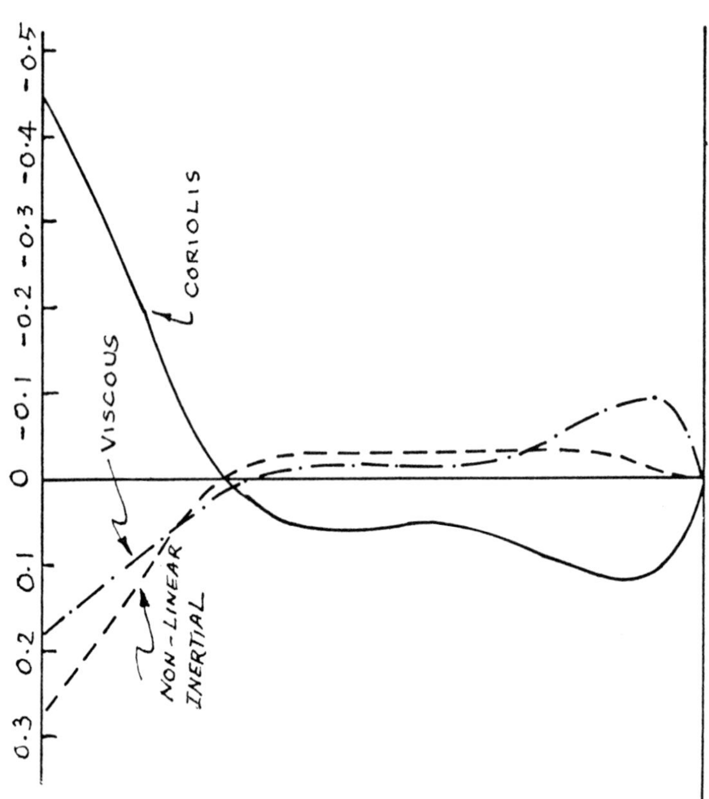

Figure 4(b). Balance of forces in y-direction at the same location.

5. REFERENCES

1. BIGELOW, H.B. - Studies of the waters on the continental shelf, Cape Cod to Chesapeake Bay, I, The cycle of temperature. Pap. Phys. Oceanogr. Meteorol., 2(4), 135 pp., 1933.

2. BIGELOW, H.B. and M. SEARS - Studies of the waters on the continental shelf, Cape Cod to Chesapeake Bay, II, Salinity. Pap. Phys. Oceanogr. Meteorol., 4(1), 95 pp., 1935.

3. BUMPUS, D.F. - A description of the circulation on the continental shelf of the east coast of the United States. Progr. Oceanogr., 6, 111-157, 1973.

4. BEARDSLEY, R.C., W.C. BOICOURT, and D.V. HANSEN - Physical oceanography of the Middle Atlantic Bight. Limnol. Oceanogr. Spec. Symp. Ser. 2, 20-34, 1976.

5. STOMMEL, H. and A. LEETMA - The circulation on the continental shelf. Proc. Nat. Acad. Sci., U.S., 69, 3380-3384, 1972.

6. PIETRAFESA, L.J. and G.S. JANOWITZ - On the effects of buoyancy flux on continental shelf circulation. J. Phys. Oceanogr., 9, 911-918. 1979.

7. KAO, T.W., H.-P. PAO, and C. PARK - Surface intrusions, fronts and internal waves: A numerical study. J. Geophys. Res., 83, 4641-4650, 1978.

8. TORRANCE, K.E. and J.A. ROCKET - Numerical study of natural convection in an enclosure with localized heating from below-creeping flow to the onset of laminar instability, J. Fluid Mech., 36, 33-54, 1969.

9. PAO, H.P. and T.W. KAO - Dynamics of establishment of selective withdrawal of a stratified fluid from a line sink. Part 1, Theory. J. Fluid Mech., 65, 657-688, 1974.

10. KAO, T.W., C. PARK, and H.P. PAO - Inflows, density currents, and fronts. Phys. Fluids, 21, 1912-1922, 1978.

11. KAO, T.W. - The dynamics of oceanic fronts. Part I: The Gulf Stream, J. Phys. Oceanogr., 10, 483-492, 1980.

DETERMINATION OF TIDAL RESIDUAL CURRENTS IN WIDE ESTUARIES, USING FINITE ELEMENT METHODS

A D Jenkins[(i)]

Time-independent finite element methods are used to solve linearised two-layer uniform depth tidal equations. Assuming that the mean (residual) flow is due to Stokes drift, the accuracy of the calculated residual current depends upon the accuracy to which the tidal velocity and its derivatives can be determined. Two Galerkin-type finite element methods are discussed:-

(a) A solution of three first-order equations in the tidal elevation and velocity components, using quadratic interpolation over 6-noded triangular elements.
(b) A solution of a single second-order equation, using quintic interpolation over 3-noded triangular elements.

Finite element solutions to the following problems with known analytical solutions are obtained:-

(a) A single Kelvin wave propagating along a straight-sided canal.
(b) G I Taylor's solution to the tidal equation in a rectangular basin.
(c) A known solution to the tidal equations in a wedge of angle $\pi/3$.

1. INTRODUCTION

Dyke [2] has argued that significant residual currents will arise in estuaries from a combination of tidal motions and density differences, and if friction is ignored, the residual flow will be largely due to Stokes drift

(i) Research Associate, Department of Offshore Engineering, Heriot-Watt University, Riccarton, Currie, Edinburgh EH14 4AS.

$$\underline{u}_s = \overline{\int_{t_o}^{t} \underline{u}(t')dt' . \nabla \underline{u}(t)} \quad , \tag{1.1}$$

where the overbar denotes an average over a complete tidal period. In this paper, two finite element methods, which can be used to evaluate the Stokes drift by solving two-layer time-independent tidal equations, are tested for accuracy. The first method uses quadratic interpolation for the tidal velocities and amplitudes over triangular elements, and the second method uses quintic interpolation for just the tidal amplitudes, also over triangular elements. The Galerkin method for finite element approximation is used in both cases.

2. THE TIDAL EQUATIONS

The following complex modal equations describe linear sinusoidal two-dimensional tidal motions in a two-layer sea [1,2] :

$$i\sigma \underline{u}_j + \underline{f} \wedge \underline{u}_j + g\nabla\zeta_j = 0 \tag{2.1}$$

$$i\sigma\zeta_j + h_j \nabla . \underline{u}_j = 0 \tag{2.2}$$

$$(j = 1,\ 2).$$

In these equations, which are correct to lowest order in the fractional density difference α,

$$\underline{u}_1 = (h'\underline{u}' + h''\underline{u}'')/h, \quad \underline{u}_2 = \underline{u}' - \underline{u}'' \tag{2.3}$$

$$\zeta_1 = \zeta', \quad \zeta_2 = \alpha(h''\zeta'/h - \zeta'') \ , \tag{2.4}$$

where $\underline{u}'$ and $\underline{u}''$ are the horizontal fluid velocities in the top and bottom layers, which have equilibrium thicknesses h' and h''. $(h = h' + h'')$. ζ' and ζ'' are the departures of the levels of the sea surface and the interface from equilibrium. $\underline{f}$ is directed upward and has magnitude equal to the Coriolis parameter (f), which is assumed constant.

From equations (2.1) and (2.2), we can obtain the following Helmholtz equations:

$$\nabla^2\zeta_j + k_j^{\,2}\zeta_j = 0 \ (j = 1,2) \tag{2.5}$$

where $k_j^{\,2} = (\sigma^2 - f^2)/gh_j$, with $h_1 = h$ and $h_2 = \alpha h'h''/h$.

The boundary conditions to be applied are ζ_j prescribed on the open parts of the boundary (essential boundary conditions), and $(u_n)_j = 0$ on the coastlines. From (2.1) we obtain the following expression for the $\underline{u}_j$:

$$\underline{u}_j = \frac{g}{\sigma^2 - f^2}(i\sigma\nabla\zeta_j - \underline{f}\wedge\nabla\zeta_j) \quad , \tag{2.6}$$

so the coastal boundary conditions become

$$i\sigma\frac{\partial\zeta_j}{\partial n} + f\frac{\partial\zeta_j}{\partial s} = 0 \quad , \tag{2.7}$$

where s is directed anticlockwise around the boundary.

3. RESIDUAL FLOW

We assume that the residual flow in each layer is entirely due to Stokes drift (1.1) in each of the top and bottom layers. In complex notation

$$\underline{u}_s = (1/2\sigma)\text{Im}(\underline{u}.\nabla\underline{u}^*) \quad . \tag{3.1}$$

The appropriate expression for the Stokes drift in each layer can be obtained using the relations (2.3).

It can be seen from (2.6) and (3.1) that the calculated Stokes drift is a function of the first and second spatial derivatives of the modal tidal elevations ζ_1 and ζ_2. The accuracy to which the Stokes drift can be determined from a numerical solution of the tidal equations is therefore limited by the accuracy to which the first spatial derivatives of the modal currents $\underline{u}_1$ and $\underline{u}_2$, or the second derivatives of the modal tidal elevations, can be determined. Since $k_2 >> k_1$, it is the accuracy of the solution of the mode 2 equation which will in general be the most critical.

4. FINITE ELEMENT METHODS

In order to test the accuracy of the proposed methods of solution of the tidal equation, we chose to test the methods against a simple analytical solution: a single Kelvin wave propagating from right to left in a parallel-sided canal, as shown in Figure 1 ($\partial_1 A$ is the open boundary and $\partial_2 A$ is the coast).

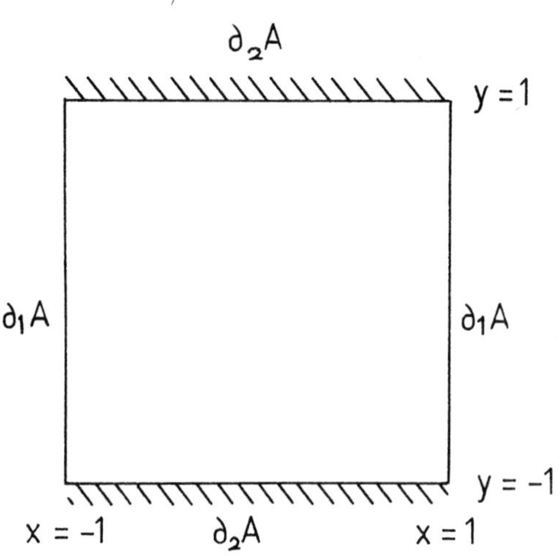

Figure 1

The analytical solution was given by

$$\zeta = \exp(fy)\exp(i\sigma x) , \qquad (4.1)$$

with $f = 0.7$, $\sigma = \sqrt{0.74}$, $g = h_j = 1$, which corresponds to a value of $k^2 = (\sigma^2 - f^2)/gh_j$ of 0.25. It can be verified from (2.6) that the y-component of $\underline{u}$ is zero everywhere, so that the coastal boundary conditions are satisfied. For both finite element methods discussed, the domain was divided into 32 triangular elements, as shown in Figure 2, and the Galerkin approach was used to produce banded systems of linear equations.

The quadratic finite element method starts with (2.1) and (2.2), and incorporates $(u_n)_j = \underline{u}_j.\underline{n} = 0$ on $\partial_2 A$ as natural boundary conditions. Essential boundary conditions are applied by constraining ζ to be equal to its analytical value at the nodes on $\partial_1 A$. 6-noded elements are used, and the complex unsymmetric global matrix has dimension 243 and bandwidth 125 for the mesh of Figure 2.

The quintic method starts with (2.5) and incorporates (2.7) on $\partial_2 A$ as natural boundary conditions. 11th degree quadrature [4,5] is used to evaluate the area integrals, and 9th degree quadrature is used for the boundary integrals, in order to calculate the elements of the global matrix. The basis functions used are the three-node 18-parameter family of Hermite interpolation polynomials described by Mitchell and Wait [3], which have as coefficients the values of ζ and

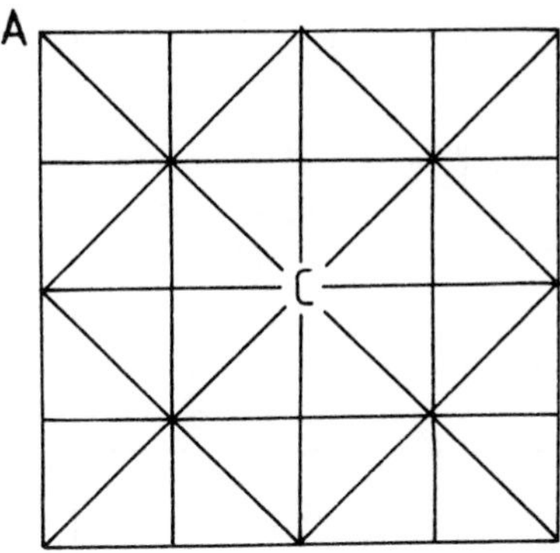

Figure 2

all its first and second derivatives at each node. Essential boundary conditions are applied by constraining

$$\zeta, \quad \frac{\partial\zeta}{\partial s} \quad \text{and} \quad \frac{\partial^2\zeta}{\partial s^2}$$

to be equal to their analytical values at the nodes on $\partial_1 A$, making "local" rotations of the global system of equations where necessary. The global matrix (also complex and unsymmetric) has dimension 150 and bandwidth 83 for the mesh of Figure 2.

5. COMPARISON OF RESULTS, FOR THE QUADRATIC AND QUINTIC FINITE ELEMENT METHODS

Table 1 shows the results of the finite element calculations, compared with the single Kelvin wave solution of (4.1). The Aberdeen University Honeywell Level 66 computer was used to run the programs, which were not optimised in respect of CPU time or core storage. (Complex quantities are represented by their moduli and their phases in degrees).

We can see from Table 1 that the quintic finite element method produces much more accurate results, especially for the velocity derivatives. It also uses less core storage, and though the program takes longer to execute, this difference in execution time is due to the time taken to evaluate the elements of the global matrix, and would be less marked if we were to increase the size of the finite element mesh.

	Quadratic Method		Quintic Method		Analytic Solution	
Core (Words)	80K		45K		-	
CPU TIME	43 s (4 s to set up global matrix)		77 s (65 s to set up global matrix)		-	
ζ at C	1.001	0	1.000	0	1.000	0
u at C	1.007	180	1.000	180	1.000	180
at A	2.019	132	2.014	131	2.014	131
v at C	0.005	-90	0.000	-90	0.000	-
at A	0.016	-42	0.001	-169	0.000	-
$\frac{\partial u}{\partial x}$ (C)	~0.9	~-90	0.859	-90	0.866	-90
(A)	~1.36	~-130	1.747	-139	1.732	-139
$\frac{\partial u}{\partial y}$ (C)	~0.8	~-170	0.701	180	0.700	180
(A)	~1.4	~ 130	1.410	131	1.410	131
$\frac{\partial v}{\partial x}$ (C)	~0.15	~	0.001	180	0.000	-
(A)	~0.3	~	0.018	-1	0.000	-
$\frac{\partial v}{\partial y}$ (C)	~0.15	~	0.002	-90	0.000	-
(A)	~0.2	~	0.001	-74	0.000	-

Table 1

6. SOLUTIONS FOR A RECTANGULAR BASIN AND FOR A WEDGE

The quintic finite element method was also used to solve two other tidal problems with known solutions: G I Taylor's solution in a rectangular basin, open at one end [6] (Figure 3); and a solution for a wedge of angle $\pi/3$ [7,8] (Figure 4).

Tables 2 and 3 summarise the respective results. The physical constants take the same values as for the single Kelvin wave problem in the rectangular basin, but for the wedge (a mode 2 solution), they are:

$f = 1.209 \times 10^{-4}s^{-1}$, $\sigma = 1.405 \times 10^{-4}s^{-1}$, $g = 9.81m/s$,
$h = 3.75 \times 10^{-2}m$.

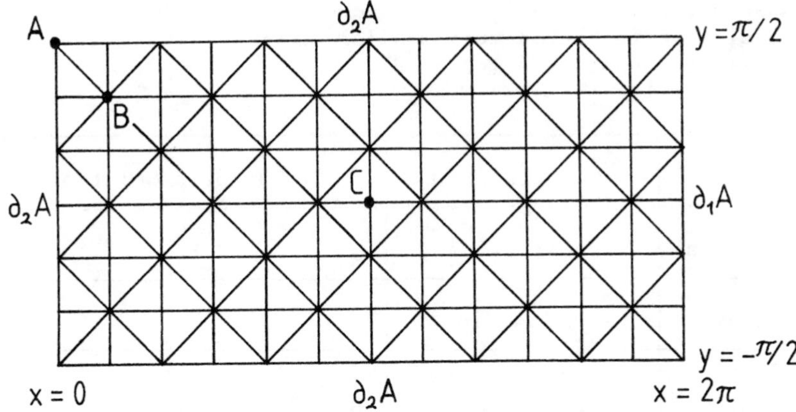

Figure 3: Finite element mesh for G I Taylor's problem

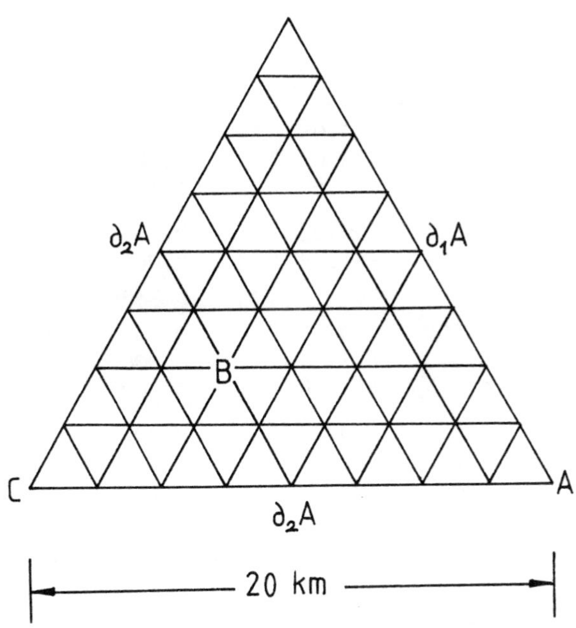

Figure 4: Finite element mesh for wedge of angle $\pi/3$

Tables 2 and 3 show that the quintic finite element method produces excellent results for ζ, and very good results for $\underline{u}$, in both cases, except at the corners of the rectangular basin, near which very large velocity gradients occur. The values for $\nabla \underline{u}$ are good away from the corners for both cases, and acceptable at the corners of the wedge (the worst case for the wedge is at the ends of the open boundary). The unacceptable errors in $\nabla \underline{u}$ at the corners of the rectangular basin become much smaller as one moves to the diagonally adjacent node (point B).

		Quintic f.e.m.		G I Taylor's Solution	
Core (words)		134K		-	
CPU time		6.4 min			
ζ	(C)	1.064	-90.0°	1.068	-90.0°
	(A)	2.268	138.0°	2.271	137.4°
	(B)	1.768	137.4°	1.777	137.4°
u	(C)	0.030	0.1°	0.032	0.0°
	(A)	0.080	-49.0°	0.000	-
	(B)	0.733	-2.4°	0.735	-2.6°
v	(C)	0.124	-90.0°	0.125	-90.0°
	(A)	0.080	-35.2°	0.000	-
	(B)	0.653	-75.5°	0.653	-75.3°
$\frac{\partial u}{\partial x}$	(C)	0.918	180.0°	0.921	180.0°
	(A)	2.803	-23.0°	0.000	-
	(B)	1.209	2.6°	1.179	4.9°
$\frac{\partial u}{\partial y}$	(C)	0.853	90.0°	0.855	90.0°
	(A)	0.857	-81.3°	0.000	-
	(B)	0.630	-16.6°	0.614	-21.6°
$\frac{\partial v}{\partial x}$	(C)	0.108	90.0°	0.108	90.0°
	(A)	0.855	177.2°	0.000	90.0°
	(B)	0.714	114.7°	0.702	119.2°
$\frac{\partial v}{\partial y}$	(C)	0.003	0.0°	0.002	0.0°
	(A)	2.742	122.9°	0.000	-
	(B)	1.070	100.1°	1.030	97.9°

Table 2: Finite element solution for rectangular basin

		Quintic f.e.m.		Analytic solution	
Core		86K			
CPU time		$3\frac{1}{2}$ min			
$\frac{\zeta}{10^{-3}m}$	B:	0.785	29.7°	0.785	29.7°
	C:	4.666	29.7°	4.666	29.7°
$\frac{u}{10^{-2}m/s}$	A:	3.942	152.9°	3.944	152.8°
	B:	2.936	-60.6°	2.936	-60.6°
	C:	0.001	-117.0°	0.000	-
$\frac{v}{10^{-2}m/s}$	A:	0.002	22.0°	0.000	-
	B:	1.696	-59.2°	1.696	-59.2°
	C:	0.002	16.6	0.000	-
$\frac{\partial u}{\partial x} / 10^{-6}s^{-1}$	A:	9.236	65.1°	9.350	65.6°
	B:	1.140	-162.5°	1.140	-162.4°
	C:	15.77	-3.9°	15.69	-4.1°
$\frac{\partial u}{\partial y} / 10^{-6}s^{-1}$	A:	8.401	-27.1°	8.332	-26.6°
	B:	3.027	131.6°	3.028	131.6°
	C:	15.12	-150.5°	15.04	-150.3°
$\frac{\partial v}{\partial x} / 10^{-6}s^{-1}$	A:	0.138	5.1°	0.000	-
	B:	3.524	86.9°	3.525	86.9°
	C:	0.063	144.1°	0.000	-
$\frac{\partial v}{\partial y} / 10^{-6}s^{-1}$	A:	0.594	14.8°	0.497	16.3°
	B:	3.370	-41.0°	3.371	-41.0°
	C:	15.81	-116.4°	15.69	-116.4°

Table 3: Finite element solution for wedge of angle $\pi/3$

REFERENCES

1. RATTRAY M, Jr - Time-dependent motion in an ocean; a unified two-layer, beta-plane approximation. Studies on Oceanography, pp19-29, 1964.

2. DYKE P P G - On the Stokes' drift induced by tidal motions in a wide estuary. Estuarine and Coastal Marine Science, Vol 11, pp17-25, 1980.

3. MITCHELL A R and WAIT R - The Finite Element Method in Partial Differential Equations, John Wiley & Sons, 1977.

4. HAMMER P C, MARLOWE O J and STROUD A H - Numerical integration over simplexes and cones. Math Tables and Other Aids to Computation, Vol 10, pp130-136, 1956.

5. FISHMAN H - Numerical integration constants. Math Tables and Other Aids to Computation, Vol 11, pp1-9, 1957.

6. TAYLOR G I - Tidal oscillations in gulfs and rectangular basins. Proc Lond math Soc, Vol 20, pp148-181, 1920.

7. PACKHAM B A and WILLIAMS W E - The diffraction of Kelvin waves at a corner. J Fluid Mechanics, Vol 34, pp517-529, 1968.

8. DYKE P P G - Residual currents in a wide estuary - a mathematical model. Proceedings of the 1980 European Conference on Environmental Pollution, Ed Bhatnagar V M, Elsevier.

SOME VARIANTS OF THE ICE-TECHNIQUE

JOACHIM BENNER

Kernforschungszentrum Karlsruhe
Institut für Reaktorentwicklung
7500 Karlsruhe, Fed.Rep.Germany

ABSTRACT

Some modifications of Harlow and Amsden's ICE-Technique are discussed. They are selected to reduce the numerical damping inherent in the ICE-method. The modifications are tested by three different problems: 1) one-dimensional shock-tube; 2) two-dimensional shallow water flow; 3) two-dimensional, incompressible flow over a backward facing step. The problems are described and the results are compared. We found, that a lower "degree of implicitness" gives the best results in almost all cases.

1 INTRODUCTION

For simulation of the depressurization process in a pressurized water reactor in case of a hypothetical blowdown accident many numerical codes exist. Some of them use the ICE (Implicit-Continuous-Eulerian) Technique, developed by Harlow and Amsden in [1] and [2]. Recently some theoretical investigations [3] and experimental results [4] have shown, that this method suffers from strong numerical damping. We therefore tried to modify the ICE-Technique slightly in order to reduce damping and to improve numerical performance.

2 DIFFERENTIAL EQUATIONS

The system of differential equations we want to solve are the Navier-Stokes-Equations for compressible, viscous flow:

(1) $$\frac{\partial \rho}{\partial t} + \text{div}\,(\rho\vec{u}) = 0$$

(2) $$\frac{\partial(\rho\vec{u})}{\partial t} + \text{div}\,(\rho\vec{u} \bullet \vec{u}) = -\,\text{grad}\,p + \nu\Delta\vec{u}$$

(3) $$\rho\,(\frac{\partial e}{\partial t} + \text{div}\,(e\vec{u}) - e\,\text{div}\,\vec{u}) = p\cdot\text{div}\,\vec{u}$$

(4) $$p = p\,(\rho, e)$$

Here e denotes the specific internal energy and $\vec{u} = (u_i, u_j)$ the velocity vector. "$\bullet$" is the dyadic product.

3 DISCRETIZATION

There are several Finite-Difference Methods to solve (1) - (4). Frequently, explicit schemes are used which all have one disadvantage. To ensure stability the maximum time step Δt and the space discretisation Δx are related by the CFL-condition [5] (c = sound velocity):

(5) $$(c + |\vec{u}|)\,\frac{\Delta t}{\Delta x} \leq 1$$

Fully implicit schemes have no restriction for the time step, but each step requires the solution of a set of non-linear algebraic equations.

The ICE-Technique uses a compromise: those parts of (1) - (4) which deal with fluid-convection are treated explicitly, parts envolving sound propagation are treated implicitly. We therefore have the stability condition [6]:

(6) $$\beta = |\vec{u}|\,\frac{\Delta t}{\Delta x} < \frac{1}{2}$$

which normally is less restrictive then (5). β is called the "Courant-Number".

But this partly implicit formulation is one reason for numerical damping. The "degree of implicitness" therefore should be as small as possible without changing the stability condition. Here we do this by taking pressures and densities with equal weights at the old and new time step (called a "half-implicit-scheme").

For space discretization a "staggered grid" is used in ICE, where different variables are calculated at different positions in the grid. For products of these variables we therefore need some interpolation-formulae. To ensure stability, ICE employs the "donor-cell-interpolation" (also called "upwind-differencing"). This is of first order accuracy and another reason for numerical damping. We therefore tried to use second-order central-differences in space and time (the "Leapfrog-scheme" in contrast to the usual "Euler-scheme", where time-discretization is one-sided). It is well known [7] that the Leapfrog-scheme produces high-frequency numerical oscillations. In order to suppress them we took some kind of avarage among Leapfrog- and Euler-scheme every N_o time step, as it was done in [8].

A scheme which incorporates all these variants is the following:

$$(7)\quad \rho^{n+1} = \rho^n + \Delta t\cdot\{FC\cdot DIV(I_o(\rho\vec{u}))^{n+1} + (1-FC)\cdot DIV(I_o(\rho\vec{u}))^n\}$$

$$(8)\quad (\rho\vec{u})^{n+1} = f_1(d,n)\cdot(\rho\vec{u})^n + f_2(d,n)\cdot(\rho\vec{u})^{n-1} - f_3(d,n)\cdot \Delta t\cdot\{DIV(I_d(\vec{u}\bullet\vec{u}))^n + FM\cdot GRAD(p)^{n+1} + (1-FM)\cdot GRAD(p)^n + \nu\cdot DELTA(\vec{u})^{n-d}$$

$$(9)\quad e^{n+1} = f_1(d,n)\cdot e^n + f_2(d,n)e^{n-1} - f_3(d,n)\cdot\Delta t\cdot \{DIV(I_d(\vec{u}\cdot e))^{n+1-d} - e^n\, DIV(\vec{u})^{n+1-d} - (p^{n+1-d}/\rho^{n+1-d})\cdot DIV(\vec{u})^{n+1-d}\}$$

$$(10)\quad p^{n+1} = p(\rho^{n+1}, e^n)$$

Here we used the notations:

d = 0 : If upwind-differences are used

d = 1 : If central-differences are used

I_d with d = 0,1 interpolation-operators:

I_o Interpolation with upwind-differences

I_1 Interpolation with central-differences

$0.5 \leq FC, FM \leq 1$

a^n denotes the quantity a taken at the nth time step.

$$\left.\begin{aligned} f_1(o,n) &= 1\\ f_2(o,n) &= 0\\ f_3(o,n) &= 1 \end{aligned}\right\} \text{ for all } n \qquad \left.\begin{aligned} f_1(1,n) &= 1\\ f_2(1,n) &= 0\\ f_3(1,n) &= 1 \end{aligned}\right\} \text{ for } n = 1$$

$$\left.\begin{aligned} f_1(1,n) &= 0\\ f_2(1,n) &= 1\\ f_3(1,n) &= 2 \end{aligned}\right\} \begin{array}{l}\text{for}\\ n \not\equiv 0 \bmod(N_0)\end{array} \qquad \left.\begin{aligned} f_1(1,n) &= 0.5\\ f_2(1,n) &= 0.5\\ f_3(1,n) &= 1.5 \end{aligned}\right\} \begin{array}{l}\text{for}\\ n \equiv 0 \bmod (N_0)\end{array}$$

DIV, GRAD and DELTA are the usual difference approximations for the operators div, grad and Δ.

In the case of discontinuous solution functions the variants of ICE with reduced damping produce numerical oscillations. To suppress them, here an artificial viscosity was used, similar to a proposition of Lapidus /¯9_7. Artificial damping is proportional to the velocity gradient and is only used in regions with fluid-compression.

4 TEST PROBLEMS

4.1 One-dimensional, compressible flow: Shock-Tube

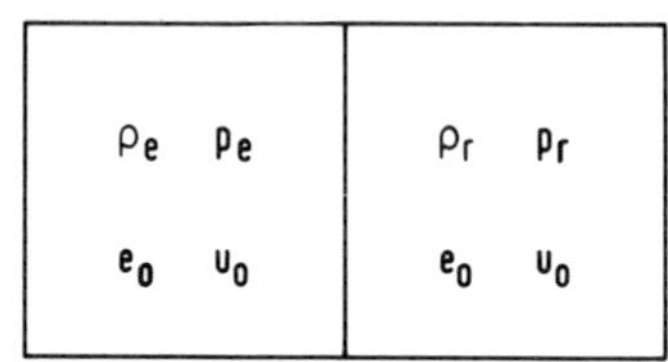

Fig. 1: Shock tube

A cylinder is devided into two semi-infinite sections by a diaphragm and filled with an ideal gas. The initial pressure and density in the two sections are different, initial velocity and specific internal energy are equal (Fig. 1). At t = 0 the diaphragm is removed, then a shock front and a contact discontinuity propagate to one side, a rarefaction wave to the other. At any time the solution can be calculated analytically /¯10_7, in the following section the numerical results are compared with this analytical solution.

In order to see the effect of fluid convection two cases are distinguished. First, we consider the case where the initial velocity of the system is zero. Then the maximum convective velocity after removing the diaphragm is 20 % of the sound velocity. Second we assume that the total configuration has a constant initial velocity which takes a value of 70 % of the sound speed. This corresponds to a Galilei-Transformation and does not change the analytical solution except for a general shift.

4.2 Two-dimensional, compressible flow: Shallow water equations

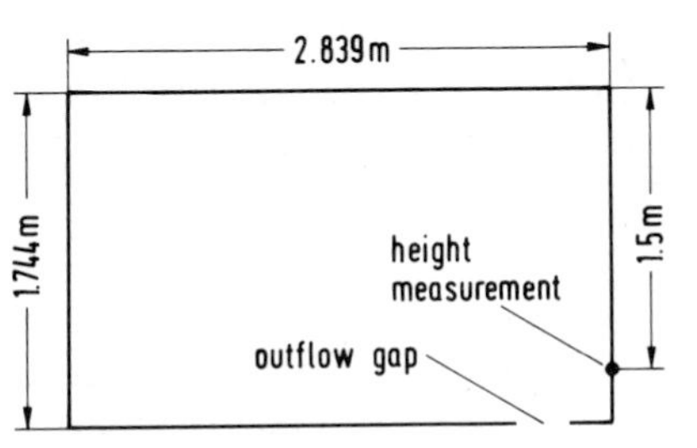

Fig. 2: Water-bed

Under some conditions /̄11_7 the equations describing the two-dimensional flow of shallow water in a plane bed with vertical boundaries under gravity (Fig. 2) are formally identical with (1), (2) and (4). For this "hydraulic analogy" we have to identify the quantities as shown in Table 1. (11) gives an "equation of state" which is independent of the energy density (similar to acoustic media). Hence, the energy equation (3) is superflows.

Ideal gas	Shallow water
ρ: Density	h: Water surface height
$\vec{u}$: Velocity	$\vec{u}$: Velocity
t: Time	t: Time
p: Pressure	$\frac{1}{2}\,gh^2$ (11)
c: Sound-velocity	$u = \sqrt{g \cdot h}$ speed of small disturbances

Table 1: Hydraulic-Analogy

In an experiment /̄11_7 the variation in time of the water surface height at several points was measured. Here the measurement at one point is used to compare the different schemes, Fig. 2 shows the location of the point and the geometry of the basin.

4.3 Two-dimensional, incompressible flow: Backward-facing step

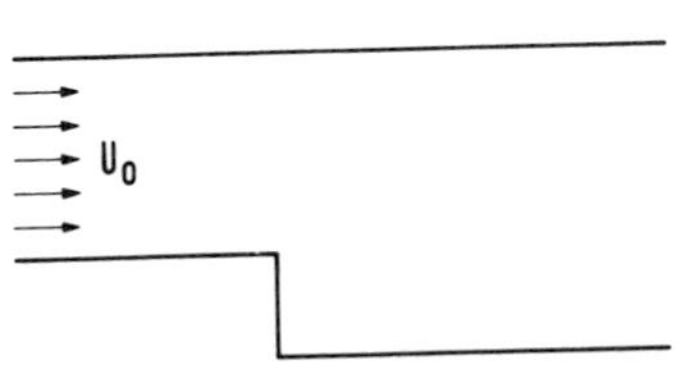

Fig. 3: Backward-facing step

In order to investigate the behaviour of the difference equations in the limiting case $c \rightarrow \infty$ we studied the flow over a backward-facing step (Fig. 3). We used very simple boundary-conditions: Constant inflow velocity at the left-hand side and free outflow at the other. We also neglected boundary-layers at the walls. This is justified because our favourite aim is to compare different numerical schemes, not to solve the test-problem exactly.

5 RESULTS

5.1 Shock-Tube

In all cases now discussed we chose a Courant-Number of 0.12. We also investigated other discretizations and found that the principal differences among the schemes did not change.

In a first set of calculations the gas was initially at rest. Fig. 4 shows the density-distribution calculated with ICE (FC = FM = 1, d = 0). Here we see strong numerical damping. By a "half-implicit upwind scheme" damping is reduced but the wellknown / 5 / numerical oscillations occur. They vanish if we use an artificial viscosity (Fig. 5).

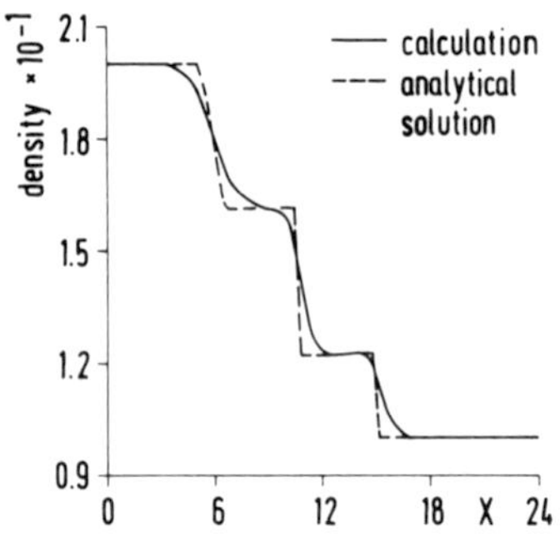

Fig. 4: FC = FM = 1, d=0

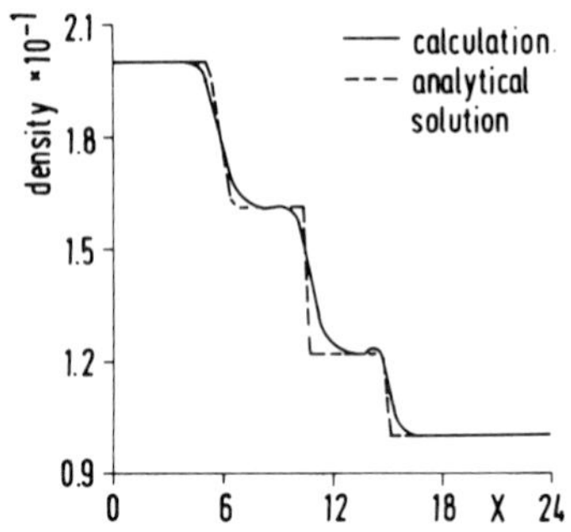

Fig. 5: FC = FM = 0.5, d=0, artificial viscosity

The results are similar if we use central differences (Fig. 6: fully-implicit, Fig. 7: half-implicit). But now the calculation-times are longer.

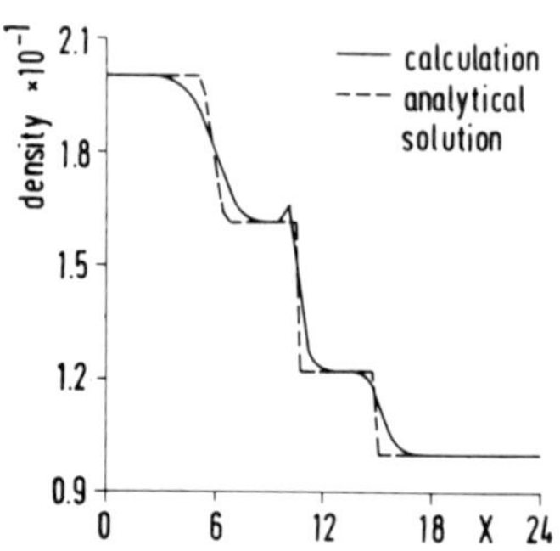

Fig. 6: FC = FM = 1, d=1

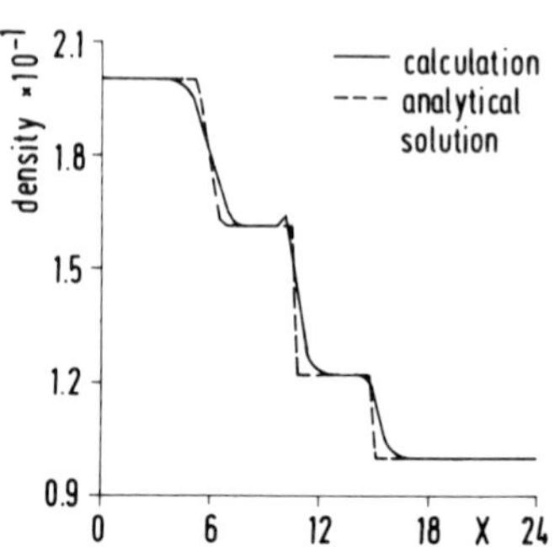

Fig. 7: FC = FM = 0.5, d=1, artificial viscosity

With large convective velocities the accuracy decreases. Full- and half-implicit upwind-schemes now give the same bad results (Fig. 8), central differences hardly improve them and produce numerical oscillations which cannot be damped (Fig. 9).

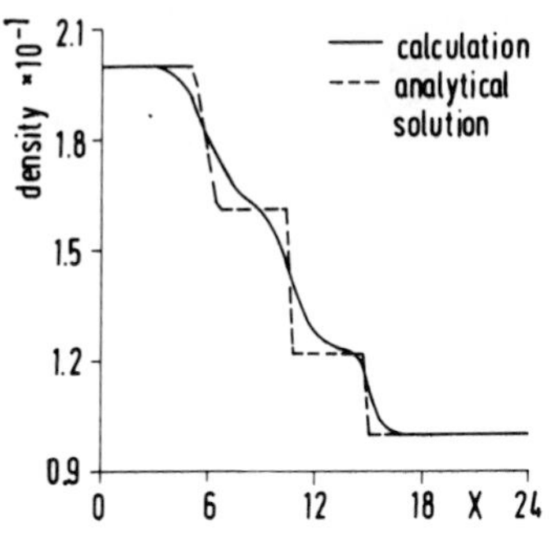

Fig. 8 : FC = FM = 0.5, d = 0

Fig. 9: FC = FM = 0.5, d = 1, artificial viscosity

For this problem we saw that the half-implicit scheme using upwind-differences is preferable to all other schemes, as long as the convective velocities are small. For problems with high velocities the numerical damping is not reduced by the variants of the ICE-Technique discussed here.

5.2 Shallow water equations

Here a discretization was selected with a Courant-Number of 0.3.

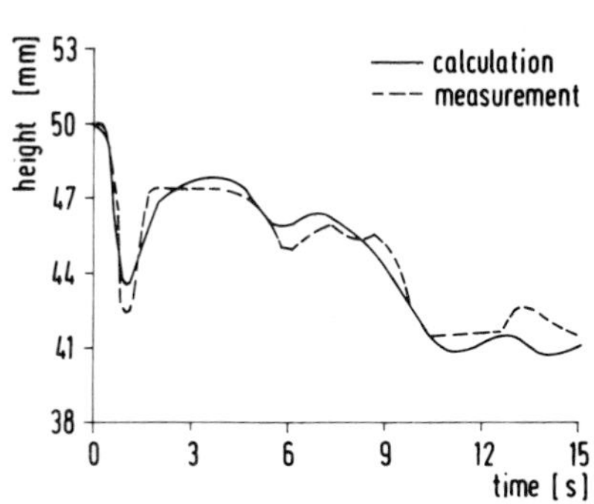

Fig. 10: Variation of water surface height, FC=FM=1, d=0

Generally we have the same results as in the last section. Fig. 10 shows measured and calculated height, here we used the ICE-Technique. Again we notice numerical smearing. The half-implicit scheme using donor-cell interpolation (Fig. 11) gives a better correspondence of experiment and calculation. The difference among both schemes is still larger at points more distant from the outflow.

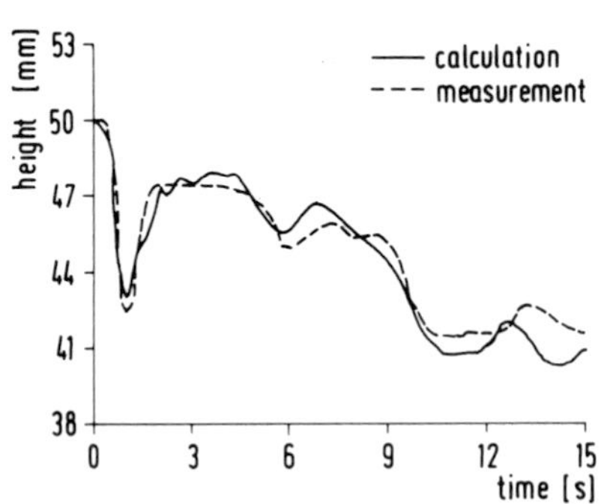

Fig. 11: Variation of water surface height,

FC= FM = 0.5, d=0

Leapfrog schemes have no advantages. So the choice of N_0 was difficult, the optimal value turned out to be very small. The use of artificial viscosity was necessary although no shock occur. The curves with full- and half-implicit calculation (Fig. 12 and Fig. 13) are similar to those obtained by upwind differences.

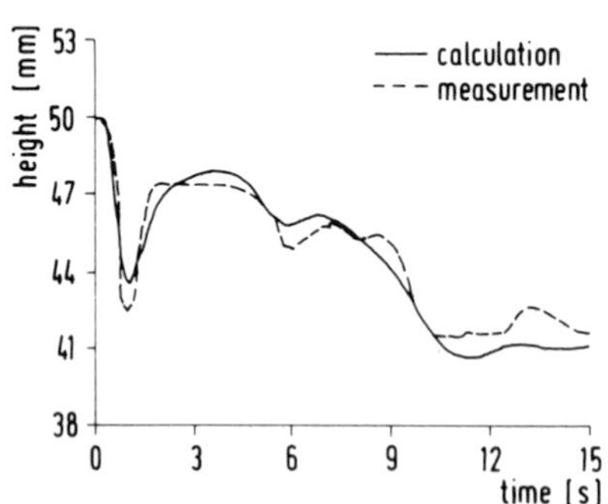

Fig. 12: Variation of water surface height,

FC=FM= 1, d=1, N_0=5

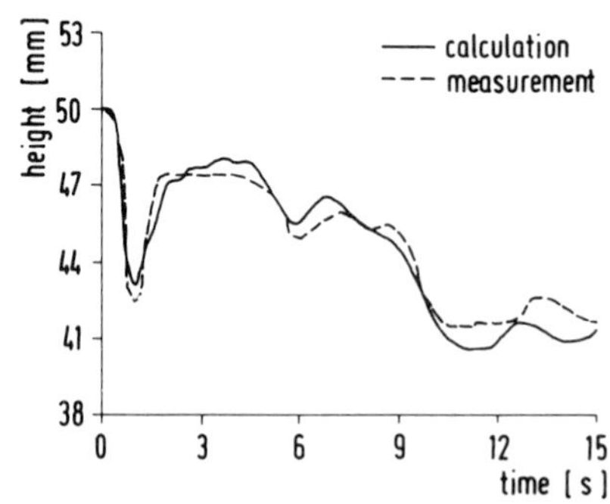

Fig. 13: Variation of water surface height,

FC= FM= 0.5, d=1, N_0=5

Scheme	time /sec/
Upwind, full-impl. FC=FM=1,d=0	262
Upwind, half-impl. FC=FM=0.5,d=0	251
Leapfrog, full-impl. FC=FM=1,d=1,N_0=5	765
Leapfrog, half-impl. FC=FM=0.5,d=1,N_0=5	517
Leapfrog, full-impl. FC=FM=1,d=1,N_0=1	686
Leapfrog, half-impl. FC=FM=0.5,d=1,N_0=1	477

Table 2: Calculation times

In termes of calculation times, the half-implicit scheme using donor-cell interpolation is again the best (Table 2). It is interesting that the Leapfrog-scheme with N_0 = 1 is the best if one wants to use central differences.

The times in Table 2 are valid for 600 time steps, calculations were performed on a IBM 3033.

We conclude that also for two-dimensional problems a half-implicit formulation improves the results of ICE. For calculations with the Leapfrog-scheme the correct choice of the avaraging-parameter N_0 is very important. We are not able to proof whether the case N_0 = 1 always is optimal.

5.3 Backward facing step

Here the numerical results lead us to different conclusions compared to the two proceeding sections. We found that half-implicit schemes are not well suited for this type of problem. The velocity distribution, calculated by central differences (Fig. 14) shows a higher velocity in the recirculating eddy then in the case of upwind-differences (Fig. 15). The stagnation point in the first case lies further downstream.

This indicates that in the incompressible limiting case central differences can reduce numerical damping. Now also the computational efford of the Leapfrog-scheme is smaller. Again it is optimal to take the avarage of Euler- and Leapfrog-scheme every time step.

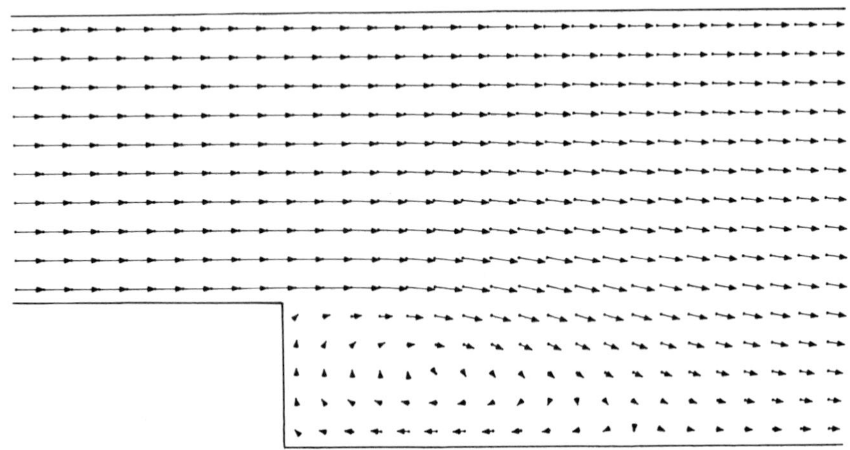

Fig. 14: Backward facing step, FC=FM=1, d=0

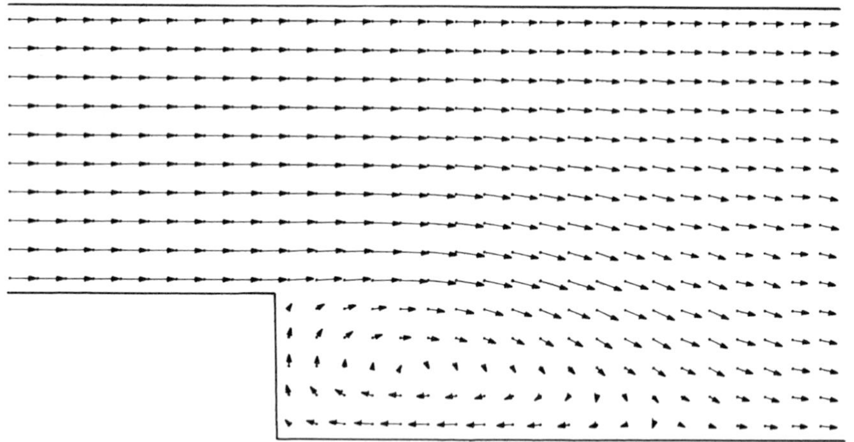

Fig. 15: Backward facing step, FC=FM=1, d=1

6 CONCLUSIONS

In order to reduce numerical damping, for compressible flow it is highly recommended to change the classical fully-implicit ICE-Technique to a half-implicit version. At the same time the "donor-cell-interpolation" should be kept. If discontinuous solutions can occur, an artificial viscosity should be used to damp numerical oscillations. The only disadvantage of the half-implicit scheme is a higher amount of computer storage.

For incompressible flow central differences appear to be preferable. All these schemes show large numerical damping for problems with very high convective velocities.

7 REFERENCES

/¯1_7 HARLOW, F.H., AMSDEN, A.A. - Numerical Calculation of Almost Incompressible Flow, J.Comp.Phys. 3, pp 80-93, 1968

/¯2_7 HARLOW, F.H., AMSDEN, A.A. - A Numerical Fluid Dynamics Calculation Method for All Flow Speeds, J.Comp.Phys. 8, pp 197-213, 1971

/¯3_7 SCHUMANN, U. - Effektive Berechnung dreidimensionaler Fluid-Strukturwechselwirkung beim Kühlmittelverluststörfall - FLUX, KfK 2645, 1979

/¯4_7 KEDZIUR, F. - Untersuchung einer Zweiphasen-Düsenströmung und Überprüfung verschiedener Rechenprogramme anhand experimenteller Ergebnisse, KfK 2946, 1980

/¯5_7 RICHTMYER, R.D., MORTON, K.W. - Difference Methods for Initial-Value Problems, Interscience, New York, 1967

/¯6_7 SCHUMANN, U. - Linear Stability of Finite Difference Equations for Three-Dimensional Flow-Problems, J.Comp.Phys. 18, pp 465-470, 1975

/¯7_7 LILLY, D.K. - On the Computational Stability of Numerical Solutions of Time Dependent Non-Linear Geophysical Fluid Dynamics Problems, Monthly Wether Review, Vol. 93, No. 1, pp 11-26, 1965

/¯8_7 SCHUMANN, U. et al. - Direct Numerical Simulation of Turbulence, Prediction Methods for Turbulent Flow, Ed. Kollmann, W., McGraw-Hill, 1980

/¯9_7 LAPIDUS, A. - A Detached Shock Calculation by Second-Order Differences, J.Comp.Phys. 2, pp 154-177, 1967

/¯10_7 HARLOW, F.H., AMSDEN, A.A. - Fluid Dynamics, a LASL Monograph, LA 4700, 1971

/¯11_7 KEDZIUR, F. et al. - PWR-Depressurization and its Hydraulic Analogy, Nucl.Eng.Des. 47, pp 25-34, 1978

SECTION 6
TURBOMACHINERY
AND
AIR FOIL FLOW

COMPUTATION OF THREE-DIMENSIONAL HYPERSONIC VISCOUS FLOWS OVER LIFTING BODIES AT HIGH ANGLES OF ATTACK

K. Y. Szema* and Clark H. Lewis**

Virginia Polytechnic Institute and State University
Blacksburg, Virginia 24061 USA

SUMMARY

The viscous shock-layer equations for three-dimensional hypersonic flow over lifting bodies at high angles of attack have been developed. For the complex three-dimensional reentry vehicle geometries of interest, the resulting equations are written in a nonorthogonal, body-oriented coordinate system, and the three velocity components are defined in the nonorthogonal coordinate directions. Since the viscous shock-layer governing equations are parabolic in both the streamwise and crossflow directions, the equations are solved by a highly efficient finite-difference scheme. The principal advantages of this technique are (i) the numerical method can be used to predict the flowfield about arbitrary geometries in both subsonic and supersonic regions, (ii) the solution is direct, and (iii) the effects of inviscid-viscous interactions are included within a single set of governing equations which are uniformly valid throughout the shock layer. Numerical solutions have been obtained for a 1:1.4 and a 1:2 ellipsoid at zero and 10-degrees angle of attack. Comparisons of surface pressure distributions were made with inviscid solutions, and the agreement is good in all the cases.

LIST OF SYMBOLS

C_p^*	specific heat at constant pressure
$\hat{e}_i$	unit vector of a general orthogonal coordinate system
$\vec{g}_1, \vec{g}_2, \vec{g}_3$	vectors in ξ_1 (streamwise), ξ_2 (normal) and ξ_3 (circumferential) directions

*Research Associate, **Professor, Department of Aerospace and Ocean Engineering.

g_{ij}	metrics $g_{ij} = \frac{\partial z}{\partial \xi_i} \frac{\partial z}{\partial \xi_j} + \frac{\partial r}{\partial \xi_i} \frac{\partial r}{\partial \xi_j} + r^2 \frac{\partial \phi}{\partial \xi_i} \frac{\partial \phi}{\partial \xi_j}$
g	$\det(g_{ij}) = g_{11}\, g_{22}\, g_{33} - g_{22}\, g_{13}^2$
h	static enthalpy, h^*/U_∞^{*2}
$\vec{N}$	shock-normal vector
p	pressure
Pr	Prandtl number
R_n	body nose radius
T	temperature
T^*_{ref}	reference temperature, U_∞^{*2}/C_p^*
$\vec{T}$	shock-tangent vector
u, v, w	streamwise, normal and crossflow velocity components nondimensionalized by the freestream velocity U_∞^*
U_N	velocity component normal to the shock
U_T	velocity component tangent to the shock
z, r, ϕ	cylindrical coordinates
Y_s	shock stand-off distance
α	angle of attack
γ	ratio of specific heats
ε	Reynolds number parameter $\varepsilon^2 = \mu^*_{ref}/\rho^*_\infty U^*_\infty R^*_n$
μ	viscosity, μ^*/μ^*_{ref}
μ_{ref}	reference viscosity, $\mu^*(T_{ref})$
ξ_1, ξ_2, ξ_3	computational coordinates
ρ	density, ρ^*/ρ^*_∞
$\left\{ {i \atop j\ k} \right\}$	Christoffel symbol of the 2nd kind $\frac{1}{2} \sum_m g^{im} \left[\frac{\partial g_{mk}}{\partial \xi_j} + \frac{\partial g_{mj}}{\partial \xi_k} - \frac{\partial g_{jk}}{\partial \xi_m} \right]$

<u>Subscript</u>

∞	dimensional freestream conditions

<u>Superscript</u>

$*$	dimensional quantity

1. INTRODUCTION

Various analyses are available to investigate the flow fields and wall-measurable quantities for hypersonic flow past an axisymmetric blunt body at different angles of attack. However, the effects of the noncircular body geometry can be significant, and the main purpose of this study is to develop a technique to predict hypersonic viscous flows over a noncircular body at typical planetary entry conditions.

Recently a numerical method for 3-D laminar, transitional and/or turbulent hypersonic flows of perfect gas over a blunt body, used for planetary entry probes, has been investigated [1] at different angles of attack by using an implicite finite-difference viscous shock-layer analysis. However, this analysis is valid for axisymmetric bodies only. For the complex three-dimensional lifting reentry vehicle geometries of interest, the resulting equations are written in a nonorthogonal body-oriented coordinate system, and the three velocity components are defined in the nonorthogonal coordinate directions. This procedure is different from writing the equations in an orthogonal coordinate system and explicity performing a coordinate transformation. Since the viscous shock-layer governing equations are parabolic in both the streamwise and crossflow directions, the equations were solved by a highly efficient finite-difference scheme [1,2]. The principle advantages of this technique are (i) the numerical method can be used to predict the flowfield about arbitrary geometries in both subsonic and supersonic regions, (ii) the solution is direct, and (iii) the effects of inviscid-viscous interactions are included within a single set of governing equations which are uniformly valid throughout the shock layer.

The basic formulation of the problem is presented in Section 2. Boundary conditions at the body surface and immediately behind the shock are given in Section 3. Coordinates and method of solution are presented in Sections 4 and 5, respectively. The results are discussed in Section 6, and the conclusions are presented in Section 7.

2. BASIC FORMULATION

The basic equations are derived from the steady Navier-Stokes equations in general body-oriented tensor form (Fig. 1). One of the coordinates, ξ_1, is chosen in the general axial direction, and another, ξ_2, in a direction normal to the body, and the third, ξ_3, around the body. The coordinate system requires orthognality only at the body surface. The normal velocity v and normal coordinate ξ_2 are assumed to be of order ε, and all terms which are of higher order than ε are neglected. The methods of obtaining these equations are discussed in detail in [2,3]. The nondimensional form of the viscous shock-layer equations that are applicable in the present case can be written as:

Continuity Equation:

$$\frac{\partial}{\partial \xi_1} (\rho u g^{\frac{1}{2}}) + \frac{\partial}{\partial \xi_2} (\rho v g^{\frac{1}{2}}) + \frac{\partial}{\partial \xi_3} (\rho w g^{\frac{1}{2}}) = 0 \qquad (1)$$

ξ_1-Momentum:

$$\rho u \frac{\partial u}{\partial \xi_1} + \rho v \frac{\partial u}{\partial \xi_2} + \rho w \frac{\partial u}{\partial \xi_3} + \rho \left[u^2 \left\{ {1 \atop 1\ 1} \right\} + 2uv \left\{ {1 \atop 1\ 2} \right\} + 2uw \left\{ {1 \atop 1\ 3} \right\} + v^2 \left\{ {1 \atop 2\ 2} \right\} + 2vw \left\{ {1 \atop 2\ 3} \right\} + w^2 \left\{ {1 \atop 3\ 3} \right\} \right]$$

$$+ \frac{g_{22}}{g} \left[g_{33} \frac{\partial p}{\partial \xi_1} - g_{13} \frac{\partial p}{\partial \xi_3} \right]$$

$$= \varepsilon^2 \left[\frac{1}{g_{22}} \left(\frac{\partial \mu}{\partial \xi_2} \frac{\partial u}{\partial \xi_2} + \mu \frac{\partial^2 u}{\partial \xi_2^2} \right) \right] \qquad (2)$$

ξ_2-Momentum:

$$\rho u \frac{\partial v}{\partial \xi_1} + \rho v \frac{\partial v}{\partial \xi_2} + \rho w \frac{\partial v}{\partial \xi_3} + \rho \left[u^2 \left\{ {2 \atop 1\ 1} \right\} + 2uv \left\{ {2 \atop 1\ 2} \right\} + 2uw \left\{ {2 \atop 1\ 3} \right\} + v^2 \left\{ {2 \atop 2\ 2} \right\} + 2vw \left\{ {2 \atop 2\ 3} \right\} + w^2 \left\{ {2 \atop 3\ 3} \right\} \right]$$

$$+ \frac{1}{g_{22}} \frac{\partial p}{\partial \xi_2} = 0 \qquad (3)$$

ξ_3-Momentum:

$$\rho u \frac{\partial w}{\partial \xi_1} + \rho v \frac{\partial w}{\partial \xi_2} + \rho w \frac{\partial w}{\partial \xi_3} + \rho \left[u^2 \left\{ {3 \atop 1\ 1} \right\} + 2uv \left\{ {3 \atop 1\ 2} \right\} + 2uw \left\{ {3 \atop 1\ 3} \right\} + v^2 \left\{ {3 \atop 2\ 2} \right\} + 2vw \left\{ {3 \atop 2\ 3} \right\} + w^2 \left\{ {3 \atop 3\ 3} \right\} \right] + \frac{g_{22}}{g} \left[g_{11} \frac{\partial p}{\partial \xi_3} - g_{13} \frac{\partial p}{\partial \xi_1} \right] = \varepsilon^2 \left[\frac{1}{g_{22}} \left(\frac{\partial \mu}{\partial \xi_2} \frac{\partial w}{\partial \xi_2} + \mu \frac{\partial^2 w}{\partial \xi_2^2} \right) \right] \qquad (4)$$

Energy Equation:

$$\rho u \frac{\partial h}{\partial \xi_1} + \rho v \frac{\partial h}{\partial \xi_2} + \rho w \frac{\partial h}{\partial \xi_3} - \left[u \frac{\partial p}{\partial \xi_1} + v \frac{\partial p}{\partial \xi_2} + w \frac{\partial p}{\partial \xi_3} \right] = \varepsilon^2 \left[\frac{1}{g_{22}} \left(\frac{\partial}{\partial \xi_2} \left(\frac{\mu}{Pr} \right) \frac{\partial h}{\partial \xi_2} + \frac{\mu}{Pr} \frac{\partial^2 h}{\partial \xi_2^2} \right) \right]$$

$$+ \varepsilon^2 \left\{ \mu \left[\frac{g_{11}}{g_{22}} \left(\frac{\partial u}{\partial \xi_2} \right)^2 + 2 \frac{g_{13}}{g_{22}} \frac{\partial u}{\partial \xi_2} \frac{\partial w}{\partial \xi_2} + \frac{g_{33}}{g_{22}} \left(\frac{\partial w}{\partial \xi_2} \right)^2 \right] \right\} \tag{5}$$

Equation of State:

$$\rho = \gamma p \left[(\gamma - 1) h \right]^{-1} \tag{6}$$

where $\left\{ {i \atop j\ k} \right\}$ are Christoffel symbols of the second kind, and the matrix g_{ij} can be obtained numerically from the grid generation.

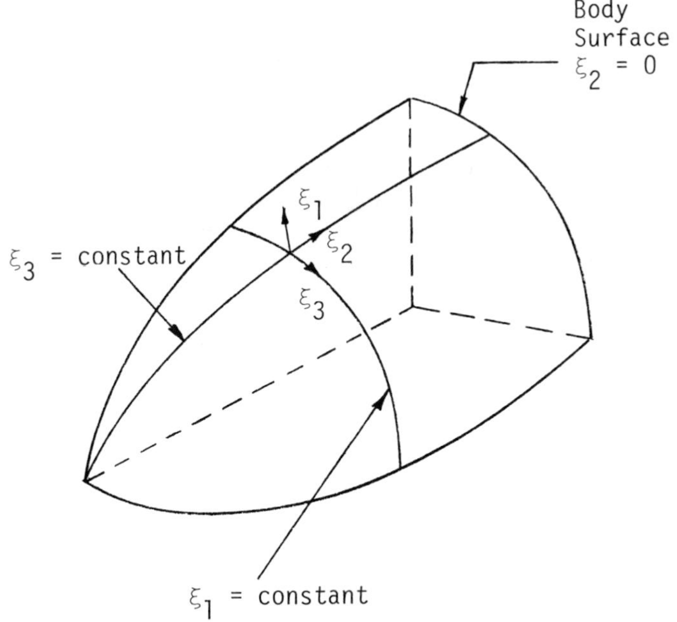

Figure 1. Coordinate System

3. BOUNDARY CONDITIONS

In order to solve the above set of governing equations, it is essential to specify appropriate boundary conditions at the body surface and at the shock. At the body surface (wall), no-slip and no-temperature-jump conditions were used. Consequently, $u_w = v_w = w_w = 0$, and the wall temperature or heat-transfer rate was specified. The conditions immediately behind the shock were obtained by using the Rankine-Hugoniot relations. However, before those equations may be applied, the tangential and normal components of the velocity must be found. These components are obtained by the method of Rakich [4]. An orthogonal set of vectors with $\vec{N}$ a unit vector normal to the shock surface and $\vec{T}$ a tangent vector is considered. The N-T plane is parallel to the direction of freestream $\vec{U}_\infty$. Then the two-dimensional Rankine-Hugoniot relations are used to calculate the conditions behind the shock. The total velocity can be written as

$$\begin{aligned}
\vec{U} &= U_T \vec{T} + U_N \vec{N} \\
&= \left[U_T\left(\frac{U_z - bN_z}{a}\right) - U_N\left(\frac{\sin\sigma}{c}\right)\right]\hat{e}_z \\
&+ \left[U_T\left(\frac{U_r - bN_r}{a}\right) + U_N\left(\frac{\cos\sigma}{c}\right)\right]\hat{e}_r \\
&+ \left[U_T\left(\frac{U_\phi - bN_\phi}{a}\right) - U_N\left(\frac{\cos\sigma \operatorname{Tan}\delta}{c}\right)\right]\hat{e}_\phi
\end{aligned} \tag{7}$$

where a, b, c, σ and δ are given in [4]. After the velocity behind the shock has been calculated, it is necessary to rotate these components into a body-normal nonorthogonal coordinate system. This is done by expressing the vectors $\hat{e}_z$, $\hat{e}_r$, $\hat{e}_\phi$ in terms of $\vec{g}_1$, $\vec{g}_2$, $\vec{g}_3$ in the ξ_1, ξ_2, ξ_3 directions

$$\begin{bmatrix} \hat{e}_z \\ \hat{e}_r \\ \hat{e}_\phi \end{bmatrix} = \begin{bmatrix} \frac{\partial z}{\partial \xi_1} & \frac{\partial r}{\partial \xi_1} & r\,\frac{\partial \phi}{\partial \xi_1} \\ \frac{\partial z}{\partial \xi_2} & \frac{\partial r}{\partial \xi_2} & r\,\frac{\partial \phi}{\partial \xi_2} \\ \frac{\partial z}{\partial \xi_3} & \frac{\partial r}{\partial \xi_3} & r\,\frac{\partial \phi}{\partial \xi_3} \end{bmatrix}^{-1} \begin{bmatrix} \vec{g}_1 \\ \vec{g}_2 \\ \vec{g}_3 \end{bmatrix} \tag{8}$$

Using Eq. 8, the shock velocity components in the body-normal nonorthogonal coordinate system are determined.

4. COORDINATES

Based on the general curvilinear coordinate governing equations, a body-oriented nonorthogonal coordinate system is constructed. This is first done on the surface of the body, where $\xi_2 = 0$, and then extended to the points away from the surface of the body. The coordinate system requires orthogonality only at the body surface, and the ξ_2 coordinate is always orthogonal to ξ_1 and ξ_3. The approach used is an extension of that presented by Blottner [5] and a detailed discussion of a similar procedure can be found in [3].

5. METHOD OF SOLUTION

Davis [6] presented an implicit finite-difference method to solve the viscous shock-layer equations for axially symmetric flows. Murray and Lewis [2] extended the method of solution to three-dimensional high angle of attack conditions. The present method of solution is identical to that of Murray and Lewis. Therefore, only an overview of the solution procedure is presented.

The solution begins on the blunt nose by obtaining an approximate stagnation solution in the wind-fixed coordinate system. The 3-D solution begins on the windward plane and marches around the body obtaining a converged solution in each ϕ sweep. After completing a sweep in the ϕ direction, the procedure then steps downstream in ξ_1 and begins the next ϕ sweep. At each point the equations are solved in the following order (i) ϕ-momentum, (ii) energy, (iii) ξ_1-momentum, (iv) integration of continuity for Y_s, and (v) coupled continuity and normal momentum equations.

6. RESULTS AND DISCUSSION

Numerical solutions of the three-dimensional nonorthogonal shock-layer equations were obtained for a 1:1.4 and a 1:2 ellipsoid at zero and 10-deg angle of attack. It should be mentioned here that the wind-fixed coordinate system is used in the 1:2 ellipsoid case. The freestream conditions for these two cases are given in Table 1. Comparisons are made for shock standoff distance and surface pressure distribution with existing inviscid results. Other data, either numerical or experimental, are not available for comparison.

Shock-standoff distance as a function of the coordinate along the body surface at different ϕ planes is illustrated in Figs. 2 and 3. The inviscid solutions from Marconi and Yaeger [7] are also presented and are in very good agreement with the present nonorthogonal viscous shock-layer calculations. It is noted that the shock standoff distance increases more rapidly between $\phi = 0$ and 30-deg for the 1:2 ellipsoid case than for the 1:1.4 case. However, this would be expected because the body curvature in the transverse direction is smaller for the 1:2 ellipsoid case in this region.

CASE	1	2
Ellipsoid Ratio	1:1.4	1:2
T_∞ (R)	81.5	203.5
ρ_∞ (slug/ft^3)	4.118E-6	2.340E-4
M_∞	10.0	5.0
T_w (R)	470	366.0
Re_∞ (1/ft)	3.056E5	4.993E6
Computational Time* IBM 370/3022(Min)	6	12

Table 1. Freestream Conditions

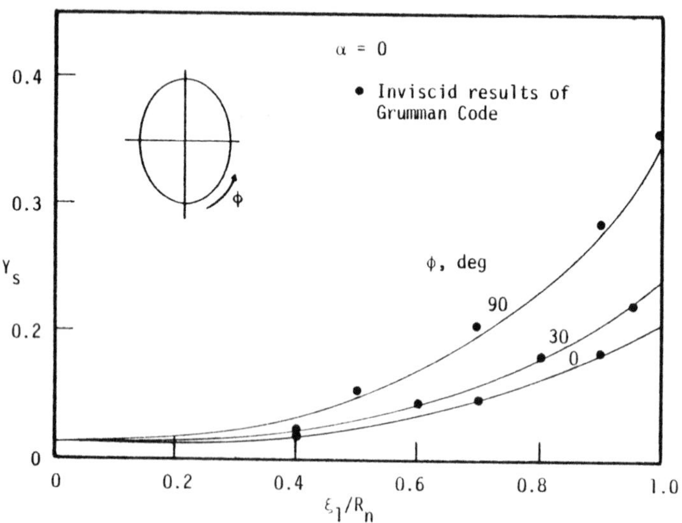

Figure 2. Shock Standoff Distance of 1:1.4 Ellipsoid

*It takes approximately two hours CDC 7600 time for a sphere nose by using time dependent method [8].

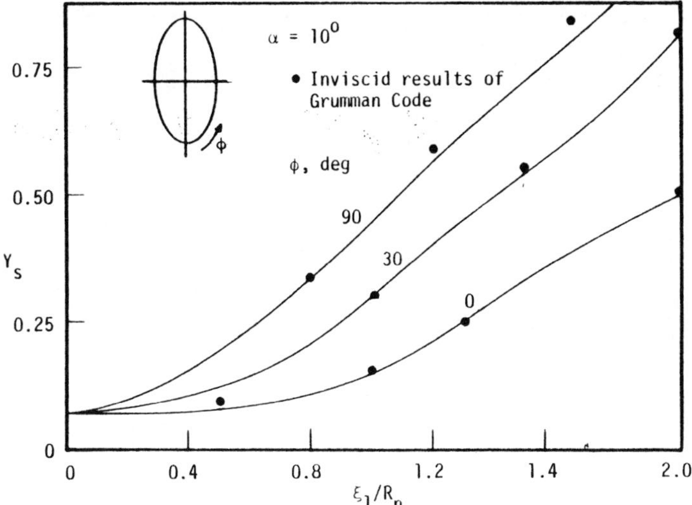

Figure 3. Shock Standoff Distance for 1:2 Ellipsoid

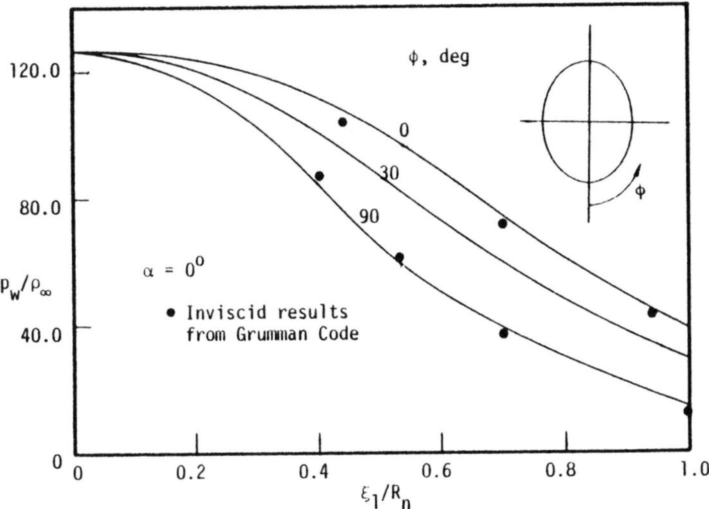

Figure 4. Surface pressure for 1:1.4 Ellipsoid

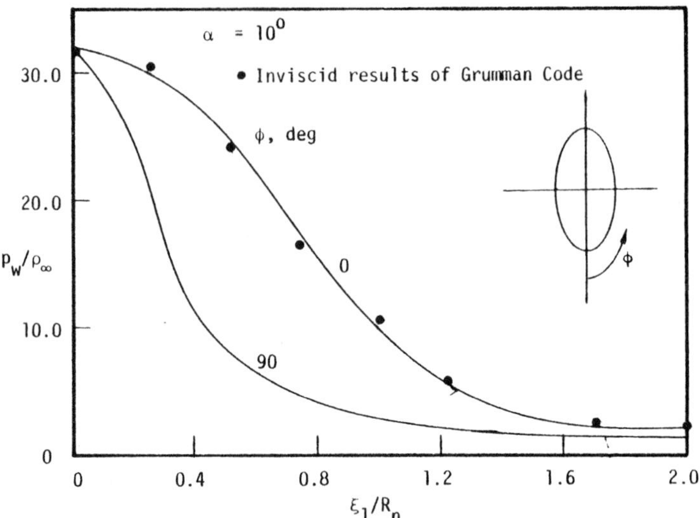

Figure 5. Surface Pressure for 1:2 Ellipsoid

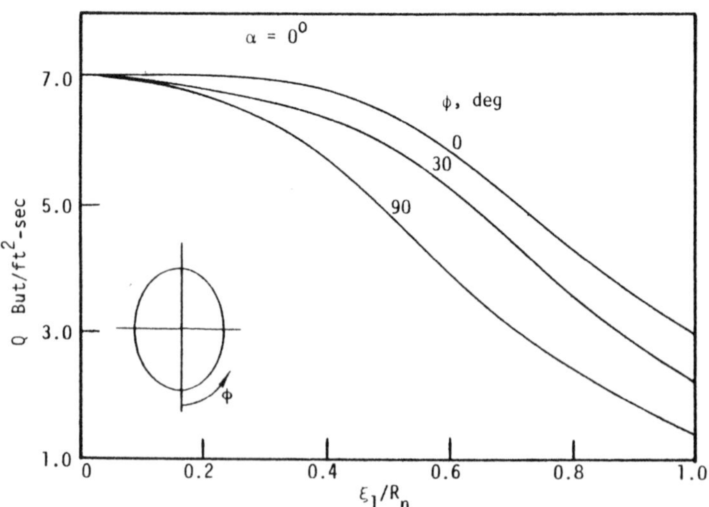

Figure 6. Convective Transfer to 1:1.4 Ellipsoid

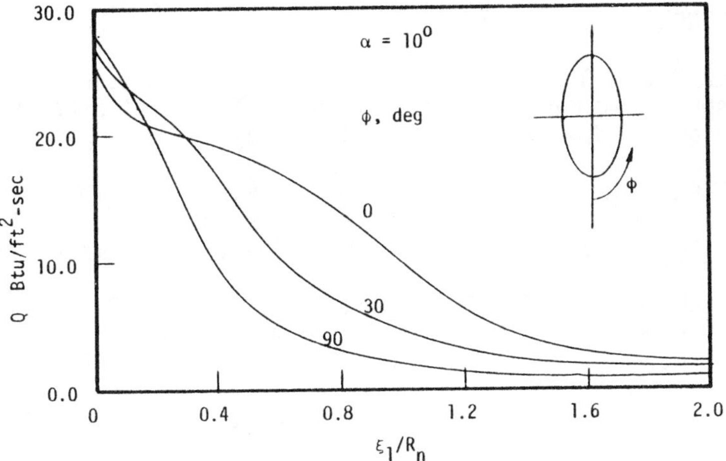

Figure 7. Convective Heat Transfer to 1:2 Ellipsoid

Figures 4 and 5 present the surface pressure distribution along the body at different ϕ-planes for these two cases. Some inviscid pressure results are also presented. It is noted that the inviscid data are in very good agreement with present results.

Figures 6 and 7 show the convective heating rate at different ϕ planes along the body surface. Since no other data are available for lifting bodies, only the present results are presented here. Because of the streamwise body curvature effects, the heating rate is higher at ϕ = 90 deg and ϕ = 30 deg than ϕ = 0 deg near the stagnation region for the 1:2 ellipsoid case. Except in the stagnation region, as was expected, the convective heat-transfer rate decreases with increasing ϕ .

7. CONCLUSIONS

The main objective of this study is to investigate the influence of the noncircular body on the shock-layer flow phenomena within reasonable computing times. A general nonorthogonal shock-layer analysis is used. The present surface pressure and shock standoff distance results are in very good agreement with existing inviscid solutions. The viscous shock-layer technique requires much less computing time than the time-dependent method [8] (Table 1). In the future we plan to include turbulent flow and equilibrium chemically reacting gas effects in the technique.

REFERENCES

1. SZEMA, K. Y. and LEWIS, C. H.: "Three-Dimensional Hypersonic Laminar, Transitional and/or Turbulent Shock Layer Flow." AIAA Paper No. 80-1457, July 1980.

2. MURRAY, A. L. and LEWIS, C. H.: "Hypersonic Three-Dimensional Viscous Shock-Layer Flow over Blunt Bodies," AIAA J., Vol. 16, No. 12, pp. 1279-1286, December 1978.

3. HELLIWELL, W. S., DICKINSON, R. P. and LUBARD, S. C.: "Viscous Flow over Arbitrary Geometries at High Angle of Attack." AIAA Paper No. 80-0064, January 1980.

4. RAKICH, J. V.: "A Method of Characteristics for Steady 3-D Supersonic Flow with Application to Inclined Bodies of Revolution." NASA TN D-5341, October 1969.

5. BLOTTNER, F. G. and ELLIS, M.: "Three-Dimensional Incompressible Boundary Layer on Blunt Bodies." Sandia Laboratories, SLA-73-0366, April 1973.

6. DAVIS. R. T.: "Numerical Solution of the Hypersonic Viscous Shock Layer Equations." AIAA J., Vol. 8, No. 5, pp. 843-851, May 1970.

7. MARCONI, F. and YAEGER, L.: "Development of a Computer Code for Calculating the Steady Super/Hypersonic Inviscid Flow Around Real Configurations." NASA CR-2675, April 1976.

8. KUTLER, P., PEDELTY, J. A. and PULLIAM. T. H.: "Supersonic Flow over Three Dimensional Ablated Nosetips Using an Unsteady Implicit Numerical Procedure. AIAA Paper No. 80-0063, January 1980.

A FINITE ELEMENT SOLUTION OF COMPRESSIBLE FLOW THROUGH CASCADES OF TURBOMACHINES

Ahmet Ş. ÜÇER*, İmdat YEĞEN**, Tahsin ÇETİNKAYA+

The solution of compressible flow through a cascade of arbitrary shape is considered in this paper. Flow through both stationary and rotating cascades may be solved. Computations may be performed on a stream sheat of revolution of any shape, allowing the solution of flow in radial, mixed, or axial flow machines.

Stream function is chosen as the field variable. The independent variables of the problem are the meridional coordinate and tangential coordinate. The finite element equations are obtained using Galerkin approach. The complicated shapes of blades are constructed accurately using 8 node isoparametric quadrilateral elements. Biquadratic shape functions are used, and element matrices are integrated using Gaussion quadrature technique. The system of equations is solved using Gauss elimination method.

The correct outlet angle is obtained through successive calculations using closure conditions. The solution technique is tested using the analytical solution of an incompressible cascade flow. Various test cases are used to establish the capabilities of the program. A transonic turbine cascade is used for deviation angle prediction. Solution is also tested on the axial flow compressor and radial in flow turbine cascades.

Convergence is generally obtained in three or four iterations. It is concluded that the computational technique gives satisfactory results for the solution of compressible flow through arbitrary cascades.

* Associate Professor, Mechanical Engineering Dept., METU, Ankara, Turkey.
** Instructor, Mechanical Engineering Dept., METU, Ankara,Turkey.
+ Teaching Assistant, Mechanical Eng.Dept., METU, Ankara,Turkey.

1. INTRODUCTION

The purpose of the paper is to outline a finite element solution technique which has been used successfully in calculating the compressible flow through arbitrary cascades.

The general formulation of the turbomachine internal flow problem has been done by C.H. Wu [1]. However, only after the recent developments in the high speed computers, solutions to the equations have become possible. The formulation may be made using stream function or the potential function as the field variable. In this investigation stream function formulation is used. The potential function formulations allows to obtain three-dimensional solution at once. However, due to the complexity of the three-dimensional solution together with the problem of computer capacity most of the research is focused on the two-dimensional analysis.

The time marching method [2,3] is a relaxation technique which is applied timewise to unsteady formulation of the cascade flow. The hyperbolic initial value problem makes the transonic flow calculations possible. But the solution needs a long computing time. In the matrix methods [4,5], the equation in terms of stream function is rewritten as a finite difference equation and the solution is performed on an orthogonal mesh.

Recently finite element methods are used for cascade flows. Habashi [6] used potential function as the field variable, psendovariational integral for the solution, and employed 8 node quadrilateral elements. D. Adler *et al.*[7] used linear triangular elements and adjusted upstream and downstream stagnation points and periodicity iteratively in the solution of Wu's equations. Hirsch *et al.*[8] used Galerkin's approach with stream function formulation and solved the equations in their quasi-harmonic form for axial flow turbomachine cascades. Thompson[9] applied triangular elements, with cubic interpolation for velocity potential, and stream function and solved linear compressor cascade flows.

The present work describes a finite element solution of flow on an arbitrary blade to blade surface using stream function as a field variable. Governing principal equation is solved in its Poisson form using 8 node, isoparametric quadrilateral elements. The solution gives the effect of any cascade configuration on the distribution of flow properties for the given flow conditions at far upstream and far downstream of the cascade.

2. FORMULATION

2.1. Principal Equation

The flow through an arbitrary turbomachine cascade may be

formulated using the equation of motion, conservation of mass and the following assumptions.

a) Flow is steady relative to the blade row.
b) Flow is isentropic.
c) Blade to blade surface is a surface of revolution.
d) The only forces are those due to accelaration and pressure gradient.

An arbitrary stream surface of revolution is represented in Figure 1. The geometry of the flow field is expressed in terms of meridional coordinate m and tangential coordinate θ. The radial coordinate must be specified as a function of m.

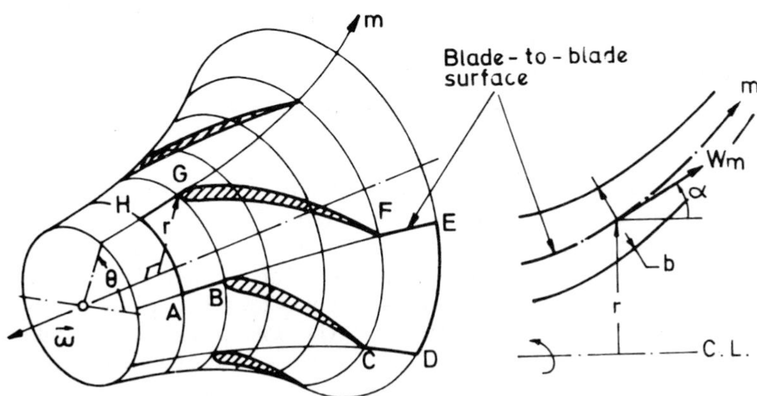

Figure 1. The Solution Domain.

The resultant principal equation describing the flow in the blade to blade surface, relative to the blade row, is expressed in terms of stream function as follows.

$$\frac{1}{r^2}\frac{\partial^2\psi}{\partial\theta^2} + \frac{\partial^2\psi}{\partial m^2} = \frac{1}{\rho r^2}\frac{\partial\rho}{\partial\theta}\frac{\partial\psi}{\partial\theta} - \left[\frac{\sin\alpha}{r} - \frac{1}{\rho b}\frac{\partial(\rho b)}{\partial m}\right]\frac{\partial\psi}{\partial m} + \frac{2\rho b\omega\sin\alpha}{\dot{m}} \qquad (1)$$

ω in equation (1) is the rotational speed of the cascade and the stream function is defined as

$$\frac{\partial\psi}{\partial m} = -\frac{\rho b W_\theta}{\dot{m}} \quad ; \quad \frac{\partial\psi}{\partial\theta} = \frac{\rho b r\ W_m}{\dot{m}} \qquad (2)$$

where W_θ and W_m are the components of relative velocity, α is the half apex angle of tangent cone and b is the thickness of the streamchannel normal to the streamsurface. $\dot{m}$ is the mass flow rate through the streamchannel between two blades. The principal equation (1) is quasi-linear and of elliptic type. It is strongly non-linear especially for high subsonic flows. This equation can be solved by iteration in its Poisson's form until the required convergence criterion is satisfied.

2.2. Boundary Conditions

The boundary condition should be given over the entire boundary of the domain of solution shown in Figure 1 for the present subsonic flow problem. Along the blade surfaces constant stream function values are specified (ψ=1 along BC and ψ=2 along GF). The stream surface is sufficiently extended to upstream and downstream. Assuming uniform flow Neumann boundary conditions are specified as

$$\frac{\partial\psi}{\partial n} = \frac{\tan\beta_{in}}{(sr)_{in}} \tag{3}$$

at the boundary AH and

$$\frac{\partial\psi}{\partial n} = -\frac{\tan\beta_{out}}{(sr)_{out}} \tag{4}$$

at the boundary DE. s in the above equations is the spacing between the blades (the pitch), and n is the direction normal to the boundaries.

Due to periodicity the flow pattern along AB is the same as that along HG and the flow pattern along CD is the same as that along FE. This property is introduced into the solution in the form of a periodic boundary condition. The solution is forced so that at each meridional station the difference between the stream function values on the two periodic boundaries is the same as that between the stream function values on the two blade surfaces. Mathematically, if ψ=a(m) along AB and ψ=b(m) along CD then ψ=1+a(m) along HG and ψ=1+b(m) along FE.

3. METHOD OF SOLUTION

3.1. Finite Element Formulation

Galerkin's weighted residual method is chosen to derive the finite element equations. The method does not require a variational statement and the mathematical form of the problem statement is conformable for this approach. In this method the weighting functions are taken equal to the interpolation functions within the solution domain and at the natural boundaries, s_2, but zero at the forced boundaries, s_1. Then equation (1) may be written as

$$\iint_\Omega \left[\frac{1}{r^2}\frac{\partial^2\psi}{\partial\theta^2} + \frac{\partial^2\psi}{\partial m^2} + q\right] v_j d\Omega + \int_{s_2} \left(f - \frac{\partial\psi}{\partial n}\right) v_j . ds = 0 \tag{5}$$

where q is the negative of the right hand side of equation (1) f corresponds to the right hand sides of equations (3) or (4), and v_j is the weighting function.

For an element with k nodes, the field variable ψ is

represented by k nodal values ψ_i and interpolation functions N_i as

$$\psi = \sum_{i=1}^{k} N_i \psi_i \tag{6}$$

Combining equations (5) and (6) and using $v_j = N_j$ (j=1,k) as weighting functions k equations for k unknowns ψ_i are obtained for each element.Applying Gauss theorem and noting that $\partial\psi/\partial m = \sum_i \partial N_i/\partial_m \psi_i$ and $\partial\psi/\partial\theta = \sum_i \partial N_i/d\theta \psi_i$, the set of equations at the element level becomes

$$[S]^e \{\psi\}^e = \{F\}^e \tag{7}$$

where,

$$[S]^e = \iint_{\Omega^e} [\frac{1}{r^2} \frac{\partial[N]}{\partial\theta} \frac{\partial[N]^T}{\partial\theta} + \frac{\partial[N]}{\partial m} \frac{\partial[N]^T}{\partial m}] \, d\Omega \tag{8}$$

and

$$\{F\} = \iint_{\Omega^e} [N] \, q d\Omega + \int_{s_2^e} [N] \, f \, ds \tag{9}$$

Two dimensional elements are used for the problem requiring the continuity of streamfunction at the element interfaces. Eight noded isoparametric quadrilateral elements are used to reconstruct the shape of the turbomachinery blades satisfactorily in the computer memory. Figure 2 shows a typical finite element mesh for Mc Donald [10] turbine cascade. It is well known that for isoparametric formulation the interpolation functions of the unknown field variable are also used to map the parent square elements into the physical coordinates [11].

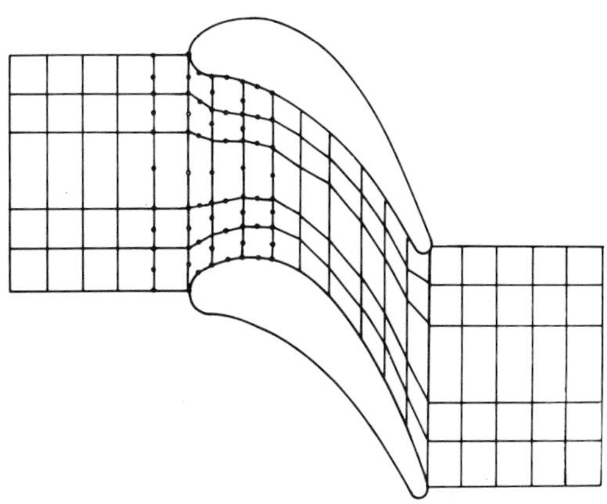

Figure 2. Finite Element Mesh.

The element stiffness matrix is formed as

$$[S]^e = \int_{-1}^{+1} \int_{-1}^{+1} [B'(\xi,\eta)]|J| \, d\xi \, d\eta \tag{10}$$

where $|J|$ is the Jacobian and $[B'(\xi,\eta)]$ is the integrand of equation (8) transformed into ξ,η plane.

It is impossible to obtain the above integral in closed form. The integration is performed using two point Gaussian Quadrature.

The first integral term in equation (9) is calculated likewise. The second term is not calculated at the element level since it disappears at the internal boundaries.

The essential boundary conditions are applied by forcing the field variable to the known values using standard techniques. The natural boundary conditions are imposed on the elements at the natural boundaries by evaluating the second integral of equation (9).

The periodic boundary condition is applied by imposing that $\psi_i - \psi_j = 1$ where i and j are the respective points of the upper and lower periodic boundaries at the same m coordinate. Application of this boundary condition distorts the symmetry of the stiffness matrix. A number of manipulations are necessary to restore the symmetry at the expense of some increase in the half-band-width.

The final system of equations is solved by Gauss elimination method. Initial solution is performed by assuming incompressible flow taking q equal to zero. The solution obtained is used to calculate the velocity and density which are used to form q for the next iteration. Iteration is continued until a convergence criterion

$$\mathrm{Max}|(\psi_{n+1}-\psi_n)/\psi_n| < 0.005 \tag{11}$$

is satisfied.

3.2. Calculation of Density and Velocity

The relative velocity is calculated from

$$W_p^2 = \frac{\dot{m}^2}{b^2\rho^2}\left[\left(\frac{1}{r}\frac{\partial\psi}{\partial\theta}\right)^2 + \left(\frac{\partial\psi}{\partial m}\right)^2\right] \tag{12}$$

where ρ is taken from the previous iteration.

Since, rothalpy I which is defined as

$$I = h + \frac{W^2-(\omega r)^2}{2} \tag{13}$$

remains constant throughout the solution domain the temperature and density at any location can be calculated from,

$$T = \frac{1}{c_p} (I- \frac{W^2-(\omega r)^2}{2} , \quad (14)$$

$$\rho = \rho_{01}(T/T_{01})^{\frac{1}{\gamma-1}} \quad (15)$$

I, T_{01}, ρ_{01} are known from the inlet conditions.

Since global iteration is performed in calculating ψ no inner iteration is used between equations (12) and (15) for density calculation.

4. RESULTS AND DISCUSSION

The computer program developed is applied to six different geometries of turbomachines including axial and radial turbines and compressors. The present method of solution is tested against existing calculations or experimental results. The effect of outlet flow angle on the solution and the response of the solution to the changes in various program parameters are investigated.

Well prepared data with clearly specified blade geometry and operating conditions is essential to obtain reliable results for comparison. The calculated and/or measured flow properties must also be given in sufficient detail. Unfortunately only a few cases satisfying the above requirements exist in open literature.

Gostelow's exact solution of flow through an axial compressor cascade [12] is generally accepted as a standard of comparison for incompressible potential flow calculations. The present program was run with Gostelow's data for blade configuration and inlet and outlet flow angles at a very low flow rate to approach incompressible flow conditions. The result is represented in figure 3 in the form of variation of pressure along the blade surfaces.

The pressure coefficient in figure 3 is defined as $C_p = 2(p-p_1)/\rho V_1^2$ where p_1 and V_1 are the inlet pressure and velocity respectively and p is the local pressure. The present calculations are in good agreement with those of Gostelow. The minor discrepancy near the leading edge is due to the fact that the high curvature of the blade surface in this region can not be accurately represented with the relatively coarse mesh of nodal points.

Figure 4 shows the variation of Mach number along the VKI-LS59-2 gas turbine cascade [13] obtained for an outlet Mach number of 0.75. With this condition the predicted flow remains subsonic throughout.

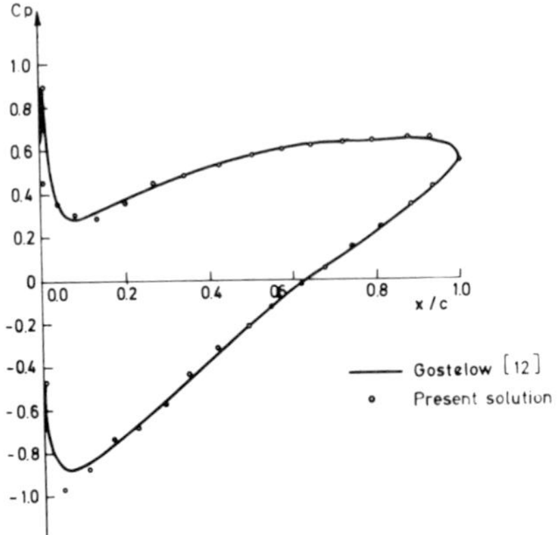

Figure 3. Comparison of Finite Element Solution with Gostelow's Exact Solution.

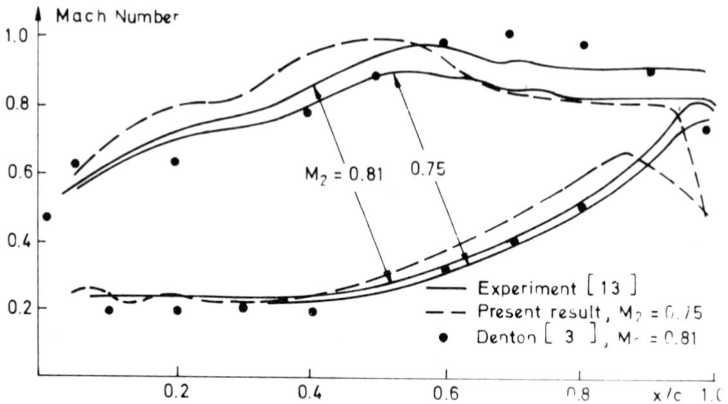

Figure 4. Flow Through VKI-LS59-2 Turbine Blades.

The present predictions are generally in agreement with measurements. The discrepancies between the two results can be partly explained by the viscous effects which were neglected in calculations.

A similar comparison between the measured and calculated results is shown in figure 5 for a DCA 48 (double circular arc 48° cambered) transonic compressor cascade [14] for an inlet Mach number of 0.60. There is a good agreement between the measured and predicted Mach numbers along the blade surfaces.

Katsanis [5] has performed hub-to-shroud and blade to blade solutions in a radial inflow gas turbine using a quasi-orthogonal method. His hub to shroud solution was used to define the stream channel thickness to be used in a blade to blade solution at the mitheight between hub and shroud. Unfortunately

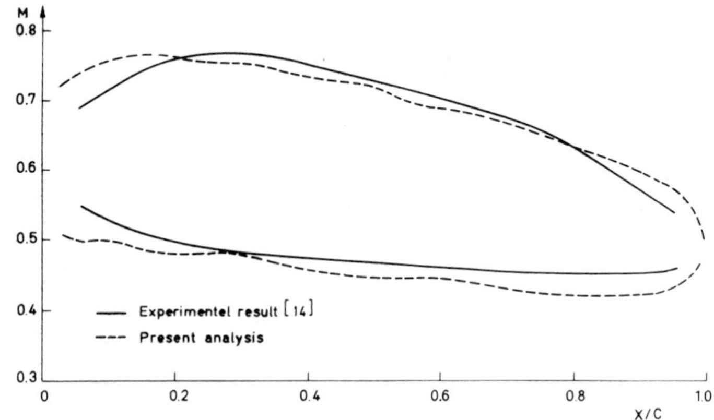

Figure 5. Flow Through DCA 48 Compressor Blades.

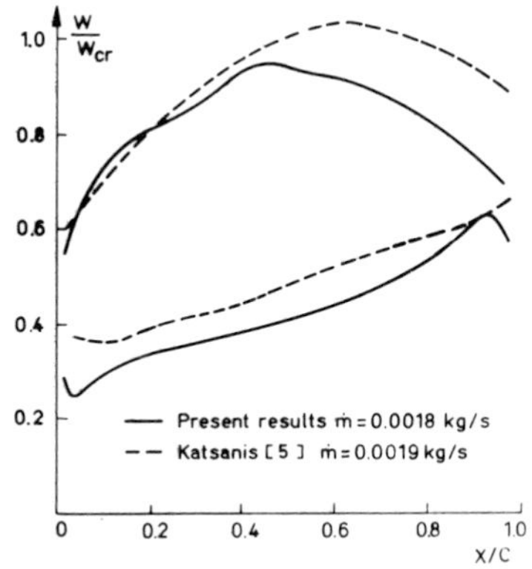

Figure 6. Flow Through a Radial Flow Turbine [5].

Katsanis' blade-to-blade solution involved regions of supersonic flow. Therefore, the present solution was performed at a lower flow rate. Consequently the two solutions are not directly comparable. Calculations were performed essentially to test the behavior of the computer program in a radial flow machine. The general trends of the two solutions are in agreement as observed in figure 6, where the nondimensional velocity distribution, W/W_{cr}, on the blade surface is given. W_{cr} is the critical velocity. The x and the chord length c are measured along the coordinate m.

The elliptic nature of equation (1) necessitates the boundary conditions to be specified over the complete boundary for a unique solution. Therefore the outlet flow angle must be specified before the solution is performed, whereas physically the outlet flow angle is determined by the inlet conditions and blade geometry. The physically correct solution is the one which satisfies a certain requirement known as Kutta condition. The condition implies that the rear stagnation point be located at the trailing edge and also that the local velocity just behind the trailing edge be tangent to the camberline.

In principle it is possible to find the correct fluid angle at the outlet by successively altering β_2 and checking the Kutta condition. The difficulty arises due to the fact that the flow can not be accurately calculated near the rounded trailing edge where the velocity varies very severely, and the exact location of the rear stagnation point can not be found. Therefore an indirect method must be used.

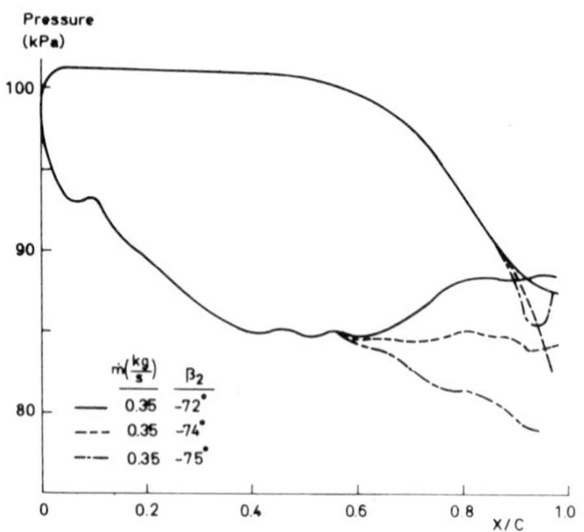

Figure 7. Effect of β_2 on Pressure Distribution.

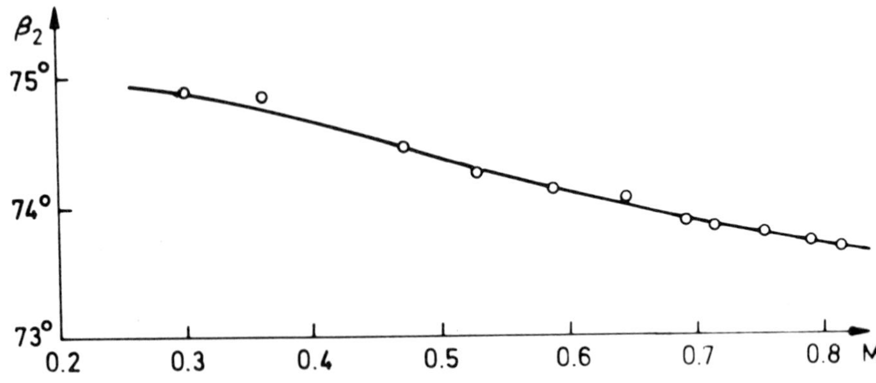

Figure 8. Prediction of Outlet Flow Angles.

Figure 7 shows the pressure distributions around a transonic turbine blade [10] at different outlet angles. It is observed that the solution near the trailing edge is very sensitive to small changes of the outlet flow angle. A pressure diagram which tends to close approximately at the trailing edge indicates a proper outlet flow angle.

According to Gostelow's Linear closure hypothesis [15] the pressure curves of the suction and pressure surfaces are linearly extrapolated from 85% chord position. Extrapolated curves closing at the trailing edge indicate the correct solution. Figure 8 shows the outlet flow angles obtained using this hypothesis for the transonic turbine blade of [10].

The solution was found to depend on the number of elements only slightly, if the nodes are chosen properly on the blade surfaces. The test cases that are presented are calculated on 150 elements and 500 nodes, but comparable results are obtained only with aproximately 100 elements and 300 nodes.

The convergence of the program is found to be very good. The program converges without any underrelaxation very rapidly. Only 2 or 3 iterations are enough for low speed linear cascade calculations. For radial cascades and high speed calculations convergence is obtained in 3 to 4 iterations. The computing time recorded was between 1 and 3 minutes on an IBM 170/145 digital computer for the reported test cases. The detailes of the computer program and results are reported in reference [16].

5. CONCLUSIONS

Isoparametric finite elements are found to be successful in simulating the complicated shapes of blades accurately. It is observed that the present computer program shows a very good converging behavior without any need of underrelaxation. It is concluded that the method gives rapid and satisfactory results that compete favorably with the existing methods in the literature.

ACKNOWLEDGEMENTS

The authors would like to acknowledge the contributions of Assoc.Prof.Dr. Hasan Akay of Purdue University, Indianapolis, in the finite element formulation.

REFERENCES

1. WU, C.H.- A General Theory of Three Dimensional Flow in Subsonic and Supersonic Turbomachines of Axial, Radial, and Mixed Flow Types, U.S. NACA TN 2604, 1952.
2. Mc. DONALD, P.W.- The Computation of Transonic Flow Through Two-Dimensional Gas Turbine Cascades, ASME paper 71-GT.89, 1971.

3. DENTON, J.D.- A Time Marching Method for Two-and Three-Dimensional Blade to Blade Flows, Aeronautical Research Council, R8M No. 3775, London, 1974.

4. SMITH, D.J.L. and FROST, D.H.- Calculation of Flow Past Turbomachine Blades, Thermodynamics and Fluid Mechanics Convention 1970, Instn. Mech. Engrs., paper no. 27, pp. 219-232.

5. KATSANIS, T.- Fortran Program for Calculating Transonic Velocities on a Blade-to-Blade Stream Surface of a Turbomachine, U.S. NASA TN D-6766, 1972.

6. HABASHI, W., DUECK, E.G. and KENNEY, D.P.- Finite Element Approach to Compressor Blade-to-Blade Cascade Analysis, AIAA Journal Vol.A, No.7, 1979.

7. ADLER, D., and KRIMMERMANN, Y.- Calculation of the Blade-to-Blade Compressible Flow Field in Turbo Impellers Using Finite Element Method, J. Mech. Eng. Sci., Vol.19, No.3, 1977, pp. 108-112.

8. HIRSH, Ch. and WARZEE, G.- Quasi 3D Finite Element Computation of Flows in Centrifugal Compressors, ASME Symp., 22nd Annual Gas Turbine Conf., New Orleans, 1980, pp. 69-75.

9. THOMPSON, D.S.- Flow Through a Cascade of Airfoils, Finite Element Methods in Flow Problems, Ed. Oden J.T., Zienkiewics, O.C., Gallaher R.H. and Taylor, C. UAH Press 1974.

10. AGARD-PEP, W.G.12, Blade-to-Blade Data, 1980.

11. ZIENKIEWICS, O.C.- The Finite Element Method, Third Edition, Mc Graw-Hill (UK), Maidenhead, Berkshire, 1977.

12. GOSTELOW, J.P.- Potential Flow Through Cascades, A comparison Between Exact and Approximate Solutions, Aeronautical Research Council, C.P. No. 807, London, 1965.

13. SIEVERDING, C.- Base Pressure Measurements in Transonic Turbine Cascades, Von Karman Institute for Fluid Dynamics, Lecture Series 84, Brussels, 1976.

14. BREUGELMANS, F.A.E.- The Compressor Blade Definition, Von Karman Institute for Fluid Dynamics, Lecture Series 59, Brussels, 1973.

15. MILLER, M.C. and SEROVY, G.K.- Deviation Angle Estimation for Axial Flow Compressors Using Inviscid Flow Solutions, Journal of Engineering for Power, Trans. ASME, Vol. 97, No.4, pp. 163-172, 1975.

16. ÇETİNKAYA, T.A.- Computation of Subsonic Compressible Flow Through The Cascades of Turbomachines, M.Sc. Thesis, METU, 1980.

SEMI-ELLIPTIC COMPUTATION OF AXI-SYMMETRIC TRANSONIC FLOWS

M.A. Serag-Eldin[I]

SUMMARY

The paper presents a numerical solution algorithm for the computation of axi-symmetric, transonic, duct flows. The algorithm solves steady state equations for the primitive variables. It is based on a specially derived pressure-correction equation which expresses a kinematic compatability condition. The algorithm is first derived and then employed to predict the flow field inside an annular convergent-divergent nozzle, featuring exceptionally small radii of curvature of the walls at the throat section.

1. INTRODUCTION

The paper presents a computational algorithm for the numerical prediction of steady, axi-symmetric, compressible, duct flows. The flows predicted may be subsonic, supersonic or transonic; but must be forward marching (non-recirculating). So far only inviscid flows have been predicted; however, boundary-layer corrections may be readily introduced.

An important asset of the present algorithm is its ability to express complicated boundary geometries accurately. This is achieved through the adoption of a grid-system in which the streamlines form one of the grid co-ordinates.

The predictions obtained give the local value of the Mach-number, static pressure, density, velocity and flow direction, at all internal points inside the calculation domain. It should thus prove useful in the design of axi-symmetric: nozzles, diffusers, various turbomachinary channels, wind tunnel sections, and many other important axi-symmetric ducts.

I. Lecturer, Mechanical Engineering Department, Cairo University.

2. CHARACTERIZATION and ORIGIN

The computational algorithm described here is a finite-difference, floating-grid, iterative, space-marching one, that solves steady-state equations for primitive variables with the aid of a pressure-correction equation. The dependent variables are:

(i) the streamwise velocity, u.
(ii) the radial velocity component, v.
(iii) the static pressure, p.
(iv) the density of the fluid, ρ.

Since viscous effects are neglected and recirculation is absent, downstream effects are presumed to influence upstream ones only by means of the pressure gradient terms. Therefore, upwind-differencing is generally employed, except for the pressure-gradient terms, for which central-differencing is adopted. The algorithm is thus designated Semi-Elliptic.

In the present algorithm, only one-dimensional storage is required for all variables, except pressure, for which two-dimensional storage is required. It combines some of the most useful features of the two-dimensional parabolic algorithms [1], and the elliptic ones [2]. The algorithm presented here is the axi-symmetric counterpart of a previous one developed for two-dimensional plane flows [3]. The latter is closely related to that of [4] but features a slightly modified pressure-correction equation.

3. DERIVATION OF ALGORITHM

3.1 Grid System

(a) Co-ordinates: A non-orthogonal, axi-symmetrical, finite-difference grid is adopted, in which the co-ordinates in a radial plane are:

(i) the stream-lines ψ = constant, where ψ is the Stokes' stream function.
(ii) a family of straight, parallel z = constant lines, where z is the distance measured along the axis of symmetry.

Grid spacings may be non-uniform in either direction, as demonstrated in Fig. 1 for an annular convergent-divergent nozzle. It is remarked that the radial position r of the internal grid nodes (located at the intersection of the grid lines) is unknown prior to solution; r is an outcome of the solution. Along a z = constant line, the radial distance $\Delta r_{i+\frac{1}{2}} \equiv r_{i+1} - r_i$ between two neighbouring streamlines ψ_{i+1}

and ψ_i, is calculated according to the continuity equation, as follows:

$$(r.\Delta r)_{i+\frac{1}{2}} = \frac{\psi_{i+1}-\psi_i}{(\rho u \cos\alpha)_{i+\frac{1}{2}}} \qquad (1)$$

where the subscript $i+\frac{1}{2}$ indicates the average value in the region between r_{i+1} and r_i, $(r.\Delta r)_{i+\frac{1}{2}} \equiv r_{i+\frac{1}{2}}.\Delta r_{i+\frac{1}{2}}$ is the flow area between r_{i+1} and r_i, and α is the inclination of the streamline to the axis of symmetry. It is calculated from the velocity components as follows:

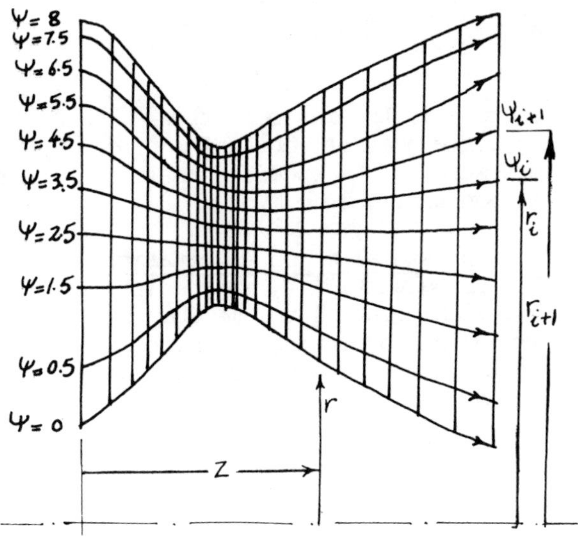

Fig. 1 Non-Orthogonal Grid.

$$\alpha = \sin^{-1}(v/u) \qquad (2)$$

(b) Staggered Grid Notation: A staggered grid distribution is adopted in which the density grid nodes coincide with those of the pressure, and the velocity grid nodes lie midway between them. This practice was first originated by [5].

Fig. 2 reveals the staggered-grid arrangement and the convention adopted to refer to the grid node locations. The subscripts n, s, e and w refer to neighbouring velocity grid nodes lying at the immediate north, south, east and west, respectively of the considered pressure node P; whereas the subscripts N, S, E and W refer to the corresponding pressure and density nodes.

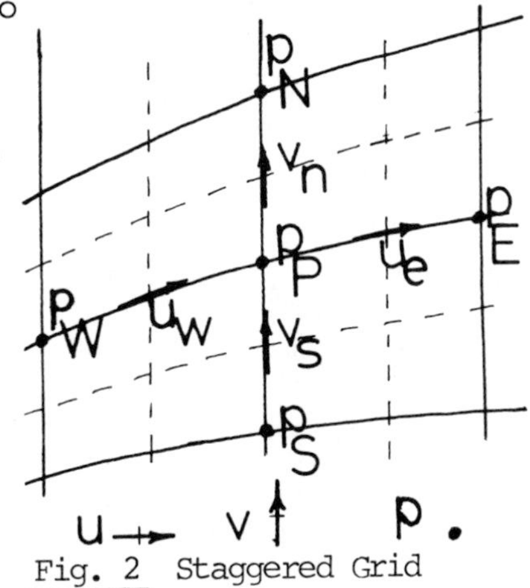

Fig. 2 Staggered Grid

3.2 Governing Equations

(a) Streamwise momentum equation: Referring to Fig. 3 which displays a typical control-volume for u, integration of Euler's equation along the central streamline and employing upwind values of u yields:

$$u_e^2 - u_w^2 = \frac{2\,\gamma}{\gamma-1}\left(\frac{p_P}{\rho_P} - \frac{p_E}{\rho_E}\right) \qquad (3)$$

where γ is the ratio of the specific heat at constant pressure to the specific heat at constant volume.

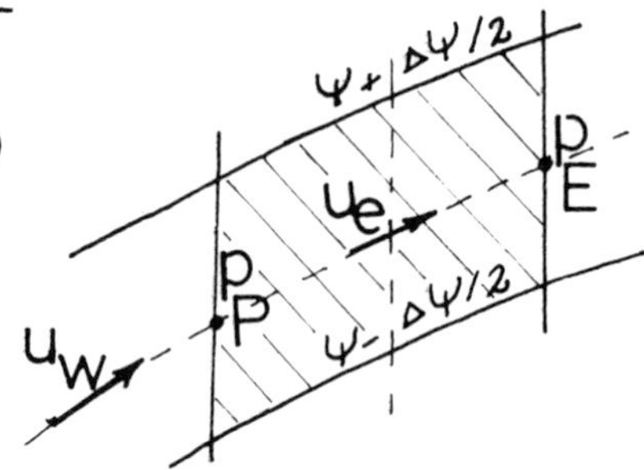

Fig. 3 u control-volume

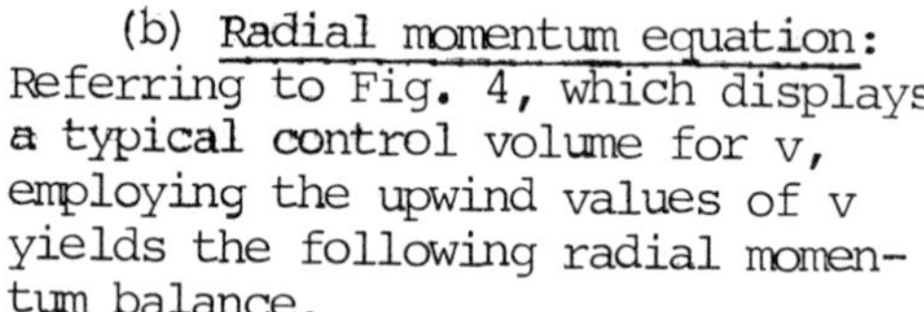

(b) Radial momentum equation: Referring to Fig. 4, which displays a typical control volume for v, employing the upwind values of v yields the following radial momentum balance.

$$\Delta\psi(v_n - v_{nw}) = (p_P - p_N)\Delta x . r_n \qquad (4)$$

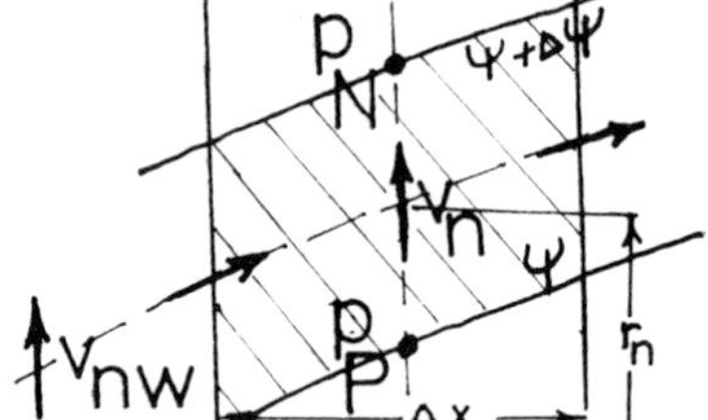

Fig. 4 v control-volume

(c) Kinematic compatability equation: It expresses a geometric requirement of the streamlines; it replaces the continuity equation in a fixed-grid system. This equation is derived here from area compatability considerations, rather than line compatability considerations as in [3] for two-dimensional flow.

Referring to the control volume for p displayed in Fig. 5, the projected areas of the control-surfaces on any z = constant plane must satisfy the following compatability relation:

$$(r.\Delta r)_e - (r.\Delta r)_w + \Delta x\,(\tan\alpha_s . r_s - \tan\alpha_n . r_n) = 0 \qquad (5)$$

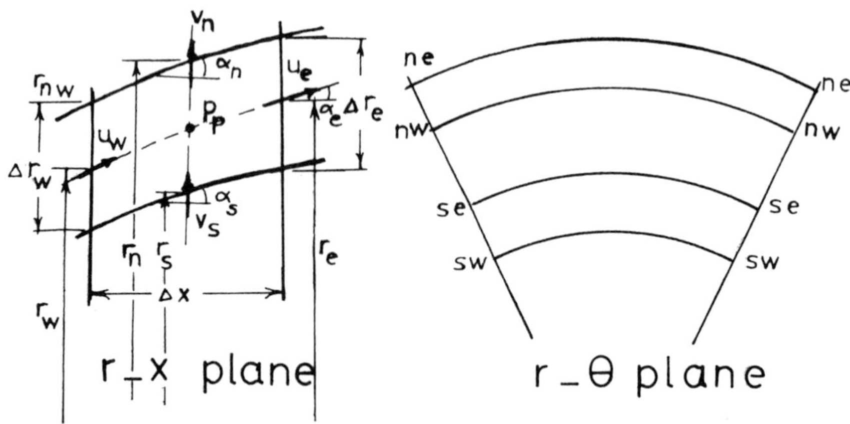

Fig. 5 The control volume for p

(d) Thermodynamic Equation

Since the flow is isentropic, the thermodynamic relations give:

$$\rho = \rho_o \, (p/p_o)^{1/\gamma} \tag{6}$$

where the subscript "o" refers to stagnation conditions.

3.3 Calculation of Mass Velocity

The elliptic and hyperbolic natures of the subsonic and supersonic regions are expressed through the differing ways of computing the mass velocity (ρu), as follows:

$$\begin{aligned} (\rho u)_e &= \rho_E \, u_e \quad \text{for} \quad M_e < 1 \\ (\rho u)_e &= \rho_P \, u_w \quad \text{for} \quad M_e > 1 \end{aligned} \tag{7}$$

where the subscripts refer to the grid-node locations displayed in Figures 2 and 5. The main implications of this practice are:

(a) In a supersonic region $(\rho u)_e$ is independent of the downstream pressure p_E, and thus downstream influences may not travel upstream.

(b) In the flow regions where the Mach-numbers are close to or greater than unity, the large density changes between successive iteration provoke no numerical instability, for, in the product $\rho_P \, u_w$ or $\rho_E \, u_e$, velocity is considered at an upstream point relative to the location of density; therefore for a pressure change at the location of the density, the considered ρ and u always vary in opposite directions. The mass-velocity is employed in the derivation of the pressure field in § 3.4.

3.4 The Pressure Correction Equation

Equation (5) expresses a kinematic compatability requirement that can only be satisfied if the pressure field is correct. This is because the areas $(r.\Delta r)$ and the angles α are calculated from densities and velocities which themselves depend on the pressure field. Thus, at each grid node P, the values of the dependent variables prevailing right after the solution of the momentum equations will result in a residual error ε_p of equation (5) which is expressed by:

$$\varepsilon_p = (r.\Delta r)_e^* - (r.\Delta r)_w^* + \Delta x \, [(\tan \alpha_s . r_s)^* - (\tan \alpha_n . r_n)^*] \tag{8}$$

where the star superscripts indicate that these are the values prevailing right after the solution of the momentum equations.

Hence, we now seek to increment r* and tan α* with corrections r' and tan α' that satisfy:

$$(r.\Delta r)'_e - (r.\Delta r)'_w + \Delta x\ [(\tan\alpha_s . r_s)' - (\tan\alpha_n . r_n)'] = -\varepsilon_p \quad (9)$$

so that the corrected values of r and tan α will satisfy equation (5). There are an infinite set of corrections that would satisfy equation (5), the set chosen is the one that would be the most likely to introduce minimum residuals in the finite-difference form of the momentum equations. This is achieved by requiring that the corrections, in addition to satisfying equation (5), would also simultaneously satisfy linearized forms of the momentum equations. For linearization purposes, it is also assumed that $(r.\Delta r)'_e$ is affected by the pressure-corrections difference $(p'_P - p'_E)$ only, $(r.\Delta r)'_w$ by $(p'_W - p'_P)$ only, $(r_n \tan\alpha_n)'$ by $(p'_P - p'_N)$ only, and $(r_s \tan\alpha_s)'$ by $(p'_S - p'_P)$ only.

Referring to Fig. 5, equation (1) gives:

$$(r.\Delta r)_e = \frac{(\Delta\psi)_e}{(\rho u)_e \cos\alpha_e} \quad (10)$$

Differentiation of equation (10) to derive $(r.\Delta r)'_e$ for pressure-corrections p'_P and p'_E yields:

$$\frac{(r.\Delta r)'_e}{(r.\Delta r)_e} = -\left[\frac{(\rho u)'_e}{(\rho u)_e} + \frac{\cos\alpha'_e}{\cos\alpha_e}\right] \quad (11)$$

Since $\cos\alpha_e = \left(\frac{\sqrt{u^2 - v^2}}{u}\right)_e$, then:

$$\cos\alpha'_e = \frac{\partial \cos\alpha_e}{\partial u_e} . u'_e = \left[\frac{v^2}{u^2} \frac{1}{\sqrt{u^2 - v^2}}\right]_e u'_e \quad (12)$$

In the last equation the term $\frac{\partial \cos\alpha_e}{\partial v_e} . v'_e$ was omitted because v'_e is presumed independent of $(p'_P - p'_E)$.

Substituting for $\cos\alpha'_e$ from equation (12) into equation (11) yields:

$$\frac{(r.\Delta r)'_e}{(r.\Delta r)_e} = -\left[\frac{(\rho u)'_e}{(\rho u)_e} + \tan^2\alpha_e . \frac{u'_e}{u_e}\right] \quad (13)$$

Similarly:

$$\frac{(r.\Delta r)'_w}{(r.\Delta r)_w} = -\left[\frac{(\rho u)'_w}{(\rho u)_w} + \tan^2\alpha_w . \frac{u'_w}{u_w}\right] \quad (14)$$

Differentiation of $r_n \tan \alpha_n$ yields:

$$(r_n \tan \alpha_n)' = r_n \tan \alpha_n' + \tan \alpha_n . r_n' \tag{15}$$

From Fig. 5 it is apparent that:

$$r_n = r_{nw} + \Delta x \tan \alpha_n/2 \tag{16}$$

Since the solution progresses from west to east, the value of r_{nw} is calculated ahead of r_n, and consequently is generally more correct at the time of application of equation (16); therefore only r_n is corrected for a correction of α_n, i.e.:

$$r_n' = \Delta x . \tan \alpha_n'/2 \tag{17}$$

Since $\tan \alpha_n = [v/\sqrt{u^2 - v^2}]_n$, therefore differentiation of $\tan \alpha_n$ gives:

$$\tan \alpha_n' = [\frac{v}{\sqrt{u^2 - v^2}}]_n' = \frac{u_n^2}{\sqrt{u_n^2 - v_n^2}} \cdot \frac{v_n'}{u_n^2 - v_n^2} \tag{18}$$

Substitution of the results of equations (17, 18) into equation (15) yields:

$$(r_n . \tan \alpha_n)' = \frac{u_n^2}{(u_n^2 - v_n^2)^{3/2}} (r_n + \Delta x \tan \alpha_n/2) \; v_n' \tag{19}$$

Similarly:

$$(r_s . \tan \alpha_s)' = \frac{u_s^2}{(u_s^2 - v_s^2)^{3/2}} (r_s + \Delta x . \tan \alpha_s/2) . v_s' \tag{20}$$

The dependence of u' and v' on p' is derived from the following linearized forms of the momentum equations:

$$u_e' = \frac{1}{(\rho u)_e} (p_P' - p_E') \tag{21}$$

$$u_w' = \frac{1}{(\rho u)_w} (p_W' - p_P') \tag{22}$$

$$v_n' = \frac{\Delta x}{(\Delta \psi)_n} (p_P' - p_N') r_n \tag{23}$$

$$v_s' = \frac{\Delta x}{(\Delta \psi)_s} (p_S' - p_P') r_s \tag{24}$$

with $p_E' = p_W' = 0$

The dependence of $(\rho u)'$ on p' is obtained by differentiating equation (7). This gives:

$$\frac{(\rho u)'_e}{(\rho u)_e} = [-\frac{1}{\rho_p . u_e^2} \cdot (1-\delta) + \frac{1}{\gamma p_P} \cdot \delta] p'_P \qquad (25)$$

$$\frac{(\rho u)'_w}{(\rho u)_w} = - \frac{1}{\rho_P \, u_w^2} [1 - M_w^2] \; (1-\delta) \; p'_P \qquad (26)$$

where: $\delta = 0$ for $M < 1$, $\delta = 1$ for $M > 1$

Substitution of the results of equations (21-26) into equations (13,14,19,20) and employing upwind values for ρ yields:

$$(r.\Delta r)'_e = - A_E . p'_P \qquad (27a)$$

$$(r.\Delta r)'_w = A_W . p'_P \qquad (27b)$$

$$\Delta x.(r_n . \tan\alpha_n)' = A_N . (p'_P - p'_N) \qquad (27c)$$

$$\Delta x(r_s . \tan\alpha_s)' = - A_S . (p'_P - p'_S) \qquad (27d)$$

where A_E, A_W, A_N and A_S are defined by the following relations:

$$A_E = (r.\Delta r)_e \; [\frac{1}{\rho_P \, u_e^2} \; (1-\delta) + \frac{1}{\gamma p_P} \delta + \frac{\tan^2\alpha_e}{\rho_P \, u_e^2}] \qquad (28a)$$

$$A_W = (r.\Delta r)_w \; [\frac{1}{\rho_P \, u_w^2} (1-M_w^2)(1-\delta) + \frac{\tan^2\alpha_w}{\rho_W \, u_w^2}] \qquad (28b)$$

$$A_N = \frac{(\Delta x)^2 . r_n}{(\Delta\psi)_n} \; \frac{u_n^2}{(u_n^2 - v_n^2)^{3/2}} \cdot (r_n + \frac{\Delta x}{2} \cdot \tan\alpha_n) \qquad (28c)$$

$$A_S = \frac{(\Delta x)^2 . r_s}{(\Delta\psi)_s} \; \frac{u_s^2}{(u_s^2 - v_s^2)^{3/2}} \cdot (r_s + \frac{\Delta x}{2} \cdot \tan\alpha_s) \qquad (28d)$$

Finally, substituting for $(r.\Delta r)'$ and $\Delta x.(r \tan\alpha)'$ in equation (9) by the right-hand sides of equation (27) yields the required pressure-correction equation:

$$(A_E + A_W + A_N + A_S) \; p'_P = A_N \; p'_N + A_S p'_S + \varepsilon_p \qquad (29)$$

This equation is solved by the Thomas algorithm for tri-diagonal matrices, to yield the value of p' at all internal grid nodes on the considered cross-stream plane.

3.5 Overall Adjustments

3.5.1 Correction of Flow Area

The solution of eauation (29) gives pressure-corrections which lead to computed radial distances satisfying the kinematic compatability equation at the individual pressure nodes. However, there is no imposed condition that the computed area of the flow A_{flow} corresponds to the area confined between the boundaries of the integration domain A_{duct}. This is satisfied by superimposing on the local pressure corrections, a block pressure-difference correction $\bar{p}'$ acting on the velocity cross-stream plane considered. $\bar{p}'$ is derived from:

$$\bar{p}' = \frac{A_{flow} - A_{duct}}{\Sigma_i A_{E,i}} \qquad (30)$$

where Σ_i designates the summation over all the individual grid nodes in the cross-stream plane considered. This pressure-difference is introduced by block correcting the pressure-field immediately upstream and immediately downstream the velocity plane considered, by increments p'_U and p'_D, respectively; computed as follows:

$$p'_U = M_m^2 \bar{p}' \qquad (31a)$$

$$p'_D = -(1-M_m^2)\bar{p}' \qquad (31b)$$

where M_m is the maximum value of the local Mach-numbers, with an upper limit of 1.

3.5.2 Correction of Flow Rate

If the flow is subsonic-supersonic, it will be choked at the cross-section of minimum area. Thus at this section only, block correction is not applied, and the difference between A_{flow} and A_{duct} is corrected by incrementing M_{in} the inlet Mach-number (and consequently the mass flow rate $\dot{m}$) at the start of each iteration sweep.

The increment M'_{in} is derived from a linearized one-dimensional analysis yielding:

$$M'_{in} = M_{in} \frac{(1 + \frac{\gamma-1}{2} M_{in}^2)}{1 - M_{in}^2} \cdot \frac{(A_{duct} - A_{flow})}{A_{flow}} \cdot \dot{m} \qquad (32)$$

4. SOLUTION PROCEDURE

Initially, the values of u,v,ρ and p are specified at the first cross-stream plane (inlet plane), and the static pressure is specified at the exit plane. Also, the internal static pressure field is assigned guessed values. The solution starts from the second downstream plane and progresses as follows:

(i) Equations (6), (3) and (4) are solved for ρ,u and v, respectively.

(ii) Equations (1) and (2) are employed to derive new values of Δr and α, using the new values of ρ, u and v.

(iii) Block corrections are calculated from equations (30,31), local pressure-corrections are obtained by solving equation (29), and block corrections are again calculated from equations (30,31). After each step, the pressure field is updated, u and v are corrected according to equations (21,23), ρ is recalculated from equation (6), and Δr and α are computed from equations (1) and (2).

(iv) The next downstream section is considered and steps (i) - (iii) are repeated. Finally when all downstream planes have been traversed, thus completing one iteration cycle, a check is made on the degree of convergence of the solution[I], and if this is not satisfactory, M_{in} is corrected according to equation (32) and a new iteration is initiated employing the latest pressure field. This sequence is repeated until a properly converged solution is attained.

5. DEMONSTRATION CASE

In order to demonstrate the ability of the computational procedure, the flow field is predicted in a convergent-divergent nozzle featuring an exceptionally small radius of curvature of the walls at the throat section, Fig. 6, similar to the one considered by [6].

Fig. 1 displays the predicted streamlines, whereas Fig. 7 displays the predicted Mach-number contours. The predictions reveal that the flow pattern deviates considerably from one-dimensionality. In particular, the Mach-number contours have steep

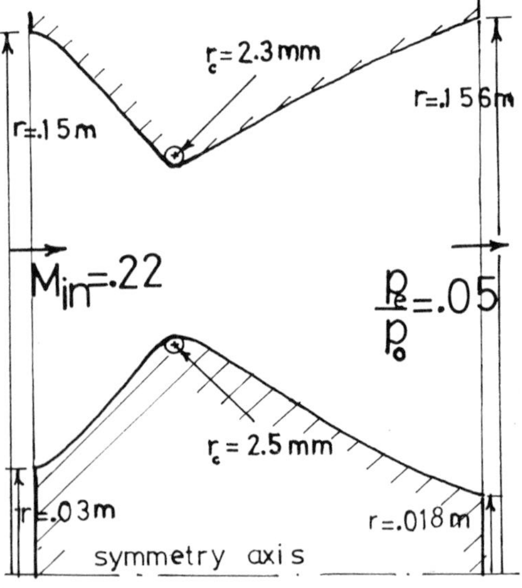

- Fig.6 Demonstration Case

I Indicated by $\Sigma_i \varepsilon_p$, and stability of predictions radial gradients near the throat section and downstream of it. As a result, the flow-coefficient is predicted to be only 0.93. The predictions also show that, although the exit section is considerably far from the throat section, yet the flow at exit is substantially non-uniform, varying from M = 2.0 to M = 2.8 , the higher Mach-numbers occurring at the larger radii.

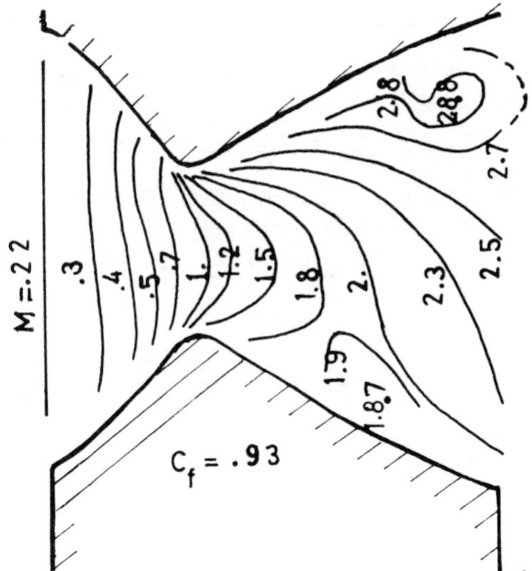

Fig. 7 Predicted Mach-numbers

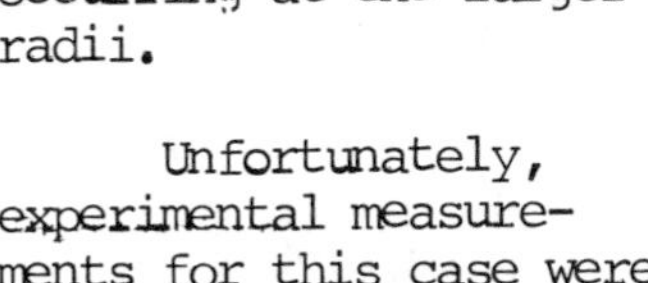

Unfortunately, experimental measurements for this case were not available for comparison; however it may be argued that since the predicted solution satisfies the momentum, continuity and thermodynamic equations for inviscid, shock-free flow, it is expected to be correct.

A total of 100 iterations was required to attain a well-converged solution.

6. COMPARISON AGAINST OTHERS

In comparison with time dependent methods, e.g. [7], the present algorithm requires considerably less storage requirements, since it requires only one-dimensional storage for all variables except pressure. Moreover, whereas the time-dependent predictions of [7] required 301 time-steps to attain a well converged solution for a simpler geometry and employing only 21 x 8 grid nodes, the present predictions required only 100 iterations employing 27 x 10 grid nodes; the computational time for each time step and each iteration should be comparable.

In comparison with relaxation methods, e.g. [8], the present algorithm has the advantage of employing primitive variables directly.

7. CONCLUSION

The paper presents a new computational algorithm for the prediction of axi-symmetric, transonic, duct flows. The algorithm is economical in both storage and computational time requirements, and is capable of predicting flows in complicated boundary geometries.

8. REFERENCES

1. SPALDING, D.B. - Genmix: A General Computer Program for Two-Dimensional Parabolic Phenomena. Heat Transfer Section Report No. HTS/75/17, Imperial College, London, 1975.

2. PUN, W.M. & SPALDING, D.B. - A General Computer Program for Two-Dimensional Elliptic Flows. Heat Transfer Section Report No. HTS/76/2, Imperial College, London, 1976.

3. SERAG-ELDIN, M.A. - A Computational Procedure for Axial Turbomachinary Flows. Proceedings of the Fifth Annual Operations Research Conference, Zagazig University, Egypt, pp 185-201, 1978.

4. ASHOK, K.S. - Flow in Axial Turbomachinary Cascades. Ph.D. Thesis, London University, 1977.

5. HARLOW, F.H. & WELCH, J.E. - Numerical Calculation of Time-Dependent Viscous Incompressible Flow of Fluid with Free Surface. Physics Fluids 8, pp 2182-2189, 1965.

6. HOPKINS, D.F. & HILL, D.E. - Transonic Flow in Unconventional Nozzles, AIAA Journal, 6, No. 5, pp 838-842, 1968.

7. CLINE, M.C. - Computation of Steady Nozzle Flow by a Time-Dependent Method. AIAA Journal, 12, No. 4, pp 419-420, 1974.

8. DODGE, P. - Transonic Relaxation. Von Karman Institute for Fluid Mechanics, Lecture Series 84, Vol. I, 1976.

HYPERSONIC VISCOUS FLOWS IN A STREAMLINE COORDINATE SYSTEM

M. D. Kim*, R. R. Thareja* and C. H. Lewis**

Virginia Polytechnic Institute and State University
Blacksburg, Virginia 24061 USA

ABSTRACT

A method of predicting the three-dimensional laminar hypersonic viscous flowfield over a body at very high angles of attack has been developed using the parabolized Navier-Stokes equations (PNS). At very high angles of attack, the primary flow direction deviates substantially from the conventional body-generator coordinate direction. Hence, in the present study the viscous flow equations are parabolized along the surface streamline direction instead of body-generator coordinate direction, which almost eliminates the restriction on the angle of attack of a flight vehicle. The streamline traces on the body surface are evaluated from a three-dimensional inviscid flowfield solution. Both the present streamline approach (SLPNS) and the conventional body-generator PNS method (PNS) have been applied to a test case for the purpose of comparison, and the computational results and computing times compare well with each other.

1. INTRODUCTION

In recent years the three-dimensional viscous shock-layer approach (VSL3D) and the parabolized Navier-Stokes method (PNS) have been applied to various problems of predicting hypersonic viscous flowfields over reentry vehicles [1-4]. The viscous shock-layer equations (VSL3D) are parabolic in both streamwise and crossflow directions and are solved by efficient methods which require substantially less computing time than the PNS method. The VSL3D method can be applied to general geometries to obtain the flowfield solution over the entire body when the angle of attack is not too high. However,

---

*Graduate Student, **Professor, Department of Aerospace and Ocean Engineering.

if the angle of attack becomes very high and a crossflow-separated region appears on the leeward side of the body, the method cannot treat that region. The PNS equations are parabolic in the streamwise direction and elliptic in the crossflow direction which makes it possible to solve the crossflow separated region.

However, when the angle of attack becomes extremely high (e.g. over 45 deg) which often can be encountered during reentry flight, those two methods may not work even in the windward region of the vehicle for the following reason. Both conventional methods take the marching direction as the first coordinate direction of the body-generator coordinate system, but at very high angles of attack, the primary flow direction deviates substantially from the body-generator axis direction, which may result in the failure or very poor results of the flowfield solution.

In the present approach (SLPNS), the same set of PNS equations are solved in a streamline coordinate system, which means that the governing equations are parabolized along the actual streamline direction, to be able to treat the cases of extremely high angle of attack. For the present analysis, the HYTAC code which has been developed by Helliwell et al. [4,5] is modified primarily in the parts related to the coordinate and metric generation. In order to generate the third coordinate ($\xi 3$) to be normal to the streamlines on the body surface, an iterative numerical scheme has been developed. The second coordinate ($\xi 2$) can be constructed by the same procedure as the one used by Helliwell [4,5] in the HYTAC code.

Streamlines on the body surface diverge rapidly from the neighboring streamlines on the windward side and converge rapidly to the neighboring streamlines on the leeward side in the cases of high angle of attack. Hence, the marching solution cannot be continued far downstream in a fixed streamline coordinate system in such cases. Both SLPNS and PNS codes have been applied to a test case at moderate angle of attack for the purpose of comparison.

2. ANALYSIS

2.1 Generation of Streamlines on a Body Surface

One of the most important steps in the present approach is to generate surface-streamline traces which will constitute the first coordinate lines ($\xi 1$) in the streamline-coordinate system. For the approximate calculation of the locations of streamlines on the body surface, an inviscid flowfield solution can be used. The inviscid flowfield can be obtained by using a set of three-dimensional inviscid flowfield programs developed by Marconi et al. [9] for complex geometries, or by using another inviscid solver [10] for sphere-cone geometries.

From the inviscid surface pressure and surface flow velocity distributions over a body, the streamline traces can be computed using the method developed by Hamilton [6].

2.2 Computational Grid

In the present approach the streamlines on the body surface constitute the first coordinate ($\xi 1$) in the curvilinear coordinate system. The third coordinate lines ($\xi 3$) are constructed to be normal to the $\xi 1$ coordinate on the body surface. An iterative numerical scheme has been developed using the geometrical relationships to accurately generate the $\xi 1$, $\xi 3$ curvilinear coordinates which are normal to each other on the body surface. Accurate streamline slopes along each streamline with respect to the reference cylindrical coordinate axes are used in the iteration scheme. An orthogonal grid on the body surface makes it easier to integrate forces and moments and to include slip and/or mass-transfer effects at wall.

The second coordinate ($\xi 2$) is constructed to be normal to both $\xi 1$ and $\xi 3$ coordinate lines in the region between the body and the shock surfaces. This procedure is the same as the one which has been used by Helliwell et al. [4] in the HYTAC code. Since the shock surface is taken as a $\xi 1$, $\xi 3$ coordinate surface ($\xi 2 = 1$), the resulting $\xi 2$ coordinate lines will be both body-normal and shock-normal. In this coordinate system, the application of the shock boundary condition becomes simple, because the $\xi 2$ coordinate is normal to the shock surface. A

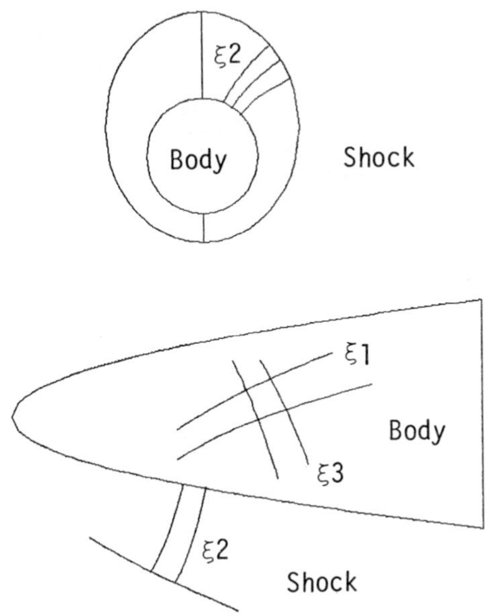

Fig. 1. Body-Normal, Shock-Normal Coordinate System

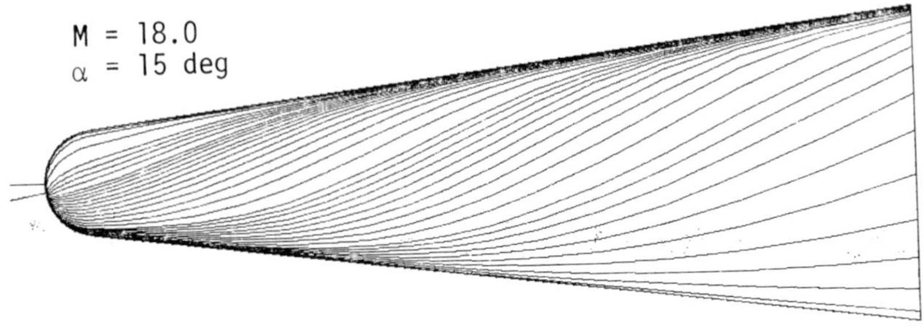

Fig. 2. Streamline Trace on a Sphere-Cone

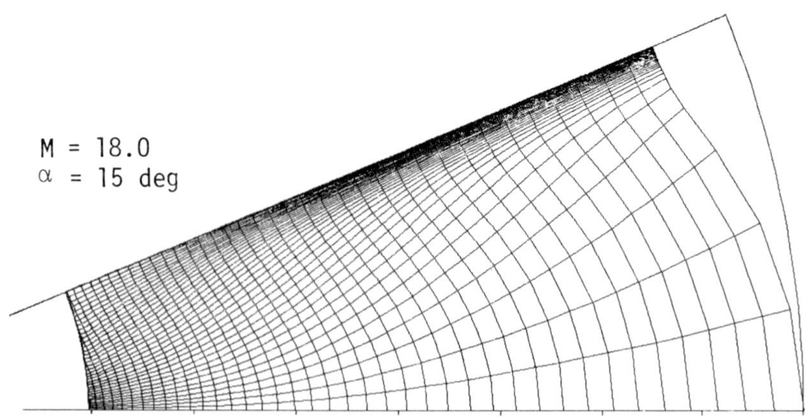

Fig. 3. $\xi 1$, $\xi 3$ Grid for a Streamline Coordinate System on a Developed Cone Surface.

body-normal and shock-normal streamline coordinate system is shown in Fig. 1, and $\xi 1$ coordinate lines (streamlines) for the test case (α = 15 deg) are shown in Fig. 2, and $\xi 1$, $\xi 3$-grid for computation is also shown in Fig. 3.

2.3 Governing Equations

In the general curvilinear coordinate system, the steady parabolized Navier-Stokes equations can be written in non-dimensionalized form with the velocity vector represented by tensor components. In the governing equations we will have the coordinate system metric tensor terms (g_{ij}), which represent the coordinate stretching and nonorthogonality. In the equations for the stress tensor components, all the derivatives with respect to the primary streamline axis ($\xi 1$) are neglected. The resulting equations are parabolic in the $\xi 1$ axis direction and elliptic in other directions. The body-normal coordinate ($\xi 2$) is normalized by the shock-layer thickness to facilitate the computation of the shock location. The complete set of the governing equations can be found in [4].

The governing equations are solved by implicit differencing in $\xi 2, \xi 3$ plane. The $\xi 1$ derivatives are approximated by backward differencing, while the $\xi 2$ and $\xi 3$ derivatives are obtained by unequably spaced three-point differencing. After differencing, the equations are linearized by the Newton-Raphson method. The linearized equations are solved by using the Gauss-Seidel iteration method. The details of the solution procedure for the equations are given by Helliwell et al. [2-4].

2.4 Boundary Conditions

At the body, the no-slip conditions are used and enthalpy is specified. To obtain the pressure at the body surface, the v-momentum equation is used, and two-point differencing in the body-normal direction is performed. At the plane of symmetry (ϕ = 0 and 180 deg) w and g13 are antisymmetric, while all other flow variables and metrics are symmetric.

At the shock, the freestream velocity vector is transformed into the computational coordinate direction, and then the Rankine-Hugoniot jump conditions are used. At this step, since the second computational coordinate $\xi 2$ is normal to the shock-surface, the conservation equations across the shock can be easily applied without further velocity vector rotations. Since the shock distance is also unknown, the continuity equation is used as a sixth equation, adding to the five conservation equations across the shock for u, v, w, p and h. The complete set of the equations for the boundary conditions can be found in [4].

2.5 Initial Data Plane

For a numerical flowfield solution which utilizes a marching scheme, the construction of an accurate initial data plane is one of the most crucial conditions for the success of the whole solution. In a previous investigation [7], the viscous shock-layer method (VSL3D) for a blunt nose was found to be able to generate a satisfactory initial data plane to start the PNS solution. Thus, the entire flowfield, including coordinates and metrics must be supplied at an initial data plane to get the SLPNS code started.

In the present study, it is noted that the streamline coordinate axes coincide with the wind axes on the spherical nose. Therefore, for the generation of an initial data plane in the streamline coordinate approach, the viscous shock-layer solution need be obtained only for zero angle of attack regardless of the actual angle of attack concerned. At the initial data plane, the coordinate and metrics for the body-normal, shock-normal coordinate system are generated from the two step body-normal data of the viscous shock-layer solution by the method described in [4].

3. RESULTS and DISCUSSION

As previously mentioned, the constructed streamline coordinate system has the following characteristics:

1. $\xi 1$ coordinate consists of the surface streamlines.
2. The body is a coordinate surface ($\xi 2 = 0$).
3. The bow shock is a coordinate surface ($\xi 2 = 1$).
4. $\xi 1$ and $\xi 3$ coordinates are necessarily orthogonal only on the body surface.
5. $\xi 2$ coordinate is always orthogonal to $\xi 1$ and $\xi 3$ coordinates.

The streamlines obtained from an inviscid flowfield on the spherical nose were found to have slight deviations from the analytic solution, and thus in the present work, the analytic streamlines were used on the nose. Hence, it was necessary to link the analytic streamlines on the sphere with the numerical solution on the conical surface through the tangent point by proper interpolations.

The shape of streamlines and the $\xi 1$, $\xi 3$-grid on a developed cone are presented in Figs. 2 and 3, respectively. Numerically constructed $\xi 1$, $\xi 3$ grids start from the sphere-cone tangency point, because in the nose region the analytic $\xi 1$, $\xi 3$-grid distribution is used. The SLPNS code produced excellent agreement with the original PNS method for a test case at $\alpha = 15$ deg., but the main purpose of the current SLPNS approach is to apply the method to the cases of extremely high angles of attack. For a sphere-cone, however, there is a limit on the angle of attack for the construction of an initial data plane on the spherical nose portion, because the inviscid flow region of the initial data plane must be supersonic. Hence, further work would be required to eliminate the limit on the angle of attack in constructing the initial data plane.

For the test case it can be observed that the streamlines on the windward side diverge rapidly from the neighboring streamlines, and the streamlines on the leeward side converge rapidly to the neighboring streamlines (see Fig. 3). Because of the rapid change in streamline pattern, the marching solution cannot be continued far downstream. For this reason, the solution for the test case has been obtained up to the station of s = 7.1 by the SLPNS approach. To restart the solution at this station, the streamlines should be re-spaced around the body, because at this station the stepsizes $\Delta\xi 3$ on the windward surface are too large, while the stepsizes $\Delta\xi 3$ on the leeward surface are too small. Of course, the use of a variable $\Delta\xi 3$ input at the initial data plane should help the marching solution be obtained further downstream. However, the restart data plane modification by respacing the streamlines around the body would be inevitable in the case of a very long body, because in the case of high angle of attack the divergence and convergence of the streamlines are remarkable.

U_∞ ft/sec (m/sec)	T ^oR (^oK)	α deg	θ_c deg	$Re_\infty x10^{-5}$ ft^{-1} (m^{-1})	T_w/T_0	ε
7223	62.8	15	7	5.447	0.127	0.055
(2202)	(34.9)			(17.871)		

TABLE I. Mach 18 Test Case Freestream Conditions

Method of Solution	s from - to	Grid Size, Number of s-steps	n-points	ϕ-planes	Time (min:sec)
PNS	1.31-10.16	50	50	19	35:24
SLPNS	1.31-7.11	31	50	19	27:16

TABLE II. Test Case Computing Times on IBM 3032

For the purpose of demonstration, a test case was chosen and both SLPNS and PNS solutions have been obtained. The results compare very well with each other. The flow properties for the test case are shown in Table I, and the computing times are presented in Table II. It can be seen that the two methods take similar computing times, and each method takes approximately 42 seconds per marching step on an IBM 3032.

The shock-layer thickness distribution along the body is shown in Fig. 4, and the slight difference between the two methods can result from the different coordinate systems, the different initial plane data and the different marching step-sizes. Figures 4 to 7 show the shock-layer thickness, wall pressure, heat-transfer rate and streamwise skin-friction distributions, and all results show that essentially identical results are produced for both methods at these modest conditions (α = 15 deg).

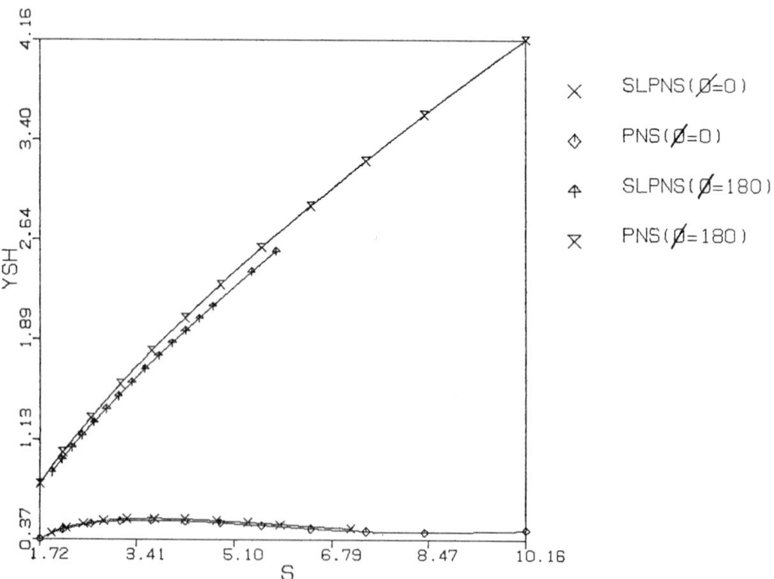

Fig. 4. Shock-Layer Thickness Distribution Along the Body

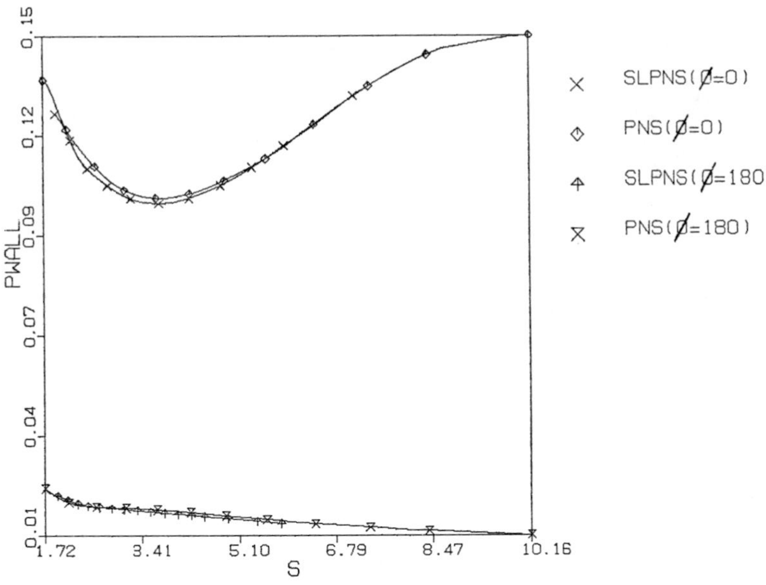

Fig. 5. Surface-Pressure Distribution Along the Body

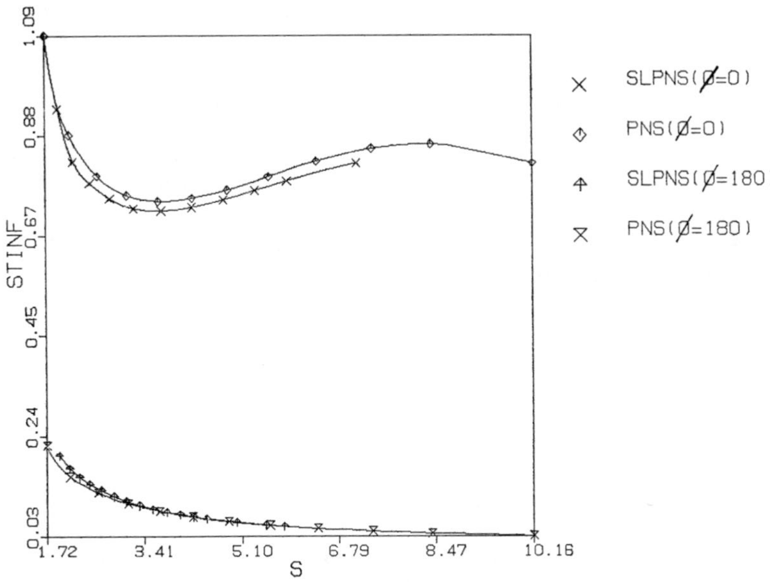

Fig. 6. Surface Heat-Transfer Distribution Along the Body

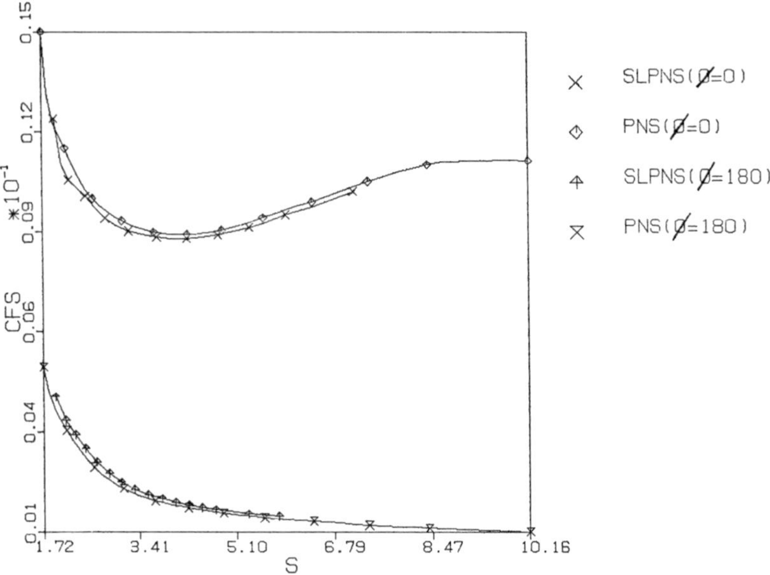

Fig. 7. Streamwise Skin-Friction Distribution Along the Body

4. CONCLUSIONS AND RECOMMENDATIONS

A procedure (SLPNS) has been developed to compute the viscous flowfield over a body at extremely high angles of attack (up to 90 deg) by employing a streamline coordinate system. Comparisons with conventional PNS results for a test case indicate that the SLPNS method can accurately solve the flowfield in comparable computing time with the body-generator PNS method.

Further work is required to develop a restarting procedure by redistributing the surface streamlines including a variable $\Delta\xi 3$ distribution around the body and to prepare the initial data plane for general bodies at extremely high angles of attack.

NOMENCLATURE

CFS	streamwise skin-friction coefficient
g_{ij}	coordinate system metric tensor; i, j = 1, 2, 3
M	freestream Mach number
p	nondimensionalized pressure, $p^*/\rho_\infty U_\infty^2$
Re_∞/ft	freestream unit Reynolds number
STINF	Stanton number
s	surface distance coordinate measured along the body, nondimensionalized by nose radius
U_∞	dimensional freestream velocity

YSH shock-layer thickness, nondimensionalized by nose radius
α angle of attack, deg
ε perturbation parameter, $[\mu(U_\infty^2/c_p)/\rho_\infty U_\infty r_n]^{1/2}$
θ_c cone half-angle, deg
$\xi 1,\xi 2,\xi 3$ streamline coordinate system

Subscript and Superscript

∞ freestream condition (dimensional quantity)
* dimensional quantity

REFERENCES

1. MURRAY, A. L. and LEWIS, C. H.: "Hypersonic Three-Dimensional Viscous Shock-Layer Flows over Blunt Bodies," AIAA J., Vol. 16, No. 12, pp. 1279-1286, December 1978.

2. LUBARD, S. C. and HELLIWELL, W. S.: "Calculation of the Flow on a Cone at High Angle of Attack," AIAA J., Vol. 12, No. 7, pp. 965-974, July 1974.

3. AGOPIAN, K., COLLINS, J., HELLIWELL, W. S., LUBARD, S. C. and SWAN, J.: "NASA Viscous 3-D Flowfield Calculations," R&D Associates, RDA-TR-6100-007, October 1975.

4. HELLIWELL, W. S., DICKINSON, R. P., and LUBARD, S. C.: "HYTAC Phase I Report: Viscous Flow over Arbitrary Geometries at High Angle of Attack," Arete Associates, AR-79-046-TR, April 24, 1979.

5. HELLIWELL, W. S., DICKINSON, R. P. and LUBARD, S. C.: "HYTAC User's Manual," Arete Associates, AR-80-207-TR, July 31, 1980.

6. HAMILTON, H. H., II: "Calculation of Heating Rates on 3-D Configurations," Thesis for Degree of Engr., School of Engineering, George Washington University, December 1979.

7. WASKIEWICZ, J. D. and LEWIS, C. H.: "Hypersonic Viscous Flows over Sphere-Cone at High Angles of Attack," AIAA Paper 78-64, January 1978.

8. SRIVASTAVA, B. N., WERLE, M. J. and DAVIS, R. T.: "Viscous Shock-Layer Solution for Hypersonic Sphere-Cones," AIAA Paper 77-693, June 1977.

9. MARCONI, F., SALAS, M., and YAEGER, L.: "Development of a Computer Code for Calculating the Steady Super/Hypersonic Inviscid Flow around Real Configurations," Vol. I - Computational Technique, NASA-CR-2675; Vol. II - Code Description, NASA-CR-2676, April 1976.

10. SOLOMON, J. M., CIMENT, M., FERGUSON, R. E., BELL, J. B., and WARDLAW, A. B., JR.: "A Program for Computing Steady Inviscid Three-Dimensional Supersonic Flow on Reentry Vehicles, Vol. I, Analysis and Programming," NSWC-WOL-TR-7728, February 1977.

SECTION 7

TWO PHASE FLOW

THE COMPUTATION OF THREE-DIMENSIONAL TURBULENT TWO-PHASE FLOWS IN MIXER VESSELS

R I Issa and A D Gosman

The paper presents a finite-volume method for the computation of the single- or two-phase turbulent three-dimensional flows arising in stirred vessels of the kind employed in the chemical process industry, consisting of a baffle-equipped cylindrical tank with a central axisymmetric impeller. The method employed solves the time-averaged Navier Stokes equations for the liquid phase in conjunction with some simpler equations for the gas phase. The turbulent Reynolds stresses are obtained from a turbulence model. Results of computations for single- and two- phase flows are presented and are compared, in the case of the former, with experimental data.

1. INTRODUCTION

1.1 Background

The stirred mixer vessel is one of the most common devices encountered in industry, particularly on the chemical process side. Yet despite their wide occurence and the simplicity of their construction and function, it has not hitherto been possible to accurately analyse or predict the behaviour of the flow field within them. This is because the flow in a stirred vessel exhibits many complexities, as will be outlined below.

A typical vessel of the kind employed by the chemical process industry consists of a cylindrical tank fitted on its inner periphery with radial baffles at uniform intervals (see fig 1).

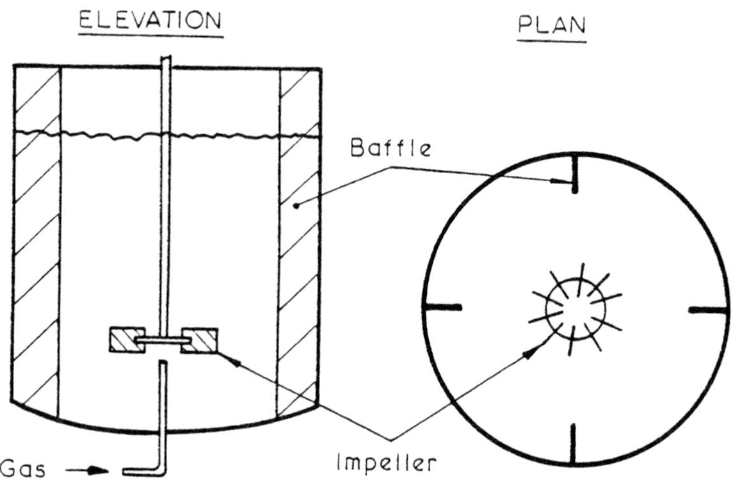

Figure 1: Typical Geometry of a Mixer Vessel.

The fluid (or mixture of fluids) within the vessel is stirred by one or more centrally-located motor-driven rotating impellers, the mixing action of which is assisted by the presence of the baffles. The impellers may be either of the propeller or turbine type. The secondary fluid , which is to be absorbed by and/or mixed with the primary fluid is introduced via external supply pipes; it is normally released into the vessel in vicinity of the impeller in order that it is dispersed rapidly by the action of the latter.

It is evident that the resulting two-phase flow is, in general, unsteady in respect of the mean motion due to the periodicity of the impeller and baffles and is three-dimensional for the same reason. Furthermore, the action of the impeller generates turbulence and this phenomenon further enhances the mixing rate.

The above complex phemonena present formidable difficulties with which a full analysis must contend. However, judicious idealisations of the real situation, such as those about to be introduced here, considerably simplify the problem and render it tractable without detracting much from the accuracy of the analysis. Firstly, the perodic unsteadiness which is imposed by the impeller blades can arguably be absorbed into time-mean characterisations of the motion which is induced in their immediate vicinity, especially if they are closely spaced. Thus the impeller can be represented by a fictitious rotating cylinder at whose porous surfaces transfer of mass, momentum and other properties can take place. Analysis of these transfer processes may then be used to provide the boundary conditions for the fluid-containing domain. In the present study, the expressions representing the transfer rates are derived from empirical correlations for the mean flow and turbulence characteristics which these devices induce.

Secondly, for a centrally-placed symmetric impeller, the flow pattern may be assumed to cyclically repeat itself in the circumferential direction, with each complete cycle taking place over the cylindrical sector contained between successive baffles. The solution domain can therefore be confined to one such sector thus minimising the computing effort. At the boundaries containing the interval, cyclic conditions are imposed. Other useful but minor idealisations that are made are: the representation of the baffles by infinitely thin sheets; the neglect of the presence of the impeller shaft; and the replacement of the secondary phase supply pipes by a fictitious source appropriately located. The resulting idealised geometry of the problem is shown in Figure 2.

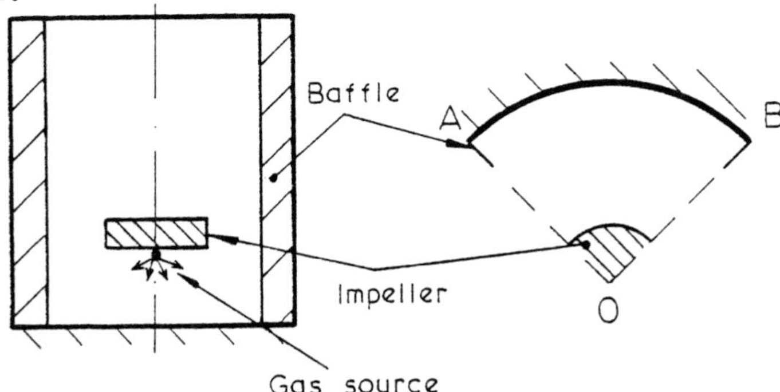

Figure 2: Idealised Mixer Vessel Geometry

A further idealisation which has been made in previous analyses (see e.g. ref. [1]) is to assume the effects of the baffles to be reproduced by a uniformly-distributed wall roughness, which has the advantage of reducing the problem to an axisymmetric one. This assumption may be expected to hold good for the case of closely spaced small baffles (failing only in the vicinity of the baffles themselves). Practical mixer vessels, however, normally employ sparsely-spaced baffles of appreciable breadth (i.e. 10% of the vessel diameter or more) making the above assumption questionable: this idealisation was therefore, not made in the present analysis.

1.2 The present contribution

The present paper describes the development and application of a finite-volume procedure for solving the equations of motion and mixing in vessels whose geometries can be idealised in the manner already discussed. Attention is focussed on the practically-important case in which the primary and secondary fluids are liquid and small (diameter of the order of a few millimetres) gas bubbles respectively. The general methodology is not, however, restricted to the particular phases chosen here.

The method employed is to solve the time-averaged Navier Stokes equations for the liquid phase in conjunction with gas momentum and continuity equations, after suitably simplifying the latter in the light of the particular two-phase regime under consideration. The turbulence Reynolds stresses which appear in the fore-mentioned equations are obtained from the 'k-ε' model and associated wall functions.

The finite-volume method employed in the solution of the equations is in many respects conventional: thus the equations are formulated in primitive-variable form, on a cylindrical-polar mesh using hybrid centred/upwind differencing and are solved iteratively. Within the iteration sequence, the variables are solved for sequentially, with pressure being calculated from its own perturbation equation, derived from the continuity requirement. The main novelties reside in the imposition of the boundary conditions. Comparisons are shown between the prediction of the method and experimental data for single-phase flow revealing that even with comparatively coarse grids the agreement is quite promising. Also presented are results of coarse grid calculations for a two- phase flow case.

2. METHODOLOGY

2.1 The governing equations

Differences of opinion exist in the literature as to what is the correct form of the transport equations for two-phase flow, as inspection of four representative works (refs [2] to [5]) will reveal. These differences extend to both the representation of the momentum exchange between the phases, and the appearance of the void fraction α (defined as the fraction of volume occupied by the particular phase) in the pressure and viscous terms. The equations used in the present work are based on the derivation of ref [3] since the arguments presented by its authors appear to be the most physically plausible. For compactness, all the equations will be stated here in orthogonal tensor notation although they are actually solved in their cylindrical polar form.

The steady-state, continuity and momentum equations for the liquid phase are respectively:

$$\frac{\partial}{\partial x_i}(\alpha^{\ell}\rho^{\ell}u_i^{\ell}) = m_{\ell g} \qquad (1)$$

and

$$\frac{\partial}{\partial x_i}(\alpha^\ell \rho^\ell u_i^\ell u_j^\ell) = -\alpha^\ell \frac{\partial p}{\partial x_j} + \frac{\partial}{\partial x_i}(\alpha^\ell \sigma_{ij}^\ell) + F_j^\ell \qquad (2)$$

where in the equations, p is the hydrostatic pressure; $m_{\ell g}$ is the rate of mass exchange between the two phases due to e.g. dissolution or chemical reaction, F_j^ℓ is the j-component of momentum exchange between the phases due to interphase drag and other effects; and σ_{ij}^ℓ is the stress tensor.

Following ocnventional single-phase flow practice, the forgoing equations are assumed also to describe turbulent flows, with the following amendments: firstly, the u_i and p are to be regarded as time-averaged properties; and secondly, the viscous stress tensor σ_{ij}^ℓ is now composed of a laminar part and a turbulent contribution (Reynolds stresses) arising from time-averaging the parent unsteady-state equations, as outlined in e.g. ref. [6]. The resulting equations, it must however be stated, are inexact, for they ignore the certain special features arising in two-phase flows, notably the effects of fluctuations in void fraction and density, and the interactions between the fluctuating velocities of the two phases. The inclusion of these effects into the analysis is currently being performed as part of an extension to the present study.

The 'effective' stresses σ_{ij}^ℓ are calculated by way of the now well-established 'k-ε' model in which k stands for the time-average kinetic energy of turbulence and ε is its dissipation rate. In this model, the concept of an eddy diffusivity is invoked, whereby the effective stress tensor is related to the strain field via the Newtonian-type relation:

$$\sigma_{ij}^\ell = \mu\left[\frac{\partial u_i^\ell}{\partial x_j} + \frac{\partial u_j^\ell}{\partial x_i} - \frac{2}{3}\delta_{ij}\left(\frac{\partial u_m^\ell}{\partial x_m} + k\right)\right] \qquad (3)$$

where μ is the effective viscosity. This latter quantity is taken to be the sum of the molecular viscosity $\mu_{\ell am}$ and a 'turbulent' viscosity μ_t:
thus:

$$\mu \equiv \mu_{\ell am} + \mu_t \qquad (4)$$

The turbulent viscosity is related to k and ε by:

$$\mu_t = C_\mu \rho k^2/\varepsilon \qquad (5)$$

where C_μ is an empirical constant given in ref. [6]. Further, the quantities k and ε are assumed to obey the following forms of transport equations, taken from ref. [6] and modified here to account for the fractional volume occupied by the liquid:

$$\frac{\partial}{\partial x_i}(\alpha^\ell \rho^\ell u_i^\ell k) = \frac{\partial}{\partial x_i}(\frac{\alpha^\ell \mu_t}{Pr_k}\frac{\partial k}{\partial x_i}) + \alpha^\ell (G - \rho^\ell \varepsilon) \tag{6}$$

$$\frac{\partial}{\partial x_i}(\alpha^\ell \rho^\ell u_i^\ell \varepsilon) = \frac{\partial}{\partial x_i}(\frac{\alpha^\ell \mu_t}{Pr_\varepsilon}\frac{\partial \varepsilon}{\partial x_i}) + \alpha^\ell (C_1 G \frac{\varepsilon}{K} - C_2 \rho^\ell \frac{\varepsilon^2}{k} \tag{8}$$

where Pr_k, Pr_ε C_1 and C_2 are additional constants also given in the cited reference, and G is a 'generation' term defined by:

$$G \equiv \sigma^\ell_{ij,t} \frac{\partial u_i^\ell}{\partial x_j} \tag{7}$$

The gas-phase continuity and momentum equations are respectively:

$$\frac{\partial}{\partial x_i}(\alpha^g \rho^g u_i^g) = - m_{\ell g} \tag{8}$$

and

$$0 = - \alpha^g \frac{\partial p}{\partial x_j} + F_j^g - \alpha^g(\rho^\ell - \rho^g) g_j \tag{9}$$

where the following relations hold between the momentum exchanges and fractional volumes:

$$F_j^g = - F_j^\ell \tag{10}$$

and

$$\alpha^\ell + \alpha^g = 1 \tag{11}$$

Equation (9) is a truncated form of the full gas momentum conservation equation, the omissions being the inertial and viscous stress terms. It can be shown that, because of the small density and diameters of the gas bubbles, the magnitudes of these terms become insignificant compared to the drag and pressure gradient contributions, thus justifying their neglect and thereby substantially simplifying the equations.

The inter-phase momentum exchange, represented by term F_j^ℓ in equation (2), must now be related to the dependent variables of the problem. Many expressions have in the past been proposed to represent this effect (see e.g. refs. [3] to [5]). The more elaborate of these models recognize different modes of such momentum exchange (such as drag due to slip, and virtual mass effect due to relative acceleration). In the present work, this exchange is assumed to arise soley from drag, since the density differential between the gas and liquid phases is very large, thus rendering other effects insignificant. Further, as the gas phase may be considered as a swarm of spherical bubbles interspersed in the liquid, the total drag force per unit volume can be evaluated as the sum of the drag forces exerted by the liquid on each individual bubble contained in that volume. The drag exerted on an individual bubble is obtained empirically from the consideration of the flow around a spherical bluff body representing a bubble. It then follows (see e.g. ref. [4]) that:

$$F_j^{\ell} = \frac{3}{8}\frac{\rho^{\ell}\alpha^{g}}{R} C_D \left(u_j^{g} - u_j^{\ell}\right) U_s \tag{12}$$

where R is the aggregate radius of the bubbles, U_s is the modulus of the slip velocity vector and C_D is a drag-coefficient to be empirically specified. The value to be ascribed to C_D varies according to the bubble Reynolds number (see e.g. ref. [2]) and empirical fomulae may be used, as is done in the present calculations, to describe this relationship.

Equations (9) can be rewritten with the aid of equations (10) and (12) to give:

$$0 = -\alpha^{g}\frac{\partial p}{\partial x_j} - \frac{3}{8}\frac{\alpha^{g}\rho^{\ell}C_D}{R}\left(u_i^{g} - u_i^{\ell}\right)U_s - \alpha^{g}\left(\rho^{\ell}-\rho^{g}\right)g_j \tag{13}$$

which serves as an equation for u_i^{g}.

Finally, as was remarked earlier, the above equations are solved on a cylindrical polar coordinate mesh which accommodates the vessel geometry. They must therefore be translated into their cylindrical-polar form; this will not, however, be done here, as such transformations are elementary and can be found in standard texts on fluid mechanics. It suffices to note that the resulting equations have the general form:

$$\frac{\partial}{\partial x}\left(\alpha^{\ell}\rho^{\ell}u^{\ell}\phi\right) + \frac{1}{r}\frac{\partial}{\partial r}\left(r\alpha^{\ell}\rho^{\ell}v^{\ell}\phi\right) + \frac{1}{r}\frac{\partial}{\partial\theta}\left(\alpha^{\ell}\rho^{\ell}w^{\ell}\phi\right)$$

$$= \frac{\partial}{\partial x}\left(\frac{\alpha^{\ell}\mu}{Pr_{\phi}}\frac{\partial\phi}{\partial x}\right) + \frac{1}{r}\frac{\partial}{\partial r}\left(\frac{r\alpha^{\ell}\mu}{Pr_{\phi}}\frac{\partial\phi}{\partial r}\right) + \frac{1}{r}\frac{\partial}{\partial\theta}\left(\frac{\alpha^{\ell}\mu}{Pr_{\phi}}\frac{1}{r}\frac{\partial\phi}{\partial\theta}\right) + S_{\phi} \tag{14}$$

where the dependent variable ϕ stands for u^{ℓ} (axial velocity), v^{ℓ} (radial velocity), w^{ℓ} (tangential velocity), k and ε, and S_{ϕ} is the source (or sink) of property ϕ. Furthermore, the continuity equations (1) and (8) may also be cast in the form of equation (14) by putting $\phi = 1$ everywhere and interchanging superscripts ℓ and g.

2.2 Boundary Conditions

The treatment will now be outlined of the types of boundaries encountered in the problem, namely: those at the surfaces of the impeller, the vessel and the free surface.

Figure 3 shows schematic representations of two typical types of impeller, namely a prepeller and a turbine.

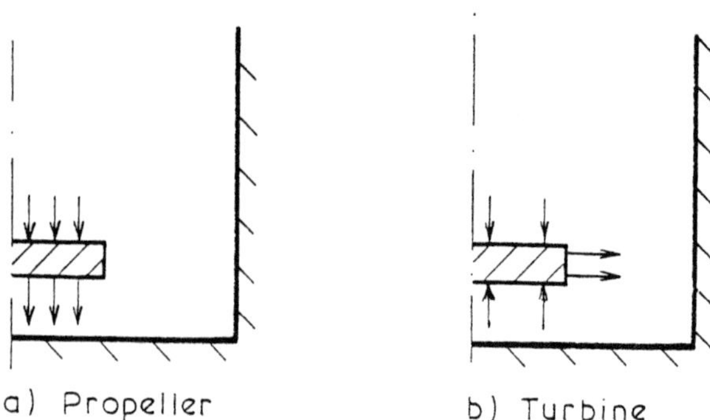

Figure 3: Schematic Representation of the Impeller

For both types, the impeller is here taken as a fictitious porous cylinder through whose surfaces fluid may enter or leave the solution domain in the directions indicated in the figure . The fluid leaving the impeller carries with it empirically-prescribed properties, such as turbulence kinetic energy and tangential momentum, appropriate to the impeller type, the outward velocities being determined from its characteristics and its discharge rate, which is dictated by the rotational speed. The fluid entering the fictitious volume carries the properties of the adjacent flow, which are determined as part of the overall solution, the magnitude of the inward velocities being determined from an overall mass balance over the impeller. Also determined from this mass balance is the void fraction of the fluid leaving the impeller, taking into account the specified gas mass flow rate from the external source, which is assumed to release the gas within the cylindrical volume.

Other boundaries that merit detailed attention are the top surface of the liquid and the meridianal planes OA and OB on figure 2 which bound the domain in the circumferential direction. At the top boundary, which because of the presence of the baffles may be assumed to be flat , all stresses vanish and the axial gradient of the void fraction is taken to be zero, while the gas escape (axial) velocity is taken to be equal to the terminal rise velocity of the bubbles. At the meridianal boundaries, cyclic conditions prevail; these are not known apriori, however, and they must be determined as part of the solution.

The remainder of the boundaries of the domain are made up of impermeable walls and an axis of symmetry, at all of which the usual appropriate constraints are applied. It should be pointed out, that the no-slip condition prevailing at walls is introduced via 'wall-functions' based on the logarithmic law of the wall as outlined in ref. [6]. Wall-functions are also invoked to describe the behaviour of k and ε at these boundaries.

2.3 Method of Solution

The equations to be solved are made up of three transport equations for the liquid phase momentum from which the liquid-phase velocity field is determined; a liquid continuity equation which is solved to determine the pressure field (details are given below); two transport equations for the turbulence quantities k and ε; three gas-phase velocity equations; and a gas-phase continuity equation which is solved to obtain the void fraction field.

The latter equation is formulated as a transport equation for the gas void fraction, which appears in it as the main dependent variable - the transport being effected solely by convection as examination of equation (8) reveals.

The finite volume solution method employed is conventional and is described in ref. [7], so it will only be outlined here. The solution domain is discretised into six-sided control volumes formed form constant-x, -r and -θ coordinate surfaces. The finite-difference equations for each variable are derived by approximate integration of the parent differential equation over each volume with the aid of assumptions about the distributions of the variable between 'nodes' centred in each volume; presently these assumed distributions are associated with upwind differencing for convection terms and centred differencing for the diffusion terms. During the process of integration due care is taken to preserve the conservation properties of the parent equation. The variables are arranged such that the velocities are located mid-way between the pressures which drive them, and the related control volumes are similarly displaced so as to remain centred about the velocity component in question. The resultant finite-difference equations connect each nodal value of property ϕ to its six nearest neighbours by linear algebraic relations. Thus for node p we have:

$$A_p\phi_p = \sum_i A_i\phi_i + S_{\phi,p} \tag{15}$$

where the A's express the combined effects of convection and diffusion, $S_{\phi,p}$ is the integrated source and the summation is over the six neighbours.

The treatment of the continuity equation also follows the practices of ref. [7] in that it is used in conjunction with the momentum finite-difference equations to derive a set of pressure perturbation equations, which are of the general form of (15) above, and which yield an updated pressure field that tends to drive the velocities in the direction of satisfying continuity.

The sets of equations such as equation (15) are solved iteratively using an ADI technique whereby the solution is effected along successive lines lying in the three coordinate directions using the Thomas algorithm for tri-diagonal matrix systems. For those lines lying in the circumferential direction, a special form of this algorithm is employed to accommodate the cyclic boundary conditions which are incorporated into the difference equations. The variables are solved for sequentially within an iteration loop, and the solution is considered converged when the residuals in the equations solved become smaller than a pre-specified tolerance value.

3. APPLICATION

3.1 Single Phase Flow

The working of the method was assessed for single-phase flows by comparison with data from the experimentally-investigated mixer of ref [8]. The liquid is water and the vessel, whose height is equal to its diameter is fitted with four equally spaced baffles. The impeller is of the turbine kind, located one third of the vessel-height above the bottom, and rotates at 1000 r.p.m. The calculations presented here were carried out with a mesh of 16 x 14 x 13 interior nodes in the x, r and θ-directions respectively. The mesh was non-uniformly distributed so as to give a greater density of nodes near the impeller and baffles than elsewhere. The computed radial profiles of the circumferential velocity on the centre plane of the impeller and of the axial velocity on the plane of the lower face of the impeller at a circumferential position 15° downstream of a baffle are shown compared with data in Figs 4 and 5 respectivly. In Figs. 6 and 7, the radial profiles of the radial and circumferential velocities on the centre plane of the impeller are compared with the data of ref. [9] for similarly-proprotioned vessels but at different impeller speeds: the shaded areas in the figures encompass the data scatter.

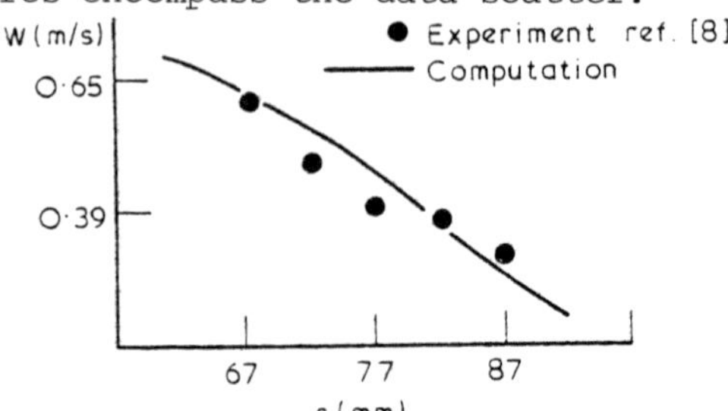

Figure 4 Radial profiles of circumferential velocity on centre-plane of impeller.

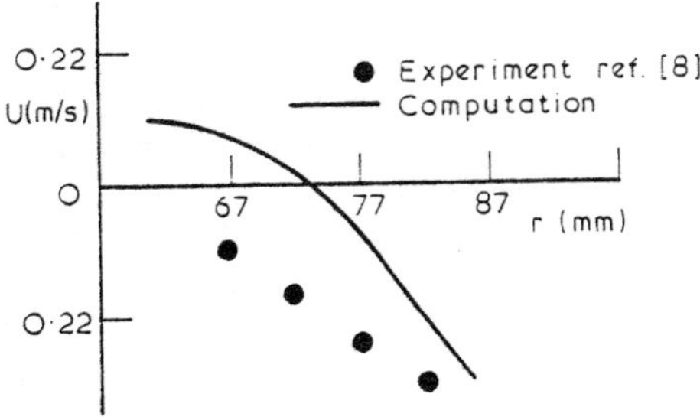

Figure 5: Radial profiles of axial velocity on plane of lower face of impeller.

In view of the many uncertainties in the computations such as empiricism in prescribing the conditions for the impeller-induced flow and the mesh coarseness, the level of agreeement between the calculations and the data is encouraging.

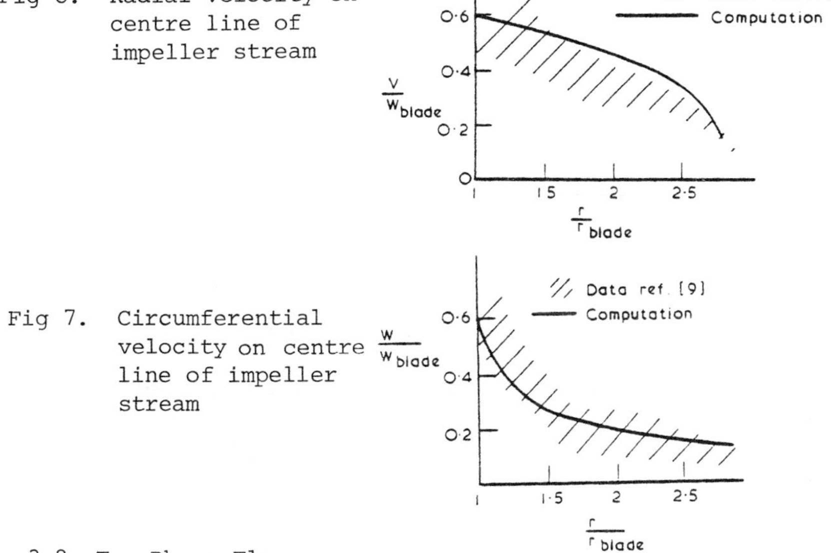

Fig 6. Radial velocity on centre line of impeller stream

Fig 7. Circumferential velocity on centre line of impeller stream

3.2 Two-Phase Flow

Calculations will now be presented for two-phase flow in a hypothetical mixer having a similar geometry to the preceding ones. In the present case, however, the impeller rotates at 180 r.p.m. and there is a constant gas discharge into the impeller region of .019 m^3/s The aggregate diameter of the gas bubbles released is .5mm, and this is assumed to be the value prevailing over the whole vessel. No experimental data are available for this case, hence the results presented can only be judged by their plausibility.

The computations were made with a very coarse grid of 8 x 5 x 7 interior nodes non-uniformly distributed to enhance resolution in the impeller region. Fig. 8 shows contours of constant gas void fractions on a constant -θ plane 12.5° downstream of a baffle. From the figure it may be deduced that the gas appears to be well dispersed by the action of the impeller with the minimum concentration occurring near the axis, as would be expected. The flow pattern prediction for this case (not shown) is qualitatively similar to that of the corresponding single-phase situation although there are differences in the magnitudes of the velocities.

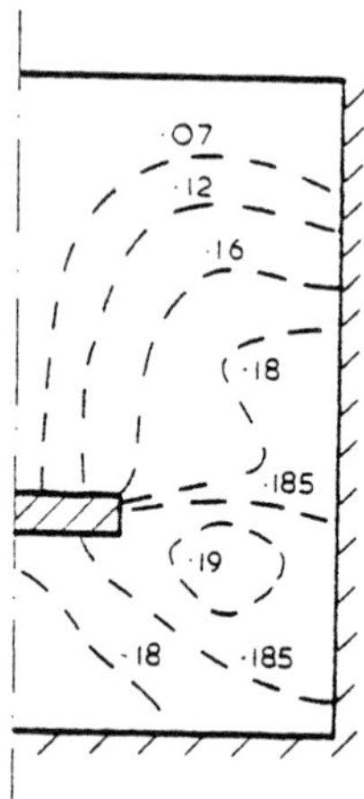

Figure 8: Constant-Void Fraction Contours

4. CONCLUSIONS

A method for the calculation of the three-dimensional, single- or two -phase flow processes occuring in mixer vessels has been described. Applications to two representative practical cases have demonstrated the capability of the method to give plausible predictions for both types of flow and the comparison with the available data for the single-phase case has been favourable. Considerable further testing and development will however be necessary before the method can be regarded as a reliable design tool. In particular, a more sophisticated and accurate representation of the impeller-induced flow is required, although it must be stressed that such a representation must remain within the realm of empiricism if the procedure is to be economically-viable. Another aspect which has hitherto not been thoroughly investigated is the modelling of turbulence in two-phase flows.

Finally, there appears to be a distinct lack of detailed and reliable experimental data, especially for the two-phase case, which makes the assessment of a computational procedure such as the present one a difficult and unsatisfying task.

References:

1. Harvey, P S
"Turbulent Flow in an Agitated Vessel" PhD Thesis, University of Bath, 1980.

2. Soo, S L
"Fluid Dynamics of Multiphase Systems", Blaisdell, 1967.

3. Harlow, H H, and Amsden, A A
"Numerical Calculation of Multiphase Fluid Flow" Journal of Comp. Phys., Vol. 17, 1975.

4. Hughes, E D, Lyczkowski, R W and McFadden, J H
"An evaluation of State-of-the-Art, Two-Velocity, Two-Phase Flow Models and their Applicability to Nuclear Reactor - Transient Analysis."
Vol. 182, Electrical Power Research Institute, 1976.

5. Anderson, T B and Jackson, R
"A Fluid Mechanical Description of Fluidized Beds - Equations of Motion"
I & EC Fundamentals vol. 6 No. 4, 1967.

6. Launder, B E and Spalding, D B
"Mathematical Models of Turbulence", Academic Press, 1972.

7. Caretto, L S, Gosman, A D, Patankar, S V and Spalding, D B
"Two Calculation Procedures for Steady, Three-Dimensional Flows with Recirculation" Proc. of the Third Int. Conf on Num. Methods in Fluid Dynamics, pp. 60-68, 1972.

8. Reed, X B Jr, Prinz, M and Hartland, S
"Laser Doppler Measurement of Turbulence in Stirred Tank". Second European Conf. on Mixing, 1977.

9. Cutter, L A
"Flow and Turbulence in a Stirred Tank" A.I.Ch.E Journal, vol, 12, No. 1, 1966.

NUMERICAL SIMULATION FOR TWO-PHASE JET PROBLEM

W. H. Lee
Los Alamos National Laboratory

V. L. Shah
Argonne National Laboratory

ABSTRACT

In the present work, we have developed a computer program TWOP for obtaining the numerical solutions of three-dimensional, transient, two-phase flow system with non-equilibrium and non-homogeneous conditions. TWOP employs two-fluid model and a set of the conservation equations formulated by Harlow and Amsden along with their Implicit Multi-Field (IMF) numerical technique that allows all degrees of couplings between the two fields. We have further extended the procedure of Harlow and Amsden by incorporating the implicit couplings of phase transition and interfacial heat transfer terms in the energy equations. Numerical results of two tested problems are presented to demonstrate the capabilities of the TWOP code. The first problem is the separation of vapor and liquid, showing that the code can handle the computational difficulties such as liquid packing and sharp interface phenomena. The second problem is the high pressure two-phase jet impinged on vertical plate, demonstrating the important role of the interfacial mass and momentum exchange.

INTRODUCTION

During loss of coolant or transient overpower accidental situations, boiling of liquid coolant in a reactor core is expected due to high temperatures of fuel pins. The fluid mixture of liquid and vapor, in such circumstances, is nonhomogeneous with both phases being in nonequilibrium thermodynamic states. It is, therefore, desirable to develop a computer code for obtaining numerical solutions of three-dimensional, transient, two-phase (gas-liquid) flow system with nonequilibrium and nonhomogeneous conditions.

The TWOP code performs three-dimensional, transient, two-phase thermal-hydraulic analyses with nonequilibrium velocity and temperature. It solves conservation of equations of mass, energy and momentum as a boundary value problem in space, and an initial value problem in time. The two-fluid model [1] and modified implicit, multifield numerical scheme are employed.

This paper summarizes the numerical results of two tested problems obtained from the TWOP code. From these results it has not only demonstrated the computational capability but also exhibited the ability of modeling the complex phenomena such as interfacial mass, momentum and energy transfers.

The next section describes the basic field equations for two-phase flow system. The following sections provide the numerical solution procedure for integrating those field equations, and then with the results of two tested problems. The results and conclusions of the present work are given in the last section.

TWO-PHASE GOVERNING EQUATIONS

When a mixture of two phases (liquid and vapor) is in nonequilibrium and nonhomogeneous condition, transfer of mass momentum and energy between the two phases occurs at the interface. This interaction between the two phases is a very complex and not well understood phenomenon. At present there are several mathematical models [1,2,3,4,5] in existence in the literature postulating the interaction between the two phases and consequently, several formulations exist that describe the governing conservation equations. In the present development, we are using the two-fluid model of Harlow and Amsden. In this model separate conservation equations are formulated for each phase and the interaction between the phases is accounted for by including evaporation (or condensation), interfacial drag and inter facial heat transfer terms in the corresponding mass, momentum and energy equations, respectively. The code, therefore, solves the conservation equations (2-continuity, 6-momentum, and 2-energy) along with the required thermodynamic constitutive relations.

The field equations for the two fluid model are

Mass Conservation:

$$\frac{\partial \rho'_\ell}{\partial t} + \nabla \cdot \rho'_\ell \vec{U}_\ell = \Gamma_c - \Gamma_e \tag{1}$$

for liquid phase, and

$$\frac{\partial \rho'_g}{\partial t} + \nabla \cdot \rho'_g \vec{U}_g = \Gamma_e - \Gamma_c \tag{2}$$

for vapor phase.

Momentum Conservation:

$$\frac{\partial \rho_\ell' \vec{U}_\ell}{\partial t} + \nabla \cdot (\rho_\ell' \vec{U}_\ell \vec{U}_\ell) = \Theta_\ell \nabla P + \vec{W}_\ell + \Gamma_c \vec{U}_g - \Gamma_e \vec{U}_\ell + \rho_\ell' \vec{g}$$

$$+ K(\vec{U}_g - \vec{U}_\ell) + \nabla \cdot (\Theta_\ell \vec{\sigma}_\ell) \quad (3)$$

for the liquid phase, and

$$\frac{\partial \rho_g' U_g}{\partial t} + \nabla \cdot (\rho_g' \vec{U}_g \vec{U}_g) = - \Theta_g \nabla P + \vec{W}_g - \Gamma_c \vec{U}_g + \Gamma_e \vec{U}_\ell + \rho_g' \vec{g}$$

$$- K(\vec{U}_g - \vec{U}_\ell) + \nabla \cdot (\Theta_g \vec{\sigma}_g) \quad (4)$$

for the vapor phase.

Energy Conservation:

$$\frac{\partial \rho_\ell' H_\ell}{\partial t} + \nabla \cdot (\rho_\ell' H_\ell \vec{U}_\ell) = \Theta_\ell \frac{\partial P}{\partial t} + \Theta_\ell \vec{U}_\ell \cdot \nabla P$$

$$+ \Theta_\ell \left[K + \left(\frac{\Gamma_e + \Gamma_c}{2} \right) \right] (\vec{U}_\ell - \vec{U}_g)^2$$

$$+ (\Gamma_c - \Gamma_e) H_g + R(T_g - T_\ell)$$

$$+ \Theta_\ell \vec{\sigma}_\ell \cdot \nabla \vec{U}_\ell + \nabla \cdot k_\ell \Theta_\ell \nabla T_\ell$$

$$+ Q_\ell \quad (5)$$

for the liquid phase, and

$$\frac{\partial \rho_g' H_g}{\partial t} + \nabla \cdot (\rho_g' H_g \vec{U}_g) = \Theta_g \frac{\partial P}{\partial t} + \Theta_g \vec{U}_g \cdot \nabla P$$

$$+ \Theta_g \left[K + \left(\frac{\Gamma_e + \Gamma_c}{2} \right) \right] (\vec{U}_g - \vec{U}_\ell)^2 + (\Gamma_e - \Gamma_c) H_g$$

$$+ R(T_\ell - T_g) + \Theta_g \vec{\sigma}_g \cdot \nabla \vec{U}_g$$

$$+ \nabla \cdot k_g \Theta_g \nabla T_g + Q_g \quad (6)$$

for the vapor phase.

Since the velocity vector is $\vec{U} = ui + vj + wk$ (7)
where u, v, and w are the velocity components in x, y, and z directions, respectively, one can have three momentum equations for each phase. In the TWOP code, there are six momentum equations all together.

The volume fractions of liquid and vapor phases are defined as

$$\Theta_\ell + \Theta_g = 1 \quad . \tag{8}$$

In order to close the set of equations, one needs the additional information, i.e., three interface jump conditions (interfacial mass, momentum, and heat transfers), equation of state for both phases, and the additional boundary restrictions, namely, wall friction and the wall heat sources.

The above mentioned 2-mass, 6-momentum, 2-energy and Eq. (8) are solved as an initial and boundary value problem to obtain Θ_g, Θ_ℓ, P, u_ℓ, v_ℓ, w_ℓ, u_g, v_g, w_g, H_ℓ, and H_g. Then the densities of ρ_ℓ and ρ_g are provided through the use of the equation of state.

In general, the drag function K is dependent on the flow regime, local void fraction, vapor and liquid density, Reyenolds' number, and phase velocity. There are two models of K function used in the TWOP code. One is

$$K = Cf(\Theta_g) \tag{9}$$

where C is some coefficient, and

$$f(\Theta_g) = \begin{cases} 1 + 10^{2000(0.01-\Theta_g)} & \text{when} \quad \Theta_g \leqslant 0.01 \\ 1 & \text{when} \quad 0.01 < \Theta_g < 0.99 \\ 1 + 10^{2000(\Theta_g-0.99)} & \text{when} \quad \Theta_g \geqslant 0.99 \end{cases} \quad . \tag{10}$$

A more elaborate form of K, suggested by Rivard[6], is also coded in the TWOP program as another option. The expression of K is

$$K = 0.375(\rho_g' + \rho_\ell')\left\{C_d|\vec{U}_g - \vec{U}_\ell| + \frac{12\left[\Theta_g \nu_g + (1 - \Theta_g)\nu_\ell\right]}{r_p}\right\} A(\Theta_g, N) \tag{11}$$

where

$$A = \begin{cases} \Theta_g^{2/3} \left(\frac{4\pi N}{3}\right)^{1/3} & \text{, when } \Theta_g \leqslant 1/2 \\ (1 - \Theta_g)^{2/3} \left(\frac{4\pi N}{3}\right)^{1/3} & \text{, when } \Theta_g > 1/2 \end{cases} \quad (12)$$

$$r_p = \begin{cases} \left(\frac{3\Theta_g}{4\pi N}\right)^{1/3} & \text{, when } \Theta_g \leqslant 1/2 \\ \left[\frac{3(1 - \Theta_g)}{4\pi N_g}\right]^{1/3} & \text{, when } \Theta_g > \end{cases} \quad (13)$$

The evaporation and condensation rates, Γ_e and Γ_c are determined from

$$\Gamma_e = \lambda_e \rho'_\ell (T_\ell - T_s)/T_s \quad \text{, for } T_\ell \geqslant T_s \quad (14)$$

$$= 0 \quad \text{, for } T_\ell > T_s$$

$$\Gamma_c = \lambda_c \rho'_g (T_s - T_g)/T_s \quad \text{, for } T_g \leqslant T_s \quad (15)$$

$$= 0 \quad \text{, for } T_g > T_s$$

where λ_g and λ_ℓ are time relaxation parameters with unit 1/sec. For the tested problem described in this paper, λ_g and λ_ℓ are set equal to 0.1.

For the dispersed flow regime, the interfacial heat transfer coefficient, R_ℓ and R_g are calculated from

$$R_\ell = 8.067 \frac{k_\ell}{r_p} \quad (16)$$

for the liquid phase and

$$R_g = \frac{1}{r_p} (1 + 0.37 \, Re^{0.5} \, Pr_g^{\,0.33}) \quad (17)$$

for the vapor phase, where

$$R_e = \frac{2\rho_g |\vec{U}_\ell - \vec{U}_g| r_p}{\mu_g} \quad \text{and } Pr = \frac{C_{p_g} \mu_g}{k_g} \ .$$

The correlations R_ℓ and R_g are obtained from Ref. [7] with r_p defined by Eq. (13).

NUMERICAL SOLUTION PROCEDURE

The finite difference equations of mass, momentum, and energy are solved as an initial value problem in time and boundary value problem in space. The procedure is iterative and the main steps necessary to determine the flow conditions at time $t+\Delta t$ from those at time t are as follows:

1. The pressure distribution at time $t+\Delta t$ is guessed.
2. The momentum equations are solved to compute first approximation of the three velocity components of each of the two phases.
3. Partial energy equations are solved for approximating enthalpies of both phases.
4. Densities and temperatures of both phases are calculated using equation of state and estimated enthalpies and pressures.
5. Liquid void fraction θ_ℓ is estimated from the continuity equation of liquid. Gas void fraction θ_g is then calculated from the relation $\theta_g = 1 - \theta_\ell$.
6. It is now checked if the gas continuity equation is satisfied. If the equation is not satisfied, then pressure is corrected, the amount of correction being dependent on the mass residual, and steps (2) to (6) are repeated. This iterative procedure is continued until the mass residual is less than the specified value.
7. Energy equations are solved for enthalpies. Temperatures and densities are then calculated using the equations of state.
8. A new timestep is chosen and steps (1) to (7) are repeated.

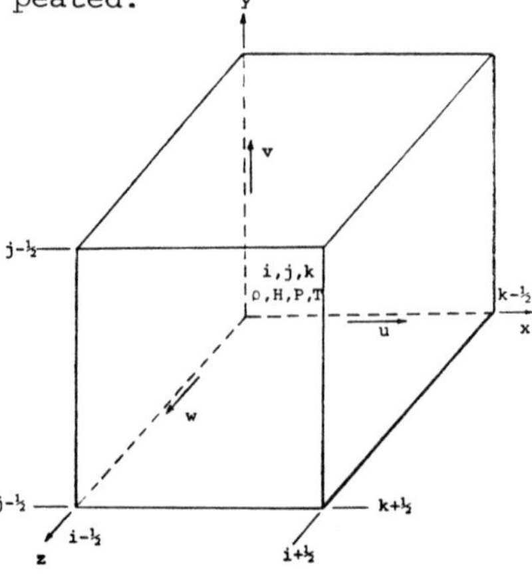

Fig. 1. Finite Difference grid notations and locations of variables; ρ,H,P,T are located at cell center, and velocities u,v,w and flux terms are at cell boundaries.

The numerical scheme used in the pressure calculation is

$$P^{r+1} = P^r - \omega \frac{D_g^r}{\left(\frac{\partial D_g}{\partial P}\right)^r} \qquad \text{when } \theta_g \geqslant \theta^* \tag{18}$$

and

$$P^{r+1} = P^r - \omega \frac{D_\ell^r}{\left(\frac{\partial D_\ell}{\partial P}\right)^r} \qquad \text{when } \theta_g < \theta^* \tag{19}$$

where θ^* is an input constant, i.e., $\theta^*=0.001$. In the pressure loop Eqs. (18) and (19) are used. The momentum exchange terms, $K(\vec{U}_g-U_\ell)$ and $K(\vec{U}_\ell-U_g)$ in the liquid and vapor momentum equations, respectively, are neglected in calculating Eqs. (18) and (19). It is found that these formulations result in greater stability for the pressure calculations for the problems that have been investigated so far. The expressions of these two derivatives with the finite difference grid shown in Fig. 1 are

$$\frac{\partial D_g}{\partial P} = \frac{1}{\left(\frac{\partial P}{\partial \rho_g'}\right)_{i,j,k} \delta t} + \frac{\delta t\left[\left(\theta_g\right)_{i+\frac{1}{2},j,k} + \left(\theta_g\right)_{i-\frac{1}{2},j,k}\right]}{(\delta x)^2} + \frac{\delta t\left[\left(\theta_g\right)_{i,j+\frac{1}{2},k} + \left(\theta_g\right)_{i,j-\frac{1}{2},k}\right]}{(\delta y)^2} + \frac{\delta t\left[\left(\theta_g\right)_{i,j,k+\frac{1}{2}} + \left(\theta_g\right)_{i,j,k-\frac{1}{2}}\right]}{(\delta z)^2} \tag{20}$$

and

$$\frac{\partial D_\ell}{\partial P} = \frac{1}{\left(\frac{\partial P}{\partial \rho_\ell'}\right)_{i,j,k} \delta t} + \frac{\delta t\left[\left(\theta_\ell\right)_{i+\frac{1}{2},j,k} + \left(\theta_\ell\right)_{i-\frac{1}{2},j,k}\right]}{(\delta x)^2} + \frac{\delta t\left[\left(\theta_\ell\right)_{i,j-\frac{1}{2},k} + \left(\theta_\ell\right)_{i,j+\frac{1}{2},k}\right]}{(\delta y)^2} + \frac{\delta t\left[\left(\theta_\ell\right)_{i,j,k-\frac{1}{2}} + \left(\theta_\ell\right)_{i,j,k+\frac{1}{2}}\right]}{(\delta z)^2} \tag{21}$$

The enthalpy equations were divided into two groups. The first group, including the heat transfer due to mass transfer, the interfacial sensible heat transfer, and the energy dissipation due to interfacial friction, is solved inside the pressure iteration loop. The second group, which contains the pressure compression work, shear stress dissipation energy and the thermal diffusion, is solved outside the pressure iteration loop. A more detailed description of the numerical solution procedure is given by Lee.[8]

RESULTS AND CONCLUSIONS

The objectives of this paper are: (1) to demonstrate the computational capabilities of the TWOP code, and (2) to assess the range of magnitudes of interfacial coupling terms and evaporation and condensation rates for the two-phase jet problem. It is our deliberate intent that models for the interfacial coupling, evaporation, and condensation be kept to very simple forms so that their effects on the system under consideration can readily be delineated. With these objectives in mind we have analysed the following two problems with TWOP.

1. Separation of Steam and Water

This is a transient problem. At time t=0, we have an isothermal uniform mixture of steam (40%) and water (60%) in a rectangular closed box. As time proceeds due to gravity, vapor starts moving up and liquid starts moving down. The separation continues until all vapor occupies the upper 40% of the box and all liquid occupies the lower 60% of the box. In order to determine the effect of interfacial drag on separation, we have carried out numerical computation for interfacial drag coefficient $K=10^3$, 10^4, and 10^5 (Kg/M^3/sec). In Fig. 2, we have shown the variations of void fraction with time in the top and the bottom cells of the vertical box. We can see that the rate of separation increases with the decrease in the value of interfacial drag coefficient.

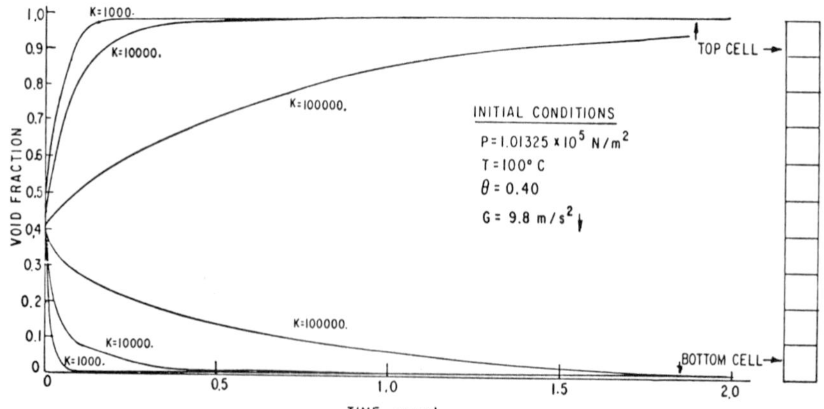

Fig. 2. Effect of interfacial drag coefficient K on the variations of void fraction in top and bottom cells.

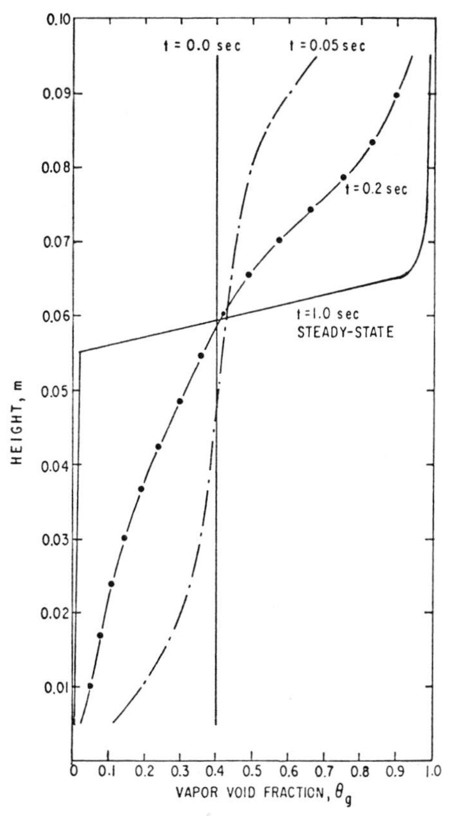

The transient and steady state vapor void fraction, θ_g, are shown in Fig. 3 for interfacial drag coefficient $K=10^4$.

It may be noted here that although this problem is physically simple, it is a very difficult problem in terms of obtaining numerical solution because of the phase disappearing phenomena.

Fig. 3. Variations of void fraction with time for drag coefficient $K=10^4$.

2. High Pressure Jet Impingement[9]

This is also a transient problem. At time t=0, a high pressure jet containing a mixture of steam (67%) and water (33%) enters into a stagnant atmosphere and impinges on a vertical wall. The experimental setup and its initial operating conditions are shown in Fig. 4. Figure 5 presents a comparison between the calculated results obtained by TWOP and the steady-state pressure measurements on the impinged wall, and corresponding sensitivity study of variations of interfacial drag coefficient (K) and evaporation rate (Γ). Note that a good agreement is reached between the experimental data on pressure distribution and the calculated results with $K=2.0 \times 10^{12}$ ($Kg/M^3/sec$) and $\Gamma=0.01$ ($Kg/M^3/sec$).

Figure 6 shows the effects of evaporation rate on the steady-state pressure distributions on the impinged wall with $K=2.0 \times 10^{12}$ ($Kg/M^3/sec$). The variations of pressure and velocity at mid-plane between the exit of nozzle and impinged wall (i.e., z=2.5 mm) are shown in Figs. 7 and 8 respectively. Figure 9 presents the velocity profiles at various axial positions (i.e., z=2,3, and 4 mm) when time $t=7.2 \times 10^{-5}$ seconds. It is noted that both K and Γ are sensitive to the calculated results. Furthermore, K and Γ are dependent on flow regime, particle size and local thermodynamic properties.

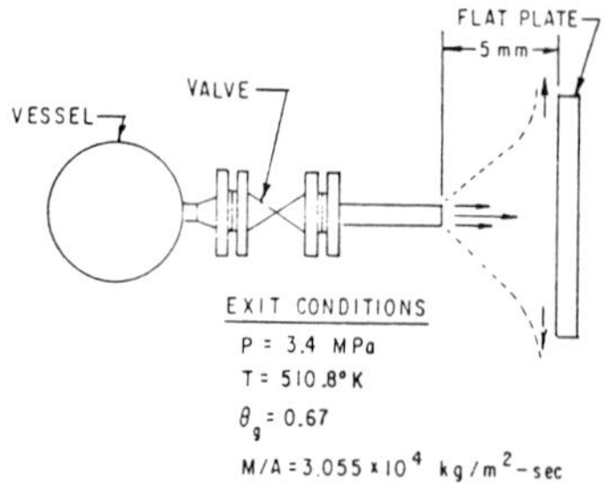

Fig. 4. Schematic layout of two-phse jet impingement experiment.

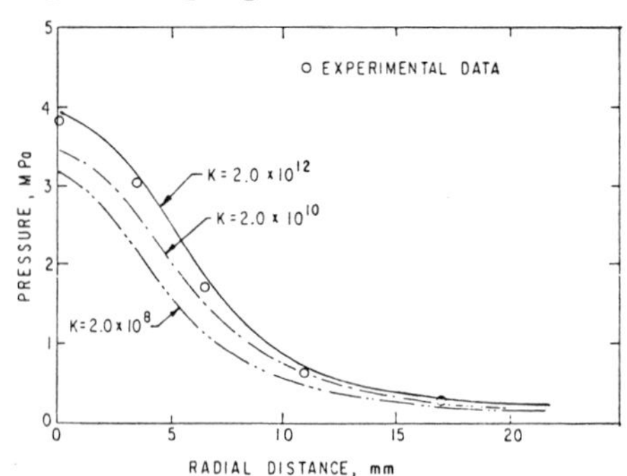

Fig. 5. Effect of K on the steady state pressure distribution over the impinged wall.

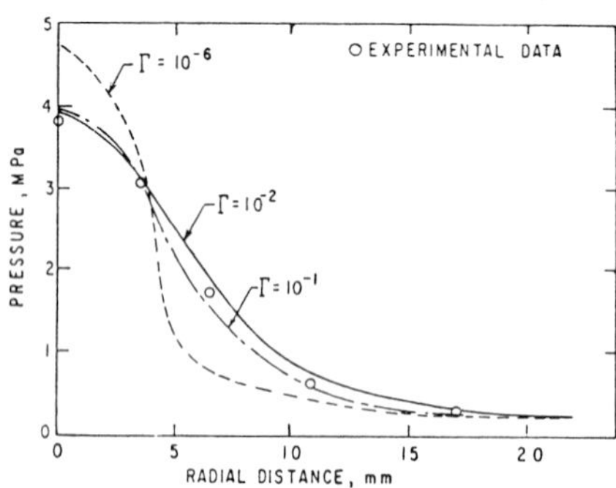

Fig. 6. Effects of evaporation rate on the steady state pressure distributions over the impinged wall.

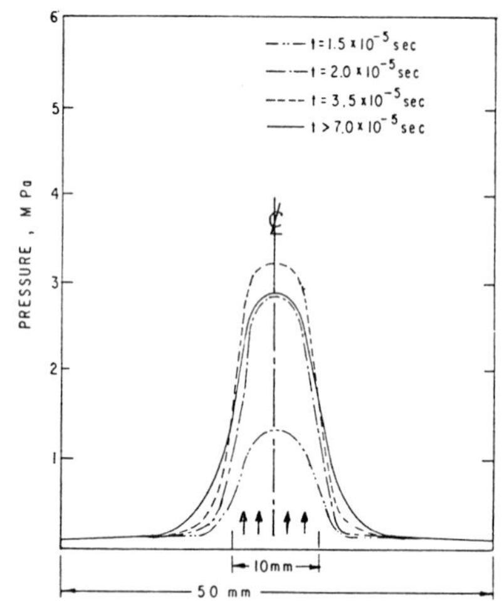

Fig. 7. Variation of pressure at mid-plane between the exit of nozzle and the impinged wall.

Based on this study, the numerical results obtained by the TWOP code thus far is very encouraging. It has not only demonstrated the computational capability but has also exhibited the ability to model the complex phenomena of the jet impingement problem via very simple interfacial drag and evaporation models.

NOMENCLATURE

Dimensionless Numbers

C_d = drag coefficient

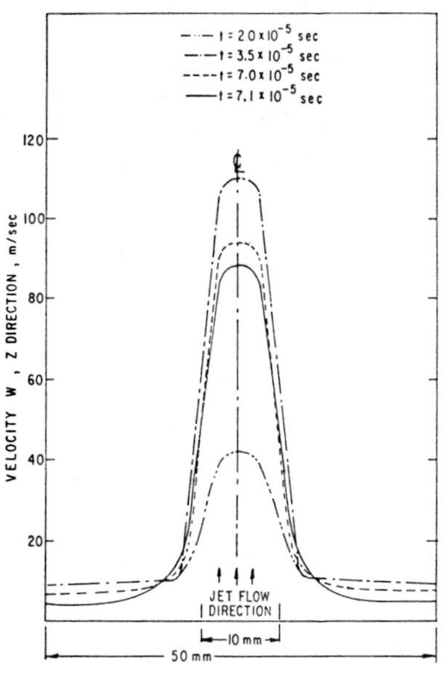

Fig. 8. Variation of velocity distribution at mid-plane between the exit of nozzle and the impinged wall.

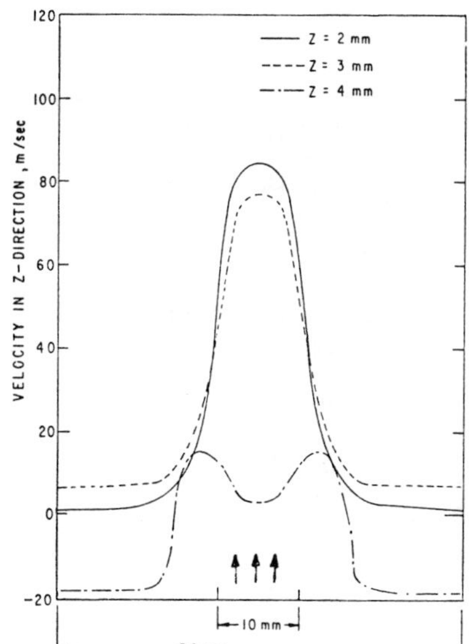

Fig. 9. Velocity profiles at various axial positions when l=7.20x10^{-5} sec. (Z-distance from the exist of nozzle)

Roman Letters

C_p = specific heat

D = $\nabla \cdot \vec{U}$

$\vec{g}$ = acceleration of gravity

H = specific enthalpy

k = heat conduction coefficient

K = drag function

N = number of particle per unit volume

P = pressure

Pr = Prandtl's number

Q = heat sources due to mass transfer and wall heat transfer

r_p = radius of particle

R = exchange function describing heat transfer between fields

Re = Reynold's number

t = time

$\vec{w}$ = wall friction

T = temperature

$\vec{U}$ = velocity with components u,v,w in the x,y,z direction, respectively

u = velocity in x direction

v = velocity in y direction

w = velocity in z direction

Greek Letters

Γ = interfacial mass transfer

Θ = void fraction

λ = time relaxation parameter

μ = dynamic viscosity

ν = Kinematic material viscosity

ρ = microscopic material density

ρ' = macroscopic material density, e.g., $\rho' = \Theta\rho$

ω = relaxation factor

Subscripts

ℓ = droplet or liquid phase

i = indicates cell number in x direction

j = indicates cell number in y direction

k = indicates cell number in z direction

s = at sturation point

g = vapor

Superscripts

r = right hand superscript indicates the iteration steps

REFERENCES

1. Harlow, F. H. and Amsden, A. A., Flow of Interpenetrating Material Phases, J. Comput. Physics, 18, 440-464 (1975).

2. Soo, S. L., On the Nature of Equations of Multiphase Multidomain Mechanics, Paper presented at 2nd Multiphase Flow and Heat Transer Symposium Workshop, Miami Beach, Florida, April 16-18, 1979.

3. Gidaspow, Dimitri, Hyperbolic Compressible Two-Phase Flow Equations Based on Stationary Principles and the Fick's Law, Two-Phase Transport and Reactor Safety, 1, 283-297, Veziroglu and Kakac, Eds., Hemisphere Publ. Corp., Washington, 1978.

4. Rudinger, G. and Chang, A., Analysis of Non-Steady Two-Phase Flow, Phys. of Fluids, 7, 1747-1754, 1964.

5. Hancox, W. T., Ferch, R. L., Liu, W. S., and Nieman, R. E., On One-Dimensional Models for Transient Gas-Liquid Flow in Ducts, Paper presented at the EPRI Workshop on Two=Phase Flow Modeling, Tampa, Florida, February, 1979.

6. Rivard, W. C. and Torrey, M. D., K-Fix: A Computer Program for Transient Two-Dimensional, Two-Fluid Flow, Los Alamos National Laboratory document, LA-NUREG-6623, 1977.

7. Solbrig, C. W., Heat Transfer and Friction Correlations Required to Describe Steam-Water Behavior in Nuclear Safety Studies, Peper presented at 15th National Heat Transfer Conference, San Francisco, California, August 1975.

8. Lee, W. H., A Pressure Iteration Scheme for Two-Phase Flow Modeling, Ibid, April 16-18, 1979.

9. Schweickert, H., Untersuchung der Vorage in einem mehrfach unterteilten containment beim Bruch einer Kuhlmittelleitung wassergekuhlter Reaktoren, Technischer Bericht BFR 50-32-C12-1, Kraftwerk Union, Germany, 1976.

CONDENSATION HEAT TRANSFER IN TWO-PHASE THERMOSYPHONS

Spendel, T.

Institut für Kernenergetik und Energiesysteme (IKE)
University of Stuttgart, Stuttgart, Federal Republic of Germany

SUMMARY

An analysis of the effects of vapor pressure variation and shear stress at the liquid-vapor interface on the local heat transfer in the condensing section of closed two-phase thermosyphons is presented. The complete two-dimensional Navier-Stokes equations are solved using a finite difference method. The equations for the liquid film flow are simplified taking into consideration the assumption of Nusselt's theory for laminar film condensation. The results show that the influence of pressure variation does not play a significant role and is dominated by the effect of vapor shear stress. Depending on the condensation rates, the exact theory yields local and average Nusselt numbers which are smaller than those predicted by Nusselt's theory. The differences are about zero to 9 % and about 2 to 5 % respectively.

1. INTRODUCTION

Heat transfer devices operating on the principle of closed two-phase thermosyphons are used in many applications, e.g. for waste heat recovery, solar energy utilization, permafrost stabilization, etc. Detailed information concerning the internal heat and mass transfer is needed for the design of these devices. The work presented here deals with the heat transfer in the condenser section of closed two-phase thermosyphons. Contrary to the work of Lee and Mital [1] the influence of forces due to vapor shear and vapor pressure drop are accounted for in the analysis.

The physical model is shown in Fig. 1. Vapor entering the condenser section of the thermosyphon is condensed at the wall. Under the influence of gravity the condensate film flows down the tube wall.

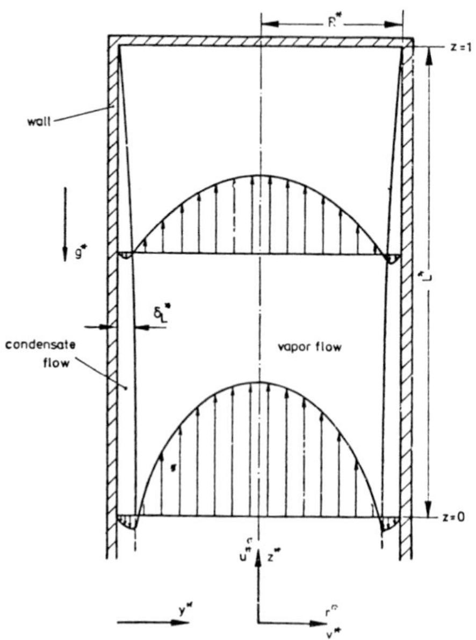

Fig. 1:
Physical model and coordinates for the condenser section of the thermosyphon

For the analysis the following assumptiones were made:

1. Liquid and vapor flow are both laminar
2. The wall temperature is constant along the condenser section
3. Vapor density variations can be neglected
4. The liquid film thickness remains negligibly small compared to the tube radius.

2. VAPOR FLOW

The vapor flow in the condenser section of a two-phase thermosyphon can be considered to be a steady axial-symmetric motion of an incompressible medium in a circular cylindrical cavity as depicted in Fig. 1. This can be described [2] by the equations for the conservation of momentum

$$v^* \frac{\partial v^*}{\partial r^*} + u^* \frac{\partial v^*}{\partial z^*} = -\frac{1}{\rho}\frac{\partial p^*}{\partial z^*} + \nu \left(\frac{\partial^2 v^*}{\partial r^{*2}} + \frac{1}{r^*}\frac{\partial v^*}{\partial r^*} - \frac{v^*}{r^{*2}} + \frac{\partial^2 v^*}{\partial z^{*2}}\right) \tag{1}$$

$$v^* \frac{\partial u^*}{\partial r^*} + u^* \frac{\partial u^*}{\partial z^*} = -\frac{1}{\rho}\frac{\partial p^*}{\partial z^*} + \nu \left(\frac{\partial^2 u^*}{\partial r^{*2}} + \frac{1}{r^*}\frac{\partial u^*}{\partial r^*} + \frac{\partial^2 u^*}{\partial z^{*2}}\right) + g^* \tag{2}$$

and by the equation for the conversation of mass

$$\frac{\partial v^*}{\partial r^*} + \frac{v^*}{r^*} + \frac{\partial u^*}{\partial z^*} = 0 \tag{3}$$

The asterices represent dimensional quantities. The dependent variables u* and v* are transformed to the stream function ψ^* and the vorticity ζ^*. The stream function is defined by

$$u^* \equiv -\frac{1}{r^*}\frac{\partial \psi^*}{\partial r^*}$$

$$v^* \equiv -\frac{1}{r^*}\frac{\partial \psi^*}{\partial z^*} \tag{4}$$

and the vorticity is defined as

$$\zeta^* \equiv \frac{\partial v^*}{\partial z^*} - \frac{\partial u^*}{\partial r^*} \tag{5}$$

The equations are made dimensionless by introducing the tube radius R* as characteristic length and the average velocity $\bar{u}_o^*$ at the condenser entrance as characteristic velocity. The definitions for stream function and vorticity are inserted in equations (1-3) to yield a set of elliptic differential equations

$$\frac{\partial(v\cdot\zeta)}{\partial r} + \frac{\partial(u\cdot\zeta)}{\partial z} = \frac{1}{Re}\left(\frac{\partial^2\zeta}{\partial r^2} + \frac{1}{r}\cdot\frac{\partial\zeta}{\partial r} - \frac{\zeta}{r^2} + \frac{\partial^2\zeta}{\partial z^2}\right) \tag{6}$$

and

$$\frac{1}{r}\frac{\partial^2\psi}{\partial r^2} - \frac{1}{r^2}\frac{\partial\psi}{\partial r} + \frac{1}{r}\frac{\partial^2\psi}{\partial z^2} = \zeta \tag{7}$$

where $Re = U_o^*\cdot R^*/\nu$ is the vapor Reynolds nomber.

To solve the set of equations (6-7) the boundary values for the stream function and the vorticity must be known. The boundary conditions for the stream function are

$$\psi(0,z) = \psi(r,L) = 0 \tag{8}$$

$$\psi(1,z) = \int_0^z V_i(z)\,dz \tag{9}$$

where the subscript i refers to the vapor-liquid interface and L is the nondimensional length of the condenser section. At the condenser entrance the vapor flow velocity profile is assumed to be parabolic. This yields the following stream function distribution

$$\psi(r,0) = -r^2\cdot\left(1-\frac{r^2}{2}\right) \tag{10}$$

Boundary values for the vorticity at the vapor-liquid interface and at the condenser entrance are obtained using a finite difference representation of equation (5). According to equation (5) the value for the vorticity at the symmetry axis is zero. At the top of the thermosyphon equations (6-7) are used to derive the vorticity. In a region adjacent to the top wall we can neglect the radial variation of ζ and ψ compared to the axial variation. The result of this derivation is

$$\zeta_w = 3 \frac{\psi_z \; \psi_w}{r \Delta z^2} + \frac{1}{2} \zeta_z \tag{11}$$

where the subscripts w and z refer to the wall and the adjacent grid point, respectively. This formula was developed by Gosman et al. [3].

The pressure gradient at the vapor-liquid interface plays an important role for the behaviour of the liquid film. It can be derived from equation (2).

$$\left.\frac{dp}{dz}\right|_i = - \frac{1}{Re} \left(\frac{\partial \zeta}{r} + \frac{\zeta}{r} \right) - v_i \left.\frac{\partial u}{\partial r}\right|_i - u_i \left.\frac{\partial u}{\partial z}\right|_i \tag{12}$$

3. LIQUID FLOW

In principle the analytical treatment of the liquid flow should start with the complete conservation equations. For the case of laminar film condensation, however, the simplifications made by Nusselt [4], i.e. neglection of the convective terms in the equation of motion and in the energy equation, are generally accepted. The valadity of this simplification was worked out by several authors. A synopsis can be found in the work of Lucas [5].

The simplified momentum and energy equations are

$$\mu_L \frac{d^2 u^*_L}{dy^{*2}} - \rho_L \, g^* \frac{dp_L^*}{dz^*} = 0 \tag{13}$$

and

$$\frac{d^2 T^*}{dy^{*2}} = 0 \tag{14}$$

where the subscript L refers to the liquid.

This system of equations is subject to the following boundary and coupling conditions:

$$(y^* = 0): \quad u^*_L = 0 \; i \; T^*_L = T^*_w \tag{15}$$

$$(y^* = \delta_L^*): \quad u_L^* = u^* \tag{16}$$

$$T_L^* = T^* \tag{17}$$

$$\dot{M}_L^* = -\dot{M}^* \tag{18}$$

$$\tau_L^* = \tau^* = \mu \frac{\partial u^*}{\partial y^*} \tag{19}$$

$$\frac{dp_L^*}{dz^*} = \frac{\partial p^*}{\partial z^*}\Big|_i \tag{20}$$

Equation (17) represents a mass balance in the thermosyphon. Equations (19-20) express the influence of the counter-current vapor flow. The discussed condensation problem is governed by five parameters: the vapor Reynolds number Re, the Froude number $Fr = \bar{u}_o{}^{*2}/(g^*R^*)$, the vapor to liquid viscosity ratio μ/u_L, the vapor to liquid density ratio ρ/ρ_L and an energy parameter E defined as

$$E = \frac{K_L(T_i^* - T_w^*)}{\mu_L h_{fg}} \tag{21}$$

where h_{fg} is the enthalpy of evaporation. Equation (13) can be integrated together with the boundary conditions to yield the velocity profile of the liquid folm in dimensionless form

$$u_L = \frac{\mu_L}{\mu} Re \left[\left(\frac{\rho}{\rho_L} \frac{1}{Fr} + \frac{dp}{dz} \right) (y^2 - y \cdot \delta_L) - \tau_i y \right] \tag{22}$$

The energy balance at the liquid-vapor interface can be written as

$$-K \frac{\partial T^*}{\partial y^*}\Big|_i + \frac{K_L (T_i^* - T_w^*)}{\delta^*} = \rho \cdot v_i^* \cdot h_{fg} \tag{23}$$

The first term on the left side of equation (23) can be neglected as the temperature in the vapor is assumed to be constant. The second term results from equation (14) and is valid for constant thermal conductivity of the liquid K_L. This term expresses Nusselt's main assumption, viz. that the enthalpy of condensation of the condensing vapor is removed by thermal conductivity only. With equation (18) and (23) equation (22) can be integrated to yield

$$\left(\frac{1}{Fr} \frac{\rho_L}{\rho} + \frac{dp}{dz}\right) \frac{\delta_L^3}{3} - \tau_i \frac{\delta_L^2}{2} = E \left(\frac{\mu_L}{\mu}\right)^2 \frac{\rho}{\rho_L} \frac{1}{Re^2} \int_z^L \frac{dz}{\delta_L(z)} \tag{24}$$

This equation describes the axial distribution of the liquid film thickness δ_L influenced by gravity and the forces due to vapor shear and due to pressure gradient. Neclecting these influences we obtain Nusselt's solution for the liquid film thickness

$$\delta_L = \left[4 \cdot E \cdot \frac{Fr}{Re^2} \cdot \left(\frac{\mu_L}{\mu}\right)^2 \cdot (L-z) \right]^{\frac{1}{4}} \tag{25}$$

4. NUMERICAL PROCEDURE

To solve flow equations (6-7) they are substituted by finite difference equations. The convection terms in equation (6) are represented by the method of upwind differencing. The set of algebraic equations is iteratively solved by the AD1-Method which shows good numerical stability and a fairly rapid convergence. Some divergence problems arose with the use of equation (5) for the vorticity at the liquid-vapor interface. Therefor ζ_i was calculated by weigthing the results of the old and new iteration step. A uniform rectangular mesh was chosen to represent to vapor flow field. The convergence of the numerical procedure was tested using a criterion recommended by Roache [6].

The calculation is started with equation (24) having set dp/dz and τ_{i*} equal to zero. This yields the boundary values for the stream function at the vapor-liquid interface. Then the vapor flow field is calculated, and the results are than inserted in equation (23). Thereby the local liquid film thickness is obtained and hence the local Nusselt number in the condenser section of the thermosyphon. Thereafter new boundary values for the stream function are calculated and the numerical procedure is repeated until the relative difference of the results between two iteration steps is less than a small preassigned value.

5. RESULTS AND DISCUSSION

For numerical analysis we assume condensation conditions which are valid for a thermosyphon containing water as working fluid with T = 100 °C, p = 1 bar, i.e. $\rho/\rho_L = 6.23 \cdot 10^{-4}$ and $\mu/\mu_L = .43$, Re = 1250 and Fr = 516. This corresponds to a thermosyphon with a tube radius of 5 mm and a performance of about 500 W. The energy parameter E was varied from 0.002 to 0.05. Figs. (2-3) show the distribution of dp/dz and τ_i with the length of z/L. In the middle of the condenser section dp/dz has a maximum for E = 0.05 and a minimum for all other cases. The first case corresponds to a relatively high condensation rate which causes pressure rise due to momentum drop, of the condensing vapor. This effect is compensated by

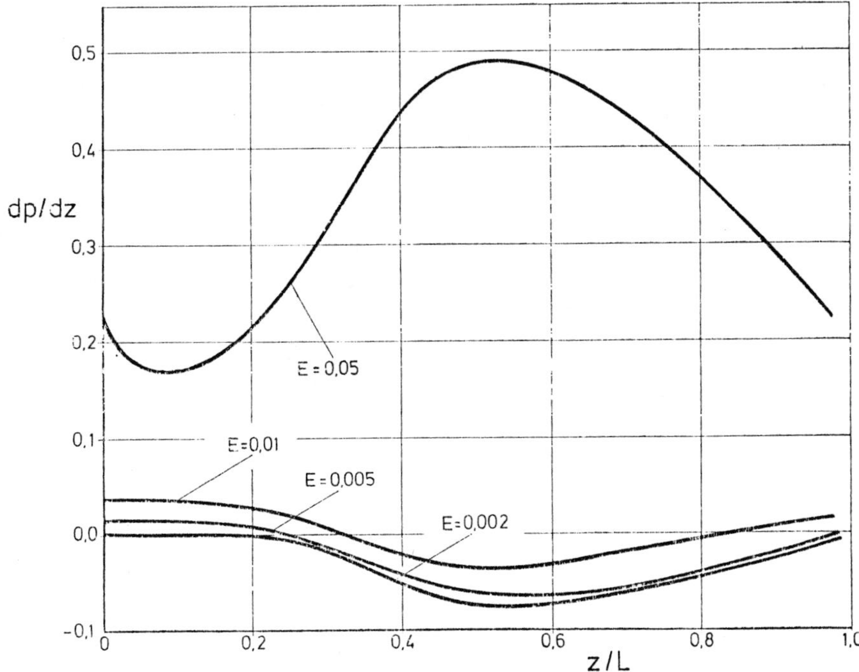

Fig. 2: Distribution of pressure gradient dp/dz along the condenser section

viscous pressure losses for relatively small condensation rates (E = 0.002, 0.005, 0.01). Fig. 3 shows the shear stress distribution at the vapor-liquid interface. τ_i has a maximum for all values of E in the middle of the condenser due to a change of the velocity profile and approaches zero at the condenser end because the axial velocity tends to zero.

Fig. 4 shows the relative deviation of the local Nusselt number of the present analysis from Nusselt's approximation solution, equation (25).

Due to the behavior of dp/dz and τ_i, the maximum deviation occurs in the middle of the condenser section. From equation (24) it is obvious that positive values of dp/dz and τ_i have an opposite influence on the film thickness and hence on the heat transfer rate. Since the local Nusselt numbers are smaller than those according to Nusselt's theory, the vapor shear influence on the film thickness must predominate over the influence of the vapor pressure gradient. Only for the case of E = 0.05, i.e. where dp/dz is relatively high due to a high condensation rate, an influence of the pressure gradient on the film thickness is noticeable.

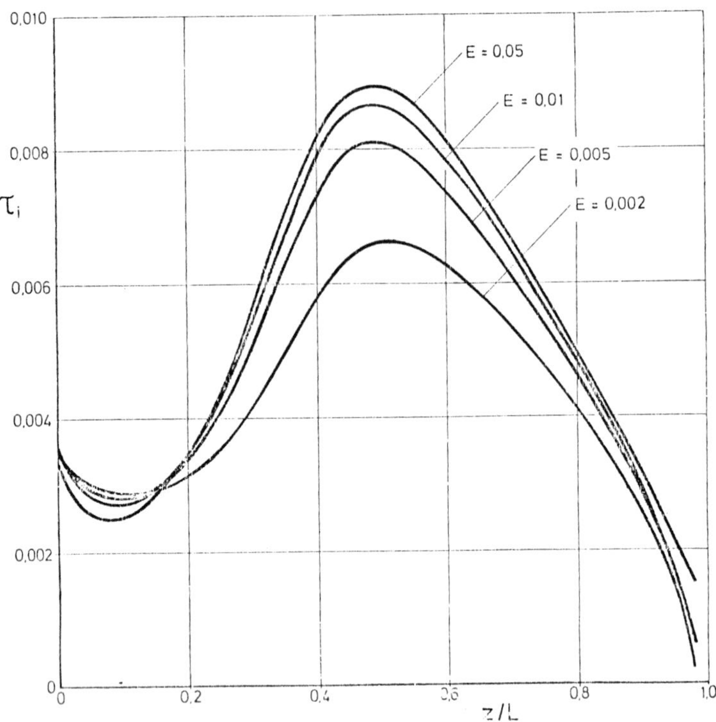

Fig. 3: Distribution of shear stress τ_i along condenser section

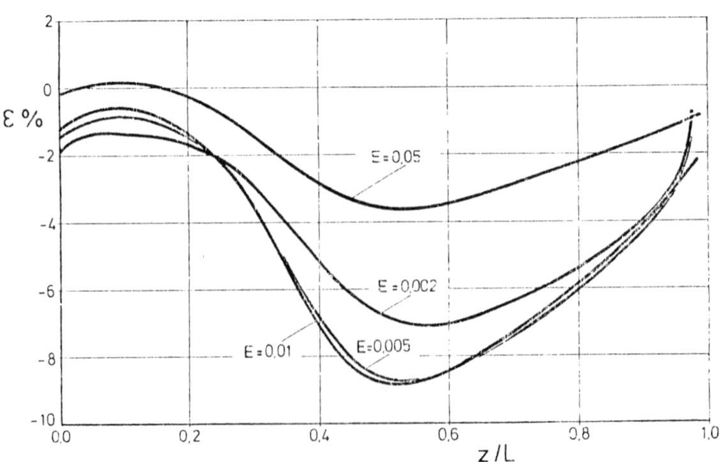

Fig. 4: Deviation of the local Nusselt number according to the present analysis from Nusselt's approcimation solution

REFERENCES

1. LEE, Y. and U. MITAL - A Two-Phase Closed Thermosyphon. J. Heat Mass Transfer, Vol. 15, pp. 1695-1707, 1972

2. MUELLER, T.J. - Numerical and Physical Experiments in Viscous Separated Flows, Porgress in Numerical Fluid Dynamics, Ed. Wirz, H., Springer Verlag, Berlin-Heidelberg-New York, 1975

3. GOSMANN, A.D., W.M. PUN, A.K. RUNCHAL, D.B. SPALDING and M. WOLFSHTEIN, Heat and Mass Transfer in Recirculating Flows, Academic Press, London and New York, 1969

4. NUSSELT, W.: Die Oberflächenkondensation des Wasserdampfes, Zeitschr. VDI, Vol. 27, pp. 541-546, 1916

5. LUCAS, K.: Die laminare Filmkondensation binärer Dampfgemische, Habilitationsschrift, Universität Bochum, 1974

6. ROACHE, P.J., Computitional Fluid Dynamics, Hermosa Publishers, Albuquerque, 1976

The numerical solution of simple one-dimensional multi-phase flows in shock tubes

I.P. Jones* and A.V. Jones+

*Computer Science and Systems Division, Atomic Energy Research Establishment, Harwell, Didcot, Oxon OX11 ORA, United Kingdom.

+Joint Research Centre, Ispra Establishment, 21020 Ispra (Varese), Italy.

ABSTRACT

This paper discusses numerical results for multi-phase flows obtained with the computer code SIMMER-II which uses the implicit multi-field method. The code is applied to two simple multi-phase problems: the flow in a shock tube containing a dusty gas and the propagation of shock waves in a mixture of water and air bubbles. The numerical solutions remain stable for low and moderate dust volume fractions and are in good agreement with the results of a simple analytic theory for the velocity and pressure behind the shock wave. The results also show that the droplet breakup model has a stabilising effect and prevents large differences in the phase velocities.

For the case of shock propagation in a bubbly liquid the SIMMER results for the speed of the shock wave are in good agreement with the experimental and theoretical results of Campbell and Pitcher.

1. INTRODUCTION

In this paper results are discussed for some simple one-dimensional multi-phase flows. These results were obtained with the computer code SIMMER-II, which uses an implicit multi-field method. The use of such methods, in which each phase is represented as a continuum, is currently the subject of some controversy. The reason for this is that for some models the underlying time dependent equations may not be hyperbolic and, in consequence, may be mathematically ill-posed. The use of donor cell differencing and implicit time differencing however renders the calculational procedures stable for sufficiently large temporal and spatial step lengths [1-3]. The program SIMMER-II [4,5] incorporates such a 'two-fluid' model and is widely used for the study of hypothetical core disruptive accidents in liquid metal cooled fast breeder reactors. Two phase flows in shock tubes have often been suggested as a means of assessing the predictive capabilities of such multi-phase models and are also of interest in their own right. Two sample cases are discussed here. The first is the flow in a shock tube containing dust particles or liquid droplets, and the second is the propagation of a shock wave through a bubbly liquid.

2. THE CALCULATION OF SHOCK WAVES IN DUSTY GASES

Within the framework of the basic multi-phase model used the treatment of droplets or dust is essentially the same except that perhaps different values of the drag coefficient are appropriate and droplets may fragment under the action of the shock wave. For this reason the terms dust and droplets are regarded as being equivalent except where droplet fragmentation is being discussed.

2.1 Single phase shock tube calculations

The starting point for this investigation was the calculation of a shock tube containing only gas. This was intended as a simple introduction to SIMMER-II the goal being to reproduce the results of Blewitt [6], also obtained using SIMMER-II. This in itself was an interesting exercise and gave much insight into the working of SIMMER. The parameter values for the cases of dusty shocks are also based upon those of the single phase calculations and so the single phase calculations provide a useful guide to the influence of the dust. Full details of the analytic solution for the flow in a shock tube are to be found in Liepmann and Roshko [7].

Blewitt [6] solved the shock tube problem with the parameters as given in Table 1. Figure 1 shows a comparison between the results of SIMMER-II and the analytical solution. The problem times and grid are different from those used by Blewitt [6] but the agreement between the numerical and analytical results is comparable. The numerical results illustrate the numerical diffusion introduced into the system by using donor cell differencing and the shock is smoothed out over several grid points. There is however a slight overshoot in the pressure profile behind the shock.

2.2 The effect of dust particles

The weak shock case just described was used as the basic case for an investigation into shock waves in dusty gases by increasing the dust concentration. In the following results

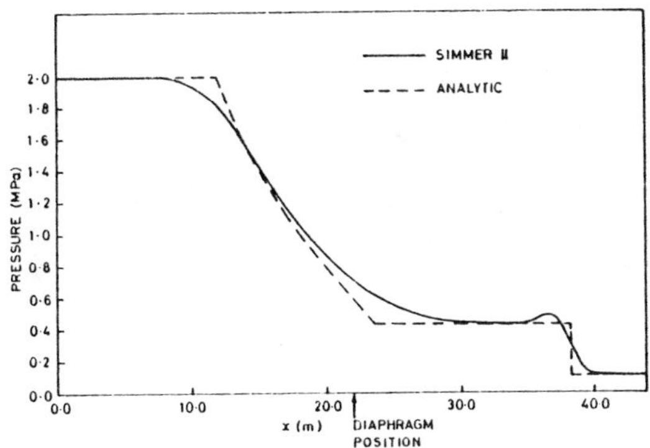

FIG.1 WEAK SHOCK, PRESSURE AS A FUNCTION OF DISTANCE x, $t = 0{\cdot}02$ secs

the parameters were as in Table 1 and any different values used in individual runs will be quoted where appropriate.

For shock tubes containing dusty gases the analytical theory is very similar to that for the single phase shock tube. As before there are undisturbed low and high pressure regions in front and behind the shock. Across a shock wave itself the gas is assumed to behave as if no droplets were present and the usual Rankine-Hugoniot relations hold. Because of their inertia the droplets remain for a while at the temperature and velocity of the undisturbed gas downstream of the shock. This is often called the 'frozen' flow since the properties of the droplets are those downstream of the shock. There then follows a 'relaxation' zone where the flow properties relax to an equilibrium region where the phase velocities and temperatures are equal, Carrier [8], Rudinger [9] and Kriebel [10]. In this equilibrium region the gas may be treated as a perfect gas with modified physical properties [9]. This theory is referred to as the phase equilibrium theory. It gives modified Rankine-Hugoniot conditions to be satisfied between the two equilibrium regions on either side of the shock. This theory may also be used for the flow everywhere when the inter-phase drag is so high that the particles and gas move with the same velocity and the medium may be treated as a continuum.

2.3 Small dust particles

The form of the momentum exchange coefficient in SIMMER-II is the same as that used by Harlow and Amsden [11]. For fixed physical parameters, drag coefficient and velocity difference, the drag increases as the particle radius decreases. Thus it is possible to get equal phase velocities by having either a large drag coefficient or very small particles. Comfort and Crowe [12] have used an approximate analysis to indicate the particle radius below which the phases move together as a homogeneous two phase flow. Such a case has been examined here using a particle radius r_p of 10^{-5} metres.

Figure 2 shows a graph of SIMMER predictions for the pressure and velocity behind a shock wave for various concentrations of dust. In these cases the high and low pressure sections had the same dust concentrations.

The results show that there is very good agreement between the SIMMER results and the analytical phase equilibrium theory, even for quite large dust volume fractions. The results also demonstrate the effect that even small amounts of dust can have.

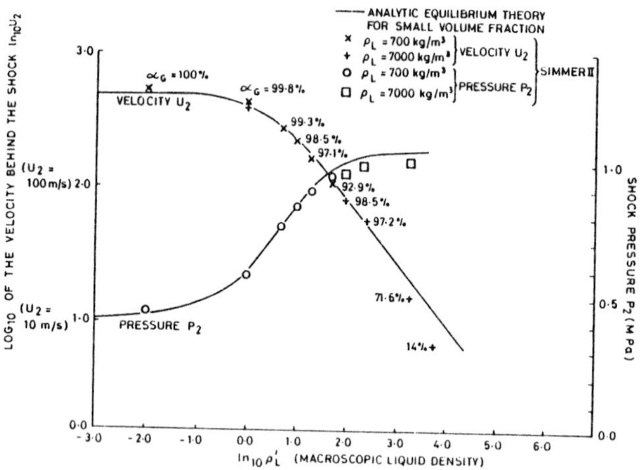

FIG 2 GRAPH OF SPEED AND PRESSURE BEHIND A SHOCK IN A SHOCK TUBE CONTAINING DROPLETS

Figures 3-4 give the graph of the various flow profiles for different dust concentrations. These show the typical shock tube profiles seen earlier for the single phase shock tube calculations and illustrate the slower speed of shock propagation. Because the gas and liquid velocities have virtually the same value the relaxation zone behind the shock is extremely thin.

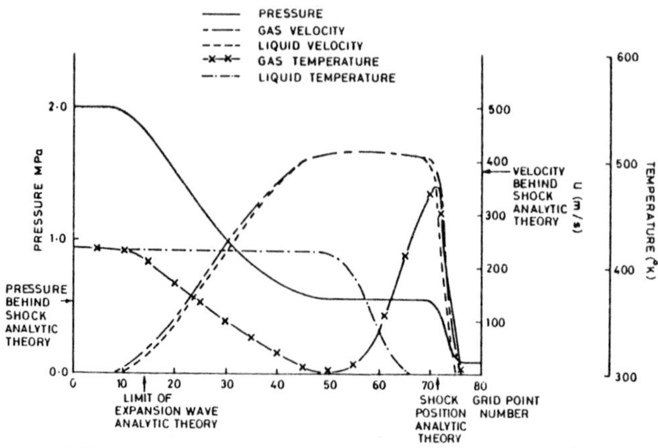

FIG 3 GRAPH OF FLOW VARIABLES AGAINST DISTANCE, $t = 0.05$ secs $\rho'_L = 1.0$ kg/m^3 $r_p = 10^{-5}$ m

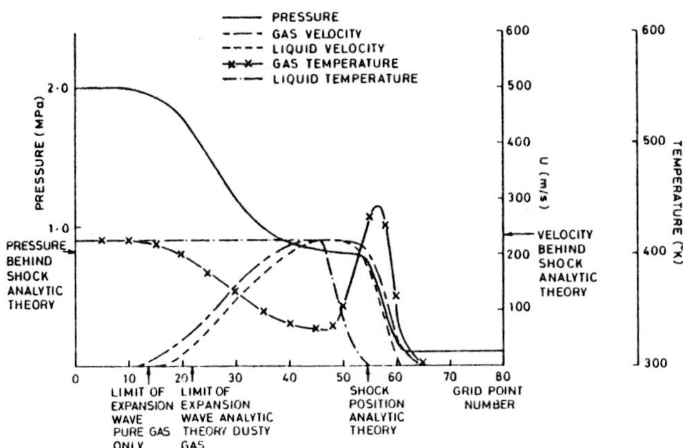

FIG 4 GRAPH OF FLOW VARIABLES AGAINST DISTANCE, $t = 0{\cdot}05$ secs. $\rho_L' = 10\,kg/m^3$, $r_p = 10^{-5}\,m$

For high volume fractions of dust the results exhibit oscillations behind the shock. These oscillations are not due to the ill-posedness of the mathematical model since equal phase velocities give rise to real characteristics [1-3]. Instead these oscillations are probably due to small changes in the liquid volume fraction causing large relative changes in the gas volume fraction and, through the gas equation of state, the pressure.

2.4 Larger dust particles

For larger dust particles the drag is less and there is more inter-phase slip. In this case the shock structure is as described by Carrier [8] with a gas shock for which the particle temperatures and velocities remain unchanged. The inter-phase drag then increases the particle velocity until equilibrium is reached some way behind the shock. If the relaxation zone is quite long, then it is not seen in the shock tube calculations since the reflected wave off the right hand wall destroys the structure of the equilibrium region long before it has a chance to get established. In this case the flow is similar to that of a single phase gas and the particle velocities remain small. For short relaxation regions the picture is one of a shock in the mixture, as in Figures 3 and 4. In Figures 5 and 6 the profiles of the variables are graphed for particle radii 10^{-4} and 10^{-3}m and a drag coefficient of 0.44. The macroscopic liquid density is 10 kg/m^3 and hence it is the same case as Figure 4.

Because of the different speeds of propagation of shock waves in the gas and in the mixture the macroscopic density of the droplets has been kept low. This is to prevent reflections of the pressure pulses in the gas from the end wall affecting the equilibrium flow region before it has had a chance to get established.

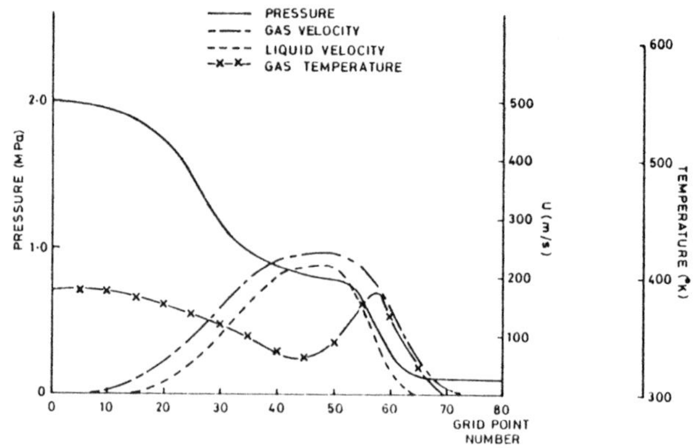

FIG.5 GRAPH OF FLOW VARIABLES AGAINST DISTANCE, $t = 0{\cdot}05$ secs, $\rho_L' = 10\ kg/m^3$, $r_p = 10^{-4}m$

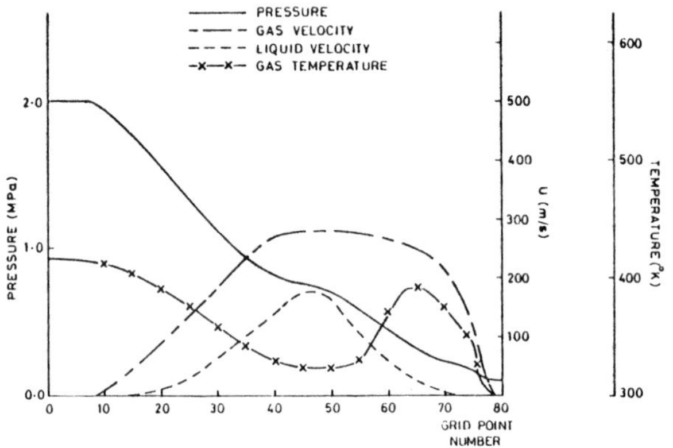

FIG 6 GRAPH OF FLOW VARIABLES AGAINST DISTANCE, $t = 0{\cdot}05$ secs, $\rho_L' = 10\ kg/m^3$, $r_p = 10^{-3}$ m

In Figure 5 the solution is very similar to that in Figure 4, except that the pressure profile near the undisturbed low pressure end is much flatter and is travelling faster. Kriebel [10] noted that it is possible for the gas 'shock' to be supersonic with respect to the mixture sound speed but subsonic with respect to the gas. In that case the initial disturbance is not a shock but just a smooth pressure rise. In Figure 6, the particle radius is larger and hence the drag exerted on the gas is a lot less. The initial disturbance propagating into the low pressure region now seems to have the form more appropriate to a gas shock with a sharp rise in temperature and particle velocity. It is unlikely that it is a true gas shock since it is travelling slightly slower than the sonic speed of the gas. If it were a shock the Rankine-Hugoniot conditions would be applicable. It does not however seem possible to relate these conditions for the shock in a simple way to the properties of the equilibrium region so that the shock speed and shock pressure ratio could be determined.

For higher particle radii the picture is one of a shock in a pure gas since the time scale for relaxation to equilibrium is much greater than the time scale for the shock propagation. For this reason the results are not presented here. Similar results to those presented here have been discussed by Otterman and Levine [13].

2.5 Droplet breakup

SIMMER-II [4] contains a droplet breakup model based on a critical Weber number. The Weber number is defined as

$$We = \frac{\rho_G |u_G - u_L|^2 \; r_p}{\sigma} \tag{1}$$

where r_p is the droplet radius, σ the surface tension between the liquid and gas, ρ_G the gas density and u_G and u_L the gas and liquid phase velocities. The criterion adopted is for the droplet to break up when the Weber number exceeds a critical value set by the user. The recommended value for this is 22.0 [4]. The mechanisms of droplet breakup under the action of a shock for single droplets are discussed by Simpkins and Bales [14], and Hanson, Domich and Adams [15].

Figures 7 and 8 show the profiles of the variables for 2 different macroscopic densities of droplets where now the droplet radius is calculated automatically by the program. The initial droplet radius is 10^{-2} metres. It does not matter what the actual value is provided it is large enough to trigger off the droplet breakup mechanism. The results are now quite different in character from the small droplet radius cases and have more in common with Figures 5 and 6. The equilibrium region behind the shock is still present and the pressure in this region is about the same. There is an appreciable velocity difference between the phases and it has roughly the same value everywhere. Signals now propagate into the undisturbed low pressure region faster than for the high drag cases and are much more smoothed out. There is also a gradual rise in pressure for a long way ahead of the equilibrium region. This pressure pulse is travelling very much faster than the equilibrium theory would give and represents either a shock wave in the gas or an acoustic wave, as for the lower drag cases.

The reason for the behaviour is quite straightforward. The situation now corresponds to the classical theory for shock propagation in dusty gases, as discussed in the previous section. There may now be a shock wave in the gas across which the droplet behaviour is undisturbed. This is followed by a relaxation region where the flow changes from its frozen state to that given by the equilibrium theory. The big difference now is that droplet breakup happens immediately behind the shock since the SIMMER model gives instantaneous breakup of drops. This has the effect of substantially

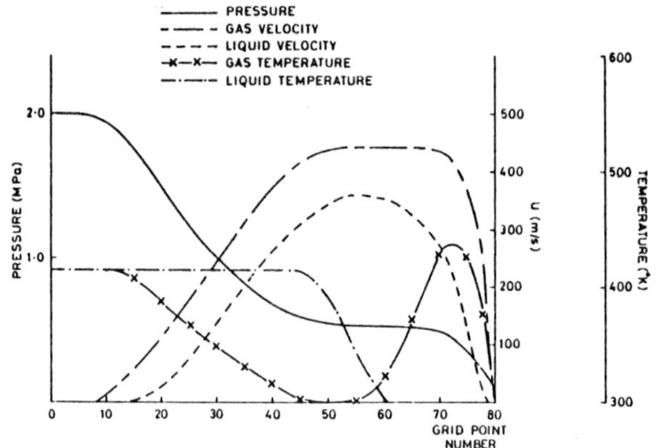

FIG.7 FLOW QUANTITIES, VARIABLE DROPLET SIZE, $\rho_L' = 1\ kg/m^3$, $t = 0{\cdot}05$

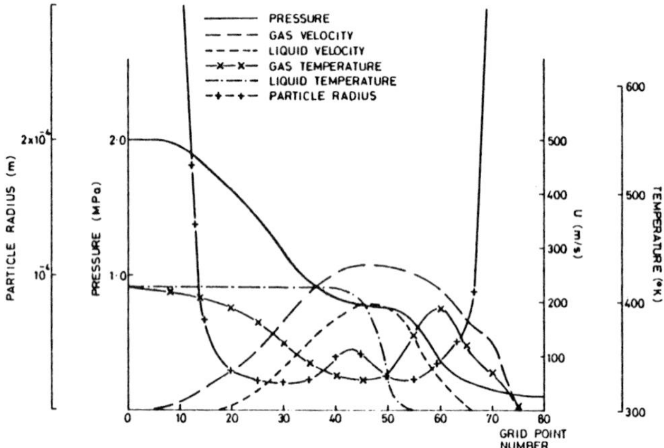

FIG.8 FLOW QUANTITIES, VARIABLE DROPLET SIZE, $\rho_L' = 10\ kg/m^3$, $t = 0{\cdot}05$ secs

increasing the drag and reducing the phase relative velocity until the Weber number criterion is satisfied. This limits the gas shock so that the velocity difference across the shock is governed by the Weber number criterion

$$u_2^2 \simeq \sigma\ We/\rho_G r_p \ . \qquad (2)$$

This will be quite a weak shock therefore and may even only be a pressure wave as with the present results. Behind the pressure wave the momentum exchange is quite large and hence the behaviour of the mixture is more like that found for very small particles. That is, it conforms quite closely to the equilibrium theory results except that there is a roughly constant velocity difference between the phases. In practice this velocity difference will gradually disappear as phase relaxation takes place but this is a much slower process. At the same time the droplets will begin to grow with the droplet coalescence model in SIMMER-II and the equilibrium process will therefore take longer to achieve than it would have without particle coalescence. The particle radius calculated by SIMMER is also shown in Figure 8 and shows the breakup across the shock and in the expansion wave, and the coalescence in the central region. Figure 8 also demonstrates the effect of the increased drag causing an inflection in the gas velocity profile.

The droplet breakup mechanism acts therefore as a stabilising mechanism since it prevents the velocity differences becoming too large. This is a physical stabilising mechanism and can also act against any unphysical instabilities arising from the model.

Figure 9 gives a graph of relative velocity in two places against the critical Weber number specified. One of these, grid point 40 is the position of the contact surface at t=0.0 and the other, grid point 50, is nearer to the shock. The results show that the effect of the critical Weber number is indeed to reduce the velocity differences and that this parameter effectively controls the maximum phase relative velocity.

This is an example of a stabilising mechanism which has not been taken into account in the theoretical analyses of the model equations. It does not really affect the question of

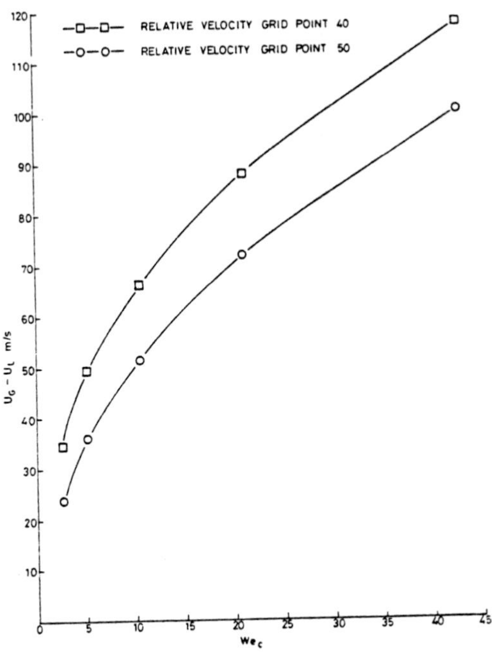

FIG.9. GRAPH OF RELATIVE VELOCITY AGAINST CRITICAL WEBER NO.

complex characteristics since no new additional differential terms are introduced into the drag terms [1-3]. It can however limit the differences in phase velocity by automatically increasing the interphase drag. This increased drag greatly aids the stability of the numerical calculations and the damping of perturbations.

3. THE CALCULATION OF SHOCK WAVES IN BUBBLY LIQUIDS

SIMMER-II is designed to handle only one two phase flow regime in its current version, that of disperse droplet flow. It is nevertheless able to treat bubbly flows to the extent that they may be considered as disperse flows with very high liquid volume fractions. Inertial effects associated with the radial (pulsatory) motion of the bubbles cannot, of course, be treated in this approximation. The same applies to effects deriving from the physical size of the bubble and the fact that the volumes of gas contained in them are not intercommunicating.

The calculations reported in this section did not, in fact use the drag model in SIMMER, except to impose a zero-slip condition between the phases for comparison with the experimental results available in the literature for the propagation of shock-waves in bubbly liquids, mainly water/glycerine mixtures. The shock waves are often generated in the experiment by having an air chamber above a column of bubbly liquid. In the chamber the air is below atmospheric pressure and at time t=0.0 a diaphragm is burst and the pressure in the chamber becomes atmospheric creating a shock wave in the bubbly liquid. This was modelled directly using SIMMER-II by modelling the air chamber and also by imposing a pressure boundary condition onto a column of bubbly fluid. The physical parameters specified were those appropriate to water containing air bubbles.

Figure 10 shows the results for the propagation velocity of the shock u_s for various pressure ratios. These agree well with the equation

$$u_s^2 = \frac{p_1}{\rho_L \alpha_G \alpha_L} \tag{3}$$

derived by Chapman and Pitcher [16] for shock propagation for an isothermal gas. This equation was in very good agreement with their experimental results and Eddington [17] has found similar agreement in his experiments. Note that the velocity of propagation is independent of the pressure downstream of the shock. This independence is also observed in the SIMMER results.

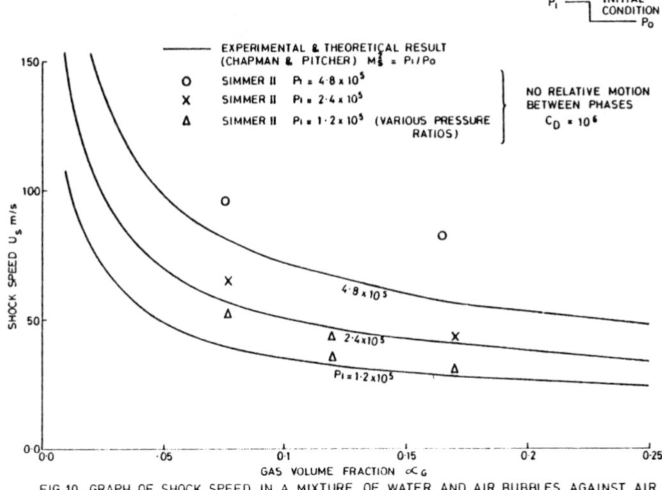

FIG.10. GRAPH OF SHOCK SPEED IN A MIXTURE OF WATER AND AIR BUBBLES AGAINST AIR VOLUME FRACTION

4. ACKNOWLEDGEMENTS

The bulk of this work was carried out whilst one of us (IPJ) was a Visiting Scientist at the Joint Research Centre, Ispra Establishment. This author would like to express his gratitude to the Commission of the European Communities for this opportunity.

5. REFERENCES

[1] RAMSHAW, J.D. and TRAPP, J.A. Characteristics, stability and short wave-length phenomena in two-phase flow equation systems. Nucl. Sci. and Eng. 66, pp.93-102, 1978.

[2] STEWART, H.B. Stability of two-phase flow calculation using two-fluid models. J. Comp. Phys. 33, pp.259-270, 1979.

[3] LYCZKOWSKI, R.W., GIDASPOW, D., SOLBRIG, C.W. and HUGHES, E.D. Characteristics and stability analyses of transient one-dimensional two-phase flow equations and their finite difference approximations. Nucl. Sci. and Eng. 66, pp.378-396, 1978.

[4] SMITH, L.L. SIMMER-II: A computer program for LMFBR disrupted core analyses, NUREG/CR-0453, LA-7515-M, 1978.

[5] SMITH, L.L. The SIMMER-II code and its applications. Int. Meeting on Fast Reactor Safety Technology, Seattle, August 19-23 1979.

[6] BLEWITT, P.J. SIMMER-II, calculation of shock tube problem. NUREG/CR-0385, LA-7481-PR, April 1 - June 30 1978.

[7] LIEPMANN, H.W. and ROSHKO, A. Elements of gas dynamics, Wiley 1960.

[8] CARRIER, G.F. Shock waves in a dusty gas. J. Fluid Mech. 4, pp.376-382, 1958.

[9] RUDINGER, G. Some properties of shock relaxation in gas flow carrying small particles. Phys. Fluids 7, pp.658-663, 1964.

[10] KRIEBEL,A.R. Analysis of normal shock waves in particle laden gas. J. Basic Eng.,Trans. ASME, pp.655-685, 1964.

[11] HARLOW, F.H. and AMSDEN, A.A. Numerical calculations of multiphase fluid flow. J. Comp. Phys. 17, pp.19-52, 1975.

[12] COMFORT, W.J. III and CROWE, C.T. Dependence of shock characteristics on droplet size in supersonic two-phase mixtures. J. Fluids Eng. 102, pp.54-58, 1980.

[13] OTTERMAN, B. and LEVINE, A.S. Analysis of gas-solid particle flows in shock tubes. AIAA Journal, 12, pp.579-582, 1974.

[14] SIMPKINS, P.G. and BALES, E.L. Water droplet response to sudden acceleration. J. Fluid Mech. 55, pp.629-639, 1972.

[15] HANSON, A.P., DOMICH, E.G. and ADAMS, H.S. Shock tube investigation of the breakup of drops by air blasts. Physics Fluids, 6, pp.1070-1080, 1963.

[16] CAMBELL, I.J. and PITCHER, A.S. Shock waves in a liquid containing gas bubbles. Proc. Roy. Soc. A, 243, pp.534-545, 1958.

[17] EDDINGTON, R.B. Investigation of supersonic phenomena in a two-phase (liquid gas) tunnel. AIAA Journal 8, pp. 65-74, 1970.

Table 1 SIMMER-II parameters, dusty shock tube

High pressure region

$T_4 = 413^{o}K$ temperature

$\rho_4 = 10.7\ Kg/m^3$ (microscopic density)

$p_4 = 2 \times 10^6\ N/m^2$ (2 MPa) pressure.

Low pressure region

$T_1 = 300^{o}K$ temperature

$\rho_1 = .738\ Kg/m^3$ (microscopic density)

$p_1 = 10^5\ N/m^2$ pressure.

General parameters

$\gamma_G = 1.333$ ratio of specific heats for the gas

$C_{VG} = 1355$ J/Kg.K specific heat at constant volume for the gas

$C_L = 1.3 \times 10^3$ J/Kg.K liquid specific heat

$\rho_L = 7050.0$ and $705.0\ Kg/m^3$ liquid microscopic density.

Liquid sonic velocity 2000 m/s.

Liquid vapour heat transfer correlation

$Nu = 2 + 0.37\ Re^{0.7}\ Pr^{0.38}$.

Grid increment .9144 m.

Number of grid points	Single phase shocks	48
	Two phase shocks	80

A MICROSTRUCTURAL MULTIPHASE COMPOSITE MATERIAL MODEL FOR CONCRETE

L. CONTRI* and B.A. SCHREFLER**

* Istituto di Scienza delle Costruzioni
** Istituto di Costruzioni, Ponti e Strade, University of Padova, Italy

SUMMARY

Concrete is investigated as a composite material where the dispersing medium is treated as a porous material containing also a liquid phase. This study adds the effects due to the particular confinement of the paste between the aggregates to the well observed behaviour of cement paste. The investigations are obviously valid for the common concrete mixes with no excess of paste. The numerical formulation of this composite material model is applied to the investigation of basic creep. It is shown that this model simulates and explains properly the creep behaviour immediately after loading as well as the deviation of creep recovery from the principle of superposition (softening) in young concrete specimens if the unloading occurs a few days after the loading. These phenomena have not been explained satisfactorily by other models.

INTRODUCTION

A great number of either structural or phenomenological models have been proposed to describe the time dependent and time independent behaviour of concrete and cement paste. These models are widely covered in the existing literature, see e.g. Neville and Dilger |1|, Bazant |2|, Stroeven |3|. Therefore only those relevant to the here presented study are briefly mentioned.

A statistically homogeneous and isotropic dispersion of particles in an elastic matrix has been investigated by Dantu |4|. Experiments made on a two-dimensional concrete specimen, with gravel formed by glass disks, have shown that the applied

load follows "main roads" through the packing of disks. These main roads are formed by an alignement of grains interconnected by the dispersing material.
Various arrangements of grains included in an elastic matrix have been used by Baker |5|, Pavlik |6| and Lusche |7| to explain the formation of axial cracking in concrete. The stress situation in the matrix around a single inclusion under repeated uniaxial loading has been investigated by Mehmel and Kern |8|. Single inclusions and symmetrical arrangements of four inclusions have also been studied by McGreath et al. |9| for the determination of crack mechanism. Biaxial stresses in concrete have been investigated further by Buyukozturk et al. |10| using a model composed by nine circular disks embedded in a mortar matrix.
Starting from statistical investigations of the packing of aggregates, Contri |11| has proposed a spatial structural element composed of two grains interconnected by cement paste. Experimental and analytical studies have simulated the curvatures towards the strain axis in the initial parts of the stress - strain diagrams observed during the loading of virgin concrete specimens |12| .
An extension of the model, which also takes into account a multiphase porous material |13|, has permitted a qualitative explanation of the difference of creep between cement paste and concrete as well as the apparent independence from loss of water of the total creep and shrinkage in advanced stage and the superposition of creep and shrinkage.
For concrete creep investigations in particular, the most recent models involve continuum mechanics, cement physics, physical chemistry and continuum thermodynamics. A creep mechanism based on the thermodynamics of multiphase equilibrium between absorbed water in the micropores of cement gel, capillary water and vapor, coupled with the diffusion of solids and water has been proposed as a hypothesis by Bazant |2|. The diffusion theory is applied by Bazant and Najjar |15| in their study of drying and wetting of concrete and cement paste, treated as a homogeneous multiphase material. The diffusivity and other material parameters are assumed to be dependent on pore humidity, temperature and degree of hydratation. The solution technique used applies to geometrically simple bodies and only unidimensional flow is investigated.
The intent of this paper is a quantitative formulation of the structural model proposed by Contri |11,13| via finite element

analysis of a multiphase composite material with porous matrix. The distributions of stress and pore pressure in the paste layers, confined between aligned aggregates, are investigated by means of an axisymmetric model. As will be shown, this limited quantities of cement paste have an enormous influence on the deformation of concrete in common mixes. This first model applies to time independent non-linear response and time dependent deformation (creep) of concrete at reference temperature and constant water content. Saturated conditions are required by the actual version of the model only for the internal parts of the cement paste layers between the aggregates. This occurs in almost every concrete for a major part of its life span and in particular in mass concrete which never loses an appreciable amount of water. The applied model leads to an understanding of some of the peculiarities in the behaviour of concrete which have not been explained by other models.

The numerical model will be further extended by coupling the heat and mass transfert equations with the already coupled equilibrium and mass continuity or flow balance equations to thus also account for temperature and humidity.

For the experimental and numerical investigations spherical aggregates have been used, but a slight change of the physical model may also account for angular aggregates.

THE STRUCTURAL MODEL

In this paper concrete is regarded as a structure of large aggregates (gravel) bound by mortar, which is a structure of fine aggregates (sand) bound by cement paste. Cement paste itself is a multiphase porous material whose solid part consists of hydrated cement and unhydrated cement grains, made up mostly of cristalline components |16|.

Inasmuch as our purpose here is the illustration of the numerical formulation of the structural model, only the necessary results of the investigations regarding the physical model will be stated. For additional details the interested reader is referred to references |11,12|.

The behaviour of mortar (simple packing |17| of sand grains, bound by cement paste) is also supposed to prevail in the analogous phenomena of concrete, even if, there, the arrangement of grains is more complex (multiple aggregations). Here only simple packings of the dispersed medium are studied and the introduced simplifications are justified by the following observations.

Common sand mixes for mortar show a prevailing number of smaller elements of similar size. For instance in a Fuller type distribution with a lower bound given by grains of a characteristical size d, more than 80% of all the elements have a size between d and 2d. A simple aggregation can therefore be assumed for the mortar.
The aggregates are taken to be adequately presented by a model composed of spherically shaped grains. Various possible packings of spherical grains had previously been investigated. If the spheres are assumed to be equal in size, the densest and most probable packings are the pyramidal and the tetrahedral ones |17|, with a porosity of 25,9%. In these packings each sphere is in contact with 12 neighbouring ones. The particularly stiff lines connecting the centres of these spheres in contact form the edges of 6 tetrahedra and respectively pyramids with a square base. The mean value of the angle between a line chosen at random and the closest of the above mentioned stiff edges is only about 20°. This is also true in particular for the direction of the principal stress in an uniaxial stress state in an equivalent homogeneous material. The pyramidal packing shows rows of spheres which are completely aligned, while in the tetrahedral one the alignment in only slightly imperfect. Sand packings, with limited variability of grain size, will be similar to one of the previous ones. Porosity of sand equal or less than 30% has in fact frequently been observed.
Keeping in mind the experiment by Dantu |4|, which has shown the existence of "main roads" for applied forces, the physical model proposed is that of a pyramidal packing of grains, loaded along the direction of one row of aligned grains. These rows are composed of a series of "basic structural elements" formed by two grains connected by cement paste. The force transmitted by the basic structural elements formed by two different grains, is near to that of an element formed by two identical grains of the smaller size |11|. Therefore only elements with identical grains will be studied.
The connection of the loaded rows to the neighbouring ones, by means of cement paste, is here disregarded. This part of cement paste will undergo shear deformation. But its importance is limited since the force is transmitted by normal stress in inhomogeneous materials with limited shear resistance at the contact surface.
The basic structural elements will now be studied in detail. The distance between the two grains is of paramount importance

for the behaviour of mortar and concrete. As will be shown, the smaller distances are much more important for our investigations since they produce a major deviation from the behaviour of a specimen formed by cement paste only and are the ones most likely to occur. The distance δ between grains is formed by the sum of the thickness of the coated layers of the matrix (cement paste) on the particles.
These distances are assumed to follow an expotential distribution $e^{-(\delta - \delta_0)}$
which presents a rather suitable model concentrating the probability on the smaller distances. The value δ_0 corresponding to the minimum free energy of the system |18|, is the mode of the distribution.
If R indicates the radius of the grains, $\alpha = 1 + \delta/2R$ is defined as a typical parameter of an element. The values of α = 1.005, 1.025, 1.05, 1.10, 1.15, 1.20 are studied. A rough estimation of the behaviour of a mortar specimen may be obtained from a structural element with the mean value $\alpha = 1.05$.

THE NUMERICAL MODEL

The dispersed particles are treated as linearly elastic materials while the dispersing medium is here an elastic or elasto-plastic saturated porous material.
The behaviour of a saturated porous medium, taking into account the compressibility of the fluid and of the solid particles |19|, is governed by the following coupled equations:

Equilibrium equations

$$\left| D_{ijkl}\, \dot{u}_{(k,l)} \right|_{,i} + \delta_{ij}\, \dot{p}_{,i} - \left(D_{ijkl}\, \delta_{kl} \frac{\dot{p}}{3k_s}\right)_{,i} + \rho F_j = 0 \tag{1}$$

Continuity equations

$$\left| k_{ij}\, (p_{,j} + \rho_2 F_j) \right|_{,i} + \dot{u}_{i,i} - \frac{1}{3k_s}\, \delta_{ij}\, D_{ijkl}\, \dot{u}_{(k,l)}$$

$$- \frac{\dot{p}}{k_f} + \frac{1}{(3k_s)^2}\, \delta_{ij}\, D_{ijkl}\, \delta_{kl}\, \dot{p} = 0 \tag{2}$$

where u_i, F_j, D_{ijkl}, k_{ij} denote the Cartesian components, respectively, of the displacement vector, the body force vector, the effective stress tangent modulus tensor

($D_{ijkl} = D_{jikl} = D_{ijlk} = D_{klij}$) and the permeability tensor; p denote the pore water pressure, ρ the mass density of the saturated soil and ρ_2 the mass density of the pore fluid. k_s denotes the bulk modulus of solid grains and k_f denotes the combined compressibility of the fluid and the solid. δ_{ij} is the Kronecker delta. A superposed dot indicates differentiation with respect to the time variable.
The associated boundary conditions are of the following four types: prescribed displacements $\bar{u}_i$, prescribed tractions $\bar{T}$, prescribed pore pressures $\bar{p}$, prescribed flow (normal to the boundary) $\bar{Q}_i$.
Finite element discretisation in space leads to the following matrix equation |19|

$$\begin{vmatrix} \underset{\sim}{0} & \underset{\sim}{0} \\ \underset{\sim}{0} & \underset{\sim}{H} \end{vmatrix} \begin{Bmatrix} \bar{\underset{\sim}{u}} \\ \underset{\sim}{p} \end{Bmatrix} + \begin{vmatrix} \underset{\sim}{K}_T & -\underset{\sim}{L} \\ -\underset{\sim}{L}^T & -\underset{\sim}{S} \end{vmatrix} \frac{d}{dt} \begin{Bmatrix} \bar{\underset{\sim}{u}} \\ \bar{\underset{\sim}{p}} \end{Bmatrix} = \begin{Bmatrix} \dfrac{d\underset{\sim}{f}}{dt} \\ \underset{\sim}{f} \end{Bmatrix} \tag{3}$$

where $\bar{\underset{\sim}{u}}$ denotes the nodal displacements, $\bar{\underset{\sim}{p}}$ the nodal values of the pore pressure, $\underset{\sim}{f}$ stands for changes of the external forces due to boundary or body force loading, $\underset{\sim}{K}_T$ denotes the tangential stiffness matrix, $\underset{\sim}{H}$ the flow matrix. $\underset{\sim}{S}$ represents the compressibility of the fluid and skeleton. $\underset{\sim}{L}$ denotes the coupling matrix representing the influence of the pore pressure in the force equilibrium and in the transposed form, the influence of the solid phase volume change upon the nodal point flux.
The solution of the simultaneous transient equations (3) may be accomplished by single step integration in the time domain. Equations (3) contain the nearly incompressible solid formulation |20|, which can be used for instantaneous load

$$\begin{vmatrix} \underset{\sim}{K}_T & -\underset{\sim}{L} \\ -\underset{\sim}{L}^T & -\underset{\sim}{S} \end{vmatrix} \begin{Bmatrix} \Delta\underset{\sim}{u} \\ \Delta\underset{\sim}{p} \end{Bmatrix} = \begin{Bmatrix} \Delta\underset{\sim}{f} \\ \underset{\sim}{0} \end{Bmatrix} \tag{4}$$

The yield criteria used in case of an elasto-plastic model are a Mohr-Coulomb type criterion and the von Mises criterion. Associated plasticity assures a symmetric tangential stiffness matrix $\underset{\sim}{K}_T$.
The finite element discretisation of the structural element investigated is shown in figure 1. Quadrilateral isoparametric elements are used along with quadratic interpolation for pore pressure and for components of displacement.

APPLICATION TO CREEP

The composite material model of concrete in the actual version allows for modelling basic creep effects for moderately stressed concrete where non-linearities, due to microcracking, are not relevant.
A mechanism which takes into account the microcracks occuring chiefly in the interface between the aggregate and the mortar or cement paste, can be introduced in the numerical model. This microcracking is important roughly beyond 0.4 strength |2|.
The effects of aging, due to cement hydratation |2| can be taken into account using a variable permeability scheme.
The time dependent deformation of concrete is known to originate chiefly in the hardened cement paste. Homogeneous material models, most common, do not explain properly all the experimentally observed behaviour. Creep curves plotted in logarithms of the time and the deformation show a change in the behaviour from the early to the later period. This is evident for instance in the experimental diagrams obtained by Le Camus, reproduced from ref. |21| and shown in figures 2 and 3. The first phase, also called short-time creep |22| is here linear while the behaviour of the second period, "long-time creep" is similar to that of a Voigt-Kelvin model.
In a semi-logarithmic coordinate system the short-time creep is curvilinear and the long-time creep linear. This is again shown in reference |21| and has been obtained by Contri experimentally with a physical model based on the composite material concept |13|
The results of this experiment as well as the model employed are shown in figure 4. The behaviour of a specimen of neat cement paste is shown in figure 5 as a comparison. For paste specimens the Voigt-Kelvin model fits the behaviour almost from the beginning, while for a concrete specimen the short-time creep is much more extended.
This creep behaviour of concrete specimens was simulated by Pfefferle |22| adding to the Voigt-Kelvin model an inertia term. This author tries to explain the short-time creep with the loading velocity, which in our opinion is not an adequate explanation.
The here presented numerical model of a composite material allows an adequate simulation of the short-time and long-time creep: short-time creep is due to the particular time transient pore water pressure distribution in the layers between the dispersed grains.

Due to the moderate stress level cement paste is here assumed to be linearly elastic and the following material properties are used:
Young modulus for cement paste: 30000 MPa , Poisson ratio: 0.157, permeability (isotropic in this case):1×10^{-12} cm/sec |15,23| , external load applied: 22 kN .
The material properties produce a ratio of two between instantaneous deformation and creep for a specimen with $\alpha = 1.05$
The deformation versus time obtained numerically for a series of structural elements is shown in figure 6 in a bi-logarithmic coordinate system. A comparison with similar results of a paste specimen shows yet again the different behaviour, as seen in figure 7. An opposite sign curvature can be observed in this case at the beginning of the diagram.
The difference can be seen even more clearly in the figures 8 and 9 where a semilogarithmic coordinate system has been used. In this case the initial curvature is practically inexistent in case of a paste specimen.
The two observed periods reveal a completely different distribution of the excess pore pressure in the paste layers. This distribution is shown in figures 10 and 11 for an element with $\alpha = 1.05$ along a radius at the centre of the paste layer and for different time values. The short-time period shows a strongly variable pressure gradient with peak values after the loading while the second period is characterised by a constant pressure gradient, which justifies the analogy with a Voigt-Kelvin model.
In the first period a redistribution of the pore pressure within the paste layer prevails while in the second period takes place a slow expulsion of water to the nearby macropores.
This expulsion occurs with a constant pressure gradient and transfert of the loading from the pore fluid to the skeleton. The quantity of water involved is very small and this explains why no external effects of load upon loss of water were observed |24|.
The present model also explains the peculiar behaviour during creep recovery of young concrete specimens with a high water content. Even if creep recovery is almost independent of age and linearly dependent on the stress drop |2|, the above mentioned specimen show a different behaviour according to the moment at which the unloading occurs. This can be clearly seen, in figure 12, reproduced from reference

|25|, where the behaviour obtained through the principle of superposition is drawn with a solid line. If the stress drop occurs a few days after the loading (e.g. 6 days in figure 12), the recovery is even higher than that predicted by the principle of superposition. (During unloading at later periods the recovery is less due to aging). This larger recovery has not been explained by other models.

The same behaviour has been simulated with the presented model, considering a stress drop in three different periods: respectively 4, 20 and 40 days after loading. The results drawn in figure 13 with full lines, are compared with the recovery predicted when the principle of superposition is applied. (The instantaneous deformation has been omitted in the drawing). In the case of unloading after 4 days a remarkable difference can be seen according to the observed behaviour, while in the other cases the predicted and calculated values are the same. The difference in behaviour in the first case is obviously connected to the distribution of pore pressure. This distribution, after unloading, differs in the first case from the latter ones: for instance, the drop of the pore pressure in the centre of the model, due to unloading, results as larger in the first case than in the last two, where it is almost equal (see figures 14 and 15). The higher deformability of the first case is due to this different pore pressure distribution, where the negative excess pore pressure is limited to a smaller, central part, with no excess pore pressure at the outer part. These effects are only a result of the confinement of the paste between the aggregates.

This also explains why "as compared with the prediction of the principle of superposition pulsating loads considerably accelerate creep of concrete even at low stress levels (cyclic creep). When pulsation occurs after a long period under constant load, cyclic creep is negligible as compared with virgin specimen. In cement paste at low stress, cyclic creep is not observed" | 2 |.

CONCLUSION

A numerical formulation of a composite material model for concrete is presented. Interpretation of test data is aided by the numerical investigation of the stress and pore pressure state in the confined paste layers between the aggregates. The model has been applied in particular to basic creep and the following has been explained:

- the existence of short-time creep after the loading of concrete specimen.
- The deviation from principle of superposition in creep recovery in the early post-loading stage observed in young concrete.
- Creep acceleration due to pulsating loads (cyclic creep) in young concrete.

The model is promising for the investigation and explanation of further observed phenomena in time-dependent and time-independent behaviour of concrete.

REFERENCES

|1| NEVILLE,A.M., and DILGER,W., Creep of Concrete: Plain, Reinforced,Prestressed, North-Holland, Amsterdam 1970 .

|2| BAZANT,Z.P., "Theory of Creep and Shrinkage in Concrete Structures: A Précis of Recent Developments", Mechanics Today, Vol.2, Pergamon Press, New York 1975, pp.1-93.

|3| STROEVEN,P., Some aspects of the micromechanics of Concrete, Univ.of Delft, Stevin Lab., 1973

|4| DANTU,P., "Etude des Contraintes dans les Milieux Heterogènes. Application au Béton", A.I.T.B.T.P., Jan.1958, pp.53-98.

|5| BAKER,A.L.L., "An analysis of deformation and failure characteristics of concrete", Mag.Concr.Res., n°11, 1959, p.33.

|6| PAVLIK,A., "Strength and deformation of crystalline building materials with special regard to concrete", Rep. Techn. and Test.Inst. for the Build.Ind. in Prague, Sept. 1971.

|7| LUSCHE,M., "Beitrag zum Bruchmechanismus von auf Druck beanspruchtem Normal- und Leichtbeton mit geschlossenem Gefüge", Schriftreihe der Zementindustrie, Beton-Verlag, n° 39, 1972.

|8| MEHMEL,A., KERN,E., "Elastiche und plastiche Stauchungen von Beton infolge Druckschwell- und Standbelastung", D.A.f. St., Heft 153, Berlin 1962.

|9| Mc.GREATH,D.R., NEWMAN,J.B., NEWMAN,K., "The influence of aggregate particles on the local strain distribution and fracture mechanism of cement paste during drying shringkage and loading to failure", Bull.RILEM, n°2, 1969, p.7.

|10| BUYUKOZTURK,O., NILSON,A.H., and SLATE F.O., "Stress-Strain Response and Fracture of a Concrete Model in Biaxial Loading", J.Am.Concr.Inst., 68, 1971, pp.590-599.

|11| CONTRI,L., "Di una particolarità notevole del diagramma tensioni-deformazioni del calcestruzzo e della sua interpretazione in base alle caratteristiche meccaniche e statistiche della struttura del materiale", Istituto di Costruzioni,Ponti e Strade dell'Università di Padova, Comunicazioni, Studi e Ricerche, n°189, July 1965.

|12| CONTRI,L., "Distribuzioni caratteristiche di curvatura nei tratti iniziali dei diagrammi tensioni-deformazioni delle malte cementizie e dei calcestruzzi, loro ricostruzione in base a modelli strutturali". Costruzioni in cemento armato, Studi e Rendiconti, Vol.6, 1969.

|13| CONTRI,L., "Ritiro e deformabilità differita dei conglomerati cementizi nell'interpretazione offerta da un nuovo modello strutturale". Costruzioni in cemento armato, Studi e Rendiconti, Vol.8, 1971, pp.3-19.

|14| McHENRI,D., "A New Aspect of Creep in Its Application to Design", Proc. ASTM, 1943, Vol.43, pp.1069-1087.

|15| BAZANT,Z.P.,NAJJAR,L.J., "Nonlinear water diffusion in non-satured concrete", Materials and Structures (RILEM), Vol.5, n°25, 1972, pp.3-20.

|16| BAZANT,Z.P., "Costitutive equation for concrete creep and shrinkage based on thermodynamics of multiphase systems", Materials and Structures (RILEM), Vol.3, n°13, 1970, pp.3-36.

|17| DERESIEWICZ,H., "Mechanics of Granular Matter", Advances in Applied Mechanics, Academic Press, Vol.V, New York 1958.

|18| BARES,R., et.al., "Some basic features in mechanics of inhomogeneous materials", Mechanical Behaviour of Materials, The Soc.of Mat.Science, Japan, Vol.V, 1972, pp.42-53.

|19| LEWIS,R.W., SCHREFLER,B., "A fully coupled consolidation model of the subsidence of Venice". Report C/R/278/1976, Univ.Col.Swansea, see also Water Resources Research, Vol. 14, n°2, 1978, pp.223-230.

|20| ZIENKIEWICZ,O.C., NORRIS,V.A., WINNICKI,L.A., NAYLOR, D.J. and LEWIS,R.W., "A Unified Approach to the Soil Mechanics Problems of Offshore Foundation", Numerical Methods in Offshore Engineering, J.Wiley & Sons, Chichester, 1978, pp.361-412.

|21| L'HERMITE,R., "Que Savons Nous de la Déformation Plastique et du Fluage du Béton?", A.I.T.B.T.P., Sept. 1957, pp.777-809.

|22| PFEFFERLE;R., "Das Kriechen des Betons, eine kritisch gedämpfte Schwingung", Beton und Stahlbetonbau, 12/1979, pp.296-301.

|23| POWERS,T.C., COPELAND,L.E., HAYES, J.C. and MANN,H.M., "Permeability of portland cement paste", American Concrete Institute Journal, Vol.51, Nov.1954, pp.285-298, (PCA Bulletin 53).

|24| L'HERMITE,R., MAMILLAN,M., "Retrait et Fluage des Bétons", A.I.T.B.T.P., Sept.1968, pp.1317-1337.

|25| BÄCKSTRÖM,S., "Creep and Creep Recovery of Cement Mortar", Proc.of the V. Congrès des Ponts et Charpentes, Publ. Prélim., 1956, pp.76-80.

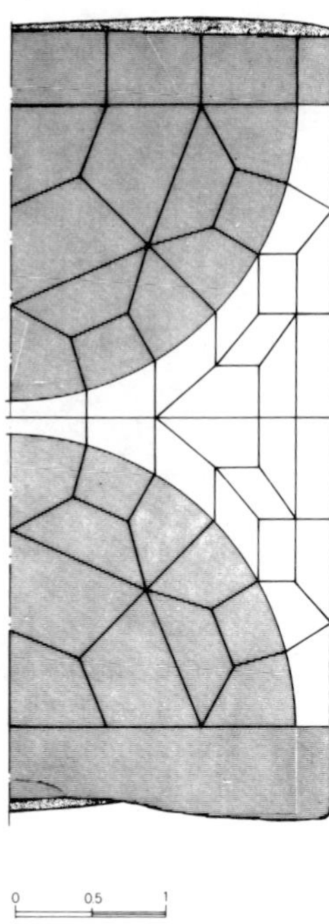

1 • Finite element discretisation of the structural element.

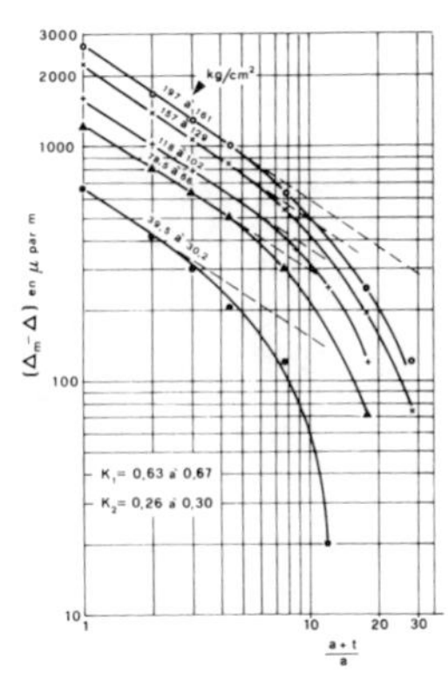

2 • Creep data reproduced from reference / 21 /.

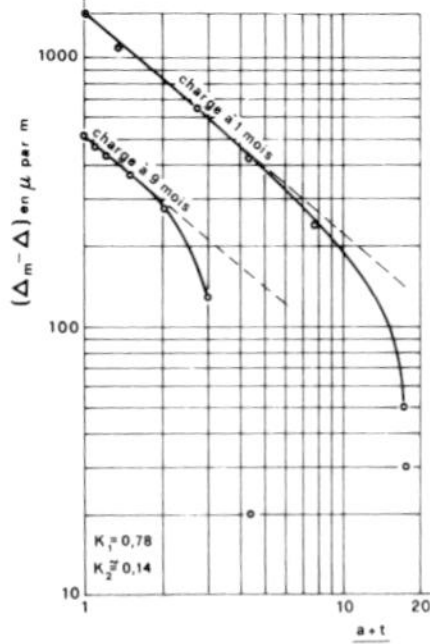

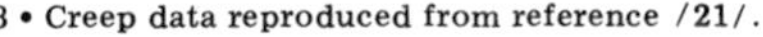

3 • Creep data reproduced from reference /21/.

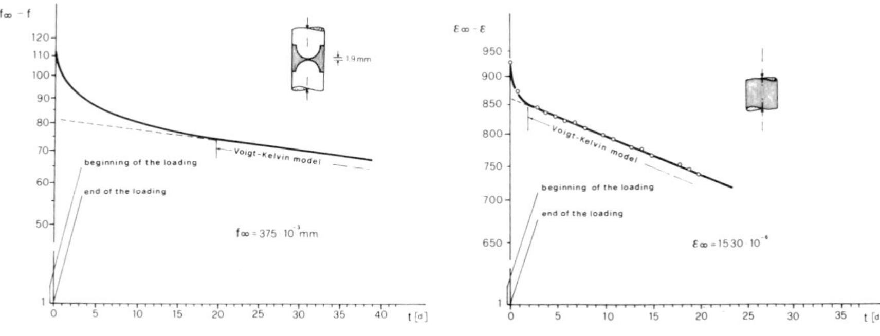

4 • Experimentally observed creep data /13/. 5 • Experimentally observed creep data for cement paste /13/.

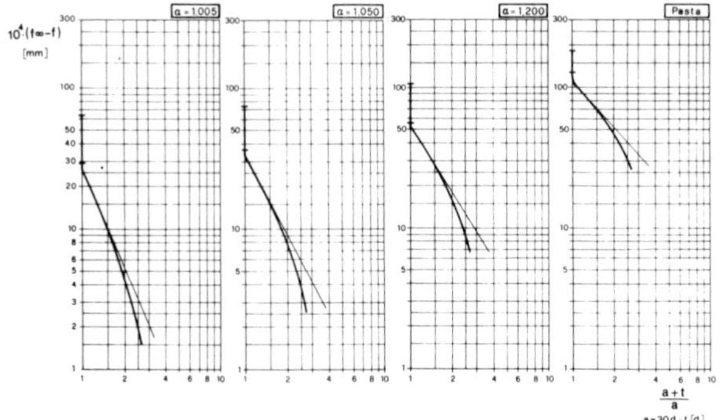

6 • Numerically obtained creep curves

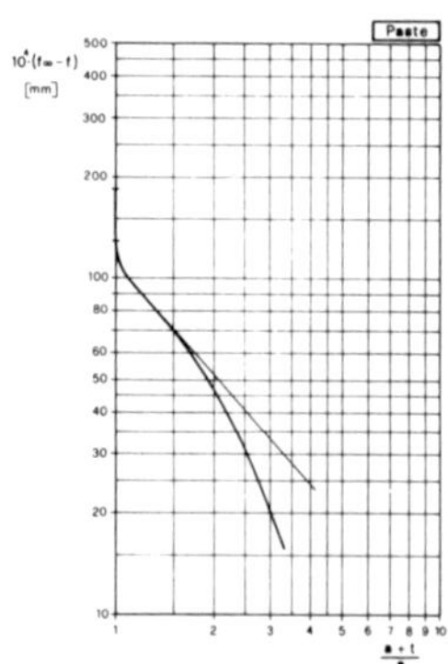

7 • Creep curves for a paste specimen.

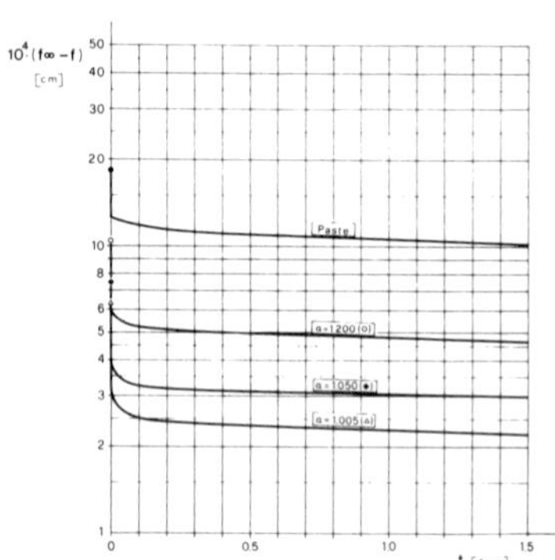

9 • Numerically obtained creep curves.

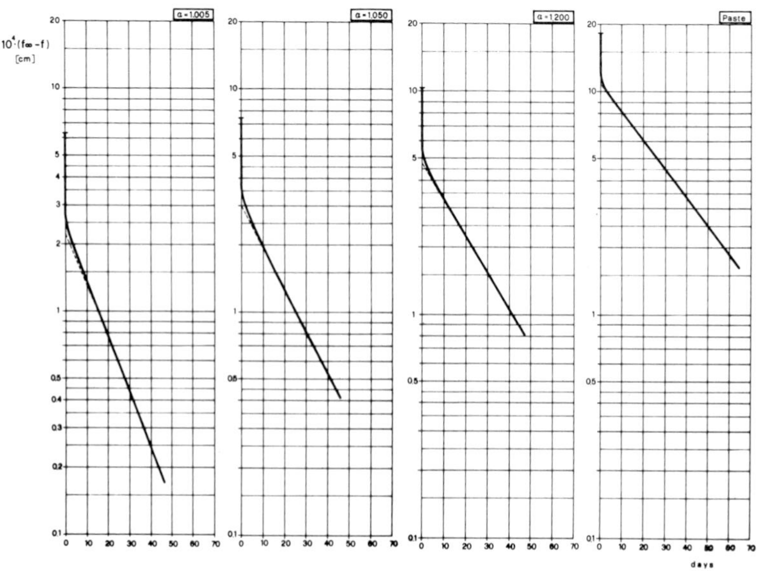

8 • Numerically obtained creep curves.

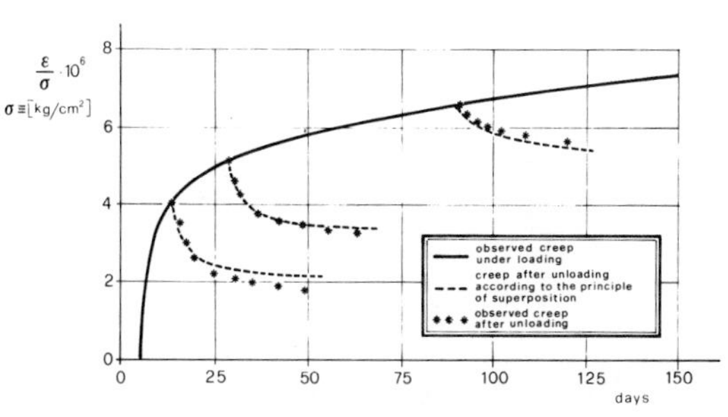

12 • Creep behaviour after unloading / 25/.

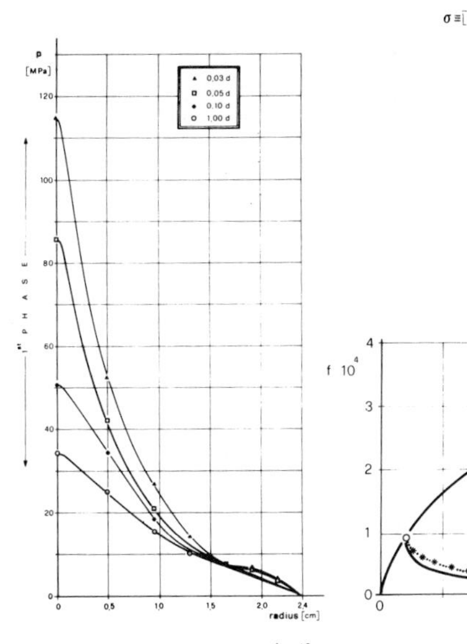

10 • Excess pore pressure in the confined paste layer.

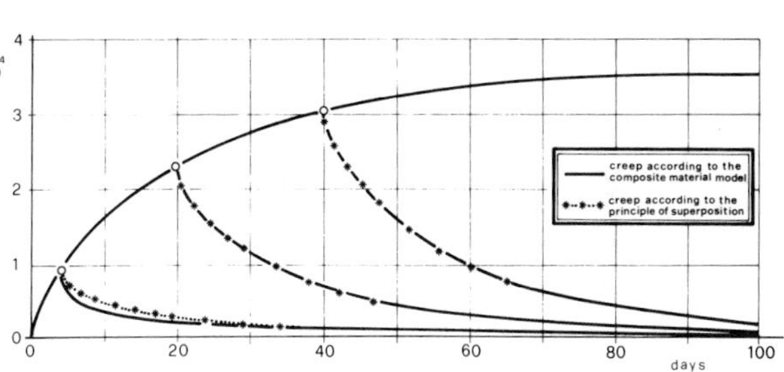

13 • Creep behaviour after unloading

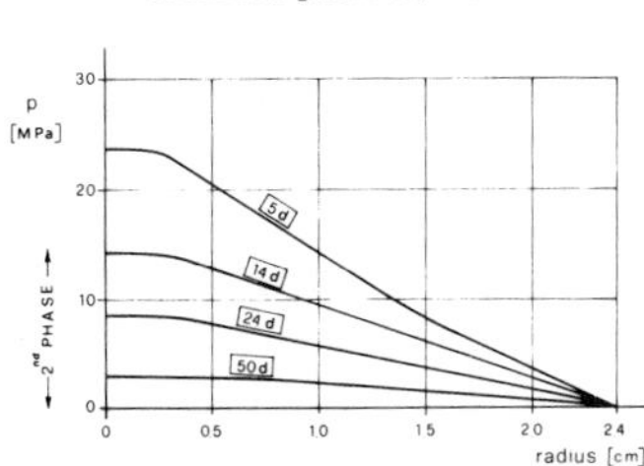

11 • Excess pore pressure in the confined paste layer.

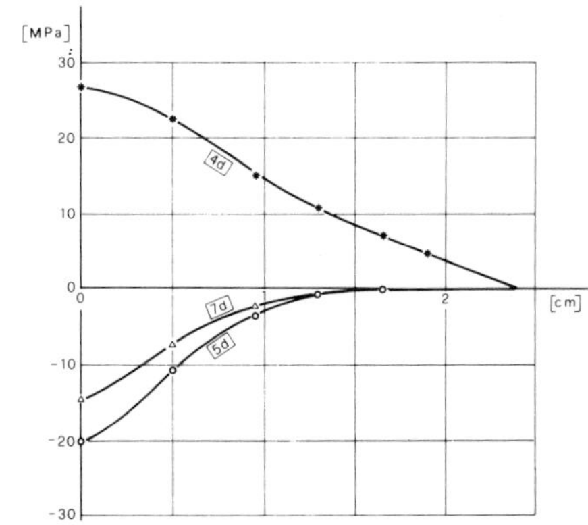

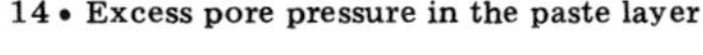

14 • Excess pore pressure in the paste layer before and after unloading (4 days).

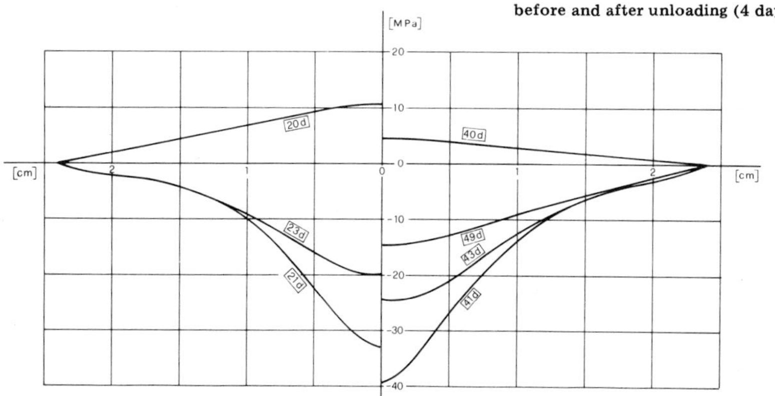

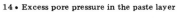

15 Excess pore pressure in the paste layers before and after unloading (at 20 and 40 days).

A NUMERICAL STUDY OF BARBOTAGE

L. G. NHAN
C.S.I.R.O. Div. of Textile Physics, Ryde, Australia, 2112
and
G. DE VAHL DAVIS
University of N.S.W., Kensington, Australia, 2033.

SUMMARY

A numerical study is described of the growth of a bubble at an orifice in a surface under a shear flow.

1. INTRODUCTION

Many physical and chemical processes involve dispersions of gas bubbles in liquids, in order to bring about efficient mass or heat transfer between the two phases. One of the common methods of forming such dispersions is barbotage, i.e., gas bubbling through orifices submerged in a flowing liquid. The phenomena involved in this process are extremely complicated, since many parameters influence the size and shape of the bubbles thus formed. Therefore, a somewhat simplified starting point in studies of this field has been the dynamics of a bubble from a single orifice, which excludes the mutual influence of bubbles formed in neighbouring orifices. In this work, a computing technique was used to simulate the dynamics of a single bubble formed on a horizontal surface in a uniform shear flow in order to study the effects of surface tension, viscosity and liquid shear flow rate on the shape of a bubble up to and at the moment of departure.

2. THE BASIC NUMERICAL METHOD

The dynamics of gas bubbling into a flowing liquid has been studied by numerically integrating the equations of fluid motion with the boundary conditions appropriate to such a system. The basic numerical techniques are those of the Marker-and-Cell (MAC) method [1], a computing scheme developed to study transient flow problems involving an incompressible viscous fluid, and noted particularly for its ability to handle free surfaces. A detailed discussion of the MAC technique has been given by Welch et al. [2], and a number of applications have appeared in the literature (e.g. Harlow and Shannon [3]; Daly and Pracht [4]).

The MAC method solves the finite-difference form of the Navier-Stokes equations, with the velocity components and pressure defined over a staggered Eulerian mesh. The primitive form of the equations are used with the primary advantage that boundary conditions are easier to implement. A Lagrangian system of marker particles is defined; these markers are moved through the grid at interpolated local fluid speeds, behaving much like dye particles in actual experiments. As time progresses, the positions of these marker particles serve to specify the location of the fluid surface and, hence, can define in which computing cells the surface boundary conditions should be applied. An extension has been made to the original MAC method in order to include the effect of surface tension, which plays an important part in the dynamics of barbotage.

Due to limitations of machine storage and computing time, the model which was constructed is that of a two-dimensional bubble growing from an infinite slit. The results can thus be expected to be only of qualitative value. They nevertheless yield useful information on the behaviour of a growing bubble in a flowing liquid.

The finite difference equations are derived from the non-dimensional differential equations

$$\frac{\partial u}{\partial t} = -\frac{\partial u^2}{\partial x} - \frac{\partial uv}{\partial y} - \frac{\partial p}{\partial x} + \nu\nabla^2 u \tag{1}$$

$$\frac{\partial v}{\partial t} = -\frac{\partial uv}{\partial x} - \frac{\partial v^2}{\partial y} - \frac{\partial p}{\partial y} + \nu\nabla^2 v + g \tag{2}$$

$$0 = \frac{\partial u}{\partial x} + \frac{\partial v}{\partial y} \tag{3}$$

in which d (the orifice width), $(dg)^{\frac{1}{2}}$, ρdg and $(d^3g)^{\frac{1}{2}}$ have been used as scaling factors for length, velocity, pressure and kinematic viscosity respectively; p is pressure/density. The origin is at the centre of the orifice; x is parallel and y is normal to the surface.

3. THE INCLUSION OF SURFACE TENSION STRESS

The surface tension effect was included in the boundary conditions at the fluid interface as an external stress. In general, the surface tension stress is expressed in terms of the two principal radii of curvature (only one in the two-dimensional case) as $\sigma(1/R_1 + 1/R_2)$. For a two-dimensional problem the non-dimensional surface tension stress is

$$\phi_{ST} = A \frac{y''}{[1 + (y')^2]^{3/2}} \tag{4}$$

where A is a non-dimensional surface tension coefficient: $\sigma/\rho g d^2$; σ is the surface tension.

To use equation (4) requires a detailed knowledge of the orientation of the bubble surface: for a given abscissa, the ordinate, the slope and the curvature of the free surface must be known in order to determine the magnitude and direction of the surface tension stress. The general technique used is patterned after the work of Daly and Pracht [4], and Daly [5]. A sequence of marker particles is laid out in order along the free surface. The kinematic free surface condition (see for example, Langlois [6]) stipulates that a particle initially on the free surface always remains on it. Hence the particles thereafter mark the free surface as they move at the local fluid velocity. The surface curvature is not determined from the surface particle positions directly, but rather from the orientation of an interpolation spline curve passed through the surface particle array.

Given a monotone sequence of points, together with appropriate end conditions, the spline fit can be used to join these points in such a way that the slopes and curvatures are continuous at the junction points. Details were described by Walsh et al. [7]. In this work, it was found that at the end point, the Neumann type of boundary condition which prescribes parabolic end segments produces satisfactory results; i.e. $y''_i = y''_{i-1}$ at the end particle i.

If the sequence of particles is not monotonic in x or y, then it must be broken into sub-sequences that are monotonic. In each succeeding sub-sequence the roles of abscissa and ordinate are interchanged, but each sequence can be monotonic increasing or decreasing in the abscissa. A greatly distorted bubble may require up to five sub-sequences to represent its surface. Boundary conditions at break points that divide the sub-sequences are critical as these sections of curve should fit together as smoothly as possible. The most successful boundary condition used was obtained by assuming that the radius of curvature is constant for the last two particles at and near the break point:

$$\frac{y''_i}{[1 + (y'_i)^2]^{3/2}} = \frac{y''_{i-1}}{[1 + (y'_{i-1})^2]^{3/2}} \tag{5}$$

To further reduce the error inherent at the break points, the spline fit was extended to the next two points after the break point, so that the boundary condition is actually applied to particles i+1 and i+2; however, values from the spline curve only up to particle i were used in the calculations of surface tension stress. This transfers any possible surface "roughness" away from the region where curvatures are actually evaluated.

The spacing of particles on the free surface is also important. Erroneous curvatures may be produced from small fluctuations in particle positions if the spacing is too close. On the other hand, the surface may be incorrectly resolved if the particles are too far apart. Different particle spacings were tested, from $0.4\delta x$ as recommended by Daly [5] to as far as $2.25\delta x$ as used by Foote [8]. It was found that the minimum particle spacing should be $1.0\delta x$ to eliminate fictitious surface noise and that the particles could be spread as far as $2.25\delta x$.

Although the surface particles were evenly spaced initially, after some period of calculation parts of the surface may be closely packed with particles while in another section of the surface the particles may be widely separated. Thus it becomes necessary to delete and add particles.

The coordinates of an added particle were first set to be the average of those of the two neighbouring particles and were then adjusted by fitting a least-squares parabola to five points, namely the points marking the position of the new particle and the positions of the two nearest particles on either side. The ordinate of the new particle was then adjusted to lie on the parabola.

Particle smoothing was also used if particles became irregularly placed so that they caused flexures in the surface line of such short wavelength that they were unlikely to be resolved in the surface tension computation. If $\overline{w}$ is the average value of the second derivative of y of particles i-1 and i+1, then particle i is smoothed if

$$|y_i'' - \overline{w}| \Big/ (|y_i''| + |\overline{w}|) > 0.4 \ .$$

The smoothing routine used is the same as that for a newly created particle. Detailed descriptions of these routines are presented by Nhan [9].

4. NUMERICAL INSTABILITY

The computational stability of the general MAC method has been discussed by Hirt [10] and Daly and Pracht [4]. Their analyses involve expanding each of the terms in the MAC difference equations in a Taylor series to reveal two effective diffusion coefficients which are required to be positive for stability. Hirt, therefore, proposed the following criteria:

$$\nu > \tfrac{1}{2}\,\delta t\,\overline{u}^2 \qquad (6)$$

$$\nu > \tfrac{1}{2}\,\delta x^2 \frac{\partial \overline{u}}{\partial x} \tag{7}$$

where $\overline{u}$ is the average maximum fluid speed and $\partial\overline{u}/\partial x$ is the average maximum velocity gradient in the direction of flow.

In addition to these conditions, an analysis of the linearized equations leads to the usual conditions on the time increment, δt, namely

$$\delta t \leqslant \delta x^2/4\nu \tag{8}$$

$$\delta t < \delta x/c \tag{9}$$

where c is the maximum fluid speed. The speed of propagation of plane capillary gravity waves is given by

$$u = \left(\frac{2\pi\sigma}{\rho\eta} + \frac{g\eta}{2\pi}\right)^{\frac{1}{2}}$$

where η is the wavelength. Since shorter waves travel fastest, the velocity u is used as an approximation for c in equation (9) when the wavelength η has the smallest value resolvable by the finite difference equation, viz. 2 δx, so that

$$c = \left[\frac{\pi\sigma}{\rho\delta x} + \frac{g\delta x}{\pi}\right]^{\frac{1}{2}}$$

and condition (9) becomes

$$\left[\frac{\pi\sigma}{\rho\delta x} + \frac{g\delta x}{\pi}\right]^{\frac{1}{2}} \delta t < \delta x\ . \tag{10}$$

Foote [8] proposed a similar condition but he neglected gravity; moreover he used the group velocity instead of c to obtain

$$\frac{3}{2}\left(\frac{\pi\sigma}{\rho\delta x}\right)^{\frac{1}{2}} \delta t < \delta x\ .$$

This is more stringent than equation (10) but was found unnecessary for the present work.

Equation (10) can be non-dimensionalized to obtain

$$\delta t < \left(\frac{\delta x^3}{\pi A + \delta x^2/\pi}\right)^{\frac{1}{2}}\ .$$

In most cases, δx^2 is much smaller than π so that $\pi A >> \delta x^2/\pi$ and one can simplify the restriction to

$$\delta t < \left(\frac{\delta x^3}{\pi A}\right)^{\frac{1}{2}} \tag{11}$$

It is necessary to observe strictly conditions (8) and (11). Within the range of values of ν used in this study, their relative importance depends on the surface tension coefficient A. When A is less than unity, condition (8) is more stringent than condition (11). Conversely, if A is much bigger than unity, then the right-hand side of (11) varies inversely with the square root of A and this condition becomes more restrictive.

The situation with respect to stability conditions (6) and (7) is not clear. These state that the numerical scheme will be unstable for an inviscid flow problem but our test run with zero surface tension and zero viscosity posed no difficulty at all. However when the surface tension was included, the transient solution behaved well up to the time when the bubble started necking rapidly prior to detachment. At that moment high velocities developed near the neck and the free surface describing the neck became extremely irregular and physically unacceptable. This instability persisted even if the time step was reduced to as small as 1/50th of the original time step. The large velocity magnitude and gradient in the instability region suggest that this could be due to violation of either (6) or (7) but insensitivity to the time step narrows our attention to the nonlinear stability condition (7) which is not dependent on the time step. Indeed, better results were obtained with identical parameters except for a higher viscosity and the calculation was able to proceed to a much later time until instability eventually developed. These facts lend support to Foote's explanation [8] that the instability comes from the initial surface noise originating in the MAC approximations of the free surface boundary conditions. Without the surface tension contribution, the surface noise remains bounded and a complete solution is attainable. On the other hand, the inclusion of the surface tension stress to the free surface boundary condition, when coupled with the large cell-to-cell velocity fluctuations, provides the exponential amplification causing termination of the run. The better results obtained with larger viscosities are an outcome of the greater amount of cell-to-cell velocity smoothing, which tends to keep the surface particles in regular alignment and thus reduces the presence of spurious surface curvatures.

5. THE EFFECT OF SHEAR INFLOW

Figure 1 shows particle plots for a bubble growing in a uniform shear flow. As the liquid near the wall approaches the bubble, it impinges on the left boundary of the bubble and changes its direction to flow downward along the bubble

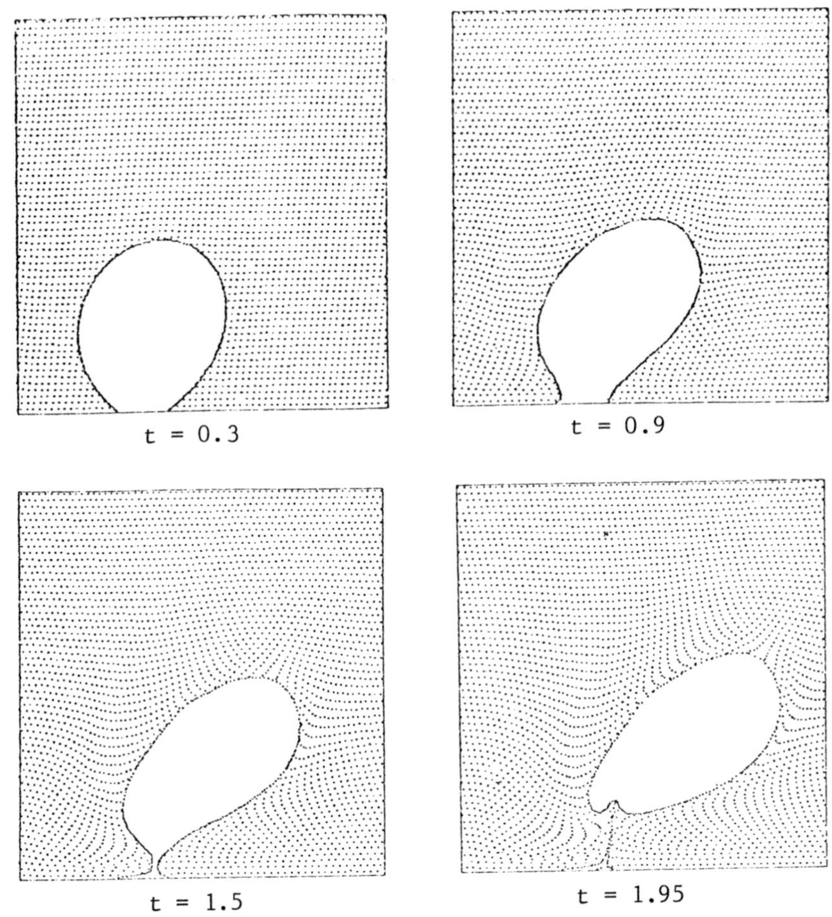

Figure 1 Particle plots showing successive stages of bubble formation and departure.

boundary and toward the bubble neck. On the downstream side of the bubble, the nearby liquid is drawn toward the neck as the bubble rises toward detachment. The oncoming shear flow gradually distorts the bubble shape, shifting its upper part sideways in the direction of flow while the bubble base remains attached to the orifice. At the same time, the bubble elongates in the vertical direction due to buoyancy; necking starts at time $t = 0.9$ and develops quite quickly up to the moment of detachment, $t = 1.5$. As the bubble leaves the surface, the elongated portion contracts rapidly into the bubble and the surrounding liquid near its base is drawn up in the wake of the bubble as it follows the liquid stream.

Figure 2 shows the shapes of two bubbles near departure for two different shear inflows. It is evident that the departure time of the bubble depends on the magnitude of the shear inflow such that a higher shear flow forces the bubble

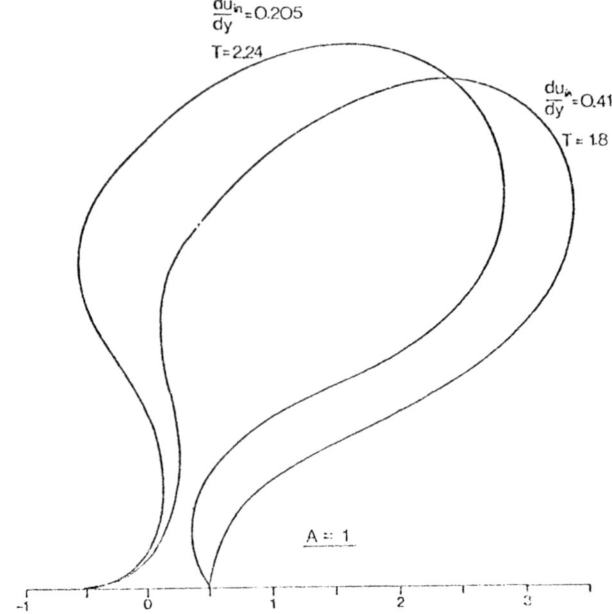

FIGURE 2 Bubble profiles just before detachment at different rates of shear.

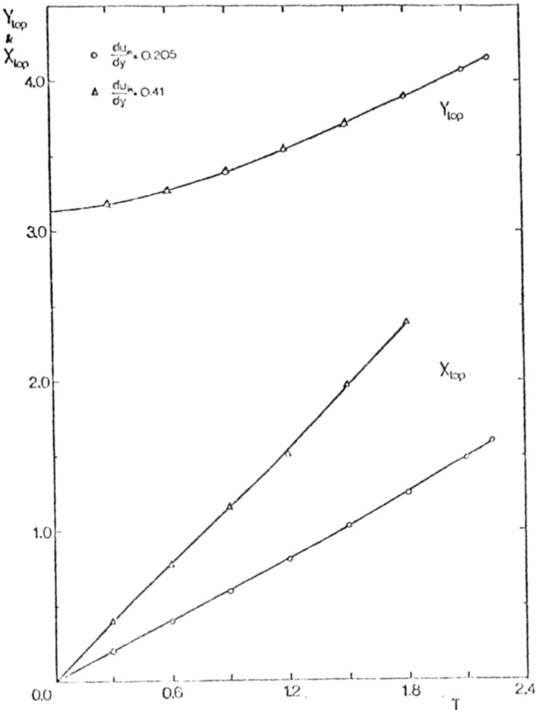

Figure 3 Time histories of coordinates of bubble top at different rates of shear.

to depart at an earlier time by hastening the necking process. Also the bubble is more tilted and narrower in a higher shear field. This is in agreement with Aziz Ul Huq's experiment [11] on air bubbles at a submerged orifice in a flowing liquid. The reduction in bubble departure size connotes a decrease in bubble departure time that could be seen in Aziz Ul Huq's plots of bubble diameter versus time and flow velocity. The vertical displacement of the bubble top, Y_{top}, however, is not significantly affected by the shear flow as shown in Figure 3, while curves of X_{top}, the horizontal displacement of the bubble top, show a similar trend of behaviour but are separated from one another by the shear ratio.

6. THE EFFECT OF SURFACE TENSION

Figure 4 shows the effects on the shape of a bubble of variation of the surface tension coefficient A. Each figure is a plot of bubble profiles at various times during the growing process. All input parameters are constant (δx=δy=0.125, δt = 0.015, ν = 0.1, g = -1 and the shear inflow du_{in}/dy = 0.205) except the surface tension coefficient, which is 0, 1 and 2 for Figures 4a, b and c respectively. In the absence of surface tension (A=0), as the bubble is raised due to buoyancy and shifted sideways by the liquid drag force, there is no force to restrain the development of a convex curvature on the bubble surface; hence it necks quickly near its base and departs with an egg-like shape that inclines toward the direction of fluid flow (Figure 4a). The growth with surface tension is different. While the buoyancy elongates the bubble vertically and thus forces it to form a neck near its base, surface tension always opposes the development of a convex curvature on the bubble boundary and in this way hinders the necking mechanism. At the same time, the liquid drag pushes the bubble sideways. The net effect is that, in the necking region, the downstream surface of the bubble moves more slowly in the direction opposite to the shear flow as the bubble continues to elongate toward detachment (see Figure 4c). Near the point of departure, the bubble is seen to be connected to the orifice by a long neck, which inclines downstream. The higher the surface tension, the longer the neck, the more tilted is the bubble and the longer is the time taken to reach detachment.

Figure 5 shows the curves of the coordinates of the position of the bubble top which indicate that the surface tension has a negligible effect on the bubble height (Y_{top}). On the other hand, curves of X_{top} for various A are clearly different from each other, indicating that the contribution of surface tension to the horizontal displacement of a bubble is more significant. At the same time level, bubbles with small surface tension are displaced further downstream. However, these low surface tension bubbles neck quickly and depart at

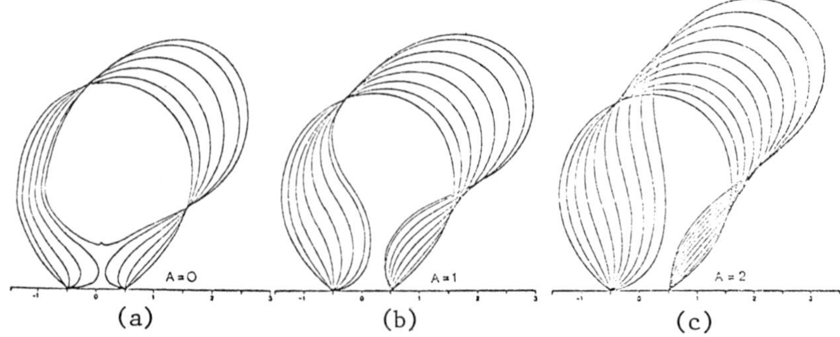

Figure 4 Sequences of bubble profiles for varying surface tensions.

an earlier time while the high surface tension bubbles continue to displace downstream and depart with much more oblique angles.

7. THE EFFECT OF VISCOSITY

Figure 6 shows bubble profiles for two different viscosity coefficients. It can be seen that the bubble shape is very similar at different viscosities, the variation of viscosity coefficient only affecting the bubble neck: the lower the viscosity coefficient, the quicker is the necking process. In addition, the angle of inclination of the bubble to the orifice is the same regardless of viscosity coefficient. This implies that the dynamic drag exerted by the flowing liquid on the bubble is the same. In other words, the dynamic drag is less sensitive to changes in viscosity than is the necking process. When considering necking as a consequence of the Helmholtz hydrodynamic instability, the above statement is in agreement with Lamb's conclusion [12] that the effect of viscosity cannot remove an instability but it does decrease the instability growth rate.

8. CONCLUSION

The use of the MAC technique for the study of barbotage has provided a useful supplement to analytic and experimental investigations. The effect of surface tension has been successfully included by employing a sequence of marker particles along the free surface to track its curvature. However, since the model is two dimensional, the results, while qualitatively plausible, cannot be quantitatively correct. There is a potential difficulty with numerical instability due to the inclusion of the surface tension effect and the following conditions must be satisfied to ensure a stable solution:

$$\delta t \leqslant \frac{\delta x^2}{4\nu} \qquad \text{and} \qquad \delta t < \left(\frac{\delta x^3}{\pi A}\right)^{\frac{1}{2}} .$$

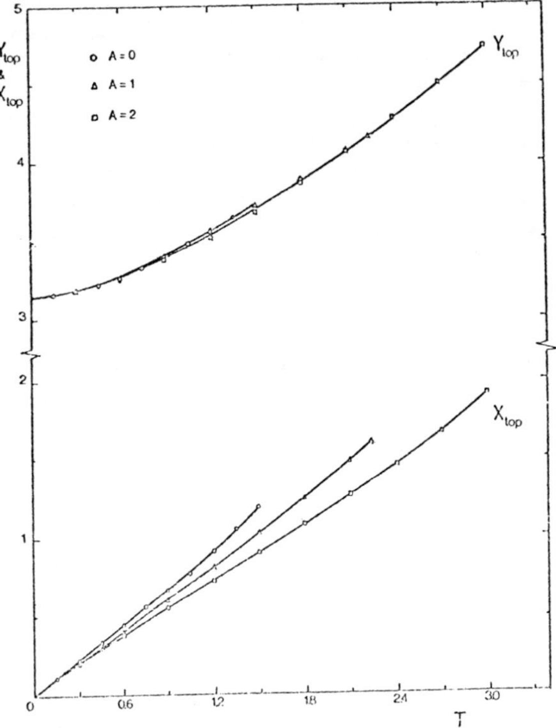

Figure 5 Time histories of coordinates of bubble top at varying surface tension.

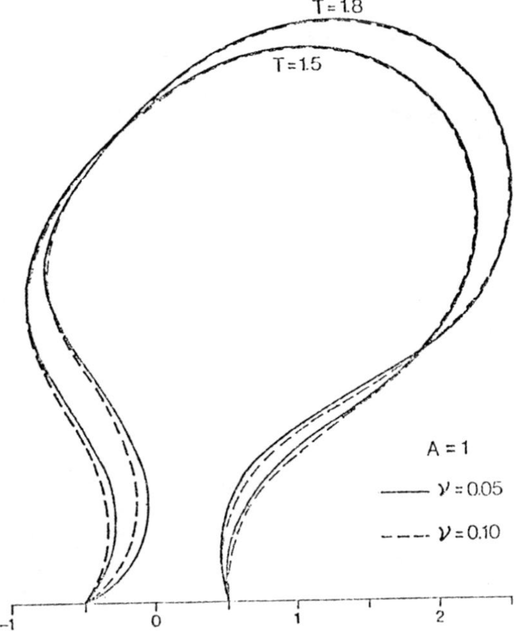

Figure 6 Bubble profiles at varying viscosity.

In a flowing liquid, liquid drag combines with buoyancy to detach the bubble and the resulting bubble shape is tilted downstream; a neck develops near the bubble base before the bubble departs. The higher the magnitude of the shear flow, the faster is the necking process and the shorter is the departure time. A larger shear flow is required to detach a smaller bubble. Increasing the surface tension retards the necking process and lengthens the bubble departure time while lowering viscosity hastens the necking process and brings about an early detachment.

9. REFERENCES

1. HARLOW, F.H. and WELCH, J.E. - Numerical Calculation of Time-Dependent Viscous Incompressible Flow of Fluid with Free Surface. Phys. Fluids, Vol. 8, 12, pp2182-2189, 1965.

2. WELCH, J.E., HARLOW, F.H., SHANNON, J.P. and DALY, B.J. - The MAC Method. A Computing Technique for Solving Viscous Incompressible, Transient Fluid Flow Problems Involving Free Surfaces. Los Alamos Scientific Laboratory, LA-3425, 1965.

3. HARLOW, F.H. and SHANNON, J.P. - The Splash of a Liquid Drop. J. Appl. Physics, Vol. 38, p3855, 1967.

4. DALY, B.J. and PRACHT, W.E. - Numerical Study of Density Current Surges. Phys. Fluids, Vol. 11, p15, 1968.

5. DALY, B.J. - A Technique for Including Surface Tension Effects in Hydrodynamic Calculations. J. Comput. Phys., Vol. 4, p97, 1969.

6. LANGLOIS, W.E. - Slow Viscous Flow, MacMillan Co., N.Y., 1964.

7. WALSH, J.L., AHLBERG, J.H. and WILSON, E.N. - Best Approximation Properties of the Spline Fit. J. Math. Mech., Vol. 11, p225, 1962.

8. FOOTE, G.B. - A Numerical Method for Studying Liquid Drop Behaviour: Simple Oscillation. J. Comput. Phys., Vol. 11, p507, 1973.

9. NHAN, L.G. - A Numerical Study of Bubble Growth. Ph.D. Thesis, University of New South Wales, 1979.

10. HIRT, C.W. - Heuristic Stability Theory for Finite-Difference Equations. J. Comput. Phys., Vol. 2, pp339-355, 1968.

11. AZIZ UL HUQ, A.M. - A Study of Bubble Formation and Departure. Ph.D. Thesis, University of New South Wales, 1972.

12. LAMB, H. - Hydrodynamics, University Press, Cambridge, 1932.

NUMERICAL STUDY ON FLOW BEHAVIOUR AND HEAT TRANSFER IN THE VICINITY OF STARTING POINT OF TRANSPIRATION

S.Kieda, K.Suzuki and T.Sato
Department of Mechanical Engineering
Kyoto University, Kyoto 606, Japan

Summary

A finite-difference scheme has been applied to a transpired boundary layer. Special attention has been paid to a local non-uniform pressure field existing in the vicinity of the starting point of transpiration, by solving the finite-difference analogues of elliptic type of differential equations. A special way adopted in this study of treating the boundary condition at the downstream end is effective to reduce the size of computation domain and enables to calculate the flow and the heat transfer accurately. The non-uniform pressure field affects the skin-friction coefficient remarkably but not so much the heat transfer rate.

1. INTRODUCTION

This paper is concerned with an incompressible, transpired laminar boundary layer on a flat plate. Flow behaviour and heat transfer characteristics in the boundary layer are studied numerically, paying special attention to those in the vicinity of starting point of transpiration.

Related with the flame propagation over oil surface or other combustible stuffs and with the fog formation from river or channel, knowledge of the flow and the heat and mass transfer characteristics in the vicinity of the front edge of evaporating surface is very important. Some basic knowledge may be obtainable through a study on a transpired boundary layer. Transpired boundary layers, both laminar and turbulent, have been studied extensively, but there is found few studies treating the vicinity of the starting point of transpiration.

Sastri and Hartnett [1] studied the heat transfer in a

laminar boundary layer on a permeable flat plate preceded by a non-transpiring part of plate by making use of boundary layer theory. The stream line lying on the wall in the non-transpiring part starts to be lifted up from the wall at the upstream edge of the transpired part of the wall. Stream lines neighbouring the wall would be displaced too, but not so much as the former stream line. Thus, pressure is expected to rise near the wall around the starting point of transpiration x=L, and this in return may decelerate the flow near the wall locally there and even upstream of x=L. Such non-uniformity of pressure and its related distortion of velocity field cannot be accounted well with the boundary layer theory. It may thus be interesting to see how these can affect the heat transfer to the wall.

Hirano and Kanno [2] found in their combustion experiment that the velocity profile has a maximum in a boundary layer with transpiration of combustible gas in the vicinity of x=L. This is deduced to be caused by a pressure non-uniformity possibly existing around x=L. In a flow of variable density with pressure gradient, local acceleration of the flow occurs most sensibly in low density region [3] so that a remarkable velocity over-shoot may have been observed in their experiment. This also might be a cause of velocity over-shoot found at the location further downstream [6]. Smith and Stewartson [4] and Smith [5] studied the flow behaviour of a compressible boundary layer around x=L, applying an elaborate theory called "the triple-deck method", and Napolitano [7] worked out recently some analyses on a compressible laminar boundary layer transpired over a port of finite size employing the similar method. The authors, however, are unaware of any theoretical study on heat and mass transfer in the similar flow situation.

One of the purposes of this paper is to study in detail the heat transfer to the flat plate in the vicinity of the starting point of transpiration. The thermal field as well as the velocity field is calculated by solving numerically the governing equations of elliptic type of differential equations. Heat transfer results to be obtained are compared with the counterparts obtained from numerical computation of boundary layer equations. The second purpose of this study is to show the capability of finite difference scheme in handling such a problem like the present one. Success of computation may prove the effectiveness of the present way [8] of treating the boundary condition at the downstream end of the computation domain. The scheme adopted presently may easily be extended to a form applicable to the more practical but complicated problems cited before.

2.COMPUTATIONAL PROCEDURE

The Navier-Stokes equation and energy equation may be written in the following forms in terms of the stream function ψ, the vorticity ζ and the enthalpy h for subsonic, two-dimensional laminar flow of constant properties.

$$\frac{\partial^2\psi}{\partial x^2} + \frac{\partial^2\psi}{\partial y^2} = -\rho\zeta \tag{1}$$

$$\frac{\partial}{\partial x}(\zeta\frac{\partial\psi}{\partial y}) - \frac{\partial}{\partial y}(\zeta\frac{\partial\psi}{\partial x}) = \mu(\frac{\partial^2\zeta}{\partial x^2} + \frac{\partial^2\zeta}{\partial y^2}) \tag{2}$$

$$\frac{\partial}{\partial x}(h\frac{\partial\psi}{\partial y}) - \frac{\partial}{\partial y}(h\frac{\partial\psi}{\partial x}) = \frac{\lambda}{C_p}(\frac{\partial^2 h}{\partial x^2} + \frac{\partial^2 h}{\partial y^2}) \tag{3}$$

where x and y are the streamwise and normal coordinates, and the properties ρ, μ, λ and C_p the density, the viscosity, the thermal conductivity and the specific heat at constant pressure of the fluid respectively. The origin of the coordinate system is located at the leading edge of the plate.

The above governing equations in conservation form have been transformed into a set of finite difference equations following the method of Gosman et al [9]. The final form of the finite difference analogues of Eqs.(1) through (3) can be expressed in the following common form.

$$\phi_P = C_E\phi_E + C_W\phi_W + C_N\phi_N + C_S\phi_S + D, \quad (\phi = \psi,\ \zeta \text{ or } h) \tag{4}$$

where the suffixes E, W, N and S denote the grid points east, west, north and south of a grid point P. The coefficients C's include the contributions of convection and diffusion, and the second upwind scheme [10] has been used for the convection term.

The boundary condition adopted in the present computation are outlined in Figure 1. At the upstream end of computation

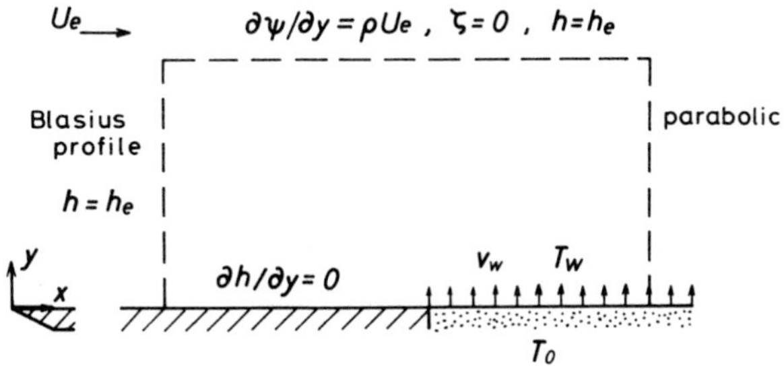

Figure 1. The computation domain and the boundary conditions

domain, the Blasius solution is used to give the inlet conditions for ψ and ζ, and the enthalpy is assumed to have uniform profile. At the free stream boundary, the following relationships are assumed.

$$\frac{\partial \psi}{\partial y} = \rho U_e, \quad \zeta = 0, \quad h = h_e \qquad (5)$$

where U_e is the free stream velocity and h_e the enthalpy assumed at the upstream end mentioned just above. On the wall, the stream function is zero upstream of x=L and it is calculated in the permeable part of the wall from the normal blowing velocity v_w distributing uniformly over $x \geqq L$. The vorticity at the wall is related to the vorticity at the grid point neighbouring the wall, assuming its linear distribution near the wall. For the enthalpy, the wall is treated adiabatic so that the normal gradient of h is set equal to zero at y=0 in the impermeable part. For x>L,

$$\rho C_p v_w (T_w - T_0) - \lambda \left(\frac{\partial T}{\partial y}\right)_{y=0} = 0 \qquad (6)$$

where T_0 is the blown gas temperature in reservoir. At the downstream end of the computation domain, all the governing equations to be solved are assumed to obey the boundary layer approximation. The boundary layer equations for ζ, ψ and h can easily be transformed into corresponding finite difference equations. They are of the following forms [8].

$$-A_j \zeta_{IN,j+1} + \zeta_{IN,j} - B_j \zeta_{IN,j-1} = C_j \qquad (7)$$

$$-\psi_{IN,j+1}/\Delta y^+ + (1/\Delta y^+ + 1/\Delta y^-)\psi_{IN,j} - \psi_{IN,j-1}/\Delta y^-$$
$$= \rho(\Delta y^+ + \Delta y^-)\zeta_{IN,j}/2 \qquad (8)$$

$$-A_j h_{IN,j+1} + h_{IN,j} - B_j h_{IN,j-1} = C_j \qquad (9)$$

$$A_j = (4\Gamma\Delta x/\Delta y^+ + P_j/2)/D_j, \quad B_j = (4\Gamma\Delta x/\Delta y^- - Q_j/2)/D_j$$

$$C_j = [\{2(\psi_{IN-1,j+1} - \psi_{IN-1,j-1}) + (P_j - Q_j)/2\}\zeta_{IN-1,j}$$
$$+ (P_j \zeta_{IN-1,j+1} - Q_j \zeta_{IN-1,j-1})/2]/D_j$$

$$D_j = 2(\psi_{IN,j+1} - \psi_{IN,j-1}) - (P_j - Q_j)/2$$
$$+ 4\Gamma\Delta x(1/\Delta y^+ + 1/\Delta y^-)$$

$$P_j = \psi_{IN,j+1} + \psi_{IN,j} - \psi_{IN-1,j+1} - \psi_{IN-1,j}$$

$$Q_j = \psi_{IN,j} + \psi_{IN,j-1} - \psi_{IN-1,j} - \psi_{IN-1,j-1}$$

$$\Delta y^+ = y_{j+1} - y_j, \quad \Delta y^- = y_j - y_{j-1}, \quad \Delta x = x_{IN} - x_{IN-1}$$

$$\Gamma = \mu \text{ for Eq.(7)}, \quad \lambda/C_p \text{ for Eq.(9)}$$

where IN specifies the grid line at the downstream end, j the j-th grid point on a normal grid line and JN the grid line at the free stream boundary. The calculated values of $\psi_{IN,j}$, $\zeta_{IN,j}$ and $h_{IN,j}$ from Eqs.(7) through (9) are used as the boundary condition at the downstream end in the next step of iterative computation procedure of Eq.(4). This method has successfully been used in calculations of a flow around a flat plate of finite length at intermediate Reynolds number [11] and the heat transfer problem in the inlet region of a circular tube [8,12]. Iterative procedure is carried out by using a successive substitution method [9].

The computation is carried out in a domain specified below.

$$0.6 \leqq x = \frac{x}{L} \leqq 1.25, \quad 0 \leqq y = \frac{y\sqrt{R_e}}{2L} \leqq 20$$

To check the reasonableness of the free stream end location, some preliminary calculations were performed with different maximum values of $\tilde{y}$, but no remarkable difference has been found among the results obtained. Thus, the size of the computation domain is believed to be large enough. In the above domain, grid points of totally 3600 (60×60) are allocated. The minimum grid space is $10^{-3}L$ both streamwisely and normally. The streamwise grid space is arranged finest around x=L, and the normal grid space is made fine in the wall region.

For specification of flow condition, blowing parameter B_0 is defined as follows.

$$B_0 = \frac{v_w}{U_e}\frac{2}{C_{f0}} \tag{10}$$

where C_{f0} is the skin friction coefficient expected at x=L when $v_w=0$ but at the same Reynolds number $R_e=(\rho U_e L/\mu)$. Then, $C_{f0}=0.664/\sqrt{R_e}$. The present calculations are mainly performed at two different values of R_e, namely at $R_e=2.5\times10^3$ and 10^4, and at two values of $B_0=0.181$ and 1.81. Skin friction coefficient C_f and Nusselt number N_u to be obtained are plotted against x/L in forms multiplied by $\sqrt{Re_x}$, where $Re_x=\rho U_e x/\mu$. Although any fluid can be taken as the working fluid, air at normal condition (Prandtl number $P_r=0.7$) is adopted as the test fluid in the present calculations.

3.RESULTS AND DISCUSSION

The skin friction coefficient C_f obtained is shown in Figures 2 and 3. For comparison, two other results are also plotted in the figures. One of them is for the case without transpiration $B_0=0$. The skin friction coefficient of this case is given by the Blasius solution and $C_f\sqrt{Re_x}=0.664$. Another one is for the case with transpiration but is calculated numerically downstream of x=L by solving the boundary layer equations, taking the Blasius profile as the inlet condition at x=L for velocity. This particular computation was carried out employing the scheme by Blottner [13].

The value of C_f obtained from Eqs.(1) through (3) starts to decrease from some location well upstream of x=L, especially at larger blowing parameter B_0. Another striking feature of C_f distribution is a narrow but deep dent in the distribution around x=L. In Figure 4 are plotted the isobar contours calculated from the numerical integration of momentum equation.

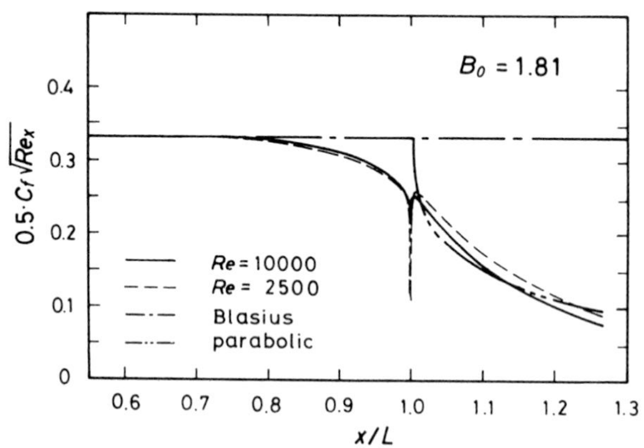

Figure 2. The distribution of C_f.

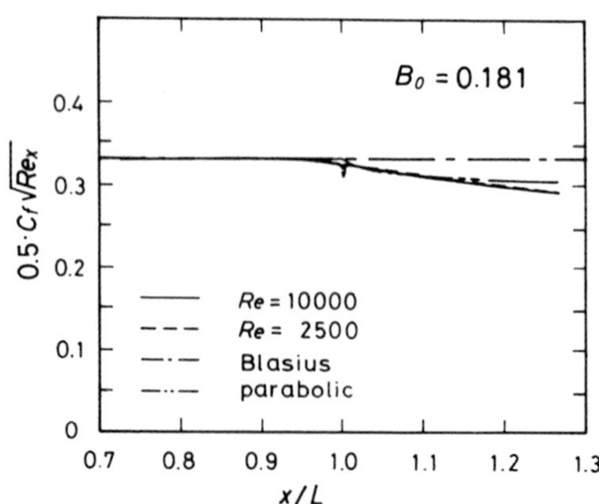

Figure 3. The distribution of C_f.

The pressure level is shown in the form of C_p defined below.

$$C_p = \frac{P - P_e}{\rho U_e^2} \tag{11}$$

where P_e is the pressure at the upstream end of computation domain and the momentum equation was integrated toward the wall from the free stream boundary. This direction of integration has been proved better to reduce the accumulation of error accompanied in the integration process than the direction parallel to the wall [11]. In Figure 5 is replotted the pressure on the wall surface. These two figures show a peaked distribution of pressure around x=L, therefore, a gradual increase of pressure well upstream of x=L and its decrease finally after x=L. This is the non-uniformity of pressure referred to before and explains why C_f starts to decrease from a point at x<L and why a sharp dent exists in the distribution of C_f. These two features cannot be predicted at all with any boundary layer theory. But it may be worthwhile to note

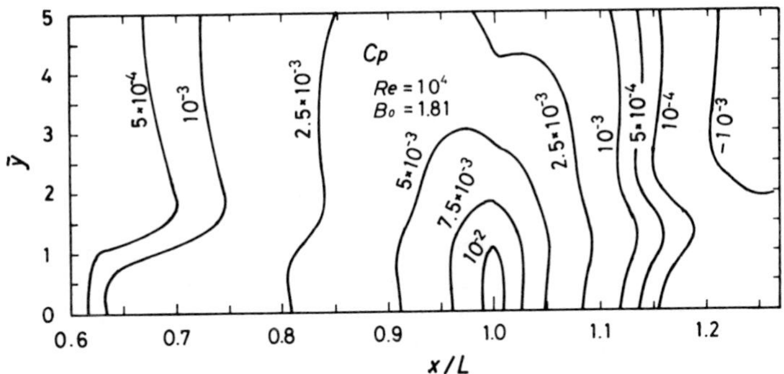

Figure 4. Isobar contours.

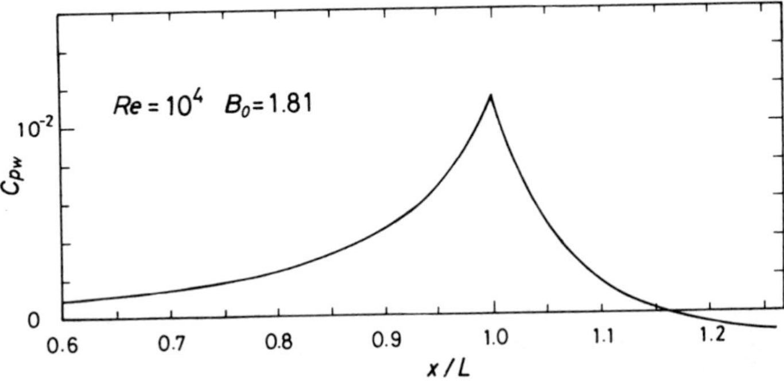

Figure 5. Streamwise distribution of wall pressure

here that the boundary layer approximation is not so bad at x>L except for a small region just downstream of x=L.

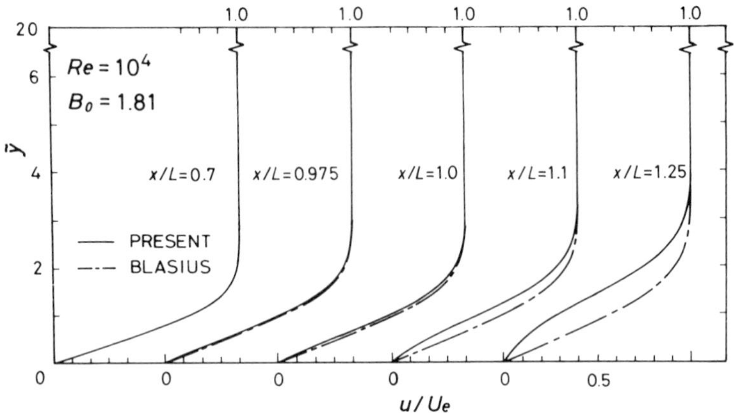

Figure 6. Velocity Profiles at typical positions.

Figure 6 shows velocity profiles at five typical positions. At x/L=0.7, the velocity profile coincides almost exactly with the Blasisus profile. This matches the result of C_f shown in Figure 4. At x/L=0.975, the profile is found to deviate from the Blasius one and the deviation becomes much more conspicuous as x/L increases further.

The deep dent of C_f curve found around x=L in Figure 3 suggests a possibility that, if the blowing rate is increased further, a local separation can occur near x=L. To see if this is actually possible, another calculation has been made at a larger blowing rate B_0=11.6. The Reynolds number has been adjusted to match the experimental condition of Smith [14]; $R_e=2\times10^5$. The computed result of C_f is plotted in Figure 7 and the wall pressure distribution obtained is compared in Figure 8 with Smith's experiment. In Figure 7 the existence of local separation can clearly be noticed just upstream of x=L. For the computation of this type of flow, the present numerical method is effective and any calculation employing the boundary layer equations cannot predict this kind of separation or whatever occurring upstream of x=L. Incidentally, at this blowing condition, the numerical computation of boundary layer equations has predicted the blow-off of the transpired boundary layer from the wall just downstream of x=L and the marching numerical integration of the equations could not go through after that point. In Smith's experiment, the flow is strongly accelerated downstream due to the blowing into a narrow duct, but this experimental situation could not be accounted in the present calculation. It may still be interesting to find a pressure peak around x=L in his result. While his paper does not say about such a local separation found in Figure 7, he has reported unsteady behaviour of the flow around x=L which may have a relation with the local separation.

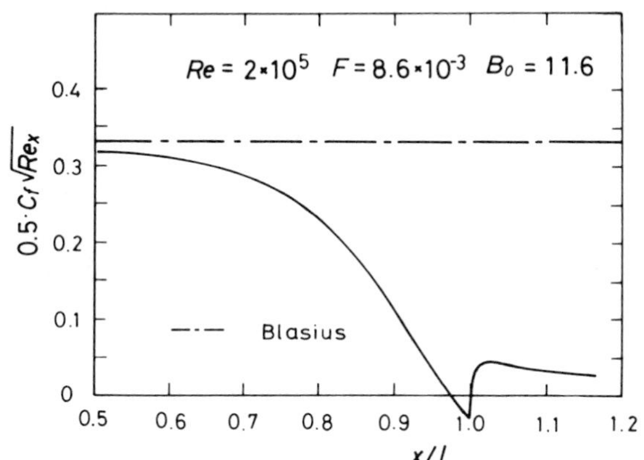

Figure 7. The distribution of C_f at B_0=11.6

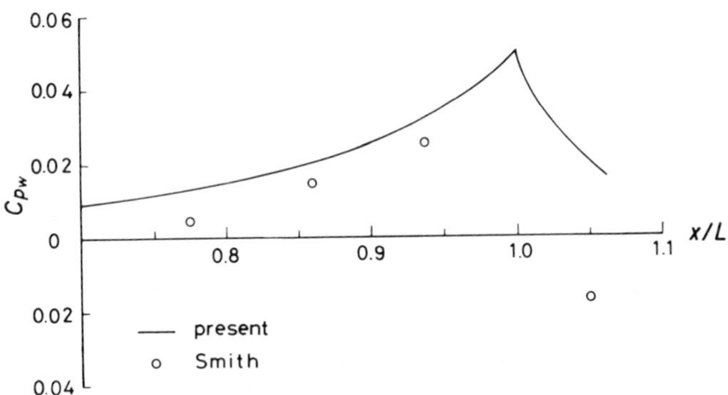

Figure 8. The distribution of wall pressure at B_0=11.6

Figures 9 and 10 compare respectively the stream lines and the equi-vorticity lines with their counterparts obtained from boundary layer theory. Figure 9 shows the displacement of stream lines due to transpiration even upstream of x=L. In connection with this, Figure 10 shows the decrease of vorticity near the wall well upstream of x=L. These results match well the results of C_f in Figure 3 and pressure distribution in Figure 5.

The Nusselt number calculated at two Reynolds number for two values of blowing parameter is plotted in Figure 11 and compared with its counterpart calculated numerically from the

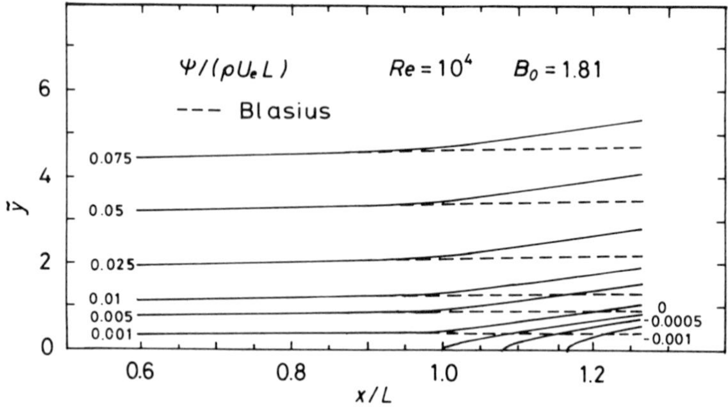

Figure 9. The displacement of stream lines.

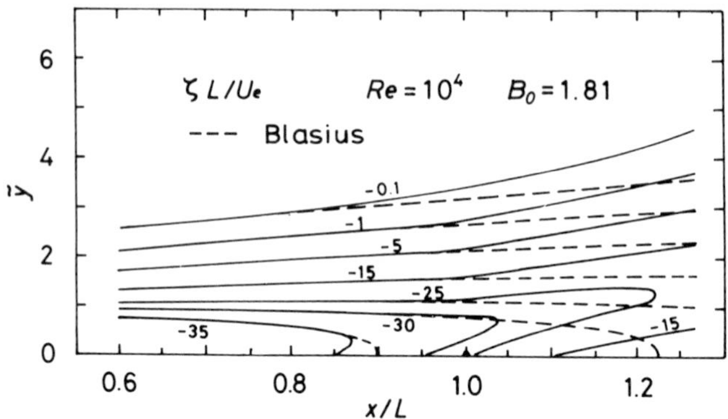

Figure 10. Contours of equi-vorticity lines.

boundary layer equations. In the Figure is also shown an additional result of Nusselt number at $R_e=2\times10^5$ and $B_0=11.6$. The result of boundary layer theory should be infinitely large at x=L, but the Nusselt number calculated from Eqs.(1) through (3) is finite there as expected. Just downstream of x=L, the difference between the both is noticeable, but the difference becomes small after short distance from x=L. In the latter region, the boundary layer theory can give a good approximation of the exactly calculated value with Eqs.(1) through (3).

In Figure 12 is plotted the cooling effectiveness. Cooling of the wall starts to occur upstream of x=L, but is limited only in a small region close to x=L. Moreover, the wall temperature there is different just a little from the free stream temperature. The local non-uniformity of pressure

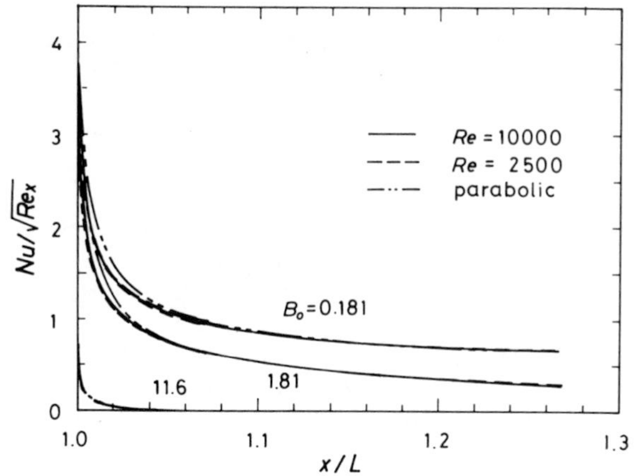

Figure 11. The distribution of Nusselt number

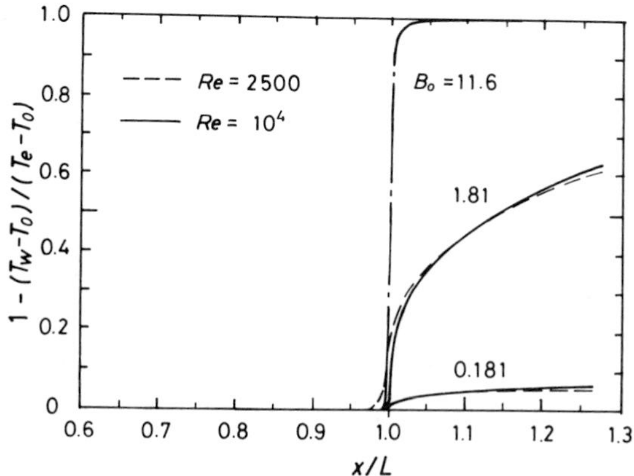

Figure 12. The cooling effectiveness.

and the related distortion of velocity field affect the skin friction coefficient remarkably, but not so much the heat transfer. At B_0=11.6 and $R_e=2\times10^5$, the skin friction coefficient has been found small downstream of x=L. This means the boundary layer there is almost in a blown-off condition. Related with this, the value of Nu is very close to zero and the wall temperature almost coincides with the temperature T_0 after x=L except for a small region just downstream of x=L.

4.CONCLUDING REMARKS

The way of treating the boundary condition at the downstream end adopted in the present study is effective to reduce the size of computation domain so that it enables to calculate the flow and thermal fields accurately around the starting point of transpiration x=L. The results obtained show that the non-uniform pressure field occurs around x=L. This can affet the skin friction coefficient remarkably but not so much the heat transfer.

5.REFERENCES

[1] SASTRI,V.M.K. and HARTNETT,J.P. Effect of an Unheated Solid Starting Length on Skin Friction and Heat Transfer in a Transpired Boundary Layer, Progr. Heat and Mass Transfer, Vol.2, Pergamon Press, pp. 213-223, 1969.

[2] HIRANO,T. and KANNO,Y., Aerodynamic and Thermal Structures of the Laminar Boundary Layer over a Flat Plate with a Diffusion Flame, 14th Symp. on Combustion, The Combustion Institute, pp.391-398, 1973.

[3] JONES,J.W. *et al*, A Turbulent Boundary Layer with Mass Addition, Combustion and Pressure Gradients, AIAA Journal, Vol.9, No.9, pp.1762-1768, 1971.

[4] SMITH,F.T. and STEWARTSON,K., On slot Injection into a Supersonic Laminar Boundary Layer, Proc. Roy. Soc. London, Ser.A, Vol.332, pp.1-22, 1973.

[5] SMITH,F.T., Boundary Layer Flow near a Discontinuity in Wall Conditions, J. Inst. Math. Appl., Vol.13, pp.127-145, 1973.

[6] SENDA,M. *et al*, Study on Turbulent Boundary Layer with Injection and Combustion, Mem. Fac. Engng Kyoto Univ., Vol.38, No.1, pp.21-36, 1976.

[7] NAPOLITANO,M., Numerical Study of Strong Slot Injection into a Supersonic Laminar Boundary Layer, AIAA Journal, Vol.18, No.1, pp.72-77, 1980.

[8] SUZUKI,K. *et al*, Numerical Study of Combined Convective Heat Transfer with Variable Fluid Properties in the Inlet Region of a Circular Tube, Proc. Int. Conf. Numerical Methods in Thermal Problems, 1981. (to appear)

[9] GOSMAN.A.D. *et al*, Heat and Mass Transfer in Recirculating Flows, Academic Press, 1969.

[10] ROACH,P.J., Computational Fluid Dynamics, Hermosa Publ., 1972.

[11] KIEDA,S. and SUZUKI,K., Numerical Study of the Flow passing a Flat Plate of Finite Length, Trans.JSME, Vol.46, No.409, pp.1655-1661, 1980. (in Japanese)

[12] KIEDA,S. *et al*, Forced-Natural Combined Heat Transfer in the Inlet Region of a Vertical Pipe, 17th National Heat Transfer Symposium of Japan, pp.19-24, 1980. (in Japanese)

[13] BLOTTNER,F.G., Investigation of Some Finite-Difference Techniques for Solving the Boundary Layer Equation, Comp. Meth. Appl. Mech. Engng, Vol.6, No.1, pp.1-30, 1975.

[14] SMITH,F.T., On Strong Blowing into an Incompressible Air-stream, J. Fluid Mech., Vol.60, Part 2, pp.241-255, 1973.

AN APPLICATION OF THE METHOD OF CHARACTERISTICS FOR THE SIMULATION OF SINGLE- AND TWO-PHASE FLOW

Timo Siikonen

Technical Research Centre of Finland
Nuclear Engineering Laboratory

SUMMARY

Several finete difference methods for the solution of the flow equations in a characteristic form are studied. The methods are compared by calculating two common test cases. The first one concerns pressure pulse propagation and the second one critical flow rate from a reservoir. Because the discretization in a cracteristic form does not conserve mass the errors in the mass balance calculations are also compared.

1. INTRODUCTION

The method of characteristics (MOC) is a widely used technique for solving one-dimensional flow problems [1-3]. Among the advantages of the method are the simplicity in general and the flexible treatment of the boundary conditions.

The method can be applied principally in two ways: using either a "wave tracing" method [1] with an unregular computational grid, or a fixed computational grid. The first method is accurate but time-consuming and difficult to apply to geometrically complex problems. The latter method is computationally more flexible and extensively used in computer codes. Linear interpolation is usually utilized between the mesh points to find out the values of pressure, velocity etc. where the characteristics and the computational grid intersect on the z-t space. The interpolation can be done either explicitly [2] or implicitly [1,3].

In computer codes based on MOC with explicit interpolation, the interpolated values are usually calculated in one subroutine. This makes it very easy to use higher-order interpolation schemes to improve the accuracy of the code. By changing the interpolation routine the numerical method may be changed to a more suitable one for the actual problem, for example to Leap-frog, Lax or Lax-Wendroff- type methods [4].

The purpose of the present study is to find the most suitable interpolation formulas for an existing computer code [2]. The interpolation methods are compared by calculating two well- known problems in reactor technology. The first problem concerns pressure pulse propagation after a rapid closure of a valve (A comparison of several numerical methods for problems of this type can also be found in ref. [5]). In the second problem the critical two-phase flow rate from a constant pressure reservoir is calculated.

2. METHOD OF CHARACTERISTICS

The mass, momentum and energy conservation laws for one dimensional two-phase flow may be written in the following form [3]

$$\frac{\partial F(U)}{\partial t} + \frac{\partial G(U)}{\partial z} = b \tag{1}$$

where the size and form of the vectors G and F depend on the approximations used. In the complete one-dimensional two-fluid model [8] there are six conservation equations, three for each phase (gas and liquid). Assuming equal velocities and thermodynamic equilibrium the number of equations is reduced to three (homogeneous model).

After defining the vector of the dependent variables as U, the equation (1) can be changed by differentation into the primitive form

$$J_1\frac{\partial U}{\partial t} + J_2\frac{\partial U}{\partial z} = c \tag{2}$$

where the matrices J_1 and J_2 are the Jacobians, with respect to the variables in U, of G and F respectively. Equation (2) can be rewritten as

$$\frac{\partial U}{\partial t} + A\,\frac{\partial U}{\partial z} = d \tag{3}$$

The choice of the dependent variables can be done in numerous ways. An extensive collection of homogeneous flow models can be found in ref. [4]. In the present study the vector U is either

$$U^T = [u,p,h] \tag{4}$$

for the homogeneous model or

$$U^T = [u,p,\rho,h_\ell,h_g] \tag{5}$$

for the five-equation (non-equilibrium) two-phase flow model. Above p stands for pressure, u for velocity, ρ for density, h_ℓ for liquid enthalpy and h_g for gas enthalpy.

The choice of vectors (4) and (5) results in the simplest alternative for the characteristic equations. For example the use of mass-flux, ρu, instead of velocity would perhaps be preferrable in steady state critical flow calculations, but the resulting equations would be much more complicated.

The characteristic form can be obtained by pre-multiplying eq. (3) by the left eigenvector ℓ_i of matrix A corresponding to the eigenvalue λ_i:

$$\ell_i\left(\frac{\partial U}{\partial t} + A\frac{\partial U}{\partial z}\right) = \ell_i d = e_i \tag{6}$$

which, since by definition

$$\ell_i A = \lambda_i \ell_i \tag{7}$$

may be written as

$$\ell_i\left(\frac{\partial U}{\partial t} + \lambda_i\frac{\partial U}{\partial z}\right) = e_i \tag{8}$$

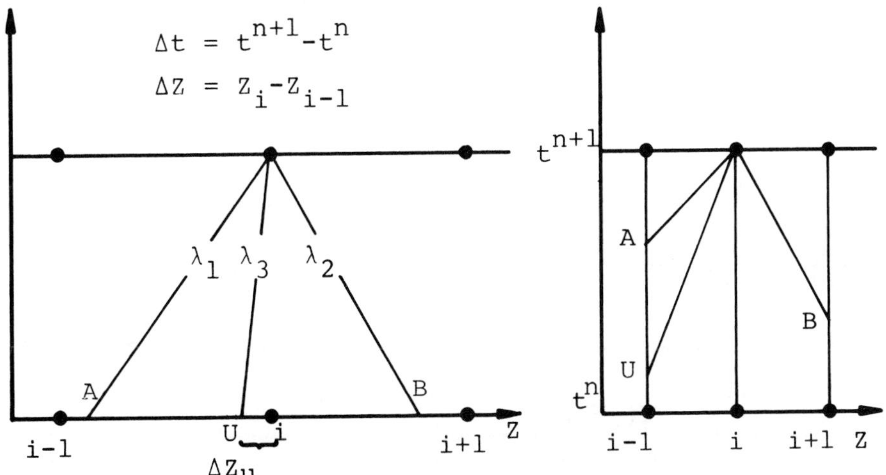

Fig. 1. Characteristics in case of explicit and implicit discretization.

For the homogeneous flow model the eigenvalues are u and $u \pm c$, where c is the velocity of sound in the mixture, and for the five equation model $u \pm c$ and u (three times). The bracketed term in eq. (8) is the directional derivative, where U is differentiated in the direction of the ith characteristic. Eq. 8 can be written as

$$\ell_i dU = e_i dt, \text{ along } dz = \lambda_i dt \tag{9}$$

For a homogeneous flow model there are three equations corresponding to three eigenvalues and for a non-equilibrium model five equations, which are integrated along their characteristics (see Fig. 1). Because the intersections of the characteristics and the previous time level are outside the fixed mesh points, some approximative technique (see chapter 3) has to be applied in order to find out the values of the dependent variables at the points A, B and U of Fig. 1.

The discretized characteristic equations corresponding to eq. (9) for the homogeneous flow (no heat addition) are

$$u_i^{n+1} - u_A + \frac{1}{\rho c}(P_i^{n+1} - P_A) = e_1 dt \text{ along } dz = (u+c)dt \quad (10)$$

$$u_i^{n+1} - u_B - \frac{1}{\rho c}(P_i^{n+1} - P_B) = e_2 dt \text{ along } dz = (u-c)dt \quad (11)$$

$$h_i^{n+1} - h_U - \frac{1}{\rho}(P_i^{n+1} - P_U) = 0 \quad \text{along } dz = udt \quad (12)$$

For frictionless flow also $e_1 = e_2 = 0$.

When the non-equilibrium model is used, e_1 and e_2 are functions of wall friction and evaporation (or condensation) and eq. (12) is replaced by three equations

$$\rho_i^{n+1} - \rho_U - \frac{1}{c^2}(P_i^{n+1} - P_U) = e_3 dt \quad (13)$$

$$h_{\ell i}^{n+1} - h_{\ell U} - \frac{1}{\rho_\ell}(P_i^{n+1} - P_U) = e_4 dt \quad (14)$$

$$h_{gi}^{n+1} - h_{gU} - \frac{1}{\rho_g}(P_i^{n+1} - P_U) = e_5 dt \quad (15)$$

where e_3-e_5 take into accout the effect of phase change.

For the solution of Eqs (10)-(12) or (10), (11) and (13)-(15) one still needs to specify the equation of state and for the five-equation model mass transfer laws, which in turn, determine the right-hand sizes of the characteristic equations (10-15).

Among the advantages of the characteristic presentation is the simplicity of eqs. (11)-(15) in comparison with the original equations of the conservation law form and the treatment of the boundary conditions. At the pipe boundaries the equations used correspond to inward pointing characteristics and the number of boundary conditions to be specified is the same as the number of outward pointing characteristics. Furthermore, the use of a separate interpolation subroutine makes it possible to easily modify the numerical sheme. The drawbacks of the characteristic method using present dependent variables are the sensitiveness for the changes of the eigenvalues and the fact that mass and energy are not conserved.

3. INTERPOLATION METHODS

The unknown values of the dependent variables in eqs. (10)-(15) at the points A, B and U are obtained by "interpolation methods". After a suitable choice of the interpolation formula some well-known difference methods can be used:

1) LINEAR INTERPOLATION. Let the distances of the points A, B and U from the point i be defined as:

$$\Delta Z_A = -(u_i^n + c_i^n)\Delta t \tag{16}$$

$$\Delta Z_B = -(u_i^n - c_i^n)\Delta t \tag{17}$$

$$\Delta Z_u = -u_i^n \Delta t \tag{18}$$

Applying linear interpolation between the points i+1 and i, and i and i-1 the following equations are obtained for the dependent variables U

$$U_A = U_i^n + \frac{\Delta Z_A}{\Delta Z}(U_i^n - U_{i-1}^n) \tag{19}$$

$$U_B = U_i^n + \frac{\Delta Z_B}{\Delta Z}(U_{i+1}^n - U_i^n) \tag{20}$$

$$U_U = U_i^n + \frac{\Delta Z_u}{\Delta Z}(U_i^n - U_{i-1}^n) \text{ if } u_i^n > 0 \tag{21}$$

$$U_U = U_i^n + \frac{\Delta Z_u}{\Delta Z}(U_{i+1}^n - U_i^n) \text{ if } u_i^n < 0 \tag{22}$$

This method forms the basis of several computer codes using the characteristic equations. Unfortunately its accuracy is of the first order only, which is insufficient for many flow problems.

2) QUADRATIC INTERPOLATION (Leith's method [7]). If a quadratic polynomical fit is used to interpolate over (i-1), (i) and (i+1), the following equation for point A is obtained.

$$U_A = U_i + \frac{\Delta Z_A}{\Delta Z}(U_{i+1} - U_{i-1}) + \frac{1}{2}\left(\frac{\Delta Z_A}{\Delta Z}\right)^2(U_{i+1} - 2U_i + U_{i-1}) \tag{23}$$

where the superscripts are dropped for simplicity. The corresponding equations are obtained for points B and U.

3) LAX'S METHOD. If the differences in eqs (19)-(22) are replaced by the central differences (between the points (i-1) and (i+1)), the method is unstable. If the following approximation is used for U_i, the method of Lax is obtained.

$$U_A = \frac{1}{2}(U_{i+1} + U_{i-1}) + \frac{\Delta Z_A}{\Delta Z}(U_{i+1} - U_{i-1}) \tag{24}$$

4) LEAP-FROG METHOD is obtained when

$$U_A = U_i^{n-1} + \frac{\Delta Z_A}{\Delta Z}(U_{i+1}^n - U_{i-1}^n) \tag{25}$$

In this case also the vector e (eqs. 11-15) must be multiplied by 2dt instead of dt.

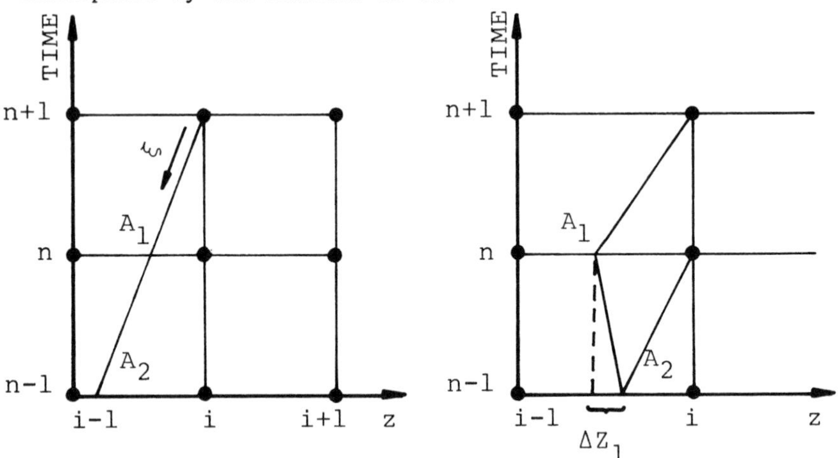

Fig. 2. Illustration of methods 5 and 6.

5) Let ξ be the independent variable in the direction of the characteristic (ξ=0 at the point (i, n+1)). Assuming quadratic fit for U we obtain

$$U(x) = U_i^{n+1} + C_1\xi + C_2\xi^2 \tag{26}$$

Thus the derivative (dU)/(dξ) corresponding to eq. 9 at the point (i, n+1) is C_1. The characteristic intersects the grid lines at the points A_1 and A_2, where $U=U_{A1}$ and $U=U_{A2}$ respectively. Thus we obtain two equations from which coefficients C_1 and C_2 can be estimated. The values U_{A1} and U_{A2} can be obtained by linear interpolation (eqs. 19-22). In the case of Fig. 2

$$U_A = \frac{4}{3}U_{A1} - \frac{1}{3}U_{A2} \tag{27}$$

Here the vector e has to be multiplied by (2/3)dt instead of dt. If the characteristic intersects the grid lines between the points (i-1, n-1) and (i-1,n) the resulting equation is more complicated than eq. (27).

6) In this method points A, B and U are treated as auxiliary mesh points between the points (i-1) and (i) (or between the points (i) and (i+1). Central differences are used in the primitive form as

$$U_{A1} = U_{A2} - \frac{\Delta t}{\Delta Z}\{\theta A^{n-1}_{i-\frac{1}{2}}(U^{n-1}_i - U^{n-1}_{i-1}) + (1-\theta)A^n_{i-\frac{1}{2}}(U^n_i - U^n_{i-1})\}$$
$$- \frac{\Delta Z_1}{\Delta Z}\{\theta(U^{n-1}_i - U^{n-1}_{i-1}) + (1-\theta)(U^n_i - U^n_{i-1})\} \tag{28}$$

where θ varies between 0 and 1.

It is also possible to use the conservation law form in this discretization and try to combine mass- and energy conservation with the method of characteristics. The drawbacks of this method are the sensitiveness for sudden disturbances in values of λ_i, and the need for more computer memory.

7) LAX-WENDROFF-type methods. In these methods the interpolation is performed using auxiliary mesh points as

$$U_A = U^n_i + \frac{\Delta Z_A}{\Delta Z}(U^{n+\frac{1}{2}}_{n+\frac{1}{2}} - U^{n+\frac{1}{2}}_{i-\frac{1}{2}}) \tag{29}$$

In the original Lax-Wendroff-method auxiliary values are obtained from (method 7A)

$$U^{n+\frac{1}{2}}_{i-\frac{1}{2}} = \frac{1}{2}(U^n_i + U^n_{i-1}) - \frac{\Delta t}{2\Delta t} A^n_{i-\frac{1}{2}}(U^n_i - U^n_{i-1}) \tag{30}$$

This method is the same as method 2. Eq. (30) can be replaced by leapfrog approximation for the mid points (method 7B)

$$U^{n+\frac{1}{2}}_{i-\frac{1}{2}} = U^{n-\frac{1}{2}}_{i+\frac{1}{2}} - \frac{\Delta t}{\Delta Z} A^n_{i-\frac{1}{2}}(U^n_i - U^n_{i-1}) \tag{31}$$

Winters [7] suggested two schemes, in the first of which eq. (29) is replaced by (method 7C)

$$U_A = U^n_i + \frac{\Delta Z_A}{\Delta Z}(U^{n+1}_{i+\frac{1}{2}} - U^{n+1}_{i-\frac{1}{2}}) \tag{32}$$

and eq. (30) by

$$U^{n+1}_{i-\frac{1}{2}} = \frac{1}{2}(U^n_i + U^n_{i-1}) - \frac{\Delta t}{\Delta Z} A^n_{i-\frac{1}{2}}(U^n_i - U^n_{i-1}) \tag{33}$$

In the other scheme eq. (33) is used and eq. (32) is replaced by (method 7D)

$$U_A = \frac{1}{2}(U^n_{i+\frac{1}{2}} + U^n_{i-\frac{1}{2}}) + \frac{\Delta Z_A}{\Delta Z}(U^{n+1}_{i+\frac{1}{2}} - U^{n+1}_{i-\frac{1}{2}}) \tag{34}$$

The matrix A is evaluated as

$$A_{i+\frac{1}{2}} = (A_i + A_{i+1})*.5 \tag{35}$$

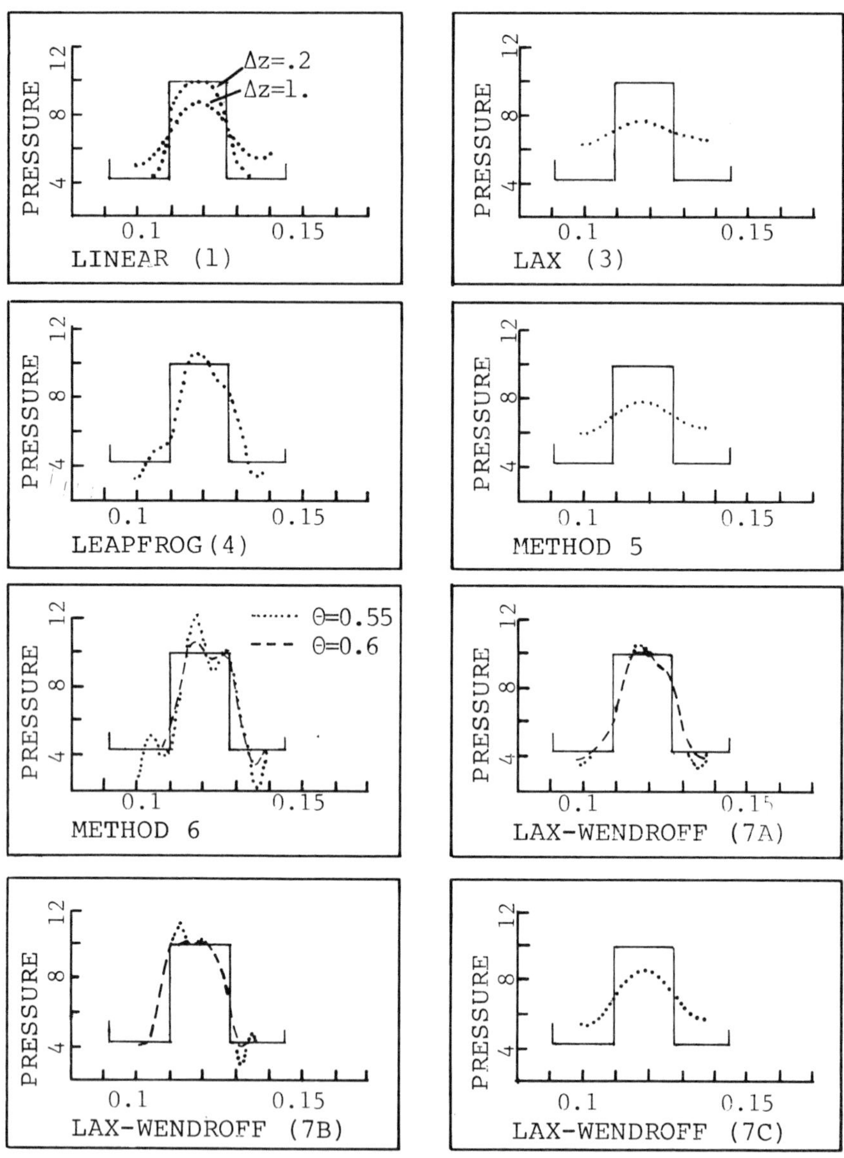

Figs. 3-10. Pressure histories at the valve between 0.1 and 0.15 secs after the valve closure. ····· undamped solution, ----- damped solution.

Hence no tedious calculations are needed for $A_{i+\frac{1}{2}}$. It is also possible to try to use the conservation law form for equations (30), (31) and (33) and to improve the conservation of mass and energy as in method no. 6. In order to maintain the second-order accuracy for Lax-Wendroff-type methods some extra boundary conditions are needed at the pipe boundaries [6]. However, because the use of different boundary conditions didn't improve the results significantly in the calculated test cases, the boundaries were finally calculated using linear interpolation only.

4. COMPARISON BETWEEN DIFFERENT INTERPOLATION METHODS

4.1 Pulse propagation

The ability of the methods to handle fast transients is tested by analyzing the hydraulic response of an instantaneous closure of a valve at the end of a 10 m long pipe with flow area of 0.01 m^2. The other end is connected to a reservoir of constant 7MPa pressure. The initial velocity, density and sound velocity are assumed to be 3.5 m/s, 750 kg/m^3 and 1100 m/s. The enthalpy is assumed to be constant. Consequently only eqs. (10) and (11) are needed to predict the transient. The time step was kept at 0.5 ms and 11 axial nodes (Δz=1 m) were used, which results in a value of .55 for the ratio $\lambda_{max}\Delta t/\Delta z$. The valve closure and the reflection of the pressure pulse from the reservoir result in a square-wave pressure oscillation at the valve. In Figs. 3-10 the pressures at the valve as calculated by different methods are compared. The second-order methods with 11 nodes are almost as accurate as the first order methods with 51 nodes, but produce oscillations around the actual solution. These oscillations can be damped either by adding a second derivative term of

$$\varepsilon \frac{c^2\Delta t^2}{\Delta z^2} (U_{i+1} - 2U_i + U_{i-1}) \tag{36}$$

where = 0.05 was used, to eqs. (10) and (11), or by increasing θ (in method 6). To insure stability for method 6, $\theta >$ 0.5 had to be used. Since method 7d has a stability criterion of $\Delta t \leq .5\lambda_{max}/\Delta z$, and it is also very dissipative, it was not included in this comparison.

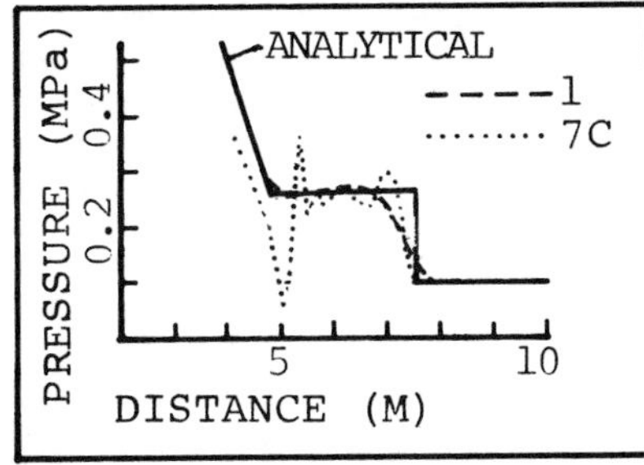

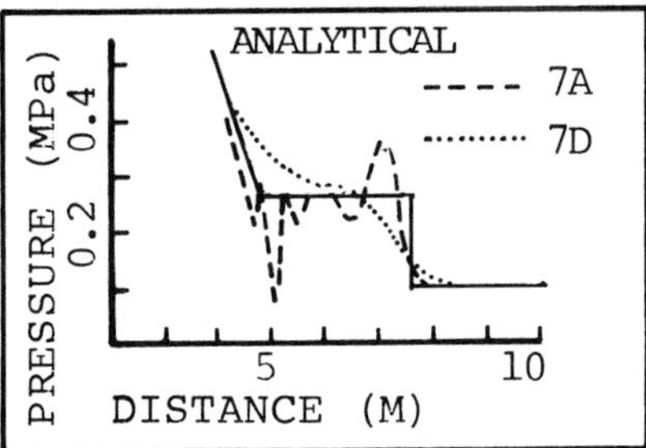

Figs. 11-12. Comparison of pressure distributions at the time 5 ms. Shock tube example.

Eq. (12) was included in the analysis of the shock tube [7].

The test case is for a 10 m long shock tube containing ideal gas with a uniform enthalpy of 2.6 10^5 J/kg and specific heat raito of 1.4. At time zero a membrane at the centre of the tube separated the high pressure gas on the left (791 kPa) from the low pressure gas on the right (101 kPa). The value of 0.25 was used for the ratio of λ_{imax} $\Delta t/\Delta z$ and the section length Δz was 0.2 m. Figs. 11-12 show comparison of the methods, which were able to calculate this problem, at the time 5 ms into the transient.

4.2 Critical Flow Example

The nonequilibrium model was used to calculate the critical flow (u=c at the valve location) out from the vessel described in chapt. 4.1. The vessel and the pipe are initially filled with saturated water at the pressure of 7 MPa. The transient was calculated until a steady flow was established in the pipe.

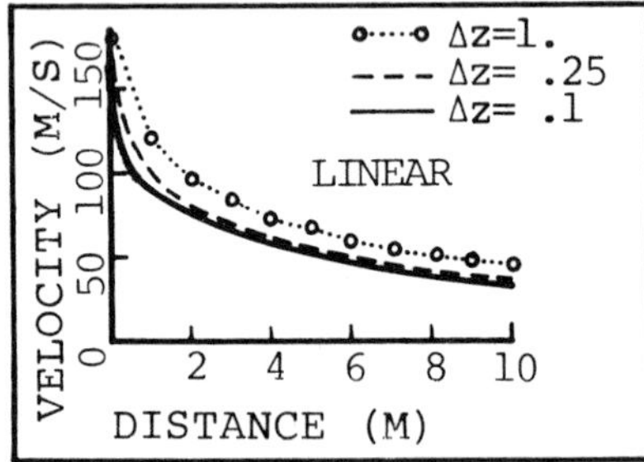

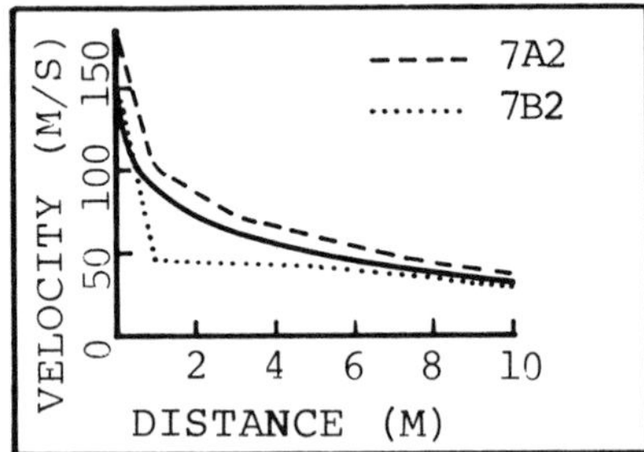

Figs. 13-14. Steady state velocity distributions in a pipe. For methods 7A2 and 7B2 Δz=1. was used.

If linear interpolation is used, a very fine mesh is needed in order to get sufficiently accurate results (Fig. 13). The erroneous velocity profile leads to overestimation of the critical flow (Fig. 15). Although the solution with 100 nodes is not yet fully converged, it was kept as a reference solution (solid lines in Figs. 13-16). The solution with quadratic method (2 and 7a) gives lower mass flux. Unfortunately the mass flux-profile contains "wiggles" near the break location, which were present with all types of boundary conditions tested. To reduce them eq. (30) was used in the conservation law form for density and mass flux (method 7A1). This choice of variables leads to slightly oscillatory behaviour, when steady state is approached. However, steady state is achieved if linear interpolation is used for U_u (method 7A2).

When method 7B was used, pressure fell below zero at the midpoints. This was due to the insufficient approximation (eq. 35) for the matrix A and vector e. It was possible to use eq. (31) for density and eq. (30) for other variables (method 7B1) or eq. (31) for velocity and pressure and eq. (30) for other variables (method 7B2).

Method of Lax and method 7D gave completely unsatisfactory results for this problem and the leapfrog-method was possible to utilize only 0.1 seconds into the transient. When steady state was approached, leapfrog-method became unstable.

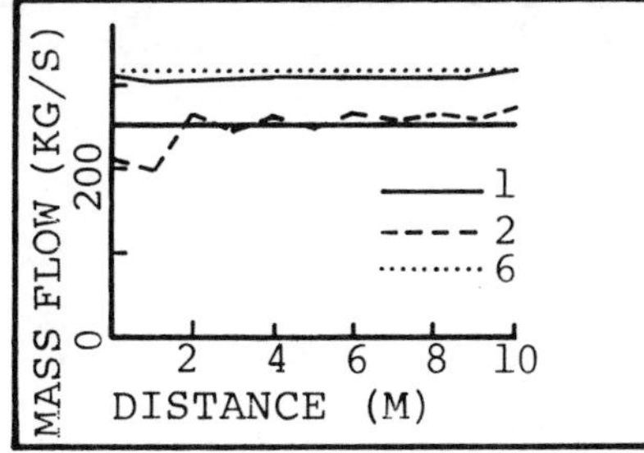

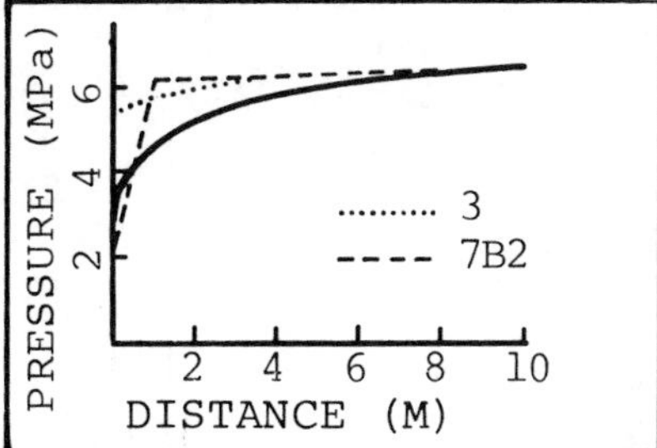

Figs. 14-15. Mass flow and pressure distributions in a pipe. Δz=1. m. Heavy solid line = reference solution.

4.3 Conservation of Mass

When the method 7B1 is used, the mass is fully conserved, which is not typical when using the characteristic form. The error in the mass balance (=(escaped mass + present mass)/initial mass) can be used for the estimation of the accuracy as illustrated by the depressurisation and mass balance histories of the 10 m long pipe closed at the upstream end (Fig. 16).

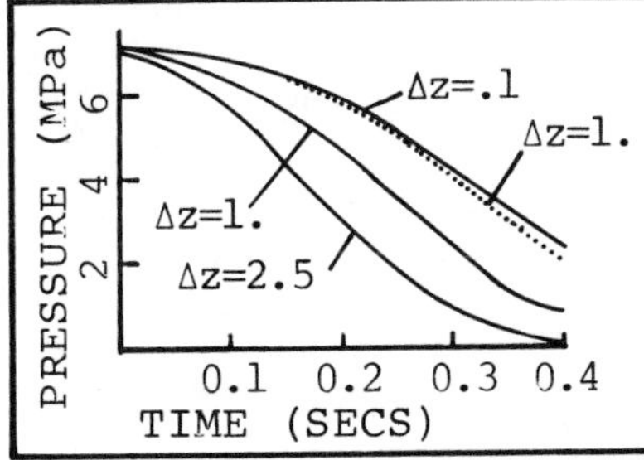

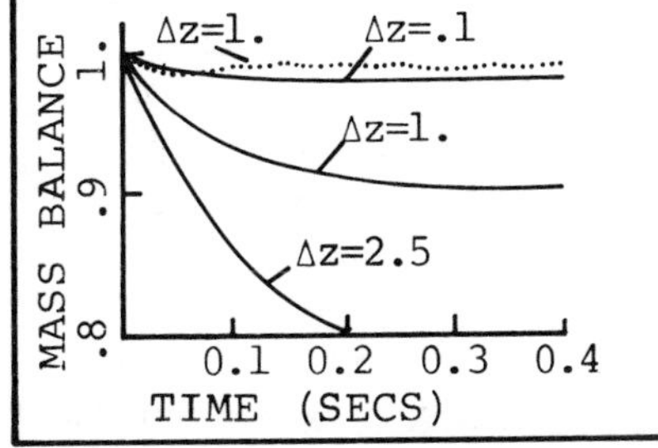

Figs. 16-17. Pressure and mass balance histories during depressurization of a pipe. To insure accurate solution the error in mass balance should be less than 5 % in this case. ——— method 1, ····· method 7A2.

5. CONCLUSIONS

The suitability of various finite difference methods for the characteristic form of flow equations has been studied. The linear interpolation method was applicable for all transients tested. However, the accuracy can be rather poor if two-phase flow is analyzed. More accurate solutions can be achieved by some Lax-Wendroff type-methods. However, these should be used carefully because of the possible anomalies in the solutions. The best result was achieved using methods 7A1 and 7A2. The various methods are summarized in Table 1.

No.	Error	Relative Comp.time	Steady state mass flow(Δz=1.)	Shock profile (Δz=.2)
1	$O(\Delta t, \Delta z)$	1.	315	good
3	$O(\Delta t, \Delta z^2, \Delta z^2/\Delta t)$	1.03	510	poor
4	$O(\Delta t^2, \Delta z^2)$	1.03	n.a	n.a
5	$O(\Delta t, \Delta z)$	1.29	not used	n.a
6	$O(\Delta t, \Delta z)$	1.31	330	n.a
2,7A	$O(\Delta t^2, \Delta z^2)$	1.17	215	poor
7A1	"	"	~215-270	not used
7A2	"	"	250	" "
7B	"	1.16	n.a	poor
7B1	"	"	~210-230	not used
7B2	"	"	~215-270	" "
7C	$O(\Delta t, \Delta z^2)$	1.17	~215-270	poor
7D	$O(\Delta t, \Delta z^2, \Delta z^2/\Delta t)$	"	320	good

Table 1. Comparison of the finite difference methods (n.a=not applicable).

6. REFERENCES

1. HANCOX, W.T., BANERJEE, S. -Numerical Standards for Flow-Boiling Analysis. Nucl. Sci. and Eng. Vol. 64, No. 1. pp. 106-123. 1977.

2. SIIKONEN, T. -Computer Program TMOC for Calculation of Pressure Transients in the Fluid Filled Piping Networks. <u>ENS/ANS Topical Meeting on Nuclear Power Reactor Safety.</u> Brussels, 1978.

3. McDONALD, B.H., HANCOX, W.T., MATHERS, W.G. -Numerical Solution of the Transient Flow Boiling Equations. <u>OECD/CSNI Specialists Meeting on Transient Two-Phase Flow</u>, Paris, 1978.

4. RICHTMYER, R.D., MORTON K.W. -Difference Methods for Initial Value Problems. John Wiley&sons, 1967.

5. STUHMILLER, J.H., FERGUSON, R.E. -Comparison of Numerical Methods for Fluid Flows. NP-1236 EPRI, Palo Alto, California 1979.

6. ROACHE, P.J. -Computational Fluid Dynamics, Hermosa Publishers, 1972.

7. WINTERS, W.S., MERTE, M. JR. -Non-equilibrium Effect in Pipe Blowdown with R-12. <u>Two-Phase Transport and Reactor Safety</u> Veziroglu, T.M., Kakac, S., Hemisphere publishing Co, 1978.

8. ISHII, M., Thermo-Fluid Dynamic Theory of Two-Phase Flow, Eyrolles, Paris, 1975.

SIMULATING INTERACTIONS BETWEEN TURBULENCE AND PARTICLES IN EROSIVE FLOW AND TRANSPORT(I)

Alfred C. Buckingham(II) Wigbert J. Siekhaus(III)

We describe a numerical procedure for simulating fluctuating gas motions and resulting forces on gas-borne particulates. We include some of the particle influences which significantly alter the driving turbulent field. The coupling apparent between dispersed phase particles and continuous phase turbulent flow is introduced with the help of observations from complementary two-phase combustion wind tunnel experiments. In the simulations, particle-to-gas relative velocities and resulting forces on the particles are obtained at incremental time steps for statistically representative particle groups using a random selection method. The fluctuating velocity components are constrained so that their ensemble averages correspond to the pre-determined local mean turbulent energy. The influence of gas-borne particles on turbulent transport, combustion kinetics and subsequent wall surface erosion/corrosion is the central issue of this research.

1. BACKGROUND

We are investigating the coupling and interaction between gas-borne particles and turbulence in internal combustion flows. Our specific interest is in describing the influences of prescribed or determined particle populations and size distributions on turbulent combustion, associated heat transfer and resulting erosion/corrosion of

---

(I)Work performed under the auspices of the U.S. Department of Energy by Lawrence Livermore National Laboratory under contract #W-7405-Eng-48 and supported by the U.S. Army Research Office under contract No. 15812-MS.

(II)Physicist, Fluid Physics, H-Division and (III)Surface Scientist, Chemistry and Materials Science Department, Lawrence Livermore National Laboratory.

container wall surfaces.

For diffusion sensitive flow regions, Marble's [1,2] theoretical investigations suggest dimensional analysis rules for estimating the effects of dispersed phase particle loading on the viscous continuous phase. Similarly, Carrier's [3] observations about fine particulate dust influences on shock tube flows suggest dimensional rules for analysis of the alteration of compressible wave propagation phenomena in the dust-laden carrier gas. These cited analyses are specifically oriented to laminar viscous compressible flow fields in which small concentrations of uniform size and mass spherical, non-colliding particles are introduced. In contrast, we examine particle influences which may be expected to develop in an intense, eddy dominated, turbulent combustion flow field where compressibility effects may emerge as a result of thermo-chemical combustion reactions. Experiments on particle laden turbulent flow field [4,5,6] suggest that there are strong, influential coupling effects which act to modify significantly the turbulent transport, convection and energy dissipation.

In our present work, we emphasize the effectiveness of solid phase additives, their size scale, their chemical compositions and their concentration on potential reduction of wall surface erosion/corrosion in internal combustion flow processes such as those developing in gas turbines, rocket nozzles and internal ballistics propellant combustion. Use is made of Crowe's self-consistent continuum equations describing the interaction between viscous continuum gas-phase and dispersed-phase particle [7] and droplet [8] quasi-continuous dispersed cloud phases.

Since our attention is on the back influence exerted by the particle field on the turbulence, we avoid direct imposition of pre-selected model probability density functions to represent the turbulent interactions in the continuum two-phase equations; an otherwise useful procedure suggested by Westbrook's detailed analysis and computations on spray combustion equations [9]. Instead we simulate the random statistical interaction between particle clusters and localized, modeled, turbulence. In this method, interactive properties for dispersed phase cloud and continuous fields are developed as ensemble averages of a statistically significant number of random fluctuations.

While the influence of the particles on the large scale features of the turbulence are explicitly simulated in space and time, the small scale (subgrid) features, particularly influences on turbulent length and time scale distributions are modeled. We develop kinetic energy, dissipation closure model differential equations of the form suggested by Danon, Wolfstein and Hestroni [10] from the experiments of Hestroni and Sokolov [5].

The explicitly computed interactions between dispersed and continuous phases are an outgrowth of the numerical treatments which have been applied to meteorological scale dynamics or relatively low Reynolds flow with sparse dispersed phase concentrations. Watson and Barr [11], for example, develop a Monte-Carlo prediction method for simulating atmospheric cloud turbulence water droplet dispersal. For smaller size scale influences, the explicitly computed channel flow coupling of particles and turbulence are merged with fine (sub-grid) scale dissipation in the two-dimensional simulations of Peskin and Kau [12]. In these examples and also in some previous computations of one of the present authors [13,14,15], exclusively one way interaction (suspended phase dispersal) is considered. This substantially reduces the computational complexity. The resulting one-way interaction, in which the turbulent continuous phase is unaffected by the suspended phase, is appropriate where particle density is sparse and where there exists a large disparity between the size scales of the large energy bearing turbulent structures and the aerosols undergoing dispersal.

In later work, preliminary development of the two-way back influence particle/turbulence model has been presented by Buckingham and Siekhaus [16]. In addition, back influence coupling has been modeled by Dukowicz [17] using a partially interactive statistical method in which droplet volume and inertial influences on continuous phase convection are included but the influences on turbulent fluctuational structure are not.

The present work describes an extension of the authors' previous investigations. We include consideration of more substantial particle concentrations and their effects on turbulent structural features together with the influences on the associated macroscopic turbulent processes: production, diffusion, convection and dissipation.

2. TECHNICAL DISCUSSION

2.1 Experimental Observations

To examine features of the interaction between particles and turbulence in a controlled situation, an experimental combustion wind tunnel facility was configured for laser Schlieren and LDV diagnostics. Current tests include low speed, steady mean flow with developing turbulent eddy structure at a rearward facing step [16]. Preselected particle concentrations with controlled size and mass distribution are injected upstream of the step into a low speed (10 to 20 m/s) mean flow of premixed combustion gas.

The current measurements are restricted to laser Schlieren diagnostics in cold (non-combustion) gas flow at about one atmosphere static pressure. LDV and laser optics will be combined for combustion, flame spread, mixing and eddy diffusion measurements in a subsequent series of planned experiments. We are currently changing the test configuration to accommodate these hot flows. The rearward facing step will be replaced by a splitter plate separating the reactant and oxidizer upstream of the eddy mixing combustion zone. While these results are not available at the time of this writing, we anticipate including some of the splitter plate hot flow results in our talk.

To determine whether or not systematic particle influences are evident, we trace the entire eddy history through shedding and downstream convection. We ensemble averaged the observations for a large number of eddies. The results were digitized frame to frame observations from fast framing (5000 s^{-1}) motion picture exposures of dual pass laser Schlieren eddy field images. We arrange these results in the histogramatic sequence of distributions given in Fig. 1. The statistical distributions for particle-free gas are monopolar with a bias to the lower shedding frequencies (larger, more energetic eddies). Adding particles produces a distinctly bipolar distribution and enhanced population of smaller (less energetic) high frequency eddies which increases with particle mass loading. The results suggest that particles of sufficient mass and concentration deplete the turbulent energy during the eddy production phase by breaking the eddy field into smaller, less energetic eddies which are dissipated more rapidly over a smaller flow region. Subsequently, we will show some theoretical results which indicate they dynamics of this partition of energy between particle and turbulent fields.

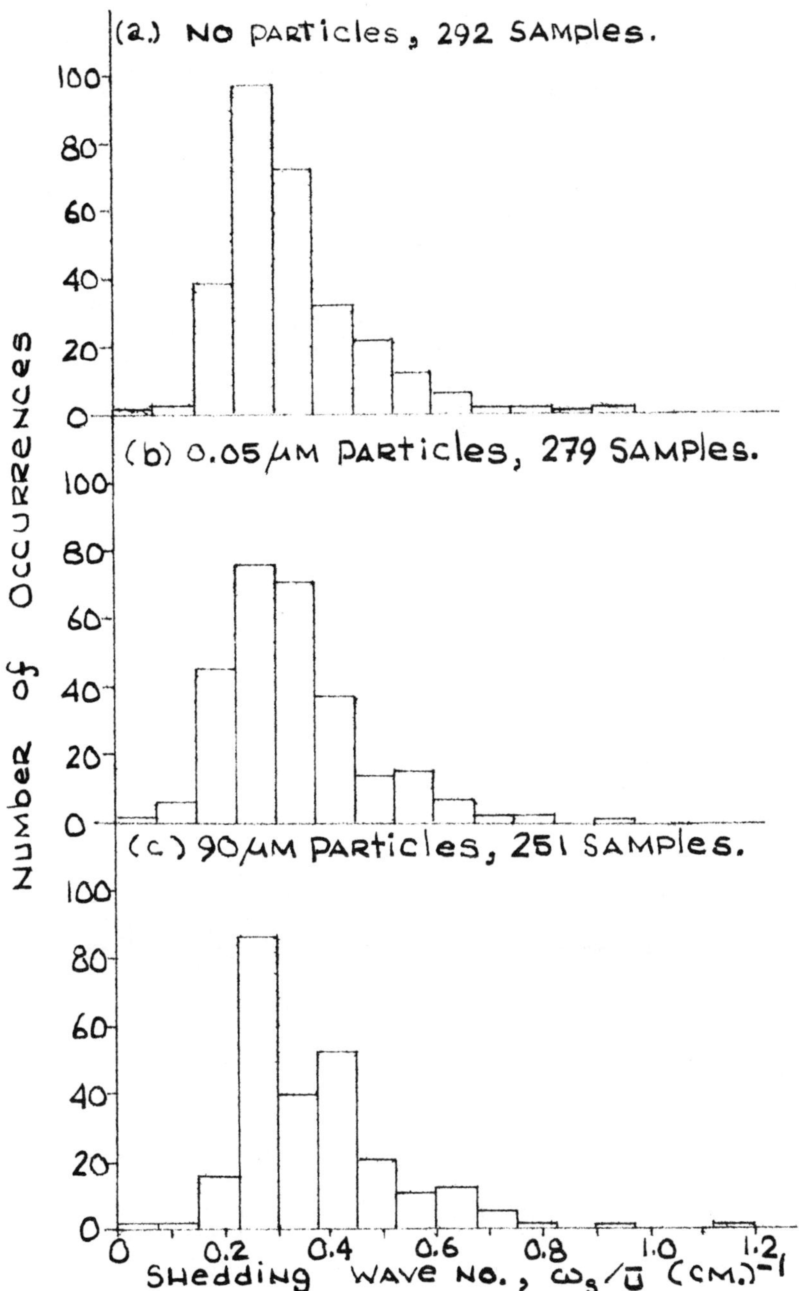

Fig. 1 Distribution of eddy shedding wave no. for indicated particle sizes and loading. Re(cm^{-1}) = 9092, U = 13.28 m/s, T = 298 K.

2.2 Continuum Turbulent Field

For the continuous phase representation we use a combined Reynolds-Favre' averaged form of unsteady conservation equations [14,15,16]. These are modified for particle volumetric displacement through the void fraction,

$$\theta \equiv \frac{V_g}{V_g + V_p} , \tag{1}$$

where V_p is the particle volume and V_g is the vapor volume and $\theta(x,y,z,t)$ is determined from the particle dispersal phase of the general solution.[I] The void fraction modifies the continuous phase mass density and particle density, respectively,

$$\bar{\rho} \rightarrow \theta\bar{\rho} \quad , \quad \bar{\rho}_p \rightarrow (1-\theta)\, \bar{\rho}_p . \tag{2}$$

In addition, as discussed by Crowe [8] in his macroscopic development of the two-phase continuum equations, consistency demands that the stress terms must be adjusted for the vapor surface area reduction associated with particle intrusion. We replace,

$$\vec{\nabla} p \rightarrow \theta \vec{\nabla} p \quad \text{and} \quad \vec{\nabla}\cdot\bar{\sigma} \rightarrow \theta\vec{\nabla}\cdot\vec{\sigma} \tag{3}$$

In the same sense, in the flux divergence terms such as those representing the flow of heat or particle concentration diffusion, the void fraction corrections are explicitly applied to the governing differential equations. In addition, the momentum equation for the continuous phase includes the frictional drag of the particles while the energy equation is supplemented by the rate of change of total energy for the particle cloud.

In the present approach, primary importance is assigned to the application of continuum turbulence closure equations which reflect the interaction of the particle loading and void fraction on the development, transport, convection and dissipation of turbulence energy. Two partial differential equations are developed from the time-averged momentum, energy and total energy continuum relations to close the system with solutions for turbulence kinetic energy, k, and dissipation, ε. The k equation is formed by extension from

---

[I] The incompressible form of the turbulence terms and closure relations is retained by applying time-averaging to the variables; ρ, p, τ, q and mass (Favre') averaging to the variables; u,v, w, h, H.

the suggestions of Danon, Wolfstein and Hestroni [11] for a developing round turbulence jet. We obtain, following order-of-magnitude simplification and reduction [16] for the turbulence energy:

$$\frac{\partial k}{\partial t} + \bar{u}_j \frac{\partial k}{\partial x_j} = \frac{\partial}{\partial x_j}\left(\frac{\mu_T}{\bar{\rho} N_k}\frac{\partial k}{\partial x_j}\right) + (1 + F_1(\ell,\theta))\,\frac{\mu_T}{\bar{\rho}}\left(\frac{\partial \bar{u}_i}{\partial x_j} + \frac{\partial \bar{u}_j}{\partial x_i}\right)$$

$$- (1 + G_1(\ell,\theta))\, k^{3/2} / F_1(\ell,\theta) \tag{4}$$

$$+ g_1(\mu, d^2, \ell)\quad \overline{[u'_{p,i} - u'_i)\, u'_i \theta\,]},$$

where terms of order Re^{-1} and smaller are neglected. In a similar fashion, the closure relations for the dissipation function is developed.

$$\frac{\partial \varepsilon}{\partial t} + \bar{u}_j \frac{\partial \varepsilon}{\partial x_j} = \frac{\partial}{\partial x_j}\left(\frac{\mu_T}{\bar{\rho} N_\varepsilon}\frac{\partial \varepsilon}{\partial x_j}\right) + (1 + F_2(\ell,\theta))\,\frac{\mu_T}{\bar{\rho}}\left(\frac{\partial \bar{u}_i}{\partial x_j} + \frac{\partial \bar{u}_j}{\partial x_i}\right)$$

$$+ (1 + G_2(\ell,\theta))k^2 / F_2(\ell,\theta)^2 \tag{5}$$

$$+ g_2\,(\mu, d^2, \ell)\quad \overline{[(u'_{p,i} - u'_i)\, u'_i\, \theta\,]},$$

where the terms on the right hand side of both equations (4) and (5) are, in order, diffusion, production, conventional dissipation and additional dissipation due to particle influence [16]. We include the additional influence of the spectral turbulence interaction length scale, ℓ, and note that the added dissipation influences of the particles vanish with particle concentration, i.e., g_1, g_2, F_1, F_2, G_1, G_2, $\rightarrow 0$ as $\theta \rightarrow 1$.

Turbulent diffusive exchange processes, other than the Reynolds stress components (for example, turbulent heat flux, turbulent mass diffusion, etc.) are modeled as linearly proportional to the Reynolds stress through the computed eddy viscosity coefficient by assumed constant Prandtl-No. like proportionalities. For example, the averaged correlation of a fluctating scalar variable, f', is approximated

$$\overline{\rho u'_j f'} = \frac{\mu_T}{\bar{\rho} N_f}\, \frac{\partial \bar{f}}{\partial x_j} \tag{6}$$

We assume a constitutive relationship for the Reynolds stress of the form,

$$-\overline{\rho u_i' u_j'} = 2/3\ \bar{\rho} k - 2\mu_T \ [\overline{S_{ij}} - 1/3 \frac{\partial \bar{u}_k}{\partial x_k} \delta jk] \quad ,$$

where we have introduced the mean rate-of-strain tensor,

$$\overline{S_{ij}} \equiv (1/2 \ \frac{\partial \bar{u}_i}{\partial x_j} + \frac{\partial \bar{u}_j}{\partial x_i}). \tag{7}$$

The turbulent dynamic eddy viscosity is given by the closure variables k and ε, by the relationship,

$$\mu_T \equiv C_1 \bar{\rho} k^2 \varepsilon^{-1} \quad \text{, where } C_1 \text{ is a model constant.} \tag{8}$$

2.3 Particle Forces and Influences

Assuming monodisperse particle size distributions, the quasi-continuous particle cloud equations are conviently represented in a Lagrangian form, moving with the mean, convective continuous phase velocity, $\bar{u}_i$. The particle momentum equation, for example, is solved and stored in a separate Lagrangian tape record for post-calculational processing and has the form:

$$\frac{\partial (\overline{\rho_p u_{p,i}})}{\partial t} + \frac{\partial}{\partial x_j} (\overline{\rho_p u_{p,i} u_{p,j}}) = \overline{F_{p,i}} \quad . \tag{9}$$

The forces imposed on the particle cloud by the turbulence, $F_{p,i}$, are functions of the relative velocity between the local particle velocities and turbulent root-mean squared fluctuational velocities in the mean flow Lagrangian system. We consider: drag, virtual mass, Bassett viscous development, Saffman lift, and particle spin contributions to the force [14,15,16].

Volumetric fluctuations also develop from the pressure - acoustical interactions emanating from the fluctuation induced pressure perturbations. These are modeled by an acoustical coupling approximation based on Lighthill's monopole distribution concept, [19], which has the form, in terms of the radiated noise intensity, I_n,

$$I_N(x,y,z,t) = \frac{\bar{\rho}}{4\pi c} \quad (\overline{\frac{d^2 V}{dt^2}}) \tag{10}$$

where c is the mean local sound speed of the continuous phase and the mean volumetric fluctions are given by the computed kinetic energy and dissipation distributions, since by means of dimensional analysis we find the approximation,

$$\frac{d^2V}{dt^2} = C_2 k^{5/2}/\varepsilon, \text{ where } C_2 \text{ is a model constant} \tag{11}$$

The equations for a right-circular turbulent jet geometry, Fig. 2 are solved using a second-order accurate space and time finite difference adaptation of the explicit McCormack method with alternating sweeps [13].

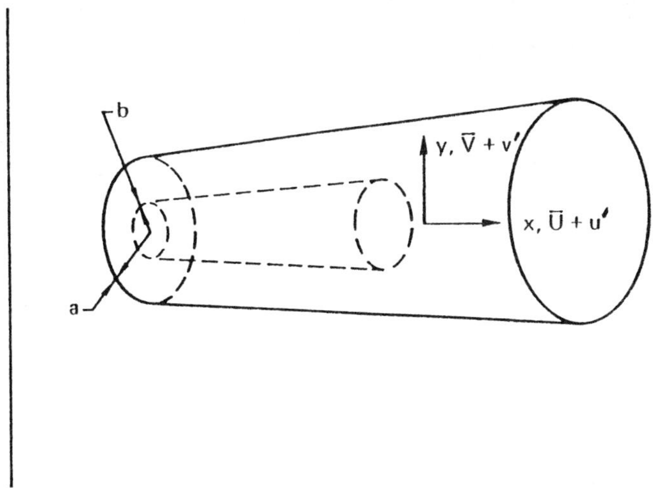

Fig. 2 Computation model round jet flow configuration, coordinates, velocity components.

2.4 Statistical Interaction

The interactive forces on the particles are associated with the random relative velocity perturbations between continuous phase and particle phases. A Monte-Carlo randomization and integral selection approximation is applied to produce the occurrences used for ensemble averaged correlation and moment determination in the current simulations [14,15,16]. Importance sampling procedures are used to reduce the sampling population and storage requirements 18 .

The current results, at the time of the preparation (January, 1981) utilize the McCormack discretization. However, we are currently using a quasi-spectral procedure based on the Stanford University concept separating the explicit large-scale eddies from the dissipative scales through spectral filtering [20,21].

The procedure commences with a calculation of the unsteady, mean Navier-Stokes flow and supplemental macroscopic turbulence closure relations for an assumed particle distribution, $\theta(x,y,z,t_0)$. This is followed by a calculation of the particle-turbulence interaction leading to a dispersal and corrected distribution pattern. This second step is basically a Monte-Carlo procedure with incremental particle-turbulence interaction steps simulating the interaction between a statistically significant number of particle-turbulent random motions (as many as 1100 particle groups are sampled to generate the representative particle clusters comprising the total particle field in the computational region). The particles are redistributed, the particle-gas phase continuum is recomputed and the altered turbulence properties are predicted on both a continuum and ensemble-averaged basis.

The partitioning of the energy and particle influences on the turbulence are illustrated by the computed velocity auto correlation coefficients for the continuous (vapor) phase shown in Fig. 3 for the longitudinal (axial) direction.

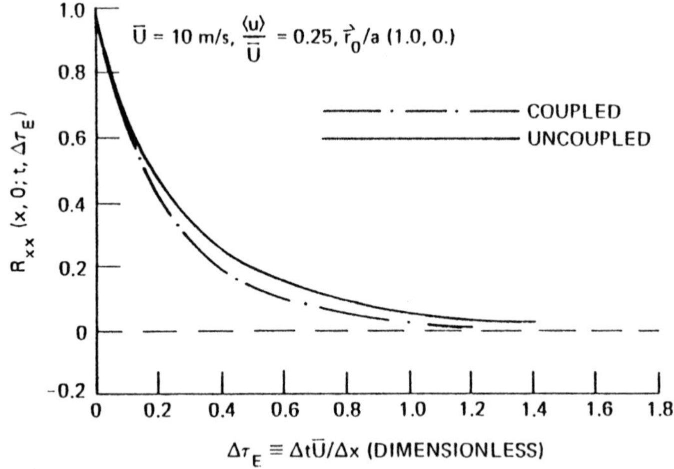

Fig. 3 Eulerian gas phase autocorrelation coefficient in the presence of 40 μm particles in longitudinal direction.

The reduction in turbulent influence is graphically represented by the shortened range of the apparent correlation when particles are added. The particles gain a residual and persistent random motion which is predicted by a post-processor numerical reduction of the difference between mean-averaged and total Lagrangian motions on a separately stored Lagrangian tape. Correlations are shown for the particles in Fig. 4 which illustrate the relative persistence of the particle fluctuations (compared to those of the gas phase) associated with the large disparity in particle to gas inertia.

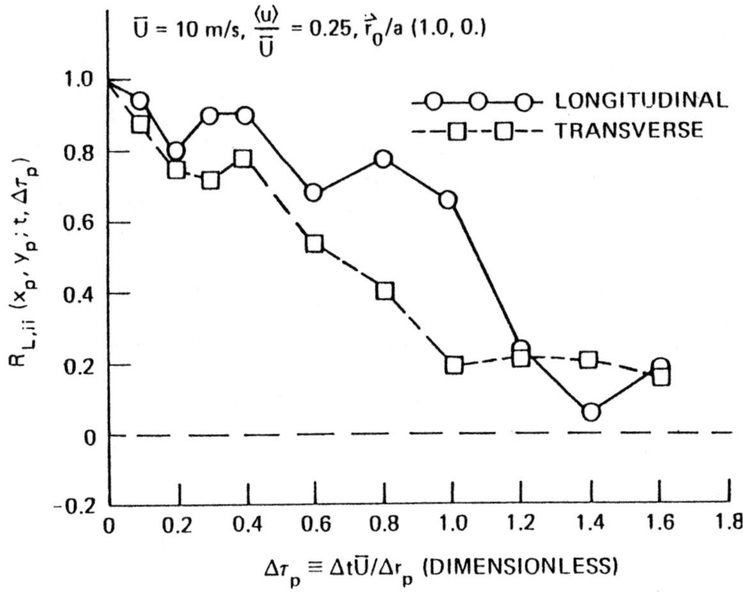

Fig. 4 Lagrangian particle phase longitudinal and transverse autocorrelation coefficients for 40 μm particles.

3. REFERENCES

1. MARBLE, F. E., "Dynamics of a Gas Containing Small Particles", in Combustion and Propulsion (5th AGARD Colloquium) Pergamon Press, NY, p. 175, (1963).
2. MARBLE, F. E., "Dynamics of Dusty Gases", Ann. Reviews of Fluid Mech. 2, 397 (1970).
3. CARRIER, G. F., J. Fluid Mech. 4, 376 (1955).
4. FULMER, R. D. and D. P. WIRTZ, "Measurement of Individual Particle Velocities in a Simulated Rocket Exhaust", AIAA J. 3, 8, 1506 (1965).
5. HESTRONI, G., and M. SOKOLOV, "Distribution of Mass, Velocity and Intensity of Turbulence in a Two Phase Turbulent Jet", J. Appl. Mech. 38, 315 (1971).
6. YUU, S., N. YASUKOUCHI and J. TOMOSADA, "Particle Turbulent Diffusion in a Dust Laden Round Jet", AIChE J. 24, 3, 508 (1978).
7. CROWE, C. T., and D. T. PRATT, "Gas Particle Flow in Combustion Chambers", WSCI 72-5 in Spring Meeting Western States Combustion Institute (April 24-25, 1972).
8. CROWE, C. T., "Vapor-Droplet Flow Equations", Univ. of California Lawrence Livermore Laboratory UCRL-51877 (August 18, 1975).
9. WESTBROOK, C. K., "Three Dimensional Numerical Modeling of Liquid Fuel Sprays", in Sixteenth Symp. (International) on Combustion, The Combustion Inst., Pittsburgh (1976).
10. DANON, H., M. WOLFSTEIN and G. HESTRONI, "Numerical Calculations of Two-Phase Turbulent Round Jet", Int. J. Multiphase Flow 3, 223 (1977).

11. WATSON, C. W., and S. BARR, "Monte Carlo Simulation of Turbulent Transport of Airborne Contaminants", Los Alamos Scientific Laboratory, Univ. of California LA-6103 (January 1976).
12. PESKIN, R. L., and C. J. KAU, "Numerical Simulation of Particulate Motion in Turbulent Gas-Solid Channel Flows", J. Fluids Engr. 101, 319 (Sept., 1979).
13. BUCKINGHAM, A. C., "Turbulent Dusty Gas Motions with Weak Statistical Coupling", Univ. of California Lawrence Livermore Laboratory UCRL-82131 and AIAA paper #79-1484, AIAA 12th Fluid and Plasmadynamics Conf. (Williamsburg, VA, July 24-26, 1979).
14. BUCKINGHAM, A. C., "Dusty Gas Influences on Transport in Turbulent Erosive Propellant Flow", Univ. of Calif. Lawrence Livermore Laboratory UCRL-82876, Rev. 1 (October 24, 1980) and to appear in AIAA Journ. Vol. 19 (March or April, 1981).
15. BUCKINGHAM, A. C., "Modeling Propellant Combustion Interacting With an Eroding Solid Surface", Univ. of California Lawrence Livermore Laboratory UCRL-83724 and AIAA paper #80-1259 in AIAA-SAE-ASME 16th Joint Propulsion Conference (Hartford, Conn. June 30-July 2, 1980).
16. BUCKINGHAM, A. C. and W. J. SEIKHAUS, "Interaction of Moderately Dense Particle Concentrations in Turbulent Flow", Univ. of Calif. Lawrence Livermore Laboratory UCRL-84559 and AIAA Paper No. 81-0346 in AIAA 19th Aerospace Sciences Meeting (St. Louis, MO January 12-15, 1981).
17. DUKOWICZ, J. K., "A Particle-Fluid Numerical Model for Liquid Sprays", J. of Comp. Phys. 335, 229 (1980).
18. FAIST, M. B., and J. T. MUCKERMAN, "Importance Sampling and Histogrammic Representations of Reactivity Functions and Product Distributions in Monte Carlo Quasiclassical Trajectory Calculations", J. Chem. Phys. 69, 9 (1978).
19. LIGHTHILL, M. J., Proc. Royal Society, (London) A, 222 (1954).
20. KWAK, D., W. C. REYNOLDS, and J. H. FERZIGER, "Three-Dimensional Time-Dependent Computation of Turbulent Flow", Stanford University Thermosciences Division Rept. TF-5 (May 1975).
21. MANSOUR, N. N., J. FERZIGER, and W. C. REYNOLDS, "Large Eddy Simulation of a Turbulent Mixing Layer", Stanford University Thermosciences Division Rept. TF-11 (April 1978).

SECTION 8
MASS TRANSPORT
AND
CONVECTION

THE SELFADAPTIVE SOLUTION OF NONLINEAR 2-D BOUNDARY VALUE PROBLEMS IN A RECTANGULAR DOMAIN*

W.Schönauer, K.Raith, G.Glotz+

Summary: A selfadaptive difference method for the solution of nonlinear systems of 2-D elliptic PDE's is presented. The nonequidistant grid and the orders of the difference star in the two coordinate directions are optimized for a given relative tolerance. The method is applied to the convective flow in a square cavity.

1. INTRODUCTION

Variable order/variable step size methods are common practice for the solution of ODE's. In this paper a similar tool for the solution of 2-D elliptic equations will be presented. The key to the method is the estimate of the discretization error from the difference of difference quotients of families of difference formulae. The knowledge of the error is used to choose the optimum grid and difference star and to adapt all the errors to a given relative tolerance, including the error of the Newton-Raphson iteration. But in order to determine the grid we must know the solution. So only repeated solution will lead the required accuracy which is the natural way for a BVP.

2. ESTIMATE OF THE DISCRETIZATION ERROR

We use the family of nonequidistant centralized difference formulae of Fig. 2.1 for the discretization of u_{xx} with arbitrary order q, and we use the same points for u_x. If u_ϕ stands for a derivative u_x, u_{xx}, u_y, u_{yy}, and $u_{\phi,d}$ for a difference formula with a discretization error $\bar{d}_\phi$, and q is the actual order then

---

*This research was supported by the Stiftung Volkswagenwerk.

+Rechenzentrum Universität Karlsruhe, D-7500 Karlsruhe 1

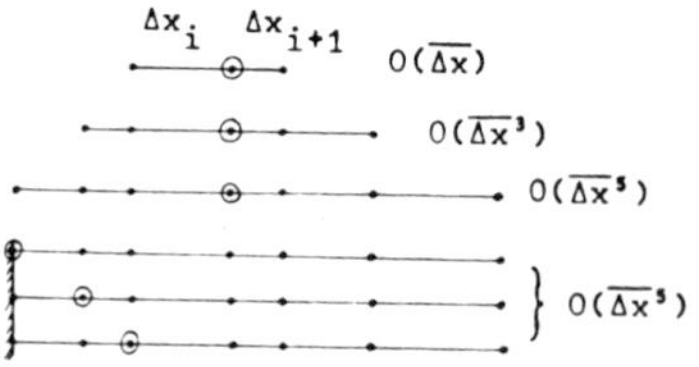

$$u_\phi = u_{\phi,d}(q) + \bar{d}_\phi(q) \qquad (2.1)$$
$$= u_{\phi,d}(q+2) + \bar{d}_\phi(q+2).$$

Fig.2.1 Examples of difference formulae $u_{xx,d}$ for u_{xx}

If we resolve for $\bar{d}_\phi(q)$ and neglect $\bar{d}_\phi(q+2)$ we get an estimate $d_\phi(q)$ for the exact error $\bar{d}_\phi(q)$:

$$d_\phi(q) := u_{\phi,d}(q+2) - u_{\phi,d}(q). \qquad (2.2)$$

We call this the principle of the difference of difference quotients. In [1,2] the generation of difference formulae of arbitrary order and the quality of the error estimate (2.2) is discussed in detail. The estimate (2.2) is much better than we ever had expected. For the difference method we replace derivatives in the following way:

$$u_\phi \Leftarrow u_{\phi,d}(q) + d_\phi(q). \qquad (2.3)$$

3. THE ERROR EQUATION

We treat the system of n nonlinear 2-D elliptic PDE's which we write in operator notation

$$Pu \equiv P(x,y,u,u_x,u_y,u_{xx},u_{yy}) = 0 , \qquad (3.1)$$

where P and u have n components. For the boundary conditions we assume a nonlinear operator

$$Gu \equiv G(x,y,u,u_x,u_y,u_{xx},u_{yy}) = 0 , \qquad (3.2)$$

with x,y on the boundary of a rectangular domain. Mixed derivatives are treated by auxiliary variables, e.g. $v = u_x$; then $u_{xy} = v_y$.

In order to linearize (3.1),(3.2) we introduce the Newton-Raphson-ansatz with iteration index ν

$$u \Leftarrow u^{(\nu+1)} = u^{(\nu)} + \Delta u^{(\nu)} . \qquad (3.3)$$

We linearize in the Newton correction $\Delta u^{(\nu)}$, drop the index ν and get from (3.1)

$$Q(x,y,\Delta u,\Delta u_x,\Delta u_y,\Delta u_{xx},\Delta u_{yy}) := \qquad (3.4)$$
$$:= -\frac{\partial P}{\partial u}\Delta u - \frac{\partial P}{\partial u_x}\Delta u_x - \ldots - \frac{\partial P}{\partial u_{yy}}\Delta u_{yy} = P(x,y,u,\ldots,u_{yy}).$$

For a system of n PDE's $\partial P/\partial u$, $\partial P/\partial u_x$,... are $n \times n$ matrices. We abbreviate (3.4) by the operator notation

$$Q\Delta u = Pu, \qquad (3.5)$$

where Q is a linear operator and (3.5) is a linear system of n PDE's for the Newton correction Δu.

For the solution of (3.5) by the difference method we replace u, Δu by their discrete values $u_d, \Delta u_d$ on the grid and we replace derivatives by (2.3) with different orders qx and qy in the x- and y-direction. On the left hand side we neglect all error terms as small of second order and on the right hand side we linearize in the discretization errors. The resulting error terms are

$$D_x = \frac{\partial P}{\partial u_x} d_x + \frac{\partial P}{\partial u_{xx}} d_{xx} \; , \; D_y = \frac{\partial P}{\partial u_y} d_y + \frac{\partial P}{\partial u_{yy}} d_{yy} \qquad (3.6)$$

If we order the resulting system for the Δu_d we get

$$Q_d \Delta u_d := (Pu)_d + D_x + D_y \; , \qquad (3.7)$$

where Q_d is a block band matrix with n×n blocks. We assume that the boundary conditions (3.2) are included in (3.7). The formal solution of (3.7) is the error equation

$$\Delta u_d = \Delta u_{Pu} + \Delta u_{D_x} + \Delta u_{D_y} = Q_d^{-1}[(Pu)_d + D_x + D_y]. \qquad (3.8)$$

The terms in the brackets are the defect and the x- and y-discretization errors on the level of the equation. Multiplication by Q_d^{-1} yields errors on the level of the solution, and the overall error Δu_d has been split up into the corresponding parts. If we want to compute efficiently we must well balance the terms in the brackets because the largest error determines the accuracy and the other smaller errors only waste computation time.

For the Newton-Raphson iteration we solve and compute

$$Q_d \Delta u_{Pu} = (Pu)_d \text{ (a)} \; , \; u_d^{(\nu+1)} = u_d^{(\nu)} + \omega \Delta u_{Pu} \text{ (b)}, \qquad (3.9)$$

where $\omega = 1$ normally, but ω is reduced if $\|(Pu)_d\|_{\nu+1} > \|(Pu)_d\|_\nu$ (nonconvergence), with $\|\cdot\|$ as the maximum norm, and reset to $\omega = 1$ as soon as possible. The iteration is stopped if

$$\|(Pu)_d\| < 0.1 \, \|D_x + D_y\| \qquad (3.10)$$

holds. Thus the defect is adapted to the discretization error and the decision is made before Δu_d is computed. If $\|(Pu)_d\|_{\nu+1} < 0.1 \, \|(Pu)_d\|_\nu$ we compute with fixed matrix Q_d (simplified Newton-Raphson).

4. SOLUTION METHOD

We assume the x,y-grid and the orders qx and qy to be prescribed. Fig. 4.1 shows a difference grid with a 9-point difference star (qx=qy=3) as an example and Fig. 4.2 shows the structure of the matrix Q_d for this star. If we want to solve (3.9a) by the Gauß method the fill-in leads to unrealistic storage requirements for problems of practical interest. Therefore we solve (3.9a) by a semiiterative method: we solve directly (with fill-in) for the blocks A,B,C of Fig.4.2 and

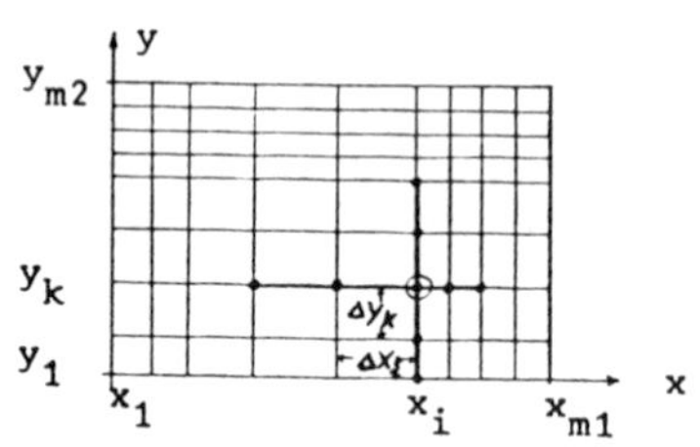

Fig.4.1 Difference grid and example of a difference star

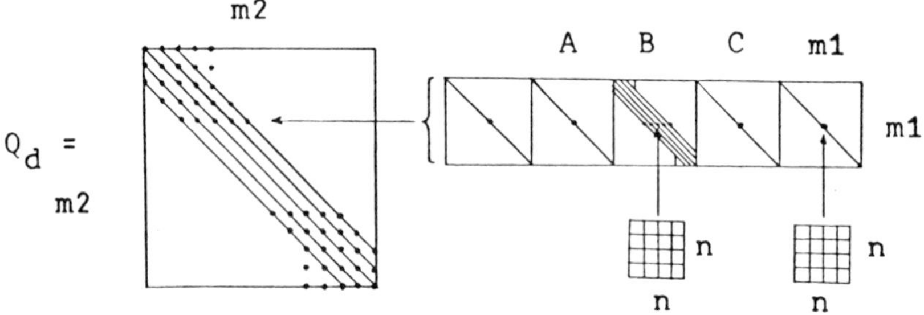

Fig.4.2 Structure of the matrix Q_d for the 9-point star of Fig. 4.1

iterate to the full matrix Q_d. For the example of Fig. 4.2 we would express the 5-point formula in the y-direction by a 3-point formula and a remainder formula

5-point formula = 3-point formula + remainder formula . (4.1)

We solve directly for

this star and iterate to .

Thus Q_d is split up into Q_d^* (direct solution) and N (remainder) and the solutions for Q_d and Q_d^* differ only by the difference of the discretization errors.

Now we have a nested iteration: the outer Newton iteration (index ν) and the inner iteration with Q_d^* (index μ)

$$\Delta u_{Pu}^{(\nu,\mu)} = \underbrace{\left(Q_d^{*(\nu)}\right)^{-1}(Pu)_d^{(\nu)}}_{\text{direct solution}} - \underbrace{\left(Q_d^{*(\nu)}\right)^{-1} N\ \Delta u_{Pu}^{(\nu,\mu-1)}}_{\text{iteration}} . \quad (4.2)$$

For the stopping of the inner iteration (4.2) we use the fact that the quadratic convergence of the Newton iteration yields doubling of the significant digits at each iteration step. We stop the μ-iteration if

$$\frac{\| Q_d^{(\nu)} \Delta u_{Pu}^{(\nu,\mu)} - (Pu)_d^{(\nu)} \|}{\| (Pu)_d^{(\nu)} \|} \leq \varepsilon^* = 0.1 \left(\frac{\| \Delta u_{Pu}^{(\nu-1)} \|}{\| u_d^{(\nu-1)} \|} \right)^2 \quad (4.3)$$

holds, but restrict ε^* by $10^{-4} \leq \varepsilon^* \leq 0.1$ and put $\varepsilon^* = 0.1$ for $\nu = 1$. For details see [2]. The elements of Q_d^* and N are stored in a "packed" form with index and element value for nonzero elements. For the elimination step (with the fill-in) only a "window" of the active part of the matrix is kept in the main store and pushed in a wave front technique over the matrix Q_d^* (resulting in many i/o operations). The speed of convergence is augmented by a selfadapted relaxation factor. Finally a new solution $u_d^{(\nu+1)}$ is computed from (3.9b) and the Newton iteration is stopped if (3.10) holds.

5. DETERMINATION OF THE OPTIMUM ORDER AND GRID

For the determination of the optimum order and grid we use onedimensional methods for each coordinate direction which we now explain for the x-direction. A relative tolerance tol (maximum error to maximum solution) is prescribed. For a system of n equations u_d has components $u_{d,1},\ldots,u_{d,n}$. We use a global, a relative local and a relative global norm:

$$\|u_d\| = \max_{j,i,k} |u_{d,j}(x_i,y_k)| ,$$

$$\|\Delta u_d\|_{rel,i} = \max_j \frac{\max_k |\Delta u_{d,j}(x_i,y_k)|}{\max_{i,k} |u_{d,j}(x_i,y_k)|} , \quad (5.1)$$

$$\|\Delta u_d\|_{rel} = \max_i \|\Delta u_d\|_{rel,i} .$$

We execute the following steps:

1. Computation of a "basic solution" for prescribed initial grid and initial orders qx_{in}, qy_{in}. If

$$\|\Delta u_{D_x}\|_{rel} < tol \quad \text{and} \quad \|\Delta u_{D_y}\|_{rel} < tol \qquad (5.2)$$

holds we have finished; if (5.2) does not hold, we go on.

2. Computation of the error norm

$$\|\Delta u_{D_x}(qx)\|_{rel} \simeq \|Q_d^{-1}(qx)\|_{rel} \cdot \|D_x(qx)\| \qquad (5.3)$$

for the orders $qx = 1,3,5,\ldots$ from the basic solution, using an estimate for $\|Q_d^{-1}(qx)\|_{rel} \simeq$ $\simeq \|Q_d^{-1}(qx_{in})\|_{rel} \simeq \|\Delta u_{D_x}(qx_{in})\|_{rel} / \|D_x(qx_{in})\|$.

3. Computation of the step size for tol/3 (equidistant error law)

$$\Delta x_{new} = [(tol/3) \,/\, \|\Delta u_{D_x}(qx)\|_{rel}]^{1/qx} \, \Delta x \qquad (5.4)$$

for increasing orders qx until we find a value qx_{opt} for which Δx_{new} does no longer (essentially) increase. Then $qx = qx_{opt}$ is fixed. Now $\Delta u_{D_x}(qx) \simeq$ $\simeq Q_d^{-1}(qx_{in}) D_x(qx)$ is computed.

4. Computation of local step sizes by

$$\Delta x_i = F \cdot 0.5(x_{i+1} - x_{i-1}) \;, \quad F = \sqrt{F_i \cdot F_{min,i}} \;,$$

$$F_i = [(tol/3)/\|\Delta u_{D_x}(qx)\|_{rel,i}]^{1/qx} \;, \qquad (5.5)$$

$$F_{min,i} = \min_{j \in E_i} F_i \;,$$

with E_i set of all indices j whose difference formulae and error formulae contain the point i. Equ (5.5) is the geometric mean of a purely local and a global error consideration.

5. Putting together the Δx_i that never a larger step size protrudes into the domain of a smaller one.

6. Now the steps 2 to 5 are executed in the same way for the y-direction, resulting in qy and a new y-grid

7. Linear interpolation of the old values u_d for the new x,y-grid, computation of a new solution and error check. If (5.2) does not hold, determination of a new order (only in second trial) and of a new grid and solution until (5.2) holds.

In these steps "computation of a solution" follows the method of section 4.

6. TESTING TECHNIQUES

While "usual" difference methods ask for the accuracy of the solution, we ask for the accuracy of the error estimate. Because there is no exact solution for a technical problem we prescribe a solution $\bar{u}(x,y)$, put it into (3.1),(3.2) and get instead of zero the values $P\bar{u}$, $G\bar{u}$. Now we solve instead of (3.1),(3.2)

$$Pu - P\bar{u} = 0 \ , \quad Gu - G\bar{u} = 0 \ , \tag{6.1}$$

which has the solution $\bar{u}$. This means changing the differential operators only by absolute (known) terms. We then may compare the exact and estimated relative errors

$$\|\bar{u}-u_d\|_{rel} \quad \text{and} \quad \|\Delta u_{D_x}+\Delta u_{D_y}\|_{rel} \tag{6.2}$$

or we may compare locally (instead of the norms) and thus gain confidence in the program. If we then drop the extra terms we have a completely tested program.

7. CONVECTION IN A SQUARE CAVITY

Merely as an illustrative example we treat the natural convection in a square cavity. The governing equations are nondimensionalized as in [3]. The nondimensional velocity in x,y-direction, voticity and temperature difference are u,v, $\zeta = v_x-u_y$, T and the differential equations are

$$P\hat{u} \equiv \begin{cases} u_{xx}+u_{yy}+\zeta_y = 0 \ , & (7.1a) \\ v_{xx}+v_{yy}-\zeta_x = 0 \ , & (7.1b) \\ u\zeta_x+v\zeta_y-P_R(\zeta_{xx}+\zeta_{yy}) - R_A P_R T_X = 0 \ , & (7.1c) \\ uT_x+vT_y-(T_{xx}+T_{yy}) = 0 \ , & (7.1d) \end{cases}$$

with $\hat{u} = (u,v,\zeta,T)$. The equations (7.1a,b) follow from the continuity equation and the definition of ζ, (7.1c) follows from the momentum equation with the Boussinesq approximation and (7.1d) follows from the energy equation. The boundary conditions are shown in Fig. 7.1, the corresponding operator $G\hat{u}$ is not written down here.

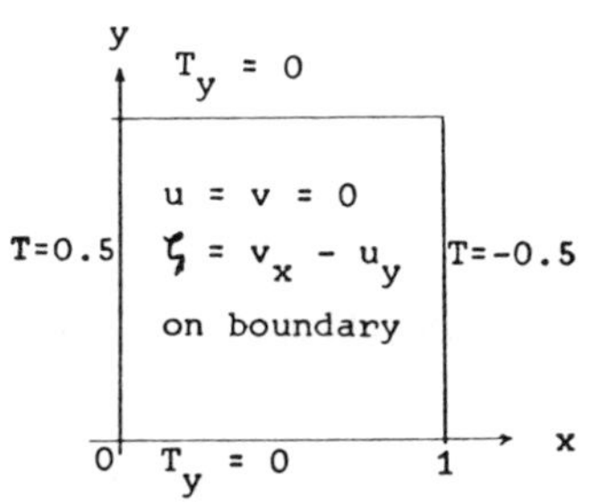

Fig. 7.1 Boundary conditions for (7.1)

Following section 6 we prescribed for the test of the program a solution $\hat{u}$ with

$$\bar{u} = \sin x \cos y \ , \ \bar{v} = -\cos x \sin y \ , \ \bar{\zeta} = 2\sin x \sin y \ , \qquad \bar{T} = -\cos x \tag{7.2}$$

and solved (7.1) for $P_R=0.71$, $R_A=10^3$, $tol=10^{-4}$. We started with an equidistant 10×10 grid and orders $qx_{in}=qy_{in}=1$ and initial profiles which were interpolated from the defect of the boundary conditions for the zero solution. The result of the selfadaptation process of section 5 was, with nx, ny the number of grid points in x,y-direction:

order qx	grid points nx	order qy	grid points ny	
1	140 343	1	1449	(estimate for equidistant grid)
3	55	3	47	
5	25	5	23	

For a 10×10 grid the orders qx, qy are limited by 5. For these orders a nonequidistant 14×11 grid was generated and the result had an exact max. relat. error of $1.7 \cdot 10^{-7}$, the error estimate was $1.8 \cdot 10^{-7}$. The CPU-time was 31 min on a UNIVAC 1106 (≈ 1/6 IBM 370/168). Fig. 7.2 a) shows the exact and Fig 7.2 b) the estimated relative error for the (most critical) vorticity

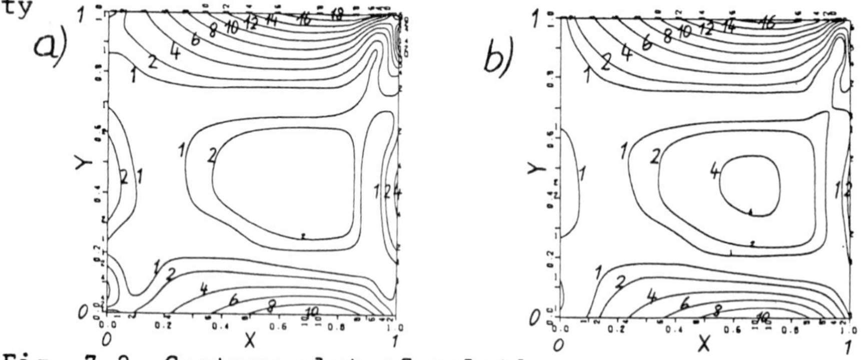

Fig. 7.2 Contour plot of relative error $10^8 \cdot |\Delta\zeta| / \|\zeta\|$, a) exact, b) estimated for the test solution (7.2)

For the solution of the real problem (7.1) with the same P_R, R_A, but $tol = 10^{-3}$ we started with a nonequidistant 15×15 grid (finer mesh near wall) with orders $qx_{in} = qy_{in} = 5$ and the same type of initial profiles. The basic solution had a max. relat. error of $2.4 \cdot 10^{-3}$ and the result of the selfadaptation process was

qx	nx	qy	ny		
1	185	1	184	assumes equidistant grid	=> nx = ny = 16 for nonequidistant grid with qx=qy=11.
3	30	3	29		
5	20	5	20		
7	17	7	17		
9	16	9	16		
11	15	11	15		

The orders were limited to 11. The max. relat. error was $1.3 \cdot 10^{-3}$ and the selfadaptation gave

qx	nx	qy	ny		
1	559	1	632	assumes equidistant grid	=> nx = ny = 18 for nonequidistant grid with qx=qy=11.
3	50	3	49		
5	27	5	27		
7	21	7	21		
9	19	9	19		
11	18	11	18		

The error was $1.2 \cdot 10^{-3}$ because the continuously finer grid near the wall revealed the larger error in the "boundary layer". Because we restarted the program with this solution still a third optimization took place with the following result

qx	nx	qy	ny		
1	762	1	620	assumes equidistant grid	=> nx = ny = 18 for nonequidistant grid with qx=qy=11.
3	57	3	56		
5	30	5	30		
7	23	7	23		
9	20	9	20		
11	19	11	19		

The max. relat. error was $0.89 \cdot 10^{-3}$ for onedimensional analysis, but with $\|\Delta\hat{u}_{D_x} + \Delta\hat{u}_{D_y}\|_{rel} = 0.48 \cdot 10^{-3}$ because of balancing of signed errors. The CPU-time for the whole process was 209 min on UNIVAC 1106. The results for v and ζ and their relat. errors are shown in Fig. 7.3. The errors of ζ are concentrated in the corners. Fig.7.4 shows the values for u and T. While the errors of u and v assume their maxima in the corners, the maximum error of T is in the middle of the cavity. The stream function ψ has been computed by the same program from

$$\psi_{xx} + \psi_{yy} - \zeta = 0 \ , \quad \psi = 0 \quad \text{on boundary} \qquad (7.3)$$

which follows from the definitions of the stream function and of the vorticity. The contour plot of ψ is shown in Fig. 7.5.

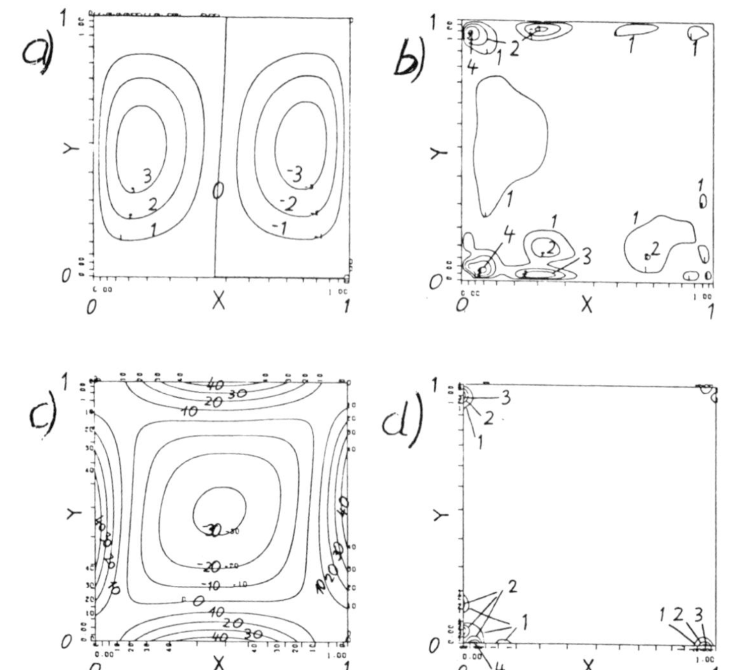

Fig. 7.3 Contour plots of a) v, b) $10^5 \cdot |\Delta v| / \| v \|$, c) ζ, d) $10^4 \cdot |\Delta \zeta| / \| \zeta \|$. The "scale" shows the selfadapted grid lines

8. CONCLUSION

The whole solution method is part of a general purpose program package (SLDGL) for the solution of elliptic and parabolic nonlinear PDE's. For other problems only the operators Pu and Gu must be interchanged. The error estimate from the difference of difference quotients opens the door for the development of reliable engineering software. We consider it as a necessary prerequisite to furnish to the engineer together wich the solution an estimate for the accuracy of this solution.

9. REFERENCES

1. SCHÖNAUER,W., GLOTZ,G., RAITH,K. - The Solution of the Laminar Boundary Layer Equations by a Variable Order Selfadaptive Difference Method, Recent Advances in Numerical Methods in Fluids, Ed. TAYLOR,C and MORGAN,K., Pineridge Press, Swansea, 1980
2. SCHÖNAUER,W., RAITH,K., GLOTZ,G. - The Principle of the Difference of Difference Quotients as a Key to the Selfadaptive Solution of Nonlinear Partial Differential Equations, to be published

3. OERTEL,H,jun. - Thermische Zellularkonvektion, Habilitationsschrift Universität Karlsruhe (1979)

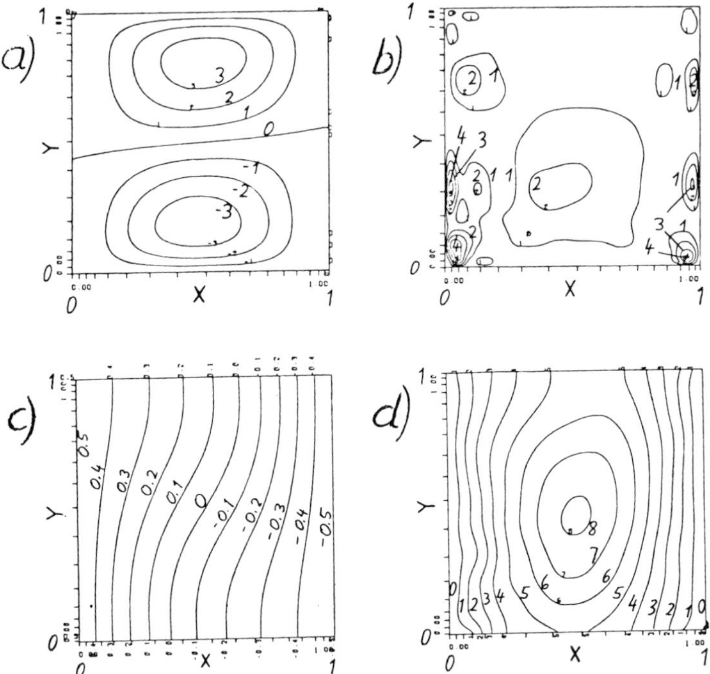

Fig. 7.4 Contour plots of a) u, b) $10^5 \cdot |\Delta u| / \|u\|$, c) T, d) $10^6 \cdot |\Delta T| / \|T\|$. The "scale" shows the selfadapted grid lines.

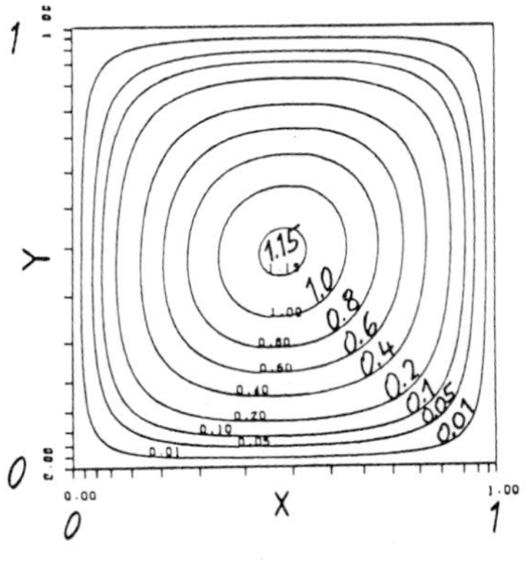

Fig. 7.5 Contour plot of the stream function ψ. The "scale" shows the grid lines.

NUMERICAL SOLUTION OF THE MOMENTUM EQUATIONS IN UNSTEADY INCOMPRESSIBLE FLOW

Y. Kronzon, I. Partom, and M. Wolfshtein

Department of Aeronautical Engineering
Technion-Israel Institute of Technology
Haifa, Israel

SUMMARY

Time dependent flow problems are studied. Solutions are obtained using standard finite-difference methods for time-dependent problems together with a velocity correction which ensures mass conservation. The method is applied to three-dimensional periodic non-viscous flow inside a cylinder, and to two-dimensional transient viscous free convection. Tests of stability and accuracy gave satisfactory results.

1. INTRODUCTION

Incompressible fluid flow is governed by four coupled simultaneous equations: the three momentum equations and the continuity equation. These govern four variables: the three velocity components and the pressure. Still, the matching between the number of equations and the number of variables is not sufficient to ensure easy solution of the problem. As each momentum equation contains second derivatives of a single velocity component and a first derivative of the pressure these equations should be solved for the velocity components. Unfortunately, the remaining equation, the continuity equation, does not contain the presssure at all for uniform density flows. Thus, this equation takes the form of a "compatibility condition" between an assumed pressure field and a calculated velocity field which satisfy the momentum equations. Out of an infinite number of such pressure-velocity fields only such fields which satisfy the continuity equation as well are solutions of the fluid flow problem.

The problem of matching a pressure field to the continuity has intrigued many investigators as reported by

Roache [1]. Some, like Thom [2], Barakat and Clark [3], or Gosman et al [4], eliminated the pressure by solving for the vorticity and the stream function in two-dimensional problems. This approach may be extended to three dimensions if the vector-potential is used, as suggested by Aziz and Hellums [5] or de Vahl Davis and Wolfshtein [6]. However, such a formulation is cumbersome when small density changes (say due to temperature differences) are to be accounted for, and formulation of boundary conditions is not always easy. Therefore, pressure-velocity (or "primitive variable") formulations have been sought for many years. Such methods were suggested by Lilly [7] or Harlow and Welch [8]. Typically, such methods are based on the solution of the momentum equations using given old pressure fields, and then seeking a new pressure field so as to ensure that the continuity equation is satisfied. Such formulations cannot satisfy all the equations exactly without iterations. The MAC method [8] for instance does not satisfy exactly any equation, but the velocity divergence remains bounded and this is enough to ensure a stable solution, accurate to the first order in the time.

Similar problems are encountered in the solution of developing duct flows. Vorticity methods (e.g. Berger and Wolfshtein [9], or Dodge [10]) are not always convenient. Primitive variables methods (e.b. Patankar and Spalding [11] or Briley [12]) require a special "pressure equation", and the governing equations are never exactly satisfied, until the solution becomes stationary with respect to the axial direction. The similarity between the "parabolized" duct flows and the parabolic unsteady flows is obvious and we propose to regard all the methods applicable to any such flow within a single framework. Thus the MAC method keeps the velocity divergence bounded, Patankar and Spalding's method keep the divergence bounded and apply small corrections to the velocity and pressure in an attempt to satisfy the equations, while Briley's method satisfies the continuity (but not the momentum) equation exactly.

In the present paper we apply Briley's method to unsteady flows. We believe that this is better procedure due to the exact (to truncation errors) conservation of mass by satisfying the continuity equation. The method is applicable to viscous and non-viscous flows, and is suitable for the solution of flows with non-uniform densities at low Mach numbers (as free convection flows). We present three-dimensional potential, and rotational non-viscous solutions, and two-dimensional laminar solutions, but we believe that the method can handle turbulent flows without special difficulties.

2. THEORETICAL PRESENTATION

2.1 Presentation of the algorithm

The governing equations for an unsteady three-dimensional viscous flow with gravitational body forces and neglecting the dissipation and pressure work, are:

$$\partial\rho/\partial t + \nabla(\rho\overline{V}) = 0 \qquad (1)$$

$$\rho[\partial\overline{V}/\partial t + (\overline{V}\cdot\nabla)\overline{V}] = -\nabla p + \mu\nabla^2\overline{V} + \rho\overline{g} \qquad (2)$$

$$\rho(\partial T/\partial t) + (\overline{V}\cdot\nabla)T = \mu/\sigma\ \nabla^2 T \qquad (3)$$

where $\overline{g}$ is the gravity force vector, and the viscosity μ and the Prandtl number σ are assumed to be constant.

Equations (1), (2) and (3) form a system of coupled, nonlinear equations. However, if they are investigated separately, it is easy to see that equations (2) and (3) are parabolic partial differential equations. Many numerical schemes are available for the solution of such equations by advancing them in the time-direction. For example we may use explicit, or alternating directions implicit methods to advance the three velocity components and the temperature from time $t^n = n\Delta t$ to time t^{n+1}. However, the pressure cannot be advanced in such a way, nor is the continuity equation satisfied by such a procedure. Indeed, there is, in general, only one pressure distribution p^{n+1} (up to a constant) which will yield such a solution of eq. (2) which satisfies the continuity eq. (1) as well. If we denote the solution to eq. (2) using a guessed pressure distribution by $\overline{V}_p^{n+1}$, we can define a correction, $\Delta\overline{V}^{n+1}$, by

$$\overline{V}^{n+1} = \overline{V}_p^{n+1} + \Delta\overline{V}^{n+1} \qquad (4)$$

and by substitution of (4) in (1) we obtain

$$\nabla(\rho^{n+1}\Delta\overline{V}^{n+1}) = -S_p^{n+1} = -(\partial\rho/\partial t)^{n+1} - \nabla(\rho^{n+1}\overline{V}_p^{n+1}) \qquad (5)$$

If eq. (5) is solved for $\Delta\overline{V}^{n+1}$ and $\overline{V}^{n+1}$ is calculated from (4), the new pressure may now be calculated from eq. (2) by substitution of ρ^{n+1}, $\overline{V}^{n+1}$. Thus, it is easily seen that an iterative scheme may be defined, comprised of the following steps:

(a) Eq. (2) and (3) are advanced from ρ^n, $\overline{V}^n$, p^n, t^n to $\overline{V}_p^{n+1}$, T_p^{n+1}.

(b) A new density is calculated, $\rho_p^{n+1} = \rho(T_p^{n+1}, \rho^n)$.

(c) Eq. (5) is solved for $\Delta\overline{V}^{n+1}$ as a function of ρ_p^{n+1}, $\overline{V}^{n+1}$. Further, $\overline{V}^{n+1}$ is calculated using (4).

(d) Eq. (2) is solved for p^{n+1} as a function of p_p^{n+1}, $\overline{V}^{n+1}$.

Now it is theoretically necessary to start a second iteration by going back to stage (a). However, in practice the iterations are often unnecessary and need not be performed.

The algorithm thus defined is complete. However, it does not define a unique velocity vector $\overline{V}^{n+1}$, as eq (5) does not have a unique solution. Yet $\overline{V}^{n+1}$ satisfies the conservation of mass (eq. (1)), and does not allow diverging solutions due to ever growing mass sources. In this sense this method is comparable to the MAC method. Eq. (1), (3) are satisfied by the solution to round off or truncation errors, while eq. (2) is satisfied only approximately. The same method may be applied to two or three-dimensional, laminar or turbulent, uniform and variable property flows, at any coordinate system.

When non-viscous flows are studied the same method is still applicable, although the differential equations are not parabolic, as the second-order terms vanish. Obviously, the method is applicable but superfluous for potential flows. It may still be useful to solve such problems in order to test the performance of the method.

2.2 The solution of the velocity correction, $\Delta\overline{V}$

The velocity correction vector $\Delta\overline{V}$ is governed by eq. (5). However, it is not convenient to calculate it from this equation, and its properties are not easy to understand. In order to study this vector we regard it as a sum of a gradient of a scalar ϕ and a rotor of a vector $\overline{\psi}$,

$$\Delta\overline{V} = \nabla\phi + \nabla \times \overline{\psi}$$

and by substitution in (5) we obtain

$$\nabla(\rho\nabla\phi) + \nabla(\rho\nabla \times\overline{\psi}) = -S$$

which may be rearranged to

$$\rho\nabla^2\phi + \nabla\rho\cdot\Delta\overline{V} = -S \qquad (6a)$$

or

$$\nabla(\rho\nabla\phi) + \nabla\rho\cdot(\nabla \times \overline{\psi}) = -S \qquad (6b)$$

The two formulations of (6) are equally valid, but none may be used to solve the problem completely, as ψ has three components, and its solution requires a solution of the momentum equations (which may be obtained if we use iterations, for instance). This is clearly illustrated if we examine the uniform density case, $\nabla\rho=0$, where eq. (6a) reduces to

$$\rho\nabla^2\phi = -S \qquad (7)$$

and $\overline{\psi}$ does not influence ϕ at all (but $\overline{\psi}$ is not necesarily negligible).

When the density is not uniform it appears better to solve (6b) with some given ψ distribution. In the present work we assumed $\nabla \times \psi = 0$ in such cases, as in the original Briley's method, although this procedure is not necessarily the only, or even the best one.

Boundary conditions for the velocity correction should not allow any mass flux into the control volume, as all this flux has been accounted for already. Therefore the boundary conditions for the correction potential ϕ are

$$\partial\phi/\partial n = 0 \tag{8}$$

where n is the normal to the boundary. Thus the problem reduces to a solution of a Poisson equation (6a) or the somewhat more complicated eq. (6b)) with Neumann boundary conditions, which is a fairly straightforward problem, provided the following integral constraint is satisfied

$$-\iiint_V s\,dV = \iint_A (\rho\nabla\phi)_n\,dA = 0 \tag{9}$$

2.3 Calculation of the pressure

As mentioned above, the pressure gradients may be obtained by substitution of $\overline{V}^{n+1}$ in the momentum equation (2)

$$\nabla p^{n+1} = \mu\nabla^2\overline{V}^{\,n+1} + \rho^{n+1}\,\overline{g} - \rho^{n+1}[(\partial\overline{V}/\partial t)^{n+1} + (\overline{V}^{n+1}\cdot\nabla)\overline{V}^{n+1}] \tag{10}$$

Yet this procedure may cause serious difficulties, because $\overline{V}^{n+1}$ is accurate to within truncation (or at best round-off) errors only. Therefore it may so happen that the condition

$$\partial/\partial x[\partial/\partial y\ p(\overline{V}^{n+1},\rho^{n+1})] = \partial/\partial y[\partial/\partial x\ p(\overline{V}^{n+1},\rho^{n+1})] \tag{11}$$

is not satisfied, and then the resulting pressure depends on the integration paths of (10). This problem is avoided if a Poisson equation for the pressure is obtained by taking the divergence of (10) as follows:

$$\nabla^2 p^{n+1} = \partial/\partial x[\partial/\partial x\ p(\overline{V}^{n+1},\rho^{n+1})] + \partial/\partial y[\partial/\partial y\ p(V^{n+1},\rho^{n+1})] \tag{12}$$

and the differentiation in the r.h.s. of (12) is performed numerically.

The boundary conditions for (12) are gradient boundary conditions obtained from (10). Thus the pressure is governed by a Poisson equation with Neumann boundary conditions which must satisfy a compatibility condition similar to (9).

The method outlined above is applicable to non-viscous flows as much as to viscous flows. The solutions obtained satisfy the continuity equation to the accuracy of the numerical method used to solve eq. (2) and (6), but has an error which may reach the order of $\nabla(p^{n+1}-p^n)$ in the momentum conservation, unless iterations are used. It appears, therefore, that both accuracy and stability require a good initial pressure guess, and relatively small steps. In the actual computations this is not a serious limitation, and the continuity constraint appears to control the solution satisfactorily, and prevent serious instabilities.

2.4 Outline of the algorithm

To conclude the section the complete numerical scheme is outlined here. The important steps are as follows:

(a) The momentum and energy equation (2), (3) are solved for $\overline{V}_p^{n+1}$, T^{n+1}, using p^n. The solution method of the equations does not appear to influence the method.

(b) The new density ρ^{n+1} is calculated **from** T^{n+1}, using a suitable equation of state.

(c) The mass source term S^{n+1} is calculated using eq. (5) from $\overline{V}_p{}^{n+1}$, ρ^{n+1}.

(d) The velocity correction potential ϕ^{n+1} is calculated using eq. (6) with boundary condtion (8). Minor adjustments of S^{n+1} in order to satisfy the integral constraint (9) are required, and allowed. However, if the required adjustments are large the solution is probably wrong due to too large a time step.

(e) The corrected velocity $\overline{V}^{n+1}$ is calculated using the relation

$$\overline{V}^{n+1} = \overline{V}_p{}^{n+1} + \nabla\phi^{n+1} \tag{13}$$

(f) The pressure gradient ∇p is calculated by substitution of ρ^{n+1} and $\overline{V}^{n+1}$ in eq. (2). It should be noted that the velocity time derivatives $\partial V/\partial t$ should be calculated at this point, using a backward derivative.

(g) The source term for the pressure Poisson equation, S^{n+1}, is calculated by numerical differentiation of the pressure gradients in all the internal points. For Cartesian coordinates this is done as in eq. (12). For other coordinate systems the appropriate equation for the numerical differentiation of the divergence of the velocity should be used.

(h) The pressure source term should be adjusted as to satisfy the constraint

$$\iiint_V S^{n+1} dV = \iint_A (\partial p/\partial n)^{n+1} dA$$

(i) The pressure Poisson equation

$$\nabla^2 p^{n+1} = S^{n+1}$$

is solved now, with gradient boundary conditions as calculated in step (f) above.

With step (i) a complete time step has been finished. If the pressure distribution is not required steps (g)-(i) are not required. If an iterative scheme is desired for each time step, each iteration cycle should be concluded with a solution of the momentum eq. (2) using $\overline{V}^{n+1}$ and p^{n+1} as completed in step (i) to compute the momentum correction to $\overline{V}^{n+1}$.

3. THREE DIMENSIONAL NON-VISCOUS FLOW

3.1 Description of the problem

The case under consideration is that of a circular cylinder filled with an incompressible and non-viscous fluid, such as fuel tanks. The cylinder is free to move at an angular velocity $\overline{\omega}$ and a translational acceleration $\overline{A}_o$. Using a coordinate system attached to the cylinder (fig. 1), the problem is governed by

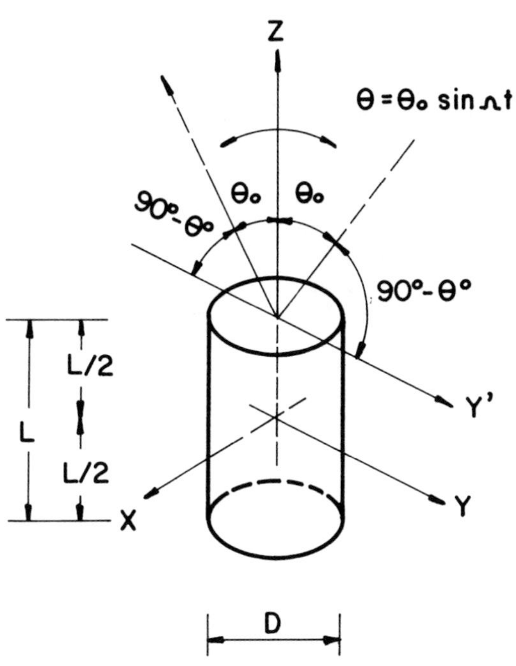

Fig. 1: The geometry of the rotating cylinder.

$$\nabla \cdot \overline{V} = 0 \tag{14}$$

$$\partial \overline{V}/\partial t + (\overline{V} \cdot \nabla)\overline{V} = -1/\rho \, \nabla p - \overline{A}_b - 2(\overline{\omega} \times \overline{V}) \tag{15}$$

$$\overline{A}_b = \overline{A}_o + (d\overline{\omega}/dt) \times \overline{R} + \overline{\omega} \times (\overline{\omega} \times \overline{R}) \tag{16}$$

where $\overline{R}$ is the radius vector.

When the divergence of (15) is calculated we get

$$\nabla^2 p = S_p(\overline{V}) = -\rho\nabla[(\overline{V} \cdot \nabla)\overline{V} + \overline{A}_b + 2(\overline{\omega} \times \overline{V})] \tag{17}$$

Eq. (15) - (17) should be solved, subject to the boundary condition

$$\overline{V}_n = 0 \tag{18}$$

The problem is solved by an explicit finite-difference method. A staggered mesh is used. The Poisson equations for the correction potential and the pressure are solved by successive-over-relaxation of the Gauss-Seidell method. The boundary conditions for the potential are eq. (8), and for the pressure they are calculated from (15).

On the axis of the cylinder, r=0, the cylindrical coordinate system becomes singular. Therefore a small cylindrical control volume surrounding the axis was defined, in which the cartesian velocity components were calculated. We did not encounter stablility problems, and the accuracy was good. Some examples will be given in the folloiwng sections. The runs were performed for

$\underline{L}/D = 2.86$

$\underline{\omega} = (\omega_0 \cos \Omega t,\ 0,\ 0)$

$\overline{A}_0 = 0$

The situation is that of a cylinder subject to periodic sinusoidal fluctuations around a lateral axis (say x).

3.2 Richardson extrapolation

The convergence of the method was tested on a solution of the flow inside a cylinder with L/D = 2.86 rotating around the x axis, at an angular velocity of 1 rad/S. The time step was 0.01S and the meshes used were (M+1)x(N+1)x(L+1) in the r, ϕ, and z respectively. Four mesh combinations were used with (M,N,L) equal to (3,8,20), (4,12,30), (5,12,30), and (6,16,40). At the point where r/D=0.214; z/L=.05 and ϕ=0 the velocity component in the r direction, u_r, is zero. Computed results for this velocity indeed approach zero as the mesh size is reduced as shown in fig. 2 (the point does not lie exactly on all meshes and the

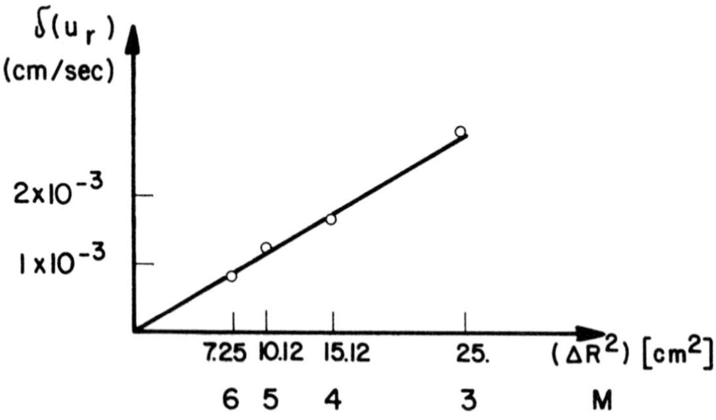

Fig. 2: Richardson's extrapolation for the rotating cylinder

values were calculated by interpolation between the mesh points). Second order convergence is easily identified.

3.3 Comparison with potential solution

As the flow in the cylinder starts from rest it is a potential flow and may be obtained by solving the potential equation. We compared such a potential solution with a solution obtained by the present method for the case

L/D = 2.86, Ω = 10 rad/S, ω_0 = $\pi/12$ with a mesh of M=3; N=8; L=20, and $\Delta t = \pi/400$. The total moment of the fluid around the x axis is

$$M_x = \int_A p dA \, \hat{n} \times \hat{r}$$

with A the area of the cylinder, $\hat{n}$ the normal to A, and $\hat{r}$ the radius vector. M_x is shown in fig. 3 and given in table 1 below.

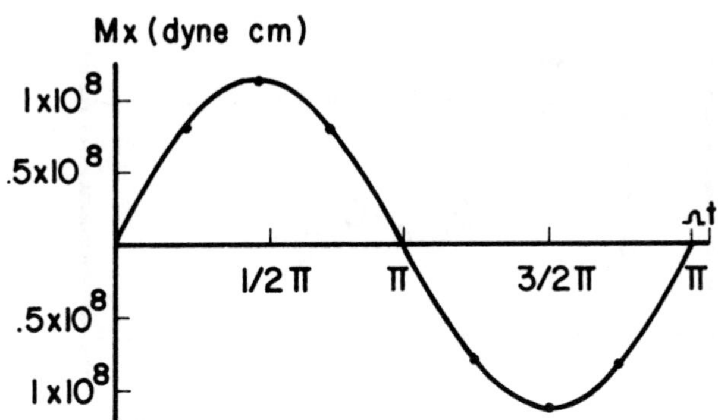

Fig. 3: The moment around the x-axis for a rotating cylinder The points are from the potential solution.

t [sec]	M_x [dyne cm] Present method	Potential
0	0	0
1/4 π/Ω	8.055×10^7	8.053×10^7
1/2 π/Ω	1.139×10^8	1.139×10^8
3/4 π/Ω	8.053×10^7	8.053×10^7
π/Ω	6.398×10^4	0
5/4 π/Ω	-8.042×10^7	-8.053×10^7
3/2 π/Ω	-1.139×10^8	-1.139×10^8
7/4 ρ/Ω	-8.066×10^7	-8.053×10^7
2 π/Ω	-1.648×10^5	0

Table 1: The moment around the x-axis

Comparison between the two methods is very good.

4. VISCOUS FREE CONVECTION IN A RECTANGULAR BOX

4.1 General description

This case is concerned with free convection in a box, similar to a solar pool. The problem is simplified by the following assumptions and restrictions:

(1) The fluid is compressible, but the density variation depends on the temperature distribution only as given by

the Boussinesq approximation $\rho=\rho_0\ [1-\beta(T-T_0)]$, where ρ_0 is a reference density at the reference temperature T_0 and β is the volumetric expansion coefficient.
(2) The flow is laminar.
(3) The Prandtl number is one.
(4) The flow is two-dimensional.
The flow occurs in a rectangular chamber without a top, as described in Fig. 4.

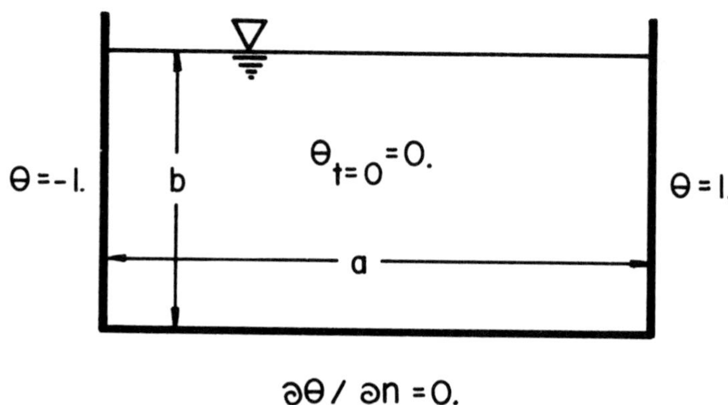

Fig. 4: The control volume for the free convection problem

Its aspect ratio, which is defined as the ratio of its height, b, to its basis, a, is equal to 0.5. The problem is solved in a cartesian coordinate system. Initially, the fluid of density, ρ_0, is at rest and at a constant uniform temperature, i.e., $u=v=0$ and $T=T_0$. At time $t=0$ the left wall is suddenly cooled to a temperature $T_0-\Delta T$ and the right wall is heated to a temperature $T_0+\Delta T$. The bottom of the chamber is adiabatic, i.e. $\partial T/\partial y=0$. The top of the chamber is a free surface defined by $v=0$, $\partial T/\partial y=0$.

If $T(x,y,t)$ is the temperature of the fluid in a certain point (x,y) at the time t, it is convenient to define a dimensionless temperature by

$$\theta = (T-T_0)/\Delta T$$

Thus, at $t\geq 0$ the dimensionless temperature of the fluid on the left wall of the mixing chamber is $\theta=-1$, and of the right one is $\theta=1$. The Boussinesq approximation is then defined by $\rho=\rho_0(1-\beta\Delta T\theta)$.

In the following test case the Grashof number is Gr=10 and 100 where $Gr=g\rho_0^2 b^2\beta\Delta T/\mu^2$. The volumetric expansion coefficient is $\beta=10^{-4}$ and the temperature difference is $\Delta T=20^\circ K$.

4.2 Numerical details

The finite difference grid is staggered having 24 and 12

mesh points in x and y direction respectively. All other details are identical to those described in section 3 above. The extension of Briley's method to the present variable density problem is made by defining a potential function, ϕ, for correcting the velocity components as follows:

$$\rho u = \partial\phi/\partial x \quad , \quad \rho v = \partial\phi/\partial y$$

The residue of the continuity equation is defined as

$$\text{Res.} = \partial\rho/\partial t + \partial(\rho u)/\partial x + \partial(\rho v)/\partial y$$

where u,v are the corrected terms of the velocity components. In this computation the residue was always less than $10-^{-5}$, when the time derivative was calculated to first order and the spatial derivatives were calculated to second order. The time step was given by

$$\Delta T = 0.01b(g\beta\Delta T)^{-1/2}$$

4.3 Results

Fig. 5 shows the stream function at different times for Gr=100. A large eddy is formed immediately upôn application of the boundary conditions, but later small secondary eddies are formed at the corners between the side walls and the bottom. The flow is nearly homogeneous, and from the hot wall to the cold wall, in most of the box. Very near to the bottom a very thin boundary layer is formed. The strength of the main eddy measured by the maximum value of the stream function is shown in fig. 6 as a function of the time. The computation was not sufficiently long to make ψ_{max} constant, but it appears to level off, and we assume that it has nearly reached such condition.

5. CONCLUSIONS

We have formulated a general method for the solution of time-dependent fluid flow problems. The method ensures exact mass conservation, but momentum conservation is satisfied only approximately. The few examples illustrate the power of the method, and its capabilities. The efficiency of the method depends on the efficiency of the marching algorithm (in the time). In the present runs explicit algorithms were used, and the computer time requirements were relatively large. However, application of implicit schemes is likely to minimize the computer time. Another saving may be accomplished if direct Poisson solvers are employed.

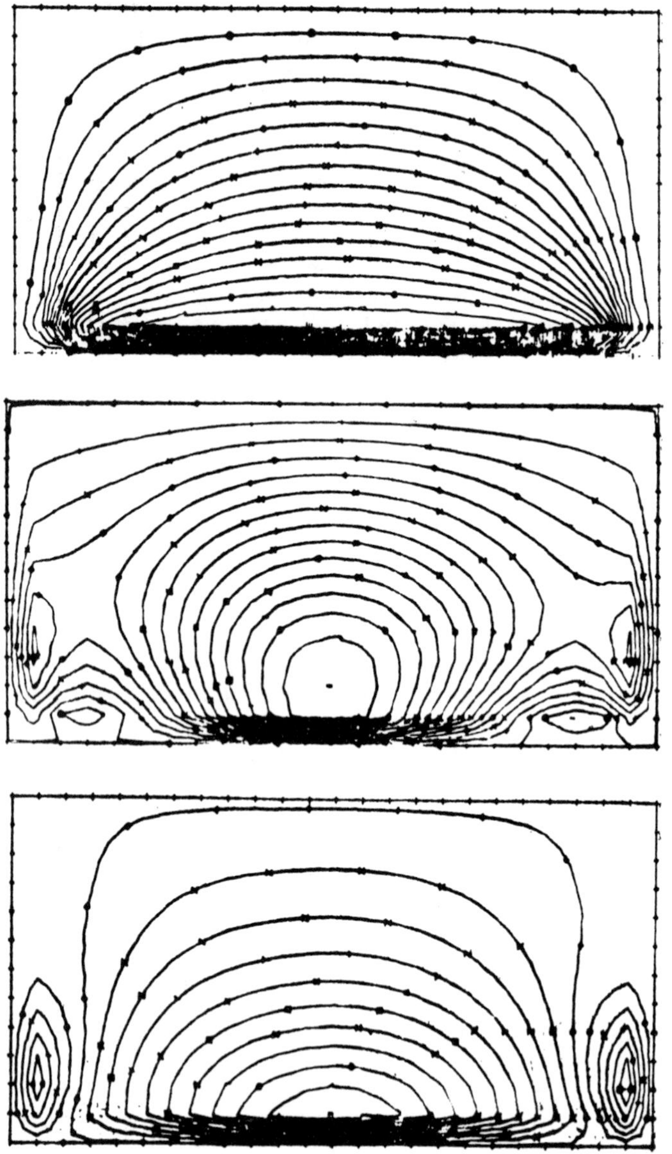

Fig. 5: Stream function for the free convection problem at: (a) t = 0.1S; (b) t = 9S; (c) t = 11.1S

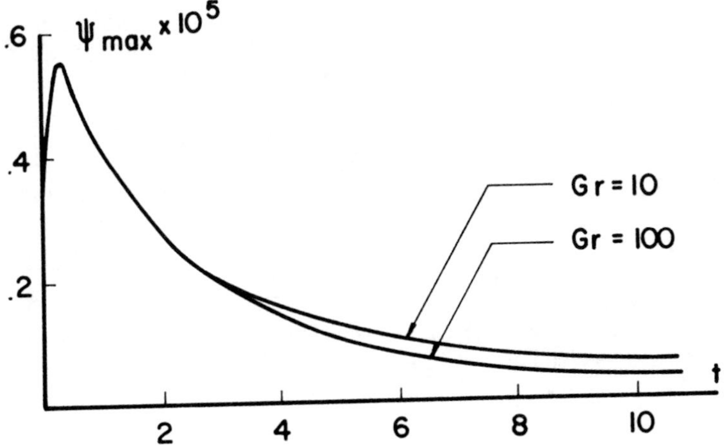

Fig. 6: The maximum stream function for the free convection problem

REFERENCES

1. ROACHE, P.J. - Computational Fluid Dynamics, Hermosa, 1972.
2. THOM, A. - The Flow Past Circular Cylinders at Low Speeds. Proc. Roy. Soc. London, A141, pp. 651-666, 1933.
3. BARAKAT, H.Z., and CLARK, J.A. - Analytical and experimental Study of the Transient Laminar Natural Convection Flows in Partially Filled Liquid Containers, Proc. 3rd Int. Heat Transfer Conf., Vol. II, p. 152, Chicago, 1966.
4. GOSMAN, A.D., PUN, W.M., RUNCHAL, A.K., SPALDING, D.B., and WOLFSHTEIN, M. - Heat and Mass Transfer in Recirculating Flows, Academic Press, London, 1969.
5. AZIZ,K., and HELLUMS, J.D. - Numerical Solution of the Three-Dimensional Equations of Motion for Laminar Natural Convection, Physics of Fluids, Vol. 10, NO. 2, pp. 314-324, 1967.
6. DE VAHL DAVIS, G., and WOLFSHTEIN,M. - A Study of a Three-Dimensional Free Jet Using the Vorticity Vector Potential Method, Proc. 4th Int. Conf. on Numerical Methods in Fluid Dynamics, Ed. Ehlers, J., Hepp, K., Weidenmuller, H.A., Springer, Berlin, 1976.
7. LILLEY, D.K. - On the Computational Stability of Numerical Solutions of Time-Dependent Non-Linear Geophysical Fluid Dynamics Problems, U.S. Weather Bureau Monthly Weather Review, Vol. 93, No. 1, pp. 11-26, 1965.
8. HARLOW, F.H., AND WELCH, J.E. - Numerical Calculations of Time-Dependent Viscous Incompressible Flow of Fluids with Free Surface, Physics of Fluids, Vol. 8, No. 12, pp. 2182-2189, 1965.

9. BERGER, M., and WOLFSHTEIN, M. - Numerical Solution of Convection in Rectangular Ducts, Physiochemical Hydrodynamics, Ed. Spalding, D.B., Advanced Publications, 1977.
10. DODGE, P.R., Numerical Method for 2D and 3D Viscous Flows, AIAA J., Vol. 15, No. 7, pp. 961-965.
11. PATANKAR, S.V., and SPALDING, D.B. - A Calculation Procedure for Heat Mass and Momentum Transfer in Three-Dimensional Parabolic Flows, J. Heat and Mass Transfer, Vol. 15, pp. 1787-1806, 1972.
12. BRILEY, W.R. - Numerical Methods for Predicting Three Dimensional Steady Flows in Ducts, J. Computational Physics, Vol. 14, pp. 8-28, 1974.

NATURAL CONVECTION BETWEEN CONCENTRIC (HORIZONTAL) CIRCULAR CYLINDERS BY A PENALTY-FINITE ELEMENT METHOD

A. Satake[1] and J. N. Reddy[2]

SUMMARY

The paper presents numerical results obtained by a penalty-finite element model for natural convection between concentric horizontal cylinders in the presence of temperature gradient between the inner and outer cylinders. The results are compared with experimental results as well as other approximate solutions for moderately high Rayleigh numbers, R_A. The agreement is found to be very good. Finite-element results are also presented for natural convection in a half-circular annulus for $R_A = 10^3$ and 10^5. These results should serve as test cases for other investigators.

1. INTRODUCTION

Study of natural convection between concentric horizontal cylinders and spheres dates back to early 1930's (see Beckmann [1]). A complete history of the many investigations made for horizontal circular cylinders can be found in the work of Kuehn and Goldstein [2]. Due to the complex nature of the governing equations of the phenomena, relatively few theoretical studies have been made that deal with approximate analytical as well as numerical solutions to natural convection in concentric circular cylinders. The first numerical solution for natural convection between horizontal concentric cylinders was due to Crawford and Lemlich [3], who examined

[1]Visiting Assistant Professor, Department of Mathematics, University of Oklahoma, Norman, OK 73019, USA.

[2]Professor, Engineering Science and Mechanics Department, Virginia Polytechnic Institute and State University, Blacksburg, VA 24061, USA.

the steady two-dimensional flow for a Prandtl number of 0.74 and for radius ratios of 2, 8 and 57; Abbot [4] considered radius ratios close to unity. Mack and Bishop [5] used a power-series expansion and Hodnett [6] used a perturbation method for natural convection between isothermal concentric cylinders, and Huetz and Petit [7] considered the case of constant heat flux on one wall, the other remaining isothermal. Existence of secondary phenomena in the form of small rolls turning in the opposite direction of the main cells was reported (for $R_A = 6 \times 10^3$, $R = 1.2$ and $P_R = 0.7$) by Powe, Carley, and Carruth [8]. Kuehn and Goldstein [2] carried finite-difference, numerical (as well as experimental) investigation of natural convection within a horizontal annulus. Custer and Shaughnessy [9] considered small Prandtl numbers by using a double perturbation expansion in powers of the Grashof and Prandtl numbers. Recently, Charrier-Mojtabi, Mojtabi and Caltagirone [10] solved the problem using the implicit alternating direction scheme on the vorticity-stream function formulation for Rayleigh number between 100 and 50,000, (P_R = 0.7 or 0.02) and radius between 1.2 and 5, and Jischke and Farschi [11] used a boundary-layer approximation to solve the problem for large Rayleigh number and large Prandtl numbers.

The present study is concerned with the numerical solution of the problem by the finite element method. The finite-element model is based on a velocity-temperature formulation without pressure. The pressure is eliminated by posing the problem as one of constrained problem in which the incompressibility (i.e., continuity of mass) condition is treated as a constraint on the velocity field. The constraint is then included in the variational formulation of the problem by means of the penalty function method of Courant [12].

2. THEORETICAL FORMULATION

Consider a steady, hydrodynamically fully developed, but thermally developing laminar flow of a viscous incompressible Newtonian fluid between two concentric circular cylinders. In writing the governing equations of the flow, we assume that Boussinesq approximation (i.e., physical properties of the fluid are constant; the density is considered to be constant except for its variation in the buoyancy term) holds. In order to cast the equations in terms of non-dimensionalized variables, we introduce the following normalization:

$$x = x'/d,\ y = y'/d,\ u = u'd/\alpha,\ v = v'd/\alpha,\ T = (T' - T_o')/(T_n' - T_c'),\ p = p'd^2/\rho_o\alpha^2,\ \alpha = K/\rho_o C_\rho \qquad (1)$$

Here quantities with primes denote the dimensional variables; (u, v) are the velocity components along (x,y)-coordinates, p is the pressure, T is the temperature (T_c' and T_h' denote the temperatures of the cold and hot walls, respectively, and T_o'

is the reference temperature), ρ_o is the reference value of the density, K is the thermal conductivity, C_p is the specific heat at constant pressure, and d is the difference of the outer radius r_o to the inner radius r_i. The governing equations of the problem can be expressed in the nondimensional form.

Conservation of mass:

$$\frac{\partial u}{\partial x} + \frac{\partial v}{\partial y} = 0 \tag{1}$$

conservation of linear momentum:

$$A(u) = -\frac{\partial p}{\partial x} + P_R[2\frac{\partial^2 u}{\partial x^2} + \frac{\partial}{\partial y}(\frac{\partial u}{\partial y} + \frac{\partial v}{\partial x})] - P_R R_A T \tag{2}$$

$$A(v) = -\frac{\partial p}{\partial y} + P_R[s\frac{\partial^2 v}{\partial y^2} + \frac{\partial}{\partial x}(\frac{\partial u}{\partial y} + \frac{\partial v}{\partial x})] \tag{3}$$

conservation of energy:

$$A(T) = \nabla^2 T \tag{4}$$

wherein P_R is the Prandtl number, R_A is the Rayleigh number, and $A(\cdot)$ is a differential operator,

$$P_R = \nu/\alpha,\ R_A = g\beta d^3(T'_n - T'_c)/\nu\alpha,\ A(\cdot) = (u\frac{\partial}{\partial x} + v\frac{\partial}{\partial y})(\cdot) \tag{5}$$

Here ν is the kinematic viscosity, β is the coefficient of thermal expansion, and g is the acceleration due to gravity.

Equations (1)-(4) must be solved subject to appropriate boundary conditions. At any given point on the boundary Γ of the fluid region Ω, the boundary conditions are either of essential type or natural type:

essential natural

$$T = (\hat{T}'-T'_o)/(T'_h-T'_c) \text{ on } \Gamma_1,\quad \frac{\partial T}{\partial n} = q'\frac{d}{\alpha}(T'_h-T'_c) \text{ on } \Gamma_2 \tag{6}$$

$$u = \hat{u}'d/\alpha \text{ on } \Gamma_{1u},\quad t_x = \hat{t}'_x(\frac{d}{\alpha})^2 \text{ on } \Gamma_{1s} \tag{7}$$

$$v = \hat{v}'d/\alpha \text{ on } \Gamma_{2u},\quad t_y = \hat{t}'_y(\frac{d}{\alpha})^2 \text{ on } \Gamma_{2s} \tag{8}$$

wherein $\hat{T}'$, $\hat{q}'$, $(\hat{u}', \hat{v}')$ and $(\hat{t}'_x, \hat{t}'_y)$ denote specified (dimensional) temperature, heatflux, velocities and boundary stresses, respectively, $n = (n_x,n_y)$ denotes the unit normal to the boundary, and Γ_1, Γ_2 etc. denote portions of boundary such that

$$\Gamma_1 \cup \Gamma_2 = \Gamma, \; \Gamma_1 \cap \Gamma_2 = 0; \; \Gamma_{iu} \cup \Gamma_{is} = \Gamma, \; \Gamma_{iu} \cap \Gamma_{is} = 0. \qquad (9)$$

Equation (9), implies that, for example, Γ_1 and Γ_2 are disjoint portions whose (direct) sum is the total boundary Γ. In any practical problem, only one of the two boundary conditions, in each of eqns. (6)-(8), is specified.

3. PENALTY FUNCTION FORMULATION

Equations (1)-(9) cannot be solved in closed-form without some simplifying assumptions. Alternatively, these equations can be solved by numerical methods of approximation. In the finite-element formulation of these equations any one of the several alternate forms (e.g., velocity-stream function, velocity-pressure, etc.) of the governing equations can be employed. The present investigation utilizes the penalty function method of Courant [12] to solve the momentum and energy equations with the continuity equation as a constraint among the velocity components. The penalty function method is a computational device to approximate solutions to constrained variational problems. Since the introduction of the concept into the finite-element analysis of incompressible fluid flow by Zienkiewicz [13], there have been several papers dealing with the application of the method to the analysis of natural convection in rectangular enclosures (see, for example, [14-16]). The present study is concerned with the natural convection in the annulus between two concentric cylinders.

Toward constructing a penalty-finite element model of eqns. (1)-(9), we give a weak formulation of the equations using the penalty function method. To this end consider the constrained variational problem associated with eqns. (1)-(9): Find (u,v,T) such that

$$\delta I_o(u,v,T) = 0, \qquad (10)$$

where

$$\delta I_o(u,v,T) = \int_\Omega \{[A(u) - P_R R_A T]\delta u + P_R[2\frac{\partial u}{\partial x}\frac{\partial \delta u}{\partial x} + 2\frac{\partial v}{\partial y}\frac{\partial \delta v}{\partial y} + (\frac{\partial u}{\partial y} + \frac{\partial v}{\partial x})(\frac{\partial \delta u}{\partial y} + \frac{\partial \delta v}{\partial x})] + A(v)\delta v + A(T)\delta T + \frac{\partial T}{\partial x}\frac{\partial \delta T}{\partial x} + \frac{\partial T}{\partial y}\frac{\partial \delta T}{\partial y}\} dxdy - \int_{\partial\Omega_{1s}} \hat{t}_x \delta u \, ds - \int_{\partial\Omega_{2s}} \hat{t}_y \delta v \, ds - \int_{\partial\Omega_1} \hat{q}\delta T \, ds, \qquad (11)$$

and u and v satisfy the divergence-free condition (1). Note that the pressure does not appear in eqn. (11). This is because the weak formulation of the momentum equations (2) and (3) give rise to the term,

$$\int_\Omega -\left(\frac{\partial p}{\partial x}\,\delta u + \frac{\partial p}{\partial y}\,\delta v\right)dxdy = \int_\Omega p\left(\frac{\partial \delta u}{\partial x} + \frac{\partial \delta v}{\partial y}\right)dxdy$$

$$+ \int_\Gamma p(\delta u n_x + \delta v n_y)ds, \quad (12)$$

in which the first term on the right-hand side vanishes in view of the constraint eqn. (1), and the second term is included in $\hat{t}_x$ and $\hat{t}_y$.

The penalty function method replaces the constrained variational form (10) by an equivalent (but approximate) unconstrained variational problem in which the original form (10) is modified by the addition (in least-squares sense) of the residual in the constraint equation (see [16]):

$$\delta I_p(u,v,T) \equiv \delta I_o(u,v,T) + \lambda\delta R(u,v) = 0, \quad (13)$$

where R(u,v) is the quadratic form, called penalty functional, given by

$$R(u,v) = \frac{1}{2}\int_\Omega \left(\frac{\partial u}{\partial x} + \frac{\partial v}{\partial y}\right)^2 dxdy, \quad (14)$$

and λ is the <u>penalty parameter</u> (preassigned). An approximation to the pressure is given by (see [17,18]),

$$p_\lambda = -\lambda\left(\frac{\partial u_\lambda}{\partial x} + \frac{\partial v_\lambda}{\partial y}\right). \quad (15)$$

Presumably, as $\lambda \to \infty$, the constraint is satisfied exactly and the solution to the penalty problem (13) converges to the solution of the original problem (10).

4. PENALTY-FINITE ELEMENT MODEL

Consider a finite-element discretization, Ω_h, of Ω. Over a typical element Ω_e, the dependent variables u, v, and T are interpolated by

$$u = \Sigma u_i N_i(x,y),\ v = \Sigma v_i N_i(x,y),\ T = \Sigma T_i N_i(x,y) \quad (16)$$

where N_i are the finite-element interpolation functions, and u_i, v_i and T_i are the nodal values of u, v, and T, respectively. Substituting (16) into the variational form associated with a typical element, we obtain the following pair of coupled nonlinear equations,

$$[K^e]\{\Delta^e\} = \{F^e\},\ [C^e]\{T^e\} = \{Q^e\}, \tag{17}$$

where

$$[K^e] = \left[\begin{array}{c|c} [H^x] + \lambda[S^x] & P_R[S^{xy}]^T + \lambda[S^{xy}] \\ \hline P_R[S^{xy}] + \lambda[S^{xy}]^T & [H^y] + \lambda[S^y] \end{array}\right]$$

$$C^e_{ij} = A_{ij} + S^x_{ij} + S^y_{ij}\ ,\ A_{ij} = \int_{\Omega_e} N_i A(N_j)\,dxdy,$$

$$S^{\xi\eta}_{ij} = \int_{\Omega_e} N_{i,\xi} N_{j,}\ dxdy\ ,\ (\xi,\eta = o,x,y)\ ,\ S^{\xi\xi}_{ij} = S^{\xi}_{ij},\ \text{etc.}$$

$$H^x_{ij} = A_{ij} + P_R(2\ S^x_{ij} + S^y_{ij})\ ,\ H^y_{ij} = A_{ij} + P_R(2S^y_{ij} + S^x_{ij}),$$

$$\{F^e\} = \{F^x_i, F^y_i\}^T,\ \{\Delta^e\} = \{u_i, v_i\}^T,\ \{T^e\} = \{T_i\}^T,$$

$$F^x_i = P_R R_A \int_{\Omega_e} TN_i\ dxdy + \int_\Gamma t_x N_i\ ds,$$

$$F^y_i = \int_\Gamma t_y N_i\ ds,\ Q^e_i = \int_\Gamma qN_i\ ds. \tag{18}$$

The element equations in (17) are assembled in the usual manner to obtain the associated global matrices. However, these matrix equations are nonlinear and require iterative procedures. The following iterative procedure is employed in the present study. At the beginning of the first iteration the matrix coefficients are computed assuming that the velocity vector is zero. Then the temperature equation is solved for {T}. Using the computed temperature field, the velocity vector {Δ} is computed, completing one cycle of iteration. Using the velocity field obtained in the previous iteration, matrix coefficients for the next iteration are computed and the procedure is repeated until the Euclidean norm of the difference of the solutions at any two successive iterations becomes sufficiently small (say, $< 10^{-4}$). In the present study we used, to accelerate the convergence, a weighted sum of the variables in computing the matrix coefficients for the next iteration. For example, at the end of r-th iteration we have, $0 < \rho_1,\ \rho_2 < 1$,

$$\begin{aligned} \{u\}^* &= \rho_1\{u\}_r + (1 - \rho_1)\{u\}_{r-1}, \\ \{\theta\}^* &= \rho_2\{T\}_r + (1 - \rho_2)\{T\}_{r-1}. \end{aligned} \tag{19}$$

The algebraic complexity and the nonlinear nature of the matrices in equation (18) forces one to use numerical integration to evaluate various matrix coefficients. Another reason which necessitates the use of numerical integration is the "reduced integration" required by the penalty method. That is, the coefficients of λ are numerically evaluated using one less number of Gauss points (in each direction) than usual (see [19,20]). In the present study the four-node isoparametric element is employed, with 1 x 1 Gauss rule for the penalty terms.

5. DISCUSSION OF THE RESULTS

All of the results presented here were obtained on an IBM 370/158 computer using the double precision arithmatic. The penalty parameter, λ, was selected to be equal to 10^8.

Two different geometries were considered. First one is an annulus between two concentric circular cylinders, and the second one is a half annulus (horizontal). We shall denote by r_i and r_o the inner and outer radii of the annulus. In the present study it is assumed that $r_o/r_i = 2.6$. Due to existence of symmetry with respect to vertical centerline of the problem, only half of the annulus was considered in the finite-element analysis. The inner cylinder is maintained at temperature T_h', and the outer cylinder is maintained at temperature T_c'.

Two Rayleigh numbers, $R_A = 10^3$ and 4.7×10^4, with $P_R = 0.706$ were used in the present study. Plots of stream lines and isotherms at Rayleigh numbers of 10^3 and 4.7×10^4 (not shown due to the space limitation) compare, visually, very well with those of Charrier-Mojtabi, et al. [10] and Kuehn and Goldstein [2]. For $R_A = 10^3$, the heat transfer is mainly by conduction, whereas for $R_A = 10^4$ convection is dominant.

A plot of the temperature and velocity along the horizontal cross section through the center of the annulus is shown in Fig. 1. The plot for the temperature agrees closely with the experimental results of Kuehn and Goldstein [2]. In Fig. 2, a plot of the local equivalent conductivity, K_{eq}, versus the angular distance θ is shown (for $R_A = 10^3$ and 4.7×10^4). Here K_{eq} is given by

$$K_{eq} = Nu(\theta)\cdot\ln(r_o/r_i),\ Nu(\theta) = \frac{\partial T}{\partial n}(\theta),$$

where Nu is the Nussett number. The Nussett number was calculated at the Gauss points closest to the walls. The present result is compared with the analytical solution of Jischke and Forshchi [11] and the numerical solution of Kuehn and Goldstein [2]. It should be noted that the solution in [11] is based on infinite Prandtl number.

Next, results are presented for an half annulus. Plots of isotherms and stream lines are shown for $R_A = 10^5$ in Fig. 3

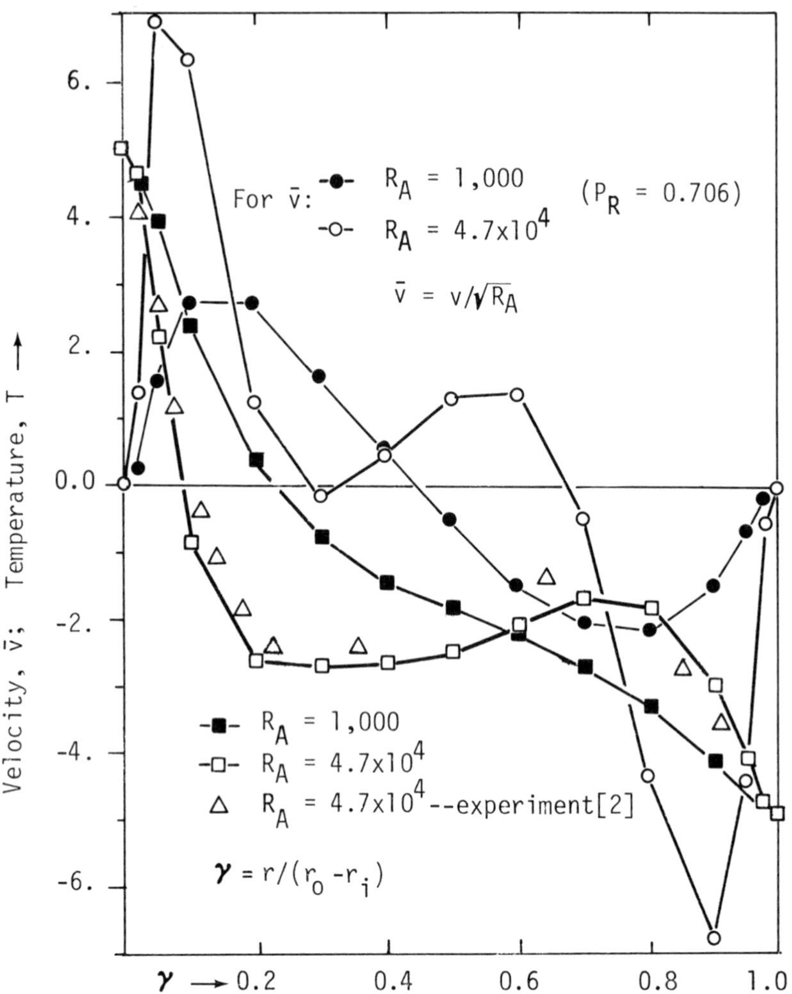

Figure 1 Vertical velocity and nondimensionalized temperature distribution along horizontal cross-section for the cylindrical annulus.

No numerical or experimental results exist in the open literature for this problem. As expected, the plots are symmetric about the vertical centerline. On the basis of the good agreement between the present results and the results of other investigators for the concentric cylinders case, it is reasonable to assume that the results for the half annulus case are also good.

6. CONCLUSIONS

Numerical solutions are presented for natural convection in concentric cylindrical annulus and half annulus using a penalty-finite element model. The finite-element solutions

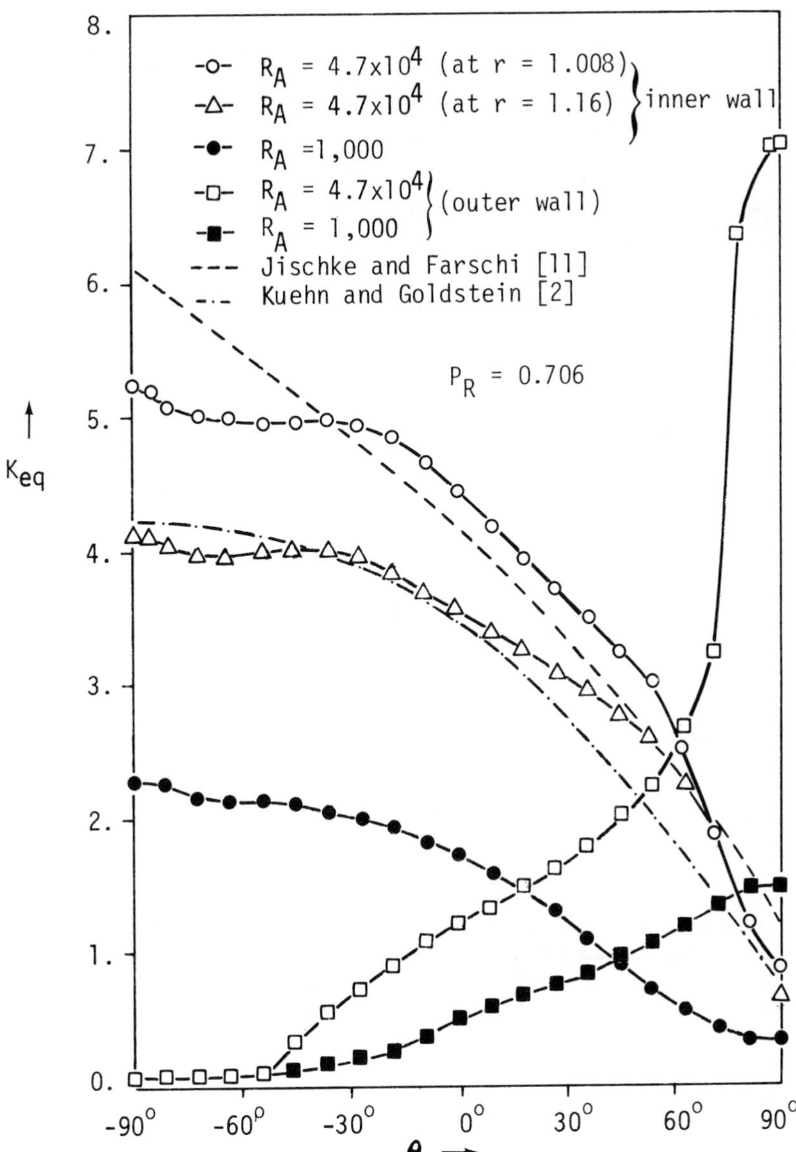

Figure 2 Plot of local equivalent conductivity versus the angular distance for cylindrical annulus.

are in good agreement with other numerical and experimental solutions for concentric cylindrical annulus.

Due to the space limitation, numerical results and plots for other boundary conditions are not included in the present paper. A more detailed discussion of the results will appear elsewhere.

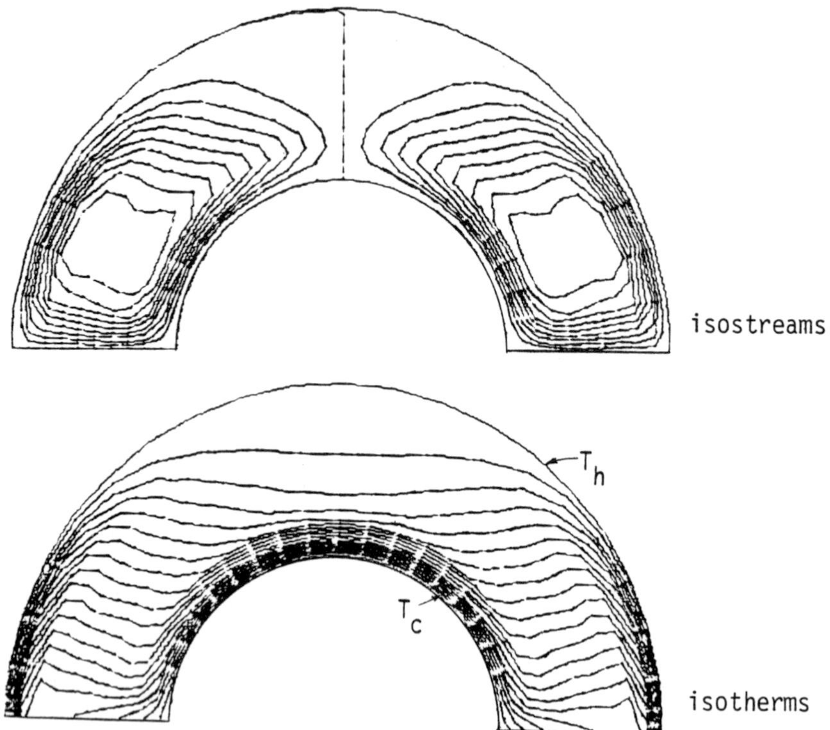

Figure 3 Plot of isotherms and streamlines for half annulus placed horizontally ($R_A = 10^5$, $P_R = 1.0$).

Three-dimensional penalty-finite element model that incorporates a turbulence mechanism is of great interest in many problems of engineering interest, and therefore must be developed. It should be also pointed out that the pressure obtained by eqn. (15) is oscillatory and alternate means of computing the pressure must be considered (for example, solve the Poisson's equation for pressure).

ACKNOWLEDGEMENT

The authors gratefully acknowledge the computer time granted by the Merrick Computer Center at the University of Oklahoma.

7. REFERENCES

1. BECKMANN, W - Die Wärmeübertragung in Zylindrischen Gasschichten bei Natürlicher Konvection. Forsch. Geb. d. Ingenieurwesen, Vol. 2, No. 5, pp. 165-178, 1931.

2. KUEHN, T. H., and GOLDSTEIN, R. J. - An Experimental and Theoretical Study of Natural Convection in the Annulus Between Horizontal Concentric Cylinders. J. Fluid Mechanics, Vol. 74, Part 4, pp. 695-719, 1976.

3. CRAWFORD, L., and LEMLICH, R. - Natural Convection in Horizontal Concentric Cylindrical Annuli. I. E. C. Fund., Vol. 1, pp. 260-264, 1962.

4. ABBOTT, M. R. - A Numerical Method for Solving the Equations of Natural Convection in a Narrow Concentric Cylindrical Annulus with a Horizontal Axis. Quart. J. Mech. Appl. Math., Vol. 17, pp. 471-481, 1964.

5. MACK, L. R., and BISHOP, E. H. - Natural Convection Between Horizontal Concentric Cylinders for Low Rayleigh Numbers. Quart. J. Mech. Appl. Math., Vol. 21, pp. 223-241, 1968.

6. HODNETT, P. F. - Natural Convection Between Horizontal Heated Concentric Circular Cylinders. J. Appl. Math. Phys., Vol. 24, pp. 507-516, 1973.

7. HUETZ, J., and PETIT, J. P. - Natural and Mixed Convection in Concentric Annular Spaces - Experimental and Theoretical Results for Liquid Metals. 5th Int. Heat Transfer Conf., Tokyo, Vol. 3, pp. 169-172, 1974.

8. POWE, R. E., CARLEY, C. T., and CARRUTH, S. L. - A Numerical Solution for Natural Convection in Cylindrical Annuli. J. Heat Transfer, Vol. 92, No. 12, pp. 210-220, 1971.

9. CUSTER, J. R., and SHAUGHNESSY, E. J. - Thermoconvective Motion of Low Prandtl Number Fluids within a Horizontal Cylindrical Annulus. J. Heat Transfer, Vol. 99, pp. 596-602, 1977.

10. CHARRIER-MOJTABI, M. C., MOJTABI, A., and CALTAGIRONE, J. P. - Numerical Solution of a Flow due to Natural Convection in Horizontal Cylindrical Annulus. J. Heat Transfer, Vol. 101, pp. 171-173, 1979.

11. JISCHKE, M. C., and FARSHCHI, M. - Boundary Layer Regime for Natural Convection Between Concentric Spheres and Circular Cylinders. Research Report No. OU-AMNE-79-5, School of AMNE, University of Oklahoma, Norman, OK, 1979.

12. COURANT, R. - Calculus of Variations and Supplementary Notes and Exercise, revised and amended by J. Moser, New York University, 1956.

13. ZIENKIEWICZ, O. C. - Constrained Variational Principles and Penalty Analysis Function Methods in Finite Elements. Lecture Notes on Mathematics, Springer-Verlag, Berlin, 1973.

14. REDDY, J. N., and MAMIDI, D. R. - Penalty Velocity-Stream Function Finite Element Models for Free Convection Heat Transfer Problems. Recent Advances in Engineering Science, Ed. Sierokoswki, R. L., University of Florida Gainesville, FL, pp. 381-386, 1978.

15. MARSHALL, R. S., HEINRICH, J. C., and ZIENKIEWICZ, O. C. - Natural Convection in a Square Enclosure by a Finite-Element Penalty Function Method Using Primitive Fluid Variables. Numerical Heat Transfer, Vol. 1, pp. 315-350, 1978.

16. REDDY, J. N., and SATAKE, A. - A Comparison of a Penalty Finite Element Model with the Stream Function - Vorticity Model of Natural Convection in Enclosures. J. Heat Transfer, Vol. 102, pp. 659-666, 1980.

17. REDDY, J. N. - On the Accuracy and Existence of Solutions to Primitive Variable Models of Viscous Incompressible Fluids. Int. J. Engineering Science, Vol. 16, pp. 921-929, 1978.

18. REDDY, J. N. - Penalty Finite Element Methods for the Solution of Advection and Free Convection Flows, Finite Element Methods in Engineering, Ed. Kabaila, A. P. and Pulmano, V. A., The University of New South Wales, Sydney, Australia, pp. 583-598, 1979.

19. ZIENKIEWICZ, O. C., and HINTON, E. - Reduced Integration, Function Smoothing and Non-Conformity in Finite Element Analysis. J. Franklin Institute, Vol. 302, pp. 443-461, 1976.

20. REDDY, J. N. - On Penalty Function Methods in the Finite-Element Analysis of Flow Problems: I. Theory Int. J. Numer. Meth. Fluids, to appear.

LAMINAR NATURAL CONVECTION ALONG VERTICAL CORNERS AND RECTANGULAR CHANNELS

K. Ramakrishna*, P.K. Khosla** and S.G. Rubin**

SUMMARY

Laminar natural convection along a vertical right angle corner and in rectangular ducts has been analyzed. Solutions for isothermal and constant heat flux walls have been obtained by a coupled strongly implicit procedure. The effect of Prandtl number on the corner flow behavior is examined. It is also shown that rectangular ducts entrain a larger amount of mass than do square ducts of equal length.

1. INTRODUCTION

The problem of natural convection flow along vertical surfaces and enclosed ducts has many technological applications. Heat dissipation from large radiators, heat-exchangers, heat losses to cryogenic tanks, fire propagation, solar collectors, cooling of electronic equipment and handling of spent nuclear reactor fuel assemblies are some of the large number of applications where natural convection is important. The growing need to conserve depletable energy resources enhances the importance of natural convection for future energy, process and domestic applications.

A large body of literature deals with natural convection from flat plates, cylindrical and spherical surfaces. All of these geometries exhibit certain types of symmetry, so that the resulting flows are not truly three-dimensional. The problem of natural convection in three dimensions, especially in boundary regions has not been given the same consideration as the quasi-two-dimensional counterparts. Studies of natural

---

* Bell Laboratories, Naperville, Illinois.

** Dept. of Aerospace Engineering and Applied Mechanics, University of Cincinnati, Cincinnati, Ohio.

convection in corner regions, rectangular channels and tanks, though of considerable technical interest, are much more complex and involve what is known as boundary regions - a boundary layer within a boundary layer. The mathematical treatment of such flows has been facilitated in recent years by high speed digital computers and improved numerical techniques for the 'corner layer'. Incompressible forced convection flow along a corner has been extensively investigated by Rubin and his co-workers.

Natural convection flow along vertical corners has been reported by Van Leeuwen et al. [1,2] and Riley and Poots [3]. Van Leeuwen et al.'s experimental investigations concern measurement of velocity and temperature at the intersection formed by two vertical copper plates (height 0.4m and width 0.3m) assembled at right angles. The corner plates were maintained at constant temperature by passage of hot water through channels at the back of the plates. The corner plates were extended downwards by unheated cardboard planes to provide a free feeding stream into the corner. The temperature of the air in the corner was measured by thermocouples and the vertical velocity was measured by a specially developed anemometer. The reported measurements, approximately at a Grashof number of 10^8, representing laminar flow, show an increased velocity and a decreased heat transfer coefficient near the corner region when compared with the flow and heat transfer coefficient at a single vertical plane at identical Grashof number. The higher velocity manifests in the form of closed isovelocity contours, which the authors have referred to as "chimney effect". It is interesting to note that the air is not stagnant at the lower unheated cardboard extensions where the chimney effect is also evident. From a consideration of measured temperature gradients and vertical velocities in a graphical interpretation, Van Leeuwen et al. [1,2] conclude that the horizontal velocity, anywhere along the diagonal is directed towards the point of maximum vertical velocity, though its magnitude is very small - of the order of 1 cm/sec - when compared with the vertical velocity of about 35 cm/sec at a height of 0.19 meters from the leading edge.

The mathematical formulation of laminar natural convection along an isothermal vertical corner has been carried out by Riley and Poots [3] and Ramakrishna et al. [4] and is identical to the matched asymptotic expansion procedure of Rubin [5] for viscous flow along a corner. The numerical results confirm the chimney effect and the reduced heat transfer coefficient in the corner region, as observed by Van Leeuwen et al. [1]. Although the experimental observations are confirmed, the predictions for air show as much as 15% variation in the heat transfer coefficient near the corner (particularly in regions close to the leading edge) as well as large differences in the velocity and temperature profiles when compared with the measurements of Van Leeuwen et al. [1].

Laminar natural convection in vertical tubes and parallel plates have also been extensively investigated. One of the earliest investigations of flows of this type was due to Elenbass [6,7], whose experimental investigations, in particular, have been a source of comparison for several of the subsequent mathematical treatments. In his first study [6], he measured heat dissipation from vertical parallel plates of 12 x 12 cms for different spacings between the plates. He also developed an approximate simplified mathematical model by fitting a modified parabolic velocity profile and a modified exponential temperature profile for the region between the plates. His measured heat dissipation, when corrected for finite size of the plates, shows good agreement with his mathematical treatment. Elenbass [7] later extended his theory developed in [6] for vertical tubes of different cross section including circular, square, rectangular and triangular shapes.

The assumption of fully developed flow by Elenbass is far from realized in practice. The formal treatment of developing natural convection flows between vertical parallel plates and in circular tubes have been presented by Bodoia and Osterle [8], Quintiere and Mueller [9], Davis and Perona [10], Kageyama and Izumi [11] and Dyer [12]. For constant temperature surfaces, the development length of the channel (defined as the distance from the channel entrance to the point along the channel where the flow approaches a fully developed value to within an agreed upon fraction - i.e., the x-derivatives tend to zero) is dependent on the volume flow rate through the channel. The volume flow rate, in turn, depends on the height of the channel.

The pressure distribution in the channel reaches a negative maximum at some point in the channel, beyond which the pressure increases and returns to atmospheric pressure. The point at which the pressure becomes atmospheric defines the channel length for the associated volume flow rate. An increasing positive pressure in the channel is contrary to the requirement of natural convection flows and therefore the pressure in the channel is always below atmospheric. This has been proved by Kageyama and Izumi [11] for circular tubes of finite or semi-finite length. Such a condition requires that the point at which the pressure in the channel becomes atmospheric defines the channel length.

It is clear from the preceding review that three-dimensional natural convection flow has received very little attention. A satisfactory solution of the corner region problem has been presented in reference [4] by the present authors.

A formal treatment of the three-dimensional natural convection in rectangular ducts has not been attempted.

Elenbass' [6] correlation for rectangular and square ducts ignores the three dimensional effects. The natural connection in a vertical square channel has been recently reformulated by Ramakrishna et al. [14]. The new procedure, which is a direct extension of the procedure for forced convection by Rubin et al. [15], takes into consideration the three dimensionality of the flow and does not assume the flow to be fully developed. Rather, the complete solution including the entrance region merging into fully developed flow is evaluated. In the present paper, natural convection along vertical corners and square ducts is reviewed. Additional results for variable wall temperature and fluid properties are presented. In addition, natural convection in the entrance length of isothermal rectangular ducts is examined.

The governing flow and energy equations for natural convection have been nondimensionalized and the resulting equations are independent of the Grashof number.

2. GOVERNING EQUATIONS

The Navier-Stokes equations utilizing the Bousinesque approximation are appropriately nondimensionalized in the following manner:

(a) Constant wall temperature ducts

$$Gr = \frac{g\beta(T_w - T_\infty)a^4}{\ell\nu^2}, \quad \xi = \frac{x'}{Gr\cdot\ell}, \quad u = \frac{u'a^2}{Gr\cdot\nu\cdot\ell}, \quad y = \frac{y'}{a},$$

$$v = \frac{v'a}{\nu}, \quad z = \frac{z'}{b}, \quad w = \frac{w'a}{\nu}, \quad \theta = \frac{T - T_\infty}{T_w - T_\infty},$$

$$p = \frac{p'_d\ a^4}{\rho Gr^2 \nu^2 \ell^2}, \quad Ra = Pr\cdot Gr, \quad \sigma = \frac{a}{b}$$

(b) The case of uniform heat flux

$$Gr^* = \frac{g\beta q_w a^5}{\ell k\nu^2}, \quad \xi = \frac{x'}{Gr^*\ell}, \quad u = \frac{u'a^2}{Gr^*\nu\ell}, \quad \theta = \frac{(T - T_0)k}{q_w a},$$

$$p = \frac{p_d a^4}{\rho\ell^2\nu^2 GR^{*2}} \qquad (1)$$

In dimensionless form, the flow equations for either type of thermal boundary conditions can be written in a universal form. In these equations, the diffusion terms in the axial direction are an order of magnitude smaller in comparison with the diffusion terms in the crossplane. Therefore, at higher Grashof numbers the axial diffusion terms may be neglected in comparison with the secondary flow diffusion effects. This leads to the following dimensionless equations and the associated boundary conditions.

$$\frac{\partial u}{\partial \xi} + \frac{\partial v}{\partial y} + \sigma \frac{\partial w}{\partial z} = 0 \tag{2a}$$

$$u \frac{\partial u}{\partial \xi} + v \frac{\partial u}{\partial y} + \sigma w \frac{\partial u}{\partial z} = - \frac{\partial p}{\partial \xi} + \frac{\partial^2 u}{\partial y^2} + \sigma^2 \frac{\partial^2 u}{\partial z^2} + \theta \tag{2b}$$

$$u \frac{\partial v}{\partial \xi} + v \frac{\partial v}{\partial y} + \sigma w \frac{\partial v}{\partial z} = -\sigma Gr^2 \frac{\partial p}{\partial y} + \frac{\partial^2 v}{\partial y^2} + \sigma^2 \frac{\partial^2 v}{\partial z^2} \tag{2c}$$

$$u \frac{\partial w}{\partial \xi} + v \frac{\partial w}{\partial y} + \sigma w \frac{\partial w}{\partial z} = - \sigma Gr^2 \frac{\partial p}{\partial z} + \frac{\partial^2 w}{\partial y^2} + \sigma^2 \frac{\partial^2 w}{\partial z^2} \tag{2d}$$

$$u \frac{\partial \theta}{\partial \xi} + v \frac{\partial \theta}{\partial y} + \sigma w \frac{\partial \theta}{\partial z} = \frac{1}{Pr} \left[\frac{\partial^2 \theta}{\partial y^2} + \sigma^2 \frac{\partial^2 \theta}{\partial z^2} \right] \tag{2e}$$

where u, v and w are components of velocity along x, y and z directions, σ is the aspect ratio of the channel, p the normalized pressure, θ is the normalized temperature, Pr is the Prandtl number of the fluid, and Gr is the Grashof number. For an infinite corner or a square duct $\sigma = 1$.

Boundary Conditions

i) Corner flow: For a right angled corner, the boundary conditions are

$$\left.\begin{array}{lll} x \geq 0 & y = 0, & z \geq 0 \\ & y \geq 0, & z = 0 \end{array}\right. \qquad u = 0 = v = w, \quad \theta = 1$$

For $x \geq 0$, $y \geq 0$ and $z \to \infty$, the solution tends to the two dimensional free convection flow along a vertical infinite flat plate.

ii) Flow in a vertical rectangular channel: In the case of a channel, the boundary conditions for an isothermal and constant heat flux walls are given as:

Constant wall temperature

$$\left.\begin{array}{lll} \xi \geq 0, & y = 0,2, & 0 \leq z \leq 2 \\ \xi \geq 0, & 0 \leq y \leq 2, & z = 0,2 \end{array}\right\} \quad \begin{array}{l} u = v = w = 0 \\ \theta = 1 \end{array}$$

$$\begin{array}{lll} \xi = 0, & 0 \leq y \leq 2, & 0 \leq z \leq 2 \end{array} \quad \begin{array}{l} u = U_{in}, \ v = w = 0 \\ \theta = 0, \ P = 0 \text{ or } -\dfrac{U_{in}^2}{2} \end{array}$$

$\xi = L, \quad 0 \leq y \leq 2, \quad 0 \leq z \leq 2 \qquad p = 0$

Constant heat flux (3)

$$\left.\begin{array}{lll} \xi \geq 0, & y = 0,2, & 0 \leq z \leq 2 \\ \xi \geq 0, & 0 \leq y \leq 2, & z = 0,2 \end{array}\right\} \begin{array}{l} u = v = w = 0 \\ \frac{\partial \theta}{\partial y} = 1, \quad \frac{\partial \theta}{\partial z} = \frac{1}{\sigma} \end{array}$$

All other boundary conditions remain the same as for the constant temperature case.

An additional equation is provided by the fact that at any cross-section of the duct, the mass flow rate is constant. This global condition in nondimensional form becomes:

$$\int_0^1 \int_0^1 U \, dydz = U_{in} = \frac{Q}{4} \tag{4}$$

where U_{in} is the dimensionless entrance velocity of the fluid, Q is the dimensionless flow rate defined by

$$Q = \frac{Q'}{Gr \cdot \nu \cdot \ell} = 4U_{in} \tag{5}$$

Equation (4) provides the axial pressure gradient which drives the flow along the channel.

The pressure gradient terms in the y and z momentum equations may be eliminated by cross differentiation and subtraction. This leads to the vorticity transport equation, where vorticity is defined by

$$\Omega = \sigma v_z - w_y \quad , \tag{6}$$

σ is the aspect ratio of the channel cross-section. This leads to

$$u \frac{\partial \Omega}{\partial \xi} + v \frac{\partial \Omega}{\partial y} + \sigma w \frac{\partial \Omega}{\partial z} + \Omega\left(\frac{\partial v}{\partial y} + \sigma \frac{\partial w}{\partial z}\right) + \left(\sigma \frac{\partial v}{\partial \xi}\frac{\partial u}{\partial z} - \frac{\partial w}{\partial \xi}\frac{\partial u}{\partial y}\right)$$
$$= \frac{\partial^2 \Omega}{\partial y^2} + \sigma^2 \frac{\partial^2 \Omega}{\partial z^2} \tag{7}$$

By introducing the cross plane potential and stream function as:

$$v = \frac{\partial \phi}{\partial y} + \sigma \frac{\partial \psi}{\partial z}, \qquad w = \sigma \frac{\partial \phi}{\partial z} - \frac{\partial \psi}{\partial y} \quad . \tag{8}$$

The continuity and equation (5) lead to

$$\phi_{yy} + \sigma^2 \phi_{zz} = - u_\xi , \quad \psi_{yy} + \sigma^2 \psi_{zz} = \Omega. \tag{9}$$

Combining equations (7) and (8) we obtain the following Poisson equations for the secondary flow velocities

$$\frac{\partial^2 v}{\partial y^2} + \sigma^2 \frac{\partial^2 v}{\partial z^2} = - \frac{\partial^2 u}{\partial \xi \partial y} + \sigma \frac{\partial \Omega}{\partial z} , \quad \frac{\partial^2 w}{\partial y^2} + \sigma^2 \frac{\partial^2 w}{\partial z^2} = - \sigma \frac{\partial^2 u}{\partial \xi \partial z} - \frac{\partial \Omega}{\partial y} . \tag{10}$$

Equations (2b,c), (4) through (10), provide the final set of equations. These have been solved numerically for the problems under consideration. The p_ξ term in equation (2b) is zero for the flow along the corner; for the channel flow it is evaluated from the global conservation condition given by equation (3). It should be noted that these equations are independent of the Grashof number.

3. SOLUTION PROCEDURE

The governing equations have been discretized using backward differencing for the ξ-derivatives and central differences for all other derivatives. The resulting implicit algebraic system of equations have been solved iteratively using a strongly implicit procedure. ξ-momentum and the energy equations are solved in a coupled fashion while v, w and Ω equations are solved separately. For flows where secondary velocities are large, additional equations may have to be coupled in order to obtain a converged solution.

Coupled 2x2 Solution Algorithm

In reference [13] a strongly implicit procedure has been developed for the solution of the stream function-vorticity form of the Navier-Stokes equations. This algorithm has the distinct advantage of being implicit in both the Y and Z directions, as well as coupling all the boundary conditions. The method of solution is unconditionally stable and converges faster than SOR, LSOR, ADI, etc. The discretized version of the equations can be written as:

$$(\tilde{L} + P)\, \tilde{W}_{n+1} = \tilde{G} + P\tilde{W}^n$$

where

$$\tilde{W} = \begin{bmatrix} \bar{U}_{ij} \\ \bar{\theta}_{ij} \end{bmatrix} ; \quad \bar{U}_{ij} = \begin{bmatrix} U_{1j} \\ U_{2j} \\ \vdots \\ U_{nj} \end{bmatrix} ; \quad \bar{\theta}_{ij} = \begin{bmatrix} \theta_{1j} \\ \theta_{2j} \\ \vdots \\ \theta_{nj} \end{bmatrix}$$

The decomposed lower and upper triangular matrices have five (5) non-zero diagonals leading to a solution of the following form:

$$\begin{bmatrix} \bar{U}_{ij} \\ \bar{\theta}_{ij} \end{bmatrix}^{n+1} = \begin{bmatrix} GM1_{ij} \\ GM2_{ij} \end{bmatrix} + \begin{bmatrix} T1_{ij} & T3_{ij} \\ T5_{ij} & T7_{ij} \end{bmatrix} \begin{bmatrix} U_{i,j+1} \\ \theta_{i,j+1} \end{bmatrix}^{n+1}$$

$$+ \begin{bmatrix} T2_{ij} & T4_{ij} \\ T6_{ij} & T8_{ij} \end{bmatrix} \begin{bmatrix} U_{i+1,j} \\ \theta_{i+1,j} \end{bmatrix}^{n+1}$$

The evaluation of U_{ij} depends not only upon the values of U, but also upon the new values of θ. The coupling increases the storage requirement; however, it accelerates the rate of convergence.

The solution algorithm for the vorticity equation and the Poisson equations for the crossplane velocities are much simpler and have the following form

$$\hat{\Omega}^{n+1}_{jk} = GM1_{jk} + T1_{jk}\, \hat{\Omega}_{j,k+1} + T2_{jk}\, \hat{\Omega}_{j+1,k}$$

In the next section, some sample calculations for a corner flow, square duct and a rectangular duct are presented.

4. RESULTS

(a) Natural Convection Along Vertical Right Angle Corner:

A detailed analysis of the problem has been presented in reference [4] by the present authors. Similarity solutions for a power law variation of wall temperature pertaining to isothermal and constant heat flux walls have been obtained for air and water. It was shown for air that the similarity analysis does predict the chimney effect as predicted by the experiments of references [1] and [2]; although, there is considerable discrepancy in the measurements and solutions. It has been shown that the non-similar solutions with stagnant initial conditions at $\xi = 0$ provide better agreement with the experiments. This agreement is excellent for heat transfer (temperature distribution), than for the velocity distribution in the corner. Additional, more accurate, measurements of velocities in the corner region are still required. Velocity distributions along the diagonal, for an isothermal corner, are shown in Fig. 1 for air with $Pr = 0.72$. The distribution of velocity, temperature along the symmetry diagonal, and heat transfer relationship for isothermal and constant heat

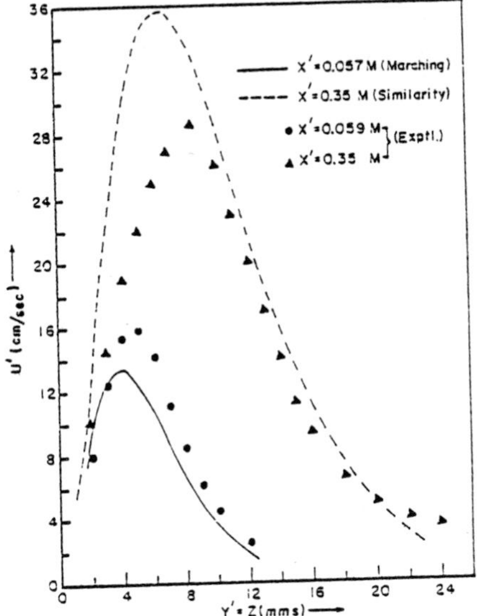

Fig. 1. Comparison of Velocity Distribution Along the Diagonal of Isothermal Corner (Pr = 0.72).

flux corners immersed in water (Pr = 4.52), are depicted in Figs. (2a,b,c). The vertical velocity along the corner is much smaller in the case of water than that for air under similar conditions. The heat transfer rate approaches the flat plate value over a shorter distance when compared with air under similar conditions. The ratio of the heat transfer for a non-isothermal wall to the heat transfer for an isothermal wall is often an important parameter for practical applications. For a corner, this ratio varies along the walls of the corner and asymptotically approaches the ratio for the vertical flat plate evaluated by Sparrow and Gregg [16]. This ratio is tabulated in Table 1 for water.

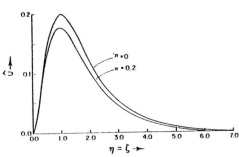

Fig. 2a. Velocity Distribution Along Diagonal (Pr = 4.52).

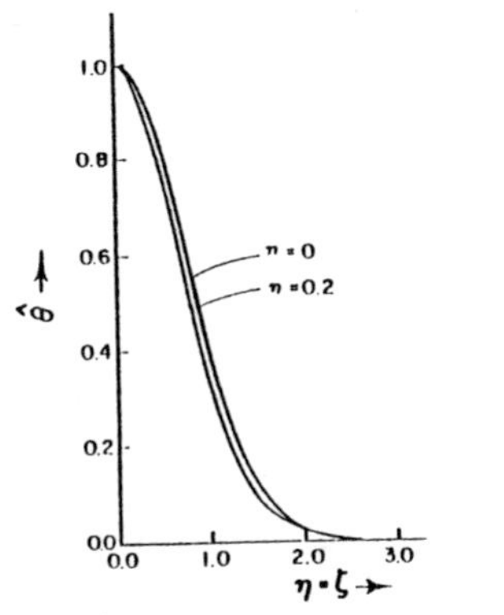

Fig. 2b. Temperature Distribution Along Diagonal (Pr = 4.52).

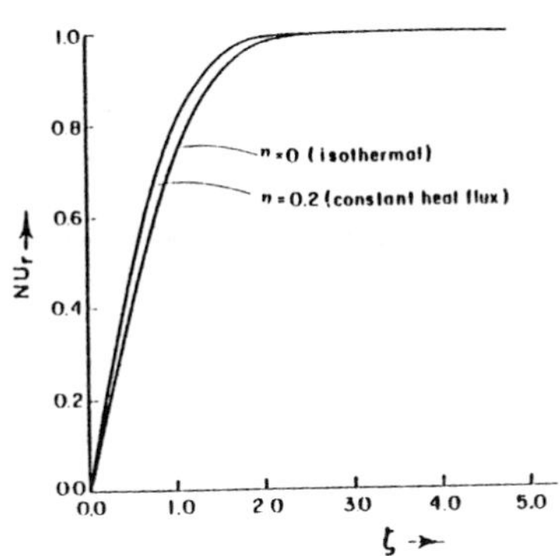

Fig. 2c. Local Nusselt Number-$NU_r = NU(0,\zeta)/NU_p$ for Water. (Pr = 4.52).

Table 1. Comparison of Heat Transfer Along Non-Isothermal and Isothermal Corners (Pr = 4.52).

ζ	$(q_{variable\ T_w}/q_{constant\ T_w})$
	n = 0.2
0.1	1.438
1.0	1.255
2.0	1.149
3.0	1.129
4.0	1.129

(b) Natural Convection Along Square Ducts

A detailed analysis and results of calculations for an isothermal and constant heat flux square duct have been presented in reference [14] by the present authors. The velocity and temperature distribution along the centerline of the square channel for isothermal walls are shown in Fig. 3. The largest ξ values depicted represent the axial coordinate of the exit plane. The isovelocity and isothermal contours for Q = 0.1 are described in Fig. 4. For short ducts, corresponding to small values of Q, the flow in a square duct is similar to the vertical corner and exhibits the "chimney effect". This chimney effect is present only twoards the exit of the channel and would indicate that this behavior is different from that of the corner. Actually if the flow is resolved within the corner region of the channel an additional chimney effect similar to the one observed in the corner flow problem should be observed. This, however, has not been considered in the present calculations. For additional results with a variety of ducts, see reference [14].

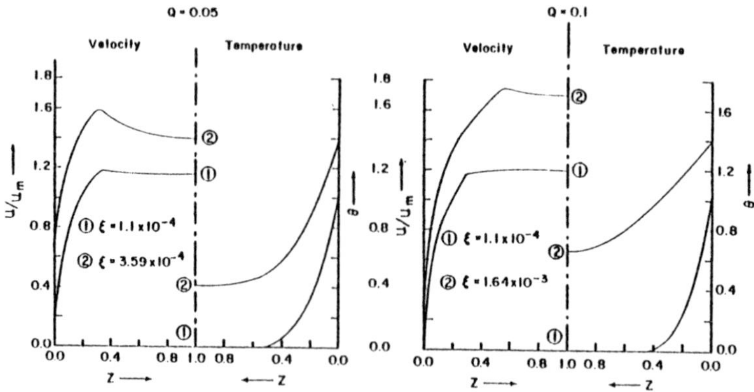

Fig. 3. Centerline Velocity and Temperature for an Isothermal Square Duct, (Pr = 0.72).

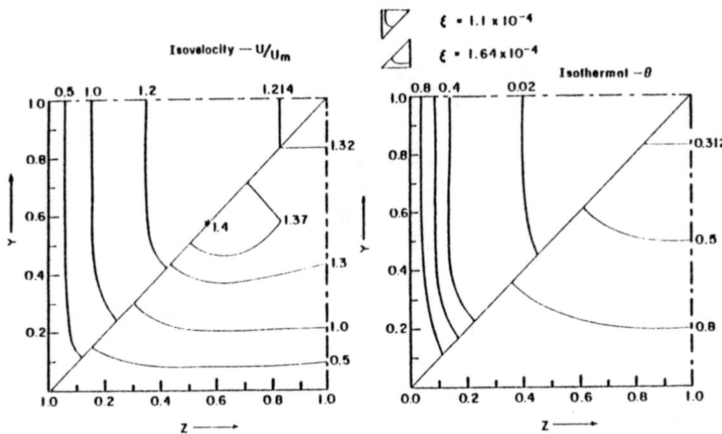

Fig. 4. Isovelocity and Isothermal Contours for Isothermal Square Duct, (Pr = 0.72, Q = 0.1).

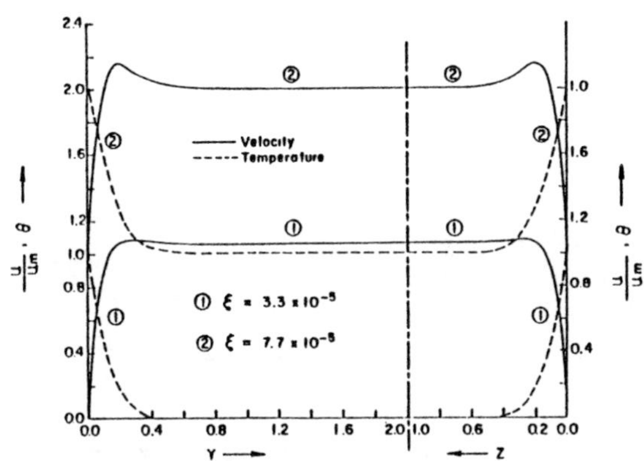

Fig. 5a. Centerline Velocity and Temperature Profiles for a Rectangular Isothermal Duct, $\sigma = 2$, $Q = 0.05$, $Pr = 0.72$.

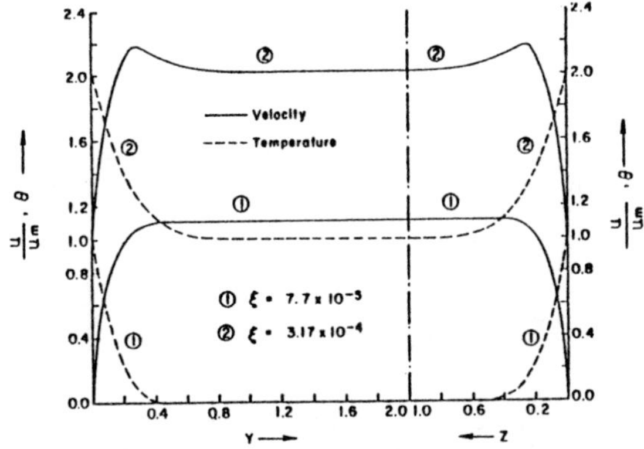

Fig. 5b. Centerline Velocity and Temperature Profiles for a Rectangular Isothermal Duct, $\sigma = 2$, $Q = 0.1$, $Pr = 0.72$.

(c) Isothermal Rectangular Channels

A limited number of calculations have been performed for a rectangular channel with the aspect ratio $\sigma = a/b = 2$. The velocity and temperature distributions along the duct centerlines, for two ducts, characterized by $Q = 0.05$ and 0.1, are presented in Figs. (5a,b). The exit plane velocity profiles at $\xi = 7.7 \times 10^{-5}$ for $Q = 0.05$ and $\xi = 3.17 \times 10^{-4}$ for $Q = 0.1$ show a chimney effect similar to the one obtained for square ducts. The isovels shown in Fig. 6 also depict the chimney effect. It should be pointed out again that the chimney flow behavior in channels is different from that in the corner region. For a calculation that resolves the corner regions both chimney effects will be obtained.

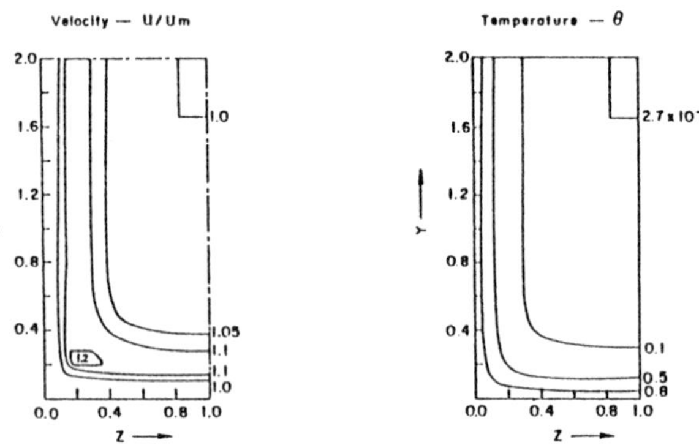

Fig. 6. Isovelocity and Isothermal Contours for an Isothermal Rectangular Duct, $\sigma = 2$, $Q = 0.05$, $Pr = 0.72$ at $\xi = 7.7 \times 10^{-5}$.

From Figs. 5 and 6, it is evident that for an equal amount of flow entrainment, the rectangular ducts are "shorter" than the square ducts. In other words, rectangular ducts of aspect ratio 2 entrain more flow than a square channel. The peripheral Nusselt numbers, defined as

$$NU_p = \frac{1}{\theta_m - 1} \left.\frac{\partial\theta}{\partial y}\right|_p , \quad NU_p' = \frac{1}{\theta_m - 1} \left.\frac{\partial\theta}{\partial z}\right|_{p'}$$

where θ_m is the dimensionless bulk mean fluid temperature, are depicted in Figs. (7a,b). The local Nusselt number defined as

$$NU_\xi = \frac{1}{(\theta_m - 1)(1+\sigma)} \left[\int_0^1 \left.\frac{\partial\theta}{\partial y}\right|_{y=0} dz + \sigma^2 \int_0^1 \left.\frac{\partial\theta}{\partial z}\right|_{z=0} dy \right]$$

is depicted in Fig. 8. The peripheral Nusselt number decreases monotonically with the axial distance everywhere along the periphery of the channel. The local Nusselt number does indicate a monotonic behavior, but a definite conclusion

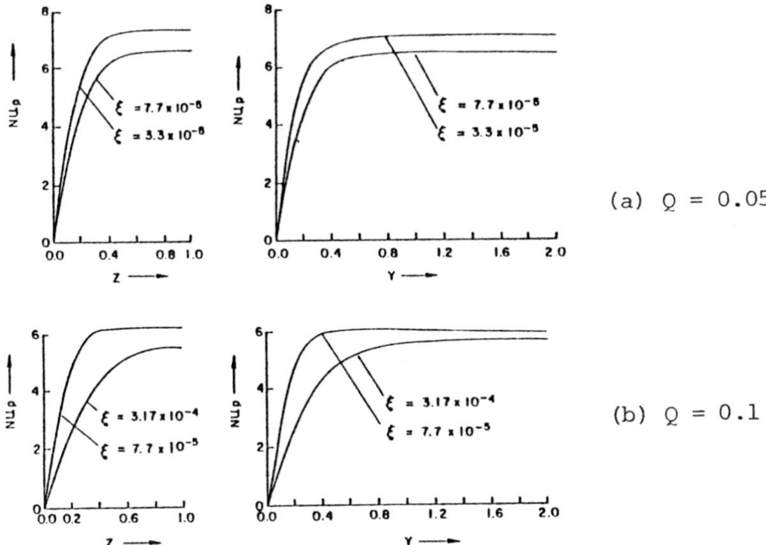

Fig. 7a,b. Peripheral Nusselt Number for an Isothermal Rectangular Duct, Pr = 0.72, σ = 2.

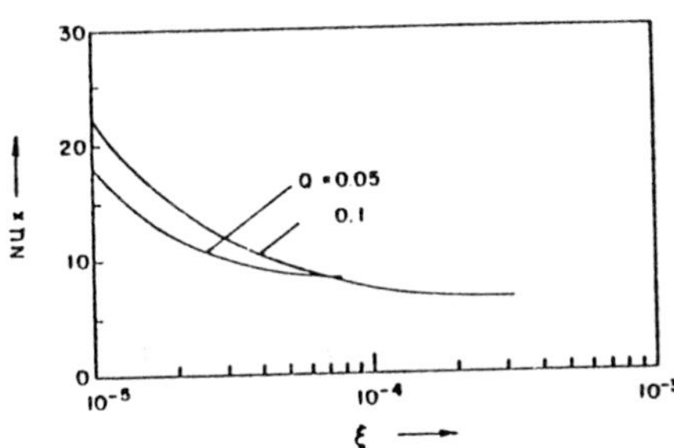

Fig. 8. Local Nusselt Number for an Isothermal Rectangular Duct (σ = 2, Pr = 0.72).

regarding the asymptotic behavior for longer channels can not be definitive from the present limited calculations.

All the computations presented in the present paper have been performed without the need for relaxation. For longer ducts, of larger aspect ratio, the present calculation procedure fails to converge. In such situations, the crossflow velocities are quite large and thereby convergence requires either underrelaxation or coupling of the secondary flow velocities with the velocity u or the temperature θ calculation.

REFERENCES

1. VAN LEEUWEN, J.H., LOOMAN, C.M., and SCHENK, J. - Int. Jl. of Heat and Mass Transfer, 14, pp. 561-564, 1971.

2. VAN LEEUWEN, J.H. and LOOMAN, C.M. - Temperature-en snelheidsveld bij vrije convectie in een hoek. Internal Reports, Technische Natuurkunde, Delft, Jan. 1979, Octo. 1969.

3. RILEY, D.S. and POOTS, G. - Quart. J. Mech. and Appl. Math., 25, pp. 401-421, 1972.

4. RAMAKRISHNA, K., KHOSLA, P.K. and RUBIN, S.G. - Proceedings of 1980 Heat Transfer & Fluid Mechanics Institute, Stanford Univ. Press, California, pp. 1-20.

5. RUBIN, S.G. - Journal of Fluid Mechanics, 26, pp. 97-110, 1966.

6. ELENBASS, W. - Physica, 9, pp. 865-873, 1942.

7. ELENBASS, W. - Physica, 9, pp. 1-28, 1942.

8. BODOIA, J.R. and OSTERLE, F.J. - Journal of Heat Transfer, 84, pp. 40-44, 1962.

9. QUINTIERE, J. and MUELLER, W.K. - Journal of Heat Transfer, 95, pp. 53-59, 1973.

10. DAVIS, L.P. and PERONA, J.J. - Int. Jl. of Heat and Mass Transfer, 14, pp889-903, 1971.

11. KAGEYAMA, J. and IZUMI, R. - Bul. J.S.M.E., 13, pp. 382-394, 1970.

12. DYER, J.R. - Int. Jl. of Heat and Mass Transfer, 18, pp. 1455-1465, 1975.

13. RUBIN, S.G. and KHOSLA, P.K. - Navier-Stokes Calculations with a Coupled Strongly Implicit Method. Part I: Finite Difference Solutions. 17th Aerospace Sciences Meeting, New Orleans, January 1979.

14. RAMAKRISHNA, K., KHOSLA, P.K and RUBIN, S.G. - Laminar Natural Convection Along Vertical Square Ducts. To appear.

15. RUBIN, S.G., KHOSLA, P.K. and SAARI, S.- Computers and Fluids, Vol. 5, pp. 151-173, 1977.

16. SPARROW, E.M. and GREGG, J.L. - Trans. ASME, 80, pp. 379-387, 1958.

NATURAL CONVECTION IN A ROTATING ANNULUS

E. LEONARDI, J.A. REIZES and G. DE VAHL DAVIS[(i)]

SUMMARY

A numerical study has been made of the steady axisymmetric velocity and temperature distributions induced in a vertical annulus by rotation, buoyancy and a combination of both.

1. INTRODUCTION

Convective flows in a rotating annulus have been extensively studied by numerical means. In general two approaches have been adopted. Williams [16] solved the set of equations written in the primitive variables, viz., velocity and pressure (as well as temperature). This approach permitted the investigation of fully three-dimensional flows; however, pressure boundary conditions, which are difficult to implement, were required and continuity was not automatically satisfied.

An alternative approach [9,10,14,15] has been to restrict studies to axisymmetric flows, requiring the solution of equations for a stream function, the circumferential velocity and the vorticity. This approach is attractive because, first, it reduces the number of equations to be solved from five to four, and second, it ensures that the continuity equation is satisfied. Moreover, the difficulties associated with pressure boundary conditions are avoided, the corresponding vorticity boundary conditions being much simpler and more easily implemented.

The problems associated with the use of primitive variables and the restriction to axial symmetry of the stream function-circumferential velocity method may be overcome by the use of a vorticity-vector potential formulation.

Although the use of such a formulation in three-

(i) University of N.S.W., Kensington, Australia, 2033.

dimensional problems is not new [1,7,8], it has so far been applied only to singly connected regions. In this paper a method which permits the numerical solution of three-dimensional flows in multiply connected regions using the vorticity-vector potential formulation is presented.

As a first step in an extensive study of three-dimensional flow in rotating annular cavities, preliminary results have been obtained for axisymmetric flows induced by rotation, buoyancy or both. These have verified the method and the program, which are now being applied to non-axisymmetric flows.

2. MATHEMATICAL FORMULATION

The annulus is assumed to be filled with a viscous heat conducting fluid, and the conditions are such that the Boussinesq approximations [2] can be made. The origin of the annulus is at the centre of the upper lid and the axes are oriented as illustrated in Figure 1. The shape of the annulus is determined by the radius ratio $RR=R_0'/R_i'$, and the aspect ratio $AR=L'/(R_0'-R_i')$.

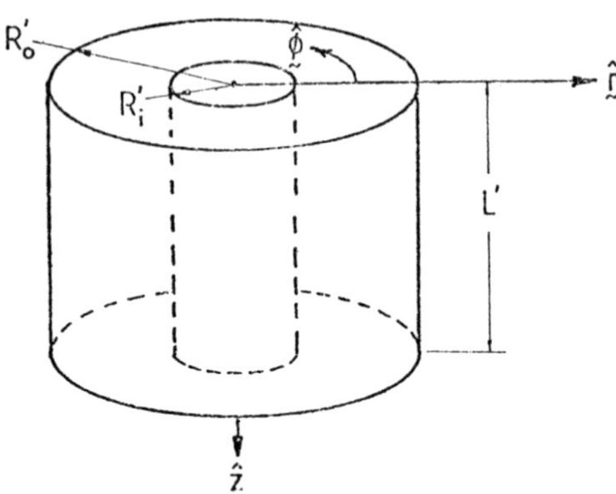

Figure 1. Definition Sketch of the System Geometry.

We use $R_0'-R_i'$, $1/\Omega'$ (where Ω' is the angular velocity of a boundary), $(R_0'-R_i')\Omega'$ and $\rho'(R_0'-R_i')^2\Omega'^2$ as scale factors for length, time, velocity and pressure; primes denote dimensional quantities. The non-dimensional temperature $\theta=(T'-T_C')/(T_H'-T_C')$, where T' denotes local absolute temperature and the subscripts H and C denote hot and cold boundaries. The equations representing the conservation of mass, momentum and energy become

$$\nabla.\underset{\sim}{V} = 0 , \tag{1}$$

$$\frac{\partial \underset{\sim}{V}}{\partial t} + \underset{\sim}{V}.\nabla\underset{\sim}{V} = -\nabla p - \frac{Ra}{Re^2 Pr}\,\theta\hat{\underset{\sim}{g}} + \frac{1}{Re}\nabla^2\underset{\sim}{V} , \tag{2}$$

$$\frac{\partial\theta}{\partial t} + \nabla.(\underset{\sim}{V}\theta) = \frac{1}{RePr}\nabla^2\theta , \tag{3}$$

where the velocity vector is $\underset{\sim}{V} = V_r\hat{\underset{\sim}{r}} + V_\phi\hat{\underset{\sim}{\phi}} + V_z\hat{\underset{\sim}{z}}$; p is the perturbation pressure from the hydrostatic value; $Ra = Pr.g'\beta'(T'_H - T'_C)(R'_0 - R'_i)^3/\nu'^2$ and $Re = (R'_0 - R'_i)^2\Omega'/\nu'$.

We recast the equations of motion into a vorticity-vector potential formulation [7]. The vorticity transport equation is

$$\frac{\partial\underset{\sim}{\zeta}}{\partial t} + \nabla\times(\underset{\sim}{\zeta}\times\underset{\sim}{V}) = -\frac{Ra}{Re^2Pr}(\nabla\times\theta\hat{\underset{\sim}{g}}) + \frac{1}{Re}\nabla^2\underset{\sim}{\zeta} , \tag{4}$$

where

$$\underset{\sim}{\zeta} = \nabla\times\underset{\sim}{V} . \tag{5}$$

The vector potential $\underset{\sim}{\xi}$ and a scalar function ψ defined as

$$\underset{\sim}{V} = \nabla\times\underset{\sim}{\xi} + \nabla\psi , \tag{6}$$

eliminate the necessity for solving the continuity equation (1). Without loss of generality, and to achieve a subsequent simplification, $\underset{\sim}{\xi}$ is chosen [13] such that

$$\nabla.\underset{\sim}{\xi} = 0 . \tag{7}$$

Since $\nabla.\underset{\sim}{V} = 0$, it follows from equation (6) that

$$\nabla^2\psi = 0 . \tag{8}$$

The value of ψ on the boundary is arbitrary, the only requirement being that equation (8) must be satisfied. As shown by Richardson and Cornish [11], the specification of the boundary conditions for $\underset{\sim}{\xi}$ is simplified by imposing

$$\frac{\partial\psi}{\partial n} = -\hat{\underset{\sim}{n}}.\underset{\sim}{V} \tag{9}$$

on ψ, where $\hat{\underset{\sim}{n}}$ is the unit vector in an outward normal direction.

For the problems considered in this paper, there is no flow through any boundary; therefore, from equations (8) and (9), ψ is identically zero and is no longer required in equation (6), which becomes

$$\underset{\sim}{V} = \nabla\times\underset{\sim}{\xi} . \tag{10}$$

A relationship between $\underset{\sim}{\zeta}$ and $\underset{\sim}{\xi}$ is obtained by taking the curl of equation (10):

$$\underset{\sim}{\zeta} = \nabla\times(\nabla\times\underset{\sim}{\xi}) = \nabla(\nabla.\underset{\sim}{\xi}) - \nabla^2\underset{\sim}{\xi}$$

$$= -\nabla^2\underset{\sim}{\xi} \quad (11)$$

Equations (3), (4), (10) and (11) yield ten component equations which must be solved for $\underset{\sim}{\zeta}$, $\underset{\sim}{\xi}$, $\underset{\sim}{V}$ and θ. The component equations are given in the Appendix.

3. BOUNDARY CONDITIONS

A wide variety of problems may be studied by varying the thermal and kinematic boundary conditions. For the problems considered here, each boundary has been assumed to be impermeable, non-slip, isothermal (different boundaries may have different temperatures) or adiabatic and either rotating at a constant angular velocity or stationary. Full details are given in Table 1. Conditions on $\underset{\sim}{\xi}$ and $\underset{\sim}{\zeta}$ are obtained from those on $\underset{\sim}{V}$.

CASE	BOUNDARY			
	$r = R_i$	$r = R_0$	$z = 0$	$z = AR$
A	$\Omega = 0$ $\theta = 1$	$\Omega = 0$ $\theta = 0$	$\Omega = 0$ Adiabatic	$\Omega = 0$ Adiabatic
B	$\Omega = 1$ $\theta = 0$	$\Omega = 0$ $\theta = 0$	$\Omega = 0$ $\theta = 0$	$\Omega = 0$ $\theta = 0$
C–E	$\Omega = 1$ $\theta = 1$	$\Omega = 0$ $\theta = 0$	$\Omega = 0$ Adiabatic	$\Omega = 0$ Adiabatic

Table 1. Boundary Conditions for Temperature, θ, and Angular Velocity, Ω.

The boundary conditions for the vector potential, as developed by Hirasaki and Hellums [5] and used by Mallinson and de Vahl Davis [7,8], cannot be applied to a multiply connected region. This was shown by Richardson and Cornish [11] who accordingly obtained different conditions. However, their derivation is very complex and the following simpler derivation is presented.

Consider the multiply connected region illustrated in Figure 2. Gauss' Divergence Theorem is applied over the region to yield

$$\iiint_{\forall} \nabla.\underset{\sim}{\xi}\, d\forall = \iint_{S_2} \underset{\sim}{\xi}.\underset{\sim}{\hat{n}}\, dS_2 - \iint_{S_1} \underset{\sim}{\xi}.\underset{\sim}{\hat{n}}\, dS_1 , \quad (12)$$

where $\forall$ is the volume of the region and S_1 and S_2 are the surface areas of the inner and outer boundaries respectively.

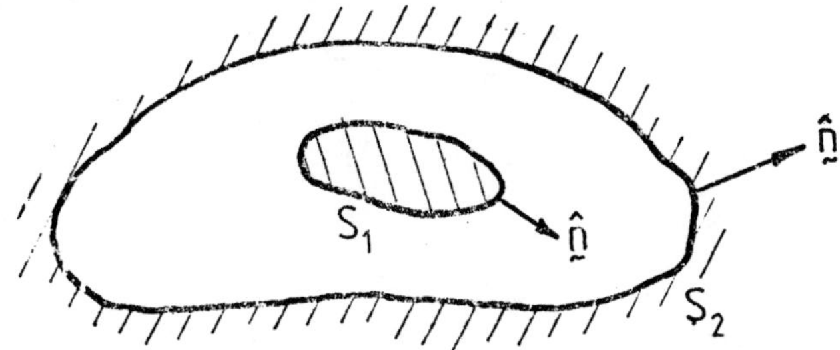

Figure 2. Cross-section of an Arbitrary Multiply Connected Region.

Since the vector potential is assumed to be solenoidal (equation (7)), equation (12) reduces to

$$\iint_{S_2} \underset{\sim}{\xi}.\underset{\sim}{\hat{n}}\, dS_2 = \iint_{S_1} \underset{\sim}{\xi}.\underset{\sim}{\hat{n}}\, dS_1 \ . \tag{13}$$

A possible solution of equation (13), for any multiply connected region, is

$$\underset{\sim}{\xi}.\underset{\sim}{\hat{n}} = \xi_n \equiv 0 \ . \tag{14}$$

It follows that for multiply connected regions the boundary conditions for $\underset{\sim}{\xi}$ are that the normal component is zero and the tangential components are related to the velocity components through equation (6) (or equation (10) if there is no flow through any boundary). For the particular geometry considered in this paper, it follows from equations (10) and (14) that the boundary conditions for $\underset{\sim}{\xi}$ are

$$\xi_z = \frac{\partial \xi_\phi}{\partial z} = 0$$

$$\text{and} \qquad \frac{\partial \xi_r}{\partial z} = V_\phi \tag{15a}$$

on the end walls ($z = 0$ and $z = AR$), and

$$\xi_r = \frac{\partial r\xi_\phi}{\partial r} = 0$$

$$\text{and} \qquad \frac{\partial \xi_z}{\partial r} = -V_\phi \tag{15b}$$

on the cylinder boundaries ($r = R_i$ and $r = R_0$).

The boundary conditions for $\underset{\sim}{\zeta}$ are obtained directly from equation (5), and are

$$\zeta_r = \frac{-\partial V_\phi}{\partial z} ,$$

$$\zeta_\phi = \frac{\partial V_r}{\partial z} \qquad (16a)$$

and

$$\zeta_z = 2\Omega$$

for the end walls, and

$$\zeta_r = 0 ,$$

$$\zeta_\phi = \frac{-\partial V_z}{\partial r} \qquad (16b)$$

and

$$\zeta_z = \frac{1}{r}\frac{\partial r V_\phi}{\partial r}$$

for the cylinder boundaries.

Since all the variables are single valued, they must be periodic in the circumferential direction.

4. METHOD OF SOLUTION

Advantage was taken of the fast convergence rate of the false transient technique of Mallinson and de Vahl Davis [7], obtained by multiplying the time derivative terms in equations (3) and (4) by appropriate factors. The finite difference approximations (FDA) of equations (3) and (4) were obtained by replacing the time derivatives with forward differences and the spatial derivatives with second order central differences. The FDAs were solved using the Samarskii-Andreyev ADI scheme [12]. FDAs for equation (11) were obtained using second order central difference approximations for the spatial derivatives and were solved using a Fourier Analysis direct method [6].

Second order FDAs were used for all derivative boundary conditions.

The overall solution procedure consisted of solving in turn the energy equation, the vorticity transport equations and the vector potential equations for one increment of the time step. The velocity field and boundary values of the vorticity were evaluated before proceeding to a new solution of the energy equation. This process was repeated until steady state was reached in accordance with a suitable convergence criterion.

5. RESULTS AND DISCUSSION

As a first step in the application of the vorticity-vector potential formulation, results have been obtained for an axisymmetric situation. In axisymmetric flows all the derivatives in the circumferential direction are zero and the equations presented in the appendix are accordingly simplified.

The particular problem discussed here consists of flow in an annulus with radius ratio RR=2, aspect ratio AR=4, and boundary conditions as outlined in Table 1. An 11×31 rectangular mesh was used; this is probably adequate for the results presented here, although a finer mesh would be desirable at higher Ra and Re. The values of Ra, Re and Pr for the cases studied are given in Table 2, together with the figure number in which the results are presented.

CASE	Ra	Re	Pr	FIGURE NO.
A	10000	no rotation (ii)	7	3
B	0	75	7	4
C	100	75	7	5
D	1000	75	7	6
E	10000	75	7	7

Table 2. Cases Studied.

The Figures 3-7 show streamlines for the motion in radial planes (i.e., with the V_ϕ component removed); in axisymmetric flow, these streamlines are contours of constant ξ_ϕ. Also shown are isotherms and contours of constant V_ϕ. The figures are the right-hand halves of vertical sections through the annulus: the axis of symmetry is to the left.

The result presented in Figures 3(a) and (b) is typical for buoyancy-induced flow in the annulus at Ra~10^4. There is a single cell with its centre slightly above the cavity mid-height as observed by de Vahl Davis and Thomas [3].

Case B is a situation leading to the formation of Taylor vortices due to the rotation of the inner cylinder and is illustrated in Figures 4(a) and (b). Four vortices are obtained. Due to end effects the two vortices near the end walls (labelled I and IV in Figure 4) are weaker and smaller than the other two vortices. Further, the centres of the vortices I and IV are nearer to the rotating cylinder than the larger vortices. Strawbridge and Hooper [14] also obtained four vortices. However in their solutions the cells nearer the end walls were larger and stronger than those in the centre of

---

(ii) A different method of non-dimensionalisation was used.

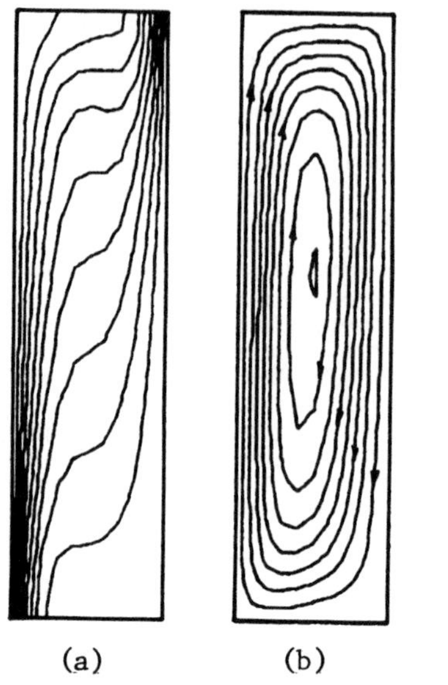

Figure 3. Isotherms (a) and Streamlines (b) for case A.

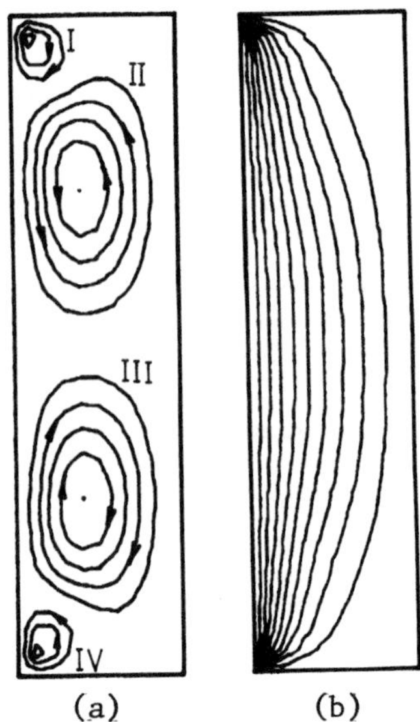

Figure 4. Streamlines (a) and Isovels of V_ϕ (b) for case B.

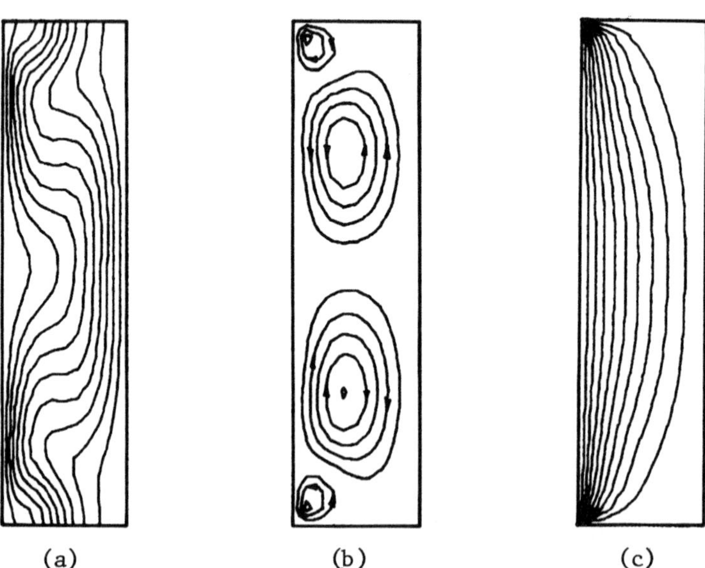

Figure 5. Isotherms (a), Streamlines (b) and Isovels of V_ϕ (c) for case C.

the annulus.

The discrepancy between these two results is attributed in part to the fact that the solutions in [14] were generated using a very coarse mesh (6×21). This was verified by generating solutions using their mesh for which four cells of equal strength were obtained. Further, they used a first order upwind differencing scheme. As discussed by de Vahl Davis and Mallinson [4], the use of upwind differences can result in a severe misrepresentation of the flow. This is particularly true for the coarse mesh used by Strawbridge and Hooper, since the distortion of the flow increases with an increase in mesh spacing.

The isotherms, streamlines and circumferential velocity contours for rotational flows with a low Rayleigh number (Ra = 100, case C) are shown in Figure 5.

The rotation about the z-axis dominates the flow. The isotherms have been severely distorted, by the Taylor vortices, when compared with the isotherms in Figure 3(a), whereas the streamlines and isovels have only been changed slightly from those in Figure 4 by the base (buoyancy driven) flow.

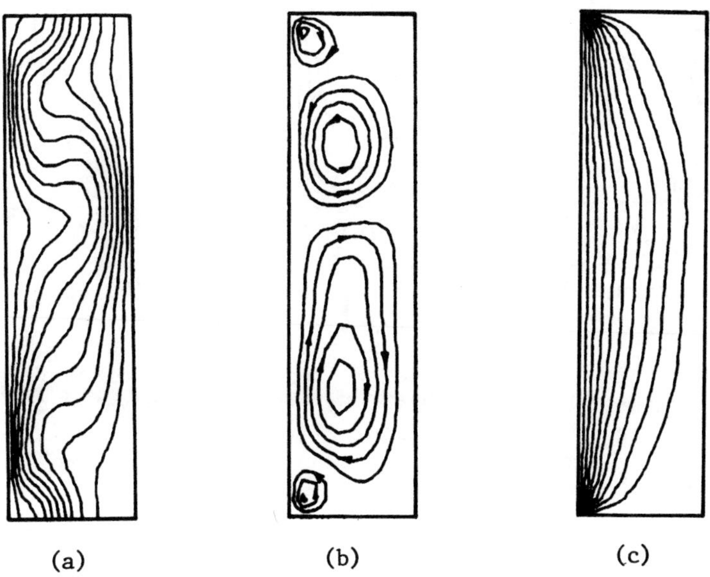

(a) (b) (c)

Figure 6. Isotherms (a), Streamlines (b) and Isovels of V_ϕ (c) for case D.

As the Rayleigh number is increased (Ra=1000, case D, Figure 6), the distortion of the isotherms is not significantly reduced. The streamlines, however, are noticeably changed. Cell I has been strengthened slightly by the base flow, whereas cells II and IV, which are rotating in the opposite direction to the base flow, have been reduced in size and strength. Cell III, which is rotating in the same direction as the base flow, has been substantially strengthened and increased in size.

When the Rayleigh number is further increased (Ra=10000, case E, Figure 7), the buoyancy-induced flow dominates. The isotherms are similar to those obtained for case A (no rotation). The Taylor vortices have now been significantly distorted. Cell I has been considerably weakened and cell II has actually reversed direction. The smaller corner cell IV has also been weakened by the dominant base flow.

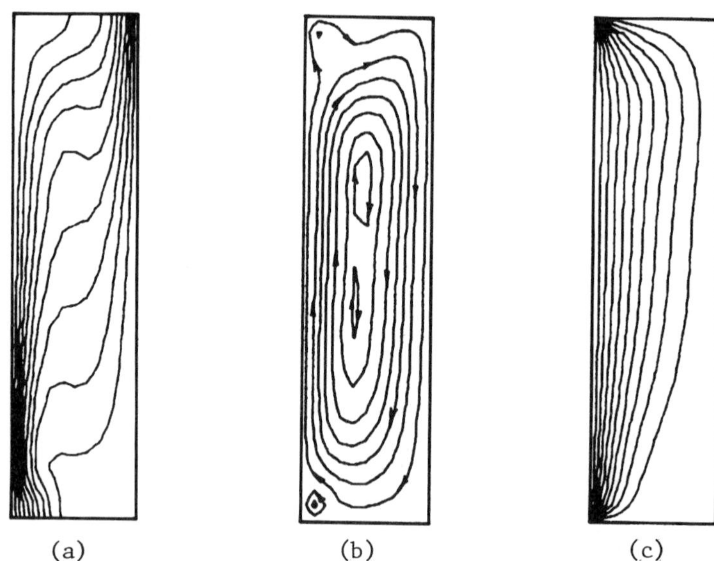

(a) (b) (c)

Figure 7. Isotherms (a), Streamlines (b) and Isovels of V_ϕ (c) for case E.

At this higher Rayleigh number the circumferential velocity has been affected by the buoyancy-induced flow (compare Figure 4(b) with Figure 7(c)) which has shifted the isovels at the top of the annulus to the right, and those at the bottom to the left.

6. CONCLUSION

The vorticity-vector potential formulation has been applied successfully to axisymmetric flows in a rotating heated annulus. Space has permitted the presentation of only a limited discussion of the interaction between the flow induced by buoyancy and rotation, fuller details of which will be published elsewhere. The extension to fully three-dimensional flows is now being made.

7. ACKNOWLEDGEMENT

We are grateful for the financial support provided under the National Energy Research, Development and Demonstration programme which is administered by the Commonwealth Department of National Development and Energy.

8. REFERENCES

1. AZIZ, K. and HELLUMS, J.D. - Numerical Solution of the Three-Dimensional Equations of Motion for Laminar Natural Convection, Phys. of Fluids, Vol. 10, pp.314-324, 1967.
2. BOUSSINESQ, J. - Theorie Analytique de la Chaleur, Gauthier-Villars, Vol. 2, p172, 1903.
3. DE VAHL DAVIS, G. and THOMAS, R.W. - Natural Convection between Concentric Vertical Cylinders, Physics of Fluids, Supplement II, Vol. 12, pp.198-207, 1969.
4. DE VAHL DAVIS, G. and MALLINSON, G.D. - An Evaluation of Upwind and Central Difference Approximations by a Study of Recirculating Flow, Computers and Fluids, Vol.4, pp.29-43, 1976.
5. HIRASAKI, G.J. and HELLUMS, J.D. - Boundary Conditions on the Vector and Scalar Potential in Viscous Three-Dimensional Hydrodynamics, Quart. of Appl. Math., Vol. 28, No. 2, pp.293-296, 1970.
6. LE BAIL, R.C. - Use of Fast Fourier Transforms for Solving Partial Differential Equations in Physics, J. Comp. Phys., Vol. 9, pp.440-465, 1972.
7. MALLINSON, G.D. and DE VAHL DAVIS, G. - The Method of the False Transient for the Solution of Coupled Elliptic Equations, J. Comp. Phys., Vol. 12, pp.435-461, 1973.
8. MALLINSON, G.D. and DE VAHL DAVIS, G. - Three-Dimensional Natural Convection in a Box: A Numerical Study, J. Fluid Mech., Vol. 83, Pt. 1, pp.1-31, 1977.
9. PIACSEK, S.A. - Numerical Experiments on Convective Flows in Geophysical Fluid Systems, 7th Symposium on Naval Hydrodynamics, Rome, pp.1-34, 1968.
10. QUON, C. - A Study of Penetrative Convection in Rotating Fluid, Trans. ASME J. of Heat Trans., Vol.101, pp.261-264, 1979.
11. RICHARDSON, S.M. and CORNISH, A.R.H. - Solution of Three-Dimensional Incompressible Flow Problems, J. Fluid Mech., Vol. 82, Pt. 2, pp.309-319, 1977.
12. SAMARSKII, A.A. and ANDREYEV, V.B. - On a High Accuracy Difference Scheme for an Elliptic Equation with Several Space Variables, U.S.S.R. Comp. Math. and Math. Phys., Vol. 3, pp.1373-1382, 1963.
13. SEDOV, L.I. - A Course in Continuum Mechanics, Vol. III, Wolters-Noordhoff Publishing, Groningen, 1972.
14. STRAWBRIDGE, D.R. and HOOPER, G.T.J. - Numerical Solutions of the Navier-Stokes Equations for Axisymmetric Flows, J. Mech. Eng. Science, Vol. 10, No. 5, pp.389-401, 1968.
15. WILLIAMS, G.P. - Thermal Convection in a Rotating Fluid Annulus; Pt. 2, J. Atmos. Sci., Vol. 24, pp.162-174, 1967.
16. WILLIAMS, G.P. - Numerical Integration of the Three-Dimensional Navier-Stokes Equations for Incompressible Flows, J. Fluid Mech., Vol. 37, Pt. 4, pp. 727-750, 1969.

APPENDIX

Equations (3), (5), (9) and (10) may be expressed as ten partial differential equations for the ten dependent variables θ, ζ_r, ζ_ϕ, ζ_z, ξ_r, ξ_ϕ, ξ_z, V_r, V_ϕ and V_z. For a cylindrical coordinate system these become

$$\frac{\partial\theta}{\partial t} = -\frac{1}{r}\frac{\partial rV_r\theta}{\partial r} - \frac{1}{r}\frac{\partial V_\phi\theta}{\partial\phi} - \frac{\partial V_z\theta}{\partial z} + \frac{1}{RePr}\left(\frac{1}{r}\frac{\partial\theta}{\partial r} + \frac{\partial^2\theta}{\partial r^2} + \frac{1}{r^2}\frac{\partial^2\theta}{\partial\phi^2} + \frac{\partial^2\theta}{\partial z^2}\right)$$

$$\frac{\partial\zeta_r}{\partial t} = \frac{1}{r}\frac{\partial V_r\zeta_\phi}{\partial\phi} - \frac{1}{r}\frac{\partial V_\phi\zeta_r}{\partial\phi} + \frac{\partial V_r\zeta_z}{\partial z} - \frac{\partial V_z\zeta_r}{\partial z} - \frac{Ra}{Re^2Pr}\frac{1}{r}\frac{\partial\theta}{\partial\phi}$$
$$+ \frac{1}{Re}\left(\frac{1}{r}\frac{\partial\zeta_r}{\partial r} + \frac{\partial^2\zeta_r}{\partial r^2} + \frac{1}{r^2}\frac{\partial^2\zeta_r}{\partial\phi^2} + \frac{\partial^2\zeta_r}{\partial z^2} - \frac{\zeta_r}{r^2} - \frac{2}{r^2}\frac{\partial\zeta_\phi}{\partial\phi}\right)$$

$$\frac{\partial\zeta_\phi}{\partial t} = \frac{\partial V_\phi\zeta_r}{\partial r} - \frac{\partial V_r\zeta_\phi}{\partial r} + \frac{\partial V_\phi\zeta_z}{\partial z} - \frac{\partial V_z\zeta_\phi}{\partial z} + \frac{Ra}{Re^2Pr}\frac{\partial\theta}{\partial r}$$
$$+ \frac{1}{Re}\left(\frac{1}{r}\frac{\partial\zeta_\phi}{\partial r} + \frac{\partial^2\zeta_\phi}{\partial r^2} + \frac{1}{r^2}\frac{\partial^2\zeta_\phi}{\partial\phi^2} + \frac{\partial^2\zeta_\phi}{\partial z^2} - \frac{\zeta_\phi}{r^2} + \frac{2}{r^2}\frac{\partial\zeta_r}{\partial\phi}\right)$$

$$\frac{\partial\zeta_z}{\partial t} = \frac{1}{r}\frac{\partial rV_z\zeta_r}{\partial r} - \frac{1}{r}\frac{\partial rV_r\zeta_z}{\partial r} + \frac{1}{r}\frac{\partial V_z\zeta_\phi}{\partial\phi} - \frac{1}{r}\frac{\partial V_\phi\zeta_z}{\partial\phi}$$
$$+ \frac{1}{Re}\left(\frac{1}{r}\frac{\partial\zeta_z}{\partial r} + \frac{\partial^2\zeta_z}{\partial r^2} + \frac{1}{r^2}\frac{\partial^2\zeta_z}{\partial\phi^2} + \frac{\partial^2\zeta_z}{\partial z^2}\right)$$

$$-\zeta_r = \frac{1}{r}\frac{\partial\xi_r}{\partial r} + \frac{\partial^2\xi_r}{\partial r^2} + \frac{1}{r^2}\frac{\partial^2\xi_r}{\partial\phi^2} + \frac{\partial^2\xi_r}{\partial z^2} - \frac{\xi_r}{r^2} - \frac{2}{r^2}\frac{\partial\xi_\phi}{\partial\phi}$$

$$-\zeta_\phi = \frac{1}{r}\frac{\partial\xi_\phi}{\partial r} + \frac{\partial^2\xi_\phi}{\partial r^2} + \frac{1}{r^2}\frac{\partial^2\xi_\phi}{\partial\phi^2} + \frac{\partial^2\xi_\phi}{\partial z^2} - \frac{\xi_\phi}{r^2} + \frac{2}{r^2}\frac{\partial\xi_r}{\partial\phi}$$

$$-\zeta_z = \frac{1}{r}\frac{\partial\xi_z}{\partial r} + \frac{\partial^2\xi_z}{\partial r^2} + \frac{1}{r^2}\frac{\partial^2\xi_z}{\partial\phi^2} + \frac{\partial^2\xi_z}{\partial z^2}$$

$$V_r = \frac{1}{r}\frac{\partial\xi_z}{\partial\phi} - \frac{\partial\xi_\phi}{\partial z}$$

$$V_\phi = \frac{\partial\xi_r}{\partial z} - \frac{\partial\xi_z}{\partial r}$$

$$V_z = \frac{1}{r}\frac{\partial r\xi_\phi}{\partial r} - \frac{1}{r}\frac{\partial\xi_r}{\partial\phi}$$

DIFFUSION CONTROLLED SEA WATER CORROSION OF OFFSHORE PIPELINES UNDER DISBONDED PROTECTIVE COATINGS

O. Kvernvold, A.K. Rastogi and T. Sontvedt

Det norske Veritas, P.O. Box 300, N1322 Hovik, Oslo, Norway

ABSTRACT

Partial differential equations for conservation of mass, momentum, energy and species concentration have been solved numerically by a finite difference solution procedure for flow occuring beneath disbonded coatings on offshore risers. Two-dimensional gaps closed either in the top or in the bottom are considered. The corrosion rate under diffusion controlled conditions are presented for different coatings and different gap widths.

1. INTRODUCTION

Insulating coatings are frequently applied on the outside of offshore risers and pipe lines to protect the metal surface from the corrosive sea water. Different kinds of thickfilm coatings such as asphalt and coal tar based coatings are most frequently used. Effectiveness of the coatings reduces strongly if there is a gap between the protecting layer and the metal surface, permitting sea water to come in contact with the metal surface. Some of the thickfilm coatings applied in the North Sea are suspected to fail from disbonding from the pipes in length up to several meters. For risers the large temperature differences between the hot oil/gas in the pipe and the cold sea water results in a convective fluid current in the crevice, which continuously transport sea water saturated with oxygen into the crevice. In these cases cathodic protection is relied upon to ensure that severe corrosion will not take place at the disbonded areas. However, recent experimental evidence [1], [2], [3] and [4] indicates that the effect of cathodic protection decreases with increasing flow velocity in the crevice. Severe corrosion problems may therefore be expected in areas with unsatisfactory cathodic protection.

In a recent paper, [5], a theoretical examination of diffussion controlled corrosion in vertical crevices open both top and bottom is reported. The numerical results obtained in [5] prove to give a reasonable values compared with available experimental results.

In the present report vertical crevices open in one end and closed in the other are examined by a finite difference method [6]. A schematic picture of the models is shown in Fig. 1 and Fig. 2. For the model shown in Fig. 1 saturated seawater enters the crevice along the hot metal wall and the largest corrosion rates are found in the inlet region. For the model shown in Fig. 2, however, the fluid enters the crevice along the outer non-reacting coating wall and corrosion is then found in larger parts of the crevice compared with the model shown in Fig. 1. In this paper four coatings (epoxy or asphalt (somastic), polyethylene and coal tar + concrete are examined for both the model shown in Fig. 1 and Fig. 2. The temperature for the oil/gas is taken to be 90°C while the sea water is assumed to have a temperature of 10°C.

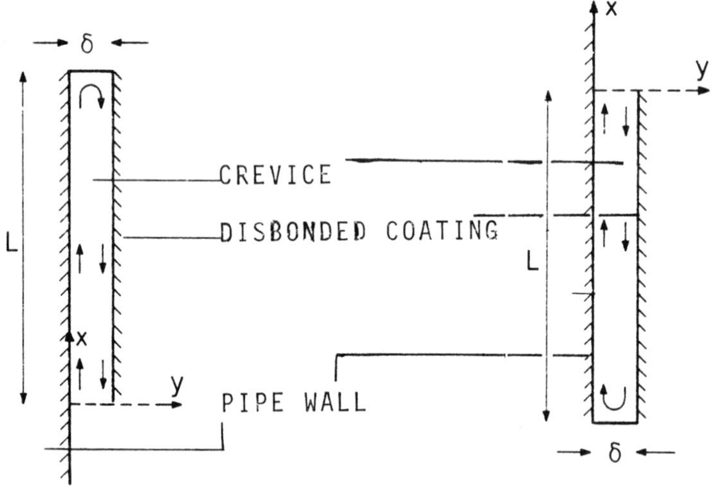

Fig. 1: Model for crevice closed in the top.

Fig. 2: Model for crevice closed in bottom.

2. GOVERNING EQUATIONS

For a crevice which is closed in one end, fluid is driven up along the hot wall and the same fluid has, due to continuity, to flow down along the cold wall (Fig. 1 and Fig. 2). The flow is buoyancy driven due to the temperature difference across the crevice. The flow is governed by the complete set of the Navier-Stokes equations. However, for the present models, with $\delta/L \ll 1$, the equations can be simplified by neglecting the pressure gradient in the y-direction, perpendicular to the wall. Then the y-component of the momentum equation can be neglected and the velocity component in that direction is found from the continuity equation. The following set of equations then has to be solved:

Continuity: $$\frac{\partial u}{\partial x} + \frac{\partial v}{\partial y} = 0 \qquad (1)$$

x-momentum: $$u\frac{\partial u}{\partial x} + v\frac{\partial u}{\partial y} = \nu\left(\frac{\partial^2 u}{\partial x^2} + \frac{\partial^2 u}{\partial y^2}\right) + g\,\beta\left(T - \frac{T_h + T_c}{2}\right) \qquad (2)$$

Energy: $$u\frac{\partial T}{\partial x} + v\frac{\partial T}{\partial y} = \frac{\nu}{Pr}\left(\frac{\partial^2 T}{\partial x^2} + \frac{\partial^2 T}{\partial y^2}\right) \qquad (3)$$

Concentration: $$u\frac{\partial c}{\partial x} + v\frac{\partial c}{\partial y} = \frac{\nu}{Sc}\left(\frac{\partial^2 c}{\partial x^2} + \frac{\partial^2 c}{\partial y^2}\right) \qquad (4)$$

The boundary conditions at the rigid walls are taken to be:

$$u = v = 0 \text{ at } y = 0,\delta\ ;\ T = T_h \text{ at } y = 0;$$
$$T = T_c \text{ at } y = \delta;\ c = 0 \text{ at } y = 0; \qquad (5)$$
$$\delta c/\delta y = 0 \text{ at } y = \delta;$$

$$\left.\begin{array}{l} u = v = 0 \\ \partial c/\partial x = \partial T/\partial x = 0 \end{array}\right\} \text{ at the closed end}$$

The boundary conditions at the closed end correspond to an insulated, impervious boundary.

Further the conditions at the open end have to be specified. They are taken to be:

$$u = \frac{g\,\beta\,(T_h - T_c)\,\delta^2}{\nu}\left(\frac{1}{12}\left(\frac{y}{\delta}\right) + \frac{1}{6}\left(\frac{y}{\delta}\right)^3 - \frac{1}{4}\left(\frac{y}{\delta}\right)^2\right) \qquad (6)$$

$$T = T_c + (T_h - T_c)\left(1 - \frac{y}{\delta}\right) \qquad (7)$$

(6) and [7] correspond to the solution of (2) and (3) far from the closed end of the crevice and are obtained by assuming $v = \partial u/\partial x = \partial T/\partial x = 0$. (6) gives the largest input velocity and thereby the largest flux of oxygen into the crevice for the above model.

The input oxygen concentration is taken to be $C = C_{in} = 0.07\ mol/m^3$ for the fluid entering the crevice. For fluid leaving the crevice the upstream value is applied, which means that $\partial c/\partial x = 0$. The same boundary condition, $\partial/\partial x = 0$, is also applied for the velocity and temperature for fluid leaving the crevice.

The material properties used in the computation, such as viscosity, Prandtl number etc. are the same as reported in [5], and are for simplicity also given in the Appendix. The boundary conditions for the temperature are obtained from the analysis in [5] and actual values listed in Table 1.

Coating / Gap width	Epoxy / Asphalt (Somastic)		Polyethylene		Coal tar + concrete	
	T_h	T_c	T_h	T_c	T_h	T_c
1.5 mm	88.4	81.2	86.0	68.0		
2.0 mm	88.4	79.1	86.3	63.1	89.3	85.4
3.0 mm	88.5	75.3	86.7	57.5		

Table 1: Temperature at the boundaries of the crevice.

3. SOLUTION PROCEDURE

Equations (2), (3) and (4) are nonlinear elliptic partial differential equations. The solution procedure of ref. [7] for partial differential equations was modified and used in the present work. This is an iterative solution procedure based on finite differences. The solution was found by first obtaining a velocity distribution from (1) and (2) for a given temperature field. This velocity field is then kept unchanged. and only the temperature equation (3) was solved to provide a new temperature field. This new temperature field was then used to give a new velocity field and so on. This iteration procedure was continued till the u, v and T values did not change anymore. The velocity field then obtained was introduced into (4). This so linearized partial differential equation was solved for oxygen concentration. The computer program which employs the above mentioned solution scheme is available at "Det norske VERITAS" under the name "DOC2" [6].

4. RESULTS

The results of the numerical calculations are shown in Table 2 and Figs. 3 - 10. Calculations reveal that the velocity distribution does not change along the crevice except in the region near the closed end, where the flow direction is turned. The velocity is further antisymmetric about $y/\delta=1/2$. In Table 2 is shown the maximum velocity for the different coatings and different gap widths considered. The magnitude of the velocity is seen to depend strongly on both the gap width and the kind of coating applied. For a given gap width the coatings which are the best insulators to heat transfer give the smallest velocities in the crevice. There are no difference in the velocity distribution in a crevice closed

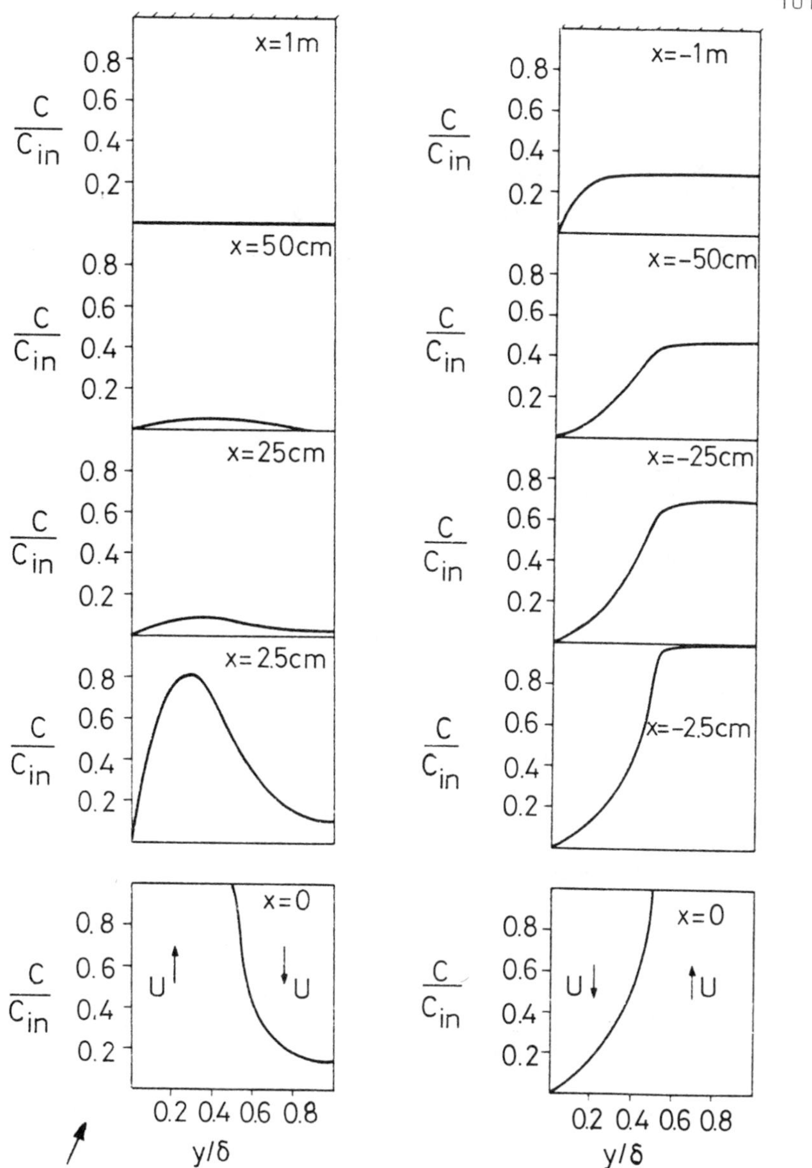

Fig. 3 and Fig. 4: The variation of oxygen distribution along crevices closed in top (Fig. 3) and closed in bottom (Fig. 4). Epoxy coating: crevice length: L = 1 m; gap width: δ = 2mm

in the top and a crevice closed in the bottom having the same gap width and the same coating. However, the oxygen distribution in the crevice is quite different in the two cases. For a crevice closed in the top the saturated sea water enters along the metal wall where the oxygen is reduced. A characteristic example for the oxygen distribution in a crevice closed in the top is shown in Fig. 3. Fig. 4 shows the oxygen distribution in a crevice with the same gap width and the same coating, but

Coating → / Gap width	Epoxy / Asphalt (Somastic)	Polyethylene	Coal tar + concrete
1.5mm	0.21	0.53	
2.0mm	0.48	1.20	0.2
3.0mm	1.55	3.48	

Table 2: Maximum convection velocity in the crevice (cm/s) for coatings referred to in Table 1.

closed in the bottom. In this case the saturated sea water enters the crevice along the non-reacting coating wall. Except for the lower part of the crevice the oxygen is transported to the metal wall by molecular diffusion only. This is a much slower transport mechanism than the convective transport. Oxygen may therefore be transported by the fluid to the bottom of the crevice along the outer coating surface before the fluid turns and goes up along the metal surface.

The oxygen distribution near the metal wall determines the corrosion rates under diffusion controlled conditions. The limiting current density is given by

$$I_{lim} = zF\ D_i\ \left(\frac{\partial c}{\partial y}\right)_{y=0} \qquad (8)$$

where I_{lim} : limiting current density
z : valency change
F : Faradays constant

In [5] it is shown that a limiting current of $1 A/m^2$ corresponds to a corrosion rate of 1.17 mm/year.

Fig. 5 shows the corrosion rates for epoxy and polyethylene coatings for crevices closed in the top. The corrosion rate at a given distance from the inlet is found to be nearly independent on the crevice length (for the same coating and the same gap width). The figure shows that the epoxy coating gives better protection than the polyethylene coating. This is due to the difference in the convective velocities (Table2).

The results displayed in Figs. 6 - 10 show the corrosion rates for crevices closed in the bottom. For comparison the corrosion rate for the corresponding crevice closed in the top is also given. The figures clearly demonstrate the difference between crevices closed in top and bottom. While the

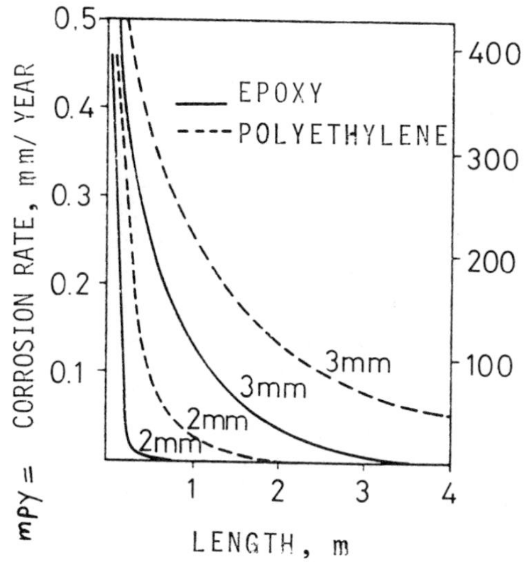

Fig. 5: The corrosion rate and limiting current density in crevices closed in the top.

corrosion rate for crevices closed in the top decreases monotonically with the distance from the inlet, crevices closed in the bottom may even have the highest corrosion rates in the bottom of the crevice (Figs. 6 and 8). Generally it is shown that crevices closed in the bottom give corrosion in a larger part of the crevice than the corresponding crevice closed in the top. The reason is, as earlier mentioned, that the saturated sea water enters the crevice along the outer non-reacting coating wall. Oxygen

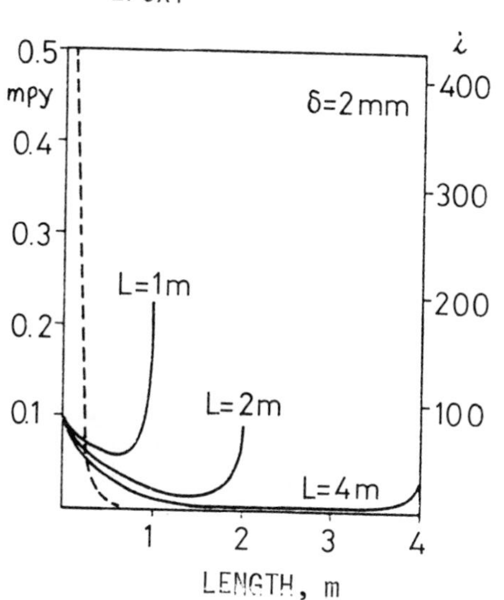

Fig. 6: The corrosion rate and limiting current density in crevices with a 2mm gap width. Epoxy coating. —— crevices closed in bottom, --- crevices closed in top

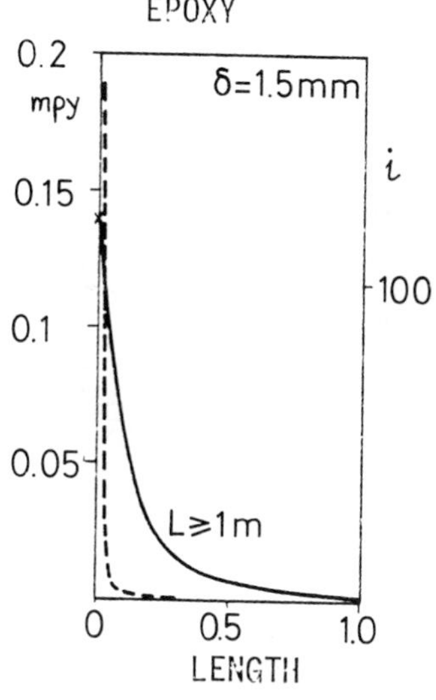

Fig. 7: The corrosion rate and limiting current density in crevices with a 1.5mm gap width. Epoxy coating. —— crevices closed in the bottom, --- crevices closed in top.

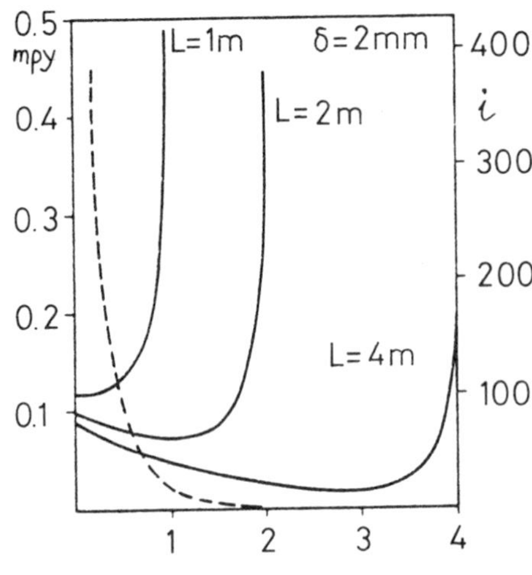

Fig. 8: The corrosion rate and limiting current density in crevices with a 2mm gap width. Polyethylene coating. —— crevices closed in bottom; --- crevices closed in top.

may therefore be transported further into the crevice before it reacts with the metal wall.

It is shown in [1], [2], [3] and [4] that cathodic protection is only effective in the first 5-10 cm from the inlet. The numerical results reveal that severe corrosion therefore may be expected in areas which is not sufficient cathodic protected if the gap width is greater than 1-2 mm depending on the coating and crevice configuration. Crevices closed in the bottom are the most critical ones because high corrosion rates are found in larger parts of the crevice compared with the corresponding crevice closed in the top.

The corrosion rates obtained for closed crevices are found to be smaller than the corresponding results for a crevice open in both ends [5]. This is expected since the input velocity is smaller for closed crevices than for open ones. In addition fluid flows into closed crevices in only half the total width (Figs. 1 and 2), while for open crevices fluid enters in the total width [5].

5. CONCLUDING REMARKS

Diffusion controlled corrosion in crevices closed either in top or bottom is examined numerically. Examples characteristic for the oil/gas production in the North Sea are examined. Calculations reveal the corrosion rate decreases monotonically with the distance from the inlet for crevices closed in the top (Fig. 5). For crevices closed in the bottom, however, the corrosion rate is found to increase in the bottom of the crevice and it may even be higher there than at the inlet depending on the coating and the crevice length (Figs. 6, 8 and 9). For a given gap width and crevice configuration, coal tar + concrete is found to give the lowest corrosion rates.

Cathodic protection is relied upon to protect the risers if the coatings fail. However, since cathodic protection is

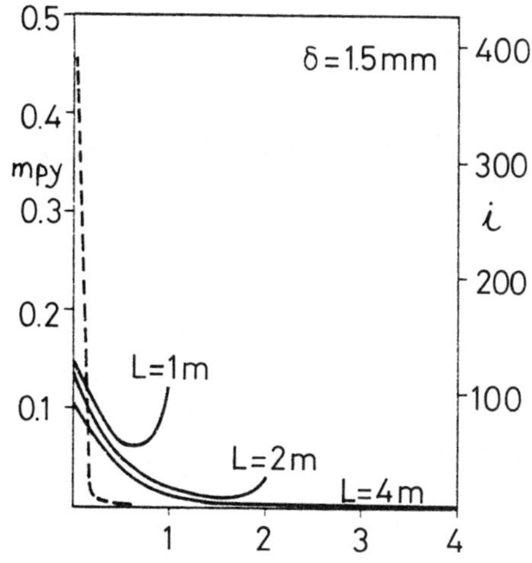

Fig. 9: The corrosion rate and limiting current density in crevices with a 2mm gap width. Polyethylene coating. —— crevices closed in bottom; --- crevices closed in top.

effective only in the first 5-10 cm from the inlet [1], [2], [3] and [4] the present results show that severe corrosion may be expected in areas which are not satisfactorily protected. For crevices closed in the top cathodic protection is expected not to give sufficient protection if the gap width is greater than 1.5 - 2 mm depending on the coating. For crevices closed in the bottom the limits are expected to be 1-1.5mm depending on the coating and also the crevice length. For comparison the limit for crevices open in both ends was found to be 0.5 mm.

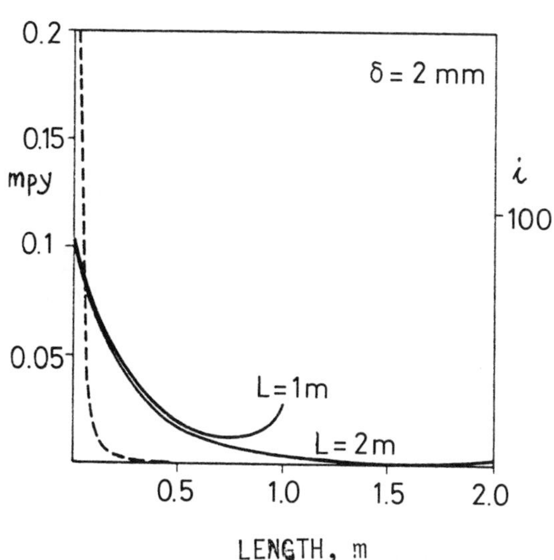

Fig. 10: The corrosion rate and limiting current density in crevices with a 2mm gap width. Coal tar + concrete coating.
—— crevices closed in bottom;
--- crevices closed in top.

6. REFERENCES

[1] VENNET, R.M. - Cathodic Protection of a Hot Riser in Cold Sea Water. Materials Performance, 18, p. 26, 1979.

[2] GROOVER, R.E. and PETERSON, M.H. - Cathodic Protection of Internal Surfaces of Pipes Containing Sea Water. Materials Performance, 13, p. 24, 1974.

[3] NTNF-Project 1830.5585 - Corrosion Control of Offshore Pipelines II. Part Activity 5.2: Effect of Cathodic Protection in Crevices. Det norske Veritas.

[4] GRATACOR, J. and PERROLLET, C. - Study of Cathodic Protection Parameters in Case of a Cracked Coating. Société Nationale Elf Aquitane (Production) Paris, 1976.

[5] RASTOGI, A.K., KVERNVOLD, O., MOLLAN, R. and SONTVEDT, T. Diffusion Controlled Corrosion under Disbonded Coatings on Offshore Risers - Theoretical Model, Part 1. VERITAS Report No. 80-0072, 1980.

[6] KVERNVOLD, O. and RASTOGI, A.K. - Computer Program Specification "DOC-2". Decay of Oxygen Concentration. VERITAS Report not yet available., 1981.

[7] GOSMAN, A.D. and PUN, W.M. - Calculation of Recirculating Flows. Imperial College Report No. HTS/74/2, 1973.

7. APPENDIX

Data for oil from the North Sea:

Thermal conductivity: $K = 0.15\ W/m^{o}C$

Heat capacity: $C_p = 1470\ Ws/kg^{o}C$

Dynamic viscosity: $\mu = 0.5 \times 10^{-3}\ ks/ms$

Density: $\rho = 800\ kg/m^{2}$

Heat transfer data for steel:

Thermal conductivity: $K = 50\ W/m^{o}C$

Wall thickness: $d = 14$ mm

Heat transfer data for the coatings:

Epoxy:
Thermal conductivity: $K = 0.35\ W/m^{o}C$

Thickness: $d = 8$ mm

Asphalt: (Somastic)
Thermal conductivity: $K = 0.61\ W/m^{o}C$
Thickness: $d = 15$ mm

Coal tar + concrete:

Coal tar:

Thermal conductivity:	$K = 0.14\ W/m^{o}C$
Thickness:	$d = 4\ mm$

Concrete:

Thermal conductivity	$K = 2\ W/m^{o}C$
Thickness:	$d = 75\ mm$

Polyethylene:

Thermal conductivity:	$K = 0.33\ W/m^{o}C$
Thickness:	$d = 2\ mm$

Data for sea water in the crevice. Temperature T ~ 80-85°C

Thermal conductivity:	$K = 0.67\ W/m^{o}C$
Dynamic viscosity:	$\mu = 0.355 \times 10^{-3}\ kg/ms$
Density:	$\rho = 972\ kg/m^{3}$
Coefficient of thermal expansion:	$\beta = 6.5 \times 10^{-4}\ {}^{o}C^{-1}$
Diffusion coefficient for oxygen:	$D_i = 7.4 \times 10^{-9}\ m^{2}/s$
The Schmidt number:	$S_c = 50$

Data for sea water outside the crevice. Temperature T ~ 10-15°C

Thermal conductivity:	$K = 0.59\ W/m^{o}C$
Dynamic viscosity:	$\mu = 1.2 \times 10^{-3}\ kg/ms$
Density:	$\rho = 1000\ kg/m^{3}$
Coefficient of thermal expansion:	$\beta = 1 \times 10^{-4}\ {}^{o}C^{-1}$

DIGITAL SIMULATION OF FORCED CONVECTION IN A ROTATING FLUID

M. Mihelčič, C. Schröck-Pauli, K. Wingerath,
H. Wenzl, W. Uelhoff, A. van der Hart

Institut für Festkörperforschung der Kernforschungsanlage Jülich, GmbH, Postfach 1913,
D-517o Jülich, West Germany

SUMMARY

We investigate the flow patterns in a rotating rigid cylindrical container filled with a homogeneous viscous liquid and stirred by a rotating disk. This rotational system is a model of the Czochralski crystal-growth arrangement, where the forced convection of melt in a crucible caused by crystal and/or crucible rotation is under consideration. The time-dependent Navier-Stokes equations governing the fluid motion are computationally solved by a finite difference procedure. We verify numerically that the presence of an obstacle (i.e. the disk) immersed in a rotating liquid gives rise to a stationary liquid cell beneath the obstacle as predicted from the Taylor-Proudman theorem. Our digital simulations also show the different effects of combined container and disk rotations in equal and opposite direction (iso- and counter-rotation). Moreover, we present the flow patterns for the periodically accelerated and decelerated iso- and counterrotational arrangement of container and disk.

1. INTRODUCTION

The aim of this paper is to investigate numerically the time-dependent convective flow patterns in a rotating rigid cylindrical container filled with a homogeneous viscous liquid and stirred by a rotating disk (see fig. 1).

We study the cases of the uniformly rotating container-disk arrangement with angular velocities in equal and opposite directions as well as the corresponding non-uniformly rotating cases, where

container and disk are periodically accelerated and decelerated.
This rotational system is a model of the Czochralski crystal-growth process in the classical and the so-called CACRT arrangement proposed in [5], where the forced convection of the (isothermal) melt in a crucible caused by crystal and crucible rotation is considered (see [2,3]).
The classification of hydrodynamic phenomena in rotating systems is assisted by the Taylor-Proudman theorem, which is strictly valid only for inviscid fluids but true as an approximation for small Ekman numbers. If an object is immersed in a rotating liquid which has a different angular velocity, the Taylor-Proudman theorem implies that the fluid isolates this obstacle by the formation of a liquid column in its vicinity, where no mixing with the outer fluid regions occurs. Such liquid columns are called Taylor-Proudman cells. In our case shown in fig. 1, the Taylor-Proudman cell is bounded by a cylindrical stagnation surface in the liquid beneath and concentric with the disk. Outside of this stagnation surface the liquid rotates essentially as a solid body provided the crucible rotation is fast enough. If the disk as well as the container rotate in the same direction ("isorotation") only a single Taylor-Proudman cell develops, whereas in the case of opposite rotational directions of container and disk ("counterrotation") the liquid column beneath the disk divides into two Taylor-Proudman cells seperated by another transverse stagnation surface.

We computationally solve the time-dependent Navier-Stokes and continuity equations in primitive variables (pressure and velocity) for axisymmetric fluid flow by the explicit finite difference procedure of the MAC-method [1,6] for a consecutive sequence of time-steps until an approximate steady-state solution is reached. We verify the emergence of Taylor-Proudman cells in two characteristic uniformly iso- and counter-rotational container-disk arrangements. Moreover, we show how it can be achieved that the non-mixing Taylor-Proudman cells vanish periodically thereby getting total stirring of the liquid.

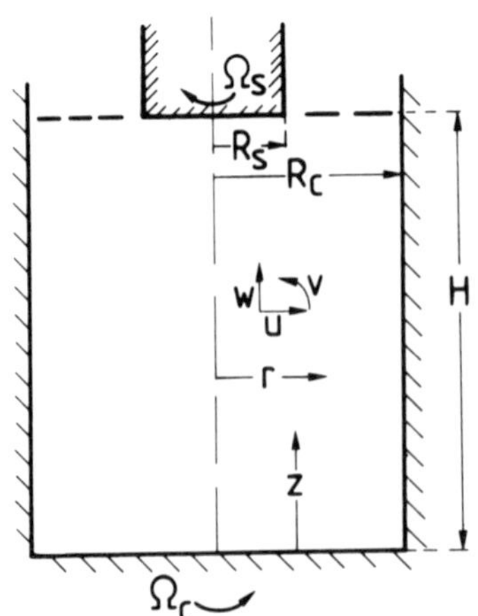

Fig. 1 : Geometry of the container-disk system.

2. THE MATHEMATICAL MODEL

2.1. The continuous equations of motion

Our calculations are based on the geometrical model of the container-disk system shown in fig. 1 . For the basic field variables in a cylindrical Eulerian coordinate system we use the fluid velocity $\vec{u} = (u,v,w)$ with radial component u , zonal component v and vertical component w , and the ratio $P = p/\rho$ of pressure p to (constant) density ρ . In addition, t denotes the elapsed time and ν the (constant) kinematic viscosity of the liquid. Then the equations governing the time-dependent incompressible viscous flow in the container are

$$\nabla \cdot \vec{u} = 0 \ , \tag{2.1}$$

$$\frac{\partial \vec{u}}{\partial t} = -(\vec{u}\cdot\nabla)\vec{u} - \nabla P + \nu\Delta\vec{u} + \vec{g} \ . \tag{2.2}$$

Eq. (2.1) is the continuity equation expressing the conservation of mass for an incompressible fluid. (2.2) consists of the well-known Navier-Stokes equations describing the local production of momentum. Using (2.1) we rewrite eq. (2.2) in the conservative form

$$\frac{\partial \vec{u}}{\partial t} = -\nabla\cdot(\vec{u}\vec{u}) - \nabla P - \nu\nabla\times(\nabla\times\vec{u}) + \vec{g} \ , \tag{2.3}$$

because in a cylindrical coordinate system (2.3) is the only differential form the finite difference analogue of which retains rigorous momentum conservation [1,4,6]. In order to satisfy the continuity equation (2.1), it is convenient to derive the well-known Poisson-equation for the pressure P :

$$\Delta P = -\frac{\partial}{\partial t}(\nabla\cdot\vec{u}) - \nabla\cdot(\nabla\cdot(\vec{u}\vec{u})) \ . \tag{2.4}$$

Obviously, in the continuous case, the first term on the right side of eq. (2.4) vanishes by (2.1), but during the discretization process we have to pay attention to $\partial(\nabla\cdot\vec{u})/\partial t$ in order to fulfil the discrete version of the mass conservation (i.e. volume conservation). Eq. (2.4) shows another advantage of the use of (2.3) instead of (2.2): In (2.4) only advective and no viscous terms appear simplifying the difference analogue, too, and thereby preventing needless accumulation of round-off errors later on.

2.2. The discretization

In our current simulations presented in section 3, we assume rotational symmetry of the flow so that it is sufficient to consider one half of a vertical cross-section of the liquid. The equations of section 2.1 are expressed in finite difference form on the grid shown in fig. 2 (see [2,3]). For $R_c = 3.4$ cm and $H = 4$ cm (see fig. 1), our discretization parameters are $\Delta r = R_c/(L-1) = h$ and $\Delta z = H/(N-1) = h$ with $h = 0.1$ cm, and $\Delta t = 0.001$ s, which is small enough to satisfy the necessary stability condition of Courant, Lewy and Friedrichs, i. e. $\Delta t < h^2/4\nu$ for $\Delta r = \Delta z$. Our discretization procedure consists in the well-known "Marker-and-Cell" - (MAC-) method in [1,6]. It is an explicit finite difference method using forward difference schemes for the first order derivatives in time and central differencing for the spatial derivatives. Starting with a certain initial state of the pressure and velocity field, a new velocity and pressure distribution is computed for a consecutive sequence of time-steps, whereby the last velocity field and a time-extrapolated pressure distribution of the last time-cycle act as initial state for the following calculations at the next time-step. Furthermore, we use vector velocity plots for showing the direction of flow and the velocity field of the fluid at the end of every time-cycle [2,3].

2.3. The boundary conditions

A rigid wall may be either of two types, no-slip or free-slip. The basic boundary conditions are simply that the normal velocity component

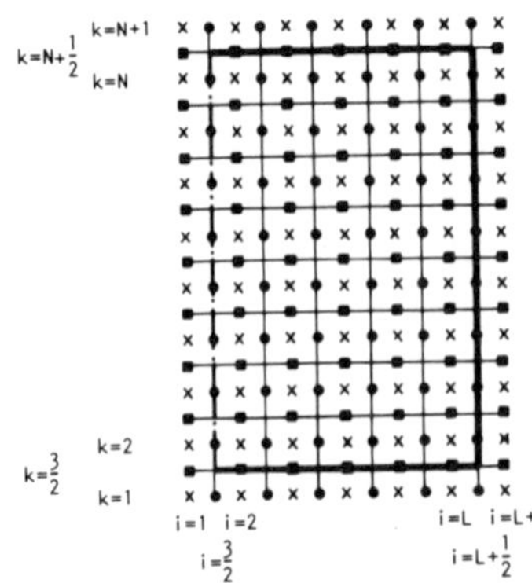

x, pressure and zonal velocity points;

•, radial velocity points;

■, vertical velocity points.

The broken line indicates the axis of symmetry; the heavy lines denote container walls and the free surface with the disk, respectively.

Fig. 2: Grid arrangement in a vertical cross section.

vanishes and, in addition, that the tangential component vanishes if no slippage is to be allowed. The decision whether to take free-slip or no-slip boundary conditions depends upon the relation between the spatial discretization parameter and the thickness of the boundary layer that would be expected to develop in the true fluid. Due to our coarse spatial discretization we choose free-slip boundary conditions for u and w at all rigid walls. Moreover, only when performing confined flow calculations by taking free-slip rigid wall boundary conditions for u and w at the free surface, too, our numerical simulations yielded correct flow patterns in the fluid [2,3]. This assumption is permitted as long as the ratio between centrifugal and gravitational forces is small, as it is the case in the present study.
The boundary conditions for the zonal velocity component v include the angular velocities Ω_c and Ω_s of the rotation of container and disk.

3. NUMERICAL RESULTS

3.1. Uniformly rotating container-disk system

In figs. 3 and 4 we present some characteristic flow patterns of our simulations in an isorotational case with $\Omega_s = 30$ rpm and $\Omega_c = 10$ rpm and in the counterrotational case with $\Omega_s = 30$ rpm and $\Omega_c = -5$ rpm. The model liquid was assumed to have the kinematic viscosity $\nu = 0.025\ cm^2/s$ of liquid silicon. We always started our simulations from the liquid at rest, switching on the appropriate crystal and crucible rotation at the time $t = 0$. The figures show the flow patterns at the time indicated, and v_{max} designates the absolute value of the maximum velocity in the vertical cross-section of the liquid shown in fig. 2.
When the disk and the container rotate in the same direction but with different angular velocities, as shown in fig. 3, two vortices set up in the beginning, one below the disk and another in the lower edge of the container. But gradually, the disk draws more fluid upwards ($\Omega_s > \Omega_c$), so that a Taylor-Proudman cell develops reaching right to the bottom of the container. The highest cell velocity occurs where the disk throws the fluid aside.
Fig. 4 shows the evolution in time of the flow patterns, where container and disk rotate in opposite directions with different angular

<u>Fig. 3 :</u> $\nu = 0.025\ cm^2/s$, $\Omega_s = 30\ rpm$

$\Omega_c = 10\ rpm$

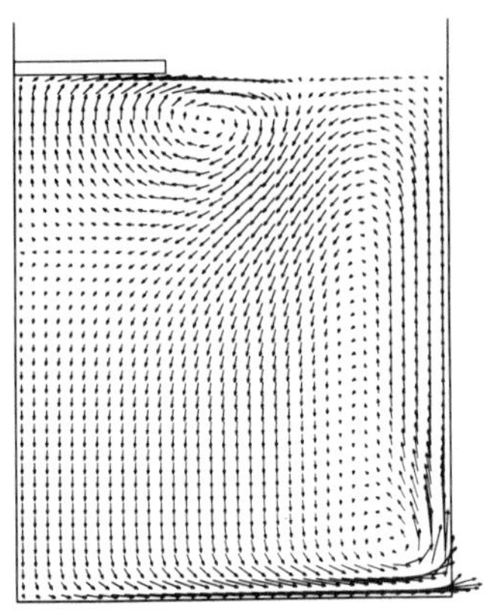

$t = 5\ s$

$v_{max} = 1.5\ cm/s$

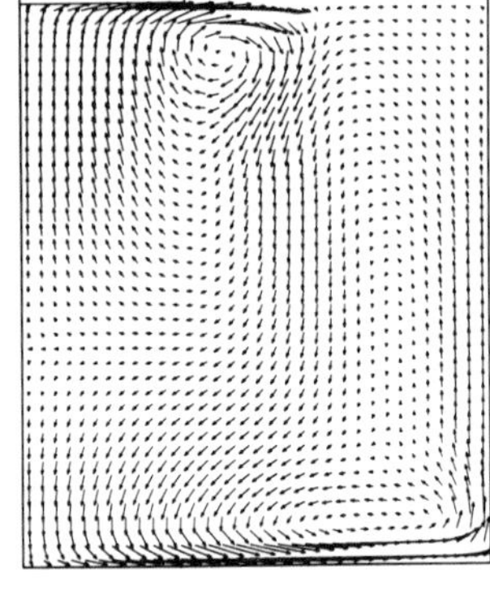

$t = 10\ s$

$v_{max} = 0.92\ cm/s$

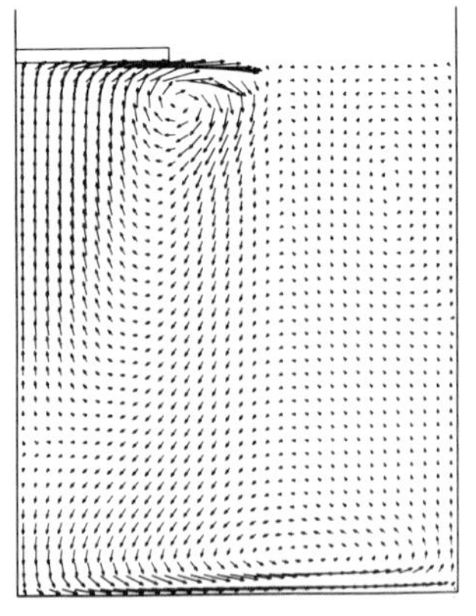

$t = 15\ s$

$v_{max} = 0.76\ cm/s$

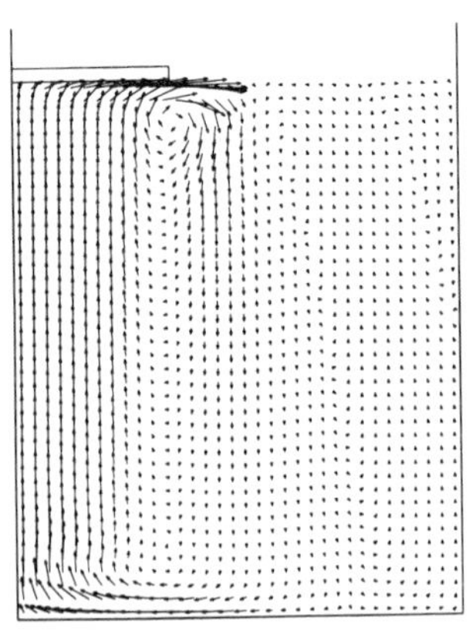

$t = 37.159\ s$

$v_{max} = 0.63\ cm/s$

Fig. 4 : $\nu = 0.025\ cm^2/s$, $\Omega_s = 30$ rpm
$\Omega_c = -5$ rpm

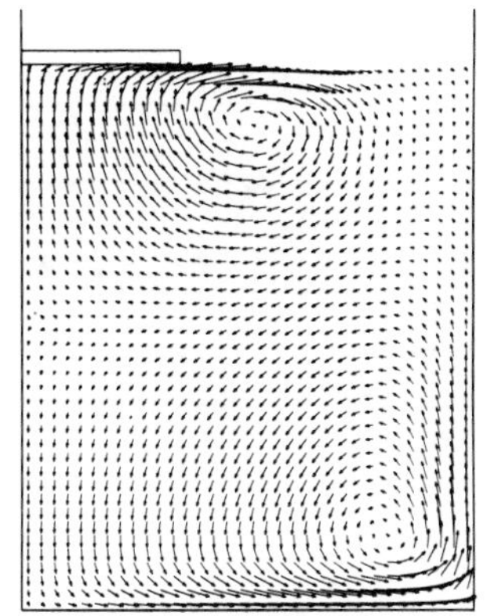

$t = 5$ s
$v_{max} = 0.86$ cm/s

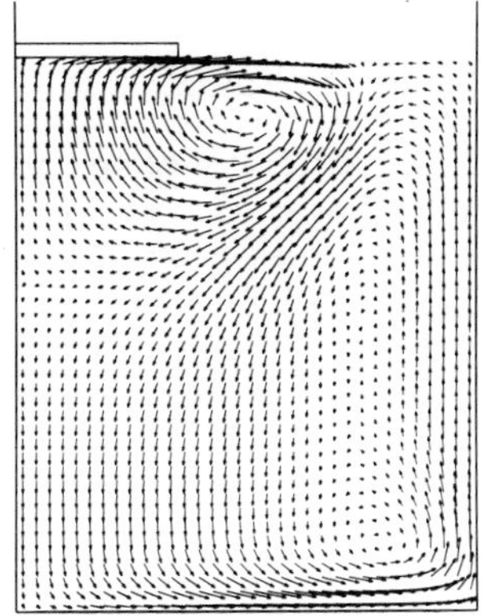

$t = 10$ s
$v_{max} = 0.85$ cm/s

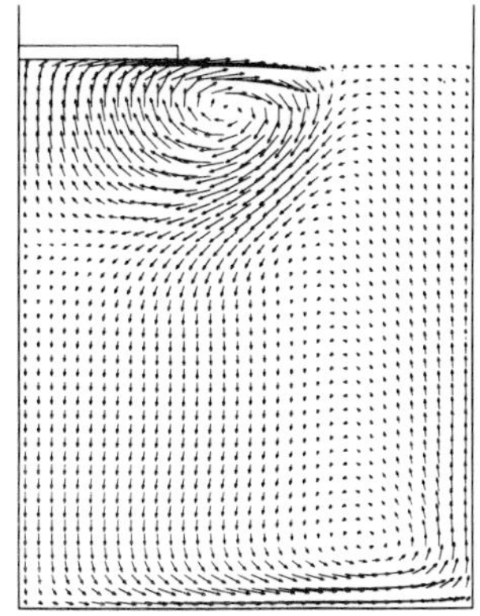

$t = 15$ s
$v_{max} = 0.85$ cm/s

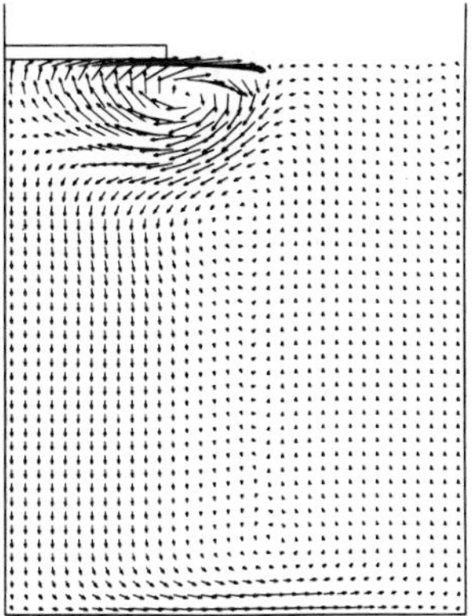

$t = 37.104$ s
$v_{max} = 0.76$ cm/s

velocities. Again two vortices occur beneath the disk and in the lower edge of the container, respectively, which gradually devolve in the two expected Taylor-Proudman cells. The one induced by the container rotation is very large due to the low angular velocity of -5 rpm . The size of the cell beneath the disk obviously depends on the relative angular velocity between container and disk. Aditionally, in this cell the maximum flow velocity appears.

3.2. Non-uniformly rotating container-disk system

On the same geometrical, physical, and numerical conditions as in section 3.1, we present the digital simulations for the periodically accelerated and decelerated rotating container-disk system in an isorotational and a counterrotational arrangement. In both cases, apart from different disk rotational directions, we vary the disk and the container rotation with a time period of 15 s by a constant acceleration from ±4o to ±8o rpm and 1o to 3o rpm , respectively, followed by a constant deceleration from ±8o to ±4o rpm of the disk and 3o to 1o rpm of the container.

In the case of counterrotation, fig. 5 shows that the originally developed Taylor-Proudman cell beneath the disk certainly changes its size nearly periodically ("pulsation effect"), but still cannot be destroyed. For equally directed disk and container rotation, however, the flow patterns in fig. 6 are completely different from the ones before. Namely, in this arrangement, the Taylor-Proudman cell beneath the disk is only formed during the acceleration phase, while deceleration causes the disappearance of the cell. Additionally, the size of the cell is in any case greater than before. For these reasons, the isorotational arrangement seems to be more advantageous in view of optimum stirring of the liquid.

4. CONCLUSIONS

The present investigations of forced convection in our rotating container-disk system filled with a model fluid of the kinematic viscosity of liquid silicon have shown that in the uniformly isorotational as well as in the uniformly counterrotational container-disk arrangement non-mixing Taylor-Proudman cells devolve which prevent total stirring of the liquid. Additionally, our calculations have exhibited the possibility to eliminate

Fig. 5 : $\nu = 0.025\ cm^2/s$, $\Omega_s = -40 : -80$ rpm

$\Omega_c = 10 : 30$ rpm

time period = 15 s

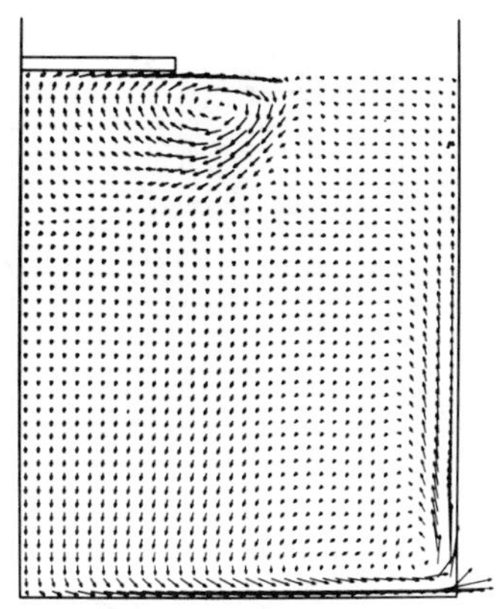

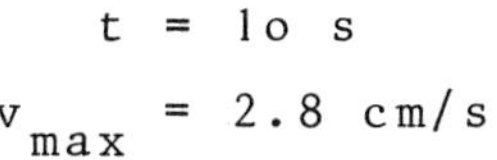

t = 10 s

$v_{max} = 2.8$ cm/s

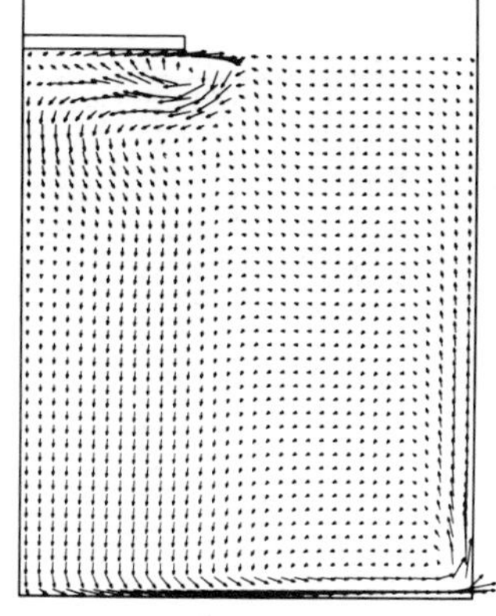

t = 15 s

$v_{max} = 2.6$ cm/s

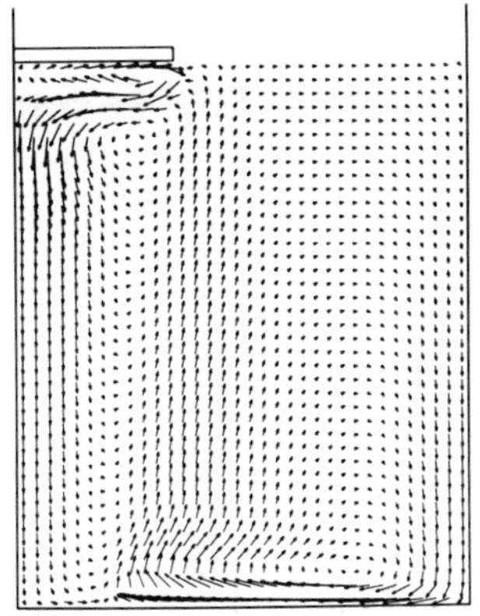

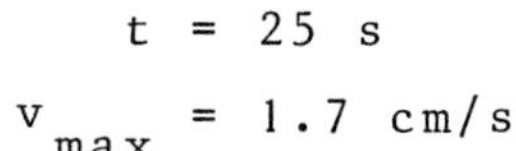

t = 25 s

$v_{max} = 1.7$ cm/s

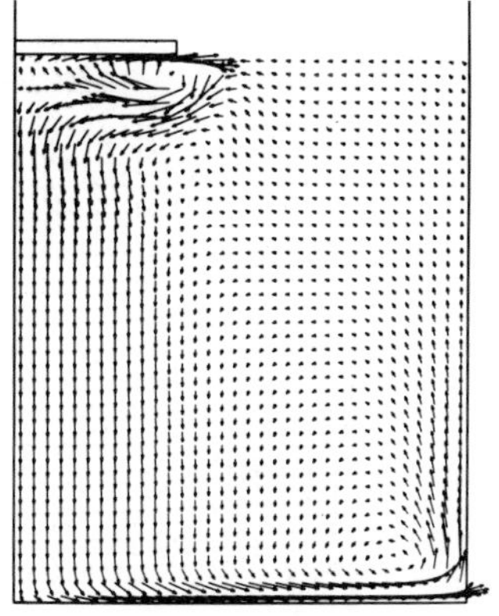

t = 35 s

$v_{max} = 1.3$ cm/s

Fig. 6 : ν = o.o25 cm^2/s , Ω_s = 4o : 8o rpm
Ω_c = 1o : 3o rpm
time period = 15 s

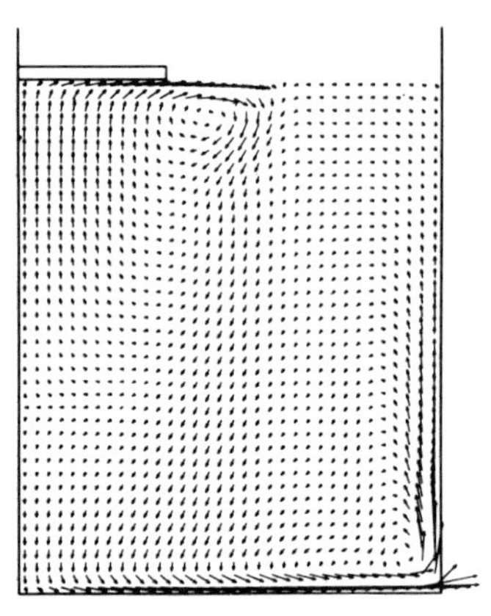

t = 1o s

v_{max} = 2.7 cm/s

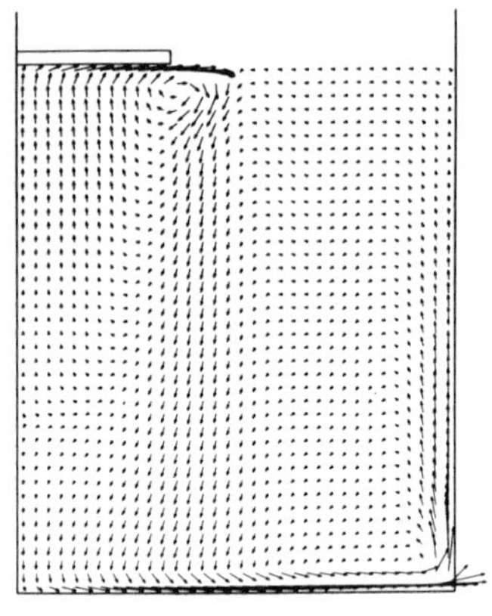

t = 15 s

v_{max} = 2.1 cm/s

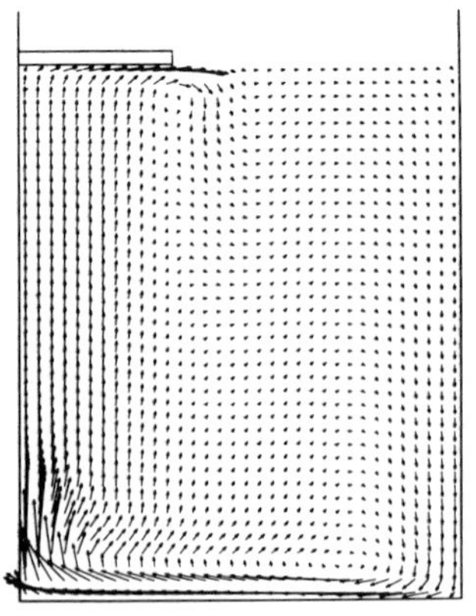

t = 25 s

v_{max} = 1.9 cm/s

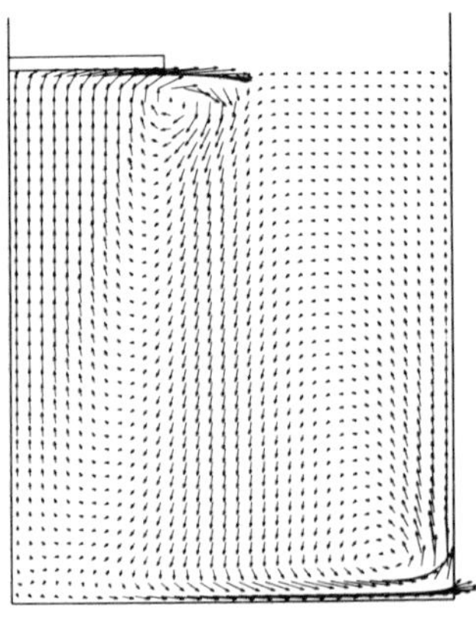

t = 35 s

v_{max} = 1.1 cm/s

these Taylor-Proudman cells periodically by the periodical acceleration and deceleration of the container and disk rotation. But this periodical disappearance of the non-mixing liquid columns only occurs in the isorotational case, despite of intuitive predictions which might have prefered the counterrotational arrangement.

5. REFERENCES

1. HARLOW, F.H./WELCH, J.E. - Numerical Calculation of Time-Dependent Viscous Incompressible Flow of Fluid with Free Surface. The Physics of Fluids, Vol. 8, pp. 2182-2189, 1965.

2. MIHELČIČ, M./SCHRÖCK-PAULI, C./WINGERATH, K./ WENZL, H./ UELHOFF, W./VAN DER HART, A. - Numerical Simulation of Convective Flow of the Melt in the Classical Czochralski Method, in ACRT and CACRT. Part 1: Simulation of Forced Convection. Bericht der Kernforschungsanlage Jülich, Jül-1682, 198o.

3. MIHELČIČ, M./SCHRÖCK-PAULI, C./WINGERATH, K./ WENZL, H./UELHOFF, W./ VAN DER HART, A. - Numerical Simulation of Forced Convection in the Classical Czochralski Method in ACRT and CACRT. J. Crystal Growth, Vol. 53, 1981 (to appear).

4. ROACHE, P.J. - Computational Fluid Dynamics, Hermosa Publ., 1972.

5. SCHEEL, H.J./MÜLLER-KRUMBHAAR, H. - Crystal Pulling Using ACRT. J. Crystal Growth, Vol. 49, pp. 291-296, 198o.

6. WELCH, J.E./HARLOW, F.H./SHANNON, P.J./DALY, B.J.- The MAC Method, Los Alamos Report, LA-3425, 1966.

7. WILLIAMS, G.P. - Numerical Integration of the Three-Dimensional Navier-Stokes Equations for Incompressible Flow. J. Fluid Mech., Vol. 37, pp. 727-75o, 1969.

A COMPARISON OF EXPLICIT INTEGRATION TECHNIQUES FOR THE ADVECTION EQUATION

R. A. Skop,[1] M. L. Morrell,[2] and G. A. Keramidas[1]

Numerical studies of two explicit, two-step time integration techniques for the one dimensional, constant velocity finite element advection equation have been conducted for both square hill and cosine hill density distributions. One of these integration techniques, the Godunov scheme, is first order accurate in time while the other, the Lax-Wendroff scheme, is second order accurate in time. The results show that, overall, the "best" numerical solutions are obtained by combining a central weighted first-step Lax-Wendroff time integration with parabolic spatial discretization either in its full or condensed [M] maxtrix form. Both the standard and central weighted first-step Godunov time integrations are found to be numerically diffusive. This diffusivity tends to override whatever spatial discretization is used. However, the positivity property possessed by the Godunov schemes can be valuable for many applications.

1. INTRODUCTION

Nonlinear hyperbolic equations of the form

$$\frac{\partial \vec{\rho}}{\partial t} + \text{div}\,(\vec{V}\,\vec{\rho}) = \vec{F}(\vec{V},\vec{\rho}) \tag{1}$$

where $\vec{\rho}$ is a vector of conserved quantities, $\vec{V}$ is the velocity field, and $\vec{F}$ is a given functional, describe flow behavior in subjects ranging from hydraulics to gas dynamics. A review of finite difference techniques for the solution of (1) can be found in Roache [1], Sod [2], and Book, et al. [3]. As it is evident from these references, finite difference solutions of (1) have reached a high level of sophistication and accuracy. The most significant criticism that can be leveled against these techniques is the difficulty encountered in treating complex flow boundaries.

A natural way of handling complex flow boundaries is through finite element discretization of the spatial derivatives of (1). In these finite element methods (see Baker [4]) the complex flow boundaries fall naturally out of the discretization as constraints on the shape function coefficients of boundary elements.

[1]Applied Mechanics Branch, Naval Research Laboratory, Washington, DC 20375
[2]Dept. of Civil Engineering, Clemson University, Clemson, SC 29631

Before blindly applying finite element methods to the solution of (1), however, it is necessary to evaluate their ability to replicate some simple flow behaviors predicted by (1). To this end, we examine in this paper numerical solutions of the simple, constant velocity advection equation

$$\frac{\partial \rho}{\partial t} + V \frac{\partial \rho}{\partial x} = 0 \tag{2}$$

that becomes, upon finite element spatial discretization,

$$[M]\,\{\dot{R}\} + V\,[K]\,\{R\} = 0. \tag{3}$$

Here, $[M]$ and $[K]$ are matrices that depend on the specific discretization, $\{R\}$ is a column matrix representing the node point values of the "density" ρ, and the dot denotes ordinary differentiation with respect to time. For practical computational purposes, two requirements can be imposed on the time integration technique used to advance equation (3):

1. It should require knowledge of $\{R\}$ only at the node points. This requirement is imposed to preserve the strongest attributes of the finite element method, namely, ease of establishing grids for complex geometric boundaries and ease of handling boundary conditions.

2. It should be explicit to ensure fast, efficient calculations. This requirement really stems from (1) where an implicit integration scheme would mandate the solution of large, nonlinear sets of algebraic equations at each time step.

The problems of concern are what combination of spatial discretization and time integration "best" models the true advection solution and whether "best" is good enough. Within bounds, the latter concern is, of course, fairly subjective.

In the remainder of this paper, we review the results of numerical experiments previously conducted by the authors [5] in connection with (3). Only the highlights are presented here. For more detailed explanations and descriptions of the numerical experiments, the reader is referred to [5].

2. PROBLEM DEFINITION

We consider the solutions of (2) and (3) on the racetrack domain $0 \leqslant x \leqslant L$ where the points $x = 0$ and $x = L$ are identical. The racetrack domain is particularly useful for studying advection since mass can neither leave nor enter the domain. Hence, all changes to the exact solution are caused purely by numerics. The exact solution of (2) is, of course, a constant motion of the initial density distribution around the track at velocity V.

For the examples studied here, the length of the track is taken as $L = 48$ (arbitrary units of distance) and the advection velocity is taken as $V = 1$ unit of distance/s. Thus, the density distribution is advected once around the track every 48 s. Two initial density distributions are considered: (a) a consine hill between x_1 and x_2,

$$\rho(x,0) = F(x) = \begin{cases} 1, & x \leqslant x_1 \text{ and } x \geqslant x_2 \\ \dfrac{3}{2} - \dfrac{1}{2}\cos\dfrac{2\pi(x - x_1)}{x_2 - x_1}, & x_1 < x < x_2 \end{cases} \tag{4}$$

and (b) a square hill also between x_1 and x_2,

$$\rho(x,0) = F(x) = \begin{Bmatrix} 1, & x \leqslant x_1 \text{ and } x \geqslant x_2 \\ 2, & x_1 < x < x_2 \end{Bmatrix}. \tag{5}$$

These initial conditions are shown in Figure 1 for $x_1 = 8$ and $x_2 = 18$.

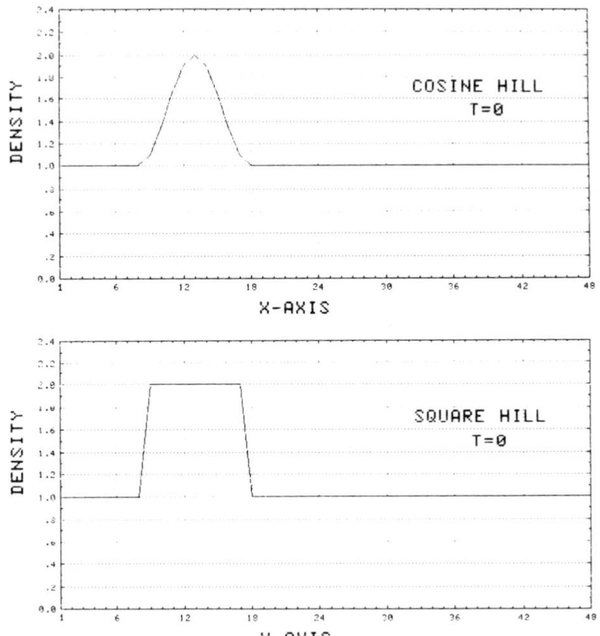

Figure 1. Initial conditions: (a) cosine hill, (b) square hill

3. CONDENSED [M] MATRIX FORMULATION

Equation (3) can be rewritten formally as

$$\{\dot{R}\} = -V\,[M]^{-1}\,[K]\,\{R\} \tag{6}$$

where $[M]^{-1}$ denotes the inverse of $[M]$. Generally, $[M]$ is a large, sparse matrix and the calculation of $[M]^{-1}$ (which is a full matrix) can be very time consuming. The process of condensing the mass matrix, or lumping the mass at the diagonal terms, is a popular procedure [4] for avoiding this time consuming inversion.

Hence, the condensed mass matrix $[M_c]$ is defined by

$$[M_c] \rightarrow \begin{cases} m_{cii} = \sum_{j=1}^{N} m_{ij}\,, & i = 1, \ldots, N \\ m_{cij} = 0, \; i \neq j & i, j = 1, \ldots, N \end{cases} \tag{7}$$

where the m_{ij} are the entres in $[M]$ and N is the rank of $[M]$. The matrix $[M_c]$ now replaces $[M]$ in (3) and (6) is replaced by

$$\{\dot{R}\} = -V\,[M_c]^{-1}\,[K]\,\{R\}. \tag{8}$$

We note that this condensation is mass preserving.

4. TIME INTEGRATION SCHEMES

Consider that the value of the nodal densities at time step n are known $\{R^n\}$. The solution at the $n+1$ time step is sought. The time increment is Δt. Then the standard two-step, explicit method can be written as:

$$\text{1st step} \qquad \{R^{n+\alpha}\} = \{R^n\} + \alpha\,\Delta t\,\{\dot{R}^n\} \tag{9a}$$

$$\text{2nd step} \qquad \{R^{n+1}\} = \{R^n\} + \Delta t\,\{\dot{R}^{n+\alpha}\} \tag{9b}$$

where α is a fraction of the time step. The time derivatives $\{\dot{R}^n\}$ and $\{\dot{R}^{n+\alpha}\}$ are determined from (6) [or (8)] using $\{R^n\}$ and $\{R^{n+\alpha}\}$ respectively.

Following the convention of Sod [2], two well-known methods are embodied in equations (9). These are the second order accurate in time Lax-Wendroff method when $\alpha = 1/2$, and the first order accurate in time Godunov method when $\alpha = 1$.

Equations (9) are referred to in this paper as the *standard* two-step methods. Because the standard two-step methods update a nodal value only as a function of its previous value, it seems plausible that introducing a weighting matrix to average the influence of adjacent nodes might improve numerical accuracy. Such averaging is used frequently in finite difference methods [1]. This weighting matrix should have a general form so as to be problem independent. One such weighting matrix is employed herein.

This weighting matrix is termed the "modified" weighting matrix $[W_m]$, the terms of which are determined from the components m_{ij} of the global mass matrix $[M]$ as

$$[W_m] \rightarrow w_{m_i j} = \frac{m_i j}{C_i} \qquad i,\ j = 1,\ \ldots,\ N \tag{10a}$$

where

$$C_i = \sum_{j=1}^{N} m_{ij}. \tag{10b}$$

Note that the sum across row 'i' of $[W_m]$ is unity so that $[W_m]$ is mass conserving.

The weighting matrix is used only in the first step of the time integration. Thus the modified two-step method is written as:

$$\text{1st step} \qquad \{R^{n+\alpha}\} = [W_m]\,\{R^n\} + \alpha\,\Delta t\,\{\dot{R}^n\} \tag{11a}$$

$$\text{2nd step} \qquad \{R^{n+1}\} = \{R^n\} + \Delta t\,\{\dot{R}^{n+\alpha}\}. \tag{11b}$$

The modified Lax-Wendroff method is obtained by letting $\alpha = 1/2$ in (11) and the modified Godunov method is obtained by letting $\alpha = 1$ in (11).

The performance of another first-step weighting matrix that gave zero weight to the node under consideration (that is $w_{ii} = 0$) was also studied in [5]. The results using this weighting were far inferior to the standard and modified methods and are not presented here.

5. NUMERICAL RESULTS

The numerically advected density distributions after 96s (or two full cycles around the track) are shown in Figures 2 through 9. The exact solution is superimposed on each figure.

For the results shown using a linear finite element spatial discretization, the track is divided into 48 elements of unit length. Taking into account that $x = 0$ and $x = L = 48$ are the same points, the number of nodal degrees of freedom $N = 48$. It is worthwhile to point out that the condensed $[M]$ formulation of the linear element discretization is identical, for one dimensional problems, to standard central differences.

For the results shown using a parabolic finite element spatial discretization, the track is divided into 24 elements of two units of length. Taking into account the mid-point degree of freedom of the parabolic element, the total number of nodal degrees of freedom is again $N = 48$.

The time step used for the solutions depicted in Figures 2 through 9 is $\Delta t = 0.2$s. As found in [5], this time step gives representative solution behavior for all time steps $\leqslant 0.33$s. For larger time steps, several of the spatial discretization, time integration combinations become numerically unstable.

In Table 1, the average absolute error, defined by

$$E = \frac{1}{N} \sum_{i=1}^{N} |\rho_i - R_i| \tag{12}$$

where ρ_i and R_i are, respectively, the exact and numerical solutions at node i, is given for each case shown in the figures. From this table and the figures, the reader is free to draw his own conclusions, or agree with those presented in the paper summary, regarding what combination(s) of spatial discretization and explicit time integration "best" models the true advection solutions.

TABLE 1
Percent errors at 96s (two cycles) with a time step $\Delta t = 0.2$s

	Linear FEM*				Linear CFM*			
	SLW*	MLW*	SG*	MG*	SLW	MLW	SG	MG
cosine hill	1	14	9	10	18	20	10	12
square hill	16	14	15	16	21	24	15	19
	Parabolic FEM				Parabolic CFM			
	SLW	MLW	SG	MG	SLW	MLW	SG	MG
cosine hill	3	1	9	9	2	4	9	9
square hill	33	8	15	15	15	12	15	15

*FEM = finite element method, CFM = condensed $[M]$ martix formulation.
*S = standard, M = modified, LW = Lax-Wendroff, G = Godunov

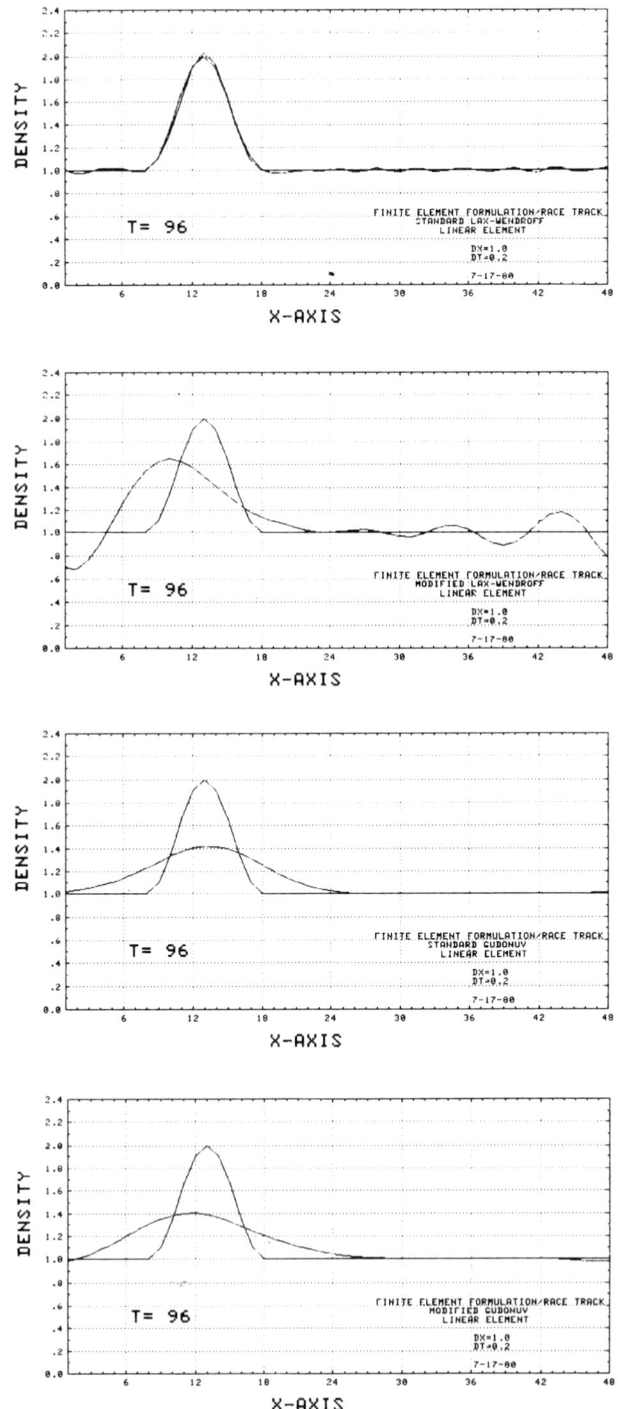

Figure 2. Advected cosine hill after 96s using linear finite elements: (a) standard Lax-Wendroff, (b) modified Lax-Wendroff, (c) standard Godunov, (d) modified Godunov.

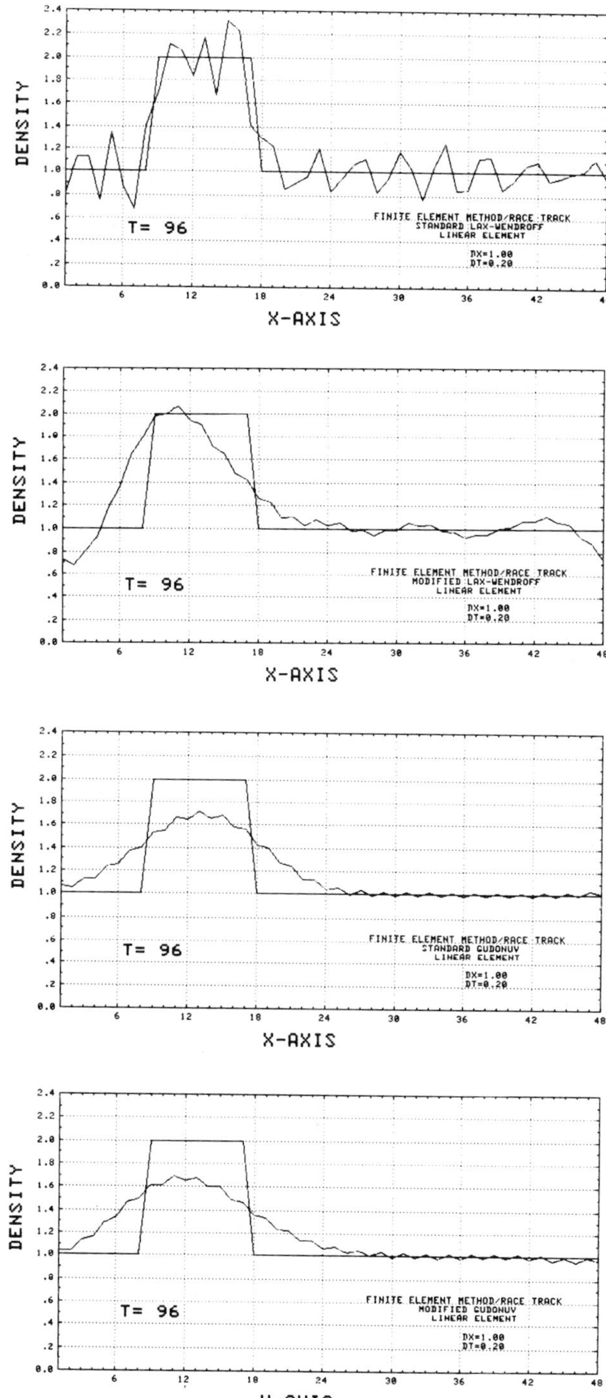

Figure 3. Advected square hill after 96s using linear finite elements: (a) standard Lax-Wendroff, (b) modified Lax-Wendroff, (c) standard Godunov, (d) modified Godunov.

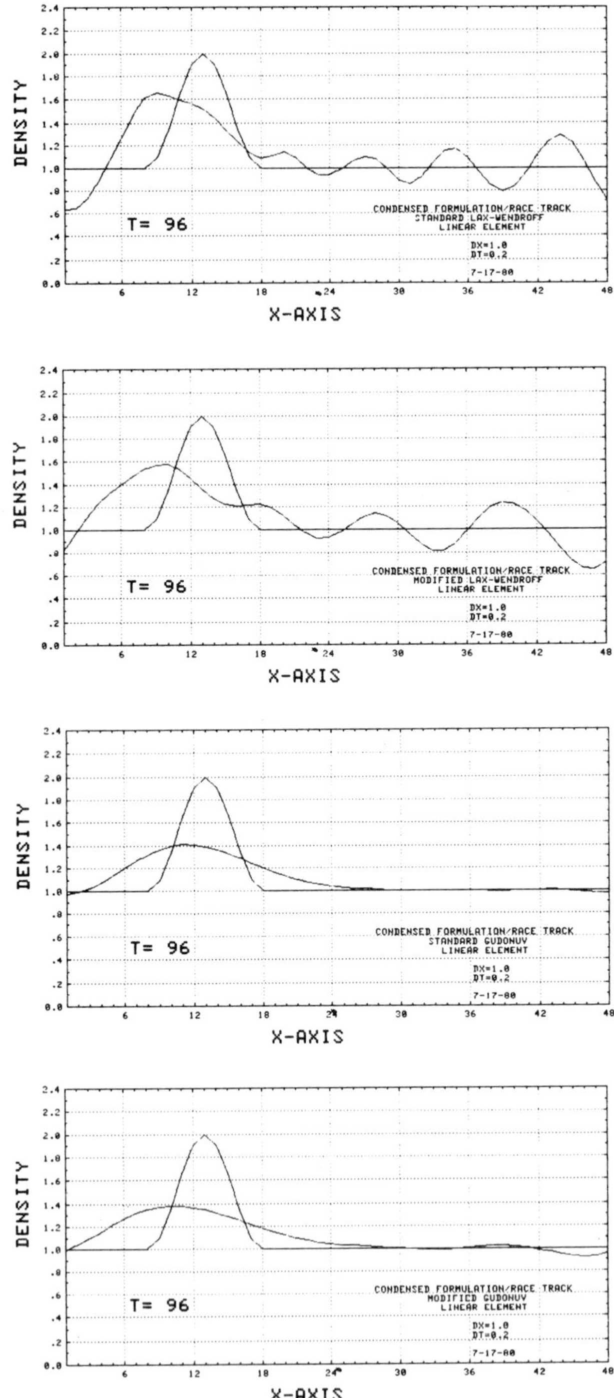

Figure 4. Advected cosine hill after 96s using linear finite elements with condensed [M] formulation: (a) standard Lax-Wendroff, (b) modified Lax-Wendroff, (c) standard Godunov, (d) modified Godunov.

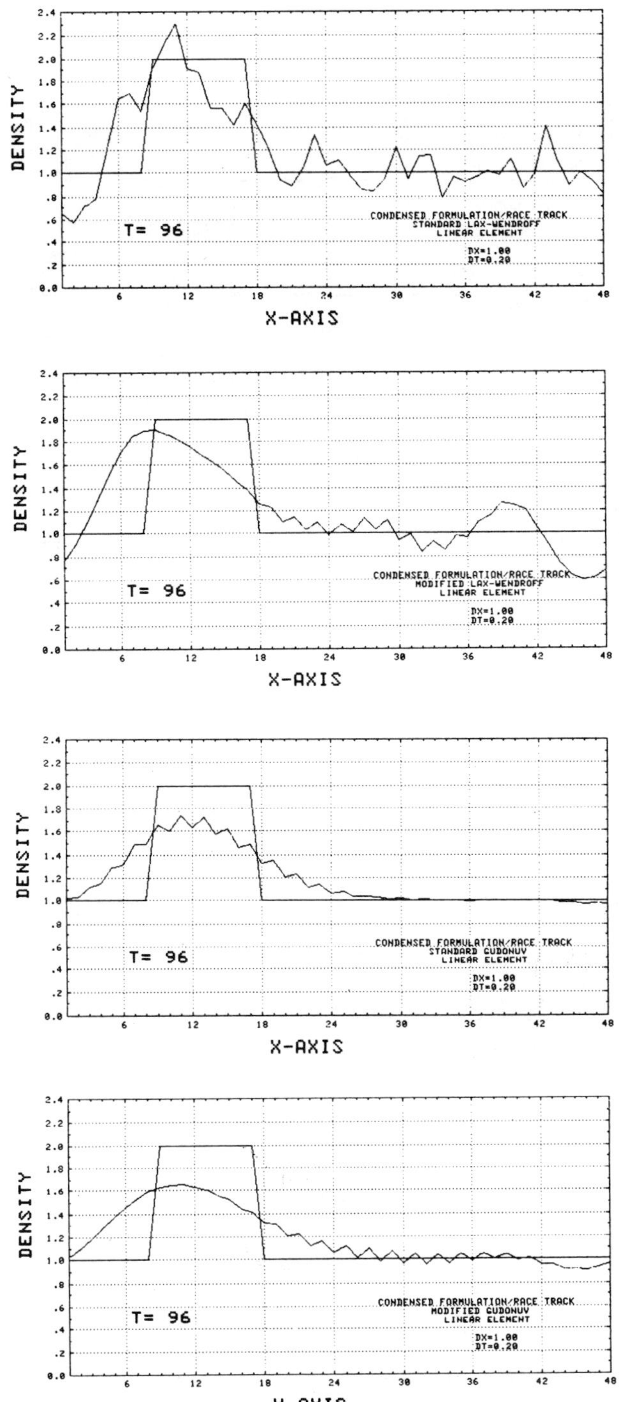

Figure 5. Advected square hill after 96s using linear finite elements with condensed $[M]$ formulation: (a) standard Lax-Wendroff, (b) modified Lax-Wendroff, (c) standard Godunov, (d) modified Godunov.

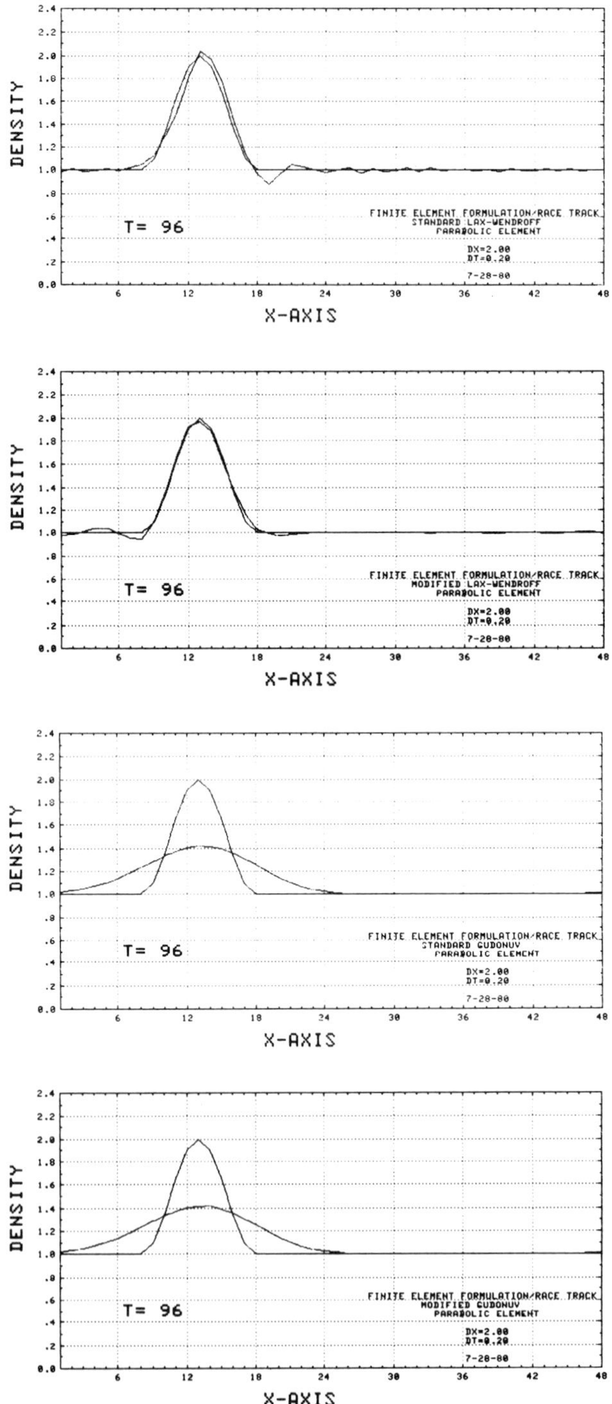

Figure 6. Advected cosine hill after 96s using parabolic finite elements: (a) standard Lax-Wendroff, (b) modified Lax-Wendroff, (c) standard Godunov, (d) modified Godunov.

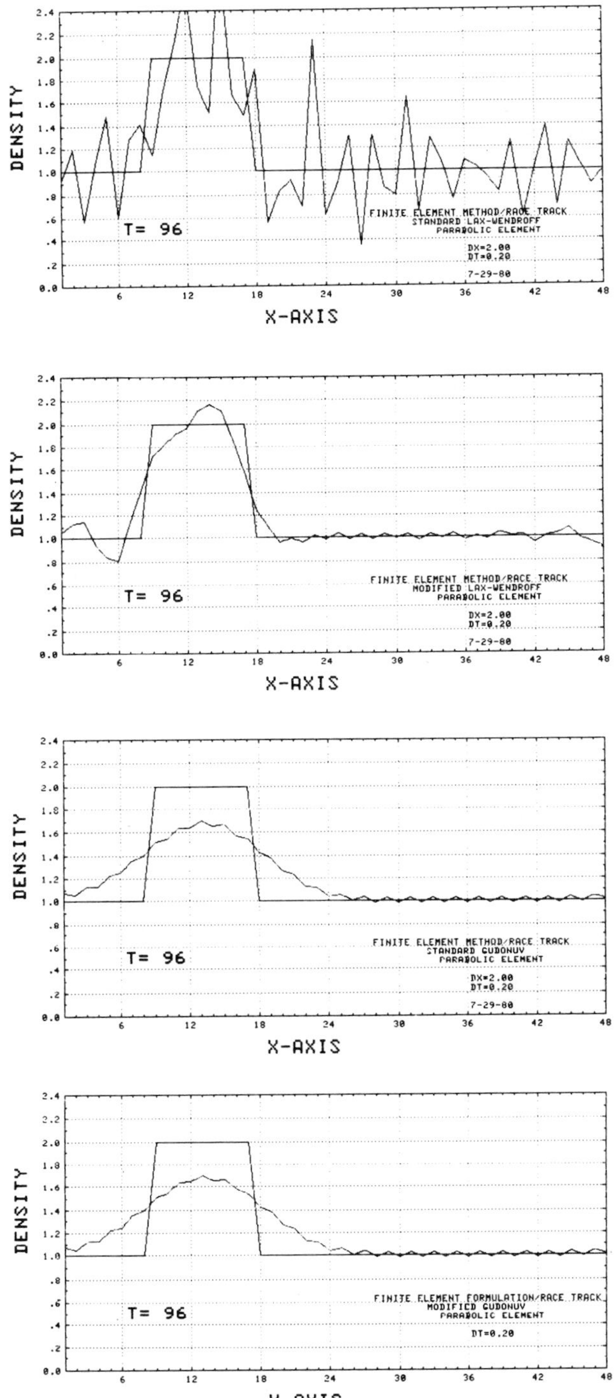

Figure 7. Advected square hill after 96s using parabolic finite elements: (a) standard Lax-Wendroff, (b) modified Lax-Wendroff, (c) standard Godunov, (d) modified Godunov.

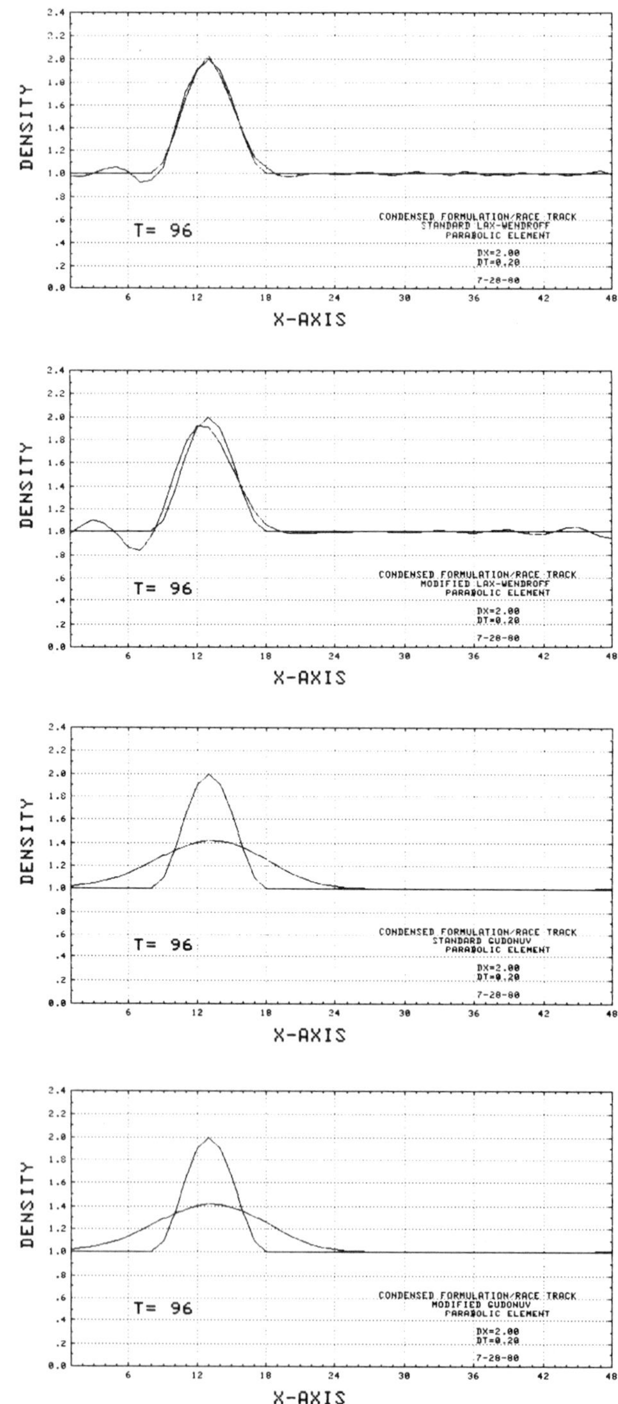

Figure 8. Advected cosine hill after 96s using parabolic finite elements with condensed $[M]$ formulation: (a) standard Lax-Wendroff, (b) modified Lax-Wendroff, (c) standard Godunov, (d) modified Godunov.

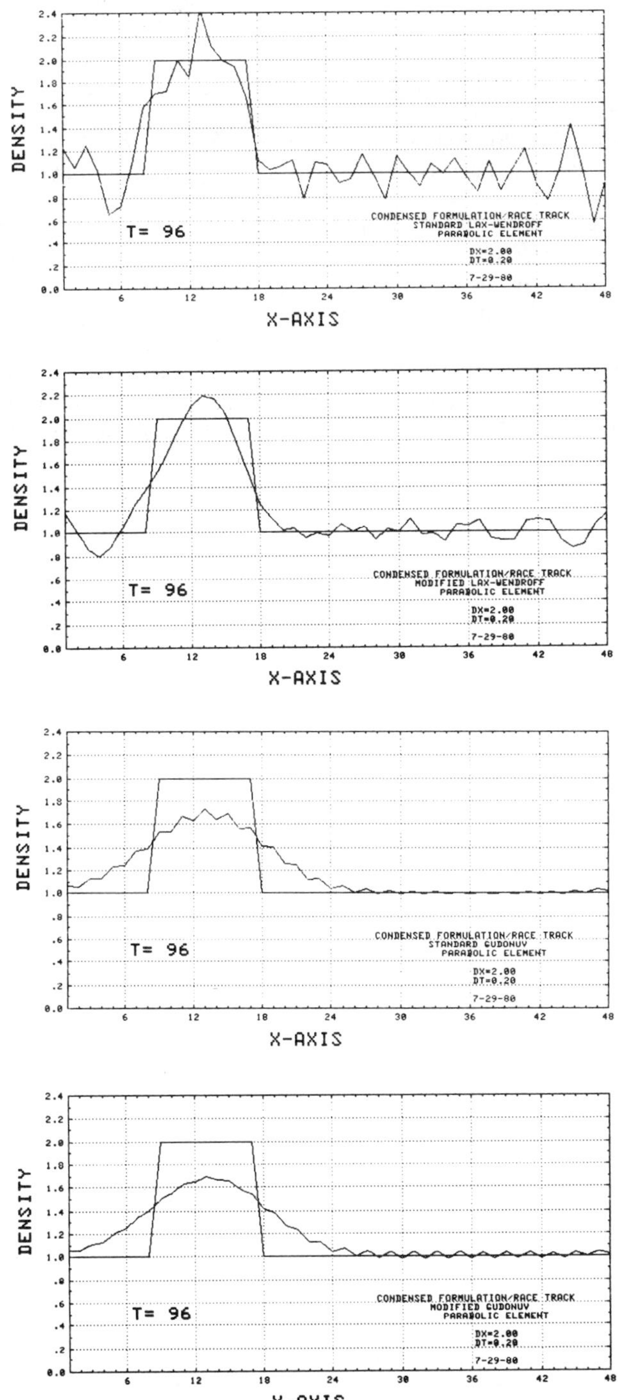

Figure 9. Advected square hill after 96s using parabolic finite elements with condensed $[M]$ formulation: (a) standard Lax-Wendroff, (b) modified Lax-Wendroff, (c) standard Godunov, (d) modified Godunov.

6. REFERENCES

[1] P.J. Roache 1972 Computational Fluid Dynamics. Albuquerque, New Mexico: Hermosa Publishers.

[2] G.A. Sod 1978 Journal of Computational Physics 27, 1-31. A survey of several finite difference methods for systems of nonlinear hyperbolic conservation laws.

[3] D.L. Book, et al. 1979 Naval Research Laboratory Memorandum Report 4095 (Washington, DC). Recent developments in computational techniques for applied hydrodynamics.

[4] A.J. Baker 1979 Finite Element Computational Fluid Mechanics. Knoxville, Tennessee: The University of Tennessee at Knoxville.

[5] M.L. Morrell, R.A. Skop, and G.A. Keramidas, 1980, Naval Research Laboratory Memorandum Report 4438 (Washington, DC). A comparison of two-step time integration schemes for the finite element advection equation.

THE EFFECT OF FORCED AND FREE CONVECTION IN THE DISCHARGE OF A PRESSURIZED GAS[(i)]

S. Paolucci and D. R. Chenoweth[(ii)]

SUMMARY

When gas issues from a pressurized container, the remaining gas expands and cools. The resulting heat flux from the walls affects the mass transfer. Numerical solutions of the two-dimensional time-dependent governing equations are presented for the resulting convective motion and mass transfer rate interaction. An explicit finite difference method is used. The region of numerical stability, as governed by the Courant-Friedrich-Lewy condition, is greatly enlarged by analytically "filtering" the acoustic waves from the equations. The resulting equations allow for solution of laminar flows in the non-Boussinesq regime. The effect of discharging the gas at different angles from the gravity vector is illustrated by several examples. In all cases, the resulting motion is characterized by three physically meaningful flow regimes. Initially forced convection dominates the flow, while as thermal equilibrium is approached free convection prevails. Between these two regimes, and in the vicinity of pressure equilibrium, the flow is controlled by the combined forced and free convection mechanisms; here the direction of discharge relative to the gravity vector is significant.

1. INTRODUCTION

During the pressurized discharge of a gas from a container there is a transient heat transfer interaction between the remaining gas and the surface on the interior of the container. Heating of the gas is accomplished in response to the rapid venting of the gas from the enclosing vessel. As the gas

(i) Work supported by U.S. Department of Energy under Contract DE-AC04-76DP00789.

(ii) Members of Technical Staff, Thermal Sciences Division, Sandia National Laboratories, Livermore, CA 94550 USA.

expands, it cools, but the container walls remain essentially at the initial temperature. As heat is transferred to the cool gas, the gas near the walls becomes warm and expands, causing a lower gas density next to the walls. Thus a potentially unstable situation develops in which a heavier gas lies adjacent to a lighter one. Whether this arrangement is stable or unstable depends, among other things, on the direction of the gravitational vector.

These various simultaneous and complicated heat and fluid flow processes produce a transient response of the temperature and pressure in the vessel. Of prime importance is to be able to predict the heat transferred to the gas, in the form of a Nusselt number, and the mass discharge rate, as functions of time. In order to accomplish this with a certain degree of accuracy and yet with minimal computer time and storage, the equations are formulated in such a way as to enable the utilization of a non-uniform coordinate mesh with capability of dynamic rezoning. However, all results presented in this paper have been obtained using a uniform computational mesh.

A significant number of numerical solutions of the compressible Navier-Stokes equations have been obtained utilizing both implicit and explicit finite difference methods. Explicit methods are appealing for a number of reasons, but primarily because they are simple, easy to program, and require comparatively few computations per time step. Also, the parallel-processing feature of new computers tend to make explicit methods attractive, and we should see an increase in their usage in the future. However, due to the small time step limitation dictated by numerical stability requirements of explicit methods, excessive computing times are required to solve practical compressible flow problems. The allowable time step is limited by the Courant-Friedrich-Lewy (CFL) stability condition, which simply states that a sound wave cannot travel more than one cell length in one time increment. Implicit methods generally have no such restriction, however, one must beware of producing computationally stable nonsense, and must be prepared to write a more complicated program, and expect a larger arithmetic operation count and more memory storage requirements.

It is evident that if acoustic waves are "filtered" out without significantly altering the physics of the problem, then an explicit solution scheme is preferable. The most popular method of accomplishing this is to automatically invoke the Boussinesq approximation, and then one is ready to do some efficient computing using explicit, schemes. Unfortunately in the present type of problem the temperature differences are large, and thus the familiar Boussinesq approximation does not hold. However, there is no need to dispair if one gives some thought on the governing physics.

In a general physical problem, there are several wave propagation speeds. It sometimes happens that one of these waves moves much faster than the others, but represents little energy in the solution. The part of the motion called sound usually carries a minute fraction of the energy present in the primary process, which is not considered to be acoustical. The above is the exact situation in our problem and many other convection dominated problems. Compressible equations in which acoustic motion is absent have been rigorously derived elsewhere [1]. The resulting approximate equations can be used to describe rather general flows for which the Mach number is small. It should be pointed out that similar equations have been previously used in the numerical work of Forester and Emery [2], and have been derived recently, albeit incompletely, by Rehm and Baum [3]. These equations are also presently being used by Humphrey and LeQuere [4] in numerical work related to convective heat losses in solar cavity receivers.

In section 2 we formulate the problem under investigation by listing and describing the appropriate governing equations and boundary conditions. This is followed in section 3 by a description of the numerical procedure used. In section 4 we present several results obtained using a uniform mesh for our computational domain. This is followed by a discussion of the results.

2. GOVERNING EQUATIONS

Consider a two-dimensional rectangular enclosure of width ℓ and height h which contains a pressurized gas. The gas is initially quiescent at a uniform temperature T_o and pressure p_o. The walls of the vessel are maintained at a temperature T_w, except at the discharge opening of size Δs, where we have adiabatic conditions. Laminar flow is assumed to exist throughout.

The problem evolves in time t and can be described in terms of the velocity components $v_i = (u,v)$ in the $x_i = (x,y)$ directions, the density ρ, temperature T, and pressure p. The governing equations are statements of conservation of mass, momentum, and energy, with the addition of a state equation describing our gas. These equations, valid under a small Mach number approximation, and given in dimensionless variables, have been derived by Paolucci and Chenoweth [1], and are given as follows:

$$\frac{\partial \rho}{\partial t} + \frac{\partial J_j}{\partial x_j} = 0, \tag{1}$$

$$\frac{\partial J_i}{\partial t} + \frac{\partial}{\partial x_j}(J_j v_i) = -\frac{\partial \pi}{\partial x_i} + \frac{1}{Fr}\rho n_i + \frac{1}{Re}\frac{\partial}{\partial x_j}\tau_{ij}, \tag{2}$$

$$\rho C_p \left(\frac{\partial T}{\partial t} + v_j \frac{\partial T}{\partial x_j}\right) - \Gamma \frac{d\bar{p}}{dt} = \frac{1}{Pe} \frac{\partial}{\partial x_j} \left(K \frac{\partial T}{\partial x_j}\right), \tag{3}$$

$$\bar{p} = \rho\, T, \tag{4}$$

where $J_j = \rho v_j$ is the mass flux density, $\pi = p/(\gamma Ma^2)$ is a reduced pressure which accounts for the hydrostatic and dynamic effects, n_i is the unit vector in the direction of gravity τ_{ij} is the viscous stress tensor given by

$$\tau_{ij} = \mu \left(\frac{\partial v_i}{\partial x_j} + \frac{\partial v_j}{\partial x_i}\right) - \frac{2}{3} \delta_{ij}\, \mu \frac{\partial v_k}{\partial x_k}, \tag{5}$$

and δ_{ij} is the Kronecker delta function. The thermal conductivity K, viscosity μ, and specific heat at constant pressure C_p are allowed to be functions of the thermodynamic variables. The pressure $\bar{p}=\bar{p}(t)$ appearing in the energy equation and the equation of state, accounts for the change in time of the static pressure. The separation of the pressure components, holding under the small Mach number approximation, is the essence of the acoustic waves "filtering" [1], however this splitting introduces $\bar{p}$ as an extra unknown. It can be shown that the equation for $\bar{p}$ is obtained by consistency with the boundary conditions [1] and is given by

$$\frac{d\bar{p}}{dt} = \frac{1}{\int_V (C_p - \Gamma)\, dV} \left[\frac{1}{Pe} \int_S K \frac{\partial T}{\partial x_j}\, dS_j - \hat{C}_p \hat{u}_i(t)\, \Delta s_i\, \bar{p} + \bar{p} \int_V v_j \frac{\partial C_p}{\partial x_j}\, dV\right], \tag{6}$$

where $\hat{C}_p$ denotes the specific heat evaluated at the discharge opening, $\hat{u}_i(t)$ is the time varying velocity boundary condition at the opening (to be specified later), and V and S denote the "volume" and "surface" of the vessel, respectively.

The independent dimensionless parameters appearing in the problem are

$$A = \frac{h}{\ell}, \qquad \Gamma = \frac{R}{C_{po}}, \qquad Ma = \frac{u_r}{c_o},$$
$$Fr = \frac{u_r^2}{(g\,\ell)}, \qquad Re = \frac{\rho_o\, u_r\, \ell}{\mu_o}, \qquad Pe = \frac{\rho_o\, C_{po}\, u_r\, \ell}{K_o}, \tag{7}$$

where c_o is the initial sound speed in the gas, u_r is a reference speed, g is the magnitude of the gravitational field, and R is the gas constant. The dimensionless parameters are the aspect ratio, a gas parameter, the Mach number, Froude number, Reynolds number, and Peclet number, respectively.

The initial and boundary conditions expressed in dimensionless form are:

$$v_i(x_j,0) = 0, \quad \bar{p}(0) = 1, \quad T(x_j, 0) = 1, \tag{8}$$

$$v_i = \begin{cases} 0 & \text{at all walls} \\ \hat{u}_i(t) & \text{at } \Delta s_i \end{cases}, \quad \begin{matrix} T = T_w \text{ at all walls} \\ m_i \dfrac{\partial T}{\partial x_i} = 0 \text{ at } \Delta s_i \end{matrix}, \tag{9}$$

where m_i is the unit outward normal to Δs_i. The problem was non-dimensionalized as follows (starred quantities are dimensional):

$$\begin{matrix} x_i^* = \ell x_i , & t^* = \dfrac{\ell}{u_r} t , & v_i^* = u_r v_i , \\ p^* = p_o p , & T^* = T_o T , & \rho^* = \rho_o \rho , \\ \mu^* = \mu_o \mu , & K^* = K_o K , & C_p^* = C_{po} C_p , \end{matrix} \tag{10}$$

where $p_o = \rho_o RT_o$, and zero subscripts denote the initial dimensional values. The value of the reference speed u_r is defined to be the initial dimensional discharge speed, in which case $\hat{u}_i(0) = 1$.

3. NUMERICAL PROCEDURE

The numerical scheme employed is based on explicit finite difference approximations and is essentially the Marker And Cell method (MAC) [5]. The governing equations (1)-(6) are differenced forward in time and central in the space dimensions. The time-step stability restriction has been greatly enlarged by "filtering" the sound waves, thus allowing the efficient use of an explicit scheme.

The finite difference grid used for the difference equations is shown in Figure 1. All the scalar variables ρ, T, and π are evaluated at the center of the cells and are differenced by using the values at the center of the four neighboring cells. The momentum flux quantities J_i=(X,Y) are evaluated at the sides of the cells. The x-momentum flux X is evaluated at the center of the sides whose normals point in the x-direction. Similarly, the y-momentum flux Y is evaluated at the center of the sides whose normals point in the y-

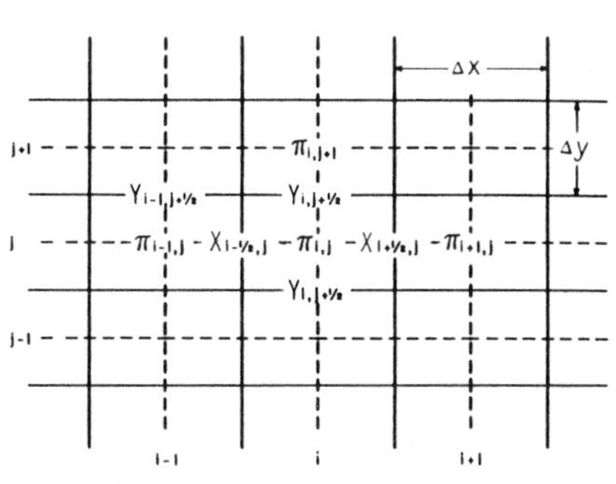

Figure 1. MAC cell structure.

direction. Their equations are similarly differenced by using the sides of neighboring cells.

The convective terms in the momentum and energy equations have been differenced using both central differencing, and upwind differencing. The upwind differencing scheme is only used when we have uniform-size cells.

The modified pressure π is governed by the elliptic equation

$$\frac{\partial^2 \pi}{\partial x_i \partial x_i} = \frac{\partial}{\partial x_i} \Lambda_i - \frac{\partial}{\partial t} \Delta , \tag{11}$$

$$\Lambda_i = - \frac{\partial}{\partial x_j} (J_j v_i) + \frac{1}{Re} \frac{\partial}{\partial x_j} \tau_{ij} + \frac{n_i}{Fr} \rho , \tag{12}$$

$$\Delta = \frac{\partial J_j}{\partial x_j} = \frac{1}{T} [-J_j \frac{\partial T}{\partial x_j} + (\frac{\Gamma}{C_p} - 1) \frac{d\bar{p}}{dt} + \frac{1}{C_p Pe} \frac{\partial}{\partial x_j} (K \frac{\partial T}{\partial x_j})]. \tag{13}$$

In order to accurately describe boundary heat fluxes in the unsteady boundary layer, especially at very small times, it is necessary most times to use a non-uniform coordinate mesh with or without dynamic rezoning. This is so since the smallest grids necessary near the boundaries can sometime be orders of magnitude smaller than necessary well outside the boundary layer. The approach used is a simple one. The grids are spaced non-uniformly in a manner to be described later. Then the spatial independent variables appearing in the difference equations are transformed to a uniform grid using the orthogonal transformation $\xi_j = \xi_j (x_j, t)$. The transformed coordinates are allowed to depend on t in addition to x_j to allow for possible rezoning as the problem progresses in time. The transformed finite difference equations for first and second order difference operators were first derived for a general orthogonal transformation by Kálnay de Rivas [6], and are given by

$$(\frac{\partial f}{\partial x})_i = \frac{f_{i+1} - f_{i-1}}{2 \Delta\xi (\frac{\partial x}{\partial \xi})_i (1 + \frac{1}{6} \delta_i)} , \tag{14}$$

$$(\frac{\partial^2 f}{\partial x^2})_i = \frac{(f_{i+1} - f_i)/(\frac{\partial x}{\partial \xi})_{i+1/2} - (f_i - f_{i-1})/(\frac{\partial x}{\partial \xi})_{i-1/2}}{(\Delta\xi)^2 (\frac{\partial x}{\partial \xi})_i (1 + \frac{5}{24} \delta_i)} , \tag{15}$$

where $\delta_i = (\Delta\xi)^2 (\frac{\partial^3 x}{\partial \xi^3})_i / (\frac{\partial x}{\partial \xi})_i$ is important when we have large grid variation. The above approximations have a truncation error of order $(\Delta\xi)^2$ for arbitrary mesh transformation. The equation

$$\left.\frac{\partial f}{\partial t}\right|_x = \left.\frac{\partial f}{\partial t}\right|_\xi + \left.\frac{\partial x}{\partial t}\right|_\xi \left.\frac{\partial f}{\partial x}\right|_t \tag{16}$$

is used to transform the time derivatives in (2) and (3) prior to the space derivative approximations.

On a study of computations using non-uniform finite difference grids, it was found by Chenoweth and Paolucci [7] that, for our type of problem, in order to compute the heat flux accurately at all times, the Roberts transformation [8]

$$\begin{aligned} \frac{\partial x}{\partial \xi} &= \frac{1}{2} H \,(1 - L^2 \tilde{x}^2) , \\ H &= L^{-1} \tanh^{-1} L, \qquad L = (1 - S)^{1/2} , \end{aligned} \tag{17}$$

where $\tilde{x} = 2x - 1$, gives best results. The values of H and L are obtained by applying the constraints $\xi = 0$ at $x = 1/2$ and $\xi = 1$ at $x = 1$, along with the definition

$$S = S(t) = \left(\frac{\partial x}{\partial \xi}\right)_{\xi=1} / \left(\frac{\partial x}{\partial \xi}\right)_{\xi=0} , \tag{18}$$

which represents a measure of grid reduction. With this transformation we then have that

$$\delta_i = 2\,(HL\Delta\xi)^2\,(3L^2\,\tilde{x}^2_i - 1) \quad . \tag{19}$$

The above transformation is nothing more than a particular exponential one.

S is still a free parameter which is used to control the grid reduction in the boundary layer as a function of time. The way S is controlled is based on boundary flux error. In the finite difference approximation S is discretized as $S(t_n)=S^n$, and an initial grid reduction value is specified, e.g. $S^o=10^{-2}$; note that $S \equiv 1$ gives us uniform grids. At the location of maximum wall heat flux, the numerical value of $(\partial T/\partial x)^n$ is evaluated there using both linear and quadratic interpolations GTL_n and GTQ_n, respectively (for details the reader is referred to Chenoweth and Paolucci [7]). If we let $\varepsilon_n = 1 - GTL_n/GTQ_n$, and ε_m be an error control parameter, then the grids can be controlled as follows: let $G^{n+1} = S^n[1+H^2\,(\varepsilon_m - \varepsilon_n)]$, then

$$S^{n+1} = \begin{cases} S^n & , \text{ if } S^n \geq G^{n+1} \\ G^{n+1} & , \text{ if } S^n < G^{n+1} \leq 1 \end{cases} \quad . \tag{20}$$

It was found [7] that if ε_m is chosen in the range $0.03 \leq \varepsilon_m \leq 0.10$, then one can be assured that several grid points remain in the boundary layer, but no more than is necessary to obtain the accuracy desired. For our particular problem, since we have two coordinate direction, we use the transformation (17) for both coordinates, however we now have $S_x = S_x(t)$ and $S_y = S_y(t)$ for grid control in the two directions, and the aspect ratio A enters in the transformation in the y-direction.

With the transformation $\xi_j = \xi_j(x_j,t)$, the resulting elliptic equation for π is separable with variable coefficients. This equation is solved at each time step using a fast Poisson solver [9]. The computational procedure is as follows:

1. Prescribe initial values.
2. Set up the computational grids on both directions.
3. Decide on time step based on stability restrictions.
4. Compute $d\bar{p}/dt$ using (6) and then compute $\bar{p}^{n+1}$ from $\bar{p}^{n+1} = \bar{p}^n + \Delta t\, d\bar{p}/dt$.
5. Evaluate T^{n+1} using (3) and ρ^{n+1} using the equation of state (4).
6. Calculate π^{n+1} from (11) using the fast Poisson solver.
7. Compute J_i^{n+1} using the momentum equations (2).
8. Evaluate the new velocities using $v_i^{n+1} = J_i^{n+1}/\rho^{n+1}$.
9. Compute maximum wall heat flux, GTL_{n+1}, and GTQ_{n+1} on both directions and repeat steps 2-8. If we have non-adapting grids, skip step 2.

4. RESULTS

The algorithm just described was applied to a problem in which air at a temperature of $T_o = 293.33°K$ and at a pressure of $p_o = 3$ atm is enclosed in a vessel of width $\ell = 3.048$ cm and height $h = 3.048$ cm. Gravity was taken to point downwards. At time t=0+, the gas is allowed to discharge at a rate of 0.372 gm/sec/cm through an opening of size 0.3048 cm. The dimensionless mass discharge rate is taken to be

$$\dot{m}(t) = (1 + t/t_r)^{\frac{1+\gamma}{1-\gamma}}, \qquad (21)$$

where $t_r = 23.52$, and $\gamma = C_{po}/C_{vo}$ is the ratio of specific heats. Equation (21) was chosen so as to have the form of choked flow discharge (the choking point is assumed to be outside the vessel), and is equivalent to specifying a condition on the velocity $\hat{u}_i(t)$.

The values of the dimensionless parameters for this particular situation are A=1, Γ=0.285, Ma=1.015×10^{-2}, Fr=42.984, Re=1.961×10^4, and Pe=1.385×10^4. In this case the gas properties μ, C_p, and K were taken to be constant, and the computational grids were chosen to be uniform. The above simplifications were applied so that we could observe the qualitative behavior of the physics, without introducing other effects or complications.

In order to illustrate some of the results, it becomes convenient to define three more dimensionless parameters:

$$\Pr = \frac{\mu_o C_{po}}{K_o} = \frac{Pe}{Re}, \quad Fo = \frac{K_o t^*}{\rho_o C_{po} \ell^2} = \frac{t}{Pe}, \quad Nu = \frac{0.5\ \ell}{(T_w - \bar{T})} \left(\overline{\frac{\partial T}{\partial x}}\right)_w, \quad (22)$$

where $\bar{T}$ is the mass averaged temperature of the gas in the vessel, and $(\overline{\partial T/\partial x})_w$ is the average temperature gradient on a wall. The first dimensionless parameter in (22) is the Prandtl number; the second is the Fourier modulus, and is essentially a scaled representation of our dimensionless time; the third is the average Nusselt number on a wall of the vessel.

Figure 2 shows the variations of $\bar{p}$ and $\bar{T}$ as functions of the Fourier modulus. Note that $\bar{p}$ can also the thought of as the mass averaged pressure since the other component of pressure, p, is of the order of Ma^2. If can be observed that the pressure $\bar{p}$ reaches a minimum of 1.4 atm at approximately a Fourier number of 2.0×10^{-3} which roughly corresponds to $t \approx t_r$. The bulk temperature $\bar{T}$, instead, reaches its minimum value of about 238°K at a Fourier number of approximately 1.0×10^{-3}, or equivalently $t \approx 0.5\ t_r$. Both $\bar{p}$ and $\bar{T}$ approach constant values at large times. The temperature asympotically approaches the dimensionless value of 1 as the gas returns to thermal equilibrium with the vessel, as expected. The slight increase in the pressure $\bar{p}$ after its minimum value needs a little explanation. This occurs because our mass discharge boundary condition is of the form of choked flow. In reality, as more and more mass is discharged from the vessel, the pressure $\bar{p}$ drops, so that eventually the flow will not be choked anymore. In order to properly simulate this process, the discharge boundary condition would have to be changed to the proper unchoked flow equation once the pressure ratio, between the interior of the vessel and the ambient, dictates it. In such case we would expect the curve for $\bar{p}$ to be monotonic in time, and reach its minimum (equal to the ambient pressure) asymptotically.

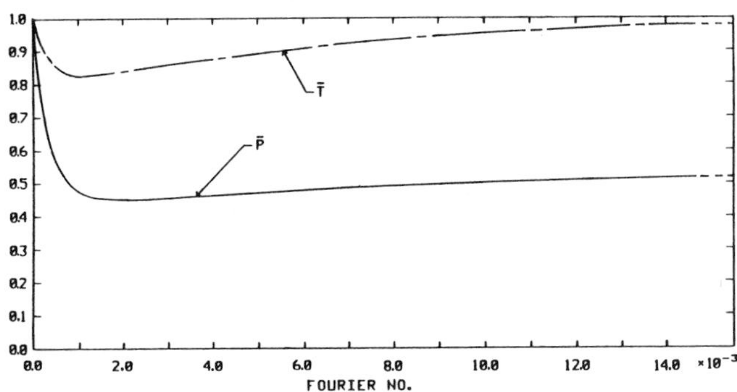

Figure 2. Average pressure and temperature as functions of the Fourier modulus.

In Figure 3 we show the Nusselt number at the top and bottom walls, and the average Nusselt number for the entire vessel, obtained by averaging the values at the four walls. The values at the side walls have not been shown for clarity, since their values almost coincide, and in turn they follow closely the average value for the container. As can be observed, all values remain essentially equal until approximately a Fourier number of 1.8×10^{-3}. Initially their values are infinite since $T_w = \bar{T}$, and they drop off sharply to $Nu \approx 4$ at $Fo \approx 0.25 \times 10^{-3}$, after which they settle into what appears to be a conduction dominated flow until $Fo \approx 1.8 \times 10^{-3}$. After this time, which occurs before the pressure minimum, the Nusselt numbers for the top and bottom sides separate, and the average value for the vessel approximately becomes the average of the top and bottom walls. This behavior is expected if, as we will shortly see, natural convection starts dominating the flow after the pressure minimum. This is so since, as the gas in the center of the vessel is colder than the gas next to the walls, the fluid will rise next to the side walls, and fall in between the side walls. Thus we have the warmer gas near the top side and colder gas near the bottom side giving rise to the Nusselt number patterns shown in Figure 3. If the simulation would have been carried out much further in time, we would expect all values of Nusselt numbers to approach a value of unity, since then the flow would be proceeding from natural convection to conduction, and eventually achieve thermal equilibrium with the walls of the container.

The results shown in Figures 2 and 3 were obtained for a vessel in which the gas was discharging from the center of the bottom wall. The corresponding velocity and temperature fields are shown in Figure 4 for different values of time. The values of time shown in this figure is the product of the Fourier modulus and the Prandtl number. From the top

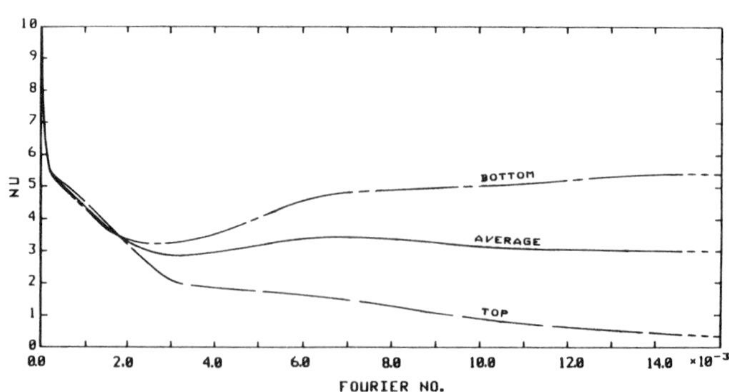

Figure 3. Nusselt numbers as functions of the Fourier modulus.

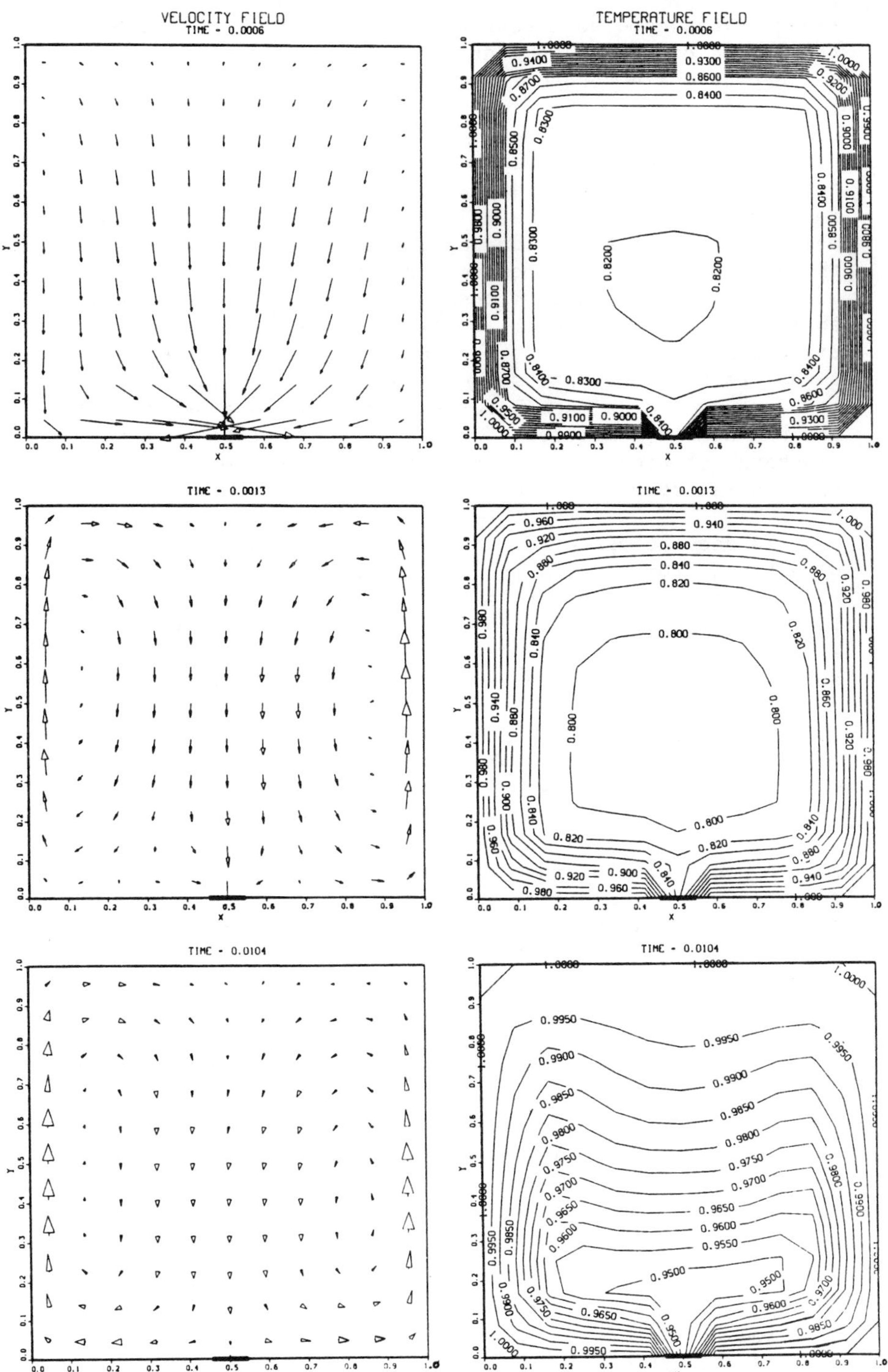

Figure 4. Sequence of velocity and temperature fields for bottom wall gas discharge.

figures it can be seen that at small times all gas rushes toward the discharge opening, and the flow is forced convection dominated, while conduction dominates the temperature field. The momentum and thermal fields appear to be uncoupled at this stage. The center figures, corresponding to a time before the pressure minimum is obtained and after that of minimum bulk temperature, show that natural convection starts playing a role in the problem. The cold gas being warmed by the side walls rise towards the top surface and falls in the center of the vessel towards the bottom surface thus combining with the momentum component due to the forced convection. Two cells appear to be setting up. The temperature field starts becoming distorted at this time reflecting the effect of the gravitational force. The last sequence shows the velocity and temperature fields much later in time. Here we see that the gas has almost completely "forgotten" the location of the discharge opening, and the flow pattern and temperature fields are natural convection dominated.

To see the effect of discharging the gas at different directions from the gravity vector, the gas was also allowed to issue from the centers of the top and left sides. The variations of $\bar{p}$, $\bar{T}$, and the different Nusselt numbers are similar to those already displayed in Figures 2 and 3, and do not quantitatively differ substantially from those. However, the effects on the velocity and temperature fields can be quite dramatic. We do not show the results here due to space limitations, however, in short the results can be summarized as follows. Regardless of the direction of the gravity vector, at early times the gas always rushes toward the discharge opening, no matter which side it is located on; the flow is forced convection dominated, and heating of the gas is by conduction. At late times, the gas does not "remember" the location of the discharge opening, no matter which side it was on, and the whole flow pattern is dominated by natural convection. In the vicinity of pressure equilibrium, and after the temperature minimum has been obtained, the relative direction of discharge with the gravity vector is important; the flow pattern can be complicated, since here we are in the mixed convection regime, where both forced and free convection are important.

5. CONCLUSION

Numerical solutions of the two-dimensional time-dependent problem of the discharging of gas from a vessel have been presented. An efficient explicit numerical scheme was used to approximate the governing compressible Navier-Stokes equations. The use of an explicit scheme was made possible by first analytically "filtering" the sound waves from the equations, without affecting the physics, thus considerably enlarging the numerical stability restriction.

The flow regimes resulting from the physics can be clearly observed, and the effect of the direction of gas discharge relative to the gravity vector has been investigated. It should be kept in mind that the amount of heating of the gas directly affects the mass discharge (especially near pressure equilibrium and after) in the more realistic problem where $\dot{m}$ is coupled to the rest of the problem by letting it depend on both $\bar{p}$ and the ambient pressure. In such case, in order to accurately determine the amount of gas discharged from the vessel, we are required to accurately determine the heat flux through the vessel walls. This can be accomplished by making use of the non-uniform adaptive grid scheme described. We are presently in the process of solving this more realistic problem in the manner described and plan to report the results in the future.

REFERENCES

1. PAOLUCCI, S. and CHENOWETH, D. R. - (To be published).
2. FORESTER, C. K. and EMERY, A. F. - A Computational Method for Low Mach Number Unsteady Compressible Free Convective Flows. J. Comp. Phys., Vol. 10, pp. 487-502, 1972.
3. REHM, R. G. and BAUM, H. R. - The Equations of Motion for Thermally Driven Buoyant Flows, J. of Res. of the N.B.S., Vol. 83, pp. 297-308, 1978.
4. HUMPHREY, J. A. C. and LE QUERE, P. - (Personal communication).
5. HARLOW, F. H. and WELCH, J. E. - Numerical Calculation of Time Dependent Viscous Incompressible Flow of Fluid with Free Surface. Phys. Fluids, Vol. 8, pp. 2182-2189, 1965.
6. KÁLNAY DE RIVAS, E. - On the Use of Nonuniform Grids in Finite-Difference Equations. J. Comp. Phys., Vol. 10, pp. 202-210, 1972.
7. CHENOWETH, D. R. and PAOLUCCI, S. - On Optimizing Nonuniform Finite-Difference Grids for Boundary Regions in Transient Transport Problems. Sandia National Laboratories Report SAND81-8204, Livermore, California, 1981.
8. ROBERTS, G. O. - Computational Meshes for Boundary Layer Problems. Proc. Second Int. Conf. on Numerical Methods in Fluid Dynamics, Lecture Notes in Physics, Ed. Holt, M., Springer-Verlag, Vol. 8, pp. 171-177, 1970.
9. SWARZTRAUBER, P. and SWEET, R. - Efficient Fortran Subprograms for the Solution of Elliptic Partial Differential Equations. NCAR Technical Note IA-109, Boulder, Colorado, 1975.

THERMOCONVECTIVE HEAT TRANSFER IN A RECTANTULAR CAVITY WITH CONSTANT WALL COOLING RATE

ROBILLARD, L., VASSEUR, P. and HUNG, N.T.

Professors
Ecole polytechnique de Montréal
Québec, CANADA, H3C 3A7

ABSTRACT

The transient natural convection in a fluid contained in a rectangular enclosure, the wall of which is maintained at a uniform temperature which changes at a steady rate, is approached by a numerical method. The time-dependent governing differential equations are solved using an alternating direction implicit (A.D.I.) finite difference method. Numerical solutions are obtained for $P_R = 0.73$ and 7.3 and a range of Rayleigh numbers $R_A = 10^2 \sim 10^8$. The transient flow and temperature fields, local and overall heat rate are presented. Although the present problem is time dependent a quasi-steady state develops if the cooling rate applied to the wall of the cavity is held constant long enough. For quasi-steady state, a generalized correlation curve for Nusselt number valid for $P_R \geqslant .73$ and $R_A \leqslant 10^7$ is presented. At relatively low Rayleigh numbers the flow is caracterized by the development of double cells with flow up the center and down the sidewalls. However it was found that the increase of the Rayleigh number leads to the development of strong secondary circulation on the axis of symmetry of the cavity near the top wall. Thus, as the Rayleigh number is increased the secondary cells grow in size. The effects of the secondary cells on the temperature fields and heat transfer coefficients are discussed. Most results are obtained in the case of a square cavity (E = 2.0) but the influence of the aspect ratio of the cavity is also studied for E = 1.0 and 4.0

1. INTRODUCTION

The present study considers the transient recirculating flows induced in a rectangular two-dimensional enclosure by decreasing the wall temperature at a constant time rate. Theoretical analysis on transient natural convection in enclosures with a initially uniform fluid temperature and a linear variation of wall temperature with time has received little attention in the litterature. Prior studies on the subject have examined theoretically [1, 2] and experimentally [3] the transient natural convection in horizontal cylinders with constant cooling rate. It was found that after an initial transient period, a quasi-steady state takes place inside the cavity for which the cooling rate of every interior point is the same as that of the cavity walls.

In this paper the transient natural convection in a rectangular cavity with wall temperature decreasing steadily and fluid initially at rest and at uniform temperature is considered. Results are presented in terms of streamline and temperature fields. The effect of cooling rate, cavity size, aspect ratio of the cavity and Prandl number are investigated. Also considered in details are representative heat transfer characteristics.

2. PROBLEM FORMULATION

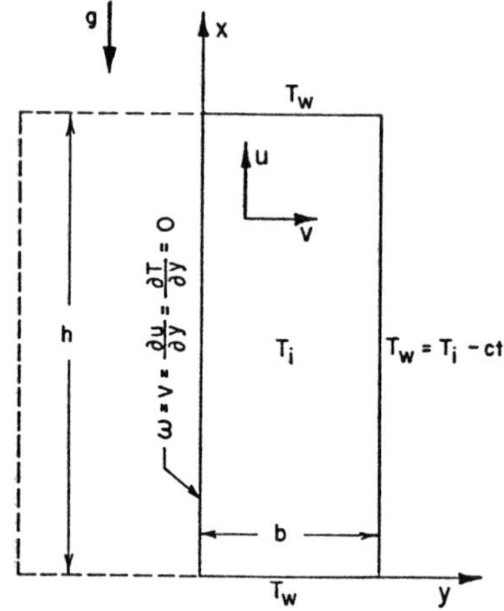

Fig. 1 Coordinate system and boundary conditions

Consider the natural convective motion of a mass of fluid contained in the closed rectangular two-dimensional cavity illustrated schematically in Figure 1. A rectangular Cartesian co-ordinate system is located in the centre of the base. Initially the water is motionless and at a uniform temperature T_i. At time t = 0 a uniform temperature $T_w = T_i - ct$, where c is a constant cooling rate, is imposed on the boundaries of the cavity. Cooling of the system is maintained long enough so that a final quasi-steady state solution, if such exists, is reached. The appropriate non dimensional equations governing the resulting transient flow of fluid in this situation are :

$$\frac{\partial \Omega}{\partial \tau} + \frac{\partial U\Omega}{\partial X} + \frac{\partial V\Omega}{\partial Y} = P_R\, R_A \frac{\partial \theta}{\partial Y} + P_R\, \nabla^2 \Omega \qquad (1)$$

$$\frac{\partial \theta}{\partial \tau} + \frac{\partial U\theta}{\partial X} + \frac{\partial V\theta}{\partial Y} = \nabla^2 \theta + 1 \qquad (2)$$

$$\nabla^2 \Psi = -\Omega \qquad (3)$$

$$U = \frac{\partial \Psi}{\partial Y} \qquad V = -\frac{\partial \Psi}{\partial X} \qquad (4)$$

with corresponding initial and boundary conditions :

$$\begin{aligned} &\tau = 0 : \quad U = V = 0 ; \quad \theta = 0, \quad \text{everywhere} \\ &\tau > 0 : \quad U = V = 0 ; \quad \theta = 0, \quad \text{on all solid boundaries} \\ &\qquad \frac{\partial U}{\partial Y} = V = 0 ; \quad \frac{\partial \theta}{\partial Y} = 0, \quad \text{on the axis of symmetry } (Y = 0) \end{aligned} \qquad (5)$$

The non dimensional variables of equations 1 to 5 are :

$$\begin{aligned} &X = x/b \qquad Y = y/b \qquad U = ub/\alpha \qquad V = vb/\alpha \\ &\tau = t\alpha/b^2 \qquad \Omega = \zeta b^2/\alpha \qquad \Psi = \psi/\alpha \qquad \theta = (T - T_w)/\Delta T \\ &R_A = gb^3\beta\Delta T/\alpha\nu \qquad P_R = \nu/\alpha \qquad \Delta T = cb^2/\alpha \\ &\alpha = K/\rho C_p \qquad \nabla^2 = \frac{\partial^2}{\partial X^2} + \frac{\partial^2}{\partial Y^2} \end{aligned} \qquad (6)$$

Inherent in the derivation of these equations is the usual Oberbeck-Boussinesq approximation [4, 5]. Also the work of compression and viscous dissipation are neglected and all fluid properties are assumed constant except for density in the buoyancy term [6].

The Rayleigh number R_A, the Prandtl number P_R and the aspect ratio $E = h/b$ are the three control parameters of the present problem. The unity ($b^2c/\Delta T\alpha = 1$) appearing on the right hand side of the energy equation can be regarded as a uniform heat source term. As a matter of fact, the present problem is known to be equivalent to transient natural convection heat transfer between a fluid with uniform internal heat sources of strength per unit time and volume $\rho c\, C_p$ and a cavity with constant wall temperature.

Equations (1) to (4) with appropriate boundary conditions are solved with a time-marching, finite-difference A.D.I. technique. The forward time, central-difference scheme, that has truncation errors of second order in X and Y, was found to be stable. Details of the technique are discussed by Roache (1976).

To expedite plotting of the results, an auxiliary computer program was written to locate points lying on specified isotherms and streamlines by linear interpolation on the computed values at the grid points. As mentioned earlier the problem under consideration is symmetrical and it was found advantageous to reproduce the computer results at a given time on a single graph with the flow pattern on the right half of the cavity and the isotherms on the left half.

3. RESULTS AND DISCUSSION

Numerical results obtained in the present study have revealed that, depending on the Rayleigh number, two different modes of convection may be observed inside the cavity.

The development of the first mode of convection (R_A = 5×10^4) is quite similar to the one obtained experimentally by Deaver and Eckert [3]. and numerically by Takeuchi and Cheng [2]. The initial stage of cooling is characterized by a pure conduction heat transfer. Due to the symmetry with respect to the vertical plane Y = 0, a pair of counterrotating vortices is formed. As the cooling progresses, the vortex pair gradually increases its strength and the convective motion progressively stratifies the core region with isotherms closely spaced near the top wall but sparely spaced near the bottom one, giving respectively large and poor heat transfer on those boundaries, Fig. 2 shows the stabilized pattern of streamlines and iostherms corresponding to the quasi-steady state situation that prevails at time τ = 0.611.

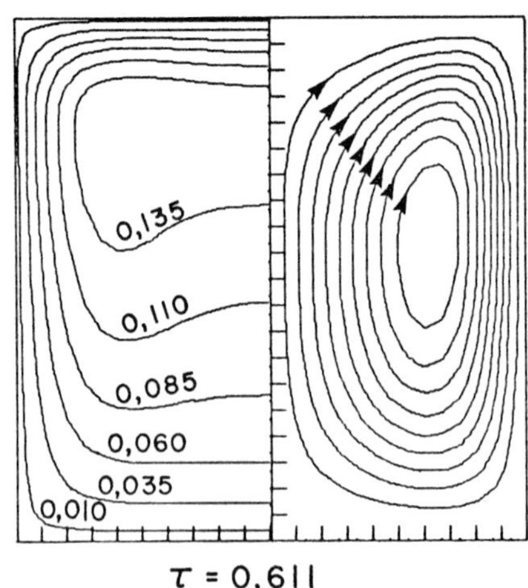

Fig. 2 Quasi-steady state streamline and isotherm fields for P_R = 7.3, R_A = 5×10^4 and E = 2 (square cavity)

The corresponding heat fluxes $\Phi^{(I)}$ are given as a function of time on Fig. 3.

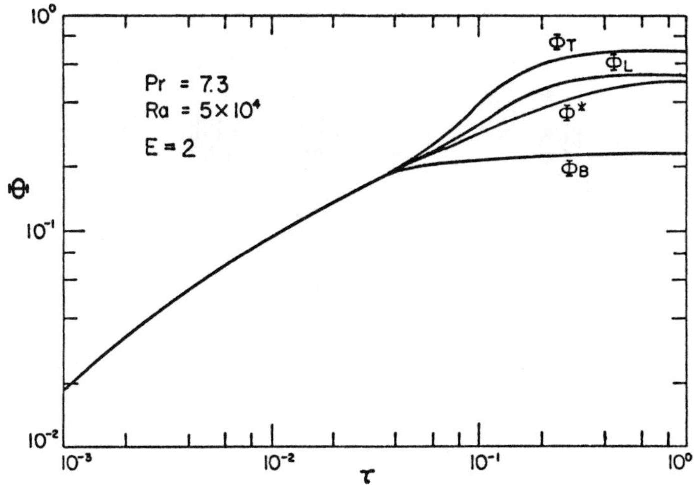

Fig. 3 Transient heat fluxes corresponding to the etablishment of the first mode

Numerical results have shown that the development of the second mode of convection occurs when R_A is greater than 8×10^4, for $P_R \geqslant 1$. At the initial stage of the cooling, the temperature and flow fields are quite similar to those observed for the first mode of convection. However, due to the higher Rayleigh number involved the flow near the side wall tend to be of the boundary layer type while the fluid in the upper central region of the cavity becomes almost stagnant. Furthermore, the fluid in this region is unstable because of the top heavy situation resulting from the particular temperature field prevailing near the top wall. The induced density field combined to the existing flow motion gives rise to an additional pair of secondary vortices symmetrically located on the axis of symmetry, near the upper boundary. The new flow field develops to the quasi-steady state situation depicted in Fig. 4 in which the secondary vortex has reached its equilibrium intensity. Its results from this particular flow pattern that the relatively cold fluid penetrates the core region not only from the bottom but also from the top of the cavity. This motion perturbs greatly the isotherm field in the centre upper region (compare Figs 4 and 2). Furthermore the second mode of convection allows the warmer core fluid to reach

(I) Φ_T, Φ_B and Φ_L are the dimensionless heat fluxes averaged over the top, bottom and side wall respectively. The subscript (*) refers to pure conduction.

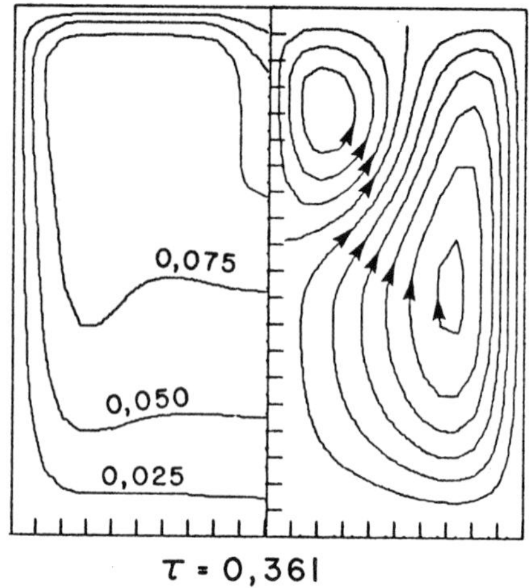

Fig. 4 Quasi-steady state streamline and isotherm fields for $P_R = 7.3$, $R_A = 3 \times 10^5$ and $E = 2$

the region near the top boundary by two paths instead of one. The noticeable bump characterizing the heat transfer Φ_T in Fig. 5 corresponds to the occurrence of the secondary motion.

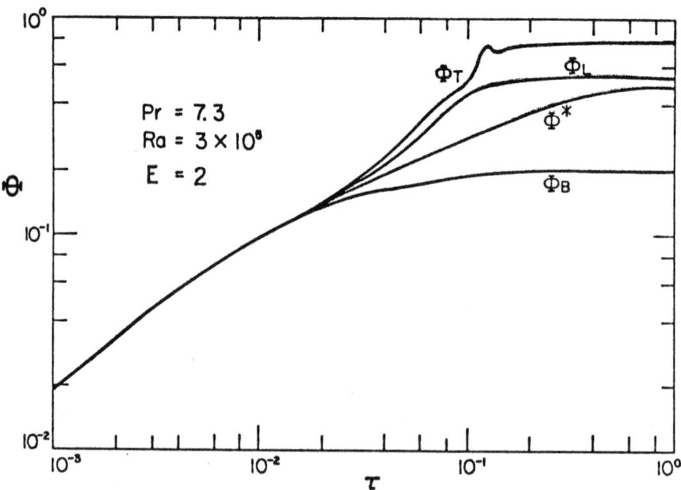

Fig. 5 Transient wall heat fluxes corresponding to the etablishment of the second mode

In order to study the effect of the Prandtl number the quasi-steady state situations of Fig. 2 and 4 were computed for a Prandtl number of .73. Results are shown on Figs 6a and 6b. It is seen that for a given Rayleigh number a decrease of P_R 1° lowers the position of the vortices center, 2° promotes the occurrence of the secondary motion (compare Figs 2 and 6a), and 3° increases the size of the secondary vortex when it is present.

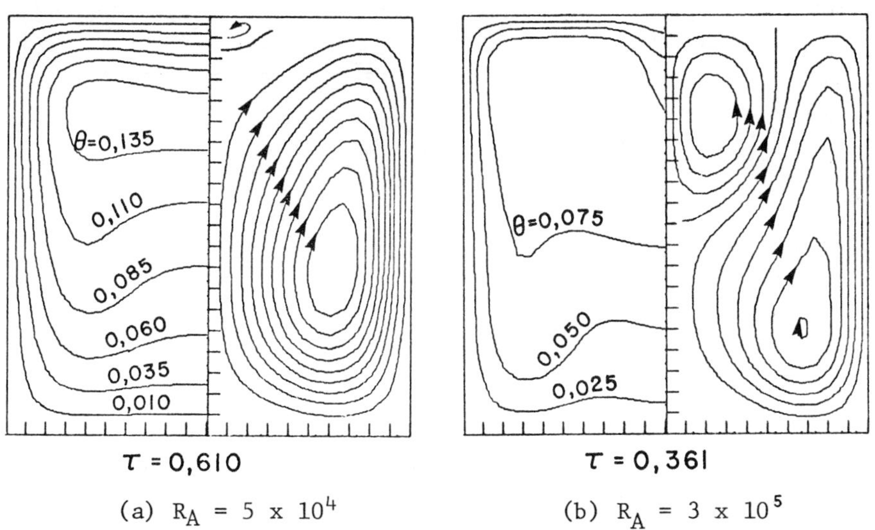

(a) $R_A = 5 \times 10^4$ (b) $R_A = 3 \times 10^5$

Fig. 6 Quasi-steady state streamline and isotherm fields for $P_R = .73$ and $E = 2$

The influence of the aspect ratio E on the flow configuration is illustrated on Figs 7a and 7b for E = 4 and 1 respectively. Quantitatively the general features of the flow and temperature field remain similar to those obtained with E = 2. However, in the case of the shallow cavity the secondary motion is enhanced. This is due to the fact that, with decreasing E, the instability zone related to the vertical density gradient near the upper horizontal boundary becomes more important whereas the stabilizing driving force generated by the horizontal density gradient near the vertical wall is reduced.

The relationship between Nusselt number N_U and the modified Rayleigh number R_A' for the quasi-steady state is of practical interest and the results obtained for cavities with E = 1, 2 and 4 are presented in Fig. 8.

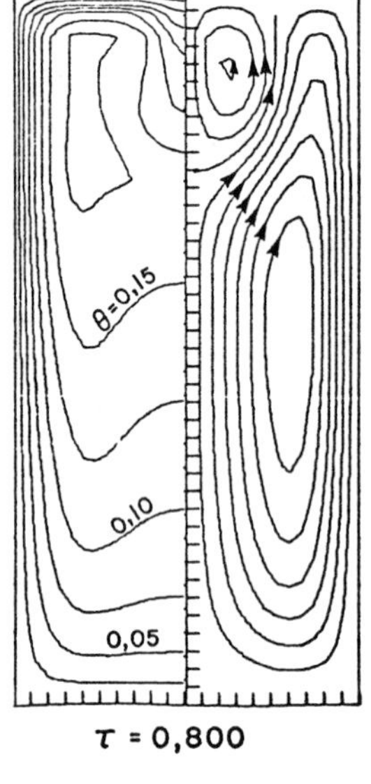

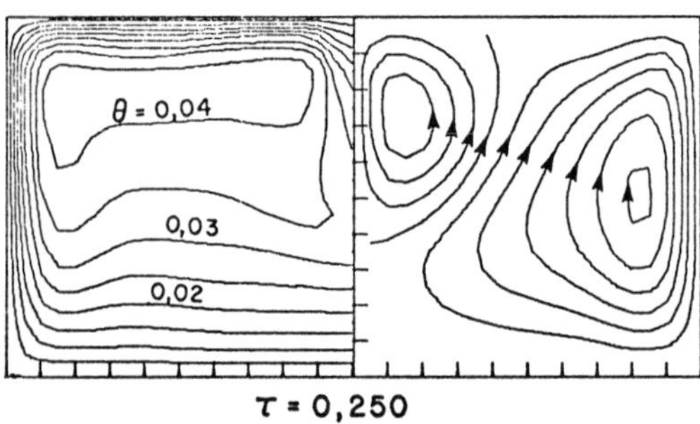

(b) $R_A = 3 \times 10^6$

$E = 1$

(a) $R_A = 7 \times 10^4$

$E = 4$

Fig. 7 Quasi-steady state streamline and isotherm fields for $P_R = 7.3$

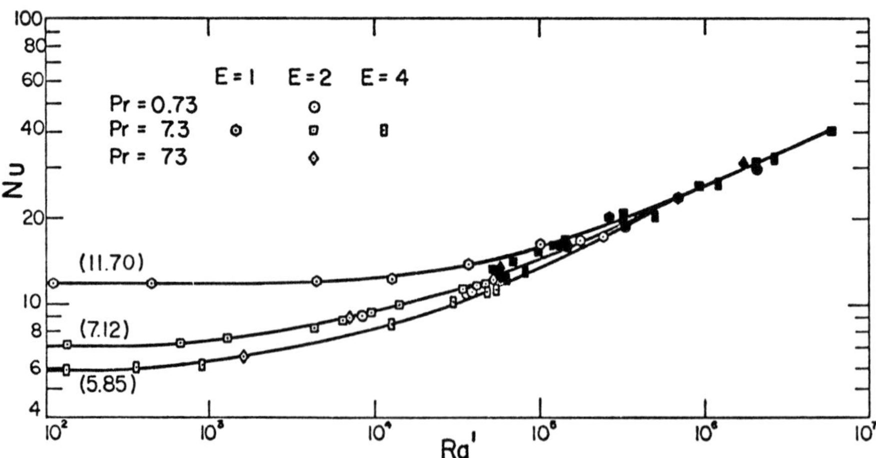

Fig. 8 Relationship between Nusselt number and modified Rayleigh number for the quasi-steady state

N_U is defined as :

$$N_U = \frac{2(\Phi_T + \Phi_B + E\Phi_L)}{(2 + E)\,\bar{\theta}} \qquad (7)$$

where $\bar{\theta} = (\bar{T} - T_w)/\Delta T$, $\bar{T}$ being the mixed mean temperature. The relationship between modified Rayleigh number R_A', based on the temperature difference between $\bar{T}$ and T_w, and the Rayleigh number R_A is given by :

$$R_A' = \frac{g(2b)^3\beta}{\nu\alpha}(\bar{T} - T_w) = 8\,\bar{\theta}\,R_A \qquad (8)$$

The limit values of N_U corresponding to the pure conduction case (i.e. $R_A' \to 0$), obtained analytically, are indicated on the graph for reference. It is seen from the curves that the pure conduction theory is valid for $R_A' < \sim 5 \times 10^2$ when $E = 4$ and $R_A' < \sim 5 \times 10^3$ when $E = 1$. Furthermore it may be observed that as R_A' increases, the three curves tend to collapse.

4. CONCLUSIONS

The thermoconvective motion taking place in a rectangular cavity with wall temperature decreasing at a constant rate, has been studied numerically for Prandtl numbers $P_R = .73$ and 7.3 aspect ratio of the cavity E = 1, 2 and 4 and a range of Rayleigh numbers $R_A = 10^2$ to 10^8. From the results obtained, the following observations may be made :

i) At a sufficiently large time, velocities, flow patterns and temperature differences between the fluid and the wall tend to become constant with time and the quasi-steady state is reached.

ii) A second mode appears for Rayleigh numbers beyond a critical value in which an additional pair of counterrotating vortices, located near the top boundary is superposed to the basic flow of the first mode, this latter consisting of a single pair of counterrotating vortices. This second mode was found to enhance the convective heat transfer near the top boundary.

iii) The effect of the Prandtl number on the present problem follows the trend already reported in past litterature. For $P_R > 1$, the flow and temperature fields and the resulting heat transfer were found to be almost independent of P_R except for very high values of the Rayleigh number. For $P_R < 1$ (.73) some effect of P_R was observed.

ACKNOWLEDGEMENTS

This work was supported by the National Research Council of Canada through grants NRC A-4197 and NRC A-9201.

REFERENCES

1. QUACK, H. - Natürliche Konvektion innerhalb eines Zylinders bei kleinen Grashof Zahlen, Wärme-Und Stoffübertragung, Vol. 3, pp. 134-137, 1970.
2. TAKEUCHI, M. and CHENG, K.C. - Transient Natural Convection in Horizontal Cylinders with Constant Cooling Rate, Wärme-Und Stoffübertragung, Vol. 9, pp. 215-225, 1976.
3. DEAVER, F.K. and ECKERT, E.R.G. - An Interferometric Investigation of Convective Heat Transfer in Horizontal Fluid Cylinder with Wall Temperature Increasing at a Uniform Rate, Heat Transfer, Vol. 4, paper NC1.1, Elsevier Publishing, Amsterdam, pp. 1-12, 1970.
4. CHANDRASKHAR, S. - Hydrodynamics and Hydromagnetic stability, Oxford University Press, 1961.
5. GRAY, D.D. and GIOGINI, A. - The Validity of the Boussinesq Approximation for Liquids and Gases, Int. J. Heat Mass Transfer, Vol. 91, pp. 545-551, 1976.
6. BOOKER, J.R. - Thermal Convection with Strongly Temperature-dependent viscosity, J. Fluid Mech., Vol. 76, Part 4, pp. 741-754, 1976.
7. ROACHE, P. - Computational Fluid Dynamics, Hermosa Publishers, 1976.

NUMERICAL CALCULATION OF THE HEAT TRANSFER BY NATURAL CONVECTION IN A CUBICAL ENCLOSURE.

S.J.M. Linthorst (*), C.J. Hoogendoorn (**)

(*) Research scientist, Dept. of Applied Physics
Delft University of Technology, The Netherlands
(**) Professor of Heat Transfer, Dept. of Applied Physics
Delft University of Technology, The Netherlands

1. SUMMARY

Numerical calculations of the heat transfer by natural convection in a cubical box are described in this paper. Two vertical opposing walls are supposed isothermal. All sidewalls are assumed to be adiabatic or perfectly conducting. The Rayleigh range considered varied between 10^4 and 10^6. We have taken air as the working fluid ($P_R = 0.71$) and considered the flow 3-dimensional, laminar and stationary. Solutions of the governing equations are obtained with a method of finite differences using a non-linear grid. A method is used to correct the results for the influence of the grid used in the numerical calculation by varying the non-linearity parameter. Application of this more accurate method showed the value of the calculated Nusselt number to be smaller than the value obtained without the correction for the grid.

2. INTRODUCTION

In literature the problem of natural convection in rectangular enclosures is mainly confined to enclosures for which a two dimensional approximation of the flow is sufficient (de Vahl Davis [1], Elder [2]). Usually applications as the heat loss in a solar collector or a double glazed window desired no further extended calculations.
Experimental investigations of Schinkel [3] showed that the flow at the midplane of the enclosure, perpendicular on the third direction, could be considered two dimensional if the depth to width ratio was beyond four.
But as shown by Mallinson and de Vahl Davis [4] the influence of the sidewalls, perpendicular on the third direction, which is neglected in a two dimensional approach, increases if the depth to width ratio decreases. Therefore it is obvious that in the case of predicting the heat transfer by natural convection in a solar collector in which a honeycomb is

inserted, a three dimensional approach is desired.

Three dimensional calculations of natural convection in enclosures are done for the horizontal orientation by Ozoe [5] and Catton [6].

For the vertical orientation Mallinson and de Vahl Davis [4] calculated the three dimensional fluid flow for several Rayleigh and Prandtl numbers.

The purpose of our work is to calculate the heat transfer by natural convection in small enclosures of which both the depth to width as the height to width ratio are less than one. In this paper only the results for a vertical cubical box are described. These calculations are used as a test for the computerprogramme. No calculations are described for honeycombs. Two vertical opposing isothermal walls are held on a certain temperature difference. All other sidewalls are assumed perfectly conducting or adiabatic. The calculations are done for air (P_R = 0.71). The Rayleigh number (R_A) considered was in the range of 10^4 to 10^6.

3. THEORY

3.1. The formulation of the problem.

The natural convection flow is described by the equations for conservation of energy, mass and momentum. If we apply the Boussinesq approximation and consider the flow three dimensional, stationary and laminar the governing equations are:

$$\frac{\partial u}{\partial x} + \frac{\partial v}{\partial y} + \frac{\partial w}{\partial z} = 0 \qquad (1)$$

$$G_R \left(\frac{\partial uu}{\partial x} + \frac{\partial vu}{\partial y} + \frac{\partial wu}{\partial z} \right) = \frac{\partial^2 u}{\partial x^2} + \frac{\partial^2 u}{\partial y^2} + \frac{\partial^2 u}{\partial z^2} - \frac{\partial p}{\partial x} + (\theta - \theta_o) \qquad (2)$$

$$G_R \left(\frac{\partial uv}{\partial x} + \frac{\partial vv}{\partial y} + \frac{\partial wv}{\partial z} \right) = \frac{\partial^2 v}{\partial x^2} + \frac{\partial^2 v}{\partial y^2} + \frac{\partial^2 v}{\partial z^2} - \frac{\partial p}{\partial y} \qquad (3)$$

$$G_R \left(\frac{\partial uw}{\partial x} + \frac{\partial vw}{\partial y} + \frac{\partial ww}{\partial z} \right) = \frac{\partial^2 w}{\partial x^2} + \frac{\partial^2 w}{\partial y^2} + \frac{\partial^2 w}{\partial k^2} - \frac{\partial p}{\partial z} \qquad (4)$$

$$G_R \left(\frac{\partial u\theta}{\partial x} + \frac{\partial v\theta}{\partial y} + \frac{\partial w\theta}{\partial z} \right) = P_R^{-1} \left(\frac{\partial^2 \theta}{\partial x^2} + \frac{\partial^2 \theta}{\partial y^2} + \frac{\partial^2 \theta}{\partial x^2} \right) \qquad (5)$$

where D is the characteristic length and $\frac{g\beta_o(T_h-T_c)D^2}{\nu_o}$,

$|g|$, $\rho_o g\beta_o(T_h-T_c)D$ are the scale factors for velocity, gravity and pressure, respectively. The non-dimensional temperature $\theta = (T - T_c) / (T_h - T_c)$, G_R the Grashof number = $G_R = g\beta_o(T_h-T_c)D^3/\nu_o^2$ and $P_R = \frac{\nu_o}{a_o}$. The Rayleigh number is defined by $R_A = G_R.P_R$.

The index "o" indicates that the value of the parameter is taken at the mid temperature: $T_o = (T_h + T_c)/2$ (For notation see fig. 1).

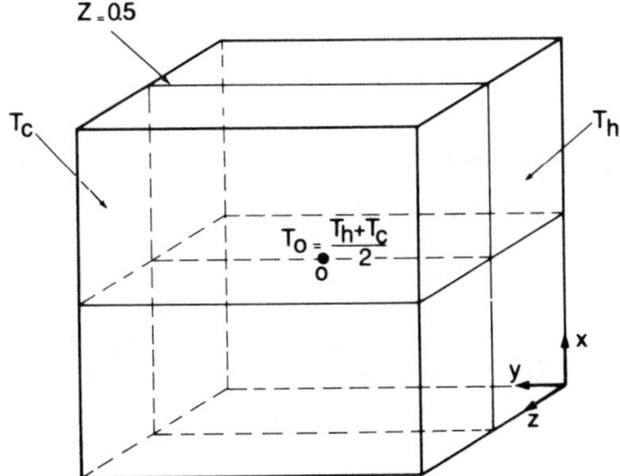

Fig. 1: A schematic view of the enclosure

The no slib condition for the velocities is imposed on all the walls:

$$u = v = w = 0 \text{ for } x = 0 \text{ and } x = 1 \quad y = 0 \text{ and } y = 1 \quad z = 0 \text{ and } z = 1 \tag{6}$$

The boundary conditions for the temperature are:
for the isothermal walls:

$$\theta = 1 \text{ for } y = 0 \qquad \theta = 0 \text{ for } y = 1 \tag{7}$$

For the sidewalls in the case of adiabatic side walls (ad):

$$\frac{\partial\theta}{\partial x} = 0 \text{ for } x = 0 \text{ and } x = 1 \qquad \frac{\partial\theta}{\partial z} = 0 \text{ for } z = 0 \text{ and } z = 1 \tag{8}$$

in the case of perfectly conducting sidewalls (pc)

$$\theta(y) = 1 - y \quad \text{for } x = 0 \text{ and } x = 1 \quad z = 0 \text{ and } z = 1 \tag{9}$$

The governing equations and the boundary conditions have two symmetry properties. First the symmetry with respect to the midplane z = 0.5, secondly the symmetry with respect to the center point 0 (0.5, 0.5, 0.5). The z - symmetry is defined by:

$$\begin{aligned} u(x,y,z) &= u(x,y,1-z) \\ v(x,y,z) &= v(x,y,1-z) \\ w(x,y,z) &= -w(x,y,1-z) \\ p(x,y,z) &= p(x,y,1-z) \\ \theta(x,y,z) &= \theta(x,y,1-z) \end{aligned} \tag{10}$$

the centro symmetry by:

$$
\begin{aligned}
u\ (x,y,z) &= -u\ (1-x,\ 1-y,\ 1-z)\\
v\ (x,y,z) &= -v\ (1-x,\ 1-y,\ 1-z)\\
w\ (x,y,z) &= -w\ (1-x,\ 1-y,\ 1-z)\\
p\ (x,y,z) &= \ \ p\ (1-x,\ 1-y,\ 1-z)\\
\theta\ (x,y,z) &= 1-\theta(1-x,\ 1-y,\ 1-z)
\end{aligned}
\qquad (11)
$$

These symmetry properties are used to reduce the computational effort required for solving the governing equations. Applying these properties it is only necessary to solve the equations in a symmetrical quarter of the enclosure.

3.2. The calculation of the heat transfer

If we consider $\alpha = -K_o \frac{1}{T_h - T_c} \left.\frac{\partial T}{\partial y'}\right|_{y'=0}$ the local heat transfer coefficient and y' the real distance. The local Nusselt number at the hot wall in non-dimensional formulation is defined by:

$$Nu\ (x,z) = -\left.\frac{\partial \theta}{\partial y}\right|_{y=0} \qquad (12)$$

The averaged Nusselt number of the hot wall:

$$\overline{Nu}_h = \int_0^1 \int_0^1 Nu\ (x,z)\ dx\ dz \qquad (13)$$

To calculate the local Nusselt number at the hot wall we fitted a cubical polynomial through values of θ at the wall and the first two grid nodes near the wall.

3.3. The numerical method

We used a finite difference method, developed by Gosman et al. [7], known under the name TEACH. This method solves the governing equations in the hydrodynamic formulation. A finite difference representation of the equations (1-5) is obtained using the so-called hybrid difference scheme. The hybrid difference scheme uses an upwind or a central difference approximation dependent on the ratio of the transport by convection and diffusion. For dominating convection the hybrid scheme uses upwind differences. This leads to five sets of linear equations which have to be solved. An iterative procedure known as the line by line method, is used to solve the equations at one line. Then the next line of grid nodes is considered. In this way a sweep through the enclosure is made for each variable. For more detailed information see Gosman [7] or Schinkel [3].

We used in our calculations a non-linear grid for all three directions. Near the wall in the boundary layer the number of grid nodes is large in comparison with the center.

The grid is symmetrical around the centerpoint O(0.5,0.5,0.5). To calculate the velocities u,v,w in the three directions we used staggered grids. The grid nodes for the different variables for a control volume is shown in fig. 2.

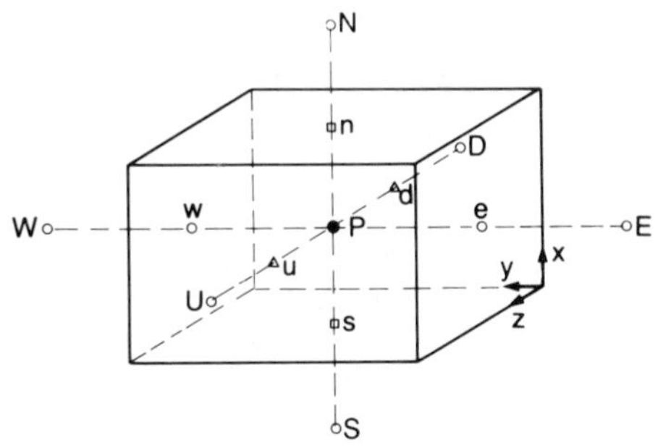

Point P: grid node for p and θ
Point s: grid node for u
Point e: grid node for v
Point d: grid node for w

Fig. 2: The position of the grid nodes for the different variables.

The non-linear grid is defined by:

$$\begin{aligned} x\ (i+1) &= 0.5\ \left(\frac{i}{k}\right)^{\alpha_x} && \text{for } i = 2,k \\ x\ (1) &= -\ x\ (2) \\ x(NI-i) &= 1 - x(i+1) && \text{for } i=1,K \\ x(NI) &= 1 - x(1) \end{aligned} \tag{14}$$

With NI, the total number of grid nodes in the x-direction, is an odd integer, $k = (NI - 1)/2$ and α_x is the non-linearity parameter. For $\alpha_x = 1$ the grid is linear. For the y- and z-direction the grid is defined in a similar way. We used 21 grid nodes each for the x- and y-direction (NI = NJ = 21). For the z-direction 11 grid nodes were used. Applying the z- and centro symmetry confined the actual number of grid nodes to 1764.
The convergence criterium defined as

$$\left| \frac{\overline{Nu}_h(i) - \overline{Nu}_h(i-1)}{\overline{Nu}_h(i-1)} \right| < \delta \tag{15}$$

where i denotes the iteration step and δ is of order 10^{-5}, was reached within 200 iterations, starting from an earlier obtained result with for example a lower R_A number or a different non-linearity parameter. To correct our results for the grid used we varied the non-linearity parameter α_y ($\alpha_x = \alpha_z = 1.5$). In the case of adiabatic sidewalls we calculated the heat transfer for each Rayleigh number with four different values of α_y; 1.5, 1.6, 1.7 and 1.8 respectively. So we obtained four Nusselt numbers at the nodes nearest to the hot wall. We fitted a second order polynomial through these ($\overline{Nu}(y)$,y)points and calculated the Nusselt-number for y = 0. This value is used as the for the grid influence corrigated $\overline{Nu}$ value. Variation of α_x and α_z had much less influence on the Nusselt number and are therefore not varied. This method is described by Schinkel [3]. The solution for a certain Rayleigh number was obtained after approximately 200 iterations, which required about 8 minutes of CPU time on a IBM 370/158 computer

4. RESULTS

We compared our results with those obtained by Mallinson and de Vahl Davis [4]. They calculated the heat transfer by natural convection mainly for enclosures with A_z = 2, adiabatic sidewalls and P_R in the range of 0.2 to 100.

Only one single case with $A_x = A_z = 1$ was calculated for air with $R_A = 5 \times 10^5$. From this single result it can be concluded that in there calculations no great difference excist between the heat transfer for an enclosure of A_z = 2 or A_z = 1. They obtained $\overline{Nu}$ = 7.43 and $\overline{Nu}$ = 7.37, respectively. This is in agreement with the few results we already obtained for A_z = 2. So comparison of our results ($A_z = A_x$ = 1) with there results is not meaningless.

In figure 3 the relations are given which we found for the adiabatic (ad) and the perfectly conducting case (pc).

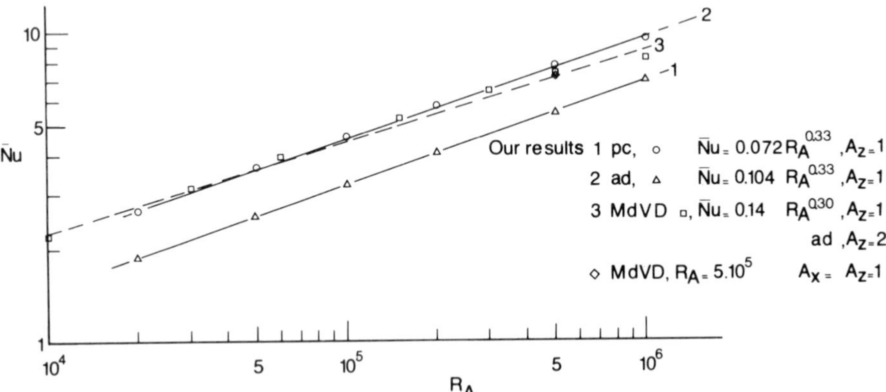

Fig. 3: The different heat transfer relations. Nusselt as function of R_A:

Both cases are calculated for $\alpha_x = \alpha_y = \alpha_z = 1.5$. Also the obtained values of Mallinson and de Vahl Davis (abbrev. as MdVD) are given for A_z = 2 and adiabatic sidewalls. Further one single value of MdVD is given for A_z = 1 and Ra = 5 10^5. First we notice that our $\overline{Nu}$-values for adiabatic sidewalls are in good agreement with those obtained by MdVD, but that the power of R_A in our obtained heat transfer relation $\overline{Nu} = 0.104\, R_A^{0.33}$ is higher than the one obtained by them

$Nu = 0.14\ R_A^{0.30}$. This is due presumably to the number of grid nodes in the y-direction; 21 and 15 respectively.
The correlations for the pc- and ad-case show almost the same power in the heat transfer relations.
In table 1 the obtained heat transfer relations are given.

	α_y	a	b	r
pc	1.5	0.072	0.330	0.9997
ad	1.5	0.104	0.330	0.9999
ad	1.6	0.109	0.324	0.9999
ad	1.7	0.116	0.317	0.9999
ad	1.8	0.122	0.313	0.9998
ad	'0'	0.148	0.294	0.9990

Table 1: The parameters of the obtained relations: $\overline{Nu}=a\ R_A^{\ b}$
r: correlation coefficient

In figure 4 the x-averaged Nusselt numbers on the hot wall are shown:

$$\langle Nu\ (z)\rangle_x = \int_0^1 Nu\ (x,z)\ dx \qquad (16)$$

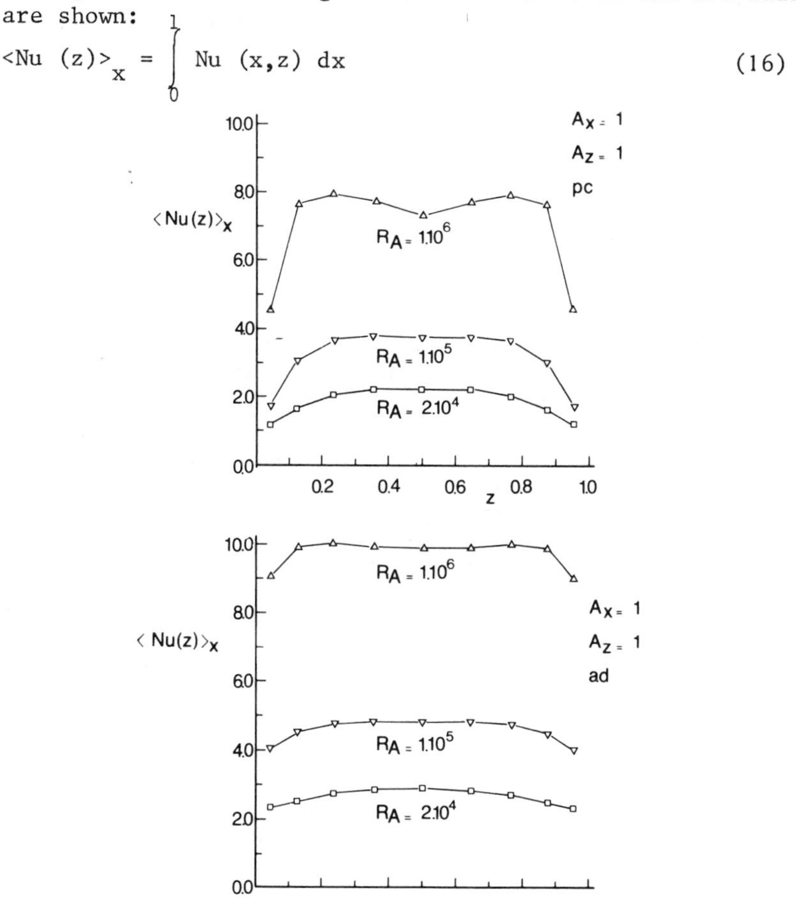

Fig. 4: The x-averaged Nusselt numbers as function of z for different R_A numbers.

The results are obtained with $\alpha_x = \alpha_y = \alpha_z = 1.5$. In comparison with the ad-case the influence of the sidewalls is larger in the pc-case, as can be seen in fig.4 on the area with almost equal heat transfer. For higher R_A number this area is increased but for $z = 0.5$ a minimum is obtained in the heat transfer. This minimum is more pronounced for increasing R_A number and is relatively greater for the pc-case than for the ad-case.

In the adiabatic case we varied the grid by changing the non-linearity parameter for the y-direction (α_y). For low R_A numbers the for the grid corrigated heat transfer was within one percent of the value obtained for $\alpha_y = 1.5$. For high R_A number ($5x10^5$, 10^6) the variation was greater ($\sim$10%). In table 1 the heat transfer relations are given for the different values of α_y and also the for the grid corrigated heat transfer relation. As can be seen in tabel 1 is the power b of the correlation for the case of '$\alpha_y = 0$' and also the regression coefficient r less than for the other grids. This is mainly due to the less accurate calculations for the high R_A numbers. For high R_A numbers more grid nodes are needed close to the walls. With a non-linear grid this is realised by varying the non-linearity parameter.

Finally in fig. 5 the isotherms and the calculated streamlines are shown at the midplane ($z = 0.5$) for three different R_A numbers in the pc-case. The development of the boundary layer flow with secondary flow is shown. The velocities calculated at the midplane varied from a maximum velocity in the x-direction of about 0.02 m/s for $R_A = 2 \times 10^4$ to a maximum of about 0.14 m/s for $R_A = 10^6$. Because of the z-symmetry property the velocity in the z-direction is zero at the midplane. So a two dimensional streamfunction could be defined at the midplane. As can be seen in figure 5 the flow is almost similarly to the calculated two dimensional cases for equal R_A number. This indicates that the influence of the velocity in the z-direction is not so large that the main flow pattern is changed. From our results we obtained that for instance for $R_A = 10^6$ in the boundary layer at the hot wall the velocity in the x-direction is up to 100 x larger than the w-velocity.

5. CONCLUSION

Results of three dimensional calculations of the heat transfer by natural convection in a cubical box are obtained. The results are in good agreement with the results of Mallinson and de Vahl Davis, but show a higher value of the power in the heat transfer relations, due to a less coars grid. The influence of the grid on the results is corrected by varying the non-linearity of the grid. The influence of the sidewalls on the heat transfer is larger for the pc-case than for the case with adiabatic sidewalls. In the depth of the enclosure the Nusselt number, averaged in the x-direction, is almost constant except near the sidewall regions and at

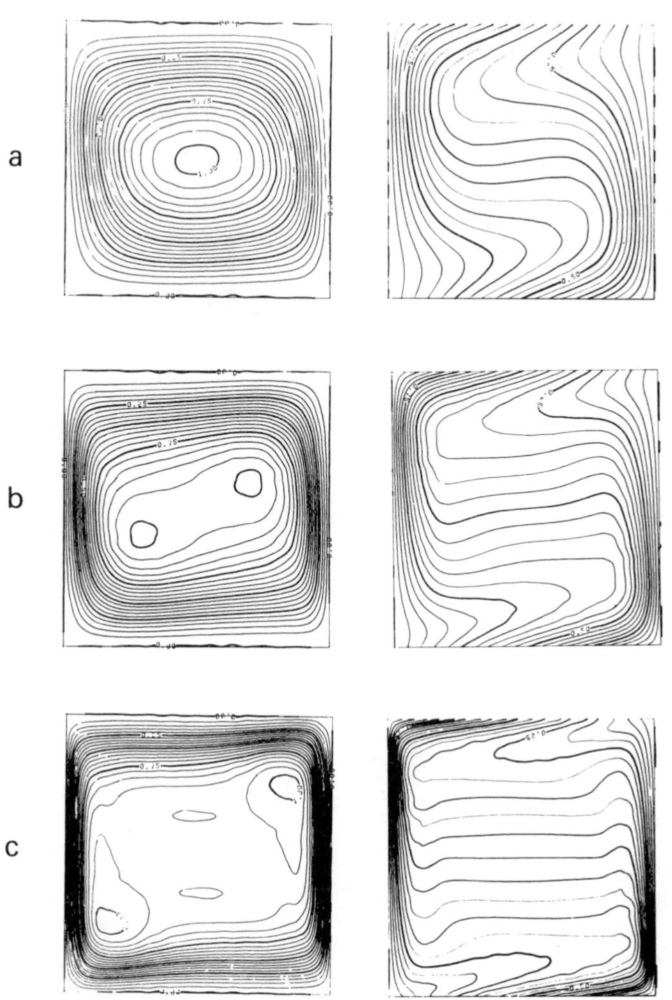

Fig. 5: Streamlines (left) and isotherms (right) calculated at the midplane ($z = 0.5$) for the case of perfectly conducting side walls.
The hot wall is on the right hand side.

a: $R_A = 2.10^4$

b: $R_A = 1.10^5$

c: $R_A = 1.10^6$

the midplane. The influence of the sidewalls decreases with increasing R_A number. However for high R_A numbers a local minimum is obtained for z = 0.5.

6. REFERENCES

[1] De Vahl Davis, G; Laminar Natural Convection In An Enclosed Rectangular Cavity; Int. J. Heat Mass Transfer; vol. 11, pp 1675-1693, 1960

[2] Elder, J.W; Numerical Experiments With Free Convection In A Vertical Slot; J. Fluid Mechanics; vol. 24, part 4, pp 823-843, 1965

[3] Schinkel, W.M.M; Natural Convection In Inclined Air Filled Enclosures; Ph.D. Thesis, Delft University of Technology, Dutch Efficiency Bureau, 1980.

[4] Mallinson, G.D; De Vahl Davis, G; Three Dimensional Natural Convection In A Box; A Numerical Study; J. Fluid Mechanics; vol. 83, part 1, pp 1-31, 1977.

[5] Ozoe, H; Yamamoto, K; Sayama, H; Churchill, S.W; Natural Convection Patterns In A Long Inclined Rectangular Box Heated From Below; Int. J. Heat Mass Transfer; vol. 20, pp 131-139, 1977.

[6] Catton, I; Effect Of Wall Conduction On The Stability Of A Fluid In A Rectangular Region Heated From Below; J. Heat Transfer, pp 446-452, november 1972.

[7] Gosman, A.D; Pun, W.M; Lecture Notes For Course Entitled "Calculation Of Recirculating Flows"; Imp. College Of Science And Technology, London, december 1973.

FINITE ELEMENTS WITH UPWINDING FOR THE DIFFUSION-CONVECTION EQUATION

P. LESAINT
Laboratoire d'analyse numérique et informatique. Faculté des Sciences, Route de Gray
25030 - BESANCON CEDEX

INTRODUCTION -

In this paper we describe several discretizations for convection diffusion problems. For the purpose of simplicity we shall only consider one spatial variable. Two main ideas will be used : the first one is now classical (ODEN, 1969, BONNEROT and JAMET 1974 and 1977) and consists in defining space-time finite elements ; the second one is to consider continuous or discontinuous finite element methods. First introduced for solving neutron transport equations (REED and HILL, 1973), this concept was then mathematically studied (LESAINT and RAVIART 1974, LESAINT, 1975) and applied to various problems such as Navier Stokes Equation (FORTIN, 1976), Stefan equation (JAMET, 1978).

The governing equation for the density ρ may be taken as

(1) $$\frac{\partial \rho}{\partial t} + u \frac{\partial \rho}{\partial x} - \varepsilon \frac{\partial^2 \rho}{\partial x^2} = 0 \qquad 0 < x < 1, \quad 0 < t < T$$

with initial and boundary conditions

(2) $$\rho(x,0) = \rho_o(x)$$

(3) $$- \varepsilon \frac{\partial \rho}{\partial x}(0,t) + u\rho\,(0,t) = 0$$

(4) $$\frac{\partial \rho}{\partial x}(1,t) = 0$$

where ε is small.

We first recall some features of already existing schemes for solving this problem. We then describe what we mean by continuous or discontinuous method on a very simple differential equation. There after we define continuous and discontinuous methods for the above equation and give some stability properties.

SOME ALREADY AVAILABLE SCHEMES -

Since ε is assumed to be small, we need a scheme which would be also valid for $\varepsilon=0$. A standard idea is to use upwinding techniques. But generally upwinding generates artificial dissipation, which should not dominate the physical one, i.e. $-\varepsilon\frac{\partial^2\rho}{\partial x^2}$. A typical finite difference approach is to write the following relations :

$$(5)\quad \frac{1}{\Delta t}(\rho_i^{n+1}-\rho_i^n)+\frac{u}{h}(\rho_i^n-\rho_{i-1}^n)-\frac{\varepsilon}{h^2}(\rho_{i+1}^n-2\rho_i^n+\rho_{i-1}^n)=0$$

It is known that for such a scheme the artificial dissipation is of order O(uh), so that we should choose h such that u h be small compared to ε.

A recent approach (PIRONNEAU, 1980) consists in solving the convective part of the equation, following the flow, i.e., the characteristics line defined by $dx=udt$. He writes :

$$(6)\quad \frac{1}{\Delta t}(\rho^{n+1}(x)-\rho^n(X^n(x))-\frac{\varepsilon}{h^2}\frac{\partial^2\rho^{n+1}}{\partial x^2}=0$$

where $X^n(x)$ is the solution at time $t_n=n\Delta t$ of the equation :

$$(7)\quad \frac{dX}{ds}=u(X,s)\quad ;\quad X(t_{n+1})=x$$

Then u^{n+1} is found by finite element technique. In this method we need to solve equation (7) backward in time for each node of the triangulation at time $t=t_{n+1}$. When using piecewise constants (resp. piecewise polynomial of degree 1) for approximating ρ^{n+1}, the accuracy is of order $h+\Delta t+\frac{h}{\Delta t}$ (resp. order $h+\Delta t+\frac{h^2}{\Delta t}$). In these two situations, the terms $\frac{h}{\Delta t}$ and $\frac{h^2}{\Delta t}$ are typically dissipative terms, arising from the interpolation procedure necessary to define $\rho^n(X^n(x))$ from $\rho^n(x)$. Our aim is to try to lower the effect of this dissipation.

CONTINUOUS AND DISCONTINUOUS METHODS -

Consider the following differential equation :

$$\frac{dv}{dt} + \sigma v = 0 \qquad 0 < t < T$$
$$v(0) = a$$

We define an approximate solution on the interval $[0,\Delta t]$ as follows : let P_1 denote the set of polynomials of degree ≤ 1.

Continuous method. Look for $v_h \in P_1$ on $[0,\Delta t]$ with :

$$\int_0^{\Delta t} \left(\frac{dv_h}{dt} + \sigma v_h\right) dt = 0$$
$$v_h(0) = a$$

we have

$$v_1 - v_o + \frac{\lambda}{2}(v_1 + v_o) = 0 \quad ,$$
$$v_o = a \qquad v_1 = v_h(\Delta t) \; , \qquad \lambda = \sigma \Delta t$$

Discontinuous method : Look for v_h P_1 on $]0,\Delta t[$ with

$$\int_0^{\Delta t} \left(\frac{dv_h}{dt} + \sigma v_h\right) w dt + (v_h(0^+) - a)\, w(0^+) = 0$$

for all $w \in P_1$ on $]0,\Delta t[$. We have

$$v_1^- = \frac{1 - \frac{\lambda}{3}}{1 + \frac{2}{3}\lambda + \frac{\lambda^2}{6}} a, \quad v_1^- = v_h(\Delta t - 0)$$

$$v_o^+ = \frac{1 + \frac{2}{3}\lambda}{1 + \frac{2}{3}\lambda + \frac{\lambda^2}{6}} a, \quad v_o^+ = v_h(0^+)$$

For this example, the continuous method is a Crank Nicolson type scheme and is second order accurate. The discontinuous méthod is third order accurate for the definition of v_1^- and has better property of stability when looking for the solution on a long interval $[0,T]$.

DEFINITION OF THE SCHEMES -

We let $\mathcal{D}_n = |0,1| \times |t_n, t_{n+1}|$, and $x_i = ih$.
Consider the space time finite elements $K_{i,n}$ on Figure 1.

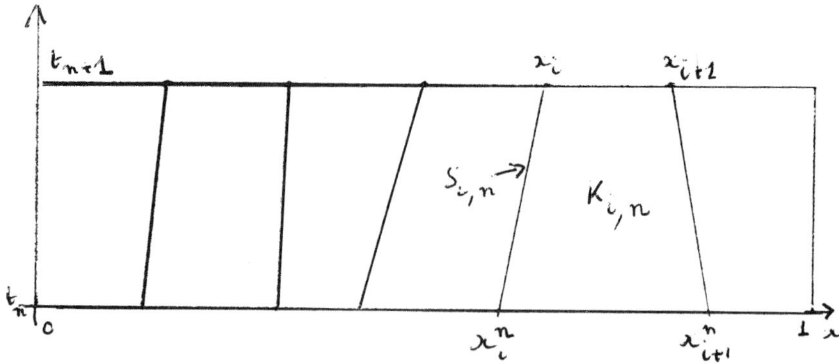

Figure 1.

The slope of the side $S_{i,n}$ is to be chosen later. We let $d_i = x_i - x_i^n$. We assume that the approximate solution is known at the time $t_n - 0$. Continuous method. We let $V_{k,l}$ be the space of continuous functions on $\mathcal{D}_n$ whose restriction to each element K is defined isoparametrically, with polynomials of degree k (resp. l) for the variable x (resp. the variable t). We look for $\rho_h \in V_{k,l}$ with :

$$\int_{\mathcal{D}_n} \left(\left(\frac{\partial \rho_h}{\partial t} + u \frac{\partial \rho_h}{\partial x}\right) w + \varepsilon \frac{\partial \rho_h}{\partial x} \frac{\partial w}{\partial x}\right) dx\, dt = 0$$

for all $w \in V_{k,l-1}$,

$$\rho_h(t_n + 0) = \rho_h(t_n - 0)$$

For $k = l = 1$, we get a generalization of a Crank Nicolson scheme, which is stable under the (weak) condition :

$$d_{i+1} - d_i \leqslant C\, \Delta t\, h \quad \text{for all } i.$$

As an example if the sides $S_{i,n}$ are approximations of the characteristic direction starting back from x_i we get roughly the condition that $\left|\frac{\partial u}{\partial x}\right|$ should be bounded.

Semi discontinuous methods. We look for $\rho_h \in V_{k,l}$, such that

$$\int_{\mathcal{D}_n} \left(\left(\frac{\partial \rho_h}{\partial t} + u \frac{\partial \rho_h}{\partial x}\right) w + \varepsilon \frac{\partial \rho_h}{\partial x} \frac{\partial w}{\partial x}\right) dx\, dt + \int_0^1 (\rho_h(t_n+0) - \rho_h(t_n-0))\, w\, (t_n+0)\, dx = 0$$

for all $w \in V_{k,l}$.

This scheme is always stable.

For k=1, l=0 we get a generalization of the purely implicit standard scheme for the heat equation. Moreover if we choose for the sides $S_{i,n}$ approximations of the characteristic direction starting back frome x_i, we get a scheme similar to scheme (6). With this new approach, we have now the possibility of playing on the scalars d_i to minimize the artificial dissipation.

Discontinuous methods. We first write equation (1) as a first order system

$$\begin{aligned} &\frac{\partial \rho}{\partial t} + u\frac{\partial \rho}{\partial x} - \sqrt{\varepsilon}\,\frac{\partial v}{\partial x} = 0 \\ &-\sqrt{\varepsilon}\,\frac{\partial \rho}{\partial x} + v = 0 \end{aligned} \qquad \text{in } \Omega = \,]0,1[\, \times \,]0,T[$$

which can also be written as

$$A\,\emptyset = A_1 \frac{\partial}{\partial t}\emptyset + A_2 \frac{\partial}{\partial x}\emptyset + A_o\,\emptyset = 0$$

where $\emptyset = (\rho, v)$ and A_o, A_1, A_2 are 2 x 2 symmetric matrices. Initial and boundary values are taken into account by using two matrices B and M defined on $\partial\Omega$, by

$$B = A_1 n_t + A_2 n_x\,, \qquad \vec{n} = (n_x, n_t) \text{ being the outer normal}$$

on $\partial\Omega$. For example :

$$B(0,t) = \begin{bmatrix} -u & \sqrt{\varepsilon} \\ \sqrt{\varepsilon} & 0 \end{bmatrix}, \quad M(0,t) = \begin{bmatrix} u & -\sqrt{\varepsilon} \\ +\sqrt{\varepsilon} & 0 \end{bmatrix}$$

so that $(B-M)(0,t)\emptyset = 0$ is equivalent to equation (3). Now assume that (ρ_h, v_h) is known for $t = t_n - 0$, we look for $\emptyset_h = (\rho_h, v_h)$ piecewise polynomials on the elements K, but not necessarily continuous on $\mathcal{D}_n$, such that

$$\sum_k \int_K (A\,\emptyset_h, W_h)\,dx\,dt - \frac{1}{2}\int_{\partial k} \left((B_K - M_K)(\emptyset_h^{in} - \emptyset_h^{out}), W_h^{in} \right) ds = 0$$

for all W_h lying in the same space as $\emptyset_h$. The matrix B_K is defined on ∂K similarly to B. The matrix M_K is to be chosen (with $M_K + M_K^* \geqslant 0$). The function $\emptyset_h^{in}$ and $\emptyset_h^{out}$ denote respectively the valeurs of $\emptyset_h$ along ∂K, inside or outside K.

Such methods are shown to be always stable and rather accurate, but they are more expensive. By choosing properly the matrices M_K and the slopes of the interelement boundaries $S_{i,n}$, the artificial dissipation can be lowered.

CONCLUSION -

Those schemes are defined for a one dimensional problem. In a next work, we will try to extend those definitions to two dimen sional situations. We could also try to use flux corrected tech nique (BORIS and BOOK, 1973) for solving the convective part of the equations.

REFERENCES -

[1] BARDOS R, BERCOVIER M, PIRONNEAU O. -
The vortex method with finite elements. Rapport INRIA n° 15, B.P. 105, 78150 LE CHESNAY - FRANCE.

[2] BONNEROT R. and JAMET P. -
A second order finite element method for the one dimensional Stefan problem. Int. Journ. Numer. Meth. Eng. 8, 811-820,(1974).

[3] BORIS J., BOOK D. -
Flux corrected transport. SHASTA, a fluid transport algorithm that works. J. of Comp. Phys. Vol. 11, 38-69, (1973).

[4] FORTIN M. -
Résolution numérique des équations de Navier-Stokes par des éléments finis de type mixte. INF LAB 7615 (1976).

[5] JAMET P. -
Galerkin-Type approximations which are discontinuous in time for parabolic equations in a variable domaine SIAM J. Numer. Anal. 15, n° 5, 912-928, (1978).

[6] LESAINT P. -
Sur la résolution des systèmes hyperboliques du premier ordre par des méthodes d'éléments finis. Thèse. Paris, (1975).

[7] LESAINT P. -
Solving the heat equation on non cylindrical domains. International Conference on Numerical Methods in Thermal Problems. Swansea, July 1979.

[8] LESAINT P. and RAVIART P.A. -
On a finite element method for solving the neutron transport equation, Mathematical aspects of finite elements in Partial Differential Equations (C. de Boor Ed) 89-123, Academic Press, New-York, (1974).

[9] PIRONNEAU O. -
On the transport diffusion algorithm and its applications to the Navier-Stokes equations. Rapport INRIA, (1980).

[10] ODEN J.T. -
A general theory of finite elements II. Applications Internat J. Numer. Methods Eng., 1, 247-259 (1969).

[11] REED W.H. and HILL T.R. -
Triangular mesh methods for the neutron transport equation LA UR 73479 LOS ALAMOS Laboratory (1973).

NUMERICAL SOLUTION FOR LAMINAR FORCED CONVECTION IN CHANNELS WITH COMBINED WALL SUCTION AND FLUID AXIAL CONDUCTION

ANTONIO CAMPO *

LOUIS C. CHOW **

1. SUMMARY

Hydrodynamic and thermal development for fluid flows in a parallel-plate channel with suction are examined. The use of two different inlet velocity profiles confirms that hydrodynamically developed solutions are nonexistent, and therefore local temperatures depend on the particular velocity profile. The energy equation, including the axial conduction term, along with the fluid flow equations are solved simultaneously by means of numerical techniques. At low Peclet numbers, it is shown that the upstream precooling due to axial conduction is significant.

2. NOMENCLATURE

D	-	half-width of the channel
ℓ	-	nondimensional length of the channel, L/D
L	-	length of the channel
Pe	-	Peclet number, $U_{be}D/\alpha$
Re_i	-	inlet Reynolds number, $U_{be}D/\nu$
Re_w	-	suction Reynolds number, $V_w D/\nu$
T	-	temperature
T_{be}	-	bulk temperature at inlet, $x = -2$
T_w	-	wall temperature
u	-	nondimensional horizontal velocity, U/U_{be}
u_b	-	nondimensional bulk horizontal velocity

* Universidad Simón Bolívar, Caracas, Venezuela

** Texas A&M University, College Station, Texas, U.S.A.

U - horizontal velocity
U_{be} - bulk horizontal velocity at inlet, $x = -2$
v - nondimensional vertical velocity, V/U_{be}
V - vertical velocity
V_w - suction velocity, equal to $-V$ at the wall
x - nondimensional horizontal coordinate, X/D
X - horizontal coordinate
y - nondimensional vertical coordinate, Y/D
Y - vertical coordinate

Greek symbols

α - thermal diffusivity
θ - nondimensional temperature $(T-T_w)/(T_{be}-T_w)$
θ_{bo} - nondimensional bulk temperature at $x = 0$
ν - kinematic viscosity
Ψ - nondimensional stream function, $u=\frac{\partial\Psi}{\partial y}$ and $v=-\frac{\partial\Psi}{\partial x}$
Ψ_i - nondimensional stream function at inlet, $x = -2$
Ω - nondimensional vorticity, $\frac{\partial v}{\partial x} - \frac{\partial u}{\partial y}$
Ω_i - nondimensional vorticity at inlet, $x = -2$

Subscripts

b - bulk
e,i - inlet, $x = -2$
o - at $x = 0$

3. INTRODUCTION

Fluid temperature calculations of internal forced convection have constituted an important and practical subject of investigation since the classical works of Graetz between 1883 and 1885 (1). Additionally, it may be added that this laminar forced convection problem under hydrodynamic developed conditions defines a fundamental case because it forms the basis for other related heat transfer problems as evidenced in the survey of (2).

In studying thermal forced convection through pipes, the axial conduction effect may become significant if the flow velocities are relatively low and/or the fluid's Prandtl number is small. This is because the axial conduction mechanism induces a streamwise temperature gradient which modifies the temperature fields and hence the heat transfer rates. This problem applied to parallel plate and cylindrical geometries with laminar flow has attracted considerable attention in recent years. The inclusion of axial conduction in the analysis is dictated by: 1) the magnitude of the Peclet number (the presence of axial conduction becomes important as the Peclet number decreases) and 2) the specified thermal boundary conditions at the heat exchange region. Papers related to this topic are examined by a number of investigators and the relevant literature has been brought together in (3-5).

Alternatively, a search of the world literature disclose that only limited consideration has been given to situations involving laminar heat transfer in ducts with wall suction. Most of the early works have been restricted to determining the fully developed velocity profiles with uniform mass suction at the walls. However, it has been demostrated in (6,7) that for strong suction, the velocity profile depends strongly on the inlet velocity shape and the fully developed solutions published previously are neither attained nor approached. On the other hand, the influence of variable suction was reported in (8) and it was concluded that it remains quite uniform along the wall. Hence, this is a valid assumption for related publications.

The present paper is concerned with a class of thermal interactions that has not yet been treated in the published literature. Attention will be focused on the interplay between forced convection flow, fluid axial conduction and wall mass suction. Specifically, the velocity development in a parallel-plate channel under the above mentioned conditions is examined. The inlet velocity profiles are assumed to be both uniform and parabolic. Another related contribution of the paper deals with the temperature development along the channel based on the associated velocity. The thermally interacting flows are analyzed here by solving the basic conservation equations. Numerical solutions were obtained by an elliptic finite-difference method discussed later.

4. GOVERNING EQUATIONS

The physical situation analyzed here is described as follows. The upstream portion of the channel ($-2 \leq x \leq 0$) has nonporous walls and is externally insulated. In the other portion ($0 \leq x < \ell$), porous walls having a uniform fluid suction are kept at a constant temperature. The governing equations in dimensionless terms for this conjugate problem are:

$$\frac{\partial(u\Omega)}{\partial x} + \frac{\partial(v\Omega)}{\partial y} = \frac{1}{Re_e}\left(\frac{\partial^2\Omega}{\partial x^2} + \frac{\partial^2\Omega}{\partial y^2}\right) \tag{1}$$

$$\frac{\partial^2\Psi}{\partial x^2} + \frac{\partial^2\Psi}{\partial y^2} = -\Omega \tag{2}$$

$$\frac{\partial(u\theta)}{\partial x} + \frac{\partial(v\theta)}{\partial y} = \frac{1}{Pe}\left(\frac{\partial^2\theta}{\partial x^2} + \frac{\partial^2\theta}{\partial y^2}\right) \tag{3}$$

$$u = \frac{\partial\Psi}{\partial y}, \quad v = -\frac{\partial\Psi}{\partial x} \tag{4}$$

where Ψ is the stream function, Ω is the vorticity and θ is the temperature.

It may be noted that the energy equation (3) contains both the radial and axial conduction terms, the latter being retained because of the focus on low Peclet number flows. An additional relation accounting for the overall conservation of mass is expressed by $U_{be}D = V_wL$.

The applicable boundary conditions for the upstream portion $(-2 \leq x \leq 0)$ are:

$$y = 0: \Psi = -1, \; \Omega = -\frac{\partial^2\Psi}{\partial y^2}, \; \frac{\partial\theta}{\partial y} = 0 \tag{5a}$$

$$y = 1: \Psi = 0, \; \Omega = 0, \; \frac{\partial\theta}{\partial y} = 0 \tag{5b}$$

$$x = -2: \Psi = \Psi_i(y), \; \Omega = \Omega_i(y), \; \theta = 1 \tag{5c}$$

where $\Psi_i(y) = y - 1$ uniform

$= -\frac{1}{2}y^3 + \frac{3}{2}y^2 - 1$ parabolic

and $\Omega_i(y) = 0$ uniform

$= 3y - 3$ parabolic

Likewise, boundary conditions for the downstream portion $(0 \leq x \leq \ell)$ are given by:

$$y = 0: \Psi = \frac{Re_w x}{Re_i} - 1, \; \Omega = -\frac{\partial^2\Psi}{\partial y^2}, \; \theta = 0 \tag{5d}$$

$$y = 1: \Psi = 0, \; \Omega = 0, \; \frac{\partial\theta}{\partial y} = 0 \tag{5e}$$

$$x = \ell: \Psi = 0, \; \Omega = 0, \; \frac{\partial\theta}{\partial x} = 0 \tag{5f}$$

This last condition indicates that at the channel exit $x = \ell$, all the fluid has been removed. In these sets of equations $\ell = \frac{L}{D} = \frac{U_{be}}{V_w}$ is a dimensionless channel length while Re_i and Re_w denote the axial and suction Reynolds number respectively.

5. SOLUTION METHODOLOGY

Solutions to the elliptic problem defined by the foregoing equations were obtained numerically by finite difference schemes. The effectiveness and limitations of the various schemes have not always been clear. Relying on a comparative study comprising the central difference scheme (CDS), the upwind difference scheme (UDS), the exponential difference scheme (EDS) and the Allen and Southwell scheme (ASS); it was finally decided to adopt the ASS because of the proved advantages in both accuracy and computation time (9).

Elliptic procedures have to be used because all the unknowns (i.e. stream function Ψ, vorticity Ω and temperature θ at the nodal points) must be solved for simultaneously. Both the stream function and temperature are usually known functions along the boundaries; however, special care is required in specifying the vorticity boundary condition. The conventional methods for handling the wall vorticities usually introduce errors that are dominant in the solution, and consequently the results are inadequate. It has been shown that the vorticity near a wall with mass suction varies exponentially at a distance away from the wall. Based on this, a new finite-difference representation for the wall vorticity has been employed whereby accurate numerical results can be obtained even with a relative coarse grid (10).

Most of the results reported in this paper are obtained with 24 and 16 grid spaces in the x and y directions respectively. Some guidance was obtained from limiting solutions for which results were available for comparison. By employing a finer grid for some critical cases, it was judged that the results are correct to within a few percent.

6. RESULTS AND DISCUSSION

The first objective of this paper is to examine the velocity development based on two different inlet velocity profiles. This is illustrated in Fig. 1 where u/u_b are plotted at various axial positions for the case of $Re_i = 300$ and $Re_w = 30$. For all cases tested, it was observed that the velocity downstream bears a strong resemblance to the corresponding inlet velocity profile corroborating the findings of (6).

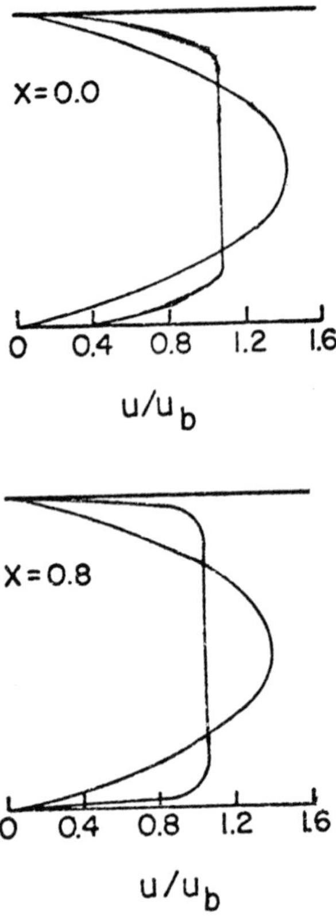

Fig. 1. Velocity development at two axial positions for Re_i = 300 and Re_w = 30.

The temperature profile at the inlet (x = -2) for all computer runs is assumed to be uniform across the parallel-plate channel. However, the temperature development downstream is strongly dependent on the inlet velocity and this relation is more significant at low Peclet numbers. In Fig. 2, the temperature profiles at x = 0 and x = 5 are drawn for Pe = 30, Re_i = 300 and Re_w = 30. At x = 0, the temperature gradient near the wall is largest for the uniform inlet velocity profile and smallest for the parabolic inlet velocity. This is because the velocity at the same location near the wall is largest for the uniform case, resulting in the largest heat convection rate. However, the opposite order for the temperature gradients occurs at locations further downstream, as evidenced in Fig. 2 for x = 5.

For small Peclet number flows through channels bounded by impermeable walls, the bulk temperature at x = 0 is lowered due to the upstream penetration of energy through the axial

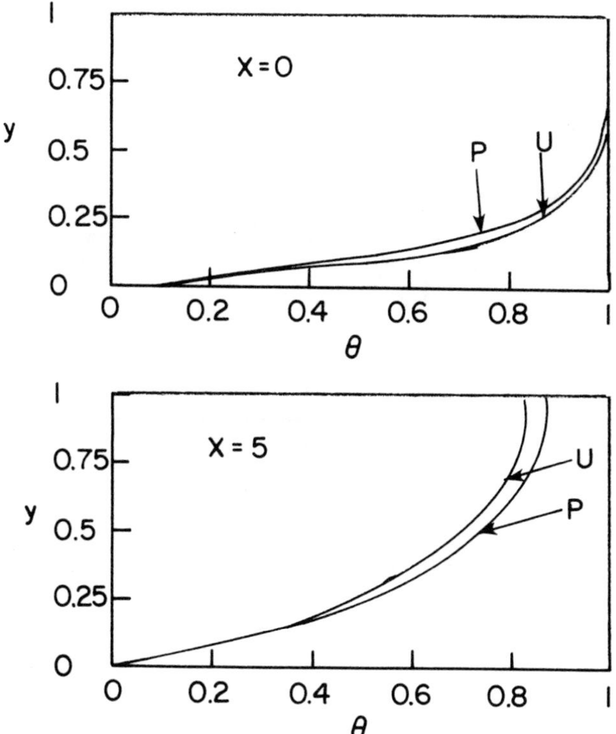

Fig. 2. Temperature development at x = 0 and x = 5 for $Re_i = 300$, $Re_w = 30$ and Pe = 30.

conduction term (3, 4). The same statement is true when there exists mass suction at the walls downstream of the origin. However, wall suction has a small effect on the bulk temperature at x = 0.

In Fig. 3 the upstream penetration depth is plotted versus the Peclet number. This depth is the nondimensional distance upstream of the origin (the location where mass suction begins) at which the bulk temperature θ_{bo} is equal to 0.99. From physical grounds, it is expected that the upstream penetration depth will depend only on the Peclet number and the inlet velocity profile. However, since the grid size chosen in the axial direction is not very fine, it is difficult to obtain this quantity accurately. Therefore, Fig. 3 represents the average values of the penetration depth without regard to the inlet velocity profile. A least squares fit of these data points yields the following correlation equation:

$$x_p = 3.71 \text{ Pe}^{-0.66} \qquad (6)$$

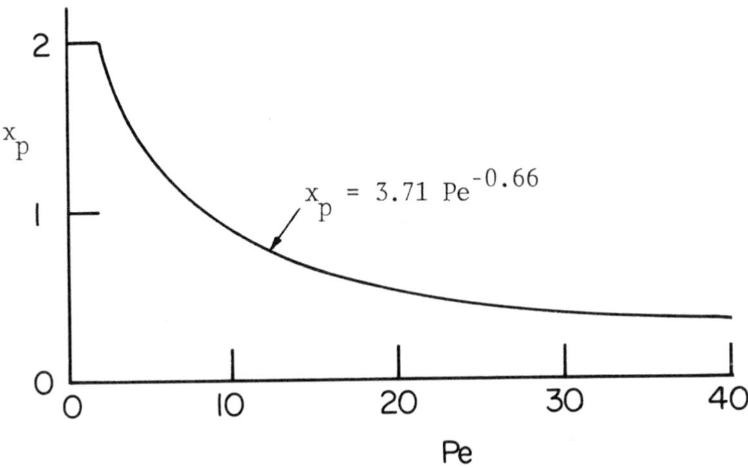

Fig. 3. Upstream penetration depth x_p versus Peclet number Pe.

7. REFERENCES

(1) GRAETZ, L. - Uber die warmeleitungsfahigkeit von flussing keiten. Annalen der Physik Chem., Vol. 18, pp. 79-84, 1883 and Vol. 25, pp. 337-357, 1885.

(2) SHAH, R.K. and LONDON, A.L. - Laminar Flow Forced Convection in Ducts, Academic Press, New York, 1978.

(3) HENNECKE, D.K. - Heat transfer by Hagen - Poiseuille flow in the thermal development region with axial conduction. Wärme-und Stoffübertragung, Vol. 1, pp. 177-184, 1968.

(4) CAMPO, A. and AUGUSTE, J.C. - Axial conduction in laminar pipe flows with nonlinear wall heat fluxes. Int. J. Heat Mass Transfer, Vol. 21, pp. 1597-1607, 1978.

(5) VICK, B., OZISIK, M.N. and BAYAZITOGLU, Y. - A method of analysis of Peclet number thermal entry region problems with axial conduction. Letters Heat and Mass Transfer, Vol. 7, pp. 235-248, 1980.

(6) RAITHBY, G.D. and KNUDSEN, D.C. - Hydrodynamic development in a duct with suction and blowing. J. Appl. Mech., Vol. 41, pp. 892-902, 1974.

(7) GUPTA, B.K. and LEVY, E.K. - Symmetric laminar channel flow with wall suction. J. Fluids Engng., Vol. 98, pp. 469-475, 1976.

(8) ARAUJO, P.M.S. and STUCKENBRUCK, S. - Laminar heat transfer in porous ducts with variable suction, ASME Paper N° 78-WA/HT-41.

(9) CHOW, L.C. and TIEN, C.L. - An examination of four differencing schemes for some elliptic-type convection equations. Numerical Heat Transfer, Vol. 1, pp. 87-100, 1978.

(10) CHOW, L.C., CHEUNG, Y.K. and TIEN, C.L. - A new finite-difference representation for the vorticity at a wall with suction. Numerical Heat Transfer, Vol. 1, pp. 417-423, 1978.

SECTION 9
NUMERICAL CONCEPTS
AND
MATHEMATICAL CONCEPTS

THE SECOND-ORDER PARTICLE-IN-CELL (PIC) COMPUTATIONAL METHOD IN THE ONE-DIMENSIONAL VARIABLE EULERIAN MESH SYSTEM

J. J. Pyun

Theoretical Applications Division
Code Development Group
Los Alamos National Laboratory
Los Alamos, New Mexico USA

As part of an effort to incorporate the variable Eulerian mesh into the second-order PIC computational method, a truncation error analysis was performed to calculate the second-order error terms for the variable Eulerian mesh system. The results show that the maximum mesh size increment/decrement is limited to be $\alpha(\Delta r_i)^2$ where Δr_i is a non-dimensional mesh size of the ith cell, and α is a constant of order one.

The numerical solutions of Burgers' equation by the second-order PIC method in the variable Eulerian mesh system were compared with its exact solution. It was found that the second-order accuracy in the PIC method was maintained under the above condition. Additional problems were analyzed using the second-order PIC methods in both variable and uniform Eulerian mesh systems. The results indicate that the second-order PIC method in the variable Eulerian mesh system can provide substantial computational time saving with no loss in accuracy.

1. INTRODUCTION

The purpose of this paper is (1) to determine the maximum mesh size change in the variable Eulerian mesh system in order to maintain the second-order accuracy in the second-order PIC computational method,[1] and (2) to verify the above result by comparing the numerical solution obtained by the second-order PIC method with the known exact solution.

It is well known that the truncation errors[2] in a finite difference equation increase if the mesh size changes rapidly. For example, consider a simple central difference equation for $\frac{du}{dr}$ at a point of $r_{i-\frac{1}{2}}$ as shown in Fig. 1. Expanding a function u in a Taylor series forward and backward from $r = r_{i-\frac{1}{2}}$ gives:

$$u_i \equiv u(r_{i-\frac{1}{2}} + \frac{\Delta r_i}{2}) = u_{i-\frac{1}{2}} + \frac{\Delta r_i}{2}\left(\frac{du}{dr}\right)_{i-\frac{1}{2}} + \frac{1}{2}\left(\frac{d^2u}{dr^2}\right)_{i-\frac{1}{2}}\left(\frac{\Delta r_i}{2}\right)^2 + 0(\Delta r_i^3) \quad (1)$$

$$u_{i-1} \equiv u(r_{i-\frac{1}{2}} - \frac{\Delta r_{i-1}}{2}) = u_{i-\frac{1}{2}} - (\frac{du}{dr})_{i-\frac{1}{2}} \frac{\Delta r_{i-1}}{2} + \frac{1}{2}(\frac{d^2u}{dr^2})_{i-\frac{1}{2}} (\frac{\Delta r_{i-1}}{2})^2 + 0(\Delta r_{i-1}^3) \quad . \tag{2}$$

The expression for $(\frac{du}{dr})_{i-\frac{1}{2}}$ is obtained by substracting Eq. (2) from Eq. (1) and dividing the resulting equation by $(\Delta r_i + \Delta r_{i-1})/2$.

$$\frac{du}{dr}_{i-\frac{1}{2}} = \frac{u_i - u_{i-1}}{0.5(\Delta r_i + \Delta r_{i-1})} + \frac{1}{4}(\frac{d^2u}{dr^2})_{i-\frac{1}{2}} (\Delta r_i - \Delta r_{i-1}) + 0(\Delta r^2) \tag{3}$$

where by $0(\Delta r^2)$ we mean the largest of $0[2\Delta r_i^3/(\Delta r_i + \Delta r_{i-1})]$, or $0[2\Delta r_{i-1}^3/(\Delta r_i - \Delta r_{i-1})]$. The finite difference Eq. (3) maintains the second-order accuracy if the value of $(\Delta r_i - \Delta r_{i-1})$ is an order of $\sim(\Delta r_i)^2$. However, the accuracy of Eq. (3) deteriates to first order if the mesh size change is greater or equal to $\sim\Delta r_i$. However, if the mesh size changes very slowly such that the fine mesh is used in the region of great interest and the coarse mesh is employed in the reminaing regions, a marked improvement in the overall computational time saving could be achieved without the expense of increasing the density of the mesh system everywhere.

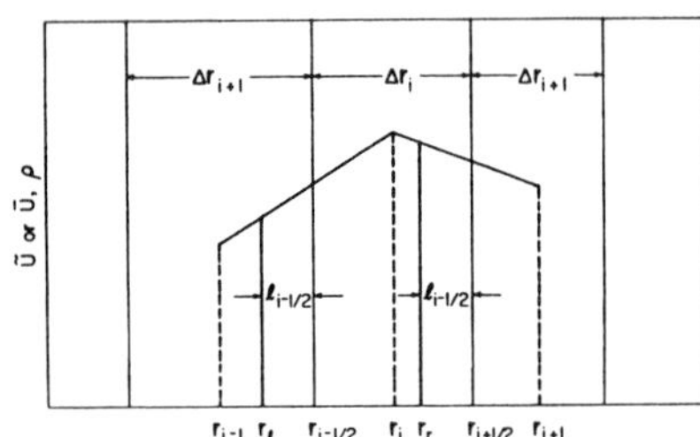

Fig. 1. Change in mesh spacing.

In Section 2 of this paper, the second-order PIC computational method is briefly reviewed. Truncation errors of the second-order PIC computational method are calculated in section 3, and the maximum mesh size change in the variable mesh system is determined based on the second-order truncation error terms in order to maintain the second-order accuracy in the second-order PIC computational method. In section 4 several sample calculations including a simple plane shock and a rarefaction wave are performed, and their numerical solutions by the second-order PIC method are compared with their known exact solutions.

2. REVIEW OF THE SECOND-ORDER PIC COMPUTATIONAL METHOD

Since the detailed descriptions of and applications of the second-order PIC computational method as well as its comparison with the first-order PIC computations method[7,8] is given in reference 1, only its brief summary will be given here.

Consider a set of one-dimensional (1D) differential equations for compressible fluid flow in a cylindrical geometry. Assuming only one material to be present in a 1D clyinder, a system of equations that we want to solve subject to initial and boundary conditions are:

$$\rho_t + u\rho' = -\rho(u' + \frac{u}{r}) \qquad \text{(mass)} \qquad (4)$$

$$u_t + uu' = -\frac{P'}{\rho} \qquad \text{(momentum)} \qquad (5)$$

$$I_t + uI' = -\frac{P}{\rho}(u' + \frac{u}{r}) \qquad \text{(energy)} \qquad (6)$$

$$P = F(\rho, I) \qquad \text{(equation of state)} \qquad (7)$$

where a subscript, t, denotes a partial derivative with respect to time and the apostrophe means a partial derivative with respective to r hereafter.

In the second-order PIC method the cylinder is divided into a cell of length Δr_i such that a total summation of Δr_i is equal to the radius of the cylinder. The initial velocity U_i, mass M_i and internal energy I_i are assigned to the ith cell such that they represent the initial conditions. Within each cell, the cell mass M_i is divided randomly among a number of "particles" that are distributed throughout the cell so that they approximate the density profile.

The second-order PIC method calculates the quantities at time $(n+1)\Delta t$ in terms of those at time $n\delta t$ (i.e., an explicit time advancement procedure) where n is a number of timesteps and Δt is a timestep. Within one timestep, the new quantities are computed in two phases:

<u>Phase 1: Lagrangian Phase</u>

(1) Calculate the cell pressure $P^{n+\frac{1}{2}}$ at $(n+\frac{1}{2})\Delta t$ timestep from the equation of state

$$P_i^{n+\frac{1}{2}} = F(\rho_i^{n+\frac{1}{2}}, I_i^{n+\frac{1}{2}}) \quad . \qquad (8)$$

Hereafter, a superscript n denotes a timestep. A subscript i indicates a cell number, and all the variables are cell-centered quantities. Half timestep quantities inside of parenthesis in Eq. (8) are calculated hereafter as below:

$$f_i^{n+\frac{1}{2}} = [f + 0.5\Delta t(f_t + uf')]_i^n \qquad (9)$$

where a quantity f denotes any state variable.

(2) The Lagrangian quantities, $\tilde{u}_i$ and $\tilde{I}_i$, at timestep $(n+1)\Delta t$ are calculated based on $(n+\frac{1}{2})\Delta t$ timestep quantities:

$$\tilde{u}_i = u_i^n - \Delta t\left(\frac{P'}{\rho}\right)_i^{n+\frac{1}{2}} \tag{10}$$

$$\tilde{I}_i = I_i^n - \Delta t\left[\frac{P}{\rho}\left(u' + \frac{u}{r}\right)\right]_i^{n+\frac{1}{2}} \tag{11}$$

(3) Temporarily assign the momentum $M_i\tilde{u}_i$ and total internal energy $M_i\tilde{I}_i$ to cell i.

Phase 2: Particle Transport and Remapping

(1) Move the particles with velocity $\bar{u}(r_r)$ where $\bar{u}(r_r)$ is obtained by linear interpolation of $\bar{u}_i$ $(=0.5u_i^n + 0.5\tilde{u}_i)$ to the particle position r_r. Along with these particles, portions of momentum and internal energy are transported across the cell boundary.

(2) The new total cell mass, momentum, and internal energy are calculated. Subsequently, the new velocity and total specific internal energy are determined.

3. TRUNCATION ERROR ANALYSIS OF THE SECOND-ORDER PIC COMPUTATIONAL METHOD IN THE VARIABLE EULERIAN MESH SYSTEM

Let us consider the velocity and density profiles in Fig. 1 along with a set of 1D differential equations, Eq. (4) through Eq. (7). For convenience, the velocity is assumed to be positive. Let $\ell_{i-\frac{1}{2}}$ and $\ell_{i+\frac{1}{2}}$ be the maximum distances between cell left and right boundaries, and the particle positions from which the particles reach the boundary in timestep Δt as shown in Fig. 1, and the particle velocity leaving a cell (i) is calculated by linear interpolation,

$$\bar{u}(r_r) = [\bar{u} - \ell(\bar{u}_r)]_{i+\frac{1}{2}} \tag{12}$$

where $$\ell_{i+\frac{1}{2}} = \bar{u}(r_r)\Delta t \tag{13}$$

$$\bar{u}_{i+\frac{1}{2}} = \frac{\bar{u}_{i+1}\,\Delta r_i + \bar{u}_i\,\Delta r_{i+1}}{\Delta r_i + \Delta r_{i+1}} \tag{14}$$

$$(\bar{u}_r)_{i+\frac{1}{2}} = \frac{\bar{u}_{i+1} - \bar{u}_i}{0.5(\Delta r_{i+1} + \Delta r_i)} \quad . \tag{15}$$

Eliminating $\bar{u}(r_r)$ from Eqs. (12) and (13) gives

$$\ell_{i+\frac{1}{2}} = \frac{\bar{u}_{i+\frac{1}{2}}\,\Delta t}{1 + \Delta t(ur)_{i+\frac{1}{2}}} \quad . \tag{16}$$

Expanding Eq. (16) by using a Taylor series gives:

$$\ell_{i+\frac{1}{2}} = (\bar{u}\Delta t - \bar{u}\bar{u}_r\Delta t^2)_{i+\frac{1}{2}} + O(\Delta t^3) \quad . \tag{17}$$

Here, we neglected the higher than the second-order terms. A similar procedure is used to derive the expression for $\ell_{i-\frac{1}{2}}$

and the result is

$$\ell_{i-\frac{1}{2}} = (\bar{u}\Delta t - \overline{uu_r}\Delta t^2)_{i-\frac{1}{2}} + O(\Delta t^3) \quad . \tag{18}$$

Let $\delta m_{i+\frac{1}{2}}$ be a mass transported from a cell (i) to (i+1), then we have,

$$\delta m_{i+\frac{1}{2}} = 2\pi \int_{r_{i+\frac{1}{2}} - \ell_{i+\frac{1}{2}}}^{r_{i+\frac{1}{2}}} \rho(r) r dr \tag{19}$$

Assuming that the density, $\rho(r)$, varies linearly, we have

$$\delta m_{i+\frac{1}{2}} = 2\pi [r\rho\ell - (r\rho_r + \rho) \frac{\ell^2}{2} + \frac{1}{3} \rho_r \ell^3]_{i+\frac{1}{2}} \tag{20}$$

$$\text{where } (\rho_r)_{i+\frac{1}{2}} = \frac{\rho_{i+1} - \rho_i}{0.5(\Delta r_{i+1} - \Delta r_1)} \quad . \tag{21}$$

Let $\delta m_{i-\frac{1}{2}}$ be a mass transported from a cell (i-1) to (i), then we could calculate $\delta m_{i-\frac{1}{2}}$ simularily

$$\delta m_{i-\frac{1}{2}} = 2\pi [r\rho\ell - (r\rho_r + \rho)\frac{\ell^2}{2} + \frac{1}{3}\rho_r \ell^3]_{i-\frac{1}{2}} \quad . \tag{22}$$

Then, the net change of a mass in a cell (i), ΔM_i, is calculated as,

$$\Delta M_i = \delta m_{i-\frac{1}{2}} - \delta m_{i+\frac{1}{2}} \quad . \tag{23}$$

Now the density change in cell (i) can be calculated as below,

$$\Delta\rho_i = \rho_i^{n+1} - \rho_i^n = \frac{\Delta M_i}{2\pi r_i \Delta r_i} \quad . \tag{24}$$

From a Taylor series expansion of $\rho_i(t + \Delta t)$, we have

$$(\rho_t)_i^n = \frac{\rho_i^{n+1} - \rho_i^n}{\Delta t} - \frac{\Delta t}{2}(\rho_{tt})_i^n - \frac{\Delta t^2}{6}(\rho_{ttt})_i^n + O(\Delta^3) \quad . \tag{25}$$

Here, we truncated the higher than second-order terms. Inserting Eq. (24) into Eq. (25) and, after complex algebraic manipulation, we have

$$\begin{aligned}(\rho_t)_i^n = &- [(\rho u)' + \frac{\rho u}{r} + \Delta t^2 A_1(r,t) + \Delta r_i^2 A_2(r,t) \\ &+ (\frac{\Delta r_{i+1} - \Delta r_i}{\Delta r_i})\Delta t A_3(r,t) - (\frac{\Delta r_i - \Delta r_{i-1}}{\Delta r_i})\Delta t A_3(r,t) \\ &+ \frac{(\Delta r_i^2 + \Delta r_i \Delta r_{i+1} + \Delta r_{i+1}^2)}{\Delta r_i} \Delta t A_4(r,t) \\ &- \frac{(\Delta r_i^2 + \Delta r_i \Delta r_{i-1} + \Delta r_{i-1}^2)}{\Delta r_i} \Delta t A_4(r,t) + O(\Delta^3)]_i^n \quad .\end{aligned} \tag{26}$$

Here, we again truncated the terms higher than the second-order terms. For simplicity, the detailed expressions for the coefficients A_1 through A_4 are not given. However, they are given in reference 9.

Next consider the momentum equation. Let $\delta(m\tilde{u})_{i+\frac{1}{2}}$ be a momentum transported along with particles from a cell (i) to (i+1), then we have

$$\delta(m\tilde{u})_{i+\frac{1}{2}} = 2\pi \int_{r_{i+\frac{1}{2}} - \ell_{i+\frac{1}{2}}}^{r_{i+\frac{1}{2}}} \rho\tilde{u}r dr \quad . \tag{27}$$

Again, assuming that the density, $\rho(r)$, and velocity, $\tilde{u}(r)$, vary linearly, we have,

$$\delta(m\tilde{u})_{i+\frac{1}{2}} = \delta m_{i+\frac{1}{2}}(\tilde{u} - \frac{\Delta t}{2}|\tilde{u}|\tilde{u}_r)_{i+\frac{1}{2}} \quad . \tag{28}$$

Let $\delta(m\tilde{u})_{i-\frac{1}{2}}$ be a momentum transported from a cell (i-1) to (i), then we could calculate $\delta(m\tilde{u})_{i-\frac{1}{2}}$ in a similar way

$$\delta(m\tilde{u})_{i-\frac{1}{2}} = \delta m_{i-\frac{1}{2}}(\tilde{u} - \frac{\Delta t}{2}|\tilde{u}|\tilde{u}_r)_{i-\frac{1}{2}} \quad . \tag{29}$$

The net change of a momentum in a cell (i), $\Delta(M\tilde{u})_i$, is calculated as

$$\Delta(M\tilde{u})_i = \delta(m\tilde{u})_{i-\frac{1}{2}} - \delta(m\tilde{u})_{i+\frac{1}{2}} \quad . \tag{30}$$

Now the velocity at timstep $(n+1)\Delta t$ can be calculated in terms of a net momentum change,

$$u_i^{n+1} = \frac{M_i\tilde{u}_i + \Delta(M\tilde{u})_i}{M_i + \Delta(M)_i} = \frac{\tilde{u}_i + \Delta(M\tilde{u})_i/M_i}{1 + \Delta(M)_i/M_i} \quad . \tag{31}$$

Expanding Eq. (31) in terms of a Taylor series results in

$$u_i^{n+1} = [\tilde{u}_i + \frac{\Delta(M\tilde{u})_i}{M_i}](1 - \varepsilon + \varepsilon^2 - \varepsilon^3) + O(\Delta^4) \tag{32}$$

where $\varepsilon = \Delta(M)_i/M_i$.

The truncation error terms higher than the third-order terms are neglected. From a Taylor series expansion of $u_i(t + \Delta t)$, we have

$$(u_t)_i^n = \frac{u_i^{n+1} - u_i^n}{\Delta t} - \frac{\Delta t}{2}(u_{tt})_i^n - \frac{\Delta t^2}{6}(u_{ttt})_i^n + O(\Delta^3) \quad . \tag{33}$$

Here, the truncation error terms higher than the second-order terms are neglected. Inserting Eq. (32) into Eq. (33) and, after complex algebraic manupulation, we have

$$(u_t)_i^n = [-uu' - \frac{P'}{\rho} + \Delta t^2 B_1(r,t) + \Delta r_i^2 B_2(r,t)$$
$$+ (\frac{\Delta r_{i+1} - \Delta r_i}{\Delta r_i}) \Delta t B_3(r,t) - (\frac{\Delta r_i - r_{i-1}}{\Delta r_{\frac{i}{2}}}) \Delta t B_3(r,t)$$
$$+ \frac{(\Delta r_i^2 + \Delta r_i \Delta r_{i+1}^2 + \Delta r_{i+1}^2)}{\Delta r_i} \Delta t B_4(r,t)$$
$$- \frac{(\Delta r_i^2 + \Delta r_i \Delta r_{i-1} + \Delta r_{i-1}^2)}{} \Delta t B_4(r,t) + 0(\Delta^3)_i^n \quad . \tag{34}$$

Again, the truncation error terms higher than the second-order terms are neglected here. Again, the detailed expressions for the coefficients B_1 through B_4 are not given. However, they are given in reference 9.

A similar algebraic manipulation could be performed for the energy equation, beginning with Eq. (30) and replacing $\tilde{u}_i$ with $\tilde{I}_i$ and the results are

$$(I_t)_i^n = [-uI' - \frac{P}{\rho} (u' + \frac{u}{r}) + \Delta t^2 D_1(r,t) + \Delta r_i^2 D_2(r,t)$$
$$+ \frac{(\Delta r_{i+1} - \Delta r_i)}{\Delta r_i} \Delta t D_3(r,t) - \frac{(\Delta r_i - \Delta r_{i-1})}{\Delta r_i} \Delta t D_3(r,t)$$
$$+ \frac{(\Delta r_i^2 + \Delta r_i \Delta r_{i+1} + \Delta r_{i+1}^2)}{\Delta r_i} \Delta t D_4(r,t)$$
$$- \frac{(\Delta r_i^2 + \Delta r_i \Delta r_{i-1} + \Delta r_{i-1}^2)}{\Delta r_i} \Delta t D_4(r,t) + 0(\Delta^3)]_i^n \quad . \tag{35}$$

The detailed expressions for the coefficients D_1 through D_4 are given in reference 9.

Equations (26), (34) and (35) show that the first-order error terms are identically cancelled out except those terms containing $\Delta t(\Delta r_{i+1} - \Delta r_i)/\Delta r_i$ and $\Delta t(\Delta r_i - \Delta r_{i-1})/\Delta r_i$. In order to maintain the second-order accuracy for the PIC method in the 1D variable Eulerian mesh system, the maximum mesh size change is limited to be $\alpha(\Delta r_i)^2$ where α is a constant of order one.

4. SAMPLE CALCULATIONS AND CONCLUSIONS

To test the second-order PIC method in the variable Eulerian mesh system, three sample problems were analyzed. The first problem is Burgers' equation and the remaining problems are the plane shock and adiabatic rarefaction wave.

4.1 Burgers' Equation

Consider Burgers' equation[3] with the following initial and boundary conditions:

$$u_t - uu_x = \nu u_{xx} \tag{36}$$

$$u(x,0) = u_0 \sin\pi x/\ell \tag{37}$$

$$u(0,t) = u(\ell,t) = 0 \quad . \tag{38}$$

The exact solution for the above equation is known and given in reference 4.

A numerical solution for Burgers' equation was obtained by the second-order PIC method in the variable mesh. A truncation error (i.e., a difference between an exact solution and numerical solution) is calculated. Figure 2 shows a ratio of velocity truncation error for doubling a mesh size. A numerical solution by the second-order PIC method is second-order accurate in time and space. Therefore, the velocity truncation error is quadrupled as the mesh size is doubled as shown in Fig. 2. However, the velocity truncation error is not quadrupled as the timestep is doubled. This is because the truncation error terms are dominated by a space error term ($\sim \Delta x^2$) as a result of a convergence criterion, $\nu\Delta t/(\Delta x_i)^2 \leqslant 0.5$. Since it is difficult to double a mesh size at the edges, the ratio of velocity truncation error was not quadrupled.

4.2 Plane Shock Wave

A plane shock wave was analyzed by using the second-order PIC method in both uniform and variable mesh system. These numerical solutions along with their exact solutions[5] in an ideal gas are shown in Figs. 3 through 6. The shock front was handled using an artifical viscosity[6] and a value of its constant coefficient used is 0.5. Figure 3 shows a spatial profile of density at 14 μs. As shown in Fig. 3, the numerical solutions for a spatial density in both uniform and variable mesh system with α=1 agree very well with their exact solutions. The maximum compression ratio for this shock is 4.0 (i.e., a strong shock in an ideal gas). In addition, a computational time saving as a result of using a variable mesh was approximately 30% without loss in accuracy over a uniform mesh case. This computational time saving is due to a smaller number of meshes used in a variable mesh system.

Figures 4 and 5 compare the numerical solutions for a spatial pressure and internal energy in both uniform and variable mesh system with their exact solutions. The mesh size increment constant, α, used here is again 1.0. Again, these numerical solutions agree very well with their exact solutions.

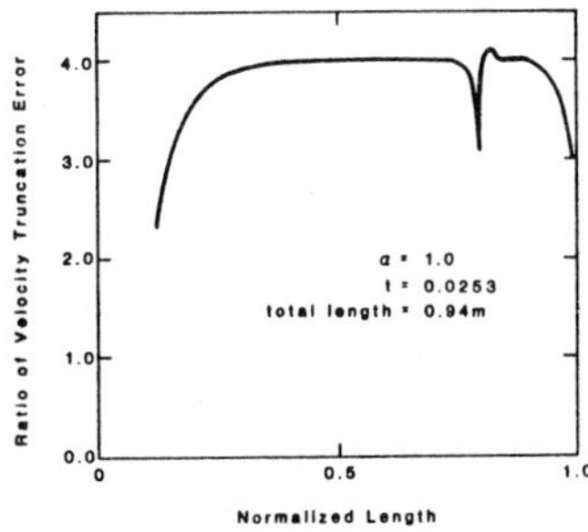

Fig. 2. Ratio of velocity truncation error for doubling mesh size.

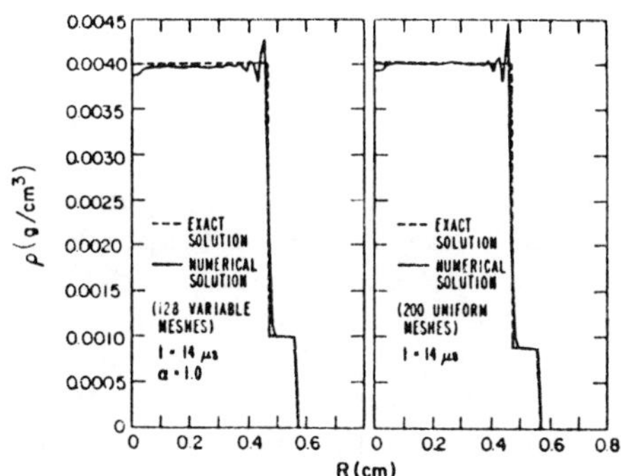

Fig. 3. Density vs R-Distance.

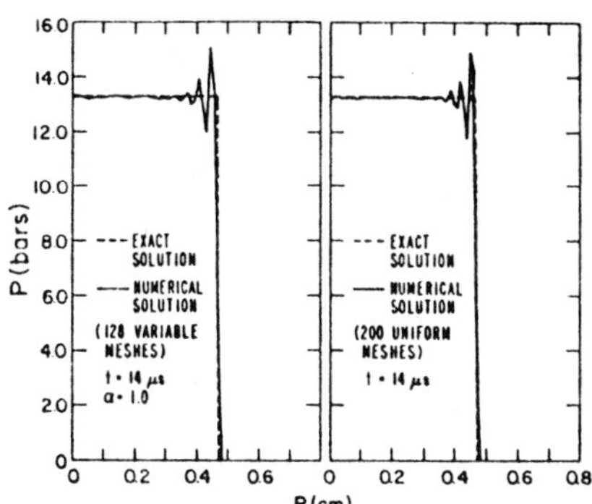

Fig. 4. Pressure vs R-Distance.

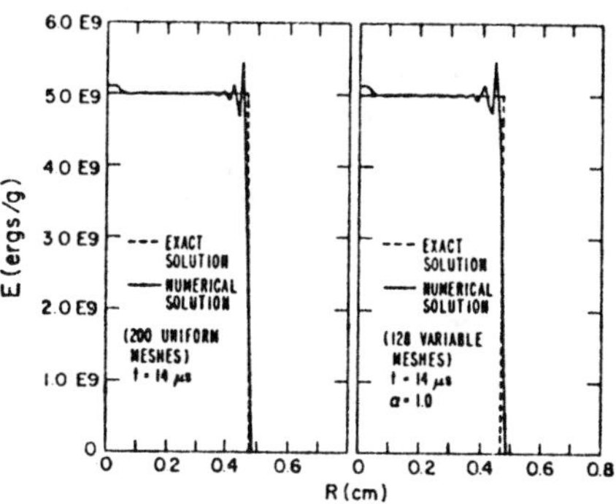

Fig. 5. Internal energy vs R-Distance.

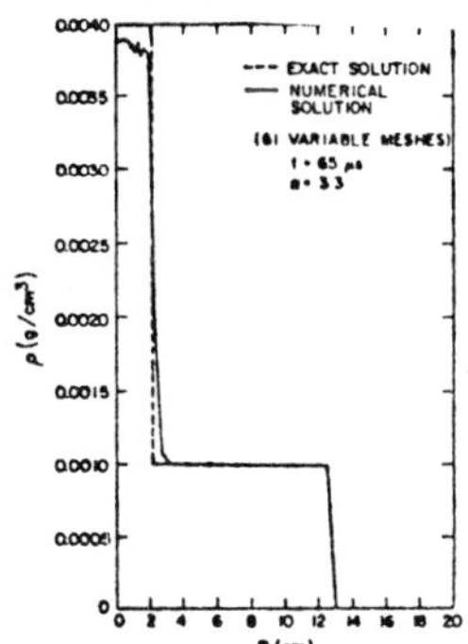

Fig. 6. Density vs R-Distance.

Figure 6 shows a comparison of a spatial density profile at 65 μs for a strong shock in an ideal gas. The variable mesh with incremental constant of 3.3 is used for this numerical solution. The maximum compression ratio if ~ 3.85 for the numerical solution whereas its exact solution is 4.0. This result indicates that the accuracy in the second-order deteriates as the mesh size rapidly increases. As shown in section 3, the maximum mesh size increase in a variable mesh system in order to maintain the second-order accuracy in the second-order PIC method is limited to be $\sim(\Delta r_i)^2$. Here we used the maximum mesh size increase of 3.3 $(\Delta r_i)^2$.

4.3 Plane Adiabatic Rarefaction Wave

A plane adiabatic rarefaction wave was analyzed using the second-order PIC method in both uniform and variable mesh system. These numerical solutions along with their exact solutions[5] in an ideal gas are shown in Figs. 7 through 9. Figures 7 and 8 show a comparison of numerical solutions for a spatial density profile at 15 μs in a uniform and variable mesh system with α=1.0 with their exact solutions. As shown in Figs. 7 and 8 the numerical solutions agree relatively well with their exact solutions. Deviation of numerical solutions from their exact solutions at the edges as shown in Figs. 7 and 8 is due to the vaccum boundary condition. In addition, a computational time saving for this problem was approximately 25% as a result of using a variable mesh over a uniform mesh case.

Figure 9 compares the numerical solution for a spatial density profile at 15 μs in a variable mesh with α=3.3 with its exact solution. The numerical solution at a vaccum boundary deteriates as shown in Fig. 9. However, a computational time saving as a result of using a variable mesh was approximately 50% over a uniform mesh case.

In conclusion, a substantial computational time saving could be achieved in the second-order PIC computational method without sacrifice in accuracy by using a variable mesh system as compared to one using a uniform mesh.

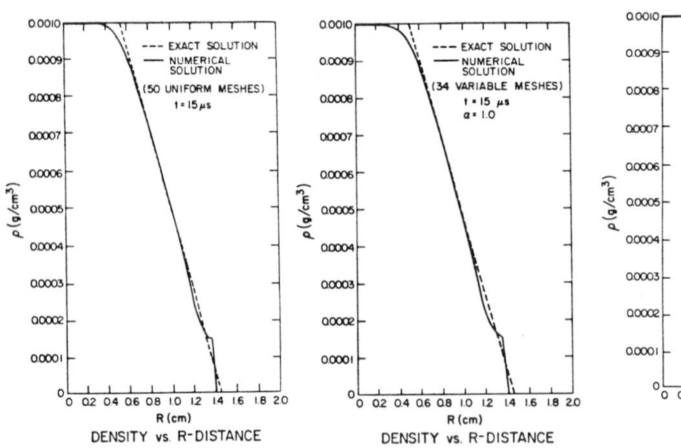

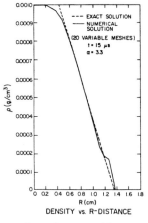

Fig. 7. Density vs R-Distance.

Fig. 8. Density vs R-Distance.

Fig. 9. Density vs R-Distance.

5. REFERENCES

1. Clark, R. A., The Second-Order Particle-In-Cell Computing Method, Accepted for publication in the Journal of Computational Physics, 1980.

2. Crowder, H. J. and Dalton, C., Errors in The Use of Non-Uniform Mesh Systems, J. Computational Physics, Vol. 7, pp.32-45, 1971.

3. Burgers, J., A Mathematical Model Illustrating the Theory of Turbulence, Advances in Applied Mechanics, Vol. 1, pp. 171-199, Academic Press, 1948.

4. Cole, J. D., On a Quasi-Linear Parabolic Equation Occurring in Aerodynamics, Quart. Appl. Math., Vol. 9, pp.225-236, 1951.

5. Harlow, F. H. and Amsden, A. A., Fluid Dynamics, Los Alamos National Laboratory document, LA-4700, 1971.

6. von Neumann, J. and Richtmeyer, R. D., A Method for the Numerical Calculation of Hydrodynamic Shocks, J. Appl. Physics, Vol. 21, pp. 232-237, 1949.

7. Evans, M. W. and Harlow, F. H., The Particle-In-Cell Method for Hydrodynamic Calculations, Los Alamos National Laboratory document, LA-2139, 1957.

8. Amsden, A. A., The Particle-In-Cell Method for the Calculation of the Dynamics of Compressible Fluids, Los Alamos National Laboratory document, LA-3466, 1966.

9. Pyun, J. J., The Second-Order Particle-In-Cell Computational Method in the 2D Variable Eulerian Mesh System, to be published as a Los Alamos National Laboratory document, 1981.

ACKNOWLEDGEMENT

The author is indebted to Dr. Robert A. Clark, Los Alamos National Laboratory, for his encouragement and invaluable guidance throughout the preparation of this paper. The theoretical aspects in particular were made possible through his assistance.

On the nature of the stationary point in finite element determination of open channel flows.

E. F. Toro[1] and M. J. O'Carroll[2]

Teesside Polytechnic, England

SUMMARY

Steady, ideal flows in open channels pose three coupled problems. The first is to determine the flow within a given region, the second is to determine the free surface boundary position and the third is to determine a critical flow rate. The second of these is non-linear and may have non-uniqueness or non-existence of solutions.

The governing variational principle covers both of the first two problems. Its second variation may become singular as critical flow is approached and this introduces numerical difficulties.

In this paper we analyse numerically the nature of the second variation for a full finite element discretisation. A qualitatively similar semidiscrete formulation gives strong analytical results. We conclude by classifying types of flows. In general those with supercritical outlet give definite variational minima and those which oscillate about subcritical (tranquil) levels become saddle-like for channel lengths beyond a transition length.

[1] Research Assistant, Department of Mathematics and Statistics
[2] Head of Department of Mathematics and Statistics

1 INTRODUCTION

The calculation of steady ideal open channel flows requires three things. Firstly the flow equations must be satisfied within the domain, secondly the free surface condition is to be met by correctly locating the boundary, and thirdly the critical or maximum steady discharge flow rate is to be achieved. Using a stream function formulation, we call these the ψ problem, the h problem and the Q problem respectively.

Several researchers have computed such flows by finite elements. Applications include flows past weirs, spillways and sluicegates [1]-[5]. All cases encounter a difficulty with the Q problem where the selection of critical discharge is based on hit-or-miss efforts such as proceeding to raise Q until iterative convergence breaks down. In the paper of Varoglu and Finn a direct approach to the Q problem is offered but is not valid. The singularity of the Hessian matrix relating to the variational formulation of the joint ψ,h problem occurs as the flow becomes critical and this corresponds to failure of (e.g.) Newton-Raphson iterations [6],[7].

In view of this it is of value to determine the definiteness or otherwise of the Hessian matrix. This paper presents a finite element calculation for the ψ,h problem together with an evaluation of the Hessian and tests for its definiteness. Algebraic expressions for the Hessian are given explicitly in terms of quadratic forms in nodal ψ values for a semi-regular family of grids fitting an arbitrary configuration. The usual functional takes a minimum with respect to ψ for fixed h as is well known but we are also concerned with its nature with respect to h. In the classical case of uniform flows of undetermined depth, the supercritical (rapid) solution is minimum-like whereas the subcritical (tranquil) solution is maximum-like [7]. Here we investigate more general flows and present the analysis of the Hessian.

2 FINITE ELEMENT FORMULATION

Take Cartesian co-ordinates (x,y) with y vertically upwards and origin at the stagnation level for two-dimensional steady ideal flow. It then turns out [7] that the ψ-problem and the h-problem are solved by making the functional

$$J = \iint (-gy + \tfrac{1}{2}(\nabla\psi)^2)\, dx\, dy, \tag{1}$$

taken over the flow domain, stationary with respect to both the volumetric stream function ψ and the free surface position.

A mesh as illustrated in Figure 1 approximates arbitrary bed and free surface shapes by polygonal curves. It consists of a set of vertical transversals at arbitrary positions

x_i in the x direction. The heights of these transversals from given bed height b_i to the unknown free surface are the unknown variables h_i. Each transversal is divided into n-1 equal intervals by n nodes including those on bed and surface. Nodal values ψ_{ij} are assigned to the nodes with $i=1(1)m$ advancing in the x direction and $j=1(1)n$ advancing in the y direction. The standard bilinear interpolation is then made for ψ within each quadrilateral element with the aid of an isoparametric transformation.

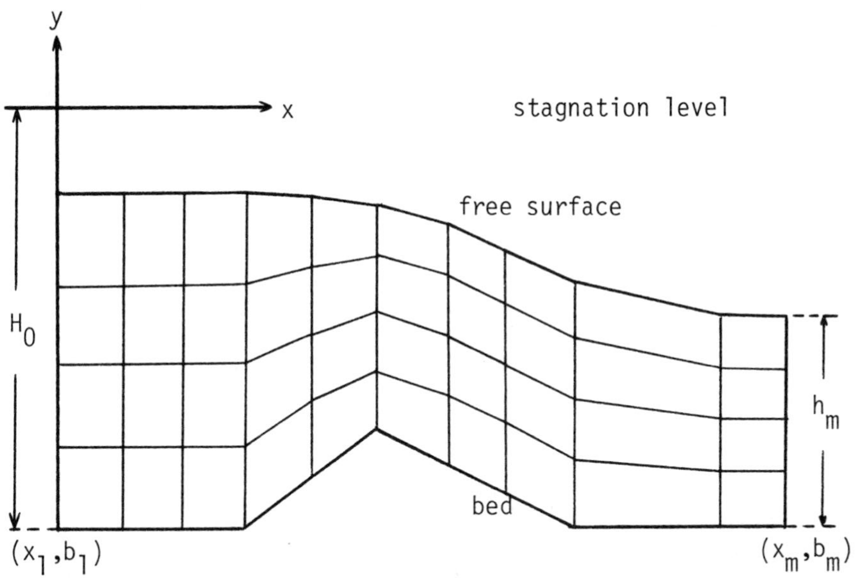

Figure 1. Typical mesh configuration

Boundary conditions require $\psi=0$ on the free surface and $\psi=Q$, the volumetric flow rate, on the bed. These are constraints on the admissible functions for (1). We define ψ so that velocity components are $(-\partial\psi/\partial y, \partial\psi/\partial x)$ giving positive flow in the x direction. The conventional definition of ψ is simply the negative of this. Boundary conditions on the ends $x=x_1$, $x=x_m$ may prescribe the ψ_{ij} or we may use the natural condition $\partial\psi/\partial n=0$ giving horizontal flow. The end values h_1, h_m may be fixed or free or may be constrained to satisfy a parallel-to-bed condition such as $h_{m-1}=h_m$.

With this discretisation the isoparametric transformation introduces rational functions and logarithmic integrals. Using third order approximation to these, the functional is given in terms of vectors ψ,h of the mn and m discrete unknowns by an explicit algebraic relation

$$J = J(\psi,h). \tag{2}$$

The lengthy algebraic detail is given in [8]. The stationary equations for this present a linear problem for ψ when h is fixed, which we solve directly. The h problem is non-linear and coupled to the ψ problem. We use the non-linear algebraic equation algorithm of Brent [9] to solve for h in terms of given ψ values. We then proceed iteratively to determine ψ at the new h values and to redetermine h at the new "frozen" ψ values. This frozen-ψ method has worked effectively and required few iterations. Aitchison [2] has built in the ψ equations as a constraint for the h-problem so that

$$J^*(h) = J(\psi(h),h) \tag{3}$$

is, at solutions of the ψ problem, a function of h alone, albeit an expensive one to compute.

3 SECOND VARIATIONS

The type of stationary solution we obtain to (1) or (2) is determined by the second variation. For example if $\delta^2 J$ is positive definite we obtain a minimum for J. The ψ-problem alone always gives a minimum for J whatever the domain is. The difficulty lies in the h-problem where purely uniform admissible flows give a minimum (maximum) for the rapid (tranquil) solutions and an inflexion at critical flow. The overall type of stationary point cannot be determined from the separate ψ and h problems and must take account of ψ-h coupling. It could be obtained from an analysis of (3) by superimposing the positive-definite nature of ψ variations but this would be complex. We will consider the nature of the frozen-ψ h problem for (2), which is the part of the computation where difficulties of critical or choked flow would be likely to occur.

The m×m Hessian matrix H with elements $\partial^2 J/\partial h_i \, \partial h_j$ is symmetric and tridiagonal. Its definiteness is determined from the signs of its m leading minors given by the recurrence relation

$$\Delta_k = H_{kk}\,\Delta_{k-1} - H^2_{k\,k-1}\,\Delta_{k-2} \tag{4}$$

with

$$\Delta_1 = H_{11},$$

$$\Delta_2 = H_{22}H_{11} - H^2_{12} \quad .$$

Detail of the elements of H is given in [8]. An algorithm based on (4) is used to determine numerically the type of stationary point for the computed finite element solutions.

4 NUMERICAL RESULTS

A range of solutions for the weir configuration of Figure 2 has been calculated for various initial heights h_1 and the condition $h_{m-1}=h_m$. The solution curves shown on the figure have

$$Q^2/(gH_0^3) = 0.16, \qquad H_0 = -b_1$$

below the flat-bed critical value of 8/27 and low enough to permit solutions over the weir. The type of stationary point with respect to h is a minimum for curves (m) and a saddle for curves (s).

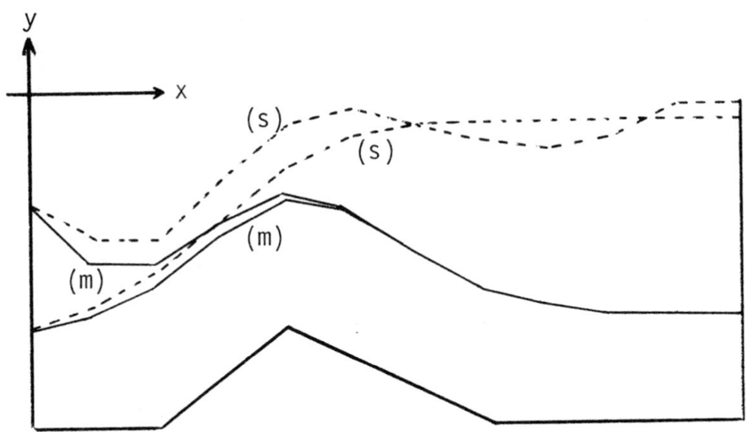

Figure 2. Minimum- and saddle-type solutions

General numerical results indicate that steady flow solutions exist only for values of Q up to a critical value depending on the bed shape with respect to the stagnation level. Below that value there may frequently be non-unique solutions. Depending on the boundary conditions set, some of the multiple "solutions" may proceed outside the permitted range $0 < h < 1$ and will therefore be excluded. There remain however many cases with non-unique solutions within the permitted range. In general those solutions near or asymptotic to the 'rapid' uniform flow are definite minima for J and are easily calculated. Solutions near to or oscillatory about the 'tranquil' uniform flow give a saddle point for J and they may be associated with nonconvergence of the iterations between h and ψ even though the non-linear h problem is solved. Where iterations have converged, and even near the critical value, convergence has been rapid (e.g. 5 iterations give an error reduction factor 10^{-6} for a net of 12×5 nodes).

As an extra check the program has been equipped with a routine to calculate the functional J at the solution and at individually perturbed values of the h_i. The results have supported the classification of minima and saddles within the arithmetic precision used.

5 A SEMIDISCRETE FORMULATION

By requiring ψ to proceed linearly in y from zero at the surface to Q at the bed the problem can be reduced to a 1-dimensional one for an unknown function h(x). This has been found to represent the qualitative behaviour of the general problem. In particular it allows a range of oscillatory solutions about the flat-bed tranquil flow and (by horizontal translation) a range of monotonic solutions asymptotic to the flat-bed rapid flow [10]. Here J reduces to

$$J = J(h(x)) \tag{5}$$

with a second variation of the form

$$\delta^2 J = \int_0^a (pv'^2 + qv^2)\ dx \tag{6}$$

for variations v and channel length a. This is amenable to analysis and full theoretical results have been obtained for flows over a flat bed [11].

These results show that there are no maxima because, as p is always positive, short wave variations are minimum-like. A range of flows asymptotic to the rapid solution always gives a minimum for J. Oscillatory solutions about the tranquil flow are minima on short enough channels with respect to purely interior variations (i.e. v(0)=v(a)=0). But longer channels can accommodate longer wavelength variations which can be maximum-like or give a singularity. The infinite-wavelength uniform flow variation about the tranquil solution is certainly maximum-like. In the special case of critical flow all internal variations are minimum-like whatever the channel lengths, while the uniform variation gives a (singular) inflexion and attendant numerical difficulty.

6 CONCLUSIONS

We have presented numerical results on the nature of full finite element solutions and have referred briefly to strong analytical results for a semidiscrete approximation to the problem.

In general flows with supercritical (rapid) outlet give a minimum for the variational principle with respect to determining the free surface. This accords with the minimum with respect to stream function and presents no special numerical difficulty. Flows which oscillate about a subcritical (tranquil) depth may generally have a transition length for the channel, beyond which potentially difficult long wave variations can introduce a singularity before passing to a strict saddle type of stationary value. The singularity of critical flow is a feature of uniform variations; internal variations are minimum-like. Some numerical difficulties may be alleviated by retaining a fixed-depth boundary condition at inlet.

7 REFERENCES

1. IKEGAWA, M, and WASHIZY, K. - Finite element method applied to analysis of flow over a spillway crest. Int. J.Numer.Meth.Engng, Vol.6, pp. 179-189, 1973.

2. AITCHISON, J. M. - A variable finite element method for the calculation of flow over a weir. Rutherford Laboratory Report, 1979.

3. McCORQUODALE, J. A. and LI, C. Y. - Finite element analysis of a sluice gate flow. Trans.Engng.Inst. Canada, 14 (C-2), I-IV, 1971.

4. VAROGLU, E. and FINN, W. D. L. - Variable domain finite element analysis of free surface gravity flow. Computers & Fluids, Vol. 6, pp. 103-114, 1978.

5. BETTS, P. L. - A variational principle in terms of stream function for free surface flows and its application to the finite element method. Computers & Fluids, Vol. 7, pp. 145-153, 1979.

6. O'CARROLL, M. J. - A variational principle for ideal flow over a spillway. International Journal for Numerical Methods, Vol. 15, pp. 767-789, 1980.

7. O'CARROLL, M. J. - Variational methods for free surfaces of cavitation, jets, open channel flows, separation and wakes. Finite Elements in Fluids, Vol. 3, Chapter 16, John Wiley & Sons, 1978.

8. TORO, E. F. - Two dimensional open channel flows part II. Mathematical Report,Teesside Polytechnic, Middlesbrough England, TPMR 80-7, December 1980.

9. BRENT, R. P. - Some efficient algorithms for solving systems of non-linear equations. SIAM Journal Numerical Analysis, Vol. 10, No.2, pp. 327-366, 1973.

10. WOODMAS, R. - Differential equation studies related to open channel flow. B.Sc. Mathematical Sciences Project Teesside Polytechnic, Middlesbrough, England, 1979-80.

11. TORO, E. F. - On the nature of solutions of two-dimensional open channel flows. Mathematical Report, Teesside Polytechnic, Middlesbrough, England, TPMR 81-1, February, 1981.

A Semi Analytic Method for Viscous Flows in the Vicinity of Singular Corners

Murli M. Gupta*

SUMMARY

Many problems of physical interest involve flows of viscous fluids in regions that contain one or more sharp corners. A typical example is the flow in a prototype cavity where the driving wall meets the cavity at two corners which are singular points of the flow. In this paper, a semi-analytic technique is introduced to solve the two-dimensional Navier-Stokes equation in small neighborhoods of the singular corners. The solutions are expressed as an asymptotic expansion in terms of the Reynolds Number R, the radial distance from the singular corner r and the angle of deviation Θ. The coefficients of the power series are obtained by numerical integration. Comparison with existing numerical solutions is also carried out.

1. Introduction

The two-dimensional Navier-Stokes equations for incompressible viscous flows have been investigated by a number of researchers and a variety of methods have been propagated to obtain the numerical solutions of these equations. Most of these methods tend to ignore the effect of corner singularities. The finite difference methods are usually designed to ensure that the computational stencil does not require the singular corner values. The finite element methods tend to smooth out the singularities by using some sort of averaging at the corner. There is increasing evidence that the corner singularities affect the fluid flows not only in the neighborhood of singular corners but also throughout the fluid flow region. It is thus important to examime the local behaviour of the Navier-Stokes equations near the singular corners.

*Department of Mathematics
The George Washington University
Washington, D.C. 20052 (USA)

The case of Stokes' flow (R=0) has been studied by several authors [1,2] who obtained analytic solutions valid in the neighborhood of singular corners. In the case of nonzero Reynolds number, Roache and Zoltani [3] modelled an interior ballistics problem by the steady state Navier-Stokes equations in a rectangular cavity which were solved on a set of successively finer meshes (11x11 to 31x31) in the corner regions of the cavity. These solutions were obtained for small R (0 to 50).

In this paper, we obtain closed form solutions of the Navier-Stokes equations in the neighborhood of singular corners for general values of R. This is done by assuming a local asymptotic form of the solutions at singular corners. The coefficients of the asymptotic series are expressed as definite integrals of certain complicated functions which are evaluated using numerical quadrature techniques.

2. GOVERNING DIFFERENTIAL EQUATIONS

The Navier-Stokes equations describing the cavity flow problem can be written in the form:

$$\nabla^4 \psi = RJ\,(\psi,\ \omega),$$
$$J(\psi,\ \omega) = (\psi_x \omega_y - \psi_y \omega_x) \tag{1}$$

where R is the Reynolds number of the flow, ψ the stream-function and ω the vorticity. The velocity components u, v are related to ψ, ω by

$$u = \psi_y\ ,\ v = -\psi_x,\ \omega = v_x - u_y = -\nabla^2 \psi$$

The equation (1) is solved in a rectangular cavity subject to the following boundary conditions:

$$\psi = 0\ ,\quad \psi_y = -1 \text{ on the sliding wall } y=1$$
$$\psi = 0\ ,\quad \psi_n = 0 \text{ on the stationary walls} \tag{2}$$

3. ASYMPTOTIC FORMULATION

We assume that the solution of equation (1) can be expressed as a power series in R, at least in the local neighborhood of singular corners.

$$\psi = \psi^{(0)} + R\psi^{(1)} + R^2\psi^{(2)} + \ldots \tag{3}$$

This yields

$$\omega = \omega^{(0)} + R\omega^{(1)} + R^2\omega^{(2)} + \ldots$$

where

$$\omega^{(k)} = -\nabla^2\psi^{(k)} \qquad k = 0, 1, 2, \ldots$$

Substituting (3) in (1) and comparing powers of R, we get

$$\left.\begin{aligned} \nabla^4\psi^{(0)} &= 0 \\ \nabla^4\psi^{(k)} &= \sum_{j=0}^{k-1} J(\psi^{(k-j-1)}, \omega^{(j)}) \quad k = 1, 2, \ldots \end{aligned}\right\} \qquad (4)$$

We introduce polar coordinates with pole at the down-stream singular corner, the sliding wall coinciding with $\theta = 0$ and the stationary wall coinciding with $\theta = -\pi/2$. The boundary conditions (2) may be rewritten as

$$\left.\begin{aligned} \psi^{(k)} &= 0 \quad \text{for} \quad \theta = 0 \ , -\pi/2 \\ \psi_\theta^{(k)} &= 0 \quad \text{for} \quad \theta = -\pi/2 \end{aligned}\right\} \quad k = 0, 1, 2, \ldots$$

$$\psi_\theta^{(0)} = -r \ , \ \psi_\theta^{(k)} = 0 \ \text{for} \ \theta = 0 \ \text{and} \ k = 1, 2, 3, \ldots \qquad (5)$$

The Stokes equation $\nabla^4\psi = 0$ admits solutions of the form [1, 2]

$$\psi^{(0)} = r\, f_o(\theta) \qquad (6)$$

with corresponding vorticity

$$\omega^{(0)} = -\nabla^2\psi^{(0)} = 0(r^{-1}) \qquad (7)$$

The value of vorticity approaches infinity as $r \to 0$. Since $J(\psi^{(0)}, \omega^{(0)}) = 0(r^{-2})$, we have from eq. (4)

$$\nabla^4\psi^{(1)} = J(\psi^{(0)}, \omega^{(0)}) = 0(r^{-2})$$

which implies $\psi^{(1)} = 0(r^2)$. Extending this dimensional argument to the remaining equations in (4), we obtain the separated form of $\psi^{(k)}$ as

$$\psi^{(k)} = r^{k+1} \quad f_k(\theta), \quad \theta = 0, 1, 2, \ldots \qquad (8)$$

From eq. (4), (5) the function $f_k(\theta)$ satisfy the following system of ordinary differential equations.

$$\left.\begin{array}{l} f_0^{iv} + 2f_0'' + f_0 = 0 \\ f_0(0) = f_0(-\pi/2) = 0 \ , \ f_0'(0) = -1 \ , \ f_0'(-\pi/2) = 0 \end{array}\right\} \tag{9}$$

For $k \geq 1$,

$$\left.\begin{array}{l} f_k^{iv} + 2(k^2+1)f_k'' + (k^2-1)^2 f_k = g_k(\theta) \\ f_k(0) = f_k(-\pi/2) = f_k'(0) = f_k'(-\pi/2) = 0 \end{array}\right\} \tag{10}$$

where

$$g_k(\theta) = \sum_{j=0}^{k-1} \Big[(j-1)f_{k-j-1}'(f_j'' + (j+1)^2 f_j) - (k-j)f_{k-j-1}(f_j''' + (j+1)^2 f_j') \Big] \tag{11}$$

The functions $g_k(\theta)$ are expressed in terms of $f_i(\theta)$ (i=0, k=1) and their derivatives. These functions become increasingly complicated as k is increased.

4. THE SEPARATED SOLUTIONS

The solution of eq. (9) is easily obtained as

$$f_0(\theta) = \frac{1}{\alpha} [\theta \cos\theta - \frac{\pi}{2} (\frac{\pi}{2} + \theta) \sin\theta], \ \alpha = \pi^2/4-1, \tag{12}$$

which agrees with the Stokes flow solutions obtained by Taylor [1] and Moffatt [2]. The solutions of eq. (11) are obtained using the method of variation of parameters. These solutions can be expressed as

$$f_k(\theta) = A_k(\theta) \cos(k-1)\theta + B_k(\theta) \cos(k+1)\theta + C_k(\theta) \sin(k-1)\theta + D_k(\theta) \sin(k+1)\theta . \tag{13}$$

where

$$A_k(\theta) = -\frac{1}{4k(k-1)} \int_0^{\theta} \sin(k-1)\,\theta\; g_k(\theta)\cdot d\,\theta + A_k(0)$$

$$B_k(\theta) = \frac{1}{4k(k+1)} \int_0^{\theta} \sin(k+1)\,\theta\; g_k(\theta)\cdot d\,\theta + B_k(0)$$

$$C_k(\theta) = \frac{1}{4k(k-1)} \int_0^{\theta} \cos(k-1)\,\theta\; g_k(\theta)\cdot d\,\theta + C_k(0) \qquad (14)$$

$$D_k(\theta) = -\frac{1}{4k(k+1)} \int_0^{\theta} \cos(k+1)\,\theta\; g_k(\theta)\cdot d\,\theta + D_k(0)\,.$$

The values of these coefficients at $\theta = 0$ are given as follows:

$$\lambda_1 = A_1(0) = -B_1(0) = \frac{1}{32} \int_0^{-\pi/2} [4\,\theta + 2\sin 2\,\theta + \pi(1 - \cos 2\,\theta)]\; g_1(\theta)\; d\,\theta \qquad (15)$$

$$\mu_1 = C_1(0) = -2D_1(0) = -\frac{1}{8} \int_0^{-\pi/2} (1 + \cos 2\,\theta)\; g_1(\theta)\; d\,\theta$$

and for $k \geqq 2$,

$$\lambda_k = \begin{cases} \dfrac{1}{8k} \displaystyle\int_0^{-\pi/2} \left(\frac{\sin(k-1)\theta}{(k-1)} + \frac{\sin(k+1)\theta}{(k+1)}\right) g_k(\theta)\,d\theta, & k \text{ odd} \\[2ex] \dfrac{1}{8k^2} \displaystyle\int_0^{-\pi/2} (\sin(k-1)\theta + \sin(k+1)\theta)\, g_k(\theta)\,d\theta, & k \text{ even} \end{cases} \qquad (16)$$

$$\mu_k = \begin{cases} -\frac{1}{8k} \int_0^{-\pi/2} (\cos(k-1)\theta + \cos(k+1)\theta) g_k(\theta) d\theta, & k \text{ odd} \\ -\frac{(k^2-1)}{8k^2} \int_0^{-\pi/2} \left(\frac{\cos(k-1)\theta}{(k-1)} + \frac{\cos(k+1)\theta}{(k+1)}\right) g_k(\theta) d\theta, & k \text{ even} \end{cases} \tag{17}$$

Once the functions $f_k(\theta)$ are known, the value of $\psi(r, \theta)$, in the neighborhood of the singular corner is obtainable from eq. (3),(8). The values of $\omega(r, \theta)$, $u(r, \theta)$, $v(r, \theta)$, can then be obtained from $\psi(r,\theta)$. The expressions for ω, u, v involve values of $f_k(\theta)$, $f_k'(\theta)$ and $f_k''(\theta)$ all of which can be expressed as linear combinations of $A_k(\theta)$, $B_k(\theta)$, $C_k(\theta)$ and $D_k(\theta)$. Consequently, there is no need to carry out any numerical differentiation.

5. ASYMPTOTIC SOLUTIONS FOR THE WALLS OF THE CAVITY

From eq. (9),(10), the value of $\psi(r, \theta) = 0$ whenever $\theta = 0$ or $-\pi/2$. Moreover,

$$u(r,0) = -1, \qquad u(r, -\pi/2) = 0$$

$$v(r,0) = 0, \qquad v(r, -\pi/2) = 0$$

Thus all boundary conditions in eq.(2) are exactly satisfied.

The values of vorticity on the sliding wall are given by

$$\omega(r,0) = -\sum_{k=0}^{\infty} r^{k-1} R^k [f_k''(0) + (k+1)^2 f_k(0)]$$

$$= \frac{\pi}{\alpha r} - 4R\lambda_1 - 8r\, R^2\lambda_2 - 4\sum_{k=3}^{\infty} k r^{k-1} R^k \lambda_k \tag{18}$$

On the stationary wall, the vorticity is given by

$$\omega(r,-\pi/2) = -\sum_{k=0}^{\infty} r^{k-1}R^k[f_k''(-\pi/2) + (k+1)^2 f_k(-\pi/2)]$$

$$= -\frac{2}{\alpha r} - 4R[\lambda_1 - \frac{\pi}{2}\mu_1 - \frac{\pi(12-\pi^2)}{64\alpha^2}] \tag{19}$$

$$+ 8rR^2[\mu_2 + \frac{1}{8}\int_0^{-\pi/2} \cos\theta\, g_2(\theta)] + O(r^2R^3) .$$

The values of λ_1, μ_1, from eq. (15) are

$$\lambda_1 = -\frac{\pi}{64}\,\frac{\pi^4 - 2\pi^2 - 88}{(\pi^2 - 4)^2} \approx 0.014\ 718\ 29508$$

$$\mu_1 = -\frac{\pi^2\,(\pi^2 - 8)}{16(\pi^2 - 4)^2} \approx -0.033\ 474\ 32447 \tag{20}$$

The values of λ_2, μ_2 obtained by numerically integrating the expressions in eq. (16), (17) are

$$\lambda_2 = 8.345648773 \times 10^{-5} \tag{21}$$

$$\mu_2 = -1.531217339 \times 10^{-3}$$

These values have relative errors smaller than 10^{-10} and absolute errors smaller than 10^{-12}.

Substituting the above values in eq. (18), (19) we obtain

$$\omega(r,o) = \frac{1}{r}[2.140922923 - 0.05887318032Rr$$
$$- 0.0006676519018\,(Rr)^2 + O(Rr)^3]; \tag{22}$$

$$\omega(r,-\pi/2) = \frac{1}{r}[-1.362953864 - 0.07493450248\,Rr$$
$$- 0.006540833896\,(Rr)^2 + O(Rr)^3] \tag{23}$$

The values of vorticity in the neighborhood of the upstream singular corner are found to be similar to the expressions above, except for a change of sign in the odd powers of Rr.

6. COMPARISON WITH NUMERAL SOLUTIONS

Using eq. (22), we obtained the values of $\omega(r,o)$ for several values of R, r and compared them with the available numerical solutions $\bar{\omega}$. In table 1 we give the values of ω, $\bar{\omega}$ as well as $(\omega - \bar{\omega})/\omega$, the relative deviation of $\bar{\omega}$ from ω. The agreement between the semi-analytic values and the numerical values is seen to be very good.

R	r	ω	$\bar{\omega}$	$(\omega-\bar{\omega})/\omega$
0	0.05	42.82	39.16	8.4%
10	0.05	42.23	38.66	8.4%
50	0.05	39.79	36.95	7.1%
100	0.05	36.60	35.23	3.7%
100	0.0625	27.95	29.85	6.8%
100	0.0201	100.49	95.49	5.0%

Table 1: Vorticity values on the sliding wall ($\theta=0$).

In table 2, we given the corresponding data on the stationary wall ($\theta= -\pi/2$). The numerical values are seen to deviate quite substantially from the predicted analytic values, even for the Stokes flow solution (R=0). The reason seems to be that the vorticity values on stationary walls are very sensitive to the inaccuracies inherent in the numerical solutions.

R	r	ω	$\bar{\omega}$	$(\omega-\bar{\omega})/\omega$
0	0.05	-27.26	-18.08	33.7%
10	0.05	-28.04	-18.54	33.8%
50	0.05	-31.82	-20.39	35.9%
100	0.05	-38.02	-22.44	41.0%
100	0.0625	-33.39	-29.96	10.3%
100	0.0201	-76.62	-59.73	22.0%

Table 2: Vorticity values on the stationary wall ($\theta= -\pi/2$)

7. CONCLUSIONS

A semi-analytic procedure has been outlined here for obtaining local asymptotic solutions of the Navier-Stokes equations in the neighborhood of singular corners. The coefficients of the expansion are obtained using numerical integration which is a stable procedure. The higher order terms can be obtained by continuing the process outlined here. This expansion would enable highly accurate calculation of streamfunction, vorticity and velocity profiles in the singular corner regions. The next step would be to subtract out the singularities and solve the resulting regular problem cheaply and more accurately as compared to the existing numerical methods.

8. REFERENCES

1. TAYLOR, G.I. - Similarity solutions of hydrodynamic problems, Aeronautics and Astronautics Ed. Hoff, N.J. and Vincenti, W.G., Pergamon Press, 1960.

2. MOFFATT, H.K. - Viscous and resistive eddies near a sharp corner. J. Fluid Mech., vol. 18, pp. 1-18, 1964.

3. ROACHE, P.J. and ZOLTANI, C.K. - A preliminary investigation of the singular behaviour of fluids near a sliding corner, A.R.O. Report No. 79-3, 1979.

ON A DEFERRED-CORRECTION PROCEDURE FOR DETERMINATION OF CENTRAL-DIFFERENCE SOLUTIONS TO THE NAVIER-STOKES EQUATIONS

Markku Lindroos*

SUMMARY

The convergence conditions of an iterative method for determination of central-difference solutions to the vorticity – stream-function formulation of the Navier-Stokes equations subject to Dirichlet boundary conditions are studied. It is found that if the vorticity values are equal in the numerical solution, and if the starting values are sufficiently close to the solution, then the iteration converges if the stream-function and vorticity iterations involved converge separately. This prompts consideration of the vorticity iteration for fixed values of the stream function. Analytical treatment of the problem becomes possible when the vorticity equation is replaced by a constant-coefficient model problem in a rectangle. The range of the relaxation parameters which yield convergence is investigated, and some comparisons with the standard SOR method are made.

1. INTRODUCTION

This paper is concerned with the numerical solution of the Navier-Stokes equations which describe the steady two-dimensional flow of a viscous incompressible fluid. The equations are expressed in terms of the stream function, ψ, and the vorticity, ζ, in the nondimensional form

$$\frac{\partial^2\psi}{\partial x^2} + \frac{\partial^2\psi}{\partial y^2} = -\zeta \,, \tag{1}$$

$$-\frac{\partial\psi}{\partial y}\frac{\partial\zeta}{\partial x} + \frac{\partial\psi}{\partial x}\frac{\partial\zeta}{\partial y} + \frac{1}{Re}\left(\frac{\partial^2\zeta}{\partial x^2} + \frac{\partial^2\zeta}{\partial y^2}\right) = 0 \,, \tag{2}$$

in which Re is the Reynolds number. The velocity components of the flow in the x and y directions are given by $u = \partial\psi/\partial y$ and $v = -\partial\psi/\partial x$, respectively. Equation (1) is referred to as the stream-function equation and equation (2) as the vorticity equation.

*Institute of Mathematics, Helsinki University of Technology, SF-02150 Espoo 15, Finland

Our objective is to study the iterative procedure proposed by Dennis and Chang [1] for determination of central-difference solutions to equations (1) and (2). This procedure is based on the idea of deferred corrections (or difference corrections), originally introduced by Fox [2].

Assume that the calculation domain is a rectangle with sides parallel to the coordinate axes. Assume further that the rectangle is covered by a uniform grid with mesh spacings Δx and Δy, and that the nodes of the grid are identified by the indices i and j, with i running in the x direction from 0 to M, and j running in the y direction from 0 to N. Equation (1) is replaced by $\Delta_h\psi_{ij} = -\zeta_{ij}$, in which Δ_h represents the usual five-point analogue of the Laplacian operator. The second-order central-difference approximation to (2) is written symbolically as $L\zeta_{ij} = 0$. If the abbreviations

$$p_{ij} = u_{ij}\Delta x\mathrm{Re} = (\psi_{i,j+1} - \psi_{i,j-1})\Delta x\mathrm{Re}/(2\Delta y), \tag{3}$$

$$q_{ij} = v_{ij}\Delta y\mathrm{Re} = -(\psi_{i+1,j} - \psi_{i-1,j})\Delta y\mathrm{Re}/(2\Delta x) \tag{4}$$

are introduced, the approximations can be written as

$$\psi_{i+1,j}+\psi_{i-1,j}+\gamma(\psi_{i,j+1}+\psi_{i,j-1})-2(1+\gamma)\psi_{ij}+(\Delta x)^2\zeta_{ij} = 0 , \tag{5}$$

$$(1-p_{ij}/2)\zeta_{i+1,j}+(1+p_{ij}/2)\zeta_{i-1,j}+\gamma[(1-q_{ij}/2)\zeta_{i,j+1}+(1+q_{ij}/2)\zeta_{i,j-1}] -2(1+\gamma)\zeta_{ij} = 0 , \tag{6}$$

in which $\gamma = (\Delta x/\Delta y)^2$. The boundary conditions are assumed to be of the Dirichlet type, that is, the values of ψ and ζ at the boundary nodes are considered to be known in advance. This is hardly ever met in practice, and hence limits the range of applicability of the results. Nevertheless, this paper guides the choice of iteration schemes and relaxation parameters even in cases where the boundary values must be determined iteratively as part of the solution.

In the method suggested by Dennis and Chang, the discrete vorticity equation is written in the form $L\zeta_{ij} \equiv M\zeta_{ij}+C\zeta_{ij} = 0$, in which M is the difference operator pertaining to upwind differences, and C is the operator which corrects the accuracy to that of central differences. (Of course, upwind differences are employed only for the convection terms of the vorticity equation; the diffusion terms are approximated by the central differences.) The upwind-difference solution is first determined from the set of equations

$$M\zeta^0_{ij} = 0, \quad \Delta_h\psi^0_{ij} = -\zeta^0_{ij} , \tag{7}$$

and this is improved according to the following iterative law:

$$M\zeta^{k+1}_{ij} + C\zeta^k_{ij} = 0, \quad \Delta_h\psi^{k+1}_{ij} = -\zeta^{k+1}_{ij} , \quad k = 0, 1, 2, \cdots . \tag{8}$$

If this process converges, the limit solution satisfies the central-difference equations

$$L\zeta_{ij} = 0, \quad \Delta_h\psi_{ij} = -\zeta_{ij} . \tag{9}$$

Thus the solution procedure consists of the outer iteration, the steps of which are counted by index k, and of the inner iterations which are required for determination of the outer iterates from the nonlinear system which has to be solved for each fixed k. The method of successive over-relaxation (SOR) can be employed successfully for the inner iterations, as the coefficient matrix associated with the operator M is diagonally dominant. The most recent outer iterates provide natural starting points for the inner iterations. Relaxation factors can also be employed to hasten convergence of the outer iteration. It is not necessary to determine the upwind-difference solution to start the iteration process. In our analyses, we will assume that the correction terms are present from the very beginning.

In practice, it is neither possible nor advisable to carry out the inner iterations to ultimate convergence. If too many inner iteration sweeps are performed for a step of outer iteration, a great deal of time may be wasted without any essential improvement being obtained in the numerical solution. However, if the inner sweeps are too few in number, the overall process of the nested iterations may diverge. The question thus arises of how the iterations should be arranged in order to obtain a convergence which is as rapid as possible.

Two cases are of special interest, since they represent the extreme situations, and are also the most amenable to mathematical analysis. The first involves only one inner iteration for each step of the outer iteration (this variant has been employed in [4], [5], and [9]), and the second involves an infinite number of them (i.e. the set of equations for the new outer iterates is solved exactly). These extreme cases of the deferred-correction procedure will be referred to as DC(1) and DC(∞) respectively.

We employ here a general theory of the iterative solution of nonlinear equations (see Ortega and Rheinboldt [8]) for study of the convergence conditions in the two extreme situations. It is found that if the vorticity is constant in the numerical solution, and if the initial estimate is sufficiently close to the solution, then the iteration converges provided that the vorticity and stream-function iterations converge separately. (In paper [6], a similar result was obtained for the ordinary SOR method which does not involve the decomposition of operator L into the sum of M and C.) This, together with the fact that the stream-function iteration has good convergence characteristics, inspires consideration of the vorticity iteration when the values of ψ are held fixed. To permit calculation of the eigenvalue spectrum of the iteration matrix of the remaining system, a constant-coefficient partial differential equation is chosen to model the vorticity equation. (This is also the primary reason for the restriction to a rectangular region.) In the case of the model problem thus generated, a theorem is stated for each of the two extreme cases of iteration.

2. ANALYSIS OF CONVERGENCE FOR THE NONLINEAR SYSTEM

This section is devoted to study of the iterative solution of the full nonlinear system of 2(M-1)(N-1) algebraic equations in 2(M-1)(N-1) variables. We present a theorem which concerns the convergence conditions of the deferred-correction procedure in the two extreme cases defined above. In the first case, single SOR sweeps of the stream-function equation are performed in alternation with single relaxation sweeps of the vorticity equation. A relaxation parameter r_ψ is employed in the ψ iteration, and a parameter r_ζ in the ζ iteration. The correction terms $C\zeta^k_{ij}$ are updated after each complete iteration cycle, consisting of the single ψ and ζ sweeps.

Assume that the unknowns are revised in each vertical grid line progressively from bottom to top, and that the grid lines are treated in order from left to right. It is convenient to introduce a vector α with components α_i denoting the nodal unknowns in the order in which they are revised:

$$\psi_{11}\ ,\ \psi_{12}\ ,\ \ldots,\ \psi_{1,N-1}\ ,\ \psi_{21}\ ,\ \ldots,\ \psi_{M-1,N-1}\ ,$$
$$\zeta_{11}\ ,\ \zeta_{12}\ ,\ \ldots,\ \zeta_{1,N-1}\ ,\ \zeta_{21}\ ,\ \ldots,\ \zeta_{M-1,N-1}\ .$$

The nonlinear system to be solved can then be written as

$$f_i(\alpha_1\ ,\ \alpha_2\ ,\ \ldots,\ \alpha_{2(M-1)(N-1)}) = 0,\ i = 1,\ \ldots,\ 2(M-1)(N-1). \quad (10)$$

The first half of these equations consists of equations (5), and the second half of equations (6). Next we divide the functions f_i into sums of two functions, $f_i = f_{1i} + f_{2i}$. In the equations which are approximations to the vorticity equation, f_{1i} represents terms associated with the operator M, and f_{2i} terms associated with the operator C. In the equations which are discrete analogues of the stream-function equation, we choose $f_{1i} = f_i$ and $f_{2i} = 0$. If we introduce the functions

$$g_i(\alpha,\beta) = f_{1i}(\alpha_1\ ,\ \ldots,\ \alpha_{i-1}\ ,\ \beta_i + (\alpha_i - \beta_i)/r_i\ ,\ \beta_{i+1}\ ,\ \ldots,\ \beta_{2(M-1)(N-1)})$$
$$+ f_{2i}(\beta_1\ ,\ \beta_2\ ,\ \ldots,\ \beta_{2(M-1)(N-1)})\ , \quad (11)$$

in which $r_i = r_\psi$ for $i = 1,\ \ldots,\ (M-1)(N-1)$ and $r_i = r_\zeta$ for $i = (M-1)(N-1)+1,\ \ldots,\ 2(M-1)(N-1)$, and define the column vector

$$G(\alpha,\beta) = (g_1(\alpha,\beta),\ g_2(\alpha,\beta),\ \ldots\ ,\ g_{2(M-1)(N-1)}(\alpha,\beta))^T, \quad (12)$$

then our iteration can be written as

$$G(\alpha^{k+1}, \alpha^k) = 0\ , \quad k = 0,\ 1,\ 2,\ \cdots\ . \quad (13)$$

We assume that our algebraic system has a solution α^* . The hypotheses of a theorem given by Ortega and Rheinboldt [8, p. 325] are then satisfied. On the basis of this theorem, we conclude that if

$$\rho(-\partial_1 G(\alpha^*,\alpha^*)^{-1}\partial_2 G(\alpha^*,\alpha^*)) < 1\ , \quad (14)$$

and if the initial approximation is sufficiently close to α^* , then iteration (13) converges to α^* . Here $\partial_1 G$ and $\partial_2 G$ are the derivatives (the Jacobian matrices) of G with respect to the vector variables α and β, and ρ stands for the spectral radius of the matrix involved. The eigenvalues, λ, of $-\partial_1 G(\alpha^*,\alpha^*)^{-1}\partial_2 G(\alpha^*,\alpha^*)$ are the roots of the characteristic equation

$$\det(\lambda\,\partial_1 G(\alpha^*,\alpha^*) + \partial_2 G(\alpha^*,\alpha^*)) = 0\,. \tag{15}$$

Let the Jacobian matrices of f_{1i} and f_{2i}, $i = 1, \ldots, 2(M-1)(N-1)$, be partitioned into the block forms

$$\left(\frac{\partial f_{1i}}{\partial \alpha_j}\right) = \begin{pmatrix} L_1 + D_1 + U_1 & (\Delta x)^2 I \\ R & L_2 + D_2 + U_2 \end{pmatrix}, \quad \left(\frac{\partial f_{2i}}{\partial \alpha_j}\right) = \begin{pmatrix} 0 & 0 \\ S & T \end{pmatrix}, \tag{16}$$

where all of the matrices occurring in the blocks are square matrices of order $(M-1)(N-1)$. L_1 and L_2 are strictly lower-triangular, D_1 and D_2 are diagonal, U_1 and U_2 are strictly upper-triangular, and I is the unit matrix. Equation (15) can now be written as (L_1, D_1, and U_1 are independent of α^*)

$$\begin{vmatrix} \lambda L_1 + (1 + r_\psi^{-1}(\lambda - 1))D_1 + U_1 & (\Delta x)^2 I \\ \lambda R(\alpha^*) + S(\alpha^*) & \lambda L_2(\alpha^*) + (1 + r_\zeta^{-1}(\lambda - 1))D_2(\alpha^*) + U_2(\alpha^*) + T(\alpha^*) \end{vmatrix} = 0\,. \tag{17}$$

If the nodal values of ζ are all equal in α^*, then $R(\alpha^*) = 0$ and $S(\alpha^*) = 0$. Equation (17) then decomposes into the two separate equations

$$\det(\lambda L_1 + (1 + r_\psi^{-1}(\lambda - 1))D_1 + U_1) = 0\,, \tag{18}$$

$$\det(\lambda L_2(\alpha^*) + (1 + r_\zeta^{-1}(\lambda - 1))D_2(\alpha^*) + U_2(\alpha^*) + T(\alpha^*)) = 0\,. \tag{19}$$

The first is the characteristic equation of the ψ iteration matrix when the values of ζ are held fixed, and the second is the characteristic equation of the ζ iteration matrix when the values of ψ are fixed at the stream-function values in α^*. The asymptotic rate of convergence is in this case determined by the iteration which yields the largest eigenvalue. The result can be given as the following theorem, valid also for regions other than rectangles:

Theorem 1. If the nodal values of ζ are all equal in the numerical solution α^*, and if the initial approximation is sufficiently close to α^*, then iteration of the coupled system converges to α^* if r_ψ and r_ζ are chosen so that the ψ and ζ iterations converge separately.

The theorem concerns a very specific situation, but may also give hints for treatment of the problems in which the vorticity depends weakly on x and y.

The theorem applies to the second extreme case, too, if r_ψ and r_ζ denote the relaxation parameters of the outer iteration. The proof is very similar to the one given above, and can be omitted. Clearly, if the inner iterations are continued until convergence, then the relaxation parameters used in them do not affect the overall convergence.

3. THE MODEL PROBLEM AND TWO USEFUL IDENTITIES

In the following sections, the convergence characteristics of the deferred-correction procedure will be considered when the stream-function iteration is neglected altogether. Moreover, to allow a more complete analytical treatment of the problem, the vorticity equation is replaced by the model equation

$$-u\frac{\partial\zeta}{\partial x} - v\frac{\partial\zeta}{\partial y} + \frac{1}{Re}\left(\frac{\partial^2\zeta}{\partial x^2} + \frac{\partial^2\zeta}{\partial y^2}\right) = 0 , \qquad (20)$$

in which u and v are constant. Occasionally a further simplification is made by a restriction to the one-dimensional counterpart of equation (20),

$$-u\frac{\partial\zeta}{\partial x} + \frac{1}{Re}\frac{\partial^2\zeta}{\partial x^2} = 0 . \qquad (21)$$

With the ordering of unknowns defined previously, the use of central differences in (20) results in a matrix equation, the $(M-1)(N-1)\times(M-1)(N-1)$ coefficient matrix of which has the block-tridiagonal structure

$$\begin{pmatrix} A & B & & & \\ C & A & B & & \\ & \ddots & \ddots & \ddots & \\ & & C & A & B \\ & & & C & A \end{pmatrix}, \text{ where } A = \begin{pmatrix} a & d & & & \\ e & a & d & & \\ & \ddots & \ddots & \ddots & \\ & & e & a & d \\ & & & e & a \end{pmatrix}, \; B = bI, \text{ and } C = cI .$$

The square matrices A, B, C, and I (the unit matrix) are of order N-1. The scalars appearing are given by (cf. equation (6))

$$a = -2(1+\gamma),\; b = 1-p/2 ,\; c = 1+p/2 ,\; d = \gamma(1-q/2),\; e = \gamma(1+q/2), \qquad (22)$$

in which $p = u\Delta x Re$, $q = v\Delta y Re$, and $\gamma = (\Delta x/\Delta y)^2$. By means of the identity

$$\begin{vmatrix} A & B & & \\ C & \ddots & \ddots & \\ & \ddots & \ddots & B \\ & & C & A \end{vmatrix} = \prod_{m=1}^{M-1}\prod_{n=1}^{N-1}\left(a - 2\sqrt{bc}\cos\frac{m\pi}{M} - 2\sqrt{de}\cos\frac{n\pi}{N}\right) , \qquad (23)$$

valid for any real or complex numbers a, b, c, d, e and any integers $M \geq 2$, $N \geq 2$, it is easy to prove that the matrix equatio always has a unique solution. Identity (23) is a generalization of the factorization which Muir [7, p. 401] has given for the determinant of a tridiagonal matrix with constant diagonals.

As will be seen below, the evaluation of the eigenvalues, λ, of the iteration matrix for the two extreme cases considered hinges on the existence of the factorization (23). To obtain the eigenvalues, the determinant of a matrix which has the block-tridiagonal structure given above has to be set equal to zero. The scalars a, b, c, d, and e are in these cases polynomials of degree 1 or 0 in λ. The eigenvalues can thus be determined by setting the factors of the double product separately to zero. By combining the factors which correspond to $m = m'$, $m = M-m'$ and $n = n'$, $n = N-n'$, the right-hand side of (23) can be reduced to a form which does not include any square-root signs. (If M is even, there is no pair for $m = M/2$, but the corresponding square-root term vanishes due to the sine function. An analogous situation arises if N is even.) The reduction is based on the identity

$$(\alpha+\sqrt{\beta}+\sqrt{\gamma})(\alpha+\sqrt{\beta}-\sqrt{\gamma})(\alpha-\sqrt{\beta}+\sqrt{\gamma})(\alpha-\sqrt{\beta}-\sqrt{\gamma}) = (\alpha^2-\beta-\gamma)^2-4\beta\gamma \ . \qquad (24)$$

This gives us a method of factorizing the characteristic polynomial of the iteration matrix into a product of polynomials which are of degree four or less in λ.

4. THE DC(1) PROCEDURE

Consider now the numerical solution of the model problem by the deferred-correction procedure. The first extreme case, in which the correction terms $C\zeta_{ij}^k$ are updated between the successive iteration sweeps, is treated in this section. If the values obtained at the single sweeps are as such accepted as the new outer iterates (that is, no acceleration parameter is used in the outer iteration), then the following iteration formula is obtained:

$$\begin{aligned}
&-r_\zeta\left(1+\frac{p+|p|}{2}\right)\zeta_{i-1,j}^{k+1}-r_\zeta\gamma\left(1+\frac{q+|q|}{2}\right)\zeta_{i,j-1}^{k+1}+[|p|+2+\gamma(|q|+2)]\zeta_{ij}^{k+1} = \\
&-r_\zeta\frac{|p|}{2}\zeta_{i-1,j}^{k}+r_\zeta\left(1-\frac{p}{2}\right)\zeta_{i+1,j}^{k}+[|p|+2+\gamma(|q|+2)-2r_\zeta(1+\gamma)]\zeta_{ij}^{k} \\
&-r_\zeta\gamma\frac{|q|}{2}\zeta_{i,j-1}^{k}+r_\zeta\gamma\left(1-\frac{q}{2}\right)\zeta_{i,j+1}^{k} \ . \qquad (25)
\end{aligned}$$

If λ is an eigenvalue of the iteration matrix and e_{ij}, $i=1, \dots, M-1$, $j=1, \dots, N-1$, are the components of the corresponding eigenvector, then

$$\begin{aligned}
&-r_\zeta\{[1+(p+|p|)/2]\lambda-|p|/2\}e_{i-1,j}-r_\zeta(1-p/2)e_{i+1,j} \\
&-r_\zeta\gamma\{[1+(q+|q|)/2]\lambda-|q|/2\}e_{i,j-1}-r_\zeta\gamma(1-q/2)e_{i,j+1} \\
&+\{[|p|+2+\gamma(|q|+2)]\lambda-[|p|+2+\gamma(|q|+2)-2r_\zeta(1+\gamma)]\}e_{ij} = 0, \qquad (26) \\
&\qquad i=1, \dots, M-1, \quad j=1, \dots, N-1,
\end{aligned}$$

when the notational convention is made that $e_{i0}=e_{iN}=0$ for $i=1, \dots, M-1$, and $e_{0j}=e_{Mj}=0$ for $j=1, \dots, N-1$. A nontrivial solution $\{e_{ij}\}$ exists if and only if the determinant of the coefficient matrix vanishes. By means of the factorization (23), the evaluation of the eigenvalues can be reduced to solution of the equations

$$\begin{aligned}
&(|p|+2)\lambda-|p|-2(1-r_\zeta)-r_\zeta\sqrt{[(2+p+|p|)\lambda-|p|](2-p)}\cos\frac{m\pi}{M} + \\
&\gamma\{(|q|+2)\lambda-|q|-2(1-r_\zeta)-r_\zeta\sqrt{[(2+q+|q|)\lambda-|q|](2-q)}\cos\frac{n\pi}{N}\}=0, \qquad (27) \\
&\qquad m=1, \dots, M-1, \quad n=1, \dots, N-1.
\end{aligned}$$

When the square-root signs are removed, fourth-degree polynomial equations are obtained. The removing process corresponds to the combination of four different factors in accordance with the identity (24). A method due to Descartes (see Fröberg [3, p. 19]) can be employed for solution of these quartic equations.

Figure 1 presents the convergence region in the (p,q) plane for $M=10$, $N=10$, $\gamma=1$, $r_\zeta=1$. The region comprises the first quadrant in its entirety and parts of the other quadrants.

For comparison, the convergence region of the ordinary SOR method for $r_\zeta = 1$ (the Gauss-Seidel method) is given. (See Russell [10] for an account of the convergence properties of SOR when applied to the solution of our model problem.) The convergence region of DC(1) can be enlarged by under-relaxation, but the iteration necessarily remains slowly convergent if p and q are negative and large in absolute value. As r_ζ increases from 1, the convergence region becomes smaller. It shrinks towards the point $p = 2$, $q = 2$ as r_ζ approaches 4, and is empty for $r_\zeta \geqq 4$.

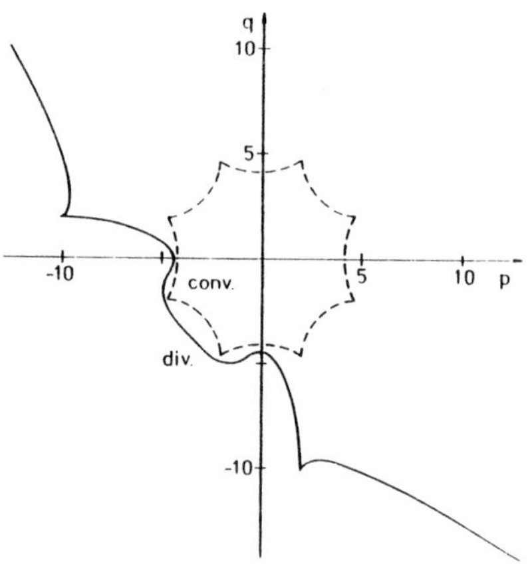

Figure 1. The convergence region for $M = 10$, $N = 10$, $\gamma = 1$, $r_\zeta = 1$: ——, DC(1); - - - , SOR.

The difficulties encountered with DC(1) for negative p and q can be avoided, and the rate of convergence improved, by reversing the order in which the mesh points are scanned. This is a consequence of the fact that the spectral radius of the iteration matrix remains unaltered if the order of scanning is reversed at the same time as the signs of the velocity components are changed. If no relaxation parameter is employed in the outer iteration, the following theorem can be established:

<u>Theorem 2.</u> If the inner iteration proceeds in the direction of the velocity components, and if the relaxation parameter used in the inner sweep is in the range $0 < r_\zeta \leqq 2/\sqrt{3}$, then the DC(1) iteration for solving the model Dirichlet problem converges for any u, v, Re, Δx, Δy, M, and N.

We sketch the idea of the proof by considering the case $p \geqq 0$, $q \geqq 0$, in which the mesh points should be scanned in the order defined previously. To begin with, we define the function

$$f(z,r,p,\phi) = (p+2)z-p-2(1-r)-r\sqrt{[2(1+p)z-p](2-p)}\ \phi . \qquad (28)$$

Equation (27) can then be written as

$$f(\lambda,r_\zeta , p,\cos\frac{m\pi}{M}) + \gamma\, f(\lambda,r_\zeta , q,\cos\frac{n\pi}{N}) = 0 . \qquad (29)$$

For fixed values of M, N, γ, and r_ζ , the roots of equation (29) are continuous functions of p and q. As a consequence, if the roots are within the unit circle of the complex plane for some $p \geqq 0$, $q \geqq 0$, and if they cannot be on the unit circle for any $p \geqq 0$, $q \geqq 0$, then they are within this circle for all $p \geqq 0$, $q \geqq 0$.

When $p = 2$ and $q = 2$, all of the roots of equations (27) coincide and are equal to $(2-r_\zeta)/2$. In the range considered, $0 < r_\zeta \leqq 2/\sqrt{3}$, these are inside the unit circle. So it remains to show that $z = e^{i\theta}$, where θ is real and i is the imaginary unit, cannot be a root of (29) for any $p \geqq 0$, $q \geqq 0$. Moreover, as the complex roots occur in conjugate pairs, it is sufficient to consider the range $0 \leqq \theta \leqq \pi$.

Equation (29) is satisfied if and only if the two terms on the left are equal in modulus and of the opposite sign, that is, if and only if they lie symmetrically with respect to the origin of the complex plane. It is therefore natural to consider, for fixed values of r_ζ and $z = e^{i\theta}$, the image of the set

$$S = \{(p,\phi) \mid p \geqq 0\ ,\ -1 \leqq \phi \leqq 1\} \tag{30}$$

in the mapping defined by f. As ϕ varies between -1 and 1, $f(z,r_\zeta, p,\phi)$ assumes values on the line segment which connects the points $f(z,r_\zeta, p,-1)$ and $f(z,r_\zeta, p,1)$. As p traverses the interval $[0,\infty)$, this line segment sweeps the image of the set S. To discover the image of S for fixed θ and r_ζ, it is accordingly sufficient to determine the loci of the points $f(e^{i\theta},r_\zeta, p,-1)$ and $f(e^{i\theta},r_\zeta, p,1)$ as p varies from 0 to ∞. (Due to the double-valuedness of the square-root function in (28), these loci may be interpreted as the graphs of the two branches of $f(e^{i\theta},r_\zeta,p,1)$ or, equally well, those of $f(e^{i\theta},r_\zeta, p,-1)$.)

For fixed values of θ and r_ζ, the smallest cone including the image of S and having the vertex at the origin may be drawn. Let us denote this cone by $C(\theta,r_\zeta)$. If the vertex angle, say $\alpha(\theta,r_\zeta)$, of the cone is less than or equal to π for any $\theta \in [0,\pi]$, then, for the particular value of r_ζ in question, equation (29) cannot have roots on the unit circle for any $p \geqq 0$, $q \geqq 0$, γ, M, and N.

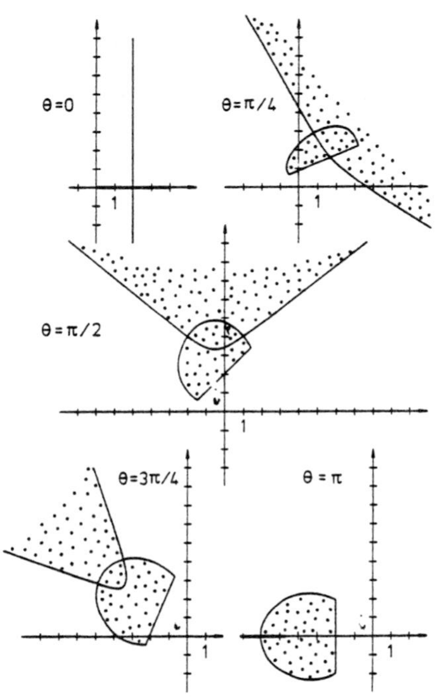

Figure 2. Image of S for $r_\zeta = 1$.

The shaded regions in Figure 2 show the image of S for $r_\zeta = 1$ at some selected values of θ. The cone $C(\theta,r_\zeta)$ has not been drawn, but it is easily understood how it rotates counterclockwise about the origin as θ is increased from 0 to π. The vertex angle $\alpha(\theta,r_\zeta)$ simultaneously changes in size; the functional relation has been presented graphically in Figure 3, which also includes this relation for some other values of r_ζ.

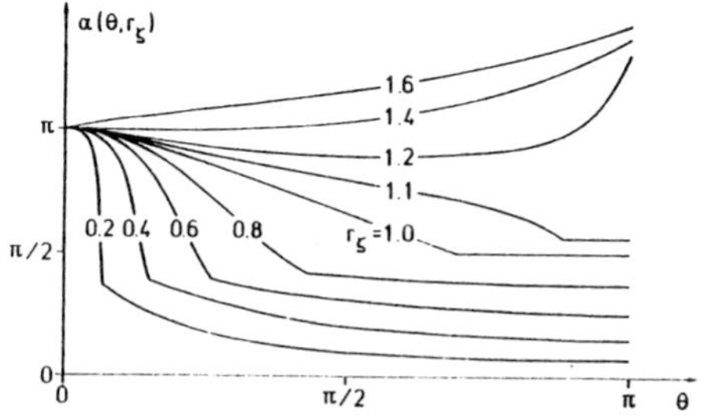

Figure 3. The vertex angle $\alpha(\theta,r_\zeta)$.

It is quite easy to prove that $\alpha(\theta,r_\zeta)$ is a nondecreasing function of r_ζ (for $r_\zeta \geq 0$). Therefore, if the vertex angle is less than or equal to π for some positive value of r_ζ , then it will be so for any smaller positive value. Figure 3 suggests, and it can be verified by a closer study, that the value $\theta = \pi$ has a decisive role as far as the determination of the critical value of r_ζ is concerned. By setting $\theta = \pi$, it is easy to see that the critical value is $r_\zeta = 2/\sqrt{3} \simeq 1.1547$. This completes the proof of Theorem 2.

The upper bound of r_ζ in Theorem 2 is determined by the most restrictive choice of p, q, γ, M, and N. For other choices, convergence may be achieved with parameter values far higher than $2/\sqrt{3}$. The value of r_ζ which minimizes the spectral radius of the iteration matrix is called the optimum relaxation parameter, and is denoted by r^*_ζ . It gives the best asymptotic rate of convergence. As a rule, r^*_ζ has to be determined numerically by varying r_ζ and evaluating the spectral radius from the solutions of the associated quartic equations.

To gain some insight into good choices of r_ζ and the achievable rates of convergence, the iterative solution of the one-dimensional model problem may be considered. This facilitates graphical representation, since the number of parameters involved is reduced. Figure 4 presents the variation of r^*_ζ as a function of $p = u\Delta x Re$ for $M = 10$. The supremum, $\bar{r}_\zeta$, of the relaxation parameters for which convergence is obtained with an arbitrary initial approximation is also shown. The respective quantities of the ordinary SOR are included for comparison. Figure 5 shows the spectral radii of the iteration matrices for the two methods under comparison, when $r_\zeta = 1$ and when $r_\zeta = r^*_\zeta$ (each method has its own optimum parameter). For an arbitrary choice of p and M, both

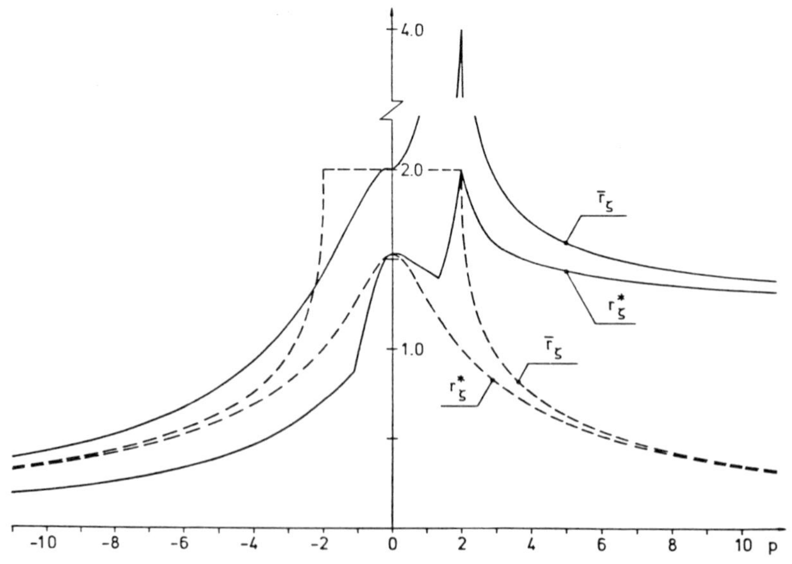

Figure 4. r^*_ζ and $\bar{r}_\zeta$ for $M = 10$: ——, DC(1); - - -, SOR.

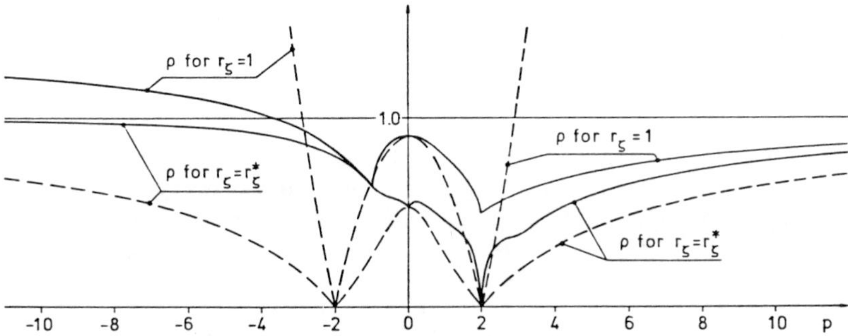

Figure 5. The spectral radius of the iteration matrix for $M=10$, when $r_\zeta=1$ and when $r_\zeta=r^*_\zeta$: ——— , DC(1) ; - - - , SOR.

methods can be made convergent by choosing a sufficiently small relaxation parameter. Numerical calculations indicate that if M is even, then for any p the spectral radius of the optimum SOR method is smaller than or equal to the spectral radius of the optimum DC(1) procedure. If M is odd, this seems to be true on an interval $(-\infty, p_M]$, where p_M depends on M, and is smallest for $M=3$ ($p_3 \simeq 6$). As the work per iteration cycle is smaller for SOR, and as SOR requires less storage, the result of our comparison is quite unfavourable for the DC(1) procedure.

5. THE DC(∞) PROCEDURE

Now we turn our attention to the second extreme case, in which the inner iteration is repeated until convergence at each step of the outer iteration. In its basic form, the iteration scheme is then as follows:

$$[1+(|p|+p)/2]\zeta^{k+1}_{i-1,j}+[1+(|p|-p)/2]\zeta^{k+1}_{i+1,j}-[|p|+2+\gamma(|q|+2)]\zeta^{k+1}_{ij}$$
$$+\gamma[1+(|q|+q)/2]\zeta^{k+1}_{i,j-1}+\gamma[1+(|q|-q)/2]\zeta^{k+1}_{i,j+1}-|p|(\zeta^{k}_{i-1,j}+\zeta^{k}_{i+1,j})/2$$
$$-\gamma|q|(\zeta^{k}_{i,j-1}+\zeta^{k}_{i,j+1})/2+(|p|+\gamma|q|)\zeta^{k}_{ij} = 0\,. \qquad (31)$$

The case in which a relaxation parameter r_ζ is employed to hasten the convergence is obtained from (31) by replacing ζ^{k+1}_{ij} by $[\zeta^{k+1}_{ij}-(1-r_\zeta)\zeta^{k}_{ij}]/r_\zeta$, and making an analogous replacement in the other terms which possess the superscript k+1. The eigenvalues of the iteration matrix for the accelerated case can be obtained from those of the unaccelerated case by multiplying them by r_ζ , and then adding the term $1-r_\zeta$. This relation, which is not valid for DC(1), would be useful in the numerical determination of the optimum relaxation parameter (it would suffice to solve the quartic equations for $r_\zeta=1$), but in the proof of Theorem 3 below basically the same pattern as was employed in proving Theorem 2 must obviously be followed. Here we content ourselves with pointing out that the eigenvalues of the iteration matrix are again obtainable from equations (29) if the function f is defined as

$$f(z,r,p,\phi) = (2+|p|)(z+r-1)-r|p|-\sqrt{[(2+|p|)(z+r-1)-r|p|]^2-p^2(z+r-1)^2\phi}\,. \qquad (32)$$

<u>Theorem 3.</u> If the inner iterations are carried out to convergence and if the relaxation parameter of the outer iteration is within the interval (0,2), then the overall iteration for solving the model Dirichlet problem converges for any u, v, Re, Δx, Δy, M, and N

The upper bound for the relaxation parameter of the outer iteration is again based on the worst-case analysis. For fixed values of p, q, γ, M, and N, the upper bound can usually be somewhat relaxed. In the case of the one-dimensional model problem, the range of convergence is given by

$$0 < r_\zeta < \bar{r}_\zeta = 2\,\frac{p^2 + (4 + 2|p| - p^2)\sin^2(\pi/M)}{p^2 + (4 - p^2)\sin^2(\pi/M)} \quad . \tag{33}$$

Figure 6 presents $\bar{r}_\zeta$ as a function of $|p|$ when $M = 10$. Also included are the optimum relaxation parameter r_ζ^* and the spectral radii for $r_\zeta = 1$ and $r_\zeta = r_\zeta^*$. All of these quantities can be expressed in analytical form for any M. The expressions are dependent on whether M is even or odd, and must be given piecewise wit respect to $|p|$. The spectral radii increase with $|p|$, and hence the rates of convergence decrease with $|p|$. On the other hand, in the inner iterations the rate of convergence increases as $|p|$ increases. The asymptotically best relaxation parameter r_ζ^* and the corresponding spectral radius can be evaluated by the general theory of the SOR method [11]; for $M = 10$, they are shown graphically in Figure 6. It should be stressed that r_ζ^* is not optimal in practice, when a finite (possibly quite small) number of sweep is performed in each inner iteration.

If M is even and $p \geq 2$, then the spectral radii of DC(∞) with $r_\zeta = 1$ and DC(1) with $r_\zeta = 1$ coincide and are equal to $p/(p+2)$. Th in this case at least, there would be little point in performing very many inner sweeps. The same is obviously true more generally for positive p. For negative p, with $|p|$ large, the situation is different.

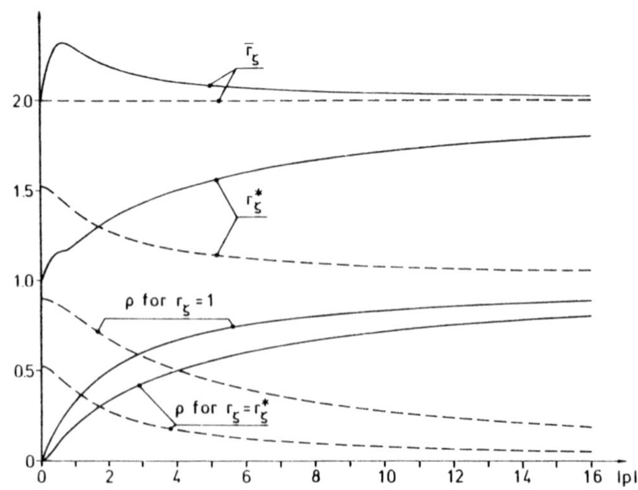

Figure 6. r_ζ^*, $\bar{r}_\zeta$, and the spectral radii for $r_\zeta = 1$ and $r_\zeta = r_\zeta^*$ when $M = 10$: ———, DC(∞); – – –, the inner SOR iteration.

6. CONCLUSION

We have derived some theoretical results which aid in evaluating the efficiency of the deferred-correction procedure proposed by Dennis and Chang. Consideration of the model Dirichlet problem provided useful information on how the vorticity iteration should be arranged. However, care must be exercised when applying the results to problems in which the assumptions made are not valid; in complex flow problems the model equation may be an over-simplification of the true nonlinear equations.

Comparisons with the SOR method suggested that, as a rule, the SOR method is superior to the correction procedure, provided that the relaxation parameter of the vorticity iteration can be evaluated properly. One alternative is to vary it from node to node in conformity to local values of the cell Reynolds numbers $|p_{ij}|$ and $|q_{ij}|$. Russell [10] has proposed a relatively simple formula for this purpose. If the problem is a difficult one, and if divergent behaviour is to be avoided at any price, it may be safer to work with the correction procedure and a large number of inner sweeps. If there is a single dominant direction of flow, one inner sweep may be sufficient to guarantee convergence even at high cell Reynolds numbers, provided that the vorticity iterations proceed in this direction.

REFERENCES

1. DENNIS, S.C.R., and CHANG, G.-Z. — Numerical integration of the Navier-Stokes equations for steady two-dimensional flow. The Physics of Fluids Supplement II, Phys. Fluids, Vol. 12, pp. II-88 - II-93, 1969.

2. FOX, L. — Some improvements in the use of relaxation methods for the solution of ordinary and partial differential equations. Proc. Roy. Soc. London A, Vol. 190, pp. 31-59, 1947.

3. FRÖBERG, C.-E. — Introduction to Numerical Analysis, 2nd edition, Addison-Wesley, Reading, Massachusetts, 1970.

4. LAINE, S.K. — A theoretical study of the effect of a step in a flat plate upon the laminar boundary layer. Helsinki University of Technology, Research Papers 41, Otaniemi, 1972.

5. LINDROOS, M. — Numerical integration of the Navier-Stokes equations for steady flow past a wavelike bulge on a flat plate. Helsinki University of Technology, Laboratory of Aerodynamics, Report No. 78-A2, Otaniemi, 1978.

6. LINDROOS, M. — On the convergence of iterative methods for solving the steady-state Navier-Stokes equations by finite differences. Paper presented at the Seventh International Conference on Numerical Methods in Fluid Dynamics, Stanford University and the NASA/Ames Research Center, June 23-27, 1980.

7. MUIR, T. — The Theory of Determinants in the Historical Order of Development, Vol. 4, Dover, New York, 1960.

8. ORTEGA, J.M., and RHEINBOLDT, W.C. — Iterative Solution of Nonlinear Equations in Several Variables, Academic Press, New York, 1970.

9. RICHARDS, C.W., and CRANE, C.M. — The accuracy of finite difference schemes for the numerical solution of the Navier-Stokes equations. Appl. Math. Modelling, Vol. 3, pp. 205-211, 1979.

10. RUSSELL, D.B. — On obtaining solutions to the Navier-Stokes equations with automatic digital computers. Aeronautical Research Council, Reports and Memoranda No. 3331, 1962.

11. YOUNG, D.M. — Iterative Solution of Large Linear Systems, Academic Press, New York, 1971.

An experimental investigation of the dispersion of a gas jet in a coflowing stream of air.

E.C.P. Ransom, J.H. Barnes and P.D. Phipps

School of Mechanical, Aeronautical and Production Engineering
Kingston Polytechnic

Synopsis

This paper describes the manner in which an experimental investigation, involving the use of a tracer gas technique coupled to a numerical method of analysis, can be used to measure molecular and momentum diffusion levels in laminar and turbulent flow. The basic, three dimensional equations of continuity for laminar flow, are developed which relate rates of change of tracer gas concentration level with functions of flow velocity and molecular diffusion coefficient. Similar equations are developed for turbulent flow by additionally incorporating terms for eddy mass diffusivity.

A numerical method is then explained which is based on brief forms of Taylors expansions. This enables the experimentally determined changes of centre line concentration level to be used:–

a. to construct mathematical models of radial and axial concentration levels in axisymmetric flow

and

b. to compute values of diffusion coefficient and eddy mass diffusivity and diffusion coefficient, for laminar and turbulent flow conditions respectively.

1. Introduction

The use of flow tracing elements for the study of fluid flow can be a most powerful technique. However the major criticisms of such methods are that invariably there is some order of incompatibility of viscous and inertial effects between the tracer and the primary fluid. Consequently it is by no means certain that the tracing agent precisely follows the same path as the fluid under investigation. Indeed, the presence of a tracer may modify the fluid properties to such an extent that the flow system then becomes unrepresentative. For this reason it is essential to reduce to a minimum the proportion of tracer added to the fluid. Clearly, accuracy of measurement is greater if the fluid dynamic properties of the tracer approach those of the fluid under investigation.

Tracer gas is, for these reasons, best for the study of air and gas flows. If the tracer gas selected has a molecular weight near to that of air or the flowing gas under investigation, it may be assumed that the tracer will follow closely the path and accelerations of the air. It is however important to consider the effect of molecular and momentum diffusion particularly in the radial direction if results are to be interpreted correctly.

This paper describes the results of a programme of work on gaseous diffusion in both laminar and turbulent flow.

2. Theory

2.1 Diffusion in laminar flow.

The law governing the mass transfer of one gas in another in a stationary system is Fick's law.[1]

The diffusion mass flux in x direction

$$\left(\frac{\dot{m}_d}{A}\right)_x = -D\,\frac{\partial C}{\partial x} \tag{1}$$

For a flow system comprising of a mixture of gases a, b, c etc., resulting in a stream of mass density ρ and mean velocity U, the total mass flux in the x direction is given by the continuity equation.

Flow mass flux in x direction,

$$\left(\frac{\dot{m}f}{A}\right)_x = \rho U = (CaUa)_x + (CbUb)_x + (CcUc)_x + \ldots\text{etc.} \tag{2}$$

where Ca, Cb, Cc etc. are the mass concentrations (or partial densitites) of components a, b and c, and $(Ua)_x$, $(Ub)_x$, $(Uc)_x$ are the actual velocities of components a, b, and c in the x direction.

Similar equations can be written for flow in the x and y directions.

It is possible to relate the mean velocity $\bar{U}$ with the actual velocities (Ua), (Ub) etc. by incorporating the effects of diffusion. This is best illustrated by considering a flow system having a mixture of two gases only (that is a binary gas mixture comprising component a and b respectively).

The diffusion mass flux of gas a in gas b is

$$\frac{\dot{m}a}{A} = -Da\left(\frac{\partial Ca}{\partial x}\right)$$

where Da is the diffusion coefficient for gas a in a field of gas b.

Hence in the x direction:

$$(CaUb) = CaU - \left(Da\,\frac{\partial Ca}{\partial x}\right) \tag{3}$$

Under laminar flow conditions the diffusion process is entirely by molecular diffusion and hence Da is equal to the molecular diffusion coefficient.

Derivation of the Unsteady Three Dimension Differential Equation of Continuity for Gas component a.

Consider a stationary elemented control volume with sides of length Δx, Δy, and Δz, through which the diffusing gas mixture is flowing.

The rate at which gas component a is entering the control volume in the x direction is:–

$$(\dot{m}a)_x - (\dot{m}a)_{x+\Delta x} = \left(\frac{\partial ma}{\partial t}\right)_x = \frac{-\partial}{\partial x}(CaUa)\,\Delta x\,\Delta y\,\Delta z \tag{4}$$

Similarly for the y and z directions:—

$$\left(\frac{\partial \dot{m}a}{\partial t}\right)_y = -\frac{\partial}{\partial y} (CaVa) \quad \Delta x \ \Delta y \ \Delta z \tag{5}$$

$$\left(\frac{\partial \dot{m}a}{\partial t}\right)_z = -\frac{\partial}{\partial x} (CaWa) \quad \Delta x \ \Delta y \ \Delta z \tag{6}$$

Hence the net rate of increase in the mass of gas a is given by

$$\left(\frac{\partial \dot{m}_a}{\partial t}\right) = \frac{\partial Ca}{\partial t} \ \Delta x \ \Delta y \ \Delta z \tag{7}$$

Assuming that the process is purely a mass transfer phenomena (without chemical reaction) then:—

$$\frac{\partial Ca}{\partial t} = \frac{\partial}{\partial x} (CaUa) - \frac{\partial}{\partial y} (CaVa) - \frac{\partial}{\partial x} (CaWa) \tag{8}$$

Substituting from equation (3)

$$\frac{\partial Ca}{\partial t} = -\frac{\partial}{\partial x}\left(CaU - \frac{D\partial Ca}{\partial x}\right) - \frac{\partial}{\partial y}\left(CaV - \frac{D\partial Ca}{\partial x}\right) - \frac{\partial}{\partial x}\left(CaW - \frac{D\partial Ca}{\partial x}\right) \tag{9}$$

but $U = \frac{1}{\rho}\left\{CaVa + CbUb\right\}$ for a binary system (from (2))

∴ Substituting in (10) gives:

$$\frac{\partial Ca}{\partial t} = -\frac{\partial}{\partial x}\left\{Ca\frac{1}{\rho}(CaVa + CbUb) - D\frac{\partial Ca}{\partial x}\right\} - \frac{\partial}{\partial x}\left\{Ca\frac{1}{\rho}(CaVa + CbVb) - D\frac{\partial Ca}{\partial y}\right\}$$

$$-\frac{\partial}{\partial z}\left\{Ca\frac{1}{\rho}(CaWa + CbWb) - D\frac{\partial Ca}{\partial y}\right\} \tag{10}$$

Assuming that the density of the gas mixture is constant, equation (10) can be written in the following form

$$\frac{\partial}{\partial x}\left(D\frac{\partial Ca}{\partial x}\right) + \frac{\partial}{\partial y}\left(D\frac{\partial Ca}{\partial y}\right) + \frac{\partial}{\partial z}\left(D\frac{\partial Ca}{\partial z}\right) - \frac{\partial Ca}{\partial t} = \frac{\partial}{\partial x} Ca \frac{(CaUa + CbUb)}{\rho}$$

$$+ \frac{\partial}{\partial y} Ca \frac{(CaVa + CbVb)}{\rho} + \frac{\partial}{\partial z} Ca \frac{(CaWa + CbWb)}{\rho} \tag{11a}$$

$$= \frac{\partial}{\partial x} Ca\,(CaU) + \frac{\partial}{\partial y} Ca\,(CaV) + \frac{\partial}{\partial z}(CaW) \tag{11b}$$

$$= U\frac{\partial Ca}{\partial x} + V\frac{\partial Ca}{\partial y} + W\frac{\partial Ca}{\partial z} \tag{11c}$$

The final form (11c) of this equation being achieved for steady flow conditions by using equation (2) with the fact that

$$\frac{\partial}{\partial x}(CaU) = Ca\frac{\partial U}{\partial x} + U\frac{\partial Ca}{\partial x} = Ca\frac{\partial U}{\partial x}$$

since with incompressible steady flow $\frac{\partial U}{\partial x} = 0$

Equation 11 may be transposed to cylindrical coordinates. By assuming that the system is axially symmetric, the tangential and radial velocities are equal to zero and the diffusion coefficient is constant:—

$$D\left\{\frac{1}{r}\frac{\partial Ca}{\partial r}+\frac{\partial^2 Ca}{\partial r^2}+\frac{\partial^2 Ca}{\partial z^2}\right\} = W\cdot\frac{\partial Ca}{\partial z}+\frac{\partial Ca}{\partial t} \qquad (12)$$

2.2 Diffusion in Turbulent Flow

The instantaneous velocity at a point in turbulent flow is normally defined as the sum of a mean velocity, known as a temporal mean velocity, and a fluctuating velocity.

Thus, for all the velocities in the x, y and z directions:

$$U = \bar{U} + U'$$
$$V = \bar{V} + V'$$
$$W = \bar{W} + W'$$

Similarly the instantaneous concentration of gas at any point in the system may be defined by:—

$$Ca = \bar{Ca} + Ca'$$

Since the principle of the conservation of mass also holds for turbulent flows, equation (11b) may be used.

The time average of this equation may be obtained by averaging each term.

$$\frac{\partial}{\partial x}(CaU) = \frac{\partial}{\partial x}\left\{(\bar{Ca}+Ca')(\bar{U}+U')\right\}$$

$$\overline{\frac{\partial (CaU)}{\partial x}} = \frac{\partial}{\partial x}\left\{\bar{\bar{Ca}}\bar{\bar{U}} + \overline{\bar{Ca}U'} + \overline{Ca'\bar{U}} + \overline{Ca'U'}\right\}$$

$$\bar{U'} = 0;\ \ \bar{Ca'} = 0$$
$$\overline{\bar{Ca}U'} = 0;\ \ \overline{Ca'\bar{U}} = 0$$
$$\overline{Ca'U'} \neq 0$$

$$\therefore\ \frac{\partial}{\partial x}(CaU) = \frac{\partial}{\partial x}\left\{\bar{Ca}\bar{U} + \overline{Ca'U'}\right\} \qquad (13)$$

Similar results apply to the y and z directions and equation (11b) may be rewritten in the form:—

$$\frac{\partial}{\partial x}\left\{D\frac{\partial \bar{Ca}}{\partial x} - \overline{U'Ca'}\right\} + \frac{\partial}{\partial y}\left\{D\frac{\partial \bar{Ca}}{\partial y} - \overline{V'Ca'}\right\} + \frac{\partial}{\partial z}\left\{D\frac{\partial \bar{Ca}}{\partial z} - \overline{W'Ca'}\right\}$$
$$= \bar{U}\frac{\partial \bar{Ca}}{\partial x} + \bar{V}\frac{\partial \bar{Ca}}{\partial y} + \bar{W}\frac{\partial \bar{Ca}}{\partial z} + \frac{\partial \bar{Ca}}{\partial t} \qquad (14)$$

Examination of equation (17) below shows that a close similarity of equation (14) to equation (11c) can be achieved by introducing an additional diffusion coefficient (the mass eddy diffusivity ϵ_D) which takes into account the extra diffusion (or mixing) which occurs due to the turbulent nature of the flow.[2]

This is defined for the x direction as:—

$$-\epsilon_D\cdot\frac{\partial Ca}{\partial x} = \overline{U'Ca'} \qquad (15)$$

Clearly this with turbulent mixing as well as molecular diffusion:—

Total mass flux in the x direction:—

$$\left(\frac{\dot{m}}{A}\right)_{molecular} + \left(\frac{\dot{m}}{A}\right)_{eddy} = -(D+\epsilon_D)\frac{\partial \bar{Ca}}{\partial x} \qquad (16)$$

Similar expressions can be written for the y and z directions and making the assumption that the eddy diffusivity has the same magnitude in all directions, substitution of type (15) equation into (14) gives:–

$$\frac{\partial}{\partial x}\left\{(D + \epsilon_D)\frac{\partial \bar{C}a}{\partial x}\right\} + \frac{\partial}{\partial y}\left\{(D + \epsilon_D)\frac{\partial \bar{C}a}{\partial y}\right\} + \frac{\partial}{\partial z}\left\{(D + \epsilon_D)\frac{\partial \bar{C}a}{\partial z}\right\}$$

$$= \bar{U}\frac{\partial \bar{C}a}{\partial x} + \bar{V}\frac{\partial \bar{C}a}{\partial y} + \bar{W}\frac{\partial \bar{C}a}{\partial z} + \frac{\partial \bar{C}a}{\partial t} \qquad (17)$$

On transposing into cylindrical coordinates, and assuming that radial and tangential velocities are zero:–

$$(D + \epsilon_D)\left\{\frac{1}{r}\frac{\partial \bar{C}a}{\partial r} + \frac{\partial^2 Ca}{\partial r^2} + \frac{\partial^2 \bar{C}a}{\partial z^2}\right\} = W\frac{\partial \bar{C}a}{\partial z} + \frac{\partial \bar{C}a}{\partial t} \qquad (18)$$

3. Solution of the Steady State Diffusion Equation

3.1 Finite Difference Equations.

Equations (12) and (18) are first non dimensionalised in terms of the tracer gas concentration at the point of injection, Co and the maximum radius of the containing tube ro. This then gives:–

$$\frac{\partial^2 C}{\partial R^2} + \frac{\partial^2 C}{\partial Z^2} + \frac{1}{R}\frac{\partial C}{\partial R} - \frac{W}{D}\cdot ro\frac{\partial C}{\partial Z} = 0 \qquad (19)$$

$$\frac{\partial^2 C}{\partial R^2} + \frac{\partial^2 C}{\partial Z^2} + \frac{1}{R}\frac{\partial C}{\partial R} - \frac{W.ro}{(D + \epsilon_D)}\frac{\partial C}{\partial Z} = 0 \qquad (20)$$

where $C = Ca/Co$
$R = r/ro$
$Z = z/ro$

A finite difference method is then used tosolve the equations. Since the system is axisymmetric, the complete solution can be achieved by considering any radial plane bounded by the centre line and an axial line on the tube wall. The axial origin of the plane is taken to be one tube radius upstream of the injector exit, and the plane terminates at the exit of the tube containing the flow. The radial distance is divided into twenty equally dispersed mesh points of pitch H. The mesh pitch in the axial direction, K, is determined by the need to create suitable conditions for the numerical solution to converge, i.e. [4]

$$K \leqslant \frac{D}{W\ .\ ro} \qquad (21)$$

Taylor series three term expansions for both radial and axial directions are then used to obtain the tracer gas concentration at mesh point, i, j, in terms of the concentrations at adjacent mesh points. The resulting Finite Difference equation which corresponds to equation (19) is:–

$$C_{(i-j)} = \frac{H^2}{2(H^2+K^2)}\Bigg[\frac{K^2}{H^2}\left\{C_{(i+1,j)} + C_{(i-1,j)}\right\} + C_{(i,j+1)} + C_{(i,j-1)}$$
$$+ \frac{1}{R}\cdot\frac{K^2}{2H}\left\{C_{(i+1,j)} - C_{(i-1,j)}\right\} - \frac{Wro}{D}\cdot\frac{K}{2}\Big\{C_{(i,j+1)}$$
$$- C_{(i,j-1)}\Big\}\Bigg] \qquad (22)$$

where $R = i.H$ and $Z = j.K$

The finite difference equation corresponding to equation (20) is the same as equation (22) but with the diffusion coefficient D replaced by the sum of the diffusion coefficient and eddy diffusivity $(D + \epsilon_D)$. It is assumed that the eddy diffusivity is uniform across a section of flow normal to the axis of symmetry.

3.2 Boundary Conditions

For the case of a stream of gas at the centre of a coflowing stream of air contained in a solid walled tube, where diffusion occurs both in the axial and radial directions, the boundary conditions are:–

a. Far upstream of the injector the gas concentration is zero at all radii

$$C = 0 \text{ at } Z = 0 \text{ for } 0 < R < 1$$

Application to the finite difference equation (22) gives

$$C_{(i,\,0)} = 0$$

b. The pipe wall is impervious to the diffusion of gas

$$\therefore \quad \left(\frac{\partial C}{\partial R}\right)_{R\,=\,1} = 0 \qquad \text{for } 0 < Z < 4.125$$

The Taylor series gives:–

$$\left(\frac{\partial C}{\partial R}\right)_{R\,=\,1} = 0 = \frac{1}{2K}\left\{C_{(i+1,\,j)} - C_{(i-1,\,j)}\right\}$$

Application to equation (22) gives

$$C_{(i+1,\,j)} = C_{(i-1,\,j)}$$

c. Far downstream of the injector at the end of the duct, it is assumed that the change of concentration in the axial direction is at a constant rate.

$$\left(\frac{\partial C}{\partial Z}\right)_{Z\,=\,4.125} = \text{constant for } 0 < R < 1$$

The Taylor series gives:–

$$\left(\frac{\partial^2 C}{\partial Z^2}\right)_{Z\,=\,4.125} = 0 = \frac{1}{K^2}\left\{C_{(i,\,j+1)} - 2C_{(i,\,j)} + C_{(i,\,j-1)}\right\}$$

$$\therefore \quad C_{(i,\,j+1)} = 2C_{(i,\,j)} - C_{(i,\,j-1)}$$

d. In laminar flow the centre line variation of concentration with distance from the injector is predicted by considering the diffusion of tracer gas radially through a thin disc of air. It is assumed that this disc moves from the injector with a uniform velocity and that there is no net diffusion in the axial direction. This leads to the equation

$$\left\{\frac{1}{r}\frac{\partial C}{\partial r} + \frac{\partial^2 C}{\partial r^2}\right\} D = \frac{\partial C}{\partial t} \tag{23}$$

The boundary conditions for this are

$$C = 1, \text{ at } t = 0 \text{ and } R = 0$$

When the disc is enclosed by a solid boundary at radius Ro the solution is[3]

$$C_{(r,\,t)} = \sum_{n=1}^{\infty} AnJo\,(\lambda nr)e^{-D\lambda n^2 t} \tag{24}$$

$$\text{where } An = \frac{2}{r^2 o[Jo(\lambda nro)]^2} \int_0^{rj} rCoJo(\lambda nr) \tag{25}$$

rj = injector radius

and λn are the roots of the equation

$$J_1(\lambda n r o) = 0 \tag{26}$$

In the case of turbulent flow, as explained earlier, the diffusion coefficient D may be replaced by the sum of the diffusion coefficient and the mass eddy diffusivity $(D + \epsilon_D)$

$$C_{(r,\,t)} = \sum_{n=1}^{\infty} AnJo(\lambda nr)_e{}^{-(D + \epsilon_D)\lambda n^2 t} \tag{27}$$

Equation (24) has been evaluated for each of the test conditions, using the effective radius of the injected stream of tracer gas. Values of gas concentration along the centre line of the system were then assigned to the centre line mesh point of the relaxation network.

e. Values for the velocity of flow at end mesh point were assigned according to the type of flow.

3.3 Numerical Solution

The Gauss-Seidal method was used for the solution of the finite different equation (22) for all points in the mesh. The calculations commenced with the mesh point farthest upstream and in the line of points adjacent to the axis of symmetry. Successive points were evaluated in sequence moving along this line towards the end of the test zone. The adjacent line of points radially outwards from the centre line was then calculated and this was repeated until all points in the network were solved. The procedure was repeated for each iteration. In order to increase the rate of convergence successive over relaxation was used with an acceleration factor of 1.4

Convergence of the solution occurred within 150 iterations for a typical laminar flow case in a region bounded by the axis of symmetry, the ends of the test zone and a line joining the mid points of each radius. To reduce computing time the boundaries of the relaxation mesh were readjusted after 150 iterations and the solution allowed to proceed.

The results of this calculation are shown in Fig. 1. which plots the concentration of tracer gas throughout the test region. Two features are noted from the solution.

a. The tracer gas diffuses upstream of the injector against the flow direction of the air.

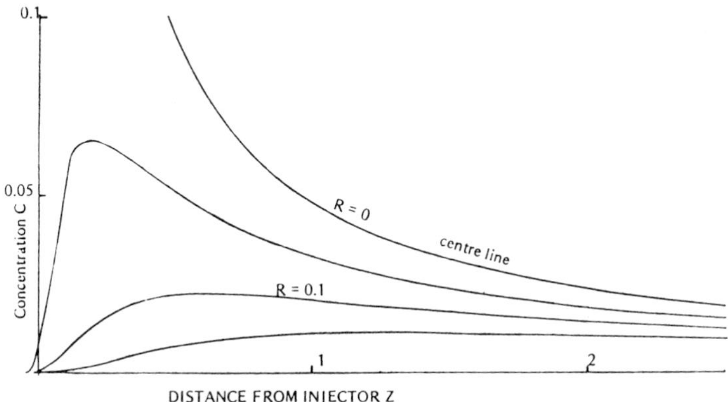

Fig. 1. Concentration of tracer gas

b. Close to the wall, in the region of very low velocity tracer gas diffuses upstream from a region of relatively high concentration, to a region of lower concentration. The rate of diffusion is great enough to cause the curve of non dimensional concentration against radius to reverse slope close to the wall (Fig. 2.), indicating that recirculation is occurring.

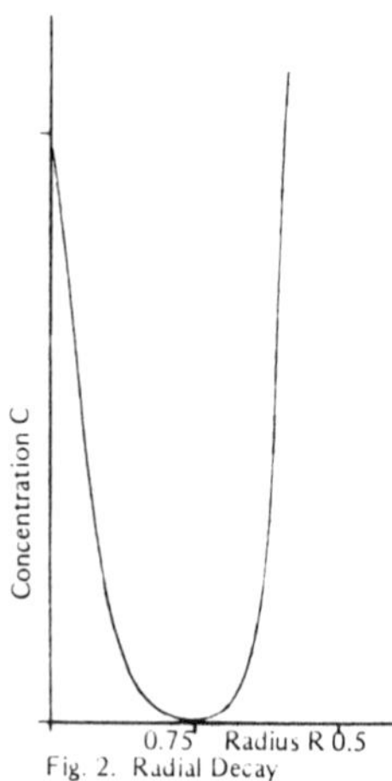

Fig. 2. Radial Decay

4. Apparatus

4.1 Test Equipment

The apparatus shown in Fig. 3. was designed for studying the radial and axial dispersion of a small diameter stream of tracer gas into a coflowing laminar or turbulent stream of air. A small variable speed centrifugal fan supplies air to a large vessel which damps out flow pulsations and fan borne turbulence. A bell mouthed intake with wire mesh screens directs the air from the vessel through a venturi meter to the test section.

The tracer gas is supplied from storage bottles through a pressure relief valve and a needle valve to the injector which is located is axially in the test sections. Previous calibration of the injector allowed the flow rate to be controlled from a knowledge of the pressure and temperature immediately upstream of the injector.

A three axis traversing gear enabled a test probe to be located at any desired situation downstream of the injector.

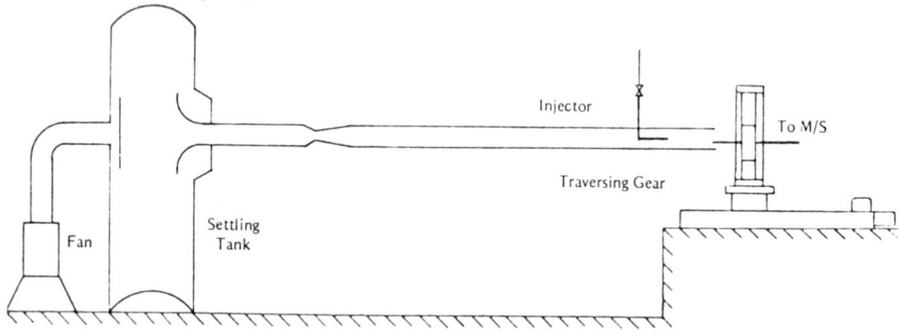

Fig. 3. Test Equipment

4.2 Quadrupole Mass Analyser

The concentration level of tracer gas in air is measured by a Quadrupole Mass Analyser whose mode of operation is briefly described below.

In order to achieve as rapid response as possible, a fast inlet system is used as shown in Fig. 4., whereby air entering the suction probe is pumped rapidly past the mass spectrometer inlet chamber. This contains a controlled leak which permits the induction of a small amount of the sampled mixture.

The heart of the V.G. Micromass Mass Analyser is Quadrupole Analysing Head which is attached to a two stage vacuum system so as to keep its interior under clean high vacuum conditions of around 10^{-7} torr. This head with its accompanying electronic equipment performs three basic functions:–

1. On admission to the analyser the gas mixture – which has been throttled from atmospheric pressure levels to high vacuum – is bombarded with a beam of electrons in order to achieve some ionisation. In fact approximately 1 molecule in 1,000 is ionised, most of these receiving a single positive charge.

2. Ionised gas is then induced to flow between an array of four long rods placed so that their axes form the long edges of a rectangular prism. Opposite pairs of rods or electrodes are connected together and to one pair is applied a potential consisting of a D.C. component and a sinusoidal R.F. voltage. The other pair of electrodes are supplied with the same voltage but of opposite sign.

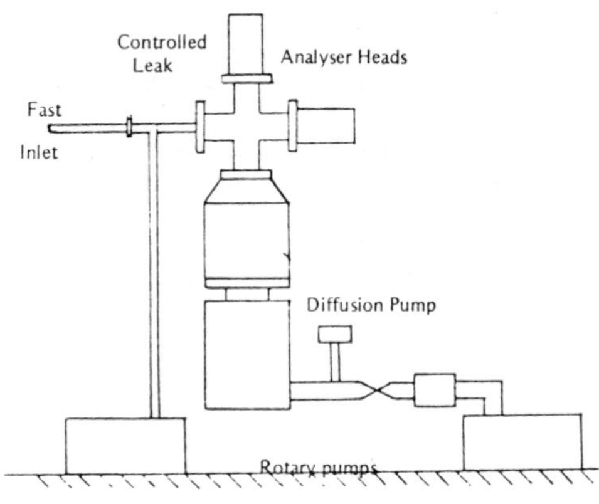

Fig. 4. Mass Spectrometer

When an ion is injected at one end of the assembly in a direction parallel to the axis of the rods, the R.F. and D.C. fields cause the ion to undergo transverse motion. Two types of motion are possible. In the first the ion oscillates about the axis and ultimately emerges from the opposite end of the array whilst in the second the ion diverges from the centreline eventually to be neutralised by striking one of the four electrodes.

The type of motion is determined by the choice of D.C. and R.F. voltages. By selection of these voltages only ions of a given mass to charge ratio have stable motion and emerge from the rod assembly, all other ions have unstable motion and strike an electrode. Thus the device acts as an ion filter on the basis of mass to charge ratio.

Orderly variation in the D.C. and R.F. voltages allows ions of different mass to charge ratios to be transmitted sequentially through the filter.

3. Ions which pass through the mass filter are directed on to a collector plate which generates a current according to the number of ions striking it. This collector is often called a Faraday Plate Collector which indicates that a means has been included to suppress the emission of secondary electrons from the surface, as these will cause a false indication of the ion current being monitored.

An analysis of a gas mixture is carried out by allowing a sample to enter the analyser head of the mass spectrometer. By varying the D.C. and R.F. voltages ions of increasing mass to charge ratio are allowed to pass through the mass filter and to strike the collector plate. A plot of the magnitude of the ion current generated at the collector plate for each mass to charge ratio provides a measure of the composition of the mixture.

Quantitative analysis of gas mixtures may be undertaken on the basis of a peak height measurement provided that a prior calibration of the instrument is carried out using standard mixtures.[5]

5. Results

5.1 Centre line decay in Laminar Flow

The experimental investigation was primarily concerned with establishing the validity of the centre line decay of concentration as predicted by equation 24. This was important since it formed one of the boundary values for the solution of equation 19 and 20.

There are no experimental values for the molecular coefficient of diffusion for argon in air and it is necessary to refer to kinetic theory to provide data. The value for the coefficient of diffusion for argon diffusing in air as determined by the kinetic theory at the conditions prevailing during the experiment is $0.000017 m^2/s$.

In order to obtain accurately measurable concentrations of gas in the mixing zone, it was found necessary to inject the gas at a greater rate than required by consideration of injector area and velocity of flow at the centre line of the duct. Within a very short distance downstream of the injector it was apparent that the stream line of injected gas experienced a reduction in velocity to match the prevailing centre line velocity. Continuity effects resulted in the injected stream line having a corresponding increase in flow area.

Using the earlier stated value for the coefficient of diffusion of argon in air and the effective injector area, calculated and experimental decay curves have been plotted (Fig. 5.). Although agreement between predicted and measured data is good a most significant variable is time of flight, t. The scatter of measured data is most probably due to the change in axial velocity with axial distance created by the presence of an injector in the flow, and not by a variation in diffusion rate. Some investigation of interference of the injector on radial velocity profile has been attempted using a hot wire anemometer, with air as the injected gas. A gradual acceleration of the centre line velocity downstream of the injector was recorded. Undoubtedly a similar acceleration exists when argon is injected but hot wire anemometry is unsuitable where the properties of the flowing fluid change from point to point.

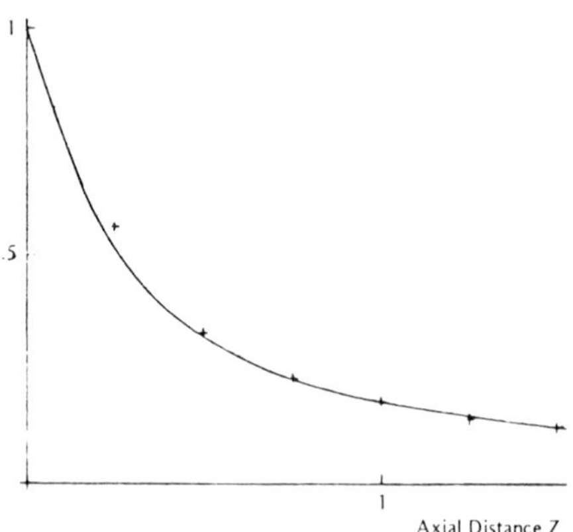

Fig. 5. Concentration Decay in Laminar Flow

All the data obtained in the laminar flow tests are found to collapse on to a single curve of non dimensional concentration against non dimensional time, provided the ratio of duct area to equivalent injector area is sufficiently large (Fig. 6).

The close agreement between theory and measurement suggests that all the centre line axial diffusion of the gas is of little importance, and that in laminar flow diffusion is governed by Fick's Law. The assumption of uniform flow has little effect upon the calculations of centre line decay.

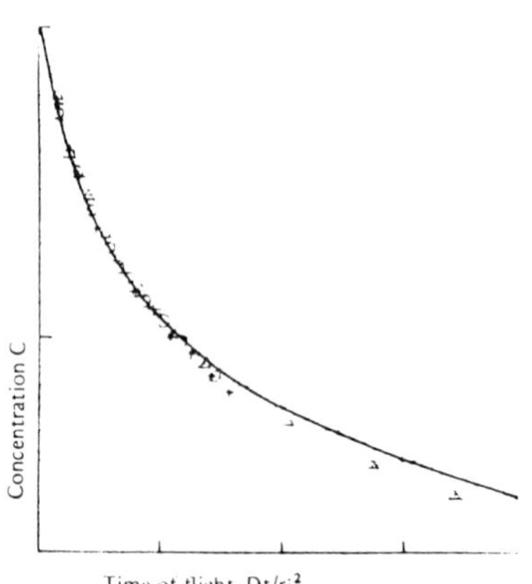

Fig. 6. Axial Decay of Concentration in Laminar Flow

5.2 Centre line decay in Turbulent Flow

As would be expected, in turbulent flow the decay in concentration of tracer gas along the axial centre line is much more rapid than calculated on the basis of a constant molecular diffusion coefficient. Also it is not surprising to find that a constant diffusion coefficient of any magnitude does not fit the measured data.

The manner in which the apparent coefficient of diffusion of argon in air was found to vary with time of flight from the injector was revealed by plotting measured data on to a family of decay curves computed on the basis of a range of constant diffusion coefficients (Fig. 7.). The intersections of the experimental curve with grid values enabled a curve of diffusivity against time of flight to be plotted. Then, by a curve fitting process, the variation of diffusivity with time of flight can be obtained.

The equation is of the form

$$(D + \epsilon_D) = a_0 + a_1 t + a_2 t^2 + a_3 t^3 \tag{28}$$

where a_0, a_1, a_2, a_3 are constants.

This equation intersects the zero time ($t = 0$) axis at a value which approximates to the coefficient of diffusion. At this instant the gas is at the point of injection. As the element of tracer gas moves downstream the diffusivity increases at an ever increasing rate for the first few jet diameters and then at an almost linear rate to the end of the test section. There are indications in all results that the rate of increase in diffusivity tends to reduce towards the exit of the test section.

Employing the derived values for the diffusivity in equation (27) a curve to fit the measured data may be produced (Fig. 8.).

In order to remove the effect of molecular diffusion the computer programme for equation (27) was rerun with the constant a_0 equal to zero. The dashed curve shows the 'self diffusion' decay of the jet, i.e. the effect of injecting air into air.

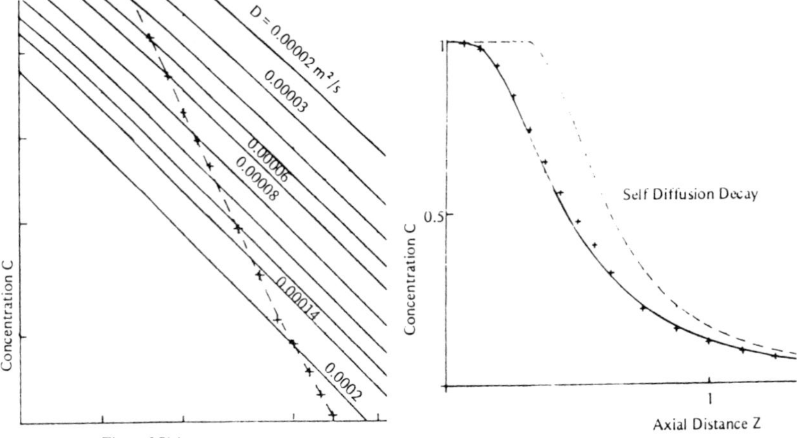

Fig. 7. Decay in Turbulent Flow

Fig. 8. Decay in Turbulent Flow

Conclusion

A useful flow tracing technique has been developed which is suitable for the study of both laminar and turbulent flows. In laminar flows the spread of the tracer gas has been shown to be due to molecular diffusion, and computer programmes have been written to predict concentrations throughout the mixing region. In turbulent flows the method has enabled eddy diffusivity to be measured.

REFERENCES

1. CRANK, J. — The Mathematics of Diffusion
Clarendon Press, Oxford 1970

2. ROHSENOW, W.H. & CHOI, H. — Heat Mass and Momentum Transfer
Prentice Hall, Inc. 1961

3. CHORLTON, F. — Boundary Valve Problems in Physics and Engineering
Van Nostrand Reinhold 1969

4. SMITH, G.D. — Numerical Solution of Partial Differential Equations: Finite Difference Methods.
Clarendon Press, Oxford 2nd Edition 1978

5. Quadrupole Mass Spectrometry,
Technical Information Leaflet No. 02.439.
V.G. – Quadrupoles Ltd,
Winsford, Cheshire.

ERROR AND STABILITY ANALYSIS OF THE FINITE ELEMENT SOLUTION FOR THE TRANSPORT EQUATION

G. A. Keramidas[1] and T. S. Papatheodorou[2]

ABSTRACT

Partial differential equations containing both convection and diffusion terms are usually the governing equations for flow processes. These have the form of second order partial differential equations and one example is the transport equation.

Several numerical techniques have been developed for the solution of the transport equation and they are based either on the finite difference or the finite element method.

It is of great importance to know whether a developed numerical scheme predicts accurately the amplitude and phase of a propagating wave or merely approximates the dynamics of the system. For modeling the transport equation, by the finite element method, one should investigate not only the efficiency of the numerical model but also its consistency for convergence and stability.

The present paper addresses these points of concern for the general in nature transport equation. The numerical model is subjected to an error analysis and the rate of convergence is investigated. Numerical results, obtained for this study, are based on a linear finite element approximation and they are used to verify the analytical formulation for error behavior.

INTRODUCTION

The solution of the transport equation by numerical methods has received considerable attention in recent years. A review and comparison of available methods can be found in [1], [2] and [3]. Comparisons between the approximate solutions and the available analytical ones have confirmed that the accuracy of the approximate methods is acceptable for engineering applications. The error involved in a numerical solution depends on a number of parameters and it is important to understand how these parameters affect the error and its behavior.

---

[1]Applied Mechanics Branch, Naval Research Laboratory
[2]Department of Computer Sciences, Clarkson College of Technology

Mathematical analyses and the approximation theory have established the theoretical foundations for error estimates and rates of convergence. Since for a numerical method both the size of the error and its rate of convergence are of interest, then an investigation based on concepts of the approximation theory should be a part of any numerical solution.

In the first part of this paper a review of the displacement variational formulation is given for the transport equation. This formulation leads to a Lagrangian type of equations by introducing the concept of generalized coordinates. The derived equations can be used for solving numerically the transport equation. In the second part of this paper the finite element method is applied to the previously derived formulation for solving numerically a boundary value problem. Futhermore, an energy function is introduced which is related to the energy function involved in the Lagrangian equations. The error of the numerical solution is estimated on this energy function and a relation is derived for the rate of convergence. The third part of the paper is focused on numerical results. The numerical solution is compared to the analytical one in order to obtain values for the error on the energy function and through numerical experimentations the rate of convergence is investigated.

Numerical results verify the approximation theory and justify the need for such an investigation. An extension of this investigation to other types of approximate solutions is of interest for future work.

VARIATIONAL FORMULATION

Consider an incompressible medium in a flow field with the intensive property to be transported at a uniform concentration C_0. The concentration C_0 will be referred to as the reference concentration and the state at this concentration will be referred to as the reference state. The value of the intensive property C at any other state will be the instantaneous concentration and the difference C-C_0 defines the instantaneous relative concentration, which is a function of the space and time coordinates. Let

$$\phi = \frac{C - C_0}{C_0} = \frac{\Delta C}{C_0} \tag{1.1}$$

be defined as the variable change per unit concentration C_0 or the instantaneous relative concentration. In the following it will be referred to as the concentration ϕ. Other properties of the medium can be defined in a similar way. The transport of the variable ϕ is governed by the equation

$$\frac{\partial \phi\,(x_k,t)}{\partial t} + \frac{\partial}{\partial x_i}\left(V_i \phi\,(x_k,t)\right) - \frac{\partial}{\partial x_i}\left(k_{ij}\frac{\partial \phi\,(x_k,t)}{\partial x_j}\right) = 0 \tag{1.2}$$

where V_i is the velocity field in the coordinate system x_k and time t, and k_{ij} is the diffusion coefficient. Equation (1.2) can be written in a different form by defining a displacement field $H_i\,(x_k,t)$ as follows

$$\phi\,(x_k,t) = \frac{\partial H_i\,(x_k,t)}{\partial x_i} = H_{i,i}(x_k,t). \tag{1.3}$$

With this definition of H_i, ϕ represents a deformation or strain analogous to mechanical strain and (1.3) is considered as a constraint in the sense of classical mechanics. In terms of the displacement H_i the governing equation (1.2) is written in a form similar to the equation of motion in mechanics as follows

$$\frac{d H_i}{d t} - k_{ij} \frac{\partial \phi}{\partial x_j} = 0 \tag{1.4}$$

where the time derivative is the total derivative given by

$$\frac{d H_i}{d t} = \frac{\partial H_i}{\partial t} + V_i \phi . \tag{1.5}$$

Equations (1.3) and (1.4) represent the displacement formulation for the transport equation and together with the appropriate boundary conditions provide a complete formulation for convective and diffusive transport phenomena.

Following the usual procedure of the principle of virtual work and virtual displacement of mechanics, a variational formulation can be derived for (1.4). The details of the derivation of this formulation can be found in [3]. Furthermore, taking the displacement field to be a function of the generalized coordinates q_k in the form

$$H_i(x_j,t) = H_i\,(q_k,\, x_j,\, t) \tag{1.6}$$

the variational formulation leads to the Lagrangian type of equations of classical mechanics given by

$$\frac{\partial D}{\partial \dot{q}_k} + \frac{\partial P}{\partial q_k} = Q_k . \tag{1.7}$$

Here D is defined as a dissipation function, P as a potential energy function; the Q_k are the boundary forces due to the distribution of ϕ at the boundary, i.e.

$$D = \frac{1}{2} \int_v \lambda_{ij}\, \dot{H}^*_i\, \dot{H}^*_j\, dv \tag{1.8}$$

$$P = \frac{1}{2} \int_v \phi^2\, dv \tag{1.9}$$

$$\dot{H}^*_i = \dot{H}_i + V_i \phi \tag{1.10}$$

and

$$Q_k = \int_s \phi\, n_i \frac{\partial H_i}{q_k}\, dS. \tag{1.11}$$

The term λ_{ij} is the inverse of k_{ij} and n_i is the unit vector normal to the boundary surface S of volume v. The vector $\dot{H}^*$ in (1.8) represents the total convective and diffusive rate of flow and the function D defines the total dissipation in the transport of ϕ.

2. APPROXIMATE SOLUTION

An approximate solution to (1.7) can be obtained by considering the displacement field to be given as a linear combinations of a set of generalized coordinates as follows

$$H_i\,(x_j,t) = q_k\,(t)\; f_{ki}\,(x_j) \qquad \begin{matrix} k = 1,n \\ j = 1,2,3 \end{matrix} \tag{2.1}$$

so that

$$\phi\,(x_j,t) = q_k(t)\; f_{ki,i}\,(x_j). \tag{2.2}$$

The generalized coordinates q_k represent degrees of freedom and the function f_{ki} specifies the extent to which q_k participates in the function H_i. In the finite element analysis, for example, equation (2.1) represents the distribution function of the displacement field where q_k can be taken as nodal displacements and/or nodal deformations. Introducing the approximations (2.1) into (1.7) one obtains

$$d_{ij}\,\dot{q}_j + p_{ij}\,q_j = Q_i \tag{2.3}$$

where

$$d_{ij} = \int_v \lambda_{mn}\,[f_{mi}\,f_{nj} + V_m\,f_{im}\,f_{jk,k}]\,dv$$

$$p_{ij} = \int_v f_{im,m}\,f_{jn,n},\,dv \tag{2.4}$$

and

$$Q_i = \int_s \phi\;n_j\,f_{ij}\,dS \tag{2.5}$$

Equations (2.3) constitute a system of n ordinary differential equations for the unknown field parameters q_i $(i = 1,n)$, which may represent the displacement field H_i. This system of equations can be solved together with the appropriate boundary and initial conditions by any numerical scheme to produce an approximate solution to the transport equation.

3. ERROR ANALYSIS

In order to obtain an approximate solution to the transport equation and to investigate error behavior, the finite element method is applied to the foregoing formulation. For the one dimensional case the approximation (2.1) yields

$$H\,(x,t) = \sum_{m=0}^{n} a_m x^m \tag{3.1}$$

and

$$\phi\,(x,t) = \sum_{m=1}^{n} m a_m x^{m-1}$$

where n is the order of the approximation. For example, for $n = 1$ we obtain the linear element, for $n = 2$ the quadratic element and $n = 3$ the cubic element approximations. The matrix equations are derived by introducing (3.1) into (2.3), (2.4) and (2.5). These derivations can be found in [3].

A total energy function E, associated to the displacement $H\,(x,t)$, can be defined from (1.7) or (2.3) as follows

$$E\,(H) = \frac{1}{2}\int_0^L H\,(x,t)^2 dx + \frac{\theta}{2}\int_0^t \left\{ \int_0^L \phi\,(x,\bar{t})^2\,dx \right\} d\bar{t} \tag{3.2}$$

where, for brevity, the dependence on t is occasionally suppressed, θ is any constant $0 < \theta < 1$ and L is a characteristic length representing the integration domain. The first term in (3.2) is related to the dissipation D and the second to the potential P. In order to evaluate the error involved in the approximate solution, it is of interest to estimate the speed with which $E\,(H - H_n)$ converges to zero. The error ϵ_n in the energy E is due to $H - H_n$, where H_n is the approximate displacement produced by the finite element method. Here n is the total number of degrees of freedom. If NE is the number of subintervals (elements) used, then $n = NE + 1$ for the linear and $n = 2\,(NE + 1)$ for the quadradic case or in general $n = m\,(NE + 1)$, $m = 1, 2, 3, \ldots$, for first, second and third, cases respectively.

The best possible speed of convergence to zero of $E(H - H_n)$ is achieved when H_n is the best approximation to H from among all sets of the form $q_i f_i$. Let U_n represent the best approximation to H, then

$$U_n = \alpha_i f_i. \tag{3.3}$$

It is a well known result of the approximation theory that the best approximation U_n exists. Then there exist constants $C_0(H)$, $C_1(H)$, $C_t(H)$, which depen- d on H and some of its derivatives but are independent of n, such that for any final time t_f

$$\sup_{0 \leqslant t \leqslant t_f} \int_0^L \left[(H - U_n)(x,t) \right]^2 dx \leqslant C_0(H)\, h^{2m+2};$$

$$\sup_{0 \leqslant t \leqslant t_f} \int_0^t \left[\frac{\partial}{\partial x} (H - U_n)(x,t) \right]^2 dx \leqslant C_1(H)\, h^{2m}; \tag{3.4}$$

$$\sup_{0 \leqslant t \leqslant t_f} \int_0^t \left[\frac{\partial}{\partial t} (H - U_n)(x,t) \right]^2 dx \leqslant C_t(H)\, h^{2m+2};$$

where $h = \Delta x = 1/NE$. Note that $h \to 0$ if and only if $n \to \infty$. Introducing (3.4) into (3.2) we find

$$\left\{ \sup_{0 \leqslant t \leqslant t_f} [E(H - U_n)](t) \right\}^{1/2} \leqslant \hat{C}(H)\, h^m \tag{3.5}$$

where $\hat{C}(H)$ is also independent of h and n. Thus, as $n \to \infty$ the best error in E converges to zero with speed h^m. For the case of the finite element method the same best speed is attained by the error convergence provided that the initial and boundary conditions are also approximated with sufficient accuracy. This is justified by the following theorem;

Theorem: If there exists a constant $\bar{C}(H)$ depending on H and its derivatives, and it is independent of h (and n), such that both

$$\int_0^L [(H - H_n)(x,0)]^2\, dx \leqslant \bar{C}(H)\, h^{2m} \tag{3.6}$$

and

$$\sup_{0 \leqslant t \leqslant t_f} \left\{ \int_0^t [(H - H_n)^2 (0,\bar{t}) - (H - H)^2 (L,\bar{t})]\, d\bar{t} \leqslant \bar{C}(H)\, h^{2m} \right.$$

are true, then there exists a constant $C(H)$ depending on H and some of its derivatives and independent of h (and n) such that

$$\left\{ \sup_{0 \leqslant t \leqslant t_f} [E(H - H_n)(t)] \right\}^{1/2} \leqslant C(H)\, h^m. \tag{3.7}$$

In order to investigate the error behavior of the approximate solution some numerical experimentation is required which will provide us with information on the rate of convergence. Details of the foregoing analysis can be found in [4].

4. NUMERICAL RESULTS AND DISCUSSION

The following problem is considered here for the transport equation

$$\frac{\partial H}{\partial t} + V_0 \phi - \frac{\partial \phi}{\partial x} = 0 \tag{4.1}$$

with the following initial and boundary conditions

$$\phi\,(x,0) = 0.0 \quad 0 \leqslant x \leqslant L$$

$$\phi\,(0,t) = 1.0 \quad 0 < t \leqslant t_f$$

$$\phi\,(L,t) = 0.0 \quad 0 \leqslant t \leqslant t_f$$

where L is the characteristic length approximating infinity. The exact solution to the above problem is given by

$$\phi\,(x,t) = \frac{1}{2}\left[\exp\,(V_0 x)\cdot erfc\left(\frac{x + V_0 t}{2\sqrt{t}}\right) + erfc\left(\frac{x - V_0 t}{2\sqrt{t}}\right)\right]$$

The numerical solution is obtained from the system of ordinary differential equations given by (2.3). A third order backward finite difference scheme is used for the time integration, and the displacement H is a linear approximation of the x coordinate. Two sets of numerical results are presented here. For both of these sets the characteristic length L was taken equal to five and the number of elements (NE) ranges from 5 to 50, thus the length of its element Δx ranges from 1.0 to 0.1 respectivelly. Each set of results was obtained for a constant value of $\Delta t/\Delta x$ and they are presented for three different times for $0 \leqslant t \leqslant 5.0$. The error ϵ_n of the numerical solution is evaluated according to (3.5) and this relation can be translated to the following equation

$$\epsilon_{NE} = C\,(H)\,(\Delta x)^m$$

where ϵ_{NE} is the error norm corresponding to (NE) elements of length Δx.

Figure 1 shows the error norm as a function of (NE). One can observe the uniform convergence of the solution for both sets of results and that the solution attains a minimum error well before $NE = 50$. For the larger value of the ratio $\Delta t/\Delta x$ the error is larger as we should expect.

Figure 1 is necessary for understanding the error behavior but in order to show the validity of the error analysis and to investigate the speed of convergence additional results are needed. From (4.1) we derive the following

$$\log\,(\epsilon_n) = m\,\log\left(\frac{n_1}{n_2}\right) \qquad (4.2)$$

where ϵ_n is the ratio of the error corresponding to n_i degrees of freedom over the error corresponding to n_2 degrees of freedom. Results obtained by (4.2) are presented in figure 2. There $\log\,(\epsilon_n)$ is given as a function of $\log\,(NE/5)$ where NE takes values from 5 to 50. Again two sets of results are presented which correspond to the two previous sets of results of figure 1. For each case the speed of convergence of the numerical solution, can be determined from figure 2 or from (4.2), as the slope m at each point of the given lines. The slope m is presented in figure 3, for each case, as a function of (NE). As one can observe from this figure the slope m or the rate of convergence attains a constant value for (NE) larger than 25. This is justified by the results for the error of the numerical solution presented in figure 1. Furthermore, from all figures one can observe that there is only a very small improvement of the numerical solution when (NE) is increased beyond the value of 25.

In order to understand and evaluate the error behavior and stability of the numerical solution all three figures should be used together. For example, the

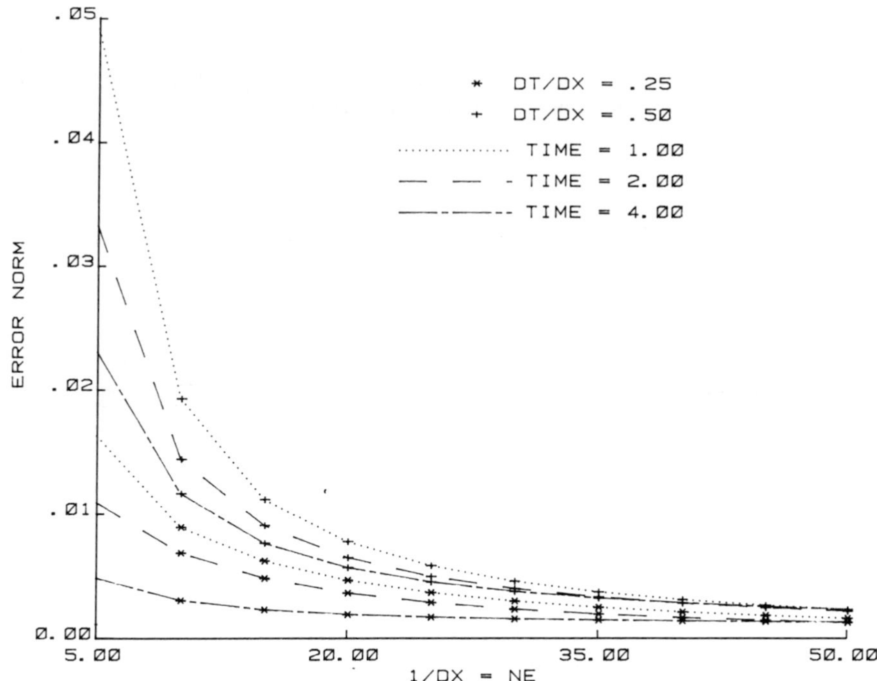

Fig. 1. — Error histories of the numerical solution.

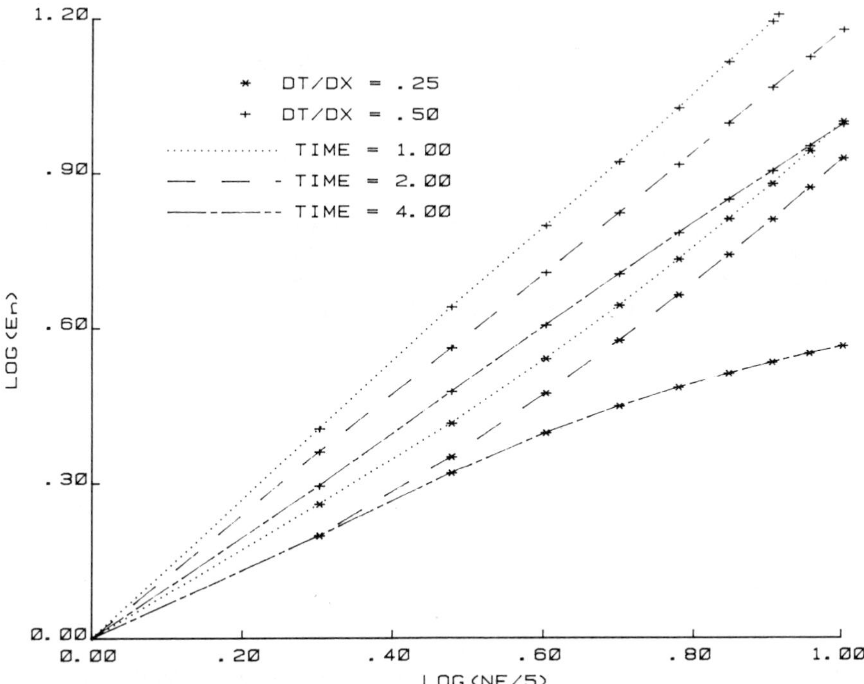

Fig. 2. — Normalized error as a functions of the normalized number of elements (NE).

values of m that are greater than one in figure 3 can be justified if the error values in figure 1 for the corresponding cases are taken into consideration. Comparing the three figures one can evaluate not only the size of the error but also how fast the solution converges by increasing the number of elements.

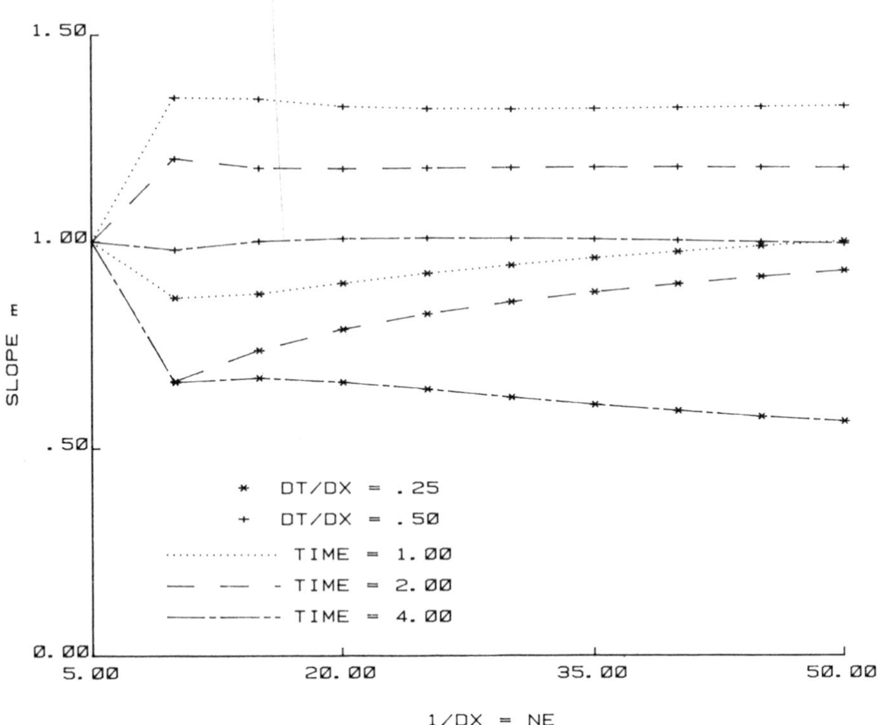

Fig. 3. — Rate of convergence of the numerical solution.

According to the approximation theory the best speed of convergence is equal to one for the linear element approximation. This can be verified by the obtained results and by taking into account the relation given by (3.5). Moreover, the results obtained verify the error analysis for the approximate solution introduced previously.

In conclusion, one should point out the importance of the error analysis for any approximate solution. Such analysis should include both a theoretical foundation, relative to the approximate solution, and numerical experimentation. Finally, to further verify the approximation theory for error analysis and stability, some higher order finite element approximations will be the subject for future investigations.

REFERENCES

1. Lee R., et al, "A comparative study of certain finite element and finite difference methods for advection-diffusion", 1975 Summer Computer Simulation Conference, Washington, DC (July 1975).

2. Smith, I.M. et al, "Raleigh-Ritz and Galerkin finite elements for diffusion-convection problems", Water Resources Research, Vol. 9, 3, (1973).

3. Keramidas, G.A., "Finite Element Modeling of Convection-Diffusion Problems", NRL Memorandum Report 4225, 1980.

4. Keramidas, G.A., Papatheodorou, T.S., "Error analysis of the finite element solution for the transport equation", NRL Memorandum Report (to be published).

SOLUTION OF FLUID FLOW PROBLEMS BY A DIRECT SPLINE INTERPOLATION METHOD

J. Häuser, D. Eppel
Institut für Physik, GKSS-Research Center
2054 Geesthacht, West-Germany

F. Tanzer
1. Phys. Institut, Universität Giessen
6300 Giessen, West-Germany

ABSTRACT

The present paper reports on the investigation and results of a direct cubic spline procedure for the numerical approximation of first and second spatial derivatives and its application to fluid flow problems. The method developed, utilizing a uniform mesh, does not demand the solution of a system of linear equations, rather the spline coefficients are determined by analytical formulas depending on the boundary values and the eigenvalues of a 2 x 2 matrix. The method is applied to the solution of the nonlinear one-dimensional Burgers equation, the atmospheric boundary layer equations, and to the diffusion equation, where the medium moves at a constant and uniform velocity. The numerical results are compared with the corresponding (stationary) analytical solutions.

The partial differential equations are reduced to a weakly coupled system of ordinary differential equations, first order in time. Using the spline approximation for the spatial derivatives, the system is integrated by a fourth-order Runge-Kutta method.

A detailed truncation error analysis for the spline derivatives is outlined, including boundary effects. It is found that the method is of fourth order accuracy for the first derivative and of second order accuracy for the second derivative, provided the approximation of the first derivatives at the boundaries is fourth order accurate. Furthermore, a stability analysis for Burgers equation, based on singular perturbation theory, is presented giving an improved estimate for the Courant and cell Reynolds number.

INTRODUCTION

In the following a description of the spline approximation to the spatial derivatives is given. Since in fluid problems only first and second derivatives occur, the construction of spline functions being twice continuously differentiable is sufficient. Let us take the two-dimensional Navier-Stokes equation without pressure terms as an example for a nonlinear parabolic partial differential equation (PDE)

$$\frac{\partial v_i}{\partial t} + v_k \frac{\partial}{\partial x_k} v_i = \mu \frac{\partial^2 v_i}{\partial x_k \partial x_k} \quad ; \; i = 1, 2 \tag{1}$$

where v_i denotes the velocity of the fluid. In cases of practical relevance analytic solutions of Eqs. (1) are not possible, hence numerical methods must be employed to approximate this PDE. For that purpose a mesh is spanned over the region of interest and function values are known at grid points only. To evaluate the first and second derivatives in Eqs. (1) finite differences [1] are being used extensively. In some cases, however, the solutions so obtained are not satisfactory because of insufficient accuracy, spurious oscillations, and large computer time or storage problems. Recently a series of papers were published by Rubin, Graves and Khosla [2, 3, 4] using polynomial splines. Their numerical procedure results in the solution of a tridiagonal or 2 x 2 block-tridiagonal system of linear equations.

The present paper describes a direct cubic spline method, and also avoids the solution of a linear system. For the sake of simplicity a rectangular solution area is used. Let $a \le x_o < x_1 \ldots < x_n = b$ and $c \le y_o \le y_1 < \ldots < y_m = d$ be two partitions of the intervals [a, b] and [c, d] in the x and y direction where $h = x_{i+1} - x_i$ and $k = y_{j+1} - y_j$ denote the constant grid spacings. From hereon the notation $\vec{v} = (v_1, v_2) = (u, v)$ and $\vec{x} = (x_1, x_2) = (x, y)$ is used. Then we take the splinefunction S(t) (t stands for x or y), constructed (see Sec. 1) from the corresponding u or v values of the different rows or columns and make the ansatz

$$\begin{aligned} u(x,y_j) &= S_j^u(x);\; v(x,y_j) = S_j^v(x);\; x \in \{x_o, x_1, \ldots, x_n\};\; j = 1(1)m \\ u(x_i,y) &= S_i^u(y);\; v(x_i,y) = S_i^v(y);\; y \in \{y_o, y_1, \ldots, y_m\};\; i = 1(1)n \end{aligned} \tag{2}$$

Since the splinefunctions have continuous derivatives of up to second order the required derivatives are obtained by differentiating $S_j(x)$, $S_i(y)$ and inserting these approximate values into Eqs. (1). Thus any finite-difference discretization is avoided. Eqs. (1) are reduced to a system of ordinary differential equations of the form

$$\frac{\partial}{\partial t} u_{i,j} = f(u_{i,j}, v_{i,j}, \frac{\partial}{\partial x}\left(S_j^u(x)\right), \frac{\partial}{\partial y}\left(S_i^u(y)\right), \frac{\partial^2}{\partial x^2}\left(S_j^u(x)\right), \frac{\partial^2}{\partial y^2}\left(S_i^u(y)\right), t)$$

$$\frac{\partial}{\partial t} v_{i,j} = f\left(u_{i,j}, v_{i,j}, \frac{\partial}{\partial x}\left(S_j^v(x)\right), \frac{\partial}{\partial y}\left(S_i^v(y)\right), \frac{\partial^2}{\partial x^2}\left(S_j^v(x)\right), \frac{\partial^2}{\partial y^2}\left(S_i^v(y)\right), t\right) \quad (3)$$

where $i = 1(1)n$; $j = 1(1)m$. This system of equations is integrated in time by a fourth order Runge-Kutta method [5, 6]. Other methods for the time integration are given in [7, 8]. Predictor-corrector codes [9] which are faster than the Runge-Kutta method require substantially more storage, so that their use is not recommended for large systems of equations. Since Eqs. (1) are reduced to a system of first-order equations initial conditions are necessary . The prescribed boundary conditions are incorporated in the spline formulation. The spline method is also applicable to multiconnected areas provided a sufficient number of successive base points in every row and column are available. The direct spline method presented in this paper is restricted to uniform meshes. The application of the direct method to nonequidistant meshes requires a parameter representation [10] or a tridiagonal system of linear equations has to be solved [2, 3, 4].

To demonstrate the method the one-dimensional nonlinear Burgers, the diffusion equation and the Ekman layer equations are calculated.

1. SPLINE FORMULATION

It is known that interpolation polynomials of higher order can exhibit strong oscillations between the tabular points while the function varies smoothly. Hence, the numerical evaluation of derivatives is normally not possible by differentiating the corresponding collocation polynomial. This problem is overcome when piecewise low-order interpolating polynomials are used, representing the function between successive base points and fulfilling additional smoothness conditions at the end of each subinterval. Functions with a maximal degree of smoothness are called splines [11]. The following gives the definition for cubic splines and the method for the direct calculation of the spline coefficients is derived.

Let the interval $I = [a,b]$ be divided into n subintervals $a = x_o < x_1 < \dots < x_n = b$ of equal length $h = x_{i+1} - x_i$. Consider a function $u(x)$ (to be approximated) such that at the mesh points x_i the values $u_i = u(x_i)$ are taken. A spline $S(x; m,k)$ is a function of degree m having m-k continuous derivatives (that is, $S(x; m, k)$ is in the set $C^{m-k}[a,b]$), coincides with a polynomial on each subinterval $I_i = [x_i, x_{i+1}]$ (k is called the deficiency of the spline) and interpolates data points u_i at mesh points x_i. In this paper we are only concerned with splines $S(x; 3, 1) \equiv S(x)$.

The spline coefficients are determined by the following four properties

(i) $S(x_i) = u_i$; $i = 0(1)n$,

(ii) $S'(x)$, $S''(x)$ are continuous on I,

(iii) $S(x)$ is a cubic polynomial in each subinterval $[x_i, x_{i+1}]$; $i = 0(1)n-1$,

(iv) boundary conditions at x_o and x_n for either the first or second derivatives are specified (see below).

For brevity we use $S'(x_i) = m_i$ and $S''(x_i) = s_i$. Properties (i) and (iii) are satisfied by the ansatz ($S_i(x)$ denotes the polynomial for $x \in [x_i, x_{i+1}]$):

$$S_i(x) = u_i + m_i(x-x_i) + 1/2\, s_i(x-x_i)^2 + (u_{i+1} - u_i - m_i h - 1/2\, s_i h^2)\left(\frac{x - x_i}{h}\right)^3 \tag{4}$$

$x_i \leq x \leq x_{i+1}$ and $0 \leq i \leq n - 1$. From this ansatz the relations follow

$$S_i(x_i) = u_i \; ; \quad S_i(x_{i+1}) = u_{i+1} \; ; \quad S_i'(x_i) = m_i \; ; \quad S_i''(x_i) = s_i .$$

Differentiating Eq.(4) gives

$$\begin{aligned} S_i'(x) &= m_i + s_i(x-x_i) + 3(u_{i+1} - u_i - m_i h - 1/2\, s_i h^2)\;(x-x_i)^2/h^3 \\ S_i''(x) &= s_i + 6(u_{i+1} - u_i - m_i h - 1/2\, s_i h^2)\;(x - x_i)/h^3 \end{aligned} \tag{5}$$

Inserting $x = x_{i+1}$ into Eq.(5) and using condition (ii) yields two formulas for the first and second derivatives

$$\begin{aligned} m_{i+1} &= m_i + s_i h + 3(u_{i+1} - u_i - m_i h - 1/2\, s_i h^2)\;/h \\ s_{i+1} &= s_i + 6(u_{i+1} - u_i - m_i h - 1/2\, s_i h^2)\;/h^2 \end{aligned} \tag{6}$$

Rewriting these equations in matrix representation one has

$$\begin{pmatrix} m_{i+1} \\ s_{i+1} \end{pmatrix} = \begin{pmatrix} -2 & -h/2 \\ -\frac{6}{h} & -2 \end{pmatrix} \begin{pmatrix} m_i \\ s_i \end{pmatrix} + \frac{3}{h}\,(u_{i+1} - u_i) \begin{pmatrix} 1 \\ \frac{2}{h} \end{pmatrix} \tag{7}$$

where $A = \begin{pmatrix} -2 & -h/2 \\ -6/h & -2 \end{pmatrix}$ has the eigenvalues $\lambda_1 = -2 + \sqrt{3}$, $\lambda_2 = -2 - \sqrt{3}$ and $\lambda_1 \lambda_2 = 1$. If m_o and s_o are known all other derivatives can be determined from the recursion formula Eq.(7). The above matrix formulation, however, is not used for numerical computations since this can lead to round-off errors for small h.

Calculating the matrix A^i one obtains

$$A^i = \begin{pmatrix} 1/2\,(\lambda_1^i + \lambda_2^i) & -h/2\,\dfrac{\lambda_1^i - \lambda_2^i}{\lambda_1 - \lambda_2} \\ -6/h\,\dfrac{\lambda_1^i - \lambda_2^i}{\lambda_1 - \lambda_2} & 1/2\,(\lambda_1^i + \lambda_2^i) \end{pmatrix} \tag{8}$$

To express the first and second derivatives at point x_i solely by their corresponding values at $x = x_o$; i.e., m_o and s_o, the following definitions are introduced

$$\vec{d}_o := \begin{pmatrix} 0 \\ 0 \end{pmatrix};\ \vec{d}_i := \begin{pmatrix} d_i^{(1)} \\ d_i^{(2)} \end{pmatrix} = \frac{3}{h} \sum_{\ell=0}^{i-1} (u_{\ell+1} - u_\ell) A^{i-1-\ell} \begin{pmatrix} 1 \\ \dfrac{2}{h} \end{pmatrix}; \tag{9}$$

$i = 1(1)\,n$

Combining Eqs. (7,8,9) results in

$$\begin{pmatrix} m_i \\ s_i \end{pmatrix} = A^i \begin{pmatrix} m_o \\ s_o \end{pmatrix} + \vec{d}_i \quad ; \quad i = 1(1)\,n \tag{10}$$

For i = n the following relations hold

$$m_n = 1/2(\lambda_1^n + \lambda_2^n)\, m_o - h/2\,\frac{\lambda_1^n - \lambda_2^n}{\lambda_1 - \lambda_2}\, s_o + d_n^{(1)} \tag{11}$$

$$s_n = -6/h\,\frac{\lambda_1^n - \lambda_2^n}{\lambda_1 - \lambda_2}\, m_o + 1/2\,(\lambda_1^n + \lambda_2^n)\, s_o + d_n^{(2)}$$

Solving for the second derivatives

$$s_o = \frac{\lambda_1 - \lambda_2}{h(\lambda_1^n - \lambda_2^n)} \left\{ (\lambda_1^n + \lambda_2^n)\, m_o - 2\, m_n + 2\, d_n^{(1)} \right\}$$

$$s_n = \frac{\lambda_1 - \lambda_2}{h(\lambda_1^n - \lambda_2^n)} \left\{ 2\, m_o - (\lambda_1^n + \lambda_2^n)\,(m_n - d_n^{(1)}) \right\} + d_n^{(2)} \tag{12}$$

Eqs.(11) or (12) are two equations for four variables. The simplest way to procure two additional relations is the approximation of either m_o, m_n or s_o, s_n by finite differences. The accuracy of the spline method so obtained is dependent on the accuracy of the finite difference approximation at least near boundaries (see Sec. 2). Eqs.(12) are now used to eliminate s_o and s_n from Eqs.(10). Thus, if m_o and m_n are known all derivatives m_i and s_i can be calculated. Hence, Eqs.(10) take the form

$$m_i = \frac{m_o\,\lambda^i(1-\lambda^{2(n-i)}) + (1 - \lambda^{2i})\,\lambda^{n-i}(m_n - d_n^{(1)}) + (1-\lambda^{2n})\, d_i^{(1)}}{1 - \lambda^{2n}} \tag{13}$$

$$s_i = \frac{-2\sqrt{3}\,m_o(1+\lambda^{2(n-i)})\lambda^i + 2\sqrt{3}\,(1+\lambda^{2i})\lambda^{n-i}(m_n - d_n^{(1)}) + h(1-\lambda^{2n})d_i^{(2)}}{h(1-\lambda^{2n})}$$

In order to avoid round-off errors $\lambda_2 = -2-\sqrt{3}$ ($|\lambda_2|>1$) has been eliminated in Eqs.(13) and $\lambda = \lambda_1$ is used for brevity. These errors can occur when A^i is calculated and in particular in the calculation of the vector $\vec{d}_i$ because of the repeated use of A^i. Next, to simplify the above equations, all terms which are independent of m_o and m_n are evaluated.

$$A^i \begin{pmatrix} 1 \\ \frac{2}{h} \end{pmatrix} = (\sqrt{3}+1) \begin{pmatrix} \frac{\lambda_1^{i+1}}{2\sqrt{3}} & \frac{\lambda_2^i}{2\sqrt{3}} \\ \frac{\lambda_1^{i+1}}{h} & \frac{\lambda_2^i}{h} \end{pmatrix} \tag{14}$$

From the definition of vector $\vec{d}_i$ and using Eq.(14) one obtains

$$\vec{d}_i = -\frac{3(\sqrt{3}-1)}{h} \sum_{\ell=o}^{i-1} (u_{\ell+1} - u_\ell) \begin{pmatrix} \frac{-\lambda^{i-1-\ell}}{2\sqrt{3}} & \frac{\lambda^{\ell-i}}{2\sqrt{3}} \\ \frac{\lambda^{i-1-\ell}}{h} & \frac{\lambda^{\ell-i}}{h} \end{pmatrix} \tag{15}$$

To get a more compact form of Eqs.(15) the following quantities are introduced

$$G_i = \sum_{\ell=0}^{i-1} (u_{\ell+1}-u_\ell)\lambda^\ell \; ; \; H_i := \sum_{\ell=0}^{i-1} (u_{\ell+1}-u_\ell)\lambda^{i-1-\ell} ;$$

$$R_i := \sum_{\ell=i}^{n-1} (u_{\ell+1}-u_\ell)\lambda^{\ell-i} \tag{16}$$

where $G_o = 0$, $H_o = 0$ and $R_n = 0$.

Hence Eqs.(15) assume the form

$$\vec{d}_i = -\frac{3(\sqrt{3}-1)}{h} \begin{pmatrix} -\frac{1}{2\sqrt{3}} H_i + \frac{1}{2\sqrt{3}\,\lambda^i} G_i \\ \frac{H_i}{h} + \frac{G_i}{h\lambda^i} \end{pmatrix} \tag{17}$$

Inserting the relations

$$G_n = G_i + \lambda^i R_i = R_o$$

$$\frac{G_i}{\lambda^i} = \frac{R_o}{\lambda^i} - R_i \tag{18}$$

into Eq.(17) the final form for the vector components of $\vec{d}_i$ is obtained

$$d_i^{(1)} = \frac{3(\sqrt{3}-1)}{2\sqrt{3}\,h}\left(H_i + R_i - \frac{R_o}{\lambda^i}\right)$$

$$d_i^{(2)} = \frac{3(\sqrt{3}-1)}{h^2}\left(-H_i + R_i - \frac{R_o}{\lambda^i}\right) \tag{19}$$

In particular we have

$$d_n^{(1)} = \frac{3(\sqrt{3}-1)}{2\sqrt{3}\,h}\left(H_n - \frac{R_o}{\lambda^n}\right);\quad d_n^{(2)} = -\frac{3(\sqrt{3}-1)}{h^2}\left(H_n + \frac{R_o}{\lambda^n}\right) \tag{20}$$

Inserting the equations for $d_n^{(1)}$ and $d_n^{(2)}$ into the equations for the first and second derivatives(Eqs.(13)) eventually leads to

$$m_i = \frac{\lambda^i(1-\lambda^{2(n-i)})}{1-\lambda^{2n}}\left(m_o - \frac{1-\lambda}{2h}R_o\right) + \frac{(1-\lambda^{2i})\,\lambda^{n-i}}{1-\lambda^{2n}}\left(m_n - \frac{1-\lambda}{2h}H_n\right) + \frac{1-\lambda}{2h}(R_i + H_i)$$

$$\frac{hs_i}{2\sqrt{3}} = \frac{-\lambda^i(1+\lambda^{2(n-i)})}{1-\lambda^{2n}}\left(m_o - \frac{1-\lambda}{2h}R_o\right) + \frac{\lambda^{n-i}(1+\lambda^{2i})}{1-\lambda^{2n}}\left(m_n - \frac{1-\lambda}{2h}H_n\right) + \frac{1-\lambda}{2h}(R_i - H_i) \tag{21}$$

Thus, for known derivatives m_o and m_n all derivatives m_i and s_i can be calculated from Eqs.(21). For the numerical calculations these equations can be substantially reduced, since for a prescribed accuracy e_r, the exponent $\ell = \log_{|\lambda|} e_r$ gives $|\lambda|^\ell \leq e_r$, e.g. for $e_r = 10^{-6}$ we find $\ell = 11$. Since the truncation error e_t is normally much larger than a relative error $e_r = 10^{-6}$ this error will be negligible in the calculation of the derivatives. Therefore all terms having exponents greater than ℓ can be dropped in Eqs.(21). Moreover, the calculation of the sums H_i and R_i is simplified, provided enough base points are available. From Eqs.(21) we know

$$\frac{hs_o}{2\sqrt{3}} = -\frac{1+\lambda^{2n}}{1-\lambda^{2n}}\left(m_o - \frac{1-\lambda}{2h}R_o\right) + \frac{2\lambda^n}{1-\lambda^{2n}}\left(m_n - \frac{1-\lambda}{2h}H_n\right) - \frac{1-\lambda}{2h}R_o \tag{22}$$

$$\frac{hs_n}{2\sqrt{3}} = -\frac{2\lambda^n}{1-\lambda^{2n}}\left(m_o - \frac{1-\lambda}{2h}R_o\right) + \frac{1+\lambda^{2n}}{1-\lambda^{2n}}\left(m_n - \frac{1-\lambda}{2h}H_n\right) + \frac{1-\lambda}{2h}H_n \tag{23}$$

Defining the new variables $\bar{m}_o = m_o - \frac{1-\lambda}{2h} R_o$; $\bar{m}_n := m_n - \frac{1-\lambda}{2h} H_n$, we obtain from Eqs.(22)

$$\frac{h}{2\sqrt{3}}((1+\lambda^{2n})\, s_n - 2\lambda^n s_o) = \frac{(1+\lambda^{2n})^2 - 4\lambda^{2n}}{1-\lambda^{2n}} \bar{m}_n + \frac{1-\lambda}{2h}((1+\lambda^{2n}) H_n + 2\lambda^n R_o)$$

$$\frac{h}{2\sqrt{3}}((1+\lambda^{2n})\, s_o - 2\lambda^n s_n) = \frac{4\lambda^{2n} - (1+\lambda^{2n})^2}{1-\lambda^{2n}} \bar{m}_o - \frac{1-\lambda}{2h}(2\lambda^n H_n + (1+\lambda^{2n}) R_o) \tag{24}$$

If we assume $n > 6$ (for $n \leq 6$ the terms for the spline coefficients can be written down explicitly) then $\lambda^{2n} \leq 9.8 \times 10^{-9}$. Thus λ^{2n} is neglected resulting in

$$\bar{m}_o = \frac{h}{2\sqrt{3}}(2\lambda^n s_n - s_o) + \frac{1-\lambda}{2h}(2\lambda^n R_o + H_n)$$

$$\bar{m}_n = \frac{h}{2\sqrt{3}}(s_n - 2\lambda^n s_o) - \frac{1-\lambda}{2h}(R_o + 2\lambda^n H_n) \tag{25}$$

Eqs.(25) show that approximations of the second derivatives at the boundary points $x = x_o$ and $x = x_n$ are necessary. If one wishes to apply the common three-point discretization formula for s_o and s_n, extrapolation to fictitious grid points x_{-1} and x_{n+1} must be used. This extrapolation can be avoided if corresponding formulas for s_1 and s_n are chosen, leading to

$$\begin{pmatrix} \bar{m}_o \\ \bar{m}_n \end{pmatrix} = -\frac{\lambda}{\lambda^2 - \lambda^{2(n-1)}} \begin{pmatrix} 1+\lambda^{2(n-1)} & 4\lambda^{n-1} \\ -4\lambda^{n-1} & -1+\lambda^{2(n-1)} \end{pmatrix} \times \begin{pmatrix} \frac{h}{2\sqrt{3}} s_1 - \frac{1-\lambda}{2h}(R_1 - H_1) \\ \frac{h}{2\sqrt{3}} s_{n-1} - \frac{1-\lambda}{2h}(R_{n-1} - H_{n-1}) \end{pmatrix} \tag{26}$$

Neglecting all terms containing $\lambda^{2(n-1)}$ (i.e., the relative error has the magnitude $\lambda^{2(n-1)-1}$), we get

$$\bar{m}_o = \frac{1}{\lambda}\left\{\frac{h}{2\sqrt{3}}(4\lambda^{n-1} s_1 + s_{n-1}) - \frac{1-\lambda}{2h}(4\lambda^{n-1}(R_1 - H_1) + R_{n-1} - H_{n-1})\right\}$$

$$\bar{m}_n = -\frac{1}{\lambda}\left\{\frac{h}{2\sqrt{3}}(s_1 + 4\lambda^{n-1} s_{n-1}) - \frac{1-\lambda}{2h}(R_1 - H_1) + 4\lambda^{n-1}(R_{n-1} - H_{n-1})\right\} \tag{27}$$

For large n Eqs.(26) can still be further simplified. Since the end point derivatives are not normally known the 'not-a-knot' condition [12] can be used, i.e., the first and last interior knots are not active. However, incorporating this condition in the spline formulation it turns out to be equivalent to the approximation of the second derivative by the familiar three-point discretization formula.

2. TRUNCATION ERROR ANALYSIS

The truncation error analysis is performed for interior points and for the two boundary points showing the influence of the approximation at these points on the overall truncation error. Introducing the translation operator $u(x+h) = e^{hD}u(x)$; $D = \frac{\partial}{\partial x}$ the following two operators are defined

$$T_i^H := (1-e^{hD}) \frac{1-\lambda^i e^{-ihD}}{1-\lambda e^{-hD}} e^{ihD}$$
$$T_i^R := (e^{hD}-1) \frac{1-\lambda^{n-i} e^{(n-i)hD}}{1-\lambda e^{hD}} e^{ihD} \tag{28}$$

The construction of these operators follows directly from the definitions of H_i and R_i (Eq. (16)). Furthermore $u_{\ell+1} - u_\ell = \Delta u_\ell = (e^{hD}-1)\, u_\ell$ (Δ denotes the forward difference operator) was used. From the construction used it is obvious that

$$H_i = T_i^H(u_o) \quad ; \quad R_i = T_i^R(u_o) \tag{29}$$

$$T_n^H = (e^{hD}-1)\frac{e^{nhD}-\lambda^n}{e^{hD}-\lambda} \quad ; \quad T_n^H(u_o) = H_n$$
$$T_o^R = (e^{hD}-1)\frac{1-\lambda^n e^{nhD}}{1-\lambda e^{hD}} \quad ; \quad T_o^R(u_o) = R_o \tag{30}$$

Writing Eqs.(21) in matrix form yields

$$\begin{pmatrix} m_i \\ \frac{hs_i}{2\sqrt{3}} \end{pmatrix} = \frac{1}{1-\lambda^{2n}} \begin{pmatrix} \lambda^i(1-\lambda^{2(n-1)}) & (1-\lambda^{2i})\lambda^{n-i} \\ -\lambda^i(1+\lambda^{2(n-i)}) & \lambda^{n-i}(1+\lambda^{2i}) \end{pmatrix} \times$$
$$\begin{pmatrix} m_o - \frac{1-\lambda}{2h} R_o \\ m_n - \frac{1-\lambda}{2h} H_n \end{pmatrix} + \frac{1-\lambda}{2h} \begin{pmatrix} H_i + R_i \\ H_i - R_i \end{pmatrix} \tag{31}$$

Eqs.(31) are now expanded in terms of 1, e^{ihD} and e^{nhD}. To this end the following definitions are introduced

$$\Delta_1 = \frac{e^{hD}-1}{1-\lambda e^{hD}} \; ; \; \Delta_2 = \frac{e^{hD}-1}{e^{hD}-\lambda} \; ; \; \Delta = e^{hD}-1 \; ; \; \Delta^* = 1+4e^{hD}+e^{2hD} \tag{32}$$

Then Eqs.(28) and (30) take the form

$$\begin{pmatrix} T_i^H \\ T_i^R \end{pmatrix} = \begin{pmatrix} 0 & -\lambda^{n-i}\Delta_1 \\ -\lambda^i\Delta_2 & 0 \end{pmatrix} \begin{pmatrix} 1 \\ e^{nhD} \end{pmatrix} + \begin{pmatrix} \Delta_1 \\ \Delta_2 \end{pmatrix} e^{ihD} \tag{33}$$

$$\begin{pmatrix} T_n^H \\ T_o^R \end{pmatrix} = \begin{pmatrix} \Delta_1 & -\lambda^n\Delta_1 \\ -\lambda^n\Delta_2 & \Delta_2 \end{pmatrix} \begin{pmatrix} 1 \\ e^{nhD} \end{pmatrix} \tag{34}$$

We express now the last term of the right-hand side of Eq.(31) by the operators T_i^R and T_i^H

$$\frac{1-\lambda}{2h} \begin{pmatrix} R_i + H_i \\ R_i - H_i \end{pmatrix} = \frac{1-\lambda}{2h} \begin{pmatrix} T_i^R + T_i^H \\ T_i^R - T_i^H \end{pmatrix} (u_o) \tag{35}$$

Using the representation for T_i^H and T_i^R from Eqs.(33) gives

$$\frac{1-\lambda}{2h} \begin{pmatrix} T_i^R + T_i^H \\ T_i^R - T_i^H \end{pmatrix} = \frac{1-\lambda}{2h} \begin{pmatrix} -\lambda^i\Delta_2 & -\lambda^{n-i}\Delta_1 \\ \lambda^i\Delta_2 & -\lambda^{n-i}\Delta_1 \end{pmatrix} \begin{pmatrix} 1 \\ e^{nhD} \end{pmatrix} + \frac{1-\lambda}{2h} \begin{pmatrix} \Delta_1 + \Delta_2 \\ \Delta_1 - \Delta_2 \end{pmatrix} e^{ihD} \tag{36}$$

$$= \frac{1-\lambda}{2h} \begin{pmatrix} -\lambda^i\Delta_2 & -\lambda^{n-i}\Delta_1 \\ \lambda^i\Delta_2 & -\lambda^{n-i}\Delta_1 \end{pmatrix} \begin{pmatrix} 1 \\ e^{nhD} \end{pmatrix} + \begin{pmatrix} \dfrac{\Delta}{h} \dfrac{1+\dfrac{\Delta}{2}}{1+\Delta+\dfrac{\Delta^2}{6}} \\ \dfrac{\Delta^2}{2\sqrt{3}h} \dfrac{1}{1+\Delta+\dfrac{\Delta^2}{6}} \end{pmatrix} e^{ihD} \tag{37}$$

By taking into account Eq.(36), the equations for the derivatives, Eqs.(21), can be written thus

$$\begin{pmatrix} m_i \\ \dfrac{hs_i}{2\sqrt{3}} \end{pmatrix} = M \left\{ \begin{pmatrix} m_o \\ m_n \end{pmatrix} - \frac{1-\lambda}{2h} \left[\begin{pmatrix} \Delta_1 & -\lambda^n\Delta_1 \\ -\lambda^n\Delta_2 & \Delta_2 \end{pmatrix} \begin{pmatrix} 1 \\ e^{hD} \end{pmatrix} - M^{-1} \right.\right.$$

$$\left[\begin{pmatrix} -\lambda^{i}\Delta_2 & -\lambda^{n-i}\Delta_1 \\ \lambda^{i}\Delta_2 & -\lambda^{n-1}\Delta_1 \end{pmatrix}\begin{pmatrix} 1 \\ e^{hD} \end{pmatrix}\right] u_o \Bigg\} + \frac{1-2}{2h}\left[\begin{pmatrix} \Delta_1 + \Delta_2 \\ \Delta_1 - \Delta_2 \end{pmatrix} e^{ihD}\right] u_o \quad (38)$$

where

$$M^{-1} = \frac{1}{\lambda^{2n}} \begin{pmatrix} \lambda^{n-i}(1+\lambda^{2i}) & -\lambda^{n-i}(1-\lambda^{2i}) \\ -\lambda^{i}(1+\lambda^{2(n-i)}) & \lambda^{i}(1-\lambda^{2(n-i)}) \end{pmatrix} \quad (39)$$

$$M^{-1}\begin{pmatrix} -\lambda^{i}\Delta_2 & -\lambda^{n-i}\Delta_1 \\ \lambda^{i}\Delta_2 & \end{pmatrix} = \begin{pmatrix} -\lambda_2 & -\lambda^{n}\Delta_1 \\ -\lambda^{n}\Delta_2 & -\Delta_1 \end{pmatrix} \quad (40)$$

With Eq.(40) the right-hand side of Eq.(38) is of the form

$$\begin{pmatrix} m_i \\ \frac{hs_i}{2\sqrt{3}} \end{pmatrix} = M\left(\begin{pmatrix} m_o \\ m_n \end{pmatrix} - \frac{1-\lambda}{2h}(\Delta_1+\Delta_2)\begin{pmatrix} 1 \\ e^{hD} \end{pmatrix}(u_o)\right) + \frac{1-\lambda}{2h}\begin{pmatrix} \Delta_1+\Delta_2 \\ \Delta_1-\Delta_2 \end{pmatrix} e^{ihD}(u_o) \quad (41)$$

The final form of Eq.(41) is obtained by expanding the terms $\Delta_1 + \Delta_2$ and $\Delta_1 - \Delta_2$ using $\Delta^* = 6 + 6\Delta + 6\Delta^2$

$$\Delta_1 = \frac{\Delta}{\Delta^*}\left(1 - \frac{e^{hD}}{\lambda}\right) = -\frac{\Delta}{\lambda\Delta^*}(1 - \lambda + \Delta)$$
$$\Delta_2 = \frac{\Delta}{\Delta^*}\left(e^{hD} - \frac{1}{\lambda}\right) = -\frac{\Delta}{\lambda\Delta^*}(1 - \lambda - \lambda\Delta) \quad (42)$$

From these relations it follows

$$\Delta_1 + \Delta_2 = -\frac{1-\lambda}{\lambda}\frac{\Delta}{\Delta^*}(\Delta + 2) \quad ; \quad \Delta_1 - \Delta_2 = -\frac{1+\lambda}{\lambda}\frac{\Delta^2}{\Delta^*} \quad (43)$$

Expansion of these terms leads to

$$\frac{1-\lambda}{2h}(\Delta_1+\Delta_2) = \frac{\Delta}{h}\,\frac{1+\Delta/2}{1+\Delta+\Delta^2/6} =: D\sum_{n=o}^{\infty} a_n h^n D^n \quad (44)$$

resulting in a recursion formula for the unknown coefficients a_n:

$$a_o = 1$$
$$a_n = \frac{2^n}{(n+1)!} - \frac{1}{6}\sum_{\ell=1}^{n} a_{n-\ell}\frac{(4+2^{\ell})}{\ell!} \quad ; \quad n = 1, 2, \ldots \quad (45)$$

Hence we have

$$a_o = 1;\ a_1 = a_2 = a_3 = 0\ ;\ a_4 = -\frac{1}{180}\ ;\ a_5 = 0\ ;\ a_6 = \frac{1}{1512}$$

The same calculations are performed for $\Delta_1 - \Delta_2$

$$\frac{1-\lambda}{2h}(\Delta_1-\Delta_2) = \frac{\Delta^2}{2\sqrt{3}\,h}\;\frac{1}{1+\Delta+\Delta^2/6} =: \frac{1}{2\sqrt{3}} D^2 \sum_{n=0}^{\infty} b_n h^n D^n \qquad (46)$$

resulting in the recursion formula :

$$b_o = 1$$

$$b_n = \frac{2^{n+2}-2}{(n+2)!} - \frac{1}{6}\sum_{\ell=1}^{n} b_{n-\ell}\frac{(4+2^{\ell})}{\ell!} \quad ; \quad n = 1, 2, \ldots \qquad (47)$$

and

$$b_o = 1 \;;\; b_1 = 0 \;;\; b_2 = -1/12 \;;\; b_3 = 0 \;;\; b_4 = 1/360.$$

Denoting the exact values of the derivatives by u'_o, u'_1, ..., u'_n and u''_o, u''_1, ..., u''_n and restricting ourselves to expansion terms of up to sixth order, one obtains from Eqs.(41)

$$\begin{pmatrix} m_i \\ \\ \frac{hs_i}{2\sqrt{3}} \end{pmatrix} = M \left\{ \begin{pmatrix} m_o - u'_o \\ \vdots \\ m_n - u'_n \end{pmatrix} + \frac{h^4}{180}\begin{pmatrix} u_o^{(5)} \\ \vdots \\ u_n^{(5)} \end{pmatrix} - \frac{h^6}{1512}\begin{pmatrix} u_o^{(7)} \\ \vdots \\ u_n^{(7)} \end{pmatrix} \right\}$$

$$+ \begin{pmatrix} u'_i - \frac{h^4}{180}u_i^{(5)} + \frac{h^6}{1512}u_i^{(7)} \\ \frac{h}{2\sqrt{3}}\left(u''_i - \frac{h^2}{12}u_i^{(4)} + \frac{h^4}{360}u_i^{(6)}\right) \end{pmatrix} + \begin{pmatrix} O(h^7) \\ O(h^5) \end{pmatrix} \qquad (48)$$

Eqs.(48) show that is important for the order of the truncation error near boundaries that the approximation of m_o and m_n is of fourth order; i.e., only then we have $m_i = u'_i + O(h^4)$ and $s_i = u''_i + O(h^2)$. Disregarding boundary effects the truncation error analysis leads to the same results as the work of Rubin et al [2, 3, 4] since the same splines are used.

3. STABILITY ANALYSIS

For the linear Burgers equation

$$u_t + \bar{u}\,u_x = \nu u_{xx} \quad ; \quad \nu, \bar{u} > 0 \qquad (49)$$

stability is assessed with the discrete perturbation theory [1]. Since the Runge-Kutta method, used for the integration of the initial value problem, Eqs.(3), employs four time levels for one time step integration, its stability analysis would be too costly. However, since the Runge-Kutta method is a higher-order accurate modification of the Euler-Cauchy procedure, the stability criteria derived for the simple Euler time integration are definitely sufficient for the Runge-Kutta method. In the formulation of the discrete perturbation theory it is assumed that up to any certain time t the actual computed velocities $\tilde{u}_i(t)$ are equal to the exact solution $u_i(t)$ for all i. At the arbitrary time step t an error $\delta u_i(t)$ is introduced only at one arbitrary base point $x = x_i$ such that

$$u_i(t) = \tilde{u}_i(t) + \delta u_i(t) \quad ; \quad \tilde{u}_\ell(t) = u_\ell(t); \; \ell = 0, 1, \ldots, i-1, i+1, \ldots, n \tag{50}$$

Hence the differential equation (49) ceases to be exact and it takes the form for the approximate solution $\tilde{u}_i(t)$

$$(\tilde{u}_i)_t + \bar{u}\,(\tilde{u}_x)_i = \nu (u_{xy})_i \tag{51}$$

Noting that $u_i(t)$ satisfies the differential equation for the exact solution. Eq.(51) reduces to

$$\frac{\partial(\delta u_i)}{\partial t} = -\bar{u}\,\delta(u_x)_i + \nu\,\delta(u_{xx})_i \tag{52}$$

where $\delta(u_x)_i := (u_x)_i - (\tilde{u}_x)_i$; $\delta(u_{xx})_i := (u_{xx})_i - (\tilde{u}_{xx})_i$ represent the error introduced in the first and second spatial derivative due to the discrete perturbation error δu_i. To obtain the error $\delta u_i(t+\Delta t)$ for the next time step, the time derivative is approximated by the forward time operator, that is

$$\delta u_i\,(t+\Delta t) = \delta u_i(t) + \Delta t\,(-\bar{u}\delta(u_x)_i + \nu\delta(u_{xx})_i \tag{53}$$

The condition for the suppression of all error growth requires that the perturbation error must not increase when t is advanced to t + Δt, which implies for the magnitude of the error

$$\left|\frac{\delta u_i(t+\Delta t)}{\delta u_i(t)}\right| = \left|1 + \frac{\Delta t}{\delta u_i(t)}\,(-\bar{u}\,\delta(u_x)_i + \nu\delta(u_{xx})_i)\right| \leq 1 \tag{54}$$

This stability condition is necessary and sufficient only for the first computational time step. Eq.(54) can also be expressed as

$$-1 \leq \frac{\delta u_i(t+\Delta t)}{\delta u_i(t)} = 1 + \frac{\Delta t}{\delta u_i(t)}\,(-\bar{u}\delta(u_x)_i + \nu\delta(u_{xx})_i) \leq 1 \tag{55}$$

The right-hand inequality is referred to as static instability and guarantees that the error is not amplified in its own direction and the left-hand inequality is called the dynamic-stability condition which assures that the overshoot is not larger than the magnitude of the error itself. Since the error propagation process in the course of time becomes extremely difficult, a more restrictive condition for the dynamic stability is required, that is, no overshoot of the error between successive time steps is allowed. Hence, the −1 of the left-hand side of Eq.(55) is replaced by O. In [1] it is shown that this condition corresponds in many practical cases to the necessary and sufficient stability requirements of the more rigorous methods (v. Neumann, Hirt etc.). From the modified Eq.(55) then, the following two stability conditions are obtained

$$\frac{1}{\delta u_i}\,(-\bar{u}\,\delta(u_x)_i + \nu\delta(u_{xx})_i) < 0 \qquad \text{static stability condition}$$

$$\Delta t \leq \frac{\delta u_i(t)}{\bar{u}\delta(u_x)_i - \nu\delta(u_{xx})_i} \qquad \text{dynamic stability condition} \tag{56}$$

The static stability condition is a necessary requirement to assure the boundedness of the computed solution. The dynamic stability condition specifies the maximum allowable time step size Δt for stable solutions. It should be emphasized that both the static stability and the modified dynamic stability are necessary but not sufficient conditions to guarantee a stable solution. This, however, is also valid for all other stability analysis methods. The disturbances $\delta(u_x)_i$ and $\delta(u_{xx})_i$ which result from the introduction of δu_i can be directly evaluated from the spline representation using δm_i and δs_i. From the definitions for R_i and H_i, Eq.(16), the errors δR_i and δH_i introduced by δu_i become

$$\delta R_o = R_o - \tilde{R}_o = (\lambda^{i-1} - \lambda^{i})\,\delta u_i \;\; ; \;\; \delta H_n = H_n - \tilde{H}_n = (\lambda^{n-i} - \lambda^{n-1-i})\,\delta u_i \tag{57}$$

and

$$\delta H_i = \delta u_i \;\; ; \;\; \delta R_i = -\delta u_i \tag{58}$$

Thus

$$\delta(R_i + H_i) = 0 \;\; ; \;\; \delta(R_i - H_i) = -2\delta u_i \tag{59}$$

Substituting Eqs.(57) in Eqs.(21) and rearranging, the error introduced in the first derivative becomes

$$\delta m_i = -\frac{1-\lambda}{2h(1-\lambda^{2n})}(\lambda^{i}(1-\lambda^{2(n-i)})(\lambda^{i-1}-\lambda^{i})+\lambda^{n-i}(1-\lambda^{2i}) \times (\lambda^{n-i}-\lambda^{n-1-i}))\,\delta u_i \tag{60}$$

The expression in brackets can be reduced to

$(\lambda^{2i} - \lambda^{2(n-i)})\ (\frac{1}{\lambda} - 1)$.

Using $\frac{\lambda-1}{2}\ (\frac{1}{\lambda} - 1) = 3$, one obtains the final result

$$\delta m_i = \frac{3}{h}\,\frac{\lambda^{2i} - \lambda^{2(n-i)}}{1 - \lambda^{2n}}\,\delta u_i \tag{61}$$

For the error introduced in the second spatial derivative, it follows from Eqs.(21)

$$\delta s_i = \frac{2\sqrt{3}}{h^2}\left(\frac{1-\lambda}{2}\left(\frac{\lambda^{i}(1+\lambda^{2(n-i)})}{1-\lambda^{2n}}\,\delta R_o - \frac{\lambda^{n-i}(1+\lambda^{2i})}{1-\lambda^{2n}}\,\delta H_n + \delta(R_i - H_i)\right)\right) \tag{62}$$

Similarly, substituting Eqs.(58) and the second of Eqs.(59) and rearranging the λ expressions, we find

$$\delta s_i = \frac{6\sqrt{3}}{h^2(1-\lambda^2)}\left(2\lambda\,\frac{1-\lambda^{2n-1}}{1-\lambda} - \lambda^{2i} - \lambda^{2(n-i)}\right)\delta u_i \tag{63}$$

Substituting the expressions for δm_i and δs_i in the modified Eq.(54) establishes the necessary computional criterion for the selection of the time step Δt

$$0 \le \left|\frac{\delta u_i(t+\Delta t)}{\delta u_i(t)}\right| = \left|1 + \frac{\Delta t}{h^2}\,\frac{3}{1-\lambda^{2n}}\left(-h\bar{u}(\lambda^{2i} - \lambda^{2(n-i)}) + \right.\right.$$

$$+ 2\sqrt{3}\ \nu \frac{2\lambda(1-\lambda^{2n-1})}{1-\lambda} - \lambda^{i} - \lambda^{2(n-i)} \Bigg)\Bigg| \leq 1 \tag{64}$$

Introducing the stability parameters $\beta = \nu \frac{\Delta t}{h}$; $c = \bar{u} \frac{\Delta t}{h}$ (Courant number) and $R_c = \frac{c}{\beta} = \frac{h\bar{u}}{\nu}$ (cell Reynolds number), Eq.(64) can be written in the form

$$0 \leq \left| \frac{\delta u_i(t+\Delta t)}{\delta u_i(t)} \right| = \Bigg| 1 + \beta \frac{6\sqrt{3}}{1-\lambda^{2n}} \ \mathrm{x} \tag{65}$$

$$\left(2\lambda \frac{1-\lambda^{2n-1}}{1-\lambda} - \lambda^{2i} - \lambda^{2(n-i)} - c\, \frac{3(\lambda^{2i}-\lambda^{2(n-i)})}{1-\lambda^{2n}}\right)\Bigg| \leq 1$$

The signs of the coefficients in Eq.)65) are as follows

$$\mathrm{sign}\,(2\lambda(1-\lambda^{2n-1}) - \lambda^{2i} - \lambda^{2(n-i)}) = -1 \qquad (\lambda < 0)$$

$$\mathrm{sign}(\lambda^{2i} - \lambda^{2(n-i)}) = \begin{cases} -1 & \text{for } n/2 < i \leq n \\ +1 & \text{for } 0 \leq i \leq n/2 \end{cases}$$

and

$$\left| \frac{\lambda^{2i} - \lambda^{2(n-i)}}{1-\lambda^{2n}} \right| \leq 1.$$

We define

$$A_i := \left| \frac{2\lambda(1-\lambda^{2n-1})}{1-\lambda} - \lambda^{2i} - \lambda^{2(n-i)} \right| \frac{6\sqrt{3}}{1-\lambda^{2n}}$$

$$B_i := 3\, \frac{|\lambda^{2i} - \lambda^{2(n-i)}|}{1-\lambda^{2n}} \tag{66}$$

Inserting these definitions in Eq.(65) yields

$$-\frac{B_i}{A_i} c \leq \beta \leq \frac{1 - B_i c}{A_i} \tag{67}$$

From Eq.(66) it is found

$$A_i > -\frac{2\lambda}{1-\lambda} 6\sqrt{3} \quad ; \quad \frac{1}{A_i} < \frac{1+\sqrt{3}}{12} \tag{68}$$

Now the quotient $\left|\frac{B_i}{A_i}\right|$ is investigated

$$\left|\frac{B_i}{A_i}\right| \leq \frac{1-\lambda^{2n}}{2\sqrt{3}\ (1+\lambda^{2n} - 2\lambda \frac{1-\lambda^{2n-1}}{1-\lambda})} < \frac{1}{2\sqrt{3}\,(1-\frac{2\lambda}{1-\lambda})} = \frac{\sqrt{3}-1}{6} \tag{69}$$

Using these two estimates, one has

$$\frac{\sqrt{2}-1}{6} c \leq \beta \leq \frac{\sqrt{3}+1}{12} - \frac{\sqrt{3}-1}{6} c \tag{70}$$

Hence we finally obtain the stability relations

$$c \le \frac{2+\sqrt{3}}{4} \quad ; \quad R_c \le 3\,(\sqrt{3}+1) \text{ and } \beta \le \frac{1+\sqrt{3}}{24} \tag{71}$$

It is found that the stability limits obtained by the present direct method are substantially improved in comparison with the stability parameters of the method of Rubin and Khosla[3]. Of course the above stability limits are valid for any cubic spline formulation. The direct method allows only a more accurate estimation of the Courant and the cell Reynolds number.

4. RESULTS

Three model problems, namely the nonlinear Burgers equation, the Ekman boundary layer equations and the classical convection-diffusion equation have been used to demonstrate the cubic spline collocation procedure presented herein. For each of these problems an analytic solution is available for comparison purposes. The cubic splines are used to approximate the first and second spatial derivatives. All problems are time-integrated using a fourth order Runge-Kutta method.

The nonlinear Burgers equation utilized is of the form

$$u_t + (u - 1/2)u_x = \nu u_{xx} \tag{72}$$

with ν constant and the boundary conditions $u \to 1$ as $x \to -\infty$ and $u \to 0$ as $x \to +\infty$. The stationary solution of (72) is [3]

$$u = (1 - \tanh(x/4\nu))/2 \tag{73}$$

The Ekman layer equations describe the friction region [13] in geophysical fluid dynamics and follow directly from the Navier-Stokes equations for a rotating fluid where the rigid wall is at $z = 0$ and a horizontally uniform flow u_g, v_g is specified far from the wall. With the boundary conditions $u = u_g$, $v = v_g$, $w = 0$ for $z \to \infty$ and $u = v = w = 0$ for $z = 0$ (no slip) it follows from the Navier-Stokes equations that

$$u = u_g = -\frac{1}{\rho f}\frac{\partial p}{\partial y} \quad z \to \infty \text{(f is the Coriolis parameter)};$$

$$u = u(z),\ v = v(z)$$

and $w \equiv 0$. This, along with the fact that u and v are functions of z alone, results in the equations

$$\begin{aligned} u_t - f v &= (K(z)\,u_z)_z - f v_g \\ v_t + f u &= (K(z)\,v_z)_z + f u_g \end{aligned} \tag{74}$$

For constant K and $v_g = 0$ these equations have the analytic solutions

$$\begin{aligned} u &= u_g(1 - \exp(-z/\delta_E)\cos(z/\delta_E)) \\ v &= u_g \exp(-z/\delta_E)\sin(z/\delta_E) \end{aligned} \tag{75}$$

where $\delta_E = \left(\frac{K}{f/2}\right)^{1/2}$ is the so called Ekman layer thickness.

The classical convection diffusion equation

$$c_t + \bar{u}\, c_x = \nu\, c_{xx} \quad ; \quad \bar{u} = \text{constant} \;;\; c = \text{concentration} \left(\frac{\text{kg}}{\text{m}}\right) \qquad (76)$$

is, for example, a model in which the medium moves at a constant and uniform velocity, describing a source in uniform wind. For $\bar{u}$ = constant this equation has the analytic solution

$$c = \frac{Q}{\sqrt{2\pi}\,\sigma} \exp\left(-(x - \bar{u}t)^2/2\sigma^2\right) \qquad (77)$$

where $\sigma = (2\nu t)^{1/2}$ and $Q = \int_{-\infty}^{+\infty} c\,dx$.

The following tables show the comparison of the computed and the exact analytic solutions for the above problems.

NONLINEAR BURGERS EQUATION			
X	SPLINE		EXACT
	B.C. 2. ORDER	B.C. 3. ORDER	
0	.50000	.50000	.50000
-0.2	.69247	.69105	.68997
-0.4	.83322	.83227	.83202
-0.6	.91859	.91807	.91683
-0.8	.96278	.96253	.96083
-1.0	.98356	.98345	.98201
-1.2	.99286	.99281	.99184
-1.4	.99692	.99690	.99632
-1.6	.99868	.99867	.99834
-1.8	.99943	.99943	.99925
-2.0	.99976	.99976	.99966

Table 1:

Solution of nonlinear Burgers equation $\nu = 1/8$, $h = -0.2$ and number of base points $n = 31$ (see also [3]). B.C. 2. order means that the second derivatives at the left and the right boundary are approximated second order accurate. The third column then reflects the influence of the boundary values on the spline solution.

EKMAN-LAYER EQUATIONS				
HEIGHT	SPLINE		EXACT	
	U	V	U	V
0.	0.000	0.000	0.000	0.000
100.	8.532	5.668	8.450	5.691
200.	15.121	6.925	15.077	6.887
300.	19.466	5.901	19.405	5.849
400.	21.832	4.010	21.748	4.070
500.	22.788	2.367	22.693	2.370
600.	22.909	1.062	22.813	1.090
700.	22.637	.252	22.545	.284
800.	22.255	- .140	22.168	- .132

Table 2:

Solution of Ekman-layer equations with only 10 base points. Hence, the somewhat reduced accuracy. $\nu = 3.5$, $u_g = 21.4$.

CONVECTION-DIFFUSION EQUATION		
X	SPLINE	EXACT
0.0	.16084	.16120
0.2	.18100	.18137
0.4	.20025	.20065
0.6	.21783	.21828
0.8	.23298	.23349
1.0	.24502	.24560
1.2	.25338	.25401
1.4	.25765	.25832
1.6	.25760	.25832
1.8	.25322	.25401
2.0	.24468	.24560

Table 3:

Solution of the convection-diffusion equation after 20 time steps. Step size h = 0.2, n = 35 and Q = 1. Boundary conditions are second order accurate. $\bar{u} = 0.2$, $\Delta t = 0.5$. The exact solution is the convected and diffused distribution for $\nu = 1/8$ and t = 2 of Eq. 77, initially centered at x = 0. Table 3 shows base points not influenced by boundary conditions.

CONCLUSIONS AND OUTLOOK

The preceding discussion of the last several sections has been focussed on a direct spline method for the approximation of spatial gradients. The equations derived for the evaluation of the first and second spatial derivatives are dependent only on the eigenvalue $\lambda = -2+\sqrt{3}$ and the first derivatives at the left and right boundary. These equations can be substantially simplified for large n, that is, for $n > \ell$ where $|\lambda^\ell| < e_r$ (e.g. $\ell = 11$ for $e_r = 10^{-6}$). The value e_r is the prescribed error bound for the approximate calculation of the spline coefficients. Normally e_r is chosen so that its value is an order of magnitude smaller than the truncation error of the numerical procedure.

The truncation error analysis, including the boundary effects, showed that the cubic spline formulation is fourth order accurate for the first derivative and second order accurate for the second derivative if the first derivatives are of fourth order accuracy at boundary points. The stability analysis for the linear Burgers equation, using the discrete perturbation theory, gave substantially improved values for the Courant number and the Reynolds cell number in comparison with the commonly employed spline formulation.

The technique was illustrated by applying it to three examples. First, the nonlinear Burgers equation was solved and compared to the analytic solution. Second, the Ekman-layer equations for constant viscosity were calculated and finally, the method was applied to the classical convection-diffusion equation.

Investigations for an improved truncation error and the extension to nonuniform meshes are being carried out [10]. At the same time attention is given to the super-step method [8] to find out whether this procedure is a definite candidate for the time integration in fluid flow problems.

ACKNOWLEDGEMENTS

We would like to thank A. Müller, GKSS-Research Center, for a number of valuable suggestions and discussions. Furthermore, we are grateful to the staff of the theory department of the Max-Planck-Institut für Plasmaphysik, Garching, for helpful suggestions. In particular, the critical remarks of A. Schlüter and D. Pfirsch are gratefully acknowledged.

REFERENCES

1. ROACHE, P. - Computational Fluid Dynamics. Hermosa Publishers, 1976.

2. RUBIN, S.G.; GRAVES, R.A.: Viscous Flow Solutions with a Cubic Spline Approximation. Computers and Fluids, Vol. 3, pp. 1-36, 1975.

3. RUBIN, S.G.; KHOSLA, P.K.: Higher Order Numerical Solutions Using Cubic Splines. AIAA, J. 14, pp. 851-858, 1976.

4. RUBIN, S.G.; KHOSLA, P.K.: Polynomial Interpolation for Viscous Flow Calculations.J.of Comp.Physics 24, pp.217-244,1977.

5. SHAMPINE, L.F.; ALLEN, R.C.: Numerical Computing. W.B. Saunders, 1973.

6. FEHLBERG, E.: Klassische Runge-Kutta-Formeln vierter und niedrigerer Ordnung mit Schrittweiten Kontrolle und ihre Anwendung auf Wärmeleitungsprobleme. Computing 6, 61-71,1970.

7. VERWER, J.G.: An Implementation of a Class of Stabilized Explicit Methods for the time integration of Parabolic Equations. ACM Transactions on Mathematical Software, Vol. 6, No. 2, pp. 188-205, 1980.

8. GENTZSCH, W.: Über ein verbessertes explizites Einschrittverfahren zur Lösung parabolischer Differentialgleichungen. DFVLR - TS, 1980.

9. SHAMPINE, L.F.; GORDON, M.K.: Computer Solution of Ordinary Differential Equations. W.H. Freeman, 1975.

10. HÄUSER, J.; EPPEL, D.; TANZER, F.: A direct spline method for nonuniform meshes. (in preparation).

11. PRENTER, P.M.: Splines and Variational Methods. Wiley-Interscience, 1975.

12. BOOR, de Carl: A Practical Guide to Splines. Springer, 1978.

13. PEDLOSKY, J.: Geophysical Fluid Dynamics. Springer, 1979.

FURTHER u,v-FORMULATIONS FOR IDEAL FLUID FLOW

E-M. SALONEN*, P. LEHTONEN* and A. PRAMILA**

SUMMARY

Two-dimensional incompressible irrotational inviscid fluid flow is analysed by the finite element method employing the velocity components as basic unknowns. The conventional least squares method with C^0 continuous elements has been found to perform poorly in this context. Numerical results from three test problems using a modified least squares method are given. The modified form is a combination of the conventional least squares and the subdomain collocation method. The effect of the number of subdomains is studied when using eight-noded isoparametric quadrilateral elements. A remarkable increase in accuracy compared with the conventional least squares method is obtained by employing only one subdomain after which the results improve more slowly. The most practical procedure is to take each element as a subdomain. It seems probable that the proposed procedure could be useful also in other applications of the least squares method.

1 INTRODUCTION

This study is a continuation of the work described in Reference [1]. Two-dimensional incompressible irrotational inviscid fluid flow is considered. The governing equations are in the domain A

$$\frac{\partial u}{\partial x} + \frac{\partial v}{\partial y} = 0 \tag{1}$$

*Associate Professor, Student, Institution of Mechanics, Department of General Sciences, Helsinki University of Technology, 02150 Espoo 15, Finland
**Associate Professor, Institution of Engineering Mechanics, Department of Mechanical Engineering, The University of Oulu, 90570 Oulu 57, Finland

$$\frac{\partial v}{\partial x} - \frac{\partial u}{\partial y} = 0 \ , \qquad (2)$$

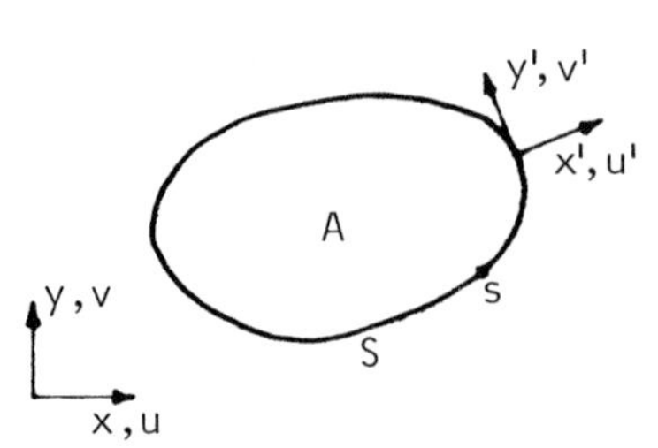

Figure 1. Some notations

on the boundary $S_{u'}$

$$u' \equiv n_x u + n_y v = \bar{u}' \qquad (3)$$

and on the boundary $S_{v'}$

$$v' \equiv -n_y u + n_x v = \bar{v}' . \qquad (4)$$

Here (Figure 1) u and v are the velocity components in the x and y directions, respectively, n_x and n_y are the components of the outward unit normal vector to the boundary and u' and v' are the normal and tangential velocity components at the boundary. Parts $S_{u'}$ and $S_{v'}$ form the whole boundary. A bar refers to a given quantity. Usually, the possible $S_{v'}$ part of the boundary arises from symmetry reasons, when for instance only a half of a symmetric domain is analysed. The given tangential velocity component $\bar{v}'$ is then zero.

Results from numerical experiments using the conventional least squares functional

$$\Pi(u,v) = \frac{1}{2}\int_A \left[\left(\frac{\partial u}{\partial x} + \frac{\partial v}{\partial y}\right)^2 + \left(\frac{\partial v}{\partial x} - \frac{\partial u}{\partial y}\right)^2\right] dA \qquad (5)$$

were given in [1]. The C^0 continuous finite element approximation is written in the form

$$\underset{\sim}{u} = \underset{\sim}{N}\, \underset{\sim}{a} \ , \qquad (6)$$

where

$$\underset{\sim}{u} = [u,v]^T \ , \qquad (7)$$

$\underset{\sim}{N}$ is the shape function matrix and $\underset{\sim}{a}$ is the column matrix consisting of the nodal velocity components u and v. At boundary nodes components u' and v' in a rotated x'y' coordinate system are used. Boundary condition (3) or (4) is satisfied at the nodes thus simply by assigning the given value $\bar{u}'$ or $\bar{v}'$ to the corresponding nodal parameter. In the program this is achieved by using penalty terms (c.f. equation (9) in [1]).

It is easy to show that the Euler equations corresponding to functional (5) are the Laplace equations for u and v. Similarly, the natural boundary condition is found to be equation (2) on $S_{u'}$ and equation (1) on $S_{v'}$ when the admissible u and v are required to satisfy boundary conditions (3) and (4). Thus also the discrete finite element equations

$$\frac{\partial \Pi(\underset{\sim}{a})}{\partial \underset{\sim}{a}} = \underset{\sim}{0} \tag{8}$$

are obviously not direct approximations to the original field equations (1) and (2) (This is of course a general feature of the least squares method. It is well known that the corresponding Euler equations are obtained from the original field equations by operating with the adjoint operator on them.). This fact probably explains the observed large violations in the mass conservation and zero circulation conditions

$$\int_A \left(\frac{\partial u}{\partial x} + \frac{\partial v}{\partial y}\right) dA = \int_S (n_x u + n_y v) ds = \int_S u' ds = 0 \tag{9}$$

$$\int_A \left(\frac{\partial v}{\partial x} - \frac{\partial u}{\partial y}\right) dA = \int_S (-n_y u + n_x v) ds = \int_S v' ds = 0 \tag{10}$$

when functional (5) was used in the numerical calculations in [1]. See also the comments on volume conservation in Reference [2] in a somewhat similar situation.

A remedy for the poor performance was proposed in [1] and some promising early results were obtained. The procedure is now studied in more detail and results from three test problems are reported in the following.

2 MODIFIED LEAST SQUARES METHOD

We take the modified least squares functional

$$\Pi(u,v;\lambda_1,\ldots,\lambda_{2m}) = \frac{1}{2}\int_A \left[\left(\frac{\partial u}{\partial x} + \frac{\partial v}{\partial y}\right)^2 + \left(\frac{\partial v}{\partial x} - \frac{\partial u}{\partial y}\right)^2\right] dA + \\ + \sum_{k=1}^{m} \left[\lambda_{2k-1} \int_{A_k} \left(\frac{\partial u}{\partial x} + \frac{\partial v}{\partial y}\right) dA + \lambda_{2k} \int_{A_k} \left(\frac{\partial v}{\partial x} - \frac{\partial u}{\partial y}\right) dA\right] \tag{11}$$

as the basis for the calculations. Notation A_k refers to the k'th subdomain of the whole domain A, m is the total number of subdomains used and the λ's are Lagrange multipliers. It is to be noted that u(x,y) and v(x,y) are unknown functions but $\lambda_1,\lambda_2,\ldots,\lambda_{2m}$ are unknown constants. The idea with the Lagrange multiplier terms is to inject more information about the original field equations (1) and (2) into the formulation. We introduce constraint equations

$$\int_{A_k} \left(\frac{\partial u}{\partial x} + \frac{\partial v}{\partial y}\right) dA = 0 \ , \tag{12}$$

$$k = 1,2,\ldots,m$$

$$\int_{A_k} \left(\frac{\partial v}{\partial x} - \frac{\partial u}{\partial y}\right) dA = 0 \tag{13}$$

by subdomain collocation and adjoin them to functional (5) via Lagrange multipliers. If m = 0, we obtain the conventional least squares method. In [1] the effect of using only one subdomain (i.e. the whole domain) was considered. Here the number of subdomains is gradually extended to the extreme case in which each element is finally a subdomain. Constraints (12) and (13) are not transformed to line integrals around the periphery of the subdomain (c.f. equations (9) and (10)) as was done in [1], since the programming is much easier when using the area integral form.

If the column matrix of Lagrange multipliers is denoted by $\underset{\sim}{\lambda}$, the discrete finite element equations are now

$$\left\{\begin{matrix} \dfrac{\partial \Pi(\underset{\sim}{a};\underset{\sim}{\lambda})}{\partial \underset{\sim}{a}} \\ \dfrac{\partial \Pi(\underset{\sim}{a};\underset{\sim}{\lambda})}{\partial \underset{\sim}{\lambda}} \end{matrix}\right\} = \underset{\sim}{0} \ , \qquad (14)$$

from which the unknowns $\underset{\sim}{a}$ and $\underset{\sim}{\lambda}$ are determined.

Looking at equations (12) and (13) one could always ask why to bother at all about the least squares method and instead obtain all the necessary equations by collocation. That has in fact been done in [3] using Galerkin, subdomain and point collocation; however, the results were not very encouraging. The system equations are unsymmetrical and have usually zeroes on the diagonal. The velocity profiles often oscillated wildly about the right values and in some cases the banded solver failed to give an answer. In the least squares method the system equations are symmetrical and no difficulty in solving them appears. The velocity profiles are smooth but the mass conservation and zero circulation conditions are violated in an intolerable way with reasonable meshes. Thus formulation (11) is a compromise between least squares and collocation trying to retain best features from both of them.

3 NUMERICAL RESULTS

To avoid "leaking" on curved boundaries it is preferable to have elements which can have curved sides. The isoparametric eight-noded quadrilateral Serendipity element (16 degrees of freedom) with 3x3 Gauss integration rule has been used to obtain the results which are reported here. The example problems are the same as in [1] and the detailed boundary conditions used are not repeated to save space.

3.1 First example

The radial flow due to a line source of strength Q per unit length at point 0 was calculated in the domain ABCD (Figure 2). The exact solution for the radial velocity is

$Q/(2\pi r)$ where r is the radius.

The numerical results with 2x2 = 4 elements and 21 nodes gave practically exact values with about three significant digits without any Lagrange multiplier terms (m = 0). This

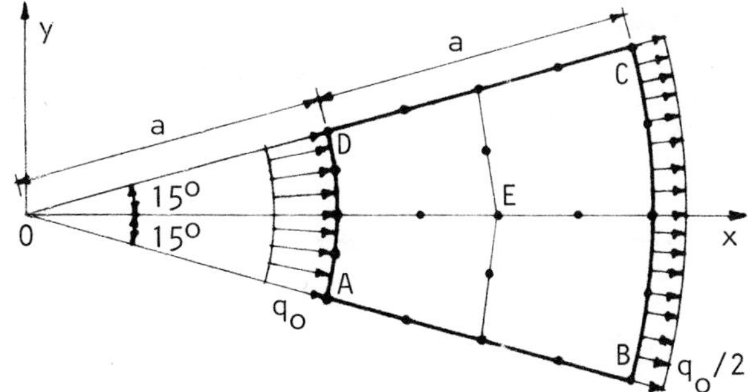

Figure 2. The first test example

good behaviour is to be attributed to the utmost simplicity of the problem. However, the results with one subdomain (m = 1) - the whole area - were at first very bad. At point E the error in radial velocity was about -24 %. This is explained as follows. The boundary is totally of type S_u, in this example. The information about the total mass conservation is thus already included through the given boundary nodal normal velocities. When we give the same information again by the Lagrange multiplier term we end up with a singular system of equations. Due to discretization and roundoff errors the computer has still obtained a solution. It is easy to correct the situation by just leaving any one of the known nodal normal velocities as not given. This was done at point x = 2a, y = 0 after which no numerical difficulties were observed when using different number (maximum was m = 4) of subdomains. The results were practically the same in all the cases. If one could devise an example, in which the boundary would be totally of type S_v, a similar procedure with respect to one tangential nodal velocity would be needed.

3.2 Second example

The flow around a cylinder between parallel walls (Figure 3) is considered. From symmetry reasons it suffices to take domain ABCDE only under consideration. A 2x4 = 8 element mesh with 37 nodes is employed.

In Figure 4 some results are given. Q is the calculated volume flow at section CD (exact value is $2q_oa$) and Γ is the circulation around boundary ABCDE (exact value is zero). As the exact solution is not known, the results obtained by Chan and Larock [4] with the velocity potential formulation and 180 six-noded triangular elements are employed for reference.

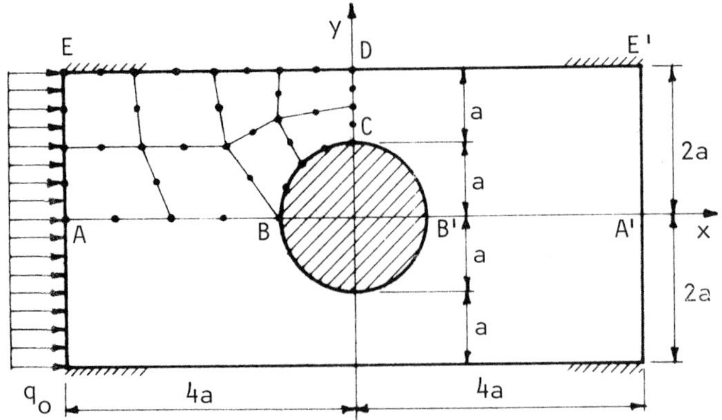

Figure 3. The second test example

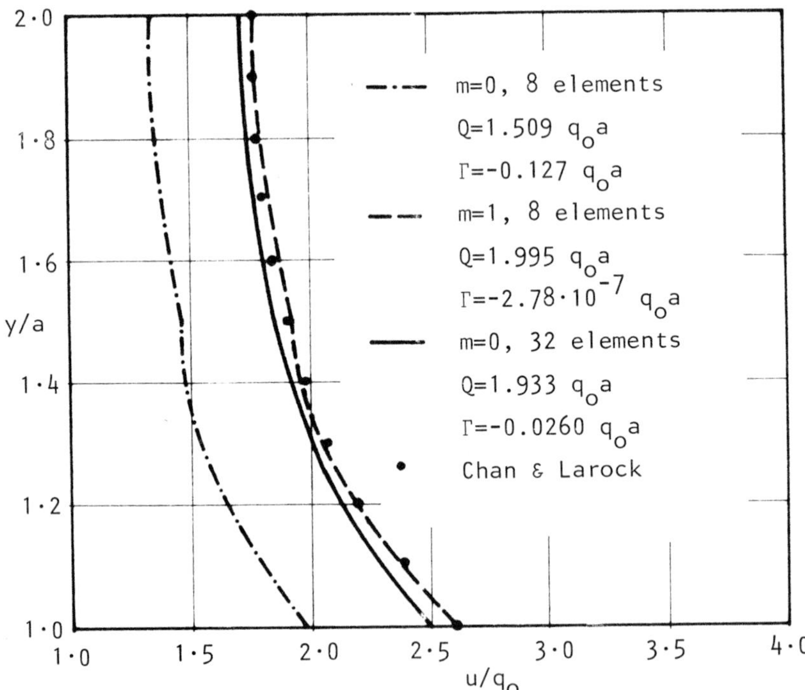

Figure 4. Horizontal velocity u at vertical center line CD

The values are reproduced by measurements from a small-scale figure and can therefore be slightly in error. Calculations were made also by a similar 4x8 = 32 element mesh having 121 nodes. It is quite remarkable that by taking just two additional unknowns (λ_1, λ_2) by employing only one subdomain (m = 1) in the modified method gives more accurate results than the conventional method (m = 0) with the finer mesh, which has

2x121-2x37 = 168 more unknowns. It is thus obvious that the least squares method is in practice useless here for economical reasons without the modification.

The number and way of selection of the subdomains has very small influence on u or Q at section CD after $m \geq 1$. As is to be expected, the influence can be still felt in the interior of the domain. For instance the value of the volume flow at a cross section at about midway between sections AE and CD is changed from value 1·591 $q_o a$ to 1·884 $q_o a$ when m is changed from 0 to 1. When m = 2 and the cross section in question is taken to be the boundary of the two subdomains, we naturally obtain the exact value 2 $q_o a$ (of course some small errors appear due to numerical integration and roundoff). Thus for instance for m = 1 we have at section CD the values 1·995 $q_o a$ and 1·9997 $q_o a$ for the volume flow with the coarser and finer mesh, respectively.

3.3 Third example

The last test example is the case of a flow around a circular cylinder with circulation (Figure 5). The stagnation

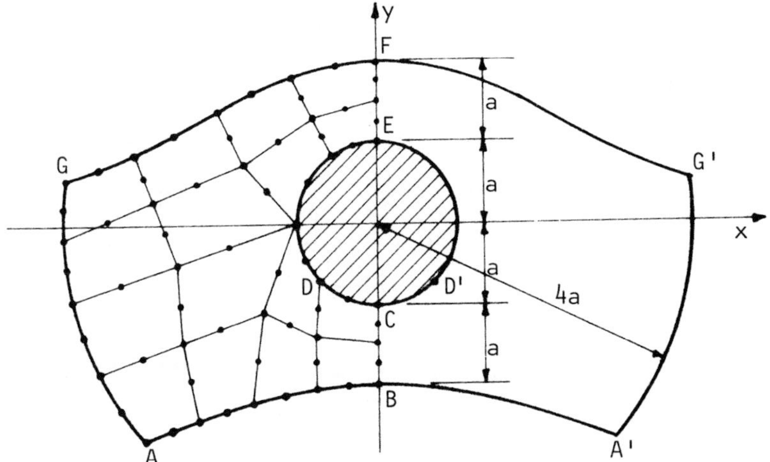

Figure 5. The third test example

points are taken to be at D and D'. The corresponding value of circulation around the cylinder is $\Gamma = -2\sqrt{2}\pi q_o a$, where q_o is the free stream velocity. From symmetry reasons only domain ABCDEFG is considered. A 4x4 = 16 element mesh with 69 nodes is used. The velocity distribution at section BC and EF for m = 0 and m = 16 is shown in Figure 6. The relative error in maximum speed (at point E) is -27·3 %, -4·9 % and -1·5 % when correspondingly m = 0, m = 1 and m = 16.

The effect of the number and selection of the subdomains

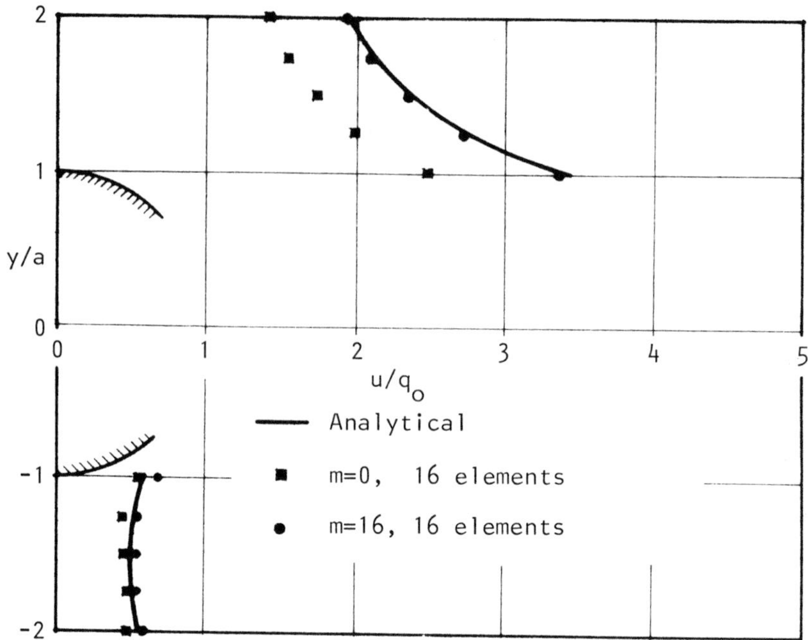

Figure 6. Horizontal velocity u at center lines BC and EF

is shown in Table 1 for q and Γ, where q is the speed at the node (x = -1·75 a, y = -0·25 a) marked in the Table and Γ is the circulation around the cylinder. Subscript a refers to

m	$\frac{q-q_a}{q_a} \cdot 100$	$\frac{\Gamma-\Gamma_a}{\Gamma_a} \cdot 100$
0	-22·45	-30·52
1	-8·80	-11·62
2	-4·71	-8·82
2	-6·69	-10·55
4	-2·59	-6·72

Table 1 continued

8	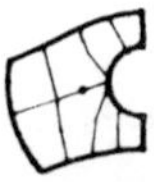 −1·63	−5·04
8	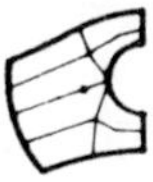 −1·87	−5·80
16	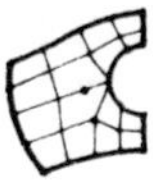 −0·53	−3·83

Table 1. Percentage error of q and Γ against the number of subdomains m

the analytical value ($q_a \approx 0{\cdot}9107\ q_o$, $\Gamma_a = -2\sqrt{2}\pi q_o a \approx -8{\cdot}886\ q_o a$). These results are typical. The most drastic improvement happens with one subdomain. The best results are nearly always obtained by taking each element as a subdomain.

4 DISCUSSION AND CONCLUSIONS

We have also performed numerical experiments using a functional, which is obtained from (11) by transforming the terms $\int(\partial u/\partial x \cdot \partial v/\partial y - \partial v/\partial x \cdot \partial u/\partial y)dA$ into a line integral along the boundary. The curvature of the boundary appears in the line integral. The numerical results were, however, generally considerable less accurate than those obtained by formulation (11) and are therefore not discussed here. Similarly a formulation based on Kelvin's principle and the penalty function method [5] has been experimented with without any great success.

Thus from the numerical experience obtained this far it is quite clear that the best and most straightforward method to solve two-dimensional ideal fluid flow with the velocity components as basic unknowns is to employ the functional

$$(u,v;\lambda_1^1,\lambda_2^1,\ldots) = \frac{1}{2}\int_A\left[\left(\frac{\partial u}{\partial x}+\frac{\partial v}{\partial y}\right)^2 + \left(\frac{\partial v}{\partial x}-\frac{\partial u}{\partial y}\right)^2\right]dA \, +$$

$$+ \sum_e\left[\lambda_1^e \int_{A^e}\left(\frac{\partial u}{\partial x}+\frac{\partial v}{\partial y}\right)dA + \lambda_2^e\int_{A^e}\left(\frac{\partial v}{\partial x}-\frac{\partial u}{\partial y}\right)dA\right] \qquad (15)$$

(superscript e refers to element e and the summation is over the number of elements). By taking each elements as a subdomain we are relieved from the difficulty of selection of the subdomains and from the corresponding additional data input. A more important advantageous feature is, that the mass conservation and zero circulation conditions around any closed reducible loop consisting of element edges are exactly (barring errors from numerical integration and roundoff) satisfied. The rather large number of Lagrange multipliers (two per each element) is naturally a drawback. However, with higher order elements the ratio of the number of the velocity degrees of freedom to the number of Lagrange multipliers remains favourable. With the eight-noded quadrilateral this ratio is for a large mesh about 6/2 = 3. For low order elements as the three-noded triangular element, expression (15) is not recommended - somewhat larger subdomains must be used - since one can easily obtain more constraints than what there are velocity degrees of freedom.

A solver taking into account the double-banded nature of the system of equations (14) was used. No difficulties were met in the solution of the equations without pivoting in spite of the fact that the coefficient matrix is no more positive definite due to the Lagrange multipliers. We had not anticipated that the constraints could be brought down to the element level. As this is the case, two interesting further modifications appear possible.

The constraint equations

$$\int_{A^e}\left(\frac{\partial u}{\partial x} + \frac{\partial v}{\partial y}\right)dA = 0 \qquad (16)$$

$$e = 1,2,\ldots$$

$$\int_{A^e}\left(\frac{\partial v}{\partial x} - \frac{\partial u}{\partial y}\right)dA = 0 \qquad (17)$$

are now local (i.e. only few adjacent nodal parameters appear in them). By adding for book-keeping purposes one internal node to each element (the position of the node is of no significance) and by associating parameters λ_1^e and λ_2^e with that node we now obtain by the conventional assembly process a banded coefficient matrix instead of the double-banded one. Thus a conventional banded solver can be used of course providing that the new ordering does not make the system much more vulnerable to a solver without pivoting. We have not yet made any experiments with this option.

The second, more exciting possibility is to use the penalty function method or "least squares inside least squares" and to write

$$\Pi(u,v) = \frac{1}{2}\int_A\left[\left(\frac{\partial u}{\partial x} + \frac{\partial v}{\partial y}\right)^2 + \left(\frac{\partial v}{\partial x} - \frac{\partial u}{\partial y}\right)^2\right]dA +$$

$$+ \frac{\alpha}{2}\sum\left\{\left[\int_{A^e}\left(\frac{\partial u}{\partial x} + \frac{\partial v}{\partial y}\right)dA\right]^2 + \left[\int_{A^e}\left(\frac{\partial v}{\partial x} - \frac{\partial u}{\partial y}\right)dA\right]^2\right\}, \qquad (18)$$

where the penalty number has been selected to be the same for each constraint for simplicity. The penalty function method was rejected in [1] as it totally destroys the possible sparsity and bandedness of the original system if the constrainst are global (i.e. many nodal parameters associated with nodes far from each other appear in them). Form (18), however, now seems to be very promising. Positive definite matrix with no increase in the number of unknowns or in bandwidth is obtained. When $\alpha = 0$, we again have the conventional least squares formulation. First trials with the second test problem gave identical results with those obtained by formulation (15) with three significant digits in the range $\alpha a^2 = 10^3 \ldots 10^5$. The calculations were in single precision by a computer with an accuracy of 7...8 digits. At the time of the Conference more results should be available.

It is interesting to note that if we are using the three-noded triangular element, the derivatives $\partial u/\partial x$, $\partial u/\partial y$ etc. are constants in each element and it is seen that formulation (18) is in fact nearly equivalent with the conventional form (5) and no increase in accuracy can be expected (without taking larger subdomains). Thus again the method should be more useful and practical with higher order elements.

It may be mentioned, that the least squares method has been applied for ideal fluid flow with C^1 continuous elements in References [6] and [7]. In [6] it is even claimed that the trial functions must have C^1 continuity in this case. However, for physical reasons the solution does not always have C^1 continuity; for instance at a solid boundary, when the curvature of the boundary has a jump. Additionally the C^1-elements are rather complicated, cannot follow easily curved boundaries, are difficult to extend into three dimensions and according to the results obtained in [7] cannot compete with C^0-elements when the modified least squares method is used.

The drawback of the least squares method in general is the fact that it is approximating "the wrong equations". It seems probable that the modified least squares method presented here for ideal fluid flow could work - when generalized in an obvious way - beneficially also in other applications.

REFERENCES

1. PRAMILA, A., SALONEN, E-M. - A u,v-Formulation for Ideal Fluid Flow. Numerical Methods in Laminar and Turbulent Flow, Ed. Taylor, C., Morgan K. and Brebbia, C.A., Pentech Press, pp. 971-982, 1978

2. ROACHE, P.J. - Computational Fluid Dynamics, Hermosa Publishers, p. 207, 1976.

3. PRAMILA, A., SALONEN, E-M. - Weighted Residual u,v Finite Element Formulations for Ideal Fluid Flow, Proc. International Conference on Computer Applications in Civil Engineering, October 23-25, 1979. University of Roorkee, Nem Chand & Bros, India.

4. CHAN, S.T.K., LAROCK, B.E. - Flow around Cylinder between Parallel Walls, Proc. Am. Soc. Civ. Eng., Vol. 98, No. EM5, pp. 1317-1322, 1972.

5. SALONEN, E-M., PRAMILA, A., LEHTONEN, P. - A Finite Element Method for Ideal Fluid Flow Employing Kelvin's Principle and the Penalty Function Method, Report No. 7, Institution of Mechanics, Helsinki University of Technology, Finland, 1979.

6. de VRIES, G., LABRUJERE, T.E., NORRIE, D.H. - A Least Squares Finite Element Solution for Potential Flow, Report No. 86, Department of Mechanical Engineering, the University of Calgary, Canada, 1976.

7. TUOMALA, M., PRAMILA, A. - A Combined Least Squares-Point Collocation Method for Ideal Fluid Flow, Report No. 6, Institution of Mechanics, Helsinki University of Technology, Finland, 1979.

NUMERICAL TECHNIQUES IN A WEDGE FLOW OF A POWER LAW FLUID

Dr. N. L. Kalthia

Assistant Professor
Department of Mathematics
S.V. Regional College of Engg. and Tech., Surat-395007 - India.

1. The problem : an introduction :

During the last three decades a large number of fluids of great commercial importance have been found whose flow behavior can not be explained fully by the solutions of Naviers stokes equation. These fluids do not have a linear relationship between shear stress tensor component and rate of strain tensor components. Many models for describing non-Newtonian behavior have been developed from molecular theories. Meter [1] lists 21 different models for describing the non-Newtonian viscosity far back in 1965. The literature study reveals that, by now, there is a lot of progress in the constitutive theories to study complex rheological phenomena.

Among the non-Newtonian fluids power law fluid has been found of great interest in some processing industries. Its flow properties have been discussed in detail in the literature [2],[3]. It is characterised by the rheological equation [4].

$$\tau_{ij} = k \left| \sum_{n=1}^{3} \sum_{l=1}^{3} e_{ln}\, e_{nl} \right|^{\frac{(n-1)}{2}} e_{ij} \qquad (1.1)$$

where
$$e_{ij} = \frac{\partial u_i}{\partial x_j} + \frac{\partial u_j}{\partial x_i} \qquad (1.2)$$

Usually k and n are known as constancy and flow behavior index respectively.

Here we shall investigate the laminar flow of such an incompressible fluid over a wedge of an angle $\pi\theta$ as shown in the figure 1.1 . (given in end)

2. Mathematical analysis :

The governing boundary layer equations for steady, two dimensional flow past a surface oriented above, under pressure gradient, following Acrivos et al [5] are :

$$\frac{\partial u}{\partial x} + \frac{\partial v}{\partial y} = 0 \tag{2.1}$$

$$u\frac{\partial u}{\partial x} + v\frac{\partial u}{\partial y} = \nu_p \frac{\partial}{\partial y}\left[\left| \frac{\partial u}{\partial y} \right|^{n-1} \frac{\partial u}{\partial y} \right] + U\frac{\partial U}{\partial x} \tag{2.2}$$

The boundary conditions to be satisfied are

$$\left.\begin{aligned} u(x,\ 0) &= v(x,\ 0) = 0 \\ u(x,\ \infty) &= U(x) \end{aligned}\right\} \tag{2.3}$$

We try to reduce the equations (2.1), (2.2) to a set of ordinary (nonlinear) differential equation by the group theoretic method of similarity analysis. This method is based on the concepts derived from the theory of transformation of groups. The main advantage of this method over the others (like free parameter method, separation variable method) is that it always allows the determination of velocity profile of the external flow not a priori but as an outcome of the theory itself. Kalthia [6], [7] has obtained transformation via this method and has reduced the equation of motion to Falkner Skam type equation as below.

Introducing the similarity variable

$$\eta = y\, x^{-\frac{1}{n+1}} \left(\frac{U}{c_o} \right)^{\frac{2-n}{n+1}} \tag{2.4}$$

and

$$\left.\begin{aligned} p &= C_o^{2-n} \left[\frac{a(2n-1)-1}{n-2} \right] \\ a &= \frac{n(n-2)+1}{n+1} \\ \beta &= \frac{\theta(n+1)}{2[1+\theta(n-1)]} \end{aligned}\right\} \tag{2.5}$$

Equation (2.2) simplifies to the nonlinear differential equation

$$n(f'')^{n-1} f''' + ff'' + \beta(1 - f'^2) = 0 \qquad (2.6)$$

Here f is a function of η , $n \neq 2$, and $\partial u/\partial y$ is positive. The axial velocity u, obtained through group transformations is given by

$$u = U f'(\eta) \qquad (2.7)$$

where

$$U = C_o^{\frac{a(n+1)-1}{(n-2)}} x \qquad (2.8)$$

While the boundary conditions in (2.3) reduce to

$$\left.\begin{array}{llll} \text{at} & \eta = 0 & & f = f'' = 0 \\ & \eta \to \infty & & f' = 1 \end{array}\right\} \qquad (2.9)$$

Case for n = 2 was treated separately and equation governing the motion turns out to be

$$2ff''' + ff'' + \frac{1}{2}(1 - f'^2) = 0 \qquad (2.10)$$

with the boundary conditions as above.

In what follows, we shall study this flow over a wedge in general and in particular

(a) The stagnation point flow ($\beta = (n+1)/2n$)

(b) Flow past a right angled wedge ($\beta = 0.5$)

(c) Flows for which $\beta = 1$.

3. Numerical solutions :

Now our task is to solve a nonlinear differential equation (2.6) with the boundary conditions (2.9) for different values of n and β. Also one of the condition is to be satisfied at the edge of a boundary layer. Such a F.S. type equation has been solved partly by Shah [8] with the traditional hit and shoot method. The method has some short falls like (i) relative large value of independent variable (ii) initial guess and sensitivity aspect etc. Nachtsheim et al. [9] were the first to remove above difficulties to a large extent. The method is based on least-squares convergence criterian. It is very less sensative to initial guesses and converges to the solution quickly. Also the method is applicable to study the relation between the properties of the solution and a value of parameter.

Here we have attempted to adopt the Nachtsheim et al [9] technique with the necessary extension to solve our problem. The equation (2.9) contains two parameters n and β and the term of f" is raised to a power (n-1). We hope to find out the shear stress at wall for arbitrary n and β.

4. Analysis of numerical method :

The method adopted for the numerical work is described into three parts :

(A) Calculations for sheat stress at wall [≡f'(0)] and velocity profiles :

Consider the equations (2.6) and (2.9). The main task is to deal successfully with the satisfaction of asymptotic boundary condition. Also we are interested in finding out the values of f" (0), such that, for a given n, when $\eta \to \eta_{edge}$(i.e. edge of the boundary layer; in other words $\eta \to \infty$)$f' \to 1$. Clearly f' is a function of f" (0). Mathematically, we should find a solution of

$$f'_{edge}\left[f''(0) \right] = 1,$$

$$\text{Or} \quad f'_{edge}[x] = 1, \text{ where } x \equiv f''(0) \qquad (4.1)$$

Expanding left hand side, to the first approximation one gets

$$f' + \frac{\partial f'}{\partial x}\Delta x = 1. \qquad (4.2)$$

The necessary correction to the assumed value of x can be obtained from this equation provided we know f'_x $(f'_x \equiv \frac{\partial f'}{\partial x})$ at $\eta = \eta_{edge}$. This partial derivative can be obtained from the perturbation equation formed by differentiating the equation (2.6). It will be

$$n(n-1)(f'')^{n-2} f''_x f''' + n(f'')^{n-1} f'''_x + f f''_x + f_x f'' - 2\beta f' f'_x = 0 \qquad (4.3)$$

The initial conditions now will come to

$$\text{at } \eta = 0 \; ; \; f_x = f'_x = 0 \; , \; f''_x = 1. \qquad (4.4)$$

The question of where to stop integration in such problems is well discussed by Nachtsheim et al [9]. Satisfaction of a single boundary condition $f' = 1$ at $\eta = \eta_{edge}$ does not lead to a unique solution but along with this if the condition $f'' = 0$ at $\eta = \eta_{edge}$ is imposed (which is also physically meaning full), asymptotic boundary condition will be satisfied uniquely. Hence, instead of choosing Δx from (4.2), it will be so taken that equations

$$f' + \frac{\partial f'}{\partial x} \Delta x = 1 \; ; \; f'' + f''_x \; \Delta x = 0, \qquad (4.5)$$

are satisfied at $\eta = \eta_{stop}$. This is, in general, impossible because of one unknown and two equations. However, the satisfactory solution can be obtained with the idea of least-square method. We define :

$$\delta_1 = f'_x \, \Delta x + f' - 1 \; ; \; \delta_2 = f''_x \Delta x + f'' \; ,$$

and attempt to find Δx such that $\delta_1^2 + \delta_2^2$ is minimum. To minimize the sum, we differentiate it with respect to Δx and equate to zero. The calculations yield the value of Δx as

$$\Delta x = \frac{f'_x \, (1-f') - f''_x f''}{f'^2_x + f''^2_x} \qquad (4.6)$$

Along with the descrepancies δ_1 and δ_2, consider the sum of deviations of the computed quantities from its asymptotic values, the error

$$E = (1 - f')^2 + f''^2 \qquad (4.7)$$

The magnitude of E will give us an idea about the satisfaction of asymptotic boundary condition. The value of x that gives $\Delta x = 0$ will correspond to the minimum with respect to x of E [can be verified from (4.6,7)]. Once E minimum is attained, no change is required in the initial conditions. This ensures the satisfaction of the asymptotic boundary condition at a finite value of η.

Scheme: For the assumed value of x, equations (2.6)

and (4.3) along with its respective boundary conditions were integrated by Runge Kutta's method or Adams - Moulton's predictor corrector method using one correction per step. Throughout the integration the step size was taken to be 0.05. Integration was carried to the specific value of η. Δx was determined at this stage. The process was repeated till the relative changes in Δx was less than small preassigned value (taken as 10^{-8}). When this was attained E was tested. If this is not achieved η was increased and the whole process was repeated. Calculations were stopped on satisfaction of the test for E. It is a matter of experience that once the first test is put through, achieving of second test was not much a problem. The whole scheme was carried over in the case of three chief flows mentioned previously i.e., for n = 0.2, 0.5, 0.8, 1, 1.2, 1.5, 2 . (Numerical calculations can be extended to cover $n \geq 2$ but boundary layer assumptions are no longer true here Acrivos et al [5]). We have included these values of n with two fold purposes (i) to compare with some available results, (ii) to study the behavior of a fluid when it is far from Newtonian model. Of course the calculations were carried for the values of n ranging from 0.1 to 2 with increment of 0.1 .

(B) Variation of shear stress at the wall with respect to β. :

The differential equation of our problem contains two parameters α and β. Given a value of n, for each β, we can find $f''(0)$ and study the effect of mutual variations, but this will be little cumborsome and also more time taking. Such variations can be obtained directly by modifying the above discussed method with the help of steepest descent as explained by Tompkins [10]. It consists of finding the curve of solutions $f''(0)$ against β in the $[\beta, f''(0)]$ plane. The points that lie on the curve of solutions are the values that allow equation (2.6) to satisfy the boundary conditions. Briefly the method is as follows :

Since f' is a function of β and $f''(0)$ i.e. x, we are interested in solving an implicit equation

$$f'_{edge}\,[\beta, x] = 1 \qquad (4.8)$$

The necessary corrections to the first approximation

can be found by solving the equations

$$f' + f'_\beta \Delta\beta + f'_x \Delta x = 1 \,, \qquad (4.9)$$

$$- f'_x \Delta\beta + f'_\beta \Delta x = 0 \,. \qquad (4.10)$$

Direction of the tangent to the curve will be given by

$$f'_x \frac{dx}{d\beta} + f'_\beta = 0 \qquad (4.11)$$

In order to obtain the solution for $(\Delta x, \Delta\beta)$, we need f'_x and f'_β . f'_x can be obtained through (4.3), while for f'_β, the equation (2.6) will be differentiated partially with respect to β. It amounts to

$$n(n-1)(f'')^{n-2} f''_\beta f''' + n(f'')^{n-1} f'''_\beta$$

$$+ f_\beta f'' + ff''_\beta - (f'^2-1) - 2\beta f'f'_\beta = 0 \qquad (4.12)$$

Here the initial conditions are :

$$\text{at } \eta = 0 \,, \; f_\beta = f'_\beta = f''_\beta = 0. \qquad (4.13)$$

Like case (A), here, also we shall impose the condition $f'' = 0$ at $\eta = \eta_{edge}$,

i.e. $f'' \,|\beta, x| = 0.$

Now Δx, $\Delta\beta$ can be found by solving the equations

$$f'' + f''_\beta \Delta\beta + f''_x \Delta x = 0, \qquad (4.14)$$

$$- f''_x \Delta\beta + f''_\beta \Delta x = 0. \qquad (4.15)$$

The direction of tangent can be obtained from

$$f''_x \frac{dx}{d\beta} + f''_\beta = 0. \qquad (4.16)$$

The required values of $\Delta\beta$, Δx for the asymptotic approach are obtained from the least-square solutions of equations (4.9, 10, 14, 15). It results to:

$$\Delta\beta = - \frac{f'_\beta (f'-1) + f''_\beta f''}{f'^2_\beta + f'^2_x + f''^2_\beta + f''^2_x} \qquad (4.17)$$

and

$$\Delta x = - \frac{f'_x (f'-1) + f''_x f''}{f'^2_\beta + f'^2_x + f''^2_\beta + f''^2_x} \qquad (4.18)$$

Also, the least-square solution of (4.11, 16) gives the direction of tangent

$$\frac{dx}{d\beta} = - \frac{f'_x f'_\beta + f''_x f''_\beta}{f'^2_x + f''^2_X} \qquad (4.19)$$

Scheme: Equations (2.6), (4.3,4.12) were integrated along with necessary boundary conditions. Necessary corrections were obtained by (4.17) and (4.18) . Relative change in Δx and $\Delta\beta$ were tested. After this test was over E was checked. As earlier, all these values were checked with 10^{-8}. When convergence was realized, equation (4.19) was applied to predict a new solution. Calculations were carried over till β attained 2.4 to 2.6. Results were procured for different values of n.

(C) Variation of f'' (0) with respect to n :

Entirely, as in (B), an algorithm was developed to study the variations of f'' (0) with respect to n, for a fixed β . The program made for calculating the variations in β was used with some changes. Here, since the parameter n occurs as an index also, it makes the parametric differentiation and hence the calculations little complicated. The process was carried out for β = 0.0, 0.5, 1.0 .

Theoretically, in the boundary layer f'' is never negative but while calculating, sometimes, for some values of n and β (both > 1), when f' is near to 1, f'' becomes too small or negative. Since f'' is raised to the power (n-1), the computer give the error of overflow/underflow or negative raised to positive fractional power etc.. Even otherwise this weakned the convergence of f' to unity. These difficulties were overcome either by adjusting the initial guess or feeding $|f''|^{n-1}$ instead of $(f'')^{n-1}$. (In fact, this is so as per the conditions in (2.6).

5. Results and discussion :

First of all let us compare the values of f''(0), an importent outcome of the problem with the available results. This is done in the table (5.1). They are in good agreement with that of Acrivos et al. [5]. As expected Roys [11] results are good enough near n = 1 only.

Table 5.1

Shear stress at the wall for various n.

β	n→	0.2	0.5	1.0	1.5	2.0
	A	-	1.1826	0.9419	0.8942	0.8844
0.5	R	-	1.0790	0.9277	0.9094	1.0243
	P.R.	2.10810	1.16135	0.92898	0.885520	0.87846
	A	-	2.209	1.246	1.081	0.981
1.0	R	-	1.8202	1.2326	1.1346	1.5263
	P.R.	3.45398	1.71326	1.23402	1.09278	1.04766
	A	8.219	2.209	1.246	-	0.981
$\frac{n+1}{2n}$	S	8.151	2.187	1.233	-	0.973
	P.R.	8.15652	2.18906	1.23402	1.03674	0.97476

A = Acrivos et al. R = Roy, S = Shah, P.R.= Present results.

Figures (5.1), (5.2), (5.3) represent velocity profiles, for the flows having $\beta = 0.5$, $\beta = 1$ and $\beta = (n+1)/2n$. It can be observed that $f' \to 1$ at a larger value of η for $n \leq 1$. No sooner $n > 1$, $f' \to 1$ is attained at $n = 2$ to 4. Profiles for $\beta = 0.5$ and $\beta = 1$ are more or less similar. Though profiles in case of $\beta = (n+1)/2n$ are similar to earlier, these are much steeper and f' reaches rapidly to unity.

Considering β as a parameter, the values of shear stress at the wall are calculated and represented in the figure (5.4) for different values of n. It can be seen that as β increases $f''(0)$ increases for all n. Rate of increase in $f''(0)$ is faster in case of smaller n. The rate is slowed down after n = 1. Also these graphs will be able to give correct initial guess for searching more correct value of $f''(0)$ at a particular values of β. For example, this can be verified in case of the values

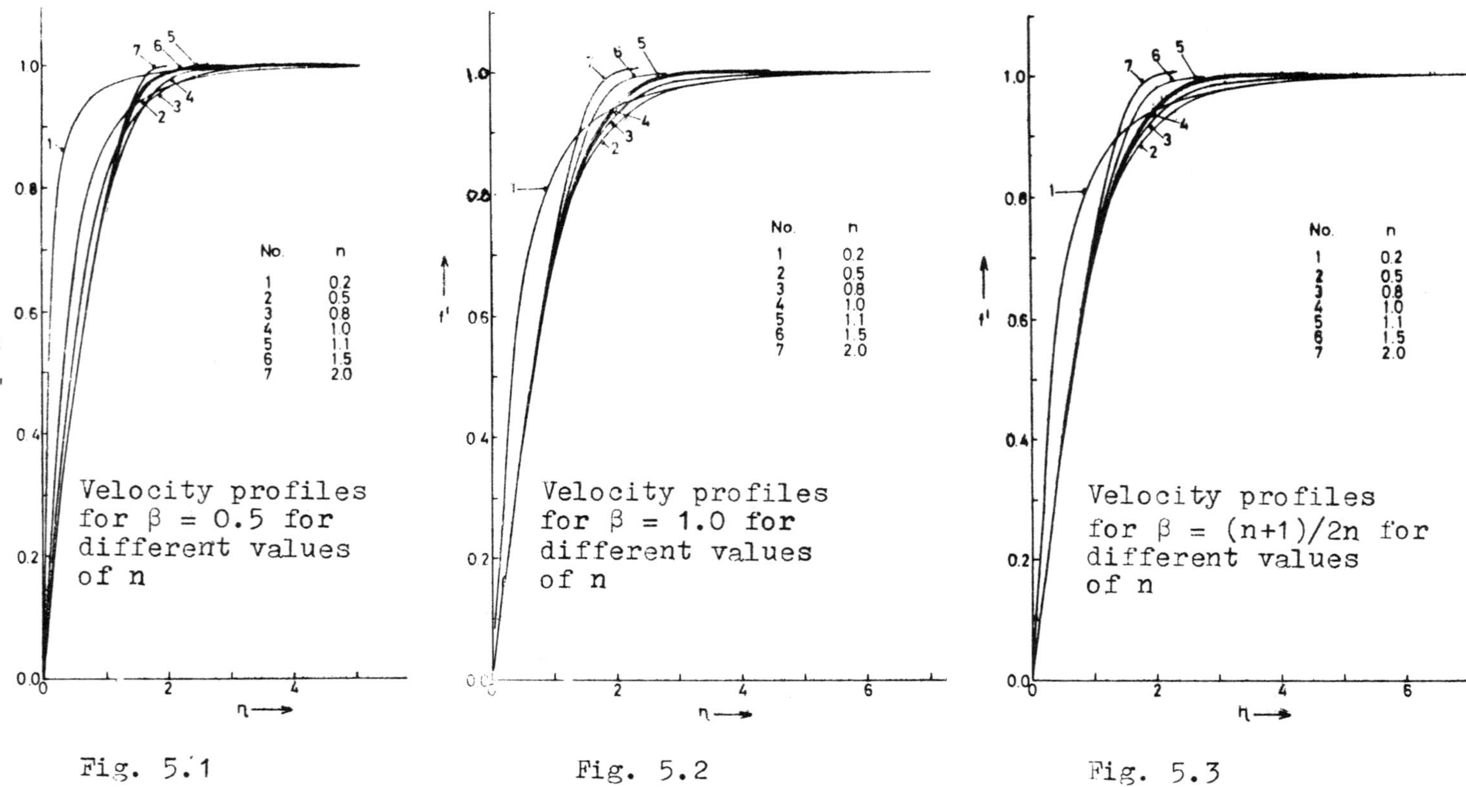
Velocity profiles for β = 0.5 for different values of n
Velocity profiles for β = 1.0 for different values of n
Velocity profiles for β = (n+1)/2n for different values of n
No. n
1 0.2
2 0.5
3 0.8
4 1.0
5 1.1
6 1.5
7 2.0
f'
η
0.0
0.2
0.4
0.6
0.8
1.0
0
2
4
6

Fig. 5.1

Fig. 5.2

Fig. 5.3

given in Table (5.2)

Table 5.2

Shear stress for n = 1 at different values of β

β	f''(0) Hartree [12]	f''(0) Acrivos et al. [5].	f''(0)read from Fig.
1.0	1.233	1.246	1.21
0.8	1.120	1.134	1.13
0.5	0.928	0.944	0.94
0.3	0.775	0.788	0.78
0.2	0.687	0.698	0.69
0.1	0.587	0.595	0.60

In figure 5.5, f''(0) versus n is plotted for β = 0.0, 0.5, 1.0. From this figure it is observed that f''(0) becomes nearly constant after n has attained the value 1.1. Also, in general, as n increases f''(0) decreases. This is not true in case of β = 0. The trend changed at n = 0.5. Rate of decrease is faster till n reaches to 1.1. If one wants to obtain f''(0) exactly, for any n, at β = 0.0, 0.5, 1.0, he can do so by easily having the initial guess with the help of graphs.

6. Asymptotic convergence :

In course of numerical computations, one important aspect was to satisfy the boundary condition $f' \to 1$ as $\eta \to \infty$, (i.e. at a large value of η). For $n \leq 1$ (Pseudoplastic fluids) the convergence of f' to unity was quite rapid and normally it was achieved for η = 5 to 7. Convergence was so strong that even if the initial guess of f''(0) differed from the correct value by a number lying between −1 and +2, it did converge within 4 to 5 runs. Indeed, one thing was observed that, for the first integration i.e. at smaller values of η, values of f', f'' tend to deviate much from the right solution. After n attains the value 1.1, (Dilatant fluids) $f' \to 1$ was attained at relatively small values of η, i.e. at η = 2 to 4. Convergence was poor in comparison to $n \leq 1$. One has to be very careful about the initial guess. In some case f' did not converge to unity till the initial guess was correct upto 4 decimal places. This behavior of dilatant fluids is common irrespective of value of β. Acrivos et al. [5], Fox, Serth and others who have worked on boundary layer flows have experienced a difficulty in fulfilling the boundary condition at infinity, especially in case of dilatant fluids having $n \geq 2$, though some of them have extended the calculations for the sake of completeness. Acrivos et al. also have shown that the boundary layer assumptions breakdown for $n \geq 2$ and hence in stricter sense the differential equation (4.16) will no longer true. Probably this is the reason why the convergence becomes poor near and after $n = 2$.

7. Conclusion :

The flow over a wedge of a power-law fluid is completely solved by numerical techniques. Velocity profiles for different parameters are obtained. It is observed that, for a fixed β, increase in n causes decrease in the shear stress at the wall ; while, for a fixed n, increase in β has always resulted to an increase in $f''(0)$.

References :

1. METER, D. N.- Ph.D. Thesis, Uni Wisconsin, Medson.
2. WILKINSON, W. L.- Non Newtonian Fluids, Pergamon Press, 1960.
3. METENER, A. N.- Hand book of Fluid dynamics ed. by STREETER, V.L. McGraw Hill New York, 1960.
4. BIRD, R. B., LIGHTFOOT, R. N., STEWART, V.L. - Transport phenomena, John Wiley, New York, 1960.
5. ACRIVOS, A.- A theoretical analysis of natural convection heat transfer to non-Newtonian fluids. A.I.Ch.E. Jl. 6, pp. 984-990.
6. KALTHIA, N.L.- Similarity solutions of the boundary layer equations for non-Newtonian fluids. Ind. J. Engg. Maths, 2, pp. 17-21.
7. KALTHIA, N.L.- Ph.D. Thesis, Ind. Inst. Tech., Kanpur.
8. SHAH, M. J.- Ph.D. Thesis, Uni. California, Berkely.
9. NACHTSHEIM, P. R., and SWIGERT, I.- Satisfaction of

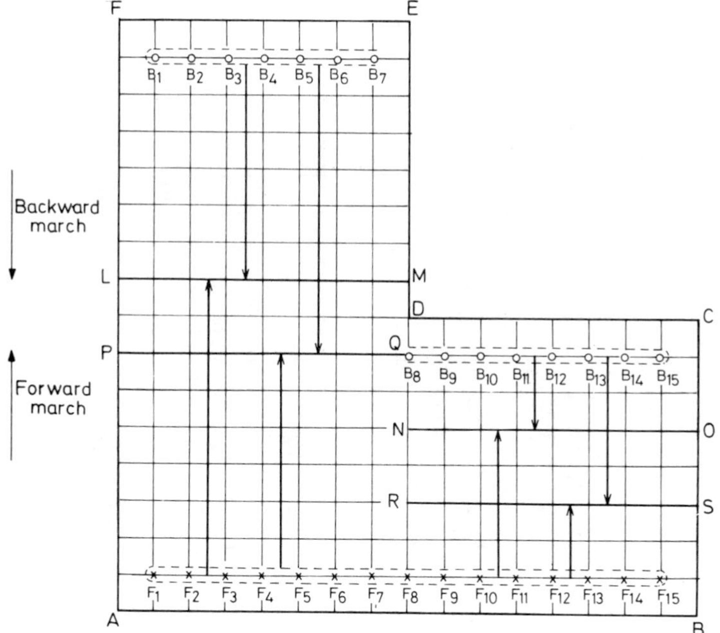

FIGURE 1(a)

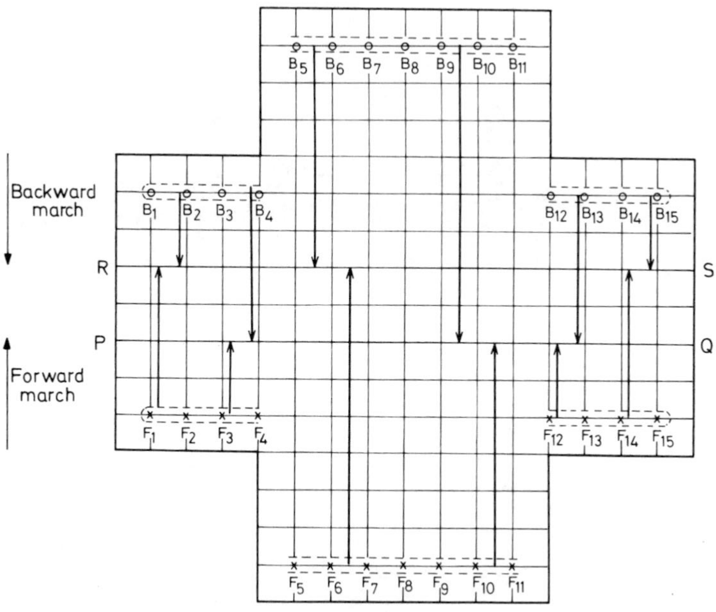

FIGURE 1 (b)

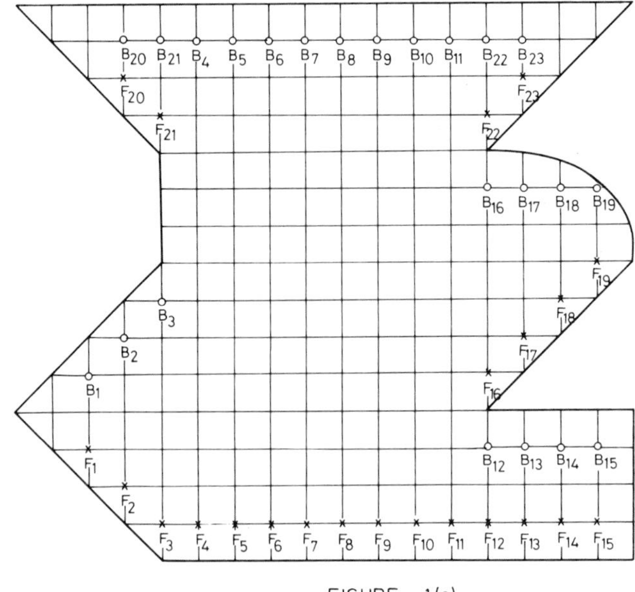

FIGURE 1(c)

FIGURE 1(a)-(c) SMT FOR IRREGULAR GEOMETRIES F_1, F_2--ETC COMPONENTS OF THE INITIAL VECTOR IN THE FORWARD MARCH B_1, B_2---ETC COMPONENTS OF THE INITIAL VECTOR IN THE BACKWARD MARCH

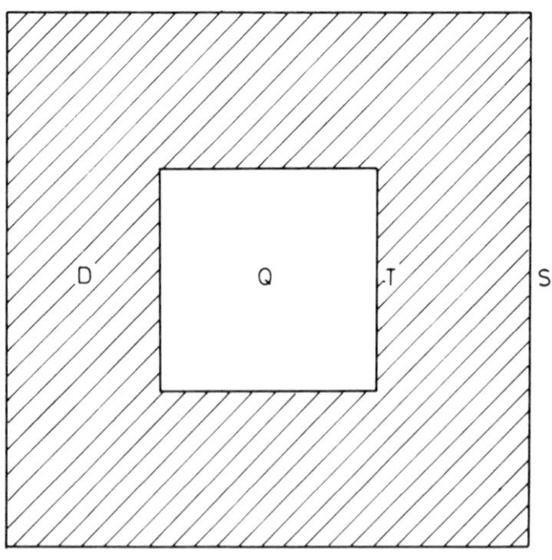

FIGURE 2 RECTANGULAR REGION WITH AN INTERIOR RECTANGLE REMOVED

asymptotic boundary conditions of system of nonlinear equations of boundary type. NASA - IND-3004.

10. TOMPKINS, C.B.- Modern Mathematics for Engineers. ed. by Backenbech, B.P. McGraw Hill, 1960.

11. ROY,S.- A perturbation solution for the wedge flow in power law fluid. Ind. J.Phys., 47, pp. 445-452.

12. HARTREE,D.R.- On the equation occuring in Falkner-Skan's approximate treatment. Proc. Camb. Phil. Soc. 33(II), pp. 223-239.

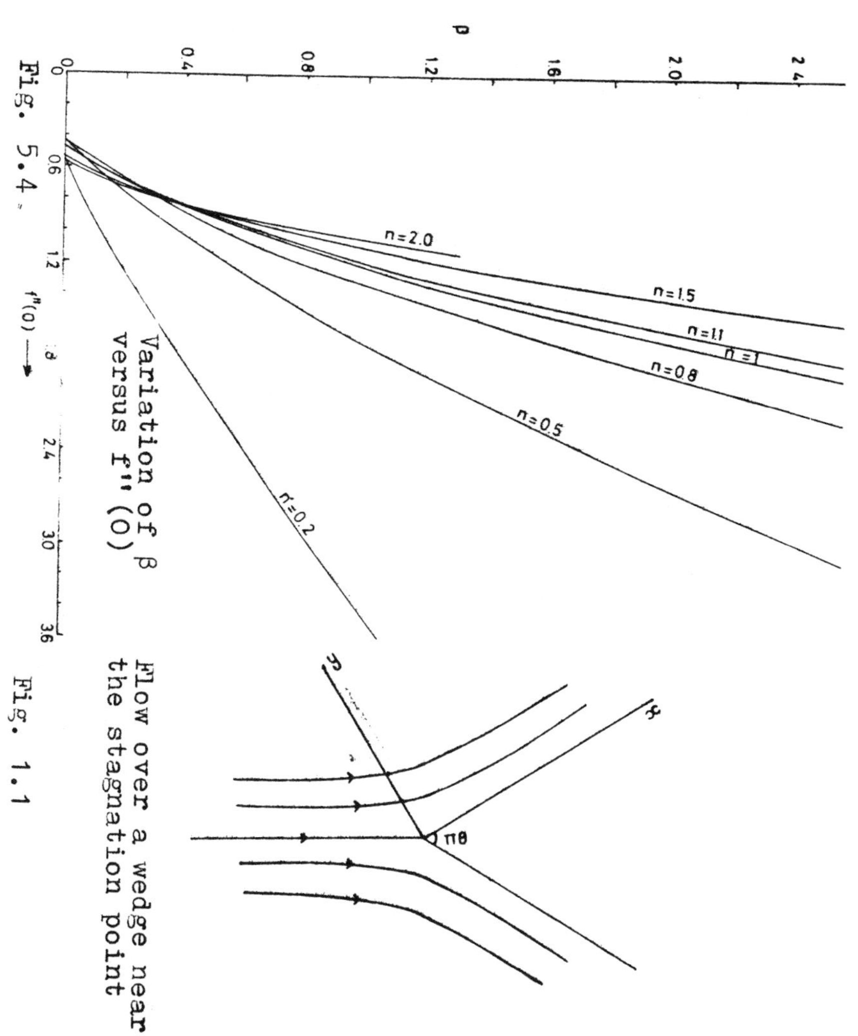

Fig. 5.4. Variation of β versus f''(O)

Fig. 1.1 Flow over a wedge near the stagnation point

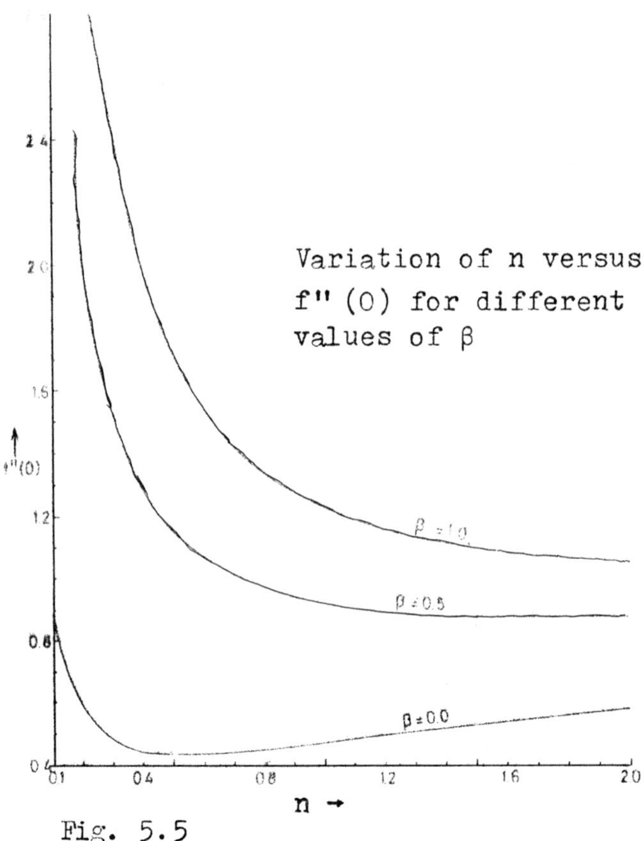
Variation of n versus
f" (0) for different
values of β
2.4
2.0
1.6
1.2
0.8
0.4
f"(0)
β = 1.0
β = 0.5
β = 0.0
0.1
0.4
0.8
1.2
1.6
2.0
n →

Fig. 5.5

THE USE OF BESSEL FUNCTION AND JACOBI POLYNOMIAL IN RADIAL VIBRATIONS OF A GAS IN AN INFINITE CYLINDERICAL TUBE

S. D. Bajpai,
Dean, Faculty of Science,
Rivers State University of Science & Technology,
Port Harcourt, Nigeria.

SUMMARY

In this paper, we have employed the Bessel function of the first kind and the Jacobi polynomial to obtain the solution of the fundamental differential equation of the radial vibrations of a gas in an infinite cylinderical tube.

1. Introduction. The use of special functions in applied sciences is of great importance in technical applications.

In this paper we consider the application of the Bessel function of the first kind and the Jacobi polynomial in the radial vibrations of a gas in an infinite cylinderical tube.

Suppose that we have an immovable tube that is so long that we may consider it as extending infinitely far in both directions. We denote the radius of any cross-section of the tube by R.

Let us suppose that the tube is filled with a gas that performs small amplitude vibrations about its equilibrilim position. Let us investigate these small amplitude vibrations, confining ourself to radial vibrations, where the velocity potential u depends only on the distance r of the vibrating particle of the gas from the z-axis (which is the axis of the cylinder) and on the time t.

In this case the wave equation [2, p.196, (33)], written in cylinderical coordinates r, ϕ, and z acquires a simpler form:

$$(1.1)\quad \frac{\partial^2 u}{\partial r^2} + \frac{1}{r}\frac{\partial u}{\partial r} = \frac{1}{a^2}\frac{\partial^2 u}{\partial t^2}.$$

Obviously, we can solve the problem on the small amplitude vibrations of the gas if we find the solution of (1.1) satisfying the initial conditions

$$(1.2)\quad u\Big|_{t=0} = f(r),\quad \frac{\partial u}{\partial t}\Big|_{t} = F(r),$$

and the boundary condition

$$(1.3)\quad \frac{\partial u}{\partial r}\Big|_{r=R} = 0,$$

In this paper we shall take

$$(1.4)\quad f(r) = \left\{1-\frac{r^2}{R^2}\right\}^{\nu/2} J_\nu\left\{\lambda\sqrt{\left(1-\frac{r^2}{R^2}\right)}\right\},$$

and

$$(1.5)\quad F(r) = \left\{1-\frac{r^2}{R^2}\right\}^{\beta} P_\nu^{(0,\beta)}\left\{1-\frac{2r^2}{R^2}\right\}.$$

The following formulae are required in the proof:

$$(1.6)\quad \int_0^R r\left\{1-\frac{r^2}{R^2}\right\}^{\nu/2} J_\nu\left\{\lambda\sqrt{\left(1-\frac{r^2}{R^2}\right)}\right\} J_0\left(\frac{r}{R}\mu_k\right)dr$$

$$= \frac{R^2\lambda^\nu}{\{\lambda^2+\mu_k^2\}^{(\nu+1)/2}} J_{\nu+1}\left\{\lambda\sqrt{(\lambda^2+\mu_k^2)}\right\},\quad \mathrm{Re}\,\nu > -1,$$

which follows from [4, p. 299, (26)].

$$(1.7)\quad \int_0^R r\left\{1-\frac{r^2}{R^2}\right\}^{\beta} P_n^{(0,\beta)}\left\{1-\frac{2r^2}{R^2}\right\} J_0\left(\frac{r}{R}\mu_k\right)dr$$

$$= \frac{2^\beta R^2\,\Gamma(\beta+n+1)}{n!\,\mu_k^{\beta+1}} J_{\beta+2n+1}(\mu_k),\quad \mathrm{Re}\,\beta > -1,$$

which follows from [1, p.189].

2. Solution of the problem

The solution of the problem to be obtained is

$$(2.1)\ u\ (r,t) = \sum_{K=1}^{\infty}\left[\frac{2\lambda^{\nu} J_{\nu+1}\sqrt{(\lambda^2+\mu_K^2)}}{\{\lambda^2+\mu_K^2\}^{(\nu+1)/2} J_0^{-2}(\mu_K)} \cos\frac{at}{R}\mu_K + \frac{2^{\beta+1} R\,\Gamma(\beta+n+1)}{n!\,a\mu_K J_0^2(\mu_K)} J_{\beta+2n+1}(\mu_K)\sin\frac{at}{R}\mu_K\right] J_0\left(\frac{r}{R}\mu_K\right)$$

where $Re\,\nu > -1$, $Re\,\beta > -1$.

<u>Proof</u>: The solution of the problem on setting u (r,t) = W (r) T (t) can be written as [2, p.197, (46)]

$$(2.2)\ u\ (r,\ t) = \sum_{K=1}^{\infty}\left[a_K \cos\frac{at}{R}\mu_K + b_K \sin\frac{at}{R}\mu_K\right] J_0\left(\frac{r}{R}\mu_K\right)$$

If t = o, then by virtue of (1.4), we have

$$(2.3)\left\{1-\frac{r^2}{R^2}\right\}^{\nu/2} J_{\nu}\left\{\lambda\sqrt{\left(1-\frac{r^2}{R^2}\right)}\right\} = \sum_{K=1}^{\infty} a_K J_0\left(\frac{r}{R}\mu_K\right).$$

Multiplying both sides of (2.3) by r $J_0\ (\frac{r}{R}\mu_j)$ and integrating with respect to r from o to R, we get

$$(2.4)\int_0^R r\left\{1-\frac{r^2}{R^2}\right\}^{\nu/2} J_{\nu}\left\{\lambda\sqrt{\left(1-\frac{r^2}{R^2}\right)}\right\} J_0\left(\frac{r}{R}\mu_j\right)dr = \sum_{K=1}^{\infty} a_K\int_0^R r J_0\left(\frac{r}{R}\mu_K\right) J_0\left(\frac{r}{R}\mu_j\right)dr.$$

Now using (1.6) and the orthogonolity property of Bessel functions [3, p.130, (5.14.9)], we have

$$(2.5)\ a_K = \frac{2\lambda^{\nu} J_{\nu+1}\{\sqrt{(\lambda^2+\mu_K^2)}}{\{\lambda^2+\mu_K^2\}^{(\nu+1)/2} J_0^{2}(\mu_K)}$$

Similarly by virtue of the relation (1.7), we get (2.6)

$$(2.6)\ b_K = \frac{2^{\beta+1} R\,\Gamma(\beta+n+1)}{n!\,a\mu_K J_0^{2}(\mu_K)} J_{\beta+2n+1}(\mu_K),$$

with the help of (2.2), (2.5) and (2.6) the solution (2.1) is obtained.

<u>Note 1</u>. Since u is the velocity potential, we have discarded the term $a_0 + b_0 t$, because the velocity distribution in the vibrating gas is not changed by this term.

Note 2. Any function which is bounded and has a finite number of maxima and minima can be represented by a series of Bessel functions, hence our f (r) is of a general character and hence may be of great physical interest. In view of the expansion property of Jacobi polynomials [5, 5, p.24], our F (r) may also encompass several cases of interest.

Note 3. The values of $\mu_1, \mu_2, \mu_3, ---$ can be obtained as [2, p.192].

Note 4. The solution (2.1) may be useful for computing different values of u (r, t).

REFERENCES

1. BHONSLE, B. R. - on some results involving Jacobi polynomials, Journ. Ind. Math. Soc. 26, pp. 187-190, 1962

2. KOSHLYAKOV, N. S.; SMIRNOV, M. M. and GLINER, E. B.- Differential Equotions of mathematical physics, North Holland Publishing Co., Amsterdam - London, 1951.

3. LEBDEV, N. N. - Special functions and their applications, Printice Hall, Inc., Englewood cliffs, N. J., 191965.

4. LUKE, Y. L. - Integrals of Bessel functions, McGraw-Hill, New York, 1962.

5. SZEGŐ, G. - Orthogonal polynomials, American Mathematical Society, 1959.

"Symmetric Marching Technique (SMT) for the Efficient Solution of Discretized Poisson Equation on Non-Rectangular Regions"

Mohan K.Kadalbajoo and K.K.Bharadwaj*
Department of Mathematics
Indian Institute of Technology, Kanpur 208016
(INDIA)

ABSTRACT : In this paper, two methods based on the symmetric marching technique (SMT) have been presented for the solution of discretized Poisson equation on non-rectangular regions. The method I illustrates the direct adaptation of SMT to irregular geometries. In method II, an efficient implementation of the capacitance matrix method has been considered using SMT. The favourable properties of SMT, to solve Poisson equation subject to several right hand side functions and different boundary conditions without extra computational effort, have been exploited for the fast generation of the capacitance matrix. Numerical results, of the model problems solved, have been presented.

---

*B.N.College, Udaipur, Rajasthan.

INTRODUCTION :

During the last decade, it has become increasingly popular to solve the discretized Poisson equation by direct rather than iterative methods and several very fast methods have been developed for the direct solution of the quite special system of linear algebraic equations which arise when Poisson's or Helmholtz's equation is solved by standard finite difference methods on rectangular region. The best known of these methods are the Hockney's FACR method [13], the block-cyclic reduction of Buzbee, Golub and Nielson [9] and the popular algorithm of Bunemann [6]. A survey of the development of these methods is available in Dorr [10]. Recently Bank and Rose [4] have developed marching techniques, based on a special block LU decompositions and appropriate block back solution. A Fourier- Toeplitz method has been developed by Fischer et.al [11] and recently discussed by Proskurowski and Widlund [17].

Most of these direct methods can be regarded as efficient computer implementations of the separation of variables method. For a continuous problem, the separation of variables can be used only for regions, which after a possible change of independent variables, are rectangular and differential operator of a special form. The discrete problem has also been imposed with the similar restrictions and a special structure of the coefficient matrix is needed for the use of fast direct techniques . As such most of the fast direct methods, except a few [1,2,3,18], are suitable for problems on rectangular regions. However even for problems on non-rectangular regions which do not allow for separation of variables, the fast direct methods can be used provided the differential operator allows for separation of variables on some different appropriate region. The idea is to imbed the given non-rectangular bounded region in a rectangle or another region for which the problem can be solved by separation of variables. Several attempts, based on this idea, have been made and the methods developed are known as capacitance matrix methods.

Early work on capacitance matrix method was carried out by Hockney [14,15] based on an idea from potential theory suggested to him by Oscar Buneman. At about the same time, Buzbee et.al. [8] used Woodbury formula to develope a similar method and gave a formulation in matrix language in which the solution of a matrix equation Au = v is obtained by solving another equation Bu = w, where B is chosen so that the equation Bu = w can be solved by fast direct techniques. Buzbee and Dorr [7] have considered an efficient implementation of this technique using the direct method of matrix decomposition.

George [12] has developed certain so-called iterative imbedding algorithms. The author, instead of generating the capacitance matrix explicitly, formulated the problem as a minimization problem and the conjugate gradient or the Davidon-Fletcher-Powell method has been employed.

Recently, Proskurowski and Widlund [17] considered the capacitance matrix for the Dirichlet and Neumann problems for Helmholtz's equation and it has been shown, that by an appropriate choice of the fast solver, the capacitance matrix can be generated quite inexpensively. The authors employed Fourier-Toeplitz method for the fast generation of the capacitance matrix and used the conjugate gradient method for solving the corresponding capacitance matrix equations efficiently. A survey of the previous work on capacitance matrix methods has also been included. Very recently, the capacitance matrix approach, has been extended to Helmholtz's equation on general three-dimensional region by O'Leary and Widlund [16].

In this paper two methods, based on the symmetric marching technique (SMT) developed in [5], have been presented for the discretized Poisson equation on non-rectangular regions. The method I, which is an extension of the idea given by Roache [18], illustrates the direct adaptation of SMT to irregular geometries. In method II, an efficient implementation of the capacitance matrix method, developed in [14,15], has been considered using SMT. The favourable properties of SMT, to solve Poisson equation subject to several right hand side functions and different boundary conditions without extra computational efforts, have been exploited for the fast generation of the capacitance matrix. Numerical results, of the model problems solved, have been presented.

2. EXTENSION OF SMT TO POISSON EQUATION ON NON-RECTANGULAR REGIONS :

In this section, two methods have been described for Dirichlet problem for Poisson equation over non rectangular regions. In method I, we have demonstrated the simple adaption of SMT to non-rectangular regions and in method II, an efficient implementation of a capacitance matrix approach [14,15] using SMT has been presented.

<u>METHOD I</u> : In order to demonstrate simple and direct adaptation of SMT to the problems over non-rectangular regions, we have considered different irregular geometries (cf. Figure 1(a) - (c)). The main feature of the method is the definition of the initial vectors in the forward and backward march and the matching levels. In the Figures 1(a)-(c), $F_1, F_2, \ldots$ denote component of the initial vector in the forward march and correspondingly $B_1, B_2, \ldots$ denote the components of the initial vector in the backward march. As illustrated in Figure 1(a), for the L-shaped region, the final vectors in the forward and backward march will be matched at the levels {LM, NO} and {PQ,RS}.

Similarly for the region in Figure 1(b) the matching levels would be RS and PQ. Computational results, corresponding to the regions in Figure 1(a) and Figure 1(b), have been presented and a comparison of SMT with Roache approach has also been included.The problems on more complex geometries, such as given in Figure 1(c) can also handled similarly using SMT. It is to be noted here that the indexing of F and B are not unique.

<u>METHOD II</u> : For definiteness, we shall confine our construction to the region given in Fig. 2 i.e. a rectangular region with an interior rectangle removed.

We consider the following problem :

$\Delta u = f$ in D ...(1)

$u = g$ on SUT ...(2)

A uniform grid is superimposed on the rectangle S and it is asumed, for simplicity, that T lies on grid points and the lines adjoning adjacent grid points. Using the usual five point approximation and writing out an equation for every grid point in S, we get an N x N system of equations

$$Av = w \, , \qquad ...(3)$$

where the vectors v and w are defined on the grid and N is the total number of grid points in the rectangle S. We write (3) in the partition form

$$\begin{bmatrix} A_{11} & A_{12} & \\ A_{21} & A_{22} & A_{23} \\ & A_{32} & A_{33} \end{bmatrix} \begin{bmatrix} v_D \\ v_T \\ v_Q \end{bmatrix} = \begin{bmatrix} w_D \\ w_T \\ w_Q \end{bmatrix} \qquad ...(4)$$

where the vector partitions with subscripts D,Q and T contain elements corresponding to grid points in D,Q and on T, respectively. If w_T and w_Q are assigned values so that the solution to (4) satisfies the boundary conditions i.e. v_T has correct values, then v_D will be correct discrete solution to our given problem. A method, closely connected to the discrete Green's function, has been described by Hockney [14,15] to handle the above problem. This approach has also been discussed by George [12] in connection with some iterative imbedding algorithms. We shall first describe briefly, the Hockney's approach and then an efficient implementation of the same will be presented using SMT.

Inverting the partitioned matrix in (4), we obtain

$$\begin{bmatrix} v_G \\ v_T \\ v_Q \end{bmatrix} = \begin{bmatrix} B_{11} & B_{12} & B_{13} \\ B_{21} & B_{22} & B_{23} \\ B_{31} & B_{32} & B_{33} \end{bmatrix} \begin{bmatrix} w_G \\ w_T \\ 0 \end{bmatrix} \quad \ldots(5)$$

and solving (5) for $B_{22}\, w_T$ gives

$$B_{22}\, w_T = v_T - B_{21}\, w_G \quad \ldots(6)$$

Since B_{22}, being non-singular, is positive definite, w_Q has been set to zero. A fast direct method can be used to solve the system

$$\begin{bmatrix} A_{11} & A_{12} & \\ A_{21} & A_{22} & A_{23} \\ & A_{32} & A_{33} \end{bmatrix} \begin{bmatrix} \bar{v}_G \\ \bar{v}_T \\ \bar{v}_Q \end{bmatrix} = \begin{bmatrix} w_G \\ 0 \\ 0 \end{bmatrix} \quad \ldots(7)$$

to obtain $B_{21}\, w_G$ as $\bar{v}_T$.

Having got $\bar{v}_T$, the w_T is then obtained by solving

$$B_{22} w_T = v_T - \bar{v}_T \quad \ldots(8)$$

The method thus requires B_{22}, which means that we need p (no. of grid points on T) corresponding columns of the inverse of the coefficient matrix A. This method, therefore requires solving p + 2 systems of the form (3), and the solution of the p linear equations (8).

We now describe the computational procedure, for the efficient implementation of the above technique, using SMT.

Step 1 : SMT and the fast generation of the capacitance matrix B_{22} (p x p) :

In order to generate the p-columns of B_{22}, corresponding to p-grid points on T, p calls of SMT, applied to every grid point in S, will be required. For the kth ($1 \leq k \leq p$) call of SMT the following finite difference formulas will be used :

$$u_{i,j+1} = 4u_{i,j}-u_{i+1,j}-u_{i-1,j}-u_{i,j-1}, \forall \text{ points } \in GUQUT - \{\text{kth grid point on T}\} \qquad ...(9)$$

$$u_{i,j+1} = 4u_{i,j}-u_{i-1,j}-u_{i+1,j}-u_{i,j-1}-1 \quad \text{for the kth grid point on T} \qquad ...(10)$$

The error propagation equation corresponding to the formulas (9) and (10) would be

$$e_{i,j+1} = 4e_{i,j}- e_{i-1,j} - e_{i+1,j} - e_{i,j-1} \quad \text{for all points} \in GUQUT \qquad ...(11)$$

Using the formulas (9), (10) and (11) the SMT as described in [5] is applied on the rectangle S and the components of the kth column of the matrix B_{22} are then the solution components corresponding to the p-grid points on T.

It is to be noted that the error propagation (11) remains fixed for all the p calls of SMT and as such the influence coefficient matrix is to be computed only once and the same will be employed for the subsequent calls of SMT. Since the major computational effort in SMT is the construction of influence coefficient matrix and as such once the influence coefficient matrix is constructed and it LU decomposition is stored, the subsequent calls of SMT will require very little extra computational effort.

Step 2 : Construction of the vector $\bar{v}_T(p\times 1)$

The finite difference formulas for SMT to construct $\bar{v}_T$ will be

$$u_{i,j+1} = 4u_{i,j}-u_{i-1,j}-u_{i+1,j}-u_{i,j-1}+\Delta^2 y g_{i,j}, \quad \text{for all points} \in D \qquad ...(12)$$

and

$$u_{i,j+1} = 4u_{i,j} - u_{i-1,j} - u_{i+1,j} - u_{i,j-1},$$

for all points $\in T \cup Q$ ...(13)

Again the error propagation equation for this case would be the same as (11) and as such the influence coefficient matrix of step 1 would be employed. The construction of the vector $\bar{v}_T$ as the SMT solution components, corresponding to the p-grid points on T, requires little computational effort.

Step 3 : Computation of the vector w_T (pxl) :

Having computed the vector $\bar{v}_T$ in Step 2, the vector w_T is computed by solving the system.

$$B_{22} w_T = v_T - \bar{v}_T$$

of p linear equations using Gausion elimination.

It is to be noted that B_{22} may not be in general symmetric, but whenever the inner irregular boundary T is symmetrical w.r.t. the outer rectangle S the capacitance matrix is symmetric and the Cholesky factorization may be prefered over Gausian elimination.

Step 4 : Computation of the final solution in the region D.

This final call of SMT will require the following finite difference formulas :

$$u_{i,j+1} = 4u_{i,j} - u_{i-1,j} - u_{i+1,j} - u_{i,j-1} + \Delta^2 y g_{i,j},$$

for all points $\in$ D ...(14)

and

$$u_{i,j+1} = 4u_{i,j} - u_{i-1,j} - u_{i+1,j} - u_{i,j-1} - w_T,$$

for all points $\in$ T ...(15)

Once again the error propagation equation for this case would be the same as (11) and the influence coefficient matrix of Step 1 would work for this case too.

3. SOLUTION ON FINER GRIDS :

The solution on the finer grids is computed using the mesh refinement technique described in [5].

4. NUMERICAL RESULTS AND DISCUSSIONS :

In this section the numerical results of the model problems solved are given.

In the Tables given below we use the following abbreviations for convenience

a.e - average error

m.e - maximum error

t - run time (in sec.)

(a) - SMT

(b) - Refining SMT solution (mesh size 1/16) to finer grids

(c) - Roache method

Problem 1. $\Delta u = o$ w.r.t. the region given in Figure 1(a),

$u(x,y) = x^2-y^2$ on the boundary of the region given in Figure 1(a), which has the exact solution $u(x,y) = x^2-y^2$.

Problem 2. $\Delta u = 2(x+y)$ w.r.t. the region given in Figure 1(a),

$u(x,y) = xy(x+y)$ on the boundary of the region given in Figure 1(a),

which has the exact solution $u(x,y) = xy(x+y)$.

Problem 3. Problem 1 w.r.t. the region given in Figure 1(b).

Problem 4. Problem 2 w.r.t. the region given in Figure 1(b).

Problem 5. $\Delta u = o$ in D (cf. Fig.2)

$$u(x,y) = x^2-y^2 \text{ on } TUS,$$

which has the exact solution $u(x,y) = x^2-y^2$.

Problem 6. $\Delta u = 2(x+y)$ in D (cf. Fig.2)

$$u(x,y) = xy(x+y) \text{ on } TUS,$$

which has the exact solution $u(x,y) = xy(x+y)$.

The computational results in Table 1-4 for the problem 1-4, using method I, show that SMT works on grids with mesh size 1/32, whereas Roache method works on grids with mesh size 1/16 and fails to converge on finer grids. Further, the solution computed by SMT are of greater accuracy as compared to Roache method. However the computational time required in SMT is greater than Roache method as indicated in the results presented. The solution on finer grids with h = 1/32 and 1/64 has been computed by refining SMT solution on coarser grid (h=1/16). As it is apparent from the results, the refinement to the finer grids does not increase the error beyond the SMT solution on coarser grid (h = 1/16).

From the results given in Table 5, for the Problem 5-6 using method II, it is clear that the method works well on grids with mesh size h = 1/16 and the refinement on the finer grids (h=1/32 ; 1/64) does not increase the error beyond the SMT solution on coarser grid (h = 1/16). The computational time required for the construction and factorization of capacitance matrix was 2.464 sec. Since the capacitance matrix depends only on the geometry of the region and not on the boundary data, the capacitance matrices for Problems 5 and 6 comes out to be identical. As such the time given for Problem 6 is the time for solving an additional problem (excluding the computational time for the construction and decomposition of the capacitance matrix).

5. CONCLUSION :

The adaptability of SMT to irregular regions has been successfully shown through the methods I and II. The method I has been shown to cover all irregular geometries, except the regions with a hole.

Method II, based on capacitance matrix has been shown to cover any irregular bounded region. The capacitance matrix depends only on the geometry of the problem and not on the boundary data and as such the Method II is specially suitable when many problems with the same geometry are to be solved. Method I is clearly more efficient than method II and whenever it works, is preferable to method II.

REFERENCES :

1. ANGEL,E. - Discrete Invariant Imbedding and Elliptic Boundary Value Problems Over Irregular Regions, J.Math.Anal. Applic. , Vol.23, pp. 471-484, 1968.

2. - Dynamic Programming and Linear Partial Differential Equations, J. Math. Anal. Applic., Vol.23, pp. 628-638, 1968.

3. - A Building Block Technique for Elliptic Boundary Value Problems Over Irregular Regions, J.Math.Anal. Applic., Vol.26, pp. 75-81,1969.

4. BANK,R.E. and ROSE,D.J. Marching Algorithms for Elliptic Boundary Value Problems. I : The Constant Coefficient Case, SIAM J. Numer. Anal., Vol.14, pp. 792-829, 1977.

5. BHARADWAJ, K.K., KADALBAJOO, M.K., and SANKAR,R. A Symmetric Marching Technique (SMT) for Discretized Poisson Equation, Submitted for publication.

6. BUNEMAN,O. A Compact Non-Iterative Poisson Solver, Rep. 294, Stanford Univ., Institute for Plasma Research, Stanford, C A., 1969.

7. BUZBEE,B.L. and DORR,F.W. The Direct Solution of the Biharmonic Equation on Rectangular Regions and the Poisson Equation on Irregular Regions; SIAM J.Numer. Anal.,Vol.11, pp. 753 - 763, 1974.

8. BUZBEE, B.L., DORR, F.W., GEORGE, J.A. and GOLUB, G.H. — The Direct Solution of the Discrete Poisson Equation On Irregular Regions, SIAM J. Numer. Anal., Vol.8, pp. 722-736, 1971.

9. BUZBEE, B.L., GOLUB, G.H., and NIELSON, C.W. — On Direct Methods for Solving Poisson Equation, SIAM J. Numer. Anal. Vol.7, pp. 627-656, 1970.

10. DORR, F.W. — The Direct Solution of the Discrete Poisson Equation on a Rectangle, SIAM Rev., Vol.12, pp. 248-263, 1970.

11. D. FISCHER, G. GOLUB, O. MALD, C. LEIVA, and O. WIDLUND — On Fourier Toeplitz Method for Separable Elliptic Problems, Math. Comp., Vol. 28, pp. 349-368, 1974.

12. GEORGE, J.A. — The Use of Direct Methods for the Solution of Discrete Poisson Equation on Non-rectangular Region, Computer Science Department Report 159, Stanford Univ., 1970.

13. HOCKNEY, R.W. — The Potential Calculation and Some Application, in "Methods of Computational Physics", Vol.9, pp.135-211, Academic Press, New York 1969.

14. - — Formation and Stability of Virtual Electrodes in a Cylinder J.Appl. Phys., 39. (1968), pp. 4166-4170.

15. - — POT 4 - A Fast Direct Poisson Solver for the Rectangle allowing Some Mixed Boundary Conditions and Internal Electrodes, IBM Research, R.C. 2870, 1970.

16. O'LEARY, D.P. and O. WIDLUND — Capacitance Matrix Methods for the Helmohtz Equation on General Three Dimensional Regions, Math. Comp., 33, pp. 849-879, 1979.

17. PROSKUROWSKI, W. and O. WIDLUND — On the Numerical Solution of Helmoltz's Equation by the Capacitance Matrix Method, Math. Comp., 30(1976) pp. 433-468.

18. ROACHE., P.J. — A New Direct Method for the Discritized Poisson Equation. Lecture Notes in Phys., Springer Verlag., 8(1971), pp. 48-53.

Table 1

Results for Problem 1 ; $h = \Delta x = \Delta y$

Mesh size (h)		1/8	1/16	1/32	1/65
(a)	a.e.	4.6E-18	7.2E-14	2.9E-3	
	m.e.	3.2E-17	1.2E-13	6.0E-2	
	t	.076	.445	3.671	
(b)	a.e			8.6E-14	2.1E-14
	m.e	-	-	1.4E-13	1.4E-13
	t			.546	.825
(c)	a.e	1.1E-17	4.9E-12		
	m.e	3.1E-16	1.7E-11	Does not converge →	
	t	.025	.149		

Table 2

Results for Problem 2 ; $h = \Delta x = \Delta y$

Mesh size (h)		1/8	1/16	1/32	1/65
(a)	a.e.	1.3E-18	3.8E-14	1.6E-3	
	m.e	5.9E-17	5.6E-13	4.7E-2	
	t	.085	.454	3.685	
(b)	a.e			7.4E-14	5.2E-14
	m.e	-	-	5.8E-13	5.8E-13
	t			.558	.882
(c)	a.e.	6.4E-17	1.6E-12		
	m.e	1.0E-16	6.6E-11	Does not converge →	
	t	.027	.152		

Table 3

Results for Problem 3 ; h = Δx y= Δy

Mesh size (h)		1/8	1/16	1/32	1/64
(a)	a.e	2.3E-18	1.2E-15	7.9E-08	
	m.e	1.7E-17	1.9E-14	1.1E-6	
	t	.089	.390	2.522	
(b)	a.e			4.6E-15	4.8E-15
	m.e	-	-	2.3E-14	2.3E-14
	t			.501	.782
(c)	a.e	1.9E-16	2.2E-10		
	m.e	5.6E-16	5.7E-10	Does not converge →	
	t	.030	.151		

Table 4

Results for Problem 4 , h = Δx = Δy

Mesh size (h)		1/8	1/16	1/32	1/64
(a)	a.e	1.9E-18	1.4E-15	6.5E-08	
	m.e	1.7E-16	1.5E-14	9.1E-7	
	t	.088	.376	2.531	
(b)	a.e			2.8E-15	3.2E-15
	m.e	-	-	1.9E-14	1.9E-14
	t			.512	.796
(c)	a.e	2.2E-17	1.4E-11		
	m.e	2.4E-16	3.6E-11	Does not converge →	
	t	.033	.153		

Table 5

Results for Problem 5 and 6 ; h = Δx = Δy ;
Dimension of capacitance matrix B_{22} is 24x24

Mesh size (h)		SMT	Refining SMT solution to finer grids	
		1/16	1/32	1/64
Prob. 5	a.e	6.5E-15	8.2E-15	8.7E-15
	m.e	3.2E-13	6.3E-13	6.3E-13
	t	3.428	3.564	3.835
Prob. 6	a.e	5.3E-10	8.9E-10	9.3E-10
	m.e	2.6E-08	7.2E-08	7.2E-8
	t*	1.015	1.145	1.264

*The capacitance matrix for Prob.6 is identical to that of Prob.5 and the time given is the time for solving an additional problem.

NUMERICAL STUDY OF A JET IN A STRATIFIED FLUID*

by

Roger PEYRET and Bernard REBOURCET
Department of Mathematics, University of Nice
C.E.A. - C.E. Limeil

1. INTRODUCTION.- The present paper concerns the unsteady laminar jet flow due to the horizontal intrusion of a fluid with uniform temperature into a stratified fluid at rest in a semi-infinite plane channel. The initial stratification of the fluid at rest is stable and is due to thermal effects. This problem is studied by a finite-difference solution of the unsteady Navier-Stokes equations in primitive variables within the Boussinesq approximation. The main numerical difficulty lies in the derivation of suitable conditions to be imposed at the downstream boundary of the computational domain. The treatment of this artificial boundary must allow the fluid to leave the domain freely without perturbing the upstream region nor creating numerical instabilities. This problem, which is already delicate for homogeneous fluids because of the velocity-pressure formulation of the Navier-Stokes equations, is enhanced here by the stratification of the fluid and the propagation of internal waves which is associated with it. In such flows, the disturbances are felt very far from their source. Various downstream conditions were tested and their results compared in order to select a technique having the required properties.

2. THE PROBLEM.-

The physical domain in which the flow takes place is shown in figure 1 where AX, A'X' and AB, A'B' are respectively isothermal and adiabatic walls. The fluid initially at rest in the channel is stably stratified by thermal effects :

$$\tilde{T}_s(\tilde{y}) = (\tilde{T}_2 - \tilde{T}_1)\,\tilde{y}/(2H) + (\tilde{T}_2 + \tilde{T}_1)/2 , \qquad \tilde{T}_2 > \tilde{T}_1 .$$

---

* This work was partly performed at the "Laboratory of Theoretical Mechanics", Univ. P. & M. Curie, Paris.

The density $\tilde{\rho}$ and the temperature $\tilde{T}$ are connected through the state law

$$\tilde{\rho} = \tilde{\rho}_0 \left[1 - \beta(\tilde{T} - \tilde{T}_0) \right], \quad \tilde{T}_0 = (\tilde{T}_2 + \tilde{T}_1)/2,$$

where β is the volume coefficient of thermal expansion. So, to the initial distribution $\tilde{T}_s(\tilde{y})$ corresponds the density $\tilde{\rho}_s(\tilde{y})$ with $\tilde{\rho}_0 = \tilde{\rho}_s(0)$, and the pressure $\tilde{p}_s(\tilde{y})$ in the fluid at rest is hydrostatic.

From the time $\tilde{t} = 0$, a fluid of same physical properties that the fluid at rest is injected with a velocity of maximal value U_0 through the slot BB'. The temperature of the injected fluid is $\tilde{T}_0$ and its density is $\tilde{\rho}_0$, i.e. the corresponding values of the stratified fluid at $\tilde{y} = 0$. Because of this local equilibrium, the intruded fluid will propagate in the horizontal direction.

Such a problem has been studied, mainly experimentally, by MAXWORTHY [1] and by MANINS [2]. Previous numerical studies in various geometrical configurations have been done by PEYRET [3], IMBERGER et al. [4] and by KAO et al. [5]. In particular, in ref. [3] the jet was assumed to be emitted into a finite basin.

Let be $\tilde{\pi}$ and $\tilde{\theta}$ the perturbations with respect to $\tilde{p}_s$ and $\tilde{T}_s$, the dimensionless variables are defined by :

$$x = \tilde{x}/(2D), \quad y = \tilde{y}/(2D), \quad t = \tilde{t}\, U_0/(2D),$$
$$\vec{U} = (u, v) = (\tilde{u}/U_0, \tilde{v}/U_0), \quad \pi = \tilde{\pi}/(\tilde{\rho}_0 U_0^2), \quad \theta = \eta \tilde{\theta}/(\tilde{T}_2 - \tilde{T}_1),$$

with $\eta = H/D$ where 2H and D are respectively the heights of the channel and of the aperture.

The Navier-Stokes equations within the Boussinesq approximation are :

$$\frac{\partial \vec{U}}{\partial t} + (\vec{U}.\nabla)\vec{U} + \nabla \pi = Re^{-1} \nabla^2 \vec{U} + Ri\, \theta\, \vec{j},$$
$$\frac{\partial \theta}{\partial t} + \vec{U}.\nabla\theta + \vec{U}.\vec{j} = (Re\, Pr)^{-1} \nabla^2 \theta,$$
$$\nabla.\vec{U} = 0,$$

where the Reynolds number Re , the Richardson number Ri and the Prandtl number Pr are defined by :

$$Re = \frac{2D\tilde{\rho}_0 U_0}{\mu}, \quad Ri = \beta g \left(\frac{2D}{U_0}\right)^2 \frac{\tilde{T}_2 - \tilde{T}_1}{2H}, \quad Pr = \frac{\mu C_p}{K}.$$

The initial conditions at $t = 0$ are $u = v = \theta = 0$, and the boundary conditions are :

- On BB' : $u = \varphi(t)(1 - 16y^2)$, $v = 0$, $\theta = -y$ with

$$\varphi(t) = \begin{cases} t^2[t^2 - 2(t_0 + t_0^{-3})t + t_0(t_0 + 3t_0^{-3})] & \text{if } t \leqslant t_0 . \\ 1 & \text{if } t > t_0 . \end{cases}$$

- On AB, B'A' : $u = v = \partial\theta/\partial x = 0$.
- On AX, A'X' : $u = v = \theta = 0$.

The domain of computation is artificially bounded downstream by CC' located at a distance L from the entry. The manner in which the flow quantities are computed at CC' will be described in section 3.2.

Moreover, preliminary numerical experiments have shown that for the values of the physical parameters considered here, the flow remains symmetrical with respect to the axis $y = 0$; so that in calculations reported here a condition of symmetry with respect to this axis is assumed and the computations are done only in a half-domain.

3. THE NUMERICAL SOLUTION.-

3.1. The method [3]. The discretization with respect to time is carried out by using the second order accurate Crank-Nicholson scheme where the pressure is evaluated at the mid-time $(n + 1/2)k$ with k = time step, let :

$$\frac{1}{k}(\vec{U}^{n+1} - \vec{U}^n) + \frac{1}{2}(\vec{A}^{n+1} + \vec{A}^n) + \nabla\pi^{n+1/2} - \frac{Ri}{2}(\theta^{n+1} + \theta^n)\vec{j} = 0$$

$$\frac{1}{k}(\theta^{n+1} - \theta^n) + \frac{1}{2}(B^{n+1} + B^n) + \frac{1}{2}(v^{n+1} + v^n) = 0,$$

$$\nabla . \vec{U}^{n+1} = 0,$$

where

$$\vec{A} = (\vec{U}.\nabla)\vec{U} - Re^{-1}\nabla^2\vec{U},$$

$$B = \vec{U}.\nabla\theta - (Re\, P_r)^{-1}\nabla^2\theta.$$

The discretization in space makes use of the staggered MAC mesh (Harlow and Welch [6]) of size h . The derivatives are approximated with centered differences. Second order accurate differences are considered except for the convective term $\vec{U}.\nabla\theta$ which is discretized by means of fourth-order accurate differences in order to minimize the associated truncation error compared to the diffusive term $(RePr)^{-1}\nabla^2\theta$. Finally non centered differences are used for the velocity component derivatives near each wall on which this component is not defined (fig. 2).

The discretization leads to a nonlinear algebraic system for the unknown values $\vec{U}_h^{n+1}$, $\pi_h^{n+1/2}$ and θ_h^{n+1} at each discretization point. The finite-difference equations are written in the symbolic form :

$$\vec{M}(\vec{U}_h^{n+1}, \pi_h^{n+1/2}, \theta_h^{n+1}) = 0, \quad E(\vec{U}_h^{n+1}, \theta_h^{n+1}) = 0, \quad D(\vec{U}_h^{n+1}) = 0,$$

where $\vec{M}$, E and D correspond respectively to the discretized momentum, temperature and continuity equations. The above system is solved by an iterative procedure based upon the artificial compressibility method [7], [8] :

$$\vec{U}_h^{n+1,m+1} - \vec{U}_h^{n+1,m} + \alpha \vec{M}(\vec{U}_h^{n+1,m}, \pi_h^{n+1/2,m}, \theta_h^{n+1,m}) = 0,$$

$$\theta_h^{n+1,m+1} - \theta_h^{n+1,m} + \chi E(\vec{U}_h^{n+1,m+1}, \theta_h^{n+1,m}) = 0,$$

$$\pi_h^{n+1/2,m+1} - \pi_h^{n+1/2,m} + \lambda D(\vec{U}_h^{n+1,m+1}) = 0,$$

where m is the index of the iterative procedure which is initialized by using the values at the previous time step. The parameters α, χ and λ are chosen in order to insure convergence of the procedure. Necessary conditions of convergence are obtained by carrying out a linearized analysis of the Von Neumann stability of the above system where gravitation are nonlinear terms are neglected. The conditions are found to be

$$\alpha > 0, \quad \chi > 0, \quad \lambda > 0,$$

$$\frac{2\alpha}{h^2}\left(\frac{1}{Re} + \frac{h^2}{4k} + \lambda\right) \leq 1, \qquad \frac{2\chi}{h^2}\left(\frac{1}{Re\,Pr} + \frac{h^2}{4k}\right) \leq 1.$$

In fact, in the above iterative procedure, the various quantities are used as soon as they are computed (Gauss-Seidel technique) ; this technique simplifies the programming and improves the convergence.

The form of the convergence criteria suggests to use $a = \alpha/h^2$, $b = \chi/h^2$ and $c = \lambda/h$ rather than α, χ and λ. Typical values of a, b, c leading to a good convergence (determined from numerical tests) of the iterative procedure are given in the Table :

Re	h	k	a	b	c
10	1/10	1/50	0.40	0.40	10.50
100	1/16	1/64	0.75	0.75	9.50
250	1/16	1/64	0.82	0.82	8.70

The convergence of the iterative procedure is assumed obtained when $\max\{|M_u|, |M_v|, |E|, |D|\} \leq 0.25\times10^{-2}$. The maximum is generally given by the v equation, the divergence equation is then satisfied at less than 10^{-3}.

The number N of iterations needed to get the convergence is about 30 in the established regime. The evolution of N during the phase of establishment $t \leq t_0$ is shown in figure 3. In this figure, the effect of the choice of $\varphi(t)$ and t_0 is illustrated.

3.2. The treatment of the downstream boundary. As it was already said, the major numerical difficulty is related to the unboundness of the physical domain in which the flow takes place. To solve this question two techniques are generally used :

(i) Map the semi-infinite domain into a bounded domain by means of a coordinate transform. Hence, the disturbances which propagate downstream will take an infinite time to reach the end and therefore no reflection of internal waves can affect the global numerical solution. This technique was used in [4]. Here, we consider the second method.

(ii) Bound the domain with an artificial boundary CC' (fig. 1) and determine the flow quantities on this boundary by suitable equations allowing the fluid to leave freely the domain. Due to stratification effects the flow is highly unsteady near CC' ; moreover the presence of eddies makes it very different from a Poiseuille type flow with $v = 0$. Therefore, the solution of CC' located at a distance L is determined here from the continuity equation

$$\text{(A)} \qquad \frac{\partial u}{\partial x} + \frac{\partial v}{\partial y} = 0,$$

and an approximate momentum equation:

$$\text{(B)} \qquad \frac{\partial \pi}{\partial y} = - \frac{\partial v}{\partial t} - v \frac{\partial v}{\partial y} + \frac{1}{Re} \frac{\partial^2 v}{\partial y^2} + Ri\ \theta,$$

associated with two other equations for which various possibilities respectively for v and θ have been tested :

$$\text{(a)} \quad \frac{\partial v}{\partial x} = 0, \qquad \text{(b)} \quad \frac{\partial^2 v}{\partial x^2} = 0,$$

$$(\alpha) \quad \frac{\partial^2 \theta}{\partial x^2} = 0, \quad (\beta) \quad \frac{\partial \theta}{\partial x} = 0, \quad (\gamma) \quad \frac{\partial \theta}{\partial t} + v \frac{\partial \theta}{\partial y} - \frac{1}{Re Pr} \frac{\partial^2 \theta}{\partial y^2} + v = 0.$$

The figure 4 shows the effect of these conditions on the pressure derivative $\partial \pi / \partial y$ and the velocity v which have been found to be the more sensitive to the downstream treatment. The final choice consists to equations (A), (B), (a) and (α). So, the equation (a) gives v on CC', then eq. (A) gives u , eq. (α) gives θ, and π is determined from eq. (B). The figure 5 explains the discretization.

4. SOME NUMERICAL RESULTS.

Here, we give only few typical numerical results. A more complete analysis of the flow can be found in [9] and [10].

The figure 7 shows the instantaneous streamlines in

the case where $Re = 250$, $Pr = 10$ and $Ri = 64$, i.e. the stratification is relatively large. It can be seen how two eddies are created near the entrance, then are propagated downstream ; while an other pair of eddies, having an opposite rotation to the first ones, appears near the slot. This process is continuously repeated so that two rows of eddies are located on both sides of a central region in which the streamlines are nearly horizontal. The striking effect of stratification upon the shape of streamlines is evident if the figure 7c is compared with the figure 6 corresponding to a neutral stratification $Ri = 0$.

The evolution of the velocity on the axis $y = 0$ is shown in figure 8. The wavy shape of the curves is a characteristic of such a flow and can be related to the presence of eddies with alternate rotation so that the streamlines are periodically expanded or constricted in the jet region as shown in fig. 7c . We can observe the existence of maxima much larger than the entry velocity ; the location of these overshoots corresponding to the local constriction of the jet.

Finally, a picture of the intruded fluid given by markers is presented in fig. 9. Here again, the difference between a neutral stratification $(Ri = 0)$ and a large one $(Ri = 64)$ is very pronounced. In the first case, the advance of the tongue is strongly blocked ; the mushroom shape is due to the succession of two effects : inertia near the entrance, diffusion farther. On the other hand, in the second case, the gravity prevents the lateral expansion of the homogeneous intruded fluid. The inertial effects are sensitive near the entrance while farther the gravitational effects are preponderant and induce the sharp appearance of the tongue.

REFERENCES

[1] MAXWORTHY, T. "Experimental and theoretical studies of horizontal jets in a stratified fluid". Int. Symp. on stratified flows, Novosibirsk, 1972.

[2] MANINS, R.C. "Intrusion into a stratified fluid". J. Fluid Mech., vol. 74, P. 547-560, 1976.

[3] PEYRET, R. "Unsteady evolution of a horizontal jet into a stratified fluid". J. Fluid Mech., vol. 78, p.49-63, 1976.

[4] KAO, T.W., PARK, C. and PAO, H.P. "Inflows, density currents and fronts". Phys. of Fluids. Vol. 21, p.1912-1922, 1978.

[5] IMBERGER, J., THOMPSON, R. and FANDRY, C. "Selective withdrawal from a finite rectangular tank". J. Fluid Mech., vol. 78, p. 489-512, 1976.

[6] HARLOW, F.H. and WELCH, J.E. "Numerical calculation of time-dependent viscous incompressible flow of fluid with free surface". Phys. of Fluids, vol. 8, p.2182-2189, 1965.

[7] CHORIN, A.J. " A numerical method for solving incompressible viscous flow problems". J. Comp. Phys., vol. 2, p. 12-26, 1967.

[8] FORTIN, M., PEYRET, R. et TEMAM, R. " Résolution numérique des équations de Navier-Stokes pour un fluide incompressible". J. Méca., Vol. 10, p. 357-390, 1971.

[9] REBOURCET, B. "Etude numérique de jets instationnaires en fluides stratifiés". Thèse 3ième cycle, Mécanique théorique. Université P. et M. Curie, Paris, 1980.

[10] PEYRET, R. et REBOURCET, B. "Développement d'un jet en fluide stratifié". Submitted to J. Méca. Théor. Appl.

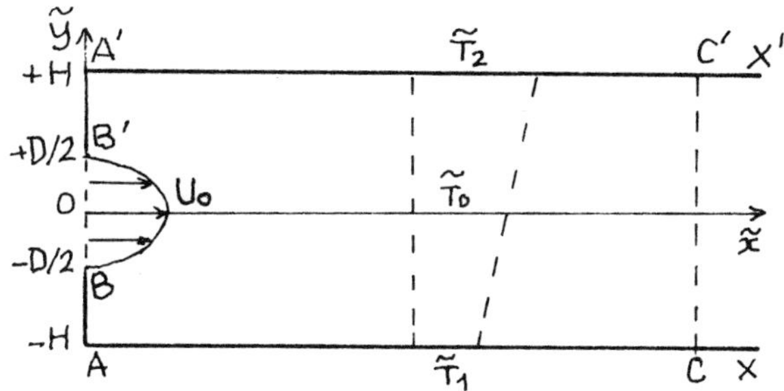

Fig. 1. Geometrical Configuration

Fig. 2. Mesh

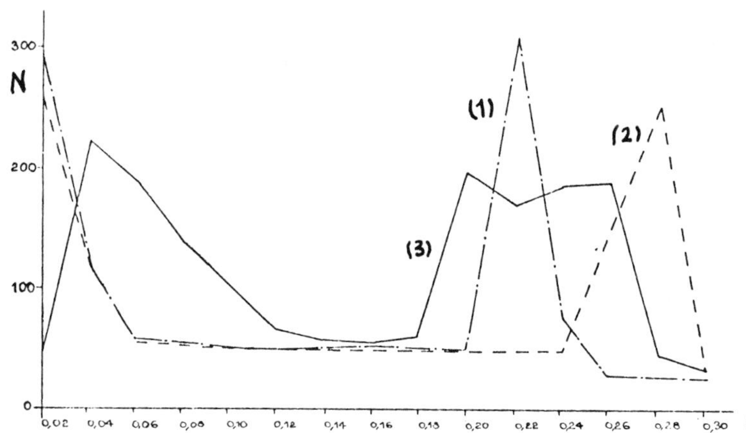

Fig. 3. Number of iterations by time step during the phase of establishment : Re = 10, Ri = 100, Pr = 10, h = 0.1, k = 0.02

(1) $\varphi(t) = t/t_0$, $t_0 = 0.20$

(2) $\varphi(t) = t/t_0$, $t_0 = 0.25$

(3) $\varphi(t) = t^2[t^2 - 2(t_0 + t_0^{-3})t + t_0(t_0 + 3t_0^{-3})]$, $t_0 = 0.25$

$[\varphi(t) = 1, \; t > t_0]$

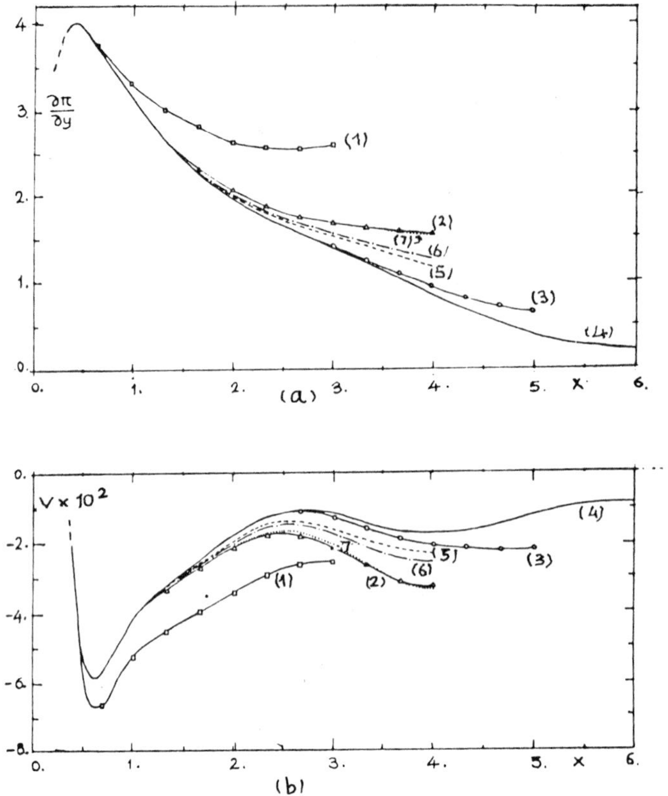

Fig. 4. Effect of the downstream boundary treatment on (a) $\frac{\partial\pi}{\partial y}(x,\ 0.85,\ 1.50)$ and (b) $v(x,\ 0.85, 1.50)$. Re = 10., Ri = 100., Pr = 1., h = 0.1, k = 0.02, η = 2.3125

(1) —□— Eqs. (a) & (β), L = 3. (5) - - - - Eqs. (a) & (α), L = 4.
(2) —△— Eqs. (a) & (β), L = 4. (6) — · — Eqs. (a) & (γ), L = 4.
(3) —○— Eqs. (a) & (β), L = 5. (7) ········ Eqs. (b) & (β), L = 4.
(4) ——— Eqs. (a) & (β), L = 6.

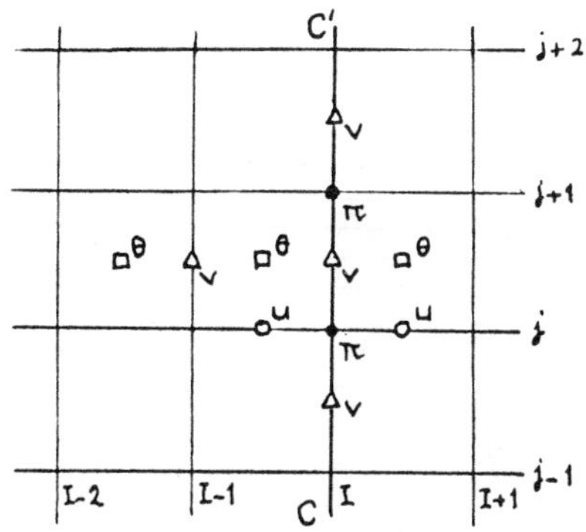

$$v^{n+1}_{I,j+1/2} = v^{n+1}_{I-1,j+1/2}$$

$$u^{n+1}_{I+1/2,j} = u^{n+1}_{I-1/2,j} - \left(v^{n+1}_{I,j+1/2} - v^{n+1}_{I,j-1/2} \right)$$

$$\theta^{n+1}_{I+1/2,j+1/2} = 2\,\theta^{n+1}_{I-1/2,j+1/2} - \theta^{n+1}_{I-3/2,j+1/2}$$

$$\begin{cases} \pi^{n+1/2}_{I,j+1} = \pi^{n+1/2}_{I,j} - \dfrac{h}{k}\left(v^{n+1}_{I,j+1/2} - v^{n}_{I,j+1/2} \right) - \dfrac{h}{2}\left(\Phi^{n+1}_{I,j+1/2} + \Phi^{n}_{I,j+1/2} \right) \\ \pi^{n+1/2}_{I,1} = 1. \end{cases}$$

$$\Phi_{I,j+1/2} = v_{I,j+1/2}\left(v_{I,j+3/2} - v_{I,j-1/2} \right)/h - \left(v_{I,j+3/2} - 2v_{I,j+1/2} - v_{I,j-1/2} \right)/(Re\, h^2)$$
$$- Ri \left(\theta_{I+1/2,j+1/2} + \theta_{I-1/2,j+1/2} \right)/2 .$$

Fig. 5. Discretization of the downstream boundary equations.

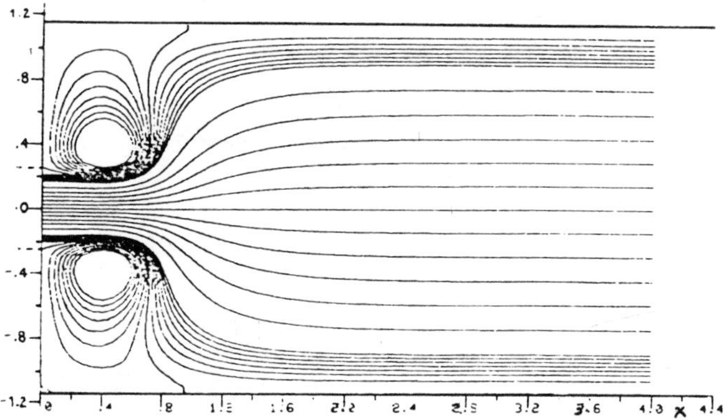

Fig. 6. Instantaneous streamlines at $t = 2.00$ in the case without stratification, $Re = 250.$, $Ri = 0.$, $\eta = 2.3125$, $L = 5.$

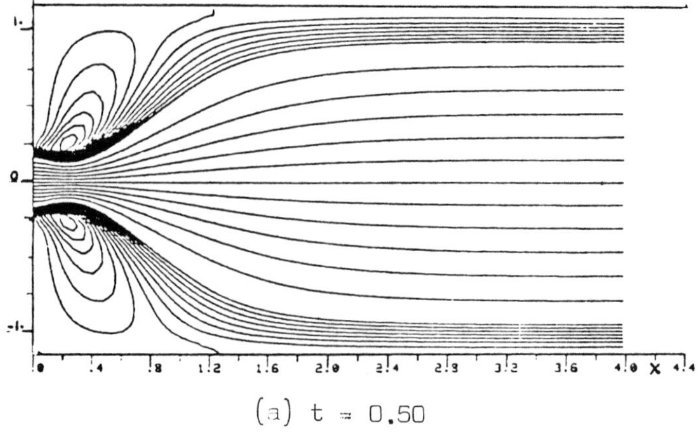

(a) t = 0.50

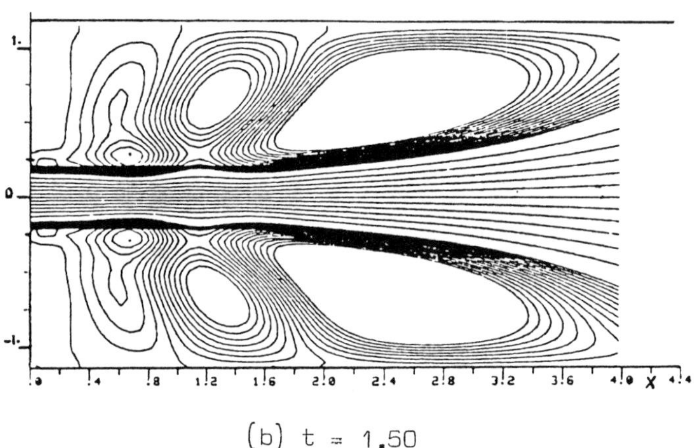

(b) t = 1.50

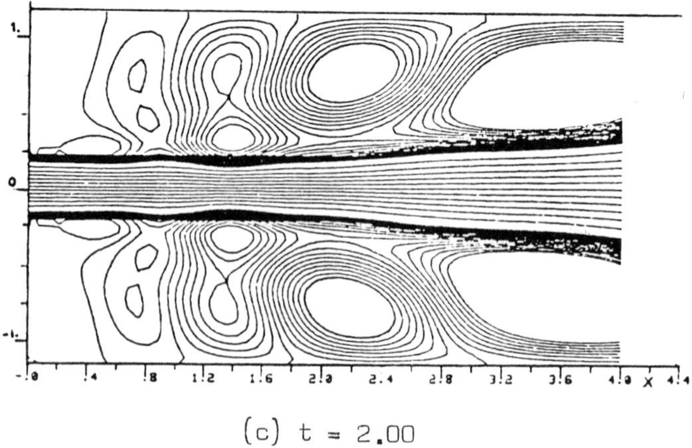

(c) t = 2.00

Fig. 7. Instantaneous streamlines, Re = 250, Ri = 64, Pr = 10, η = 2.3125, L = 5.

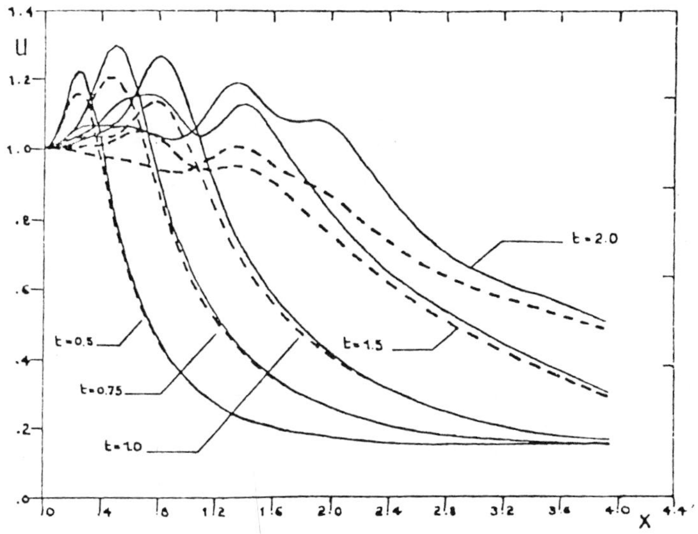

Fig. 8. Velocity u on the axis $y = 0$; $Ri = 64$, $Pr = 10$
$\eta = 2.3125$, $L = 5$; - - - - Re = 100. , ——— Re = 250.

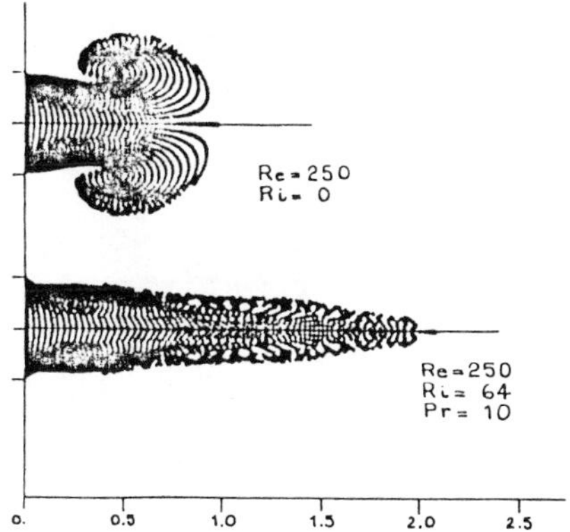

Fig. 9. Visualization of the intruded fluid at $t = 2$.

FINITE ELEMENT ANALYSIS OF MIXED CONVECTION APPLIED TO THE STORAGE OF SOLAR ENERGY

Denis AUBRY (+) Denis ZANDVLIET (++)

1. SUMMARY

In this paper, we consider the numerical computations of transient mixed convection in a fluid : the Navier Stokes equations are coupled with the equation of conservation of energy because the density is dependent on the temperature.

The finite element method is used for the space discretisation. The incompressibility condition is approximately satisfied by using the penalisation technique together with selective reduced integration. A special A_o stable two step method is used for the time discretisation as we are concerned with long time prediction and would like to use large steps. The method initially proposed by Zlamal (1) for parabolic problem involves one parameter : the influence of this parameter in the choice of the time step, and the stability is illustrated by numerical examples and compared to the more classical method of Crank Nicholson. It is shown that the order of accuracy is increased by one and this seems to be important in the present case, as suggested by Raviart et al (10)

Finally we apply these techniques to the transient flow of a fluid inside a storage tank of solar energy. It is interesting to control the evolution of such a flow because we are looking for the efficiency of the storage tank, and the global behaviour of a solar system including the collectors is primarily related to the stratification of the temperature in the tank.

(+) Ecole Centrale des Arts et Manufactures, 92290 Chatenay (F)

(++) Centre Scientifique des Techniques du Batiment, Paris.

2. FORMULATION OF THE PROBLEM

The present investigation is limited to a two dimensional analysis. Let Ω be a bounded open region in R^2 with piecewise smooth boundary Γ. We employ the summation convention on repeated indices i, j and k only. A comma is used to denote partial differentiation and a comma followed by a subscript t is used to denote partial differentiation wrt. the time variable.

2.1 Basic equations

Let $\vec{u}(x,t)$ be the velocity field for $(x,t) \in \Omega \times R^+$
The momentum equation associated with the transient flow of viscous fluid is :

(2.1) $\rho\,(u_{i,t} + u_i . u_{i,j}) - \sigma_{ij,j} \quad - \rho g_i \beta\,(\Theta - \Theta r) = 0$

(2.2) with : $\sigma_{ij} = -p\delta_{ij} + \mu(u_{i,j} + u_{j,i})$

where ρ, σ, $\vec{g}$, β, p, μ, θ, θ_r denote respectively the fluid density, the stress tensor, the gravity vector, the coefficient of volume expansion, the pressure, the viscosity, the temperature and reference temperature.

The condition of fluid incompressibility is enforced through the equation : $u_{i,i} = 0$ (2.3)

The transport of thermal energy in the fluid is described by:

(2.4) $\rho C\,(\Theta_{,t} + u_j . \Theta_{,j}) - k\,\Theta_{,j,j} = 0$

k is the thermal diffusivity, C the heat capacity.

We complete the formulation of the problem with usual boundary conditions :

(2.5) $\begin{cases} \vec{u} = \bar{\vec{u}} & \text{on } \Gamma_u \\ \sigma_{ij} n_j = 0 & \text{on } \Gamma_t \end{cases}$

where Γu and Γt correspond to a given partition of the boundary Γ for the fluid problem, and :

(2.6) $\begin{cases} \Theta = \bar{\Theta} & \text{on } \Gamma_\Theta \\ -k.\ \Theta_{,i}\, n_i = 0 & \text{on } \Gamma_\psi \end{cases}$

and Γ_Θ, Γ_ψ correspond to another partition of Γ for the heat transfert problem. Note that n_i is the outward unit normal to the boundary.

These equations must be augmented by initial conditions :

(2.7) $\vec{u}\,(x,o) = \vec{u}_o(x)$; $\Theta\,(x,o) = \Theta_o$ for $x \in \Omega$

Equation (2.1) to (2.6) form a complete set to determine the temperature field in a moving fluid. All the coefficients but the density are independant on the temperature. In this manner, the velocity field is coupled to the temperature field

2.2 Penalty function formulation

We eliminate the pressure variable by the use of the well-known penalisation technique.
We replace the pressure in the constitutive equation (2.2) by :

(2.8) $p^{\lambda} = - \lambda u_{i,i}$

If λ is large enough the continuity equation (2.3) is approximately satisfied.

2.3 Weak formulation

For $\vec{u}$, $\vec{v}$, $\vec{w}$, we define the following forms :

(2.9) $a_1 (\vec{u},\vec{v}) = (\lambda\, u_{i,i}, v_{j,j}) + \mu\, ((u_{i,j}+u_{j,i}), v_{i,j})$

(2.10) $a_2 (\vec{w} ; \vec{u},\vec{v}) = (\rho\, w_j\, u_{i,j}, v_i)$

(2.11) $a_0 (\vec{u},\vec{v}) = (\rho\, u_i, v_i)$

(2.12) $a_3 (\Theta,v) = (\rho\beta(\Theta - \Theta_r)\, g_i, v_i)$

where $(.,.)$ denotes the usual scalar product in $L^2(\Omega)$

Similary for Θ,τ we define :

(2.13) $b_1 (\Theta,\tau) = (k\, \Theta_{,i}, \tau_{,i})$

(2.14) $b_2 (\vec{u} ; \Theta, \tau) = (\rho c u_j\, \Theta_{,j}, \tau)$

(2.15) $b_0 (\Theta, \tau) = (\rho c\, \Theta, \tau)$

Let us finally define the space of admissible velocities and temperature :

(2.16) $\begin{cases} V = \{ \vec{v} \,/\, \vec{v} = o \text{ on } \Gamma_u \} \\ H = \{ \Theta \,/\, \Theta = o \text{ on } \Gamma_\Theta \} \end{cases}$

and let us assume that there exists $\vec{u}^*$ and Θ^*defined in Ω such that :

(2.17) $\begin{cases} \vec{u}^* = \vec{\bar{u}} \text{ on } \Gamma_u \\ \Theta^* = \overline{\Theta} \text{ on } \Gamma_\Theta \end{cases}$

then the weak form of Boussinesq equations reads :

Find $(\vec{u}, \Theta)$ such that :

$$(2.18)\begin{cases} \frac{d}{dt} a_o(\vec{u},\vec{v})+a_1(\vec{u},\vec{v})+a_2(\vec{u};\vec{u},\vec{v}) -a_3(\Theta,\vec{v}) = o, \forall \vec{v} \varepsilon V \\ \frac{d}{dt} b_o(\Theta,\tau)+b_1(\Theta,\tau)+b_2(\vec{u};\Theta,\tau) = 0, \quad \forall \tau \varepsilon H \\ \vec{u}(o) = \vec{u}_o, \quad \Theta(o) = \Theta_o \end{cases}$$

where from now on $\vec{u}$ (t) (resp Θ(t)) denotes the mapping $x \rightarrow u(x,t)$ for each t .

3. FINITE ELEMENT METHOD

3.1 Approximation

We associate to V and H two finite dimensional spaces V_h, H_h with basis functions $\vec{w}_i$, ζ_i respectively.

These functions are computed using the classical steps of the finite element method. Then the approximate Boussinesq equations consists in finding $(\vec{u}_h, \Theta_h)$

Solutions of :

$$(3.1)\quad \frac{d}{dt}a_o(\vec{u}_h,\vec{v}_h)+a_1(\vec{u}_h,\vec{v}_h)+a_2(\vec{u}_h;\vec{u}_h,\vec{v}_h)-a_3(\Theta_h,\vec{v}_h) = 0 \qquad \forall \vec{v}_h \varepsilon V_h$$

$$(3.2)\quad \frac{d}{dt}b_o(\Theta_h,\tau_h)+b_1(\Theta_h,\tau_h)+b_2(\vec{u}_h;\Theta_h,\tau_h) = 0 \qquad \forall \tau_h \varepsilon H_h$$

$$(3.3)\quad \vec{u}_h(0) = \vec{u}_{oh}, \quad \Theta_h(o) = \Theta_{oh}$$

where $\vec{u}_{oh}$, Θ_{oh} denotes spatial approximations to the initial values. It is well known that the above equations are non linear because of the presence of the convection terms in the forms a_2 (.,.) and b_2 (.,.).

Numerical integration is used to compute numerically the different forms and it is recalled more particularly that a_1 is selectively integrated.(Cf. Hughes (7)).

$$(3.4)\quad a_1(\vec{u}_h,\vec{v}_h) = a_1^{\lambda}(\vec{u}_h,\vec{v}_h) + a_1^{\mu}(\vec{u}_h,\vec{v}_h)$$

with obvious notations. When λ gets larger and larger, the term a_1^{λ} is predominant and the form should not be definite positive on V_h. The reduced integration on a_1^{λ} is a good technique to introduce the so-called rank deficiency condition.

3.2 Elements

We use the Serendip family element for the mapping of the domain Ω. The construction of the shape functions is easy :

$$w_I(J) = \delta_{IJ}$$

where I and J are two nodes of the triangulation.

These elements are correct regarding the penalty formulation as selective reduced integration is used (see Hughes (7), Malkus (9)).

The orders of integration for the three different terms are not equal ; for a four node element they would be :

Terms	Penalty	Viscosity	Convection
Number of Integration points (Gauss method)	1	2 x 2	3 x 3

In fact, we use the same number of integration points for viscosity and convection terms because cost consideration. This rule does not seem to be wrong.

4. TWO STEP METHOD

In this paragraph, we focus our attention on a two-step method and we study only the discretization with respect to the time variable because the presentation is simpler. Obviously the discretization with respect to the space and time variable should be done simultaneously.

4.1

Aspointed out by Raviart (10) , in order to achieve second order accuracy, which is important when large time steps are necessary ; it is interesting to resort to multi step methods. We intend to use here the two step method proposed by Zlamal (1). Let $\Delta t = T/N$ be the time step, where N is a positive integer, and $t_n = n\Delta t$.

We shall denote by $\vec{u}^n$ an approximation of $\vec{u}(t_n)$ and consider the following problem :

Find $(\vec{u}^q, \Theta^q), (\vec{u}^{q+1}, \Theta^{q+1}), \ldots\ldots$
such that :

$$(4.1)\quad \begin{cases} \frac{1}{\Delta t} \sum_{i=o}^{2} \alpha_i \; a_o(\vec{u}^{n+i},\vec{v}) + \sum_{i=o}^{2} \beta_i \{a_1(\vec{u}^{n+i},\vec{v}) + a_2(\vec{u}^{n+i};\vec{u}^{n+i},\vec{v}) - a_3(\Theta^{n+i},\vec{v})\} = 0 \\ \frac{1}{\Delta t} \sum_{i=o}^{2} \alpha_i \; b_o(\Theta^{n+i},\tau) + \sum_{i=o}^{2} \beta_i \{b_1(\Theta^{n+i},\tau) + b_2(\vec{u}^{n+i};\Theta^{n+i},\tau)\} = 0 \end{cases}$$

starting from the given functions $\vec{u}^o, \vec{u}^1 ; \Theta^o, \Theta^1$

and where : for $0 < \gamma \leq 1$

$$(4.2)\quad \begin{array}{lll} \alpha_2 = 1/2 + \gamma & ; & \beta_2 = 1/4\,(1+\gamma)^2 \\ \alpha_1 = -2\gamma & ; & \beta_1 = 1/2\,(1-\gamma)^2 \\ \alpha_o = 1/2 + \gamma & ; & \beta_o = 1/4\,(1-\gamma)^2 \end{array}$$

The method is of order ≥ 2 according to the choice of γ. It is A_o stable (Dahlquist ((2), Cryer (3)).

The solution of the above equations for $\{\vec{u}^{n+2}, \Theta^{n+2}\}$ is difficult because, it is non linear. It is solved by a predictor corrector method : let us call $\vec{u}^{\bar{n}}$ a linear combination of $\vec{u}^n$, $\vec{u}^{n+1}$ and $t_{\bar{n}} \in (t_n, t_{n+2})$ where $\bar{n}$ is chosen carefully so that the order of the predictor is at least two . Then we replace the above equations by :

Find $(\vec{u}^q, \Theta^q), (\vec{u}^{q+1}, \Theta^{q+1}), \ldots$

$$(4.3)\quad \begin{cases} \frac{1}{\Delta t} \sum_{i=o}^{2} \alpha_i a_o(\vec{u}^{n+i},v) + \sum_{i=o}^{2} \beta_i \{a_1(\vec{u}^{n+i},\vec{v}) + a_2(\vec{u}^{\bar{n}};\vec{u}^{n+i},\vec{v}) - a_3(\Theta^{\bar{n}},\vec{v})\} = 0 \\ \frac{1}{\Delta t} \sum_{i=o}^{2} \alpha_i b_o(\Theta^{n+i},\tau) + \sum_{i=o}^{2} \beta_i \{b_1(\Theta^{n+i},\tau) + b_2(\vec{u}^{\bar{n}};\Theta^{n+i},\tau)\} = 0 \end{cases}$$

(4.4) where $\vec{u}^{\bar{n}}$ is given by :

$$\vec{u}^{\bar{n}} = (2\beta_2+\beta_1)\vec{u}^{n+1} + (\beta_o - \beta_2)\,\vec{u}^n$$

for the predictor.

4.2 Solution procedure

An iterative scheme is used to solve (4.3). If we set :

$$(4.5)\quad \begin{cases} \delta \vec{u}^{\,n+2,k} = \vec{u}^{n+2,k} - \vec{u}^{n+2,k-1} \\ \delta \Theta^{n+2,k} = \Theta^{n+2,k} - \Theta^{n+2,k-1} \end{cases}$$

where k denotes the iteration number and if we define the following form :

(4.6)

$$\begin{cases} a(\vec{u}^{n+2,k},\theta^{n+2,k},\vec{v}) = \frac{1}{\Delta t}\alpha_2 a_o(\vec{u}^{n+2,k},\vec{v})+\beta_2\{a_1(\vec{u}^{n+2,k},\vec{v}) \\ \qquad + a_2(\bar{\vec{u}}^{n},\vec{u}^{n+2,k},\vec{v}) - a_3(\theta^{n+2,k},\vec{v})\} \\ \qquad + \frac{1}{\Delta t}\sum_{i=o}^{1}\alpha_i a_o(\vec{u}^{n+i},\vec{v}) \\ + \sum_{i=o}^{1}\beta_i\ \{a^1(\vec{u}^{n+i},\vec{v})+a_2(\vec{u}^{n+i};\vec{u}^{n+i},\vec{v})-a_3(\theta^{n+i},\vec{v})\} \end{cases}$$

and use a similar definition for b $(\vec{u}^{n+2,k},\theta^{n+2,k},\tau)$, then the iterative scheme is as follows :

- Step n+2, iteration number k , find $(\vec{u}^{n+2,k},\theta^{n+2,k})$ such that :

(4.7)

$$\begin{cases} \frac{1}{\Delta t}\alpha_2\ a_o(\delta\vec{u}^{n+2,k},\vec{v}) + \beta_2\ a_1(\delta\vec{u}^{n+2,k},\vec{v}) = \\ \qquad - a(\vec{u}^{n+2,k-1},\theta^{n+2,k-1},\vec{v}) \\ \frac{1}{\Delta t}\alpha_2\ b_o(\delta\theta^{n+2,k},\tau) + \beta_2\ b_1(\delta\theta^{n+2,k},\tau) = \\ \qquad - b(\vec{u}^{n+2,k-1},\theta^{n+2,k-1},\tau) \end{cases}$$

Thus at each iteration, we have to solve a symmetric linear system.

5. NUMERICAL RESULTS

Experience from numerical experiments conducted Zlamal (1) to propose to start the scheme by the one step θ-method with θ=1/3, and then to use the above two steps method with the coefficient γ given by (4.2) equal to 1/3. From our own numerical experiments in the linear case (diffusion equation) we found these choices to be efficient and we resorted the same technique in the full nonlinear case.

5.1 One dimensionnal transport of a temperature front

initially, the fluid is isothermal (Θ_o=0.). The temperature of the left point is suddenly increased by ΔΘ=1. The velocity is $u=u_o$. The dimensionless parameter Pe (Peclet number) is given by:

$$Pe = u\ D_h\ \rho\ C_p\ /\ K$$

From a numerical point of view, the characteristic dimension is the length of an element.

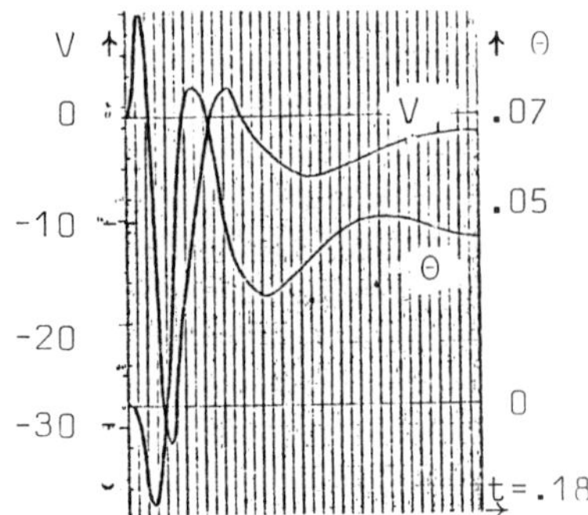

fig. 1 : Vertical velocity and temperature history at point (x=.73 , y=.5)

-Thermal Driven Cavity-

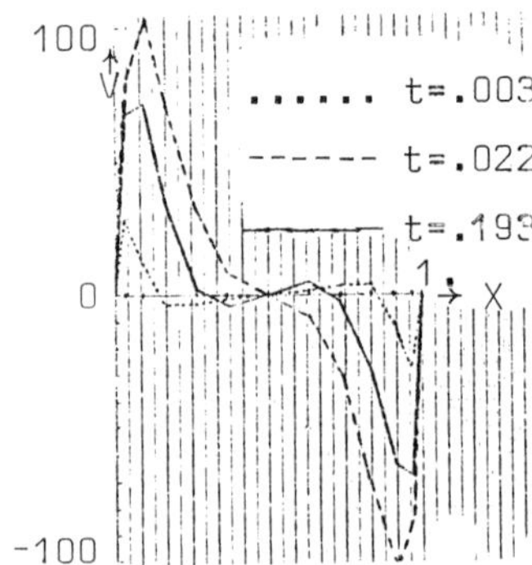

fig. 2 : Vertical velocity profil history at y=.5

-Thermal Driven Cavity-

Let Δt_c be the time step imposed by the convection term for an explicit scheme. Then :

$$\Delta t_c = h/|u|$$

Let $\Delta t_d = \rho C_p\ h^2/K$ be the corresponding time step for the diffusion term. Then the ratio :

$$\Delta t_d/\Delta t_c = Pe_h$$

could be viewed as a local Peclet number in an element.
We found that the computer error$|\Theta\text{-}\Theta h|/\Theta$ at a node is about 10%, and decreases in time, when the time step is $\Delta t_c/2$ and when the value of Pe_h is 1500. It is obvious that the diffusion term smoothes out the solution of a such transport problem. However if the solution is not smooth, we note that the error becomes greater and greater and is transported in the flow even though the algorithm converges at each timestep. When Pe_h is not too large, we get a very good approximation of the solution. We found also that the optimum value of the parameter Υ is about 1/3.

5.2 Thermally driven cavity

The top and the bottom of the cavity are perfectly adiabatic. The temperature of the left and the right wall is suddenly increased by $\Delta\Theta/2 = 0.5$ and $\Delta\Theta/2=-0.5$ respectively while the fluid was initially isothermal ($\Theta_o=0.$). The Rayleigh and Prandtl numbers Ra and Pr are respectively equal to :

$$Ra= 10^5 \text{ and } Pr=1.$$

which corresponds to a classical bench mark test frequently presented in the litterature.

We use a graded mesh (13x13) with four nodes element and 169 nodes.

We compared our results with those obtained by Gresho(8) from Lawrence Livermore Laboratory. The concordance between the velocity and the temperature field is fairly good. We distinguish too the same three stages in the transient

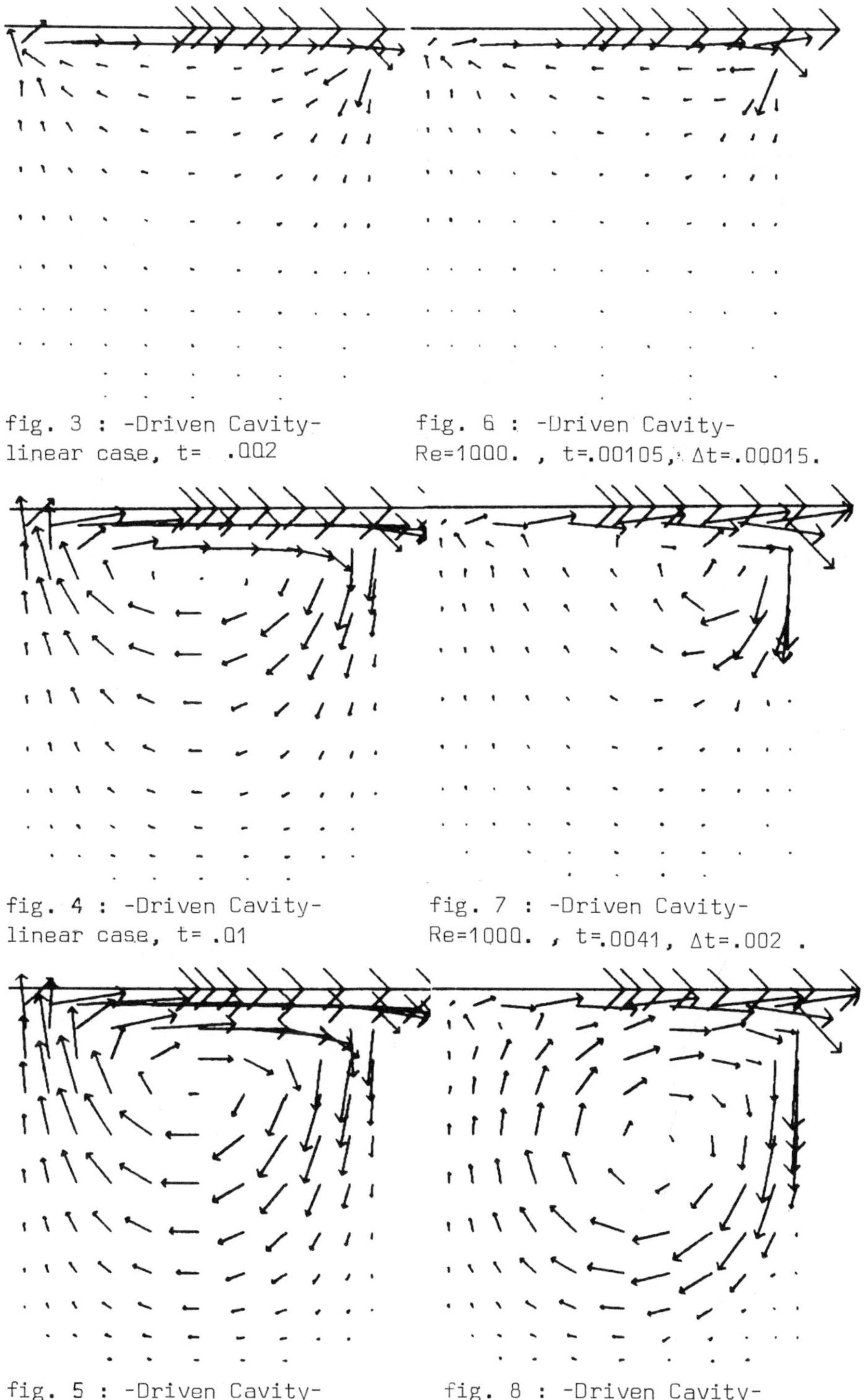

fig. 3 : -Driven Cavity-
linear case, t= .002

fig. 6 : -Driven Cavity-
Re=1000. , t=.00105, Δt=.00015.

fig. 4 : -Driven Cavity-
linear case, t= .01

fig. 7 : -Driven Cavity-
Re=1000. , t=.0041, Δt=.002 .

fig. 5 : -Driven Cavity-
linear case, t= .05

fig. 8 : -Driven Cavity-
Re=1000. , t= .0281, Δt=.002

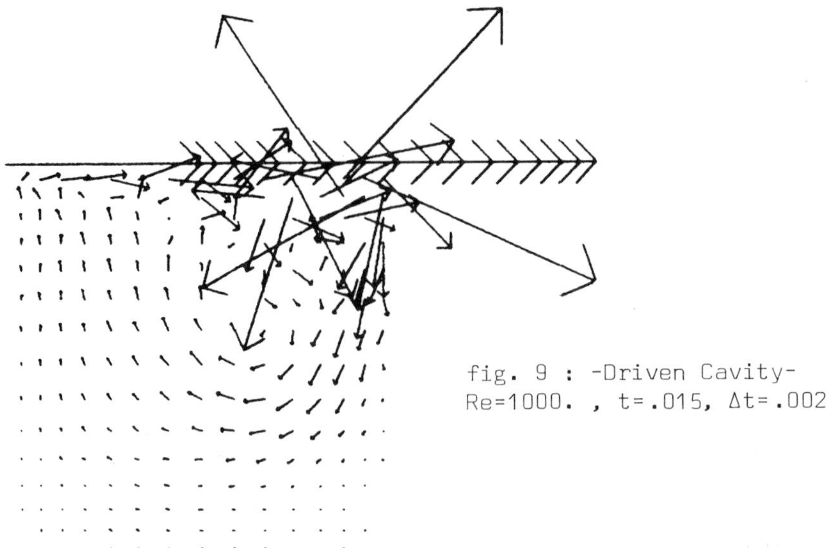

fig. 9 : -Driven Cavity-
Re=1000. , t=.015, Δt=.002

solution: conduction stage from t=0. to ≃0.01, "overshoot" stage from t≃ .01 to ≃0.05, and the reco very stage from t ≃0.05 to ≃0.15 (see fig. 1 and 2)

5.3 Driven cavity

we use the same mesh as for thermally driven cavity. The fluid is now isothermal and initially in a quiescent state. The velocity of the top boundary is suddenly set to 1. The Reynolds number governing the flow is Re=1000.

From the figure 6 to 8 it is possible to distinguish two stages: a viscous stage from t=0. to ≃0.001 where the velocity field seems to be linearly distribued, a convection stage where we note that a vortex is growing in the right top corner and moving towards the center of the cavity.

The comparison between the linear and nonlinear cases is presented on figure 3 to figure 8.

Figure 9 shows what happens when the time step is too large: though the algorithm still converges, a substential error appears in top boundary and grows(there the mesh is 17x17).

5.4 Storage of solar energy

We consider a perfectly adiabatic tank. The inlet is at the top left corner, and the outlet is at the right bottom corner. The fluid is initially isothermal (Θ_0=0.). The inlet is suddenly increased by $\Delta\Theta$=10. and velocity by u=.001 . The value of the Prandtl number and the Peclet number are :

$$Pr=1. \qquad Pe=15.$$

In the first example (fig. 10),.. the Rayleigh number

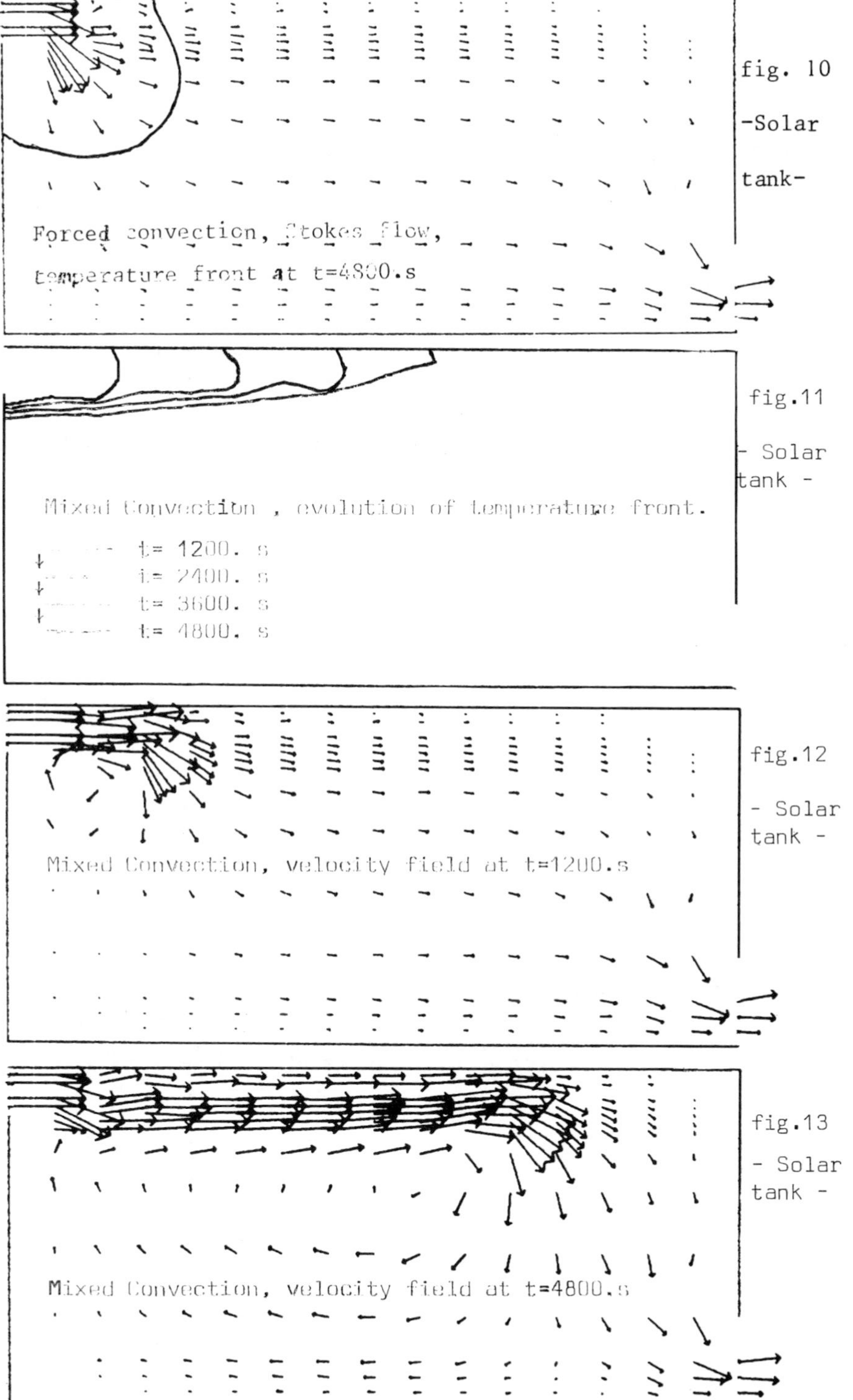
Forced convection, Stokes flow,
temperature front at t=4800.s
fig. 10
-Solar
tank-
Mixed Convection , evolution of temperature front.
t= 1200. s
t= 2400. s
t= 3600. s
t= 4800. s
fig.11
- Solar
tank -
Mixed Convection, velocity field at t=1200.s
fig.12
- Solar
tank -
Mixed Convection, velocity field at t=4800.s
fig.13
- Solar
tank -

is zero value; in the second example (fig 11 to fig.13), the value of Ra is : $Ra=2.5\ 10^4$
The time step is constant and equal to 120s . This simulation needs 40 steps and required about 23 minutes on UNIVAC 1110 .

The effect of the buoyancy term is obvious. A vortex appears and grows in the Stokes flow and the heat volume is deformed towards the top of the tank.

REFERENCES

1. ZLAMAL M.-Finite element methods in heat conduction problems, The mathematics of finite element and applications II, J. Whiteman, Academic Press, 1977.
2. DAHLQUIST G. -A special stability problem for linear multistep method, BIT 3, pp 27-43, 1963.
3. CRYER C.W.-A new class of highly-stable methods, BIT 13, pp 153-159, 1973.
4. TAYLOR C., IJAM A. -Coupled convective/conductive heat transfert including velocity field evaluation, The Math. of F. E. and applications II, J. Whiteman, Academic Press, 1977.
5. HUGHES Th. J.R., TAYLOR R.L., LEVY J-F. -High Reynolds number steady, incompressible flows by a F.E.M., The Math. of F.E. and applications II, J. Whiteman, Academic Press, 1977.
6. ZIENKIEWIEZ O.C., GALLAGHER R.H., HOOD P. -Newtonian and non-newtonian viscous incompressible flow. Temperature induced flows Finite element solutions. , The Math. of F.E. and applications J. Whiteman, Academic Press, 1997.
7. HUGHES Th.J.R., LIU W.K., BROOKS A. -Review of finite element analysis of incompressible viscous flows by the penalty formulation, Journal of Computational Physics, Vol. 30, 1979.
8. GRESHO Ph.M., LEE R.L., CHAN S.T., SANI R.L., -Solution of the time dependent incompressible Navier-Stokes and Boussinesq equations using the Galerkin F.E.M., Approximations Methods for Navier-Stokes Problems, R.Rautmann, Springer-Verlag, 1979.
9. MALKUS D.S., Int. J. Solids Struct., 12, pp 731-738, 1976.
10. GIRAULT V., RAVIART R.A., -Finite element approximation of the Navier-Stokes equations, Springer Verlag Lecture Notes Math., 749.

° °
°

FINITE ELEMENT SOLUTION OF SUBSONIC AND TRANSONIC CASCADE FLOWS

W.G. HABASHI*
Concordia University, Montreal, Canada
P.L. KOTIUGA**
Pratt & Whitney Aircraft of Canada Ltd.

ABSTRACT

In this work a study of the application of the finite element method for transonic flows in axial turbomachines is undertaken.

Solution techniques capable of accurately predicting flows from the incompressible regime up to the establishment of shocks in the transonic regime are presented. In the subsonic and shockless transonic regimes a local linearization method is utilized. In the full transonic regime the artificial compressibility method of Eberle [1] is employed to introduce artificial viscosity in the supersonic regions. The two approaches can be combined in a unified package and appropriate switches used to select the relevant method in each flow regime.

1. GOVERNING EQUATIONS AND BOUNDARY CONDITIONS

For the steady, two-dimensional, inviscid, compressible flow of a fluid in a cascade, one can write in terms of the velocity potential

$$\frac{\partial}{\partial x}\left(\frac{\rho b}{\rho_o}\ \frac{\partial \Phi}{\partial x}\right) + \frac{\partial}{\partial y}\left(\frac{\rho b}{\rho_o}\ \frac{\partial \Phi}{\partial y}\right) = 0 \qquad (1)$$

Figure 1 is a sketch of the computational domain adopted. The appropriate boundary conditions become

* Associate Professor, Department of Mechanical Engineering, and Consultant, Pratt & Whitney Aircraft of Canada Ltd.

** Aerodynamicist, Pratt & Whitney Aircraft of Canada Ltd. Longueuil, Quebec, Canada.

ON AB: $\frac{\partial \Phi}{\partial n} = a_o[M \cos\beta]_{in}$ i.e. only the axial speed is specified. (2a)

ON CD: $\frac{\partial \Phi}{\partial n} = a_o[M \cos\beta]_{ex}$ i.e. only the axial speed is specified. (2b)

ON AE: $\Phi(x) = \Phi_1(x) + a_o[rsM \sin\beta]_{in}$, a periodicity condition (2c)

where Φ_1 is the potential at the corresponding x point on BG

ON FC: $\Phi(x) = \Phi_2(x) + a_o[rsM \sin\beta]_{ex}$, a periodicity condition (2d)

where Φ_2 is the potential at the corresponding x point on HD.

ON EF, GH: the no-penetration condition ($\partial\Phi/\partial n = 0$) is enforced, where n is the outward normal direction to the blade surface at any point. (2e)

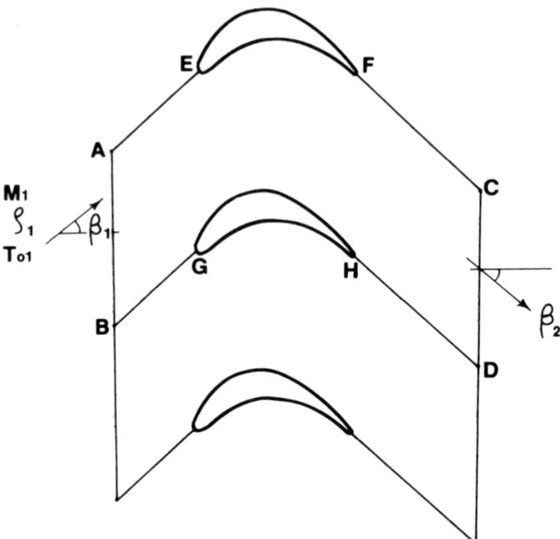

Fig. 1. Numerical Solution Domain.

Alternatively, if one selects a stream function formulation, the governing equation becomes

$$\frac{\partial}{\partial x}\left(\frac{\rho_o}{\rho b}\ \frac{\partial \psi}{\partial x}\right) + \frac{\partial}{\partial y}\left(\frac{\rho_o}{\rho b}\ \frac{\partial \psi}{\partial y}\right) = 0 \tag{3}$$

The equation is usually normalized by defining

$$\frac{\partial \psi}{\partial x} = -\frac{\rho b}{\dot{m}}\, v\ ;\ \frac{\partial \psi}{\partial y} = \frac{\rho b}{\dot{m}}\, u \tag{4}$$

The boundary conditions in this case become

ON AB: $\psi = y/rs$ (5a)

CD ÷ again only the axial speed is specified, giving in conjunction with mass continuity

$$\left(\frac{\rho}{\rho_o} \frac{M\cos\beta}{br}\right)_{ex} = \left(\frac{\rho}{\rho_o} \frac{M\cos\beta}{br}\right)_{in} \quad (5b)$$

BG: $\psi = \psi_1(x)$; unknown (5c)

HD: $\psi = \psi_2(x)$; unknown (5d)

AE and BG are periodic boundaries along which

AE: $\psi(x) = \psi_1(x) + 1$; FC: $\psi(x) = \psi_2(x) + 1$ (5e)

GH: $\psi = 0$; EF: $\psi = 1$ (5f)

The governing equations (1) or (3) are complemented by the isentropic relationship

$$\frac{\rho}{\rho_o} = \left[1 - \frac{\gamma-1}{2} \frac{\nabla\phi\cdot\nabla\phi}{a_o^2}\right]^{\frac{1}{\gamma-1}} \quad (6)$$

2. SUBSONIC SOLUTIONS

For the subsonic regime, the governing equation is elliptic and to speed up convergence a local linearization scheme appropriate to finite elements [2,3] is adopted. Our initial field is normally the solution of the Laplace equation. By determining, in each element, a local velocity direction ξ_e and a normal to it η_e, say at the centroid, one can show [2] that the appropriate variational integral for ψ is

$$I(\psi) = \sum_{e=1}^{N} \frac{1}{2\rho_e b} \iint_{A_e} \left[(1-M_e^2)\psi_{\xi_e}^2 + \psi_{\eta_e}^2\right] d\xi_e d\eta_e - \oint_C \frac{1}{\rho b} \psi\left(\frac{\partial\psi}{\partial n}\right) ds \quad (7)$$

where C is the outer contour of the calculation domain and N the total number of elements in the domain.

Similarly, for the velocity potential Φ

$$I(\Phi) = \sum_{e=1}^{N} \frac{\rho_e b}{2} \iint_{A_e} \left[(1-M_e^2)\Phi_{\xi_e}^2 + \Phi_{\eta_e}^2 + 2V_e M_e^2 \Phi_{\xi_e}\right] d\xi_e d\eta_e - \oint_C \rho b \Phi\left(\frac{\partial\Phi}{\partial n}\right) ds \quad (8)$$

Equations (7) and (8) provide a physical algorithm in which the change in the type of governing equation at transonic speeds shows up naturally in the integral and contributes to rapid convergence at these speeds.

3. TRANSONIC SOLUTIONS

For the transonic regime the proper influences have to be accounted for and downstream influences excluded at supersonic points. While finite difference relaxation schemes are highly developed for this problem, finite element methods have proved difficult and cumbersome for transonic flows. Recently, however, Eberle [1] and Hafez et al [4] have developed the artificial compressibility (AC) method for finite elements and finite differences, respectively. By replacing the density ρ at supersonic points by an artificial density $\tilde{\rho}$ calculated at a point slightly upstream an artificial viscosity is introduced into equation (1), and shock waves evolve naturally in the solution of transonic flows. It is not clear in Eberle's work exactly how to retard the artificial density $\tilde{\rho}$. Hafez's scheme, however, is systematic and assigns at a supersonic point P the density $\tilde{\rho}$ such that

$$\tilde{\rho} = \rho - \mu \left(\frac{\partial \rho}{\partial s}\right) \Delta s \tag{9}$$

where

$$\frac{\partial \rho}{\partial s} \Delta s \simeq \frac{u}{V}\left(\frac{\partial \rho}{\partial x}\right) \Delta x + \frac{v}{V}\left(\frac{\partial \rho}{\partial y}\right) \Delta y \tag{9a}$$

μ being a multiplier of artificial compressibility and s is the streamline direction. If in a finite element solution, the grid is aligned approximately with the streamlines one can safely adopt Hafez's approach of calculating the artificial density using the density gradient between the centroids of adjacent elements.

4. FINITE ELEMENT DISCRETIZATION

Within each element, assume

$$\psi(x,y) = \sum_{i=1}^{n} N_i(\xi,\eta)\psi_i \tag{10}$$

$$x = \sum_{i=1}^{n} N_i(\xi,\eta)x_i; \quad y = \sum_{i=1}^{n} N_i(\xi,\eta)y_i$$

where in the present work n has been taken as 3,4 or 8, i.e. as simple constant-derivative triangular elements or isoparametric 4 and 8 noded quadrilaterals. Upon minimization one obtains at the element level, for the stream function formulation,

$$[K]^{(e)} \{\psi\}^{(e)} = \{R\}^{(e)} \tag{11}$$

$$k_{ij} = \iint_{A_e} \frac{\rho_o}{\rho_e b}\left[(1-M_e^2)\frac{\partial N_i}{\partial \xi_e}\frac{\partial N_j}{\partial \xi_e} + \frac{\partial N_i}{\partial \eta_e}\frac{\partial N_j}{\partial \eta_e}\right] d\xi_e d\eta_e \tag{12}$$

$R_i = \oint \frac{N_i \rho_o}{\rho_e b}\left(\frac{\partial \psi}{\partial n}\right) ds$, for elements having a boundary on AB or CD, i.e.

$$R_i = \left[\frac{\rho_o}{\rho_{ex}} \frac{\tan\beta_{ex}}{s} + \frac{\rho_o}{\rho_{in}} \frac{\tan\beta_{in}}{s}\right] \int_o^s N_i d\theta \tag{13}$$

Since all shape functions N_i, are in terms of the local coordinates (ξ,η) one proceeds as follows:

$$\begin{Bmatrix} \dfrac{\partial N_i}{\partial x} \\ \\ \dfrac{\partial N_i}{\partial \theta} \end{Bmatrix} = [J]^{-1} \begin{Bmatrix} \dfrac{\partial N_i}{\partial \xi} \\ \\ \dfrac{\partial N_i}{\partial \eta} \end{Bmatrix} \tag{14}$$

where $[J]^{-1}$ can be explicitly calculated for each element. A further transformation is needed to obtain the shape function derivatives with respect to the local flow direction σ in each element

$$\begin{Bmatrix} \dfrac{\partial N_i}{\partial \xi_e} \\ \\ \dfrac{\partial N_i}{\partial \eta_e} \end{Bmatrix} = \begin{bmatrix} \cos\sigma & \sin\sigma \\ \\ -\sin\sigma & \cos\sigma \end{bmatrix}^{(e)} \begin{Bmatrix} \dfrac{\partial N_i}{\partial x} \\ \\ \dfrac{\partial N_i}{r d\theta} \end{Bmatrix} \tag{15}$$

After substituting (15) into (12), the influence matrix is formed for each element, directly or by Gaussian integration, and assembled in the usual fashion to yield

$$[K]\ \{\psi\} = \{R\} \tag{16}$$

The variational integral is identical in the transonic case with $M_e = 0$ in equation (12) and with the element density replaced by $\tilde{\rho}$. Only 4 node bilinear elements have been used for transonic solutions.

5. DOMAIN DISCRETIZATION

The construction of appropriate computational grids for cascades is always a challenging problem. In an industrial situation where such an analysis program would be used in a "black-box" form, any mesh generation scheme should be fully automated. Several schemes have been presented in the literature for the generation of such grids (5,6,7).

In the present work we start by splining the coordinates of the blade. Local splines are found necessary for high turning turbine blades. Leading and trailing edge circles are then fitted at both ends of compressor blades. Several element layers resembling the blade shape could be added. The remainder of the grid is generated isoparametrically between the last element layers (Fig. 2). Upstream and downstream

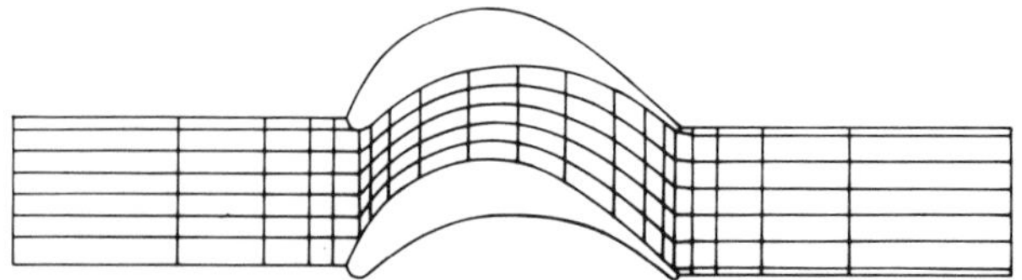

Fig. 2. Finite Element Automated Grid.

sections are also constructed in an expanding fashion. For the case of 8-node isoparametric elements, the location of the exact mid-points is an important factor in the solution but has received scant attention in the literature. On the blade, for example, one proceeds by connecting the two corner nodes of each element by a straight line. By constructing a normal to this line, at its mid-point, and finding the intersection of this normal with the spline describing the blade, one would have located the exact mid-point of the parabola that replaces the spline in the analysis. If the analysis is for triangular elements, the same grid is used and each parabolic element broken internally into 6 triangular elements.

6. FEATURES OF THE SOLUTION

The program tested contained the following options that could be specified interactively by the user:

a) Stream function solution.
b) Velocity potential solution.
c) Local-linearization.
d) Triangular, 4-node and 8-node quadrilateral elements.
e) Arbitrary radius and height of the axisymmetric stream tube analyzed in the blade-to-blade plane.
f) Complete freedom in specification of number of points and their distribution on the blade, across the pitch and in the upstream/downstream sections. The inlet/exit sections are built in an expanding fashion and can also be slanted to match the flow angles.

During each iteration, the matrix [K] of equation (16) is assembled in the "Skyline" form described in [8]. Essentially the element topology is first scanned and the bandwidth of each row predetermined. The variable bandwidth matrix [K] is then stored in vector form, with a vector of pointers to the addresses of diagonal elements. The scheme is efficient, especially if element layers are generated around the blade and the bandwidth of only a few rows increases substantially. The symmetric matrix is solved by LLT decomposition using the program described in the previous reference.

For subsonic flow, convergence to an accuracy of 0.01% in both density and stream function (or velocity potential), is reached in about 4 to 6 iterations and global solution times are below 7 seconds on a Cyber 175 computer. An important comment to be added concerns the extraction of Mach numbers from the ψ or ϕ solution. Although for 8-node isoparametric elements the Mach number distribution is linear within an element, it is discontinuous across the elements. The method of [9] has proven innately accurate in obtaining corner values of the Mach number from the accurate values obtainable at the Gaussian points. Large non-physical velocity peaks would usually result at leading and trailing edges without the use of this method.

For the artificial compressibility method we use the subsonic solver for 3 to 4 iterations to provide an initial good guess to the transonic solution. Afterwards, a line successive overrelaxation (LSOR) method is used to implement the AC solution. Two comments are worth making here, First, at each vertical station of the inlet/exit sections, a periodicity condition must be enforced. For the symmetric test cases presented here (Figs. 9 and 10), the periodicity constant is zero. However, to enable the use of a tridiagonal solver, one can, for example assume all points on AE (Fig. 1) to be known, with periodicity respected, from the previous iteration. Solution symmetry, however, is immediately destroyed and one needs to repeat the line relaxation several times, at the same line, to recover symmetry. The physical explanation is simple since such an approach is equivalent to specifying Dirichlet conditions on AE while imposing a Neumann condition on BG during the solution. If one chooses to implicitly account for periodicity at each line the tridiagonal structure is destroyed. Using the skyline method, however, the matrix size at each line is limited and solution times are unaffected.

Secondly, it is imperative to update the first and last stations in the cascade since no ϕ values are specified on these lines. We solve for the last station, and the one preceeding it, simultaneously during each field sweep. This has the minor effect of increasing the LSOR matrix skyline only when the last line is reached. The same procedure is used for the first two stations.

7. TEST CASES AND RESULTS

Several test cases are presented here. The method's accuracy has been verified against analytical, experimental and other numerical data. In Figure 3 the stream function solution is compared to the incompressible analytical solution of Gostelow [10]. All other cases presented use the velocity potential approach.

Figures 4 and 5 compare the solutions for a flow over a compressor blade designed by Stanitz's hodograph method [11], using triangular and 8-node isoparametric elements. The superiority of the latter element is evident. Figure 5 also

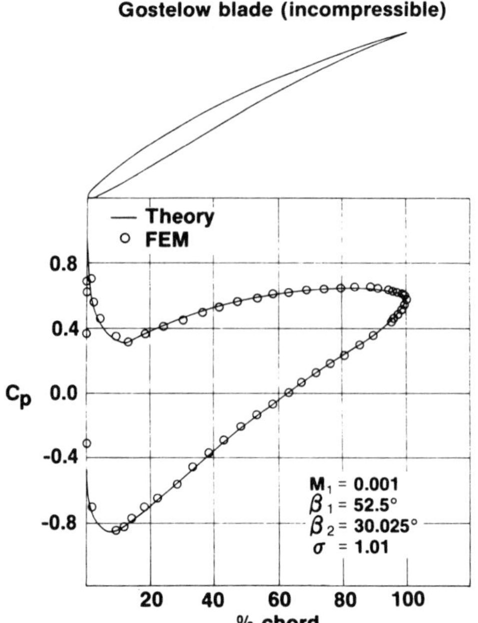

Fig. 3. Incompressible Flow Over Gostelow Cascade.

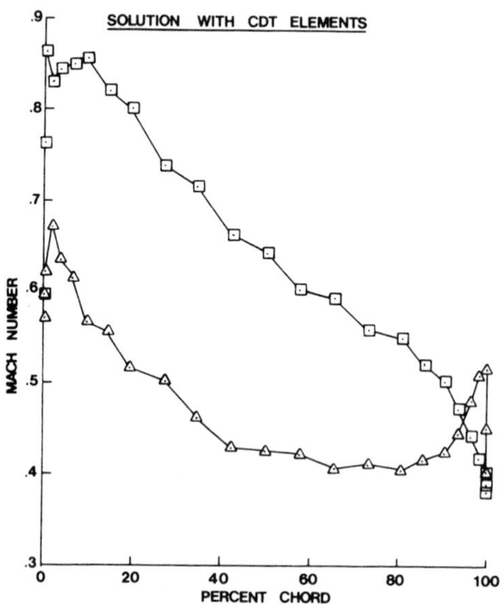

Fig. 4. Hodograph Designed Blade, Triangular Elements Solution.

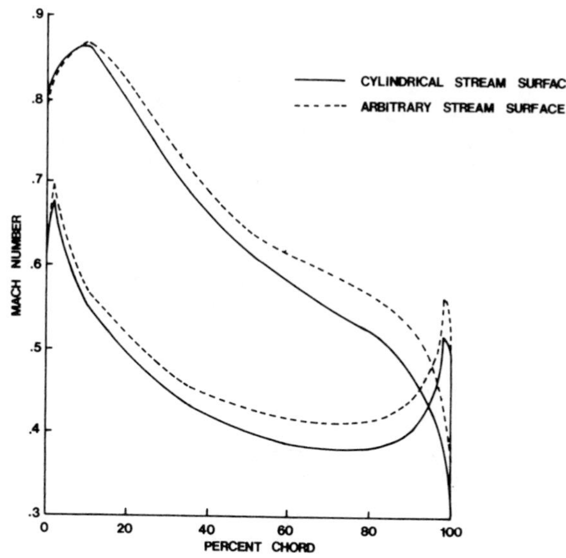

Fig. 5. Hodograph Designed Blade, Isoparametric Elements Solution.

shows the solution to the same blade but on an arbitrary surface described by

x	r	b	A/A_{LE}
-2.5523	6.6231	1.0191	.94465
0.0	7.1454	1.000	1.000
0.952	6.9550	1.0006	0.88323
1.904	6.7625	0.97235	0.92024

More points are used to describe this surface in the actual solution and R and b become additional variables in the Gaussian integration of each element. Although no experimental or other numerical data is available for comparison in this case, we estimate accuracy by integrating the torque profile around the blade and comparing against the imposed angular momentum change between inlet and exit. The error is less than 0.2%. A shockless transonic turbine blade designed by Hobson [12] is computed in Figure 6. The result is in excellent agreement with the hodograph design.

Figure 7 demonstrates the shock capturing capability of the AC method for the flow over a non-staggered cascade of NACA 0012 airfoils at $M_\infty = 0.77$; we note in this diagram the perfect symmetry of the pressure and suction surface distributions. The results compare well to the transonic small disturbance results obtainable by the method of Jones [13].

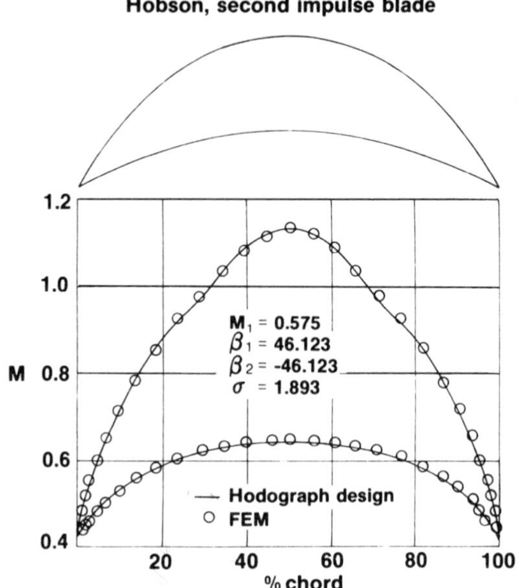

Fig. 6. Hobson Second Impulse Shockless Transonic Blade.

Fig. 7. NACA 0012 Unstaggered Cascade.

Figure 8 represents the results for a 6% parabolic arc isolated airfoil solved for in a low-solidity cascade form, at $M_\infty = 0.9$ and also compared against Jones [13]. Finally, Figures 9 and 10 show the calculated Mach numbers and contours over a staggered transonic compressor blade row.

8. CONCLUSIONS

A versatile finite element method is presented for the solution of transonic cascades. Test cases compare favorably against theory, experiments and hodograph designs in both the subsonic and transonic regimes. Solution times for the subsonic cases are very low (7 secs) while those for transonic solutions with shocks are in the other of 100-200 secs on a Cyber 175 computer.

ACKNOWLEDGEMENTS

This work has been partially supported by Grants A-3662 and PRAI 7901 of the Natural Sciences and Engineering Research Council of Canada (NSERC).

The authors would like to acknowledge the many challenging and fruitful discussions we have had with Dr. D.J. Jones of NSERC.

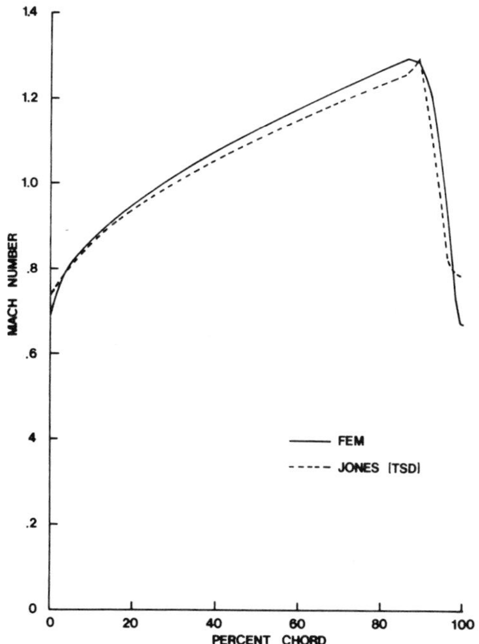

Fig. 8. 6% Parabolic Arc, Low Solidity Cascade M = 0.9.

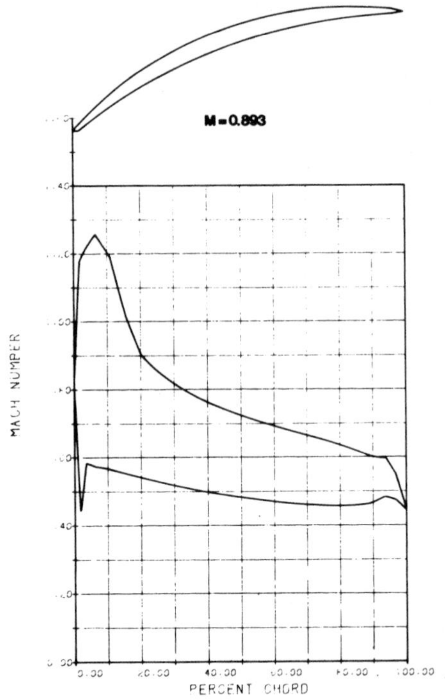

Fig. 9. Cascade of MCA Blades, Solidity = 2.04.

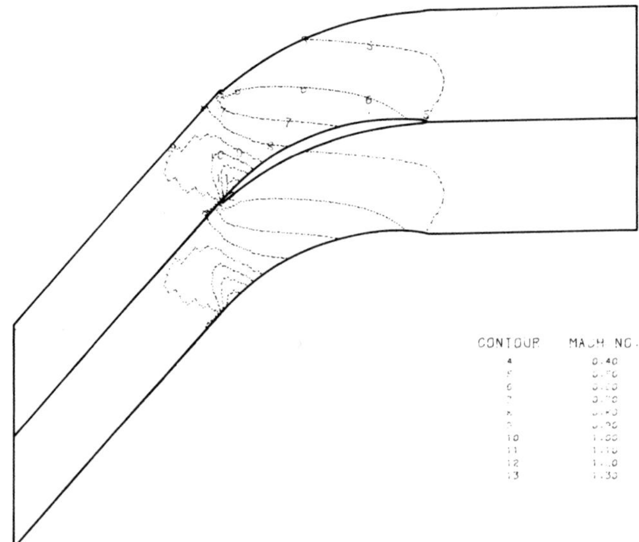

Fig. 10. Mach Number Contours for the Above Cascade.

REFERENCES

1. Eberle, A. - Transonic Potential Flow Computations by Finite Elements: Airfoil and Wing Analysis, Airfoil Optimization. Messerschmidt-Böllkow-Blohm GBH, UF 1428 (ö).

2. Shen, S.F. and Habashi, W.G. - Local Linearization of the Finite Element Method and its Applications to Compressible Flows. International Journal for Numerical Methods in Engineering, Vol. 10, pp. 565-577, 1976.

3. Habashi, W.G., Dueck, E.G. and Kenny, D.P. - Finite-Element Approach to Compressor Blade-to-Blade Cascade Analysis. AIAA Journal, Vol. 17, No. 7, pp. 693-698, 1979.

4. Hafez, M.M., South, J.C. and Murman, E.M. - Artificial Compressibility Methods for Numerical Solution of Transonic Full Potential Equations. AIAA Journal, Vol. 17, pp. 838-844, 1979.

5. Hirsch, C. and Warzee, G. - Finite Element Computation of Subsonic Cascade Flows. Proceedings of the 6th Canadian Congress of Applied Mechanics, Vancouver, B.C., 1977.

6. Ives, D.C. and Liutermoza, J.F. - Second-Order Accurate Calculation of Transonic Flow Over Turbomachinery Cascades. AIAA Journal, Vol. 17, pp. 870-876, 1979.

7. Caspar, J.R., Hobbs, D.E. and Davis, R.L. - Calculation of Two-Dimensional Potential Cascade Flow Using Finite Area Methods. AIAA Journal, Vol. 18, No. 1, pp. 103-109, 1980.

8. Bathe, K. and Wilson, E.L. - Numerical Methods in Finite Element Analysis, Prentice Hall, 1976.

9. Hinton, E., Scott, F.C. and Ricketts, R.E. - Local Least Squares Stress Smoothing for Parabolic Isoparametric Elements. International Journal for Numerical Methods in Engineering, Vol. 9, pp. 235-239, 1975.

10. Gostelow, J.P. and Smith, D.J.L. - Test Cases for Turbomachinery Flow Field Computation. Cambridge University, Engineering Department, CUED/A/TURBO/TR 48, 1973.

11. Stanitz, J.D. and Sheldrake, L.J. - Application of a Channel Design Method to High Solidity Cascades and Tests of an Impulse Cascade with 90° of Turning. NACA Rept. 1116, 1950.

12. Hobson, D.C. - Shock-Free Transonic Flow in Turbomachinery Cascades. Report CUED/A/TURBO/TR 65, 1974.

13. Jones, D.J. and Dickinson, R.G. - A Description of the NAE Two-Dimensional Transonic Small Disturbance Computer Method. NRC/Canada, Lab. Tech. Rept. LTR-HA-39, 1980.